BIRKHÄUSER

Progress in Mathematics
Volume 265

Geometry and Dynamics of Groups and Spaces

In Memory of Alexander Reznikov

Mikhail Kapranov
Sergiy Kolyada
Yuri I. Manin
Pieter Moree
Leonid Potyagailo
Editors

Birkhäuser
Basel · Boston · Berlin

Editors:

Mikhail Kapranov
Department of Mathematics
Yale University
10 Hillhouse Avenue
New Haven CT 06520, USA
e-mail: mikhail.kapranov@yale.edu

Yuri Ivanovich Manin
Pieter Moree
Max-Planck-Institut für Mathematik
Vivatsgasse 7
53111 Bonn, Germany
e-mail: manin@mpim-bonn.mpg.de
moree@mpim-bonn.mpg.de

Sergiy Kolyada
Institute of Mathematics
National Academy of Sciences of Ukraine
Tereshchenkivs'ka, 3
01601 Kyiv, Ukraine
e-mail: skolyada@imath.kiev.ua

Leonid Potyagailo
UFR de Mathématiques
Université de Lille 1
59655 Villeneuve d'Ascq Cedex, France
e-mail: Leonid.Potyagailo@univ-lille1.fr

2000 Mathematics Subject Classification: 08, 11, 18, 20, 22, 37, 51, 54, 55, 60

Library of Congress Control Number: 2007938509

Bibliographic information published by Die Deutsche Bibliothek.
Die Deutsche Bibliothek lists this publication in the Deutsche Nationalbibliografie; detailed bibliographic data is available in the Internet at http://dnb.ddb.de

ISBN 978-3-7643-8607-8 Birkhäuser Verlag AG, Basel · Boston · Berlin

Basel · Boston · Berlin
P.O. Box 133, CH-4010 Basel, Switzerland
Part of Springer Science+Business Media
Printed on acid-free paper produced from chlorine-free pulp. TCF ∞
Printed in Germany

ISBN 978-3-7643-8607-8 e-ISBN 978-3-7643-8608-5

9 8 7 6 5 4 3 2 1 www.birkhauser.ch

Contents

Preface

This book is a collection of selected papers on recent trends in the study of various branches of mathematics. Most of the authors who contributed to this volume were invited speakers at the International Conference "*Geometry and Dynamics of Groups and Spaces. In Memory of Alexander Reznikov*" which was held at the Max Planck Institute for Mathematics (Bonn), Germany, in September 22–29, 2006.

Alexander (Sasha) Reznikov (1960–2003) was a brilliant mathematician who died unfortunately very early. This conference in his remembrance focused on topics Sasha made a contribution to. In particular: hyperbolic, differential and complex geometry; geometric group theory; three dimensional topology; dynamical systems.

The list of participants: Belolipetsky Mikhail; Bismut Jean-Michel; Boileau Michel; Breuillard Emmanuel; Danilenko Alexandre; Delzant Thomas; Deninger Christopher; Esnault Hélène; Franks John; Gal S.R.; Gunesch Roland; Hai Phung Ho; Kaimanovich Vadim; Kapovich Michael; Klingenberg Wilhelm; Makar-Limanov Leonid; Milman Vitali; Morishita M.; Moree Pieter; Navas Andres; Neretin Yuri; Papazoglu Panagiotis; Parker John R.; Porti Joan; Rosellen Markus; Simpson Carlos; Swenson Eric L.; Szczepanski A.; Tomanov Georges; Tsygan Boris; Verjovsky Sola Alberto; Wang Shicheng; Zakrzweski Wojtek.

Talks took place in an informal and constructive atmosphere, and it was a pleasure to see discussions taking place between groups of participants all over the Institute and at all times of day. The exceptional quality of the lectures and the great interest they generated among the conference participants gave us, the organizers of the conference, the idea of this book. The editors wish to record their thanks to the staff at the Max-Planck Institute and to the many researchers who participated for all their efforts in making this such a stimulating experience.

This volume is a collection of papers which aims at reflecting the present state of the art in a most active area of research at the intersection of several branches of mathematics. In the volume we include an unpublished manuscript "Analytic Topology of Groups, Actions, Strings and Varietes" by Sasha (this article is posted, essentially, in the form it appeared as the ArXive preprint math.DG/0001135 in January of 2000; the referee corrected obvious typos, updated and added few references, and made several comments in the form of footnotes and italicized remarks) and some short speeches/recollections about Sasha.

Topics discussed in the book include analytic topology of groups, actions, strings and varieties, category theory, homological algebra, the quantum dilogarithm, Chow groups, parabolic bundles, quantum spaces over non-archimedean fields, Nori's fundamental group scheme, Baumslag-Solitar groups, finitely generated branch groups, Serre's property (FA), hypoelliptic equations, Hodge theory, index theory and related fixed point theorems, determinants and determinant bundles, analytic torsion, Kleinian groups, hyperbolic manifolds, geometric functional analysis, Chern character and cyclic homology, Milnor invariants and l-class groups, geodesic flow, operads, algebras, inner cohomomorphisms, symmetry and deformations in noncommutative geometry, Chebyshev-Dickson polynomials, convolution operators, lattices, discrete harmonic functions, ergodic transformations and rank-one actions, hyperbolic surfaces, eigenfunctions, non-Archimedean metric spaces.

The articles collected in this volume should be of interest to specialists in such areas of mathematics as algebra, dynamical systems, geometry, group theory, functional analysis, number theory, probability theory and topology. The broad spectrum of topics covered should also present an exciting opportunity for graduate students and young researchers working in any of these areas who are willing to put their research in a wider mathematical perspective.

Alexander (Sasha) Reznikov (1960–2003)

Alexander (Sasha) Reznikov was born in Kiev (Ukraine, former USSR) on January 14, 1960. From a very early age he was fascinated with mathematics. It took him only 8 years (instead of the usual 10 years) to finish the primary and secondary schools. In 1975 (the last year of the secondary school), Sasha won the second prize at the International Mathematical Olympiad. Because of this, according to the former Soviet rules he got the right to be admitted as a student to any University of the Soviet Union without the entrance exams. For this reason, he, luckily, was able to avoid the enormous obstacles faced by other Jewish students who tried to enter prestigious Soviet universities during the 1970s and 1980s.[1] Sasha was admitted to Kiev State University at the age of 15 and was a brilliant student during his years at the University. He successfully participated in the student mathematical life and started to do research early on. Despite his success, the fact that he was a Jew effectively barred him from graduate programs and jobs at research institutes. Sasha got a job at a state planning institution which had nothing to do with mathematics. Outside of his working hours there, he visited the Kiev Institute of Mathematics and worked on his PhD thesis under the supervision of Myroslav Gorbachuk. Around that time he became interested in Jewish history and joined a small group studying the history of the state of Israel. Very soon, the Soviet secret service discovered this activity and reported Sasha to his employer. He was forced to quit his job and to leave Kiev. Sasha started to travel the country. He worked in Lithuania, Tajikistan and other remote regions, doing mostly manual work, unable to continue his research in mathematics. Fortunately, the times were changing and starting from 1988 Soviet Jews were again allowed to emigrate. Sasha emigrated to Israel in 1989 and already a year later completed his PhD thesis at the Tel Aviv University under the supervision of Vitali Milman. After spending a year as a postdoc at ICTP in Trieste, he became a lecturer at the Hebrew University in Jerusalem. He remained there until he took a chair in Durham in 1997. Sasha Reznikov, Professor of Pure Mathematics at the University of Durham, died on 5 September 2003, at the age of 43.

We will now describe briefly the most important mathematical contributions of Sasha Reznikov. He started as a classical Riemannian geometer and one of his most impressive results is a proof of the so-called weak Blaschke conjecture [Re19]. A compact Riemannian manifold is called a *Blaschke manifold* if the length of the maximal geodesic segment α starting at any point p, is independent of p and α. The problem is whether the only possible Blaschke manifolds are spheres and projective spaces over the reals, complex numbers, quaternions and Cayley numbers, equipped with their canonical metrics. Sasha proved that all Blaschke manifolds have the same volume as the spheres or projective spaces on which they are modelled.

Sasha Reznikov's most influential work is his proof of Spencer Bloch's conjecture on representations of the fundamental group of an algebraic variety. The proof

[1] This situation is described in the book by M. Shifman "You Failed Your Math Test, Comrade Einstein: Adventures and Misadventures of Young Mathematicians".

is a remarkable combination of arithmetic and analytic methods. Sasha proved that for any smooth complex projective variety, and any representation of its fundamental group into $\mathrm{SL}(2, \mathbf{C})$, the second Chern class in the Deligne cohomology of the associated holomorphic bundle is torsion [Re14]. More generally, he later showed that for all flat bundles on a smooth projective variety, all Chern classes in the Deligne cohomology, except the first one, are torsion [Re17]. These are outstanding results which opened new directions of the research in this field. C. Soulé gave a talk at the Bourbaki Séminaire devoted to this theorem of Reznikov, [So].

For these results Sasha Reznikov was awarded a sectional talk at the European Congress of Mathematics in Barcelona in 2000 [Re2].

In the middle of 1990s he became interested in the geometry and topology of 3-manifolds and geometric group theory. He attempted to prove the famous Haken–Waldhausen–Thurston conjecture that any irreducible 3-manifold with infinite fundamental group has a finite covering with positive first Betti number (such 3-manifolds are called *virtually Haken*)[2]. During these years, he wrote a series of important papers [Re4], [Re5], [Re6], [Re7] and [Re9]. In [Re9] he proved several restrictions on manifolds which are not virtually Haken. In particular, it follows from Reznikov's theorem that if the manifold M is hyperbolic and is not virtually Haken, then for every prime number p there exists a finite covering $N \to M$ such that rank $H_1(N, \mathbb{F}_p) \geq 4$. In the paper [Re4] he discussed similarities between the 3-manifold topology and the theory of number fields. These are illustrated by several interesting examples concerning Heegaard splittings of 3-manifolds. In [Re8] Sasha developed an analogy between the symplectomorphism groups and linear groups; he proved in particular that the inclusion of the compact Lie group $PSU(n+1)$ into the symplectomorphism group of the complex projective n-space is injective on the rational homology.

The final work of Sasha Reznikov "Analytic topology of groups, actions, strings and varieties" was written in 2000 and remained unpublished since then. It contains many important ideas and results, in particular it was instrumental in study of the property T (Kazhdan property) of groups of diffeomorphisms of the circle $\mathbb{S}^1$. We are very glad that this paper is included in our Proceedings and we thank the referee for his careful reading of the paper and many useful remarks and corrections. We keep the referee's comments in the form of footnotes and italicized remarks.

Sasha Reznikov wrote 34 mathematical papers. Most of them are written in a very short period during the 1990s. Their mathematical scope is enormous; the results of Sasha Reznikov belong to several different areas of mathematics: Riemannian and symplectic geometry, 3-dimensional topology, geometric group theory, algebraic geometry and dynamical systems. This diversity of Sasha's research interests is reflected by the variety of topics covered by the articles which appear in these Proceedings of the Conference dedicated to the memory of Sasha Reznikov. The Conference was held at the Max-Planck Institute of Mathematics

[2]This problem remains open.

in Bonn in 2006 from September 22 until September 29 as a satellite conference to the annual meeting of the German Mathematical Society which in 2006 took place in the period September 18–September 22 at the University of Bonn.

We are deeply thankful to Sasha's mother, Ida Reznikova, for providing us with the details of his biography which we used in the introduction of the volume.

Many thanks are due to all contributors, especially to Michael Kapovich and Boris Tsygan for their contributions and help with the preparation of the volume. We are very grateful to Birkhäuser for publishing this volume in the "Progress in Mathematics" series.

Mikhail Kapranov, Yale
Sergiy Kolyada, Kiev
Yuri Manin, Bonn/
Northwestern
Pieter Moree, Bonn
Leonid Potyagailo, Lille

References

[Re1] Reznikov, A. *The structure of Kähler groups. I. Second cohomology. Motives, polylogarithms and Hodge theory, Part II* (Irvine, CA, 1998), 717–730, Int. Press Lect. Ser., 3, II, Int. Press, Somerville, MA, 2002.

[Re2] Reznikov, A. *Analytic topology*, European Congress of Mathematics, Vol. I (Barcelona, 2000), 519–532, Progr. Math., 201, Birkhäuser, Basel, 2001.

[Re3] Reznikov, A. *Knotted totally geodesic submanifolds in positively curved spheres*, Mat. Fiz. Anal. Geom. 7 (2000), no. 4, 458–463.

[Re4] Reznikov, A. *Embedded incompressible surfaces and homology of ramified coverings of three-manifolds*, Selecta Math. (N.S.) 6 (2000), no. 1, 1–39.

[Re5] *Regulators in analysis, geometry and number theory*, Edited by Alexander Reznikov and Norbert Schappacher. Progress in Mathematics, 171. Birkhäuser Boston, Inc., Boston, MA, 2000, 324 pp.

[Re6] Reznikov, A. *Continuous cohomology of the group of volume-preserving and symplectic diffeomorphisms, measurable transfer and higher asymptotic cycles*, Selecta Math. (N.S.) 5 (1999), no. 1, 181–198.

[Re7] Reznikov, A. and Moree, P. *Three-manifold subgroup growth, homology of coverings and simplicial volume*, Asian J. Math. 1 (1997), no. 4, 764–768.

[Re8] Reznikov, A. *Characteristic classes in symplectic topology*, Appendix D by Ludmil Katzarkov. Selecta Math. (N.S.) 3 (1997), no. 4, 601–642.

[Re9] Reznikov, A. *Three-manifolds class field theory (homology of coverings for a nonvirtually b_1-positive manifold)*, Selecta Math. (N.S.) 3 (1997), no. 3, 361–399.

[Re10] Reznikov, A. *Yamabe spectra*, Duke Math. J. 89 (1997), no. 1, 87–94.

[Re11] Reznikov, A. *Fundamental group of self-dual four-manifolds with positive scalar curvature*, Israel J. Math. 97 (1997), 93–99.

[Re12] Reznikov, A. *Quadratic equations in groups from the global geometry viewpoint*, Topology 36 (1997), no. 4, 849–865.

[Re13] Reznikov, A. *Volumes of discrete groups and topological complexity of homology spheres*, Math. Ann. 306 (1996), no. 3, 547–554.

[Re14] Reznikov, A. *Rationality of secondary classes*, J. Differential Geom. 43 (1996), no. 3, 674–692.

[Re15] Reznikov, A. *Determinant inequalities with applications to isoperimetrical inequalities* Geometric aspects of functional analysis (Israel, 1992–1994), 239–244, Oper. Theory Adv. Appl., 77, Birkhäuser, Basel, 1995.

[Re16] Reznikov, A. *Simpson's theory and superrigidity of complex hyperbolic lattices*, C.R. Acad. Sci. Paris Sér. I Math. 320 (1995), no. 9, 1061–1064.

[Re17] Reznikov, A. *All regulators of flat bundles are torsion.*, Ann. of Math. (2) 141 (1995), no. 2, 373–386.

[Re18] Reznikov, A. *The volume and the injectivity radius of a hyperbolic manifold*, Topology 34 (1995), no. 2, 477–479.

[Re19] Reznikov, A. *The weak Blaschke conjecture for* $\mathbf{C}P^n$, Invent. Math. 117 (1994), no. 3, 447–454.

[Re20] Reznikov, A. *Harmonic maps, hyperbolic cohomology and higher Milnor inequalities*, Topology 32 (1993), no. 4, 899–907.

[Re21] Reznikov, A. *Symplectic twistor spaces*, Ann. Global Anal. Geom. 11 (1993), no. 2, 109–118.

[Re22] Reznikov, A. *Affine symplectic geometry. I. Applications to geometric inequalities*, Israel J. Math. 80 (1992), no. 1-2, 207–224.

[Re23] Reznikov, A. *Linearization and explicit solutions of the minimal surface equation*, Publ. Mat. 36 (1992), no. 1, 39–46.

[Re24] Reznikov, A. *Noncommutative Gauss map*, Compositio Math. 83 (1992), no. 1, 53–68.

[Re25] Reznikov, A. *Lower bounds on volumes of vector fields*, Arch. Math. (Basel) 58 (1992), no. 5, 509–513.

[Re26] Reznikov, A. *The space of spheres and conformal geometry*, Riv. Mat. Univ. Parma (4) 17 (1991), 111–130.

[Re27] Reznikov, A. *A strengthened isoperimetric inequality for simplices*, Geometric aspects of functional analysis (1989–90), 90–93, Lecture Notes in Math., 1469, Springer, Berlin, 1991.

[Re28] Reznikov, A. *Integral operators in the spaces of sections of flat bundles and volumes of cut loci*, C.R. Acad. Sci. Paris Sér. I Math. 310 (1990), no. 10, 713–717.

[Re29] Reznikov, A. *The spectrum of the Laplacian on manifolds with closed geodesics*, Ukrain. Mat. Zh. 38 (1986), no. 3, 385–388, 407.

[Re30] Reznikov, A. *Completely geodesic foliations of Lie groups*, Differentsialnaya Geom. Mnogoobraz. Figur, No. 16 (1985), 67–70, 124.

[Re31] Reznikov, A. *The weak Blaschke conjecture for* $H\mathrm{P}^n$, Dokl. Akad. Nauk SSSR 283 (1985), no. 2, 308–312.

[Re32] Reznikov, A. *The volume of certain manifolds with closed geodesics*, Ukrain. Geom. Sb. No. 28 (1985), 102–106, iv; translation in J. Soviet Math. 48 (1990), no. 1, 83–86.

[Re33] Reznikov, A. *Blaschke manifolds of the projective plane type*, Funktsional. Anal. i Prilozhen. 19 (1985), no. 2, 88–89.

[Re34] Lyubashenko, V. and Reznikov, A. *Boundary values of solutions of operator-differential equations*, Dokl. Akad. Nauk Ukrain. SSR Ser. A 1984, no. 4, 73–75.

[So] Soulé, C. *Classes Caractéristiques Secondaires des Fibres Plats*, Séminaire Bourbaki, 48ème année, 1995–96, n 819, pp. 411–424.

Short Recollections about Sasha

Ida Reznikova – (Sasha's mother)

Sasha was born in Kiev on 14 January 1960. He was a quiet child, and seemed similar in his development to our other children. He was bored in elementary school, and the teacher advised us to let him skip a grade. In the sixth grade he again became bored, and the mathematics teacher insisted that he skips grade again, so he went to the eight grade of the mathematical school N 145.

He was so fascinated with mathematics that he was invited to the International Mathematical Olympiad in the last, 10th grade. The Soviet team shared with the US one the third and fourth prizes, while Sasha personally was awarded the second prize. He wanted to enter the Moscow University. He had the right to be admitted, as the winner of the Olympiad. But since he was only 15 when he graduated from the high school, he went to the Kiev University, as we lived in Kiev.

After attending all the lectures once, Sasha realized that he has nothing to gain by going to the lectures, as he knew the material. He regretted very much that he did not enroll in Moscow University, which he could have done without entrance exams but missed the chance. Sasha remained in Kiev.

After graduating summa cum laude, he could not find a job in any research institution, although he was a star at the university. The Dean told us right away that "people such as Sasha" (i.e., Jews) will not be admitted to any research institution, so Sasha got a job in a state planning establishment which had nothing to do with mathematics.

... After a while he met a woman with background in humanities. He learned that she wants to emigrate to Israel, but he had no plans to emigrate himself, and neither did anyone in our family at that time. What he did not know was that she was followed by the KGB, because she insisted on her right to emigrate.

The KGB reported Sasha to his employer, that he keeps company of "undesirable" people. They wanted to use Sasha to uncover and punish all the people involved, so they put pressure on the administration of his institute. The administration organized a staff meeting with public criticism of Sasha's lifestyle. Among other things, he was accused of wearing jeans all the time. Jeans were considered an expensive bourgeois excess, it is hard to believe this now, but it was for real. Sasha's father was a retired Army officer (32 years of service), he had a sizable military pension while still working in his retirement, and I worked as well. So we could afford to buy our children expensive items. Anyway, Sasha was expelled from the Komsomol (Young Communist League). The local newspaper published a very negative article about Sasha. The situation at work became intolerable and he wanted to leave his job. They summoned his father, put him to shame for raising "such a son". All the service that Sasha did to the university, his excellence at the olympiads were ignored. The year before he was admitted to the Institute of Mathematics for graduate study, and the people there treated him well and wanted him to continue his studies. They were decent people who understood that he was picked at for no reason.

In August 1985 Sasha packed his things and left, without telling us where. He wrote me a letter with no return address, saying that he leaves, that he can stand up for himself and support himself if necessary.

His absence dragged on, we started a search, and when we located him, Misha, his brother, came to see him with hope of influencing him somehow. But Sasha did not return home, he worked at a meteorological station for a pittance. He was afraid to look for a better job knowing that his expulsion from the Komsomol would come out. He found a place in a small town in Lithuania. Misha said that Sasha looks OK, he is fine, does not want to return and does not want any visitors.

In the Fall of 1986 I decided to come see Sasha, although I did not know the address. Misha was strongly opposed to the idea. I knew only that Sasha lives in Turmantas (Lithuania). Not knowing how would I find him there, I did not take many things with me, and also did not take a lot of money or food. But Turmantas was a small village and I easily found Sasha there. He looked good, one could tell he spent a lot of time outdoors, I was glad to see that. But the hut where he lived, was truly horrible, it had an outhouse in the backyard. His trousers were ragged, I was walking around the hut thinking what to bring next time.

He received me well, told me about his life. I was in a good mood seeing my son so tall and handsome, with bright shining eyes.

In a couple of days, Borya (our eldest son) took a couple of days off work and went to see him with a full suitcase of necessities and a considerable sum of money, but he did not find Sasha. Borya was told that Sasha quit his job and left for an unknown destination. We could not imagine where could he be and how to find him. Someone suggested to contact the police, which I was reluctant to do so as not to get Sasha in more trouble. But there was no other option.

The police helped us, and in 1987 I learned that Sasha was in Dushanbe (Tajikistan). This was in summer. Misha who flew in to see him, had to keep all money and things and fly back. After this, Sasha went into hiding again. I did not believe that and flew to Dushanbe the next week with the hope to find him there, but he left already. I spoke with his former coworkers, they spoke with high regards of him, regretted that he left. In fact, he was offered graduate study in the area of agriculture which was related to his job. Whatever he started, he succeeded. Everyone was so fond of him, and I cried when I saw his apartment.

We decided not to bother him for a while, but I could not stand being in the dark for long. I started searching for him again and learned that he is in Tselinograd (Kazakhstan), has a decent job, although still far from his beloved mathematics. I flew to see him some time in 1988. I realized that his only choice for normal life is to leave the USSR where he was forever branded as a "renegade".

At that time several people began to emigrate, using invitations from Israel. Sasha went to Israel in 1989, as he felt that a mathematician should do creative work while he is still young, without wasting any time. This was the beginning of a very happy period of our life. Sasha resumed contact with us, was calling home. I decided that it would be better if we went to Israel as well since Sasha is there, wants us to be in his life, this was pure happiness. We emigrated with my husband to Israel in December 1990.

Sergiy Kolyada

For the first time I met Sasha in 1974. I remember that day. We both took part in one of the mathematical competitions ("mathematical battles") which were at that time so popular among high school and university students in Kiev as well as in many other cities of the former Soviet Union.

In this picture you can see Sasha who congratulates Kostya Rybasov, the captain of our winning team of Ukrainian physical-mathematical boarding high school at the Kiev State University. Here, close to Sasha, you can see Borya Tsygan. So, from that time we were close friends with Sasha.

By the way, in those years at the Ukrainian high schools there were many students devoted to mathematics. Later many of them became outstanding mathematicians. Apart from Borya Tsygan and Kostya Rybasov already mentioned let me add at least the names of Sasha Blokh and Mikhail Lyubich from those, who were of the same age as we. Sasha Goncharov, another friend of Sasha Reznikov, was one year younger.

After we finished our high school studies in 1975, Sasha, Borya Tsygan, Kostya Rybasov, me and some others became students of the Kiev State University, while Lyubich and Blokh chose the Kharkov State University.

I will not speak on the life of Sasha during the Soviet era. Let me only add that, for several reasons, at that time it was not easy for students of the Kiev State University to work seriously in mathematics, although there were of course many good mathematicians in Kiev. Being students, we were trying to participate in good scientific seminars, for instance in the seminar of Sergei Samborski at the Kiev Polytechnic Institute. I think we were second year students when Sasha wrote his first paper (on inequalities in triangles). I remember it very well, since

Sasha asked me to read his paper and tell him my opinion before submitting it to the journal "In the world of Mathematics". Though Sasha was in fact two years younger than other students in our group, he always was the life and soul of the group. He was an extraordinary person.

After graduation from the university, Sasha was a PhD student at the Institute of Mathematics of the Ukrainian Academy of Sciences (a correspondence post-graduate study). Formally, his supervisor was Prof. Myroslav Gorbachuk, but he also had very close scientific contacts with Prof. Georgii Kats, Prof. Aleksei Pogorelov and his group in Kharkov. Sasha often used to give lectures at scientific seminars in Kharkov.

Unfortunately, that time a tragic event entered his life and this badly influenced all the rest of his life. Since in those years I was not in a contact with him, I learnt about this matter too late... This is briefly described in his mother's recollection above.

Having left Kiev, Sasha interrupted contacts with all of us. The reason was that he was afraid that we could have problems with the KGB because of him. In 1991 he sent me a letter from Trieste. To illustrate the situation, let me say that the letter was not sent by post but was brought me by a Chinese mathematician and the letter was put into three (!) envelopes.

From that time on, until his death we had been exchanging letters with Sasha regularly. Unfortunately, since 1985 I could not meet him personally any more, but I heard Sasha was very active, making a name for himself in mathematics, talking to many mathematicians.

I believe that the conference and the volume in memory of Sasha Reznikov will significantly contribute to the development of Mathematics.

Boris Tsygan

I knew Sasha since sixth or seven grade. We went to different schools and lived in different parts of Kiev but we met at mathematical competitions. We went to two Ukrainian Olympiads together as members of the Kiev team. I remember that the mathematical battle that is described in this preface ended for us in a police station. We were detained for playing cards in one of the parks for which the city was famous. That particular park, or rather a forest resembling Bois de Boulogne, was on the grounds of the Exhibition of Advanced Experience. I cannot recall the policies or the logic behind the detention, but it seems that minors playing cards were frowned upon. To recall a touching sign of the times, our cards were made of perfocards that we used in our computer science program at school. In 1975 we entered the same class of the Kiev University. That Sasha did not go to Moscow was a disappointment for him but a blessing for me.

Sasha was my first mentor in advanced mathematics but he was to me much more. He was the first real mathematical talent that I met. More generally, he was the first person in my life with a talent to create intangible beautiful things from the inside of his head. To meet such a person early in one's life one has to be lucky, especially if one lives outside of a few great cities. I was duly impressed.

Sasha was always aiming high, both mathematically and otherwise. Back then, when I read biographies of famous mathematicians, I noticed qualities and traits that I recognized in Sasha (I still do). I also remember thinking that perhaps the most interesting thing I will ever write will be memoirs about him.

We had a group of several undergraduates studying math. Sasha was an undisputed leader. He would go to Moscow, then come back and tell us what he learned there from discussions with prominent mathematicians, in particular with Joseph Bernstein. We were also reading and discussing books. Our studies tended towards representation theory and geometry of manifolds, both ignored in the undergraduate University curriculum at Kiev at the time. When we started the second or third year, S.N. Samborsky and Yu.L. Daletsky at the Polytechnical Institute organized a learning seminar on geometry and topology of manifolds. We started by studying the book Analysis on Complex Manifolds, by Wells, and Langs $SL_2(\mathbf{R})$. I was entranced (for Sasha, I think, all this was interesting but not that new). Later we attended special topics courses and seminars by Yu.A. Drozd at the University and A.N. Tolpygo at the Polytechnical, or later at A.N.'s place. The topics were algebraic geometry, algebraic groups, Lie groups, etc. Looking back, I can compare these to graduate courses and graduate seminars at the best universities in the world. This was due partly to brilliant teaching and partly to Sasha's stellar presence. After the meetings we had long walks discussing poetry, history, and mathematics.

It was with Sasha that I went to my first conference, or rather to a Summer school in Kazan, Tatarstan. To make it in time, one of Sasha's exams had to be taken two days earlier. This was theoretical mechanics. Yu.L. Daletsky noted that the professor was his classmate, and phoned her from his office. She was friendly and nice and told us to come to take the exam early. When we came and the exam started, suddenly we were subjected to a frontal assault that I had experienced only once before or since, at my entrance exams at Moscow. After a brutal examination, we are told that we failed but we are more than welcome to come again when we learn the subject. Those long June days in Kiev, when we were sitting in Shevchenko Park opposite the University and discussing classical mechanics with Sasha, are among my best memories. We tried again, and failed. (Sasha: "But Arnold writes..." Professor: "Go to Arnold and ask him to pass you"). Then came the day of my (not Sasha's) regular exam. I come, enter the classroom, get under yet another frontal assault, and exit with yet another failing grade. Everybody is shocked, both me and my classmates who have never seen me fail a test. One of them can't believe that this happened, and asks me to swear by crossing myself. Dazed and confused, I oblige her, and get a strong public rebuke from Sasha: how can you, a Jew... One has to understand the atmosphere back then, in 1977, to appreciate the audacity of this (not that the sign of the Cross was wildly popular with the authorities). Two days later we come to the exam with Sasha's group, get a friendly and charming treatment and good grades, and leave for Kazan two days after the Summer school started.

When it was the time to graduate, we were told by head of the chair of Analysis that he called all the places that asked for "young specialists" and only one was willing to take a Jew. All other options were at high schools. He was sorry and really did not know what to do. This left Sasha and me in a difficult situation. Working at a place that had nothing to do with mathematics was a common thing for many, especially for Jews; arrangements would be made for the person to have enough time for research. The teaching load at high school did not allow this. I recall that there was no shadow of ill will between us. The situation was resolved when we arranged for me to work at the computing center of the factory where my father worked. My position was proudly called mathematician, the salary was slightly better than average, the supervisor was a PhD in functional analysis. I did have to show up and to work there but I got some time for research and was quite happy.

Sasha went to the place for which we were supposed to compete. Two years later I entered the graduate school at Moscow University. Sasha's life took a difficult turn. We did meet from time to time when he was still in Kiev. When he left, we lost contact for years.

We resumed contact when he was in Israel and I at Penn State. In the Spring of 1994 I visited him in Jerusalem. Apart from always exciting mathematical discussions, we talked about many other things, like in our youth. This was another first for me: for the first time in my life I had a conversation with an Israeli. I remember a calm, informed, nuanced, and intelligent view of the country's situation, a sort of talk that I heard sometimes since then from Israelis and not so often from others.

Mathematical conversations with Sasha were always a joy to me. For full disclosure, I do remember a conversation in which mathematical arguments were mixed with (my) threats of violence, but we were fourteen or fifteen at the time. In more mature age, there could be an abrasive word now and then, but my general recollection is of high intellectual charge, generosity, and a sense of high purpose. A compliment from him was not frequent and meant a lot. Often he would listen to your question or your plan of action, think hard, make a few interesting remarks, and then quietly ask: but, in reality, does anyone care? This would help me concentrate my mind on more important things.

Yan Soibelman

My recollections about Sasha are sparse. I recall meeting him for the first time in Kiev, around 1980, at the seminar of Yuri Daletsky. At that time I was a graduate student at the University of Rostov and returned to my relatives in Kiev for vacations (perhaps, it was winter). Sasha gave an "educational" talk about G-structures. I recall his attempt to visualize the latter notion on the blackboard, and then asking the audience: "What is it?". The suggested (by him) answer was "a military unit". "But in fact", said Sasha, "it is a G-structure".

At that time I was interested in papers by Krichever, Novikov and others on non-linear integrable systems. I was surprised to see how enthusiastically Sasha responded to my questions, since, in my opinion, nobody in Kiev was interested in such things. Later I realized that it was his typical reaction to any new and interesting mathematics.

I found him a very pleasant person, but he explained me that it was all due to his interest in the famous book of Dale Carnegie. “A few years ago I was a completely different person”, he told me.

We had more time for discussions four (?) years later, when I returned to Kiev as a fresh PhD, and quickly realized that my Jewish origin made it impossible to find a job there. Sasha was not much lucky for the same reason. He got an obscure job in Kiev. I recall visiting him there once. It was a typical agency, with a lot of people in one room, a lot of noise, and an almost visible atmosphere of “killing” the working time. Sasha’s work there had hardly any relation to serious mathematics. Nevertheless he was still very optimistic about his future.

Perhaps during that period (it lasted 1.5 year) we went together to Vinnica, the town of my childhood. In fact we joined a group of teachers from the school, where my mother taught Russian literature. I still have a few photos from that trip. I recall Sasha’s attempt to find a synagogue in Vinnica. Since nobody knew where it was, he started to ask people in the street. Only a person who lived in USSR at the beginning of 80’s can imagine how pedestrians reacted to that question.

Two years later I returned to Rostov-on-Don, where friends found a job for me at the university. Very soon I heard about Sasha's problems with the KGB in Kiev. He was looking for a place to escape from Kiev. I spoke to my former scientific advisor, Igor Simonenko (without telling him all the details which I did not know myself at the time), and he agreed to take Sasha as a graduate student. Perhaps, things got worse in Kiev, since very soon Sasha asked me about a "not-so-visible" position, e.g., a meteorologist at a station. Me and my wife Tanya tried to find such a position at her agricultural institute in the Rostov region, but at the time when we succeeded, Sasha left Kiev and went into hiding. First time I was sending him letters to Latvia, but then he changed the place, and I lost him for several years until we met in Bonn in 1993.

Jean-Michel Bismut

Je crois bien que c'est lors de l'une de mes premières rencontres avec Sasha qu'au hasard d'une ballade dans le Quartier Latin, nous sommes allés voir un film de Jean Renoir, 'Le déjeûner sur l'herbe'. Je l'avais certes mis en garde sur le fait que sans le secours de sous-titres, il risquait de s'ennuyer. J'ai le souvenir que dans les éclats de rire des spectateurs qui ponctuaient la projection d'une œuvre poétique et grinçante, se détachait, clair, le rire de Sasha. A la sortie, alors que je ne cachais pas ma perplexité, il devait me dire qu'il avait apprécié chaque minute de ce film.

Venu un soir chez nous, alors qu'il était à peine entré et qu'il avait fait connaissance de l'un de mes fils âgé de six ou huit ans, je ne sais plus, nous entendîmes le bruit d'une cavalcade effrénée. Avec mon fils, et sans le secours des mots, il avait improvisé un jeu de poursuite et de guerre, où mon fils avait enfin trouvé à qui parler. Ayant souvent vu passer des mathématiciens, et croyant savoir à qui ils avaient à faire d'ordinaire, mes enfants devaient souvent évoquer entre eux le passage singulier de Sasha Reznikov.

Nous étions allés au Théâtre du Châtelet, à une représentation de 'Fidelio' par le Staatsoper de Berlin, dans une mise en scène qui devait, pour une fois, être à la hauteur des espérances qu'on pouvait y mettre. Au dernier acte, dans le balcon où nous nous trouvions, les portes s'étaient brusquement ouvertes, et nous avions été bousculés par des personnages gigantesques, choristes introduits au dernier moment parmi les spectateurs, entonnant le chœur final. Bouleversés, nous achevâmes la soirée dans un bistro, où j'ai encore le souvenir de nos rires mêlés.

Ce séminaire de Sasha à Orsay me revient aussi en mémoire. Il devait y annoncer la preuve de la conjecture d'hyperbolisation. La veille, tard le soir, il m'avait appelé pour me dire qu'un trou avait été trouvé dans la preuve. Je lui avais proposé d'annuler le séminaire, ce qu'il avait refusé. Crânement, il avait fait face. Ou quelques semaines avant, lorsqu'il expliquait sa preuve, à Jerusalem, dans un contexte d'extrême tension, devant un auditoire souvent perplexe.

Et puis plus rien.

An English translation:
If I remember correctly, it was during one of my first meetings with Sasha that while strolling around the Latin Quarter, we decided to watch a movie by Jean Renoir, 'Le déjeûner sur l'herbe'. I had warned him that without the help of subtitles, he would simply be bored. I vividly remember that among the bursts of laughter coming from the audience watching this poetic and ironic movie, I could also hear the unique laugh of Sasha. At the exit, while I was somewhat perplexed by the movie, Sasha told me that he had enjoyed every minute of it.

One evening he came over to our place. He had just entered the apartment and met one of my sons, who was six or eight years old, I am no longer sure, we suddenly heard the noise of a frantic cavalcade. Without words, he and my son had invented a game of chase and war, in which my son had finally found a suitable partner. Having previously met mathematicians at home, and convinced they knew what to expect, my children would often remember their surprising meeting with Sasha Reznikov.

On another occasion, we had all gone to the Théâtre du Châtelet, to see 'Fidelio' by the Berlin Staatsoper, in a production which, for once, was at the level that one might expect. During the final act, in our balcony, the back doors opened suddenly, and we were pushed aside by gigantic characters, singers who had been introduced into the audience to sing the final choir. Still in shock, we completed the evening in a 'bistro' nearby, and I remember the common bursts of laughter.

The seminar by Sasha at Orsay also comes back to my mind. He was to announce the proof of the hyperbolisation conjecture. The day before, late in the evening, he called to say a gap had been found in the proof. I hinted he could cancel the seminar, which he bluntly refused. Or a few weeks earlier in Jerusalem, when he was explaining the proof to a somewhat skeptical audience, at a time of extreme tension outside.

And then nothing more.

Progress in Mathematics, Vol. 265, xxv–xxix

Sipping Tea with Sasha

Pieter Moree

Dedicated to the memory of Alexander Reznikov

Abstract. In this non-mathematical note I describe how my one joint paper with Sasha arose and how he was the source of inspiration for two other papers.

Mathematics Subject Classification (2000). 14N10, 11A07, 41A60.

Keywords. three-manifolds, Artin's primitive root conjecture.

1. Summer tea in 1996

One of my favorite moments in my mathematical career occurred when I met Sasha Reznikov for the very first time. It was in the summer of 1996 and at that time the arrival of new visitors at MPIM was not that frequent, so that often I would make a chat with a new visitor. In this case the visitor was Sasha. I asked him what area of mathematics he was working in. He did not like my attempt to pin him down to a particular field of mathematics and told me that he considered himself not to belong to any mathematical field in special, rather he considered himself to be an amateur over a broad range of topics. The subsequent discussion, however, made clear that mathematically I seemed to be very far removed from his mathematical interests. The usual reaction from somebody mathematically quite removed from me on hearing about my number theoretical interests would be disinterest. So I got accustomed to anticipating on this, by just mumbling one sentence and then changing topic. So I mumbled to Sasha that I was working on something called the Artin primitive root conjecture and was about to change topic in the next sentence. However, his reaction could not have been more different than what my prejudices suggested: he was immediately extremely enthousiastic:

'that is PRECISELY what I need !!'

Sasha was close to making a lot of progress on a conjecture of Lubotzky and Shalev. They had made the following conjecture:

Conjecture 1. *Let M be a hyperbolic three-manifold. Let $f(d)$ denote the number of subgroups of index d in $\pi_1(M)$. Then, for all d sufficiently large, $f(d) > \exp(C_1 d)$, where $C_1 > 0$ is a constant that depends at most on M.*

Modulo a number theoretic ingredient, Sasha [8] could prove: suppose the three-manifold M is a rational homology sphere, then there exist $\alpha, C_2 > 0$ such that for infinitely many d we have $f(d) > \exp(Cd^\alpha)$.

In order to complete his proof he needed a variation of the following celebrated result of Heath-Brown regarding Artin's primitive root conjecture [3].

Theorem 1. *Let q, r and s be three distinct primes. Then at least one of them is a primitive root for infinitely many primes.*

Namely:

Theorem 2. *Let q, r, s be three distinct primes each congruent to $3 (\mathrm{mod}\ 4)$. Then for at least one of them, say q, there are infinitely many primes p such that q is a primitive root modulo p and $p \equiv \pm 1 (\mathrm{mod}\ q)$.*

Recall that Artin's primitive conjecture (1927) states that if g is not equal to -1 or a square, then there should be infinitely many primes p such that g is a primitive root modulo p. Thus the above theorems go some way towards establishing this conjecture. (For an elementary introduction to this material, see, e.g., Moree [4] or Ram Murty [6].)

The restriction $3(\mathrm{mod}\ 4)$ on the three primes is quite essential. If $l \equiv 1 (\mathrm{mod}\ 4)$, then l is not a primitive root modulo p for any prime $p \equiv \pm 1 (\mathrm{mod}\ l)$: by quadratic reciprocity one has $(\frac{p}{l}) = (\frac{l}{p})$ and since $(\frac{p}{l}) = (\frac{\pm 1}{l}) = 1$, it follows that $(\frac{l}{p}) = 1$ and therefore, $l^{(p-1)/2} \equiv 1 (\mathrm{mod}\ p)$, and so l is not a primitive root modulo p.

It was the last condition on p in Theorem 2 that was crucial to Sasha's proof. I quickly realized that the latter result can be obtained by making some minor modifications in Heath-Brown's proof. Two days after our tea discussion I phoned Sasha and informed him that I had solved his problem. His enthusiasm was boundless and he repeatedly told me that I had saved him about a half year of work (since his experience with number theory was not that large, he said he would need a lot of time to study the literature, etc.).

Unfortunately, the original Artin's primitive root conjecture can be presently only proved assuming the Generalized Riemann Hypothesis (GRH) and so I was quite happy that what Sasha needed could be proved unconditionally. A result of the form 'Conjecture 1 is true assuming GRH' would certainly have been a bit comical!

Soon I had written down several pages of material, more or less equalling the number of pages Sasha had written. To me it looked odd to have an appendix about as large as the main text of the paper and so I proposed to Sasha to coauthor the paper, but with the alphabetical order reversed so as to suggest my minor contribution to the paper. This is what Sasha agreed on and in this way we could write a more homogeneous paper.

Sasha's paper is closely related to his paper [7] where the analogy of covers of three-manifolds and class field theory plays a big role (an analogy that was apparently first noticed by B. Mazur). Sasha and Mikhail Kapranov (at the time also at the institute) were both very interested in this analogy. Eventually, in August 1996, Kapranov and Reznikov both lectured on this (and I explained in about 10 minutes my contribution to Reznikov's proof). I was pleased to learn sometime ago that this lecture series even made it into the literature, see Morishita [5]. Meanwhile several further authors studied the analogy in further detail: Nguyen Quang Do, Sikora and Waldspurger, for example.

After the paper was finished, Sasha submitted it to a top mathematical journal. Much later, the editor informed us that he had not been able to get a referee report and could understand our frustration and would not be insulted if we wanted to publish our paper somewhere else. Ironically, with MathSciNet the paper suffered a similar fate: it does not have a reviewer. (Since H.-W. Henn with whom I discussed Sasha's contribution to the paper told me he was impressed by the range of techniques used by Sasha, I was not completely surprised by this outcome.)

Meanwhile S.-T. Yau, whom Sasha had sent a preprint, had written Sasha that he wanted to publish the paper (provided all the other editors approved it) in the journal he had recently founded: the Asian Journal of Mathematics, This is the reason why the paper eventually ended up in the Asian Journal of Mathematics. The referee of this journal described Sasha's proof as 'ingenious'.

2. Summer tea in 1997

About a year later, in the summer of 1997, I met Sasha again during tea time. Whilst I was happily eating cookies he asked me without much ado whether I knew how the drilling machines of dentists looked like in the old days. (Somehow I immediately lost my appetite for further cookies...) I told him I knew: of lots of arms that can move in all directions. So if one has arms of length $\alpha_1, \ldots, \alpha_m$, then the vertical lengths the dentists arms can make form a subset of the positive values amongst $\pm\alpha_1 \pm \alpha_2 \pm \cdots \pm \alpha_m$. He then told me and Amnon Besser, who also was present, that considerations in work he was doing with Luca Migliorini (then also at the Max-Planck) on the cobordism theory of the moduli space of polygons suggested the following to be true: Let $m \geq 3$. Let $\underline{\alpha} = (\alpha_1, \alpha_2, \ldots, \alpha_m) \in \mathbb{R}^m_{>0}$ and suppose that there is no $\underline{\epsilon} \in \{\pm 1\}^m$ satisfying $\langle \underline{\epsilon}, \underline{\alpha} \rangle = 0$ (inner product is zero). Let $1 \leq i < j \leq m$. Let $\underline{\alpha}_{i,j} \in \mathbb{R}^{m-2}_{>0}$ be the vector obtained from $\underline{\alpha}$ on deleting α_i and α_j. Let

$$S_{i,j}(\underline{\alpha}) := \{\underline{\epsilon} \in \{\pm 1\}^{m-2} : |\alpha_i - \alpha_j| < \langle \underline{\epsilon}, \underline{\alpha}_{i,j} \rangle < \alpha_i + \alpha_j\}.$$

Then Sasha suggested that the cardinality of $S_{i,j}$ should be independent of the choice of i and j. This claim seemed very implausible to Amnon and me and so we immediately set out to disprove it. Indeed, the claim turns out to be false.

However, this suggestion coming from somebody as mathematically gifted as Sasha gave me the impression that perhaps there was a 'twisted version' of his proposed variant that really would be invariant. We then considered the quantity defined by $N_{i,j}(\underline{\alpha}) = \sum_{\underline{\epsilon}\in S_{i,j}(\underline{\alpha})} \prod_{k=1}^{m-2} \epsilon_k$. We showed that in case m is odd, this quantity is indeed independent of i and j. Moreover, $(-1)^{\#S(i,j)}$ is independent of i and j (this also holds for even m). In [1] Amnon and I gave three proofs for this, all in principle accessible to talented high school students. We were proud of the fact that our submitted version of the paper did not contain any reference whatsoever. Unfortunately the referee insisted that a reference to Gradshtyn and Ryzhik be included for the easy integral

$$\frac{1}{2\pi}\int_0^{2\pi} \cot(\frac{x}{2}) \sin(\beta x) = \mathrm{sgn}(\beta),$$

with β an integer. So this is how the paper ended up having one reference...

In Gijswijt and Moree [2] (a paper dedicated to the memory of Sasha) we prove a simple combinatorial principle which has the main result of [1] has a corollary. Several other examples of the combinatorial principle are given.

I am fully aware that the mathematical importance of the two papers described in this section is far less than that of the first section, but I had comparable fun in working on them and in both cases the initial discussion over tea with Sasha is still engraved in my memory.

3. Outside tea time

A few times I had contact with Sasha outside the institute. Also in these situations his originality shone through. Once the discussion, over dinner, was about the importance of making the right choice for a PhD subject. According to Sasha this was highly comparable with marrying the right wife. After all, one has to live at least for four years with a PhD topic, so it better be suitable. Moreover, just as one should avoid the temptation to marry a woman because she is rich and drives around in a flashy yellow Lamborghini, one should avoid the temptation to go for a PhD subject just because it is a popular/fashionable/hot topic at the time. The basic point is that, in the first place, one should love his PhD topic.

It was always my impression (but based on a modest amount of time spent together) that Sasha's ideal was to be a mathematical butterfly who with ease went from mathematician to mathematician to exchange ideas and carry out cross fertilisation and who did not want be (mathematically) confined in any sense and did his best to a 'one-of-a-kind' mathematician. On some moments I observed him when he was working alone he made the impression that behind this lightness projected to the outside world there was a much deeper, serious and very hard-working side to him.

In any case, to me, being much more of 'flower type of mathematician' than 'butterfly type', Sasha was and in my memory will always be the butterfly that during two brief summers came along and made my 'flowery' existence more rosy and fruitful.

References

[1] A. Besser and P. Moree, On an invariant related to a linear inequality, *Arch. Math. (Basel)* **79** (2002), 463–471.

[2] D. Gijswijt and P. Moree, A combinatorial identity arising from cobordism theory, *Acta Math. Univ. Comenian. (N.S.)* **74** (2005), 199–203.

[3] D.R. Heath-Brown, Artin's conjecture for primitive roots, *Quart. J. Math. Oxford Ser.* (2) **37** (1986), 27–38.

[4] P. Moree, Artin's primitive root conjecture – a survey –, arXiv:math.NT/0412262, pp. 30. (An updated version discussing recent literature is in preparation.)

[5] M. Morishita, On certain analogies between knots and primes, *J. Reine Angew. Math.* **550** (2002), 141–167.

[6] M. Ram Murty, Artin's conjecture for primitive roots, *Math. Intelligencer* **10** (1988), 59–67.

[7] A. Reznikov, Three-manifolds class field theory (homology of coverings for a nonvirtually b_1-positive manifold), *Selecta Math. (N.S.)* **3** (1997), 361–399.

[8] A. Reznikov and P. Moree, Three-manifold subgroup growth, homology of coverings and simplicial volume, *Asian J. Math.* **1** (1997), 764–768.

Pieter Moree
Max-Planck-Institut für Mathematik
Vivatsgasse 7
D-53111 Bonn
Germany
e-mail: moree@mpim-bonn.mpg.de

Part I

Analytic Topology of Groups, Actions, Strings and Varieties

Alexander Reznikov

1999/2000

Progress in Mathematics, Vol. 265, 3–93

Analytic Topology of Groups, Actions, Strings and Varieties

Alexander Reznikov

Abstract. This paper is devoted to applications of Analysis to Topology. The latter is very broadly understood and includes geometric theory of finitely generated groups, group cohomology, Kazhdan groups, actions of groups on manifolds, superrigidity, fundamental groups of Kähler and quaternionic Kähler manifolds and conformal field theory. Chapters 1 and 3 treat analytical aspects of geometry of finitely generated groups. Chapter 2 uses Analysis to study groups acting on the circle. Chapter 4 studies the groups of symplectomorphisms, Chapter 5 volume-preserving actions. Chapter 6 deals with fundamental groups of Kähler and quaternionic Kähler manifolds.

Mathematics Subject Classification (2000). 20F65, 22E40, 53C26, 53C55.

Keywords. Kähler manifolds, discrete groups, Kazhdan Property (T), rigidity.

Contents

Introduction

This paper is devoted to applications of Analysis to Topology. The latter is very broadly understood and includes geometric theory of finitely generated groups, group cohomology, Kazhdan groups, actions of groups on manifolds, superrigidity, fundamental groups of Kähler and quaternionic Kähler manifolds and conformal field theory. The motivation and philosophy which led to the present research will be reflected upon in [107] and here we will merely say that we believe Analysis to be a major tool in studying finitely generated groups. An alternative viewpoint is provided by the arithmetic methods, notably by passing to the pro-p completion and using Galois cohomology. This will be described in [108].

Each of the six chapters which constitute this paper opens with a short overview. Below we describe the overall structure of the paper. Chapters 1 and 3 treat analytical aspects of geometry of finitely generated groups. Given an immersion $M \hookrightarrow N$ of negatively curved manifolds (M is compact), there is a boundary map $\partial\tilde{M} \to \partial\tilde{N}$, and it has remarkable regularity properties. Invoking the Thurston's theory, we show that the actions A of pseudo-Anosov homeomorphisms on $W_p^{1/p}(S^1)/\text{const}$ have striking properties from the viewpoint of functional analysis, namely,

$$\sum_{n\in\mathbb{Z}} \|A^n v\|^p < \infty$$

for some $v \neq 0$. We apply this to the classical problem: when a surface fibration is negatively curved and derive a strong necessary condition.

We then develop a theory of quantization for the mapping class group. The classical work on $\text{Diff}^\infty(S^1)$ suggests a two-step quantization: First, obtaining a symplectic representation in $Sp(W_2^{1/2}(S^1)/\text{const})$ with image in the restricted symplectic group [98] and then using the Shale-Weil representation. The first step meets obstacles and the second step breaks down completely for the mapping class group: First, because $\text{Map}_{g,1}$ does not act smoothly on S^1, so it is unclear why it can be represented in $Sp(W_2^{1/2}(S^1)/\text{const})$. Second, even if it can be so represented (this happens to be the case), there is no way to show that the image lies in the restricted symplectic group (it almost certainly does not). The solution comes at the price of abandoning the classical scheme and developing a theory of a new object which we call bicohomology spaces $\mathcal{H}_{p,g}$. The mapping class group Map_g acts on $\mathcal{H}_{p,g}$ and the latter shows some remarkable properties, like duality and existence of vacuum. The last property is translated into the fact that $H^1(\text{Map}_{g,1}, W_p^{1/p}(S^1)/\text{const})$ is not zero. Finally, we find Map_g-equivariant maps of the space of all discrete representations of the surface group into $PSL_2(\mathbb{C})$ to our spaces $\mathcal{H}_{p,g}$.

Chapter 2 uses Analysis to study groups acting on the circle (we need $\text{Diff}^{1,\alpha}$ regularity, so $\text{Map}_{g,1}$ is not included). Our first main theorem says roughly that Kazhdan groups do not act on the circle. Very special cases of this result, for lattices in Lie groups, were recently proved (see the references in Chapter 2). The

Hilbert transform, which played a major role in Chapter 1, is crucial for the proof of this result as well. We then develop a theory of higher characteristic classes for subgroups of $\mathrm{Diff}^{1,\alpha}$ (the first being classically known as the integrated Godbillon-Vey class). All these classes vanish on $\mathrm{Diff}^{\infty}(S^1)$. It is safe to say that the less is smoothness, the more interesting is the geometry "of the circle".

Chapter 3 brings us back to the asymptotic geometry of finitely generated groups. We propose, for a non-Kazhdan group, to study the asymptotic behavior of unitary cocycles. We prove a general convexity result which shows that an embedding of G in the Hilbert space, given by a unitary cocycle, is "uniform". We then prove a growth estimate for unitary cocycles of a surface group, using very heavy machinery from complex analysis, adjusted to our situation. A similar result for cocycles in $H^1(G, l^p(G))$ is proven in Chapter 1.

Chapter 4 studies the groups of symplectomorphisms. There is a mysterious similarity between groups acting on the circle and groups acting symplectically on a compact symplectic manifold. In parallel with the above-mentioned result in Chapter 2 we show, roughly, that transformations of a Kazhdan group acting on a symplectic manifold must satisfy a partial differential equation. An example is $Sp(2n, \mathbb{Z})$ acting linearly on T^{2n} and, very probably, Map_g acting on the space of stable bundles over a Riemann surface. (I don't know for sure if Map_g is Kazhdan.)[1]

In the dimension 2, the result is very easy and was known before. We also introduce new characteristic classes for the symplectomorphism groups, in addition to the two series of classes defined in our previous papers, and use them to express a volume of a negatively curved manifold through the Busemann functions on the universal cover.

Chapter 5 studies volume-preserving actions. We introduce a new technique into the subject, that of (infinite-dimensional, non-positively curved) spaces of metrics. We define a invariant of an action which is the infimum of the displacement in the space of metrics and show that for the action of a Kazhdan group which does not fix a $\log L^2$-metric, this invariant is positive (a weak version of this result for the special case of lattices was known before). We then turn to a major open problem, that of non-linear superrigidity and prove what seems to be first serious breakthrough after many years of effort.

Chapter 6 deals with fundamental groups of Kähler and quaternionic Kähler manifolds. The situation is exactly the opposite to the studied in Chapters 2 and 4, namely, these groups tend to be Kazhdan. We first extend our rationality theorem for secondary classes of flat bundles over projective varieties to the case of quasiprojective varieties, answering a question posed to us by P. Deligne. We then prove that a fundamental group of a compact quaternionic Kähler manifold is Kazhdan, therefore providing a very strong restriction on its topology. We also

[1] The group Map_1 is not Kazhdan since it is virtually free. The group Map_2 is not Kazhdan since it is commensurable to the mapping class group of the 6 times punctured sphere and, hence, virtually maps onto a free group. J. Andersen recently proved that Map_g is not Kazhdan for all g, see [4].

discuss polynomial growth of the group cohomology classes for Kähler groups, proved nontrivial in a previous paper.

The paper uses many different analytic techniques. Within each chapter, there is a certain coherence in the point of view adopted for study.

I started this project on a chilly evening of November, 1998 in an African café in Leipzig and finished it on a hot afternoon of July, 1999 in Jerusalem. The manuscript was written up by August, 11, 1999; I would appreciate any mentioning of a possible overlap with any paper/preprint which appeared before this date. During the long time when the paper was being typed and then polished, I found a proof of several statements which were conjectured in the paper. In particular, I constructed a cocycle, with values in $W_p^{1/p}(S^1)/\text{const}$, for the group of quasisymmetric homeomorphisms, which was conjectured in Chapter 1. The proofs will appear in a sequel to this paper.

Editor's notes. This paper is published, essentially, in the form it appeared as the ArXive preprint math.DG/0001135 in January of 2000. We corrected obvious typos, updated and added few references, and made several comments in the form of footnotes and italicized remarks.

1. Analytic topology of negatively curved manifolds, quantum strings and mapping class groups

Chapter 1 opens with observations concerning the cohomology $H^1(G, l^\infty(G))$ of a finitely-generated group. If G is amenable we produce plenty of polynomial cohomology classes in $H^*(G, \mathbb{R})$ given by an explicit formula (Theorem 1.3). Then we prove a nonvanishing Theorem 1.4 saying that if there are Euclidean-type quasi-geodesics in the Cayley graph of G, then $G/[G,G]$ is infinite.[2]

We then review some standard facts about l^p-cohomology in Sections 1.2–1.4. One defines an asymptotic invariant of a finitely generated group G, called the *constant of coarse structure* $\alpha(G)$, as the infimum of p, $1 \le p \le \infty$, such that $H^1(G, l^p(G)) \neq 0$. For all nonelementary discrete groups of motions of complete manifolds of pinched negative curvature, $\alpha(G) < \infty$. For discrete subgroups G of $SO^+(1,n)$, $\alpha(G) \le \delta(G)$, where δ is the critical exponent of the group. In Section 1.4 we review function spaces. A classical result in weighted Sobolev spaces may be reformulated as an identification of the l^p-cohomology of cocompact real hyperbolic lattices:

$$H^1(G, l^p(G)) = W_p^{(n-1)/p}(S^{n-1})/\text{const}.$$

It follows that $\alpha(G) = n - 1$.

In Section 1.5 we prove the first result in the program to classify groups according to the cocycle growth. For surface groups we show that, if

$$L_g\mathcal{F} - \mathcal{F} \in l^p(G), \forall g \in G,$$

[2]Provided that G is amenable.

then

$$|\mathcal{F}(g)| \leq \text{const} \cdot [\text{length}(g)]^{1/p'}, \forall p' > p.$$

Here $\mathcal{F} : G \to \mathbb{R}$ is any function (Theorem 1.14). This result, no doubt, generalizes to higher-dimensional cocompact lattices in simple Lie groups of rank one.

In Section 1.6 we present a new theory for boundary maps of negatively curved spaces, associated with immersions of closed manifolds. The most striking is a partial regularity result (Theorem 1.16, part 4).

As is well known, the group of quasisymmetric ($n = 2$) or quasiconformal ($n \geq 3$) homeomorphisms of S^{n-1} acts on $W_p^{\frac{n-1}{p}}(S^{n-1})$ for $p > n-1$. The action of $\mathcal{G}_1$ on $W_2^{1/2}(S^1)/\text{const}$ is in fact symplectic. We give application to the regularity of quasisymmetric homeomorphisms (Theorem 1.27). In Corollary 1.29 we prove that the unitary representation of a discrete subgroup G of $SO(1,2)$ in $W_2^{1/2}(S^1)/\text{const}$ is the same for all groups in the Teichmüller space $\mathbf{T}(G)$.

In Theorem 1.31 we establish some striking properties of invertible operators A in the Banach spaces $W_p^{1/p}(S^1)/\text{const}$, $p > 2$, induced by pseudo-Anosov maps in $\text{Map}_{g,1}$, namely

$$\sum_{k \in \mathbb{Z}} \|A^k v\|^p < \infty$$

for some $0 \neq v$. In Theorem 1.33 we find a new inequality in topological Arakelov theory, based on the work of [76]. In Theorem 1.34 we find very strong restrictions on subgroups $G \subset \text{Map}_g$, such that induced group extension $\widetilde{G}$:

$$1 \to \pi_1(\Sigma_g) \to \widetilde{G} \to G \to 1$$

is the fundamental group of a negatively curved compact manifold (this is a classical problem). In Section 1.10 we extend the theory to the limit case $p = 1$, introducing an L^1-analogue of Zygmund spaces, which we call $\mathcal{L}_{k,\alpha}$.

In Section 1.11 we start a new theory of secondary quantization of Teichmüller spaces. We introduce the bicohomology spaces $\mathcal{H}_{g,p}$ and show that Map_g acts on these spaces. We show (difficult!) that $\mathcal{H}_{2,p}$ is an infinite-dimensional Hilbert space and there is a symmetric bilinear nondegenerate form of signature (∞, m) which is Map_g-invariant. What is the value of m, we don't know at the time of writing (August, 1999). So does the secondary quantization lead to ghosts? We provide a holomorphic realization in the space of L^2-holomorphic 2-forms on $\mathbf{H}^2 \times \mathbf{H}^2/G$ and $\mathbf{H}^2 \times \overline{\mathbf{H}^2}/G$ (Theorem 1.52). In Section 1.12 we interpret $\mathcal{H}_{p,g}$ as operator spaces (Proposition 1.54), and prove the existence of vacuum (Theorem 1.57). We prove that $H^1(\text{Map}_{g,1}, W_p^{1/p}(S^1)/\text{const}) \neq 0$ for $p \geq 2$. It still may be true that $\text{Map}_{g,1}$ is Kazhdan, because the action is not orthogonal.

In Section 1.13 we construct Map_g-equivariant maps of the space of discrete representations of the surface group into $SO^+(1,3) = PSL_2(\mathbb{C})$ to our spaces $\mathcal{H}_{p,g}$ (Theorem 1.59). In Theorem 1.60 we summarize our knowledge of the functional-analytic structure coming from hyperbolic 3-manifolds which fiber over the circle.

1.1. Metric cohomology

1.1.1. Let G be a finitely generated group. Let $\mathbb{K} = \mathbb{R}, \mathbb{C}$. Let V be a locally convex topological $\mathbb{K}$-vector space which is a G-module, that is, there is a homomorphism $G \to \mathrm{Aut}(V)$. If $\{g_i, i = 1, \ldots, n\}$ is a finite generating set of G, then the evaluation map $f \mapsto \{f(g_i)\}$ establishes an injective map $Z^1(G,V) \to \prod_{i=1}^n V$ of the space of (*inhomogeneous*) 1-cocycles of G in V. One calls the induced topology on $Z^1(G,V)$ the *cocycle topology*; it does not depend on the choice of generators. The image of the coboundary map $V \to Z^1(G,V)$ is not necessarily closed. Let $\overline{B^1(G,V)}$ denote its closure; the quotient $Z^1(G,V)/\overline{B^1(G,V)}$ is called the *reduced* first cohomology space. One way to produce nontrivial cohomology classes is to consider limits of sequences of coboundaries, that is, nontrivial elements of $\overline{B^1(G,V)}/B^1(G,V)$. That amounts to considering nets $\{v_\alpha \in V\}$ such that $g_i v_\alpha - v_\alpha \to l(g_i)$ for $i = 1, \ldots, n$.[3]

If V is a Banach space and G acts isometrically without invariant vectors, then $B^1(G,V)$ is closed in $Z^1(G,V)$ if and only if there are no almost invariant vectors, that is, sequences v_j, $\|v_j\| = 1$, such that $\|g_i(v_j) - v_j\| \to 0$ for all $i = 1, \ldots, n$. This statement is an immediate consequence of the Banach theorem and is called Guichardet's lemma [48]. So if there are almost invariant vectors, then $H^1(G,V) \neq 0$, though the reduced cohomology $H^1_{\mathrm{red}}(G,V)$ may be zero.

If V is Banach and G acts isometrically, let $l \in Z^1(G,V)$ be a cocycle. Then

$$\|l(g)\| \leq \max_{i=1}^{n} \|l(g_i)\| \cdot \mathrm{length}(g),$$

where length(g) is the length of the element g in the word metric induced by $\{g_i\}$. The proof is immediate by induction, using the cocycle equation $l(gh) = gl(h) + l(g)$.

Now let $V_j, j = 1, \ldots, m$ be a collection of Banach spaces on which G acts isometrically and let $\varphi : \otimes_{j=1}^m V \to \mathbb{K}$ be a continuous map in a sense that

$$\varphi(\otimes_{j=1}^m v_j) \leq \mathrm{const} \cdot \prod_{j=1}^m \|v_j\|.$$

Let $l_j \in Z^1(G,V_j)$ and let $l \in Z^m(G,\mathbb{K})$ be the cup product $l(g_1, \ldots, g_m) = \varphi(\otimes_{j=1}^m l_j(g_j))$.

Lemma 1.1. *$l \in Z^m(G,\mathbb{K})$ has polynomial growth, more precisely,*

$$|l(g_1, \ldots, g_m)| \leq \mathrm{const} \cdot \prod_{i=1}^m \mathrm{length}(g_i).$$

Proof. The proof is immediate from the remarks made above. □

A general definition of polynomial cohomology is to be found in [24]. As we will see, Lemma 1.1 is a very powerful tool for constructing cocycles of polynomial growth in concrete situations.

---

[3] Here $l(g_i)$ is the evaluation of the limiting cocycle l.

Proposition 1.2. *G be an infinite finitely generated group. Consider the left action of G on $l^\infty(G)$. Then $H^1(G, l^\infty(G)) \neq 0$. Moreover, $H^1(G, l_0^\infty(G)) \neq 0$.*

Proof. Let $\{g_i\}$ be a finite set of generators of G, and let length(g) be the word length of the element g. Define a right-invariant word metric by $\rho(x,y) = \text{length}(xy^{-1})$. Let $x_0 \in G$ and let $F(x) = \rho(x_0, x)$. Obviously, F is unbounded. Now let $l(g) = L_g F - F$ where L_g is a left action on functions, that is, $l(g)(x) = F(g^{-1}x) - F(x)$. We find

$$|l(g)(x)| = |\rho(x_0, g^{-1}x) - \rho(x_0, x)| \leq |\rho(g^{-1}x, x)| = \rho(g^{-1}, 1).$$

So l is a cocycle of G in $l^\infty(G)$. If it were trivial, we would have a bounded function f such that $L_g F - F = L_g f - f$ that is, $F - f$ would be invariant, therefore constant, a contradiction. The second statement of the proposition will be proved later, in Section 1.3. □

1.1.2. Now let G be amenable. In this case we have a continuous map

$$\varphi : \prod_{j=1}^{m} l^\infty(G) \to \mathbb{K}$$

given by $(f_1, \ldots, f_m) \mapsto \int_G f_1 \cdots f_m$. By the integral we mean the left-invariant normalized mean of bounded functions. We obtain

Theorem 1.3. *Let G be a finitely generated amenable group, let ρ_j, $j = 1, \ldots, m$ be a collection of right-invariant word metrics on G. Fix $x_0 \in G$. Then the formula*

$$l(g_1, \ldots, g_m) = \int_G \prod_{j=1}^{m} [\rho_j(x_0, g_j^{-1}x) - \rho_j(x_0, x)]$$

defines a real-valued m-cocycle on G of polynomial growth:

$$|l(g_1, \ldots, g_m)| \leq \text{const} \cdot \prod_{j=1}^{m} \text{length}(g_j)$$

for any word length length$(\cdot)$.

Examples. Let $G = \mathbb{Z}$. If we choose the generators $\{-1, 1\}$, then length$(g) = |g|$ (the absolute value of the integer) and

$$\rho(x_0, g^{-1}x) - \rho(x_0, x) = |x_0 - x + g| - |x_0 - x| \to \pm|g|$$

as $x \to \pm\infty$ and

$$\int_{\mathbb{Z}} (|x_0 - x + g| - |x_0 - x|) = 0.$$

However, if we choose the generators $\{-1, 2\}$, then

$$\text{length}(g) = \begin{cases} |g|, & g \leq 0 \\ \frac{g}{2}, & g \geq 0 \text{ and even} \\ \frac{g+1}{2}, & g \geq 0 \text{ and odd.} \end{cases}$$

Then

$$\text{length}(x_0 + g - x) - \text{length}(x_0 - x)$$

will have limits $\frac{g}{2}$ when $x \to -\infty$ and $-g$ when $x \to \infty$; hence

$$\int_{\mathbb{Z}} [\text{length}(x_0 + g - x) - \text{length}(x_0 - x)] = -\frac{g}{4}.$$

So we obtain a cocycle $l : \mathbb{Z} \to \mathbb{R}$ given by $g \mapsto -\frac{g}{4}$. Now, if $G = \mathbb{Z}^k$, $k \geq 2$, let $\rho_j, j = 1, \dots, k$ be the word metrics defined by the sets of generators

$$\{e_1^{\pm 1}, e_2^{\pm 1}, \dots, e_j^{-1}, e_j^2, e_{j+1}^{\pm 1}, \dots, e_k^{\pm 1}\}$$

where e_s is the generator of the sth factor. If $1 \leq j_1 < j_2 < \cdots < j_m \leq k$ is a set of indices, then Theorem 1.3 provides a cocycle

$$l(g_1, \dots, g_m) = \int_{\mathbb{Z}^k} \prod_{r=1}^{m} [\rho_{j_r}(x_0, g_j^{-1}x) - \rho_{j_r}(x_0, x)].$$

If $\pi_i : \mathbb{Z}^k \to \mathbb{Z}$ is the projection to the ith factor, then

$$l(g_1, \dots, g_m) = (-\frac{1}{4})^m \cdot \prod_{r=1}^{m} \pi_{j_r}(g_r).$$

It follows that classes of cocycles, given by Theorem 1.3, generate the real cohomology space of $\mathbb{Z}^k$.

Remark. If G is amenable, ρ is a right-invariant word metric and for some $x_0, g \in G$,

$$\int_G [\rho(x_0, g^{-1}x) - \rho(x_0, x)] \neq 0,$$

then $H_1(G, \mathbb{R}) \neq 0$ and in fact $g^s \notin [G, G]$ for all $s \neq 0$. This is a direct corollary of Theorem 1.3. A more interesting structure theorem is given below.

Theorem 1.4. *Let G be a finitely generated amenable group, ρ a right-invariant word metric. For $g \in G$ assume the following convexity condition: there is some $C > 0$, such that for any $x \in G$ there exists $N \geq 0$ such that $\rho(g^k, g^{-1}x) - \rho(g^k, x) \geq C$ for $k \geq N$. Then $H_1(G, \mathbb{R}) \neq 0$ and, moreover, $g^s \notin [G, G]$ for all $s \neq 0$.*

Remark. The functions

$$f(x) := \rho(x_0, g^{-1}x) - \rho(x_0, x)$$

considered above, are *quasi-morphisms*, i.e., maps

$$G \to \mathbb{R}$$

such that there exists a constant K so that

$$|f(xy) - f(x) - f(y)| \leq K$$

for all $x, y \in G$. The same averaging procedure as above shows that every quasi-morphism f of an amenable group G is within finite distance from a homomorphism $f^* : G \to \mathbb{R}$. Compare [59, 91].

Corollary 1.5. *Let G be the Heisenberg group*

$$\langle x,y,z|[x,y]=z,[x,z]=[y,z]=1\rangle.$$

Then for any right-invariant word metric ρ, there exists $a\in G$ such that

$$\liminf_{k\to\infty}[\rho(z^k,z^{-1}a)-\rho(z^k,a)]\le 0.$$

Proof of the corollary. Since $z\in[G,G]$, the result follows from Theorem 1.4. Indeed, G is nilpotent and, therefore, amenable. □

Proof of the theorem. Consider the 1-cocycle

$$l(\gamma)(x)=\rho(x_0,\gamma^{-1}x)-\rho(x_0,x),$$

$l\in Z^1(G,l^\infty(G))$. Set $x_0=g^n$; thus

$$l_n(g)(x)=\rho(g^n,g^{-1}x)-\rho(g^n,x).$$

If for any x and sufficiently big n, $\rho(g^n,g^{-1}x)-\rho(g^n,x)>C$, then the pointwise limit $\lim_{n\to\infty}l_n(g)(x)$ exists and is $\ge C$. Since $|l_n(z)(x)|\le\rho(z^{-1},1)$, there is a subsequence n_k such that $l_{n_k}(z)$ converges pointwise for any z to a bounded function $l(z)$. One sees immediately that $l:G\to l^\infty(G)$ is a cocycle, so $z\mapsto\int_G l(z)$ is a homomorphism from G to $\mathbb{R}$. Since $l(g)(x)\ge C>0$ for all x, $\int_G l(g)\ge C>0$, so $H_1(G,\mathbb{R})\ne 0$ and $g^s\notin[G,G]$, as desired. □

1.1.3. Let $\varphi:\mathbb{R}_+\to\mathbb{R}_+$ be a smooth function such that

$$\lim_{x\to\infty}\varphi(x)=\infty,\qquad \lim_{x\to\infty}\varphi'(x)=0.$$

Let G be a finitely generated group and let ρ be a right-invariant word metric on G. Consider $F(x)=\varphi(\rho(x_0,x))$, where $x_0\in G$ is a fixed element. Since

$$\begin{aligned}|(L_gF-F)(x)| &= |F(g^{-1}x)-F(x)|\\ &= |\varphi(\rho(x_0,g^{-1}x))-\varphi(\rho(x_0,x))|\\ &\le \sup\nolimits_{t\in I}|\varphi'(t)|\cdot|\rho(g^{-1}x,x)|\\ &\le \sup\nolimits_{t\in I}|\varphi'(t)|\rho(g^{-1},1)\end{aligned}$$

where

$$I=[\min(\rho(x_0,x),\rho(x_0,g^{-1}x)),\max(\rho(x_0,x),\rho(x_0,g^{-1}x))],$$

we see that $L_gF-F\in l_0^\infty$. Therefore $H^1(G,l_0^\infty(G))\ne 0$, because the cocycle L_gF-F cannot be trivial as a cocycle valued in l_0^∞ (by the same reasons as in the proof of the first statement of Proposition 1.2). The proof of Proposition 1.2 is now complete. □

Notice that, since $\rho(u,v)=\mathrm{length}(u\cdot v^{-1})$,

$$\rho(x_0,x)-\mathrm{length}(g)\le\rho(x_0,g^{-1}x)\le\rho(x_0,x)+\mathrm{length}(g),$$

so that

$$|(L_gF-F)(x)|\le\sup_{|t-\rho(x_0,x)|\le\mathrm{length}(g)}|\varphi'(t)|\times\rho(g^{-1},1).$$

Remark. Let $S(N) = \{g | \text{length}(g) = N\}$. If

$$\lim_{N\to\infty} S(N)/S(N-1) = 1 \qquad \text{and} \qquad \lim_{N\to\infty} \sum_{k=1}^{N} S(k)/S(N) = \infty,$$

then for $p > 1$ there is a radial function $F(x) = \varphi(\rho(x))$ such that $L_g F - F \in l^p(G)$ and the cocycle $l : G \to l^p(G)$ defined by $g \mapsto L_g F - F$ is not a coboundary. Note that G is automatically amenable. On the other hand, if $S(N) \sim e^{cN}$, then such radial function does not exist. This follows at once from the Hardy's inequality. To produce classes in $H^1(G, l^p(G))$, one needs to use some more elaborate geometry than just distance function. In the next section we produce such classes for negatively curved groups/manifolds, using the visibility angles.

1.2. Constants of coarse structure for negatively curved groups

1.2.1. Throughout this section we assume that G is a finitely generated, non-amenable group, therefore $B^1(G, l^p(G))$ is closed in $Z^1(G, l^p(G))$ for $p \geq 1$.[4]

Definition 1.6. The number

$$\alpha(G) = \inf_{1\leq p\leq\infty} \{p | H^1(G, l^p(G)) \neq 0\}$$

is called the constant of coarse structure of G.

Remark. The definition makes sense since, by Proposition 1.2, we have

$$H^1(G, l^\infty(G)) \neq 0.$$

We will need a proof of the following well-known fact (see, for example [94]). The argument below is a slightly modified, from the nonpositive curvature to negative curvature, version of a classical argument of [81, 82].

Proposition 1.7. *Let M^n be a complete Riemannian manifold of negative curvature, whose fundamental group G does not have a fixed point at infinity, and satisfying $K(M) \leq -1, \text{Ric}(M) \geq -(n-1)K$. Then $\alpha(G) \leq (n-1)\sqrt{K}$.*

Proof. Let $q_0 \in \widetilde{M}$. Consider the map of G onto an orbit $\mathcal{O}$ of $q_0 : g \mapsto gq_0$; it is equivariant with respect to the left action of G on itself. Let $q \notin \mathcal{O}, s \in \widetilde{M}$ and let $v_q(s)$ be the outward pointing vector from q to s, that is, the unit vector in $T_s\widetilde{M}$, tangent to the geodesic segment connecting q and s. For $x \in G$, consider $F(x) = v_q(xq_0)$. Note that $F(x)$ takes values in $T_{xq_0}\widetilde{M}$. We can consider the restriction of $T\widetilde{M}$ to $\mathcal{O}$ as an equivariant vector bundle over $\mathcal{O}$. Pulling back to G, we obtain an left-equivariant vector bundle over G, equipped with an invariant Euclidean structure. Then F is a section of this bundle. Now consider $(L_gF - F)(x)$. Since the action of G on sections is given by $L_gF(x) = g_*F(g^{-1}x)$, where g_* is the derivative map $(g_* : T_s\widetilde{M} \to T_{gs}\widetilde{M})$, we get

$$(L_gF - F)(x) = g_*F(g^{-1}x) - F(x) = g_*v_q(g^{-1}xq_0) - v_q(xq_0) = v_{gq}(xq_0) - v_q(xq_0).$$

[4] See [49] and [16].

Hence

$$\|(L_gF - F)(x)\| = |2\sin\frac{1}{2}\sphericalangle(gq, xq_0, q)| \le \sphericalangle(gq, xq_0, q).$$

Let $E|G$ be the equivariant Euclidean vector bundle considered above (the pullback of G of $T\widetilde{M}|\mathcal{O}$). Let $L^p(E)$ be the Banach space of L^p-sections of E. We claim that $L_gF - F \in L^p(E)$ for $p > (n-1)\sqrt{K}$. Let $r(x) = \mathrm{dist}_{\tilde{M}}(q_0, xq_0)$. For g, q_0, q fixed we have

$$\sphericalangle(gq, xq_0, q) \le \mathrm{const}_1 \cdot e^{-r(x)}$$

by the standard angle comparison theorem, since $K(M) \le -1$. On the other hand, for fixed $\delta > 0$,

$$\#(x|r - \delta \le r(x) \le r + \delta) \le \mathrm{const}_2 e^{(n-1)\sqrt{K}r}$$

by the Bishop's theorem. Therefore $L_gF - F \in L^p(E)$ for $p > (n-1)\sqrt{K}$. Note we only need that G acts discretely in $\widetilde{M}$.

The map $l : G \to L^p(E)$ defined by $l(g) = L_gF - F$ is obviously a cocycle. If it were trivial, we would have an L^p-section $s \in L^p(E)$, such that $F - s$ is invariant. This means that

$$g_*(v_q(g^{-1}xq_0) - s(g^{-1}x)) = v_q(xq_0) - s(x),$$

or

$$v_{gq}(xq_0) - g_*s(g^{-1}x) = v_q(xq_0) - s(x).$$

Note that since $\|F(x)\| = 1$, $F - s$ is invariant and $\|s(g)\| \to 0$ as $\mathrm{length}(g) \to \infty$, $\|(F - s)(x)\| = 1$ for all x. In particular, $w = v_q(xq_0) - s(x)$ has norm one. Fix x and let g vary. We get

$$\|v_{gq}(xq_0) - w\| = \|g_*s(g^{-1}x)\| \to 0$$

as $\mathrm{length}(g) \to \infty$. Let P_+, P_- be the attractive and repelling fixed points of g on the sphere at infinity of $\widetilde{M}$.

Let w_+, w_- be unit vectors in $T_{xq_0}(\widetilde{M})$, tangent to geodesic rays, connecting xq_0 to P_+, P_-. Then

$$\lim_{|n|\to\infty} \|v_{g^nq}(xq_0) - w_\pm\| \to 0.$$

It follows that $w_\pm = w$. Therefore all elements of G are parabolic and, hence, have a common fixed point at infinity, which is a contradiction. So $H^1(G, l^p(E)) \ne 0$. However, $l^p(E)$ is equivariantly isometric to $l^P(G) \otimes T_{p_0}(\widetilde{M})$. Thus

$$H^1(G, l^p(E)) \simeq H^1(G, l^p(G)) \otimes T_{p_0}(\widetilde{M}).$$

We deduce that $H^1(G, l^p(G)) \ne 0$. □

The estimate of the proposition is sharp. We will see later that if G is a cocompact lattice in $SO^+(1, n)$, i.e., $K(M) = -1$, then $\alpha(G)$ is exactly $n - 1$.

Now, let G be a discrete nonamenable subgroup of $SO^+(1,n)$, or, equivalently, $K(M) = -1$. Recall that the critical exponent $\delta(G)$ is defined by

$$\delta(G) = \inf\{\lambda | \sum_{g\in G} e^{-\lambda r(g)} < \infty\}$$

where $r(g) = \mathrm{dist}_{\widetilde{M}}(p_0, gp_0)$ for some fixed $p_0 \in \widetilde{M}$. By a well-known theorem [90], $\delta(G)$ is equal to the Hausdorff dimension of the conical limit set $\dim(\Lambda_c(G)) \subset S^{n-1}$. Note that if G is geometrically finite and $\Lambda(G) \neq S^{n-1}$, then $\dim \Lambda(G) < n-1$ by [116] and [123]. We now have

Proposition 1.8. *Let G be a discrete subgroup of $SO^+(1,n)$, without a fixed point at infinity. Then $\alpha(G) \leq \delta(G)$.*

Proof. The proposition follows from the proof of Proposition 1.7. Indeed, we only need that $\sum_{g\in G} e^{-pr(g)} < \infty$ to conclude that one has a cocycle $l : G \to l^p(G)$. It has been proven already that this cocycle is not a coboundary. □

Remark. The relation of the constant of coarse structure to the "conformal dimension at infinity" is discussed in [95].

Remark. We refer the reader to [16] for a proof of a version of Proposition 1.8 in the context of group actions on $CAT(-1)$ spaces.

1.3. Function spaces: an overview

For $s \geq 0$, the integer and fractional part of s are denoted $[s]$ and $\{s\}$ respectively. The Sobolev-Slobodečky space $W^s_p(\mathbb{R}^n)$, $(p > 1)$ consists of measurable locally integrable functions f on $\mathbb{R}^n$ such that $D^\alpha f \in L^p(R^n)$ for $|\alpha| \leq [s]$ and

$$\sum_{|\alpha|=[s]} \iint \frac{|D^\alpha f(x) - D^\alpha f(y)|^p}{|x-y|^{n+\{s\}p}}\, dxdy < \infty.$$

The space of Bessel potentials H^s_p consists of functions f for which the Liouville-type operator

$$\mathcal{D}^s f = ((1+|\xi|^2)^{s/2}\hat{f}(\xi))^\wedge$$

satisfies $\mathcal{D}^s f \in L^p$. Warning: $H^s_p \neq W^s_p$ if s is not an integer. For $p = 2$ the condition is equivalent to

$$(1+\triangle)^{s/2}\hat{f} \in L^2(\mathbb{R}^n).$$

Here $f(x) \to \hat{f}(\xi)$ is the Fourier transform and $\triangle = -\sum \frac{\partial^2}{\partial x_i^2}$.

The space of BMO functions $\mathrm{BMO}(\mathbb{R}^n)$ is defined as the space of functions f for which

$$\sup_Q \frac{1}{|Q|} \int_Q |f(x) - f_Q|\, dx < \infty,$$

where Q runs over all cubes in $\mathbb{R}^n$ and

$$f_Q = \frac{1}{|Q|} \int_Q f(x)\, dx,$$

$|Q| = \int_Q 1\, dx$. One has $W_p^{n/p} \subset BMO$ for all $1 < p < \infty$, and, moreover, $H_p^{n/p} \subset H_{p_1}^{n/p_1}$ for $1 < p < p_1 < \infty$ (this follows from Theorem 2.7.1 of [120]). In some sense, BMO is the limit of $H_p^{n/p}$ as $p \to \infty$.

If $f \in W_p^1$, the restrictions of f to the hyperplanes $\{x_n = \epsilon\} \subset R^n$ (where $(x_1, \dots, x_n)$ are Euclidean coordinates) have both L^p and nontangential[5] limits a.e. on $\mathbb{R}^{n-1} = \{x | x_n = 0\}$, and the limit function $f|_{\mathbb{R}^{n-1}}$, called the trace of f, satisfies

$$f|_{\mathbb{R}^{n-1}} \in W_p^{1-1/p}.$$

By the nontangential limit we mean the following. Let $y \in \mathbb{R}^{n-1}$, $\delta > 0$ and let C_δ be a Stolz cone centered at y, that is, the set

$$\{(z, x_n) | x_n \geq \delta \cdot |z - y|\}.$$

Then a function f defined in $\mathbb{R}^n_+ = \{x_n > 0\}$ has the nontangential limit $f(y)$ at y if

$$f(x) \underset{\substack{x \to y \\ x \in C_\delta}}{\longrightarrow} f(y)$$

for all δ. Note that the points in C_δ are within a bounded distance from any geodesic of the hyperbolic metric

$$\frac{\sum_{i=1}^n dx_i^2}{x_n^2},$$

which has y as a point at infinity. The trace theorem mentioned above may be found in [120], Section 2.7.2. Notice that functions in $W_2^1(\mathbb{R}^2)$ have traces in $W_2^{1/2}(\mathbb{R}^1)$.

Now let $\Omega \subset \mathbb{R}^n$ be a bounded domain with a smooth boundary. We define $W_p^s(\Omega)$ as the space of locally integrable functions with $D^\alpha f \in L^p$ for $|\alpha| \leq s$ and such that

$$\sum_{|\alpha| = [s]} \iint \frac{|D^\alpha f(x) - D^\alpha f(y)|^p}{|x - y|^{n + \{s\} p}} < \infty.$$

Equivalently, $W_p^s(\Omega)$ is a space of restrictions of function from $W_p^s(\mathbb{R}^n)$ to Ω. See [120, Chapter 3]. One also defines $H_p^s(\Omega)$ as the space of restrictions of $H_p^s(\mathbb{R}^n)$ on Ω. For a compact smooth manifold M without boundary (in particular, for the boundary $\partial\Omega$) one easily defines the spaces $W_p^s(M)$ and $H_p^s(M)$ [120, Chapter 3] (H_p^s is $F_{p,2}^s$ in Triebel's notations).

If M is compact and g a Riemannian metric on M, let $\triangle_g$ be the corresponding Laplace–Beltrami operator. One can construct the space of Bessel potentials $(1 + \triangle)^{-s/2}(L_p(M))$. It is known [100, Theorem 1, Section 2.3.2.5], [55], that this space coincides with W_p^s (and not H_p^s). Warning: our W_p^s is called $H^{p,s}$ in [100] and in many other sources. In particular, $W_2^s(S^1)$ consists of functions

$$f = \sum_{n \in \mathbb{Z}} a_n e^{in\theta},$$

[5] usually called *conical*

such that $\sum |n|^{2s}|a_n|^2 < \infty$. We will see that $W_2^{1/2}(S^1)$ is especially important in topology.

If $f \in W_p^1(\Omega)$ then f has an L^p nontangential limit a.e. on $\partial\Omega$ and $f|_{\partial\Omega} \in W_p^{1-1/p}(\partial\Omega)$. In particular, for the unit disc $D \subset \mathbb{R}^2$, and a function $f \in W_2^1(D)$, we have $f|_{S^1} \in W_2^{1/2}(S^1)$.

We will need trace theorems for weighted Sobolev-Lorentz spaces [69, 70], [125], [71], [72, 73], [128]. Let Ω be as above and let $\rho(x) = \mathrm{dist}(x, \partial\Omega)$. Consider $L_p^1(\Omega, \rho^\alpha)$ as the space of functions f such that $\int_\Omega |\nabla f|^p \cdot \rho^\alpha \, dx < \infty$. Then f has nontangential limits a.e. on $\partial\Omega$ and

1) $f|_{\partial\Omega} = 0$ if $\alpha \le -1$,
2) $f|_{\partial\Omega} \in W_p^{\frac{p-1-\alpha}{p}}(\partial\Omega), \quad \alpha > -1.$

Moreover,

$$\|f\|_{W_p^{\frac{p-1-\alpha}{p}}} \le \mathrm{const} \cdot \int_\Omega |\nabla f|^p \rho^\alpha \, dx$$

and for any $f \in W_p^{\frac{p-1-\alpha}{p}}(\partial\Omega)$ and harmonic h, $h|_{\partial\Omega} = f$, one has

$$\int_\Omega |\nabla h|^p \rho^\alpha \, dx \le \mathrm{const} \cdot \|f\|_{W_p^{\frac{p-1-\alpha}{p}}}.$$

1.4. l^p-cohomology of cocompact real hyperbolic lattices

The following result is an immediate corollary of the Poincaré inequality in the hyperbolic space, which is equivalent to the Hardy's inequality, and the classical results on traces of functions in weighted Sobolev spaces, reviewed in the previous section. It first appeared in print, with a different proof, in [94]. We include a proof here, as many parts of the proof will be used later on.

Theorem 1.9. *Part 1. Let $G \subset SO^+(1,n)$ be a cocompact (uniform) lattice. Then there is a G-equivariant isomorphism of Banach spaces*

$$H^1(G, l^p(G)) \simeq W_p^{\frac{n-1}{p}}(S^{n-1})/\mathrm{const}$$

for $p > n-1$. For $1 < p \le n-1$, $H^1(G, l^p(G)) = 0$.

Part 2. Let G be a cocompact lattice in $SO^+(1,n)$ and let $l \in H^1(G, l^p(G))$, let $\mathcal{F} : G \to \mathbb{R}$ be a primitive function for l (unique up to a constant). Let $\partial G \approx S^{n-1}$ be the boundary of G as a word-hyperbolic group. Then for almost all points $x \in \partial G$, $\mathcal{F}(g)$ has nontangential limits as $g \to x$, and the limit function

$$\mathcal{F}|_{S^{n-1}} \in W_p^{\frac{n-1}{p}}(S^{n-1}).$$

Corollary 1.10. *The constant of coarse structure $\alpha(G)$ equals $n-1$.*

Remarks.

1) Theorem 1.9 is the first step in the program of linearization of 3-dimensional topology, which we will develop below in this chapter. The crucial fact is that $W_2^{1/2}(S^1)$ admits a natural action of the extended mapping class group $\mathrm{Map}_{g,1}$. This will be proved in Section 1.7 below.
2) Let $\mathbb{H}^n$ be the hyperbolic n-space. Since $G = \pi_1(\mathbb{H}^n/G)$, by the work of [44] we know that $H^1(G, l^p(G))$ is isomorphic to the L^p-cohomology of $\mathbb{H}^n$. So Theorem 1.9 computes the L^p-cohomology of the hyperbolic space.

    Recall that any class l in $H^1(G, l^p(G))$ has a primitive function $\mathcal{F} : G \to \mathbb{R}$ defined up to a constant, such that $l(g) = L_g F - F$. This follows from the fact that a module of all functions $\mathbb{R}^G$ is coinduced from the trivial subgroup and therefore cohomologically trivial [17].

Corollary 1.11. *If $1 < p < p_1 < \infty$, then a natural map $H^1(G, l^p(G)) \to H^1(G, l^{p_1}(G))$ is injective. In fact, for $n-1 < p < p_1 < \infty$ one has the commutative diagram*

$$\begin{array}{ccc} H^1(G, l^p(G)) & \xrightarrow{\sim} & W_p^{\frac{n-1}{p}}(S^{n-1})/\mathrm{const} \\ \downarrow & & \downarrow \\ H^1(G, l^{p_1}(G)) & \xrightarrow{\sim} & W_{p_1}^{\frac{n-1}{p_1}}(S^{n-1})/\mathrm{const} \end{array}$$

where the right vertical arrows exists by an embedding theorem of Sobolev-Slobodečki space [120, 2.7.1].

Proof of Corollary 1.11. The commutative diagram is implied by the proof of Theorem 1.9. Injectivity follows immediately. □

Proof of Theorem 1.9. Though a shorter proof of part 1 of the theorem can be given using [44], in order to prove part 2 we need to make an isomorphism $H^1(G, l^p(G)) \simeq L^p H^1(\mathbb{H}^n)$ explicit. Here $L^p H^1(V)$ is the L^p-cohomology of a complete Riemannian manifold V.

Let l be a cocycle in $Z^1(G, l^p(G))$. We have then an affine isometric action $g \overset{\pi}{\mapsto} (v \mapsto L_g v + l(g))$ of G on $l^p(G)$. Associate with this action a smooth locally trivial affine Banach bundle over

$$E = [\tilde{M} \times l^p(G)]/\text{diagonal action}$$

over the manifold $M = \mathbb{H}^n/G$. By local triviality, smooth partition of unity and affine structure on fibers one constructs a smooth section s of this bundle. It can be interpreted as an equivalent smooth map $s : \widetilde{M} \to l^p(G)$, that is, $s(g^{-1}x) = L_g s(x) + l(g)$. We note that, since M is compact, there is a constant $C > 0$ such that $\|\nabla s(x)\| < C$ for all

$$x \in \widetilde{M} \simeq \mathbb{H}^n, \quad \nabla s \in T_x^* \widetilde{M} \otimes l^p(G).$$

Now let $\mathcal{F} \in \mathbb{R}^G$ be a primitive for l, i.e., $l(g) = L_g \mathcal{F} - \mathcal{F}$. Set $\sigma(x) = s(x) + \mathcal{F}$. This is a function

$$\sigma : \widetilde{M} \to \mathbb{R}^G$$

with the same derivative as s in the sense that for all $g \in G$, $\nabla \sigma_g = \nabla s_g$, where σ_g, s_g means the gth coordinate. Next, we claim that σ is invariant, i.e., $\sigma(g^{-1}x) = L_g \sigma(x)$. In fact, $l(g) = L_g \mathcal{F} - \mathcal{F}$, hence

$$s(g^{-1}x) = L_g s(x) + L_g \mathcal{F} - \mathcal{F},$$

and $\sigma(g^{-1}x) = L_g \sigma(x)$. Therefore, for $x \in \tilde{M}$ and $g, h \in G$ we have

$$\sigma(g^{-1}x)(h) = \sigma(x)(g^{-1}h).$$

Let $f(x) = \sigma(x)(1)$, then $\sigma(x)(g) = f(gx)$. Since $\nabla \sigma(x) = \nabla s(x) \in l^p$ and is bounded in norm, we have for all $x \in \widetilde{M}$ that

$$\sum_{g \in G} |\nabla f(gx)|^p < C.$$

In particular,

$$\int_{\widetilde{M}} |\nabla f|^p = \int_{\widetilde{M}/G} \sum_{g \in G} |\nabla f(gx)|^p < C \cdot \mathrm{Vol}(M).$$

In other words, $|\nabla f| \in L^p(\mathbb{H}^n)$. Now, we can use the Poincaré disk model for the hyperbolic space, that is, the unit ball $B^n \subset \mathbb{R}^n$ with the hyperbolic metric[6]

$$g_h = \frac{g_e}{(1-r^2)^2}.$$

Let μ_e, μ_h denote the Euclidean and the hyperbolic measures respectively, $|\nabla f|_e, |\nabla f|_h$ denote the norms of the gradient of a function in the Euclidean and hyperbolic metrics respectively, $\rho(z) = 1 - r(z)$ denote the Euclidean distance to the boundary $\partial B^n \approx S^{n-1}$. Then

$$\mathrm{const}_2 \cdot \rho^{p-n} \cdot |\nabla f|_e^p \cdot \mu_e \le |\nabla f|_h^p \cdot \mu_h \le \mathrm{const}_1 \cdot \rho^{p-n} |\nabla f|_e^p \mu_e,$$

so we have $\int_{B^n} \rho^{p-n} |\nabla f|_e^p \mu_e < \infty$.

By the theorem of Kudryavcev–Vasharin–Lizorkin–Uspenski–Lions mentioned above, we find that $f|_{(1-\epsilon)S^{n-1}}$ has an L^p-limit $f|_{S^{n-1}}$ to which it converges nontangentially a.e., and, moreover,

$$f|_{S^{n-1}} \in W_p^{\frac{n-1}{p}}(S^{n-1})$$

for $p > n-1$ and $f|_{S^{n-1}} = \mathrm{const}$ for $p \le n-1$. We claim that the map $l \mapsto f|_{S^{n-1}}$ is a well-defined bounded linear operator

$$H^1(G, l^p(G)) \to W_p^{\frac{n-1}{p}}, \quad p > n-1.$$

First, we observe that since

$$s : \widetilde{M} \to l^p(G), \quad \sigma(x) = s(x) + \mathcal{F}$$

and $\sigma(x)(g) = f(gx)$, we have for almost all $x \in \widetilde{M}$, $f(gx) - \mathcal{F}(g) \in l^p(G)$ (as a function of g). In particular, $f(gx) - \mathcal{F}(g) \to 0$ as $\mathrm{length}(g) \to \infty$. This proves

[6]Here g_e is the flat metric on $\mathbb{R}^n$, $r = |x|$.

that, identifying G with an orbit of x, the function $\mathcal{F}(g)$ has a nontangential limit a.e. on the boundary $\partial G \approx S^{n-1}$ and

$$\mathcal{F}|_{S^{n-1}} = f|_{S^{n-1}} \in W_p^{\frac{n-1}{p}}.$$

In particular, $f|_{S^{n-1}}$ (mod constants) does not depend on the choice of a section s. Since changing l by a coboundary leads to an isomorphic affine $l^p(G)$-bundle, $f|_{S^{n-1}}$ (mod constants) depends only on the class $[l] \in H^1(G, l^p(G))$. So we get a well-defined operator

$$H^1(G, l^p(G)) \to W_p^{\frac{n-1}{p}}/\text{const}.$$

We claim that this operator is bounded. The affine flat bundle E was defined as

$$\widetilde{M} \underset{G}{\times} l^p(G),$$

where G acts on $l^p(G)$ by $v \mapsto L_g v + l(g)$. It is enough to show, that there is a constant C, depending only on G but not on l, such that E possesses a Lipschitz section s with $\|\nabla s\| < C \cdot \|l\|$, where $\|l\| = \sup_i \|l(g_i)\|$ for a choice of generators $g_i, i = 1, \ldots, m$. We note that l effectively controls the monodromy of the flat connection in E. The construction of s mentioned above, that is, the choice of an open covering $\cup U_\alpha = M$, flat sections s_α over U_α, a partition of unity $\sum f_\alpha = 1$ with $supp f_\alpha \subset U_\alpha$, so that $s = \sum f_\alpha s_\alpha$, gives a bound of $|\nabla s|$ in terms of monodromy, as desired.

We note that by [44],

$$H^1(G, l^p(G)) = L^p H^1(\mathbb{H}^n),$$

so with any class in $H^1(G, l^p(G))$ we have associated a function f such that df is in L^p, or, equivalently, $\int_{\mathbb{H}^n} |\nabla f|_h^p \mu_h < \infty$. What we in fact did above was an explicit construction of this correspondence between l^p- and L^p-cohomology.

So far we constructed a bounded operator

$$H^1(G, l^p(G)) \to W_p^{\frac{n-1}{p}}(S^{n-1}), \quad p > n-1.$$

We wish to show that this operator is in fact an isomorphism of Banach spaces. To this end, we will need the Poincaré inequality in hyperbolic space.

Proposition 1.12. *(Poincaré inequality in $\mathbb{H}^n$). Let f be a locally integrable measurable function with $\int_{\mathbb{H}^n} |\nabla f|_h^p \, d\mu_h < \infty$. Then:*

1) *If $p \leq n-1$, then $\int_{\mathbb{H}^n} |f - c|^p \, d\mu_h < \infty$ for some constant c.*
2) *If $p > n-1$ and $f|_{S^{n-1}}$ as an element of*

$$W_p^{\frac{n-1}{p}}(S^{n-1})$$

is zero, then

$$\int_{\mathbb{H}^n} |f|^p \, d\mu_h < \infty.$$

Proof. This is a special case of a general theorem contained in [115]. □

We now claim that $H^1(G, l^p(G)) = 0$ for $p \leq n-1$. This in fact follows immediately from $H^1(G, l^p(G)) = L^pH^1(\mathbb{H}^n)$ [44] and Proposition 1.12. Now, if $p > n-1$, then we claim that the operator

$$H^1(G, l^p(G)) \to W_p^{\frac{n-1}{p}}(S^{n-1})/\text{const}$$

constructed above is injective. In fact, if $f|_{S^{n-1}} = 0$, then by Proposition 1.12, $f \in L^p(\mathbb{H}^n, \mu_h)$, and therefore,

$$\int_M \Sigma_g |f(gx)|^p \, d\mu_h < \infty.$$

Hence, for almost all $x \in \tilde{M}$, $\sum_g |f(gx)|^p < \infty$. But $f(gx) - \mathcal{F}(g) \in l^p(G)$, so $\mathcal{F} \in l^p(G)$ and $[l] = 0$. Now, given

$$h \in W_p^{\frac{n-1}{p}}(S^{n-1})$$

we let H denote its harmonic extension to B^n. Then [128], [73],

$$\int_{B^n} \rho^{p-n} |\nabla H|^p \, d\mu_e < \|h\|_{W_p^{\frac{n-1}{p}}(S^{n-1})},$$

so dH is an L^p 1-form on $\mathbb{H}^n$. This shows that the injective operator

$$H^1(G, l^p(G)) = L^pH^1(\mathbb{H}^n) \to W_p^{\frac{n-1}{p}}(S^{n-1})/\text{const}$$

has a bounded right inverse, and, hence, it is an isomorphism by the Banach theorem. This proves Theorem 1.9. □

Corollary 1.13. *Let G be a cocompact lattice in $SO^+(1,n)$ and let $\mathcal{F} : G \to \mathbb{R}$ be such that $L_g\mathcal{F} - \mathcal{F} \in l^p(G)$, for all $g \in G$ $(p > n-1)$. Then the limit function $u = \mathcal{F}|_{S^{n-1}}$ belongs to $L^q(S^{n-1})$ for all $q > 1$. In fact,*

$$\sup_{1<q<\infty} \left(\frac{n-1}{q}\right)^{1/q} \|u\|_{L^q(S^{n-1})} < \infty$$

Moreover, u is in the linear hull of all functions f satisfying

$$\int_{S^{n-1}} \exp(|f|^p) < \infty.$$

Proof. This is an immediate corollary of Theorem 1.9 and the properties of the Orlicz space $L_\infty(\log L)_{-a}$ and the fact that

$$W_p^{(n-1)/p}(S^{n-1}) \subset L_\infty(\log L)_{-a}(S^{n-1})$$

for $a \geq 1/p$, see [30]. □

We will use this corollary in a sequel to this paper [110] in analyzing the local behavior of the Cannon-Thurston Peano curves, corresponding to fibers of the hyperbolic 3-manifolds, fibered over the circle.

1.5. Growth of primitives for l^p-cocycles on the surface group

Theorem 1.14. *Let G be a cocompact lattice in $SO(2,1)$ and let $\mathcal{F} : G \to \mathbb{R}$ be such that for all $g \in G$, $L_g\mathcal{F} - \mathcal{F} \in l^p(G)$, $p > 1$. Then for any word metric on G,*

$$|\mathcal{F}(g)| \leq \text{const} \cdot [\text{length}(g)]^{1/p'}$$

for all $p' > p$.

Proof. The result follows from Theorem 1.9 and the following lemma. □

Lemma 1.15. *Let u be a harmonic function in the unit disc such that $u|_{S^1} \in W_p^{1/p}(S^1)$. Then*

$$|u(z)| \leq \text{const} \cdot [\log(1 - |z|)]^{1/p}.$$

Proof. Here we only treat the case $p = 2$. The full proof will be given in Section 1.11. Let $u(e^{i\theta}) = \sum_{n\in\mathbb{Z}} a_n e^{in\theta}$. Since $(1 + \triangle)^{1/4}u \in L^2$, we have $\{|n|^{1/2}a_n\} \in l^2(\mathbb{Z})$, therefore for $|z| < 1$ $\quad (b_n = |a_n| + |a_{-n}|)$, we obtain

$$\begin{aligned} u(z) - a_0 &\leq \textstyle\sum_{n>0}(|a_n| + |a_{-n}|)|z|^n \\ &= \textstyle\sum |n|^{1/2} b_n \cdot \frac{1}{|n|^{\frac{1}{2}}}|z|^n \\ &\leq (\textstyle\sum |n| b_n^2)^{1/2} \cdot (\textstyle\sum \frac{1}{|n|}|z|^{2n})^{1/2} \\ &\leq \text{const} \cdot [\log(1 - |z|)]^{1/2}. \end{aligned}$$

□

1.6. Embedding of negatively curved manifolds and the boundaries of their universal covers

A problem of fundamental importance in topology is the following: Let $M^m \overset{\varphi}{\hookrightarrow} N^n$ be a smooth π_1-injective embedding of manifolds of nonpositive curvature. Let $\widetilde{\varphi} : \widetilde{M} \to \widetilde{N}$ be a lift of $\widetilde{\varphi}$. Is there a limit map

$$S^{m-1} \approx \partial\widetilde{M} \overset{\partial\widetilde{\varphi}}{\to} \partial\widetilde{N} \approx S^{n-1}$$

and if there is, how smooth it is? For instance, let N^3 be a compact hyperbolic 3-manifold, and M^2 be an incompressible embedded surface in N^3. Then there always exists a limit continuous map $S^1 \overset{\partial\widetilde{\varphi}}{\to} S^2$. Moreover, if M is not a virtual fiber of a fibration over the circle, then $\partial\widetilde{\varphi}(S^1)$ is a quasifuchsian Jordan curve. If M is a virtual fiber, then $\partial\widetilde{\varphi} : S^1 \to S^2$ is a Peano curve in a sense that its image fills S^2 [21]. This deep dichotomy follows from the result of [12].

Remark. In fact, whenever H is a finitely-generated subgroup of $G = \pi_1(N^3)$, then H is Gromov-hyperbolic and there exists a continuous equivariant map

$$f : \partial H \to \partial G.$$

The map f is an embedding unless N^3 is finitely covered by surface bundle over the circle and H is commensurable to the fundamental group of the fiber. This result is the combination of Thurston's covering theorem, solution of the Tameness Conjecture and [21].

Conjecturally, the continuous map f exists for every discrete embedding $H \hookrightarrow PSL(2,\mathbb{C})$. The proof of this conjecture was recently (2007) announced by M. Mitra.

We have the following very general theorem (the embedding condition is superfluous but makes the proof more transparent):

Theorem 1.16. *Let $M^m \overset{\varphi}{\hookrightarrow} N^n$ be a smooth π_1-injective embedding of complete Riemannian manifolds, of pinched negative curvature. Suppose M is compact. Let $\widetilde{\varphi} : \widetilde{M} \to \widetilde{N}$ be a lift of $\widetilde{\varphi}$. Let $p_0 \in \widetilde{N}$ and $\pi : \widetilde{N}\backslash\{p_0\} \to S^{n-1}(T_{p_0}\widetilde{N})$ be the radial geodesic projection of $\tilde{N}\backslash\{p_0\}$ onto the unit tangent sphere. Identify $T_{p_0}\widetilde{N}$ with $\mathbb{R}^n$. Let $q_0 \in \widetilde{M}$. Then:*

1) *For almost all unit tangent vectors $v \in T_{q_0}(\widetilde{M})$, the restriction of $\pi\widetilde{\varphi}$ to the geodesic $\gamma(q_0, v)$ starting at q_0 and having the tangent vector v, has an L^1-derivative as a map $\widetilde{\varphi}|_{\gamma(q_0,v)} : \mathbb{R}_+ \to \mathbb{R}^n$.*
2) *For almost all v there exists a limit $\lim_{t\to\infty} \pi\widetilde{\varphi}[\gamma(q_0, v)(t)]$.*
3) *The resulting measurable map $\partial\widetilde{M} \approx S^{m-1} \overset{\partial\widetilde{\varphi}}{\to} S^{n-1} \approx \partial\widetilde{N}$ does not depend on the choice of p_0, q_0.*
4) *If both M, N are (real) hyperbolic, then for any $p > n - 1$, $\partial\widetilde{\varphi}$ induces a bounded linear operator*
$$\partial\widetilde{\varphi}_* : W_p^{\frac{n-1}{p}}(S^{n-1}) \to W_p^{\frac{m-1}{p}}(S^{m-1}).$$
5) *If M is hyperbolic and $-K \le K(N) \le -1$, then for $p > (n-1)\sqrt{K}$, $\partial\widetilde{\varphi}$ induces a bounded linear operator*
$$\partial\widetilde{\varphi}_* : C^\infty(S^{n-1}) \to W_p^{\frac{m-1}{p}}(S^{m-1})$$
for $p > (n-1)\sqrt{K}$.

Theorem 1.17. *Let N^3 be a compact oriented hyperbolic three-manifold, let $M^2 \overset{\varphi}{\hookrightarrow} N^3$ be an incompressible immersed surface, and let x_1, x_2, x_3 be Euclidean coordinates on $S^2 \approx \partial N^3$. Then:*

1) *If $\partial\widetilde{\varphi}$ is quasifuchsian, then $x_i \circ \partial\widetilde{\varphi} : S^1 \to \mathbb{R}$ are in $W_p^{1/p}$ for $p \ge 2$.*
2) *If M^2 is a virtual fiber, then $x_i \circ \partial\widetilde{\varphi} : S^1 \to \mathbb{R}$ are in $W_p^{1/p}$ for $p > 2$ (but probably not in $W_2^{1/2}$).*

Proof of Theorem 1.16. We will assume $-k \le K(M) \le -1, -K \le K(N) \le -1$. For $x \in \widetilde{N}$ let $r(x) = \rho(p_0, x)$.

Lemma 1.18. *For $r_0 > 0$ and $r(x) > r_0$, $|\nabla\pi(x)| \le \mathrm{const}(r_0)e^{-r(x)}$, where we view π as a map $N\backslash\{p_0\} \to \mathbb{R}^n$.*

Proof. The proof is an immediate application of the comparison theorem, mentioned above in the proof of Proposition 1.7. □

Lemma 1.19.
$$\int_{\widetilde{N}\setminus B(p_0,r_0)} |\nabla\pi(x)|^p < \infty \ \textit{for}\ p > (n-1)\sqrt{K}.$$

Proof. The proof repeats the argument in the proof of Proposition 1.7. □

Now consider a tubular neighborhood of M in N. There exists an embedding $M \times I \to N$ where $I = [-1, 1]$. Moreover, the restriction of the Riemannian metric g_N on N to $M \times I$ is equivalent to the product metric $g_M + dx^2$. (We say two metrics are equivalent if each one is bounded above by another one times a constant.) It follows that there is an embedding

$$\widetilde{M} \times I \xrightarrow{\Phi} \widetilde{N}$$

such that $g_{\widetilde{N}}|\widetilde{M} \times I$ is equivalent to $g_{\widetilde{M}} + dx^2$. Since φ is π_1-injective, for any $r_0 > 0$ there is $r_1 > 0$ such that if $\rho_M(q_0, z) > r_1$, then $\rho_N(p_0, \Phi(z,t)) > r_0$ for $t \in [-1, 1]$. It follows that

$$\iint_{\widetilde{M}\setminus B(q_0,r_1)\times I} |\nabla\pi \circ \Phi|^p \ dVol(\widetilde{M})dt < \infty$$

Therefore for almost all $t_0 \in I$,

$$\int_{\widetilde{M}\setminus B(q_0,r_1)} |\nabla(\pi \circ \Phi(z,t_0))|^p \ dVol(\widetilde{M}) < \infty$$

Fix such t_0 and let $f = \pi \circ \Phi(z, t_0) : \widetilde{M}\setminus B(q_0, r_1) \to \mathbb{R}^n$. We know that

$$\int_{\widetilde{M}\setminus B(q_0,r_1)} |\nabla f|^p \ dVol(\widetilde{M}) < \infty.$$

Expressing the integral in polar coordinates and taking into account that $K(M) \leq -1$ we have

$$\int_{S^{m-1}(T_{q_0}\widetilde{M})} dv \int_{r_1}^{\infty} e^{(m-1)t} |\nabla f|^p \ dt < \infty.$$

In particular, for almost all $v \in S^{m-1}(T_{q_0}\widetilde{M})$,

$$\int_{r_1}^{\infty} e^{(m-1)t} |\frac{\partial f}{\partial t}|^p \ dt < \infty.$$

In other words, for such v,

$$|\frac{\partial f}{\partial t}| \cdot e^{\frac{(m-1)}{p}t} \in L^p[r_1, \infty],$$

therefore

$$|\frac{\partial f}{\partial t}| \in L^1[r_1, \infty],$$

since

$$e^{-\frac{m-1}{p}t} \in L^{p'}[r_1, \infty].$$

This proves 1). The statements 2) and 3) follow directly.

Now suppose $K(M) = K(N) = -1$. Let $u \in W_p^{\frac{n-1}{p}}(S^{n-1})$, $p > n-1$. Then the harmonic extension g of u satisfies

$$\int_{\widetilde{N}} |\nabla g|^p < \infty.$$

as we know from [72], [128] and the proof of Theorem 1.14. By the argument above, there is a $t_0 \in I$, such that the composite function $g \circ \Phi(z, t_0)$ satisfies

$$\int_{\widetilde{M}} |\nabla (g \circ \Phi(z, t_0)|^p < \infty$$

But then the trace $g \circ \Phi(z, t_0)|_{\partial \widetilde{M}}$ lies in $W_p^{\frac{m-1}{p}}(S^{m-1})$. This proves part 4) of Theorem 1.16. The proof of part 5) is identical. Theorems 1.16 and 1.17 (part 2) are now proved. To prove part 1 of Theorem 1.17, we observe that the restriction of any function $u \in W_2^1(S^2)$ to a quasicircle belongs to the class $W_2^{\frac{1}{2}}(S^1)$. This follows immediately from the invariance of $W_2^1(S^2)$ under quasiconformal homeomorphisms, and the fact that functions from $W_2^1(B^2)$ have traces in $W_2^{\frac{1}{2}}(S^1)$. (Note that the Dirichlet energy of a function of two variables is an invariant of the conformal class of a metric.) □

As the reader has, probably, noticed, we could assume that $\pi_1(M) = \pi_1(N)$, so that $\pi_1(M)$ acts discretely in $\widetilde{N}$ and $N = \widetilde{N}/\pi_1(M)$. On the other hand, the proof does not use the fact that M is embedded, so Theorem 1.16 stays true for (finite-to-one) immersions in N.

We will outline now, having in mind the applications in the sequel to this paper, how to study the limit maps from the point of view of ergodic theory. The results thus obtained are weaker then those proved above, but apply to non-discrete representations. Our treatment can be seen as a development of the vague remark of [121, 6.4.4]. Let M^m be a compact hyperbolic manifold, $\widetilde{N} = \mathbb{H}^n$ and $\rho : \pi_1(M) \to \mathrm{Iso}(\widetilde{N})$ a discrete faithful representation. Let $N = \widetilde{N}/\rho(\pi_1(M))$. We would like to study a boundary map $\partial\widetilde{\varphi} : \widetilde{M} \to \widetilde{N}$ where φ is a smooth map $M \to N$, inducing ρ.

Lemma 1.20. *There exists a $\pi_1(M)$-equivariant measurable map ψ from $\partial \widetilde{M} = S^{m-1}$ to the space of probability measures on $\partial \widetilde{N} = S^{n-1}$.*

Proof. For any compact Riemannian manifold M, any compact metric space X and any representation $\rho : \pi_1(M) \to \mathrm{Homeo}(X)$, there is a $\pi_1(M)$-equivariant harmonic function from $\widetilde{M}$ to the affine space of charges on X, taking values in probability measures. This simple fact in various degrees of generality was proved in [35], [38], [63]. If M is hyperbolic, then the Poisson boundary of $\widetilde{M}$ is $\partial \widetilde{M}$, and the result follows. □

Now let $\psi_0 + \psi_c$ be the decomposition of ψ into atomic and non-atomic parts. Obviously, ψ_c is also $\pi_1(M)$-equivariant. We claim that $\psi_c = 0$. First, $\int \psi_c$ is a $\pi_1(M)$-invariant function on $\partial\widetilde{M} = S^{n-1}$, whence is constant, since $\pi_1(M)$ acts on S^{n-1} ergodically. Thus, if $\psi_c \neq 0$ we may assume ψ_c is a probability measure. Second, let G be the center of gravity map from the nonatomic measures on $\partial\widetilde{N}$ to N [36]. Then $G \circ \phi_c$ is a $\pi_1(M)$-equivariant map from $\partial\widetilde{M}$ to N. In particular, $\rho(G \circ \psi_c(x), G \circ \psi_c(y))$ is a $\pi_1(M)$-invariant function on $\partial\widetilde{M} \times \partial\widetilde{M}$ whence a constant by [54] and [118]. It follows easily that $G \circ \psi_c = \text{const}$ which is impossible since ρ is discrete. So $\psi_c = 0$.

We deduce that ψ is atomic, $\psi(z) = \sum_{i=1}^{\infty} m_i \delta(\psi_i(z)), m_1 \geq m_2 \geq \cdots$. Though $\psi_i(z) : \partial\widetilde{M} \to \partial\widetilde{N}$ are not uniquely defined, $m_i : \partial\widetilde{M} \to \mathbb{R}$ are. It follows that m_i are $\pi_1(M)$-invariant, whence constant. Since $\sum m_i = 1$, there is some i such that $m_{i+1} < m_1$. Choose the first such i. Then $m_1 = \cdots = m_i$ and we get a measurable equivariant map

$$\partial\widetilde{M} = S^{m-1} \to \underbrace{S^{n-1} \times \cdots \times S^{n-1}}_{i} / S_i,$$

where S_i is the symmetric group in i letters.

So far we did not use the fact that ρ is discrete, but only that $\rho(\pi_1(M))$ does not have fixed points in $\widetilde{N} = \mathbb{H}^m$. So:

Proposition 1.21. *Let M^m be a compact hyperbolic manifold and let $\rho : \pi_1(M) \to SO^+(1,n)$ be such that $\rho(\pi_1(M))$ does not have fixed points in $\mathbb{H}^n$. Then there exists a $\pi_1(M)$-equivariant measurable map*

$$S^{m-1} = \partial\widetilde{M} \xrightarrow{\psi} (\textit{subsets of cardinality } i \textit{ of } S^{n-1} = \partial\mathbb{H}^n)$$

for some $i \geq 1$.

Remark. This proposition is a very special case of a general existence theorem for equivariant measurable maps between boundaries of symmetric spaces, see, for instance, to [134].

Using cross-ratios and ergodicity of the action of $\pi_1(M)$ on $\partial\widetilde{M} \times \partial\widetilde{M}$, one can easily show that $i = 1$. Now to any $x \in \widetilde{M}$ one associates the Poisson measure μ_x on S^{m-1}. Its pushforward $\psi_* \mu_x$ is a probability measure on S^{n-1}. The pushforward of a measure by a multivalued map is defined by

$$\int_{S^{n-1}} f \, d[\psi_* \mu] = \int_{S^{m-1}} \sum_{y \in \psi(x)} f(y) \, d\mu,$$

where $f \in C(S^{n-1})$.

Now under some natural conditions $\psi_* \mu_x$ does not have atoms and using the barycenter map G in $\mathbb{H}^n$ one can define $s(x) = G(\psi_* \mu_x)$. This can easily be shown to be continuous equivariant map $\widetilde{M} \xrightarrow{s} \mathbb{H}^n$, again under some natural assumption on ρ. The multivalued map ψ should be regarded as a weak radial limit of s, but we will not pursue this point any further.

1.7. The action of quasisymmetric and quasiconformal homeomorphisms on $W_p^{(n-1)/p}$

The well-known result [99] characterizes quasiconformal maps between domains D_1, D_2 in $\mathbb{R}^n, n > 2$ as those which induce an isomorphism of the Banach spaces $BMO(D_1)$ and $BMO(D_2)$. We will see now that this result in the case $D_1 = D_2 = \mathbb{R}^n$ is a limit as $p \to \infty$ of the following result which establishes a quasiconformal invariance of fractional Sobolev spaces $W_p^{n/p}$. Of special importance is the fact that the result holds for $n = 1$ and quasisymmetric homeomorphisms of S^1. The proof of the following lemma is "almost" contained in the remarks made in [94, 95].

Lemma 1.22. *Let $\mathcal{G}_{n-1}, n \geq 2$ be the group of quasisymmetric ($n = 2$) or quasiconformal ($n \geq 3$) homeomorphisms of S^{n-1}. Then for any $p > 1 \quad (n = 2)$ or $p \geq n-1 \quad (n \geq 3)$, $\mathcal{G}_{n-1}$ leaves invariant the Sobolev-Slobodečki space $W_p^{n-1/p}(S^{n-1})$. For any $\Phi \in \mathcal{G}_{n-1}$, the corresponding map*

$$\Phi_* : W_p^{n-1/p}(S^n) \to W_p^{n-1/p}(S^{n-1})$$

is an automorphism of the Banach space $W_p^{n-1/p}(S^{n-1})$.

Theorem 1.23. *There exists for any $n \geq 2$ a (nondegenerate) bounded antisymmetric multi-linear map*

$$\underbrace{W_n^{\frac{n-1}{n}}(S^{n-1})/\mathrm{const} \times \cdots \times W_n^{\frac{n-1}{n}}(S^{n-1})/\mathrm{const}}_{n} \to \mathbb{R}\ ,$$

defined on the smooth functions by $f_1, \ldots, f_n \to \int_{S^n} f_1\, df_2 \cdots df_n$, which is invariant under $\mathcal{G}_{n-1}$.

In particular, we have

Corollary 1.24. *There exists a representation*

$$\mathcal{G}_1 \to Sp(W_2^{1/2}(S^1)/\mathrm{const}),$$

defined by $\Phi(f) = f \circ \Phi^{-1}$.

Proof of Lemma 1.22. We will need a result, proved for $n = 2$ in [2], for $n = 3$ in [22] and for $n \geq 4$ in [124]:

Theorem 1.25. *Let $\phi : S^{n-1} \to S^{n-1}$ be quasisymmetric ($n = 2$) or quasiconformal ($n \geq 3$). Then there exists an extension $\widetilde{\phi}$ of ϕ as a homeomorphism of B^n, which is a quasiisometry of the hyperbolic metric:*

$$\mathrm{const}_2 \cdot g_h \leq \widetilde{\phi}_* g_h \leq \mathrm{const}_1 \cdot g_h.$$

Now let $f \in W_p^{\frac{n-1}{p}}$ $(p > n-1)$. Let u be a harmonic function in B^n, extending f. We know that

$$\int |\nabla u|_h^p \, d\mu_h \leq \text{const}_3 \|f\|_{W_p^{\frac{n-1}{p}}(S^{n-1})}.$$

It follows that

$$\int |\nabla(u \circ \widetilde{\phi})|_h^p \, d\mu_h \leq \text{const}_4 \|f\|_{W_p^{\frac{n-1}{p}}} < \infty,$$

and by the trace theorem,

$$\|u \circ \widetilde{\phi}\|_{W_p^{\frac{n-1}{p}}} \leq \text{const}_5 \|f\|_{W_p^{\frac{n-1}{p}}},$$

which proves the theorem for $p > n-1$. For $p = n-1, n \geq 3$, the result is standard. □

Proof of Theorem 1.23. Let $f_1, \ldots, f_n \in W_n^{\frac{n-1}{n}}(S^{n-1})$. Let u_i be a harmonic extension of f_i. The result follows at once from the formula

$$\int_{S^{n-1}} f_1 \, df_2 \cdots df_n = \int_{B^n} du_1 du_2 \cdots du_n.$$

Since $\int |\nabla u_i|_h^n \, du_h < \infty$, the integral $\int_{B^n} du_1 \cdots du_n$ is finite by the Hölder inequality. The invariance is obvious. □

Proof of Corollary 1.24. The formula

$$\langle f_1, f_2 \rangle = \int_{S^1} f_1 \, df_2$$

gives $W_2^{1/2}/\text{const}$ the structure of a symplectic Hilbert space. This means that the map

$$W_2^{1/2}/\text{const} \to (W_2^{1/2}/\text{const})^*$$

given by $f \mapsto \langle f, \cdot \rangle$ is an isomorphism (not isometry) of Hilbert spaces. By $Sp(W_2^{1/2}/\text{const})$ we mean the group of invertible bounded operators which leave this symplectic form invariant. The result now follows from Lemma 1.22 and Theorem 1.23. □

1.8. Boundary values of quasiconformal maps and regularity of quasisymmetric homeomorphisms

Proposition 1.26. *Let ϕ be a quasiconformal map, defined in a neighborhood of the unit ball B^n. Then $\phi|_{S^{n-1}}$ as a map $S^{n-1} \to \mathbb{R}^n$ belongs to the class $W_n^{\frac{n-1}{n}+\delta}$ for some $\delta > 0$. In particular, if $n = 2$ and*

$$\phi(e^{i\theta}) = \sum_{n \in \mathbb{Z}} a_n e^{in\theta},$$

then

$$\sum |n|^{1+\delta} |a_n|^2 < \infty.$$

If ϕ is only defined in B^n then for almost all $\alpha \in S^{n-1}$ there exists a limit

$$\lim_{r\to 1} \phi(rx) \qquad \text{and} \qquad \phi|_{S^{n-1}} \in W^{\frac{n-1}{n}}.$$

In particular, for $n = 2$ and

$$\phi(e^{i\theta}) = \sum_{n\in\mathbb{Z}} a_n e^{in\theta}, \quad \sum |n||a_n|^2 < \infty.$$

Remark. The last statement for conformal maps is the "Flächensatz".

Proof. Since ϕ as a map $B^n \to \mathbb{R}^n$ belongs to W^1_n, the last statement follows immediately from the trace theorem. To prove the first, recall that ϕ is locally in $W^1_{n+\delta'}, \delta' > 0$ [13], [39]. Therefore

$$\phi|_{S^{n-1}} \in W_n^{\frac{n-1}{n}+\delta},$$

again by the trace theorem. □

Theorem 1.27. *Let $\varphi : S^1 \to S^1$ be a quasisymmetric homeomorphism. Then, as a map $S^1 \to \mathbb{R}^2$, φ belongs to $W_p^{1/p+\delta(p)}, \delta(p) > 0$, for all $p > 1$. If $\varphi(e^{i\theta}) = \sum_{n\in\mathbb{Z}} a_n e^{in\theta}$, then*

$$\sum_{n\in\mathbb{Z}} |n|^{p'/p+\delta} |a_n|^{p'} < \infty$$

for all $1 < p \leq 2$.

Proof. Let $\Phi : D^2 \to D^2$ be a quasiisometry of the hyperbolic plane, extending φ. We know that Φ, Φ^{-1} are Hölder in the Euclidean metric. Let f be a smooth function defined in a neighborhood of D^2. Then for $p > 1$

$$\int_{D^2} |\nabla f|_h^p \rho_e^{-\epsilon}(x, \partial D^2) \cdot \, d\mu_h < \infty$$

for $\epsilon > 0$ small enough (one needs $\epsilon < p - 1$).

Since Φ is a quasiisometry for the hyperbolic metric and bi-Hölder for the Euclidean metric, we have for $g = f \circ \Phi$:

$$\int_{D^2} |\nabla g|_h^p \rho_e^{-\beta}(y, \partial D^2) \, d\mu_h < \infty$$

for some $\beta > 0$. Rewriting in Euclidean terms, we have

$$\int_{D^2} |\nabla g|_e^p \cdot [\rho(y, \partial D^2)]^{p-\beta-2} < \infty,$$

therefore

$$g|_{S^1} \in W_p^{\frac{1}{p}+\delta}$$

by the trace theorem for weighted Sobolev spaces. Letting f be a Euclidean coordinate function, we get

$$\varphi \in W_p^{\frac{1}{p}+\delta}.$$

The last statement follows from the Young-Hausdorff theorem. □

Remark. It was a famous problem in the 1950s if φ is absolutely continuous (that is, in W_1^1). Though the answer is well known to be negative, we see that φ is as close to be absolutely continuous as one wishes. We will use Theorem 1.27 in a sequel to this paper to prove the existence of the vacuum vector for quantized moduli space for $p > 1$. We also notice that the argument above together with the proof of Theorem 1.16 shows the following: if $\varphi : M \to N$ is an π_1-injective immersion of hyperbolic manifolds, M compact, such that for $g \in \pi_1(M)$ and some fixed $z_0 \in \tilde{N}$

$$\rho(z_0, \varphi_*(g)z_0) \geq \text{const} \cdot \text{length}(g),$$

then $\partial\tilde{\varphi}$ is of class $W_p^{(m-1)/p+\delta}$ and therefore is Hölder continuous. This does not apply, however, in the case of the Cannon-Thurston curve. See [110] for the further study.

Remark. The above assumption on φ is equivalent to the assumption that $\tilde{\varphi}$ is a quasi-isometric embedding. It, hence, follows that $\partial\tilde{\varphi}$ is a quasi-symmetric embedding.

1.9. Teichmüller spaces and quantization of the mapping class group

We denote by Map_g the mapping class group of the closed oriented surface Σ^g of genus g and $\text{Map}_{g,1}$ the extended mapping class group, i.e., the mapping class group of the surface Σ^g with one point removed. Let $\Gamma_g = \pi_1(\Sigma^g)$, then $\text{Map}_{g,1}$ is isomorphic to $\text{Aut}(\Gamma_g)$ and one has the exact sequence

$$1 \to \Gamma_g \to \text{Map}_{g,1} \to \text{Map}_g \to 1,$$

see, e.g., [9].

Theorem 1.28. *(Quantization of the moduli space) For any $p > 1$ there exists a (nontrivial) representation*

$$\text{Map}_{g,1} \stackrel{\pi_p}{\to} \text{Aut}(W_p^{1/p}(S^1)/\text{const})$$

given by the formula

$$\pi_p(\varphi)(f) = f \circ \Phi^{-1},$$

where Φ is a quasisymmetric homeomorphism of S^1, induced by φ and a choice of a hyperbolic structure in Σ^g. For $p = 2$ the representation

$$\pi_2 : \text{Map}_{g,1} \to \text{Aut}(W_2^{1/2}(S^1)/\text{const})$$

is symplectic, that is, $\pi_2(\text{Map}_{g,1}) \subset Sp(W_2^{1/2}(S^1)/\text{const})$.

Proof. Fix a hyperbolic structure on $\Sigma = \Sigma^g$. Then, by the classical theorem of Nielsen, one gets an embedding $\text{Map}_{g,1} \to \mathcal{G}_1$. The theorem now follows from Theorem 1.23. $\square$

Now let $G \xrightarrow{\pi_0} PSL_2(\mathbb{R})$ be a Fuchsian group, possibly infinitely generated. We recall that the Teichmüller space $\mathbf{T}(G)$ is defined as follows: Points of $\mathbf{T}(G)$ are equivalence classes of discrete representations

$$G \xrightarrow{\pi} PSL_2(\mathbb{R})$$

which are quasiconformally conjugate to π_0, that is, there is a quasisymmetric homeomorphism Φ of S^1 such that $\pi = \Phi \circ \pi_0 \circ \Phi^{-1}$. Two representations are equivalent if they differ by conjugation by an element of $PSL_2(\mathbb{R})$. Note that this definition is equivalent to the standard one by [27].

Corollary 1.29. *Let π_0, π be two discrete representation of a group G. Then, if π lies in the Teichmüller space of π_0, then the unitary representations*

$$G \overset{\pi_0}{\to} PSL_2(\mathbb{R}) \overset{\beta}{\to} U(W_2^{1/2}(S^1)/\mathrm{const})$$

and

$$G \overset{\pi}{\to} PSL_2(\mathbb{R}) \overset{\beta}{\to} U(W_2^{1/2}(S^1)/\mathrm{const})$$

are unitarily equivalent.

Remarks.

1) The fact that $PSL_2(\mathbb{R})$ acts in $W_2^{1/2}(S^1)/\mathrm{const}$ by unitary operators (with respect to the complex structure given by the Hilbert transform) is well known [87]. In fact, this unitary representation belongs to the discrete series and may be realized in L^2 holomorphic 1-forms in B^2.
2) This result should be contrasted with the rigidity theorem from [10] in the case of representations which are not in the discrete series. In fact, a proof of Corollary 1.29 is contained in the introductory remarks in Section 8 of [10].

Proof. Since $\pi = \Phi \circ \pi_0 \circ \Phi^{-1}$ and $\mathcal{G}_1$ act in $W_2^{1/2}(S^1)/\mathrm{const}$, we get an invertible operator A such that $\beta \circ \pi = A\, \beta \circ \pi_0\, A^{-1}$. By the polar decomposition $A = UP$ where P is positive self-adjoint, U is unitary, P commutes with $\beta \circ \pi_0$ and U intertwines $\beta \circ \pi_0$ and $\beta \circ \pi$, as desired. □

The following special case is very important. Let $\pi_0 : G \to PSL_2(\mathbb{R})$ be a Fuchsian group corresponding to a Riemann surface of finite type (that is, a torsion-free lattice in $PSL_2(\mathbb{R})$). Let $\Sigma = \mathbb{H}^2/G$ and let $\varphi \in \mathrm{Map}(\Sigma, x_0)$, $x_0 \in \Sigma$. Let Φ be a quasisymmetric homeomorphism of S^1 which is the boundary value of a quasiconformal homeomorphism Ψ of (Σ, x_0), representing φ. Then

$$\pi_0 \circ \varphi^{-1} = \Phi \pi_0 \Phi^{-1}.$$

Let $A_\varphi : W_2^{1/2}/\mathrm{const} \to W_2^{1/2}/\mathrm{const}$ be the invertible operator, representing φ. Let $P_\varphi^2 = A_\varphi^* A_\varphi$. Then P_φ commutes with $\beta \circ \pi_0$. We obtained the following

Theorem 1.30. *Let $\pi_0 : G \to PSL_2(\mathbb{R})$ be a torsion-free lattice. Let $\Sigma = \mathbb{H}^2/G, x_0 \in \Sigma, \varphi \in \mathrm{Map}(\Sigma, x_0)$, Ψ a quasiconformal homeomorphism inducing φ, Φ the trace*

of its lift to $\mathbb{H}^2$ *on* S^1, A_φ *the invertible operator in* $W_2^{1/2}(S^1)/\text{const}$ *given by* $A_\varphi(f) = f \circ \Phi^{-1}$. *Then the self-adjoint bounded operator*

$$P_\varphi^2 = A_\varphi^* A_\varphi$$

commutes with $\beta \circ \pi_0$. *Moreover, if* $P_\varphi^2 = \int \lambda \, dE_\lambda$ *is the spectral decomposition, then* E_λ *commute with* $\beta \circ \pi_0$.

Remarks.

1) If G is cocompact, then we know that $W_2^{1/2}/\text{const} \approx H^1(G, l^2(G))$, so $W_2^{1/2}/\text{const}$ is a Hilbert module over the type II factor defined by G of dimension $\dim_G W_2^{1/2}/\text{const} = L^2 b_1(G) = 2g - 2$.
2) In practice, finding A_φ is difficult. The reason is that Φ is not a diffeomorphism, so the explicit formulae of Chapter 2 do not make sense. Moreover, Φ is given in a very implicit way as the boundary value of a quasiconformal map, defined by a quadratic differential on $\mathbb{H}^2$ which is G-invariant!

We will now show that for $p > 2$ the operator A_φ shows very unusual properties, from the point of view of functional analysis.

Theorem 1.31. *Let* G *be the fundamental group of a closed hyperbolic surface* Σ^g. *Let* $\varphi \in \text{Map}_{g,1}$ *be such that its image in* Map_g *is represented by a pseudo-Anosov homeomorphism* Ψ. *Let* $A = A_\varphi$ *be the operator, representing* φ *in* $W_p^{1/p}(S^1)$, $p > 2$. *Then there is an element* $v \in W_p^{1/p}(S^1) \setminus \{0\}$ *such that*

$$\sum_{k \in \mathbb{Z}} \|A^k(v)\|^p < \infty.$$

Proof. Let M be the mapping torus of Ψ, that is, $\mathbb{R} \times \Sigma/\mathbb{Z}$ where $1 \in \mathbb{Z}$ acts by $(t, x) \to (t+1, \Psi(x))$. Then M is hyperbolic [122], [92]. We will view M as a fibration over the circle $\mathbb{R}/\mathbb{Z}$ with coordinate t, $0 \le t < 1$; the fiber over t will be called Σ_t. We can trivialize $M \xrightarrow{\psi} \mathbb{R}/\mathbb{Z}$ over $I = [0, 1/2]$ so that (t, x_0), $0 \le t \le 1/2$ will be a horizontal curve. Let g be the hyperbolic metric on M and g_0 be some hyperbolic metric on Σ, then g and $g_0 + dt^2$ are equivalent on $\Sigma \times [0, 1/2] \simeq \psi^{-1}([0, 1/2]) \subset M$. Lifting to $\widetilde{M} = \mathbb{H}^3$, we get a fibration $\mathbb{H}^3 \xrightarrow{\widetilde{\psi}} \mathbb{R}$ with $\widetilde{\psi}^{-1}(t) = \widetilde{\Sigma}_t$. Let $\gamma \in \pi_1(M) \subset PSL_2(\mathbb{C})$ be the monodromy element, corresponding to φ. Let $f : \mathbb{H}^3 \to \mathbb{R}$ be such that

$$\int_{\mathbb{H}^3} |\nabla f|^p \, d\mu_h < \infty.$$

We then have

$$\sum_{k \in \mathbb{Z}} \int_{\gamma^k(\widetilde{\psi}^{-1}[0,1/2])} |\nabla f|^p \, d\mu_h \le \int_{\mathbb{H}^3} |\nabla f|^p \, d\mu_h < \infty;$$

on the other hand the left-hand side is

$$\sum_{k\in\mathbb{Z}} \int_{\widetilde{\psi}^{-1}[0,1/2]} |\nabla(f\circ\gamma^k)|^p \, d\mu_h \geq \text{const} \cdot \int_0^{\frac{1}{2}} dt \int_{\widetilde{\Sigma}_t} \sum_{k\in\mathbb{Z}} |\nabla(f\circ\gamma^k)|^p \, dVol(g_0).$$

It follows that for some t_0,

$$\int_{\widetilde{\Sigma}_{t_0}} \sum_{k\in\mathbb{Z}} |\nabla(f\circ\gamma^k)|^p \, dVol(g_0) < \infty.$$

Since g_0 is a hyperbolic metric, for any function F on $\widetilde{\Sigma}$

$$\int_{\widetilde{\Sigma}} |\nabla F|^p \, dVol(g_0) = \text{const} \cdot \|F|\partial\widetilde{\Sigma}\|^p_{W_p^{1/p}(S^1)/\text{const}},$$

actually, we may let the LHS be the definition of the norm in $W_p^{1/p}(S^1)/\text{const}$, making the constant equal one. So

$$\sum_{k\in\mathbb{Z}} \|f\circ\gamma^k|\partial\widetilde{\Sigma}_{t_0}\|^p_{W_p^{1/p}(S^1)/\text{const}} < \infty.$$

We will now identify $f\circ\gamma^k|\partial\widetilde{\Sigma}_{t_0}$. We have a boundary map

$$\partial\widetilde{\Sigma}_{t_0} = S^1 \xrightarrow{\alpha} S^2 = \partial\mathbb{H}^3.$$

We know that $\gamma^k\circ\alpha = \alpha\circ\varphi^{-k}$, so $f\circ\gamma^k = A_\varphi^k f$ and finally

$$\sum_{k\in\mathbb{Z}} \|A_\varphi^k(f|\partial\widetilde{\Sigma}_t)\|^p_{W_p^{1/p}(S^1)/\text{const}} < \infty.$$

Now, for any $u \in W_p^{2/p}(S^2)$ we let f be its harmonic extension. In particular, any smooth function u will do. Since $\alpha : S^1 \to S^2$ is continuous and nonconstant, we can take u such that $V = u\circ\alpha$ is nonconstant. Then

$$\sum_{k\in\mathbb{Z}} \|A_\varphi^k v\|^p_{W_p^{1/p}(S^1)/\text{const}} < \infty,$$

as desired. □

We remark that $\sum_{k\in\mathbb{Z}} \|A_\varphi^k v\|^p < \infty$ will hold for all v which are in the image of the bounded operator

$$W_p^{2/p}(S^2) \to W_p^{1/p}(S^1),$$

induced by $\partial\widetilde{\Sigma} \to \partial\mathbb{H}^3$.

Corollary 1.32. *Suppose that the space of fixed vectors of A_φ acting in $W_p^{1/p}/\text{const}$ possesses a complementary invariant subspace W. Then the specter of A_φ in W satisfies*

$$\sigma(A_\varphi|W) \cap S^1 \neq \phi.$$

Proof. Suppose the opposite, then $W = W_+ \oplus W_-$ such that $A_\varphi^k|W_+$ and $A_\varphi^{-k}|W_-$ are strict contractions for some $k > 0$. But then

$$\sum_{k\in\mathbb{Z}} \|A_\varphi^k v\|^p = \infty$$

for all $v \in W_p^{1/p}/\mathrm{const}$. □

We now turn to a generalization. Let $\widetilde{G} \subset \mathrm{Map}_{g,1}$ be a subgroup, which contains $\pi_1(\Sigma_g)$, so that we have an extension

$$1 \to \pi_1(\Sigma_g) \to \widetilde{G} \to G \to 1.$$

Notice that $G \subset \mathrm{Map}_g$. A well-known problem in the hyperbolic topology is: when there exists a fibration

$$\begin{array}{ccc} \Sigma & \to & Q \\ & & \downarrow \\ & & T \end{array}$$

with $\pi_1(Q) = \widetilde{G}$ such that Q is a compact manifold of negative curvature. In case T is a closed surface, Corollary F.3 to Theorem F.1 of [109] provided some necessary condition. This condition is unfortunately void, as we will show now.

Theorem 1.33. *Let $\Sigma^{g_1} \to Q \to \Sigma^{g_2}$ be a surface fibration over a surface (Σ^{g_i} are hyperbolic and oriented). Let Σ be a section of this fibration. Then*

$$|\Sigma \cap \Sigma| \le 2g_2 - 2.$$

Proof. Let ξ be the vertical tangent bundle[7] for Σ, $e(\xi)$ its Euler class, then $\Sigma \cap \Sigma = (e(\xi), [\Sigma])$. We have the natural homomorphism $\pi_1(Q) \to \mathrm{Map}_{g_1,1}$ and the composite homomorphism

$$\pi_1(\Sigma) \to \pi_1(Q) \to \mathrm{Map}_{g,1},$$

which we call φ. The inclusion

$$\mathrm{Map}_{g_1,1} \to \mathcal{G}_1$$

induces the Euler class ϵ in $H^2(\mathrm{Map}_{g,1})$ coming from the action of $\mathcal{G}_1$ on S^1. By [77], $\varphi^{-1}\epsilon = e(\xi)$. Moreover, it is well known (and obvious) that ϵ is a bounded class (see [78]). In fact, for any homomorphism $\pi_1(\Sigma^g) \xrightarrow{\varphi} \mathrm{Homeo}(S^1)$, the Milnor-Wood inequality implies that

$$|(\varphi^*\epsilon, [\Sigma])| \le 2g - 2.$$

This proves the theorem. □

[7] I.e., the restriction to Σ of the 2-plane bundle over Q tangent to the fibers of the fibration.

Remark. If the fibration $Q \to \Sigma^{g_2}$ is holomorphic and the action of $\pi_1(\Sigma^{g_2})$ on $H_1(\Sigma^{g_1}, \mathbb{R})$ is simple, then a famous inequality of Arakelov [5] reads $\Sigma \cap \Sigma < 0$ for all holomorphic sections. By Theorem 1.33,

$$-(2g_2 - 2) < \Sigma \cap \Sigma < 0.$$

We now have the following result, which seems to be a very strong restriction on G.

Theorem 1.34. *Let $1 \to \pi_1(\Sigma^g) \to \widetilde{G} \to G \to 1$ be an extension. Suppose $\widetilde{G}$ is the fundamental group of a compact manifold Q^n $(n = 4)$ of negative curvature*

$$-K \leq K(Q^n) \leq -1.$$

Then for $p > (n-1)\sqrt{K}$ there is a vector

$$\text{const} \neq v \in W_p^{1/p}(S^1),$$

such that

$$\sum_{g \in G} \|A_g v\|^p_{W_p^{1/p}/\text{const}} < \infty. \tag{$*$}$$

Proof. Since the proof is essentially identical to the proof of Theorem 1.31, we will only indicate the differences. Let $q_0 \in \widetilde{Q}$ and let $u : S^{n-1}(T_{q_0}\widetilde{Q}) \to \mathbb{R}$ be a smooth function. Composing with the geodesic projection $\widetilde{Q}\backslash\{0\} \to S^{n-1}(T_{q_0}\widetilde{Q})$ we arrive to a function $f : \widetilde{Q}\backslash B(q_0, r) \to \mathbb{R}$ with $\int_{\widetilde{Q}} |\nabla f|^p \, dVol < \infty$ for $p > (n-1)\sqrt{K}$. Since Σ is embedded in Q, one has a limit map $\partial\widetilde{\Sigma} = S^1 \to S^{n-1} = \widetilde{Q}$ by Theorem 1.16. Let $v = u \circ \alpha$, where we identify $\partial\widetilde{Q}$ and $S^{n-1}(T_{q_0}\widetilde{Q})$. Then $v \in W_p^{1/p}(S^1)$ by Theorem 1.17. As in Theorem 1.31, we have the inequality $(*)$. Finally, if $v = \text{const}$, for any choice of u, then α is almost everywhere a constant map, say to $z \in S^{n-1}$. Since α is equivariant, it follows that $\pi_1(\Sigma_g)$ stabilizes z. This is obviously impossible. □

1.10. Spaces $\mathcal{L}^{(n-1)}_{k,\alpha}$ and cohomology with weights

In this section we will describe a limit form of Theorem 1.9 when $p = 1$, and discuss l^{n-1}-cohomology with weights of cocompact lattices in $SO^+(1,n)$.

Let G be a finitely generated group, $w : G \to \mathbb{R}_+$ a function such that $w(g) \to \infty$ as length$(g) \to \infty$. Consider the space

$$l^p(G, w) = \{f : \sum_g |f(g)|^p w^{-1}(g) < \infty\}.$$

Suppose

$$\frac{L_g w}{w} = O(1), \forall g \in G$$

and the same for $\frac{R_g w}{w}$. Then $l^p(G, w)$ becomes a G–bimodule.

Examples.

1. If $r(g) = \text{length}(g)$, then consider $w(g) = r^\alpha(g),\ \alpha > 0$ or

$$w(g) = r(g)^\alpha \log r(g) \log\log r(g) \cdots \underbrace{\log\log\cdots\log}_{k} r(g),\ \alpha > 0.$$

2. Consider $w(g) = e^{\alpha r(g)}$, $\alpha > 0$.

Now let G be a cocompact lattice in $SO^+(1,n)$, We know by Theorem 1.14, that $H^1(G, l^p(G)) \neq 0$ exactly for $p > n-1$. In particular, $H^1(G, l^{n-1}(G)) = 0$. However, introduction of weights changes the situation.

Theorem 1.35. *Let G be a cocompact lattice in $SO^+(1,n)$, then for any $k \geq 1$ and $\alpha > 0$,*

$$H^1(G, l^p(G, w)) \neq 0$$

for $p = n-1$ and $w = r(g) \log r(g) \cdots (\underbrace{\log\log\cdots\log}_{k} r(g))^\alpha,\ \alpha > 1, k \geq 1$.

Proof. The proof essentially repeats the arguments of Proposition 1.7. Let $u : S^{n-1} \to \mathbb{R}$ be any smooth function and denote again by u its harmonic extension in B^n. We have $|\nabla u|_e <$ const, therefore

$$|\nabla u|_h(z) < \text{const} \cdot \rho_e(z, S^{n-1})^{-1}.$$

Let $\mathcal{F}(h) = u(h^{-1}z_0)$, then a direct computation shows that $L_g\mathcal{F} - \mathcal{F} \in l^{n-1}(G, w)$ and $\mathcal{F} - \text{const} \neq l^{n-1}(G, w)$ so $l(g) = L_g\mathcal{F} - \mathcal{F}$ is a nontrivial cocycle if u is one of the coordinate functions on S^{n-1}, as in Proposition 1.7. □

We would like to compute $H^1(G, l^{n-1}(G, w))$. The construction in Theorem 1.9 produces from any class in $H^1(G, l^p(G, w))$ a function in $L^1_w(\mathbb{H}^n)$. The latter space is defined as the space of locally integrable functions f with distributional derivatives such that

$$\int_{\mathbb{H}^n} |\nabla f|^{n-1} \cdot w^{-1}(z) < \infty \qquad (*)$$

where

$$w(z) = \rho_h(z_0, z) \log \rho_h(z_0, z) \cdots (\underbrace{\log\log\cdots\log}_{k} \rho_h(z_0, z))^\alpha.$$

Definition 1.36. The space $\mathcal{L}^{(n-1)}_{k,\alpha}$ is defined as the Banach space of traces of $L^1_w(\mathbb{H}^n)$ on S^{n-1}. The norm in $\mathcal{L}^{(n-1)}_{k,\alpha}$ is defined as the infimum of integrals $(*)$ taken over the set of all functions f with the given trace.

Remark. The norm just defined depends on z_0. Therefore the natural action of $SO^+(1,n)$ in $\mathcal{L}^{(n-1)}_{k,\alpha}$ is not isometric.

We will describe $\mathcal{L}^1_{k,\alpha}$ as a Zygmund-type space. One can analogously describe $\mathcal{L}^{(n-1)}_{k,\alpha}$ for $n > 2$, of course, but we will not need it.

Theorem 1.37. *$\mathcal{L}^1_{k,\alpha}$ consists of all function $u : S^1 \to \mathbb{R}$ for which, whenever $a > 0$, wc have*

$$\int_0^a dh \int_0^{2\pi} \frac{|u(x+h)-u(x)|}{h^2 \log h \cdots \underbrace{\log \cdots \log^\alpha h}_{k}} < \infty.$$

Proof. The proof is the word-by-word repetition of the Uspenski's argument in [128]. One does not need to use Hardy's inequality, since $p = 1$. □

Theorem 1.38. *$\mathcal{G}_1$ acts on $\mathcal{L}^1_{k,\alpha}$ by*

$$A_\Phi u(x) = u \circ \Phi^{-1}.$$

Corollary 1.39. *If $\Phi : S^1 \to S^1$ is quasisymmetric, then, as a function $S^1 \to \mathbb{R}^2$, ϕ belongs to $\mathcal{L}^1_{k,\alpha}$.*

We suggest the reader to compare this result to [23] and [37].

Proof. Let $\psi : B^2 \to B^2$ be a quasiisometry of the hyperbolic metric, extending Φ. If u satisfies $(*)$ then $u \circ \Phi^{-1}$ satisfies $(*)$ as well, whence the result. □

Embedding $\mathrm{Map}_{g,1} \subset \mathcal{G}_1$ we obtain a representation

$$\mathrm{Map}_{g,1} \to \mathrm{Aut}(\mathcal{L}^1_{k,\alpha}),$$

which is a limiting case of Theorem 1.28.

1.11. Bicohomology and the secondary quantization of the moduli space

We will now introduce a very important notion of bicohomology spaces which, to some extent, linearize 3-dimensional topology.

Definition 1.40. Let G be a finitely generated group. For $p > 1$ define

$$\mathcal{H}_p(G) = H^1(G_r, H^1(G_l, l^p(G)),$$

where r and l stand for the right and left action, respectively.

Proposition 1.41. *A group $Out(G)$ of outer automorphism of G acts naturally in $\mathcal{H}_p(G)$.*

Proof. By definition, $Out(G) = \mathrm{Aut}(G)/(G/Z(G))$. Obviously $\mathrm{Aut}(G)$ acts on $H^1(G_l, l^p(G))$ extending the right action of G, so $\mathrm{Aut}(G)/(G/Z(G))$ will act on $H^1(G_r, (H^1(G_l, l^p(G)))$. □

For a surface group $\pi_1(\Sigma_g)$ we write $\mathcal{H}_{p,g} = \mathcal{H}_p(G)$.

Theorem 1.42. *There exists a natural representation*

$$\mathrm{Map}_g \to \mathrm{Aut}(\mathcal{H}_{p,g}).$$

Moreover, for $p > 1$, $\mathcal{H}_{p,g}$ is a nontrivial Banach space. For $p = 2$, $\mathcal{H}_{p,g}$ is an infinite-dimensional Hilbert space. There is a pairing

$$\mathcal{H}_{p,g} \times \mathcal{H}_{p',g} \to \mathbb{R}\ ,$$

which is Map_g*-invariant. For* $p = p' = 2$ *this pairing is a nondegenerate symmetric bilinear form. Therefore, we obtain a representation*

$$\mathrm{Map}_g \xrightarrow{\psi} O(\infty, m),\ 0 \le m \le \infty,$$

which we call the secondary quantization of the moduli space of Riemann surfaces.

Proof. The proof of this theorem will occupy the rest of this section.

For a compact oriented manifold M let $\Omega^{1/p}$ be the space of measurable $1/p$-powers of densities such that for $\omega \in \Omega^{1/p}$

$$\int_M |\omega|^p < \infty.$$

Then $\Omega^{1/p}$ is a Banach space, and for $p = 2$, a Hilbert space. Let G be a finitely generated group acting in M.

Lemma 1.43. *Suppose that every element* $g \in G \setminus \{1\}$ *has finitely many repelling points, say* $x_1^-, \dots, x_n^-$*, and finitely many attractive points, say* $x_1^+, \dots, x_m^+$*, such that for any set of neighborhoods* U_i^-, U_+^+ *of* $x_i^\pm$*, there is* N *such that for* $k \ge N$*,*

$$g^k(M \backslash \bigcup_i U_i^-) \subset \bigcup_i U_i^+.$$

Suppose there are $g_1, g_2, g_3, g_4 \in G \setminus \{1\}$ *such that the sets*

$$\bigcup_i U_{i,s}^- \cup U_{i,s}^+$$

are disjoint for different $s = 1, 2, 3, 4$*. Then the action of* G *in* $\Omega^{1/p}$ *does not have almost invariant unit vectors.*

Proof. Suppose the opposite, then there is a sequence $\omega_j \in \Omega^{1/p}$, $\|\omega_j\| = 1$ and $\|g_s^k \omega_j - \omega_j\| \underset{j\to\infty}{\longrightarrow} 0$ for all s, k. Choose $k_s, U_{i,s}^\pm$ such that

$$g_s^{k_s}(M \backslash \bigcup_i U_{i,s}^-) \subset \bigcup_i U_{i,s}^+$$

and $\bigcup_i U_{i,s}^-$ (respectively $\bigcup_i U_{i,s}^+$) do not intersect for different s. Let ω be such that $\|\omega\| = 1$ and

$$\|g_s^{k_s}(\omega) - \omega\| < (2/3)^{1/p} - (1/3)^{1/p}.$$

For $E \subset M$ define $C(E, \omega) = \int_E |\omega|^p$. We claim that

$$C(M \backslash (\bigcup U_{s,i}^- \cup \bigcup U_{s,i}^+), \omega) < 2/3.$$

Suppose the opposite, then, by the invariance of the density $|\omega|^p$,

$$\begin{aligned} &C(M \backslash (\bigcup U_{s,i}^- \cup \bigcup U_{s,i}^+), \omega \circ g_s^{k_s}) \\ &= C(g_s^{k_s}(M \backslash (\bigcup U_{s,i}^- \cup \bigcup U_{s,i}^+)), \omega) \\ &\le C(g_s^{k_s}(M \backslash \bigcup U_{s,i}^-), \omega) \le 1/3. \end{aligned}$$

It follows that

$$\begin{aligned}
&[\textstyle\int_{M\setminus(\bigcup U^-_{s,i}\cup\bigcup U^+_{s,i})} |\omega - \omega\circ g_s^{k_s}|^p]^{1/p}\\
&\geq |[\textstyle\int_{M\setminus(\bigcup U^-_{s,i}\cup\bigcup U^+_{s,i})}|\omega^p|]^{1/p} - [\int_{M\setminus(\bigcup U^-_{s,i}\cup\bigcup U^+_{s,i})} |\omega\circ g_s^{k_s}|^p]^{1/p}|\\
&\geq (2/3)^{1/p} - (1/3)^{1/p},
\end{aligned}$$

a contradiction.

Thus $C(\cup U^-_{s,i},\omega) + C(\cup U^+_{s,i},\omega) \geq 1/3$. Since $\cup U^{\pm}_{s,i}$ are disjoint for different s, we get

$$1 \geq \sum_{s=1}^{4} C(\bigcup U^-_{s,i},\omega) + C(\bigcup U^+_{s,i},\omega) \geq 4/3,$$

a contradiction. This proves the lemma. □

Corollary 1.44. *Let $G \subset SO^+(1,n)$ be a cocompact lattice. Then the natural isometric action of G in $W_p^{(n-1)/p}(S^{n-1})$ does not have almost invariant vectors. In particular, $H^1(G, W_p^{(n-1)/p}(S^{n-1}))$ is Banach for $p > (n-1)$.*

Proof. For $u \in W_p^{(n-1)/p}(S^{n-1})/\mathrm{const}$, let f be a harmonic extension of u so that

$$\|u\| = \int_{\mathbb{H}^n} |\nabla f|^p.$$

Since the energy density $|\nabla f|^p d\mu_h$ is invariant under the isometries of $\mathbb{H}^n$, the proof of the Lemma 1.43 applies directly. □

Corollary 1.45. *The space $\mathcal{H}_{p,g}$ is Banach (Hilbert for $p = 2$).*

Proof. $H^1(G_l, l^p(G)) = W_p^{1/p}(S^1)/\mathrm{const}$. □

We now describe the pairing

$$\mathcal{H}_{p,g} \times \mathcal{H}_{p',g} \to \mathbb{R}\ .$$

This pairing is given by the cup-product in cohomology

$$H^1(G_r, H^1(G, l^p(G))) \times H^1(G_r, H^1(G_l, l^{p'}(G)))$$

$$\to H^2(G_r, H^2(G_l, l^p(G) \otimes l^{p'}(G)))$$

followed by the duality $l^p(G) \times l^{p'}(G) \to \mathbb{R}$ and evaluating twice on the fundamental cycle in $H_2(G,\mathbb{R})$. We have also an analytic description, namely a pairing

$$W_p^{1/p}(S^1)/\mathrm{const} \times W_{p'}^{1/p'}(S^1)/\mathrm{const} \to \mathbb{R}$$

is given on smooth function by $f, g \to \int_{S^1} f dg$ and then extended as a bounded bilinear form. This induces a pairing

$$H^1(G, W_p^{1/p}/\mathrm{const}) \times H^1(G, W_{p'}^{1/p'}/\mathrm{const}) \to \mathbb{R}\ .$$

Lemma 1.46. *(Cf. [65]) Let G be a finitely presented group which is realized as fundamental group of a compact Riemannian manifold M. Let $\rho : G \to O(\mathcal{H})$ be an orthogonal representation, which does not have almost invariant vectors. Let $[l] \in H^1(G, \mathcal{H})$. Let E be the flat vector bundle with fiber $\mathcal{H}$ over M, corresponding to ρ. Then there is a harmonic 1-form $\omega \in \Omega^1(M, E)$, corresponding to $[l]$.*

Proof. This is a reformulation of [65]. □

Corollary 1.47. *Let M be Kähler. Then, for ρ as in the previous lemma, we have*

1) *There is a natural complex structure in $H^1(G, \mathcal{H})$, making it a complex Hilbert space.*
2) *The pairing*
$$H^1(G, \mathcal{H}) \times H^1(G, \mathcal{H}) \to \mathbb{R}\ ,$$
given by $[l_1], [l_2] \to ([\omega]^{n-1}([l_1], [l_2]), [M])$ where $[\omega]$ is a Kähler class, $[M]$ is the fundamental class and $([l_1], [l_2]) \in H^2(G, \mathbb{R})$ is the cup-product composed with the scalar product $\mathcal{H} \times \mathcal{H} \to \mathbb{R}$, is a non-degenerate symplectic structure on $H^1(G, \mathcal{H})$.

Proof. The proof is the same as for finite-dimensional $\mathcal{H}$, once we have the Hodge theory by the previous lemma. □

We are now ready to prove that the symmetric pairing
$$\mathcal{H}_{2,g} \times \mathcal{H}_{2,g} \to \mathbb{R}$$
is nondegenerate. Realize G as a lattice in $SO(1, 2)$. Then $H^1(G, l^p(G)) = W_p^{1/p}(S^1)/\text{const}$. Recall that *the Hilbert transform H* is a bounded operator
$$H : L^p(S^1)/\text{const} \to L^p(S^1)/\text{const} \quad (p > 1)$$
defined as follows. For $u \in L^p(S^1)$ let f be its harmonic extension to the unit disk and g the conjugate harmonic function; then $Hu = g|S^1$. Since
$$\int_{\mathbb{H}^2} |\nabla f|^p = \int_{\mathbb{H}^2} |\nabla g|^p.$$
H restricts to $W_p^{1/p}(S^1)$ as an isometry.

Now, the symplectic duality $\int f dg$ in $W_2^{1/2}(S^1)/\text{const}$ is simply equal to (Hf, g). Moreover, H is $SO(1, 2)$-invariant. Then Corollary 1.47 implies that the pairing of Theorem 1.42 is also nondegenerate.

We still have to prove that $\mathcal{H}_{g,p} \neq 0$ and for $p = 2$ is infinite dimensional. We first describe an element of $\mathcal{H}_{g,p}$ associated to a given realization $G \hookrightarrow SO(1, 2)$ as a cocompact lattice, which we will call a principal state.

Given a smooth compact oriented manifold M, let $\text{Diff}^1(M)$ denote the group of orientation-preserving C^1-smooth diffeomorphisms of M. Then one has a cocycle l in $Z^1(\text{Diff}^1(M), C^0(M))$ defined as [15]
$$l(\phi) = \log \frac{\phi_* \mu}{\mu},$$

where μ is any smooth density on M, and $\phi_*\mu$ a left action. The class $[l] \in H^1(\mathrm{Diff}^1(M), C^0(M))$ does not depend on μ. For $r \geq 1$ one similarly gets a class in $H^1(\mathrm{Diff}^r(M), C^{r-1}(M))$. Now, let $M = S^{n-1}$ and consider the standard conformal action of $SO^+(1,n)$ on S^{n-1}. We get the class

$$[l]_p \in H^1(SO^+(1,n), W_p^{(n-1)/p}(S^{n-1})/\mathrm{const})$$

for all $p > 1$ simply because $C^\infty(S^{n-1}) \subset W_p^{(n-1)/p}(S^{n-1})$. We claim that $[l]_p \neq 0$ for $n = 2$ and $p > n-1$. Since the action is isometric, it follows from the following lemma (we prove and use it only for $n = 2$).

Lemma 1.48. *Fix $z_0 \in B^n$ and let $r(g) = \rho_h(z_0, g^{-1}z_0)$. Then for any fixed μ,*

$$\lim_{r(g)\to\infty} \|l(g)\|_{W_p^{(n-1)/p}(S^{n-1})/\mathrm{const}} = \infty.$$

Proof. (Only for $n = 2$) We choose for μ the harmonic (Poisson) measure μ_0, associated with z_0. Then $l(g) = \log \frac{g_*\mu_0}{\mu_0}$. For $\beta \in S^{n-1}$,

$$l(g)(\beta) = B_\beta(z_0, gz_0)$$

where $B_\beta(z_0, \cdot)$ is the Busemann function of B^n corresponding to $\beta \in \partial B^n$ and normalized at z_0, that is, $B_\beta(z_0, z_0) = 0$ (see, for example, [18]).

We will make the computation only for $n = 2$. Let $z_0 = 0$, $gz_0 = w$, then

$$B_\beta(0, w) = \log\left((1 - |w|^2) \cdot |w - \beta|^{-2}\right).$$

Notice that $\log |\frac{\beta - w}{1 - \bar{w}\beta}| = 0$, since $|\beta| = 1$, so

$$B_\beta(0, w) = \log(1 - |w|^2) - 2\log|1 - \bar{w}\beta| = -2\log|1 - \bar{w}\beta| (\mathrm{mod\ const}).$$

Notice that $\log|1 - \bar{w}z|$ is defined and is harmonic in $|z| \leq 1$, so

$$\begin{aligned} \|B_\beta(0,w)\|^p_{W_p^{1/p}(S^1)/\mathrm{const}} &= 2^p \textstyle\int_{B^2} [\nabla(\log|1 - \bar{w}z)|)]_h^p \, d\mu_h \\ &= 2^p \textstyle\int_{B^2} \frac{|w|^p}{|1 - \bar{w}z|^p} \frac{1}{(1-|z|^2)^{2-p}} \, dz d\bar{z}. \end{aligned} \qquad (*)$$

Sublemma. *The integral* $(*)$ *grows as* $\log(1 - |w|)$ *as* $|w| \to 1$.

Proof. Computing in the polar coordinates, we obtain

$$\int_0^1 dr \frac{1}{(1 - r^2)^{2-p}} \int_0^{2\pi} \frac{d\theta}{|1 - r|w|e^{i\theta}|^p}.$$

It is elementary to see that the inner integral grows as $\frac{1}{(1 - r|w|)^{p-1}}$, so we arrive at

$$\int_0^1 dr \frac{1}{(1-r)^{2-p}} \frac{1}{(1 - r|w|)^{p-1}} \sim \int_0^a \frac{ds}{s^{2-p}(A + s)^{p-1}}$$

where $a > 0$ is fixed and $A = 1 - |w|$. Furthermore, we have $(s = At)$

$$\int_0^{a/A} \frac{dt}{t^{2-p}(1+t)^{p-1}} \sim \int_0^{a/A} \frac{dt}{t} \sim \log|A|,$$

which proves the sublemma. □

Finally,

$$\|B_\beta(0,w)\|_{W_p^{1/p}(S^1)/\mathrm{const}} \sim [\log(1-|w|)]^{1/p},$$

where $\sim$ means that the ratio converges to a constant. □

The proof for $n>2$ will be given elsewhere.

Note that for $p=2$ we have (for $n=2$)

$$\|l(g)\|_{W_2^{1/2}(S^1)/\mathrm{const}} \sim \|g\|^{1/2}$$

where $\|g\|$ is the hyperbolic length of the (pointed) geodesic loop, representing g. This exponent in the RHS is the maximal possible. We will prove later a general theorem (Theorem 3.3) showing that for any orthogonal or unitary representation of $G=\pi_1(\Sigma)$ in a Hilbert space $\mathcal{H}$ and any cocycle $l \in Z^1(G,\mathcal{H})$,

$$\|l(g)\| \le \mathrm{const}\cdot \mathrm{length}(g)^{1/2} \log\log \mathrm{length}(g)$$

as g converges nontangentially to almost all $\theta \in S^1 = \partial G$.

Returning to the principal states $[l]_p \in H^1(G, W_p^{(n-1)/p}(S^{n-1})/\mathrm{const})$, let E be a flat affine bundle over $M=\mathbb{H}^n/G$ with fiber $W_p^{(n-1)/p}(S^{n-1})$, associated with the affine action

$$g \mapsto R_g + l(g).$$

Note that

$$s : z \mapsto \log\frac{\mu(z)}{\mu(z_0)}$$

is a G-equivariant section of the lift of E to $\widetilde{M}=\mathbb{H}^n$, or, equivalently, defines a section of E. We claim that this section is harmonic. This immediately reduces to the statement that $B_\beta(z_0,z)$ is harmonic mod constants as a function of z. In the upper half-plane model this simply means that $(x,y)\mapsto \log y$ is harmonic mod constants. The harmonic section just defined does not lift to a harmonic section of the flat affine bundle with fiber $W_p^{(n-1)/p}(S^{n-1})$. For $n=2$ we can say more. Let

$$H : W_p^{1/p}(S^1)/\mathrm{const} \to W_p^{1/p}(S^1)/\mathrm{const}$$

be the Hilbert transform. It makes $W_p^{1/p}(S^1)/\mathrm{const}$ into a complex Banach space. Then a direct inspection shows that the section of E defined above is (anti)holomorphic (depending on the choice of a sign of H). The will be used later. Equivalently, ds is an (anti)-holomorphic one-form on $\mathbb{H}^2/G$, valued in E. Again, this holomorphic form does not lift to a d and δ-closed form of a flat bundle with fiber $W_p^{1/p}(S^1)$ even for $p=2$. This latter bundle is a flat bundle with fiber a Hilbert space, but whose monodromy is not orthogonal. The Hodge theory of [65] and [60] does not apply and, in fact, not every cohomology class is represented by a d and δ-closed form. We will discuss these subtle obstructions to the Hodge theory in a sequel to this paper [110].

As an application of the computation made above, we will complete the proof of Lemma 1.15 for $p > 1$. Let $u \in W_p^{1/p}(S^1)/\text{const}$ and let $f : B^2 \to \mathbb{R}$ be the harmonic extension of u. We claim that

$$|f(w)| \leq c \cdot [\log(1 - |w|)]^{1/p'}.$$

Since the Hilbert transform is invertible in $W_p^{1/p}(S^1)/\text{const}$, we can assume that the Fourier coefficients $\hat{u}(n) = 0$ for $n < 0$, so that f is holomorphic:

$$\begin{aligned}
|f(w)| &= \left|\tfrac{1}{2\pi i} \int_{S^1} \tfrac{u(\xi)d\xi}{\xi - w}\right| = \tfrac{1}{2\pi}\left|\int_0^{2\pi} \tfrac{u(e^{i\theta})e^{i\theta}}{e^{i\theta}-w} d\theta\right| \\
&= \tfrac{1}{2\pi}\left|\int_0^{2\pi} \tfrac{u(e^{i\theta})d\theta}{1-w\cdot e^{-i\theta}}\right| = \tfrac{1}{2\pi}\left|\int_0^{2\pi} \tfrac{u(e^{-i\theta})d\theta}{1-w\cdot e^{i\theta}}\right| \\
&= \left|\tfrac{1}{2\pi i} \int_0^{2\pi} \tfrac{[u(e^{-i\theta})\cdot e^{-i\theta}]ie^{i\theta}\, d\theta}{1-we^{i\theta}}\right| \\
&= \left| - \tfrac{1}{2\pi i w} \int_0^{2\pi} [u(e^{-i\theta}) \cdot e^{-i\theta}][\log(1 - we^{i\theta})]' d\theta \right| \\
&= \left| - \tfrac{1}{2\pi i w} < u(e^{-i\theta}) \cdot e^{-i\theta}, \log(1 - we^{i\theta}) > \right| \\
&\leq \tfrac{1}{2\pi|w|} \|u(e^{-i\theta})e^{-i\theta}\|_{W_p^{1/p}(S^1)/\text{const}} \cdot \|\log(1 - we^{i\theta})\|_{W_{p'}^{1/p'}/\text{const}} \\
&\leq c\|u\|_{W_p^{1/p}(S^1)/\text{const}} |\log(1 - |w|)|^{1/p'}.
\end{aligned}$$

It is very plausible that the result is also true for $u \in W_p^{\frac{n-1}{p}}(S^{n-1})/\text{const}$ and $n \geq 3$. Our proof obviously does not work in this case.

We now start to prove that $\mathcal{H}_{2,g}$ is infinite dimensional. Let M_0, M_0' be factors generated by the left (respectively, right) actions of G in $l^2(G)$ [85]. Notice that $H^1(G_l, l^2(G))$ can be viewed as the cohomology of the complex

$$l^2(G) \xrightarrow{d_0} \bigoplus_{i=1}^{2g} l^2(G) \xrightarrow{d_1} l^2(G), \tag{$*$}$$

computed from the standard CW-decomposition of Σ^g with one zero-dimensional cell, $2g$ one-dimensional cells and one two-dimensional cell. Notice that d_0, d_1 are given by matrices with entries in $\mathbb{Z}[G]$, acting on $l^2(G)$ from the left. Letting $\Delta_l = d_0 d_0^* + d_1^* d_1$ we can view $H^1(G, l^2(G))$ as $\text{Ker}\Delta_l$. Notice that $\Delta_l \in M_0$. It follows that $H^1(G, l^2(G))$ is a module over M_0'. Now, since M_0 is type II, there is a decomposition

$$W_2^{1/2}(S^1)/\text{const} = H^1(G_l, l^2(G)) = \text{Ker } \Delta_l = \bigoplus_{j=1}^{m} H_m,$$

for any $m \geq 1$ where H_m are isomorphic right G-modules. Since we know already that $H^1(G_r, W_2^{1/2}(S^1)/\text{const}) \neq 0$, and H_j are all isomorphic, it follows that $H(G_r, H_j) \neq 0$ for all j, therefore $\dim H^1(G, W_2^{1/2}(S^1))/\text{const} \geq m$. This concludes the proof of Theorem 1.42. □

There are natural invariant von Neumann algebras acting in $\mathcal{H}_{2,g}$. Indeed, let M_1' be the double commutant of M_0' in $H^1(G, l^2(G)) = \text{Ker}\Delta_l$ and M_1 be the commutant of M_0'. We could define M_1' as a von Neumann algebra, generated by the right action of G in $H^1(G, l^2(G))$ and M_1 as the commutant of M_1'. It follows that M_1, M_1' do not depend on the choice of the complex $(*)$ and therefore $\mathcal{M}ap_{g,1} = \mathrm{Aut}(G)$ acts in $H^1(G, l^2(G))$ leaving M_1', M_1 invariant. Now consider $\mathcal{H}_{2,g} = H^1(G_r, H^1(G_l, l^2(G))$. Then

$$\mathcal{H}_{2,g} = \mathrm{Ker}\ \Delta_r : \bigoplus_{i=1}^{2g} H^1(G_l, l^2(G)) \to \bigoplus_{i=1}^{2g} H^1(G_l, l^2(G))$$

where the right Laplacian is defined exactly as the left one. It follows that $\mathcal{H}_{2,g}$ is a module over M_1. Let M_2 be the double commutant of M_1 and M_2' be its commutant. We proved the following theorem, except for the last statement.

Theorem 1.49. *There are infinite-dimensional von Neumann algebras M_2, M_2' acting in $\mathcal{H}_{2,g}$, which are invariant under the action of* Map_g. *Moreover, there is an involution τ of $\mathcal{H}_{2,g}$ which commutes with the* Map_g*-action and permutes M_2, M_2'.*

Proof. Everything is already proved except for the last statement. Observe that there is an involution $\tau : l^2(G) \to l^2(G)$ defined by $\tau f(g) = f(g^{-1})$. The Lyndon-Serre-Hochschild spectral sequence of the extension

$$1 \to G \to G \times G \to G \to 1$$

shows that $\mathcal{H}_{2,g} = H^2(G \times G, l^2(G))$. Let σ be the involution of $G \times G$ defined by $\sigma(g,h) = (h,g)$. Then one has $\tau[(g,h)v] = (\sigma(g,h))\tau(v)$ where $g, h \in G$ and $v \in l^2(G)$. It follows that τ induces an involution, which we also call τ, in $\mathcal{H}_{2,g}$, which obviously commutes with Map_g-action and permute M_2 and M_2'. This completes the proof of Theorem 1.49. □

Note that, since the unitary representation of G in

$$H^1(G_l, l^2(G)) = W_2^{1/2}(S^1)/\mathrm{const}$$

extends to an irreducible representation of $PSL_2(\mathbb{R})$, the commutant M_1 of G in $W_2^{1/2}(S^1)/\mathrm{const}$ possesses a faithful trace defined by

$$tr(a) \cdot Id = \int_{PSL_2(\mathbb{R})/G} gag^{-1} dg.$$

Proposition 1.50. *Let $\widetilde{\mathcal{H}}_{2,g}$ be the completion of M_1 under the norm $tr\ xx^*$. Then $\widetilde{\mathcal{H}}_{2,g}$ is a Hilbert space and there is a representation*

$$\tilde{\rho} : \mathrm{Map}_g \to \mathrm{Aut}(\widetilde{\mathcal{H}}_{2,g}),$$

leaving invariant the nondegenerate form $x \mapsto tr\ x^2$.

I do not know at the time of writing if $\widetilde{\mathcal{H}}_{2,g}$ is isomorphic to $\mathcal{H}_{2,g}$ as a Map_g-module.

We now turn to the holomorphic realization of $\mathcal{H}_{2,g}$. Fix a realization of G as a cocompact lattice in $SO^+(1,2)$, then $\mathcal{H}_{2,g} = H^1(G, W_2^{1/2}(S^1)/\mathrm{const})$. Recall that G commutes with the Hilbert transform in $W_2^{1/2}(S^1)/\mathrm{const}$. Let $S = \mathbb{H}^2/G$, then S is a hyperbolic Riemann surface, homeomorphic to Σ^g. For any element $w \in$

$H^1(G, W_2^{1/2}(S^1)/\text{const})$ we have, by Lemmata 1.43 and 1.46, a unique harmonic form in the flat Hilbert bundle E with fiber $W_2^{1/2}(S^1)/\text{const}$, associated with the action of G.

Uniqueness should be explained. We have the following general fact.

Lemma 1.51. *Let M be a compact Riemannian manifold and $\rho : \pi_1(M) \to O(H)$ an orthogonal representation on a real Hilbert space, without fixed vectors, $[\omega] \in H^1(\pi_1(M), H)$. Then there is at most one harmonic form, $\omega \in \Omega^1(M, E)$, representing $[\omega]$.*

Proof. If $\omega_1 \neq \omega_2$ are two such forms, then $\omega_1 - \omega_2$ is the derivative of a harmonic section of M. But the standard Bochner vanishing theorem shows that such section should be parallel, hence ρ has a fixed nonzero vector. Contradiction. □

Observe that H makes $W_2^{1/2}(S^1)/\text{const}$ into a complex Hilbert space. Then $\frac{1}{2}(\omega - H(\omega \circ J))$, where J is a complex structure on S, will be a holomorphic 1-form in E, whereas $\frac{1}{2}(\omega + H(\omega \circ J))$ will be an anti-holomorphic 1-form. Let $\mathcal{H}_{2,g}^{\pm}$ be the spaces of holomorphic (respectively, anti-holomorphic) 1-forms in E, then $\mathcal{H}_{2,g} = \mathcal{H}_{2,g}^{+} \bigoplus \mathcal{H}_{2,g}^{-}$. Now, $W_2^{1/2}(S^1)/\text{const}$ is identified with the space of exact L^2-harmonic 1-forms in the hyperbolic plane $\mathbb{H}^2$. The latter is isomorphic as a complex Hilbert space (with the complex structure, defined by the Hodge star operator) to the space of exact L^2-holomorphic 1-form in $\mathbb{H}^2$. Thus, any element in $\mathcal{H}_{2,g}^{+}$ defines a holomorphic 1-form on S valued in the bundle with fibers L^2-holomorphic 1-forms on $\mathbb{H}^2$. In other words, let G act diagonally in $\mathbb{H}^2 \times \mathbb{H}^2$ and

$$Q = \mathbb{H}^2 \times \mathbb{H}^2/G,$$

then we have an L^2-holomorphic 2-form on Q. The space $\mathcal{H}_{2,g}^{+}$ therefore is identified with the space of L^2-holomorphic 2-forms on Q. Similarly, $\mathcal{H}_{2,g}^{-}$ is identified with the space of L^2-holomorphic 1-form on

$$Q' = \mathbb{H}^2 \times \overline{\mathbb{H}^2}/G,$$

where $\overline{\mathbb{H}^2}$ is obtained from $\mathbb{H}^2$ by reversing the complex structure (i.e., $\bar{J} = -J$). Notice that as complex surfaces, Q and Q' are not biholomorphic: Q contains a compact curve (the quotient of the diagonal) whereas Q' does not. We proved the following:

Theorem 1.52. *(Holomorphic realization of the quantum moduli space). Fix an embedding $G \hookrightarrow SO^+(1,2)$ as a cocompact lattice. Then $\mathcal{H}_{2,g}$ splits as $\mathcal{H}_{2,g}^{+} \oplus \mathcal{H}_{2,g}^{-}$, where $\mathcal{H}_{2,g}^{+}$ (respectively, $\mathcal{H}_{2,g}^{-}$) is identified with the space of L^2-holomorphic 2-forms on $Q = \mathbb{H}^2 \times \mathbb{H}^2/G$ (respectively, $Q' = \mathbb{H}^2 \times \overline{\mathbb{H}^2}/G$). Moreover, the splitting is orthogonal with respect to the canonical symmetric scalar product in $\mathcal{H}_{2,g}$ and the restriction of this scalar product on $\mathcal{H}_{2,g}^{\pm}$ is positive (respectively, negative).*

Example. The principal state $[l]_2$ lies in $\mathcal{H}_{2,g}^{-}$. We do not know at the time of writing if $\mathcal{H}_{2,g}^{+} = 0$.

1.12. $\mathcal{H}_{p,g}$ as operator spaces and the vacuum vector

In this section we will develop an algebraic and an analytic theory of $\mathcal{H}_{p,g}$ as spaces of operators between $W_q^{1/q}(S^1)/\mathrm{const}$, which commute with the action of G. We use rather rough estimates of matrix elements, so the ranges of indices for which the action is established is certainly not the best possible. We start with the following

Lemma 1.53. *Let $u \in W_p^{1/p}(S^1)$ and $a \in l^q(G)$. Then $\sum a(g)R_g u \in W_r^{1/r}(S^1)$ where $\frac{1}{p}+\frac{1}{q}-1=\frac{1}{r}$.*

Remark. R_g means the action of G in $W_p^{1/p}(S^1)$ (a reminiscent of the actions from the right in $l^2(G)$).

Proof. Let f be a harmonic extension of u, so that

$$\int_{\mathbb{H}^2} |\nabla f|^p \, d\mu_h < \infty.$$

By the Young–A. Weil inequality [53], $l^p * l^q \subset l^r$, so if $h = \sum a_g R_g f$, we have

$$\begin{aligned}
\textstyle\int_{\mathbb{H}^2} |\nabla h|^r \, d\mu_h &= \textstyle\int_{\mathbb{H}^2/G} d\mu_h(z) \sum_g |\nabla h(g^{-1}z)|^r \\
&\le \textstyle\int_{\mathbb{H}^2/G} d\mu_h(z) \|\nabla h(g^{-1}z)\|_{l^r} \\
&\le c \cdot \textstyle\int_{\mathbb{H}^2/G} d\mu_h(z) \|a(g)\|_{l^q(G)} \|\nabla f(g^{-1}z)\|_{l^p} \\
&\le c \cdot \|a(g)\|_{l^q(G)} \textstyle\int_{\mathbb{H}^2} |\nabla f|^p d\mu_h.
\end{aligned}$$

The result follows from the estimate

$$\|\sum a(g)R_g u\|_{W_r^{1/r}(S^1)/\mathrm{const}} \le c \cdot \|a(g)\|_{l^q(G)} \|u\|_{W_p^{1/p}(S^1)/\mathrm{const}}. \qquad \square$$

Now recall that we have a canonical pairing

$$B : W_r^{1/r}(S^1)/\mathrm{const} \times W_{r'}^{1/r'}(S^1)/\mathrm{const} \to \mathbb{R},$$

so that the formula

$$(u, v) \mapsto (a \mapsto B(\sum a(g)R_g u, v))$$

defines a map

$$W_p^{1/p}(S^1)/\mathrm{const} \times W_{r'}^{1/r'}/\mathrm{const} \to l^{q'}(G).$$

Now an element of $W_{r'}^{1/r'}(S^1)/\mathrm{const}$ defines an element of

$$\mathrm{Hom}_G(W_p^{1/p}(S^1)/\mathrm{const}, l^{q'}(G))$$

and the induced map

$$\begin{aligned}
&\mathrm{Hom}(H^1(G, W_p^{1/p}(S^1)/\mathrm{const}, H^1(G, l^{q'}(G)) \\
&= \mathrm{Hom}(H^1(G, W_p^{1/p}(S^1)/\mathrm{const}, W_{q'}^{1/q'}(S^1)/\mathrm{const}).
\end{aligned}$$

In other words, we have a map

$$H^1(G, W_p^{1/p}(S^1)/\mathrm{const}) \to \mathrm{Hom}_{\mathbb{R}}(W_{r'}^{1/r'}(S^1)/\mathrm{const} \to W_{q'}^{1/q'}(S^1)/\mathrm{const}),$$

and it is immediate to check that the image lies in Hom_G. Thus we have

Proposition 1.54. *The construction above defines a* Map_g*-equivariant map*

$$\mathcal{H}_{p,g} \to \mathrm{Hom}_G(W_{r'}^{1/r'}(S^1)/\mathrm{const}, W_{q'}^{1/q'}(S^1)/\mathrm{const})$$

for p, q', r' *satisfying* $\frac{1}{p} \geq 1 + \frac{1}{q'} - \frac{1}{r'}$.

The induced map in $H^1(G, \cdot)$ produces a bounded Map_g-equivariant product

$$\mathcal{H}_{p,g} \times \mathcal{H}_{r',g} \to \mathcal{H}_{q',g}.$$

We again stress that the range of indices for which this product is defined should be improved. We will see that viewing $\mathcal{H}_{p,g}$ as an operator space helps to understand Map_g-action. We turn now to an analytic description of the above. Let $l \in Z^1(G, W_p^{1/p}(S^1)/\mathrm{const})$. The construction of Theorem 1.9 produces a smooth map

$$F : \mathbb{H}^2 \to W_p^{1/p}(S^1)/\mathrm{const},$$

satisfying $F(g^{-1}z) = R_g F(z) + l(g), g \in G$, In particular, $g^*(\nabla F)(g^{-1}z) = R_g(\nabla F)(z)$. Now let $v \in W_{r'}^{1/r'}(S^1)/\mathrm{const}$, where $r \geq p$. Then we have a scalar function

$$\langle F, v\rangle : \mathbb{H}^2 \to \mathbb{R},$$

where $\langle\cdot,\cdot\rangle$ is the pairing

$$W_r^{1/r}(S^1)/\mathrm{const} \times W_{r'}^{1/r'}(S^1)/\mathrm{const} \to \mathbb{R}$$

defined in Theorem 1.23. Since $r \geq p$, $W_p^{1/p}(S^1) \subset W_r^{1/r}(S^1)$, so $\langle F, v\rangle$ is defined. Without further assumption one can only assert that

$$|\nabla\langle F, v\rangle| \leq \mathrm{const}.$$

However, if we assume $r > p$, say $\frac{1}{p} = 1 + \frac{1}{q'} - \frac{1}{r'}$, then $\langle F, v\rangle$ will satisfy

$$\sum_g |\nabla\langle F, v\rangle(gz)|^{q'} < \mathrm{const}$$

for all $z \in \mathbb{H}^2$. Integrating over $\mathbb{H}^2/G$, we get

$$\int_{\mathbb{H}^2} |\nabla\langle F, v\rangle|^{q'} < \infty,$$

thus there exists $\langle F, v\rangle|S^1 \in W_{q'}^{1/q'}(S^1)$. This defines the desired map

$$H^1(G, W_p^{1/p}(S^1)/\mathrm{const}) \to \mathrm{Hom}_G(W_{r'}^{1/r'}(S^1)/\mathrm{const}, W_{q'}^{1/q'}(S^1)/\mathrm{const}).$$

We will use this description now to compute the operator associated with the principal state

$$[l]_p \in H^1(G, W_p^{1/p}(S^1)/\mathrm{const}).$$

Proposition 1.55. *For $p, r', q' > 1$, $\frac{1}{p} \geq 1 + \frac{1}{q'} - \frac{1}{r'}$, the operator in*

$$\mathrm{Hom}_G(W_{r'}^{1/r'}(S^1)/\mathrm{const}, W_{q'}^{1/q'}(S^1)/\mathrm{const}),$$

associated to the principal state $[l]_p$ is proportional to the Hilbert transform

$$H : W_{r'}^{1/r'}(S^1)/\mathrm{const} \to W_{r'}^{1/r'}(S^1)/\mathrm{const},$$

followed by the embedding $W_{r'}^{1/r'}(S^1) \hookrightarrow W_{q'}^{1/q'}(S^1)$.

Proof. First, we observe that the Hilbert transform acts as an isometric operator in $W_p^{1/p}(S^1)/\mathrm{const}$ for all $p > 1$. This follows at once from the definition of the norm as

$$\|u\| = \int_{\mathbb{H}^2} |\nabla f|^p d\mu_h,$$

where $\Delta f = 0$ and $f|S^1 = u$ (mod const). We will prove the proposition by a direct unimaginative computation. Let

$$g(z) = \frac{z + z_0}{1 + \bar{z}_0 z}, \qquad |z_0| < 1, |z| < 1,$$

so that $g(0) = z_0$. Then the Jacobian of g on the unit circle is

$$\frac{1 - |z_0|^2}{|z - z_0|^2},$$

so

$$l(g) = \log(1 - |z_0|^2) - \log|z - z_0|^2.$$

Let $\varphi : S^1 \to \mathbb{R}$ be smooth. Then

$$\langle \varphi, l(g) \rangle = \int_{S^1} \varphi'(\theta) \cdot [\log(1 - |z_0|^2) - \log|e^{i\theta} - re^{i\varphi}|^2] d\theta,$$

where $z_0 = re^{i\varphi}$. Obviously,

$$\int_{S^1} \varphi'(\theta) \log(1 - |z_0|^2) = 0,$$

so

$$\begin{aligned} \langle \varphi, l(g) \rangle &= -\int_{S^1} \varphi(\theta) \cdot [\log|e^{i\theta} - re^{i\varphi}|^2]' \\ &= -\int \varphi(\theta) \cdot \frac{2r\sin(\theta - \varphi)}{1 + r^2 - 2r\cos(\theta - \varphi)} . \end{aligned}$$

As $z_0 = re^{i\varphi} \underset{r \to 1}{\to} e^{i\varphi}$, this integral converges to

$$-v.p. \int \varphi(\theta) \frac{2\sin(\theta - \varphi)}{2 - 2\cos(\theta - \varphi)} = v.p. \int \varphi(\theta) \cdot \frac{1}{\tan\frac{\theta - \varphi}{2}} = \pi H\varphi(\theta)$$

almost everywhere on S^1. The proposition is proved, since smooth functions are dense in $W_{r'}^{1/r'}(S^1)$. □

Notice that since H commutes with the action of $SO^+(1,2)$, for any cocycle $m \in Z^1(G, W_p^{1/p}(S^1)/\text{const})$, Hm is also a cocycle. In particular, $H[l]_p \in H^1(G, W_p^{1/p}(S^1)/\text{const})$. We wish to compute the corresponding operator

$$\mathrm{Hom}(W_{r'}^{1/r'}(S^1)/\text{const} \to W_{q'}^{1/q'}(S^1)/\text{const}).$$

Let F, as above, be a smooth map

$$F : \mathbb{H}^2 \to W_p^{1/p}(S^1)/\text{const},$$

satisfying $F(g^{-1}z) = R_g F(z) + l_p(g)$. For $v \in W_{r'}^{1/r'}(S^1)$ we need to find a limit on the boundary of $\langle HF, v\rangle$. But H respects the pairing $\langle \cdot, \cdot \rangle$ and $H^2 = -1$, so $\langle HF, v\rangle = -\langle F, Hv\rangle$, whose limit on S^1 is $\pi H(-Hv) = \pi v$. We proved the following lemma:

Lemma 1.56. *For $p, r', q' > 1$, $\frac{1}{p} \geq 1 + \frac{1}{q'} - \frac{1}{r'}$, the operator in*

$$\mathrm{Hom}_G\left(W_{r'}^{1/r'}(S^1)/\text{const}, W_{q'}^{1/q'}(S^1)/\text{const}\right),$$

associated with $\frac{1}{\pi}H[l]_p$, is the identity.

Theorem 1.57.

A. *The element $v = H[l]_2 \in \mathcal{H}_{2,g}$ does not depend on the choice of the lattice $G \hookrightarrow SO^+(1,2)$.*
B. *The action of* Map_g *in $\mathcal{H}_{2,g}$ fixes v.*

Remark. The theorem is, beyond doubt, true for all $p > 1$ and not only $p \geq 2$, however I cannot prove this at the moment of writing this paper (July, 1999). (**Added January, 2000**). This is in fact true. The proof will appear in [110]).

The vector v is called a vacuum vector.

Proof. Consider two embeddings $i_1, i_2 : G \to SO^+(1,2)$ as cocompact lattices and let v_1, v_2 be the corresponding vacuum vectors. We view v_1, v_2 as elements of $H^1(G_r, H^1(G_l, l^2(G))$. Let A_1, A_2 be the associated operators

$$A_1, A_2 : H^1(G_l, l^{r'}(G)) \to H^1(G_l, l^{q'}(G)).$$

We know that $A_1 = A_2 = id$. It follows that the operator, associated with $v_1 - v_2$ is zero. We are going to show that $v_1 - v_2$ is zero. Since

$$v_1 - v_2 \in H^1(G_r, V),$$

where

$$V := H^1(G_l, l^2(G)) \simeq W_2^{1/2}(S^1)/\text{const},$$

by the result of [65] cited above (Lemma 1.46), there exists a harmonic section F of the affine Hilbert bundle over $M = \mathbb{H}^2/G$ with fiber V and the monodromy

$$g \mapsto R_g(\cdot) + m(g).$$

Here $m(g)$ is any cocycle, representing $v_1 - v_2$. Let $v \in W_{r'}^{1/r'}(S^1)/\text{const}$, then, denoting by F again the lift of this section on $\widetilde{M} = \mathbb{H}^2$, we see that $\langle F, v\rangle$ is a harmonic function such that

$$\int_{\mathbb{H}^2} |\nabla(\langle F, v\rangle)|^{q'}\, d\mu_h < \infty,$$

and the trace of $\langle F, v\rangle$ on S^1 is constant. It follows that $\langle F, v\rangle$ is constant itself, therefore (since v is arbitrary) $F = w = \text{const}$ and

$$m(g) = R_g w - w,$$

so $v_1 - v_2 = 0$. This proves A. Now, if $\phi \in \mathrm{Map}_{g,1} = \mathrm{Aut}(\pi_1(\Sigma^g))$, simply apply A to i_1 and $i_1 \circ \phi$. □

We wish to compute v. Recall that $[l]_2$ is given by the cocycle

$$g \mapsto -2\log|\beta - w|,$$

$\beta \in S^1$, $w = g(0)$. This is equal to $2\log|1 - \bar{w}\beta|$. The latter function is the real part of $2\log(1-\bar{w}z)$ which is holomorphic in the disk $\{|z| \le 1\}$. Hence the Hilbert transform of $2\log|1-\bar{w}\beta|$ is $2\mathrm{Arg}(1-\bar{w}\beta)$. This means that the cocycle

$$m(g)(\beta) = 2\mathrm{Arg}(W - \beta) \pmod{\text{const}}$$

where $W = 1/\bar{w}$, $w = g(0)$, represents v.

Theorem 1.58. *$H^1(\mathrm{Map}_{g,1}, H^1(G_l, l^p(G))) \neq 0$ for $p \ge 2$.*

Proof. We embed G as a lattice in $SO^+(1,2)$ and identify $H^1(G_l, l^p(G))$ and $W_p^{1/p}(S^1)/\text{const}$. We know that

$$H^0(\mathrm{Map}_g, H^1(G_r, W_p^{1/p}(S^1)/\text{const}) \ni v \neq 0.$$

Notice that $H^0(G_r, W_p^{1/p}(S^1)/\text{const}) = 0$ since any G-invariant harmonic 1-form in $\mathbb{H}^2$ has infinite p-energy. So in the spectral sequence

$$\begin{aligned} E_{i,j}^2 &: H^i(\mathrm{Map}_g, H^j(G_r, W_p^{1/p}(S^1)/\text{const})) \\ &\Longrightarrow H^{i+j}(\mathrm{Map}_{g,1}, W_p^{1/p}(S^1)/\text{const}) \end{aligned}$$

the second differential

$$\begin{aligned} d_2 &: H^0(\mathrm{Map}_g, H^1(G_r, W_p^{1/p}(S^1)/\text{const}) \\ &\Longrightarrow H^2(\mathrm{Map}_g, H^0(G_r, W_p^{1/p}(S^1)/\text{const}) \end{aligned}$$

must be zero. Therefore the vacuum vector v survives in E^∞. □

It is plausible that, in fact,

$$H^1(\mathcal{G}_1, W_p^{1/p}(S^1)/\text{const}) \neq 0 \qquad (p > 1)$$

for the group $\mathcal{G}_1$ of quasisymmetric homeomorphisms. (**Added January, 2000**. This is in fact true. The formula

$$\Phi \to \mathrm{Arg}\Phi^{-1}(\beta) - \mathrm{Arg}(\beta) \bmod \text{const}$$

defines a cocycle of $\mathcal{G}_1$ in $W_p^{1/p}(S^1)/\text{const}$ for any $p > 1$. The proof will appear in [110]).

We will now give an improved version of the theory for $p = 2$. Recall that $W_2^{1/2}(S^1)/\text{const}$ is an irreducible unitary (with aspect to the complex structure given by the Hilbert transform) $SO^+(1,2)$-module lying in the discrete series. It follows that it is L^2-integrable, that is, Lemma 1.53 holds with $p = q = r = 2$. Therefore Proposition 1.54 holds for $p = q' = r' = 2$. In other words, we have a Map_g-equivariant map

$$\mathcal{H}_{2,g} \to \text{End}_G(W_2^{1/2}(S^1)/\text{const})$$

A left inverse to this map is given by the evaluation on the vacuum vector. Therefore $\mathcal{H}_{2,g}$ is a direct summand in $\text{End}_G(W_2^{1/2}(S^1)/\text{const})$, which is M_1 in the terminology of Section 1.11.

1.13. Equivariant mappings of the Teichmüller Space, the space of quasifuchsian representations and the space of all discrete representations, into $\mathcal{H}_{p,g}$

Theorem 1.59. *The map which associates to a discrete, faithful cocompact representation*

$$G \to SO^+(1,2)$$

its principal state

$$[l]_p \in H^1(G_r, H^1(G_l, l^p(G))$$

is an Map_g*-equivariant map of the Teichmüller space* $\mathbf{T}_g$ *to* $\mathcal{H}_{p,g}$ *for all* $p > 1$.

A. *Let* $\varphi : G \to SO^+(1,3)$ *be a discrete and faithful representation. Let* $\alpha_\varphi : S^1 \to S^2$ *be the limit map of the boundaries*

$$S^1 = \partial\widetilde{\Sigma} \to \partial\mathbb{H}^3 = S^2,$$

defined in Section 1.6, *associated to* φ. *For* $p > 2$ *let*

$$[l]_p \in H^1(SO^+(1,3), W_p^{2/p}(S^2)/\text{const})$$

be the principle state. Define the map

$$\mu : \varphi \mapsto A_\varphi \varphi^*[l]_p \in H^1(G_r, H^1(G_l, l^p(G))),$$

by first pulling back $[l]_p$ *to* $\varphi^*[l]_p \in H^1(G, W_p^{2/p}(S^2)/\text{const})$ *and then applying the operator*

$$A_\varphi : W_p^{2/p}(S^2)/\text{const} \to W_p^{1/p}(S^1)/\text{const},$$

induced by α_φ *and defined in Section* 1.6. *Then* μ *is a* Map_g*-equivariant map*

$$\text{Hom}_{\text{discrete}}(G, SO^+(1,3))/SO^+(1,3) \to \mathcal{H}_{p,g}$$

for all $p > 2$.

B. *The image of the restriction of* μ *to*

$$\text{Hom}_{\text{quasifuchsian}}(G, SO^+(1,3))/SO^+(1,3),$$

is contained in $\mathcal{H}_{2,g}$.

Proof. The proof is already contained in Sections 1.6–1.12. We observe that from the operator viewpoint, the map of A sends any realization of G as a lattice in $SO^+(1,2)$ to the Hilbert transform of $W_p^{1/p}(S^1)/\text{const}$, followed by the identification

$$H^1(G_l, l^p(G)) \simeq W_p^{1/p}(S^1)/\text{const},$$

which depends on the lattice. In other words, fix one lattice embedding

$$\beta_0 : G \to SO^+(1,2).$$

Then any other lattice embedding

$$\beta : G \to SO^+(1,2)$$

can be written as

$$\beta(g) = \Phi_{\beta_0,\beta}\beta_0(g)\Phi^{-1}_{\beta_0,\beta},$$

where $\Phi_{\beta_0,\beta} \in \mathcal{G}_1$ is a quasisymmetric map. Then the operator, associated with β is

$$\Phi_{\beta_0,\beta} H \Phi^{-1}_{\beta_0,\beta} \in \text{Aut}(W_p^{1/p}(S^1)/\text{const}).$$

This gives a Map_g-equivariant map

$$\mathbf{T}_g \to \text{Aut}_G(W_p^{1/p}(S^1)/\text{const}).$$

For $p = 2$ one gets a map

$$\mathbf{T}_g \to Sp_G(W_2^{1/2}(S^1)/\text{const})$$

because the Hilbert transform and $\mathcal{G}_1$-action are symplectic (Section 1.7), which can be described as follows. First, one embeds $\mathbf{T}_g$ in the universal Teichmüller space

$$\mathbf{T} = \mathcal{G}_1/SO^+(1,2).$$

Then, using the representation,

$$\mathcal{G}_1 \to Sp(W_2^{1/2}(S^1)/\text{const})$$

defined in Section 1.7, one defines an embedding to Sp/U:

$$\mathbf{T} \to Sp(W_2^{1/2}(S^1)/\text{const})/U$$

where U is the group of operators in $Sp = Sp(W_2^{1/2}(S^1)/\text{const})$ which commutes with H, the latter is regarded as a complex structure on the space $W_2^{1/2}(S^1)/\text{const}$. Finally, one uses the Cartan embedding $Sp/U \to Sp$. □

Theorem 1.60. *(Linearization of pseudo-Anosov automorphisms). Let* $\phi \in \text{Map}_{g,1} = \text{Aut}(\pi_1(\Sigma^g))$ *be a pseudo-Anosov automorphism. Then for any* $p > 1$ *there exists a nontrivial element* $S_p \in \mathcal{H}_{p,g}$ *with the following properties:*

1) *For* $p_1 < p_2$*,* S_{p_2} *is the image of* S_{p_1}*, under the natural map* $\mathcal{H}_{p_1,g} \to \mathcal{H}_{p_2,g}$.
2) S_p *is invariant under* $\bar{\phi} \in \text{Map}_g$.

3) *There is a cocycle* $\tilde{l}_p \in Z^1(G, W_p^{1/p}(S^1)/\mathrm{const})$, *representing* S_p, *such that for any* $g \in G$

$$\sum_{n\in\mathbb{Z}} \|\tilde{l}_p(g) \circ \Phi^n\|_{W_p^{1/p}(S^1)/\mathrm{const}} < \infty$$

where $\Phi : S^1 \to S^1$ *is a quasisymmetric homeomorphism, associated with* ϕ. *In other words,*

$$\sum_{n\in\mathbb{Z}} \|A_\varphi^m \tilde{l}_p(g)\| < \infty$$

where $A_\varphi \in \mathrm{Aut}(H^1(G_l, l^p(G))$ *is induced by* ϕ.

Proof. This theorem is an immediate corollary of Theorems 1.59, 1.28 and 1.31 combined with [122] (see [92] for a complete proof), which shows that the mapping torus of any homeomorphism $\Psi : \Sigma \to \Sigma$, representing φ is a hyperbolic 3-manifold. □

It is plausible that such S_p is unique up to a multiplier. Knowing S_p is essentially equivalent to knowing the hyperbolic volumes of all ideal simplices with vertices on the limit curve $S^1 \to S^2$.

2. Theory of groups acting on the circle

Our first main result in this chapter is Theorem 2.7 which says, roughly, that a Kazhdan group cannot act on the circle. This general theorem is a culmination of many years of study and various special results concerning the actions of lattices in Lie groups, see [130], [31], [42]. One can see here a historic parallel with a similar, but easier, general theorem of [3] and [129] concerning Kazhdan groups acting on trees, which also followed a study of the actions of lattices. Our technique is absolutely different from the cited papers and uses the fundamental cocycle, introduced and studied in Section 2.1. We also use standard facts from Kazhdan groups theory [52].

In Sections 2.2, 2.3 we quantize equivariant maps between boundaries of universal covers, studied in Section 1.6. Our main tool is the theory of harmonic maps into infinite-dimensional spaces, as developed in [65], see also [60]. In Section 2.4 we review some facts about Banach–Lie groups and regulators. In Section 2.5 we describe a series of higher characteristic classes of subgroups of $\mathrm{Diff}^{1,\alpha}(S^1)$. We present two constructions. One uses an extension to a restricted linear group of a Hilbert space of classes originally defined in [33] for infinite Jacobi matrices. The other construction uses the action of a restricted symplectic group $Sp(W_2^{1/2}(S^1)/\mathrm{const})$ on the infinite-dimensional Siegel half-space. In both constructions we use an embedding of $\mathrm{Diff}^{1,\alpha}$ into the restricted linear group by the unitary and symplectic representation of Diff, respectively. Using the geometry of the Siegel upper half-space, we prove that our classes have polynomial growth.

There is a striking similarity between the theory of this chapter and the theory of symplectomorphism group, see Chapter 4, [102] and [104]. Note that the extended mapping class group action is not $C^{1,\alpha}$-smooth, so the results of this chapter do not apply to this group. On the other hand, Map_g does act symplectically on a smooth compact symplectic manifold (i.e., the $2g$-dimensional torus).

2.1. Fundamental cocycle

Let $\mathrm{Diff}^{1,\alpha}(S^1)$ denote the group of orientation-preserving diffeomorphisms with derivative in the Hölder space $C^\alpha(S^1)$, which consists of functions f such that

$$|f(x) - f(y)| < c|x - y|^\alpha.$$

There is a series of unitary representations of $\mathrm{Diff}^{1,\alpha}(S^1)$ in $L^2_{\mathbb{C}}(S^1, d\theta)$ given by

$$(\pi(g)(f))(x) = f(g^{-1}x) \cdot [(g^{-1})'(x)]^{\frac{1}{2}+i\beta}, \quad \beta \in \mathbb{R}.$$

We will mostly consider $\beta = 0$, in which case one has an orthogonal representation in $L^2_{\mathbb{R}}(S^1, d\theta)$. An invariant meaning is, of course, a representation in half-densities on S^1. Now consider the Hilbert transform H as an operator in $L^2_{\mathbb{R}}(S^1, d\theta)$ given by the usual formula

$$Hf(\varphi) = \frac{1}{\pi} v.p. \int_{S^1} \frac{f(\theta)}{\tan \frac{\varphi-\theta}{2}} d\theta.$$

We wish to consider $[H, \pi(g^{-1})]$. This is a bounded operator in $L^2(S^1, d\theta)$ given by an integral kernel which we are going to compute. Observe that

$$\frac{1}{\tan \frac{\varphi-\theta}{2}} = \frac{2}{\varphi - \theta} + \text{smooth kernel}.$$

A computation from [98] shows that

$$H[\pi(g)f](\varphi) = \frac{2}{\pi} v.p. \int_{S^1} \frac{d\theta}{\varphi - \theta} f(g^{-1}(\theta))[(g^{-1}(\theta))']^{1/2} + \text{smooth kernel} \circ \pi(g),$$

so

$$(\pi(g^{-1})H\pi(g)f)(\varphi) = [g'(\varphi)]^{1/2} \cdot \frac{2}{\pi} v.p. \int_{S^1} \frac{d\theta f(g^{-1}(\theta))[(g^{-1}(\theta))']^{1/2}}{g(\varphi) - \theta} + \pi(g^{-1}) \circ \text{smooth kernel} \circ \pi(g).$$

Letting $\theta = g(\eta)$ we have

$$\begin{aligned}
&(\pi(g^{-1})H\pi(g)f)(\varphi) \\
&= [g'(\varphi)]^{1/2} \frac{2}{\pi} v.p. \int_{S^1} \frac{f(\eta) \cdot [g'(\eta)]^{1/2}}{g(\varphi) - g(\eta)} d\eta + \pi(g^{-1}) \circ \text{smooth kernel} \circ \pi(g) \\
&= \frac{2}{\pi} v.p. \int_{S^1} \frac{[g'(\varphi)g'(\eta)]^{1/2}}{g(\varphi) - g(\eta)} f(\eta) d\eta + \pi(g^{-1}) \circ \text{smooth kernel} \circ \pi(g).
\end{aligned}$$

Finally,

$$[(\pi(g^{-1})H\pi(g)-H)](\varphi)$$
$$=\frac{1}{\pi}\int_{S^1}\frac{[g'(\varphi)g'(\eta)]^{1/2}(\varphi-\eta)-(g(\varphi)-g(\eta))}{(g(\varphi)-g(\eta))(\varphi-\eta)}f(\eta)d\eta \tag{1.1}$$
$$+\pi(g^{-1})\circ \text{smooth kernel}\circ\pi(g)+\text{smooth kernel}.$$

For a Hilbert space $\mathcal{H}$ and $p\geq 1$ we denote by $J_p(\mathcal{H})$ the Shatten class of operators such that the sum of the pth powers of their singular numbers converges. By $J_{p+}(\mathcal{H})$ we mean the intersection of all $J_q(\mathcal{H})$ with $q>p$.

Now recall that $g\in \mathrm{Diff}^{1,\alpha}(S^1)$. The following proposition sharpens that of [98] for $\mathrm{Diff}^\infty(S^1)$:

Proposition 2.1. A. *For* $\alpha>1/2$, $\pi(g^{-1})H\pi(g)-H\in J_2(L^2(S^1,d\theta))$.
B. *For* $\alpha>0$, $\pi(g^{-1})H\pi(g)-H\in J_{1/\alpha+}(L^2(S^1),d\theta)$.

Proof. As $\varphi-\eta\to 0$,

$$\frac{[g'(\varphi)g'(\eta)]^{1/2}(\varphi-\eta)-(g(\varphi)-g(\eta))}{(g(\varphi)-g(\eta))(\varphi-\eta)}<\text{const}\cdot(\varphi-\eta)^{\alpha-1},$$

so the kernel in (1.1) is in $L^2(S^1\times S^1,d\theta\otimes d\theta)$ for $\alpha>1/2$. This proves A.

To prove B we notice that by [96], the estimate on the kernel implies that the operator lies in $\mathcal{J}_{1/\alpha+}$. Strictly speaking, the conditions of [96] require C^∞ smoothness off the diagonal, whereas we have only the Hölder continuity, but the result stays true. □

Observe that $GL(L^2(S^1,d\theta))$ acts in J_p by conjugation. We deduce the following

Proposition 2.2. *The map*

$$l:g\mapsto\pi(g)H\pi(g^{-1})-H$$

is a 1-cocycle of $\mathrm{Diff}^{1,\alpha}(S^1)$ *in* $J_p(L^2(S^1,d\theta))$ *for* $p>1/\alpha$. *In particular,* l *is a 1-cocycle of* $\mathrm{Diff}^{1,\alpha}(S^1)$ *in* J_2 *for* $\alpha>1/2$.

We will call l the fundamental cocycle of $\mathrm{Diff}^{1,\alpha}(S^1)$.

Now let G be a subgroup of $\mathrm{Diff}^{1,\alpha}(S^1)$. We obtain a class in

$$H^1(G,J_p(L^2(S^1,d\theta))$$

by restricting l to G. We are going to show that this class is never zero, except for completely pathological actions of G on S^1.

Proposition 2.3. *Let* G *be a subgroup of* $\mathrm{Diff}^{1,\alpha}(S^1)$, $0<\alpha<1$. *Suppose* $p>1/\alpha$. *If* $[l]\in H^1(G,J_p)$ *zero, then the unitary action of* G *in* $L^2_{\mathbb{C}}(S^1,d\theta)$ *is reducible. Moreover, if* $H^1(G,J_p)=0$ *then* $L^2_{\mathbb{C}}(S^1,d\theta)$ *a direct sum of countably many closed invariant subspaces.*

Proof. If $[l]=0$ then there is $A\in J_p$ such that

$$\pi(g)H\pi(g^{-1})-H=\pi(g)A\pi(g^{-1})-A.$$

Hence $[\pi(g),H-A]=0$. Since H has two different eigenvalues with infinitely-dimensional eigenspaces, $H-A\neq \text{const}\cdot Id$, so the action of G in $L^2_{\mathbb{C}}(S^1,d\theta)$ is reducible.

Next, consider the operator R in $L^2(S^1,d\theta)$ with the kernel

$$K(\varphi,\eta)=\frac{1}{|\tan\frac{\varphi-\eta}{2}|}.$$

One sees immediately that R is a self-adjoint unbounded operator. Repeating the computation above, we deduce that $\pi(g)R\pi(g^{-1})-R\in J_p$, so $\tilde{l}(g)=\pi(g)R\pi(g^{-1})-R$ is another cocycle. If this cocycle is trivial, then we get an unbounded self-adjoint operator $R-A$ which commutes with the action of G. An application of the spectral theorem shows that $L^2(S^1,d\theta)$ is a countable sum of invariant subspaces. □

Corollary 2.4. *The restriction of $l,\tilde{l}$ to $SO^+(1,2)$ is not zero, for all $\alpha>0$.*

Proof. $SO^+(1,2)$ acts in $L^2_{\mathbb{C}}(S^1,d\theta)$ as a representation of the principal series, which are irreducible. □

We now specialize to $\alpha=1/2$ and $p=2$. Since $[\tilde{l}]\in H^1(SO^+(1,2),J_2)$ is nonzero, $\|\tilde{l}(g)\|_{J_2}$ is unbounded as a function of g [52]. In fact, one has the following

Proposition 2.5. *Let $\pi: SO^+(1,2)\to U(\mathcal{H})$ be a unitary representation and let $l: SO^+(1,2)\to\mathcal{H}$ be a continuous cocycle. Suppose $[l]\neq 0$. Then:*

A. *For any cocompact lattice $G\subset SO^+(1,2)$, $[l]|G\neq 0$.*
B. *$\|l(g^n)\|$ is unbounded as $n\to\infty$ for any hyperbolic g.*
C. *$\|l(\gamma^n)\|$ is unbounded as $n\to\infty$ for any parabolic $\gamma\neq 1$.*

Proof. Let $V\subset SO^+(1,2)$ be compact and such that $V\cdot G=SO^+(1,2)$. For $v\in V, g\in G$ we have

$$l(vg)=\pi(v)l(g)+l(v),$$

so $\|l(vg)\|\leq\|l(g)\|+\|l(v)\|$. If $l|G$ is bounded, then so is l. This proves A. Next, let P be the image of $SO^+(1,2)/K$ under the Cartan embedding, where K is a maximal compact subgroup. For the same reason as above, $l|P$ is unbounded. Let $S^1\subset P$ be a nontrivial orbit of K in $P\approx\mathbb{H}^2$. Notice that P is closed under raising into an integral power and there is a compact $V\subset SO^+(1,2)$ such that

$$P\subseteq\bigcup_{n\geq 1}(S^1)^n\cdot V$$

where $(S^1)^n$ is the image of S^1 under raising to the nth power. We deduce that $l|\cup_{n\geq 1}(S^1)^n$ is unbounded. Let $\gamma\in S^1$. Then any element in $(S^1)^n$ is of the form $k\gamma^n k^{-1}$, $k\in K$, so

$$\|l(k\gamma^n k^{-1})\|\leq\|l(k)\|+\|l(k^{-1})\|+\|l(\gamma^n)\|.$$

Thus $\|l(\gamma^n)\|$ is unbounded. Since γ can be any hyperbolic element, B follows. Notice that we proved that $\|l(g_k)\|$ is unbounded for any sequence $g_k \in P$, which escapes all compact sets. Now let $g \in SO^+(1,2)$ be parabolic $\neq 1$, and let τ be the involution fixing K. Then $\tau(g^n) \cdot g^{-n} \in P$ and escapes all compact sets, so $\|l[(\tau g^n) \cdot g^{-n}]\|$ is unbounded. It follows that either $\|l(\tau g^n)\|$ or $\|l(g^{-n})\|$ is unbounded. But all parabolics are conjugate in $SO^+(1,2)$, so C follows. □

Proposition 2.6. *Let $G \subset \mathrm{Diff}^{1,\alpha}(S^1)$, $\alpha > 1/2$. Suppose that G contains an element g which is conjugate in $\mathrm{Diff}^{1,\alpha}(S^1)$ to a hyperbolic or a nontrivial parabolic fractional-linear transformation. Then $[l]|G \neq 0$ in $H^1(G, J_2)$.*

Proof. Any such g is conjugate in $\mathrm{Diff}^{1,\alpha}(S^1)$ to an element $g' \in SO^+(1,2)$ for which $\|l(g'^n)\|$ is unbounded, so $\|l(g^n)\|$ is unbounded as well. □

We are ready to formulate the main result of this section.

Theorem 2.7. *Let $G \subset \mathrm{Diff}^{1,\alpha}(S^1)$, $\alpha > 1/2$. Suppose that one of the following holds:*

1) *The natural unitary action ($\beta = 0$) of G in $L^2(S^1, d\theta)$ given by*

$$\pi(g)(f)(\varphi) = f(g^{-1}(\varphi)) \cdot [(g^{-1}(\varphi))']^{1/2},$$

is irreducible or is a direct sum of finitely many irreducible factors.

2) *G contains an element, conjugate in $\mathrm{Diff}^{1,\alpha}(S^1)$ to a hyperbolic fractional-linear transformation.*
3) *G contains an element conjugate in $\mathrm{Diff}^{1,\alpha}(S^1)$ to a parabolic ($\neq 1$) fractional-linear transformation.*
4) $$\sup_{g\in G} \int_{S^1}\int_{S^1} \left[\frac{\sqrt{g'(\varphi)g'(\eta)}(\varphi-\eta) - (g(\varphi)-g(\eta))}{(g(\varphi)-g(\eta))(\varphi-\eta)}\right]^2 d\varphi d\eta = \infty.$$

Then G is not Kazhdan.

Proof. The proof follows from the formula (1.1) on page 55 and Propositions 2.3, 2.5 and 2.6. □

Remark. The above proof shows that $H^1(G, J_2) \neq 0$. In order to conclude that G is not Kazhdan, one also has to introduce a G-invariant structure of a Hilbert space on J_2. We refer the reader to [88, 89] for a generalization of Reznikov's ideas.

2.2. Construction of $N = 2$ quantum fields with lattice symmetry

It is possible that the physical space-time is discrete. Accordingly, in the axiomatic quantum field theory it is possible that the fields are invariant not under the whole Poincaré group, but only under a lattice in it. See [7] in this respect. We are going to construct mathematical objects, which yield such invariance on one hand, and quantize the equivariant measurable maps considered in Chapter 1, Section 1.6, on the other. Below H is the Hilbert transform.

Theorem 2.8. *Let G be a cocompact lattice in $SO^+(1,2)$. Let $\mathcal{H} = L^2_{\mathbb{R}}(S^1, d\theta)$ with the orthogonal action π, corresponding to $\beta \in \mathbb{R}$. Then there exists a measurable map to the space of bounded operators*

$$S^1 \xrightarrow{\rho} \mathcal{B}(\mathcal{H})$$

with the following properties:

1) *Equivariance: for $s \in S^1$ and $g \in G$*

$$\rho(gs) = \pi(g)\rho(s)\pi(g^{-1})$$

almost everywhere on S^1.

2) *One has*

$$\int_{S^1} (\rho(s) - H)\psi(s)ds \in J_2$$

for $\psi \in C^\infty(S^1)$.

3) *There exists $J \in J_2(\mathcal{H})$ such that $\rho(s)$ is a weak nontangential limit*

$$\rho(s) = \lim_{g \to s} \pi(g)(H + J)\pi(g^{-1})$$

as $g \in G$ converges nontangentially to $s \in S^1 = \partial G$ a.e. on S^1.

Proof. As a Hilbert space with orthogonal G-action,

$$J_2 = L^2(S^1 \times S^1, d\theta \otimes d\theta).$$

By the proof of Lemma 1.43 in Chapter 1, G does not have almost invariant vectors in J_2. Let $\Sigma = \mathbb{H}^2/G$ and let E be a flat affine vector bundle over Σ with a fiber J_2 and monodromy

$$g \mapsto Ad\pi(g) + l(g).$$

Then by a result of [65], and [60, Lemma I.11.6], there exists a harmonic map

$$\tilde{f} : \mathbb{H}^2 \to J_2$$

satisfying

$$\tilde{f}(gx) = \pi(g)\tilde{f}(x)\pi(g^{-1}) + l(g)$$

Consider $f(x) = \tilde{f}(x) + H$. Then

$$f(gx) = \pi(g)f(x)\pi(g^{-1}),$$

in particular, $\|f(x)\|$ is bounded in the operator norm. The operator version of the Fatou theorem, see [86] and references therein, shows that f has nontangential limit values a.e. on S^1, say $\rho(s)$. Obviously, ρ is G-invariant. On the other hand, $\tilde{f}$ is a Bloch harmonic J_2-valued function, that is,

$$\sup_{x \in \mathbb{H}^2} \|\nabla \tilde{f}\|_{J_2} < \infty.$$

It follows that $\|\tilde{f}(w)\|_{J_2} < c\cdot\log(1-|w|)$, $w \in B^2 = \mathbb{H}^2$. This implies by a standard argument (see, e.g., [45] that $\tilde{f}$ has a limit on S^1 as an element of $\mathcal{D}'(S^1, J_2)$. So for $\psi \in C^\infty(S^1)$,

$$\int_{S^1} (\rho - H)\psi \in J_2,$$

which proves the theorem. □

Remarks. 1) As was mentioned above, the invariant meaning of the representation π is that $L^2(S^1, d\theta)$ should be regarded as the space of half-densities. Accordingly, an integral operator is defined by a kernel which is a half-density on $S^1 \times S^1$ of the type $K(\varphi, \eta)(d\varphi d\eta)^{1/2}$. If $K(\varphi, \eta)$ is smooth and has a zero of the second order on the diagonal $\Delta \subset S^1 \times S^1$, then one has an invariant definition of its residue or second derivative, which is a quadratic differential. A direct computation which we leave to the reader shows that for $g \in \mathrm{Diff}^\infty(S^1)$:

i) $l(g) = \pi(g)H\pi(g^{-1}) - H$ is given by a kernel which has a zero of second order on Δ.

ii) The corresponding residue $S(g)$ is the Schwarzian of g.

This shows that $l(g)$ is a quantization of the Schwarzian cocycle. The operator field $\rho(s)$ of Theorem 2.8 seems therefore to be related to objects axiomatized, but not constructed, in [8].

2) The theorem and its proof stay valid for any representation

$$\varphi : G \to \mathrm{Diff}^{1,\alpha}(S^1),$$

$\alpha > 1/2$, such that the action on $S^1 \times S^1$ satisfies the very mild conditions of Lemma 1.43 in Chapter 1.

2.3. Construction of $N = 3$ quantum fields with lattice symmetry

The theory developed have for $\mathrm{Diff}(S^1)$ does not generalize to $\mathrm{Diff}(S^n)$, for $n \geq 2$. The reason is that the action of $\mathrm{Diff}(S^1)$ on S^1 is conformal. There are two ways to generalize various aspects of the theory to higher dimensions, by either considering $SO^+(1,n)$ acting on S^{n-1} or, very surprisingly, the group of symplectomorphisms of a compact symplectic manifold M (see Chapter 4). Here we consider the action of $SO^+(1,3) \simeq PSL_2(\mathbb{C})$ on S^2. We let $d(x,y)$ denote the spherical distance on S^2. Let $d\theta$ denote the spherical measure and let $\mathcal{H} = L^2(S^2, d\theta)$. For $g \in SO^+(1,3)$ let $\mu_g(x)$ denote the conformal factor, that is $\mu_g^2(x)$ is the Jacobian of g with respect to $d\theta$. The formula

$$\pi(g)f(x) = f(g^{-1}(x)) \cdot \mu_{g^{-1}}^{1+i\beta}(x), \quad \beta \in \mathbb{R},$$

defines a unitary representation of $SO^+(1,3)$ in $\mathcal{H}$. Now we introduce the operator H with the kernel

$$K(\varphi, \theta) = \frac{1}{d^2(\varphi, \eta)}.$$

This operator is self-adjoint and unbounded. Our goal is to compute

$$\pi(g)H\pi(g^{-1}) - H = l(g).$$

Proposition 2.9. *$l(g) \in J_2$ for all $g \in SO^+(1,3)$ and $\beta = 0$.*

Proof. The proof is a direct computation. One needs to show that as $d(x,y) \to 0$,

$$d^2(g(x), g(y)) - \mu_g(x)\mu_g(y)d^2(x,y)$$

is of order $d^4(x,y)$. In other words, for a fractional-linear transformation g of $\mathbb{C}$ one needs to show that as $x \to y$, $Im\ x, Im\ y > 0$, $g(x) = x$,

$$\left|\frac{g(x)-g(y)}{g(x)-\overline{g(y)}}\right|^2 - |g'(x)||g'(y)|\frac{Im\ y}{Im\ g(y)}\left|\frac{x-y}{x-\bar{y}}\right|^2$$

is of order $|x-y|^4$. This verifies the result for the hyperbolic metric instead of spherical metric, which is of course equivalent. One computes directly using the Taylor series for the holomorphic function g. □

Now arguing as in Section 2.2 we arrive at the following result:

Theorem 2.10. *Let G be a cocompact lattice in $SO^+(1,3)$. Let $\mathcal{H} = L^2_{\mathbb{R}}(S^2, d\theta)$ with the orthogonal action of G corresponding to $\beta = 0$. Then there exists a harmonic map*

$$\mathbb{H}^3 \xrightarrow{\psi} J_2(\mathcal{H})$$

with the property that $z \mapsto \psi(z) + H$ is equivariant:

$$\psi(gz) + H = \pi(g)(\psi(z) + H)\pi(g^{-1})$$

for all $g \in G$ and $z \in \mathbb{H}^3$.

Since H is unbounded, the boundary value of $\psi(z) + H$ does not exist as a measurable map to the space of bounded operators. It is possible that there is a more clever choice of a conformally natural singular integral operator which is bounded, but I don't know how to do it. Note in this respect that there is a very different realization of the orthogonal representation of $SO^+(1,3)$ in the space of functions on S^2, discovered in [101]. Namely, look at the natural action of $SO^+(1,3)$ on the smooth half co-densities, that is, sections of $\sqrt{\Lambda^2 TS^2}$. Using the spherical metric, we can identify this space with $C^\infty(S^2)$. Then the above-mentioned action leaves invariant a nonnegative quadratic form

$$Q(f) = \int_{S^2} ((\Delta f)^2 - 2|\nabla f|^2)dArea$$

whose kernel consists of constants and linear functions. It is possible that there are G-equivariant quantum fields valued in operators acting on the associated Hilbert space.

2.4. Banach-Lie groups and regulators: an overview

A Banach–Lie group is a Banach manifold with a compatible group structure. The usual Lie theory largely extends to this case. In particular, if $\mathcal{G}$ is a Banach–Lie group and $\mathfrak{g}$ is its Banach-Lie algebra, then a continuous n-cocycle on $\mathfrak{g}$ defines a left-invariant closed form on $\mathcal{G}$, so that one has a homomorphism

$$H^n_{\text{cont}}(\mathfrak{g}, \mathbb{K}) \to H^n_{\top}(\mathcal{G}, \mathbb{K})$$

where $H^*_{\top}$ is the cohomology of a topological space. In [102] we defined $\mathbb{K}$-homotopy groups of a Lie algebra, so that there is a map

$$\pi_i(\mathcal{G}) \otimes \mathbb{K} \to \pi_i(\mathfrak{g})$$

which in the case $\mathcal{G} = SL_n(C^\infty(M))$, M a compact manifold, $n \gg 1$, reduces to the Chern character

$$K_i^{\top}(M) \to HC_i(C^\infty(M)) = \Omega^i(M)/d\Omega^{i-1}(M) \oplus H^{i-2}(M, \mathbb{K}) \oplus \cdots$$

($\mathcal{G}$ is not a Banach-Lie group but a Fréchet-Lie group in this case). More interesting is a secondary class (= regulator) map. Define the algebraic K-theory of $\mathcal{G}$ as

$$K_i^{\text{alg}}(\mathcal{G}) = \pi_i((B\mathcal{G}^\delta)^+)$$

and the augmented K-theory as a kernel of the map $K_i^{\text{alg}} \to K_i^{\top}$:

$$0 \to \overline{K}_i^{\text{alg}}(\mathcal{G}) \to K_i^{\text{alg}}(\mathcal{G}) \to \pi_i(B\mathcal{G}) = \pi_{i-1}(\mathcal{G}).$$

Then the regulator map is a homomorphism

$$r : \overline{K}_i^{\text{alg}}(\mathcal{G}) \to \text{coker}(\pi_i(\mathcal{G}) \otimes \mathbb{K} \to \pi_i(\mathfrak{g})).$$

Lifting this map to the cohomology, that is, constructing a map

$$H^*_{\text{cont}}(\mathfrak{g}, \mathbb{K}) \to H^*(\mathcal{G}^\delta, \mathbb{K}),$$

meets obstructions described in the van Est spectral sequence. If $\mathcal{K} \subset \mathcal{G}$ is a closed subgroup such that $\mathcal{G}/\mathcal{K}$ is contractible, then these obstructions vanish and one gets a map

$$H^*_{\text{cont}}(\mathfrak{g}, \mathfrak{k}) \to H^*(\mathcal{G}^\delta)$$

given explicitly by a Dupont-type construction [28]. This is essentially the same as the geometric construction of the secondary classes of flat $\mathcal{G}$-bundles, described in [103]. In case $\mathcal{G} = SL_n(C^\infty(M))$ this gives the usual regulator map in the algebraic K-theory. However, for various diffeomorphism groups one construct new interesting classes. For the symplectomorphism groups, two series of classes, mentioned in the Introduction to Chapter 4, were constructed in [102] and [104], and a new class associated to a Lagrangian submanifold, will be constructed in Chapter 4. The symmetric spaces for $\text{Sympl}(M)$, used in [102] are sort of continuous direct products of finite-dimensional Siegel upper half-spaces. On the other hand, the symmetric space which we will use in this chapter to construct classes in $H^*(\text{Diff}^{1,\alpha}(S^1))$ is an infinite-dimensional Siegel half-space. The trouble is, however, that, for a compact manifold Y, (say, S^1) the group of diffeomorphisms of finite smoothness, like $\text{Diff}^k(Y)$, is not a Banach-Lie group: the multiplication from the right is not a diffeomorphism (the multiplication from the left is). This is neatly explained in [1]. Luckily, to construct secondary classes we only use the fact that the multiplication from the left is a diffeomorphism.

2.5. Characteristic classes of foliated circle bundles

It is well known that the continuous cohomology of $\mathrm{Diff}^\infty(S^1)$ is generated by the Euler class and by the integrated Godbillon-Vey class, see [40], [34] and references therein. Moreover, the square of the Euler class is zero. This already shows that the degree of smoothness is crucial. For if one considers the action of the extended mapping class group

$$\mathrm{Map}_{g,1} \hookrightarrow \mathcal{G}_1 \hookrightarrow \mathrm{Homeo}(S^1),$$

then the pull-back of the Euler class has nonzero powers to a degree which goes to infinity with g [80], [77], [84]. It appears that the scarcity of the cohomology of $\mathrm{Diff}(S^1)$ is a consequence of the (artificial) restriction of excessive degree of smoothness. Notice that the proofs in [34] depend hopelessly on C^∞- smoothness. We will give two constructions of a series of new classes in $H^*(\mathrm{Diff}^{1,\alpha}(S^1))$, $0 < \alpha < 1$, using both the unitary representation in $L^2(S^1, d\theta)$ and the symplectic representation in $Sp(W_2^{1/2}(S^1)/\mathrm{const})$. As in the case of the powers of the Euler class, nonvanishing of these classes is an obstruction to smoothability, i.e., to a conjugation to a subgroup of $\mathrm{Diff}^\infty(S^1)$. We will also prove that our classes are of polynomial growth if $\alpha > 1/2$. A related result (but not the argument) for C^∞ Gelfand-Fuks cohomology in all dimensions is to be found in [20]. Both in spirit and technology, the construction of the classes in $H^*(\mathrm{Diff}^{1,\alpha}(S^1))$ resembles our construction of a series of classes in $H^k_{\mathrm{cont}}(\mathrm{Sympl}(M), \mathbb{R})$, $k = 2, 6, 10, \ldots$, where M is a compact symplectic manifold and $Sympl(M)$ is its symplectomorphism group [104].

We start with the construction using the unitary representation. By Proposition 2.1, $\pi(g)H\pi(g^{-1}) - H \in J_p$ where $g \in \mathrm{Diff}^{1,\alpha}(S^1)$, $\quad p > 1/\alpha$, π is a unitary action in $L^2_{\mathbb{C}}(S^1, d\theta)$, and H is the complexification of the Hilbert transform. In other words, $H(e^{in\theta}) = \mathrm{sgn}(n)\cdot e^{in\theta}$. The subgroup of $\Phi \in GL(\mathcal{H})$, $\mathcal{H} = L^2_{\mathbb{C}}(S^1, d\theta)$ such that $\Phi H \Phi^{-1} - H \in J_p$ will be denoted $GL_{J_p}(\mathcal{H})$, following [98]. The unitary subgroup of $GL_{J_p}(\mathcal{H})$ is denoted $U_{J_p}(\mathcal{H})$. Let $\mathcal{H}_+, \mathcal{H}_-$ be the eigenspaces of H with the eigenvalues $+1$ and -1 respectively. Let $Gr_{J_p}(\mathcal{H})$ denote the restricted Grassmanian $U_{J_p}/U(\mathcal{H}_+) \times U(\mathcal{H}_-)$. Then $Gr_{J_p}(\mathcal{H})$ is a Banach manifold, modelled on the Banach space J_p. The Banach-Lie group $GL_{J_p}(\mathcal{H})$ acts smoothly on $Gr_{J_p}(\mathcal{H})$. On the other hand, though $\mathrm{Diff}^{1,\alpha}(S^1)$ is a group and a Banach manifold, it is not a Banach-Lie group [1]. However, multiplication from the left $L_g(h) = gh$ is a diffeomorphism (but not the multiplication from the right). The embedding

$$\mathrm{Diff}^{1,\alpha}(S^1) \to U_{J_p} \to GL_{J_p}(\mathcal{H})$$

is not continuous. However, the induced action of $\mathrm{Diff}^{1,\alpha}(S^1)$ on $Gr_{J_p}(\mathcal{H})$ is smooth [98].

We will introduce a series of U_{J_p}-invariant differential forms on $Gr_{J_p}(\mathcal{H})$. These forms induce cohomology classes in the Lie algebra cohomology

$$H^*(\mathrm{Lie}(U_{J_p})),$$

extending the classes introduced in [33] for the Lie algebra of Jacobi matrices. Notice that the tangent space to the origin of $Gr_{J_p}(\mathcal{H})$ can be identified with the space of matrices of the form

$$C = \begin{pmatrix} 0 & B \\ A & 0 \end{pmatrix}$$

where $A \in J_p(\mathcal{H}_+, \mathcal{H}_-)$ and $B \in J_p(\mathcal{H}_-, \mathcal{H}_+)$. Let

$$C_1, \ldots, C_{2k}, \quad (k \text{ odd})$$

be a collection of such matrices. Define

$$\mu_k(C_1, \ldots, C_{2k}) = \sum_{\sigma \in S_{2k}} \mathrm{sgn}(\sigma) P_k(\rho(C_{\sigma(1)}, C_{\sigma(2)}), \ldots, \rho(C_{\sigma(2k-1)}, C_{\sigma(2k)}) \qquad (\dagger)$$

where P_k is the kth invariant symmetric functions of k matrices, which is a polarization of $tr\ A^k$ (not an elementary symmetric polynomial, as in [34]). Now, $\rho(C_1, C_2)$ is defined as follows: let $\pi(C)$ be the left upper corner of C, i.e., an operator in $\mathcal{B}(\mathcal{H}_+)$. Then

$$\rho(C_1, C_2) = \pi([C_1, C_2]) - [\pi(C_1), \pi(C_2)].$$

The "meaning" of π is that of a connection of a principal bundle on something like the classifying space of the Lie algebra $\mathrm{Lie}(GL_{J_p})$, and of ρ is that of the curvature of this connection. Then μ_k becomes a characteristic class, somewhat analogous to the characteristic classes in the standard Chern–Weil theory. Note that μ_k is defined for all $k \geq [1/\alpha] + 1$. In [33], $\rho(C_1, C_2) \in \mathfrak{gl}(\infty, \mathbb{K})$ and μ_k is defined for all k. The form μ_2 defines the famous "Japanese cocycle", [126].

Lemma 2.11. *μ_k is U_{J_p}-invariant and closed.*

Proof. The invariance is obvious. The proof of closedness is standard and left to the reader, see the remarks above and [33]. □

Pulling back to $\mathrm{Diff}^{1,\alpha}(S^1)$ (this is possible by the remarks made above) we obtain a left-invariant closed differential form on $\mathrm{Diff}^{1,\alpha}(S^1)$. Pulling back to the universal cover $\widetilde{\mathrm{Diff}}^{1,\alpha}(S^1)$, we obtain a left-invariant closed differential form $\tilde{\mu}_k$ on $\widetilde{\mathrm{Diff}}^{1,\alpha}(S^1)$. The next theorem follows:

Theorem 2.12. *The secondary characteristic class, corresponding to $\tilde{\mu}_k$ is a well-defined class $r(\tilde{\mu}_k)$ in $H^{2k}([\widetilde{\mathrm{Diff}}^{1,\alpha}]^\delta, \mathbb{R})$.*

Proof. The group $\widetilde{\mathrm{Diff}}^{1,\alpha}(S^1)$ is contractible. □

Observe that for $\alpha > 1/2$, the class $\mu_1 \in H^2(\mathrm{Diff}^{1,\alpha}(S^1))$ is just the integrated Godbillon-Vey class.

Our second construction uses the symplectic action. For simplicity, we only treat the case $\alpha > 1/2$. Recall (Corollary 1.24 in Chapter 1) that $\mathcal{G}_1$ acts symplectically in $V = W_2^{1/2}(S^1)/\mathrm{const}$. Restricting to $\mathrm{Diff}^{1,\alpha}(S^1)$, we obtain a representation

$$\mathrm{Diff}^{1,\alpha}(S^1) \xrightarrow{\pi} Sp(V).$$

Let H be the Hilbert transform in V, normalized so that $H^2 = -1$. Denote by Sp_{J_p} the subgroup of operators $A \in Sp(V)$ such that $[A, H] \in J_p$. Let $U = U(V)$ denote the unitary group of A such that $[A, H] = 0$. Denote by

$$X = Sp_{J_p}/U$$

the restricted Siegel half-space. This is a Banach contractible manifold [93]. For $p = 2$ this is a Hilbert manifold with the canonical Sp_{J_2}-invariant Riemannian metric of nonpositive curvature. The metric is defined as follows. The tangent space $T_H(X)$ is identified with the space of operators A such that $A \in \mathrm{Lie}(Sp_{J_2})$ and $AH = -HA$. It follows that $A \in J_2$, and $A = A^*$. Then the metric is defined as trA^2. This definition is dimension-free and so the proof that the curvature is nonpositive follows from the explicit formulae, as in finitely-dimensional case.

Lemma 2.13. *For $\alpha > 1/2$, $\pi(\mathrm{Diff}^{1,\alpha}(S^1)) \subset Sp_{J_2}(V)$.*

Proof. We will use the computation of [113]. Let $g \in \mathrm{Diff}^{1,\alpha}(S^1)$. We need to show that

$$S = \sum_{n,m>0} \frac{m}{n} \left| \int_{S^1} e^{i(ng(\theta)+m\theta)}\, d\theta \right|^2 < \infty.$$

As in [113] we have, using a trick of Kazhdan,

$$S = \sum_{N=1}^{\infty} \sum_{m=1}^{N-1} \frac{m}{n} \left| \int_{S^1} e^{iN\varphi} \cdot [g_\beta^{-1}]'(\varphi) d\varphi \right|^2,$$

where $\beta = \frac{n}{N}$, $n = N - m$, $g_\beta(\theta) = \beta g(\theta) + (1-\beta)\theta$, $\theta \in S^1 = \mathbb{R}/2\pi\mathbb{Z}$. For $0 \le \beta \le 1$, g_β^{-1} are uniformly in $\mathrm{Diff}^{1,\alpha}(S^1)$ with $\alpha > 1/2$, so

$$\int_{S^1} e^{iN\varphi} [g_\beta^{-1}]'(\varphi)\, d\varphi \le \mathrm{const} \cdot N^{-\alpha} \cdot c_N$$

with $\sum_{N=1}^{\infty} c_N^2 < \infty$. Since $\sum_{m=1}^{N-1} \frac{m}{n} \sim \log N$, we have

$$S \le \mathrm{const} \cdot \sum_{N=1}^{\infty} N \log N N^{-2\alpha} \cdot c_N^2 < \infty. \qquad \square$$

Now let k be odd, $A_1, \dots, A_{2k} \in T_H(X)$, we define ν_k to be the form μ_k defined in (†) restricted to the Lie subalgebra of symplectic matrices. It necessarily vanishes for even k, so only odd k remain.[8]

Lemma 2.14. *ν_k is closed and Sp_{J_2}-invariant.*

Proof. The proof is identical to the finite-dimensional case, see [14]. □

Theorem 2.15. *The secondary characteristic class corresponding to ν_k, defines an element $r(\nu_k)$ in $H^{2k}_{\mathrm{cont}}(Sp_{J_2}(V))$ and in $H^{2k}([\mathrm{Diff}^{1,\alpha}(S^1)]^\delta, \mathbb{R})$, $\alpha > 1/2$. All these classes are of polynomial growth.*

Proof. Only the last statement needs a proof. For points $x_0, \dots, x_s \in X$ define the *geodesic span* $\sigma(x_0, \dots, x_s)$ inductively as follows:

$\sigma(x_0, x_1)$ is the geodesic segment joining x_0 and x_1.

$\sigma(x_0, \dots, x_s)$ is the union of geodesic segments joining x_0 and points of $\sigma(x_1, \dots, x_s)$.

By the standard comparison theorems

$$\mathrm{Vol}_s(\sigma(x_0, \dots, x_s)) \leq \mathrm{const} \cdot [\max_{0 \leq i \leq j \leq s} \rho(x_i, x_j)]^s,$$

where $\rho(\cdot, \cdot)$ is the distance function (this is where we use non-positive curvature). By [28], $r(\nu_k)$ can be represented by a cocycle

$$g_1, \dots, g_{2k} \mapsto \int_{\sigma(x_0, g_1 x_0, g_1 g_2 x_0, \dots, g_1, g_2 \cdots g_{2k} x_0)} \nu_k$$

where $g_i \in Sp_{J_2}$ and $x_0 \in X$ is fixed. Since ν_k is uniformly bounded, the result follows. □

We will give an independent proof of polynomial growth of

$$\mu_2 \in H^2(\mathrm{Diff}^{1,1}(S^1)).$$

Let $\mathrm{Var}(S^1)$ be a space of functions of bounded variation on S^1 mod constants. Then for $f_1, f_2 \in \mathrm{Var}(S^1)$,

$$\int_{S^1} f_1 \cdot d\, f_2 \leq c \|f_1\|_{\mathrm{Var}} \cdot \|f_2\|_{\mathrm{Var}}.$$

Now, $\mathrm{Homeo}(S^1)$ acts isometrically in $\mathrm{Var}(S^1)$ and there is a cocycle $\psi \in H^1(\mathrm{Diff}^{1,1}(S^1), \mathrm{Var})$ given by $g \mapsto \log(g^{-1})'$. By an formula of Thurston, μ_2 can be represented as

$$\int_{S^1} \psi(g_1)\, d\psi(g_2).$$

The result now follows from Lemma 1.1 in Chapter 1. For μ_2 as a class in $H^2(\mathrm{Diff}^\infty(S^1))$ see also [20]. □

[8]The Editors are grateful to Boris Tsygan for clarifying the definition of the form ν_k which was absent in the original version of the paper.

2.6. Examples

A typical example of a subgroup of $\mathrm{Diff}^{1,\alpha}(S^1)$ is the following. Let $K \subset \mathbb{R}$ be a subfield (i.e., a number field). Let $S^1(K)$ denote the set of K-rational points of $S^1 \subset \mathbb{R}^2$. Define G_K as the group of C^1-diffeomorphisms g such that there are points $x_0, \ldots, x_n = x_0 \in S^1(K)$ in this order such that $g_k = g|_{[x_k, x_{k+1}]}$ is the restriction of an element of $PSL_2(K)$. The C^1-condition simply means that $g'_k(x_{k+1}) = g'_{k+1}(x_{k+1})$. Then automatically $G_K \subset \mathrm{Diff}^{1,1}(S^1)$. Groups of this type, or rather their obvious analogues which act by piecewise-affine transformations on S^1 viewed as $\mathbb{R}/\mathbb{Z}$, appeared in [119], [46], [19], etc., where various properties were studied. The "proper" Thompson group can be smoothed, that is, embedded in $\mathrm{Diff}(S^1)$ [43] so that the Theorem 2.7 applies. However, it also acts on a tree so it is not Kazhdan already by the result of [3], [129]. Generally speaking, subgroups of $\mathrm{Diff}^{1,\alpha}(S^1)$, like the ones described above, do not have any obvious action on a tree and one needs our Theorem 2.7 to show that they are not Kazhdan. A parallel theorem for the symplectomorphism groups will be given in Chapter 4. Note also that the proof that our characteristic classes constructed in Section 2.5 are in polynomial cohomology agrees with a recent result on the growth of the Dehn function of the Thompson group [50].

3. Geometry of unitary cocycles

In this chapter we return to the asymptotic geometry of finitely generated groups. If G is not Kazhdan, then an orthogonal cocycle $l \in Z^1(G, \mathcal{H})$ should be viewed as a way to linearize the geometry of G. Our first result is a convexity theorem 3.1 which says that the map of G into the Hilbert space $\mathcal{H}$ given by l coarsely respects the geometry in a sense that inner points of big "domains" in G are mapped inside the convex hull of the image of boundary points.

We have seen in Chapter 1 that primitive functions $\mathcal{F} : G \to \mathbb{R}$ of cocycles in $Z^1(G, l^p(G))$ of a surface group satisfy

$$|\mathcal{F}(g)| < c \cdot \mathrm{length}(g)^{1/p}.$$

Here, we start a general study of cocycle growth. We show in Theorem 3.3 that for any orthogonal cocycle $l : G \to \mathcal{H}$,

$$\|l(g)\| < c(\theta)[\mathrm{length}(g) \log\log\ \mathrm{length}(g)]^{1/2}$$

for almost all $\theta \in \partial G \simeq S^1$ and g converging to θ nontangentially. We use in the proof a modified version of Makarov's law of iterated logarithm. The result extends to all complex hyperbolic cocompact lattices of any dimension.

Using another deep result of Makarov, we show the following in Theorem 3.5. Let G be a surface group, $\beta : G \to \mathbb{Z}$ a surjective homomorphism and $G_0 = \mathrm{Ker}\beta$. Then the conical limit set of G_0 has Hausdorff dimension 1, in particular, the critical exponent $\delta(G_0) = 1$. We do not know if this set has a full Lebesgue measure (it is certainly a doable problem).

Remark. More generally, if G is a discrete group of isometries of a $CAT(-1)$ space, $H \subset G$ is a normal subgroup such that G/H is amenable, then $\delta(H) = \delta(G)$, see [111].

Notice that the proof of Lemma 1.48 in Chapter 1 shows that the estimate on $\|l(g)\|$ is essentially sharp. It also shows that this estimate does not hold in other Banach spaces. However, imposing various restrictions on a Banach space, one still hopes to get an estimate, reflecting a fine structure of G.

3.1. Smooth and combinatorial harmonic sections

Let G be a finitely generated group, $\pi : G \to O(\mathcal{H})$ an orthogonal representation without almost invariant vectors and $l : G \to \mathcal{H}$ a nontrivial cocycle (with respect to π). If M is a compact Riemannian manifold with $\pi_1(M) = G$ (so that G is finitely presented) then one forms a flat affine bundle E over M with fiber $\mathcal{H}$ and monodromy

$$g \mapsto (v \mapsto \pi(g)v + l(g)).$$

A result of [65] and [60, Lemma I.11.6] states that there is a harmonic section f of E. If M is Kähler then there is another cocycle $m : G \to \mathcal{H}$ so that the complex affine bundle $E \otimes \mathbb{C}$ with the monodromy

$$g \mapsto (v + iw \mapsto \pi(g)v + i\pi(g)w + l(g) + im(g))$$

admits a holomorphic section. Our first result is a combinatorial version of this theorem.

Let $\{\gamma_i\}$ be a finite set of generators for G. Let V be the space of "sections", that is, G-equivariant maps

$$f : G \to \mathcal{H}.$$

This simply means that $f(g^{-1}x) = \pi(g)f(x) + l(g)$. Obviously, every such map is determined by $f(1) \in \mathcal{H}$. Therefore, $V \approx \mathcal{H}$. The combinatorial Laplacian is defined as

$$\triangle f(x) = \sum_i f(\gamma_i x) + f(\gamma_i^{-1}x) - 2f(x).$$

Theorem 3.1. *There exists an equivariant map $f : G \to \mathcal{H}$ with $\triangle f = 0$.*

Proof. Let $v = f(1)$, then $f(x^{-1}) = xv + l(x)$. Therefore

$$\begin{aligned}
\triangle f(x^{-1}) &= \sum f(\gamma_i x^{-1}) + f(\gamma_i^{-1}x^{-1}) - 2f(x^{-1}) \\
&= \sum x\gamma_i^{-1}v + l(x\gamma_i^{-1}) + x\gamma_i v + l(x\gamma_i) - 2xv - 2l(x) \\
&= \sum x(\gamma_i^{-1}v + \gamma_i v - 2v) + \sum xl(\gamma_i^{-1}) + l(x) + \\
&\quad +xl(\gamma_i) + l(x) - 2l(x) \\
&= x\sum(\gamma_i^{-1} + \gamma_i - 2)v + x\sum[l(\gamma_i^{-1}) + l(\gamma_i)],
\end{aligned}$$

so that we need only to solve the equation

$$\sum(\gamma_i^{-1} + \gamma_i - 2)v = -\sum[l(\gamma_i^{-1}) + l(\gamma_i)].$$

Notice that $\widetilde{\triangle} : \mathcal{H} \to \mathcal{H}$ defined by $v \mapsto \sum(\gamma_i^{-1} + \gamma_i - 2)v$ is self-adjoint. Moreover, since $\widetilde{\triangle} = -\sum(\pi(\gamma_i) - 1)^*(\pi(\gamma_i) - 1)$, $\widetilde{\triangle}$ is nonpositive and if $0 \in spec(\widetilde{\triangle})$, then

$\pi : G \to O(\mathcal{H})$ has almost invariant vectors. Therefore, $\widetilde{\triangle}$ is invertible and the result follows. $\square$

3.2. A convexity theorem

We keep the notation of 3.1. Any cocycle $l : G \to \mathcal{H}$ can be seen as an "embedding" of G in the Hilbert space. If $\|l(g)\| \to \infty$ as length$(g) \to \infty$, then this embedding is uniform in the sense that $\|l(g) - l(h)\| \to \infty$ as $\rho(g,h) \to \infty$ for any left-invariant word metric on G.[9]

For instance, Proposition 1.7 in Chapter 1 implies that any group G acting discretely (but possibly not cocompactly) on an Hadamard manifold of pinched negative curvature, admits a uniform embedding into $l^p(G)$ for some $p \in (1, \infty)$. We are, however, interested in a finer geometry of the cocycle embeddings.

For a finite $A \subset G$ and $C > 0$, the C-interior $\mathrm{int}_C(A)$ is defined as $\{x | \rho(x,y) < C \Rightarrow y \in A\}$. The C-boundary $\partial_C(A)$ is defined as $A \backslash \mathrm{int}_C(A)$.

Theorem 3.2. *Let $\pi : G \to \mathcal{O}(\mathcal{H})$ be an orthogonal representation without almost invariant vectors. Let $l : G \to \mathcal{H}$ be a cocycle for π. Then there are constants $C_1, C_2(l) > 0$ such that for any finite $A \subset G$ and any $x \in A$,*

$$\mathrm{dist}_{\mathcal{H}}(l(x) - \overline{\mathrm{conv}}(l(\partial_{C_1} A))) \leq C_2. \qquad (*)$$

Proof. Let $f : G \to \mathcal{H}$ be an equivariant harmonic map of Theorem 3.1. Since $\|f(x^{-1}) - l(x)\| = \|f(1)\| = \text{const}$, we can replace $(*)$ by the condition

$$\mathrm{dist}_{\mathcal{H}}(f(x) - \overline{\mathrm{conv}} f(\partial_{C_1}(A)) \leq C_2',$$

where however, one uses the right-invariant word metric on G in the definition of $\partial_C(A)$. This result follows from the maximum principle for harmonic functions. Indeed, let $x \in \mathrm{int}_{C_1}(A)$ be such that $\mathrm{dist}_{\mathcal{H}}(f(x) - \overline{\mathrm{conv}} f(\partial_{C_1}(A)))$ is maximal possible (and $> C_2$) (the choice of C_1, C_2 will be made later). Let v be a unit vector, such that

$$(f(x) - y, v) = \mathrm{dist}_{\mathcal{H}}(f(x) - \overline{\mathrm{conv}} f(\partial_{C_1}(A))$$

for some $y \in \overline{\mathrm{conv}} f(\partial_{C_1}(A))$. Let $h(z) = (f(z) - y, v)$. Then $h(x) > C_2$ and $h(\partial_{C_1}(A)) \subset (-\infty, 0]$. Moreover, $\widetilde{\triangle} h = 0$ and $h(z) \leq h(x)$ for $z \in \mathrm{int}_{C_1}(A)$. It follows that $h(\gamma_i x) = h(x)$ for all i. Replacing x by $\gamma_i x$ and continuing until we hit $\partial_{C_1} A$, we arrive to a contradiction with $C_1 = 2$, $C_2 = 2\|f(1)\| + 1$. $\square$

3.3. Cocycle growth for a surface group

In this section we continue, for general representations, the discussion started in Chapter 1, Section 5.2. Recall that, for any group G, any primitive function $\mathcal{F} : G \to \mathbb{R}$ of a class in $H^1(G, l^p(G))$ satisfies

$$|\mathcal{F}(g)| \leq \text{const} \cdot \text{length}(g)$$

[9] Existence of such cocycles is a strong negation of the Kazhdan property T, called the *Haagerup property* or *a-T-menability*.

at least if G is finitely presented. However, if $G = \pi_1(\Sigma)$, a surface group, then one has much finer estimate, established in Theorem 1.14 in Chapter 1:

$$|\mathcal{F}(g)| \leq \text{const} \cdot \text{length}(g)^{1/p'}$$

for all $p' > p$.

Theorem 3.3. *Let $G = \pi_1(\Sigma)$ be a surface group. Let $\pi : G \to O(\mathcal{H})$ be an orthogonal representation without almost-invariant vectors and let $l : G \to \mathcal{H}$ be a cocycle. Then for almost all $\theta \in S^1 \approx \partial G$,*

$$\|l(g)\| \leq \text{const}(\theta)[\text{length}(g) \cdot \log\log\ \text{length}(g)]^{1/2} \qquad (*)$$

as g converges nontangentially to θ. Here "almost all" refers to the Lebesgue measure on ∂G, identified with S^1 under some lattice embedding $G \hookrightarrow SO^+(1,2)$.

Remark. Nontangential convergence of points of B^2 to $\theta \in \partial B^2$ is an invariant of quasi-conformal homeomorphism (since the latter are quasi-isometries of the hyperbolic metric). Therefore $(*)$ is $\text{Map}_{g,1}$-invariant. Let $A \subset S^1$ be an exceptional set where $(*)$ does not hold. It follows that

$$\text{meas}\ \varphi(A) = 0$$

for all $\varphi \in \text{Map}_{g,1}$, considered as a quasisymmetric homeomorphism of S^1. Here meas is the Lebesgue measure.

Proof. Complexifying, we find a holomorphic section of an affine bundle $E_{\mathbb{C}}$ as in Section 3.1. Lifting to $\mathbb{H}^2$, we obtain an equivariant holomorphic map (we replace $\mathcal{H}$ by $\mathcal{H} \otimes \mathbb{C}$)

$$\widetilde{f} : \mathbb{H}^2 \to \mathcal{H} \otimes \mathbb{C}.$$

Notice that $\widetilde{f}$ is a Bloch function, that is, $\|\nabla \widetilde{f}\| \leq$ const. The result now follows from a version of the Makarov law [74] of iterated logarithms for Hilbert-space-valued Bloch functions. □

Proposition 3.4. *Let $\mathcal{H}$ be a complex Hilbert space. Let $\psi : B^2 \to \mathcal{H}$ be holomorphic and $\|\nabla\psi\|_h \leq$ const. Then for almost all $\theta \in S^1$,*

$$\limsup_{z \to \theta} \frac{\|\psi(z)\|}{\sqrt{\log(1-|z|)\log\log\log(1-|z|)}} < \infty.$$

Proof. We will simply note which changes should be made in a proof for complex–valued functions [97]. The Hardy's identity [97, page 174] holds in the following form. Let S be a Riemannian surface, $z_0 \in S$, $g : S \to \mathcal{H}$ a holomorphic function, (x, y) normal coordinates in the neighborhood of z_0. Let n be a positive integer. Then

$$\frac{\partial}{\partial x}(g, g)^{n+1} = (n+1)(g, g)[(g'_x, g) + (g, g'_x)],$$

$$\frac{\partial^2}{\partial x^2}(g,g)^{n+1} = n(n+1)(g,g)^{n-1}[g'_x,g)+(g,g'_x)]^2 \\ + (n+1)(g,g)^n[2(g'_x,g'_x)+(g''_x,g)+(g,g''_x)]$$

and the same for $\frac{\partial^2}{\partial y^2}$. Summing up, we have

$$\begin{aligned}\triangle(g,g)^{n+1} &= (\frac{\partial^2}{\partial x^2}+\frac{\partial^2}{\partial y^2})(g,g)^{n+1} \\ &= n(n+1)(g,g)^{n-1}\cdot 4|(g',g)|^2 + (n+1)(g,g)^n \cdot 2(g',g'),\end{aligned}$$

because $\triangle g = 0$ and $g'_y = \sqrt{-1}g'_x$. If S is a unit disc then in polar coordinates $z = re^{it}$

$$\triangle = \frac{\partial^2}{\partial r^2}+\frac{1}{r}\frac{\partial}{\partial r}+\frac{1}{r^2}\frac{\partial^2}{\partial t^2} = \frac{1}{r}\frac{\partial}{\partial r}(r\frac{\partial}{\partial r})+\frac{1}{r^2}\frac{\partial^2}{\partial t^2} = \frac{1}{r^2}[(r\frac{\partial}{\partial r})^2+\frac{\partial^2}{\partial t^2}].$$

So

$$\frac{1}{r^2}((r\frac{\partial}{\partial r})^2+\frac{\partial^2}{\partial t^2})(g,g)^{n+1} = 4n(n+1)(g,g)^{n-1}|(g',g)|^2 + 2(n+1)(g,g)^n|g'|^2.$$

Integrating over $0 \le t \le 2\pi$ and using the Cauchy-Schwartz inequality, we arrive at the inequality of [97, Theorem 8.9]. The rest of the proof will go unchanged once we know the Hardy-Littlewood maximal theorem for $(g,g)^n$, which is used in [97, page 187]. Let

$$g^*(s,\xi) = \max_{0\le r\le 1-e^{-s}} |g(r\xi)|, e \le s < \infty, \xi \in S^1.$$

Since $g : B^2 \to \mathcal{H}$ is holomorphic, it is also harmonic and yields the Poisson formula. Then a proof of the Hardy-Littlewood maximal theorem given in [68] applies, since it reduces it to the Hardy-Littlewood inequality for the maximal function of $|g|$. □

Remark. Theorem 3.3 holds for complex hyperbolic cocompact lattices[10]. This is because Makarov's law of iterated logarithms holds for the complex hyperbolic space, as we can see by passing to totally geodesic spaces of complex dimension 1. It is plausible that a version of Theorem 3.3 holds for real hyperbolic lattices (but not quaternionic and Cayley, as these are Kazhdan, see a new proof in Chapter 4). On the other hand, another deep result of [75] saying that Bloch functions are nontangentially bounded for a limit set of Hausdorff dimension one, fails for Hilbert space valued functions. In fact, we have shown in Chapter 1 that there are unitary cocycles on a surface group such that $\|l(g)\| \to \infty$ as length$(g) \to \infty$.

If G is any finitely generated group, and we are given an orthogonal representation $\pi : G \to O(\mathcal{H})$ and a cocycle $l \in Z^1(G,\mathcal{H})$ with a control on $\|l(g)\|$ from below, then for any embedding of the surface group $\pi_1(\Sigma)$ into G we immediately have a comparison inequality between the word lengthes of elements of $\pi_1(\Sigma)$ in $\pi_1(\Sigma)$ and in G. To get a nontrivial result, we need a low bound on $\|l(g)\|$ better

[10]Provided that $\tilde{f}$ can be chosen pluriharmonic.

then $[\text{length}(g) \log\log \text{length}(g)]^{1/2}$. To find such groups and cocycles seems to be a very attractive problem.

We will now use similar ideas to estimate the Hausdorff dimension of limit sets of some infinite index subgroups of $G = \pi_1(\Sigma)$.

Theorem 3.5. *Let $\beta : G \to \mathbb{Z}$ be a surjective homomorphism and let $G_0 = \mathrm{Ker}\beta$. Let A be the conical limit set of G_0. Then* $\dim A = 1$.

Proof. Let $[\beta] \in H^1(G, \mathbb{Z})$ be the induced class. Realize G as a cocompact lattice in $SO^+(1,2)$ so that $S = \mathbb{H}^2/G$ is a hyperbolic surface. Let ω be a holomorphic 1-form on S such that $Re[\omega] = [\beta]$ and let $\widetilde{\omega}$ be a lift of ω on $\mathbb{H}^2$. Let $f : \mathbb{H}^2 \to \mathbb{C}$ be holomorphic with $df = \omega$. Then f is a Bloch function. By a result of [75], there is a set $B \subset S^1$ with $\dim B = 1$ such that

$$\limsup_{z \to \theta} |f(z)| < \infty$$

for any $\theta \in B$ and nontangential convergence. Notice that $f(gz) = f(z) + ([\omega], [g])$ where $g \in G$ and $[g]$ is the image of g in $H_1(G, \mathbb{Z})$. Now it is clear that $B \subseteq A$, so $\dim A = 1$.

Remark. This result does not contradict a theorem of [116] and [123] because G_0 is infinitely generated.

In the opposite direction we have the following. Let Σ_1, Σ_2 be two closed surfaces and let $\psi : \Sigma_1 \to \Sigma_2$ be a smooth ramified covering. Let $G_i = \pi_1(\Sigma_i)$ and let $G_0 = \mathrm{Ker}(\psi_*) : G_1 \to G_2$. Let $G_1 \hookrightarrow SO^+(1,2)$ be a realization of G_1 as a lattice. Then for any $z \in B^2$,

$$\sum_{g \in G_0} |1 - gz| < \infty.$$

In other words, either $\delta(G_0) < 1$ or $\delta(G_0) = 1$ and G_0 is of convergence type. In the latter case, the Patterson-Sullivan measure of the conical limit set of G_0 is zero. To see this, notice that we can find hyperbolic structures on Σ_i, $i = 1, 2$ so that ψ is holomorphic. Let $\widetilde{\psi}$ be a lift of ψ as a map $\widetilde{\psi} : B^2 \to B^2$. Since $\widetilde{\psi}$ is a bounded holomorphic function, $\widetilde{\psi}$ has limit values almost everywhere on S^1. By Chapter 1, Section 6.5., $|\widetilde{\psi}|S^1| = 1$ almost everywhere. So $\widetilde{\psi}$ is an inner function. Let $C \subset B^2$ be a countable set of zeros of $\widetilde{\psi}$. We claim that C is a finite union of orbits of G_0. First, it is clear that C is G_0-invariant. Let $Q \subset B^2$ be compact which contains a fundamental domain for G_1. Then $\widetilde{\psi}(Q)$ is compact so there is a finite set $R \subset G_2$ such that $g(0) \notin \widetilde{\psi}(Q)$ if $g \notin R$. Let $T \subset G_1$ be finite and such that $\psi_*(T) \supseteq R$. Let $Q_1 = \bigcup_{g \in T^{-1}} gQ$ so that Q_1 is compact and therefore $C \cap Q$ is finite. Let $x \in C$, then $x = gy$ with $y \in Q$. So $0 = \widetilde{\psi}(x) = \psi_*(g)\widetilde{\psi}(y)$, i.e., $\psi_*(g^{-1})(0) = \widetilde{\psi}(y) \in \widetilde{\psi}(Q)$. This means $\psi_*(g^{-1}) \in R$ so $g^{-1} \in TG_0$, and $g \in G_0T^{-1}$, say $g = g_0t^{-1}$, $g_0 \in G_0$, $t \in T$. Then $t^{-1}y \in C$ and $t^{-1}y \in Q_1$, so there are finitely many options for $t^{-1}y$.

We deduce that there are $x_1, \dots, x_n$ such that $C = \bigcup_{i=1}^n G_0 x_i$. The decomposition formula for inner functions implies that

$$\widetilde{\psi}(z) = c \cdot \prod_{\substack{g_0 \in G_0 \\ 1 \le i \le n}} \frac{\overline{g_0 x_i}}{g_0 x_i} \frac{z - g_0 x_i}{1 - \overline{g_0 x_i} z}$$

which gives an explicit formula for holomorphic maps between hyperbolic Riemann surfaces (one still needs to find x_i). By a well-known result on zeros of a bounded holomorphic function [68, IV: B, Theorem 1],

$$\sum_{g_0 \in G_0} (1 - |g_0 x_i|) < \infty.$$

The rest follows from [90]. □

4. A theory of groups of symplectomorphisms

We already observed the intriguing similarity between groups acting on the circle and groups acting symplectically on a compact symplectic manifold. The two leading topics studied in Chapter 2, namely, (non-) Kazhdan groups acting on S^1 and characteristic classes, have exact analogues for $\mathit{Sympl}(M)$. In fact, a theory of characteristic classes parallel to 2.5, was already presented in [102] and [104]. In the second cited paper, we noticed that the Kähler action of $\mathrm{Sympl}(M)$ on the twistor variety allows us to define a series of classes in $H^{2k}_{\mathrm{cont}}(\mathrm{Sympl}(M), \mathbb{R})$, k odd, which are highly non-trivial. In the first cited paper, we introduced bi-invariant forms on $\mathrm{Sympl}(M)$ and the classes in $H^{\mathrm{odd}}_{\mathrm{T}}(\mathrm{Sympl}(M))$ and $H^{\mathrm{odd}}(\mathrm{Sympl}(M)^\delta, \mathbb{R}/A)$ (cohomology of a topological space and a discrete group) where A is the group of periods of the above-mentioned forms. Here we present a fundamental class in $H^1(\mathrm{Sympl}(M), L^2(M))$ whose nontriviality on a subgroup $G \subset \mathrm{Sympl}(M)$ implies that G is not Kazhdan, similarly to the situation in $\mathcal{D}iff^{1,\alpha}(S^1)$. From the nature of our class it is clear that its vanishing imposes severe restriction on the symplectic action, roughly, the transformations of G should satisfy a certain PDE. We give an explicit formula for our class in the case of a flat torus.

We then introduce a characteristic class in $H^{n+1}(\mathrm{Sympl}^\delta(M^{2n}), \mathbb{R})$ associated with an immersed compact Lagrangian submanifold. This class is a symplectic counterpart, and a generalization, of the Thurston-Bott class [15]. We use this class to give a formula for the volume of compact negatively curved manifold through Euclidean volumes of "Busemann bodies" (the images of the manifold under Busemann functions).

4.1. Deformation quantization: an overview

Let F be a field and $A|F$ a (commutative) algebra. A deformation of A is an algebra structure of $A[[\hbar]]$ over $F[[\hbar]]$, extending that A, so that if $x, y \in A$, then

$$x * y = x \cdot y + b_1(x,y)\hbar + b_2(x,y)\hbar^2 + \cdots$$

where $x \cdot y$ is the multiplication in A and $x * y$ is the deformed multiplication.

If $F = \mathbb{R}$, $A = C^\infty(M)$, where M is a symplectic manifold, then a deformation quantization is a deformation of A with $b_1(f,g) = \{f,g\}$, the Poisson bracket. A deformation quantization always exists by a result of [79], [127], [6], [32]. For any algebra $A|F$ one defines the Hochschild cohomology $HH^k(A) = \mathrm{Ext}^k_{A\otimes A}(A,A)$. There is a natural Lie superalgebra structure in $HH^*(A)$, see [41]. There exists a simple explicit complex, computing $HH^k(A)$ with

$$C^k(A) = \mathrm{Hom}_F(\bigotimes_{i=1}^k A, A).$$

In particular, b_1 above is a cocycle (for any deformation). If $F = \mathbb{R}$ and A is a topological algebra, one modifies the definitions to obtain topological Hochschild cohomology. If M is a smooth manifold and $A = C^\infty(M)$ with the pointwise multiplication, then

$$HH^k(A) = \Gamma(M, \Lambda^k TM),$$

a space of (anti-symmetric) poly-vector fields. The Lie superalgebra structure coincides with the classical bracket of poly-vector fields.

We will need an explicit form of the cocycle condition for a 2-cocycle $b : A \otimes A \to \mathbb{R}$:

$$xb(y,z) - b(xy,z) + b(x,yz) - b(x,y)z = 0.$$

4.2. A fundamental cocycle in $H^1(\mathrm{Sympl}(M), L^2(M))$

Let (M^{2n}, ω) be a compact symplectic manifold. Fix a deformation quantization

$$f * g = f \cdot g + \{f,g\}\hbar + \sum_{i=2}^\infty c_i(f,g)\cdot \hbar^i.$$

Let $\Phi : M \to M$ be symplectic and let

$$\begin{aligned} f\tilde{*}g &= (f\circ\Phi^{-1} * g\circ\Phi^{-1})\circ\Phi \\ &= f\cdot g + \{f,g\}\hbar + \textstyle\sum_{i=2}^\infty c_i'(f,g)\cdot\hbar^i. \end{aligned}$$

Lemma 4.1. *Let $A|F$ be an algebra and let*

$$f * g = f\cdot g + c_1(f,g)\hbar + \cdots + c_{k-1}(f,g)\hbar^{k-1} + \sum_{i=k}^\infty c_i(f,g)\cdot\hbar^i$$

and

$$f\tilde{*}g = f\cdot g + c_1(f,g)\hbar + \cdots + c_{k-1}(f,g)\hbar^{k-1} + \sum_{i=k}^\infty c_i'(f,g)\cdot\hbar^i$$

be two deformations, which coincide up to the order $\hbar^{k-1}$. Then

$$c_i - c_i' : A\otimes A \to A$$

is a Hochschild cocycle.

Proof. $(f * g) * p - (f \tilde{*} g) \tilde{*} p$

$$= c_k(f,g) \cdot p + c_k(f \cdot g, p) - c'_k(f,g) \cdot p - c'_k(f \cdot g, p) \pmod{\hbar^{k+1}}.$$

Similarly,

$$f * (g * p) - f \tilde{*} (g \tilde{*} p)$$
$$= f \cdot c_k(g,p) + c_k(f, g \cdot p) - f c'_k(g,p) - c'_k(f, g \cdot p) \pmod{\hbar^{k+1}}.$$

Thus, for $c = c_k - c'_k$,

$$f \cdot c(g,p) + c(f, g \cdot p) - c(f,g)p - c(f \cdot g, p) = 0,$$

which means that c is a 2-cocycle. □

Lemma 4.2. *The formula*

$$\Phi \mapsto \left[(f,g) \mapsto c_2(f \circ \Phi^{-1}, g \circ \Phi^{-1}) \circ \Phi - c_2(f,g)\right]$$

defines a smooth cocycle of $\mathrm{Sympl}(M)$ *in the space* $Z^2(C^\infty(M), C^\infty(M))$ *of Hochschild 2-cocycles for* $C^\infty(M)$.

Proof. Follows from Lemma 4.1. □

Passing to the Hochschild cohomology, we obtain a 1-cocycle of $Sympl(M)$ in

$$HH^2(C^\infty(M)) = \Gamma(M, \Lambda^2 TM).$$

Using the symplectic structure, we identify $\Gamma(M, \Lambda^2 TM)$ with $\Omega^2(M)$, the space of 2-forms on M. Multiplying by ω^{n-1} we obtain a cocycle

$$\mu \in H^1(Sympl(M), C^\infty(M)).$$

4.3. Computation for the flat torus and the main theorem

If M is a coadjoint orbit of a compact Lie group, one can find an explicit formula for the deformation quantization $f * g$. The classical case $M = T^{2n}$, the flat torus, is due to H. Weyl.

Proposition 4.3. *One has the following deformation quantization on* T^{2n}:

$$f * g = \sum_{k=0}^{\infty} \frac{1}{k!} \left(-\frac{i\hbar}{2}\right)^k \sigma^{i_1 j_1} \cdots \sigma^{i_k j_k} \frac{\partial^k f}{\partial y_{i_1} \cdots \partial y_{i_k}} \frac{\partial^k g}{\partial y_{j_1} \cdots \partial y_{j_k}}$$

where σ^{ij} *are entries of the matrix inverse to the matrix* (σ_{ij}) *of a (constant) symplectic form, and the "repeated indices" summation agreement is applied.*

Now, since our definition of a fundamental cocycle is completely explicit, one can derive an explicit formula for μ in this case. We give an answer for T^2 (the formula for T^{2n} is completely analogous). The computation is tedious (takes several pages) but straightforward and is left to reader. Here is the formula for T^2:

$$\Phi \mapsto \frac{\partial^2 \Phi_2}{\partial y_1^2} \frac{\partial^2 \Phi_1}{\partial y_2^2} - \frac{\partial^2 \Phi_1}{\partial y_1^2} \frac{\partial^2 \Phi_2}{\partial y_2^2}$$

where $\Phi = (\Phi_1, \Phi_2)$ a symplectomorphism of the torus T^2.

Summing up, we have:

Theorem 4.4. *Let M^{2n} be a compact symplectic manifold, let* Sympl(M) *be its symplectomorphism group, acting orthogonally on the Hilbert space $L^2(M)$. There exists a cocycle*

$$\mu \in Z^1(\mathrm{Sympl}(M), L^2(M)),$$

defined canonically by the given deformation quantization of $C^\infty(M)$ and satisfying the following properties:

A. *Let*

$$f \tilde{*} g = (f \circ \Phi^{-1} * g \circ \Phi^{-1}) \circ \Phi = f \cdot g + \{f, g\}\hbar + c_2'(f, g) \cdot \hbar^2 + \cdots,$$
$$f * g = f \cdot g + \{f, g\}\hbar + c_2(f, g) \cdot \hbar^2 + \cdots$$

and let us identify the class of the Hochschild cocycle $c_2' - c_2$ with a section ν of $\Lambda^2 TM$. Let $\hat{\nu}$ be a 2-form obtained from ν by lifting the indices using the symplectic form. Then

$$\mu(\Phi) \cdot \omega^n = \hat{\nu} \cdot \omega^{n-1}.$$

B. *$\mu(\Phi)$ depends only on the second jet of Φ.*

C. *For $M = T^2$ and the Weyl deformation quantization, $\Phi = (\Phi_1, \Phi_2)$,*

$$\mu(\Phi) = \frac{\partial^2 \Phi_2}{\partial y_1^2}\frac{\partial^2 \Phi_1}{\partial y_2^2} - \frac{\partial^2 \Phi_1}{\partial y_1^2}\frac{\partial^2 \Phi_2}{\partial y_2^2}.$$

D. *If G is a Kazhdan subgroup of* Sympl(M)*, then*

$$\|\mu(\Phi)\|_{L^2} < \mathrm{const} \qquad \forall \Phi \in G.$$

Examples.

1) $M = T^{2n}$, $G = Sp(2n, Z)$ (this group is Kazhdan for $n \geq 2$). Then μ is identically zero.
2) Let Γ be a surface group, and let M be a component of

$$\mathrm{Hom}(\Gamma, SO(3))/SO(3),$$

consisting of representations with nontrivial Stiefel-Whitney class. Then M is a compact symplectic manifold and Map_g acts symplectically on M. We do not know if part D of Theorem 4.4 holds in this case and if Map_g is Kazhdan or not. There is a "Teichmüller structure" on M defined by a holomorphic map of the Teichmüller space into the twistor variety of M, described in [104], see also Chapter 5.

Remark. The case of two-dimensional M^2 is much easier, simply because $SL_2(\mathbb{R})$ is not Kazhdan. If Sympl(M, x_0) denotes the subgroup of Sympl(M) fixing $x_0 \in M$ then one gets a nontrivial unitary cocycle on Sympl(M, x_0) by pulling back from $SL_2(\mathbb{R})$ under the tangent representation. Using the measurable transfer (i.e., Shapiro's lemma) one constructs a cocycle of Sympl(M). See [131] for details.

4.4. Invariant forms on the space of Lagrangian immersions and new regulators for symplectomorphism groups

In this section we will "symplectify" the Thurston-Bott class in the cohomology of diffeomorphism groups. Let M be any (possibly noncompact) symplectic manifold, and let $L_0 \hookrightarrow M$ be a Lagrangian immersion of a compact oriented manifold L_0. Let $\mathrm{Lag}(L_0, M)$ be the space of Lagrangian immersions of L_0 into M which can be jointed to L_0 by an exact Lagrangian homotopy. This means the following. If $f_t : L_0 \to M$ is a smooth family of Lagrangian immersions, then $\frac{d}{dt} f_t|_{t=0}$ is a vector field along L_0. Projecting to the normal bundle NL_0 and accounting that NL_0 is canonically isomorphic to T^*L_0, we get a 1-form on L_0 which is immediately seen to be closed. A Lagrangian homotopy f_t is exact, if this form is exact for all t. There is therefore a well-defined function F (modulo constants) on L which can be seen as a tangent vector of such deformation.

Definition 4.5. The canonical $(n+1)$-form ν on $\mathrm{Lag}(L, M)$ is defined by

$$\nu(F_0 \cdots F_n) = \int_L F_0 dF_1 \cdots F_n = \mathrm{Vol}_{n+1}(\widetilde{Q}) \qquad (*),$$

where $\widetilde{Q}$ is any chain in $\mathbb{R}^{n+1}$ spanning $(F_0, \dots, F_n)(L)$.

Proposition 4.6. *The form ν is closed.*

Proof. This proof is left as an exercise for reader. □

Let $\mathrm{Sympl}_0(M)$ be the group of Hamiltonian transformations of M. Then $\mathrm{Lag}(L, M)$ is invariant under $\mathrm{Sympl}_0(M)$. The following is obvious:

Proposition 4.7. *The form ν is $\mathrm{Sympl}_0(M)$-invariant.*

The standard theory of regulators [103], [102] implies that:

First, one has an induced class in $H^{n+1}(\mathfrak{g}, \mathbb{R})$, where

$$\mathfrak{g} = \mathrm{Lie}(\mathrm{Sympl}_0(M)) = C^\infty(M)/\mathrm{const},$$

which is given by $(*)$, where now $F_i \in C^\infty(M)$.

Second, one has a class in

$$\mathrm{Hom}(\pi_{n+1}(B\mathrm{Sympl}_0^\delta(M)^+, \mathbb{R}/A)) \qquad (n+1 \geq 5),$$

where A is the group of periods of ν on maps $\Sigma^{n+1} \to \mathrm{Sympl}_0(M)$ of homology spheres to $\mathrm{Sympl}_0(M)$. This class often lifts to a class in

$$H^{n+1}(\mathrm{Sympl}_0^\delta(M), \mathbb{R})$$

under suitable conditions on the topology of $\mathrm{Sympl}_0(M)$ (see discussion in the papers cited above).

For instance, let Q be a compact oriented simply connected manifold, $M = T^*Q$ and $L_0 = Q$, the zero section. Then we obtain a class $[\nu]$ in

$$H^{n+1}(\mathrm{Sympl}_0(T^*Q), \mathbb{R}).$$

Note that the restriction of this class to

$$\mathrm{Diff}(Q) \hookrightarrow \mathrm{Sympl}_0(T^*Q)$$

is zero, as $\mathrm{Diff}(Q)$ fixes the zero section. However, our class is an extension of the Thurston-Bott class [15] in $\mathrm{Diff}(Q)$ by means of the following construction. Let $G \subset \mathrm{Sympl}_0(T^*Q)$ be the subgroup of symplectomorphisms of the form

$$p_x \mapsto \phi^* p_x + df(x),$$

where $f \in C^\infty(Q)$, $\phi \in \mathrm{Diff}(Q)$, $x \in Q$, $p_x \in T_x^*Q$. Then G is the extension

$$0 \to C^\infty(Q)/\mathrm{const} \to G \to \mathcal{D}iff(Q) \to 1.$$

Any 1-cocycle $\psi \in Z^1(\mathcal{D}iff(Q), C^\infty(Q)/\mathrm{const})$ induces a splitting of this exact sequence:

$$S_\psi : \mathcal{D}iff(Q) \to G.$$

Now let μ be a smooth density on Q then $\psi = \frac{\phi_* \mu}{\mu}$ is a 1-cocycle, so it defines such a splitting. The pull-back $S_\psi^*([\nu]|G)$ of our class to $\mathcal{D}iff(Q)$ is precisely the Thurston-Bott class.

We summarize this discussion:

Theorem 4.8. A. *The formula*

$$\nu(F_0, \dots, F_n) = \int_L F_0 dF_1 \cdots dF_n = \mathrm{Vol}_{n+1}(\widetilde{Q})$$

defines an $\mathrm{Sympl}_0(M)$*-invariant closed* $(n+1)$*-form in* $\mathrm{Lag}(L, M)$. *It induces a class* $[\nu] \in H^{n+1}(\mathrm{Lie}(\mathrm{Sympl}_0(M), \mathbb{R})$ *and a regulator*

$$[\nu] : \pi_{n+1}(B\mathrm{Sympl}_0^+(M)) \to \mathbb{R}, \qquad n+1 \geq 5,$$

which lifts to a class

$$[\nu] \in H^{n+1}(\mathrm{Sympl}_0^\delta(M), \mathbb{R})$$

if $\widetilde{H}_i(\mathrm{Lag}(L, M), \mathbb{R})) = 0, \quad 0 \leq i \leq n+1.$

B. *In particular, if Q is a smooth oriented simply-connected closed manifold, then*

$$[\nu] \in H^{n+1}(\mathrm{Sympl}^\delta(T^*Q), \mathbb{R})$$

pulls back to the Thurston-Bott class under any splitting

$$\mathcal{D}iff(Q) \to C^\infty(Q)/\mathrm{const} \rtimes \mathcal{D}iff(Q),$$

coming from a smooth density on Q.

4.5. A volume formula for negatively-curved manifolds

This section is ideologically influenced by [51] and discussions with G. Besson (Grenoble, 1996). Let N^n be an Hadamard manifold, i.e., a complete, simply-connected nonpositively curved Riemannian manifold. Let CN be the space of oriented geodesics in N, which is a symplectic manifold of dimension $2n-2$. Any point $x \in N$ defines a Lagrangian sphere $S_x \subset CN$ of geodesics passing through x.

Lemma 4.9. *The pull-back $S^*\nu$ of the form $\nu \in \Omega^n(CN)$ to N is the Riemannian volume form on N times a constant.*

Proof. An exercise in Jacobi fields. □

Now, if G acts discretely, freely and cocompactly on N, preserving orientation, we have

$$\langle S^*\nu, [N/G]\rangle = c \cdot \mathrm{Vol}(N/G).$$

Here $[N/G]$ is the fundamental class of N/G.

Corollary 4.10. *Under the above assumptions, $[\nu] \neq 0$ in $H^n(\mathrm{Sympl}^\delta(N), \mathbb{R})$.*

We now assume that the curvature of N is strictly negative and, moreover, the induced action of G on the sphere at infinity S_∞ is of the class $C^{1,\frac{n-1}{n}}$. For $n = 2$ this is always the case [56], whereas for $n \geq 3$ it seems to require a pinching of the curvature. Notice that the map

$$s_+ : CN \to S_\infty,$$

sending any geodesic $\gamma(t)$ to $\gamma(\infty)$, is a Lagrangian fibration. Therefore, if we fix a Lagrangian section of s_+, we will have a symplectomorphism $CN \simeq T^*(S_\infty)$. Fix $p_0 \in N$, then S_{p_0} is such a section. Notice that the induced homomorphism $G \to \mathrm{Sympl}(T^*S_\infty)$ is given by,

$$g \mapsto (z \mapsto \pi(g)z + dF(p_0, g^{-1}p_0, \theta)),$$

where $g \in G$, $z \in T^*_\theta S_\infty$, $\pi : G \to \mathcal{D}iff^{1,\frac{n-1}{n}}(S_\infty) \to \mathrm{Sympl}(T^*S_\infty)$ is induced by the action of G on S_∞ and $B(p_0, g^{-1}p_0, \theta))$ is the Busemann function. Our assumption imply that $B(p_0, p_1, \cdot) \in C^{\frac{n-1}{n}}(S_\infty) \subset W_n^{\frac{n-1}{n}}(S_\infty)$. Recall that for $F_1, \ldots, F_n \in W_n^{\frac{n-1}{n}}(S_\infty)$ we have the n-form

$$\int_{S_\infty} F_1 dF_2 \cdots dF_n = \int_{B^n} du_1 \cdots du_n,$$

where u_i is a harmonic extension of F_i.

We derive

Corollary 4.11. *Let N^n/G be a compact negatively curved manifold such that the induced action of G on S_∞ is of class $C^{1,\frac{n-1}{n}}$. If the fundamental class of G is*

$$\sum_i [g_1{}^{(i)} \cdots g_n{}^{(i)}],$$

then the following volume formula holds:

$$\mathrm{Vol}(N/G) = c(n) \cdot \sum_i \int_{S_\infty} F_1{}^{(i)} dF_2{}^{(i)} \cdots dF_n{}^{(i)},$$

where $F_k{}^{(i)}(\theta) = B\left(p_0, \left(g_k{}^{(i)}\right)^{-1} p_0, \theta\right)$.

One can say that the volume of a negatively curved manifold is a sum of Euclidean volumes of Busemann bodies in $\mathbb{R}^n$ bounded by $(F_1, \ldots, F_n)(S_\infty)$.

Replacing the Busemann cocycle by a Jacobian cocycle $\frac{g*\mu}{\mu}$, where μ is a smooth density on S_∞, we arrive to a similar formula for the Godbillon-Vey-Thurston-Bott invariant of N/G, under the same regularity assumptions. This seems to have been also accomplished in a preprint [57] cited in [56], though I was unable to obtain this paper from its author. The case $n = 2$ is, however, covered in [56].

5. Groups of volume-preserving diffeomorphisms and the nonlinear superrigidity alternative

In this chapter, we shift the focus from linear functional analytic techniques to nonlinear PDE, notably harmonic maps into nonlocally compact spaces, a theory recently developed in [65] and [60]. The main idea is to use twistor varieties, which were in the center of the characteristic classes construction of [104], for a deeper study of volume-preserving actions of groups. We introduce an invariant Λ of a volume-preserving action, which is a sort of a $\log L^2$-version of the *sup*-displacement studied in [132]. Our first main result, Theorem 5.6, states that if G is a Kazhdan group acting on a compact manifold M preserving volume, then either $\Lambda > 0$ or G fixes a $\log L^2$-metric. A much weaker analogue of this result for the special case of lattices in Lie groups and *sup*-displacement was known before [132, Theorem 4.8].

We then apply our technique to a major open problem in the field, that of the nonlinear superrigidity of volume-preserving actions of lattices in Lie groups. From a nonlinear version of Margulis theorem given in [133] one knows that a volume preserving action of a lattice in a semisimple Lie group of rank ≥ 2 on a low dimensional (with respect to the group) manifold fixes a **measurable** Riemannian metric. Since measurable metrics do not define a geometry on a manifold, one wishes, of course, to prove a much stronger result: that the action preserves a smooth metric. Zimmer proved (see [132] and references therein) that such stronger result would follow if one is able to find an invariant metric whose dilations with respect to any smooth metric are in the class L^2_{loc}. The central question of how to find such a "bounded" invariant metric was left completely open. We present a completely new approach to the problem which leads to Theorem 5.7. It states that if a cocompact lattice in a Lie group acts on M preserving volume, then either it **nearly** preserves a $\log L^2$-metric, or a sort of a G-structure. This theorem, though constituting a clear progress towards the main problem is still less than what one wants in two respects: First, we deal with $\log L^2$-metrics, not L^2-metrics, second, we leave open a very delicate situation when an action nearly preserves a $\log L^2$-metric, but does not exactly preserve such a metric. This situation is purely infinite-dimensional: If an action on a proper space of nonpositive curvature nearly preserves a point, it actually fixes a point either in the space or at infinity. As already said, we use a heavy machinery: harmonic maps into twistor varieties and vanishing results of [83] and [25]. These results will also be applied in the next chapter to study quaternionic Kähler groups.

It is well-known that the original Kolmogorov's definition of entropy used extremum over all partitions and only became computable after it was realized by Kolmogorov and Sinai that certain partitions realize entropy. In a way of a pleasant similarity, we show how to compute our invariant Λ for $G = \mathbb{Z}$ in case G leaves a geodesic in the twistor space invariant, like a hyperbolic element of $SL(n, \mathbb{Z})$ acting on T^n. This clearly shows the advantage of the $\log L^2$-displacement over the *sup*-displacement.

5.1. $\log L^2$-twistor spaces

C^∞-twistor varieties were used in [104] to define secondary characteristic classes for volume-preserving and symplectic actions. More specifically, we defined, for a compact oriented manifold M equipped with a volume form ν, a series of classes in $H^*_{\text{cont}}(\mathcal{D}iff_\nu(M))$ of dimension $5, 9, 13, \ldots$ (where $\mathcal{D}iff_\nu(M)$ is the group of volume-preserving diffeomorphisms). Likewise, for a compact symplectic manifold M we defined classes in $H^*_{\text{cont}}(Sympl(M))$ of dimensions of $2, 6, 10, \ldots$. For the purposes of the present paper, we will need to work with a $\log L^2$-version of the twistor varieties, defined below.

Remark. I would like to use an opportunity to note that for some strange reasons I overlooked the integrated Euler class in $H^n_{\text{cont}}(\mathcal{D}iff_\nu(M^n))$. The definition is exactly like that in [104] for classes in dimensions $5, 9, \ldots$, if one realizes that there exists an n-form on the twistor variety for M, which is $\mathcal{D}iff_\nu$-invariant. Alternatively, if $\mathcal{D}iff_\nu(M, p_0)$ denotes the subgroup fixing a point p_0, then one pulls back the Euler class of $SL_n(\mathbb{R})$ under the tangent representation

$$\mathcal{D}iff_\nu(M, p_0) \to SL_n(\mathbb{R}),$$

and then applies the measurable transfer (see the above cited paper). The just defined class, viewed as a class in $H^n(\mathcal{D}iff^s_\nu(M))$, is bounded. This follows from the fact that the Euler class is bounded [117] exactly in the same manner as in [104].

We now define the $\log L^2$-twistor variety X for (M, ν). First, one defines a bundle $\mathcal{P}$ of Riemannian metrics with the volume form ν as an $SL(n)/SO(n)$-bundle, associated with a principal $SL(n)$-bundle, defined by ν. Fix a smooth section (i.e., a Riemannian metric with the volume form ν) g_0 of this bundle. For any other measurable section g of $\mathcal{P}$ define

$$\rho^2(g_0, g) = \int_M \rho_x^2(g_0, g)d\nu, \qquad (*)$$

where ρ_x is the distance in $\mathcal{P}_x$ induced by the (fixed once and forever) $SL(n)$-invariant metric on $SL(n)/SO(n)$. Now the twistor variety X consists of $\log L^2$-metrics, that is, measurable metrics g for which

$$\rho(g_0, g) < \infty.$$

Alternatively, let A_x be a self-adjoint (with respect to $(g_0)_x$) operator such that $g_x = g_0(A_x\cdot, \cdot)$. Then $(*)$ can be written as

$$\int_M \|\log A_x\|^2 d\nu < \infty.$$

The crucial fact about $\mathcal{P}$ is a following

Proposition 5.1. *$\mathcal{P}$ is a complete Hilbert–Riemannian manifold with nonpositive curvature operator. The action of $\mathcal{D}iff_\nu(M)$ on $\mathcal{P}$ is isometric.*

Proof. We will only define the metric, leaving all routine checks to the reader. The tangent space at g_0 consists of L^2-sections of S^2T^*M, with trace identically zero. If A is such a section (so that A_x is g_0-self-adjoint for all $x \in M$) then we define the square of the length of A as

$$\int_M trA^2 d\nu.$$

This metric is invariant under the $SO(n)$-valued gauge transformations. Now we define a $\log L^2$ $SL(n)$-gauge group as the group of measurable sections A of $\mathrm{Aut}(TM)$ such that, with respect to g_0,

$$\int_M \|\log(A^*A)\|^2 d\nu < \infty.$$

Then $\mathcal{P}$ is a homogeneous space under the action of this group. We define a Riemannian metric on $\mathcal{P}$ as the unique invariant metric, which agrees at g_0 with the metric just defined.

Now let (M^{2n}, ω) be a compact symplectic manifold. Let $\mathfrak{T}$ be the twistor bundle, that is, the $Sp(2n)/U(n)$-bundle, associated with the principal $Sp(2n)$-bundle, defined by ω. A smooth section of $\mathfrak{T}$ is exactly a tamed almost-complex structure. One then defines a space Z of $\log L^2$-sections of $\mathfrak{T}$ as above (the C^∞-version was used in [104]). □

Proposition 5.2. *The spaces X and Z are nonpositively curved in the sense of Alexandrov.*

Proof. The proofs are standard. □

5.2. A new invariant of smooth volume-preserving dynamical systems

Let (M, ν) be a compact oriented manifold with the volume form ν. Let G be a finitely generated group which acts on M by smooth transformations, preserving ν. We are going to define a new dynamical invariant which we call Λ. This is a nonnegative real number. Though Λ depends on the choice of a system of generators of G, positivity of Λ is independent of the generating set. This relates our Λ to the Kolmogorov's entropy. The invariant Λ is highly nontrivial already for $G = \mathbb{Z}$, that is, as a new invariant of volume-preserving diffeomorphisms. It is also an invariant under conjugation in $\mathcal{D}iff_\nu(M)$. The central result of this section is Theorem 5.6 below stating that if G is a Kazhdan group then either $\Lambda > 0$ or G fixes a $\log L^2$-Riemannian metric (again this connects Λ to the Kolmogorov's entropy).

Let $g_1, \ldots, g_n$ be a system of generators for G. Let $X = \mathcal{P}$ be the twistor variety for (M, ν). Let ρ be the distance function for X, introduced in Section 5.1. We define Λ as the displacement of the G-action:

$$\Lambda = \inf_{z \in X} \max_i \rho(g_i z, z).$$

Proposition 5.3. *Λ is invariant under conjugation in $\mathcal{D}iff_\nu(M)$.*

Proof. ρ is $\mathcal{D}iff_\nu$-invariant. □

Proposition 5.4. *Let $M = (T^n, can)$ and let $G = \mathbb{Z}$ act by iterations of a hyperbolic element of $SL(n, \mathbb{Z})$. Then $\Lambda > 0$.*

Proof. The proof is based on an observation about Alexandrov non-positively curved spaces and a trick from [104].

Lemma 5.5. *Let X be an Alexandrov non-positively curved space and let $\phi : X \to X$ be an isometry which leaves invariant a geodesic γ of X. Then the displacement of ϕ is realized on the points of γ, that is, for $y \in \gamma$,*

$$\rho(y, \phi y) = \min_{x \in X} \rho(x, \phi x).$$

Proof. For $x \in X$ let $y \in \gamma$ be a point which realizes the distance from x to γ. Then $\rho(y, \phi y) \leq \rho(x, \phi x)$. □

Now let X be the twistor space of T^n and let $Y \subset X$ be the space of metrics, invariant under the action of T^n (we view T^n as a Lie group). Then Y is totally geodesic in X, because it is the manifold of fixed points of a family of isometries. As a Riemannian manifold, $Y \simeq SL(n)/SO(n)$. Any hyperbolic matrix ϕ, by definition leaves invariant a geodesic in Y. The result follows. □

The main result in the theory of the invariant Λ is as follows:

Theorem 5.6. *Let G be a Kazhdan group acting on a compact oriented manifold (M, ν) preserving a volume form ν. Then either $\Lambda > 0$ or G fixes a $\log L^2$-metric on M.*

Proof. Consider an isometric action of G on X. If the displacement function $\sup_i \rho(g_i z, z)$ is not bounded away from zero, then either there is a fixed point $z_0 \in X$ for G, or G is not Kazhdan, by a result of [65]. The theorem follows. □

5.3. Non-linear superrigidity alternative

Theorem 5.7. *Let G be either a semisimple Lie group of rank≥ 2, or $Sp(n,1)$ or* $\mathrm{Iso}(\mathbb{C}a\mathbb{H}^2)$. *Let $\Gamma \subset G$ be a cocompact lattice. Let (M^n, ν) be a compact oriented manifold, on which Γ acts preserving the volume form ν. Then either*

a) *Γ preserves a $\log L^2$-metric on M, or*
b) *there exists a sequence $g_0, g_1, \dots$ of smooth Riemannian metrics on M with the volume form ν such that*

$$\int_M \|\log A_i\|^2_{g_0} d\nu \to \infty,$$

where $g_i = g_0(A_i \cdot, \cdot)$ and

$$0 < \mathrm{const}_1 < \sup_j \int_M \|\log B_{ij}\|^2_{g_i} d\nu < \mathrm{const}_2, \qquad (i \to \infty),$$

$\gamma_j^* g_i = g_i(B_{ij}\cdot, \cdot)$, $\{\gamma_j\}$ is a fixed finite set of generators for Γ, or
c) *there is a nonconstant totally geodesic Γ-invariant map*

$$\Psi : G/K \to X,$$

where K is a maximal compact subgroup of G.

Remarks.

1) In case b) we say that Γ nearly fixes a $\log L^2$-metric on M.
2) The case c) implies, for G simple, that $\dim G/K \leq \dim SL(n)/SO(n)$, a so-called Zimmer conjecture.
3) For $G = SL(m,\mathbb{R}), m \geq 3$ and $n = m$, one deduces in case c) the existence of a measurable frame field $\hat{e}(x), \hat{e} = (e_1, \dots, e_n)$, such that for almost all $x \in M$,

$$\pi(\gamma)_*[\hat{e}(x)] = \gamma\hat{e}(\pi(\gamma)x)$$

where $\gamma \in \Gamma$ and $\pi(\gamma)$ is an action of γ on M.
4) Conversely, the standard action of $\Gamma = SL(n,\mathbb{Z})$ on T^n does not satisfy a) (which is well known) and b). To see this, we notice that $SL(n,\mathbb{Z})$ leaves invariant a totally geodesic space Y introduced in the proof of Proposition 5.4. The argument of this proof implies that it is enough to show that the displacement function of the action of Γ on Y diverges to ∞ as one escapes all compact subsets of Y. This follows from the fact that Y is a Riemannian symmetric space of non-compact type and Γ does not fix a point at infinity of Y.
5) The statement of Theorem constitutes a definite progress in the nonlinear superrigidity problem. There is still a mystery in the option b) where one would prefer the statement that Γ fixes a “point at infinity” of the space of metrics X, perhaps a measurable distribution of k-dimension planes, $k \leq n$.

At the time of writing this chapter (August, 1999) I am unable to make such a reduction.

Proof. The proof follows a long-established tradition [112], [25], [83], see also a treatment of [61], in a new infinite-dimensional target context. If neither a) or b) holds then, accounting that Γ is Kazhdan, we deduce that the displacement function of Γ tends to infinity as one escapes all bounded sets in X. Let $F \to \Gamma \setminus G/K$ be a flat fibration with the fiber X, corresponding to the action of Γ in X. A theorem of [65], or [60] implies that there is a harmonic section of F. By Proposition 5.1 and the main theorem of [25] and [83], this section must be totally geodesic. The result follows. □

In the case of a symplectic action of a lattice Γ on a compact symplectic manifold (M, ω), we have a completely similar theorem, as follows.

Theorem 5.8. *Let G be either a semi-simple Lie group of rank ≥ 2, or $Sp(n, 1)$ or $\mathrm{Iso}(\mathbb{C}a\mathbb{H}^2)$, Γ a cocompact lattice in G which acts symplectically on a compact symplectic manifold (M^{2n}, ω). Then either*

a) *Γ fixes a $\log L^2$ tamed almost-complex structure J, or*

b) *there exists a sequence of tamed smooth almost-complex structures $J_i \in Z$ with $\rho(J_0, J_i) \to \infty$ and*

$$0 < \mathrm{const}_2 < \sup_j \rho(\gamma_j J_i, J_i) < \mathrm{const}_1,$$

or

c) *there is a Γ-invariant totally geodesic map*

$$\Psi : G/K \to Z.$$

Proof. The proof is exactly the same as above. □

In case c) and G simple it follows that $\dim G/K \leq \dim Sp(2n)/U(n)$. If $M = (T^{2n}, can)$, $G = Sp(2n, \mathbb{R})$ and case c) one deduces an existence of a measurable symplectic frame $\hat{e}(x) = (e_1, \ldots, e_{2n}(x))$, such that for $\gamma \in \Gamma$,

$$\pi(\gamma)_* [\hat{e}(x)] = \gamma \hat{e}(\pi(\gamma)(x)).$$

6. Kähler and quaternionic Kähler groups

In a letter to the author [26] P. Deligne asked if one can extend the author's theorem on rationality of secondary characteristic classes of a flat bundle over a projective variety to quasiprojective varieties. In 1994 the author was able to answer this question positively in the special case of noncompact ball quotients using the analytic technique of [47] and the scheme of the original proof for projective varieties. Here we present a full answer to Deligne's question, Theorem 6.1, using the analytic technique of [62], who produced harmonic maps of infinite energy but controlled growth.

We then turn to a well-known open problem of finding restrictions on topology of compact quaternionic Kähler manifolds. In the case of positive scalar curvature the situation is well understood, but in the case of negative scalar curvature the twistor spaces of [114] are not Kähler and its technique fails. The only result known was a theorem of [25] stating that the fundamental group does not have infinite linear representations unless the manifold is locally symmetric. Our result, Theorem 6.4, states that the fundamental group is Kazhdan. This is, of course, a severe restriction (Kazhdan groups are rare). As a by-product of our technique, we obtain a new proof of a classical theorem, stating that the lattices in semisimple Lie groups of rank ≥ 2, $Sp(n,1)$ $(n \geq 2)$ and $\mathrm{Iso}(\mathbb{C}a\mathbb{H}^2)$ are Kazhdan. We also show, using Section 1.1, that the nontrivial classes in the second cohomology of a Kähler non-Kazhdan group, constructed in [106], are of polynomial growth. This again is very rare for "just a group", as polynomial growth in cohomology is connected to a polynomial isoperimetric inequality in the Cayley graph, which needs special reasons to hold. This means that Kähler groups are rare, too.

6.1. Rationality of secondary classes of flat bundles over quasiprojective varieties

A rationality theorem for the secondary classes of flat bundles over compact Kähler manifolds (previously known as the Bloch conjecture [11]) was proved in 1993 in [103] and [105]. In a letter to the author [26] P. Deligne asked if one can prove such a statement for local systems with logarithmic singularities over a quasiprojective variety. It turns out that the answer is positive:

Theorem 6.1. *Let X be a quasiprojective variety, $\rho : \pi_1(X) \to SL(n,\mathbb{C})$ a representation. Let $b_i(\rho)$ be the imaginary part and $ChS_i(\rho)$ be the $\mathbb{R}/\mathbb{Z}$-part of the secondary class $c_i(\rho) \in H^{2i-1}(X,\mathbb{C}/\mathbb{Z})$ of the flat bundle with monodromy ρ. Then*

A. $b_i(\rho) = 0 \quad (i \geq 2)$ *(the Vanishing Theorem).*
B. $ChS_i(\rho) \in H^{2i-1}(X,\mathbb{Q}/\mathbb{Z})$ *(the Rationality Theorem).*

Proof. For any smooth manifold, A implies B, as explained in the above cited papers. So we only prove A. As explained in the above cited papers, we may assume that ρ is irreducible. Then, by the recent result [62], the associated $SL(n,\mathbb{C})/SU(n)$ flat bundle over X possesses a pluriharmonic section s which satisfies the Sampson degeneration condition. This means the following. The derivative Ds_x, $x \in X$ can be viewed as an $\mathbb{R}$-linear map to the space P of Hermitian matrices. Let

$$(Ds_x)^{\pm}_{\mathbb{C}}(Y) = (Ds_x(Y) \pm \sqrt{-1}Ds_x(\sqrt{-1}Y))$$

be the corresponding map of TX to $P \otimes \mathbb{C}$. Then the image of $(Ds_x)^{\pm}_{\mathbb{C}}$ consists of commuting matrices. Now the first proof of the Main Theorem in [105] applies word-by-word and the result follows. □

6.2. Property T for Kähler and quaternionic Kähler groups

There are two ways to geometrize the group theory. One approach (the *time geometry* in the terminology of [107]) is to consider finitely generated groups which act on a (usually compact) space with some structure (a volume form, a symplectic form, a tree, a circle, a conformal structure, etc). An amazing phenomenon, amply demonstrated in the previous chapters is that these groups tend to be not Kazhdan. Another approach (the *space geometry*) is to consider groups which are fundamental groups of a compact (or close to compact) manifold with some structure (like Kähler). It happens that these groups tend to be Kazhdan. Therefore these two families of "geometric" groups are essentially disjoint. The following is the main result of [106].

Theorem 6.2. *Let G be the fundamental group of a compact Kähler manifold. If G is not Kazhdan, then $H^2(G,\mathbb{R}) \neq 0$. Moreover, if H is not Kazhdan and $\psi : G \to H$ is surjective, then the map*

$$\psi^* : H^2(H,\mathbb{R}) \to H^2(G,\mathbb{R})$$

is nonzero.

I would like to mention an important property, which I overlooked in [106]:

Proposition 6.3. *Under the conditions of the above theorem, there is a nontrivial class of polynomial growth in $H^2(G,\mathbb{R})$.*

Proof. It was proven in [106] that there exists a unitary representation $\rho : G \to U(\mathcal{H})$ and a class $l \in H^1(G,\mathcal{H})$ such that the class γ in $H^2(G,\mathbb{R})$ given by $\langle l, l\rangle$ is nonzero. Here $\langle l, l\rangle$ is obtained by taking the imaginary part of the image of l under the cup-product followed by the scalar product in $\mathcal{H}$. Now the result follows from Lemma 1.1 in Chapter 1. □

It is extremely rare for a finitely generated group to have nonzero polynomial cohomology.

We now turn to quaternionic Kähler manifolds. If the scalar curvature is positive, then the topology is very well understood due to the work of Solomon [114]. To the contrary, if the scalar curvature is negative, the only known result (due to Corlette) is that the fundamental group satisfy the geometric superrigidity [25]. This means that if $\pi_1(X)$ admits a Zariski-dense representation in an algebraic Lie group, then $\pi_1(X)$ is a lattice, and X a symmetric space of a certain type. However, it is a rare occasion for a group to have any finite-dimensional linear representations with infinite image. Using a combination of ideas of [25], [106] and [65], we now prove a very strong structure theorem:

Theorem 6.4. *Let X be a quaternionic Kähler manifold of negative scalar curvature. Then $\pi_1(X)$ is Kazhdan.*

Proof. Suppose not. Then by [65] there exists an affine flat Hilbert bundle E over X with a nonparallel harmonic section. By the vanishing result of [25], this section must be totally geodesic. Then X is covered by a flat torus, which is a contradiction. □

Remark. The same argument provides a new proof of the classical theorem [64], [67], that the (cocompact) lattices in semisimple Lie groups of rank ≥ 2, $Sp(n,1)$ and $\mathrm{Iso}(\mathbb{C}a\mathbb{H}^2)$ are Kazhdan. One uses a vanishing result of [83] (see also the treatment in [61]) for lattices in semisimple Lie groups of rank ≥ 2, and the above-mentioned result of [25] for $Sp(n,1)$ and $\mathrm{Iso}(\mathbb{C}a\mathbb{H}^2)$. Once established for cocompact lattices, the result follows for all lattices because a Lie group is Kazhdan if and only if all its lattices are Kazhdan.

References

[1] M. Adams, T. Ratiu, R. Schmid, *The Lie group structure of diffeomorphism groups and Fourier integral operators with applications*, in: Infinite dimensional Lie groups with applications, V. Kac, Editor, Springer, 1985.

[2] L. Ahlfors, A. Beurling, *The boundary correspondence under quasiconformal mappings*, Acta Math., **96** (1956), 125–142.

[3] R. Alperin, *Locally compact groups acting on trees and property T*, Monatsh. Math., **93** (1982), 261–265.

[4] J. Andersen, *Mapping Class Groups do not have Kazhdan's Property (T)*, Preprint, arXiv:0706.2184, 2007.

[5] S. Arakelov, *Families of algebraic curves with fixed degeneracies*, (Russian) Izv. Akad. Nauk SSSR Ser. Mat. 35 (1971), 1269–1293.

[6] F. Bayen, M. Flato, C. Fronsdal, A. Lichnerowicz, D. Sternheimer, *Deformation theory and quantization*, Ann. Phys. **III** (1978), 61–110.

[7] A.A. Belavin, *Discrete groups and integrability of quantum systems*, Funct. Anal. Appl, **14** (1980), 18–26 (Russian).

[8] A.A. Belavin, A.M. Polyakov, A.B. Zamolodchikov, *Infinite conformal symmetry in two-dimensional quantum field theory*, Nucl. Phys., **B241** (1984), 333–380.

[9] J. Birman, *Braids, links, and mapping class groups.* Annals of Mathematics Studies, No. 82. Princeton University Press, Princeton, N.J.; University of Tokyo Press, Tokyo, 1974.

[10] C. Bishop, T. Steger, *Representation-theoretic rigidity in* $\mathrm{PSL}(2,\mathbb{R})$, Acta Math. 170 (1993), no. 1, 121–149.

[11] S. Bloch, *Applications of the dilogarithm functions in algebraic K-theory and algebraic geometry*, in: Proc. Int. Symp. Alg. Geom, Kyoto, Kinokumiya, 1977, 103–114.

[12] F. Bonahon, *Bouts des variétés hyperboliques de dimension* 3. Ann. of Math. (2) **124** (1986), no. 1, 71–158.

[13] B. Bojarski, *Generalized solutions of a system of differential equations of the first order with discontinuous coefficients*, Math. Sbornik, **43** (1957), 451–503 (Russian).

[14] A. Borel, *Stable real cohomology of arithmetic groups*, Ann. Sci. Ecole Norm. Sup. **7** (1974), 235–272 (1975).

[15] R. Bott, *On the characteristic classes of groups of diffeomorphisms*, Enseign. Math. **23** (1977), 209–220.

[16] M. Bourdon, F. Martin, A. Valette, *Vanishing and non-vanishing for the first L_p-cohomology of groups*, Comment. Math. Helv. 80 (2005), no. 2, 377–389.

[17] K.S. Brown, *Cohomology of Groups*, Springer Verlag, 1982.

[18] G. Besson, G. Courtois, S. Gallot, *Entropies et rigidités des espaces localement symétriques de courbeure strictement négative*, GAFA **5** (1995), 731–799.

[19] K.S. Brown, R. Geoghegan, *An infinite dimensional torsion free FP_∞ group*, Invent. Math. **77** (1984), 367–381.

[20] A. Connes, M. Gromov, H. Moscovici, *Group cohomology with Lipschitz control and higher signatures*, GAFA **3** (1993), 1–78.

[21] J. Cannon, W. Thurston, *Group invariant Peano curves*, Geom. Topol. **11** (2007), 1315–1355.

[22] L. Carleson, *The extension problem for quasiconformal mappings*, in: Contributions to Analysis, AP, 1974, 39–47.

[23] L. Carleson, *On mappings, conformal at the boundary*, J. Analyse Math. **19** (1967), 1–13.

[24] A. Connes, H. Moscovici, *Cyclic cohomology, the Novikov conjecture and hyperbolic groups*, Topology **29** (1990), 345–388.

[25] K. Corlette, *Archimedean superrigidity and hyperbolic geometry*, Ann. Math. **135** (1992), 165–182.

[26] P. Deligne, *A letter to the author*, 1994.

[27] A. Douady, C.J. Earle, *Conformally natural extensions of homeomorphisms of the circle*, Acta Math., **157** (1986), 23–48.

[28] J.L. Dupont, *Simplicial de Rham cohomology and characteristic classes of flat bundles*, Topology, **15** (1976), 233–245.

[29] D.E. Edmunds, B. Opic, *Weighted Poincaré and Friedrichs inequalities*, J. London Math. Soc., **47** (1993), 79–96.

[30] D.E. Edmunds, H. Triebel, *Logarithmic Sobolev spaces and their applications to spectral theory*, Proc. London Math. Soc. **71** (1995), 333–371.

[31] B. Farb, P. Shalen, *Real-analytic action of lattices*, Invent. Math. **135** (1999), 273–296.

[32] B. Fedosov, *Index theorems.* In "Partial Differential Equations, VIII", 155–251. Encyclopaedia Math. Sci., 65, Springer, Berlin, 1996.

[33] B.L. Feigin, B.L. Tsygan, *Cohomology of Lie algebras of generalized Jacobi matrices*, Funct. Anal. Appl. **17** (1983), 86–87 (Russian).

[34] D.B. Fuks, *Cohomology of Infinite-Dimensional Lie Algebras*, Contemporary Soviet Mathematics. Consultants Bureau, New York and London, 1986.

[35] H. Furstenberg, *A Poisson formula for semi-simple Lie groups*, Ann. Math., **77** (1963), 335–386.

[36] H. Furstenberg, *Boundary theory and stochastic processes on homogeneous spaces*, Proc. Symp. Pure Math. **26** (1973), 193–233.

[37] F.P. Gardiner, D. Sullivan, *Symmetric structures on a closed curve*, Amer. J. Math. **114** (1992), 683–736.

[38] L. Garnett, *Foliations, the ergodic theorem and Brownian motion*, J. Funct. Anal., **51** (1983), 285–311.

[39] F. Gehring, *The L^p-integrability of the partial derivatives of a quasiconformal mapping*, Acta Math., **130** (1973), 265–277.

[40] I.M. Gelfand, D.B. Fuks, *Cohomology of Lie algebras of vector fields on the circle*, Funct. Anal. Appl. **2** (1968), 92–93 (Russian).

[41] M. Gerstenhaber, *The cohomology structure of an associative ring*, Ann. of Math. (2),**78** (1963), 267–288.

[42] E. Ghys, *Actions de réseaux sur le cercle*, Invent. Math. **137** (1999), 199–231.

[43] E. Ghys, V. Sergiescu, *Sur un groupe remarquable de difféomorphismes du cercle*, Comment. Math. Helv. **62** (1987), 185–239.

[44] V.M. Goldshtein, V.I. Kuzminov, I.A. Shvedov, *On a problem of Dodziuk*, Trudy Mat. Inst. Steklov, **193** (1992), 72–75.

[45] V.I. Gorbachuk, M.L. Gorbachuk, *Boundary value problems for operator differential equations*, Mathematics and its Applications **48**, Kluwer Academic Publishers Group, Dordrecht, 1991.

[46] P. Greenberg, V. Sergiescu, *An algebraic extension of the braid group*, Comment. Math. Helv. **66** (1991), 109–138.

[47] M. Gromov, R. Schoen, *Harmonic maps into singular spaces and p-adic rigidity for lattices in groups of rank one*, Publ. Math. IHES, **76** (1992), 165–246.

[48] A. Guichardet, *Cohomologie des groupes topologiques et des algèbres de Lie*, Cedic/Fernand Nathan, 1980.

[49] A. Guichardet, *Sur la cohomologie des groupes topologiques II*, Bull. Sci. Math. **96** (1972), 305–332.

[50] V.S. Guba, *Polynomial upper bounds for the Dehn function of R. Thompson group F*, Journ. Group Theory **1** (1998), 203–211.

[51] U. Hämenstadt, *A lecture at Leipzig conference "Perspectives in Geometry"*, 1998.

[52] P. de la Harpe, A. Valette, *La propriété (T) de Kazhdan pour les groupes localement compactes*, Astérisque **175** (1989).

[53] E. Hewitt, K.A. Ross, *Abstract Harmonic Analysis*, Springer, 1970.

[54] E. Hopf, *Statistik der geodatischen Linien in Mannigfaltigkeiten negativer Krümmung*, Ber. Verb. Sachs. Akad. Wiss. Leipzig **91** (1939), 261–304.

[55] L. Hörmander, *Linear partial differential operators*, Die Grundlehren der mathematischen Wissenschaften, Bd. 116, Academic Press Inc., Publishers, New York,1963.

[56] S. Hurder, A. Katok, *Differentiability, rigidity and Godbillon-Vey classes for Anosov flows*, Publ. Math. IHES, No. 72 (1990), 5–61 (1991).

[57] S. Hurder, D. Lehmann, *Homotopy characteristic classes of foliations*, Ill. J. Math. **34**, (1990) No. 3, 628–655.

[58] N. Ivanov, V. Turaev, *The canonical cocycle for the Euler class of a flat vector bundle*, Dokl. Akad. Nauk SSSR 265 (1982), no. 3, 521–524.

[59] B.A. Johnson, *Cohomology in Banach algebras*, Memoirs of the American Mathematical Society, No. 127. American Mathematical Society, 1972.

[60] J. Jost, *Equilibrium maps between metric spaces*, Calc. Var. Partial Dif. Eq.,**2** (1994), 173–204.

[61] J. Jost, S.T. Yau, *Harmonic maps and superrigidity*, Proc. Symp. Pure Math., **54** (1993), 245–280.

[62] J. Jost, K. Zuo, *Harmonic maps of infinite energy and rigidity results for representations of fundamental groups of quasiprojective varieties*, J. Diff. Geom., **47** (1997), 469–503.

[63] V.A. Kaimanovich, A.M. Vershik, *Random walks on discrete groups: boundary and entropy*, Ann. Probab. **11** (1983), 457–490.

[64] D. Kazhdan, *On the connection of the dual space of a group with the structure of its closed subgroups*, (Russian) Funkcional. Anal. i Priložen., **1** (1967) 71–74.

[65] N. Korevaar, R. Schoen, *Global existence theorems for harmonic maps to non-locally compact spaces*, Comm. Geom. Anal., **5** (1993), 333–387.

[66] N. Korevaar, R. Schoen, *Sobolev spaces and harmonic maps for metric space targets*, Comm. Geom. Anal., **1** (1993), 561–659.

[67] B. Kostant, *On the existence and irreducibility of certain series of representations*, Lie Groups and Their Representations, Halsted, NY, 1975, 231–329.

[68] P. Koosis, *Introduction to H_p Spaces*, Cambridge UP, 1998.

[69] L.D. Kudryavcev, *Habilitationsschrift*, Steklov Math. Inst., 1956.

[70] L.D. Kudryavcev, *Direct and inverse imbedding theorems. Applications to the solutions of elliptic equations by variational methods,* Trudy Steklov Math. Inst. **55** (1959).

[71] J.L. Lions, *Théorèmes de trace et d'interpolation*, I, Ann. Schuola Norm. Super. Pisa, **13** (1959), 389–403.

[72] P.I. Lizorkin, *Boundary values of functions from "weight" classes*, Sov. Math. Dokl. **1** (1960), 589–593.

[73] P.I. Lizorkin, *Boundary values of a certain class of functions*, Dokl. Anal. Nauk SSSR **126** (1959), 703–706 (Russian).

[74] N.G. Makarov, *On the distortion of boundary sets under conformal mappings*, Proc. London Math. Soc. **51** (1985), 369–384.

[75] N.G. Makarov, *On the radial behavior of Bloch functions*, Soviet Math. Dokl. **40** (1990), 505–508.

[76] S. Matsumoto, S. Morita, *Bounded cohomology of certain groups of homeomorphisms*, Proc.Amer. Math. Soc. **94** (1985), 539–544.

[77] S. Morita, *Characteristic classes of surface bundles*, Invent. Math., **90** (1987), 551–577.

[78] S. Morita, *Characteristic classes of surface bundles and bounded cohomology*, in: A Fête of Topology, AP, 1988, 233–257.

[79] J.E. Moyal, *Quantum mechanics as a statistical theory*, Proc. Cambridge Philos. Soc.**45** (1949), 99–124.

[80] E.Y. Miller, *The homology of the mapping class group*, J. Diff. Geom. **24** (1986), 1–14.

[81] A.S. Mishchenko, *Infinite-dimensional representation of discrete groups and higher signatures*, Math. USSR. Izv. **8** (1974), 85–111.

[82] A.S. Mishchenko, *Hermitian K-theory, the theory of characteristic classes and methods of functional analysis*, Russian Math. Surveys **31** (1976), 71–138.

[83] N. Mok, Y.-T. Siu, S.-K. Yeung, *Geometric superrigidity*, Invent. Math. **113** (1993), 57–83.

[84] D. Mumford, *Towards an enumerative geometry of the moduli space of curves*, in: Arithmetic and geometry, Progress in Math. **36**, Birkhäuser, 1983, 271–328.

[85] F.J. Murray, J. von Neumann, *On rings of operators*, Ann. Math **37** (1936), 116–129, TAMS **41** (1937), 208–248, Ann. Math., **41** (1940), 94–161, **44** (1943), 716–808.

[86] S.N. Naboko, *Nontangential boundary values of operator-valued R-functions*, Leningrad Math. Journ. **1** (1990), 1255–1278.

[87] S. Nag, *The complex analytic theory of Teichmüller spaces*, Canadian Mathematical Society Series of Monographs and Advanced Texts, A Wiley-Interscience Publication,John Wiley & Sons Inc., New York, 1988.

[88] A. Navas, *Actions de groupes de Kazhdan sur le cercle*, Ann. Sci. Ecole Norm. Sup. (4) 35 (2002), no. 5, 749–758.

[89] A. Navas, *Quelques nouveaux phénomènes de rang* 1 *pour les groupes de difféomorphismes du cercle*, Comment. Math. Helv. 80 (2005), no. 2, 355–375.

[90] P.J. Nicholls, *A measure on the limit set of a discrete groups*, in: Ergodic Theory, Symbolic Dynamics and Hyperbolic Spaces, T. Bedford, M. Keane, C. Series, eds., Oxford UP, 1991, 259–296.

[91] G.A. Noskov, *Bounded cohomology of discrete groups with coefficients*, (Russian) Algebra i Analiz 2 (1990), no. 5, 146–164; translation in Leningrad Math. J. 2 (1991), no. 5, 1067–1084.

[92] J.-P. Otal, *The hyperbolization theorem for fibered* 3*-manifolds.* Translated from the 1996 French original by Leslie D. Kay. SMF/AMS Texts and Monographs, 7. American Mathematical Society; Société Mathématique de France, Paris, 2001.

[93] R. Palais, *On the homotopy type of certain groups of operators*, Topology **3** (1965), 271–279

[94] P. Pansu, *Cohomologie L^p des variétés à courbure négative, cas du degré un,* Rend. Sem. Mat. Torino (1989), 95–120.

[95] P. Pansu, *Differential forms and connections adapted to a contact structure, after M. Rumin*, in: Symplectic Geometry, D. Salamon, Ed., Cambridge UP (1993), 183–195.

[96] A. Pietsch, *Operator Ideals*, VEB, Berlin, 1978.

[97] C. Pommerenke, *Boundary Behaviour of Conformal Maps*, Springer, 1992.

[98] A. Pressley, G. Segal, *Loop Groups*, Clarendon Press, Oxford, 1986.

[99] H.M. Reimann, *Functions of bounded mean oscillation and quasiconformal mappings*, Comment. Math. Helv. **49** (1974), 260–276.

[100] S. Rempel, B.-W. Schulze, *Index Theory of Elliptic Boundary Problems*, Academie-Verlag, Berlin, 1982.

[101] A. Reznikov, *The space of spheres and conformal geometry*, Riv. Math. Un. Parma **17** (1991), 111–130.

[102] A. Reznikov, *Characteristic classes in symplectic topology*, Sel. Math. **3** (1997), 601–642.

[103] A. Reznikov, *Rationality of secondary classes*, J. Diff. Geom. **43** (1996), 674–682.

[104] A. Reznikov, *Continuous cohomology of volume-preserving and symplectic diffeomorphisms, measurable transfer and higher asymptotic cycles*, Sel. Math. **5** (1999), 181–198.

[105] A. Reznikov, *All regulators of flat bundles are torsion*, Ann. Math. **141** (1995), 373–386.

[106] A. Reznikov, *The structure of Kähler groups. I. Second cohomology*, in: Motives, polylogarithms and Hodge theory, Part II (Irvine, CA, 1998), 717–730, Int. Press Lect. Ser., 3, II, Int. Press, Somerville, MA, 2002.

[107] A. Reznikov, *Analytic topology*, in: European Congress of Mathematics, Vol. I (Barcelona, 2000), 519–532, Progr. Math., 201, Birkhäuser, Basel, 2001.

[108] A. Reznikov, *Arithmetic topology of units, ideal classes and three and a half-manifolds*, in preparation.

[109] A. Reznikov, *Harmonic maps, hyperbolic cohomology and higher Milnor inequalities*, Topology **32** (1993), 899–907.

[110] A. Reznikov, *Analytic Topology II*, in preparation.

[111] T. Roblin, *A Fatou theorem for conformal densities with applications to Galois coverings in negative curvature*, Isr. J. Math. 147, (2005) 333–357.

[112] Y.T. Siu, *The complex-analyticity of harmonic maps and strong rigidity of compact Kähler manifolds*, Ann. Math. **112** (1980), 73–111.

[113] G.B. Segal, *Unitary representations of some infinite dimensional groups*, Comm. Math. Phys. **80** (1981), 301–342.

[114] S. Salamon, *Quaternionic Kähler manifolds*, Invent. Math. **67**(1982), 143–171

[115] R. Strichartz, *Improved Sobolev inequalities*, Trans. Amer. Math. Soc.**279** (1983), 397–409.

[116] D. Sullivan, *Entropy, Hausdorff measures old and new, and limit sets of geometrically finite Kleinian groups*, Acta Math. **153**(1984), 259–277

[117] D. Sullivan, *A generalization of Milnor's inequality concerning affine foliations and affine manifolds*, Comment. Math. Helv. 51 (1976), no. 2, 183–189.

[118] D. Sullivan, *On the ergodic theory at infinity of arbitrary discrete group of hyperbolic motions*, in: Riemann Surfaces and Related Topics, I. Kra and B. Maskit, Editors, Up (1981), 465–496.

[119] J. Thompson, Unpublished.

[120] H. Triebel, *Theory of Function Spaces*, Birkhäuser, 1983.

[121] W. Thurston, *Three-dimensional geometry and topology*, Princeton Mathematical Series, **35**. Princeton University Press, Princeton, NJ, 1997.

[122] W. Thurston, *Hyperbolic structure on 3-manifolds, II: surface groups and 3-manifolds which fiber over the circle*, Preprint, 1986.

[123] P. Tukia, *The Hausdorff dimension of the limit set of a geometrically finite Kleinian group*, Acta Math. **152** (1989), 127–140.

[124] P. Tukia, J. Väisälä, *Quasiconformal extension from dimension n to $n+1$*, Ann. Math., **115** (1982), 331–348.

[125] A.A. Vasharin, *The boundary properties of functions having a finite Dirichlet integral with a weight*, Dokl. Anal. Nauk SSSR **117** (1957), 742–744.

[126] J.-L. Verdier, *Les répresentations des algèbres de Lie affines: applications à quelques problèmes de physique (d'après E. Date, M. Jimbo, M. Kashivara, T. Miwa*, Séminaire Bourbaki, Exposé 596 (1981–1982), 1–13.

[127] J. Vey, *Déformation du crochet de Poisson sur une variété symplectique*, Comment. Math. Helv., **50** (1975), 421–454.

[128] S.V. Uspenskiĭ, *Imbedding theorems for weighted classes*, Trudy Math. Inst. Steklov **60** (1961), 282–303 (Russian) (Amer. Math. Soc. Trans. **87** (1970)).

[129] Y. Watatani, *Property* (T) *of Kazhdan implies property* (FA) *of Serre*, Math. Japon. **27** (1981), 97–103.

[130] D. Witte, *Arithmetic groups of higher* Q*-rank cannot act on* 1*-manifolds*, Proc. Amer. Math. Soc. **122** (1994), 333–340.

[131] R. Zimmer, *Kazhdan groups acting on manifolds*, Invent. Math., **75** (1984), 425–436.

[132] R. Zimmer, *Lattices in semisimple groups and invariant geometric structures on compact manifolds*, Discrete Groups in Geometry and Analysis, Progress in Math. **67** (1987), 152–210.

[133] R. Zimmer, *Strong rigidity for ergodic actions of semisimple Lie groups*, Annals of Math., **112** (1980), 511–529.

[134] R. Zimmer, *Ergodic theory and semisimple groups.* Monographs in Mathematics, 81. Birkhäuser Verlag, Basel, 1984.

Part II

Research Articles

Progress in Mathematics, Vol. 265, 97–111

Jørgensen's Inequality for Non-Archimedean Metric Spaces

J. Vernon Armitage and John R. Parker

Dedicated to the memory of Alexander Reznikov

Abstract. Jørgensen's inequality gives a necessary condition for a non-elementary group of Möbius transformations to be discrete. In this paper we generalise this to the case of groups of Möbius transformations of a non-Archimedean metric space. As an application, we give a version of Jørgensen's inequality for $\mathrm{SL}(2, \mathbb{Q}_p)$.

Mathematics Subject Classification (2000). 20H20, 11E95.

Keywords. Discrete group, p-adic numbers.

1. Introduction

In [6] Jørgensen proved a famous inequality giving a necessary condition for a non-elementary subgroup of $\mathrm{SL}(2, \mathbb{C})$ to be discrete. Intuitively, this inequality says that if two elements of $\mathrm{SL}(2, \mathbb{C})$ generate a non-elementary discrete group, then they cannot both be very close to the identity. Jørgensen's theorem both makes this statement precise and gives explicit uniform bounds.

The methods used to prove this inequality have been generalised to a wide variety of different contexts but, generally, the statements look rather different from that given by Jørgensen. For example, a geometrical interpretation says there is always an embedded tubular neighbourhood of a very short geodesic in a hyperbolic manifold and that this neighbourhood, or "collar", has volume uniformly bounded away from zero. Hence handles in hyperbolic manifolds cannot be both short and thin.

In [7] Markham and Parker gave a general formulation of Jørgensen's inequality for Möbius transformations on a metric space which recovers many known versions as special cases. In these examples the one-point compactification of the metric space in question is the boundary of a rank one symmetric space of non-compact type, that is one of real, complex or quaternionic hyperbolic spaces or the octonionic hyperbolic plane. Additionally, this result applies when the metric space is a field, for example the p-adic numbers $\mathbb{Q}_p$ in which case $\text{Möb}(\mathbb{Q}_p) = \text{PSL}(2, \mathbb{Q}_p)$. In the main result of this paper, Theorem 3.1, we show that for non-Archimedean metric spaces one obtains a better inequality than Theorem 2.4 of [7]. In the case of $\mathbb{Q}_p$ this improved version of Jørgensen's inequality looks very similar to the original statement given by Jørgensen in [6]; see Theorem 4.2. We interpret this theorem geometrically in terms of the action of our group on an infinite, regular $p+1$ valent tree.

In the final section, we consider function field spaces. There is a strong analogy between these spaces and the p-adic numbers. It is possible to give a version of Theorem 4.2 in this case, but we leave details to the reader.

We would like to thank the referee for his/her valuable comments. Also, we would like to thank Guyan Robertson for his help, in particular for telling us about reference [3].

2. Non-Archimedean Möbius transformations

Let X be a non-empty set. A distance or *metric* on X is a function ρ from pairs of elements (x, y) to the real numbers satisfying:

(i) $\rho(x, y) \geq 0$ with equality if and only if $x = y$;
(ii) $\rho(x, y) = \rho(y, x)$;
(iii) $\rho(x, y) \leq \rho(x, z) + \rho(z, y)$ for all $z \in X$.

The inequality in (iii) is called the triangle inequality. A metric is said to be *non-Archimedean* if the triangle inequality is replaced with the following stronger inequality, called the *ultrametric inequality*:

(iv) $\rho(x, y) \leq \max\{\rho(x, z), \rho(z, y)\}$ for all $z \in X$.

A simple consequence of the ultrametric inequality is the fact that every triangle in a non-Archimedean metric space is isosceles:

Lemma 2.1. *Suppose that ρ is a non-Archimedean metric on a space X. If x, y and z are points of X so that $\rho(x, y) < \rho(x, z)$, then $\rho(x, z) = \rho(y, z)$.*

Proof. We have

$$\rho(y, z) \leq \max\Bigl\{\rho(x, y), \rho(x, z)\Bigr\} = \rho(x, z)$$

by hypothesis. Likewise,

$$\rho(x, z) \leq \max\Bigl\{\rho(x, y), \rho(y, z)\Bigr\} = \rho(y, z)$$

since otherwise we would have $\rho(x, z) \leq \rho(x, y)$ which would be a contradiction. Therefore, we have

$$\rho(y, z) \leq \rho(x, z) \leq \rho(y, z)$$

and hence these quantities are equal. □

Many metrics arise from valuations on a ring. Let R denote a non-trivial ring. An *absolute value* (or *valuation* or *norm*) on R is a real-valued function $x \longmapsto |x|$ on R satisfying:

(i) $|x| \geq 0$ with equality if and only if $x = 0$;
(ii) $|xy| = |x|\,|y|$;
(iii) $|x+y| \leq |x| + |y|$.

Once again, a valuation is said to be *non-Archimedean* if the inequality in (iii) is replaced with the stronger inequality:

(iv) $|x+y| \leq \max\{|x|, |y|\}$.

Given a valuation $|\ |$ on a ring R we may define a metric on R by:

$$\rho(x,y) = |x-y|.$$

Examples.

(i) The standard absolute value on $\mathbb{R}$ or $\mathbb{C}$, which gives rise to the Euclidean metric.
(ii) Fix a prime number p and let $r \in \mathbb{Q}$ be non-zero. Write $r = p^f u/v$ where $f \in \mathbb{Z}$ and u, v are coprime integers both of which are also coprime to p. Then define a valuation $|\ |_p$ on $\mathbb{Q}$ by:

$$|r|_p = p^{-f}, \qquad |0|_p = 0. \tag{1}$$

One can then show that $|r+s|_p \leq \max\{|r|_p, |s|_p\}$. This valuation is called the *p-adic valuation*.

Let X be a complete non-Archimedean metric space with metric ρ. Following [7], we now define the Möbius transformations on X. Let $\mathrm{Aut}(X) \subset \mathrm{Isom}(X)$ be a group of isometries of X. This may be either the full isometry group or a sufficiently large subgroup that preserves some extra structure on X. We will suppose that $\mathrm{Aut}(X)$ acts transitively on X. The metric ρ induces a topology on X and we give $\mathrm{Aut}(X)$ the corresponding compact-open topology. Let o be a distinguished point of X. (Since $\mathrm{Aut}(X)$ acts transitively, in fact we may take o to be any point of X.) Suppose that the stabiliser of o in $\mathrm{Aut}(X)$ is compact. We make some more assumptions about X that allow us to extend $\mathrm{Aut}(X)$ to the group of Möbius transformations on X.

Given $d \in \mathbb{R}_+$, a *dilation* with *dilation factor* $d^2 \in \mathbb{R}_+$ is a map $D_d : X \longrightarrow X$ with $D_d o = o$ and for all $z, w \in X$ we have

$$\rho(D_d z, D_d w) = d^2 \rho(z,w). \tag{2}$$

(It may seem more natural to have taken d rather than d^2. However that would have introduced square roots into our formulae, such as (7) below.) Note that if $d \neq 1$, then D_d has a unique fixed point in X.

Let $X \cup \{\infty\}$ be the one-point compactification of X. Suppose that there is an involution R interchanging o and ∞ and so that if $z,\, w \in X - \{o\}$, then

$$\rho(Rz, o) = \frac{1}{\rho(z,o)}, \tag{3}$$

$$\rho(Rz, Rw) = \frac{\rho(z,w)}{\rho(z,o)\rho(w,o)}. \tag{4}$$

We may think of R as reflection in the unit sphere of centre $o \in X$.

Let Möb(X) be the group generated by Aut(X), D_d and R for all d in some multiplicative subgroup of $\mathbb{R}_+$. We call Möb(X) the group of *Möbius transformations* of X. There is a natural topology on $X \cup \{\infty\}$ induced from the metric ρ (so neighbourhoods of ∞ are the exteriors of compact subsets of X). This enables us to define the compact-open topology for continuous functions from $X \cup \{\infty\}$ to itself. We will be interested in discrete subgroups of Möb(X) with respect to this topology.

Proposition 2.2 (Proposition 2.1 of [7]). *Let X be a metric space and* Möb(X) *be the group generated by* Aut(X), D_d *and R satisfying* (2), (3) *and* (4).

(i) *Let A be any element of* Möb(X) *for which $A\infty = \infty$. Then there exists a positive number d_A so that for all $z,\, w \in X$*

$$\rho(Az, Aw) = {d_A}^2 \rho(z,w).$$

(ii) *Let B be any element of* Möb(X) *for which $B\infty \neq \infty$. Then there exists a positive number r_B so that for all $z,\, w \in X - \{B^{-1}\infty\}$*

$$\rho(Bz, Bw) = \frac{{r_B}^2 \rho(z,w)}{\rho(z, B^{-1}\infty)\rho(w, B^{-1}\infty)},$$

$$\rho(Bz, B\infty) = \frac{{r_B}^2}{\rho(z, B^{-1}\infty)}.$$

The intuition behind Proposition 2.2(ii) is that B is like reflection in a sphere of radius r_B followed by an isometry. Also, we see that for all $B \in$ Möb(X) with $B\infty \neq \infty$ we have

$$\frac{\rho(Bz, z)}{\rho(Bz, B\infty)} = \frac{\rho(z, B^{-1}z)}{\rho(B^{-1}z, B^{-1}\infty)}. \tag{5}$$

Lemma 2.3 (Lemma 2.2 of [7]). *Let X be a metric space. If* Aut(X) *acts transitively on X, then* Möb(X) *acts 2-transitively on $X \cup \{\infty\}$. That is, given any two pairs $x_1,\, y_1;\, x_2,\, y_2$ of points in $X \cup \{\infty\}$, there exists $B \in G$ so that $B(x_2) = x_1$ and $B(y_2) = y_1$.*

Lemma 2.4 (Lemma 2.3 of [7]). *Let X be a metric space. Suppose that $B \in$* Möb(X) *fixes distinct points $x,\, y \in X \cup \{\infty\}$. Then B is conjugate to $A \in$* Möb(X) *with fixed points o and ∞. Moreover, the dilation factor ${d_A}^2$ of A is independent of the conjugating map.*

Define the *cross-ratio* of quadruples of points in $X \cup \{\infty\}$ by

$$\begin{aligned}\mathbb{X}(z_1, z_2; w_1, w_2) &= \frac{\rho(w_1, z_1)\rho(w_2, z_2)}{\rho(w_2, z_1)\rho(w_1, z_2)},\\ \mathbb{X}(z_1, \infty; w_1, w_2) &= \frac{\rho(w_1, z_1)}{\rho(w_2, z_1)}.\end{aligned}$$

Using Proposition 2.2 it is not hard to show that the cross-ratio of four points is preserved by the action of Möb(X). Also, the cross-ratios satisfy the following property that resembles the ultrametric inequality

Proposition 2.5. *Let X be a non-Archimedean metric space. Let z_1, z_2, w_1, w_2 be four distinct points in $X \cup \{\infty\}$. Then*

$$\begin{aligned}\mathbb{X}(w_2, z_2; w_1, z_1) &\le \max\Bigl\{1,\ \mathbb{X}(z_1, z_2; w_1, w_2)\Bigr\},\\ 1 &\le \max\Bigl\{\mathbb{X}(w_2, z_2; w_1, z_1),\ \mathbb{X}(z_1, z_2; w_1, w_2)\Bigr\}.\end{aligned}$$

Proof. When $z_2 = \infty$ we have

$$\mathbb{X}(z_1, \infty; w_1, w_2) = \frac{\rho(w_1, z_1)}{\rho(w_2, z_1)}, \quad \mathbb{X}(w_2, \infty; w_1, z_1) = \frac{\rho(w_1, w_2)}{\rho(w_2, z_1)}$$

and the result follows directly from

$$\rho(w_1, w_2) \le \max\Bigl\{\rho(w_1, z_1),\ \rho(w_2, z_1)\Bigr\}, \quad \rho(w_2, z_1) \le \max\Bigl\{\rho(w_1, z_1),\ \rho(w_1, w_2)\Bigr\}.$$

Now using the invariance of the cross-ratio under Möb(X) we get the result for general quadruples of points. □

Let A be an element of Möb(X) fixing x, $y \in X \cup \{\infty\}$ with dilation factor ${d_A}^2$ which may be 1 (see Lemma 2.4). Suppose that m_A is a positive number so that for all points $z \in X \cup \{\infty\} - \{x, y\}$ we have

$$\mathbb{X}(x, Az; y, z) \le d_A m_A. \tag{6}$$

This is a conjugation invariant statement of the following inequality in the special case when $x = o$ and $y = \infty$:

$$\rho(z, Az) \le d_A m_A \rho(o, z). \tag{7}$$

Observe that combining (7) with Proposition 2.2 gives

$$\rho(z, A^{-1}z) \le {d_A}^{-1} m_A \rho(z, o)$$

and so $m_{A^{-1}} = m_A$. The number m_A gives a quantitative measure of how near A is to the identity: if A is close to the identity, then the distance from z to Az should be small and hence m_A must be small. We remark that such an m_A always exists. For example using $Ao = o$ and the ultrametric inequality, we obtain

$$\rho(z, Az) \le \max\{\rho(o, z),\ \rho(o, Az)\} = d_A \max\{d_A,\ 1/d_A\}\rho(o, z). \tag{8}$$

Thus one may always take $m_A = \max\{d_A,\ 1/d_A\} \ge 1$.

Lemma 2.6. *Let X be a non-Archimedean metric space. Suppose that $A \in \text{Möb}(X)$ is conjugate to a dilation with $d_A \neq 1$. If m_A is any positive number satisfying* (6)*, then $m_A \geq \max\{d_A,\, 1/d_A\} > 1$.*

Proof. Assume A fixes o and ∞ and that m_A is any positive number satisfying (7). Since $d_A \neq 1$, then $\rho(z, o) \neq \rho(o, Az)$. Hence, using Lemma 2.1, we have equality in (8). In other words,

$$\rho(z, Az) = d_A \max\{d_A,\, 1/d_A\}\rho(o, z)$$

and so if m_A satisfies (7) we have

$$d_A m_A \rho(o, z) \geq \rho(z, Az) = d_A \max\{d_A,\, 1/d_A\}\rho(o, z)$$

and $m_A \geq \max\{d_A,\, 1/d_A\} > 1$ as claimed. □

The intuition behind Lemma 2.6 is that, when $d_A \neq 1$, the map A is uniformly bounded away from the identity. For example, when A fixes o and ∞ we have

$$\rho(z, Az) = d_A \max\{d_A,\, 1/d_A\}\rho(z, o) \geq \rho(z, o)$$

and so $\rho(z, Az)$ is bounded below by a number depending on z but independent of A.

3. The main theorem

The main result of this paper is:

Theorem 3.1. *Let X be a complete non-Archimedean metric space and suppose that $\text{Aut}(X)$ is a group of isometries of X that acts transitively on X with compact stabilisers. Suppose that $\text{Möb}(X)$, the group of Möbius transformations on X, satisfies hypotheses* (2), (3) *and* (4)*. Let A be an element of $\text{Möb}(X)$ with exactly two fixed points, which we denote by x and y. Let m_A be a positive number satisfying* (6)*. If Γ is a discrete subgroup of $\text{Möb}(X)$ containing A, then for all $B \in \Gamma$ so that $\{Bx,\, By\} \cap \{x,\, y\} = \emptyset$ we have*

$$m_A^2 \max\Big\{1,\, \mathbb{X}(Bx, y; x, By)\Big\} \geq 1. \tag{9}$$

Using Lemma 2.3, since $\text{Aut}(X)$ acts transitively on X we see that $\text{Möb}(X)$ acts 2-transitively on $X \cup \{\infty\}$. Thus, without loss of generality, in what follows we shall suppose that A fixes $x = o$ and $y = \infty$. Then the cross-ratio in (9) becomes:

$$\mathbb{X}(Bo, \infty; o, B\infty) = \frac{\rho(o, Bo)}{\rho(B\infty, Bo)}.$$

We now begin the proof of Theorem 3.1. This will broadly follow Section 2.3 of [7]. The main difference will come from the fact that we are working with a non-Archimedean metric. Our strategy is to assume that the hypothesis (9) fails. In particular, we must have $m_A < 1$ and so $d_A = 1$, using Lemma 2.6. (Recall, that as we saw above if $d_A \neq 1$, then A is uniformly bounded away from the identity in the sense that $\rho(z, Az) \geq \rho(z, o)$.) We construct a sequence B_n for $n = 0,\, 1,\, \ldots$

as follows. Let B_n be defined by $B_0 = B$ and $B_{n+1} = B_n A B_n^{-1}$. Let $x_n = B_n o$ and $y_n = B_n \infty$ be the fixed points of B_{n+1}. Let r_n denote r_{B_n}. We shall show that when the hypothesis (9) is not true, then the B_n form a sequence of distinct elements of Γ that tend to the identity as n tends to infinity. This contradicts our hypothesis that Γ is discrete.

We begin by supposing that $x_n, y_n \notin \{o, \infty\}$ for all n. We then show that x_n tends to o and y_n tends to ∞ as n tends to infinity, Corollary 3.8. This immediately implies that the B_n are distinct.

Lemma 3.2. *Let A be a dilation fixing o and ∞ with $m_A < 1$. With the above notation,*

$$\begin{aligned} \rho(o, x_{n+1}) &\leq m_A \rho(x_{n+1}, y_n) \frac{\rho(o, x_n)}{\rho(x_n, y_n)}, \\ \frac{1}{\rho(x_{n+1}, y_{n+1})} &\leq \frac{m_A}{\rho(x_{n+1}, y_n)} \frac{\rho(o, y_n)}{\rho(x_n, y_n)}. \end{aligned}$$

Proof. This follows the proof of Lemma 2.5 of [7]. Using Lemma 2.6, since $m_A < 1$ we have $d_A = 1$. Using Proposition 2.2 and (7) we have

$$\begin{aligned} \rho(o, x_{n+1}) &= \rho(o, B_n A {B_n}^{-1} o) \\ &= \frac{{r_n}^2 \rho({B_n}^{-1} o, A {B_n}^{-1} o)}{\rho(A {B_n}^{-1} o, {B_n}^{-1} \infty) \rho({B_n}^{-1} o, {B_n}^{-1} \infty)} \\ &\leq \frac{m_A {r_n}^2 \rho(o, {B_n}^{-1} o)}{\rho(A {B_n}^{-1} o, {B_n}^{-1} \infty) \rho({B_n}^{-1} o, {B_n}^{-1} \infty)} \\ &= \frac{m_A \rho(B_n A B_n^{-1} o, B_n \infty) \rho(o, B_n o)}{\rho(B_n o, B_n \infty)} \\ &= m_A \rho(x_{n+1}, y_n) \frac{\rho(o, x_n)}{\rho(x_n, y_n)}. \end{aligned}$$

We have used (5) on the penultimate line. Similarly, we have

$$\begin{aligned} \frac{1}{\rho(x_{n+1}, y_{n+1})} &= \frac{1}{\rho(B_n A {B_n}^{-1} o, B_n A {B_n}^{-1} \infty)} \\ &= \frac{\rho(A {B_n}^{-1} o, {B_n}^{-1} \infty) \rho(A {B_n}^{-1} \infty, {B_n}^{-1} \infty)}{{r_n}^2 \rho(A {B_n}^{-1} o, A {B_n}^{-1} \infty)} \\ &\leq \frac{m_A \rho(A {B_n}^{-1} o, {B_n}^{-1} \infty) \rho(o, {B_n}^{-1} \infty)}{{r_n}^2 \rho({B_n}^{-1} o, {B_n}^{-1} \infty)} \\ &= \frac{m_A \rho(o, B_n \infty)}{\rho(B_n A B_n^{-1} o, B_n \infty) \rho(B_n o, B_n \infty)} \\ &= \frac{m_A}{\rho(x_{n+1}, y_n)} \frac{\rho(o, y_n)}{\rho(x_n, y_n)}. \end{aligned}$$

□

Suppose that

$$\mathbb{X}_n = \mathbb{X}(B_n o, \infty; o, B_n\infty) = \frac{\rho(o, B_n o)}{\rho(B_n o, B_n\infty)} = \frac{\rho(o, x_n)}{\rho(x_n, y_n)}.$$

We may rewrite our hypothesis that (9) fails as

$$m_A < 1 \quad \text{and} \quad m_A^2 \mathbb{X}_0 < 1.$$

We shall show, first, that if the hypothesis (9) fails to hold, then there is an $N \geq 1$ so that $\mathbb{X}_N \leq 1$ and, secondly, that this implies that $\mathbb{X}_n$ tends to zero as n tends to infinity.

Lemma 3.3. *Suppose that $\mathbb{X}_n > 1$, then $\mathbb{X}_{n+1} \leq m_A^2 \mathbb{X}_n^2$.*

Proof. Since $\mathbb{X}_n > 1$ we have $\rho(x_n, y_n) < \rho(o, x_n)$. Therefore, using Lemma 2.1, we see that $\rho(o, y_n) = \rho(o, x_n)$. This means that

$$\mathbb{X}_{n+1} = \frac{\rho(o, x_{n+1})}{\rho(x_{n+1}, y_{n+1})} \leq m_A^2 \frac{\rho(o, x_n)\rho(o, y_n)}{\rho(x_n, y_n)^2} = m_A^2 \left(\frac{\rho(o, x_n)}{\rho(x_n, y_n)} \right)^2 = m_A^2 \mathbb{X}_n^2. \quad \square$$

Lemma 3.4. *Suppose that $\mathbb{X}_n \leq 1$, then $\mathbb{X}_{n+1} \leq m_A^2 \mathbb{X}_n$.*

Proof. If $\mathbb{X}_n \leq 1$, then $\rho(o, x_n) \leq \rho(x_n, y_n)$. Thus

$$\rho(o, y_n) \leq \max\Big\{\rho(o, x_n),\ \rho(x_n, y_n)\Big\} = \rho(x_n, y_n).$$

This means that

$$\mathbb{X}_{n+1} = \frac{\rho(o, x_{n+1})}{\rho(x_{n+1}, y_{n+1})} \leq m_A^2 \frac{\rho(o, x_n)\rho(o, y_n)}{\rho(x_n, y_n)^2} \leq m_A^2 \frac{\rho(o, x_n)}{\rho(x_n, y_n)} = m_A^2 \mathbb{X}_n. \quad \square$$

Lemma 3.5. *Suppose that $m_A^2 \mathbb{X}_0 < 1$. Then there exists $N \geq 0$ so that $\mathbb{X}_N \leq 1$.*

Proof. If $\mathbb{X}_0 \leq 1$, then we choose $N = 0$. Suppose that $\mathbb{X}_k > 1$ for all $0 \leq k \leq n-1$. Then, using Lemma 3.3, we have

$$m_A^2 \mathbb{X}_n \leq \left(m_A^2 \mathbb{X}_{n-1}\right)^2 \leq \cdots \leq \left(m_A^2 \mathbb{X}_0\right)^{2^n}.$$

Since $m_A^2 \mathbb{X}_0 < 1$, this sequence is eventually at most m_A^2. Therefore we can only have $\mathbb{X}_n > 1$ for finitely many values of n. Hence there exists N with $\mathbb{X}_N \leq 1$. $\square$

Lemma 3.6. *Suppose that $\mathbb{X}_N \leq 1$, then $\mathbb{X}_n \leq m_A^{2n-2N}$ for all $n \geq N$. In particular, if $m_A < 1$, then $\mathbb{X}_n$ tends to zero as n tends to infinity.*

Proof. We use induction. We have $\mathbb{X}_N \leq 1 = m_A^0$. Suppose that $\mathbb{X}_n \leq m_A^{2n-2N}$ for some $n \geq N$. Then, using Lemma 3.4, we have $\mathbb{X}_{n+1} \leq m_A^2 \mathbb{X}_n \leq m_A^{2n+2-2N}$. The result follows. $\square$

We now use the fact that $\mathbb{X}_n$ tends to zero as n tends to infinity to show that $\rho(o, x_n)$ tends to zero and $\rho(x_n, y_n)$ tends to infinity as n tends to infinity.

Lemma 3.7. *Suppose that $m_A < 1$ and $\mathbb{X}_N \leq 1$. Then for all $n \geq N$ we have*

$$\rho(o, x_n) \leq m_A^{n-N} \rho(o, x_N) \quad \text{and} \quad \frac{1}{\rho(x_n, y_n)} \leq \frac{m_A^{n-N}}{\rho(x_N, y_N)}.$$

In particular, $\rho(o, x_n)$ tends to zero and $\rho(x_n, y_n)$ tends to infinity as n tends to infinity.

Proof. Using Lemma 3.6, we see that $m_A \mathbb{X}_n \leq m_A^{2n+1-2N} \leq m_A < 1$ for all $n \geq N$. Thus

$$\rho(o, x_{n+1}) \leq m_A \mathbb{X}_n \rho(x_{n+1}, y_n) < \rho(x_{n+1}, y_n),$$

and so, using Lemma 2.1, we see that $\rho(o, y_n) = \rho(x_{n+1}, y_n)$. As we already know that $\rho(o, y_n) \leq \rho(x_n, y_n)$, this means

$$\rho(x_{n+1}, y_n) = \rho(o, y_n) \leq \rho(x_n, y_n).$$

Using Lemma 3.2, we have

$$\rho(o, x_{n+1}) \leq m_A \rho(o, x_n) \frac{\rho(x_{n+1}, y_n)}{\rho(x_n, y_n)} \leq m_A \rho(o, x_n).$$

Using induction, we see that $\rho(o, x_n) \leq m_A^{n-N} \rho(o, x_N)$ as claimed.

Similarly, from the second part of Lemma 3.2, we have

$$\frac{1}{\rho(x_{n+1}, y_{n+1})} \leq \frac{m_A}{\rho(x_{n+1}, y_n)} \frac{\rho(o, y_n)}{\rho(x_n, y_n)} \leq \frac{m_A}{\rho(x_n, y_n)}.$$

Again, we use induction to get

$$\frac{1}{\rho(x_n, y_n)} \leq \frac{m_A^{n-N}}{\rho(x_N, y_N)}.$$ □

Corollary 3.8. *The points x_n tend to o and the points y_n tend to ∞ as n tends to infinity.*

We claim that the B_n lie in a compact subset of Möb(X). Hence (a subsequence of) the B_n tend to the identity. Since the B_n are distinct, we see that $\langle A, B \rangle$ is not discrete. This will prove the main theorem in the case where x_n, $y_n \neq o, \infty$.

In order to verify the claim, observe that we may choose D_n lying in a compact subset of Möb(X) so that $D_n B_n {D_n}^{-1}$ fixes both o and ∞. Secondly, since B_n is conjugate to A, using Lemma 2.4 we see that the dilation factor of $D_n B_n {D_n}^{-1}$ is 1. Thus for all z, $w \in X$ we have

$$\rho(D_n B_n {D_n}^{-1} A^{-1} z, D_n B_n {D_n}^{-1} A^{-1} w) = \rho(A^{-1} z, A^{-1} w) = \rho(z, w).$$

Hence $D_n B_n {D_n}^{-1} A^{-1}$ is in Aut(X) and fixes o. By hypothesis the stabiliser of o in Aut(X) is compact. Hence B_n lies in a compact subset of Möb(X) as claimed.

We need to treat the case where there is an $N \geq 0$ for which either x_N or y_N is o or ∞, and so $x_{N+1} = o$ or $y_{N+1} = \infty$. Without loss of generality, suppose $y_{N+1} = \infty$ and hence $y_n = \infty$ for all $n \geq N + 1$.

Suppose $x_n \neq o$ for all n. We will not use (9) but only the fact that $\langle A, B \rangle$ is discrete. (Note that taking $N = 0$ this shows that if $\langle A, B \rangle$ is discrete, then

$\{Bo, B\infty\} \cap \{o, \infty\}$ cannot be just one point.) Consider the sequence B_n as defined above. By construction, B_n is conjugate to A and fixes ∞ for $n \geq N+1$ and so $d_{B_n} = d_A = 1$. In other words, B_n is an isometry of X for $n \geq N+1$. Hence for $n \geq N+1$ we have

$$\begin{aligned}
\rho(x_{n+1}, o) &= \rho(B_n A B_n{}^{-1} o, o) \\
&= \rho(A B_n{}^{-1} o, B_n{}^{-1} o) \\
&\leq m_A \rho(B_n{}^{-1} o, o) \\
&= m_A \rho(B_n o, o) \\
&= m_A \rho(x_n, o) \\
&\leq m_A^{n-N} \rho(x_{N+1}, o).
\end{aligned}$$

Therefore x_n tends to o as n tends to infinity and, arguing as above, B_n is a sequence of distinct elements of $\langle A, B\rangle$ converging to the identity. Again, $\langle A, B\rangle$ cannot be discrete.

Finally, suppose $x_{N+1} = o$ and $y_{N+1} = \infty$ for some $N \geq 0$. Thus B_{n+1} fixes both o and ∞ for all $n \geq N+1$. Again we will not use (9), but this time we only use the fact that $\{Bo, B\infty\} \cap \{o, \infty\} = \emptyset$. Since A has precisely two fixed points, if $B_{n+1} = B_n A B_n{}^{-1}$ fixes both o and ∞, then B_n either fixes both o and ∞ or interchanges them. Without loss of generality, suppose that N is the smallest index for which $x_{N+1} = o$ and $y_{N+1} = \infty$. Since $\{B_0 o, B_0 \infty\} \cap \{o, \infty\} = \emptyset$, we may assume that $N \geq 1$. Then $B_N o = \infty$ and $B_N \infty = o$ and we see that B_N has an orbit of size 2. Thus B_N^2 fixes points that B_N does not. Since B_N is conjugate to A, this is a contradiction. This proves the theorem.

4. The p-adic numbers

In this section we consider the case where $X = \mathbb{Q}_p$, the p-adic numbers, that is, the completion of $\mathbb{Q}$ with respect to the p-adic valuation (1). We show that Möb(X) is then the matrix group $\mathrm{PSL}(2, \mathbb{Q}_p) = \mathrm{SL}(2, \mathbb{Q}_p)/\{\pm I\}$ acting on $\mathbb{Q}_p \cup \{\infty\}$ by Möbius transformations. Discrete subgroups of $\mathrm{SL}(2, \mathbb{Q}_p)$ have been considered by Ihara [5] and Serre in Chapter II.1 of [8], in particular page 84. Our main theorem gives a necessary condition for a subgroup of $\mathrm{SL}(2, \mathbb{Q}_p)$ to be discrete, Theorem 4.2. This is very similar to the standard version of Jørgensen's inequality, [6]. In [4] Gromov and Schoen considered more general p-adic representations of lattices in non-compact, semisimple Lie groups. Our main result should apply in many of these cases.

The construction of the p-adic numbers and their properties in terms of non-Archimedean spaces is well known; see Artin [1], Cassels [2], and Serre [8], for example. We recall that a *p-adic integer* is any p-adic number α with $|\alpha|_p \leq 1$. Thus, the ring of p-adic integers, denoted $\mathbb{Z}_p$, is the p-adic unit ball in $\mathbb{Q}_p$. Each

p-adic integer α has an expansion

$$\alpha = \sum_{n=0}^{\infty} a_n p^n \tag{10}$$

where $a_n \in \{0, 1, \ldots, p-1\}$ and so $\mathbb{Z}_p$ is compact; see Lemma 2 on page 10 of Cassels [2]. Likewise, a *p-adic unit* is any $u \in \mathbb{Q}_p$ so that $u \in \mathbb{Z}_p$ and $u^{-1} \in \mathbb{Z}_p$. That is, u has the form (10) with $a_0 \neq 0$. Since the set of units is the intersection of two compact subsets of $\mathbb{Q}_p$, we see that it is compact.

We now show how to define a tree T whose boundary is $\mathbb{Q}_p \cup \{\infty\}$. This idea is due to Serre, [8], but our treatment will follow Figà-Talamanca [3]. The closed balls in $\mathbb{Q}_p$ are the vertices of T, that is

$$V = \left\{ x + p^k\mathbb{Z}_p \ : x \in \mathbb{Q}_p,\ k \in \mathbb{Z} \right\}.$$

Two vertices $x + p^k\mathbb{Z}_p$ and $y + p^j\mathbb{Z}_p$ are joined by an edge of T if and only if either $k = j + 1$ and $x - y \in p^j\mathbb{Z}_p$ or else $j = k + 1$ and $x - y \in p^k\mathbb{Z}_p$; see page 8 of [3]. In other words, $|j - k| = 1$ and one of the balls is contained in the other. Notice that each ball $x + p^k\mathbb{Z}_p$ of radius p^{-k} is contained in exactly one ball $x + p^{k-1}\mathbb{Z}_p$ of radius p^{-k+1} and contains exactly p balls $x + yp^k + p^{k+1}\mathbb{Z}_p$ of radius p^{-k-1} where $y = 0, 1, \ldots, p-1$. Hence each vertex has exactly $p+1$ edges emanating from it. Therefore the graph T we have just constructed is an infinite, regular $p+1$ tree.

We now find the boundary of T; see [3]. We consider geodesic paths through T. In other words, such a path is a (possibly infinite) sequence of vertices v_j so that for all j the vertices v_j, v_{j+1} are joined by an edge and $v_{j-1} \neq v_{j+1}$, that is there is no back tracking. The semi-infinite geodesic path $p^{-k}\mathbb{Z}_p$ for $k = 0, 1, 2, \ldots$ identifies a point of the boundary denoted by ∞. Every other semi-infinite geodesic path starting at the vertex $\mathbb{Z}_p$ eventually consists of a sequence of nested, decreasing balls $x + p^k\mathbb{Z}_p$ for $k = K, K+1, K+2, \ldots$. The limit of this sequence is the point x of $\mathbb{Q}_p$. Choosing a starting point other than $\mathbb{Z}_p$ makes only finitely many changes to these paths. Hence the boundary of T is $\mathbb{Q}_p \cup \{\infty\}$.

Any two distinct points z, w in $\mathbb{Q}_p \cup \{\infty\}$ are the end points of a unique doubly infinite geodesic path through T. We denote this path by $\gamma(z, w)$. The cross-ratio $\mathbb{X}(z_1, z_2; w_1, w_2)$ has the following interpretation in terms of T.

Lemma 4.1. *Suppose that z_1, z_2, w_1, w_2 are four distinct points of $\mathbb{Q}_p \cup \{\infty\}$. Let $\gamma(z_1, w_2)$ and $\gamma(z_2, w_1)$ be the geodesics joining z_1, w_2 and z_2, w_1. If $\mathbb{X}(z_1, z_2; w_1, w_2) = p^d > 1$, then the shortest path in T from $\gamma(z_1, w_2)$ to $\gamma(z_2, w_1)$ has d edges. If $\mathbb{X}(z_1, z_2; w_1, w_2) \leq 1$, then $\gamma(z_1, w_2)$ and $\gamma(z_2, w_1)$ intersect.*

Proof. Without loss of generality we suppose that $w_1 = o$ and $z_2 = \infty$. Then we have $\mathbb{X}(x, \infty; o, y) = \rho(o, x)/\rho(x, y)$. The geodesic $\gamma(o, \infty)$ passes through vertices $p^j\mathbb{Z}_p$ for $j \in \mathbb{Z}$.

Suppose first that the first few terms in the expansion of x and y are the same. In other words, we have $x = p^j(a + bp^k)$ and $y = p^j(a + cp^k)$ where $k > 0$ and a, b, c are units with $b \neq c$. Then $\rho(o, x) = p^{-j}$ and $\rho(x, y) = p^{-j-k}$ and thus we have $\mathbb{X}(x,\infty;o,y)=p^k>1$. Every vertex on the geodesic $\gamma(x,y)$ has the form $p^j(a+bp^k+p^l\mathbb{Z}_p)$ or $p^j(a+cp^k+p^l\mathbb{Z}_p)$ where $l\geq k>0$. The points of $\gamma(o,\infty)$ and $\gamma(x,y)$ closest to each other are $p^j\mathbb{Z}_p$ and $p^j(a+p^k\mathbb{Z}_p)$. The geodesic segment joining them has k edges and passes through the $k+1$ vertices $p^j(a+p^l\mathbb{Z}_p)$ for $l=0,\,1,\,\ldots,\,k$. This proves the first part of the lemma.

Suppose now that the first few terms of z_1 and w_2 are not the same. That is, we have $z_1=ap^j$ and $w_2=bp^k$ where a and b are units and either $j\neq k$ or, if $j=k$, then $a-b$ is a unit. Then $\rho(o,z_1)=p^{-j}$, $\rho(w_2,z_1)=p^{-\min(j,k)}$ and $\mathbb{X}(z_1,\infty;o,w_2)=p^{\min(0,k-j)}\leq 1$. Then the geodesic joining z_1 and z_2 passes through $p^j\mathbb{Z}_k$, which also lies on the geodesic joining o and ∞. □

We claim that Möb($\mathbb{Q}_p$) is PSL$(2,\mathbb{Q}_p)$ acting on $\mathbb{Q}_p$ via Möbius transformations. Let Aut($\mathbb{Q}_p$) be the collection of maps $Ax=(ax+b)a$ where a is a unit in $\mathbb{Q}_p$ and b is any element of $\mathbb{Q}_p$. For $d=p^{-m}$ the dilation D_d is defined by $D_dx=p^{2m}x$ and satisfies (2):

$$\rho(D_dx,D_dy)=|p^{2m}z-p^{2m}w|_p=|p^{2m}|_p|z-w|_p=d^2|z-w|_p=d^2\rho(z,w).$$

The inversion R is given by $Rx=-1/x$ and clearly satisfies (3) and (4):

$$\begin{aligned}\rho(Rx,o) &= \left|\frac{-1}{x}\right|_p = \frac{1}{|x|_p} = \frac{1}{\rho(x,o)},\\ \rho(Rx,Ry) &= \left|\frac{-1}{x}-\frac{-1}{y}\right|_p = \frac{|x-y|_p}{|x|_p|y|_p} = \frac{\rho(x,y)}{\rho(x,o)\rho(y,o)}.\end{aligned}$$

As elements of SL$(2,\mathbb{Q}_p)$ these three maps are given by

$$A=\begin{pmatrix}a & b\\ 0 & a^{-1}\end{pmatrix},\quad D_d=\begin{pmatrix}p^m & 0\\ 0 & p^{-m}\end{pmatrix},\quad R=\begin{pmatrix}0 & -1\\ 1 & 0\end{pmatrix}.$$

The maps A, D_d and R also act on T. Consider the vertex $v=p^j(x+p^k\mathbb{Z}_p)$ where x is a unit and $k\geq 0$. The action of isometries and dilations is straightforward; see pages 9 and 10 of [3]:

$$A\big(p^j(x+p^k\mathbb{Z}_p)\big)=p^j(a^2x+p^k\mathbb{Z}_p)+ba,\qquad D_d\big(p^j(x+p^k\mathbb{Z}_p)\big)=p^{j+2m}(x+p^k\mathbb{Z}_p).$$

The action of R is slightly more complicated. Let y be the unit with $xy=-1$. Then

$$R\big(p^j(x+p^k\mathbb{Z}_p)\big)=p^{-j}(y+p^k\mathbb{Z}_p).$$

if $k>0$ and $R(p^j\mathbb{Z}_p)=p^{-j}\mathbb{Z}_p$. One may easily check that R preserves the structure of T.

Clearly Aut($\mathbb{Q}_p$) acts transitively on $\mathbb{Q}_p$: For any $b\in\mathbb{Q}_p$ the map $A(x)=x+b$ sends o to b. Notice that the stabiliser of o in Aut($\mathbb{Q}_p$) comprises those maps $A(x)=a^2x$ where a is a unit. Since the units form a compact subset of $\mathbb{Q}_p$, we see that Aut(X) acts with compact stabilisers. This means that the hypotheses of Theorem 3.1 are satisfied in this case. In fact, we can restate Theorem 3.1 in a more familiar form:

Theorem 4.2. *Let A be a an element of* $\mathrm{SL}(2,\mathbb{Q}_p)$ *conjugate to a diagonal matrix. Let B be any element of* $\mathrm{SL}(2,\mathbb{Q}_p)$ *so that, when acting on* $\mathbb{Q}_p\cup\{\infty\}$ *via Möbius transformations, B neither fixes nor interchanges the fixed points of A. If* $\Gamma = \langle A,B\rangle$ *is discrete, then*

$$\max\left\{\left|\mathrm{tr}^2(A)-4\right|_p,\ \left|\mathrm{tr}[A,B]-2\right|_p\right\}\geq 1.$$

Proof. Suppose that

$$A=\begin{pmatrix}\lambda & 0\\ 0 & \lambda^{-1}\end{pmatrix},\qquad B=\begin{pmatrix}a & b\\ c & d\end{pmatrix},$$

where λ, a, b, c, $d\in\mathbb{Q}_p$ and $ad-bc=1$. Then $m_A=|\lambda-\lambda^{-1}|_p$ and $Bo=b/d$, $B\infty=a/c$. By hypothesis neither Bo nor $B\infty$ is either o or ∞, so

$$\mathbb{X}(Bo,\infty;o,B\infty)=\frac{|b/d|_p}{|a/c-b/d|_p}=\frac{|b|_p|c|_p}{|ad-bc|_p}=|bc|_p.$$

We can then calculate

$$\begin{aligned}\left|\mathrm{tr}^2(A)-4\right|_p &= |\lambda-\lambda^{-1}|_p^2 = m_A^2,\\ \left|\mathrm{tr}[A,B]-2\right|_p &= |\lambda-\lambda^{-1}|_p^2\,|bc|_p = m_A^2\mathbb{X}(Bo,\infty;o,B\infty).\end{aligned}$$

The result follows directly from Theorem 3.1. □

We can interpret this result geometrically in terms of the action of $\langle A,B\rangle$ on T as follows. Let A be as in the proof of Theorem 4.2 and write $\lambda=p^ja$ where $a=a_0+a_1p+\cdots$ is a unit. Then $\lambda^{-1}=p^{-j}b$ where $b=b_0+b_1p+\cdots$ is the unit with $ab=1$.

Suppose that $m_A=|\lambda-\lambda^{-1}|<1$. Then $j=0$ and p divides $a-b$, that is $a_0=b_0$. Since $ab=1$ this means that $a_0b_0=a_0^2$ is congruent to 1 (mod p); that is λ^2 is congruent to 1 (mod p). In this case, $A(z)=\lambda^2z$ fixes each vertex $p^j\mathbb{Z}_p$. In other words, A fixes $\gamma(o,\infty)$. For such maps, Theorem 3.1 states that $\mathbb{X}(Bo,\infty;o,B\infty)\geq 1/m_A>1$. Geometrically this means that $\gamma(o,\infty)$ does not intersect its image under B.

On the other hand, if A has $d_A\neq 1$, then A maps the geodesic $\gamma(o,\infty)$ to itself shifting each vertex along by a fixed number of edges (see page 77 of Serre [8]). Recall that in this case $m_A\geq 1$ and $\rho(z,Az)\geq\rho(z,o)$. This corresponds to the fact that A must translate each vertex by a whole number of edges and so cannot have arbitrarily short translation length.

5. Function field spaces

We now explain how a function field can be thought of as resembling the p-adic numbers $\mathbb{Q}_p$ as developed in Section 4. We consider a field k and the field $k(t)$ of rational functions over k. The elements of $k(t)$ are quotients of two elements of the polynomial ring $k[t]$ over k. Then $k(t)$ is analogous to $\mathbb{Q}$ and $k[t]$ to $\mathbb{Z}$. We choose an irreducible polynomial $p(t)$ in $k[t]$ which plays the role analogous to the prime p in the definition of $\mathbb{Q}_p$. We consider an element $\phi(t) \in k(t)$ and we write

$$\phi(t) = p(t)^f \, \frac{u(t)}{v(t)}$$

where $f \in \mathbb{Z}$ and $u(t)$, $v(t)$ are polynomials in $k[t]$ without common factors and so that $p(t)$ does not divide either $u(t)$ or $v(t)$. Following (1) above, we define

$$\big|\phi(t)\big|_{p(t)} = c^{-f} \tag{11}$$

where $c > 1$ and we develop the theory in a manner resembling the p-adic case.

There is another approach, which we prefer in this section; see Artin [1] or Section II.1.6 of Serre [8]. If we replace the irreducible polynomial $p(t)$ with the rational function $1/t$ (which corresponds to ∞ at $t = 0$), then the valuation corresponding to (11) is

$$\left|\frac{u(t)}{v(t)}\right| = c^{\deg(u)-\deg(v)}. \tag{12}$$

Here the polynomials $u(t)$ and $v(t)$ have no common factors and have degree $\deg(u)$ and $\deg(v)$ respectively. This valuation corresponds to the standard absolute value in the case of $\mathbb{Q}$. The valuation (12) is non-Archimedean and leads to an ultrametric space.

In number theoretic applications (for example to the function fields of curves defined over finite fields) it is natural to define the number c in (11) to be q, where k is the field $\mathbb{F}_q$ of q elements.

We are led, accordingly, to consider $u(t) = a_n t^n + a_{n-1}t^{n-1} + \cdots + a_0$ in $k[t]$ with a_n, a_{n-1}, $\ldots$, $a_0 \in k$ and $n \geq 0$. We introduce the valuation

$$\big|u(t)\big| = \begin{cases} c^{\deg(u)} = c^n & \text{if } u(t) \neq 0, \\ 0 & \text{if } u(t) = 0. \end{cases} \tag{13}$$

It follows that

$$\begin{aligned} \big|u(t)\,v(t)\big| &= \big|u(t)\big|\,\big|v(t)\big|, \\ \big|u(t) + v(t)\big| &\leq \max\Big\{\big|u(t)\big|\,\big|v(t)\big|\Big\}, \\ \big|u(t) + v(t)\big| &= \big|u(t)\big| \quad \text{if } \deg(v) < \deg(u). \end{aligned}$$

If we take $k = \mathbb{F}_q$ and $c = q$, then $\big|u(t)\big| = q^{\deg(u)}$ (for $u(t) \neq 0$) is the number of residue classes of polynomials in $\mathbb{F}_q[t]$ modulo $u(t)$, which is why $c = q$ is the natural choice. (Each residue class may be represented by a polynomial of degree less than $u(t)$. There are q choices for each of the $\deg(u)$ coefficients.)

If $k(t)$ denotes the quotient field of $k[t]$, then the the valuation defined by (13) extends in the obvious way to (12). The field $k\{t\}$ of formal Laurent series in $1/t$ consists of the series

$$\phi = \phi(t) = a_n t^n + a_{n-1}t^{n-1} + \cdots + a_0 + a_{-1}t^{-1} + a_{-2}t^{-2} + \cdots$$

which is the completion of $k(t)$ with respect to the valuation (12) and is analogous to the completion of $\mathbb{Q}$ with respect to the Archimedean valuation. For such a ϕ we have

$$|\phi| = \big|\phi(t)\big| = c^n. \tag{14}$$

We may define $\text{Möb}\big(k\{t\}\big)$ in terms of $\text{SL}\big(2, k\{t\}\big)$ acting on $k\{t\}$ via Möbius transformations and, similarly, $\text{Aut}\big(k\{t\}\big)$. The dilations D_d are given by $D_d(\phi) = t^{2m}\phi$. We can prove the analogue of Theorem 4.2 with the valuation (14) in place of the p-adic valuation.

References

[1] E. Artin, Algebraic Numbers and Algebraic Functions, Nelson 1968.

[2] J.W.S. Cassels, Lectures on Elliptic Curves. London Mathematical Society Student Texts **18**, Cambridge University Press, 1991.

[3] A. Figà-Talamanca, *Local fields and trees.* Harmonic Functions on Trees and Buildings, ed A. Korányi, A.M.S. Contemporary Mathematics **206** (1997), 3–16.

[4] M. Gromov & R. Schoen, *Harmonic maps into singular spaces and p-adic superrigidity for lattices in groups of rank one*, I.H.E.S. Publ. Math. **76** (1992), 165–246.

[5] Y. Ihara, *On discrete subgroups of two by two projective linear groups over p-adic fields.* J. Mathematical Society of Japan **18** (1966) 219–235.

[6] T. Jørgensen, *On discrete groups of Möbius transformations*, American J. Maths. **98** (1976), 739–749.

[7] S. Markham & J.R. Parker, *Jørgensen's inequality for metric spaces with applications to the octonions*, Advances in Geometry **7** (2007), 19–38.

[8] J.-P. Serre, Trees, Springer 1980.

J. Vernon Armitage and John R. Parker
Department of Mathematical Sciences
University of Durham
Durham DH1 3LE
England
e-mail: `j.v.armitage@durham.ac.uk`
`j.r.parker@durham.ac.uk`

Progress in Mathematics, Vol. 265, 113–246

The Hypoelliptic Dirac Operator

Jean-Michel Bismut

In memory of Sasha Reznikov

Abstract. In this paper, given a compact oriented spin Riemannian manifold X, we construct a deformation of the classical Dirac operator D^X into a hypoelliptic operator acting on the total space $\mathcal{X}$ of the tangent bundle TX. This construction is parallel to the deformation of the de Rham Hodge operator we had obtained in a previous work. If X is complex and Kähler, we produce this way a deformation of the Hodge theory of the corresponding Dolbeault complex.

By adapting results of Lebeau and ourselves already proved in the context of de Rham theory, we show that the deformation of the Dolbeault Hodge theory has essentially the same analytical properties as the corresponding deformation in de Rham theory.

We define hypoelliptic Quillen metrics, and we relate them to classical Quillen metrics in a formula where the Gillet-Soulé genus R appears. This formula is parallel to a formula we had proved with Lebeau for Ray-Singer metrics in the context of de Rham theory.

We develop the theory in the equivariant setting, and also for holomorphic torsion forms.

Mathematics Subject Classification (2000). 35H10, 58A14, 58J20, 58J52.

Keywords. Hypoelliptic equations, Hodge theory, index theory and related fixed point theorems, determinants and determinant bundles, analytic torsion.

Contents

The author is indebted to a referee and to Xiaonan Ma for reading the paper very carefully.

Introduction

The purpose of this paper is to show that if X is a compact Riemannian manifold, the Dirac operator D^X can be deformed into a family of first-order hypoelliptic operators $\mathcal{D}_{Y_b}, b > 0$ acting on the total space $\mathcal{X}$ of the tangent bundle, which interpolates in the proper sense between D^X and the geodesic flow on $\mathcal{X}$. If X is a complex Kähler manifold X, we obtain a deformation of the standard Dolbeault-Hodge theory into a new Hodge theory whose corresponding Laplacian is a second-order hypoelliptic differential operator on $\mathcal{X}$.

First we will give the proper perspective to this work. In [B05], a similar deformation was obtained in the context of the de Rham-Hodge theory of X. The resulting Laplacian has properties formally similar to the ones described above.

The underlying motivation for [B05] was the construction of a semiclassical version of a Witten like deformation [W82] of a non existing de Rham-Hodge theory of the loop space LX of X. The corresponding Laplacian was shown to be formally self-adjoint with respect to a Hermitian form of signature (∞, ∞). The whole construction relies formally on a path integral representation of the Witten deformation.

The program outlined in [B05] was in large part carried out in joint work with Lebeau in [BL06]. In this work, we showed that indeed the de Rham hypoelliptic Laplacian on T^*X has many of the properties of the standard elliptic Laplacian, and also that as $b \to 0$, it converges in the proper sense to the standard elliptic de Rham-Hodge Laplacian on X. The standard conclusions of Hodge theory were shown to be correct except for a discrete family of values of the parameter $b > 0$, and also to be valid near $b = 0$. Moreover, the classical elliptic Ray-Singer torsion [RS71] was proved to be equal to its hypoelliptic deformation, while the corresponding equivariant versions were related by a topological formula. A whole machinery was also developed to handle local index theoretic questions in the hypoelliptic context.

The hypoelliptic deformation of the Dirac operator which is obtained here is of a different nature from the geometric point of view, even if analytically, some aspects are very similar.

We will first explain the construction from a functional analytic point of view, and later describe the effective construction in the case of complex manifolds.

Let (X, g^{TX}) be a compact Riemannian manifold, let LX be its loop space, i.e., the set of smooth maps $S^1 \to X$. Then LX is naturally equipped with a natural L^2 Riemannian metric. Namely if $Y \in TLX$, so that Y is a smooth section of TX along the given loop $x.$, then

$$|Y|^2_{g^{TLX}} = \int_{S^1} |Y|^2_{g^{TX}} \, ds. \tag{0.1}$$

Then S^1 acts isometrically on LX by the maps $k_s x. = x_{s+\cdot}$, and the associated generating Killing vector field is the speed $K = \dot{x}$.

Now we proceed as in Atiyah [A85]. Let K' be the 1-form on LX dual to K by the metric, so that if $Y \in TLX$, then

$$K'(Y) = \int_{S^1} \langle Y, \dot{x} \rangle \, ds. \tag{0.2}$$

Set

$$d_K = d + i_K. \tag{0.3}$$

Then d_K is the equivariant de Rham operator acting on $\Omega^{\cdot}(LX)$.

Let E be the energy,

$$E(X) = \frac{1}{2} |K|^2 = \frac{1}{2} \int_{S^1} |\dot{x}|^2 \, ds. \tag{0.4}$$

Then

$$d_K K' = E + dK'. \tag{0.5}$$

Since K is Killing,

$$d_K^2 K' = 0. \tag{0.6}$$

Set

$$\alpha = \exp\left(-d_K K'/2\right). \tag{0.7}$$

From (0.6), (0.7), we get

$$d_K \alpha = 0. \tag{0.8}$$

Let $L(x,\dot{x})$ be any Lagrangian, which has an associated Hamiltonian $\mathcal{H}$. Let I be the S^1-invariant function on LX,

$$I(x) = \int_{S^1} L(x,\dot{x})\, ds. \tag{0.9}$$

Prominent among these functionals is the energy E, whose critical points are the closed geodesics. Let Φ^{TLX} be the Mathai-Quillen equivariant Thom form [MaQ86] on TLX which is associated to the Levi-Civita connection ∇^{TLX}.

In [B05], given $b > 0$, the following formal path integral is considered,

$$J_b = \int_{LX} \alpha \wedge \left(b^2 \nabla I\right)^* \Phi^{TLX}. \tag{0.10}$$

If $L(x,\dot{x}) = \frac{1}{2}|\dot{x}|^2$, we can write the integral J_b in the form

$$J_b = \int_{LX} \exp\left(-\frac{1}{2}\int_{S_1}\left(|\dot{x}|^2 + b^4 |\ddot{x}|^2\right) ds + \dots\right). \tag{0.11}$$

In (0.11), we only wrote the bosonic part of the term appearing in the exponential, the fermionic part, or differential form contribution, being ignored.

When $b \to 0$, we recover formally the classical Brownian integral, and when $b \to +\infty$, the integral should concentrate on closed geodesics. From a dynamical point of view, it was argued in [B05] that the dynamics of the path x which corresponds to the functional integral (0.11) is given by the differential equation

$$\ddot{x} = \frac{1}{b^2}\left(-\dot{x} + \dot{w}\right), \tag{0.12}$$

where w is a standard Brownian motion. Making $b = 0$ in (0.12) leads to the equation of Brownian motion, and $b = +\infty$ to the equation of the geodesics.

The integral J_b should be thought as the Lagrangian counterpart to the Hamiltonian theory which is precisely the theory of the hypoelliptic Laplacian in de Rham theory which was developed in [B05].

At this stage, let us observe that if $I = E$, the energy functional appears twice in (0.10), first in α and also in $\left(b^2\nabla E\right)^* \Phi^{TLX}$.

Now we will explain the Lagrangian version of the construction of our hypoelliptic Dirac operator. We will assume X to be even dimensional, oriented and spin. Let $\pi : \mathcal{X} \to X$ be the total space of TX. The Levi-Civita connection ∇^{TX} induces the identification $T\mathcal{X} = \pi^*(TX \oplus TX)$, the second copy of TX being the fibre of π. Given $b > 0$, we equip $T\mathcal{X}$ with the metric $g_b^{T\mathcal{X}} = g^{TX} \oplus b^4 g^{TX}$.

Let $L\mathcal{X}$ be the loop space of $\mathcal{X}$, and let $\mathcal{K}$ be the corresponding generating vector field. We can embed LX into $L\mathcal{X}$ by the map $i : x \dashrightarrow (x, K)$, so that $i_* K = \mathcal{K}$. Since $K = \dot{x}$, using the splitting above, we get $i_* K = (\dot{x}, \ddot{x})$.

Let $g_b^{TLX} = i^* g_b^{TL\mathcal{X}}$. The metric g_b^{TLX} is a H^1 type metric on LX. Namely, if $Y \in TLX$, let $\dot{Y}$ be the covariant derivativc of Y with respect to ∇^{TX} along the given loop $x.$. Then

$$|Y|^2_{g_b^{TLX}} = \int_{S^1} \left(|Y|^2_{g^{TX}} + b^4 \left|\dot{Y}\right|^2_{g^{TX}} \right) ds. \tag{0.13}$$

By (0.13), we find in particular that

$$|K|^2_{g_b^{TLX}} = \int_{S^1} \left(|\dot{x}|^2_{g^{TX}} + b^4 \left|\ddot{x}\right|^2_{g^{TX}} \right) ds. \tag{0.14}$$

We introduce a vector bundle $\left(E, g^E, \nabla^E\right)$ on X equipped with a metric and a unitary connection. Let LE be the loop space of E. Let $\mathrm{ch}\left(LE, \nabla^{LE}\right)$ be the even form on LX obtained in [B85], which lifts the Chern character form of E on X to a d_K-closed form on LX. Let $\mathcal{K}'_b$ be dual to $\mathcal{K}$ by the metric $g_b^{TL\mathcal{X}}$. In the Lagrangian formulation, the idea will be to consider a path integral of the type

$$\mathbf{J}_b = \int_{LX} \exp\left(-i^* d_{\mathcal{K}} \mathcal{K}'{}_b / 2\right) \wedge \mathrm{ch}\left(LE, \nabla^{LE}\right). \tag{0.15}$$

We can rewrite (0.15) in the form,

$$\mathbf{J}_b = \int_{LX} \exp\left(-d_K i^* \mathcal{K}'_b / 2\right) \wedge \mathrm{ch}\left(LE, \nabla^{LE}\right). \tag{0.16}$$

Let us make $b = 0$ in (0.15). We get

$$\mathbf{J}_0 = \int_{LX} \exp\left(-d_K K' / 2\right) \wedge \mathrm{ch}\left(LE, \nabla^{LE}\right). \tag{0.17}$$

Now by arguments of Witten explained in detail in [A85, B85], (0.17) is the Lagrangian counterpart to the well-known McKean-Singer heat equation formula [MKS67] for the index of the classical Dirac operator acting on spinors of X twisted by E.

The question is to know what is the Hamiltonian counterpart to the path integral in (0.15) for $\mathbf{J}_b$ for $b > 0$. Before doing this, observe that by (0.14), (0.15), as $b \to +\infty$, the integral for $\mathbf{J}_b$ will localize on the closed geodesics of X. It is useful to note the similarity of the path integrals in (0.11) and in (0.15), by simple inspection of (0.14).

The construction which is done in the present paper is not exactly the Hamiltonian counterpart to (0.15). Indeed let R^{TX} be the curvature of ∇^{TX}. If Y is the tautological section of $\pi^* TX$ on $\mathcal{X}$, we consider the 2-form $\widehat{R^{TX} Y}$ on X with values in the vertical fibre TX. Let $\widehat{R^{TX} Y}'_b$ be the corresponding three form on $\mathcal{X}$ one obtains by duality on the vertical component with respect to the vertical metric $b^4 g^{TX}$. This form is of horizontal degree 2 and vertical degree 1. Note that $\pi_* \mathcal{K} = K$. Consider the odd form on $L\mathcal{X}$,

$$\mathcal{L}_b = \mathcal{K}' + \widehat{R^{TX} K}'_b. \tag{0.18}$$

Then $\mathcal{L}_b$ is the sum of a 1-form and of a 3-form. Observe that

$$i^* i_{\mathcal{K} + \widehat{R^{TX} K}} \mathcal{L}_b = \int_{S^1} \left(|\dot{x}|^2 + b^4 \left| \ddot{x} + R^{TX} \dot{x} \right|^2 \right) ds. \tag{0.19}$$

The path integral which is considered implicitly in this paper is given by

$$\mathfrak{J}_b = \int_{LX} \exp\left(-d_K i^* \mathcal{L}_b / 2\right) \wedge \mathrm{ch}\left(LE, \nabla^{LE}\right). \tag{0.20}$$

For reasons which should be obvious to the reader, the expression in (0.19) contributes to $d_K i^* \mathcal{L}_b$.

Again for $b = 0$, $\mathfrak{J}_b$ reduces to the classical expression in (0.17). When $b \to +\infty$, the integrand in (0.20) should localize near the set of closed geodesics. However elementary considerations show that there may be divergences of local currents when $b \to +\infty$, which is not the case for the de Rham deformation of [B05]. The Hamiltonian counterpart to the path integral (0.20) will precisely be our hypoelliptic Dirac operator $\mathcal{D}_{Y_b}$.

If the manifold X is complex and the metric g^{TX} is Kähler, the path integral can be written in another form. Indeed LX is a complex manifold, and K is a holomorphic vector field. Put

$$\overline{\partial}_K = \overline{\partial} + i_{K^{(1,0)}}, \qquad \partial_K = \partial + i_{K^{(0,1}}, \tag{0.21}$$

so that

$$\overline{\partial}_K^2 = 0, \qquad \partial_K^2 = 0, \qquad d_K = \overline{\partial}_K + \partial_K. \tag{0.22}$$

Let ω^X be the Kähler form on X which is associated to the metric g^{TX}, and let $\widehat{\omega}^{TX}$ be the fibrewise Kähler form associate to the metric $\widehat{g}^{TX}$ on the fibres of TX. Note that we do not assume any more that $\widehat{g}^{TX}$ is related to g^{TX}, nor that $\widehat{g}^{TX}$ induces a Kähler metric on X. From the holomorphic Hermitian connection $\widehat{\nabla}^{TX}$ on $\left(TX, \widehat{g}^{TX}\right)$, we get a horizontal subbundle $T^H \mathcal{X}$ of $T\mathcal{X}$, so that $\widehat{\omega}^X$ can be viewed as a $(1,1)$ form on $\mathcal{X}$. For $b \geq 0$, set

$$\omega_b^{\mathcal{X}} = \pi^* \omega^X + b^4 \widehat{\omega}^X. \tag{0.23}$$

Note that for $b > 0$, the form $\omega_b^{\mathcal{X}}$ on $\mathcal{X}$ is closed only if the metric $\widehat{g}^{TX}$ is flat.

The form $\omega_b^{\mathcal{X}}$ lifts to a S^1-invariant form on $L\mathcal{X}$. Then the integral (0.20) can be written in the form

$$\mathfrak{J}_b = \int_{LX} \exp\left(i \overline{\partial}_K \partial_K i^* \omega_b^{L\mathcal{X}}\right) \wedge \mathrm{ch}\left(LE, g^{LE}\right). \tag{0.24}$$

Working instead with the path integral $\mathbf{J}_b$ would mean that if y is the tautological section of $\pi^* TX$ on $\mathcal{X}$, we would replace $\widehat{\omega}^{\mathcal{X}}$ by the closed form $i \overline{\partial} \partial \left|y\right|^2_{\widehat{g}^{TX}}$ (of which $\widehat{\omega}^X$ is the vertical component). Finally note that when $b = 0$, equation

(0.24) is just the formula established in [B90a, eq. (15)] in the context of classical elliptic Hodge theory.

Let us now briefly explain the explicit construction of the hypoelliptic Dirac operator in the case where X is a compact complex manifold. Let $\left(E, g^E\right)$ be a holomorphic Hermitian vector bundle on X. We still denote by $\pi : \mathcal{X} \to X$ the total space of TX. Let y be the tautological section of $\pi^* TX$ on $\mathcal{X}$. Let $i : X \to \mathcal{X}$ be the embedding of X into $\mathcal{X}$ as the zero locus of y. The Koszul complex $(\Lambda^{\cdot}(T^*X), i_y)$ provides a resolution of $i_* \mathcal{O}_X$. Consider the Dolbeault-Koszul complex $\left(\Omega^{(0,\cdot)}(\mathcal{X}, \pi^*(\Lambda^{\cdot}(T^*X) \otimes E)), \overline{\partial}^{\mathcal{X}} + i_y\right)$. Let $\widehat{g}^{TX}$ be a Hermitian metric on the fibre TX of π, and let $\widehat{\nabla}^{TX}$ be the corresponding holomorphic Hermitian connection, from which we get a smooth identification $T\mathcal{X} = \pi^*(TX \oplus TX)$. Using this identification, we can write the operator $A''_Y = \overline{\partial}^{\mathcal{X}} + i_y$ in the form,

$$A''_Y = \nabla^{\mathbf{I}\cdot\prime\prime} + \overline{\partial}^V + i_y. \tag{0.25}$$

In (0.25), $\nabla^{\mathbf{I}\prime\prime}$ is a horizontal Dolbeault operator, and $\overline{\partial}^V$ the fibrewise Dolbeault operator.

Let ω^X be the Kähler form of a Kähler metric g^{TX} on TX. Set

$$A'_Y = e^{i\omega^X} \left(\nabla^{\mathbf{I}\cdot\prime} + \overline{\partial}^{V*} + i_{\overline{y}}\right) e^{-i\omega^X}. \tag{0.26}$$

We have the obvious,

$$A''^2_Y = 0, \qquad\qquad A'^2_Y = 0. \tag{0.27}$$

Set

$$A_Y = A''_Y + A'_Y. \tag{0.28}$$

By a result of Hörmander [Hö67], $\frac{\partial}{\partial u} - A^2_Y$ is a hypoelliptic differential operator on $\mathcal{X}$. The operator A_Y is a special case of the hypoelliptic Dirac operator.

For $b > 0$, when replacing $\widehat{g}^{TX}$ by $b^4 g^{TX}$, the Laplacian $A^2_{Y,b}$ is shown to interpolate between the classical Dolbeault-Hodge operator on X, and the geodesic flow on $\mathcal{X}$. Moreover, $A^2_{Y,b}$ is self-adjoint with respect to an explicit Hermitian form of signature (∞, ∞).

Set

$$\lambda = \det H^{(0,\cdot)}(X, E|_X). \tag{0.29}$$

Let $g^{TX\prime}$ be another Kähler metric on TX. The metrics $g^{TX\prime}, g^E$ determine a classical Quillen metric [Q85b, BGS88c] on $\| \|^2_\lambda$ on λ via the classical Dolbeault-Hodge Laplacian on X. On the other hand, by proceeding as in [BL06], the metrics $g^{TX}, \widehat{g}^{TX}, g^E$ determine a hypoelliptic generalized metric $\| \|_{\lambda,h}$ on λ.

An important result contained in this paper is a comparison formula relating the elliptic and hypoelliptic Quillen metrics. Indeed let $\widetilde{\mathrm{Td}}\left(TX, g^{TX\prime}, \widehat{g}^{TX}\right)$ be the Bott-Chern class in the sense of [BoC65, BGS88a] of TX for the couple of metrics $g^{TX\prime}, \widehat{g}^{TX}$. Let R be the Gillet-Soulé additive genus [GS92] associated to the formal power series

$$R(x) = \sum_{\substack{n \geq 1 \\ n\,\mathrm{odd}}} \left(2\frac{\zeta'(-n)}{\zeta(-n)} + \sum_{j=1}^{n} \frac{1}{j} \right) \zeta(-n) \frac{x^n}{n!}. \tag{0.30}$$

In (0.30), $\zeta(s)$ is the Riemann zeta function.

We give the comparison formula established in Theorem 10.8.

Theorem 0.1. *The generalized metric* $\| \|^2_{\lambda,h}$ *is a Hermitian metric. Moreover,*

$$\log \left(\frac{\| \|^2_{\lambda,h}}{\| \|^2_{\lambda}} \right) = - \int_X \widetilde{\mathrm{Td}} \left(TX, g^{TX\prime}, \widehat{g}^{TX}\right) \mathrm{ch}\left(E, g^E\right) + \int_X \mathrm{Td}(TX) R(TX) \mathrm{ch}(E). \tag{0.31}$$

By (0.31), we find that the hypoelliptic Quillen metric $\| \|_{\lambda,h}$ does not depend on the Kähler metric g^{TX}, and also that it has all the nice properties of classical Quillen metrics, even when $\widehat{g}^{TX}$ is not Kähler. In Subsection 10.8, we relate the appearance of the genus R to the immersion formula of [BL91] for Quillen metrics, where the genus R also appears.

We establish an analogue of Theorem 0.1 also in the equivariant case. The equivariant genus $R(\theta, x)$ of [B94] appears instead. Moreover, for values of the deformation parameter $b > 0$ small enough, we get corresponding formulas relating the elliptic holomorphic torsion forms [BK92, M00a] to their hypoelliptic analogues.

Let us also mention that for $b > 0$, the hypoelliptic Dirac operator $\mathcal{D}_{Y_b}$ is a perturbation of a Quillen superconnection over X [Q85a] by an operator of order 0. In the context of the local families index theorem of [B86], superconnections have appeared naturally as adiabatic limits of Dirac operators. Here the situation is actually the opposite. The elliptic Dirac operator appears as the limit of a family of perturbed superconnections.

When X is an oriented spin manifold of odd dimension, we extend the definition of the $\mathbf{R}/\mathbf{Z}$-valued reduced eta invariant of Atiyah-Patodi-Singer [APS75a, APS75b, APS76] to $\mathcal{D}_{Y_b}$, and we show that when the horizontal and vertical metrics coincide, this reduced eta invariant coincides with the reduced eta invariant of the corresponding elliptic Dirac operator.

Our paper is organized as follows. In Section 1, given a section Z of π^*TX on $\mathcal{X}$, we define the Dirac like operator $\mathcal{D}_Z$ on $\mathcal{X}$, and we prove that if Y is the tautological section of π^*TX, if $b > 0$ and $Y_b = Y/b^2$, the operator $\frac{\partial}{\partial u} - \mathcal{D}^2_{Y_b}$ is hypoelliptic. Also we show that $\mathcal{D}^2_{Y_b}$ interpolates in the proper sense between $D^{X,2}$ as $b \to 0$ and the geodesic flow as $b \to +\infty$.

In Section 2, we extend the above constructions in the context of families of compact manifolds. The object to be deformed is the standard Levi-Civita superconnection [B86] of the given fibration, whose curvature is elliptic along the fibres X. The deformation is still by superconnections whose curvature is now hypoelliptic along the fibres $\mathcal{X}$.

In Section 3, we consider the case of complex manifolds, and of holomorphic families. We show in particular that under the proper Kähler condition, the deformation of the elliptic Dolbeault complex is a special case of what was done in Sections 1 and 2.

In Section 4, we state known results on the local index theory for elliptic Dirac operators, and on the local families index theorem. Also we briefly recall the construction of holomorphic elliptic torsion forms.

In Section 5, we consider the case where the fibres X form a vector bundle E. If E is holomorphic, we show that corresponding hypoelliptic torsion forms can be expressed using the R-genus of [GS92]. This is a computation which is completely distinct from the one in [B90b], where the genus R appears in the evaluation of elliptic torsion forms for vector bundles equipped with a Koszul complex.

In Section 6, we prove local index results for the hypoelliptic Dirac operator.

In Section 7, by using arguments from [BL06], we show that as $b \to 0$, the hypoelliptic Dirac Laplacian converges in the proper sense to the standard Laplacian $D^{X,2}$, and we also construct holomorphic hypoelliptic torsion forms, and hypoelliptic Quillen metrics.

In Section 8, we compare the hypoelliptic torsion forms to their elliptic counterparts in the case where the vertical metric $\widehat{g}^{TX}$ is proportional to the horizontal metric g^{TX}, and as a by-product, we compare the corresponding Quillen metrics. The proof relies on two intermediate results whose proof is deferred to Section 9.

Section 9 is devoted to the proof of these intermediate results, one which consists in studying the asymptotics as $b \to 0$ of an analogue of the L^2-metric on the harmonic forms.

In Section 10, we consider the case where the horizontal and vertical metrics on $\mathcal{X}$ are unrelated, and we prove corresponding comparison formulas.

Finally in Section 11, we define the reduced eta invariant of the hypoelliptic Dirac operator, and we show that it coincides mod $\mathbf{Z}$ with the reduced eta invariant of the associated elliptic Dirac operator.

Even though the construction of our operators is substantially different from the one in [B05], the analytic arguments used in the paper are entirely taken from our joint work with Lebeau [BL06] on the de Rham complex. We have tried to make the references to [BL06] as explicit as possible.

In the whole paper, if $\mathcal{A}$ is a $\mathbf{Z}_2$-graded algebra, if $a, b \in \mathcal{A}$, $[a, b]$ denotes the supercommutator of a and b.

The results contained in this paper have been announced in [B06].

1. A Dirac operator on the total space of the tangent bundle

Let X be a compact Riemannian manifold and let $\pi : \mathcal{X} \to X$ be the total space of its tangent bundle TX. The purpose of this section is to construct a hypoelliptic Dirac operator on $\mathcal{X}$. Also we show that a family of such operators interpolates between the ordinary Dirac operator on X and the geodesic flow.

This section is organized as follows. In Subsection 1.1, we recall elementary facts on Clifford algebras and their relations to exterior algebras.

In Subsections 1.2 and 1.3, we briefly consider the spin representation in the case of even- and odd-dimensional real vector spaces.

In Subsection 1.4, we recall elementary properties of the classical elliptic Dirac operator D^X on X.

In Subsection 1.5, we construct the tautological superconnection $\mathcal{D}$ on the total space $\mathcal{X}$ of TX. The principal symbol of $\mathcal{D}$ is elliptic along the fibre, and nilpotent horizontally.

In Subsection 1.6, given a section Z of π^*TX over $\mathcal{X}$, we construct the operator $\mathcal{D}_Z$ as a perturbation of $\mathcal{D}$ by an operator of order 0.

In Subsection 1.8, we show that if our section Z is antiinvariant under the obvious involution of TX, the operator $\mathcal{D}_Z$ is formally self-adjoint with respect to a Hermitian form of signature (∞, ∞).

In Subsection 1.9, we give a Lichnerowicz formula for $\mathcal{D}_Z^2$.

In Subsection 1.10, we show that if Z is regular, the operator $\frac{\partial}{\partial u} - \mathcal{D}_Z^2$ is hypoelliptic. This includes the case where Z is the tautological section Y of π^*TX.

In Subsection 1.11, we introduce the obvious action of dilations on sections of TX over $\mathcal{X}$. We obtain a family of hypoelliptic Dirac operators $\mathcal{D}_{Y_b}$.

In Subsection 1.12, after a suitable conjugation of $\mathcal{D}_{Y_b}$, we relate the constant term in the expansion of the conjugate operator $\mathcal{E}_{Y,b}$ to the elliptic Dirac operator D^X.

In Subsection 1.13, we prove a corresponding for $\mathcal{E}_{Y,b}^2$. The results of Subsections 1.12 and 1.13 will later be used to justify the fact that as $b \to 0$, the hypoelliptic Laplacian $\mathcal{D}_{Y_b}^2$ converges in the proper sense to the elliptic Laplacian $D^{X,2}$.

In Subsection 1.14, we show that as $b \to +\infty$, the family $\mathcal{D}_{Y_b}$ converges in the proper sense to the geodesic flow on $\mathcal{X}$.

The results obtained in this section are formally related to the results we gave in [B05] in the context of de Rham theory.

1.1. The Clifford algebra

Let V be a real Euclidean vector space of dimension n. We identify V and V^* by the scalar product of V. Let $c(V)$ be the Clifford algebra of V, i.e., the algebra spanned by $1, X \in V$, with the commutation relations for $X, Y \in V$,

$$XY + YX = -2 \langle X, Y \rangle . \tag{1.1}$$

Then $c(V)$ is naturally filtered by length, and the corresponding Gr is just $\Lambda^{\cdot}(V^*)$. Moreover, $c(V)$ and $\Lambda^{\cdot}(V^*)$ are canonically isomorphic as vector spaces.

Observe that if in the right-hand side of (1.1), we replace $-2\langle X,Y\rangle$ by $2\langle X,Y\rangle$, we obtain a new Clifford algebra. When tensoring with $\mathbf{C}$, this algebra coincides with the original one. In the sequel, we will often not distinguish the two algebras.

If $e \in V$, let $e^* \in V^*$ correspond to e by the scalar product. Consider the operators acting on $\Lambda^{\cdot}(V^*)$,

$$c(e) = e^* \wedge - i_e, \qquad \widehat{c}(e) = e^* \wedge + i_e. \tag{1.2}$$

If $e, e' \in V$, then

$$[c(e), c(e')] = -2\langle e, e'\rangle, \quad [\widehat{c}(e), \widehat{c}(e')] = 2\langle e, e'\rangle, \qquad [c(e), \widehat{c}(e')] = 0. \tag{1.3}$$

By (1.3), we find that $\Lambda^{\cdot}(V^*)$ is equipped with two Clifford actions. When identifying $\Lambda^{\cdot}(V^*)$ and $c(V)$ as vector spaces, $e \to c(e)$ corresponds to the left action of $c(V)$ on itself, and $e \to \widehat{c}(e)$ induces the right action of $c(V)$ on itself, where the right action takes into account the $\mathbf{Z}_2$-grading.

1.2. The even-dimensional case

Assume that n is even so that $n = 2\ell$, and also that V is oriented. Let $S^V = S^V_+ \oplus S^V_-$ be the Hermitian vector space of spinors. The dimension of $S^V, S^V_\pm$ is respectively $2^\ell, 2^{\ell-1}$. Then $c(V) \otimes_{\mathbf{R}} \mathbf{C}$ acts on S^V and can be identified with $\operatorname{End}\left(S^V\right)$, this identification being an identification of $\mathbf{Z}_2$-graded algebras, so that

$$c(V) \otimes_{\mathbf{R}} \mathbf{C} = S^V \widehat{\otimes} S^{V*}. \tag{1.4}$$

By (1.4), we have the identification of vector spaces,

$$\Lambda^{\cdot}(V^*) \otimes_{\mathbf{R}} \mathbf{C} \simeq S^V \widehat{\otimes} S^{V*}. \tag{1.5}$$

In the sequel, we will often omit the notation $\otimes^{\mathbf{C}}_{\mathbf{R}}$, so that the real vector spaces are implicitly complexified.

Using (1.5), we obtain

$$\Lambda^{\cdot}(V^*) \widehat{\otimes} S^V \simeq S^V \widehat{\otimes} S^{V*} \widehat{\otimes} S^V. \tag{1.6}$$

Note that (1.6) is now naturally equipped with three Clifford actions of $c(V)$. If $e \in V$, the first two Clifford actions $c(e), \widehat{c}(e)$ act on $\Lambda^{\cdot}(V^*)$ and were described in (1.2). The last Clifford action operates on the second copy of S^V. If $e \in V$, it will be denoted $c(\widehat{e})$. Of course these three actions of $c(V)$ anticommute.

Note that $S^V \widehat{\otimes} S^{V*} \widehat{\otimes} S^V$ is canonically equipped with an involution i, in which the two copies of S^V are interchanged, with a corresponding sign. Under i, we get the canonical isomorphism,

$$\Lambda^{\cdot}(V^*) \widehat{\otimes} S^V \simeq S^V \widehat{\otimes} \Lambda^{\cdot}(V^*). \tag{1.7}$$

To distinguish the two copies of S^V, we will denote the second copy by $\widehat{S}^V$. Put

$$\widehat{\Lambda}^{\cdot}(V^*) = \widehat{S}^V \widehat{\otimes} S^{V*}. \tag{1.8}$$

The canonical isomorphism (1.7) is now written in the form,

$$\Lambda^{\cdot}(V^*)\widehat{\otimes}\widehat{S}^V \simeq S^V\widehat{\otimes}\widehat{\Lambda}^{\cdot}(V^*). \tag{1.9}$$

When acting on $\Lambda^{\cdot}(V^*)$, we will slightly modify the Clifford actions, by a factor $\sqrt{2}$. Indeed we will write,

$$\begin{aligned} c(e) &= e^*\wedge/\sqrt{2}-\sqrt{2}i_e, & \widehat{c}(e) &= e^*\wedge/\sqrt{2}+\sqrt{2}i_e, \\ c(\widehat{e}) &= \widehat{e}^*\wedge - i_{\widehat{e}}, & \widehat{c}(e) &= \widehat{e}^*\wedge + i_{\widehat{e}}. \end{aligned} \tag{1.10}$$

Note that instead of $\widehat{c}(e)$, we could have written instead $\widehat{c}(\widehat{e})$, the hat on e being in this case irrelevant. The notation in (1.10) is entirely consistent. For instance the operator $\widehat{c}(e)$ can be expressed in two ways, depending on which side of (1.9) it acts.

By (1.10), we get

$$e^*\wedge = \frac{1}{\sqrt{2}}\left(\widehat{c}(e)+c(e)\right), \qquad i_e = \frac{1}{2\sqrt{2}}\left(\widehat{c}(e)-c(e)\right). \tag{1.11}$$

1.3. The odd-dimensional case

Assume now that $n = 2\ell - 1$ is odd and that V is oriented. Let S^V be the corresponding Hermitian vector space of spinors. The dimension of S^V is $2^{\ell-1}$. Moreover,

$$c(V)\otimes_{\mathbf{R}}\mathbf{C} = \left(S^V\otimes S^{V*}\right)\oplus\left(S^V\otimes S^{V*}\right). \tag{1.12}$$

From (1.12), we get the identification of vector spaces,

$$\Lambda^{\cdot}(V^*) \simeq \left(S^V\otimes S^{V*}\right)\oplus\left(S^V\otimes S^{V*}\right). \tag{1.13}$$

By (1.13), we obtain

$$\Lambda^{\cdot}(V^*)\otimes S^V \simeq \left(S^V\otimes S^{V*}\otimes S^V\right)\oplus\left(S^V\otimes S^{V*}\otimes S^V\right). \tag{1.14}$$

Now exchanging the two copies of S^V in each term in the right-hand side of (1.14) produces an involution i. Equivalently i induces the canonical identification similar to (1.9),

$$\Lambda^{\cdot}(V^*)\widehat{\otimes}\widehat{S}^V \simeq S^V\widehat{\otimes}\widehat{\Lambda}^{\cdot}(V^*). \tag{1.15}$$

Even though S^V is not $\mathbf{Z}_2$-graded, the $\widehat{\otimes}$ in (1.15) indicates that if $\sigma = \pm 1, \widehat{\sigma} = \pm 1$ define the $\mathbf{Z}_2$-gradings on $\Lambda^{\cdot}(V^*), \widehat{\Lambda}^{\cdot}(V^*)$ if $e \in V$, $c(\widehat{e})$ acts like $\sigma\otimes c(\widehat{e})$ on $\Lambda^{\cdot}(V^*)\widehat{\otimes}S^V$, and $c(e)$ acts like $c(e)\otimes\widehat{\sigma}$ on $S^V\widehat{\otimes}\widehat{\Lambda}^{\cdot}(V^*)$. This ensures that the three Clifford actions on (1.15) anticommute. Also equations (1.10), (1.11) still hold.

In the sequel, whether in the even- or odd-dimensional case, the Clifford multiplication operators $c(e), \widehat{c}(e), c(\widehat{e})$ will always be considered as odd.

1.4. The elliptic Dirac operator

Let X be a compact oriented spin manifold of dimension n. Let g^{TX} be a Riemannian metric on TX, let $\nabla^{TX,L}$ be the Levi-Civita connection on TX, and let $R^{TX,L}$ be the curvature of $\nabla^{TX,L}$. Let S be the Ricci tensor, and let K be the scalar curvature of $\left(X, g^{TX}\right)$.

Let $c(TX)$ be the bundle of Clifford algebras of $\left(TX, g^{TX}\right)$. If $e \in TX$, we denote by $c(e)$ the corresponding element in $c(TX)$.

If n is even with $n = 2\ell$, let $S^{TX} = S^{TX}_+ \oplus S^{TX}_-$ be the $\mathbf{Z}_2$-graded Hermitian vector bundle of $\left(TX, g^{TX}\right)$ spinors. Then S^{TX} is of dimension 2^ℓ and $S^{TX}_\pm$ of dimension $2^{\ell-1}$. If n is odd, let S^{TX} denote the Hermitian vector bundle of $\left(TX, g^{TX}\right)$ spinors.

The Levi-Civita connection $\nabla^{TX,L}$ lifts to a Hermitian connection $\nabla^{S^{TX},L}$ on S^{TX} which preserves the $\mathbf{Z}_2$-grading when n is even.

Let $\left(E, g^E, \nabla^E\right)$ be a complex finite-dimensional Hermitian vector bundle on X equipped with a Hermitian connection, whose curvature is denoted by R^E. We denote by $\nabla^{S^{TX}\otimes E,L}$ the connection on $S^{TX} \otimes E$ which is induced by $\nabla^{S^{TX},L}$ and ∇^E.

We equip $C^\infty\left(X, S^{TX}\widehat{\otimes} E\right)$ with the obvious L^2 Hermitian product. Let $e_1, \dots, e_n$ be an orthonormal basis of TX.

Definition 1.1. Let D^X be the Dirac operator,

$$D^X = c\left(e_i\right) \nabla^{S^{TX}\widehat{\otimes} E,L}_{e_i}. \tag{1.16}$$

Then D^X is an elliptic first order differential operator acting on $C^\infty\left(X, S^{TX}\widehat{\otimes} E\right)$. Also D^X is odd and formally self-adjoint.

Let Δ^H be the Bochner Laplacian. If $e_1, \dots, e_n$ is a locally defined smooth orthonormal basis of TX, then

$$\Delta^H = \nabla^{S^{TX}\otimes E,L,2}_{e_i} - \nabla^{S^{TX}\otimes E,L}_{\nabla^{TX,L}_{e_i} e_i}. \tag{1.17}$$

The Lichnerowicz formula asserts that

$$D^{X,2} = -\Delta^H + \frac{K}{4} + \frac{1}{2} c\left(e_i\right) c\left(e_j\right) R^E\left(e_i, e_j\right). \tag{1.18}$$

1.5. The superconnection $\mathcal{D}$

By (1.9), (1.15), we have the canonical isomorphism,

$$\Lambda^{\cdot}\left(T^*X\right) \widehat{\otimes} S^{TX} \simeq S^{TX} \widehat{\otimes} \widehat{\Lambda}^{\cdot}\left(T^*X\right). \tag{1.19}$$

Set

$$F = \Lambda^{\cdot}\left(T^*X\right) \widehat{\otimes} \widehat{S}^{TX} \widehat{\otimes} E \simeq S^{TX} \widehat{\otimes} \widehat{\Lambda}^{\cdot}\left(T^*X\right) \widehat{\otimes} E. \tag{1.20}$$

Let $\pi : \mathcal{X} \to X$ be the total space of TX. If $x \in X$, Y denotes the generic element in the fibre T_xX. To distinguish the elements of the fibre T^*X dual to TX from usual 1-forms on X, we will denote them with a hat.

Let ∇^{TX} be a Euclidean connection TX with respect to g^{TX}, which is not necessarily equal to $\nabla^{TX,L}$, and let R^{TX} be its curvature.

Set

$$A = \nabla^{TX} - \nabla^{TX,L}. \tag{1.21}$$

Then A is a 1-form on X with values in antisymmetric sections of $\mathrm{End}\,(TX)$.

Let $\nabla^{S^{TX}}$ be the connection induced by ∇^{TX} on S^{TX}. We now denote by $\nabla^{S^{TX}\widehat{\otimes}E}$ the connection induced by $\nabla^{S^{TX}}, \nabla^{E}$ on $S^{TX}\widehat{\otimes}E$.

We view the curvature R^{TX} as a section of $\Lambda^{\cdot}(T^*X)\,\widehat{\otimes}\mathrm{End}\,(TX)$, so that

$$R^{TX} = \frac{1}{2} e^i \wedge e^j \wedge R^{TX}(e_i, e_j)\,. \tag{1.22}$$

The connection ∇^{TX} induces a horizontal vector bundle $T^H\mathcal{X} \subset T\mathcal{X}$, so that

$$T\mathcal{X} = T^H\mathcal{X} \oplus \pi^* TX. \tag{1.23}$$

Also π_* induces an isomorphism $T^H\mathcal{X} \simeq \pi^*TX$, so that we can write (1.23) in the form,

$$T\mathcal{X} = \pi^*(TX \oplus TX)\,. \tag{1.24}$$

If $e \in TX$, let $e^H \in T^H\mathcal{X}$ be the horizontal lift of e. Note that e^H depends explicitly on the connection ∇^{TX}. In the sequel, we will write e instead of e^H.

We denote by $\mathbf{I}$ the vector bundle $C^\infty\left(TX, \pi^*\left(\widehat{S}^{TX}\widehat{\otimes}E\right)|_{TX}\right)$ over X.

Let $e_1, \dots, e_n$ be an orthonormal basis of TX, and let $e^1, \dots, e^n$ be the corresponding dual basis of T^*X. We denote by $\widehat{e}_1, \dots, \widehat{e}_n$ the corresponding orthonormal basis of the fibre TX of $\mathcal{X}$, and by $\widehat{e}^1, \dots, \widehat{e}^n$ the dual basis.

Now we will use the notation of Subsections 1.2 and 1.3 with $V = TX$ with respect to the metric g^{TX}. In particular if $e \in TX$, $c(\widehat{e})$ acts on $\widehat{S}^{TX}\widehat{\otimes}E$.

Definition 1.2. Set

$$D^V = c(\widehat{e}_i)\,\nabla_{\widehat{e}_i}. \tag{1.25}$$

The operator D^V acts on $\mathbf{I}$.

Note that $\nabla^{\widehat{S}^{TX}\widehat{\otimes}E}$ induces a connection $\nabla^{\mathbf{I}}$ on $\mathbf{I}$. Indeed if $U \in TX$, if s is a smooth section of $\mathbf{I}$, set

$$\nabla^{\mathbf{I}}_U s = \nabla^{\widehat{S}^{TX}\widehat{\otimes}E}_{U^H} s. \tag{1.26}$$

Let $\Omega^{\cdot}(X, \mathbf{R})$ denote the vector space of smooth differential forms on X. More generally $\Omega^{\cdot}(X, \mathbf{I})$ denotes the space of smooth sections of $\Lambda^{\cdot}(T^*X)\,\widehat{\otimes}\mathbf{I}$ over X. Of course

$$\Omega^{\cdot}(X, \mathbf{I}) \simeq C^\infty(\mathcal{X}, \pi^*F)\,. \tag{1.27}$$

The operator $\nabla^{\mathbf{I}}$ acts naturally on $\Omega^{\cdot}(X, \mathbf{I})$, its action on $\Omega^{\cdot}(X, \mathbf{R})$ being just the de Rham operator d^X.

It will be useful to equip $\Lambda^{\cdot}(T^*X)$ with the connection $\nabla^{\Lambda^{\cdot}(T^*X),L}$ induced by the Levi-Civita connection $\nabla^{TX,L}$ with respect to g^{TX}. The vector bundle $F = \Lambda^{\cdot}(T^*X)\,\widehat{\otimes}\widehat{S}^{TX}\widehat{\otimes}E$ is then equipped with an Euclidean connection ∇^F, induced by the connections $\nabla^{\Lambda^{\cdot}(T^*X),L}, \nabla^{\widehat{S}^{TX}}, \nabla^E$. The canonical isomorphism in (1.20) is parallel if and only if $\nabla^{TX} = \nabla^{TX,L}$.

Proposition 1.3. *The following identity holds,*

$$\nabla^{\mathbf{I}} = e^i \wedge \nabla^F_{e_i}. \tag{1.28}$$

Proof. Since the Levi-Civita connection $\nabla^{TX,L}$ is torsion free, the de Rham operator d^X is given by

$$d^X = e^i \wedge \nabla^{\Lambda^{\cdot}(T^*X),L}_{e_i}. \tag{1.29}$$

Equation (1.28) is now a trivial consequence of (1.29). □

We use the notation,

$$c\left(\widehat{R^{TX}Y}\right) = \frac{1}{2} e^i \wedge e^j \wedge c\left(\widehat{R^{TX}(e_i, e_j)\,Y}\right). \tag{1.30}$$

Now we will use the formalism of Quillen's superconnections [Q85a]. The reader who is not familiar with superconnections can just treat them as ordinary operators.

Definition 1.4. Let $\mathcal{D}$ be the superconnection on $\mathbf{I}$,

$$\mathcal{D} = \nabla^{\mathbf{I}} + \frac{D^V}{\sqrt{2}}. \tag{1.31}$$

Let Δ^V be fibrewise Laplacian along the fibres of TX with respect to the metric g^{TX}.

Proposition 1.5. *The following identities hold,*

$$\mathcal{D}^2 = -\frac{1}{2}\Delta^V - \nabla_{\widehat{R^{TX}Y}} + \frac{1}{4}\left\langle R^{TX} e_i, e_j\right\rangle c\left(\widehat{e}_i\right) c\left(\widehat{e}_j\right) + R^E. \tag{1.32}$$

Proof. Clearly,

$$\mathcal{D}^2 = \nabla^{\mathbf{I},2} + \frac{1}{2} D^{V,2}. \tag{1.33}$$

We have the trivial,

$$D^{V,2} = -\Delta^V. \tag{1.34}$$

Moreover, one verifies easily that

$$\nabla^{\mathbf{I},2} = -\nabla_{\widehat{R^{TX}Y}} + \frac{1}{4}\left\langle R^{TX} e_i, e_j\right\rangle c\left(\widehat{e}_i\right) c\left(\widehat{e}_j\right) + R^E. \tag{1.35}$$

Equation (1.32) follows from (1.34)–(1.35). □

Remark 1.6. At this early stage, the reader may ask why we did not choose instead the Levi-Civita superconnection [B86],

$$\mathcal{D}_L = \nabla^{\mathbf{I}} + \frac{D^V}{\sqrt{2}} - \frac{c\left(\widehat{R^{TX}Y}\right)}{2\sqrt{2}}. \tag{1.36}$$

Indeed this superconnection appears as such in [B90b, B94] in relation with local index theory. We gave up writing up the present paper also for $\mathcal{D}_L$. One reason is that in the case of holomorphic manifolds, the corresponding operator is well defined only in the case where TX is nonpositive. Still in the real case, most of the arguments given for $\mathcal{D}$ also work for $\mathcal{D}_L$. It could also be that in certain cases, using $\mathcal{D}_L$ instead of $\mathcal{D}$ is more appropriate.

1.6. The operator $\mathcal{D}_Z$

Let Z be a smooth section of π^*TX over $\mathcal{X}$. Let Z^* be the section of T^*X which corresponds to Z by the metric g^{TX}. By (1.10), $\widehat{c}(Z)$ acts on $\pi^*\Lambda^{\cdot}(T^*X)$ by the formula

$$\widehat{c}(Z) = Z^* \wedge /\sqrt{2} + \sqrt{2} i_Z. \tag{1.37}$$

Then $\widehat{c}(Z)$ acts naturally on $C^\infty(\mathcal{X}, \pi^*F)$. Similarly the superconnection $\mathcal{D}$ can also be viewed as an operator acting on $C^\infty(\mathcal{X}, \pi^*F)$.

Definition 1.7. Let $\mathcal{D}_Z$ be the operator acting on $C^\infty(\mathcal{X}, \pi^*F)$,

$$\mathcal{D}_Z = \mathcal{D} + \frac{\widehat{c}(Z)}{\sqrt{2}}. \tag{1.38}$$

By (1.31), (1.37), (1.38), we get

$$\mathcal{D}_Z = \nabla^{\mathbf{I}} + \frac{D^V}{\sqrt{2}} + \frac{Z^*\wedge}{2} + i_Z. \tag{1.39}$$

Because of the presence of i_Z in the right-hand side of (1.39), $\mathcal{D}_Z$ is no longer a superconnection on $\mathbf{I}$.

Proposition 1.8. *The following identities hold,*

$$\sqrt{2}\mathcal{D}_Z = (c(e_i) + \widehat{c}(e_i)) \nabla^F_{e_i} + c(\widehat{e}_i) \nabla_{\widehat{e}_i} + \widehat{c}(Z). \tag{1.40}$$

Proof. Using (1.11), (1.25), (1.28) and (1.39), we get (1.40). □

1.7. The time parameter

For $t > 0$, if the metric g^{TX} is replaced by g^{TX}/t, the corresponding operator $\mathcal{D}^t_{tZ}$ is given by

$$\mathcal{D}^t_{tZ} = \nabla^{\mathbf{I}} + \frac{\sqrt{t}}{\sqrt{2}} D^V + \frac{Z^*}{2} \wedge + t i_Z. \tag{1.41}$$

Let N^H be the number operator which counts the degree in the variables e^i. By (1.41), we get

$$t^{N^H/2} \mathcal{D}^t_{tZ} t^{-N^H/2} = \sqrt{t} \mathcal{D}_Z. \tag{1.42}$$

A key implicit assumption in the above constructions is that the metrics on the two copies of TX in the right-hand side of (1.24) are the same. In fact the operator D^V and the morphism $\widehat{c}(Z)$ are calculated with respect to the same metric g^{TX}. This way, we also handle the case where these two metrics are proportional, with a fixed constant of proportionality. Indeed given $a \in \mathbf{R}^*$, let r_a be the dilation $(x, Y) \to (x, aY)$. The pull back of the vertical metric g^{TX} by r_a is just $a^2 g^{TX}$.

However we will temporarily relax this assumption. Given $b > 0, t > 0$, we equip $T\mathcal{X} = \pi^* (TX \oplus TX)$ with the metrics $\frac{g^{TX}}{t} \oplus \frac{b^4}{t^3} g^{TX}$. The associated superconnection $\mathcal{D}_{b,t}$ is given by

$$\mathcal{D}_{b,t} = \nabla^{\mathbf{I}} + \frac{t^{3/2}}{b^2} \frac{D^V}{\sqrt{2}}. \tag{1.43}$$

If Z is a section of TX over $\mathcal{X}$, by (1.10), the operator $\widehat{c}^t(Z)$ associated with the metric g^{TX}/t is given by

$$\widehat{c}^t(Z) = \frac{1}{t\sqrt{2}} Z^* \wedge + \sqrt{2} i_Z. \tag{1.44}$$

Set

$$\mathcal{D}_{b,t,Z} = \mathcal{D}_{b,t} + \frac{\widehat{c}^t(Z)}{\sqrt{2}}, \tag{1.45}$$

so that

$$\mathcal{D}_{b,t,Z} = \nabla^{\mathbf{I}^{\cdot}} + \frac{t^{3/2}}{b^2} \frac{D^V}{\sqrt{2}} + \frac{Z^*}{2t} \wedge + i_Z. \tag{1.46}$$

For $a > 0$, let K_a be the map $s(x, Y) \to s(x, aY)$. Observe that

$$K_{t/b^2} \mathcal{D}_{b,t,Z} K_{b^2/t} = \nabla^{\mathbf{I}^{\cdot}} + \frac{\sqrt{t}}{\sqrt{2}} D^V + \frac{K_{t/b^2} Z^*}{2t} \wedge + i_{K_{t/b^2} Z}. \tag{1.47}$$

Comparison with (1.41) shows that

$$K_{t/b^2} \mathcal{D}_{b,t,Y} K_{b^2/t} = \mathcal{D}^t_{tY/b^2}. \tag{1.48}$$

Therefore when $Z = Y$, the two above constructions are equivalent.

The point of view developed in this subsection will play an important role in Section 7.

1.8. A property of self-adjointness

We consider the operator $\mathcal{D}_Z$ as acting on $C^\infty\left(\mathcal{X}, \pi^*\left(S^{TX} \widehat{\otimes} \widehat{\Lambda}^{\cdot}(T^*X) \widehat{\otimes} E\right)\right)$, i.e., we write temporarily the operator $\mathcal{D}_Z$ in the form given in (1.40).

Consider the map $r : Y \to -Y$. This map acts naturally on the fibrewise sections of $\widehat{\Lambda}^{\cdot}(T^*X)$. Namely if $\alpha \in \widehat{\Lambda}^i(T^*X)$, then

$$r^* \alpha(x, Y) = (-1)^i \alpha(-Y). \tag{1.49}$$

Of course the involution r^* extends to the full $C^\infty(\mathcal{X}, \pi^* F)$.

Let $C^{\infty,c}(\mathcal{X}, \pi^* F)$ be the vector space of compactly supported sections of $\pi^* F$ over $\mathcal{X}$. Let $\langle \rangle_{L^2}$ be the obvious L^2 Hermitian product on $C^{\infty,c}(\mathcal{X}, \pi^* F)$. If $s, s' \in C^{\infty,c}(\mathcal{X}, \pi^* F)$, put

$$\eta(s, s') = \langle r^* s, s' \rangle_{L^2}. \tag{1.50}$$

Then η is a Hermitian form, i.e., it is nondegenerate, and moreover,

$$\eta(s', s) = \overline{\eta(s, s')}. \tag{1.51}$$

Let $\mathcal{D}_Z^\dagger$ be the adjoint of $\mathcal{D}_Z$ with respect to η, so that

$$\eta\left(\mathcal{D}_Z s, s'\right) = \eta\left(s, \mathcal{D}_Z^\dagger s'\right). \tag{1.52}$$

Let Z_- be the section of TX on $\mathcal{X}$ given by

$$Z_-\left(x, Y\right) = -Z\left(x, -Y\right). \tag{1.53}$$

Theorem 1.9. *The following identity holds,*

$$\mathcal{D}_Z^\dagger = \mathcal{D}_{Z_-}. \tag{1.54}$$

Proof. Observe that if $e \in TX$, $c\left(e\right), c\left(\widehat{e}\right)$ are skew-adjoint operators in the usual sense, while $\widehat{c}\left(e\right)$ is self-adjoint. Equation (1.54) now follows from (1.40). □

1.9. The Lichnerowicz formula for $\mathcal{D}_Z^2$

Theorem 1.10. *The following identity holds,*

$$\mathcal{D}_Z^2 = \frac{1}{2}\left(-\Delta^V + |Z|^2 + c\left(\widehat{e}_i\right)\widehat{c}\left(\nabla_{\widehat{e}^i} Z\right)\right) - \nabla_{\widehat{R^{TX}Y}} + \frac{1}{4}\left\langle R^{TX} e_i, e_j\right\rangle c\left(\widehat{e}_i\right)c\left(\widehat{e}_j\right) \\ + \frac{e^i}{\sqrt{2}}\widehat{c}\left(\nabla_{e_i}^{TX,L} Z\right) + \nabla_Z^F + R^E. \tag{1.55}$$

Proof. By (1.38), we get

$$\mathcal{D}_Z^2 = \mathcal{D}^2 + \frac{|Z|^2}{2} + \left[\mathcal{D}, \frac{\widehat{c}\left(Z\right)}{\sqrt{2}}\right]. \tag{1.56}$$

By (1.25), (1.31), by equation (1.32) in Proposition 1.5 and by (1.37), (1.56), we get (1.55). □

1.10. A hypoellipticity property

We will say that the section Z of $\pi^* TX$ is fibrewise regular if the map $Y \to Z\left(x, Y\right)$ is regular, or equivalently if $\widehat{e} \in TX \to \nabla_{\widehat{e}} Z \in TX$ is invertible. This is the case in particular if $Z = cY, c \neq 0$.

Theorem 1.11. *If Z is fibrewise regular, the operator $\frac{\partial}{\partial u} - \mathcal{D}_Z^2$ is hypoelliptic.*

Proof. This is an easy consequence of equation (1.55) and of a result of Hörmander [Hö67]. □

1.11. A rescaling on Y and the elliptic Dirac operator

Set

$$Z_b = K_{1/b^2} Z. \tag{1.57}$$

Then

$$Z_b\left(x, Y\right) = Z\left(x, Y/b^2\right). \tag{1.58}$$

Definition 1.12. For $b > 0$, set

$$\mathcal{E}_{Z,b} = K_b \mathcal{D}_{Z_b} K_b^{-1}. \tag{1.59}$$

By (1.40), (1.58), we get

$$\sqrt{2}\mathcal{E}_{Z,b} = \left(c\left(e_i\right) + \widehat{c}\left(e_i\right)\right) \nabla^F_{e_i} + \frac{1}{b}\left(D^V + \widehat{c}\left(bZ\left(x, \frac{Y}{b}\right)\right)\right). \tag{1.60}$$

Of special interest is the case where $Z(x,Y) = Y$. By (1.60), we get

$$\sqrt{2}\mathcal{E}_{Y,b} = \left(c\left(e_i\right) + \widehat{c}\left(e_i\right)\right) \nabla^F_{e_i} + \frac{1}{b}\left(D^V + \widehat{c}(Y)\right). \tag{1.61}$$

Equations (1.60), (1.61) should be put in parallel with [B05, Propositions 2.34 and 2.39], where corresponding results are proved for the de Rham complex.

1.12. The projection of the hypoelliptic Dirac operator

Now we use the fact that $F = S^{TX} \widehat{\otimes} \widehat{\Lambda}^{\cdot}(T^*X) \widehat{\otimes} E$, and also the representation of the Clifford actions in the second line of (1.10).

Recall that a result of Witten [W82], [B05, Proposition 2.40] asserts that when acting on smooth L^2 sections of $\widehat{\Lambda}^{\cdot}(T^*X)$ along the fibres TX, $\ker\left(D^V + \widehat{c}\left(\widehat{Y}\right)\right)$ is 1-dimensional and spanned by $\exp\left(-\left|Y\right|^2/2\right)$. When acting on smooth L^2 sections of F along the fibre TX, the corresponding result is that $\ker\left(D^V + \widehat{c}\left(\widehat{Y}\right)\right)$ is spanned by $S^{TX}\widehat{\otimes}E \otimes \exp\left(-\left|Y\right|^2/2\right)$.

We embed $C^\infty\left(X, S^{TX}\widehat{\otimes}E\right)$ into $C^\infty\left(\mathcal{X}, \pi^* F\right)$ by the L^2 isometric embedding $s \to \pi^* s \exp\left(-\left|Y\right|^2/2\right)/\pi^{n/4}$. Let P be the orthogonal projection operator on the image.

Evaluating P is quite easy. Indeed let $p : \widehat{\Lambda}^{\cdot}(T^*X) \to \widehat{\Lambda}^0(T^*X) = \mathbf{R}$ be the obvious projection. Then p maps $F = S^{TX}\widehat{\otimes}\widehat{\Lambda}^{\cdot}(T^*X)\widehat{\otimes}E$ into $S^{TX}\widehat{\otimes}E$. If s is a L^2 section of $\pi^* F$,

$$Ps(x,Y) = \frac{\exp\left(-\left|Y\right|^2/2\right)}{\pi^{n/4}} \int_{TX} ps\left(x,Y'\right) \exp\left(-\left|Y'\right|^2/2\right) \frac{dY'}{\pi^{n/4}}. \tag{1.62}$$

In the sequel, we identify $C^\infty\left(X, S^{TX}\widehat{\otimes}E\right)$ to its image by the above embedding.

Now we establish an analogue of [B05, Proposition 2.41].

Theorem 1.13. *The following identity of operators holds,*

$$P\left(\left(c\left(e_i\right) + \widehat{c}\left(e_i\right)\right) \nabla^F_{e_i}\right) P = D^X. \tag{1.63}$$

Proof. By (1.10), $\widehat{c}(e)$ maps $\mathbf{R} = \widehat{\Lambda}^0 (T^*X)$ into its orthogonal. In the case where $\nabla^{TX} = \nabla^{TX,L}$, (1.63) is obvious.

Now we consider the general case. By (1.21),

$$\nabla_{\cdot}^{\widehat{S}^{TX}} = \nabla_{\cdot}^{\widehat{S}^{TX},L} + \frac{1}{4} \langle A(\cdot) e_j, e_k \rangle c(\widehat{e}_j) c(\widehat{e}_k) . \tag{1.64}$$

Let $\nabla^{F,L}$ be the connection on F which is induced by the connections $\nabla^{TX,L}, \nabla^E$. By (1.64), we deduce a similar equation relating ∇^F and $\nabla^{F,L}$.

If $U \in TX$, let $U^{H,L}$ be the horizontal lift of U with respect to the connection $\nabla^{TX,L}$. By (1.21), we get

$$U^H = U^{H,L} - \widehat{A(U)Y}. \tag{1.65}$$

In (1.65), $\widehat{A(U)Y}$ denotes the vector field corresponding to $A(U)Y$. By (1.64), (1.65), we get

$$\nabla_{U^H}^F = \nabla_{U^{H,L}}^{F,L} - \nabla_{\widehat{A(U)Y}} + \frac{1}{4} \langle A(U) e_j, e_k \rangle c(\widehat{e}_j) c(\widehat{e}_k) . \tag{1.66}$$

Now we already proved (1.63) when ∇^F is replaced by $\nabla^{F,L}$. Using the fact that A takes its values in antisymmetric endomorphisms and (1.66), we still get (1.63) in full generality. □

1.13. The hypoelliptic Laplacian and its relation to the elliptic Laplacian

By Theorem 1.10, we get

$$\begin{aligned}\mathcal{E}_{Z,b}^2 = \frac{1}{2b^2} \Big(-\Delta^V + |bZ(x, Y/b)|^2 + c(\widehat{e}_i) \widehat{c}(\nabla_{\widehat{e}_i} (bZ(x, Y/b))) \Big) - \nabla_{\widehat{R^{TX}Y}} \\ + \frac{1}{4} \langle R^{TX} e_i, e_j \rangle c(\widehat{e}_i) c(\widehat{e}_j) + \frac{e^i}{\sqrt{2}b} \widehat{c}\left(\nabla_{e_i}^{TX,L} bZ(x, Y/b)\right) + \frac{1}{b} \nabla_{bZ(x,Y/b)}^F + R^E.\end{aligned} \tag{1.67}$$

We identify the canonical section Y of TX to the corresponding horizontal vector field Y^H on $\mathcal{X}$. Note that this vector field still depends on ∇^{TX}. If $\nabla^{TX} = \nabla^{TX,L}$, we will write instead $Y^{H,L}$. By (1.65),

$$Y^H = Y^{H,L} - \widehat{A(Y)Y}. \tag{1.68}$$

Using (1.66)–(1.68), we obtain,

$$\begin{aligned}\mathcal{E}_{Y,b}^2 = \frac{1}{2b^2} \Big(-\Delta^V + |Y|^2 + c(\widehat{e}_i) \widehat{c}(e_i) \Big) - \nabla_{\widehat{R^{TX}Y}} + \frac{1}{4} \langle R^{TX} e_i, e_j \rangle c(\widehat{e}_i) c(\widehat{e}_j) \\ + \frac{1}{b} \Big(\nabla_Y^{F,L} - \nabla_{\widehat{A(Y)Y}} + \frac{1}{4} \langle A(Y) e_j, e_k \rangle c(\widehat{e}_j) c(\widehat{e}_k) - \frac{e^i}{\sqrt{2}} \widehat{c}(A(e_i)Y) \Big) + R^E.\end{aligned} \tag{1.69}$$

By (1.69), we find that

$$\frac{\mathcal{E}_{Y,b}^2}{2} = \frac{\alpha}{b^2} + \frac{\beta}{b} + \gamma. \tag{1.70}$$

In (1.70), we have the obvious identities,

$$\begin{aligned}
\alpha &= \frac{1}{2}\left(-\Delta^V + |Y|^2 + c\left(\widehat{e}_i\right)\widehat{c}\left(e_i\right)\right),\\
\beta &= \nabla_Y^{F,L} - \nabla_{\widehat{A(Y)Y}} + \frac{1}{4}\left\langle A\left(Y\right)e_j, e_k\right\rangle c\left(\widehat{e}_j\right)c\left(\widehat{e}_k\right) - \frac{e^i}{\sqrt{2}}\widehat{c}\left(A\left(e_i\right)Y\right),\\
\gamma &= -\nabla_{\widehat{R^{TX}Y}} + \frac{1}{4}\left\langle R^{TX}e_i, e_j\right\rangle c\left(\widehat{e}_i\right)c\left(\widehat{e}_j\right) + R^E.
\end{aligned} \tag{1.71}$$

Note that

$$\alpha = \frac{1}{2}\left(D^V + \widehat{c}\left(Y\right)\right)^2. \tag{1.72}$$

The operator α is Witten's harmonic oscillator [W82]. By [W82], when acting on sections of $\Lambda^{\cdot}\left(T^*X\right)$ along the fibres T^*X, its kernel is spanned by $\exp\left(-|Y|^2/2\right)$. In general the kernel is spanned by $S^{TX}\widehat{\otimes}E\otimes\exp\left(-|Y|^2/2\right)$. Let $\ker\alpha^{\perp}$ be the orthogonal vector space to $\ker\alpha$ with respect to the obvious L^2 Hermitian product. We denote by α^{-1} the inverse of the restriction of α to $\ker\alpha^{\perp}$. Observe that β maps $\ker\alpha$ into $\ker\alpha^{\perp}$.

Now we establish the obvious analogue of [B05, Theorem 3.14].

Theorem 1.14. *The following identity holds,*

$$P\left(\gamma - \beta\alpha^{-1}\beta\right)P = \frac{D^{X,2}}{2}. \tag{1.73}$$

Proof. By squaring the expansion (1.61) for $\mathcal{E}_{Y,b}$, by proceeding as in [B05, Remark 3.15], and using Theorem 1.13, we can give a simple proof of (1.73). Here we give a direct proof.

By (1.10), (1.11), (1.71), we get

$$P\gamma P = \frac{1}{8}\left\langle R^{TX}\left(e_i, e_j\right)e_j, e_i\right\rangle + \frac{1}{4}c\left(e_i\right)c\left(e_j\right)R^E\left(e_i, e_j\right). \tag{1.74}$$

Let N^V be the vertical number operator, i.e., the operator counting the degree in the Grassmann variables $\widehat{e}^i$. Then

$$\alpha = \frac{1}{2}\left(-\Delta^V + |Y|^2 - n\right) + N^V. \tag{1.75}$$

In other words, α is the sum of a bosonic and of a fermionic number operator. By (1.10), (1.71) and (1.75), we get

$$\alpha^{-1}\beta P = \Big(\nabla_Y^{F,L} + \frac{1}{12}\left\langle A\left(Y\right)e_j, e_k\right\rangle \widehat{e}^j\widehat{e}^k - \frac{1}{6}\left\langle A\left(e_j\right)Y, e_k\right\rangle \widehat{e}^j\widehat{e}^k + \frac{1}{2}\left\langle A\left(e_i\right)e_i, Y\right\rangle$$
$$- \frac{1}{4}\left\langle A\left(e_j\right)Y, e_k\right\rangle c\left(e_j\right)\widehat{e}^k\Big)P, \tag{1.76}$$

$$P\beta = P\Big(\nabla_Y^{F,L} + \frac{1}{4}\left\langle A\left(Y\right)e_j, e_k\right\rangle i_{\widehat{e}_j}i_{\widehat{e}_k} - \frac{1}{2}\left\langle A\left(e_j\right)Y, e_k\right\rangle i_{\widehat{e}^j}i_{\widehat{e}^k} - \frac{1}{2}\left\langle A\left(e_i\right)e_i, Y\right\rangle$$
$$- \frac{1}{2}\left\langle A\left(e_j\right)Y, e_k\right\rangle c\left(e_j\right)i_{\widehat{e}_k}\Big).$$

Moreover,

$$\int_{\mathbf{R}} x^2\exp\left(-x^2\right)\frac{dx}{\pi^{n/2}}dx = \frac{1}{2}. \tag{1.77}$$

By (1.76) and (1.77), we get

$$-P\beta\alpha^{-1}\beta P = \frac{1}{2}\Big(-\Delta^H - \frac{1}{4}\left\langle A\left(e_i\right)e_j, A\left(e_j\right)e_i\right\rangle + \frac{1}{4}\Big|\sum_1^n A\left(e_i\right)e_i\Big|^2$$
$$- \frac{1}{2}\left\langle\nabla_{e_i}^{TX,L}A\left(e_j\right)e_j, e_i\right\rangle\Big). \tag{1.78}$$

Finally from (1.21), we obtain,

$$\left\langle R^{TX}\left(e_i, e_j\right)e_j, e_i\right\rangle - \left\langle A\left(e_i\right)e_j, A\left(e_j\right)e_i\right\rangle + \Big|\sum_{i=1}^n A\left(e_i\right)e_i\Big|^2$$
$$- 2\left\langle\nabla_{e_i}^{TX,L}A\left(e_j\right)e_j, e_i\right\rangle = K. \tag{1.79}$$

By (1.18), (1.74), (1.78), (1.79), we get (1.73). The proof of our theorem is completed. □

Remark 1.15. When $A = 0$, the proof is of course very simple. Also note that the inverse α^{-1} contributes by factors $1, 1/2, 1/3$ to $P\beta\alpha^{-1}\beta P$, contrary to what happens in [BL91, B05, BL06], where only the factor $1/2$ appears.

1.14. Interpolating between the Laplacian and the geodesic flow

Definition 1.16. Put

$$\mathcal{F}_{Z,b} = K_{b^2}\mathcal{D}_{Z_b}K_{b^2}^{-1}. \tag{1.80}$$

By (1.28), (1.38), we get

$$\mathcal{F}_{Z,b} = e^i\nabla_{e_i}^F + \frac{D^V}{\sqrt{2}b^2} + \frac{\widehat{c}\left(Z\right)}{\sqrt{2}}. \tag{1.81}$$

Put

$$\widehat{c}\left(\nabla^{TX,L}_{\cdot} Z\right) = -e^i \widehat{c}\left(\nabla^{TX,L}_{e_i} Z\right). \tag{1.82}$$

By (1.55), we get

$$\begin{aligned}\mathcal{F}^2_{Z,b} = \frac{|Z|^2}{2} - \nabla_{\widehat{R^{TX}Y}} + \frac{1}{4}\left\langle R^{TX} e_i, e_j\right\rangle c\left(\widehat{e}_i\right) c\left(\widehat{e}_j\right) \\ - \widehat{c}\left(\nabla^{TX,L}_{\cdot} Z\right)/\sqrt{2} + \nabla^F_Z + R^E + \mathcal{O}\left(1/b^2\right).\end{aligned} \tag{1.83}$$

One can directly obtain (1.83) by squaring the expansion in (1.81).

The vector field $Y^{H,L}$ is just the generator of the geodesic flow on $\mathcal{X}$. If $\nabla^{TX} = \nabla^{TX,L}$, the operator ∇^F_Y is the obvious lift of the generator of the geodesic flow. The constant term as $b \to +\infty$ in the expansion of $\mathcal{F}^2_{Y,b}$ is a perturbation of the operator ∇^F_Y.

2. A deformation of the Levi-Civita superconnection

In this section, we construct a hypoelliptic deformation of the Levi-Civita superconnection of [B86], i.e., we extend the results of Section 1 to the case of families.

This section is organized as follows. In Subsection 2.1, given a submersion $p : M \to S$ with compact fibre X, we briefly recall some aspects of the construction of the Levi-Civita superconnection.

In Subsection 2.2, if $\pi : \mathcal{M} \to M$ is the total space of TX, we construct a superconnection $\mathcal{A}$ over M, which is the obvious extension to families of manifolds of the superconnection $\mathcal{D}$ obtained in Section 1. Also we construct a superconnection $\mathcal{A}_Z$ over S, which extends the construction of $\mathcal{D}_Z$ to the case of families.

In Subsection 2.3, we show that if Z is antiinvariant, then $\mathcal{A}_Z$ is self-adjoint with respect to the Hermitian form η.

In Subsection 2.4, we give a formula for the curvature $\mathcal{A}^2_Z$ of $\mathcal{A}_Z$.

In Subsection 2.5, we give formulas which relate a family of such superconnections, which depends on $b > 0$, to the Levi-Civita superconnection.

In Subsection 2.6, we show that this family of superconnections interpolates in the proper sense between the Levi-Civita superconnection and the fibrewise geodesic flow.

For simplicity, we consider only the case where the dimension of the fibres is even.

This section is an analogue of [B05, Section 4], where similar results were established in the context of the families of fibrewise de Rham complexes as in [BLo95].

2.1. The Levi-Civita superconnection

Let $p : M \to S$ be a submersion of smooth manifolds with compact oriented fibre X, of even dimension n. We assume that TX is a spin vector bundle on M. Let g^{TX} be an Euclidean metric on TX, let $T^H M$ be a horizontal vector bundle on

M, so that

$$TM = T^H M \oplus TX. \tag{2.1}$$

Let $P^{TX} : TM \to TX$ be the projection associated to the above splitting. Let (E, g^E, ∇^E) be a complex Hermitian vector bundle on M equipped with a Hermitian connection.

Let $H = C^\infty \left(X, \left(S^{TX} \widehat{\otimes} E\right)|_X\right)$ be the infinite-dimensional vector bundle over S of smooth sections of $S^{TX}\widehat{\otimes}E$ along the fibre X. This is a $\mathbf{Z}_2$-graded vector bundle over S. Let D^X be the family of Dirac operators considered in Subsection 1.4 acting along the fibres X.

Let us now briefly describe the construction of the Levi-Civita superconnection [B86]. First $(T^H M, g^{TX})$ determines an Euclidean connection ∇^{TX} on TX. If g^{TM} is any Euclidean metric on TM which is such that the splitting (2.1) is orthogonal, then ∇^{TX} is the projection of the Levi-Civita connection $\nabla^{TM,L}$ on TX with respect to the splitting (2.1). In particular ∇^{TX} restricts to the Levi-Civita connection along the fibres. Let R^{TX} be the curvature of ∇^{TX}.

If $U \in TS$, let $U^H \in T^H M$ be the horizontal lift of U. If U is a vector field, the Lie derivative operator L_{U^H} acts naturally on the smooth sections of the tensor algebra associated to TX, and this action is also tensorial with respect to U.

In [B86, Section 1], we have constructed natural tensors which are associated to the above data. First a 2-form T on M with values in TX is defined which is such that:

- T vanishes on $TX \times TX$.
- If $U, V \in TS$,

$$T\left(U^H, V^H\right) = -P^{TX}\left[U^H, V^H\right]. \tag{2.2}$$

- If $U \in TS, A \in TX$,

$$T\left(U^H, A\right) = \frac{1}{2}\left(g^{TX}\right)^{-1}\left(L_{U^H} g^{TX}\right) A. \tag{2.3}$$

It follows from (2.3) that if $U \in TS, A, B \in TX$, then

$$\left\langle T\left(U^H, A\right), B\right\rangle = \left\langle T\left(U^H, B\right), A\right\rangle. \tag{2.4}$$

Let $e_1, \ldots, e_n$ be an orthonormal basis of TX, let $f_1, \ldots, f_m$ be a basis of TS. The corresponding dual bases are denoted the usual way.

Let dv_X be the Riemannian volume element along the fibres X. By the above, if $U \in TS$,

$$L_{U^H} dv_X = \left\langle T\left(U^H, e_i\right), e_i\right\rangle dv_X. \tag{2.5}$$

The connection ∇^{TX} induces a Hermitian connection on S^{TX}, which restricts to the Levi-Civita connection along the fibres. Let $\nabla^{S^{TX}\widehat{\otimes}E}$ be the obvious connection on $S^{TX}\widehat{\otimes}E$.

Definition 2.1. If $s \in C^\infty\left(M, S^{TX}\widehat{\otimes}E\right), U \in TS$, set

$$\nabla^H_U s = \nabla^{S^{TX}\widehat{\otimes}E}_{U^H} s + \frac{1}{2}\left\langle T\left(U^H, e_i\right), e_i\right\rangle s. \tag{2.6}$$

Then (2.6) defines a Hermitian connection on H.

Let T^H be the restriction of T to $T^HM \times T^HM$. Note that T^H is obtained via equation (2.2). Set

$$c\left(T^H\right) = \frac{1}{2} f^\alpha f^\beta c\left(T\left(f_\alpha^H, f_\beta^H\right)\right). \tag{2.7}$$

Definition 2.2. Let A be the Levi-Civita superconnection on H,

$$A = \frac{D^X}{\sqrt{2}} + \nabla^H - \frac{c\left(T^H\right)}{2\sqrt{2}}. \tag{2.8}$$

In [B86], the fibrewise connection ${}^1\nabla^{\Lambda^\cdot(T^*S)\widehat{\otimes} S^{TX}\widehat{\otimes} E}$ along the fibres X was defined,

$${}^1\nabla_\cdot^{\Lambda^\cdot(T^*S)\widehat{\otimes} S^{TX}\widehat{\otimes} E} = \nabla_\cdot^{S^{TX}\widehat{\otimes} E} + \frac{1}{2}\left\langle T\left(f_\alpha^H, e_i\right), \cdot\right\rangle f^\alpha c\left(e_i\right) + \frac{1}{4}\left\langle T^H, \cdot\right\rangle. \tag{2.9}$$

For $a > 0$, let $\psi_a \in \operatorname{End}\left(\Lambda^\cdot\left(T^*S\right)\right)$ be given by $\alpha \to a^{\deg \alpha/2}\alpha$. Set

$${}^1\nabla_{a,\cdot}^{\Lambda^\cdot(T^*S)\widehat{\otimes} S^{TX}\widehat{\otimes} E} = \psi_a^{-1}\,{}^1\nabla_\cdot^{\Lambda^\cdot(T^*S)\widehat{\otimes} S^{TX}\widehat{\otimes} E}\psi_a. \tag{2.10}$$

Let K be the scalar curvature of the fibres X. The Lichnerowicz formula given in [B86, Theorem 3.6] asserts that

$$\begin{aligned} A^2 = -\frac{1}{2}\,{}^1\nabla_{1/2,e_i}^{S^{TX}\widehat{\otimes} E,2} + \frac{K}{8} + \frac{1}{4} c\left(e_i\right) c\left(e_j\right) R^E\left(e_i, e_j\right) \\ + \frac{1}{\sqrt{2}} c\left(e_i\right) f^\alpha R^E\left(e_i, f_\alpha^H\right) + \frac{1}{2} f^\alpha f^\beta R^E\left(f_\alpha^H, f_\beta^H\right). \end{aligned} \tag{2.11}$$

2.2. A construction of superconnections and fibrewise operators on TX

Here we are inspired by a construction in [B05, Subsection 4.22]. Let $\pi : \mathcal{M} \to M$ be the total space of the vector bundle TX over M. Let $Y \in TX$ be the generic element of the fibre TX. Then $\mathcal{M}$ fibres over M with fibre TX, and moreover, TX is equipped with a connection ∇^{TX}, which in turn induces a horizontal vector bundle $T^H\mathcal{M}$ on $\mathcal{M}$. Let $\mathcal{H}$ be the vector bundle of smooth sections of $\widehat{S}^{TX}\widehat{\otimes} E$ along the fibres TX.

We denote by $\mathcal{T}$ the tensor T defined in Subsection 2.1 which is associated here to the fibration $p : \mathcal{M} \to M$. Note that by (2.2), (2.3), $\mathcal{T}$ vanishes on $TX \times T^H\mathcal{M}$, and moreover, if $U, V \in TM$,

$$\mathcal{T}\left(U^H, V^H\right) = R^{TX}\widehat{(U, V)}\,Y. \tag{2.12}$$

We define the operator D^V as in (1.25), and the connection $\nabla^{\mathcal{H}}$ as in (1.26). Let $\mathcal{A}$ be the superconnection over $\mathcal{H}$,

$$\mathcal{A} = \nabla^{\mathcal{H}} + \frac{D^V}{\sqrt{2}}. \tag{2.13}$$

Let us observe that $\mathcal{A}$ depends on T^HM only indirectly through the choice of ∇^{TX}. In fact, any other Euclidean connection could be used over TX, at the cost of extra notational complication.

Note that $\nabla^{\mathcal{H}}$ acts on smooth sections of $\Lambda^{\cdot}(T^*M)\widehat{\otimes}\mathcal{H}$, and that by construction, it acts like the de Rham operator on smooth sections of $\Lambda^{\cdot}(T^*M)$. The splitting (2.1) induces the identification,

$$\Lambda^{\cdot}(T^*M) \simeq \pi^*\Lambda^{\cdot}(T^*S)\widehat{\otimes}\Lambda^{\cdot}(T^*X). \tag{2.14}$$

The connection ∇^{TX} induces a connection $\nabla^{\Lambda^{\cdot}(T^*X)}$ on $\Lambda^{\cdot}(T^*X)$. Moreover, using (2.14), we find that the action of $\nabla^{\Lambda^{\cdot}(T^*X)}$ induces a corresponding action on the smooth sections of $\Lambda^{\cdot}(T^*M)$. By construction, the action of this operator on smooth sections of $\Lambda^{\cdot}(T^*S)$ is given by the de Rham operator of S.

Note that the operator i_T acts on sections of $\Lambda^{\cdot}(T^*M)$. A simple result is that if d^M is the de Rham operator of M, then

$$d^M = \nabla^{\Lambda^{\cdot}(T^*X)} + i_T. \tag{2.15}$$

The operator $\nabla^{\Lambda^{\cdot}(T^*X)}$ extends to an operator acting on smooth sections of $\Lambda^{\cdot}(T^*M)\widehat{\otimes}\mathcal{H}$. By (2.15), we get

$$\nabla^{\mathcal{H}} = \nabla^{\Lambda^{\cdot}(T^*X)\widehat{\otimes}\mathcal{H}} + i_T. \tag{2.16}$$

As in (1.20), set

$$F = \Lambda^{\cdot}(T^*X)\widehat{\otimes}\widehat{S}^{TX}\widehat{\otimes}E \simeq S^{TX}\widehat{\otimes}\widehat{\Lambda}^{\cdot}(T^*X)\widehat{\otimes}E. \tag{2.17}$$

Recall that $e_1,\dots,e_n$ is an orthonormal basis of TX. To fit with the notation in Section 1, we denote by $e_1,\dots,e_n$ a copy of this base of TX, and by $e^1,\dots,e^n$ the associated dual basis. In particular $\Lambda^{\cdot}(T^*X)$ is generated as an algebra by $e^1,\dots,e^n$. Note here that we specifically refer to the first copy of $\Lambda^{\cdot}(T^*X)$ which appears in the right-hand side of (2.17).

By (1.25), (2.13), (2.16), we get

$$\mathcal{A} = e^i\wedge\nabla^F_{e_i} + f^\alpha\wedge\nabla^F_{f^H_\alpha} + e^i f^\alpha i_{T(e_i,f^H_\alpha)} + i_{T^H} + \frac{c(\widehat{e}_i)}{\sqrt{2}}\nabla_{\widehat{e}_i}. \tag{2.18}$$

Observe that if S is reduced to a point, then

$$\mathcal{A} = \mathcal{D}. \tag{2.19}$$

In this case ∇^{TX} is just the Levi-Civita connection, and not any arbitrary Euclidean connection as in Subsection 1.5.

Of course, we can still use the identities in (1.10), (1.11) to reexpress $\mathcal{A}$ in terms of the Clifford variables $c(e_i), c(\widehat{e}_i), \widehat{c}(e_i), 1\le i\le n$.

Let now Z be a smooth section of π^*TX over $\mathcal{M}$, let Z^* be the smooth section of π^*T^*X which corresponds to Z by the metric g^{TX}.

Definition 2.3. Set

$$\mathcal{A}_Z = \mathcal{A} + \frac{\widehat{c}(Z)}{\sqrt{2}}. \tag{2.20}$$

By (1.37), (2.18), (2.20), we get

$$\mathcal{A}_Z = \frac{c\left(\widehat{e}_i\right)}{\sqrt{2}}\nabla_{\widehat{e}_i} + Z^* \wedge /2 + i_Z + e^i \wedge \nabla^F_{e_i} + f^\alpha \wedge \nabla^F_{f^H_\alpha} + e^i f^\alpha i_{T(e_i, f^H_\alpha)} + i_{T^H}. \quad (2.21)$$

Note that because of the term i_Z in (2.21), in general, $\mathcal{A}_Z$ is not a superconnection over M. Still it can be viewed as a superconnection over S on the vector bundle $C^\infty\left(\mathcal{X}, \pi^* F|_{\mathcal{X}}\right)$. Finally when S is a point, $\mathcal{A}_Z$ is just $\mathcal{D}_Z$.

Proposition 2.4. *The following identity holds,*

$$\begin{aligned}\mathcal{A}_Z = & \frac{1}{\sqrt{2}}\left(\left(c\left(e_i\right) + \widehat{c}\left(e_i\right)\right)\nabla^F_{e_i} + D^V + \widehat{c}(Z)\right) \\ & + f^\alpha\left(\nabla^F_{f^H_\alpha} + \frac{1}{2}\left\langle T\left(f^H_\alpha, e_i\right), e_i\right\rangle + \frac{1}{2}c\left(e_i\right)\widehat{c}\left(e_j\right)\left\langle T\left(f^H_\alpha, e_i\right), e_j\right\rangle\right) \\ & - \frac{1}{2\sqrt{2}}\left(c - \widehat{c}\right)\left(T^H\right). \quad (2.22)\end{aligned}$$

Proof. This is an obvious consequence of (1.10), (1.11), (2.4) and (2.21). □

2.3. A self-adjointness property of the superconnection $\mathcal{A}_Z$

We will now extend Theorem 1.9. We define the Hermitian form η on $C^\infty\left(\mathcal{X}, \pi^* F\right)$ as in (1.49). We define the adjoint of $\mathcal{A}^\dagger_Z$ of the superconnection $\mathcal{A}_Z$ with respect to η as in [BLo95, Section 1], by simply taking adjoints with respect to the Hermitian form η instead of taking adjoints with respect to a classical fibrewise Hermitian product. In the conventions of [BLo95], one important point is that $\left(f^\alpha\right)^\dagger = -f^\alpha$.

Theorem 2.5. *The following identities hold,*

$$\mathcal{A}^\dagger_Z = \mathcal{A}_{Z_-}. \quad (2.23)$$

Proof. We use equation (2.22) in Proposition 2.4, which leads immediately to (2.23). □

2.4. The curvature of the superconnection operator $\mathcal{A}_Z$

Now we extend Theorem 1.10, i.e., we give a formula for the curvature $\mathcal{A}^2_Z$ of $\mathcal{A}_Z$. We use the notation

$$\widehat{c}\left(\nabla^{TX}_{\cdot} Z\right) = -e^i \widehat{c}\left(\nabla^{TX}_{e_i} Z\right) - f^\alpha \widehat{c}\left(\nabla^{TX}_{f^H_\alpha} Z\right). \quad (2.24)$$

Also we define the fibrewise connections ${}^1\nabla^{\Lambda^\cdot(T^*S)\widehat{\otimes}F}_{\cdot}, {}^1\nabla^{\Lambda^\cdot(T^*S)\widehat{\otimes}F}_{a,\cdot}$ as in (2.9), (2.10).

Theorem 2.6. *The following identity holds,*

$$\begin{aligned}\mathcal{A}^2_Z = & \frac{1}{2}\left(-\Delta^V + |Z|^2 + c\left(\widehat{e}_i\right)\widehat{c}\left(\nabla_{\widehat{e}_i} Z\right)\right) - \nabla_{\widehat{R^{TX}Y}} + \frac{1}{4}\left\langle R^{TX} e_i, e_j\right\rangle c\left(\widehat{e}_i\right) c\left(\widehat{e}_j\right) \\ & - \frac{1}{\sqrt{2}}\widehat{c}\left(\nabla^{TX}_{\cdot} Z\right) + {}^1\nabla^F_{1/2, Z} + R^E. \quad (2.25)\end{aligned}$$

Proof. When $Z = 0$, the proof of (2.25) is the same as the proof of Proposition 1.5. The general case is obtained by using the formula for $Z = 0$, (1.3), (1.10), (2.4), (2.18) and also the identity,

$$\nabla^F_Z + f^\alpha \left(\frac{e^i}{2} - i_{e_i}\right) \left\langle T\left(f^H_\alpha, Z\right), e_i\right\rangle + \frac{1}{2}\left\langle T^H, Z\right\rangle = {}^1\nabla^F_{1/2,Z}. \tag{2.26}$$

□

We will say that the section Z of $\pi^* TX$ is fibrewise regular if the map $\widehat{e} \in TX \to \nabla_{\widehat{e}} Z\left(x, Y\right) \in TX$ is invertible. The obvious extension of Theorem 1.11 holds when Z is fibrewise regular.

2.5. A rescaling on Y and the elliptic Levi-Civita superconnection

We still define Z_b as in (1.57). As in (1.59), set

$$\mathcal{B}_{Z,b} = K_b \mathcal{A}_{Z_b} K_b^{-1}. \tag{2.27}$$

By (2.22), we get

$$\begin{aligned}\mathcal{B}_{Z,b} = &\frac{1}{\sqrt{2}}\left(c\left(e_i\right) + \widehat{c}\left(e_i\right)\right)\nabla^F_{e_i} + \frac{1}{\sqrt{2}b}\left(D^V + \widehat{c}\left(bZ\left(x, Y/b\right)\right)\right)\\ &+ f^\alpha\left(\nabla^F_{f^H_\alpha} + \frac{1}{2}\left\langle T\left(f^H_\alpha, e_i\right), e_i\right\rangle + \frac{1}{2}c\left(e_i\right)\widehat{c}\left(e_j\right)\left\langle T\left(f^H_\alpha, e_i\right), e_j\right\rangle\right)\\ &- \frac{1}{2\sqrt{2}}\left(c - \widehat{c}\right)\left(T^H\right).\end{aligned} \tag{2.28}$$

Now we take $Z = Y$. Let $\mathfrak{A}$ be the constant term in the expansion of $\mathcal{B}_{Y,b}$.

We identify $C^\infty\left(X, S^{TX}\widehat{\otimes}E\right)$ to its image in $C^\infty\left(\mathcal{X}, \pi^* F\right)$ as in Subsection 1.11, i.e., by the isometric embedding $s \to \pi^* s \exp\left(-\left|Y\right|^2/2\right)/\pi^{n/4}$. Recall that the operator P was defined in Subsection 1.11. Now we establish the obvious extension of Theorem 1.13.

Theorem 2.7. *The following identity holds,*

$$P\mathfrak{A}P = A. \tag{2.29}$$

Proof. We use (2.28). The first term in the right-hand side of (2.28) was already considered in Theorem 1.13. The first two terms in the second line of the same equation contribute by ∇^H to the right-hand side of (2.29), and the third term by 0. The last term in (2.28) contributes by $-c\left(T^H\right)/2\sqrt{2}$. Comparing with (2.8), we get (2.29). □

By (2.25) in Theorem 2.6, we get

$$\begin{aligned}\mathcal{B}^2_{Z,b} = &\frac{1}{2b^2}\left(-\Delta^V + \left|bZ\left(x, Y/b\right)\right|^2 + c\left(\widehat{e}_i\right)\widehat{c}\left(\nabla_{\widehat{e}_i}\left(bZ\left(x, Y/b\right)\right)\right)\right) - \nabla_{\widehat{R^{TX}Y}}\\ &+ \frac{1}{4}\left\langle R^{TX} e_i, e_j\right\rangle c\left(\widehat{e}_i\right)c\left(\widehat{e}_j\right) - \frac{1}{\sqrt{2}b}\widehat{c}\left(\nabla^{TX}_{\cdot} bZ\left(x, Y/b\right)\right) + \frac{1}{b}{}^1\nabla^F_{1/2, bZ(x,Y/b)} + R^E.\end{aligned} \tag{2.30}$$

By (2.30), as in (1.70), we get

$$\mathcal{B}^2_{Y,b} = \frac{\alpha}{b^2} + \frac{\beta}{b} + \gamma. \tag{2.31}$$

Also we have the identities,

$$\begin{aligned} \alpha &= \frac{1}{2}\left(-\Delta^V + |Y|^2 + c(\widehat{e}_i)\,\widehat{c}(e_i)\right), \\ \beta &= {}^1\nabla^F_{1/2,Y}, \\ \gamma &= -\nabla_{\widehat{R^{TX}Y}} + \frac{1}{4}\left\langle R^{TX} e_i, e_j\right\rangle c(\widehat{e}_i)\, c(\widehat{e}_j) + R^E. \end{aligned} \tag{2.32}$$

The obvious extension of Theorem 1.14 is as follows.

Theorem 2.8. *The following identity holds,*

$$P\left(\gamma - \beta\alpha^{-1}\beta\right)P = A^2. \tag{2.33}$$

Proof. Again we give two proofs. The first proof uses Theorem 2.7 and proceeds as the first proof of Theorem 1.14.

The second proof proceeds as the second proof of Theorem 1.14. We will here concentrate on the proof of the analogue of (1.74), which is

$$\begin{aligned} P\gamma P = \frac{K}{8} &+ \frac{1}{4} c(e_i)\, c(e_j)\, R^E(e_i, e_j) \\ &+ \frac{1}{\sqrt{2}} c(e_i)\, f^\alpha R^E\left(e_i, f^H_\alpha\right) + \frac{1}{2} f^\alpha f^\beta R^E\left(f^H_\alpha, f^H_\beta\right). \end{aligned} \tag{2.34}$$

First let us take the case where $R^E = 0$. With respect to (1.74), we only have to explain why there is no term containing any of the f^α. There is no term of degree 1 in the f^α because R^{TX} takes its values in antisymmetric endomorphisms, and also because $\ker\alpha$ is concentrated in degree 0. There is no term of degree 2 in the f^α for the same reason. The case of R^E is also very simple, but of course, there are now contributions of the f^α. This completes our second proof of (2.33). □

Now we replace S by $S \times \mathbf{R}^*_+$, with $b > 0$ the generic element of $\mathbf{R}^*_+$. This means that $\mathcal{A}_{Y,b}$ is replaced by $db\frac{\partial}{\partial b} + \mathcal{A}_{Y,b}$. Then

$$\begin{aligned} K_b\left(db\frac{\partial}{\partial b} + \mathcal{A}_{Y_b}\right)K_b^{-1} &= \mathcal{B}_{Y,b} + db\frac{\partial}{\partial b} - \frac{db}{b}\nabla_{\widehat{Y}}, \\ K_b\left(db\frac{\partial}{\partial b} + \mathcal{A}_{Y_b}\right)^2 K_b^{-1} &= \mathcal{B}^2_{Y,b} - 2\frac{db}{b^2}\frac{\widehat{c}(Y)}{\sqrt{2}}. \end{aligned} \tag{2.35}$$

2.6. Interpolating between the elliptic curvature and the geodesic flow

Definition 2.9. Put

$$\mathcal{C}_{Z,b} = K_{b^2}\mathcal{A}_{Z_b}K_{b^2}^{-1}. \tag{2.36}$$

As in (1.81), by (2.16), (2.21), as $b \to \infty$,

$$\mathcal{C}_{Z,b} = \nabla^{\mathcal{H}} + \frac{\widehat{c}(Z)}{\sqrt{2}} + \mathcal{O}\left(1/b^2\right). \tag{2.37}$$

By (2.37), we get the strict analogue of (1.83), i.e.,

$$\begin{aligned} \mathcal{C}^2_{Z,b} = \frac{|Z|^2}{2} - \nabla_{\widehat{R^{TX}Y}} + \frac{1}{4}\left\langle R^{TX} e_i, e_j \right\rangle c\left(\widehat{e}_i\right) c\left(\widehat{e}_j\right) \\ - \widehat{c}\left(\nabla^{TX}_{\cdot} Z\right)/\sqrt{2} + {}^1\nabla^F_{1/2,Z} + R^E + \mathcal{O}\left(1/b^2\right). \end{aligned} \tag{2.38}$$

Observe that the vector field Y^H on $\mathcal{X}$ is the generator of the fibrewise geodesic flow. Therefore as $b \to +\infty$, $\mathcal{C}^2_{Y,b}$ converges to a simple perturbation of this fibrewise generator.

3. The complex case

Let X be a compact complex Kähler manifold. In this section we construct a holomorphic Hodge theory on the total space $\mathcal{X}$ of TX, whose Laplacian is a hypoelliptic second order differential operator. This Laplacian still depends on a parameter $b > 0$. We show that this family of operators interpolates between classical holomorphic Hodge theory on X and the geodesic flow. Families are considered as well. This section should be understood as the analogue of [B05], where instead the de Rham complex was considered. Also observe that part of the results which are established in this section should be viewed as special cases of the results of Sections 1 and 2.

This section is organized as follows. In Subsection 3.1, we recall elementary results on Clifford algebras when the underlying vector space is complex.

In Subsection 3.2, we introduce our complex manifold X and the total space $\mathcal{X}$ of TX.

In Subsection 3.3, we construct the holomorphic Levi-Civita superconnection associated to the projection $\pi : \mathcal{X} \to X$.

In Subsection 3.4, given a holomorphic section z of $\pi^* TX$ over $\mathcal{X}$, we obtain an operator A_Z, which is an analogue of the operator $\mathcal{D}_Z$ constructed in Section 1.

In Subsection 3.5, by a conjugation involving classical Hodge operators, we obtain another operator B_Z.

In Subsection 3.6, we show that if z is antiinvariant, A_Z is self-adjoint with respect to a Hermitian form ϵ of signature (∞, ∞).

In Subsection 3.7, we give a Lichnerowicz formula for A^2_Z.

In Subsection 3.8, if y is the tautological section of TX on $\mathcal{X}$, we relate the component A''_Y of A_Y to the tautological Koszul complex on $\mathcal{X}$.

In Subsection 3.9, from the dilations on $\mathcal{X}$, we obtain a family of conjugate superconnections $C_{Z,b}$. We give a crucial formula which relates this family to the Hodge operator $\overline{\partial}^X + \overline{\partial}^{X*}$ on X, which suggests that along the lines of the arguments in [B05], as $b \to 0$, this family converges in the proper sense to $\overline{\partial}^X + \overline{\partial}^{X*}$.

In Subsection 3.10, we state various results on the Hodge theory of the hypoelliptic Laplacian, by showing that formally, corresponding results of [BL06], which were obtained in the context of de Rham theory, can be adapted without any change. A by-product of these considerations is that for $b > 0$ small enough, the classical conclusions of elliptic Hodge theory still hold in this case.

In Subsection 3.11, we start considering the case of families of complex manifolds. In this subsection, we recall the results of [BGS88b] on locally Kähler fibrations and on the holomorphic version of the Levi-Civita superconnection.

In Subsection 3.12, we construct a superconnection $\mathcal{A}_Z$ which extends to families the construction of A_Z.

In Subsection 3.13, when z is antiinvariant, we establish a self-adjointness property of $\mathcal{A}_Z$ with respect to a generalized Hermitian form $\underline{\epsilon}$.

In Subsection 3.14, we consider the superconnection $\mathcal{A}_{Y_b}$, and we relate the asymptotic expansion as $b \to 0$ of a conjugate version of these superconnections to the classical holomorphic Levi-Civita superconnection.

3.1. The Clifford algebra on a complex vector space

Let V be a real Euclidean vector space of even dimension $2n$. Let J be a complex structure on V, i.e., an antisymmetric endomorphism of V such that $J^2 = -1$. Let $W, \overline{W} \subset V \otimes_{\mathbf{R}} \mathbf{C}$ be the eigenspaces of J associated to the eigenvalues $i, -i$, so that $V \otimes_{\mathbf{R}} \mathbf{C} = W \oplus \overline{W}$. Then W is naturally equipped with a Hermitian product. Let ω be the 2-form on V, such that if $e, e' \in V$,

$$\omega(e, e') = \langle e, Je' \rangle . \tag{3.1}$$

Then ω is a $(1,1)$ form on V.

We use the same notation as in Subsection 1.1. Note that $W, \overline{W}$ are identified to $\overline{W}^*, W^*$ by the Hermitian product. If $e \in W$, set

$$c_{\overline{W}}(e) = \sqrt{2}e^* \wedge, \qquad c_{\overline{W}}(\overline{e}) = -\sqrt{2}i_{\overline{e}}, \tag{3.2}$$
$$\widehat{c}_W(e) = \sqrt{2}i_e, \qquad \widehat{c}_W(\overline{e}) = \sqrt{2}\overline{e}^* \wedge .$$

We extend $c_W, c_{\overline{W}}$ by linearity to $V \otimes_{\mathbf{R}} \mathbf{C}$. Then (3.2) defines actions of $c(V)$ and $\widehat{c}(V)$ on $\Lambda^{\cdot}\left(\overline{W}^*\right), \Lambda^{\cdot}(W^*)$.

As is well known [H74], we have the canonical isomorphism,

$$\Lambda^{\cdot}\left(\overline{W}^*\right) \simeq S^V \otimes (\det W)^{1/2} . \tag{3.3}$$

so that the action of $c(V)$ on S^V is precisely obtained via $c_{\overline{W}}$.

Let $w_1, \dots, w_n$ be an orthonormal basis of W, and let $w^1, \dots, w^n$ be the corresponding dual base of W^*. Set

$$L = -\sqrt{-1}w^i \wedge \overline{w}^i, \qquad \Lambda = \sqrt{-1}i_{\overline{w}_i}i_{w_i}. \tag{3.4}$$

The operator L is just multiplication by ω. Moreover, L, Λ act on $\Lambda^{\cdot}(V^*)$ and are adjoint to each other.

Let N be the number operator acting on $\Lambda^{\cdot}(V^*) \otimes_{\mathbf{R}} \mathbf{C}$. Set

$$H = \frac{1}{2}(N - n). \tag{3.5}$$

Then we have the sl_2 commutation relations,

$$[H, L] = L, \qquad [H, \Lambda] = -\Lambda, \qquad [L, \Lambda] = 2H. \tag{3.6}$$

If $U \in V$, set

$$c'(U) = \exp(-i\Lambda) c_{\overline{W}}(U) \exp(i\Lambda), \quad \widehat{c}'(U) = \exp(-i\Lambda) \widehat{c}_W(U) \exp(i\Lambda). \tag{3.7}$$

If $U \in V$, let $u \in W$ be the component of U in W, so that $U = u + \overline{u}$. Using (3.2), we get easily

$$c'(U) = \sqrt{2}(u^* \wedge -i_U), \qquad \widehat{c}'(U) = \sqrt{2}(\overline{u}^* \wedge + i_U). \tag{3.8}$$

Of course (3.8) still provides us with two anticommuting representations of $c(V)$ and $\widehat{c}(V)$. By (3.8), we find in particular that

$$\frac{1}{\sqrt{2}}(c'(U) + \widehat{c}'(U)) = U^* \wedge. \tag{3.9}$$

Note that

$$\Lambda^{\cdot}(V^* \otimes_{\mathbf{R}} \mathbf{C}) = \Lambda^{\cdot}\left(\overline{W}^*\right) \widehat{\otimes} \Lambda^{\cdot}(W^*). \tag{3.10}$$

We introduce another copy $\widehat{\Lambda}^{\cdot}\left(\overline{W}^*\right)$ of $\Lambda^{\cdot}\left(\overline{W}^*\right)$, whose elements will wear hats. By (3.10), we get

$$\Lambda^{\cdot}(V^* \otimes_{\mathbf{R}} \mathbf{C}) \widehat{\otimes} \widehat{\Lambda}^{\cdot}\left(\overline{W}^*\right) = \Lambda^{\cdot}\left(\overline{W}^*\right) \widehat{\otimes} \widehat{\Lambda}^{\cdot}\left(\overline{W}^*\right) \widehat{\otimes} \Lambda^{\cdot}(W^*). \tag{3.11}$$

As in (1.6), (3.11) is equipped with three anticommuting Clifford actions. Also the algebra in (3.11) is equipped with an involution exchanging the first two factors in the right-hand side. Set

$$\widehat{\Lambda}^{\cdot}(V^* \otimes_{\mathbf{R}} \mathbf{C}) = \widehat{\Lambda}^{\cdot}\left(\overline{W}^*\right) \widehat{\otimes} \Lambda^{\cdot}(W^*). \tag{3.12}$$

Using this involution, we have a nontrivial isomorphism,

$$\Lambda^{\cdot}(V^* \otimes_{\mathbf{R}} \mathbf{C}) \widehat{\otimes} \widehat{\Lambda}^{\cdot}\left(\overline{W}^*\right) \simeq \Lambda^{\cdot}\left(\overline{W}^*\right) \widehat{\otimes} \widehat{\Lambda}^{\cdot}(V^* \otimes_{\mathbf{R}} \mathbf{C}). \tag{3.13}$$

Note that the isomorphism in (3.13) is essentially the same as the one in (1.15).

3.2. A compact Hermitian manifold

Let X be a compact complex manifold of complex dimension n. Let $T_{\mathbf{R}}X$ be the real tangent space of X. Let J be the complex structure on $T_{\mathbf{R}}X$. We denote by $TX, \overline{TX}$ the holomorphic and antiholomorphic tangent bundles to X. Let g^{TX} be a Hermitian metric on $T_{\mathbf{R}}X$, let ω^X be the associated Kähler form as in (3.1). Let ∇^{TX} be the holomorphic Hermitian connection on TX, and let R^{TX} be its curvature. Let $\nabla^{T_{\mathbf{R}}X}$ be the connection on $T_{\mathbf{R}}X$ which is induced by ∇^{TX}. The connection $\nabla^{T_{\mathbf{R}}X}$ coincides with the Levi-Civita connection $\nabla^{T_{\mathbf{R}}X,L}$ if and only if

g^{TX} is Kähler. Let $\nabla^{\Lambda^\cdot(\overline{T^*X})}$ be the connection on $\Lambda^\cdot\left(\overline{T^*X}\right)$ which is induced by ∇^{TX}.

Let T be the torsion of ∇^{TX}. Then T maps $TX \times TX$ into TX, $\overline{TX} \times \overline{TX}$ into $\overline{TX}$ and vanishes on $TX \times \overline{TX}$.

As in (2.15), we have the identity,

$$d^X = \nabla^{\Lambda^\cdot(T^*_{\mathbf{R}}X)} + i_T. \tag{3.14}$$

From (3.14), we obtain

$$d^X \omega^X = i_T \omega^X. \tag{3.15}$$

Equation (3.15) allows us to express T in terms of $d^X \omega^X$.

Let (E, g^E) be a holomorphic Hermitian vector bundle on X, let ∇^E be the corresponding holomorphic Hermitian connection, and let R^E be its curvature. Let $\langle\rangle_{\Lambda^\cdot(\overline{T^*X})\otimes E}$ be the Hermitian product on $\Lambda^\cdot\left(\overline{T^*X}\right) \otimes E$ which is associated to the metrics g^{TX}, g^E. Let dv_X be the volume form on X which is associated to g^{TX}. Let $\nabla^{\Lambda^\cdot(\overline{T^*X})\widehat{\otimes}E}$ be the connection on $\Lambda^\cdot\left(\overline{T^*X}\right)\widehat{\otimes}E$ which is induced by $\nabla^{\Lambda^\cdot(\overline{T^*X})}, \nabla^E$.

Let $\left(\Omega^{(0,\cdot)}(X,E), \overline{\partial}^X\right)$ be the Dolbeault complex of smooth antiholomorphic forms on X with coefficients in E. We equip $\Omega^{(0,\cdot)}(X,E)$ with the L^2 Hermitian product,

$$\langle s, s'\rangle_{L^2} = (2\pi)^{-n} \int_X \langle s, s'\rangle_{\Lambda^\cdot(\overline{T^*X})\otimes E}\, dv_X. \tag{3.16}$$

Let $\overline{\partial}^{X*}$ be the formal adjoint of $\overline{\partial}^X$ with respect to the Hermitian product (3.16).

Set

$$D^X = \sqrt{2}\left(\overline{\partial}^X + \overline{\partial}^{X*}\right). \tag{3.17}$$

Then by [H74], if g^{TX} is Kähler, D^X is a Dirac operator of the kind already considered in (1.16). In the right-hand side of (1.16), $\nabla^{S^{TX}\widehat{\otimes}E}$ should be replaced by $\nabla^{\Lambda^\cdot(\overline{T}^*X)\widehat{\otimes}E}$.

3.3. The holomorphic Levi-Civita superconnection

Let $\mathcal{X}$ be the total space of TX, and let $\pi : \mathcal{X} \to X$ denote the obvious projection. The fibre of the projection π will often be denoted $\widehat{TX}$, to distinguish it from the genuine tangent fibre TX. Then $\mathcal{X}$ is still a complex manifold. The connection ∇^{TX} induces a horizontal vector bundle $T^H\mathcal{X} \subset T\mathcal{X}$, so that we have the splitting of smooth vector bundles,

$$T\mathcal{X} = T^H\mathcal{X} \oplus \pi^*\widehat{TX}. \tag{3.18}$$

If $e \in T_{\mathbf{R}}X$, let $e^H \in T^H_{\mathbf{R}}\mathcal{X}$ denote its horizontal lift. As before, we will often omit the upper script H.

Of course π_* induces the identification $T^H\mathcal{X} \simeq \pi^*TX$, so that (3.18) can be written in the form,

$$T\mathcal{X} = \pi^*\left(TX \oplus TX\right). \tag{3.19}$$

By (3.19), we get

$$\Lambda^{\cdot}\left(\overline{T^*\mathcal{X}}\right) = \pi^*\left(\Lambda^{\cdot}\left(\overline{T^*X}\right)\widehat{\otimes}\Lambda^{\cdot}\left(\overline{T^*X}\right)\right). \tag{3.20}$$

Elements of the second copy of $\Lambda^{\cdot}\left(\overline{T^*X}\right)$ in the right-hand side of (3.20) are vertical antiholomorphic forms. These forms will often be hatted to distinguish them from the elements of the first copy, which will be denoted in the standard way. In particular instead of (3.20), we will now write

$$\Lambda^{\cdot}\left(\overline{T^*\mathcal{X}}\right) = \pi^*\left(\Lambda^{\cdot}\left(\overline{T^*X}\right)\widehat{\otimes}\widehat{\Lambda}^{\cdot}\left(\overline{T^*X}\right)\right). \tag{3.21}$$

We will use the results of Subsection 3.1, with $V = T_{\mathbf{R}}X$. In particular by (3.13), we have the isomorphism,

$$\Lambda^{\cdot}\left(T^*_{\mathbf{R}}X\otimes_{\mathbf{R}}\mathbf{C}\right)\widehat{\otimes}\widehat{\Lambda}^{\cdot}\left(\overline{T^*X}\right) \simeq \Lambda^{\cdot}\left(\overline{T^*X}\right)\widehat{\otimes}\widehat{\Lambda}^{\cdot}\left(T^*_{\mathbf{R}}X\otimes_{\mathbf{R}}\mathbf{C}\right). \tag{3.22}$$

By analogy with (1.20), we now set

$$F = \Lambda^{\cdot}\left(T^*_{\mathbf{R}}X\otimes_{\mathbf{R}}\mathbf{C}\right)\widehat{\otimes}\widehat{\Lambda}^{\cdot}\left(\overline{T^*X}\right)\widehat{\otimes}E \simeq \Lambda^{\cdot}\left(\overline{T^*X}\right)\widehat{\otimes}\widehat{\Lambda}^{\cdot}\left(T^*_{\mathbf{R}}X\otimes_{\mathbf{R}}\mathbf{C}\right)\widehat{\otimes}E. \tag{3.23}$$

Let $\nabla^{\Lambda^{\cdot}(T^*_{\mathbf{R}}X),L}$ be the connection on $\Lambda^{\cdot}\left(T^*_{\mathbf{R}}X\right)$ which is induced by $\nabla^{TX,L}$. Then using the first isomorphism for F in (3.23), the connections $\nabla^{\Lambda^{\cdot}(T^*_{\mathbf{R}}X),L}, \nabla^{TX}, \nabla^{E}$ induce a corresponding connection ∇^{F} on F. This connection is compatible with the isomorphisms in (3.23) only if g^{TX} is Kähler.

Definition 3.1. Let $\mathfrak{F}$ be the vector bundle of elements of $\Lambda^{\cdot}\left(T^*_{\mathbf{R}}\mathcal{X}\otimes_{\mathbf{R}}\mathbf{C}\right)\widehat{\otimes}\pi^*E$ whose restriction to the fibres TX is of type $(0,\cdot)$.

Clearly $C^{\infty}\left(\mathcal{X},\mathfrak{F}\right)\subset\Omega^{(\cdot,\cdot)}\left(\mathcal{X},E\right)$ is stable by the Dolbeault operator $\overline{\partial}^{\mathcal{X}}$. By (3.20), we get

$$\mathfrak{F}\simeq\pi^*F. \tag{3.24}$$

In view of the above, one can then say that $\overline{\partial}^{\mathcal{X}}$ acts naturally on $C^{\infty}\left(\mathcal{X},\pi^*F\right)$.

Definition 3.2. Let $\mathbf{I}^{\cdot}$ be the vector bundle of sections of $\pi^*\left(\widehat{\Lambda}^{\cdot}\left(\overline{T^*X}\right)\widehat{\otimes}E\right)$ which are smooth along the fibre TX.

Clearly,

$$C^{\infty}\left(\mathcal{X},\pi^*F\right) = \Omega^{(\cdot,\cdot)}\left(X,\mathbf{I}^{\cdot}\right). \tag{3.25}$$

Also

$$C^{\infty}\left(\mathcal{X},\pi^*F\right) = \Omega^{(0,\cdot)}\left(\mathcal{X},\pi^*\left(\Lambda^{\cdot}\left(T^*X\right)\widehat{\otimes}E\right)\right). \tag{3.26}$$

We denote by y the generic section of TX, by Y the generic section of $T_{\mathbf{R}}X$, so that $Y = y+\overline{y}$, and $|Y|^2 = 2\,|y|^2$. Set

$$\omega^{\mathcal{X}} = i\overline{\partial}\partial\,|y|^2\,. \tag{3.27}$$

Let $\omega^{\mathcal{X},V}$ be the Kähler form along the fibres of TX, i.e., the restriction of $\omega^{\mathcal{X}}$ to the fibres TX. Put

$$\omega^{\mathcal{X},H} = i\left\langle R^{TX}y,\overline{y}\right\rangle. \tag{3.28}$$

Then a simple computation shows that

$$\omega^{\mathcal{X}} = \omega^{\mathcal{X},V}+\omega^{\mathcal{X},H}. \tag{3.29}$$

Equation (3.29) gives the splitting of $\omega^{\mathcal{X}}$ into its vertical and horizontal parts. The same equation indicates that $\pi : \mathcal{X} \to X$ is a Kähler fibration in the sense of [BGS88b]. Equivalently the horizontal vector bundle $T^H\mathcal{X}$ is just the orthogonal bundle to the vertical vector bundle TX with respect to $\omega^{\mathcal{X}}$.

Let $\overline{\partial}^V$ denote the Dolbeault operator acting on $\mathbf{I}^{\cdot}$ along the fibres of TX. Also $\nabla^{\widehat{\Lambda}^{\cdot}(\overline{T^*X})\widehat{\otimes}E}$ determines a connection on $\mathbf{I}$. Namely if $U \in T_{\mathbf{R}}X$, if $U^H \in T^H_{\mathbf{R}}\mathcal{X}$ is the horizontal lift of U, if s is a smooth section of $\mathbf{I}^{\cdot}$, set

$$\nabla^{\mathbf{I}^{\cdot}}_U s = \nabla^{\widehat{\Lambda}^{\cdot}(\overline{T^*X})\widehat{\otimes}E}_{U^H} s. \tag{3.30}$$

Since R^{TX} is of type $(1,1)$, we find easily that

$$\nabla^{\mathbf{I}'',2} = 0, \qquad \nabla^{\mathbf{I}^{\cdot}\prime,2} = 0, \qquad \left[\nabla^{\mathbf{I}^{\cdot}}, \overline{\partial}^V\right] = 0. \tag{3.31}$$

In the sequel, our operators will act on $\Omega^{(0,\cdot)}\left(\mathcal{X}, \pi^*\left(\Lambda^{\cdot}\left(T^*X\right)\widehat{\otimes}E\right)\right)$. Now we establish a special case of [BGS88b, Theorem 2.8].

Proposition 3.3. *We have the identity of operators,*

$$\overline{\partial}^{\mathcal{X}} = \nabla^{\mathbf{I}^{\cdot}\prime\prime} + \overline{\partial}^V. \tag{3.32}$$

Proof. Let $d^{\mathcal{X}}$ be the de Rham operator on $\mathcal{X}$. Let $\mathbf{J}^{\cdot}$ denote temporarily the vector bundle on X of smooth sections of the vertical exterior algebra along the fibre TX. Using a notation very similar to the notation in (2.15), we get

$$d^{\mathcal{X}} = \nabla^{\mathbf{J}^{\cdot}} + d^V + i_{\widehat{R^{TX}Y}}. \tag{3.33}$$

In (3.33), $\nabla^{\mathbf{J}}$ is the obvious connection on $\mathbf{J}$ which is induced by ∇^{TX}. Since R^{TX} is of type $(1,1)$, from (3.37), we get

$$\overline{\partial}^{\mathcal{X}} = \nabla^{\mathbf{J}^{\cdot}\prime\prime} + \overline{\partial}^V + i_{\widehat{R^{TX}y}}. \tag{3.34}$$

Now when acting on $\Omega^{(0,\cdot)}\left(\mathcal{X}, \pi^*\left(\Lambda^{\cdot}\left(T^*X\right)\widehat{\otimes}E\right)\right)$, the operator $i_{\widehat{R^{TX}y}}$ vanishes, so that we get (3.32). □

By (3.31) or by (3.32), we get

$$\left(\nabla^{\mathbf{I}^{\cdot}\prime\prime} + \overline{\partial}^V\right)^2 = 0. \tag{3.35}$$

Let $\overline{\partial}^{V*}$ be the fibrewise adjoint of $\overline{\partial}^V$ with respect to the obvious fibrewise L^2 Hermitian product. The operators $\nabla^{\mathbf{I}^{\cdot}\prime}, \overline{\partial}^{V*}$ verify identities similar to (3.31). As in (3.35), we have

$$\left(\nabla^{\mathbf{I}^{\cdot}\prime} + \overline{\partial}^{V*}\right)^2 = 0. \tag{3.36}$$

Now we follow [BGS88b, Section 2] and [B97, eq. (2.35)],

Definition 3.4. Set

$$A'' = \nabla^{\mathbf{I}^{\cdot}''} + \overline{\partial}^V, \qquad A' = e^{i\omega^X}\left(\nabla^{\mathbf{I}^{\cdot}'} + \overline{\partial}^{V*}\right)e^{-i\omega^X}, \qquad A = A'' + A'. \tag{3.37}$$

Of course,

$$A''^2 = 0, \qquad\qquad A'^2 = 0. \tag{3.38}$$

Equation (3.37) for A'', A' can be made more symmetric with respect to ω^X. Our conventions try to fit with various traditions.

When g^{TX} is Kähler, conjugation by $e^{i\omega^X}$ can be ignored. By (3.37), we get in this case,

$$A = \nabla^{\mathbf{I}^{\cdot}} + \overline{\partial}^V + \overline{\partial}^{V*}. \tag{3.39}$$

By comparing (1.31) and (3.39), we see that when g^{TX} is Kähler, A is a special case of $\mathcal{D}$. This point will be made precise in equation (3.45).

3.4. The superconnection operator A_Z

Let z be a holomorphic section of π^*TX over $\mathcal{X}$. Let $Z = z + \overline{z}$ be the corresponding section of $T_{\mathbf{R}}X$.

Let $\overline{z}^* \in T^*X$ correspond to $\overline{z} \in \overline{TX}$ by g^{TX}. Then $i_z, \overline{z}^*\wedge, i_{\overline{z}}$ act on $\Lambda^{\cdot}(T^*_{\mathbf{R}}X \otimes_{\mathbf{R}} \mathbf{C})$. Ultimately these operators act on $C^\infty(\mathcal{X}, \pi^*F)$.

Put

$$A''_Z = \nabla^{\mathbf{I}^{\cdot}''} + \overline{\partial}^V + i_z, \quad A'_Z = e^{i\omega^X}\left(\nabla^{\mathbf{I}^{\cdot}'} + \overline{\partial}^{V*} + i_{\overline{z}}\right)e^{-i\omega^X}, \qquad A_Z = A''_Z + A'_Z. \tag{3.40}$$

Note that

$$A''^2_Z = 0, \qquad\qquad A'^2_Z = 0. \tag{3.41}$$

Indeed the first identity in (3.41) is a trivial consequence of the fact that z is holomorphic, and the second one is the obvious conjugate.

Clearly,

$$A''_Z = A'' + i_z, \qquad\qquad A'_Z = A' + i_{\overline{z}} + \overline{z}^* \wedge. \tag{3.42}$$

Moreover, A_Z is not a superconnection. By (3.8), (3.39), (3.42), we get

$$A_Z = A + \frac{\widehat{c}\,(Z)}{\sqrt{2}}. \tag{3.43}$$

As we saw before, when g^{TX} is Kähler, A is a special case of $\mathcal{D}$. Similarly, when g^{TX} is Kähler, A_Z is a special case of $\mathcal{D}_Z$.

Let $e_1, \ldots, e_{2n}$ be an orthonormal basis of $T_{\mathbf{R}}X$ with respect to g^{TX}, let $\widehat{e}_1, \ldots, \widehat{e}_{2n}$ be the corresponding orthonormal basis of the vertical fibre $\widehat{T_{\mathbf{R}}X}$. Note that

$$\overline{\partial}^V + \overline{\partial}^{V*} = \frac{1}{\sqrt{2}} c_{\widehat{TX}}(\widehat{e}_i)\nabla_{\widehat{e}_i}. \tag{3.44}$$

Recall that the connection ∇^F on F was defined after (3.23). By (3.8), (3.9), (3.39), (3.43), (3.44), when g^{TX} is Kähler, we get

$$A_Z = \frac{1}{\sqrt{2}}\left(c'(e_i) + \widehat{c}'(e_i)\right)\nabla^F_{e_i} + \frac{1}{\sqrt{2}}\left(c_{\overline{TX}}(\widehat{e}_i)\nabla_{\widehat{e}_i} + \widehat{c}'(Z)\right). \tag{3.45}$$

3.5. A conjugation on A_Z

Let $\nabla^{F,h}$ be the connection on F which is induced by ∇^{TX} and ∇^E. This connection coincides with ∇^F only if g^{TX} is Kähler.

Let $w_1, \ldots, w_n$ be an orthonormal basis of TX with respect to g^{TX}, let $w^1, \ldots, w^n$ be the corresponding dual basis of T^*X. We define the Hodge operators L, Λ as in (3.4), i.e.,

$$L = -\sqrt{-1}w^i \wedge \overline{w}^i, \qquad \Lambda = \sqrt{-1}i_{\overline{w}_i}i_{w_i}. \tag{3.46}$$

Note in particular that equation (3.46) contains only horizontal variables, and no vertical hatted variables.

Definition 3.5. Set

$$B_Z = \exp(i\Lambda)A_Z\exp(-i\Lambda). \tag{3.47}$$

Let $\nabla^{\mathbf{I}^{\cdot}*}$ be the formal adjoint of the operator $\nabla^{\mathbf{I}^{\cdot}}$ with respect to the obvious L^2 Hermitian product on $C^\infty(\mathcal{X}, \pi^*F)$. Then $\nabla^{\mathbf{I}^{\cdot}*}$ splits as

$$\nabla^{\mathbf{I}^{\cdot}*} = \nabla^{\mathbf{I}^{\cdot}\prime\prime*} + \nabla^{\mathbf{I}^{\cdot}\prime*}. \tag{3.48}$$

In (3.48), $\nabla^{\mathbf{I}^{\cdot}\prime\prime*}$ decreases the horizontal antiholomorphic degree by 1, and $\nabla^{\mathbf{I}^{\cdot}\prime*}$ decreases the horizontal holomorphic degree by 1.

Theorem 3.6. *The following identity holds,*

$$\begin{aligned}B_Z = {} & \left(\overline{w}^i \wedge + i_{w_i}\right)\nabla^{F,h}_{\overline{w}_i} + \left(w^i \wedge - i_{\overline{w}_i}\right)\nabla^{F,h}_{w_i} \\ & + \frac{1}{2}\left(\overline{w}^i \wedge + i_{w_i}\right)\left(\overline{w}^j \wedge + i_{w_j}\right)i_{T(\overline{w}_i,\overline{w}_j)} - \frac{1}{2}\left(w^i \wedge - i_{\overline{w}_i}\right)\left(w^j \wedge - i_{\overline{w}_j}\right)T^*(w_i, w_j)\wedge \\ & + \overline{\partial}^V + i_z + \overline{\partial}^{V*} + \overline{z}^*\wedge. \end{aligned} \tag{3.49}$$

If the metric g^{TX} is Kähler, then

$$B_Z = \nabla^{\mathbf{I}\prime\prime} + \nabla^{\mathbf{I}^{\cdot}\prime} + \nabla^{\mathbf{I}^{\cdot}\prime\prime*} - \nabla^{\mathbf{I}^{\cdot}\prime*} + \overline{\partial}^V + i_z + \overline{\partial}^{V*} + \overline{z}^*\wedge. \tag{3.50}$$

Proof. By (3.14), (3.40), (3.42) and (3.46), we get

$$A_Z = \overline{w}^i \wedge \nabla^{F,h}_{\overline{w}_i} + w^i \wedge \nabla^{F,h}_{w_i} + i_T - T^{(1,0)*}\wedge + \overline{\partial}^V + \overline{\partial}^{V*} + \overline{z}^*\wedge + i_Z. \tag{3.51}$$

Under conjugation by $\exp(i\Lambda)$, $w^i\wedge, \overline{w}^i\wedge$ are changed into $w^i \wedge -i_{\overline{w}_i}, \overline{w}^i + i_{w_i}$ while the annihilation operators are not changed. From (3.51), we get (3.49). If the metric g^{TX} is Kähler, the terms containing T in (3.49) disappear and equation (3.50) follows easily. The proof of our theorem is completed. □

3.6. A self-adjointness property

In this subsection, we assume the metric g^{TX} to be Kähler.

Let $\mathcal{N}^{(0,1)}$ be the number operator on $\Lambda^{\cdot}\left(\overline{T^*\mathcal{X}}\right)$, let $N^{H(1,0)}$ be the number operator on $\Lambda^{\cdot}(T^*X)$. Put

$$\mathcal{N} = \mathcal{N}^{(0,1)} - N^{H(1,0)}. \tag{3.52}$$

Then $\mathcal{N}$ defines a $\mathbf{Z}$-grading on $\Omega^{(0,\cdot)}\left(\mathcal{X}, \pi^*\left(\Lambda^{\cdot}(T^*X)\widehat{\otimes}E\right)\right)$. Clearly A''_Z increases the total degree by 1, and A'_Z decreases the total degree by 1.

Let $dv_{\mathcal{X}}$ be the obvious volume form on $\mathcal{X}$. We equip $C^{\infty,c}(\mathcal{X}, \pi^*F)$ with the normalized L^2 Hermitian product,

$$\langle s, s'\rangle_{L^2} = (2\pi)^{-2n}\int_{\mathcal{X}} \langle s, s'\rangle_F \, dv_{\mathcal{X}}. \tag{3.53}$$

We still denote by r the map $Y \to -Y$. Let r^* the corresponding action on $\widehat{\Lambda}^{\cdot}(T^*_{\mathbf{R}}X \otimes_{\mathbf{R}} \mathbf{C}) \simeq \Lambda^{\cdot}(T^*X)\widehat{\otimes}\widehat{\Lambda}^{\cdot}\left(\overline{T^*X}\right)$. Of course r^* acts trivially on $\Lambda^{\cdot}\left(\overline{T^*X}\right)$. Let η be the Hermitian form on $C^{\infty,c}(\mathcal{X}, \pi^*F)$,

$$\eta(s, s') = \langle r^* s, s'\rangle_{L^2}. \tag{3.54}$$

Put

$$\epsilon(s, s') = \eta\left(e^{i\Lambda}s, e^{i\Lambda}s'\right). \tag{3.55}$$

Then ϵ is still a Hermitian form.

In the sequel, the adjoints of A''_Z, A'_Z, A_Z will be taken with respect to ϵ, and the adjoints of B''_Z, B'_Z, B_Z with respect to η. They will be denoted with an upper $\dagger$.

Theorem 3.7. *The following identities hold,*

$$A''^{\dagger}_Z = A'_{Z_-}, \qquad B''^{\dagger}_Z = B'_{Z_-}, \tag{3.56}$$
$$A^{\dagger}_Z = A_{Z_-}, \qquad B^{\dagger}_Z = B_{Z_-}.$$

Proof. By making $T = 0$ in (3.49), we obtain the fourth identity in (3.56). The third equation is an obvious consequence of the second one. Now A_Z, B_Z are the sum of their components of degree 1 and -1. The first two equations in (3.56) are now obvious. □

3.7. The Lichnerowicz formula for A^2_Z

In this subsection, we assume the metric g^{TX} to be Kähler. Let Δ^V be the fibrewise Laplacian with respect to the metric g^{TX}. Let $|Z|$ be the norm of Z with respect to g^{TX}. Let $e_1, \ldots, e_{2n}$ be an orthonormal basis of $T_{\mathbf{R}}X$, let $e^1, \ldots, e^{2n}$ be the corresponding dual basis of $T^*_{\mathbf{R}}X$.

Theorem 3.8. *The following identity holds,*

$$A^2_Z = \frac{1}{2}\left(-\Delta^V + |Z|^2 + c(\widehat{e}_i)\widehat{c}(\nabla_{\widehat{e}_i}Z)\right) - \nabla_{\widehat{R^{TX}Y}} + \frac{1}{4}\langle R^{TX}e_i, e_j\rangle c(\widehat{e}_i)c(\widehat{e}_j)$$
$$+ \frac{1}{2}\mathrm{Tr}\left[R^{TX}\right] + \frac{e^i}{\sqrt{2}}\widehat{c}\left(\nabla^{TX_{\mathbf{R}}}_{e_i}Z\right) + \nabla^F_Z + R^E. \tag{3.57}$$

Proof. As we saw in Subsection 3.4, since g^{TX} is Kähler, A_Z is a special case of $\mathcal{D}_Z$. We can then use equation (1.55) in Theorem 1.10. We claim that we get (3.57). In fact the appearance of $\frac{1}{2}\mathrm{Tr}\left[R^{TX}\right]$ in the right-hand side is a consequence of (3.3). The proof of our theorem is completed. □

3.8. The Koszul complex of $\mathcal{X}$

By equation (3.32) in Proposition 3.3, by (3.37), (3.40), we get the identity of operators acting on $\Omega^{(0,\cdot)}\left(\mathcal{X},\pi^*\left(\Lambda^{\cdot}\left(T^*X\right)\widehat{\otimes}E\right)\right)$,

$$A''_Y = \overline{\partial}^{\mathcal{X}} + i_y. \tag{3.58}$$

We identify $\Lambda^{\cdot}\left(T^*X\right)\widehat{\otimes}\widehat{\Lambda}^{\cdot}\left(\overline{T^*X}\right)$ to the vertical exterior algebra.

Let i be the embedding of X into $\mathcal{X}$ as the zero set of y. Then i^* maps $\Omega^{(0,\cdot)}\left(\mathcal{X},\pi^*\left(\Lambda^{\cdot}\left(T^*X\right)\widehat{\otimes}E\right)\right)$ into $\Omega^{(0,\cdot)}\left(X,E\right)$. Let $\mathcal{S}^{(0,\cdot)}\left(\mathcal{X},\pi^*\left(\Lambda^{\cdot}\left(T^*X\right)\widehat{\otimes}E\right)\right)$ be the Schwartz space of sections of $\pi^*\left(\Lambda^{\cdot}\left(T^*X\right)\widehat{\otimes}E\right)$ on $\mathcal{X}$. Note that the operator $\overline{\partial}^{\mathcal{X}}+i_y$ also acts on $\mathcal{S}^{(0,\cdot)}\left(\mathcal{X},\pi^*\left(\Lambda^{\cdot}\left(T^*X\right)\widehat{\otimes}E\right)\right)$.

Proposition 3.9. *The map i^* is a quasiisomorphism of $\mathbf{Z}$-graded complexes*

$$\left(\Omega^{(0,\cdot)}\left(\mathcal{X},\pi^*\left(\Lambda^{\cdot}\left(T^*X\right)\widehat{\otimes}E\right)\right),\overline{\partial}^{\mathcal{X}}+i_y\right)\to\left(\Omega^{(0,\cdot)}\left(X,E\right),\overline{\partial}^{X}\right),$$

so that the cohomology of the complex $\left(\Omega^{(0,\cdot)}\left(\mathcal{X},\pi^*\left(\Lambda^{\cdot}\left(T^*X\right)\widehat{\otimes}E\right)\right),\overline{\partial}^{\mathcal{X}}+i_y\right)$ is $H^{(0,\cdot)}\left(X,E\right)$. The map i^* induces a quasiisomorphism

$$\left(\mathcal{S}^{(0,\cdot)}\left(\mathcal{X},\pi^*\left(\Lambda^{\cdot}\left(T^*X\right)\widehat{\otimes}F\right)\right),\overline{\partial}^{\mathcal{X}}+i_y\right)\to\left(\Omega^{(0,\cdot)}\left(X,E\right),\overline{\partial}^{X}\right).$$

Proof. Note that the Koszul complex on $\left(\mathcal{O}_{\mathcal{X}}\left(\pi^*\Lambda^{\cdot}\left(T^*X\right)\right),i_y\right)$ is a resolution of the sheaf $i_*\mathcal{O}_X$. The first part of our theorem follows.

Put

$$\widehat{L}=-\sqrt{-1}w^i\wedge\widehat{\overline{w}}^i,\qquad \widehat{\Lambda}=\sqrt{-1}i_{\widehat{\overline{w}}_i}i_{w_i}. \tag{3.59}$$

In (3.59), $\widehat{L}$ is just the multiplication operator by the fibrewise Kähler form $\widehat{\omega}^{\mathcal{X},V}$. Clearly,

$$\left[\overline{\partial}^V+i_y,\overline{\partial}^{V*}+\overline{y}^*\wedge\right]=\frac{1}{2}\left(-\Delta^V+|Y|^2-i\left(\widehat{L}-\widehat{\Lambda}\right)\right). \tag{3.60}$$

By [B90b, Proposition 1.5 and Theorem 1.6], we know that the operator in (3.60) is essentially self-adjoint on $\mathcal{S}^{(0,\cdot)}\left(\widehat{TX},\pi^*\Lambda^{\cdot}\left(T^*X\right)\right)$, its spectrum is $\mathbf{N}$, and its kernel is 1-dimensional and spanned by

$$\beta=\exp\left(i\widehat{\omega}^{\mathcal{X},V}-|Y|^2/2\right). \tag{3.61}$$

Moreover, its resolvent acts on $\mathcal{S}^{(0,\cdot)}\left(\widehat{TX},\pi^*\Lambda^{\cdot}\left(T^*X\right)\right)$. Clearly,

$$\left(\overline{\partial}^V+i_y\right)\beta=0. \tag{3.62}$$

From the above, it follows that in a given fibre, the cohomology of the complex $\left(\mathcal{S}^{(0,\cdot)}\left(\widehat{TX},\pi^*\Lambda^{\cdot}(T^*X)\right),\overline{\partial}^V+i_y\right)$ is concentrated in degree 0.

Using (3.32), (3.58) and (3.62), we get

$$\left(\overline{\partial}^{\mathcal{X}}+i_y\right)\beta=0. \tag{3.63}$$

Moreover,

$$i^*\beta=1 \tag{3.64}$$

Using the above and a form of the Leray-Hirsch theorem, we get the last part of our proposition. □

Remark 3.10. In the above construction, we have assumed that the metric on the vertical fibre $\widehat{TX}$ is the same as the given metric on the tangent bundle TX coming from the metric of X. When defining A'_Z, A_Z in (3.40), we may as well assume that the two metrics are indeed distinct. This extension will be developed in full generality in Section 10. In Subsection 3.9 and in Section 7, we will already relax this assumption, by only assuming the horizontal and vertical metrics to be proportional, with a fixed constant of proportionality.

3.9. A rescaling on y and the elliptic Hodge operator on X

Given $b>0$, we define the holomorphic section z_b of TX as in (1.57).

Definition 3.11. For $b\in\mathbf{R}^*$, set

$$C_{Z,b}=K_bB_{Z_b}K_b^{-1}. \tag{3.65}$$

By (3.49), we get

$$\begin{aligned}C_{Z,b}={}&\left(\overline{w}^i\wedge+i_{w_i}\right)\nabla^{F,h}_{\overline{w}_i}+\left(w^i\wedge-i_{\overline{w}_i}\right)\nabla^{F,h}_{w_i}\\&+\frac{1}{2}\left(\overline{w}^i\wedge+i_{w_i}\right)\left(\overline{w}^j\wedge+i_{w_j}\right)i_{T(\overline{w}_i,\overline{w}_j)}-\frac{1}{2}\left(w^i\wedge-i_{\overline{w}_i}\right)\left(w^j\wedge-i_{\overline{w}_j}\right)T^*(w_i,w_j)\wedge\\&+\frac{1}{b}\left(\overline{\partial}^V+i_{bz(x,Y/b)}+\overline{\partial}^{V*}+b\overline{z}(x,Y/b)^*\wedge\right).\end{aligned} \tag{3.66}$$

Set

$$\begin{aligned}E={}&\left(\overline{w}^i\wedge+i_{w_i}\right)\nabla^{F,h}_{\overline{w}_i}+\left(w^i\wedge-i_{\overline{w}_i}\right)\nabla^{F,h}_{w_i}+\frac{1}{2}\left(\overline{w}^i\wedge+i_{w_i}\right)\left(\overline{w}^j\wedge+i_{w_j}\right)i_{T(\overline{w}_i,\overline{w}_j)}\\&-\frac{1}{2}\left(w^i\wedge-i_{\overline{w}_i}\right)\left(w^j\wedge-i_{\overline{w}_j}\right)T^*(w_i,w_j)\wedge.\end{aligned} \tag{3.67}$$

By (3.66), (3.67), we obtain,

$$C_{Y,b}=E+\frac{1}{b}\left(\overline{\partial}^V+i_y+\overline{\partial}^{V*}+\overline{y}^*\wedge\right). \tag{3.68}$$

As we already saw in the proof of Proposition 3.9, the kernel of the operator $\overline{\partial}^V+i_y+\overline{\partial}^{V*}+\overline{y}^*\wedge$ acting on $\mathcal{S}^{(0,\cdot)}\left(\widehat{TX},\pi^*\Lambda^{\cdot}(T^*X)\right)$ is 1-dimensional and spanned by $\beta = \exp\left(i\widehat{\omega}^{\mathcal{X},V} - \frac{|Y|^2}{2}\right)$. Note that

$$(2\pi)^{-n} \int_{\widehat{TX}} |\beta|^2 \, dv_{TX} = 1. \tag{3.69}$$

Let P be the orthogonal projection operator on this kernel.

We embed $\Omega^{(0,\cdot)}(X,E)$ into $\Omega(\mathcal{X}, \pi^* F)$ by the isometric embedding $\alpha \to \pi^*\alpha \wedge \beta$, so that $\Omega^{(0,\cdot)}(X,E)$ is identified with $\ker\left(\overline{\partial}^V + i_y + \overline{\partial}^{V*} + \overline{y}^* \wedge\right)$.

Theorem 3.12. *The following identity holds,*

$$PEP = \overline{\partial}^X + \overline{\partial}^{X*}. \tag{3.70}$$

Proof. We find easily that

$$\begin{aligned} PEP = \overline{w}^i \nabla^{\Lambda^{\cdot}\left(\overline{T^*X}\right)\otimes E}_{\overline{w}_i} &- i_{\overline{w}_i} \nabla^{\Lambda^{\cdot}\left(\overline{T^*X}\right)\otimes E}_{w_i} \\ &+ \frac{1}{2}\overline{w}^i \wedge \overline{w}^j \wedge i_{T(\overline{w}_i,\overline{w}_j)} - \frac{1}{2} i_{\overline{w}_i} i_{\overline{w}_j} T^*(w_i,w_j) \wedge . \end{aligned} \tag{3.71}$$

Using [B89, eqs. (1.41) and (2.21)], which gives an expression for $\overline{\partial}^X + \overline{\partial}^{X*}$, we get (3.70). □

Remark 3.13. Theorem 3.12 is remarkable. Indeed it indicates that even in the case of non Kähler metrics, the formalism of Theorem 1.13 remains valid.

3.10. The analytic and spectral properties of $A^2_{Y_b}$

In this subsection, we assume g^{TX} to be Kähler. We will describe the analytic and spectral properties of the operator $A^2_{Y_b}$. We fix $b > 0$.

Inspection of equations (1.69) and (3.57) shows that the hypoelliptic operator $A^2_{Y_b}$ has the same analytic properties as the hypoelliptic Laplacian considered in [BL06, Sections 15 and 17]. In particular it has compact resolvent and discrete spectrum which is located in the domain δ_+ to the right of a curve γ indicated in Figure 3.1. The precise description of γ is as follows. Given $b > 0$, constants $\lambda_0 > 0, c_0 > 0$ depending on $b > 0$ are defined so that

$$\gamma = \left\{\lambda = -\lambda_0 + \sigma + i\tau, \sigma, \tau \in \mathbf{R}, \sigma = c_0 |\tau|^{1/6}\right\}. \tag{3.72}$$

Note that γ depends explicitly on b. Moreover, the corresponding characteristic subspaces are finite dimensional and included in $\mathcal{S}^{(0,\cdot)}\left(\mathcal{X}, \pi^*\left(\Lambda^{\cdot}\left(T^*X\right)\widehat{\otimes}E\right)\right)$. If $\lambda \in \operatorname{Sp} A^2_{Y_b}$, let $\mathcal{S}_\lambda \subset \mathcal{S}^{(0,\cdot)}\left(\mathcal{X}, \pi^*\left(\Lambda^{\cdot}\left(T^*X\right)\widehat{\otimes}E\right)\right)$ denote the corresponding characteristic subspace. By Theorem 3.7, the operator $A^2_{Y_b}$ is self-adjoint with respect to the Hermitian form ϵ. By proceeding as in [BL06, Proposition 3.1] , we deduce that the spectrum of $A^2_{Y_b}$ is conjugation invariant.

By proceeding as in [BL06, Subsection 3.2], we find that there is a natural splitting

$$\mathcal{S}^{(0,\cdot)}\left(\mathcal{X}, \pi^*\left(\Lambda^{\cdot}\left(T^*X\right)\widehat{\otimes}E\right)\right) = \mathcal{S}_0 \oplus \mathcal{S}_*, \tag{3.73}$$

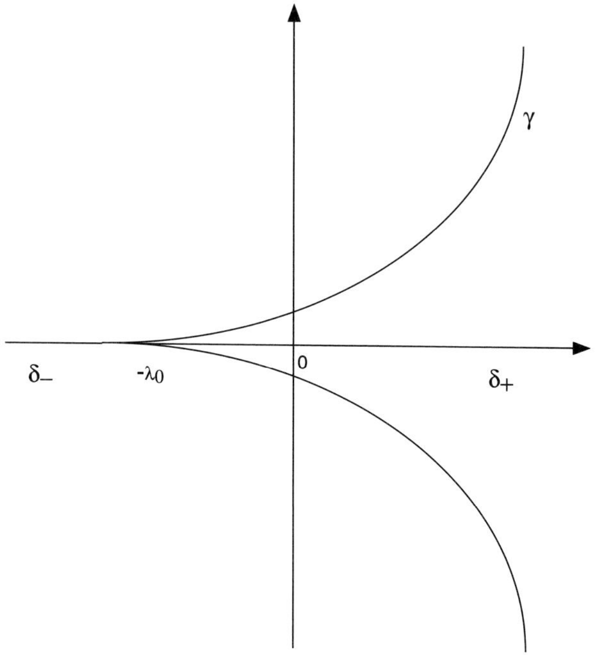

FIGURE 3.1

which is is A''_{Y_b}, A'_{Y_b} stable. In (3.73), $\mathcal{S}_*$ is just the image of $1-\mathbf{P}$, where $\mathbf{P}$ is the spectral projector on $\mathcal{S}_0$. By [BL06, Theorem 3.2],

$$\mathcal{S}_* = \operatorname{Im} A''_{Y_b}|_{\mathcal{S}_*} \oplus \operatorname{Im} A'_{Y_b}|_{\mathcal{S}_*}. \tag{3.74}$$

The splitting (3.73) is ϵ-orthogonal, so that the restriction of the Hermitian form ϵ to $\mathcal{S}_0$ and $\mathcal{S}_*$ is nondegenerate. Also the complex $\left(\mathcal{S}_*, A''_{Y_b}\right)$ is acyclic. Combining this result with Proposition 3.9, we get

$$H^\cdot\left(\mathcal{S}_0, A''_{Y_b}\right) \simeq H^{(0,\cdot)}(X,E). \tag{3.75}$$

Let us now describe the behaviour of $A^2_{Y_b}$ as $b \to 0$. We replace A_{Y_b} by its conjugate $C_{Y,b}$ defined in (3.65). Inspection of equations (1.61), (1.63), (1.70), (1.71), and (3.66)–(3.70) shows that the analysis of the operator $C^2_{Y,b}$ is formally the same as the analysis of the operator $\mathfrak{A}'^2_{\phi,\mathcal{H}_c}$ which is done in [BL06, Sections 3 and 17].

Recall that D^X was defined in (3.17), and that the projector P was defined in Subsection 3.9.

Let $\delta = (\delta_0, \delta_1, \delta_2)$ with $\delta_0 \in \mathbf{R}, \delta_1 > 0, \delta_2 > 0$. Put

$$\mathcal{W}_\delta = \left\{\lambda \in \mathbf{C}, \operatorname{Re}\lambda \leq \delta_0 + \delta_1 \left|\operatorname{Im}\lambda\right|^{\delta_2}\right\}. \tag{3.76}$$

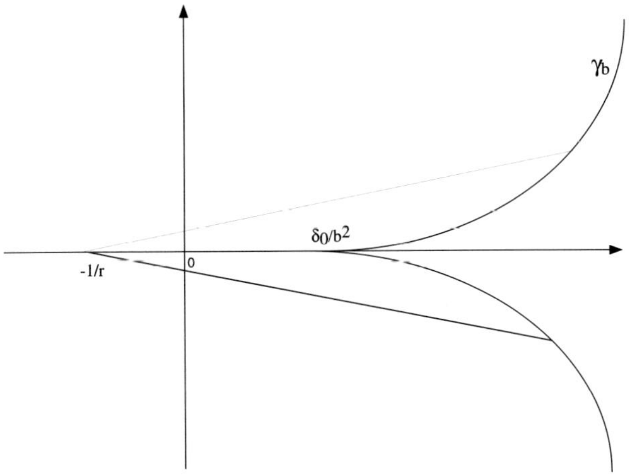

FIGURE 3.2

For $r > 0, b > 0$, set

$$\mathcal{W}_{\delta',b,r} = \left\{\lambda \in \mathcal{W}_{\delta'}/b^2, r\mathrm{Re}\,\lambda + 1 \leq |\mathrm{Im}\,\lambda|\right\}. \tag{3.77}$$

The domain $\mathcal{W}_{\delta',b,r}$ is located to the left of γ_b and outside of the cone in Figure 3.2.

Using the results of [BL06, Section 3] and Theorem 3.12, we find that given $r > 0$, there exists $\delta' = (\delta'_0, \delta'_1, \delta'_2)$ with $\delta'_0 \in]0,1], \delta'_1 > 0, \delta'_2 = 1/6$ such that for $b > 0$ small enough, for $\lambda \in \mathcal{W}_{\delta',b,r}$, for $N \in \mathbf{N}^*$ large enough, the norm of the operator

$$\left(C^2_{Y,b} - \lambda\right)^{-N} - P\left(D^{X,2}/2 - \lambda\right)^{-N} P$$

can be suitably estimated in any natural Sobolev norm on the corresponding kernels by Cb^v, with $v \in]0,1[$.

By proceeding as in the proof of [BL06, Theorem 3.5], we find that given $M > 0$, for $b > 0$ small enough,

$$\begin{aligned} &\mathrm{Sp}\, C^2_{Y,b} \cap \{\lambda \in \mathbf{C}, \mathrm{Re}\lambda \leq M\} \subset \mathbf{R}_+, \\ &\mathcal{S}_0 = \ker A''_{Y_b} \cap \ker A'_{Y_b}. \end{aligned} \tag{3.78}$$

For $b > 0$ small enough, Figure 3.2 can be replaced by Figure 3.3.

By (3.78), we find that as in [BL06], for $b > 0$ small enough, the usual conclusions of Hodge theory hold, i.e.,

$$\mathcal{S}_0 = \ker A^2_{Y_b}, \qquad \mathcal{S}_0 \simeq H^{(0,\cdot)}(X,E). \tag{3.79}$$

Finally by proceeding as in [BL06, Theorem 3.8], one can show that the set of $b > 0$ such that the usual conclusions of Hodge theory do not hold for $A^2_{Y_b}$ is discrete.

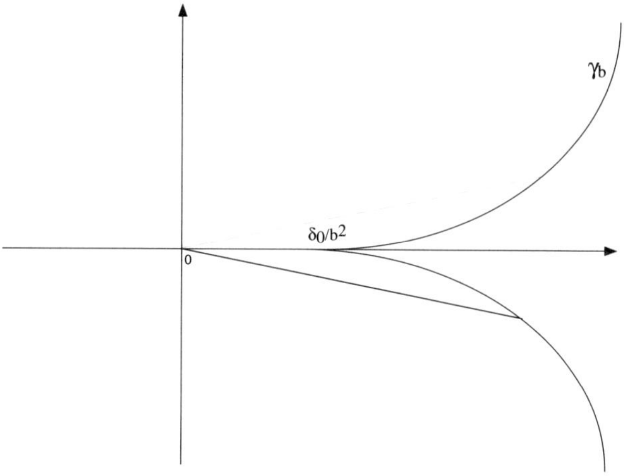

FIGURE 3.3

3.11. Kähler fibrations and the Levi-Civita superconnection

Let $p : M \to S$ be a holomorphic submersion of complex manifolds with complex fibre X of dimension n. Let ω^M be a smooth real $(1,1)$-form on M which is closed, and which restricts to the Kähler form along the fibre X associated to a fibrewise Kähler metric g^{TX} on the relative tangent bundle $TX = TM/S$. Let $T^H M \subset TM$ be the orthogonal bundle to TX with respect to ω^M. Then we have the smooth splitting,

$$TM = T^H M \oplus TX. \tag{3.80}$$

By (3.80), we deduce the smooth isomorphism,

$$\Lambda^{\cdot}(T^*M) \simeq \pi^*\Lambda^{\cdot}(T^*S)\widehat{\otimes}\Lambda^{\cdot}(T^*X). \tag{3.81}$$

By following the terminology in [BGS88b, Definition 1.4], we will say that the triple $\left(p, g^{TX}, T^H M\right)$ defines a Kähler fibration with associated $(1,1)$ form ω^M. Let $\omega^{M,V}, \omega^{M,H}$ be the restriction of ω^M to $T_{\mathbf{R}}X, T^H_{\mathbf{R}}M$, so that

$$\omega^M = \omega^{M,H} + \omega^{M,V}. \tag{3.82}$$

Note that $\omega^{M,V}$ is just the Kähler form ω^X along the fibre X which is associated with the metric g^{TX}. As in (3.16), g^{TX}, g^E induce a Hermitian product $\langle\rangle_{L^2}$, on $\Omega^{(0,\cdot)}(X,E)$.

Let ∇^{TX} be the holomorphic Hermitian connection on $\left(TX, g^{TX}\right)$, and let R^{TX} be its curvature. Let dv_X be the volume form along the fibre X which is associated to g^{TX}. Let $\nabla^{T_{\mathbf{R}}X}$ be the connection on $T_{\mathbf{R}}X$ which is induced by ∇^{TX}. By [BGS88b, Theorem 1.5], [B97, Theorems 1.1 and 2.3], $\nabla^{T_{\mathbf{R}}X}$ is exactly the connection associated to $\left(T^H_{\mathbf{R}}M, g^{T_{\mathbf{R}}X}\right)$ which was considered in Subsection 2.1. It is shown in the same references that the associated tensor T which was considered in Subsection 2.1 is of type $(1,1)$. Moreover, if $A \in TS, B \in TX$, then $T\left(A^H, \overline{B}\right) \in TX$ and $T\left(\overline{A}^H, B\right) \in \overline{TX}$. Finally by [BGS88b, Theorem 1.7],

$$d^X \omega^{M,H} + i_{T^H} \omega^{M,V} = 0. \tag{3.83}$$

Let E be a holomorphic vector bundle on M, let g^E be a Hermitian metric on E, let ∇^E be the holomorphic Hermitian connection on E, and let R^E be its curvature. Let $\nabla^{\Lambda^{\cdot}\left(\overline{T^*X}\right)\otimes E}$ be the connection on $\Lambda^{\cdot}\left(\overline{T^*X}\right) \otimes E$ which is induced by ∇^{TX}, ∇^E. If $U \in T_{\mathbf{R}}S$, let $U^H \in T^H_{\mathbf{R}}M$ be its horizontal lift.

If s is a smooth section on S of $\Omega^{(0,\cdot)}(X,E)$, if $U \in T_{\mathbf{R}}S$, set

$$\nabla_U^{\Omega^{(0,\cdot)}(X,E)} s = \nabla_{U^H}^{\Lambda^{\cdot}\left(\overline{T^*X}\right)\otimes E} s. \tag{3.84}$$

By [BGS88b, Theorem 1.14], $\nabla^{\Omega^{(0,\cdot)}(X,E)}$ is a Hermitian connection on the vector bundle $\Omega^{(0,\cdot)}(X,E)$, and its curvature is of type $(1,1)$.

By (3.80), we get the identification,

$$\Omega^{(0,\cdot)}(M,E) = \Omega^{(0,\cdot)}\left(S, \Omega^{(0,\cdot)}(X,E)\right). \tag{3.85}$$

The operators $\nabla^{\Omega^{(0,\cdot)}(X,E)\prime\prime}$ and $\overline{\partial}^X$ act on $\Omega^{(0,\cdot)}(M,E)$. By [BGS88b, Theorem 2.8], we get the identity of operators acting on $\Omega^{(0,\cdot)}(M,E)$,

$$\overline{\partial}^M = \nabla^{\Omega^{(0,\cdot)}(X,E)\prime\prime} + \overline{\partial}^X. \tag{3.86}$$

By (3.86), we obtain,

$$\left(\nabla^{\Omega^{(0,\cdot)}(X,E)\prime\prime} + \overline{\partial}^X\right)^2 = 0. \tag{3.87}$$

Similarly, by [BGS88b, Theorem 2.6],

$$\left(\nabla^{\Omega^{(0,\cdot)}(X,E)\prime} + \overline{\partial}^{X*}\right)^2 = 0. \tag{3.88}$$

Now we follow [BGS88b, Section 2].

Definition 3.14. Put

$$\begin{aligned} B'' &= e^{-i\omega^{M,H}/2}\left(\nabla^{\Omega^{(0,\cdot)}(X,E)\prime\prime} + \overline{\partial}^X\right) e^{i\omega^{M,H}/2}, \\ B' &= e^{i\omega^{M,H}/2}\left(\nabla^{\Omega^{(0,\cdot)}(X,E)\prime} + \overline{\partial}^{X*}\right) e^{-i\omega^{M,H}/2}, \\ B &= B'' + B'. \end{aligned} \tag{3.89}$$

Then

$$B''^2 = 0, \qquad B'^2 = 0. \tag{3.90}$$

It is shown in [BGS88b] that B is a special case of a Levi-Civita superconnection. In particular,

$$B = \nabla^{\Omega^{(0,\cdot)}(X,E)} + \overline{\partial}^X + \overline{\partial}^{X*} - \frac{1}{2}\left(T^{H(1,0)*} \wedge - i_{T^{H(0,1)}}\right). \tag{3.91}$$

In particular the curvature B^2 is an elliptic operator of order 2 along the fibre X. It was explicitly evaluated in [BGS88b, Theorem 2.10] using the formula of [B86, Theorem 3.6] for the curvature of a general Levi-Civita superconnection, which is also given in (2.11).

3.12. Kähler fibrations and the hypoelliptic superconnection

We make the same assumptions as in Subsection 3.11 and we use the corresponding notation. Let $\mathcal{M}$ be the total space of TX, and let $\pi : \mathcal{M} \to M, q : \mathcal{M} \to S$ be the obvious projections. We denote by $\mathcal{X}$ the fibre of π.

The connection ∇^{TX} induces a smooth splitting,

$$T\mathcal{M} = \pi^* \left(TM \oplus TX \right). \tag{3.92}$$

By (3.92), we get the identification

$$\Lambda^{\cdot} \left(\overline{T^*\mathcal{M}} \right) = \pi^* \left(\Lambda^{\cdot} \left(\overline{T^*M} \right) \widehat{\otimes} \widehat{\Lambda}^{\cdot} \left(\overline{T^*X} \right) \right). \tag{3.93}$$

We still define F as in (3.23), and $\mathfrak{F} \subset \Lambda^{\cdot} \left(T^*_{\mathbf{R}}\mathcal{M} \otimes_{\mathbf{R}} \mathbf{C} \right) \widehat{\otimes} \pi^* E$ as in Definition 3.1. Instead of (3.24), we have the smooth isomorphism,

$$\mathfrak{F} \simeq \Lambda \left(T^*_{\mathbf{R}} S \otimes_{\mathbf{R}} \mathbf{C} \right) \otimes \pi^* F. \tag{3.94}$$

Then the Dolbeault operator $\overline{\partial}^{\mathcal{M}}$ acts on $C^{\infty} \left(\mathcal{M}, \mathfrak{F} \right)$.

Also we define $\mathbf{I}^{\cdot}$ as in Definition 3.2. Then $\mathbf{I}^{\cdot}$ is a $\mathbf{Z}$-graded vector bundle over M, on which $\overline{\partial}^V$ acts naturally. The obvious analogue of equation (3.32) in Proposition 3.3 holds, i.e.,

$$\overline{\partial}^{\mathcal{M}} = \nabla^{\mathbf{I}^{\cdot}\prime\prime} + \overline{\partial}^V. \tag{3.95}$$

The obvious analogues of (3.35), (3.36) still hold.

Let y be the generic element of the fibre TX. Instead of (3.27), we now set

$$\omega^{\mathcal{M}} = i\overline{\partial}\partial \left| y \right|^2. \tag{3.96}$$

Let $\omega^{\mathcal{M},V}$ be the Kähler form along the fibres TX. Put

$$\omega^{\mathcal{M},H} = i \left\langle R^{TX} y, \overline{y} \right\rangle. \tag{3.97}$$

Then the analogue of (3.29) still holds.

The form $\pi^*\omega^M + \omega^{\mathcal{M}}$ is a closed $(1,1)$ form on $\mathcal{M}$. Let $T^H\mathcal{M} \subset T\mathcal{M}$ be the horizontal lift of $T^H M$ in $T\mathcal{M}$ with respect to the connection ∇^{TX}. Then $T^H\mathcal{M}$ is a horizontal subbundle with respect to the projection $q : \mathcal{M} \to S$. Moreover, $T^H\mathcal{M}$ is orthogonal to $T\mathcal{X}$ with respect to the form $\pi^*\omega^M + \omega^{\mathcal{M}}$.

Incidentally we do not claim that there is an associated Kähler fibration $p : \mathcal{M} \to S$, because $\pi^*\omega^M + \omega^{\mathcal{M}}$ may well be degenerate along the fibre $\mathcal{X}$. However this will be totally irrelevant in the sequel.

Definition 3.15. Set

$$\mathcal{A}'' = \nabla^{\mathbf{I}^{\cdot}\prime\prime} + \overline{\partial}^V, \qquad \mathcal{A}' = \nabla^{\mathbf{I}^{\cdot}\prime} + \overline{\partial}^{V*}, \qquad \mathcal{A} = \mathcal{A}'' + \mathcal{A}'. \tag{3.98}$$

By comparing (2.13) and (3.98), we find that $\mathcal{A}$ is a special case of the superconnection defined in (2.13).

Let z be a holomorphic section of $\pi^* TX$ on $\mathcal{M}$. Let $Z = z + \overline{z}$ be the corresponding section of $T_{\mathbf{R}} X$. Put

$$\mathcal{A}''_Z = \nabla^{\mathbf{I}^{\cdot}\prime\prime} + \overline{\partial}^V + i_z, \quad \mathcal{A}'_Z = e^{i\omega^M}\left(\nabla^{\mathbf{I}^{\cdot}\prime} + \overline{\partial}^{V*} + i_{\overline{z}}\right)e^{-i\omega^M}, \quad \mathcal{A}_Z = \mathcal{A}''_Z + \mathcal{A}'_Z. \tag{3.99}$$

Then

$$\mathcal{A}''^2_Z = 0, \qquad \mathcal{A}'^2_Z = 0. \tag{3.100}$$

Observe that since the form ω^M is closed,

$$\mathcal{A}'_Z = \nabla^{\mathbf{I}^{\cdot}\prime} + \overline{\partial}^{V*} + i_{\overline{z}} + \overline{z}^* \wedge . \tag{3.101}$$

By (3.99), (3.101), we get

$$\mathcal{A}_Z = \mathcal{A} + \frac{\widehat{c}\,(Z)}{\sqrt{2}}. \tag{3.102}$$

Using (3.8), (3.9), we find that the superconnection $\mathcal{A}_Z$ in (3.99) is a special case of the superconnection $\mathcal{A}_Z$ in (2.20).

3.13. A self-adjointness property

Recall that the Hermitian forms ϵ, η were defined in (3.54), (3.55). We define the forms $\underline{\epsilon}, \underline{\eta}$ on $C^{\infty,c}(\mathcal{X}, \pi^* F)$ with values in $\bigoplus \Lambda^{(p,p)}(T^*_{\mathbf{R}} S)$ by the formulas,

$$\underline{\epsilon}(s, s') = \epsilon\left(s, e^{-i\omega^{M,H}} s'\right), \qquad \underline{\eta}(s, s') = \eta\left(s, e^{-i\omega^{M,H}} s'\right), \tag{3.103}$$

Let $\dagger$ be the antilinear involution acting on $\Lambda^{\cdot}(T^*_{\mathbf{R}} S) \otimes_{\mathbf{R}} \mathbf{C}$ which is such that $(\alpha \wedge \alpha')^\dagger = \alpha'^\dagger \wedge \alpha^\dagger$, and moreover, if $f \in T^* S$, $f^\dagger = -\overline{f}$. Note that

$$\left(i\omega^{M,H}\right)^\dagger = i\omega^{M,H}. \tag{3.104}$$

We claim that $\underline{\epsilon}, \underline{\eta}$ are Hermitian forms, in the sense that

$$\underline{\epsilon}(s', s) = (\underline{\epsilon}(s, s'))^\dagger, \qquad \underline{\eta}(s', s) = (\underline{\eta}(s, s'))^\dagger. \tag{3.105}$$

Indeed this is a consequence of the corresponding fact for the forms ϵ, η, and also of (3.104).

By defining L, Λ as in (3.46), we can define the conjugate superconnection $\mathcal{B}_Z$ by the formula

$$\mathcal{B}_Z = \exp(i\Lambda)\,\mathcal{A}_Z \exp(-i\Lambda). \tag{3.106}$$

The operators $\mathcal{B}''_Z, \mathcal{B}'_Z$ are obtained in the same way from $\mathcal{A}''_Z, \mathcal{A}'_Z$.

Let $\underline{N}^{(1,0)}, \underline{N}^{(0,1)}$ be the number operators on $\Lambda^{\cdot}(T^*\mathcal{M}), \Lambda^{\cdot}(\overline{T^*\mathcal{M}})$. Set

$$\underline{N} = \underline{N}^{(0,1)} - \underline{N}^{(1,0)}. \tag{3.107}$$

Note that (3.52) and (3.107) are compatible. Let $\underline{N}^{H(0,1)}, \underline{N}^{H(1,0)}$ be the number operators on $\Lambda^{\cdot}(\overline{T^*M}), \Lambda^{\cdot}(T^*M)$. Let N^V be the number operator on $\widehat{\Lambda}^{\cdot}(\overline{T^*X})$.

We identify $\underline{\mathcal{N}}$ to its restriction to $\Lambda^{\cdot}(T^*_{\mathbf{R}}S)\widehat{\otimes}_{\mathbf{R}}\pi^*F$, which is given by

$$\underline{\mathcal{N}} = N^V + \underline{N}^{H(0,1)} - \underline{N}^{H(1,0)}. \tag{3.108}$$

In what follows, we consider the operators defined in (3.98)–(3.108) as operators acting on $C^\infty\left(\mathcal{M}, \Lambda^{\cdot}(T^*_{\mathbf{R}}S)\widehat{\otimes}_{\mathbf{R}}\pi^*F\right)$.

The vector space $C^\infty\left(\mathcal{M}, \Lambda^{\cdot}(T^*_{\mathbf{R}}S)\widehat{\otimes}_{\mathbf{R}}\pi^*F\right)$ is naturally $\mathbf{Z}$-graded by $\underline{\mathcal{N}}$. Given $k \in \mathbf{Z}$, we will say that an operator acting on $C^\infty\left(\mathcal{M}, \Lambda^{\cdot}(T^*_{\mathbf{R}}S)\widehat{\otimes}_{\mathbf{R}}\pi^*F\right)$ is of degree k if it increases the degree by k. Let $\nabla^{F,h}$ be the connection on F which is induced by ∇^{TX}, ∇^E.

Let $f_1, \ldots, f_m$ be a basis of TS, let $f^1, \ldots, f^m$ be the corresponding dual basis. We will establish an analogue of Theorem 3.7. Adjoints will still be denoted with a $\dagger$. The adjoint of $\mathcal{A}_Z$ will be taken with respect to $\underline{\epsilon}$, and the adjoint of $\mathcal{B}_Z$ with respect to $\underline{\eta}$.

Theorem 3.16. *The operator $\mathcal{A}''_Z$ is of total degree 1, and the operator $\mathcal{A}'_Z$ is of total degree -1. Moreover,*

$$\begin{aligned} \mathcal{A}''^{\dagger}_Z &= \mathcal{A}'_{Z_-}, & \mathcal{B}''^{\dagger}_Z &= \mathcal{B}'_{Z_-}, \\ \mathcal{A}^{\dagger}_Z &= \mathcal{A}_{Z_-}, & \mathcal{B}^{\dagger}_Z &= \mathcal{B}_{Z_-}. \end{aligned} \tag{3.109}$$

Proof. We will first establish the fourth identity in (3.109). By the analogue of (3.14), we get

$$d^M = \nabla^{\Lambda^{\cdot}(T^*_{\mathbf{R}}X)} + i_T. \tag{3.110}$$

Since ω^M is closed, using (3.110), we get

$$\nabla^V \omega^{M,H} + i_{T^H}\omega^X = 0, \qquad \nabla^H \omega^{M,H} = 0. \tag{3.111}$$

In (3.111), ∇^V, ∇^H denote vertical and horizontal differentiation respectively. The first equation in (3.111) just expresses the known fact [BGS88b, Theorem 1.7] that T^H is a fibrewise Hamiltonian vector field with Hamiltonian ω^H. By (3.99), (3.101) and (3.110), we get

$$\begin{aligned} \mathcal{B}_Z = \left(\overline{w}^i \wedge + i_{w_i}\right)\nabla^{F,h}_{\overline{w}_i} + \left(w^i \wedge - i_{\overline{w}_i}\right)\nabla^{F,h}_{w_i} + \overline{f}^\alpha \wedge \nabla^{F,h}_{\overline{f}^H_\alpha} + f^\alpha \wedge \nabla^{F,h}_{f^H_\alpha} \\ + \overline{f}^\alpha \wedge \left(w^i \wedge - i_{\overline{w}_i}\right) i_{T\left(\overline{f}^H_\alpha, w_i\right)} + f^\alpha \wedge \left(\overline{w}^i \wedge + i_{w_i}\right) i_{T(f^H_\alpha, \overline{w}_i)} + i_{T^H} \\ + \overline{\partial}^V + i_z + \overline{\partial}^{V*} + \overline{z}^* \wedge . \end{aligned} \tag{3.112}$$

By proceeding as in the proof of Theorem 3.7 and using (3.111), (3.112), the proof of the fourth identity in (3.109) follows easily. The third equation in (3.109) now follows from the fourth one. By splitting $\mathcal{A}_Z, \mathcal{B}_Z$ into their components of degree 1 and -1, we get the first two equations in (3.109). The proof of our theorem is completed. □

Remark 3.17. The observations we made in Remark 3.10 remain still valid in the families context.

3.14. Hypoelliptic and elliptic superconnections

Given $b \in \mathbf{R}^*$, we define $\mathcal{C}_{Z,b}$ from $\mathcal{B}_{Z_b}$ as in (3.65).

Set

$$\mathcal{E} = f^\alpha \wedge \nabla^{F,h}_{f_\alpha} + \overline{f}^\alpha \wedge \nabla^{F,h}_{\overline{f}_\alpha} + \left(\overline{w}^i \wedge + i_{w_i}\right) \nabla^{F,h}_{\overline{w}_i} + \left(w^i \wedge - i_{\overline{w}_i}\right) \nabla^{F,h}_{w_i}$$
$$+ \overline{f}^\alpha \wedge \left(w^i \wedge - i_{\overline{w}_i}\right) i_{T\left(\overline{f}^H_\alpha, w_i\right)} + f^\alpha \wedge \left(\overline{w}^i + i_{w_i}\right) i_{T(f_\alpha, \overline{w}_i)} + f^\alpha \wedge \overline{f}^\beta i_{T\left(f^H_\alpha, \overline{f}^H_\beta\right)}. \tag{3.113}$$

By (3.112), we get

$$\mathcal{C}_{Y,b} = \mathcal{E} + \frac{1}{b}\left(\overline{\partial}^V + i_y + \overline{\partial}^{V*} + \overline{y}^* \wedge\right). \tag{3.114}$$

We define the projector P as in Subsection 3.9. Finally recall that the superconnection B was defined in Definition 3.14.

We establish the obvious extension of Theorem 3.12.

Theorem 3.18. *The following identity holds,*

$$e^{-i\omega^{M,H}/2} P\mathcal{E}P e^{i\omega^{M,H}/2} = B. \tag{3.115}$$

Proof. By (3.113), we obtain

$$P\mathcal{E}P = \nabla^{\Omega^\cdot(X,E)} + \overline{\partial}^X + \overline{\partial}^{X*} - \overline{f}^\alpha \wedge i_{\overline{w}_i} i_{T\left(\overline{f}^H_\alpha, w_i\right)} + i_{T^{H(0,1)}}. \tag{3.116}$$

Also by (2.4),

$$i_{\overline{w}_i} i_{T\left(\overline{f}^H_\alpha, w_i\right)} = \left\langle T\left(\overline{f}^H_\alpha, w_i\right), w_j\right\rangle i_{\overline{w}_i} i_{\overline{w}_j} = 0. \tag{3.117}$$

Finally by (3.83) and (3.116), we get

$$e^{-i\omega^{M,H}/2} P\mathcal{E}P e^{i\omega^{M,H}/2} = \nabla^{\Omega^\cdot(X,E)} + \overline{\partial}^X + \overline{\partial}^{X*} - \frac{1}{2}\left(T^{H(1,0)*} \wedge - i_{T^{H(0,1)}}\right). \tag{3.118}$$

Comparing (3.91) with (3.118), we get (3.115). The proof of our theorem is completed. □

4. The local index theory for the elliptic Laplacian and the elliptic holomorphic torsion forms

In this section, we state the local families index theorem of [B86]. Also we explain the construction of the holomorphic analytic torsion forms associated to Kähler fibrations, and we state the curvature theorem for Quillen metrics on the determinant of the cohomology. These objects will be constructed using the usual elliptic Hodge theory. In the next sections, we will extend these constructions to hypoelliptic Hodge theory.

This section is organized as follows. In Subsection 4.1, we introduce the action of a compact Lie group G which preserves the various geometric structures which were considered before.

In Subsection 4.2, we state the local families index theorem.

In Subsection 4.3, we give the double transgression formulas for the Chern character forms in the case of Kähler fibrations, and we study their limit as $t \to 0$.

In Subsection 4.4, we obtain their limit as $t \to +\infty$.

In Subsection 4.5, we construct the holomorphic analytic torsion forms in the equivariant context.

In Subsection 4.6, we state the curvature theorem of [BGS88c] for Quillen metrics.

Finally in Subsection 4.7, we give a curvature theorem for equivariant Quillen metrics.

4.1. A group action

Let G be a compact Lie group. First we make the same assumptions as in Section 1. We assume that G acts isometrically on X, and that the action of G is oriented and lifts to the corresponding spin bundle. Then the action of G lifts as an isometric action on TX, and also to a unitary action S^{TX} which preserves the $\mathbf{Z}_2$-grading when n is even. Finally we assume that the action of G lifts to $\left(E, g^E, \nabla^E\right)$.

Then G acts on $C^\infty\left(X, S^{TX}\widehat{\otimes}E\right)$, so that if $g \in G, s \in C^\infty\left(X, S^{TX}\widehat{\otimes}E\right)$,

$$(gs)(x) = gs\left(g^{-1}x\right). \tag{4.1}$$

Also G commutes with the elliptic Dirac operator D^X which was constructed in Definition 1.1.

Now we make the same assumptions as in Subsection 1.5. Also we assume the connection ∇^{TX} to be G-invariant. This is the case in particular if $\nabla^{TX} = \nabla^{TX,L}$.

Recall that the vector bundle F was defined in (1.20). Then G acts on $\mathcal{X}$ and the action lifts to $C^\infty(\mathcal{X}, \pi^*F)$. Finally the operator $\mathcal{D}$ commutes with G.

Assume that Z is a G-invariant section of π^*TX over $\mathcal{X}$. Then G also commutes with $\mathcal{D}_Z$. Note here that the tautological section Y is G-invariant.

Now we consider the more general context of Section 2. We assume that G acts on M along the fibres X. Also we assume that G preserves $\left(T^HM, g^{TX}\right)$, and that the action of G lifts to S^{TX}. Finally we still suppose that the action of G lifts to E and preserves g^E, ∇^E. It should be clear that under obvious conditions, like the G-invariance of the section Z, the superconnections which were constructed in Section 2 commute with the action of G.

We work now in the context of Section 3. In the case of a single compact complex Hermitian manifold X, we assume that G acts holomorphically on X and preserves the Hermitian metric g^{TX}. Also we assume that the action of G on X lifts to a holomorphic action on E which preserves the metric g^E.

Then G acts on $\Omega^{(0,\cdot)}(X,E)$ by a formula similar to (4.1), and this action commutes with $\overline{\partial}^X, \overline{\partial}^{X*}$, and so it commutes with the operator D^X in (3.17).

Again G acts on $C^\infty(\mathcal{X}, \pi^*F)$, and this action commutes with the operators in (3.37). If the holomorphic section z of TX is G-invariant, then G also commutes with A_Z.

Finally we make the same assumptions as in Subsection 3.11. We assume that G acts holomorphically along the fibres X, and we assume that the form ω^M is G-invariant. Then G commutes with the superconnection B. Under the assumptions of Subsection 3.12, G acts on $\mathcal{M}$ and preserves $\omega^{\mathcal{M},H}$. It commutes with the superconnection $\mathcal{A}$. If the section z of TX is G-invariant, G also commutes with $\mathcal{A}_Z$.

4.2. The local families index theorem

We make the same assumptions as in Subsection 2.1, and we use the corresponding notation. Let G be a compact Lie group as in Subsection 4.1. We assume that the compatibility conditions of that subsection are verified.

Given $t > 0$, let A_t be the Levi-Civita superconnection on the $\mathbf{Z}_2$-graded vector bundle H considered in (2.8) which is attached to $T^HM, g^{TX}/t, \nabla^E$. Recall that for $a > 0$, $\psi_a \in \operatorname{End}(\Lambda^{\cdot}(T^*S))$ was defined before equation (2.10). Then one verifies easily that

$$A_t = \psi_{1/t}\sqrt{t}A\psi_t. \tag{4.2}$$

Also A_t^2 is an elliptic second order differential operator along the fibre X.

In the sequel we use the notation of Quillen [Q85a]. If L is trace class operator acting on H, let $\operatorname{Tr_s}[L]$ denote the supertrace of L.

If $\alpha_t, t \geq 0$ is a family of smooth forms on S depending on $t > 0$, we will say that as $t \to 0$, $\alpha_t = \alpha_0 + \mathcal{O}(t)$ if given a compact subset $K \subset S$, for any $k \in \mathbf{N}$, the sup over K of the norm of the derivatives of order k of $\alpha_t - \alpha_0$ can be dominated by $C_{K,k}t$. We will use a similar notation when $t \to +\infty$.

Let $\operatorname{ind}(D^X) \in K^0(S)$ be the index bundle associated to the family of operators D^X in the sense of Atiyah-Singer [AS71]. Then $\operatorname{ind}(D^X)$ is an equivariant vector bundle. If $g \in G$, let $\operatorname{ch}_g(\operatorname{ind}(D^X)) \in H^{\cdot}(S, \mathbf{C})$ be the corresponding equivariant Chern character.

Take $g \in G$. Let $M_g \in M$ be the fixed point set of g. Then M_g fibres on S with fibre X_g. Moreover, the restriction of T^HM to M_g is included in TM_g.

Let $\widehat{A}_g(TX, \nabla^{TX}), \operatorname{ch}_g(E, g^E)$ be the even closed smooth forms on M_g in Chern-Weil theory which are associated to the connections of ∇^{TX}, ∇^E on $TX|_{M_g}, E|_{M_g}$ and to the $\widehat{A}$ and ch genera as in the Lefschetz fixed point formulas of Atiyah-Bott [AB66, AB67, AB68]. Note that except when $g = 1$, there is a $\pm$ ambiguity in the definition of $\widehat{A}_g(TX, \nabla^{TX})$. This sign ambiguity will be noted explicitly.

Let φ be the endomorphism of $\Lambda^{\cdot}(T^*_{\mathbf{R}}S \otimes_{\mathbf{R}} \mathbf{C})$ such that if α is a form of degree p, then $\varphi\alpha = (2i\pi)^{-p/2}\alpha$. For $t > 0$, set

$$u_t = \varphi \operatorname{Tr_s}\left[g \exp\left(-A_t^2\right)\right]. \tag{4.3}$$

Now we state the local families index theorem in [B86, Theorems 4.12 and 4.16] and [M00a, Theorem 2.10].

Theorem 4.1. *The even forms u_t are closed, their cohomology class $[u_t]$ does not depend on t and is given by*

$$[u_t] = \mathrm{ch}_g\left(\ker D^X\right). \tag{4.4}$$

As $t \to 0$,

$$u_t = \int_{X_g} \pm \widehat{A}_g\left(TX, g^{TX}\right) \mathrm{ch}_g\left(E, g^E\right) + \mathcal{O}(t). \tag{4.5}$$

4.3. The local index theorem in the holomorphic elliptic case and the double transgression formulas

We make the same assumptions as in Subsection 3.11. Let G be a compact Lie group. We assume that the compatibility conditions of Subsection 4.1 are verified.

Let P^S be the vector space of smooth forms on S which are sums of forms of type (p,p). Let $P^{S,0} \subset P^S$ be the space of forms $\alpha \in P^S$ such that there exist smooth forms β, γ on S for which $\alpha = \overline{\partial}^S \beta + \partial^S \gamma$. We use a similar notation for forms on other manifolds.

Take $g \in G$. Then $M_g \subset M$, the fixed point set of g, is a complex submanifold of M which fibres on S with complex fibre X_g. For $g \in G$, the forms $\mathrm{Td}_g\left(TX, g^{TX}\right), \mathrm{ch}_g\left(E, g^E\right)$ are the Todd and Chern character forms on M_g which appear in the context of the holomorphic fixed point formulas of Atiyah-Bott, which are associated to the holomorphic Hermitian connections on the given vector bundles.

Let $R^{\cdot}\pi_* E$ be the direct image of E. Then G acts on $R^{\cdot}\pi_* E$. Also let $\mathrm{ch}_g\left(R^{\cdot}\pi_* E\right) \in H^{\cdot}(S, \mathbf{C})$ be the equivariant Chern character of $R^{\cdot}\pi_* E$.

For $t > 0$, let B_t be the superconnection defined in Definition 3.14 which is associated with $\left(\omega^M/t, g^E\right)$. Let N be the number operator acting on $\Omega^{(0,\cdot)}\left(X, E|_X\right)$. Set

$$u_t = \varphi \mathrm{Tr_s}\left[g \exp\left(-B_t^2\right)\right], \quad w_t = \varphi \mathrm{Tr_s}\left[g\left(N - n + i\frac{\omega^{M,H}}{t}\right)\exp\left(-B_t^2\right)\right]. \tag{4.6}$$

Theorem 4.2. *For any $t > 0$, the forms u_t, w_t lie in P^S. The forms u_t are closed, and their cohomology class does not depend on $t > 0$. More precisely,*

$$[u_t] = \mathrm{ch}_g\left(R^{\cdot}\pi_* E\right). \tag{4.7}$$

Moreover,

$$\frac{\partial}{\partial t} u_t = -\frac{\overline{\partial}\partial}{2i\pi t} w_t. \tag{4.8}$$

As $t \to 0$,

$$u_t = \int_{X_g} \mathrm{Td}_g\left(TX, g^{TX}\right) \mathrm{ch}_g\left(E, g^E\right) + \mathcal{O}(t). \tag{4.9}$$

There exists forms $C_{-1} \in P^S, C_0 \in P^S$ such that as $t \to 0$,

$$w_t = \frac{C_{-1}}{t} + C_0 + \mathcal{O}(t). \tag{4.10}$$

Moreover,

$$C_{-1} = \int_{X_g} \frac{\omega^M}{2\pi} \mathrm{Td}_g\left(TX, g^{TX}\right) \mathrm{ch}_g\left(E, g^E\right), \tag{4.11}$$
$$C_0 = -\int_{X_g} \mathrm{Td}'_g\left(TX, g^{TX}\right) \mathrm{ch}_g\left(E, g^E\right) \text{ in } P^S/P^{S,0}.$$

Proof. The fact that the form u_t is closed was established in [B86, Theorem 3.4] and in [M00a, Theorem 2.10] in the equivariant case. By [BGS88b, Theorems 2.2 and 2.9], the forms u_t, w_t lies in P^S. By [B86, Theorems 4.12 and 4.16] and by [M00a, Theorem 2.10] in the equivariant case, we get (4.9). By [BGS88b, Theorem2.16] and [M00a, Theorem 2.6], we get (4.10), (4.11). □

4.4. The limit as $t \to +\infty$ of the superconnection forms

For $s \in S$, let $\mathfrak{H}^{\cdot}_s \subset \Omega^{(0,\cdot)}_s\left(X, E|_X\right)$ be the vector space of harmonic forms on X_s, i.e.,

$$\mathfrak{H}^{\cdot}_s = \ker\left(\overline{\partial}^{X_s} + \overline{\partial}^{X_s *}\right). \tag{4.12}$$

By Hodge theory, for any $s \in S$,

$$H^{(0,\cdot)}_s\left(X, E|_X\right) \simeq \mathfrak{H}^{\cdot}_s. \tag{4.13}$$

In the sequel we assume that the $H^{(0,i)}\left(X, E|_X\right), 1 \leq i \leq n$ have locally constant dimension. They are the fibres of a holomorphic **Z**-graded vector bundle, which we denote by $H^{(0,\cdot)}\left(X, E|_X\right)$. This assumption is equivalent to $R^{\cdot}\pi_* E$ being locally free.

Then $\mathfrak{H}^{\cdot}$ is a finite-dimensional smooth subvector bundle of $\Omega^{(0,\cdot)}(X, E)$. It inherits a smooth metric $g^{\mathfrak{H}^{\cdot}}$ from the Hermitian product (3.16) on $\Omega^{(0,\cdot)}(X, E)$. Let $g^{H^{(0,\cdot)}(X,E|_X)}$ denote the corresponding metric on $H^{(0,\cdot)}\left(X, E|_X\right)$.

Recall that

$$\mathrm{ch}_g\left(H^{(0,\cdot)}\left(X, E|_X\right), g^{H^{(0,\cdot)}(X,E|_X)}\right) = \sum_0^n (-1)^i \mathrm{ch}_g\left(H^{(0,i)}\left(X, E|_X\right), g^{H^{(0,i)}(X,E|_X)}\right). \tag{4.14}$$

Set

$$\mathrm{ch}'_g\left(H^{(0,\cdot)}\left(X, E|_X\right), g^{H^{(0,\cdot)}(X,E|_X)}\right) = \sum_0^n (-1)^i i\,\mathrm{ch}_g\left(H^{(0,i)}\left(X, E|_X\right), g^{H^{(0,i)}(X,E|_X)}\right). \tag{4.15}$$

Put

$$u_\infty = \mathrm{ch}_g\left(H^{(0,\cdot)}(X,E|_X), g^{H^{(0,\cdot)}(X,E|_X)}\right),$$
$$w_\infty = \mathrm{ch}'_g\left(H^{(0,\cdot)}(X,E|_X), g^{H^{(0,\cdot)}(X,E|_X)}\right) - n\mathrm{ch}_g\left(H^{(0,\cdot)}(X,E|_X), g^{H^{(0,\cdot)}(X,E|_X)}\right). \tag{4.16}$$

The following result was established in [BeGV92, Theorem 9.19].

Theorem 4.3. *As $t \to \infty$,*

$$u_t = u_\infty + \mathcal{O}\left(1/\sqrt{t}\right), \qquad w_t = w_\infty + \mathcal{O}\left(1/\sqrt{t}\right). \tag{4.17}$$

4.5. The elliptic analytic torsion forms

We still assume that the assumption made after (4.13) holds.

Definition 4.4. For $s \in \mathbf{C}, \mathrm{Re}\, s > 1$, put

$$\zeta^1(s) = -\frac{1}{\Gamma(s)}\int_0^1 t^{s-1}(w_t - w_\infty)\,dt. \tag{4.18}$$

For $s \in \mathbf{C}, \mathrm{Re}\, s < 1/2$, set

$$\zeta^2(s) = -\frac{1}{\Gamma(s)}\int_1^\infty t^{s-1}(w_t - w_\infty)\,dt. \tag{4.19}$$

By Theorems 4.2 and 4.3, $\zeta^1(s), \zeta^2(s)$ are holomorphic functions which extend to holomorphic functions near $s = 0$.

Following [BGS88b, Definition 2.19], [BK92, Definition 3.8], and [M00a, Definition 2.11] in the equivariant case, we now define the holomorphic elliptic torsion forms.

Definition 4.5. Set

$$T_g\left(\omega^M, g^E\right) = \frac{\partial}{\partial s}\left(\zeta^1 + \zeta^2\right)(s)\,|_{s=0}. \tag{4.20}$$

Then $T_g\left(\omega^M, g^E\right)$ is an even form on S which lies in P^S. It is called an elliptic equivariant analytic torsion form. In degree 0, it coincides with the equivariant version of the classical Ray-Singer analytic torsion [RS73].

By (4.10), we get

$$T_g\left(\omega^M, g^E\right) = -\int_0^1\left(w_t - \frac{C_{-1}}{t} - C_0\right)\frac{dt}{t} - \int_1^\infty (w_t - w_\infty)\frac{dt}{t} + C_{-1} + \Gamma'(1)(C_0 - w_\infty). \tag{4.21}$$

Now we recall the basic result of [BGS88b, Theorem 2.20], [BK92, Theorem 3.9] and [M00a, Theorem 2.12] in the equivariant case.

Theorem 4.6. *The following identity holds,*

$$\frac{\overline{\partial}\partial}{2i\pi} T_g\left(\omega^M, g^E\right) = \mathrm{ch}_g\left(H^{(0,\cdot)}\left(X, E|_X\right), g^{H^{(0,\cdot)}(X,E|_X)}\right) - \int_{X_g} \mathrm{Td}_g\left(TX, g^{TX}\right)\mathrm{ch}_g\left(E, g^E\right). \quad (4.22)$$

Proof. This is an obvious consequence of Theorems 4.2 and 4.3. □

Remark 4.7. By results which were established in [BK92, Theorems 3.10 and 3.11], [B95, Theorem 2.5] and in [M00a, Theorem 2.13], the classes $T_g\left(\omega^M, g^E\right) \in P^S/P^{S,0}$ verify anomaly formulas, when ω^M, g^E vary, which can be expressed in terms of Bott-Chern classes [BGS88a]. An important consequence of those results is that the class of $T_g\left(\omega^M, g^E\right)$ only depends on the metrics g^{TX}, g^E.

In the sequel, when $g = 1$, we will write $T\left(\omega^M, g^E\right)$ instead of $T_1\left(\omega^M, g^E\right)$.

4.6. Quillen metrics and the curvature theorem

Set

$$\lambda = \det R^{\cdot}\pi_* E. \quad (4.23)$$

Then λ is a holomorphic complex line. Moreover, for any $s \in S$, we have the canonical isomorphism,

$$\lambda_s = \det H^{(0,\cdot)}\left(X_s, E|_{X_s}\right). \quad (4.24)$$

If the $H^i\left(X, E|_X\right)$ have locally constant dimension, (4.24) extends to an isomorphism of holomorphic line bundles.

Let g^{TX} be a smooth fibrewise Kähler metric on TX. We do not assume any more that g^{TX} is induced by the form ω^M.

By (4.24), the L^2 metric $g^{H^{(0,\cdot)}(X,E|_X)}$ on $H^{(0,\cdot)}\left(X, E|_X\right)$ induces a metric $|\,|_\lambda$ on the fibres λ_s.

If $\alpha \in \Lambda^{\cdot}\left(T^*_{\mathbf{R}}S\right)$, let $\alpha^{(0)}$ be the component of α which has degree 0.

Definition 4.8. Set

$$\|\,\|_{\lambda_s} = |\,|_{\lambda_s} \exp\left(\frac{1}{2} T\left(\omega^M, g^E\right)^{(0)}_s\right). \quad (4.25)$$

If $\left(\mu, \|\,\|_\mu\right)$ is a holomorphic Hermitian vector bundle on S, let $c_1\left(\mu, \|\,\|_\mu\right)$ be the real $(1,1)$ form in P^S associated to the holomorphic Hermitian connection on μ which represents the first Chern class of μ in Chern-Weil theory.

We now state a result which was established in [BGS88a, Theorem 0.1]

Theorem 4.9. *The Quillen metric $\|\,\|_\lambda$ is a smooth metric on the line bundle λ, and moreover,*

$$c_1\left(\lambda, \|\,\|_\lambda\right) = \left[\int_X \mathrm{Td}\left(TX, g^{TX}\right)\mathrm{ch}\left(E, g^E\right)\right]^{(2)}. \quad (4.26)$$

4.7. The equivariant determinant bundle

Here we follow [B95, Section 2]. For simplicity, we assume first that S is reduced to a point.

Let $\widehat{G}$ be the set of equivalence classes of complex irreducible representations of G. An element of $\widehat{G}$ is specified by a complex finite-dimensional vector space W together with an irreducible representation $\rho_W : G \to \operatorname{End}(W)$. Let χ_W be the character attached to this representation.

We have the isotypical decomposition,

$$H^{(0,\cdot)}(X, E|_X) = \bigoplus_{W\in\widehat{G}} \operatorname{Hom}_G\left(W, H^{(0,\cdot)}(X, E|_X)\right) \otimes W, \tag{4.27}$$

which is orthogonal with respect to $g^{H^{(0,\cdot)}(X,E|_X)}$.

If $W \in \widehat{G}$, set

$$\lambda_W = \det\left(\operatorname{Hom}_G\left(W, H^{(0,\cdot)}(X, E|_X)\right) \otimes W\right). \tag{4.28}$$

Put

$$\lambda = \bigoplus_{W\in\widehat{G}} \lambda_W. \tag{4.29}$$

Let $|\,|_{\lambda_W}$ be the metric on λ_W induced by $g^{H^{(0,\cdot)}(X,E|_X)}$. Set

$$\log\left(|\,|^2_\lambda\right) = \sum_{W\in\widehat{G}} \log\left(|\,|^2_{\lambda_W}\right) \frac{\chi_W}{\dim W}. \tag{4.30}$$

The symbol $|\,|^2_\lambda$ is said to be the equivariant L^2 metric on λ.

Definition 4.10. If $g \in G$, set

$$\log\left(\|\,\|^2_\lambda\right)(g) = \log\left(|\,|^2_\lambda\right)(g) + T_g\left(g^{TX}, g^E\right). \tag{4.31}$$

The symbol $\|\,\|_\lambda$ is called an equivariant Quillen metric on λ.

We go back to the case of a general base S. We make the same assumptions as in Subsections 3.3 and 4.3. Let g^{TX} be any fibrewise Kähler metric on TX. In particular do not assume that g^{TX} is induced by ω^M. Let g^E be a Hermitian metric on E. Then by the above construction, the fibres λ_s are equipped with the equivariant Quillen metric $\|\,\|^2_{\lambda_s}$.

Now we have the curvature theorem for equivariant Quillen metrics, which follows from the arguments in [BGS88b], from [B95, Theorem 2.5] and from Theorem 4.6.

Theorem 4.11. *The metric $\|\,\|^2_\lambda$ is smooth on S. Moreover, for any $g \in G$,*

$$c_1\left(\lambda, \|\,\|_\lambda\right)(g) = \left[\int_{X_g} \operatorname{Td}_g\left(TX, g^{TX}\right) \operatorname{ch}_g\left(E, g^E\right)\right]^{(2)}. \tag{4.32}$$

5. The case of a vector bundle

In this section, we apply the constructions of the previous sections to the case where the fibres X are vector bundles. Also we evaluate explicitly certain supertraces. In the case of holomorphic vector bundles, some of these supertraces will be evaluated in terms of the genera $R(x)$, $R(\theta, x)$ of [GS91, B94]. The results of this section will play a crucial role in Section 8 when proving the formula comparing the elliptic and hypoelliptic holomorphic torsion forms.

This section is organized as follows. In Subsection 5.1, we introduce the functions $\varphi(u,x)$, $\sigma(u,\eta,x)$.

In Subsection 5.2, we give a formula for the hypoelliptic curvature in the case where the fibres X form a vector bundle.

In Subsection 5.3, we evaluate certain supertraces of the corresponding heat kernels.

In Subsection 5.4, we give a formula for the hypoelliptic curvature in the case of holomorphic vector bundles.

In Subsection 5.5, we evaluate corresponding supertraces in the holomorphic case in terms of the function $\sigma(u,\eta,x)$.

Finally in Subsection 5.6, we give various formulas for the genus $R(\theta,x)$.

5.1. The functions $\varphi(u,x)$ and $\sigma(u,\eta,x)$

We introduce the functions $\varphi(u,x)$, $\sigma(u,\eta,x)$ which were defined in [B90b, Definition 6.1] and [B94, Definition 4.1],

$$\varphi(u,x) = \frac{4}{u}\sinh\left(\frac{x+\sqrt{x^2+4u}}{4}\right)\sinh\left(\frac{-x+\sqrt{x^2+4u}}{4}\right), \tag{5.1}$$

$$\sigma(u,\eta,x) = 4\sinh\left(\frac{x-2\eta+\sqrt{x^2+4u}}{4}\right)\sinh\left(\frac{-x+2\eta+\sqrt{x^2+4u}}{4}\right).$$

Clearly,

$$\varphi(u,x) = \frac{\sigma(u,0,x)}{u}. \tag{5.2}$$

By [B90b, Theorem 6.2] and [B94, Proposition 4.2],

$$\varphi(u,x) = \prod_{k=1}^{\infty}\left(1+\frac{ix}{2k\pi}+\frac{u}{4k^2\pi^2}\right)\left(1-\frac{ix}{2k\pi}+\frac{u}{4k^2\pi^2}\right), \tag{5.3}$$

$$\sigma(u,i\theta,x) = \left(\theta^2+i\theta x+u\right)\prod_{k\in\mathbf{Z}^*}\left(\frac{(\theta+2k\pi)^2+i(\theta+2k\pi)x+u}{4k^2\pi^2}\right).$$

Note that

$$\varphi(u,x) = \varphi(u,-x). \tag{5.4}$$

Recall that the $\widehat{A}(x)$ genus is given by,

$$\widehat{A}(x) = \frac{x/2}{\sinh(x/2)}. \tag{5.5}$$

Then

$$\varphi(0,x) = \widehat{A}^{-1}(x). \tag{5.6}$$

In the sequel we will use the notation,

$$\widehat{\sigma}(u,\eta,x) = \sigma(u,\eta,-x). \tag{5.7}$$

Let n be an even integer, and let A be an antisymmetric real (n,n) matrix. Set

$$\Phi(u,A) = \det{}^{1/2}\left[\varphi(u,A)\right]. \tag{5.8}$$

Similarly, if A, B are commuting antisymmetric real (n,n) matrices, set

$$\Sigma(u,B,A) = \det{}^{1/2}\left[\sigma(u,B,A)\right]. \tag{5.9}$$

Note that there is no ambiguity in taking the square root in the right-hand sides of (5.8), (5.9). We define $\widehat{\Sigma}(u,B,A)$ in a similar way, so that

$$\widehat{\Sigma}(u,B,A) = \Sigma(u,B,-A). \tag{5.10}$$

Set

$$\Phi'(u,A) = \frac{\partial}{\partial c}\Phi(u,A+c)|_{c=0}, \qquad \Sigma'(u,B,A) = \frac{\partial}{\partial c}\Sigma(u,B,A+c)|_{c=0}, \tag{5.11}$$

$$\widehat{\Sigma}'(u,B,A) = \frac{\partial}{\partial c}\widehat{\Sigma}(u,B,A+c)|_{c=0}.$$

5.2. The hypoelliptic curvature in the case of real spin vector bundles

Let S be a manifold. Let $\pi : E \to S$ be a real oriented spin vector bundle of even dimension n. Let g^E be a metric on E, let ∇^E be a metric preserving connection on E, and let R^E be its curvature. Let $S^E = S^{E_+} \oplus S^{E_-}$ be the corresponding $\mathbf{Z}_2$-graded vector bundle of (E, g^E) spinors.

We can apply the constructions of Section 2 to the projection π. Here we take the twisting bundle (denoted E in Section 2) to be trivial. The only difference with Section 2 is that the fibres E are not compact, but this will be irrelevant.

Let $\mathbf{E}$ be the total space of E. Let $\mathcal{E}$ be the total space of the tangent bundle TE to the fibre E. Note that

$$TE = E \oplus E. \tag{5.12}$$

In (5.12), the first copy represents the fibre E itself, and the second copy of E represents the tangent bundle to the fibre.

We denote by U, Y the generic sections of the first and second copies of E. As before, we will denote by $\widehat{\Lambda}^{\cdot}(E^*)$ the exterior algebra of the dual of the second copy of E. Also if $e \in E$, we denote by ∇_e a differentiation operator along the first copy of E, by $\nabla_{\widehat{e}}$ the differentiation operator along the second copy of E.

Clearly $C^\infty(E \oplus E, S^E)$ is a vector bundle on S. Let $\mathcal{A}^E_{Y_b}$ be the associated superconnection considered in (2.21). Let $\mathcal{B}^E_{Y,b}$ be the conjugate superconnection defined in (2.27).

Let Δ^V be the fibrewise Laplacian along the second copy of E. By (2.25),

$$\mathcal{A}_{Y_b}^{E,2} = \frac{1}{2}\left(-\Delta^V + \frac{|Y|^2}{b^4} + \frac{1}{b^2} c(\widehat{e}_i)\widehat{c}(e_i)\right) - \nabla_{\widehat{R^E Y}} + \frac{1}{4}\left\langle R^E e_i, e_j\right\rangle c(\widehat{e}_i) c(\widehat{e}_j) + \frac{1}{b^2}\left(\nabla_Y + \frac{1}{2}\left\langle R^E U, Y\right\rangle\right). \tag{5.13}$$

By (2.30), we obtain,

$$\mathcal{B}_{Y,b}^{E,2} = \frac{1}{2b^2}\left(-\Delta^V + |Y|^2 + c(\widehat{e}_i)\widehat{c}(e_i)\right) - \nabla_{\widehat{R^E Y}} + \frac{1}{4}\left\langle R^E e_i, e_j\right\rangle c(\widehat{e}_i) c(\widehat{e}_j) + \frac{1}{b}\left(\nabla_Y + \frac{1}{2}\left\langle R^E U, Y\right\rangle\right). \tag{5.14}$$

There are no variables $c(e_i)$ appearing in (5.14). We may as well take the second representation of the operators $c(\widehat{e}_i), \widehat{c}(e_i)$ given in (1.10), i.e., assume that these operators act on $\widehat{\Lambda}^{\cdot}(E^*)$.

The operators in (5.13), (5.14) act on $C^\infty\left(E \oplus E, \Lambda^{\cdot}(T^*S)\widehat{\otimes}\widehat{\Lambda}^{\cdot}(E^*)\right)$, which is a vector bundle on S.

5.3. The evaluation of the trace of the heat kernels

Let $P_b((U,Y),(U',Y'))$ be the smooth heat kernel associated to the operator $\exp\left(-\mathcal{A}_{Y_b}^{E,2}\right)$ with respect to the volume $\frac{dUdY}{(2\pi)^n}$.

Let g be an oriented parallel isometry of E, whose action lifts to S^E. Let E^1 be the eigenbundle of E which corresponds to the eigenvalue 1, and let $E^{1,\perp}$ be the orthogonal subbundle of E^1. Then E splits orthogonally as

$$E = E^1 \oplus E^{1,\perp}, \tag{5.15}$$

and g preserves the splitting. Let $n_1 = \dim E^1$. Let $o(E^1)$ be the orientation bundle of E^1. Let E^{-1} be the eigenbundle of g associated to the eigenvalue -1 and let n_{-1} be the dimension of E^{-1}. Then n_1, n_{-1} are even, and $E^1 \oplus E^{-1}$ is oriented. The orientation bundle of E^{-1} is just $o(E^1)$. More generally the orientation bundle of $E^{1,\perp}$ is $o(E^1)$.

We can write g in the form

$$g = e^B, \tag{5.16}$$

where B is an antisymmetric parallel section of $\operatorname{End}(E)$ which commutes with g. By using the conventions in (5.9), if A is an antisymmetric matrix which commutes with g, set

$$\Sigma_g(u,A) = \Sigma(u,B,A). \tag{5.17}$$

Similarly, we set

$$\Sigma_g\left(u,E,\nabla^E\right) = \Sigma_g\left(u, -\frac{R^E}{2i\pi}\right). \tag{5.18}$$

A similar notation will be used when replacing σ by $\widehat{\sigma}$, and Σ by $\widehat{\Sigma}$, so that

$$\widehat{\Sigma}_g\left(u,E,\nabla^E\right)=\Sigma_g\left(u,\frac{R^E}{2i\pi}\right). \tag{5.19}$$

Finally, put

$$\Sigma'_g\left(u,E,\nabla^E\right)=\Sigma'\left(u,B,-\frac{R^E}{2i\pi}\right),\quad \widehat{\Sigma}'_g\left(u,E,\nabla^E\right)=\widehat{\Sigma}'\left(u,B,-\frac{R^E}{2i\pi}\right), \tag{5.20}$$

The vector bundle $E^{1,\perp}$ is not necessarily oriented or spin. Let $g^{E^{1,\perp}}$ be the metric induced by g^E on $E^{1,\perp}$. Let $S^{E^{1,\perp}}$ be the locally defined vector bundle of spinors for $\left(E^{1,\perp},g^{E^{1,\perp}}\right)$. The action of g acts on $S^{E^{1,\perp}}$ is not necessarily well defined. However since E is oriented and spin, $\mathrm{Tr_s}^{S^{E^{1,\perp}}}[g]$ is well defined as a locally constant function on S.

Note that g acts on $\widehat{\Lambda}(E^*)$. In the sequel we will consider expressions of the type $\mathrm{Tr_s}^{S^{E^{1,\perp}}\widehat{\otimes}\widehat{\Lambda}^{\cdot}(E^*)}[gA]$, which seem to be ambiguous. However A will act trivially on $S^{E^{1,\perp}}$, so that we have an identity of the type,

$$\mathrm{Tr_s}^{S^{E^{1,\perp}}\widehat{\otimes}\widehat{\Lambda}^{\cdot}(E^*)}[gA]=\mathrm{Tr_s}^{S^{E^{1,\perp}}}[g]\,\mathrm{Tr_s}^{\widehat{\Lambda}^{\cdot}(E^*)}[gA], \tag{5.21}$$

so that there is no ambiguity left.

We will also manipulate infinite-dimensional determinants. We will use the conventions in [B90b, Section 6] and [B94, Section 5]. The operators which we will consider will be of the form $1+A$, with A trace class or A Hilbert-Schmidt. When A is trace class, the definition is obvious. When A is Hilbert-Schmidt, in the special cases which are considered here, the definition is also obvious, because by indexing the eigenvalues of A by $k\in\mathbf{Z}$, the product of the corresponding eigenvalues $\prod_{-k}^{+k}\lambda_i$ will turn out to be absolutely convergent when $k\to+\infty$. We will use the notation det for such determinants, and the notation det $'$ when only the obvious infinite terms in the product are excluded. These infinite terms appear when considering the inverse of a noninvertible operator.

Theorem 5.1. *The following identity holds,*

$$\varphi\int_{E^{1,\perp}\times E}\mathrm{Tr_s}^{S^{E^{1,\perp}}\widehat{\otimes}\widehat{\Lambda}^{\cdot}(E^*)}\left[gP_b\left(g^{-1}(U,Y),(U,Y)\right)\right]\frac{dUdY}{(2\pi)^{n-n_1/2}}=\pm\widehat{A}_g\left(E,\nabla^E\right). \tag{5.22}$$

Proof. Let $p((U,Y),(y',Y'))$ be the smooth kernel associated to the hypoelliptic heat kernel $\exp\left(\frac{\Delta^V}{2}-\frac{\nabla_Y}{b^2}\right)$ with respect to $\frac{dUdY}{(2\pi)^n}$. Given $(U,Y)\in E\oplus E$, let $S_{(g,U,Y)}$ be the probability law over $C([0,1],E\oplus E)$ of the corresponding bridge (y_t,Y_t) which starts at (U,Y) at time 0, and returns to (gU,gY) at time 1. Under $S_{g,U,Y}$,

$$\dot{U}=-\frac{Y}{b^2}. \tag{5.23}$$

By (5.23), we get

$$\frac{1}{b^2}\int_0^1 Y_t dt = -(g-1)U. \tag{5.24}$$

Using (5.13) and Itô's formula, we get

$$\begin{aligned}
P_b\left(\left(U,Y\right),g\left(U,Y\right)\right) = p\left(\left(U,Y\right),g\left(U,Y\right)\right) E^{P_{g,U,Y}}\Bigg[\exp\Bigg(-\frac{1}{2}\Bigg(\int_0^1 \frac{|Y|^2}{b^4}dt \\
+\int_0^1 \left|R^E Y\right|^2 dt - \int_0^1 \left\langle R^E Y, 2\dot{Y} + \frac{U}{b^2}\right\rangle dt\Bigg) \\
\exp\left(-\frac{1}{2b^2}c\left(\widehat{e}_i\right)\widehat{c}\left(e_i\right) - \frac{1}{4}\left\langle R^E e_i, e_j\right\rangle c\left(\widehat{e}_i\right)c\left(\widehat{e}_j\right)\right)\Bigg].
\end{aligned} \tag{5.25}$$

When E is replaced by $E' \oplus E''$, both sides of (5.22) behave multiplicatively. Therefore to establish this identity, we may and we will assume either that $E = E^1$, or that $E^1 = \{0\}$.

1) *The case where* $g = 1$. First we assume that $g = 1$. Set

$$J = \frac{d}{dt}. \tag{5.26}$$

Let $P_{(Y,Y)}$ be the probability law of the Brownian bridge on $C\left(S^1, E\right)$ which starts at Y at time 0 and returns at Y at time 1. Let $F \in \mathrm{End}\left(E\right)$ be a self-adjoint positive endomorphism. Let Q_F be the probability measure on $C\left(S^1, E\right)$ of the Gaussian process whose covariance is the operator $\left(-J^2 + F\right)^{-1}$, with periodic boundary conditions. Recall that by [B90b, eq. (7.36)] and [B94, eq. (5.25)], we have the identity of measures on $C\left(S^1, E\right)$,

$$\exp\left(-\frac{1}{2}\int_0^1 \langle FY, Y\rangle\, dt\right) P_{(Y,Y)}\frac{dY}{(2\pi)^{n/2}} = \frac{Q_F}{\det{}^{1/2}\left(F\right)\det{}'^{\,1/2}\left(1 - FJ^{-2}\right)}. \tag{5.27}$$

Set

$$h = \int_0^1 Y dt, \qquad Y_t' = Y_t - h. \tag{5.28}$$

Let $C_0\left(S^1, E\right)$ be the vector space of continuous functions $f : S^1 \to E$ such that $\int_{S^1} f dt = 0$. Note that under Q_F, h and Y' are independent, the probability law of h is a centred distribution with covariance F^{-1}, and the probability law Q_F' of Y' is the law of a Gaussian process $Y \in C_0\left(S^1, E\right)$ with covariance $\left(-J^2 + F\right)^{-1}$.

Let L^2_0 be the set of $f \in L^2([0,1],E)$ such that $\int_0^1 f dt = 0$. Let J^{-1} be the operator which is the inverse of J over L^2_0. By (5.24), (5.27), (5.28), we get

$$\begin{aligned}
&\int_E \operatorname{Tr_s}^{\widehat{\Lambda}^{\cdot}(E^*)}\left[P_b\left((0,Y),(0,Y)\right)\right]\frac{dY}{(2\pi)^{n/2}} \\
&= b^{2n}\frac{1}{\det{}'^{1/2}\left(1-J^{-2}\left(1/b^4-R^{E,2}\right)\right)} \\
&\quad\cdot E^{Q'_{1/b^4-R^{E,2}}}\left[\exp\left(-\frac{1}{2}\int_0^1\left\langle Y, R^E\left(-\frac{J^{-1}}{b^4}+2J\right)Y\right\rangle dt\right)\right] \\
&\quad\cdot \operatorname{Tr_s}^{\widehat{\Lambda}^{\cdot}(E^*)}\left[\exp\left(-\frac{1}{2b^2}c\left(\widehat{e}_i\right)\widehat{c}\left(e_i\right)-\frac{1}{4}\left\langle R^E e_i, e_j\right\rangle c\left(\widehat{e}_i\right)c\left(\widehat{e}_j\right)\right)\right].
\end{aligned} \tag{5.29}$$

We have the easy identity,

$$\begin{aligned}
&E^{Q'_{1/b^4-R^{E,2}}}\left[\exp\left(-\frac{1}{2}\int_0^1\left\langle Y, R^E\left(-\frac{J^{-1}}{b^4}+2J\right)Y\right\rangle dt\right)\right] \\
&= \det{}'^{1/2}\left(1-J^{-2}\left(1/b^4-R^{E,2}\right)\right) \\
&\quad\cdot \det{}'^{-1/2}\left[\left(1-J^{-1}R^E-J^{-2}/b^4\right)\left(1-J^{-1}R^E\right)\right].
\end{aligned} \tag{5.30}$$

By (5.29), (5.30), we get

$$\begin{aligned}
&\int_E \operatorname{Tr_s}^{\widehat{\Lambda}^{\cdot}(E^*)}\left[P_b\left((0,Y),(0,Y)\right)\right]\frac{dY}{(2\pi)^{n/2}} \\
&= b^{2n}\det{}'^{-1/2}\left[\left(1-J^{-1}R^E-J^{-2}/b^4\right)\left(1-J^{-1}R^E\right)\right] \\
&\quad\cdot \operatorname{Tr_s}^{\widehat{\Lambda}^{\cdot}(E^*)}\left[\exp\left(-\frac{1}{2b^2}c\left(\widehat{e}_i\right)\widehat{c}\left(e_i\right)-\frac{1}{4}\left\langle R^E e_i, e_j\right\rangle c\left(\widehat{e}_i\right)c\left(\widehat{e}_j\right)\right)\right].
\end{aligned} \tag{5.31}$$

By (5.3),

$$\det{}'^{1/2}\left[1-J^{-1}R^E-J^{-2}/b^4\right]=\Phi\left(1/b^4,R^E\right). \tag{5.32}$$

By using (5.6) and making $b=+\infty$ in (5.32), we obtain,

$$\det{}'^{-1/2}\left[1-J^{-1}R^E\right]=\widehat{A}\left(-R^E\right). \tag{5.33}$$

Moreover, by [B90b, Theorem 6.4],

$$\begin{aligned}
\operatorname{Tr_s}^{\widehat{\Lambda}^{\cdot}(E^*)}&\left[\exp\left(-\frac{1}{2b^2}c\left(\widehat{e}_i\right)\widehat{c}\left(e_i\right)-\frac{1}{4}\left\langle R^E e_i, e_j\right\rangle c\left(\widehat{e}_i\right)c\left(\widehat{e}_j\right)\right)\right] \\
&= b^{-2n}\Phi\left(1/b^4,R^E\right).
\end{aligned} \tag{5.34}$$

By combining (5.33) with (5.31)–(5.34), we get (5.22) when $g=1$.

2) *The case where no eigenvalue of g is equal to* 1. Now we assume that no eigenvalue of g is equal to 1. Let $P_{(Y,gY)}$ be the probability law on $C([0,1],E)$ of the Brownian bridge which starts at Y at time 0 and ends at gY at time 1. Let $C_g([0,1],E)$ be the subspace of $C([0,1],E)$ of the $Y_{\cdot}$ such that $Y_1=gY_0$. Let J_g be the skew-adjoint operator $\frac{d}{dt}$ acting on $C^\infty([0,1],E)$ with the boundary conditions $Y_1=gY_0$. Note that J_g is invertible.

If F is taken as before, let $Q_{F,g}$ be the probability law on $C_g([0,1],E)$ of the Gaussian process with covariance $\left(-J_g^2+F\right)^{-1}$. Then by [B90b, proof of Theorem 7.3] and by [BL06, eq. (7.107)], we have the identity,

$$\exp\left(-\frac{1}{2}\int_0^1 \langle FY,Y\rangle\,dt-\frac{1}{2}\left|(1-g)\,Y\right|^2\right)P_{(Y,gY)}\frac{dY}{(2\pi)^{n/2}} = \frac{Q_{F,g}}{\left|\det\left(e^{\sqrt{F}/2}-e^{-\sqrt{F}/2}g\right)\right|}. \tag{5.35}$$

Also a simple identity established in [BL06, eq. (7.113)] shows that

$$\det\left(e^{\sqrt{F}/2}-e^{-\sqrt{F}/2}g\right)=\det(1-g)\det{}^{1/2}\left(1-FJ_g^{-2}\right). \tag{5.36}$$

Moreover, under Q_F, the probability law of the random variable $\int_0^1 Y\,dt$ is a centred Gaussian with variance $\sigma^2>0$. Finally note that under $S_{g,U,Y}$, since $U_1=gU_0$, by (5.23),

$$U_\cdot=-J_g^{-1}Y_\cdot. \tag{5.37}$$

By proceeding as in (5.24), (5.26) and in (5.29) , we get

$$\begin{aligned}
&\int_E \operatorname{Tr_s}^{S^E\widehat{\otimes}\widehat{\Lambda}^\cdot(E^*)}\left[gP_b\left((U,Y),g(U,Y)\right)\right]\frac{dU\,dY}{(2\pi)^n}\\
&=\frac{1}{\det{}^2(1-g)}\frac{1}{\det{}^{1/2}\left(1-J_g^{-2}\left(1/b^4-R^{E,2}\right)\right)}\\
&\quad\cdot E^{Q_{1/b^4-R^{E,2},g}}\left[\exp\left(-\frac{1}{2}\int_0^1\left\langle Y,R^E\left(-\frac{J_g^{-1}}{b^4}+2J_g\right)Y\right\rangle dt\right)\right]\\
&\quad\cdot\operatorname{Tr_s}^{S^E\widehat{\otimes}\widehat{\Lambda}^\cdot(E^*)}\left[g\exp\left(-\frac{1}{2b^2}c(\widehat{e}_i)\widehat{c}(e_i)-\frac{1}{4}\left\langle R^Ee_i,e_j\right\rangle c(\widehat{e}_i)c(\widehat{e}_j)\right)\right].
\end{aligned} \tag{5.38}$$

By proceeding as in (5.30), we get

$$\begin{aligned}
&E^{Q_{1/b^4-R^{E,2},g}}\left[\exp\left(-\frac{1}{2}\int_0^1\left\langle Y,R^E\left(-\frac{J_g^{-1}}{b^4}+2J_g\right)Y\right\rangle dt\right)\right]\\
&=\det{}^{1/2}\left[1-J_g^{-2}\left(1/b^4-R^{E,2}\right)\right]\\
&\quad\cdot\det{}^{-1/2}\left[\left(1-J_g^{-1}R^E-J_g^{-2}/b^4\right)\left(1-J_g^{-1}R^E\right)\right].
\end{aligned} \tag{5.39}$$

By (5.38), (5.39), we obtain

$$\begin{aligned}
&\int_E \operatorname{Tr_s}^{S^E\widehat{\otimes}\widehat{\Lambda}^\cdot(E^*)}\left[gP_b\left((U,Y),g(U,Y)\right)\right]\frac{dU\,dY}{(2\pi)^n}\\
&=\frac{1}{\det{}^2(1-g)}\det{}^{-1/2}\left[\left(1-J_g^{-1}R^E-J_g^{-2}/b^4\right)\left(1-J_g^{-1}R^E\right)\right]\\
&\quad\cdot\operatorname{Tr_s}^{S^E\widehat{\otimes}\widehat{\Lambda}^\cdot(E^*)}\left[g\exp\left(-\frac{1}{2b^2}c(\widehat{e}_i)\widehat{c}(e_i)-\frac{1}{4}\left\langle R^Ee_i,e_j\right\rangle c(\widehat{e}_i)c(\widehat{e}_j)\right)\right].
\end{aligned} \tag{5.40}$$

By [B94, Proposition 4.2],

$$\det{}^{1/2}\left[1 - J_g^{-1}R^E - J_g^{-2}/b^4\right] = \frac{\Sigma_g\left(1/b^4, R^E\right)}{\det\left(1-g\right)}. \tag{5.41}$$

Moreover, by [B94, Theorem 4.5], we get

$$\mathrm{Tr_s}^{\widehat{\Lambda}^{\cdot}(E^*)}\left[g\exp\left(-\frac{1}{2b^2}c\left(\widehat{e}_i\right)\widehat{c}\left(e_i\right) - \frac{1}{4}\left\langle R^E e_i, e_j\right\rangle c\left(\widehat{e}_i\right)c\left(\widehat{e}_j\right)\right)\right] = \Sigma_g\left(1/b^4, R^E\right). \tag{5.42}$$

By [B94, eq. (5.35)],

$$\det{}^{-1/2}\left[1 - J_g^{-1}R^E\right] = \frac{\widehat{A}_g\left(-R^E\right)}{\widehat{A}_g\left(0\right)}. \tag{5.43}$$

Let us just mention here that if $\pm i\theta_j, 0 < \theta_j \leq \pi, 1 \leq j \leq n/2$ are the angles of the action of g on E, then

$$\widehat{A}_g\left(0\right) = \prod_1^{n/2}\left(2i\sin\left(\theta_j/2\right)\right)^{-1}. \tag{5.44}$$

Moreover,

$$\mathrm{Tr_s}^{S^E}\left[g\right] = \pm\overline{\widehat{A}}_g^{-1}\left(0\right). \tag{5.45}$$

By (5.40)–(5.45), we get (5.22).

The proof of our theorem is completed. □

5.4. The case of holomorphic vector bundles

Now we assume that S is a complex manifold, and that E is a complex holomorphic vector bundle on S, equipped with a Hermitian metric g^E. Let ∇^E be the holomorphic Hermitian connection on $\left(E, g^E\right)$, and let R^E be its curvature.

We can then apply the prescriptions of Section 3.1 to the total space $\mathbf{E}$ of E, which fibres on S with fibre E. The total space $\mathcal{E}$ of TE is the total space of $E \oplus E$. Let (u, y) be the generic section of $E \oplus E$.

Let $\omega^{\mathbf{E}}$ be the closed $(1,1)$ form on $\mathbf{E}$ which is the obvious analogue of the form $\omega^{\mathcal{M}}$ in (3.96), so that

$$\omega^{\mathbf{E}} = i\overline{\partial}\partial\left|u\right|^2. \tag{5.46}$$

We will denote by $\overline{\partial}^E, \overline{\partial}^{\widehat{E}}$ the $\overline{\partial}$ operators along the fibres of the first and second copies of $E \oplus E$, let $\overline{\partial}^{\widehat{E}*}$ be the fibrewise adjoint of $\overline{\partial}^{\widehat{E}}$. Then

$$\begin{aligned}
\mathcal{A}''_{Y_b} &= \nabla^{\cdot\,\prime\prime} + \overline{\partial}^E + i_{R^E u} + \overline{\partial}^{\widehat{E}} + i_y/b^2,\\
\mathcal{A}'_{Y_b} &= e^{i\omega^{\mathbf{E}}}\left(\nabla^{\mathbf{I}^{\cdot}\prime} + \partial^E + i_{R^E\overline{u}} + \overline{\partial}^{\widehat{E}*} + i_{\overline{y}}/b^2\right)e^{-i\omega^{\mathbf{E}}},\\
\mathcal{A}_{Y_b} &= \mathcal{A}''_{Y_b} + \mathcal{A}'_{Y_b}.
\end{aligned} \tag{5.47}$$

By (3.97) and (5.47), we obtain,

$$\mathcal{A}'_{Y_b} = \nabla^{\mathbf{I}^{\cdot}\prime} + \partial^E + i_{R^E\overline{u}} + +\overline{\partial}^{\widehat{E}*} + i_{\overline{y}}/b^2 + \overline{y}^* \wedge /b^2. \tag{5.48}$$

Let Δ^V be the Laplacian along the second copy of E. By (5.13), we get

$$\mathcal{A}^2_{Y_b} = -\frac{1}{2}\left(-\Delta^V + \frac{|Y|^2}{b^4} + \frac{1}{b^2}c\left(\widehat{e}_i\right)\widehat{c}\left(\widehat{e}_i\right)\right) - \nabla_{\widehat{R^E Y}} + \frac{1}{4}\left\langle R^E e_i, e_j\right\rangle c\left(\widehat{e}_i\right)c\left(\widehat{e}_j\right)$$
$$+\frac{1}{2}\mathrm{Tr}\left[R^E\right] + \frac{1}{b^2}\left(\nabla_Y + \frac{1}{2}\left\langle R^E U, Y\right\rangle\right). \tag{5.49}$$

5.5. An evaluation of supertraces in the holomorphic case

We make the same assumptions as in Subsection 5.4. Also we assume that g is a holomorphic unitary parallel section of $\mathrm{End}(E)$. Let $B \in \mathrm{End}(E)$ be a skew-adjoint locally parallel section of $\mathrm{End}(E)$ such that

$$g = e^B. \tag{5.50}$$

We will now write $\widehat{\Sigma}_g\left(E, g^E\right)$ instead of $\widehat{\Sigma}_g\left(E, \nabla^E\right)$. By (5.9), we get

$$\widehat{\Sigma}_g\left(E, g^E\right) = \det{}^E\left[\sigma\left(u, B, \frac{R^E}{2i\pi}\right)\right]. \tag{5.51}$$

Let N^V be the number operator on $\widehat{\Lambda}^{\cdot}\left(E^*\right)$.

Theorem 5.2. *The following identity holds,*

$$\varphi\int_{E^{1,\perp}\times E}\mathrm{Tr_s}^{\Lambda^{\cdot}\left(\overline{E}^{1,\perp}\right)\widehat{\otimes}\widehat{\Lambda}^{\cdot}\left(E^*_{\mathbf{R}}\otimes_{\mathbf{R}}\mathbf{C}\right)}\left[gP_b\left(g^{-1}\left(U,Y\right),\left(U,Y\right)\right)\right]\frac{dUdY}{\left(2\pi\right)^{2n-n_1}}$$
$$= \mathrm{Td}_g\left(E, g^E\right). \tag{5.52}$$

Moreover,

$$\varphi\int_{E^{1,\perp}\times E}\mathrm{Tr_s}^{\Lambda^{\cdot}\left(\overline{E}^{1,\perp}\right)\widehat{\otimes}\widehat{\Lambda}^{\cdot}\left(E^*_{\mathbf{R}}\otimes_{\mathbf{R}}\mathbf{C}\right)}\left[g\left(N^V - n\right)P_b\left(g^{-1}\left(U,Y\right),\left(U,Y\right)\right)\right]\frac{dUdY}{\left(2\pi\right)^{2n-n_1}}$$
$$= \mathrm{Td}_g\left(E, g^E\right)\left(\frac{\widehat{\Sigma}'_g}{\widehat{\Sigma}_g}\left(1/b^4, E, g^E\right) - \frac{n}{2}\right). \tag{5.53}$$

Proof. Equation (5.52) follows from Theorem 5.1. Also note that

$$N^V - n = -\frac{i}{4}\left\langle J^E e_i, e_j\right\rangle c\left(\widehat{e}_i\right)c\left(\widehat{e}_j\right) - \frac{n}{2}. \tag{5.54}$$

First we consider the case where $g = 1$. Using (5.34) and (5.54), we get

$$\mathrm{Tr_s}^{\widehat{\Lambda}^{\cdot}\left(E^*_{\mathbf{R}}\right)}\left[\left(N^V - n\right)\exp\left(-\frac{1}{2b^2}c\left(\widehat{e}_i\right)\widehat{c}\left(e_i\right) - \frac{1}{4}\left\langle R^E e_i, e_j\right\rangle c\left(\widehat{e}_i\right)c\left(\widehat{e}_j\right)\right)\right]$$
$$= b^{-4n}\left(-\Phi'\left(1/b^4, R^E\right) - \frac{n}{2}\Phi\left(1/b^4, R^E\right)\right). \tag{5.55}$$

By (5.29)–(5.34) and (5.55), we get (5.53).

Now we consider the case where no eigenvalue of g is equal to 1. By (5.42), (5.54), we get

$$\mathrm{Tr_s}^{\widehat{\Lambda}^{\cdot}(E^*_{\mathbf{R}})}\left[g\left(N^V-n\right)\exp\left(-\frac{1}{2b^2}c\left(\widehat{e}_i\right)\widehat{c}\left(e_i\right)-\frac{1}{4}\left\langle R^E e_i,e_j\right\rangle c\left(\widehat{e}_i\right)c\left(\widehat{e}_j\right)\right)\right]$$
$$=-\Sigma'_g\left(1/b^4,R^E\right)-\frac{n}{2}\Sigma_g\left(1/b^4,R^E\right). \quad (5.56)$$

Moreover,

$$\mathrm{Tr_s}^{\Lambda^{\cdot}\left(\overline{E}^*\right)}=\det\left(1-g\right). \quad (5.57)$$

By (5.40)–(5.42), (5.56), (5.57), we also get (5.53) in the case where no eigenvalue of g is equal to 1. The proof of our theorem is completed. □

Remark 5.3. We see easily that when $u\to 0$,

$$\mathrm{Td}_g\left(E,g^E\right)\left(\frac{\widehat{\Sigma}'_g}{\widehat{\Sigma}_g}\left(u,E,g^E\right)-\frac{n}{2}\right)=-\mathrm{Td}'_g\left(E,g^E\right)+\mathcal{O}\left(u\right). \quad (5.58)$$

Equation (5.58) gives the asymptotics of the form in (5.53) as $b\to\infty$.

5.6. The genus $R\left(\theta,x\right)$

Let

$$L\left(\theta,s\right)=\sum_{n=1}^{+\infty}\frac{e^{in\theta}}{n^s} \quad (5.59)$$

be the Lerch series. Let $\zeta\left(\theta,s\right)$ and $\eta\left(\theta,s\right)$ be its real and imaginary parts, so that

$$\zeta\left(\theta,s\right)=\sum_{n=1}^{+\infty}\frac{\cos\left(n\theta\right)}{n^s},\qquad \eta\left(\theta,s\right)=\sum_{n=1}^{+\infty}\frac{\sin\left(n\theta\right)}{n^s}. \quad (5.60)$$

We will introduce the genera $D\left(\theta,x\right),R\left(\theta,x\right)$ of [B94].

Definition 5.4. For $\theta\in\mathbf{R}/2\pi\mathbf{Z}$, let $R\left(\theta,x\right)$ be the formal power series

$$D(\theta,x)=\sum_{\substack{n\geq 0\\ n\,\mathrm{odd}}}\left\{\left(\Gamma'\left(1\right)+\sum_{j=1}^{n}\frac{1}{j}\right)\zeta(\theta,-n)+2\frac{\partial\zeta}{\partial s}(\theta,-n)\right\}\frac{x^n}{n!}$$
$$+\sum_{\substack{n\geq 0\\ n\,\mathrm{even}}}i\left\{\left((\Gamma'\left(1\right)+\sum_{j=1}^{n}\frac{1}{j}\right)\eta(\theta,-n)+2\frac{\partial\eta}{\partial s}(\theta,-n)\right\}\frac{x^n}{n!}, \quad (5.61)$$

$$R(\theta,x)=\sum_{\substack{n\geq 0\\ n\,\mathrm{odd}}}\left\{\sum_{j=1}^{n}\frac{1}{j}\zeta(\theta,-n)+2\frac{\partial\zeta}{\partial s}(\theta,-n)\right\}\frac{x^n}{n!}$$
$$+\sum_{\substack{n\geq 0\\ n\,\mathrm{even}}}i\left\{\sum_{j=1}^{n}\frac{1}{j}\eta(\theta,-n)+2\frac{\partial\eta}{\partial s}(\theta,-n)\right\}\frac{x^n}{n!}.$$

Set

$$\widehat{A}(\theta,x)=\widehat{A}(x)\text{ if }\theta\in 2\pi\mathbf{Z}, \qquad (5.62)$$
$$=\frac{\widehat{A}(x+i\theta)}{x+i\theta}\text{ if }\theta\notin 2\pi\mathbf{Z}.$$

Note that as $u\to 0$,

$$\frac{\partial\sigma/\partial x}{\sigma}(u,i\theta,-x)=\frac{\partial\widehat{A}/\partial x}{\widehat{A}}(\theta,x)+\mathcal{O}(u), \qquad (5.63)$$

and that as $u\to+\infty$,

$$\frac{\partial\sigma/\partial x}{\sigma}(u,i\theta,-x)=\mathcal{O}\left(1/\sqrt{u}\right). \qquad (5.64)$$

Now we proceed as in [B94, Section 6].

Definition 5.5. For $s\in\mathbf{C}, 0<\operatorname{Re}(s)<1/2, \theta\in\mathbf{R}, x\in\mathbf{C}$ and $|x|<2\pi$ if $\theta\in 2\pi\mathbf{Z}, |x|<\inf_{k\in\mathbf{Z}}|\theta+2k\pi|$ if $\theta\notin 2\pi\mathbf{Z}$, set

$$C(s,\theta,x)=\frac{1}{\Gamma(s)}\int_0^{+\infty}u^{s-1}\frac{\partial\sigma/\partial x}{\sigma}(u,i\theta,-x)\,du. \qquad (5.65)$$

By simple arguments given in [B94], $C(s,\theta,x)$ extends to a holomorphic function near $s=0$. By (5.63), (5.64), we get

$$\frac{\partial}{\partial s}C(0,\theta,x)=\int_0^1\left(\frac{\partial\sigma/\partial x}{\sigma}(u,i\theta,-x)-\frac{\partial\widehat{A}/\partial x}{\widehat{A}}(\theta,x)\right)\frac{du}{u}$$
$$+\int_1^{+\infty}\frac{\partial\sigma/\partial x}{\sigma}(u,i\theta,-x)\frac{du}{u}-\Gamma'(1)\frac{\partial\widehat{A}/\partial x}{\widehat{A}}(\theta,x). \qquad (5.66)$$

We state the result established in [B94, Theorems 7.2 and 7.8].

Theorem 5.6. *The following identity holds,*

$$\frac{\partial}{\partial s}C(0,\theta,x)=D(\theta,x). \qquad (5.67)$$

Moreover,

$$R(\theta,x)=D(\theta,x)-\Gamma'(1)\frac{\partial\widehat{A}/\partial x}{\widehat{A}}(\theta,x). \qquad (5.68)$$

Put

$$R(x)=R(0,x). \qquad (5.69)$$

The series $R(x)$ and the corresponding additive genus were introduced by Gillet and Soulé [GS91] when stating a conjectural arithmetic Riemann-Roch theorem. It reappears in the immersion formula for Quillen metrics established in [BL91], which was one important step in the proof by Gillet and Soulé of their Riemann-Roch theorem. The series $R(\theta,x)$ was introduced in [B94], and its role in an equivariant situation was made explicit in [B95]. The fact that it reappears in Sections 8 and 10 of the present paper is remarkable.

Let E be a complex holomorphic Hermitian vector bundle equipped with a unitary automorphism g as in Subsection 5.5. Then E splits as

$$E = \bigoplus_{\theta \in \mathbf{R}/2\pi\mathbf{Z}} E^{\theta}, \tag{5.70}$$

so that g acts on E^{θ} by multiplication by $e^{i\theta}$. Set

$$R_g(E) = \sum_{\theta \in \mathbf{R}/2\pi\mathbf{Z}} R\left(\theta, E^{\theta}\right). \tag{5.71}$$

Similarly we identify R to the corresponding additive genus.

6. The local index theory of the hypoelliptic Dirac operator

In this section, we establish various local index theoretic results for the hypoelliptic Dirac operator. One of the key results which is established in this section is that this operator has the same local index theory as the elliptic Dirac operator. This remarkable property will play a crucial role in the sequel. This section should be considered as an analogue of [BL06, Sections 4 and 11], where corresponding results are established for the hypoelliptic de Rham operator.

This section is organized as follows. In Subsection 6.1, we give various properties of the hypoelliptic heat kernel.

In Subsection 6.2, we recall a result given in [BL06] on the curvature of the fibrewise connection ${}^{1}\nabla_{1/4}^{\Lambda^{\cdot}(T^*S)\widehat{\otimes}F}$.

In Subsection 6.3, we construct a superconnection $\mathbb{A}_{Y,b}^{t}$ depending on the parameters $b > 0, t > 0$.

In Subsection 6.4, given $g \in G$, we state a local index theorem as $t \to 0$ for the supertraces of the heat kernel associated to $\mathbb{A}_{Y,b}^{t,2}$.

In Subsection 6.5, we show that the estimates which are needed to prove our local index theorem can be localized near M_g.

In Subsection 6.6, we prove that given $x \in X_g$, we can locally replace $\mathcal{X}$ by $T_xX \oplus T_xX$.

In Subsection 6.7, we rescale our local coordinates on X, and also we introduce a Getzler rescaling on our Clifford variables. This way, we obtain a rescaled operator $\mathfrak{P}_{b,t}$.

In Subsection 6.8, we evaluate the limit $\mathfrak{P}_{b,0}$ of $\mathfrak{P}_{b,t}$ as $t \to 0$.

In Subsection 6.9, we obtain the local trace for the heat kernel associated to the scalar part of $\mathfrak{P}_{b,0}$.

In Subsection 6.10, we compute simple finite-dimensional supertraces.

In Subsection 6.11, we evaluate the local supertrace of the heat kernel associated to $\mathfrak{P}_{b,0}$.

In Subsection 6.12, we establish our local index theorem.

Finally in Subsection 6.13, we establish local index formulas when rescaling b into $\sqrt{t}b$.

In the whole section, we make the same assumptions as in Section 2.

6.1. The hypoelliptic heat kernel

For $b > 0$, we consider the operators $\mathcal{E}^2_{Y,b}$, which have been explicitly computed in (1.69). By Theorem 1.9, these operators are formally self-adjoint with respect to the Hermitian form η in (1.50). Also by Theorem 1.11, $\frac{\partial}{\partial u} - \mathcal{E}^2_{Y,b}$ is hypoelliptic.

Let $\mathcal{S}(\mathcal{X}, \pi^* F)$ be the Schwartz space of sections of $\pi^* F$ over $\mathcal{X}$. Inspection of equation (1.69) shows that the operators $\mathcal{E}^2_{Y,b}$ are essentially of the same form as the operators which were considered by Bismut-Lebeau in [BL06].

This makes in particular that all spectral and analytic properties which were established in [BL06] remain valid in the present context. Some of these properties were summarized in Subsection 3.10, with the possible exception that the picture of the spectrum of $D^2_{Y_b}$ as $b \to 0$ remains the one given in Figure 3.2, and not the one in Figure 3.3, simply because in general, there is no cohomological interpretation for the characteristic subspace associated with the 0 eigenvalue for $D^{X,2}$ or $\mathcal{D}^2_{Y_b}$.

Besides the properties already given in Subsection 3.10, the following facts are known:

- Given $b > 0$, there exist $\lambda_0 > 0, c_0 > 0$ such that if
$$\gamma = \left\{ \lambda = -\lambda_0 + \sigma + i\tau, \sigma, \tau \in \mathbf{R}, \sigma = c_0 \left|\tau\right|^{1/6} \right\}, \tag{6.1}$$
if $\delta_\pm \subset \mathbf{C}$ is the closed domain to the right or to the left of γ, then the resolvent set of any of the above operators contains δ_-, and moreover, if $\lambda \in \delta_-$, there are uniform bounds on the standard L^2 norm of the corresponding resolvent.
- For $t > 0$, the heat operator $\exp\left(-t\mathcal{E}^2_{Y,b}\right)$ is trace class, and given by smooth kernels over $\mathcal{X}$, which lie in the obvious Schwartz space over $\mathcal{X} \times \mathcal{X}$.
- As $b \to 0$, the resolvents and the heat kernels for $\mathcal{E}^2_{Y,b}$ converge in the proper sense to the resolvent and heat kernel for $D^{X,2}$. The precise meaning of this sentence is explained in detail in [BL06, Sections 3 and 17].

The same arguments as in [BL06] allow us to conclude that under the assumptions of Section 2, similar results hold for the hypoelliptic curvatures $\mathcal{B}^2_{Y,b}$ over compact subsets of S. In particular these operators converge in the proper sense to B^2 as $b \to 0$, and this uniformly over compact subsets of S.

6.2. The curvature of ${}^1\nabla^{\Lambda^\cdot(T^*S)\widehat{\otimes}F}$

We introduce another copy of $\Lambda^\cdot(T^*X)$, which will be denoted $\underline{\Lambda}^\cdot(T^*X)$. This exterior algebra is viewed specifically as the exterior algebra along the fibre X.

Put

$$\begin{aligned}
\underline{T}^H &= \frac{1}{2}\left\langle T\left(f_\alpha^H, f_\beta^H\right), e_i\right\rangle \underline{e}^i \wedge f^\alpha \wedge f^\beta, \\
\underline{T}^0 &= f^\alpha \wedge \underline{e}^i \wedge e^j \left\langle T\left(f_\alpha^H, e_i\right), e_j\right\rangle, \\
\left|\underline{T}^0\right|^2 &= \sum_{j=1}^n \left(\sum_{\substack{1\le i\le n \\ 1\le\alpha\le m}} \left\langle T\left(f_\alpha^H, e_i\right), e_j\right\rangle f^\alpha \wedge \underline{e}^i \right)^2 .
\end{aligned} \tag{6.2}$$

Recall that for $t>0$, the fibrewise connection ${}^1\nabla_{t,\cdot}^{\Lambda^\cdot(T^*S)\widehat{\otimes} S^{TX}\widehat{\otimes} E}$ was defined in (2.10). We will denote it temporarily as ${}^1\underline{\nabla}_t^{\Lambda^\cdot(T^*S)\widehat{\otimes} F}$ instead of ${}^1\nabla_t^{\Lambda^\cdot(T^*S)\widehat{\otimes} F}$ to emphasize it is a fibrewise connection. Its curvature ${}^1\underline{\nabla}_t^{\Lambda^\cdot(T^*S)\widehat{\otimes} F,2}$ is a section of $\underline{\Lambda}^2(T^*X)\widehat{\otimes}\Lambda^\cdot(T^*S)\widehat{\otimes}\mathrm{End}(F)$. When $t=1$, we will write instead ${}^1\underline{\nabla}^{\Lambda^\cdot(T^*S)\widehat{\otimes} F}$. A straightforward computation, which is done in [BL06, eq. (11.16)], shows that

$$\begin{aligned}
{}^1\underline{\nabla}_{1/4}^{\Lambda^\cdot(T^*S)\widehat{\otimes} F,2} = {}&- \left\langle \underline{R}^{TX} e_i, e_j\right\rangle \left(e^i i_{e_j} + \frac{1}{4} c\left(\widehat{e}_i\right) c\left(\widehat{e}_j\right)\right) + \underline{\nabla}^{TX} \underline{T}^H \\
&- \underline{\nabla}^{TX} \underline{T}^0 \left(f_\alpha^H, e_i\right) f^\alpha \left(e^i - i_{e_i}\right) - \left|\underline{T}^0\right|^2 + \underline{R}^E .
\end{aligned} \tag{6.3}$$

In (6.3), $\underline{R}^E$ is just the fibrewise curvature of $\overline{\nabla}^E$.

6.3. The parameters $b>0, t>0$

We replace S by $S\times\mathbf{R}_+^*\times\mathbf{R}_+^*$. Over (b,t), we equip TX with the metric g^{TX}/t and we replace in the previous constructions b by $b/\sqrt{t}$. Let $\mathcal{A}_{Y_{b/\sqrt{t}}}^t$ be the superconnection associated to the metric g^{TX}/t and to the section $Y_{b/\sqrt{t}}$.

Put

$$\mathbb{A}_{Y,b}^t = K_{\sqrt{t}} \mathcal{A}_{Y_{b/\sqrt{t}}}^t K_{1/\sqrt{t}} . \tag{6.4}$$

Then

$$\mathbb{A}_{Y,b}^t = \mathcal{A} + \frac{\sqrt{t}}{2b^2} Y^* \wedge + \frac{t^{3/2}}{b^2} i_Y . \tag{6.5}$$

The main advantage of the superconnection in (6.5) is that the dependence on b,t has been concentrated just on the terms where $Y^*\wedge, i_Y$ appear.

Set

$$\overline{\mathbb{A}} = \mathbb{A}_{Y,b}^t + db\frac{\partial}{\partial b} + dt\frac{\partial}{\partial t} . \tag{6.6}$$

6.4. The hypoelliptic local index Theorem

Let $g\in G$. Let $M_g\subset M$ be the fixed set of g. Then M_g is a submanifold of M, which fibres of S with compact fibre $X_g\subset X$. Moreover, $T^HM|_{M_g}\subset TM_g$ is a horizontal vector bundle T^HM_g on M_g. Finally the connection ∇^{TX} induces a connection $\nabla^{TX|_{M_g}}$ on $TX|_{M_g}$. The action of g on $TX|_{M_g}$ is parallel with respect to $\nabla^{TX|_{M_g}}$. Also its restriction ∇^{TX_g} to TX_g is the connection canonically attached to the given metric on TX_g and to T^HM_g.

We have the identity,

$$\varphi \mathrm{Tr_s}\left[g\exp\left(-\overline{\mathbb{A}}^2\right)\right] = m_{b,t} + n_{b,t}\wedge dt + o_{b,t}\wedge db + p_{b,t}\wedge db\wedge dt. \tag{6.7}$$

In (6.7), $m_{b,t}, n_{b,t}, o_{b,t}, p_{b,t}$ are smooth forms on S.

We prove a hypoelliptic version of the local families index theorem of [B86].

Theorem 6.1. *The even form $\varphi \mathrm{Tr_s}\left[g\exp\left(-\mathbb{A}^2\right)\right]$ is closed on $S\times \mathbf{R}_+^*\times\mathbf{R}_+^*$. The even forms $m_{b,t}$ are closed on S, and their cohomology class does not depend on b,t. For $g=1$, these are real forms. As $t\to 0$,*

$$m_{b,t} = \int_{X_g} \pm \widehat{A}_g\left(TX|_{X_g}, \nabla^{TX|_{X_g}}\right)\mathrm{ch}_g\left(E,\nabla^E\right) + \mathcal{O}(t), \tag{6.8}$$
$$n_{b,t} = \mathcal{O}(1), \qquad o_{b,t} = \mathcal{O}(t).$$

Proof. The first part of the theorem is standard. The fact that for $g=1$ the considered forms are real is a consequence of Theorem 2.5 and of the arguments given in the proof of [BF86a, Theorem 1.5]. Subsections 6.5-6.12 will be devoted to the proof of Theorem 6.1. □

6.5. Localization of the problem

From now on, we will assume S to be compact, to ensure the proper uniformity of the estimates. If S is non compact, the estimates below will be valid uniformly over compact subsets of S.

Put

$$\mathfrak{M}_{b,t} = K_b\overline{\mathbb{A}}^2 K_{1/b}. \tag{6.9}$$

For $t>0$, set

$$c_t(U) = U^*\wedge/\sqrt{2t} - \sqrt{2t}i_U, \qquad \widehat{c}_t(U) = U^*\wedge/\sqrt{2t} + \sqrt{2t}i_U. \tag{6.10}$$

We denote by ${}^2\underline{\nabla}_{t,\cdot}^{\Lambda^\cdot(T^*S)\widehat{\otimes}F}$ the fibrewise connection ${}^1\underline{\nabla}_{t,\cdot}^{\Lambda^\cdot(T^*S)\widehat{\otimes}F}$ in (2.10), in which the $c(e_i)$ have been replaced by $c_t(e_i)$. This is in fact just the connection ${}^1\underline{\nabla}_{\cdot}^{\Lambda^\cdot(T^*S)\widehat{\otimes}F}$ associated to the metric g^{TX}/t.

By (2.30), we get

$$\begin{aligned}\mathfrak{M}_{b,t} = \frac{1}{2b^2}&\left(-\Delta^V + t^2|Y|^2 + tc(\widehat{e}_i)\widehat{c}_t(e_i)\right) - \nabla_{\widehat{R^{TX}Y}}\\ &+\frac{1}{4}\left\langle R^{TX}e_i,e_j\right\rangle c(\widehat{e}_i)c(\widehat{e}_j) + \frac{t^{3/2}}{b}\,{}^2\underline{\nabla}_{t/2,Y}^{\Lambda^\cdot(T^*S)\widehat{\otimes}F} + R^E\\ &-2\frac{db}{b^2}\left(\frac{\sqrt{t}}{2}Y^*\wedge + t^{3/2}i_Y\right) + \frac{dt}{2b}\left(\frac{Y^*}{2\sqrt{t}}\wedge + 3\sqrt{t}i_Y\right).\end{aligned} \tag{6.11}$$

Clearly,

$$\mathrm{Tr_s}\left[g\exp\left(-\overline{\mathbb{A}}^2\right)\right] = \mathrm{Tr_s}\left[g\exp\left(-\mathfrak{M}_{b,t}\right)\right]. \tag{6.12}$$

Let dx be the volume element of X, let dY be the volume element along the fibre TX. Let $\exp(-\mathfrak{M}_{b,t})((x,Y),(x',Y'))$ be the smooth heat kernel associated to the operator $\exp(-\mathfrak{M}_{b,t})$ with respect to $dxdY/(2\pi)^n$.

Clearly,

$$\mathrm{Tr_s}\left[g\exp\left(-\mathfrak{M}_{b,t}\right)\right] = \int_{\mathcal{X}} \mathrm{Tr_s}\left[g\exp\left(-\mathfrak{M}_{b,t}\right)\left(g^{-1}\left(x,Y\right),\left(x,Y\right)\right)\right] dxdY/\left(2\pi\right)^n . \tag{6.13}$$

Let d_X be the Riemannian distance along the fibres X with respect to g^{TX}. Let a_X be a lower bound for the injectivity radius of the fibres X.

Let $N_{X_g/X}$ be the orthogonal bundle to TX_g in $TX|_{X_g}$. We identify X_g to the zero section of $N_{X_g/X}$.

Given $\eta > 0$, let $\mathcal{V}_\eta$ be the η-neighbourhood of X_g in $N_{X_g/X}$. Then there exists $\eta_0 \in]0, a_X/32]$ such that if $\eta \in]0, 8\eta_0]$, the map $(x, Z) \in N_{X_g/X} \to \exp_x^X(Z) \in X$ is a diffeomorphism from $\mathcal{V}_\eta$ on the tubular neighbourhood $\mathcal{U}_\eta$ of X_g in X. In the sequel, we identify $\mathcal{V}_\eta$ and $\mathcal{U}_\eta$. This identification is g-equivariant. Let $\alpha \in]0, \eta_0]$ be small enough so that if $d_X\left(g^{-1}x, x\right) \le \alpha$, then $x \in \mathcal{U}_{\eta_0}$.

Let dv_{X_g} be the volume element on X_g, and let $dv_{N_{X_g/X}}$ be the volume element along the fibres of $N_{X_g/X}$. Let $k\left(x,y\right), x \in X_g, y \in N_{X_g/X,x}, |y| \le \eta_0$ be the smooth function with values in $\mathbf{R}_+$ such that on $\mathcal{U}_{\eta_0}$,

$$dx = k\left(x,y\right) dv_{N_{X_g/X}}\left(y\right) dv_{X_g}\left(x\right). \tag{6.14}$$

Note that

$$k\left(x,0\right) = 1. \tag{6.15}$$

Using (6.11) and comparing with the corresponding equation [BL06, eq. (4.47)], by proceeding as in [BL06, Subsection 4.7], we find that for any $\beta > 0$, there exist $C > 0, c > 0$ such that for $t \in]0,1]$, the part of the integral in (6.13) which is integrated over $\pi^{-1}\left(X \setminus \mathcal{U}_\beta\right)$ can be dominated by $C\exp\left(-c/t\right)$.

6.6. Replacing $\mathcal{X}$ by $T_xX \oplus T_xX$

Let $\gamma\left(s\right) : \mathbf{R} \to [0,1]$ be a smooth even function such that

$$\begin{aligned}\gamma\left(s\right) =& 1 \text{ if } |s| \le 1/2,\\ & 0 \text{ if } |s| \ge 1.\end{aligned} \tag{6.16}$$

If $y \in TX$, set

$$\rho\left(y\right) = \gamma\left(\frac{|y|}{4\eta_0}\right). \tag{6.17}$$

Then

$$\begin{aligned}\rho\left(y\right) =& 1 \text{ if } |y| \le 2\eta_0,\\ & 0 \text{ if } |y| \ge 4\eta_0.\end{aligned} \tag{6.18}$$

First we describe the case where S is reduced to one point, that is the case of a single fibre X.

Take $\epsilon \in]0, a_X/2]$. If $s \in S, x \in X_g$, let $B^X\left(x,\epsilon\right)$ be the geodesic ball of centre x and radius ϵ in X, and let $B^{T_xX}\left(0,\epsilon\right)$ be the open ball of centre 0 and radius ϵ in T_xX. The exponential map $\exp_x$ identifies $B^{T_xX}\left(0,\epsilon\right)$ to $B^X\left(x,\epsilon\right)$.

We fix ϵ such that $0 < \epsilon < a_X/2$. Take $x \in X_g$. We use the geodesic coordinates $y \in T_xX \in B^{TX}(0,\epsilon) \to \exp_x(y) \in \mathcal{U}_\epsilon$ to parametrize a neighbourhood of x in X. Also over $\mathcal{U}_\epsilon$, we trivialize the vector bundle TX along the geodesics based at x using the fibrewise Levi-Civita connection. This way we have identified $\pi^{-1}\mathcal{U}_\epsilon \subset \mathcal{X}$ to $B^{TX}(0,\epsilon) \times T_xX$.

Also over $\mathcal{U}_\epsilon$, we trivialize the vector bundle F along the geodesics based at x using parallel transport with respect to the connection ∇^F.

Definition 6.2. Let $g_x^{T_xX}$ be the metric on T_xX given by,

$$g_x^{T_xX} = \rho^2(y)\, g^{T_yX} + \left(1-\rho^2(y)\right) g^{T_xX}. \tag{6.19}$$

In particular the metric g_x^{TX} is just the given metric g^{TX} on $B^{T_xX}(0,2\eta_0) \simeq B^X(x,2\eta_0)$, and coincides with the flat metric g^{T_xX} outside of $B^{T_xX}(0,4\eta_0)$. Note that the above constructions are g-invariant.

Similarly let $g_x^{E_x}$ be the metric on E_x over T_xX which is given by

$$g_x^{E_x} = \rho^2(y)\, g^{E_y} + \left(1-\rho^2(y)\right) g^{E_x}. \tag{6.20}$$

Let $\mathfrak{N}_{b,t}$ be the operator of the type $\mathfrak{M}_{b,t}$ which is associated to the metrics $g_x^{T_xX}, g_x^{F_x}$. Clearly $\mathfrak{N}_{b,t}$ coincides with $\mathfrak{M}_{b,t}$ for $|y| \le 2\eta_0$.

Now we consider the case where S is not necessarily reduced to one point. We proceed as in [BL06, Subsection 4.8]. Near $x \in M_g$, there is a coordinate system which identifies an open neighbourhood $\mathcal{V}$ of x in M_g to an open ball centred at 0 in $\mathbf{R}^m \times \mathbf{R}^\ell$, such that the projection $\pi_g : M_g \to S$ coincides with the obvious projection $\mathbf{R}^m \times \mathbf{R}^\ell \to \mathbf{R}^m$. Also the vector bundle $TX|_{M_g}$ can be trivialized as an Euclidean vector bundle over $\mathcal{V}$, in such a way that the action of g on $TX|_{M_g}$ is constant. In particular the vector bundle TX is trivialized near 0 on $\mathbf{R}^m \times \{0\}$.

If $x' \in \mathcal{V}$, we use the exponential map $\exp_{x'}^X$ to identify the ball $B^{T_{x'}X}(0,4\eta_0)$ to an open ball along the fibre containing x'. The map $(s,y) \in \mathbf{R}^m \times T_xX \to \exp_s^X y \in M$ gives a coordinate chart for M near $x \in M_g$, such that the projection $\pi : M \to S$ is just $(s,y) \to s$, and moreover, $g(s,y) = (s, gY)$. From this chart, we find that the metric g^{TX} pulls back to a metric on T_xX, i.e., the metric $g^{T_{s,y}X}$ pulls back to a metric on T_xX, which we still denote $g^{T_{s,y}X}$.

We still define the metric $g_x^{T_xX}$ on T_xX as in (6.19). In particular for $|y| \ge 4\eta_0$, this new metric is a 'constant metric', which does not depend on (s,y). Note that the metric $g_x^{T_xX}$ is g-invariant.

Similarly we can choose a new g-invariant horizontal vector bundle T^HM_x which coincides with the given T^HM for $|y| \le 2\eta_0$, and is given by a 'constant' horizontal vector space for $|y| \ge 4\eta_0$.

We will trivialize $\Lambda^{\cdot}(T_{\mathbf{R}}^*S)\widehat{\otimes}F$ over $\mathcal{U}_\epsilon$ by parallel transport with respect to the connection ${}^2\underline{\nabla}_{t,\cdot}^{\Lambda^{\cdot}(T^*S)\widehat{\otimes}F}$. Using this connection will actually play an essential role only in Subsection 6.13. Let $\mathfrak{N}_{b,t}$ be the operator of the type $\mathfrak{M}_{b,t}$ which is associated to $g_x^{T_xX}, g_x^E, T^HM_x$.

We fix $b_0 > 0$. We give an analogue of [BL06, Proposition 4.19].

Proposition 6.3. *There exist $c > 0$, $C > 0$ such that for $a \in \left[\frac{1}{2}, 1\right]$, $t \in]0,1]$, $b \in \left[\sqrt{t}, b_0\right]$, $x \in X_g$,*

$$\Bigg| \int_{\pi^{-1}\left\{U \in N_{X_g/X,x}, |U| \le \eta_0\right\}} \Big(\mathrm{Tr_s}\left[g \exp\left(-a\mathfrak{M}_{b,t}\right)\left(\left(U,Y\right), g\left(U,Y\right)\right)\right] k\left(x,U\right)$$

$$- \mathrm{Tr_s}\left[g \exp\left(-a\mathfrak{N}_{b,t}\right)\left(\left(U,Y\right), g\left(U,Y\right)\right)\right]\Big) \frac{dU dY}{\left(2\pi\right)^n} \Bigg| \le C \exp\left(-c/t\right). \quad (6.21)$$

Proof. Using equation (6.11) for $\mathfrak{M}_{b,t}$ and comparing with [BL06, eq. (4.47)], the proof of our proposition is identical to the proof of [BL06, Proposition 4.19]. □

6.7. A Getzler rescaling

For $a > 0$, let I_a be the map $s\left(U,Y\right) \to s\left(aU,Y\right)$. Set

$$\mathfrak{O}_{b,t} = I_{t^{3/2}} \mathfrak{N}_{b,t} I_{1/t^{3/2}}. \quad (6.22)$$

In the sequel $e_1, \ldots, e_n$ denotes an orthonormal basis of T_xX, which is such that $e_1, \ldots, e_\ell$ is an oriented orthonormal basis of T_xX_g, and $e_{\ell+1}, \ldots, e_n$ is an oriented orthonormal basis of $N_{X_g/X,x}$.

Now we still introduce another copy of TX_g, whose elements are written in Gothic. This means that $\mathfrak{e}_1, \ldots, \mathfrak{e}_\ell$ is an orthonormal basis of this third copy of TX_g. In the sequel, $\mathfrak{e}_1, \ldots, \mathfrak{e}_\ell$ will be considered as odd Grassmann variables, which generate another copy of the exterior algebra $\Lambda^{\cdot}\left(TX_g\right)$.

Definition 6.4. Let $\mathfrak{P}_{b,t}$ be the operator obtained from $\mathfrak{O}_{b,t}$ by making the following replacements:

- For $1 \le i \le \ell$, we replace the operators i_{e_i} by $i_{e_i} + \mathfrak{e}_i/t^{3/2}$.
- For $\ell + 1 \le i \le n$, we replace $e^i\wedge$ by $\sqrt{t}e^i\wedge$, and i_{e_i} by $i_{e_i}/\sqrt{t}$.

We do not write explicitly the dependence of these operators on x.

Let $\widehat{\mathrm{Tr}}_{\mathrm{s}}$ be the linear map defined on the algebra $\mathcal{A}$ spanned by the $e^i, i_{e_i}, \mathfrak{e}_i$, for $1 \le i \le \ell$ with values in $\mathbf{R}$, which up to permutation vanishes on all the monomials except on the monomial $\prod_{i=1}^{\ell} e^i \mathfrak{e}_i$, with

$$\widehat{\mathrm{Tr}}_{\mathrm{s}} \left[\prod_{i=1}^{\ell} e^i \mathfrak{e}_i\right] = 1. \quad (6.23)$$

We extend this functional to a functional mapping

$$\Lambda^{\cdot}\left(T^*S\right) \widehat{\otimes} \mathcal{A} \widehat{\otimes} \left(\mathrm{End}\left(\widehat{S}^{TX}\right) \widehat{\otimes} \mathrm{End}\left(\Lambda^{\cdot}\left(N^*_{X_g/X}\right)\right) \widehat{\otimes} E\right)_x$$

into $\Lambda^{\cdot}\left(T^*S\right)$, by taking the classical supertrace on the last factor.

Then we will consider monomials in the $e^i, i_{e_i} + \mathfrak{e}_i/t^{3/2}, 1 \le i \le \ell$. To evaluate $\widehat{\mathrm{Tr}}_{\mathrm{s}}$ on such monomials, we first normal order them, i.e., we put the i_{e_j} to the right of any e^i. Once this is done, we eliminate any term where one of the i_{e_i} appears, and we apply the rule (6.23). We extend $\widehat{\mathrm{Tr}}_{\mathrm{s}}$ as above.

Proposition 6.5. *The following identity holds,*

$$t^{3(n-\ell)/2}\mathrm{Tr_s}\left[g\exp\left(-\mathfrak{N}_{b,t}\right)\left(g^{-1}\left(t^{3/2}U,Y\right),\left(t^{3/2}U,Y\right)\right)\right]$$
$$= (-i)^{\ell/2}\,\widehat{\mathrm{Tr_s}}\left[g\exp\left(-\mathfrak{P}_{b,t}\right)\left(g^{-1}\left(U,Y\right),\left(U,Y\right)\right)\right]. \quad (6.24)$$

Proof. To establish (6.24), it is enough to consider the case where $g=1$, so that $\ell=n$. Note that when acting on $\Lambda^{\cdot}\left(T_x^*X\right)$,

$$\mathrm{Tr_s}\left[\prod_{i=1}^{n} e^i i_{e_i}\right]=1. \quad (6.25)$$

From (6.23), (6.25), we get (6.24). □

6.8. The limit as $t\to 0$ of $\mathfrak{P}_{b,t}$

Consider the vector bundle $TX|_{M_g}$ over M_g. The total space of the tangent bundle to the fibres is just $\pi^*\left(TX|_{M_g}\oplus TX|_{M_g}\right)$. As in Section 5, we will denote with a hat elements of the second copy. Here U is the canonical section of the first copy of TX, and Y is the canonical section of the second copy. In particular ∇_Y is a vertical differentiation operator along the first copy of TX.

Definition 6.6. Put

$$\mathfrak{P}_{b,0}=\frac{1}{2b^2}\left(-\Delta^V+\sqrt{2}\sum_{i=1}^{\ell}c\left(\widehat{e}_i\right)\mathfrak{e}_i\right)-\nabla_{\widehat{i^*R^{TX}Y}}$$
$$+\frac{1}{4}\left\langle i^*R^{TX}e_i,e_j\right\rangle c\left(\widehat{e}_i\right)c\left(\widehat{e}_j\right)+\frac{\nabla_Y}{b}+i^*R^E-2\frac{db}{b^2}\sum_{1}^{\ell}\left\langle Y,e_i\right\rangle\mathfrak{e}_i. \quad (6.26)$$

Of course these operators still depend on $x\in M_g$.

In the next theorem, we make $dt=0$.

Theorem 6.7. *As $t\to 0$,*

$$\mathfrak{P}_{b,t}\to\mathfrak{P}_{b,0}. \quad (6.27)$$

More precisely,

$$\mathfrak{P}_{b,t}=\mathfrak{P}_{b,0}+\sqrt{t}\Bigg(\frac{1}{2b^2}\sum_{i=1}^{\ell}c\left(\widehat{e}_i\right)\frac{e^i}{\sqrt{2}}\wedge-i^*\sum_{\ell+1\le i\le n}e^i\wedge\nabla_{\widehat{R^{TX}(e_i,\cdot)Y}}$$
$$+i^*\sum_{\ell+1\le i\le n}e^i\left(\frac{1}{4}\left\langle R^{TX}\left(e_i,\cdot\right)e_m,e_{m'}\right\rangle c\left(\widehat{e}_m\right)c\left(\widehat{e}_{m'}\right)+R^E\left(e_i,\cdot\right)\right)$$
$$-\frac{db}{b^2}\sum_{i=1}^{\ell}\left\langle Y,e_i\right\rangle e^i\Bigg)+\mathcal{O}\left(t\left(1+|Y|\right)\right)$$
$$+\mathcal{O}\left(t^{3/2}|y|\left(1+|Y|\right)\right)+\mathcal{O}\left(\frac{t^2}{b}|y|\,|Y|\right)+\frac{db}{b^2}\left(t^{3/2}|Y|\left(1+\sqrt{t}|y|\right)\right). \quad (6.28)$$

Proof. Note that since the $c(U)$ anticommute with the $c\left(\widehat{V}\right), \widehat{c}(V)$, the tensor $\sum_{i=1}^{n} c(\widehat{e}_i)\widehat{c}(e_i)$ is parallel with respect to ${}^1\nabla^{\Lambda^{\cdot}(T^*S)\widehat{\otimes}F}$.

By (6.10), for $1 \leq i \leq \ell$, $t\widehat{c}_t(\widehat{e}_i)$ is replaced by $\sqrt{t/2}e^i \wedge +\sqrt{2}\left(i_{t^{3/2}e_i} + \mathfrak{e}_i\right)$, and for $\ell + 1 \leq i \leq n$, it is replaced by $t\widehat{c}(e_i)$. By (6.11) and the above, we find that as $t \to 0$, once the above replacements are done,

$$\sum_{i=0}^{n} tc(\widehat{e}_i)\widehat{c}_t(\widehat{e}_i) \to \sqrt{2}\sum_{i=0}^{\ell} c(\widehat{e}_i)\mathfrak{e}_i. \tag{6.29}$$

Also when $t \to 0$,

$$\nabla_{\widehat{R^{TX}Y}} \to \nabla_{i^*\widehat{R^{TX}Y}}. \tag{6.30}$$

Inspection of (6.11), (6.26) shows that to establish (6.27), we only need to show that in the given trivialization,

$$I_{t^{3/2}}t^{3/2}\,{}^2\underline{\nabla}_{t/2,Y}^{\Lambda^{\cdot}(T^*S)\widehat{\otimes}F} I_{1/t^{3/2}} \to \nabla_Y. \tag{6.31}$$

Let $\underline{\Gamma}(y)$ be the connection form for ${}^1\underline{\nabla}^{\Lambda^{\cdot}(T^*S)\widehat{\otimes}F}$ in the given trivialization of $\Lambda^{\cdot}(T^*S)\widehat{\otimes}F$. By [ABP73], near x

$$\Gamma_y = \frac{1}{2}{}^1\underline{\nabla}_x^{\Lambda^{\cdot}(T^*S)\widehat{\otimes}F,2}(U,\cdot) + \mathcal{O}\left(|y|^2\right). \tag{6.32}$$

Now using (6.3) and (6.32), we get (6.31) easily.

To get the more precise expansion in (6.28), we need to control the remainders in the above computations. We consider first the terms in (6.11) with the exception of $\frac{t^{3/2}}{b}{}^2\underline{\nabla}_{t/2,Y}^{\Lambda^{\cdot}(T^*S)\widehat{\otimes}F}$. The coefficient of $\sqrt{t}$ in the right-hand side of (6.28) is immediately obtained from (6.11). Neither $\mathfrak{P}_{b,0}$ nor none of the terms which appear in this coefficient contains annihilation operators i_{e_j}. By (2.9), (2.10),

$${}^2\underline{\nabla}_t^{\Lambda^{\cdot}(T^*S)\widehat{\otimes}F} = \nabla_{\cdot}^{S^{TX}\otimes E} + \frac{1}{2t}\left\langle T\left(f_\alpha^H, e_i\right),\cdot\right\rangle f^\alpha\left(e^i - ti_{e_i}\right) + \frac{1}{4t}\left\langle T^H,\cdot\right\rangle. \tag{6.33}$$

In view of (6.33), we find that when taking the Taylor expansion at $t = 0$ of such terms, we get a remainder of the form $\mathcal{O}(t(1+|Y|)) + \mathcal{O}\left(t^{3/2}|y|(1+|Y|)\right)$. Because of (6.33), in the considered trivialization, the contribution to the remainder of the term $-2\frac{db}{b^2}t^{3/2}i_Y$ is of the form,

$$\frac{db}{b^2}\mathcal{O}\left(t^{3/2}|Y|\left(1+\sqrt{t}|y|\right)\right), \tag{6.34}$$

which accounts for the last term in the right-hand side of (6.28). Also the term $-2\frac{db}{b^2}\frac{\sqrt{t}}{2}Y^*$ contributes to the remainder by a term of the same type.

As to the contribution to the remainder of ${}^2\nabla_{t,Y}^{\Lambda^{\cdot}(T^*S)\widehat{\otimes}F}$, using (6.3) and (6.32), we find easily that this contribution is of the type $\mathcal{O}\left(\frac{t^2}{b}|y||Y|\right)$. This concludes the proof of our theorem. □

6.9. An evaluation of the traces of certain heat kernels

We use the same notation as in Section 5. In particular E denotes a real vector bundle of even dimension n.

Let $R \in \mathrm{End}(E)$ be antisymmetric. For $b > 0$, set

$$M_{R,b} = -\frac{1}{2b^2}\Delta^V - \nabla_{\widehat{RY}} + \frac{\nabla_Y}{b}. \tag{6.35}$$

We denote by $p_{R,b}(\cdot,\cdot)$ the heat kernel associated to the operator $\exp(-M_{R,b})$ with respect to the volume element $\frac{dUdY}{(2\pi)^n}$ on $E \oplus E$.

We will assume that the eigenvalues λ of R are such that $|\lambda| < 2\pi$.

Theorem 6.8. *The following identity holds,*

$$\int_E p_{R,b}((0,Y),(0,Y))\frac{dY}{(2\pi)^{n/2}} = b^{2n}\widehat{A}^2(R). \tag{6.36}$$

Proof. Note that when $R = 0$, equation (6.36) was already established in [BL06, Proposition 4.27]. Using the multiplicativity of the integral in (6.36), we may and we will assume that R is invertible.

Let $\widehat{M}_{R,b,\xi}$ be the partial Fourier transform of $M_{R,b}$ in the variable y. Clearly,

$$\widehat{M}_{R,b,\xi} = -\frac{1}{2b^2}\Delta^V - \nabla_{\widehat{RY}} + \frac{2i\pi}{b}\langle Y,\xi\rangle. \tag{6.37}$$

Let $q_{R,b,\xi}(Y,Y')$ be the smooth kernel associated to the operator $\exp\left(-\widehat{M}_{R,b,\xi}\right)$ with respect to the volume $dY'/(2\pi)^{n/2}$. Then

$$\int_E p_{R,b}((0,Y),(0,Y))\frac{dY}{(2\pi)^{n/2}} = \int_{E\times E} q_{R,b,\xi}(Y,Y)\,dYd\xi. \tag{6.38}$$

Set

$$N_{R,b} = -\frac{1}{2b^2}\Delta^V - \nabla_{\widehat{RY}}. \tag{6.39}$$

Let $t \in \mathbf{R} \to w_t \in E$ be a standard Brownian motion. Consider the stochastic differential equation,

$$\dot{Y} = RY + \frac{\dot{w}}{b}, \tag{6.40}$$

with given a initial condition $Y_0 \in E$. Then the infinitesimal generator associated to the diffusion $Y_\cdot$ is precisely $N_{R,b}$.

Let $s_{R,b}(Y,Y')$ be the smooth kernel associated to $\exp(-N_{R,b})$ with respect to the volume $dY'/(2\pi)^{n/2}$. Given $Y \in E$, let $P^Y_{R,b}$ be the probability law of the diffusion bridge $Y_\cdot$ whose infinitesimal generator is $N_{R,b}$, starting at Y at time 0, and returning to Y at time 1. By the Feynman-Kac formula, we get

$$q_{R,b,\xi}(Y,Y) = E^{P^Y_{R,b}}\left[\exp\left(-\frac{2i\pi}{b}\left\langle\int_0^1 Y_t dt,\xi\right\rangle\right)\right] s_{R,b}(Y,Y). \tag{6.41}$$

Let J_R be the first order differential operator on $[0,1]$,

$$J_R = \frac{d}{dt} - R, \tag{6.42}$$

with periodic boundary conditions. Under the given conditions on R, the operator J_R is invertible. If Y is as in (6.40), then there is a unique $A \in E$ such that

$$Y_t = e^{tR}A + \frac{1}{b}J_R^{-1}\dot{w}, \qquad Y_0 = A + \frac{1}{b}\left(J_R^{-1}\dot{w}\right)_0. \tag{6.43}$$

We claim that the probability law of $\left(J_R^{-1}\dot{w}\right)_0$ is a nondegenerate centred Gaussian distribution on E. In fact its covariance matrix is given by the Green kernel on the diagonal associated to J_R^{-2}. By using Fourier series, it is clear that this is a non degenerate matrix. We denote the probability law of $\left(J_R^{-1}\dot{w}\right)_0$ by $m(z)\,dz/(2\pi)^{n/2}$.

By (6.43), we get

$$Y_1 = e^R\left(Y_0 - \frac{1}{b}\left(J_R^{-1}\dot{w}\right)_0\right) + \frac{1}{b}\left(J_R^{-1}\dot{w}\right)_0. \tag{6.44}$$

By (6.44), we find that

$$s_{R,b}(Y_0,Y_1) = \frac{b^n}{\det\left(1-e^R\right)}m\left(b\left(1-e^R\right)^{-1}\left(Y_1 - e^RY_0\right)\right). \tag{6.45}$$

By (6.45), we get

$$s_{R,b}(Y,Y) = \frac{b^n}{\det\left(1-e^R\right)}m(bY). \tag{6.46}$$

Let T_b be the probability law on $C\left(S^1,E\right)$ of the process $U = J_R^{-1}\dot{w}/b$. We claim that we have the identity of positive measures,

$$\int_E s_{R,b}(Y,Y)\,P_{R,b}^Y\frac{dY}{(2\pi)^{n/2}} = \frac{T_b}{\det\left(1-e^R\right)}. \tag{6.47}$$

Indeed let S_b^Z be the probability law on $C\left(S^1,E\right)$ of the process $U = \left(J_R^{-1}\dot{w}\right)$ conditional on $U_0 = Z$. By (6.43), (6.44), if $f: C([0,1],\mathbf{R}) \to \mathbf{R}, g: E \to \mathbf{R}$ are bounded measurable functions, given $Y_0 \in \mathbf{R}$,

$$E\left[f(Y)\,g(Y_1)\right] = \int_{C([0,1],E)\times E} f\left(e^{\cdot R}(Y - Z/b) + U_\cdot/b\right) \cdot g\left(e^R(Y - Z/b) + Z/b\right)dS_b^Z(U_\cdot)\,m(Z)\frac{dZ}{(2\pi)^{n/2}}. \tag{6.48}$$

We can rewrite (6.48) in the form,

$$\begin{aligned}
&E\left[f\left(Y\right)g\left(Y_1\right)\right]\\
&=\frac{b^n}{\det\left(1-e^R\right)}\int_{C([0,1],E)\times E} f\left(e^{\cdot R}\left(1-e^R\right)^{-1}\left(Y-Z\right)+U_{\cdot}/b\right)\\
&\cdot g\left(Z\right)dS_b^{b\left(1-e^R\right)^{-1}\left(Z-e^RY\right)}\left(U\right)m\left(b\left(1-e^R\right)^{-1}\left(Z-e^RY\right)\right)\frac{dZ}{\left(2\pi\right)^{n/2}}. \quad (6.49)
\end{aligned}$$

By (6.45), (6.49), we find that $P_{R,b}^Y$ is the image of dS_b^{bY} by the map $U_{\cdot} \to U_{\cdot}/b$. Equation (6.47) follows.

Under T_b, the probability law of $\int_0^1 Y_t dt$ is a centred Gaussian of variance $-R^{-2}/b^2$. By (6.38), (6.41) and (6.47), we get

$$\int_E p_{R,b}\left(\left(0,Y\right),\left(0,Y\right)\right)\frac{dY}{\left(2\pi\right)^{n/2}}=\frac{b^{2n}\det^{1/2}\left(-R^2\right)}{\det\left(1-e^R\right)}. \quad (6.50)$$

Since R is invertible, $\det R>0$. Moreover,

$$\det{}^{1/2}\left(-R^2\right)=\det\left(-R\right). \quad (6.51)$$

Now observe that

$$\frac{\det\left(-R\right)}{\det\left(1-e^R\right)}=\widehat{A}^2\left(R\right). \quad (6.52)$$

By (6.50)–(6.52), we get (6.36). The proof of our theorem is completed. □

Now we will make another computation of traces. We assume that E is even dimensional. Let $g\in\operatorname{End}\left(E\right)$ be an oriented isometry of E which commutes with R. We assume that no eigenvalue of g is equal to 1. We can then write g in the form

$$g=e^B, \quad (6.53)$$

where $B\in\operatorname{End}E$ is antisymmetric and commutes with R. We may and we will assume that the eigenvalues μ of B are such that $0<|\mu|<2\pi$.

Theorem 6.9. *The following identity holds,*

$$\int_{E\times E} p_{R,b}\left(g^{-1}\left(U,Y\right),\left(U,Y\right)\right)\frac{dUdY}{\left(2\pi\right)^n}=\frac{1}{\det\left(1-g\right)\det\left(1-g^{-1}e^R\right)}. \quad (6.54)$$

Proof. We will use the same notation as in the proof of Theorem 6.8. Clearly,

$$\begin{aligned}
&\int_{E\times E} p_{R,b}\left(g^{-1}\left(U,Y\right),\left(U,Y\right)\right)\frac{dUdY}{\left(2\pi\right)^n}\\
&=\int_{E\times E\times E} q_{R,b,\xi}\left(g^{-1}Y,Y\right)\exp\left(-2i\pi\left\langle\left(g^{-1}-1\right)U,\xi\right\rangle\right)dUdYd\xi/\left(2\pi\right)^{n/2}. \quad (6.55)
\end{aligned}$$

By (6.55), we obtain,

$$\int_{E\times E} p_{R,b}\left(g^{-1}(U,Y),(U,Y)\right)\frac{dUdY}{(2\pi)^n} = \frac{1}{\det\left(1-g^{-1}\right)}\int_E q_{R,b,0}\left(g^{-1}Y,Y\right)\frac{dY}{(2\pi)^{n/2}}. \qquad (6.56)$$

Since $M_{R,b,0} = N_{R,b}$, we have $q_{R,b,0} = s_{R,b}$. By (6.45), we get

$$q_{R,b,0}\left(g^{-1}Y,Y\right) = \frac{b^n}{\det\left(1-e^R\right)} m\left(b\left(1-e^R\right)^{-1}\left(1-g^{-1}e^R\right)Y\right). \qquad (6.57)$$

By (6.56), (6.57), since $\det g = 1$, we get (6.54). The proof of our theorem is completed. □

6.10. A computation of certain finite-dimensional supertraces

We make the same assumptions as in Subsection 6.9. Now we state a basic formula of Mathai-Quillen [MaQ86].

Proposition 6.10. *The following identity holds,*

$$\widehat{\operatorname{Tr}}_{\mathrm{s}}\left[\exp\left(-c\left(\widehat{e}_i\right)\mathfrak{e}_i/\sqrt{2}b^2 - \frac{1}{4}\left\langle Re_i,e_j\right\rangle c\left(\widehat{e}_i\right)c\left(\widehat{e}_j\right)\right)\right] = b^{-2n}\widehat{A}^{-1}(R). \qquad (6.58)$$

6.11. The generalized traces of the heat kernels associated to $\mathfrak{P}_{b,0}$

Now we make again the same assumptions as in Subsections 6.1. Recall that the functional $\widehat{\operatorname{Tr}}_{\mathrm{s}}$ was defined in Subsection 6.7.

If α is a smooth form on M_g, let $\alpha^{\max}$ be the density of the component of α which has maximal vertical degree ℓ with respect to the volume form dv_{X_g} along the fibre X_g.

Theorem 6.11. *The following identity holds,*

$$\varphi\int_{N_{X_g/X}\times TX}\widehat{\operatorname{Tr}}_{\mathrm{s}}\left[g\exp\left(-\mathfrak{P}_{b,0}\right)\left(g^{-1}(U,Y),(U,Y)\right)\right]\frac{dUdY}{(2\pi)^{n-\ell/2}} = \pm\left[\widehat{A}_g\left(TX|_{M_g},\nabla^{TX|_{M_g}}\right)\operatorname{ch}_g\left(E,\nabla^E\right)\right]^{\max}. \qquad (6.59)$$

Proof. First we make $db = 0$. We will use the notation of Subsections 6.9 and 6.10. In the operator $\mathfrak{P}_{b,0}$, we replace $TX|_{M_g}$ by E, R^E by 0 and R^{TX} by an antisymmetric matrix $R \in \operatorname{End}(E)$ commuting with g. These operators now act on the total space of $E\oplus E$. By splitting these operators according to the eigenspaces of g, we may and we will assume that either $g=1$ or that no eigenvalue of g is equal to 1.

Assume first that $g=1$. Using (6.36) and (6.58), we get

$$\int_{TX} \widehat{\mathrm{Tr_s}}\left[\exp\left(-\mathfrak{P}_{b,0}\right)\left(\left(0,Y\right),\left(0,Y\right)\right)\right] \frac{dY}{\left(2\pi\right)^{n/2}} = \widehat{A}\left(R\right). \tag{6.60}$$

By (6.60), we get (6.59).

Let us now assume that no eigenvalue of g is equal to 1. Simple character identities show that

$$\mathrm{Tr_s}^{S^E}\left[g\exp\left(-\frac{1}{4}\left\langle Re_i,e_j\right\rangle c\left(\widehat{e}_i\right)c\left(\widehat{e}_j\right)\right)\right] = \pm\left(-1\right)^{n/2}\widehat{A}_g^{-1}\left(-R\right). \tag{6.61}$$

Moreover, we have the trivial

$$\det\left(1-g^{-1}e^R\right) = \left(-1\right)^{n/2}\widehat{A}_g^{-2}\left(-R\right). \tag{6.62}$$

Also

$$\mathrm{Tr_s}^{\Lambda^{\cdot}\left(N^*_{X_g/X}\right)}\left[g\right] = \det\left(1-g\right). \tag{6.63}$$

By (6.54) and by (6.61)–(6.63), we get (6.59).

Now we extend the above to the case where db is not made equal to 0. In fact we will show that the term containing db in $\mathfrak{P}_{b,0}$ does not contribute to the limit. To do this we simply observe that if we conjugate the operator $\mathfrak{P}_{b,0}$ by the operator $\exp\left(-2\frac{db}{b}\sum_1^\ell\left\langle U,e_i\right\rangle\mathfrak{e}_i\right)$, the conjugated operator is just $\mathfrak{P}_{b,0}$, in which db has been made equal to 0. This shows that there is no term db in the right-hand side of (6.59). One can instead use the change of variables $(U,Y)\to(-U,-Y)$ to obtain the vanishing of the contribution of db by a parity argument. The proof of our theorem is completed. □

6.12. A proof of Theorem 6.1

First we make $dt=0$. By Proposition 6.5, we get

$$\begin{aligned}\int_{\left\{U\in N_{X_g/X},|U|\le\eta_0\right\}\times T_xX} &\mathrm{Tr_s}\left[g\exp\left(-\mathfrak{N}_{b,t}\right)\left(g^{-1}\left(U,Y\right),\left(U,Y\right)\right)\right]\frac{dUdY}{\left(2\pi\right)^n}\\ =\int_{\left\{U\in N_{X_g/X},|U|\le\eta_0/t^{3/2}\right\}\times T_xX}&\left(-i\right)^{\ell/2}\widehat{\mathrm{Tr_s}}\left[g\exp\left(-\mathfrak{P}_{b,t}\right)\left(g^{-1}\left(U,Y\right),\left(U,Y\right)\right)\right]\\ &\cdot k\left(x,t^{3/2}U\right)\frac{dUdY}{\left(2\pi\right)^n}.\end{aligned} \tag{6.64}$$

By proceeding as in [BL06, proof of Theorem 4.25], we find that as $t\to0$,

$$\begin{aligned}&\widehat{\mathrm{Tr_s}}\left[g\exp\left(-\mathfrak{P}_{b,t}\right)\left(g^{-1}\left(U,Y\right),\left(U,Y\right)\right)\right]\to\\ &\qquad\widehat{\mathrm{Tr_s}}\left[g\exp\left(-\mathfrak{P}_{b,0}\right)\left(g^{-1}\left(U,Y\right),\left(U,Y\right)\right)\right].\end{aligned} \tag{6.65}$$

By combining the above convergence with uniform estimates similar to the ones in [BL06, Theorem 4.25], we find that as $t \to 0$,

$$\int_{\{U\in N_{X_g/X},|U|\le\eta_0/t^{3/2}\}\times T_xX} \widehat{\mathrm{Tr_s}}\left[g\exp\left(-\mathfrak{P}_{b,t}\right)\left(g^{-1}\left(U,Y\right),\left(U,Y\right)\right)\right] \cdot k\left(x,t^{3/2}U\right)\frac{dUdY}{\left(2\pi\right)^n} \to \int_{\left(N_{X_g/X}\times TX\right)_x} \widehat{\mathrm{Tr_s}}\left[g\exp\left(-\mathfrak{P}_{b,0}\right)\left(g^{-1}\left(U,Y\right),\left(U,Y\right)\right)\right]\frac{dUdY}{\left(2\pi\right)^n}. \tag{6.66}$$

By combining the estimates obtained in Subsection 6.5 with Proposition 6.3, Theorem 6.11 and (6.66), we find that as $t \to 0$,

$$\varphi\mathrm{Tr_s}\left[g\exp\left(-\overline{\mathbb{A}}^2\right)\right] \to \int_{X_g} \pm\widehat{A}_g\left(TX|_{X_g},\nabla^{TX|_{X_g}}\right)\mathrm{ch}_g\left(E,\nabla^E\right). \tag{6.67}$$

To obtain an estimate of the remainder as in Theorem 6.1, one should proceed as in [BL06], by using in particular the expansion as $t \to 0$ of $\mathfrak{P}_{b,t}$ which was given in (6.28). If we proceed as in [BL06], we find that as $t \to 0$,

$$\varphi\mathrm{Tr_s}\left[g\exp\left(-\overline{\mathbb{A}}^2\right)\right] \to \int_{X_g} \pm\widehat{A}_g\left(TX|_{X_g},\nabla^{TX|_{X_g}}\right)\mathrm{ch}_g\left(E,\nabla^E\right)+\mathcal{O}\left(\sqrt{t}\right). \tag{6.68}$$

Let us briefly explain how to get $\mathcal{O}(t)$ instead of $\mathcal{O}\left(\sqrt{t}\right)$ in (6.68). Indeed let E be the coefficient of $\sqrt{t}$ in the right-hand side of (6.28). We will now show that

$$\int_{N_{X_g/X}\times TX} \widehat{\mathrm{Tr_s}}\left[gE\exp\left(-\mathfrak{P}_{b,0}\right)\left(g^{-1}\left(U,Y\right),\left(U,Y\right)\right)\right]\frac{dUdY}{\left(2\pi\right)^n} = 0. \tag{6.69}$$

From (6.69), we conclude easily that in the right-hand side of (6.68), $\mathcal{O}\left(\sqrt{t}\right)$ can be replaced by $\mathcal{O}(t)$.

Now we establish (6.69). The first term in E does not contribute to (6.69) because the operator $\mathfrak{P}_{b,0}$ contains only even monomials in the Clifford variables $c\left(\widehat{e}_i\right)$, and moreover, only even monomials in these Clifford variables contribute to $\widehat{\mathrm{Tr_s}}$. Moreover, $\mathfrak{P}_{b,0}$ does not contain creation or annihilation operators in the variables $e^j, \ell+1 \le j \le n$. This explains why when taking the generalized supertrace $\widehat{\mathrm{Tr_s}}$, the second kind of terms in E does not contribute to (6.69). We are left with the last term in E, which contains $\frac{db}{b^2}$. Its contribution also vanishes by a parity argument already given in the proof of Theorem 6.11. This completes the proof of the fact that $\mathcal{O}\left(\sqrt{t}\right)$ can be replaced by $\mathcal{O}(t)$. This way, we get the first and third identities in (6.8).

Now we make $db = 0$. We will show that as $t \to 0$,

$$tn_{b,t} = \mathcal{O}(t), \tag{6.70}$$

which is just the second identity in (6.8). To evaluate the asymptotics as $t \to 0$ of $tn_{b,t}$, we may as well replace dt by tdt in equation (6.11) for $\mathfrak{M}_{b,t}$. Except for irrelevant coefficients, when multiplied by t, the term containing dt in the right-hand side of equation (6.11) for $\mathfrak{M}_{b,t}$ is of the same type as the term containing db. We can then proceed as before, and we get (6.70). This completes the proof of Theorem 6.1.

6.13. The forms $m_{\sqrt{t}b,t}$ as $t \to 0$

In this subsection, we make $dt = 0$.

Theorem 6.12. *Given $b > 0$, as $t \to 0$,*

$$m_{\sqrt{t}b,t} = \int_{X_g} \pm \widehat{A}_g \left(TX, \nabla^{TX}\right) \operatorname{ch}_g \left(E, \nabla^E\right) + \mathcal{O}\left(\sqrt{t}\right), \quad \sqrt{t} o_{\sqrt{t}b,t} = \mathcal{O}\left(\sqrt{t}\right). \tag{6.71}$$

Proof. The remainder of the subsection is devoted to the proof of our theorem. □

Recall that the operator $\mathfrak{M}_{b,t}$ was defined in (6.11), and also that we make $dt = 0$. When replacing b by $\sqrt{t}b$, we will implicitly replace db by $\sqrt{t}db$. Given $b > 0$, set

$$\overline{\mathfrak{M}}_{b,t} = K_{1/\sqrt{t}} \mathfrak{M}_{\sqrt{t}b,t} K_{\sqrt{t}}. \tag{6.72}$$

By equation (6.11), we get

$$\begin{aligned} \overline{\mathfrak{M}}_{b,t} = \frac{1}{2b^2} \left(-\Delta^V + |Y|^2 + c\left(\widehat{e}_i\right) \widehat{c}_t\left(e_i\right)\right) - \nabla_{\widehat{R^{TX}Y}} \\ + \frac{1}{4} \left\langle R^{TX} e_i, e_j \right\rangle c\left(\widehat{e}_i\right) c\left(\widehat{e}_j\right) + \frac{\sqrt{t}}{b} {}^2\nabla^{\Lambda^{\cdot}(T^*S) \widehat{\otimes} F}_{t/2,Y} \\ + R^E - \sqrt{2} \frac{db}{b^2} \widehat{c}_t\left(Y\right). \end{aligned} \tag{6.73}$$

Recall that N^H is the number operator of $\Lambda^{\cdot}\left(T^*X\right)$. Put

$$\overline{\mathfrak{M}}'_{b,t} = t^{N^H/2} \overline{\mathfrak{M}}_{b,t} t^{-N^H/2}. \tag{6.74}$$

The effect of the conjugation in (6.74) is in particular to replace the $\widehat{c}_t\left(e_i\right)$ in (6.74) by $\widehat{c}\left(e_i\right)$.

Take $x \in X_g$. We choose the same coordinate system near x as in Subsection 6.6. We define an operator $\overline{\mathfrak{N}}'_{b,t}$ from $\overline{\mathfrak{M}}'_{b,t}$ by the same principle as in Definition 6.2. Put

$$\overline{\mathfrak{O}}_{b,t} = I_{\sqrt{t}} \overline{\mathfrak{N}}'_{b,t} I_{1/\sqrt{t}}. \tag{6.75}$$

We will express the operator $\overline{\mathfrak{O}}_{b,t}$ in terms of the operators $c\left(e_i\right), \widehat{c}\left(e_i\right), c\left(\widehat{e}_i\right)$. In particular the operators e^i, i_{e_i} have temporarily disappeared from the scene. But they will come back from another corner.

We define the operator $\overline{\mathfrak{P}}_{b,t}$ from the operator $\overline{\mathfrak{O}}_{b,t}$ by replacing the $c\left(e_i\right), 1 \le i \le \ell$ by $\sqrt{2}e^i \wedge /\sqrt{t} - \sqrt{t} i_{e_i}/\sqrt{2}$ for $1 \le i \le \ell$. It is here crucial that the operators $\widehat{c}\left(e_i\right), 1 \le i \le \ell$ be left untouched in this procedure.

The vector bundle $TX|_{M_g}$ on M_g verifies all the assumptions verified by the vector bundle E in Section 5. In particular we can define the operator $\mathcal{B}^{TX|_{M_g}}_{Y,b}$

acting on the fibres of $TX|_{M_g} \oplus TX|_{M_g}$ as in that section. This operator depends implicitly on $x \in M_g$.

Theorem 6.13. *As $t \to 0$,*

$$\overline{\mathfrak{P}}_{b,t} \to \overline{\mathfrak{P}}_{b,0} = \mathcal{B}_{Y,b}^{TX|_{M_g},2} + R^E - \sqrt{2}\frac{db}{b^2}\widehat{c}(Y). \tag{6.76}$$

Proof. The considerations we made at the beginning of the proof of Theorem 6.7 remain valid here. Also note that under the suggested replacements, the $c_t(e_i)$ are first changed into $c(e_i)$, and later they are changed back to $c_t(e_i)$ for $1 \leq i \leq \ell$, while remaining as they are for $\ell+1 \leq i \leq \ell$. Because of (6.74), in (6.73), the $\widehat{c}_t(e_i)$ have been replaced by $\widehat{c}(e_i)$, and the $e^i\wedge$ have been changed into $\sqrt{t}e^i\wedge$. When making the transformations indicated above, we obtain easily all the terms which appear in equation (5.14) for the right-hand side of (6.76), except maybe for the contribution of the term $\frac{\sqrt{t}}{b}{}^2\nabla_{t/2,Y}^{\Lambda^\cdot(T^*S)\widehat{\otimes}F}$ which appears in (6.73).

Under the transformations indicated in (6.74), the operator ${}^2\nabla_{t/2,Y}^{\Lambda^\cdot(T^*S)\widehat{\otimes}F}$ is changed into ${}^1\nabla_{t/2,Y}^{\Lambda^\cdot(T^*S)\widehat{\otimes}F}$. Using (6.3) and (6.32), we find that after doing the extra transformations indicated above, as $t \to 0$,

$$I_{\sqrt{t}}\sqrt{t}\,{}^1\nabla_{t/2,Y}^{\Lambda^\cdot(T^*S)\widehat{\otimes}F} I_{1/\sqrt{t}} = \nabla_Y + \frac{1}{4}\sum_{1\leq i,j\leq \ell}\left\langle \underline{R}^{TX}(U,Y)e_i, e_j\right\rangle e^i \wedge e^j$$
$$+ \frac{1}{4}\underline{\nabla}^{TX}\underline{T}^H(U,Y) - \frac{1}{2}i^*\underline{\nabla}^{TX}\underline{T}^0(U,Y) - \frac{1}{4}\left|\underline{T}^0\right|^2(U,Y) + \mathcal{O}\left(\sqrt{t}\right). \tag{6.77}$$

By [BG04, Theorem 3.26] or [B05, Theorem 4.15], we can rewrite (6.77) in the form,

$$I_{\sqrt{t}}\sqrt{t}\,{}^1\nabla_{t/2,Y}^{\Lambda^\cdot(T^*S)\widehat{\otimes}F} I_{1/\sqrt{t}} = \nabla_Y + \frac{1}{2}\left\langle i^*R^{TX}U, Y\right\rangle + \mathcal{O}\left(\sqrt{t}\right). \tag{6.78}$$

This proof of our theorem is completed. □

Now we conclude the proof of Theorem 6.12. Indeed by proceeding as in [BL06, Section 11] and using Theorems 5.1 and 6.13 and also the arguments given in the proof of Theorem 6.1, we get (6.71).

7. Double transgression formulas and hypoelliptic holomorphic torsion forms

This section is entirely devoted to the case of complex manifolds, along the lines of Section 3. In the context of families of complex manifolds considered in Subsections 3.11–3.14 and also in Section 4, for $b > 0$ small enough, we obtain hypoelliptic holomorphic torsion forms, which are the obvious analogues of the corresponding objects which were obtained in Section 4 in the elliptic case. We show in particular how to construct a hypoelliptic Quillen metric, the properties of which are very close to the properties of classical elliptic Quillen metrics.

In the whole section, we will assume for simplicity that the metrics on TX and $\widehat{TX}$ are proportional with a fixed constant of proportionality, which will be itself made to vary. This assumption will be lifted in Section 10.

This section is organized as follows. In Subsection 7.1, for $b > 0, t > 0$, we introduce the Hermitian forms $\widehat{\underline{\epsilon}}_{b,t}$, and the associated hypoelliptic superconnections $\mathcal{A}_{Y,b,t}$.

In Subsection 7.2, we establish simple commutator identities in the case of a single fibre.

In Subsection 7.3, we study the variation of our superconnections with respect to the variables b, t.

In Subsection 7.4, we introduce holomorphic hypoelliptic superconnections forms, and we establish double transgression formulas similar to the ones we gave in Section 4.

In Subsection 7.5, we study the limit of these forms as $t \to 0$.

In Subsection 7.6, we show that the limits of the hypoelliptic forms as $b \to 0$ are precisely the associated elliptic forms. Also we study their limit as $t \to +\infty$. Moreover, we establish crucial uniform estimates in the variables b, t.

In Subsection 7.7, we define the hypoelliptic holomorphic torsion forms.

In Subsection 7.8, we construct a hypoelliptic generalized Quillen metric on the determinant of the cohomology.

In Subsection 7.9, we show that the hypoelliptic Quillen metric is smooth.

In Subsection 7.10, these results are extended to the equivariant case.

Finally in Subsection 7.11, we study the dependence on b of the hypoelliptic Quillen metric.

We make the same assumptions as in Subsections 3.11–3.14 and in Subsections 4.1 and 4.4.

7.1. A metric formulation of the hypoelliptic Laplacian

We will sometimes denote by $\widehat{TX}$ the fibre of $\pi : \mathcal{M} \to M$, to distinguish it from the tangent bundle TX. For $b > 0$, let $\epsilon_b, \underline{\epsilon}_b$ be the obvious analogues of the Hermitian forms $\epsilon, \underline{\epsilon}$ in (3.55), (3.103), in which the fibrewise metric $g^{\widehat{TX}}$ along the fibre $\widehat{TX}$ is replaced by $b^4 g^{\widehat{TX}}$, while keeping fixed the metric g^{TX} along the fibres X. Clearly,

$$\begin{aligned} \epsilon_b(s, s') &= b^{4n} \epsilon\left(s, b^{-4N^V} s'\right), \\ \underline{\epsilon}_b(s, s') &= b^{4n} \underline{\epsilon}\left(b^{-2N^V} s, b^{-2N^V} s'\right) = b^{4n} \epsilon\left(s, b^{-4N^V} e^{-i\omega^{M,H}} s'\right). \end{aligned} \tag{7.1}$$

We use the same notation as in Subsection 3.11. We still define $\mathcal{A}''_Y$ as in (3.99), so that

$$\mathcal{A}''_Y = \nabla^{\mathbf{I}^{\cdot\prime\prime}} + \overline{\partial}^V + i_y. \tag{7.2}$$

Let $\mathcal{A}'_{Y,b}$ be the adjoint of $\mathcal{A}''_Y$ with respect to $\underline{\epsilon}_b$. Set

$$\mathcal{A}_{Y,b} = \mathcal{A}''_Y + \mathcal{A}'_{Y,b}. \tag{7.3}$$

Recall that N^H is the number operator on $\Lambda^{\cdot}(T^*_{\mathbf{R}}X \otimes_{\mathbf{R}} \mathbf{C})$, so that

$$N^H = N^{H(0,1)} + N^{H(1,0)}. \tag{7.4}$$

For $b > 0, t > 0$, let $\underline{\epsilon}_{b,t}$ be the Hermitian form defined as before which is associated to the Hermitian metrics $g^{TX}/t, b^4 g^{TX}/t^3$ on $TX, \widehat{TX}$. Using (3.55), (7.1), we get

$$\underline{\epsilon}_{b,t}(s,s') = \left(\frac{b}{t}\right)^{4n} \eta\left(e^{it\Lambda}s, \left(\frac{t^3}{b^4}\right)^{N^V} t^{N^H} e^{it\Lambda - i\omega^{M,H}/t} s'\right), \tag{7.5}$$

which can also be written in the form,

$$\underline{\epsilon}_{b,t}(s,s') = \left(\frac{b}{t}\right)^{4n} \eta\left(s, \left(\frac{t^3}{b^4}\right)^{N^V} t^{N^H} e^{iL/t} e^{it\Lambda - i\omega^{M,H}/t} s'\right). \tag{7.6}$$

Set

$$\widehat{\underline{\epsilon}}_{b,t}(s,s') = t^n \underline{\epsilon}_{b,t}(s,s'). \tag{7.7}$$

We denote by $\widehat{\epsilon}_{b,t}$ the fibrewise restriction of $\widehat{\underline{\epsilon}}_{b,t}$, which is obtained from (7.6), (7.7) by making $\omega^{M,H} = 0$.

In the sequel we will use $\widehat{\underline{\epsilon}}_{b,t}$ instead of $\underline{\epsilon}_{b,t}$ just for convenience. However the objects we can associate to these two forms are essentially the same.

Theorem 7.1. *The following identities hold,*

$$\widehat{\underline{\epsilon}}_{b,t}^{-1} \frac{\partial}{\partial b} \widehat{\underline{\epsilon}}_{b,t} = -\frac{4}{b}\left(N^V - n\right), \qquad \widehat{\underline{\epsilon}}_{b,t}^{-1} \frac{\partial}{\partial t} \widehat{\underline{\epsilon}}_{b,t} = \frac{1}{t}\left(3N^V - 2n + i\frac{\omega^M}{t}\right). \tag{7.8}$$

Proof. The first identity in (7.8) is obvious. To establish the second identity, we use the sl_2 commutation relations in (3.6). We represent $L, \Lambda, 2H = N^H - n$ by the matrices

$$L = \begin{pmatrix} 0 & 0 \\ 1 & 0 \end{pmatrix}, \qquad \Lambda = \begin{pmatrix} 0 & 1 \\ 0 & 0 \end{pmatrix}, \qquad 2H = \begin{pmatrix} -1 & 0 \\ 0 & 1 \end{pmatrix}. \tag{7.9}$$

Set

$$Q_t = t^{2H} e^{iL/t} e^{it\Lambda}. \tag{7.10}$$

Then one verifies easily that

$$Q_t = \begin{pmatrix} 1/t & i \\ i & 0 \end{pmatrix}. \tag{7.11}$$

By (7.11), we get

$$Q_t^{-1} \frac{\partial}{\partial t} Q_t = \begin{pmatrix} 0 & 0 \\ i/t^2 & 0 \end{pmatrix}. \tag{7.12}$$

We can rewrite (7.12) in the form,

$$Q_t^{-1} \frac{\partial}{\partial t} Q_t = i\frac{L}{t^2}. \tag{7.13}$$

By (7.6), (7.7) and (7.13), we get the second identity in (7.8). The proof of our theorem is completed. □

Definition 7.2. Let $\mathcal{A}'_{Y,b,t}$ be the adjoint of $\mathcal{A}''_Y$ with respect to $\underline{\widehat{\epsilon}}_{b,t}$. Set

$$\mathcal{A}_{Y,b,t} = \mathcal{A}''_Y + \mathcal{A}'_{Y,b,t}. \tag{7.14}$$

By proceeding as in (3.99), (3.101), we get

$$\mathcal{A}'_{Y,b,t} = \nabla^{\mathbf{I}^{\cdot\prime}} + \frac{t^3}{b^4}\overline{\partial}^{V*} + \frac{\overline{y}^*}{t}\wedge + i_{\overline{y}}. \tag{7.15}$$

For $a > 0$, we denote by ρ_a^* the obvious action of r_a^* on smooth sections of $\widehat{\Lambda}^{\cdot}\left(\overline{T^*X}\right)$. We extend the action of ρ_a^* to $C^\infty\left(\mathcal{M}, \pi^*\Lambda^{\cdot}\left(T^*_{\mathbf{R}}S\otimes_{\mathbf{R}}\mathbf{C}\right)\widehat{\otimes}F\right)$. The difference with r_a^* is that ρ_a^* has no action on $\Lambda^{\cdot}\left(T^*X\right)$.

By (7.2), (7.15),

$$\rho^*_{t^{3/2}/b^2}\mathcal{A}_{Y,b,t}\rho^{*-1}_{t^{3/2}/b^2} = \nabla^{\mathbf{I}^{\cdot}} + \overline{\partial}^V + \overline{\partial}^{V*} + \frac{\sqrt{t}}{b^2}\overline{y}^*\wedge + \frac{t^{3/2}}{b^2}i_Y. \tag{7.16}$$

When taking into account equations (1.11) and (3.8), we find that the operator in (7.16) is exactly the operator denoted $\mathbb{A}^t_{Y,b}$ in (6.5).

Now we prove an analogue of (4.2).

Proposition 7.3. *The following identity holds,*

$$t^{3/2N^V+N^H/2}K_t\mathcal{A}_{Y,b,t}K_t^{-1}t^{-3/2N^V-N^H/2} = \psi_{1/t}\sqrt{t}\mathcal{A}_{Y,b}\psi_t. \tag{7.17}$$

Proof. First note that the operator $\nabla^{\mathbf{I}^{\cdot}}$ preserves the vertical degree, is invariant under conjugation by K_t and increases the total degree by 1. Its contribution to $\mathcal{A}_{Y,b,t}$ then fits with (7.17). This is also obviously the case for $\overline{\partial}^V + i_y$ and $\frac{t^3}{b^4}\overline{\partial}^{V*} + \frac{\overline{y}^*}{t}\wedge + i_{\overline{y}}$. The proof of our proposition is completed. □

7.2. Commutator identities

Let $A''_{Y,b}, A'_{Y,b}, A_{Y,b}$ be the fibrewise operator components of $\mathcal{A}''_{Y,b}, \mathcal{A}'_{Y,b}, \mathcal{A}_{Y,b}$, i.e., their components of degree 0 in $\Lambda^{\cdot}\left(T^*_{\mathbf{R}}S\otimes_{\mathbf{R}}\mathbf{C}\right)$. Let $w_1, \ldots, w_n$ be a basis of TX and let $w^1, \ldots, w^n$ be the corresponding dual basis of T^*X. We denote with hats the corresponding bases of $\widehat{TX}, \widehat{TX}^*$. Set

$$\partial^V = w^i \wedge \nabla_{\widehat{w}_i}, \qquad \overline{\partial}^V = \overline{w}^i \wedge \nabla_{\widehat{\overline{w}}_i}. \tag{7.18}$$

Then ∂^V can be viewed as the conjugate of the operator $\overline{\partial}^V$.

Let $\widehat{Y}$ denote the fibrewise radial vector field, which generates the group of dilations r_{e^t}. Let $L_{\widehat{Y}}$ denote the corresponding Lie derivative operator, which acts naturally on $C^\infty\left(\mathcal{X}, \pi^*F\right)$.

In the sequel, † denotes the adjoint with respect to $\widehat{\epsilon}_{b,t}$. This means that the results which follow are only valid in the case where S is reduced to a point.

Now we establish an analogue of the commutator identities of [B05, Proposition 4.34].

Proposition 7.4. *The following identities hold,*

$$\partial^{V\dagger} = -\overline{\underline{\partial}}^V, \qquad i^\dagger_{\widehat{\overline{y}}} = \frac{b^4}{t^3}\widehat{y}^* \wedge . \tag{7.19}$$

Moreover,

$$L_{\widehat{Y}} = \nabla_{\widehat{Y}} + N^V + N^{H(1,0)}, \quad L^\dagger_{\widehat{Y}} = -\nabla_{\widehat{Y}} + N^V + i\frac{L}{t} - N^{H(0,1)} - n. \tag{7.20}$$

Also,

$$L_{\widehat{Y}} = \left[A''_Y, \partial^V + i_{\widehat{\overline{y}}}\right], \qquad L^\dagger_{\widehat{Y}} = \left[A'_{Y,b,t}, -\overline{\underline{\partial}}^V + \frac{b^4}{t^3}\widehat{y}^* \wedge\right]. \tag{7.21}$$

In particular,

$$\left[A''_Y, \partial^V + i_{\widehat{\overline{y}}}\right] + \left[A'_{Y,b,t}, -\overline{\underline{\partial}}^V + \frac{b^4}{t^3}\widehat{y}^* \wedge\right] = 2N^V - n + N^{H(1,0)} - N^{H(0,1)} + i\frac{L}{t}. \tag{7.22}$$

Proof. A simple computation shows that if $f \in TX$,

$$t^{N^H} e^{iL/t} e^{it\Lambda} \overline{f}^* \wedge e^{-it\Lambda} e^{-iL/t} t^{-N^H} = -i_{\overline{f}}. \tag{7.23}$$

Using (7.6), (7.18) and (7.23), we get the first identity in (7.19). A similar computation leads to the second identity in (7.19). The first identity in (7.20) is trivial. Now we use the notation in the proof of Theorem 7.1. By using the commutation relations in (3.6), we find that

$$e^{-it\Lambda} e^{-iL/t} \left(N^H - n\right) e^{iL/t} e^{it\Lambda} = n - N^H + 2iL/t. \tag{7.24}$$

By (7.6), we find that (7.24) is just the identity,

$$\left(N^H - n\right)^\dagger = \left(n - N^H\right) + 2iL/t. \tag{7.25}$$

Since L, Λ commute with $N^{H(0,1)} - N^{H(1,0)}$, we get

$$\left(N^{H(0,1)} - N^{H(1,0)}\right)^\dagger = N^{H(0,1)} - N^{H(1,0)}. \tag{7.26}$$

By (7.25), (7.26), we obtain,

$$N^{H(1,0)\dagger} = n - N^{H(0,1)} + iL/t. \tag{7.27}$$

Finally,

$$\nabla^\dagger_{\widehat{Y}} = -\nabla_{\widehat{Y}} - 2n. \tag{7.28}$$

By the first identity in (7.20) and by (7.27), (7.28), we get the second identity in (7.20).

The first identity in (7.21) is trivial, or can be proved directly using (7.20). By (7.19) and taking adjoints in the first identity in (7.21), we get the second identity in (7.21). This last identity can also be proved by a direct computation. Finally (7.22) follows from (7.20), (7.21). The proof of our proposition is completed. □

Remark 7.5. When replacing $A''_{Y,b,t}, A'_{Y,b,t}$ by $\mathcal{A}''_{Y,b,t}, \mathcal{A}'_{Y,b,t}$, the identities in (7.21), (7.22) are no longer correct. In particular equation (7.22) should be adequately modified. Extra terms involving the tensor T will appear in the right-hand side of (7.22).

Moreover, our proposition remains valid even if the horizontal metric g^{TX} and the vertical metric $g^{\widehat{TX}}$ are unrelated, as long as $\widehat{y}^*$ is calculated with respect to the vertical metric.

7.3. The variation of $\mathcal{A}'_{Y,b,t}$ with respect to b, t

From now on, we assume S to be arbitrary.

Proposition 7.6. *The following identities hold,*

$$\frac{\partial}{\partial b}\mathcal{A}'_{Y,b,t} = \left[\mathcal{A}'_{Y,b,t}, \widehat{\underline{\epsilon}}_{b,t}^{-1}\frac{\partial}{\partial b}\widehat{\underline{\epsilon}}_{b,t}\right], \qquad \frac{\partial}{\partial t}\mathcal{A}'_{Y,b,t} = \left[\mathcal{A}'_{Y,b,t}, \widehat{\underline{\epsilon}}_{b,t}^{-1}\frac{\partial}{\partial t}\widehat{\underline{\epsilon}}_{b,t}\right] \tag{7.29}$$

Proof. This is because $\mathcal{A}'_{Y,b,t}$ is the $\widehat{\underline{\epsilon}}_{b,t}$ adjoint of $\mathcal{A}''_{Y}$. □

7.4. The double transgression formulas

Set

$$\begin{aligned} u_{b,t} &= \varphi \mathrm{Tr_s}\left[g\exp\left(-\mathcal{A}^2_{Y,b,t}\right)\right], \\ v_{b,t} &= \varphi \mathrm{Tr_s}\left[gb\widehat{\underline{\epsilon}}_{b,t}^{-1}\frac{\partial}{\partial b}\widehat{\underline{\epsilon}}_{b,t}\exp\left(-\mathcal{A}^2_{Y,b,t}\right)\right], \\ w_{b,t} &= \varphi \mathrm{Tr_s}\left[gt\widehat{\underline{\epsilon}}_{b,t}^{-1}\frac{\partial}{\partial t}\widehat{\underline{\epsilon}}_{b,t}\exp\left(-\mathcal{A}^2_{Y,b,t}\right)\right], \\ \underline{w}_{b,t} &= \varphi \mathrm{Tr_s}\left[g\left(\mathcal{N} - n + i\frac{\omega^{M,H}}{t}\right)\exp\left(-\mathcal{A}^2_{Y,b,t}\right)\right]. \end{aligned} \tag{7.30}$$

In the sequel $d^{\mathbf{R}^{*2}_+}$ denotes the de Rham operator on $\mathbf{R}^{*2}_+$.

Theorem 7.7. *For any $(b,t) \in \mathbf{R}^*_+ \times \mathbf{R}^*_+$, the forms $u_{b,t}, v_{b,t}, w_{b,t}, \underline{w}_{b,t}$ are sums of forms of type (p,p). They are real for $g = 1$. The forms $u_{b,t}$ are closed, and their common cohomology class does not depend on b, t. Also,*

$$\frac{\partial}{\partial t}u_{b,t} = -\frac{\overline{\partial}\partial}{2i\pi t}w_{b,t}, \qquad \frac{\partial}{\partial b}u_{b,t} = -\frac{\overline{\partial}\partial}{2i\pi b}v_{b,t}. \tag{7.31}$$

Let a be the odd form on $S \times \mathbf{R}^*_+ \times \mathbf{R}^*_+$,

$$a = \frac{dt}{t}w_{b,t} + \frac{db}{b}v_{b,t}. \tag{7.32}$$

Then

$$d^{\mathbf{R}^{*2}_+}a \in dbdtP^{S,0}. \tag{7.33}$$

Finally,

$$w^{(0)}_{b,t} = \underline{w}^{(0)}_{b,t}. \tag{7.34}$$

Proof. First we prove that the forms in (7.30) are sums of forms of type (p,p). Clearly,

$$\mathcal{A}_{Y,b,t}^{2} = \left[\mathcal{A}_{Y}'', \mathcal{A}_{Y,b,t}'\right]. \tag{7.35}$$

By Theorem 3.16 and by (7.35), the operator $\mathcal{A}_{Y}^{2}$ is of total degree 0. Also the operators $N^{V}, \omega^{M}, \omega^{\mathcal{M},H}$ are of total degree 0. Since supertraces vanish on supercommutators, we deduce that the forms in (7.30) are sums of form of type (p,p). As we saw in (7.16), after an explicit conjugation, the superconnection $\mathcal{A}_{Y,b,t}$ is a special case of the superconnection $\mathbb{A}_{tY/b^2}^{t}$ which appears in Theorem 6.1. Therefore the forms $u_{b,t}$ are special cases of the forms $m_{b,t}$ in Theorem 6.1. This gives the fact that for $g=1$, these forms are real. Also Theorem 3.16 shows that for $g=1$, the operators appearing in (7.30) are adequately self-adjoint, and this makes the corresponding differential forms real.

By the same arguments as in [BGS88b, Theorem 2.9], we get (7.31). By proceeding as in the proof of [B97, Theorem 4.1], we easily deduce the identities in (7.33).

By (7.8), (7.30), we get

$$w_{b,t}^{(0)} - \underline{w}_{b,t}^{(0)} = \mathrm{Tr_s}\left[\left(2N^{V} - n + N^{H(1,0)} - N^{H(0,1)} + i\frac{L}{t}\right)\exp\left(-\mathcal{A}_{Y,b,t}^{2}\right)\right]. \tag{7.36}$$

By (7.22), (7.36), we get (7.34). This concludes the proof of our theorem. □

Remark 7.8. By proceeding as in [B97], the inclusion (7.33) can be turned into an explicit equality. Moreover, Remark 7.5 indicates that our definition of the form $\underline{w}_{b,t}$ is not the right one in positive degree. Indeed by adding the extra terms which appear in the right-hand side of (7.22) which involve the tensor T, we can make an analogue of (7.34) valid in any degree where the equality is now taken in $P^{S}/P^{S,0}$. Equations (7.31)–(7.33) would still be valid when replacing $w_{b,t}$ by $\underline{w}_{b,t}$.

7.5. The forms $u_{b,t}, v_{b,t}, w_{b,t}$ as $t \to 0$

Recall that

$$\mathrm{Td}\,(x) = \frac{x}{1-e^{-x}}. \tag{7.37}$$

Let $\mathrm{Td}'\,(x)$ be the derivative of $\mathrm{Td}\,(x)$. More generally set

$$\mathrm{Td}\,(x_1,\dots,x_n) = \prod_{i=1}^{n} \mathrm{Td}\,(x_i)\,. \tag{7.38}$$

Set

$$\mathrm{Td}'\,(x_1,\dots,x_n) = \frac{\partial}{\partial c}\mathrm{Td}\,(x_1+c,\dots,x_n+c)\,|_{c=0}. \tag{7.39}$$

We identify Td' with the corresponding obvious genus. Then Td'/Td is an additive genus.

By the same procedure, we can as well define the closed form $\mathrm{Td}_{g}'\left(TX, g^{TX}\right)$ on M_g.

Theorem 7.9. *For any $b > 0$, as $t \to 0$,*

$$u_{b,t} = \int_{X_g} \mathrm{Td}_g\left(TX, g^{TX}\right) \mathrm{ch}_g\left(E, g^E\right) + \mathcal{O}(t), \tag{7.40}$$
$$\frac{1}{4} v_{b,t} = \int_{X_g} \mathrm{Td}'_g\left(TX, g^{TX}\right) \mathrm{ch}_g\left(E, g^E\right) + \mathcal{O}(t).$$

As $t \to 0$, there are forms $C_{b,-1}, C_{b,0} \in P^S$ such that

$$w_{b,t} = \frac{C_{b,-1}}{t} + C_{b,0} + \mathcal{O}(t) \tag{7.41}$$

Moreover,

$$C_{b,-1} = \int_{X_g} \frac{\omega^M}{2\pi} \mathrm{Td}_g\left(TX, g^{TX}\right) \mathrm{ch}_g\left(E, g^E\right),$$
$$C_{b,0} = -3 \int_{X_g} \mathrm{Td}'_g\left(TX, g^{TX}\right) \mathrm{ch}_g\left(E, g^E\right) \tag{7.42}$$
$$+ n \int_{X_g} \mathrm{Td}_g\left(TX, g^{TX}\right) \mathrm{ch}_g\left(E, g^E\right).$$

For $b > 0$, as $t \to 0$,

$$u_{\sqrt{t}b,t} = \int_{X_g} \mathrm{Td}_g\left(TX|_{M_g}, g^{TX|_{M_g}}\right) \mathrm{ch}_g\left(E, g^E\right) + \mathcal{O}\left(\sqrt{t}\right),$$
$$\frac{1}{4} v_{\sqrt{t}b,t} = -\int_{X_g} \mathrm{Td}_g\left(TX, g^{TX}\right) \left(\frac{\widehat{\Sigma}'_g}{\widehat{\Sigma}_g}\left(1/b^4, TX, g^{TX}\right) - \frac{n}{2}\right) \tag{7.43}$$
$$\cdot \mathrm{ch}_g\left(E, g^E\right) + \mathcal{O}\left(\sqrt{t}\right).$$

Proof. As we saw in the proof of Theorem 7.7, the forms $u_{b,t}$ are special cases of the forms $m_{b,t}$ in Theorem 6.1. By using this theorem, we obtain the first identity in (7.40).

As we saw in (5.54),

$$N^V - n = -\frac{i}{4} \left\langle J^{TX} e_i, e_j \right\rangle c(\widehat{e}_i) c(\widehat{e}_j) - \frac{n}{2}. \tag{7.44}$$

Also by Proposition 6.10, we get

$$\mathrm{Tr}_s^{\widehat{\Lambda}^{\cdot}(\overline{T^*X})}\Bigg[\left(N^V - n\right) \exp\Big(-c(\widehat{e}_i)\mathfrak{e}_i/\sqrt{2}b^2 - \frac{1}{4}\left\langle R^{TX} e_i, e_j \right\rangle c(\widehat{e}_i) c(\widehat{e}_j)$$
$$- \frac{1}{2}\mathrm{Tr}\left[R^{TX}\right]\Big)\Bigg] = b^{-2n} \frac{\partial}{\partial c} \widehat{A}^{-1}\left(R^{TX} + icJ^{TX}\right) \exp\left(-\frac{1}{2}\mathrm{Tr}\left[R^{TX} + c1\right]\right)|_{c=0}$$
$$= -b^{-2n} \frac{\mathrm{Td}'}{\mathrm{Td}^2}\left(-R^{TX}\right) \exp\left(-\mathrm{Tr}\left[R^{TX}\right]\right). \tag{7.45}$$

By proceeding as in the proof of Theorem 6.11 and using (7.45), in the case where E is trivial and $g=1$, we get

$$\int_{T^*X} \widehat{\operatorname{Tr}_{\mathrm{s}}}\left[\left(N^V-n\right)\exp\left(-\mathfrak{P}_{b,0}\right)\left(\left(0,Y\right),\left(0,Y\right)\right)\right]\frac{dUdY}{\left(2\pi\right)^{n-\ell/2}}=-\operatorname{Td}'\left(-R^{TX}\right). \tag{7.46}$$

By (7.8) and (7.30), we get the second identity in (7.40) when $g=1$.

Assume now that no eigenvalue of g is equal to 1. Then by (6.61) and (7.44), or by a simple direct computation, we get

$$\begin{aligned}\operatorname{Tr_s}^{\widehat{\Lambda}^{\cdot}\left(\overline{T^*X}\right)}&\left[g\left(N^V-n\right)\exp\left(-\frac{1}{4}\left\langle R^{TX}e_i,e_j\right\rangle c\left(\widehat{e}_i\right)c\left(\widehat{e}_j\right)-\frac{1}{2}\operatorname{Tr}\left[R^{TX}\right]\right)\right]\\&=\left(-1\right)^n\frac{\partial}{\partial c}\operatorname{Td}_g^{-1}\left(-R^{TX}+c1\right)\det{}^{TX}\left[ge^{-R^{TX}}\right]\\&=-\left(-1\right)^n\frac{\operatorname{Td}_g'}{\operatorname{Td}_g^2}\left(-R^{TX}\right)\det{}^{TX}\left[ge^{-R^{TX}}\right].\end{aligned} \tag{7.47}$$

By combining (6.54) with (6.62), (6.63) and (7.47), we get the second identity in (7.40) also in this case.

Now we will establish (7.41), (7.42). Observe that by (7.8), (7.30),

$$\varphi\operatorname{Tr_s}\left[g\left(3N^V-2n\right)\exp\left(-\mathcal{A}_{Y,b,t}^2\right)\right]=-\frac{3}{4}v_{b,t}+nu_{b,t}. \tag{7.48}$$

By (7.40), we find that as $t\to0$,

$$\begin{aligned}\varphi\operatorname{Tr_s}\left[g\left(3N^V-2n\right)\exp\left(-\mathcal{A}_{b,t}^2\right)\right]&=-3\int_{X_g}\operatorname{Td}_g'\left(TX,g^{TX}\right)\operatorname{ch}_g\left(E,g^E\right)\\&+n\int_{X_g}\operatorname{Td}_g\left(TX,g^{TX}\right)\operatorname{ch}_g\left(E,g^E\right)+\mathcal{O}\left(t\right).\end{aligned} \tag{7.49}$$

By (7.8) and (7.30), to obtain the asymptotics of $w_{b,t}$ as $t\to0$, we now need to study the behaviour of $\varphi\operatorname{Tr_s}\left[gi\frac{\omega^M}{t}\exp\left(-\mathcal{A}_{Y,b,t}^2\right)\right]$ as $t\to0$. When performing the rescalings indicated in Definition 6.4, only the part of ω^M which lies in $\Lambda^{\cdot}\left(N_{X_g/X,x,\mathbf{R}}^*\otimes_{\mathbf{R}}\mathbf{C}\right)$ is rescaled by a factor $\sqrt{t}$ or t. One concludes easily that as $t\to0$,

$$\varphi\operatorname{Tr_s}\left[gi\frac{\omega^M}{t}\exp\left(-\mathcal{A}_{Y,b,t}^2\right)\right]=\frac{C_{b,-1}}{t}+\mathcal{O}\left(1\right). \tag{7.50}$$

By combining (7.8), (7.30), (7.49), (7.50), we get the first term in the asymptotic expansion of $w_{b,t}$ as $t\to0$ in (7.41).

More generally, we know that the left-hand side of (7.50) has an asymptotic expansion as $t\to0$. We will now proceed as in [BGS88b].

Indeed using the fact that ω^M is closed, we get

$$\begin{aligned}\frac{\partial}{\partial t}\mathrm{Tr_s}\left[g\omega^M \exp\left(-\mathcal{A}^2_{Y,b,t}\right)\right] &= \frac{\partial}{\partial c}\mathrm{Tr_s}\left[g\omega^M \exp\left(-\mathcal{A}^2_{Y,b,t} - c\left[\mathcal{A}''_Y, \frac{\partial}{\partial t}\mathcal{A}'_{Y,b,t}\right]\right)\right]|_{c=0} \\ &= \overline{\partial}^S \frac{\partial}{\partial c}\mathrm{Tr_s}\left[g\omega^M \exp\left(-\mathcal{A}^2_{Y,b,t} - c\frac{\partial}{\partial t}\mathcal{A}'_{Y,b,t}\right)\right]|_{c=0} \\ &\quad - \frac{\partial}{\partial c}\mathrm{Tr_s}\left[g i_y\omega^M \exp\left(-\mathcal{A}^2_{Y,b,t} - c\frac{\partial}{\partial t}\mathcal{A}'_{Y,b,t}\right)\right]|_{c=0}. \end{aligned} \tag{7.51}$$

By (7.15),

$$\frac{\partial}{\partial t}\mathcal{A}'_{Y,b,t} = \frac{3t^2}{b^4}\overline{\partial}^{V*} - \frac{\overline{y}^*}{t^2}\wedge. \tag{7.52}$$

Let $da, \overline{da}$ be odd Grassmann variables which anticommute with all the other odd operators or Grassmann variables. If $\alpha \in \Lambda^\cdot\left(T^*_{\mathbf{R}}S \otimes_{\mathbf{R}} \mathbf{C}\right) \widehat{\otimes} \mathbf{C}\left[da, \overline{da}\right]$, let $\alpha^{da\overline{da}} \in \Lambda^\cdot\left(T^*_{\mathbf{R}}S \otimes_{\mathbf{R}} \mathbf{C}\right)$ be the coefficient which appears as a factor of $da\overline{da}$ in the expansion of α. By (7.51), (7.52), we obtain

$$\begin{aligned}-\frac{\partial}{\partial c}\mathrm{Tr_s}\left[g i_y\omega^M \exp\left(-\mathcal{A}^2_{Y,b,t} - c\frac{\partial}{\partial t}\mathcal{A}'_{Y,b,t}\right)\right]|_{c=0} \\ = \mathrm{Tr_s}\left[g \exp\left(-\mathcal{A}^2_{Y,b,t} - \frac{\overline{da}}{t} i_y\omega^M - da\left(\frac{3t^3}{b^4}\overline{\partial}^{V*} - \frac{\overline{y}^*}{t}\wedge\right)\right)\right]^{da\overline{da}}. \end{aligned} \tag{7.53}$$

To handle (7.53), as suggested by (6.9) and (7.16), we conjugate the operator in the exponential in the right-hand side of (7.53) by $K_b\rho^*_{t^{3/2}/b^2}$. Note that

$$\begin{aligned}K_b\rho^*_{t^{3/2}/b^2}\left(\frac{\overline{da}}{t}i_y\omega^M + da\left(\frac{3t^3}{b^4}\overline{\partial}^{V*} - \frac{\overline{y}^*}{t}\wedge\right)\right)\rho^{*-1}_{t^{3/2}/b^2}K_{1/b} \\ = \overline{da}\frac{\sqrt{t}}{b}i_y\omega^M + da\left(\frac{3}{b}\overline{\partial}^{V*} - \frac{\sqrt{t}}{b}\overline{y}^*\wedge\right). \end{aligned} \tag{7.54}$$

Using (3.2), (3.9), we find that performing the rescaling indicated in Definition 6.4, the right-hand side of (7.54) consists either of terms which are unaffected, or which are rescaled by $\sqrt{t}$. Ultimately by defining the operator $\mathfrak{P}_{b,0}$ as in (6.26), we find that to study the limit as $t \to 0$ of (7.53), the limit operator to be considered is given by

$$\mathfrak{Q}_{b,0} = \mathfrak{P}_{b,0} + \frac{3da}{b}\overline{\partial}^{V*}. \tag{7.55}$$

By proceeding as in the proof of Theorem 6.1, using (7.51)–(7.55) and the fact that no term with $\overline{da}$ appears in $\mathfrak{Q}_{b,0}$, we find that as $t \to 0$,

$$-\frac{\partial}{\partial c}\mathrm{Tr_s}\left[g i_y\omega^M \exp\left(-\mathcal{A}^2_{Y,b,t} - c\frac{\partial}{\partial t}\mathcal{A}'_{Y,b,t}\right)\right]|_{c=0} = \mathcal{O}(t). \tag{7.56}$$

By proceeding as before and using the parity argument which was given at the end of the proof of Theorem 6.11, we find that as $t \to 0$,

$$\frac{\partial}{\partial c} \mathrm{Tr_s}\left[g\omega^M \exp\left(-\mathcal{A}^2_{Y,b,t} - c\frac{\partial}{\partial t}\mathcal{A}'_{Y,b,t}\right)\right]|_{c=0} = \mathcal{O}(t) . \qquad (7.57)$$

By (7.51), (7.56), (7.57), we find that as $t \to 0$,

$$\frac{\partial}{\partial t} \mathrm{Tr_s}\left[g\omega^M \exp\left(-\mathcal{A}^2_{Y,b,t}\right)\right] = \mathcal{O}(t) . \qquad (7.58)$$

By (7.8), (7.30), (7.49), (7.50), (7.58), we get the asymptotic expansion for $w_{b,t}$ in (7.41) and (7.42).

Now we will establish the identities in (7.43). We modify the definition of $\overline{\mathfrak{P}}_{b,0}$ in (6.76) by adding $\frac{1}{2}\mathrm{Tr}\left[R^{TX}\right]$. The first identity in (7.43) is just a consequence of Theorem 6.12. By proceeding as in the proof of Theorem 6.12, we find that given $b > 0$, as $t \to 0$,

$$\frac{1}{4} v_{\sqrt{t}b,b} = -\varphi \int_{N_{X_g/X} \times T^*X} \widehat{\mathrm{Tr_s}}\left[g\left(N^V - n\right) \exp\left(-\mathfrak{P}_{b,0}\right)\left(g^{-1}(U,Y),(U,Y)\right)\right] \cdot \frac{dU dY}{(2\pi)^{2n-\ell}} + \mathcal{O}\left(\sqrt{t}\right) . \qquad (7.59)$$

By using equation (5.53) in Theorem 5.2 and (7.59), we get the second identity in (7.43).

The proof of our theorem is completed. □

7.6. The limit of the hypoelliptic superconnection forms as $b \to 0$ and $t \to \infty$

Here we assume S to be compact. Also as in Subsection 4.4, we assume that the $H^{(0,i)}(X, E|_X), 0 \le i \le n$ have locally constant rank, so that $H^{(0,\cdot)}(X, E|_X)$ is a holomorphic **Z**-graded vector bundle on S.

By (7.16) with $t = 1$, we get

$$\rho^*_{1/b^2} \mathcal{A}_{Y,b} \rho^{*-1}_{1/b^2} = \mathcal{A}_{Y_b} . \qquad (7.60)$$

Recall that A_{Y_b} is the component of degree 0 in $\Lambda^{\cdot}(T^*_{\mathbf{R}}S \otimes_{\mathbf{R}} \mathbf{C})$ of $\mathcal{A}_{Y_b}$. By (7.60), we get

$$\rho^*_{1/b^2} A_{Y,b} \rho^{*-1}_{1/b^2} = A_{Y_b} . \qquad (7.61)$$

We take $b_0 > 0$ small enough so that the conclusions of Subsections 3.10 and 6.1 hold uniformly over the fibres $X_s, s \in S$ for $b \in]0, b_0]$. In particular for $s \in S$, the spectrum of $A^2_{Y_b}$ lies in the domain indicated in Figure 3.3, and moreover, (3.79) holds.

By the above, for $b \in]0, b_0]$, we have the canonical isomorphism,

$$\ker A^2_{Y,b} \simeq H^{(0,\cdot)}(X, E|_X) . \qquad (7.62)$$

It follows from the results of Subsection 3.10 that for $b \in]0, b_0]$, the restriction of the Hermitian form ϵ_b to $\ker A^2_{Y,b}$ is nondegenerate. Let $g_b^{H^{(0,\cdot)}(X,E|_X)}$ denote the corresponding holomorphic Hermitian form on $H^{(0,\cdot)}(X, E|_X)$ via the canonical isomorphism (7.62).

For $b \in]0, b_0]$, the holomorphic vector bundle $H^{(0,\cdot)}(X, E|_X)$ is equipped with the Hermitian form $g_b^{H^{(0,\cdot)}(X,E|_X)}$. There is a corresponding holomorphic connection $\nabla_b^{H^{(0,\cdot)}(X,E|_X)}$ on $H^{(0,\cdot)}(X, E|_X)$ preserving $g_b^{H^{(0,\cdot)}(X,E|_X)}$. We denote by $\mathrm{ch}\left(H^{(0,\cdot)}(X, E|_X), g_b^{H^{(0,\cdot)}(X,E|_X)}\right)$ the associated Chern character form on S.

The operator $\mathcal{N}$ defines the grading on $\ker A_{Y,b}^2 \simeq H^{(0,\cdot)}(X, E|_X)$.

Let $\eta > 0$ be small, and let $\delta \subset \mathbf{C}$ be the circle of centre 0 and radius η. Put

$$\mathbf{P}_b = \frac{1}{2i\pi} \int_\delta \frac{d\lambda}{\lambda - A_{Y,b}^2}. \tag{7.63}$$

As we saw in Subsection 3.10, for $b > 0$ small enough, the operator $\mathbf{P}_b$ is well defined. For $b > 0$ small enough, $\mathbf{P}_b$ is exactly the spectral projector for $A_{Y,b}^2$ which is associated to the eigenvalue 0. By (3.79), for $b > 0$ small enough, the associated characteristic subspace is just $\ker A_{Y,b}^2 \simeq H^{(0,\cdot)}(X, E|_X)$. In particular the operators $\mathbf{P}_b\left(N^V - n\right)\mathbf{P}_b, \mathbf{P}_b\left(3N^V - 3n + iL\right)\mathbf{P}_b$ act on $\ker A_{Y,b}^2$, and so they act on $H^{(0,\cdot)}(X, E|_X)$.

Definition 7.10. Set

$$\begin{aligned}
u_{b,\infty} &= \mathrm{ch}\left(H^{(0,\cdot)}(X,E), g_b^{H^{(0,\cdot)}(X,E)}\right), \\
v_{b,\infty} &= -4\varphi\mathrm{Tr_s}\left[\mathbf{P}_b\left(N^V - n\right)\mathbf{P}_b \exp\left(-\nabla_b^{H^{(0,\cdot)}(X,E),2}\right)\right], \\
w_{b,\infty} &= \varphi\mathrm{Tr_s}\left[\mathbf{P}_b\left(3N^V - 2n + iL\right)\mathbf{P}_b \exp\left(-\nabla_b^{H^{(0,\cdot)}(X,E),2}\right)\right], \\
\underline{w}_{b,\infty} &= \varphi\mathrm{Tr_s}\left[(\mathcal{N} - n)\exp\left(-\nabla_b^{H^{(0,\cdot)}(X,E),2}\right)\right].
\end{aligned} \tag{7.64}$$

Proposition 7.11. *The following identity holds,*

$$w_{b,\infty} = \underline{w}_{b,\infty}. \tag{7.65}$$

Proof. By (7.64),

$$\begin{aligned}
w_{b,\infty} - \underline{w}_{b,\infty} = \varphi\mathrm{Tr_s}\Big[\mathbf{P_b}\left(2N^V - n + N^{H(1,0)} - N^{H(0,1)} + iL\right)\mathbf{P}_b \\
\cdot \exp\left(-\nabla_b^{H^{(0,\cdot)}(X,E|_X),2}\right)\Big].
\end{aligned} \tag{7.66}$$

For $b \in]0, b_0]$, $A_Y'', A_{Y,b}'$ both vanish on $\ker A_{Y,b}^2$. By (7.22) with $t = 1$ and by (7.66), we get (7.65). □

Recall that u_t, w_t were defined in (4.6).

Definition 7.12. Set

$$u_{0,t} = u_t, \qquad v_{0,t} = 2nu_t, \qquad w_{0,t} = w_t. \tag{7.67}$$

Now we state a result which is an obvious analogue of [BL06, Theorem 5.4].

Theorem 7.13. *There exists $b_0 > 0$ such that given $t_0 > 0$, there exists $C > 0$ such that for $b \in]0, b_0], t \geq t_0$,*

$$|u_{b,t} - u_{b,\infty}| \leq \frac{C}{\sqrt{t}}, \qquad |v_{b,t} - v_{b,\infty}| \leq \frac{C}{\sqrt{t}}, \qquad |w_{b,t} - w_{b,\infty}| \leq \frac{C}{\sqrt{t}}. \tag{7.68}$$

Moreover, given $t_0 > 0, v \in]0,1[$, there exists $C_{t_0,v} > 0$ such that for $b \in]0, b_0]$, $t \geq t_0$,

$$|u_{b,t} - u_{0,t}| \leq C_{t_0,v} b^v, \quad |v_{b,t} - v_{0,t}| \leq C_{t_0,v} b^v, \qquad |w_{b,t} - w_{0,t}| \leq C_{t_0,v} b^v. \tag{7.69}$$

Proof. We claim that the arguments which were given to establish [BL06, Theorem 5.4] can be easily modified to establish our theorem.

Indeed our theorem is about uniform convergence as $t \to \infty$ of forms depending on (b,t), and about uniform convergence as $b \to 0$ of such forms. However inspection of equations (2.31), (2.32), of Theorems 2.7, 2.8, 3.18 and of (7.16), (7.17), (7.60) shows that analytically the situation is essentially the same as in [BL06]. The only point where we will give more details is when studying the convergence of the forms $u_{b,t}, v_{b,t}, w_{b,t}$ as $b \to 0$.

By Theorem 3.18, the proof of the estimate in (7.69) for $u_{b,t}$ follows the same lines as the proof of the corresponding result in [BL06]. Also $v_{b,t}, w_{b,t}$ are of the form

$$h_{b,t} = \varphi \mathrm{Tr_s}\left[E_t \exp\left(-\mathcal{A}_{Y,b,t}^2\right)\right], \tag{7.70}$$

where E_t is an endomorphism which does not depend on Y, and is invariant by conjugation by $\rho_a^*, K_a, a > 0$. Recall that the projector P was defined in Subsection 3.9. Set

$$F_t = e^{-i\omega^{M,H}/2t} P \exp(it\Lambda) E_t \exp(-it\Lambda) P e^{i\omega^{M,H}/2t}. \tag{7.71}$$

Put

$$h_{0,t} = \varphi \mathrm{Tr_s}\left[F_t \exp\left(-B_t^2\right)\right]. \tag{7.72}$$

Inspection of the arguments in [BL06] and of (3.115), (7.16), (7.17), shows that under the conditions on the parameters given before (7.69),

$$|h_{b,t} - h_{0,t}| \leq C_{t_0,v} b^v. \tag{7.73}$$

The question then arises to calculate F_t in the two remaining cases to be considered in (7.69). Note that in the case of $v_{b,t}$,

$$F_t = -4P\left(N^V - n\right)P. \tag{7.74}$$

Now one verifies easily that

$$PN^V P = n/2, \tag{7.75}$$

so that in the case of $v_{b,t}$,

$$F_t = 2n. \tag{7.76}$$

This fits with equation (7.67) for $v_{0,t}$.

In the case of $w_{b,t}$, by (7.8),

$$E_t = 3N^V - 2n + i\frac{\omega^M}{t}, \tag{7.77}$$

so that by (3.6),

$$\exp(it\Lambda) E_t \exp(-it\Lambda) = 3N^V - 3n + i\frac{\omega^M}{t} + N^{H(1,0)} + N^{H(0,1)} + it\Lambda. \tag{7.78}$$

Moreover, by (7.75), we get

$$PN^{H(1,0)}P = \frac{n}{2}. \tag{7.79}$$

Also we have the trivial,

$$P\omega^M P = \omega^{M,H}, \qquad P\Lambda P = 0. \tag{7.80}$$

By (7.75), (7.77)–(7.80), we get

$$F_t = N^{H(0,1)} - n + i\frac{\omega^{M,H}}{t}. \tag{7.81}$$

By comparing with (4.6), we find that as it should be, $h_{0,t} = w_t$. The proof of our theorem is completed. □

Remark 7.14. Obvious analogues of (7.68) and (7.69) are also valid for $\underline{w}_{b,t}$, with $\underline{w}_{0,t} = w_{0,t}$. As was pointed out in Remark 7.8, there is a better definition of $\underline{w}_{b,t}$ in positive degree. When using this better definition, one still finds that Theorem 7.13 remains valid with this new definition, the extra terms disappearing in the limit $b \to 0$ or $t \to +\infty$.

7.7. The hypoelliptic holomorphic torsion forms

We still assume the assumptions of Subsection 7.6 to be in force.

We will imitate the construction of the elliptic torsion forms in Subsection 4.5. We take $b \in]0, b_0]$.

Definition 7.15. For $s \in \mathbf{C}, \operatorname{Re} s > 1$, set

$$\zeta^1(s) = -\frac{1}{\Gamma(s)} \int_0^1 t^{s-1} (w_{b,t} - w_{b,\infty})\, dt. \tag{7.82}$$

For $s \in \mathbf{C}, \operatorname{Re} s < 1/2$, set

$$\zeta^2(s) = -\frac{1}{\Gamma(s)} \int_1^\infty t^{s-1} (w_{b,t} - w_{b,\infty})\, dt. \tag{7.83}$$

Observe that by Theorems 7.9 and 7.13, $\zeta^1(s), \zeta^2(s)$ are holomorphic on their domain of definition, and they extend to holomorphic functions near $s = 0$.

Definition 7.16. Set

$$T_{b,g}\left(\omega^{M},g^{E}\right)=\frac{\partial}{\partial s}\left(\zeta^{1}+\zeta^{2}\right)(s)\,|_{s=0}. \tag{7.84}$$

Then $T_{b,g}\left(\omega^{M},g^{E}\right)$ is an even form on S which lies in P^{S}. It is called an hypoelliptic equivariant holomorphic torsion form. When $g=1$, we will write $T_{b}\left(\omega^{M},g^{E}\right)$ instead of $T_{b,1}\left(\omega^{M},g^{E}\right)$.

The strict analogue of equation (4.21) is now

$$T_{b,g}\left(\omega^{M},g^{E}\right)=-\int_{0}^{1}\left(w_{b,t}-\frac{C_{b,-1}}{t}-C_{b,0}\right)\frac{dt}{t}-\int_{1}^{\infty}\left(w_{b,t}-w_{b,\infty}\right)\frac{dt}{t} \\ +C_{b,-1}+\Gamma'(1)\left(C_{b,0}-w_{b,\infty}\right). \tag{7.85}$$

The main result of this subsection is the obvious extension of Theorem 4.6.

Theorem 7.17. *The following identity holds,*

$$\frac{\overline{\partial}\partial}{2i\pi}T_{b,g}\left(\omega^{M},g^{E}\right)=\mathrm{ch}_{g}\left(H^{(0,\cdot)}\left(X,E|_{X}\right),g_{b}^{H^{(0,\cdot)}(X,E|_{X})}\right) \\ -\int_{X_{g}}\mathrm{Td}_{g}\left(TX,g^{TX}\right)\mathrm{ch}_{g}\left(E,g^{E}\right). \tag{7.86}$$

Proof. This is an obvious consequence of equation (7.31) in Theorem 7.7, of equations (7.40)–(7.42) in Theorem 7.9, of equation (7.68) in Theorem 7.13 and of (7.85). □

Remark 7.18. It is remarkable that the hypoelliptic torsion forms $T_{b,g}\left(\omega^{M},g^{E}\right)$ verify an equation which is the obvious analogue of the corresponding equation for the elliptic forms $T_{g}\left(\omega^{M},g^{E}\right)$ in Theorem 4.6. This makes indeed the comparison of the forms possible in $P^{S}/P^{S,0}$. By proceeding as in [BK92], one could also establish anomaly formulas for the hypoelliptic torsion forms. However since we will prove a comparison formula between the elliptic and the hypoelliptic torsion forms, and since the anomaly formulas are already known for the elliptic torsion forms, we obtain this way the corresponding anomaly formulas in the hypoelliptic case. For other considerations related to the anomaly formulas, we refer the reader to Section 10.

7.8. A generalized Quillen metric on the determinant of the cohomology

For the moment we assume that S is reduced to a point. Recall that the line λ was defined in (4.24) by

$$\lambda=\det H^{(0,\cdot)}\left(X,E|_{X}\right). \tag{7.87}$$

We will use the results of Subsection 3.10, with the slight modification that $\mathcal{S}_{0}\subset\mathcal{S}^{(0,\cdot)}\left(\mathcal{X},\pi^{*}\left(\Lambda^{\cdot}\left(T^{*}X\right)\widehat{\otimes}E\right)\right)$ now denotes the characteristic subspace associated to the eigenvalue 0 for the operator $A^{2}_{Y,b}$. By (3.75), we have the canonical isomorphism,

$$\lambda\simeq\det\mathcal{S}_{0}. \tag{7.88}$$

Let $g_b^{\mathcal{S}_0}$ be the restriction of ϵ_b to $\mathcal{S}_0$. As we saw in Subsection 3.10, the Hermitian form $g_b^{\mathcal{S}_0}$ is non degenerate.

Definition 7.19. Let $| |^2_{\det \mathcal{S}_0}$ be the generalized metric on $\det \mathcal{S}_0$ which is induced by $g_b^{\mathcal{S}_0}$. Let $| |^2_\lambda$ be the corresponding generalized metric on λ via the isomorphism (7.88).

Note that the above objects depend explicitly on $b > 0$.

In Definition 7.19, we follow the terminology in [B05, Section 1.4]. Indeed if $g_b^{\mathcal{S}_0}$ turns out to be a standard Hermitian metric, then $| |^2_\lambda$ is the corresponding Hermitian metric on λ. The above notation is just the obvious extension to the general case. Note here that the square in $| |^2_\lambda$ is not meant to indicate any positivity. To the contrary, as explained in detail in [B05, Section 1.4], the generalized metric $| |^2_\lambda$ has a definite sign $\epsilon\left(| |^2_\lambda\right)$. If $g^{\mathcal{S}_0}$ has signature (p,q), this sign is $(-1)^q$.

Let $\left(A^2_{Y,b}\right)^{-1}$ be the inverse of $A^2_{Y,b}$ on $\mathcal{S}_*$. For $g \in G, s \in \mathbf{C}, \operatorname{Re} s \gg 0$, set

$$\vartheta_g(s) = -\operatorname{Tr_s}^{\mathcal{S}_*}\left[g\left(\mathcal{N} - n\right)\left(A^2_{Y,b}\right)^{-s}\right]. \tag{7.89}$$

When $g = 1$, we will write $\vartheta(s)$ instead of $\vartheta_1(s)$.

By proceeding as in [BL06, Subsection 6.3], we claim that the nonzero real number $e^{\vartheta'(0)}$ is well defined. Indeed we use the results on the spectrum of $A^2_{Y,b}$ in Subsection 3.10 which are the strict analogues of corresponding results in [BL06], and in particular the fact that the spectrum is conjugation invariant, and also the results in (7.34) and in Theorem 7.9 on the asymptotic expansion on $\underline{w}_{b,t}$ as $t \to 0$. The sign of $e^{\vartheta'(0)}$ depends only on the finite number of negative eigenvalues of $A^2_{Y,b}$.

Definition 7.20. Put

$$S_b\left(g^{TX}, g^E\right) = e^{\vartheta'(0)}. \tag{7.90}$$

Let $\| \|^2_{\lambda,h}$ be the generalized metric on λ,

$$\| \|^2_{\lambda,h} = S_b\left(g^{TX}, g^E\right) | |^2_\lambda. \tag{7.91}$$

Let $\epsilon\left(\| \|^2_{\lambda,h}\right)$ be the sign of the generalized metric $\| \|^2_\lambda$. By (7.91), we get

$$\epsilon\left(\| \|^2_{\lambda,h}\right) = \operatorname{sign}\left(S_b\left(g^{TX}, g^E\right)\right)\epsilon\left(| |^2_\lambda\right). \tag{7.92}$$

Now we adapt Definition 7.16 to the case where S is a point, so that $M = X$. Here $T^H M = \{0\}$. For $b \in \mathbf{R}^*_+$ small enough, we have defined the real number $\mathcal{T}_b\left(g^{TX}, g^E\right)$, which is just $\mathcal{T}_{b,g}\left(g^{TX}, g^E\right)$ in the case where $g = 1$. Here the notation ω^M has been dropped.

For $b > 0$ small enough, by the results of Subsection 3.10, there is no ambiguity in the definition of $\vartheta(s)$, since the nonzero eigenvalues have positive real part.

Proposition 7.21. *For $b > 0$ small enough, the following identity holds,*

$$T_b\left(g^{TX}, g^E\right) = \frac{1}{2}\vartheta'(0). \tag{7.93}$$

Proof. By (7.34), we can replace $w_{b,t}$ by $\underline{w}_{b,t}$ in the definition of $T_b\left(g^{TX}, g^E\right)$. Our proposition is now obvious. □

By proceeding as in [BL06], i.e., by a suitable truncation procedure, there is an analogue of Proposition 7.21 for arbitrary $b > 0$.

7.9. A smooth generalized metric on the determinant bundle

We suppose again that S is arbitrary. Moreover, we do no longer assume that the $H^{(0,i)}\left(X, E|_X\right), 0 \leq i \leq n$ have locally constant rank. Let λ be the holomorphic line bundle on S which was defined in (4.23). For every $s \in S$, we have the canonical isomorphism in (4.24) which we repeat,

$$\lambda_s = \det H^{(0,\cdot)}\left(X_s, E|_{X_s}\right). \tag{7.94}$$

We replace temporarily S by $S' = S \times \mathbf{R}_+^*$. The line bundle λ lifts to S'. For $s' = (s,b) \in S'$, we can equip the fibres λ_s with the generalized metric $\| \|^2_{\lambda,h,s,b}$. For simplicity, this metric will be denoted as $\| \|^2_{\lambda,h}$.

Theorem 7.22. *The metric $\| \|^2_{\lambda,h}$ is a smooth generalized metric on the line bundle λ over S'.*

Proof. The proof of our theorem is strictly identical to the proof of [BGS88c, Theorems 1.6 and 3.14] and of [BL06, Theorem 6.14]. □

7.10. The equivariant determinant bundle

First we assume that S is reduced to a point. We use the same notation as in Subsection 4.7. In particular if $W \in \widehat{G}$, λ_W was defined in (4.28). Also λ was defined in (4.29).

Clearly G commutes with $A^2_{Y,b}$. The splitting (3.73) is preserved by G. The generalized metric $g_b^{\mathcal{S}_0}$ is also G-invariant.

Now we proceed as in [B05, Subsection 1.12]. If $W \in \widehat{G}$, set

$$\mathcal{S}_{0,W} = \operatorname{Hom}_G\left(W, \mathcal{S}_0\right) \otimes W. \tag{7.95}$$

Then we have the isotypical decomposition,

$$\mathcal{S}_0 = \bigoplus_{W \in \widehat{G}} \mathcal{S}_{0,W}. \tag{7.96}$$

By [B05, Proposition 1.24], the decomposition (7.96) is $g_b^{\mathcal{S}_0}$ orthogonal, and the restriction of $g_b^{\mathcal{S}_0}$ to each term in the right-hand side of (7.96) is nondegenerate.

The $\mathcal{S}_{0,W}$ are subcomplexes of $\mathcal{S}_0$. Set

$$\mu_W = \det \mathcal{S}_{0,W}, \qquad \mu = \bigoplus_{W \in \widehat{G}} \mu_W. \tag{7.97}$$

There are canonical isomorphisms,

$$\lambda_W \simeq \mu_W. \tag{7.98}$$

By (7.98), we have the canonical isomorphism,

$$\lambda \simeq \mu. \tag{7.99}$$

We use the notation in [B05, Definition 1.25].

Definition 7.23. Let $\log\left(|\,|^2_\mu\right)$ be the logarithm of the generalized equivariant metric on μ which is associated to $g_b^{\mathcal{S}_0}$, and let $\log\left(|\,|^2_\lambda\right)$ be the corresponding object on λ via the canonical isomorphism (7.99).

Let $\log|\,|^2_{\det \mu_W}$ be the logarithm of the generalized metric on μ_W associated to the restriction of $g_b^{\mathcal{S}_0}$ to $\mathcal{S}_{0,W}$. Recall that χ_W is the character of the representation associated to W. Then we have the identity

$$\log\left(|\,|^2_\mu\right) = \sum_{W\in\widehat{G}} \log\left(|\,|^2_{\mu_W}\right) \otimes \frac{\chi_W}{\operatorname{rk} W}. \tag{7.100}$$

As before, some care has to be given to the fact that the generalized metrics in (7.100) are not necessarily positive. Therefore each term $\log\left(|\,|^2_{\mu_W}\right)$ contains implicitly the logarithm of the sign of the corresponding metric. The logarithm of the sign is just $ki\pi$, with $k \in \mathbf{Z}$ determined modulo 2.

Recall that $\vartheta_g(s)$ was formally defined in (7.89). Note that there are still ambiguities in the definition of $\vartheta_g(s)$, due to the negative part of the spectrum. These ambiguities are lifted as before, by still splitting the spectrum of $A^2_{Y,b}$. There remain ambiguities of the type

$$\sum k_W i\pi\chi_W, \tag{7.101}$$

with $k_W \in \mathbf{Z}$ being unambiguously determined mod 2. In the case where G is trivial, we got rid of the ambiguity in Subsection 7.8 by taking the exponential.

Now we imitate equation (7.91) in Definition 7.20.

Definition 7.24. Put

$$\log\left(\|\,\|^2_{\lambda,h}\right) = \log\left(|\,|^2_\lambda\right) + \vartheta'_\cdot(0). \tag{7.102}$$

The object in (7.102) will be called the logarithm of the generalized equivariant Quillen metric on λ.

The same arguments as in Proposition 7.21 show that for $b > 0$ small enough,

$$T_{b,g}\left(g^{TX}, g^E\right) = \frac{1}{2}\vartheta'_g(0). \tag{7.103}$$

The obvious analogue of Theorem 7.22 is that the generalized equivariant Quillen metric on λ is 'smooth' over $S' = S \times \mathbf{R}^*$. In the case where G is trivial, this was already proved in Theorem 7.22. In the general case, this means in particular that the sign of the various metrics remains constant, or equivalently, that the integers which express the logarithms of these signs remain constant modulo 2. The proof is exactly the same as the proof of Theorem 7.22.

7.11. The dependence on b of the hypoelliptic Quillen metric

Proposition 7.25. *For any $g \in G$, for $b > 0$, the following identity holds,*

$$\frac{\partial}{\partial b} \log \| \|^2_{\lambda,h}(g) = \frac{4}{b} \int_{X_g} \mathrm{Td}'_g\left(TX, g^{TX}\right) \mathrm{ch}_g\left(E, g^E\right). \tag{7.104}$$

Proof. We may assume that S is reduced to a point. By Theorem 7.7, the form a is closed. By proceeding as in the proof of [BL06, Theorem 6.19], where we use Theorem 7.7 instead of [BL06, Theorem 4.3] and also equation (7.40) in Theorem 7.9 instead of [BL06, Theorem 4.8], we get (7.104). The proof of our proposition is completed. □

Remark 7.26. The fact that the hypoelliptic Quillen metric $\| \|_{\lambda,h}$ depends very simply on b is remarkable. Proposition 7.25 will be considerably extended in Section 10, where it will be shown that the dependence of the Quillen metric on the metric on the vertical fibre $\widehat{TX}$ can be expressed in terms of Bott-Chern classes.

8. A comparison formula for the hypoelliptic torsion forms and the hypoelliptic Quillen metrics

In this section, we give a formula comparing the hypoelliptic and the elliptic torsion forms, for $b > 0$ small enough. We still assume that the horizontal and vertical metrics on X are proportional with a fixed constant of proportionality. Corresponding comparison formulas are also given for Quillen metrics. From the results of this section and also using anomaly formulas, we will obtain an extension of this result to the general case in Section 10.

This section is organized as follows. In Subsection 8.1, we recall the definition of Bott-Chern classes.

In Subsection 8.2, we give the comparison formula.

In Subsection 8.3, we introduce a rectangular contour $\Gamma \subset \mathbf{R}^{*2}_{+}$. Our main result will be obtained by integrating on Γ the form a of Theorem 7.7 and by taking part of the contour Γ to infinity.

In Subsection 8.4, we state two intermediate results, whose proof is deferred to Section 9.

In Subsection 8.5, we study the asymptotics of the contribution of the four sides of Γ to the integral of a as Γ is taken to infinity.

In Subsection 8.6, we verify that the divergences in the asymptotics effectively compensate in $P^S/P^{S,0}$.

In Subsection 8.7, we prove our main identity.

In Subsection 8.8, we give a formula comparing elliptic and hypoelliptic Quillen metrics.

In the whole section, we use the notation of Sections 4 and 7. Also the assumptions of Subsections 7.6–7.7 are supposed to be in force. In particular S is assumed to be compact.

8.1. Bott-Chern classes

Let Z be a complex manifold. Let $P^Z, P^{Z,0}$ be the space of smooth forms on Z which were defined in Subsection 4.3. Let E be a holomorphic vector bundle on Z, let $g^E, g^{E\prime}$ be two Hermitian metrics on Z. Let Q be a characteristic polynomial. Then $Q\left(E, g^E\right)$ and $Q\left(E, g^{E\prime}\right)$ are closed forms in P^Z, which lie in the same cohomology class $Q(E)$.

The theory of Bott-Chern classes [BGS88a, Section 1 f)] implies the existence of a class $\widetilde{Q}\left(E, g^E, g^{E\prime}\right) \in P^Z/P^{Z,0}$ such that

$$\frac{\overline{\partial}\partial}{2i\pi}\widetilde{Q}\left(E, g^E, g^{E\prime}\right) = Q\left(E, g^{E\prime}\right) - Q\left(E, g^E\right). \tag{8.1}$$

Indeed [BGS88a, Theorem 1.29] asserts that the class $\widetilde{Q}\left(E, g^E, g^{E\prime}\right)$ is uniquely characterized by (8.1), by the fact that it vanishes for $g^{E\prime} = g^E$, and by functoriality.

Observe that the class of $\mathrm{Td}_g\left(TX, g^{TX}\right)\mathrm{ch}_g\left(E, g^E\right)$ in $P^{M_g}/P^{M_g,0}$ does not depend on g^{TX}, g^E. This class will be denoted $\mathrm{Td}_g(TX)\,\mathrm{ch}_g(E)$.

8.2. The main result

Theorem 8.1. *For $b \in]0, b_0], 0 \leq i \leq n$, the Hermitian forms $g_b^{H^{(0,i)}(X,E|_X)}$ are Hermitian metrics. For $b \in]0, b_0]$, the following identity holds,*

$$\begin{aligned} T_{b,g}\left(\omega^M, g^E\right) - T_g\left(\omega^M, g^E\right) - \widetilde{\mathrm{ch}}_g\left(g^{H^{(0,\cdot)}(X,E|_X)}, g_b^{H^{(0,\cdot)}(X,E|_X)}\right) \\ - \int_{X_g} \mathrm{Td}'_g(TX)\,\mathrm{ch}_g(E)\log\left(b^4\right) \\ - \int_{X_g} \mathrm{Td}_g(TX)\,R_g(TX)\,\mathrm{ch}_g(E) = 0 \,\mathrm{in}\, P^S/P^{S,0}. \end{aligned} \tag{8.2}$$

Proof. The remainder of the section is devoted to the proof of our theorem. □

8.3. A contour integral

Recall that the even differential form a was defined in equation (7.32). Let β, ϵ, A be such that $0 < \beta < b_0, 0 < \epsilon < 1 < A$. Let Γ be the oriented rectangular contour in $\mathbf{R}_+^{*2}$ indicated in Figure 8.1. The contour Γ is made of four oriented pieces $\Gamma_1, \ldots, \Gamma_4$. It bounds a domain Δ.

By construction, $\int_\Gamma a$ lies in P^S.

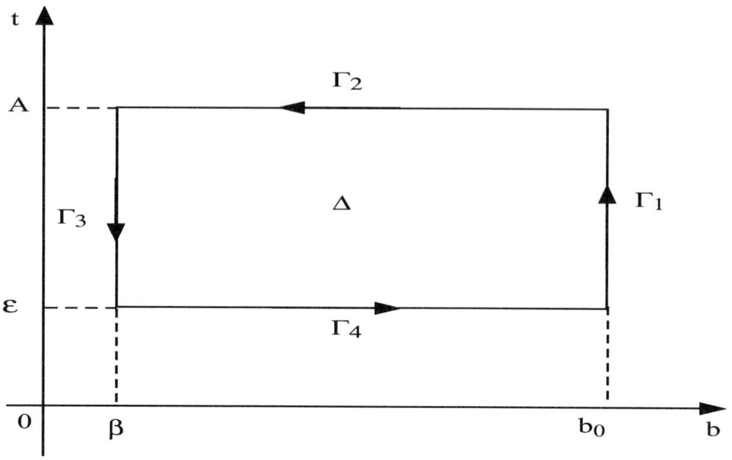

FIGURE 8.1

Proposition 8.2. *The following identity of even forms holds on S,*

$$\int_{\Gamma} a \in P^{S,0}. \tag{8.3}$$

Proof. Equation (8.3) follows from (7.33). □

As was explained in Remark 7.8, the form in (8.3) can be made completely explicit.

For $1 \leq k \leq 4$, set

$$I_k^0 = \int_{\Gamma_k} a. \tag{8.4}$$

We can rewrite equation (8.3) in the form,

$$\sum_{k=1}^{4} I_k^0 \in P^{S,0}. \tag{8.5}$$

To obtain Theorem 8.1, we will make $A \to \infty, \beta \to 0, \epsilon \to 0$ in this order in (8.5).

When S is compact and Kähler, $P^{S,0}$ is closed in P^S, so that limits can be taken in (8.5) without difficulty. In the case of a general compact manifold S, we should also control the right-hand side when taking the various limits. This can be done along arguments similar to the ones which we will use in the sequel. A similar problem was already studied in [B97]. The somewhat tedious details will be left to the reader.

8.4. Two intermediate results

In view of equation (7.40) in Theorem 7.9, we will use the notation,

$$v_{b,0} = 4 \int_{X_g} \mathrm{Td}'_g \left(TX, g^{TX}\right) \mathrm{ch}_g \left(E, g^E\right). \tag{8.6}$$

For $u > 0$, set

$$m_u = -\int_{X_g} \mathrm{Td}_g\left(TX, g^{TX}\right) \frac{\widehat{\Sigma}'_g}{\widehat{\Sigma}_g}\left(u, TX, g^{TX}\right) \mathrm{ch}_g\left(E, g^E\right). \tag{8.7}$$

Recall that $g^{H^{(0,\cdot)}(X,E|_X)}$ is the classical L^2 metric on $H^{(0,\cdot)}(X, E|_X)$ which was defined in Subsection 4.4, and that for $b \in]0, b_0]$, $g_b^{H^{(0,\cdot)}(X,E|_X)}$ is the Hermitian form on $H^{(0,\cdot)}(X, E|_X)$ introduced in Subsection 7.6.

Theorem 8.3. *For any $v \in]0,1[$, when $b \to 0$,*

$$g_b^{H^{(0,\cdot)}(X,E|_X)} = b^{2n}\left(g^{H^{(0,\cdot)}(X,E|_X)} + \mathcal{O}\left(b^v\right)\right). \tag{8.8}$$

Theorem 8.4. *There exist $C > 0, \alpha \in]0,1]$ such that for $\epsilon \in]0,1], b \in]0,1]$,*

$$\left|v_{\sqrt{\epsilon}b,\epsilon} - v_{0,\epsilon}\right| \leq C b^{\alpha}. \tag{8.9}$$

For any $b'_0 \geq 1$, there exist $C > 0$ such that for $\epsilon \in]0,1], b \in [\sqrt{\epsilon}, b'_0]$,

$$\left|v_{b,\epsilon} - v_{b,0}\right| \leq C\frac{\epsilon}{b^2}. \tag{8.10}$$

Proof. The proof of Theorems 8.3 and 8.4 is deferred to Section 9. □

Remark 8.5. By Theorem 8.3, $b \in]0, b_0]$, and for b small enough, for any $i, 0 \leq i \leq n$, $g_b^{H^{(0,i)}(X,E|_X)}$ is a Hermitian metric. Since for $b \in]0, b_0]$, the forms $g_b^{H^{(0,i)}(X,E|_X)}$ are nondegenerate, for $b \in]0, b_0]$, the signature of these Hermitian forms remains constant. Therefore for $b \in]0, b_0]$, the $g_b^{H^{(0,i)}(X,E)}$ are Hermitian metrics, which is the first part of Theorem 8.1.

By (4.9) in Theorem 4.2 and by (7.43) in Theorem 7.9, we can take the limit as $\epsilon \to 0$ in (8.9), which is a known identity because of (5.64). Similarly by replacing b by $b\sqrt{\epsilon}$ in (8.10) with $b \geq 1$, and making $\epsilon \to 0$, we obtain an inequality which itself follows from (5.63).

8.5. The asymptotics of the I_k^0

We start from identity (8.5), which asserts that

$$\sum_{k=1}^{4} I_k^0 = 0 \text{ in } P^S/P^{S,0}. \tag{8.11}$$

1) The term $\mathbf{I_1^0}$. Clearly,

$$I_1^0 = \int_{\epsilon}^{A} w_{b_0,t}\frac{dt}{t}. \tag{8.12}$$

α) $A \to \infty$. By Theorem 7.13, as $A \to \infty$,

$$I_1^0 - w_{b_0,\infty}\log(A) \to I_1^1 = \int_{\epsilon}^{1} w_{b_0,t}\frac{dt}{t} + \int_{1}^{\infty}\left(w_{b_0,t} - w_{b_0,\infty}\right)\frac{dt}{t}. \tag{8.13}$$

β) $\beta \to 0$. The term I_1^1 remains constant and equal to I_1^2.

γ) $\epsilon \to 0$. By equation (7.41) in Theorem 7.9, as $\epsilon \to 0$,

$$I_1^2 - \frac{C_{b_0,-1}}{\epsilon} + C_{b_0,0} \log(\epsilon) \to I_1^3 = \int_0^1 \left(w_{b_0,t} - \frac{C_{b,-1}}{t} - C_{b_0,0} \right) \frac{dt}{t} + \int_1^{\infty} \left(w_{b_0,t} - w_{b_0,\infty} \right) \frac{dt}{t} - C_{b_0,-1}. \quad (8.14)$$

δ) *Evaluation of* I_1^3.

Proposition 8.6. *The following identity holds,*

$$I_1^3 = -T_{b_0,g}\left(\omega^M, g^E\right) + \Gamma'(1)\left(C_{b_0,0} - w_{b_0,\infty}\right). \quad (8.15)$$

Proof. This follows from (7.85) and (8.14). □

2) The term $\mathbf{I_2^0}$. We have the identity,

$$I_2^0 = -\int_{\beta}^{b_0} v_{b,A} \frac{db}{b}. \quad (8.16)$$

α) $A \to \infty$. By Theorem 7.13, as $A \to \infty$,

$$I_2^0 \to I_2^1 = -\int_{\beta}^{b_0} v_{b,\infty} \frac{db}{b}. \quad (8.17)$$

Also using (7.8), (7.64), (8.17) and [BGS88a, Corollary 1.30], we get

$$I_2^1 = \widetilde{\mathrm{ch}}_g \left(g_{\beta}^{H^{(0,\cdot)}(X,E|_X)}, g_{b_0}^{H^{(0,\cdot)}(X,E|_X)} \right) \text{ in } P^S/P^{S,0}. \quad (8.18)$$

β) $\beta \to 0$. Clearly,

$$I_2^1 = \widetilde{\mathrm{ch}}_g \left(g^{H^{(0,\cdot)}(X,E|_X)}, g_{b_0}^{H^{(0,\cdot)}(X,E|_X)} \right) - \widetilde{\mathrm{ch}}_g \left(g^{H^{(0,\cdot)}(X,E|_X)}, g_{\beta}^{H^{(0,\cdot)}(X,E|_X)} \right) \text{ in } P^S/P^{S,0}. \quad (8.19)$$

By equation (8.8) in Theorem 8.3 and by [BGS88a, Corollary 1.30], as $\beta \to 0$, for any $v \in]0,1[$,

$$\widetilde{\mathrm{ch}}_g \left(g^{H^{(0,\cdot)}(X,E|_X)}, g_{\beta}^{H^{(0,\cdot)}(X,E|_X)} \right) = -2n\mathrm{ch}_g \left(H^{(0,\cdot)}(X, E|_X), g^{H^{(0,\cdot)}(X,E|_X)} \right) \log(\beta) + \mathcal{O}(\beta^v). \quad (8.20)$$

So by (8.19), (8.20), we find that as $\beta \to 0$,

$$I_2^1 - 2n\mathrm{ch}_g \left(H^{(0,\cdot)}(X, E|_X), g^{H^{(0,\cdot)}(X,E|_X)} \right) \log(\beta) \to I_2^2 = \widetilde{\mathrm{ch}}_g \left(g^{H^{(0,\cdot)}(X,E|_X)}, g_{b_0}^{H^{(0,\cdot)}(X,E|_X)} \right) \text{ in } P^S/P^{S,0}. \quad (8.21)$$

γ) $\epsilon \to 0$. As $\epsilon \to 0$, I_2^2 remains constant and equal to I_2^3.

3) The term $\mathbf{I_3^0}$. We have the identity,

$$I_3^0 = -\int_\epsilon^A w_{\beta,t}\frac{dt}{t}. \tag{8.22}$$

α) $A \to \infty$. By equation (7.68) in Theorem 7.13, as $A \to \infty$,

$$I_3^0 + w_{\beta,\infty}\log(A) \to I_3^1 = -\int_\epsilon^1 w_{\beta,t}\frac{dt}{t} - \int_1^\infty (w_{\beta,t} - w_{\beta,\infty})\frac{dt}{t}. \tag{8.23}$$

β) $\beta \to 0$. By equations (7.68) and (7.69) in Theorem 7.13, as $\beta \to 0$,

$$I_3^1 \to I_3^2 = -\int_\epsilon^1 w_{0,t}\frac{dt}{t} - \int_1^\infty (w_{0,t} - w_{0,\infty})\frac{dt}{t}. \tag{8.24}$$

γ) $\epsilon \to 0$. By (4.10), (7.67) and (8.24), as $\epsilon \to 0$,

$$\begin{aligned} &I_3^2 + \frac{C_{-1}}{\epsilon} - C_0\log(\epsilon) \\ &\to I_3^3 = -\int_0^1 \left(w_t - \frac{C_{-1}}{t} - C_0\right)\frac{dt}{t} - \int_1^{+\infty}(w_t - w_\infty)\frac{dt}{t} + C_{-1}. \end{aligned} \tag{8.25}$$

δ) *Evaluation of* I_3^3.

Proposition 8.7. *The following identity holds,*

$$I_3^3 = T_g\left(\omega^M, g^E\right) - \Gamma'(1)\left(C_0 - w_\infty\right). \tag{8.26}$$

Proof. Our proposition follows from (4.21) and (8.25). □

4) The term $\mathbf{I_4^0}$. Clearly,

$$I_4^0 = \int_\beta^{b_0} v_{b,\epsilon}\frac{db}{b}. \tag{8.27}$$

α) $A \to \infty$. The term I_4^0 remains constant and equal to I_4^1.

β) $\beta \to 0$. By Theorem 7.13, as $\beta \to 0$,

$$I_4^1 + v_{0,\epsilon}\log(\beta) \to I_4^2 = \int_0^{b_0}(v_{b,\epsilon} - v_{0,\epsilon})\frac{db}{b} + v_{0,\epsilon}\log(b_0). \tag{8.28}$$

γ) $\epsilon \to 0$. Take $\epsilon > 0$ small enough so that $b_0/\sqrt{\epsilon} > 1$. Set

$$J_1^0 = \int_0^1 \left(v_{\sqrt{\epsilon}b,\epsilon} - v_{0,\epsilon}\right)\frac{db}{b}, \qquad J_2^0 = \int_1^{b_0/\sqrt{\epsilon}}\left(v_{\sqrt{\epsilon}b,\epsilon} - v_{\sqrt{\epsilon}b,0}\right)\frac{db}{b}. \tag{8.29}$$

Clearly,

$$I_4^2 = J_1^0 + J_2^0 + v_{0,\epsilon} \log\left(\sqrt{\epsilon}\right) + 4\int_{X_g} \mathrm{Td}'_g\left(TX, g^{TX}\right) \mathrm{ch}_g\left(E, g^E\right)\left(\log\left(b_0\right) - \log\left(\sqrt{\epsilon}\right)\right). \tag{8.30}$$

By equation (4.9) in Theorem 4.2, by equation (7.43) in Theorem 7.9, by (7.67), by equation (8.9) in Theorem 8.4, as $\epsilon \to 0$,

$$J_1^0 \to J_1^1 = \int_0^1 4m_{1/b^4} \frac{db}{b} = \int_1^\infty m_u \frac{du}{u}. \tag{8.31}$$

By (5.58), we know that as $u \to 0$,

$$m_u = -\int_{X_g} \left(\frac{n}{2}\mathrm{Td}_g\left(TX, g^{TX}\right) - \mathrm{Td}'_g\left(TX, g^{TX}\right)\right) \mathrm{ch}_g\left(E, g^E\right) + \mathcal{O}(u). \tag{8.32}$$

By (7.43) in Theorem 7.9, by (8.6), by (8.10) in Theorem 8.4 and by (8.32), as $\epsilon \to 0$,

$$J_2^0 \to J_2^1 = \int_0^1 \left(m_u - \int_{X_g} \left(\mathrm{Td}'_g\left(TX, g^{TX}\right) - \frac{n}{2}\mathrm{Td}_g\left(TX, g^{TX}\right)\right)\mathrm{ch}_g\left(E, g^E\right)\right)\frac{du}{u}. \tag{8.33}$$

Using again (4.9) in Theorem 4.2, and also (8.29)–(8.33), we find that as $\epsilon \to 0$,

$$I_4^2 + \int_{X_g} \left(2\mathrm{Td}'_g\left(TX, g^{TX}\right) - n\mathrm{Td}_g\left(TX, g^{TX}\right)\right) \mathrm{ch}_g\left(E, g^E\right) \log(\epsilon) \to I_4^3 = J_1^1 + J_2^1 + 4\int_{X_g} \mathrm{Td}'_g\left(TX, g^{TX}\right) \mathrm{ch}_g\left(E, g^E\right) \log\left(b_0\right). \tag{8.34}$$

δ) *Evaluation of* I_4^3. Recall that the genus R_g was defined in (5.71). We define the closed form $\frac{\widehat{A}'_g}{\widehat{A}_g}\left(TX, g^{TX}\right)$ in the same way as $\frac{\mathrm{Td}'_g}{\mathrm{Td}_g}\left(TX, g^{TX}\right)$ in Subsection 7.5.

Proposition 8.8. *The following identity holds,*

$$I_4^3 = \int_{X_g} \mathrm{Td}_g\left(TX, g^{TX}\right)\left(R_g + 2\Gamma'(1)\frac{\widehat{A}'_g}{\widehat{A}_g}\right)\left(TX, g^{TX}\right) \mathrm{ch}_g\left(E, g^E\right) + 4\int_{X_g} \mathrm{Td}'_g\left(TX, g^{TX}\right) \mathrm{ch}_g\left(E, g^E\right) \log\left(b_0\right). \tag{8.35}$$

Proof. Observe that

$$\frac{\mathrm{Td}'_g}{\mathrm{Td}_g}\left(TX, g^{TX}\right) = \frac{\widehat{A}'_g}{\widehat{A}_g}\left(TX, g^{TX}\right) + \frac{n}{2}. \tag{8.36}$$

By (8.7), (8.33) and (8.36), we get

$$J_2^1 = \int_{X_g} \mathrm{Td}_g\left(TX, g^{TX}\right) \cdot \int_0^1 \left(-\frac{\widehat{\Sigma}'_g}{\widehat{\Sigma}_g}\left(u, TX, g^{TX}\right) - \frac{\widehat{A}'_g}{\widehat{A}_g}\left(TX, g^{TX}\right)\right) \frac{du}{u} \mathrm{ch}_g\left(E, g^E\right). \tag{8.37}$$

Using (5.66), equations (5.67) and (5.68) in Theorem 5.6, and (8.31), (8.34), (8.37), we get (8.35). □

8.6. Matching the divergences

Proposition 8.9. *The following identity holds,*

$$\sum_{k=1}^4 I_k^3 = 0 \text{ in } P^S/P^{S,0}. \tag{8.38}$$

Proof. We start from equation (8.11). As $A \to \infty$, by (8.13), (8.23), we have the diverging terms

$$\left(w_{\beta,\infty} - w_{b_0,\infty}\right) \log(A). \tag{8.39}$$

By (7.64) and by equation (7.65) in Proposition 7.11, it is clear that the forms $w_{b,\infty} \in P^S$ are closed and that their cohomology class does not depend on b. Using (8.39), we thus find that

$$\sum_{k=1}^4 I_k^1 = 0 \text{ in } P^S/P^{S,0}. \tag{8.40}$$

By (7.67), (8.21), (8.28), as $\beta \to 0$, we have the diverging terms,

$$2n\left(-\mathrm{ch}_g\left(H^{(0,\cdot)}\left(X, E|_X\right), g^{H^{(0,\cdot)}(X,E|_X)}\right) + u_\epsilon\right) \log(\beta). \tag{8.41}$$

By Theorems 4.2 and 4.3, (8.41) lies in $P^{S,0}$. So by (8.40), we get

$$\sum_{k=1}^4 I_k^2 = 0 \text{ in } P^S/P^{S,0}. \tag{8.42}$$

By (8.14), (8.25), (8.34), as $\epsilon \to 0$, we have the diverging terms,

$$\left(-C_{b_0,-1} + C_{-1}\right)\frac{1}{\epsilon} + \left(C_{b_0,0} - C_0 + \int_{X_g}\left(2\mathrm{Td}'_g\left(TX, g^{TX}\right) - n\mathrm{Td}_g\left(TX, g^{TX}\right)\right)\mathrm{ch}_g\left(E, g^E\right)\right) \log(\epsilon). \tag{8.43}$$

By (4.11), (7.42), the expression in (8.43) vanishes in $P^S/P^{S,0}$. By (8.42), we get (8.38). The proof of our proposition is completed. □

8.7. A proof of Theorem 8.1

We will now establish Theorem 8.1. Indeed using Proposition 8.6, (8.21), Propositions 8.7, 8.8 and 8.9, we get

$$\begin{aligned}
&-T_{b_0,g}\left(\omega^M, g^E\right) + T_g\left(\omega^M, g^E\right)\\
&+\widetilde{\mathrm{ch}}_g\left(H^{(0,\cdot)}\left(X, E|_X\right), g^{H^{(0,\cdot)}(X,E|_X)}, g_{b_0}^{H^{(0,\cdot)}(X,E|_X)}\right)\\
&+\int_{X_g} \mathrm{Td}_g\left(TX, g^{TX}\right) R_g\left(TX, g^{TX}\right) \mathrm{ch}_g\left(E, g^E\right)\\
&+\int_{X_g} \mathrm{Td}'_g\left(TX, g^{TX}\right) \mathrm{ch}_g\left(E, g^E\right) \log\left(b_0^4\right)\\
&+\Gamma'(1)\left(C_{b_0,0} - C_0 + 2\int_{X_g} \mathrm{Td}_g\left(TX, g^{TX}\right) \frac{\widehat{A}'_g}{\widehat{A}_g}\left(TX, g^{TX}\right) \mathrm{ch}_g\left(E, g^E\right)\right)\\
&+\Gamma'(1)\left(w_\infty - w_{b_0,\infty}\right) = 0 \text{ in } P^S/P^{S,0}.
\end{aligned} \tag{8.44}$$

Now we use (4.11), (7.42) as in (8.43), to conclude that the sum of the terms in the last two lines of (8.44) vanishes in $P^S/P^{S,0}$. Therefore (8.44) is just (8.2) for $b = b_0$. The proof of (8.2) for $0 < b \leq b_0$ is of course the same. We have completed the proof of Theorem 8.1.

8.8. A comparison formula for the Quillen metrics

We make the same assumptions as in Subsections 4.6 and 7.8. We may as well assume that S is reduced to a point.

First we assume that G is trivial. Put

$$\lambda = \det\left(H^{(0,\cdot)}\left(X, E|_X\right)\right). \tag{8.45}$$

Let $\| \|_\lambda$ denote the elliptic Quillen metric on λ which was defined in Subsection 4.6. For $b > 0$, let $\| \|_{\lambda,h,b}$ be the hypoelliptic generalized Quillen metric which was defined in Subsection 7.8.

Theorem 8.10. *For $b > 0$, the generalized metric $\| \|^2_{\lambda,h,b}$ is a Hermitian metric. Moreover, the following identity holds,*

$$\log\left(\frac{\| \|^2_{\lambda,h,b}}{\| \|^2_\lambda}\right) = \int_X \mathrm{Td}(TX) R(TX) \mathrm{ch}(E) + \int_X \mathrm{Td}'(TX) \mathrm{ch}(E) \log\left(b^4\right). \tag{8.46}$$

Proof. By Theorem 7.25, it is enough to prove our theorem for one given $b > 0$. For $b > 0$ small enough, by Theorem 8.1, $g_b^{H^{(0,\cdot)}(X,E|_X)}$ is a Hermitian metric. Also as we saw in Subsections 3.10 and 7.8, for $b > 0$ small enough, $\exp\left(\theta'(0)\right) > 0$. Therefore for $b > 0$ small enough, $\| \|^2_{\lambda,h,b}$ is a Hermitian metric. Finally for $b > 0$ small enough, (8.2) in degree 0 is just (8.46). The proof of our theorem is completed. □

Now we consider the case of a general Lie group G. We define λ as in (4.29). For $b > 0$, we define the elliptic equivariant Quillen metric $\| \|_{\lambda}^{2}$ as in Definition 4.10, and the generalized hypoelliptic equivariant metric $\| \|_{\lambda,h,b}$ as in Definition 7.24.

Theorem 8.11. *For $b > 0, g \in G$, the following identity holds,*

$$\log\left(\frac{\| \|_{\lambda,h,b}^{2}}{\| \|_{\lambda}^{2}}\right)(g) = \int_{X_g} \mathrm{Td}'_g(TX)\,\mathrm{ch}_g(E)\log\left(b^4\right) + \int_{X_g} \mathrm{Td}_g(TX)\,R_g(TX)\,\mathrm{ch}_g(E). \tag{8.47}$$

Proof. The proof of our theorem is the same as the proof of Theorem 8.10. □

Remark 8.12. By Theorem 8.1, for $b > 0$ small enough, (8.47) is an identity of complex numbers. For an arbitrary $b > 0$, one should remember that (7.100) and (7.102) are affected by mod 2 ambiguities, which cannot be ignored in (8.47).

9. A proof of Theorems 8.3 and 8.4

The purpose of this section is to establish Theorems 8.3 and 8.4. The methods to establish these results are closely related to the methods used in [BL06] in the context of de Rham theory.

This section is organized as follows. In Subsection 9.1, we prove Theorem 8.3. In Subsection 9.2, we prove Theorem 8.4.

9.1. A proof of Theorem 8.3

In this subsection, we first assume that S is reduced to one point.

Recall that the vector space of harmonic forms $\mathfrak{H}^{\cdot}$ on X was defined in (4.12), and that we have the canonical isomorphism in (4.13).

For $b > 0$, set

$$\beta_b = \exp\left(b^2 i\widehat{\omega}^{\mathcal{X},V} - b^2\frac{|Y|^2}{2}\right). \tag{9.1}$$

The form β_b is exactly the form β in (3.61) which is attached to the metric $b^2 g^{TX}$.

If $s \in \mathfrak{H}$, put

$$s_b = \pi^* s \wedge \beta_b. \tag{9.2}$$

Using equation (3.32) in Proposition 3.3 and (3.63), we get

$$\left(\overline{\partial}^X + i_y\right) s_b = 0, \tag{9.3}$$

and moreover,

$$i^* s_b = s. \tag{9.4}$$

By Proposition 3.9 and by (9.3), (9.4), we find that s_b is a suitable representative in $\Omega^{(0,\cdot)}\left(\mathcal{X}, \pi^*\left(\Lambda^{\cdot}(T^*X)\widehat{\otimes}E\right)\right)$ of the class $[s] \in H^{(0,\cdot)}(X, E|_X)$.

Recall that the spectral projector $\mathbf{P}_b$ on $\ker A^2_{Y,b}$ is given by (7.63). For $b > 0$ small enough, $\mathbf{P}_b s_b$ is just the canonical representative in $\ker A^2_{Y,b}$ of the class $[s] \in H^{(0,\cdot)}(X, E|_X)$.

Take $s, s' \in \mathfrak{H}^{\cdot}$. By definition,

$$\langle [s], [s'] \rangle_{g_b^{H^{(0,\cdot)}(X,E|_X)}} = \epsilon_b \left(\mathbf{P}_b s_b, \mathbf{P}_b s'_b \right). \tag{9.5}$$

By (7.60),

$$A_{Y,b} = \rho^*_{b^2} A_{Y_b} \rho^{*-1}_{b^2}. \tag{9.6}$$

Recall that the operator $C_{Y,b}$ was defined in (3.65), and is given by equation (3.68). By (3.47), (3.65),

$$A_{Y_b} = K_b^{-1} \exp(-i\Lambda) C_{Y,b} \exp(i\Lambda) K_b. \tag{9.7}$$

Moreover, we have the obvious,

$$\rho^*_{b^2} = b^{2N^V} K_{b^2}. \tag{9.8}$$

By (9.6)–(9.8), we get

$$A_{Y,b} = b^{2N^V} K_b \exp(-i\Lambda) C_{Y,b} \exp(i\Lambda) K_b^{-1} b^{-2N^V}. \tag{9.9}$$

Incidentally note that the order of the factors in the conjugating element in (9.9) is irrelevant.

Set

$$\mathbb{P}_b = \frac{1}{2i\pi} \int_\delta \frac{d\lambda}{\lambda - C^2_{Y,b}} d\lambda. \tag{9.10}$$

For $b > 0$ small enough, $\mathbb{P}_b$ is the spectral projector on $\ker C^2_{Y,b}$. By (9.9),

$$\mathbf{P}_b = b^{2N^V} K_b \exp(-i\Lambda) \mathbb{P}_b \exp(i\Lambda) K_b^{-1} b^{-2N^V}. \tag{9.11}$$

By (3.54), (3.55), (7.1), (9.5), we get

$$\langle [s], [s'] \rangle_{g_b^{H^{(0,\cdot)}(X,E|_X)}} = b^{2n} \eta \left(\mathbb{P}_b \exp(i\Lambda) \pi^* s \wedge \beta, \mathbb{P}_b \exp(i\Lambda) \pi^* s' \wedge \beta \right). \tag{9.12}$$

Recall that the projector P was defined in Subsection 3.9. Let $\mathfrak{P}$ be the orthogonal projection operator from $\Omega^{(0,\cdot)}(X, E)$ on $\mathfrak{H}^{\cdot}$. By proceeding as in [BL06, eqs. (3.67) and (10.11)], since $s \in \mathfrak{H}^{\cdot}$, for $v \in]0, 1[$, as $b \to 0$,

$$\mathbb{P}_b \exp(i\Lambda) \pi^* s \wedge \beta = \left(\mathfrak{P} \widehat{\otimes} P \right) \exp(i\Lambda) \pi^* s \wedge \beta + \mathcal{O}(b^v) \|s\|. \tag{9.13}$$

Now observe that Λ decrease the degree in $\Lambda^{\cdot}(T^*X)$ by 1. Therefore when grading forms in $\Lambda^{\cdot}(T^*X) \widehat{\otimes} \widehat{\Lambda}^{\cdot} \left(\overline{T^*X} \right)$ by $N^V - N^{H(1,0)}$, $(\exp(i\Lambda) - 1) \pi^* s \wedge \beta$ is of positive degree. Since P is a projector on forms of degree 0 and $s \in \mathfrak{H}^{\cdot}$, we find that

$$\left(\mathfrak{P} \widehat{\otimes} P \right) \exp(i\Lambda) \pi^* s \wedge \beta = \left(\mathfrak{P} \widehat{\otimes} P \right) \pi^* s \wedge \beta = \pi^* s \wedge \beta. \tag{9.14}$$

Moreover, $r^* \beta = \beta$. Using (3.69) and (9.12)–(9.14), we find that

$$\langle [s], [s'] \rangle_{g_b^{H^{(0,\cdot)}(X,E|_X)}} = b^{2n} \left(\langle s, s' \rangle_{g^{H^{(0,\cdot)}(X,E|_X)}} + \mathcal{O}(b^v) \|s\| \|s'\| \right), \tag{9.15}$$

which is just Theorem 8.3 in the case of one single fibre.

When S is compact and not reduced to a point, the above estimates can be made uniform over S, so that we still get Theorem 8.3.

9.2. A proof of Theorem 8.4

Using the above arguments, one verifies easily that the proof of Theorem 8.4 can be obtained by the same arguments as the proof of the corresponding statements in [BL06, Theorem 8.6]. Details are left to the reader.

10. Generalized hypoelliptic torsion forms and anomaly formulas

The purpose of this section is to extend the results of Sections 3, 7 and 8 in the case where the horizontal g^{TX} on TX and the vertical metric $\widehat{g}^{TX}$ on $\widehat{TX}$ are distinct. Still g^{TX} will still be assumed to be Kähler.

We show in particular that the hypoelliptic Quillen metric does not depend on g^{TX}, and moreover, that there is an explicit comparison formula with elliptic Quillen metrics which extends the formula of Theorem 8.11. Similar results are established in the equivariant context and also for hypoelliptic torsion forms.

This section is organized as follows. In Subsection 10.1, we briefly consider the Hodge theory of a Dolbeault-Koszul complex associated to two distinct metrics.

In Subsection 10.2, we extend the results of Section 3 to the more general situation which is considered here.

In Subsection 10.3, we extend certain results of local index theory of Section 7.

In Subsection 10.4, we construct corresponding hypoelliptic torsion forms.

In Subsection 10.5, we give anomaly formulas for these torsion forms.

In Subsection 10.6, we give a general formula comparing the hypoelliptic to the elliptic torsion forms.

In Subsection 10.7, we state a corresponding comparison formula for Quillen metrics.

Finally in Subsection 10.8, we relate our comparison formulas to the immersion formulas for standard Quillen metrics or elliptic torsion forms of [BL91, B95, B97].

10.1. The Dolbeault-Koszul complex on a complex vector space with two distinct metrics

Let W be a complex vector space of dimension n, let $W_{\mathbf{R}}$ be the corresponding real vector space of dimension $2n$. Let y be the tautological section of W.

The Dolbeault-Koszul complex $\left(\mathcal{S}^{(0,\cdot)}\left(W, \Lambda^{\cdot}\left(W^{*}\right)\right), \overline{\partial}^{W} + i_{y}\right)$ is naturally $\mathbf{Z}$-graded by the operator $N^{(0,1)} - N^{(1,0)}$. The arguments given in Subsection 3.8 apply to this complex. In particular its cohomology is 1-dimensional and concentrated in degree 0.

Let $g^{W}, \widehat{g}^{W}$ be two Hermitian metrics on W.

We view W as a complex manifold equipped with the metric g^W on its tangent bundle. In particular $\Lambda^{\cdot}\left(\overline{W}^*\right)$ is equipped with the metric induced by $\widehat{g}^W$. Moreover, we equip $\Lambda^{\cdot}(W^*)$ with the metric induced by g^W.

We equip $\mathcal{S}^{(0,\cdot)}(W, \Lambda^{\cdot}(W^*))$ with the L^2 Hermitian product defined as in (3.16). Here the volume element on W is calculated with respect to $\widehat{g}^W$, and the Hermitian product on $\Lambda^{\cdot}\left(\overline{W}^*\right)\widehat{\otimes}\Lambda^{\cdot}(W^*)$ incorporates the two metrics $g^W, \widehat{g}^W$.

The formal adjoint $\left(\overline{\partial}^W + i_y\right)^*$ of $\overline{\partial}^W + i_y$ is given by

$$\left(\overline{\partial}^W + i_y\right)^* = \overline{\partial}^{W*} + \overline{y}^* \wedge . \tag{10.1}$$

In (10.1), $\overline{\partial}^{W*}$ is the adjoint of $\overline{\partial}^W$ and only depends on $\widehat{g}^W$, and $\overline{y}^* \in W^*$ corresponds to $y \in W$ by the metric g^W.

Set

$$L = \left[\overline{\partial}^W + i_y, \overline{\partial}^{W*} + \overline{y}^* \wedge\right]. \tag{10.2}$$

The case where $g^W = \widehat{g}^W$ was considered in detail in [B90b, Section 1]. We will now use the results of [B90b] in this more general context.

Let $c \in \operatorname{End}(W)$ be self-adjoint positive with

$$g^W = \widehat{g}^W c^2, \tag{10.3}$$

so that if $U \in W$,

$$|U|^2_{g^W} = |cU|^2_{\widehat{g}^W}. \tag{10.4}$$

Let $\gamma_1, \dots, \gamma_n > 0$ be the eigenvalues of c. The vector space W splits into n mutually orthogonal complex lines. Using this splitting, we may as well assume that W is 1-dimensional, and that given $\gamma > 0$, then

$$g^W = \gamma^2 \widehat{g}^W. \tag{10.5}$$

In what follows, we will evaluate the Laplacian Δ and the norm $|Y|$ with respect to $\widehat{g}^W$. Let $w \in W$ be of norm 1 with respect to $\widehat{g}^W$, and let $w^* \in W^*$ be dual to w. An easy computation shows that

$$L = \frac{1}{2}\left(-\Delta + \gamma^2 |Y|^2\right) + \gamma^2 \overline{w}^* \wedge w^* - i_{\overline{w}} i_w. \tag{10.6}$$

For $a > 0$, let r_a be the map $y \to ay$. Set

$$M = r^*_{1/\sqrt{\gamma}} L r^*_{\sqrt{\gamma}}. \tag{10.7}$$

By (10.6), we get

$$M = \gamma\left(\frac{1}{2}\left(-\Delta + |Y|^2\right) + \overline{w}^* \wedge w^* - i_{\overline{w}} i_w\right). \tag{10.8}$$

Recall that β_b was defined in (9.1). By [B90b, Proposition 1.5 and Theorem 1.6] and by (10.8), we find that the spectrum of L is just $\gamma\mathbf{N}$, and also that $\ker L$

is one dimensional and generated by the form

$$\beta_{\sqrt{\gamma}} = \exp\left(\gamma w^* \wedge \overline{w}^* - \gamma |Y|^2 /2\right). \tag{10.9}$$

Clearly,

$$\left(\overline{\partial}^W + i_y\right)\beta_{\sqrt{\gamma}} = 0, \qquad \left(\overline{\partial}^{W*} + \gamma^2 \overline{y}^* \wedge\right)\beta_{\sqrt{\gamma}} = 0. \tag{10.10}$$

Also if i is the embedding of $\{0\}$ in W, then

$$i^* \beta_{\sqrt{\gamma}} = 1. \tag{10.11}$$

Then $\beta_{\sqrt{\gamma}}$ is the canonical harmonic representative of 1 viewed as the generator of the cohomology of $\{0\}$. An easy computation shows that

$$\left|\beta_{\sqrt{\gamma}}\right|^2_{L^2} = \frac{1}{\gamma}. \tag{10.12}$$

The above results extend in the obvious way to the case where W is of arbitrary dimension. The canonical harmonic representative β of the class 1 is given by the obvious extension of (10.9). By (10.12), we get

$$|\beta|^2_{L^2} = \frac{1}{\det c}. \tag{10.13}$$

10.2. Horizontal and vertical metrics and the hypoelliptic Laplacian

We make the same assumptions and we use the same notation as in Subsections 3.11 and 3.12. In particular g^{TX} denotes the metric on TX whose Kähler form $\omega^{M,V}$ is the restriction of the closed form ω^M to TX.

Let $\widehat{g}^{TX}$ be another G-invariant metric on the vertical fibre $\widehat{TX}$. We denote by $\widehat{\nabla}^{TX}$ the corresponding holomorphic Hermitian connection, and by $\widehat{R}^{TX}$ its curvature. Let $c \in \operatorname{End}(TX)$ be the $\widehat{g}^{TX}$ self-adjoint positive section such that

$$g^{TX} = \widehat{g}^{TX} c^2. \tag{10.14}$$

We still define the operator A_Z as in (3.40), and the superconnection $\mathcal{A}_Z$ as in (3.99). All the other objects are now defined with respect $\omega^M, \widehat{g}^{TX}$. There is a Lichnerowicz formula for $A^2_{Y_b}$ as in (3.57). The vertical operator which appears first in the right-hand side of (3.57) is fibrewise a harmonic oscillator similar to L in (10.6).

The L^2 Hermitian product in (3.53) which is used in the definition of ϵ in (3.54) should be modified to take into account the fact that the horizontal and vertical metrics are now different. Then we claim that these new operators have essentially the same properties as the less general operators considered in Sections 3 and 7.

Theorems 3.12 and 3.18 have to be slightly modified. Indeed the map $\alpha \in \Omega^{(0,\cdot)}(X, E) \to \pi^* \alpha \wedge \beta$ is no longer an isometry because of (10.13). The metric on $\Omega^{(0,\cdot)}(X, E)$ has to be suitably modified so that the isometry property is restored.

The adaptation of the methods and results of [BL06] to this new situation has to be done with some care. Indeed the approach of [BL06] in the context of de Rham theory uses in an essential way the fact that the fibrewise harmonic oscillator which appears there is covariantly constant. By the results of Subsection 10.1, this is no longer the case for the fibrewise harmonic oscillators which appear in $A^2_{Y_b}$. In fact the methods used in [BL06] have to be suitably modified in a nontrivial way.

To define holomorphic torsion forms in this more general context for $b > 0$ small enough, we are forced to go all the way in extending the results of [BL06], if only to know that the picture on the spectrum of $A^2_{Y_b}$ given in Figure 3.3 still remains correct.

If we are only concerned with hypoelliptic Quillen metrics or their equivariant versions, the methods used in [BL06] for fixed $b > 0$ suffice to obtain the results which are needed in this section.

10.3. The local index theorem in the general case

For $b > 0, t > 0$, consider the metrics $g^{TX}/t, b^4\widehat{g}^{TX}/t^3$ on $TX, \widehat{TX}$. We define $\underline{\epsilon}_{b,t}, \widehat{\underline{\epsilon}}_{b,t}, \widehat{\epsilon_{b,t}}$ as in Subsection 7.1. Otherwise we construct our superconnections as in Section 7. Then the obvious analogue of Theorem 7.7 still holds.

Now we prove the suitable extension of part of Theorem 7.9 to this more general situation.

Theorem 10.1. *For any $b > 0$, as $t \to 0$,*

$$u_{b,t} = \int_{X_g} \mathrm{Td}_g\left(TX, \widehat{g}^{TX}\right) \mathrm{ch}_g\left(E, g^E\right) + \mathcal{O}\left(t\right), \tag{10.15}$$

$$\frac{1}{4} v_{b,t} = \int_{X_g} \mathrm{Td}'_g\left(TX, \widehat{g}^{TX}\right) \mathrm{ch}_g\left(E, g^E\right) + \mathcal{O}\left(t\right).$$

As $t \to 0$, there are forms $C_{b,-1}, C_{b,0} \in P^S$ such that

$$w_{b,t} = \frac{C_{b,-1}}{t} + C_{b,0} + \mathcal{O}\left(t\right). \tag{10.16}$$

Moreover,

$$\begin{aligned} C_{b,-1} &= \int_{X_g} \frac{\omega^M}{2\pi} \mathrm{Td}_g\left(TX, \widehat{g}^{TX}\right) \mathrm{ch}_g\left(E, g^E\right), \\ C_{b,0} &= -3 \int_{X_g} \mathrm{Td}'_g\left(TX, \widehat{g}^{TX}\right) \mathrm{ch}_g\left(E, g^E\right) \\ &\quad + n \int_{X_g} \mathrm{Td}_g\left(TX, \widehat{g}^{TX}\right) \mathrm{ch}_g\left(E, g^E\right). \end{aligned} \tag{10.17}$$

Proof. We need to go again along the lines of the proof of Theorems 6.1 and 7.9, having a keen interest in the rescalings of Definition 6.4 and in the convergence result of Theorem 6.7. Let us just say here that we take $e_1, \ldots, e_{2n}$ to be an orthonormal basis of $T_{\mathbf{R}}X$ with respect to $\widehat{g}^{TX}$, which is such that $e_1, \ldots, e_{2\ell}$ is an orthonormal basis of $T_{\mathbf{R}}X_g$. Let $\widehat{\mathfrak{P}}_{b,0}$ be the operator in (6.26) which is associated to the metric $\widehat{g}^{TX}$.

Put

$$B = \nabla^{TX} - \widehat{\nabla}^{TX}. \tag{10.18}$$

Then B is a section of $\Lambda^{\cdot}(T^*M)\otimes \mathrm{End}(TX)$. Set

$$\mathfrak{P}_{b,0}=\widehat{\mathfrak{P}}_{b,0}+\frac{1}{b}\sum_1^{2\ell}\langle i^*BY,e_i\rangle\,\mathfrak{e}_i. \tag{10.19}$$

By proceeding as in the proof of Theorem 6.7, we find that the obvious analogue of (6.27) still holds, with $\mathfrak{P}_{b,0}$ now given by (10.19).

Now we will use the same conjugation argument as in the proof of Theorem 6.11. Namely note that

$$\exp\left(\sum_1^{2\ell}\langle i^*BU,e_i\rangle\,\mathfrak{e}_i\right)\mathfrak{P}_{b,0}\exp\left(-\sum_1^{2\ell}\langle i^*BU,e_i\rangle\,\mathfrak{e}_i\right)=\widehat{\mathfrak{P}}_{b,0}, \tag{10.20}$$

so that when evaluating the trace of the heat kernel as in Theorem 6.11, we may as well replace $\mathfrak{P}_{b,0}$ by $\widehat{\mathfrak{P}}_{b,0}$, for which the computations of Theorem 6.11 are still valid. Once we have observed this, the proof of our theorem continues as the proof of Theorem 7.9. □

Remark 10.2. By changing the trivialization, we could as well get $\widehat{\mathfrak{P}}_{b,0}$ as the limit of $\mathfrak{P}_{b,t}$ as $t\to 0$ instead of $\mathfrak{P}_{b,0}$. This is related to the fact that the operator $\left[\nabla^{\mathbf{I}},i_Y\right]$ does not depend on g^{TX}.

There are nontrivial analogues of equation (7.43) in Theorem 7.9. Still the explicit formulas are harder to get at. This is left as an inspiring exercise to the reader. Anyway we will not need this more general result.

10.4. The generalized hypoelliptic torsion forms

In the sequel, we assume S to be compact, and also that the $H^{(0,i)}(X,E|_X),0\leq i\leq n$ have locally constant rank. For $b>0$ small enough, we still denote by $g_b^{H^{(0,\cdot)}(X,E|_X)}$ the Hermitian form induced by ϵ_b on $H^{(0,\cdot)}(X,E|_X)$. This Hermitian form depends on $g^{TX},\widehat{g}^{TX},g^E$. We use otherwise the same notation as in Definition 7.10.

By arguments we already outlined, the analogue of equation (7.68) in Theorem 7.13 still hold. The extension of equations (7.69) will not be needed.

We can then define hypoelliptic holomorphic torsion forms $T_{b,g}\left(\omega^M,\widehat{g}^{TX},g^E\right)$ as in Definition 7.16. If $b=1$, we will write $T_g^h\left(\omega^M,\widehat{g}^{TX},g^E\right)$. Incidentally observe that

$$T_{b,g}\left(\omega^M,\widehat{g}^{TX},g^E\right)=T_g^h\left(\omega^M,b^4\widehat{g}^{TX},g^E\right). \tag{10.21}$$

By (10.21), we find that by suitably rescaling $\widehat{g}^{TX}$, we may as well assume that in the above statements, $b\in]0,2]$.

The obvious extension of Theorem 7.17 is the following result.

Theorem 10.3. *For $b > 0$ small enough,*

$$\frac{\overline{\partial}\partial}{2i\pi} T_g^h\left(\omega^M, b^4\widehat{g}^{TX}, g^E\right) = \mathrm{ch}_g\left(H^{(0,\cdot)}\left(X, E|_X\right), g_b^{H^{(0,\cdot)}(X,E|_X)}\right) - \int_{X_g} \mathrm{Td}_g\left(TX, b^4\widehat{g}^{TX}\right)\mathrm{ch}_g\left(E, g^E\right). \quad (10.22)$$

Proof. Using Theorem 10.1, the proof of (10.22) is the same as the proof of Theorem 7.17. □

10.5. The anomaly formulas for the hypoelliptic torsion forms

We denote by $\omega^{M\prime}, \widehat{g}^{TX\prime}, g^{E\prime}$ another triplet of data having the same properties as $\omega^M, \widehat{g}^{TX}, g^E$. We denote by $g_b^{H^{(0,\cdot)}(X,E|_X)\prime}$ the corresponding generalized metric on $H^{(0,\cdot)}\left(X, E|_X\right)$.

As in Subsection 9.1, for $b > 0$ small enough, $g_b^{H^{(0,\cdot)}(X,E|_X)}, g_b^{H^{(0,\cdot)}(X,E|_X)\prime}$ are standard Hermitian metrics. Therefore the Bott-Chern class

$$\widetilde{\mathrm{ch}}_g\left(H^{(0,\cdot)}\left(X, E|_X\right), g_b^{H^{(0,\cdot)}(X,E|_X)}, g_b^{H^{(0,\cdot)}(X,E|_X)\prime}\right)$$

is well defined.

Theorem 10.4. *For $b > 0$ small enough ,the following identity holds,*

$$\begin{aligned} T_g^h\left(\omega^{M\prime}, b^4\widehat{g}^{TX\prime}, g^{E\prime}\right) - T_g^h\left(\omega^M, b^4\widehat{g}^{TX}, g^E\right) &= \widetilde{\mathrm{ch}}_g\left(H^{(0,\cdot)}\left(X, E|_X\right), g_b^{H^{(0,\cdot)}(X,E|_X)}, g_b^{H^{(0,\cdot)}(X,E|_X)\prime}\right) \\ &\quad - \int_{X_g} \widetilde{\mathrm{Td}}_g\left(TX, b^4\widehat{g}^{TX}, b^4\widehat{g}^{TX\prime}\right)\mathrm{ch}_g\left(E, g^E\right) \\ &\quad - \int_{X_g} \mathrm{Td}_g\left(TX, b^4\widehat{g}^{TX\prime}\right)\widetilde{\mathrm{ch}}_g\left(E, g^E, g^{E\prime}\right) \text{ in } P^S/P^{S,0}. \end{aligned} \quad (10.23)$$

Proof. If $\omega^{M\prime} = \omega^M$, we get (10.23) from Theorem 10.3 by the classical technique of deformation of the given data over $\mathbf{P}^1$ explained in detail in [BGS88a, BGS90a]. In the sequel we may as well assume that $\widehat{g}^{TX}, \widehat{g}^E$ are kept fixed and that only ω^M is made to vary.

Let $c \in \mathbf{R} \to \omega_c^M$ be a smooth family of closed forms having the same properties as ω^M. Let $g_{b,c}^{H^{(0,\cdot)}(X,E|_X)}$ be the corresponding family of Hermitian forms on $H^{(0,\cdot)}\left(X, E|_X\right)$. Then by proceeding as in [BK92] and using the same local index theoretic techniques which were used in the proof of Theorem 7.9, and in particular in equations (7.51)–(7.58), one can show that for $b > 0$ small enough,

$$\frac{\partial}{\partial c} T_g^h\left(\omega_c^M, b^4\widehat{g}^{TX}, g^E\right) = \frac{\partial}{\partial c}\widetilde{\mathrm{ch}}_g\left(H^{(0,\cdot)}\left(X, E|_X\right), g_b^{H^{(0,\cdot)}(X,E|_X)}, g_{b,c}^{H^{(0,\cdot)}(X,E|_X)}\right) \text{ in } P^S/P^{S,0}. \quad (10.24)$$

By integrating (10.24), we get (10.23) in full generality.

A simpler proof will be obtained as a consequence of the anomaly formulas of [BK92] and of the general comparison formula of Theorem 10.6. This will be explained in Remark 10.7. □

Remark 10.5. The most surprising aspect of formula (10.23) is that when varying ω^M, there is no local index contribution to the variation of $T_g^h\left(\omega^M, b^4\widehat{g}^{TX}, g^E\right)$. Equivalently the dependence of the class $T_g^h\left(\omega^M, b^4\widehat{g}^{TX}, g^E\right)$ on $\omega^M \in P^S/P^{S,0}$ only appears via the metric $g_b^{H^{(0,\cdot)}(X,E|_X)}$. Nevertheless recall that when proving Theorem 10.1, we used explicitly the fact that ω^M is closed. This is already clear in the proof of Theorem 7.9.

10.6. A comparison formula for hypoelliptic torsion forms

We will now give an extension of the comparison formula Theorem 8.1 to the case of general $\widehat{g}^{TX}$. Let $\omega^M, \widehat{g}^{TX}, g^E$ be taken as before, and let $\omega^{M\prime}$ be another closed form like ω^M. Let $g^{TX\prime}$ be the Hermitian metric induced by $\omega^{M\prime}$ on TX.

Let $g^{H^{(0,\cdot)}(X,E|_X)\prime}$ be the L^2 metric on $H^{(0,\cdot)}(X, E|_X)$ associated with the metrics $g^{TX\prime}, g^E$ as in Subsection 4.4.

Theorem 10.6. *For $b > 0$ small enough, the Hermitian form $g_b^{H^{(0,\cdot)}(X,E|_X)}$ is a Hermitian metric. Moreover,*

$$\begin{aligned} T_g^h\left(\omega^M, b^4\widehat{g}^{TX}, g^E\right) - T_g\left(\omega^{M\prime}, g^E\right) \\ - \widetilde{\mathrm{ch}}_g\left(H^{(0,\cdot)}(X, E|_X), g^{H^{(0,\cdot)}(X,E|_X)}, g_b^{H^{(0,\cdot)}(X,E|_X)}\right) \\ + \int_{X_g} \widetilde{\mathrm{Td}}_g\left(TX, g^{TX\prime}, b^4\widehat{g}^{TX}\right) \mathrm{ch}_g\left(E, g^E\right) \\ - \int_{X_g} \mathrm{Td}_g(TX) R_g(TX) \mathrm{ch}_g(E) = 0 \,\mathrm{in}\, P^S/P^{S,0}. \quad (10.25) \end{aligned}$$

Proof. By using the anomaly formulas for $T_g^h\left(\omega^{M\prime}, \widehat{g}^{TX}, g^E\right)$ with respect to $\widehat{g}^{TX}$ which were established in Theorem 10.4, we only need to establish equation (10.25) for one given choice of the metric $\widehat{g}^{TX}$. Also using the anomaly formulas for the holomorphic elliptic torsion forms of [BK92, Theorem 3.10], we only need to establish our theorem for only one choice of $\omega^{M\prime}$. Therefore we may as well take $\omega^{M\prime} = \omega^M$, and also $\widehat{g}^{TX} = g^{TX}$. Our formula then reduces to equation (8.2) in Theorem 8.1. □

Remark 10.7. As a by-product of Theorem 10.6, we obtain the hard part of Theorem 10.4, which describes the variation of $T_g^h\left(\omega^M, b^4\widehat{g}^{TX}, g^E\right)$ as a function of ω^M. This way, one can avoid using any sophisticate local index techniques to obtain these formulas.

10.7. Generalized hypoelliptic Quillen metrics

We will now assume that S is reduced to a point. Let $g^{TX}, g^{TX\prime}$ be Kähler metrics on TX, let $\widehat{g}^{TX}$ be any other metric on TX, let g^E be a Hermitian metric on E.

By proceeding as in Subsections 7.8–7.10, we can define an equivariant generalized Quillen metric $\| \|^2_{\lambda,h}$ on λ, which is obtained using the construction of a more general hypoelliptic Laplacian of Subsection 10.2, which is associated with the metrics $g^{TX}, \widehat{g}^{TX}, g^E$. The main point is that we do not need to introduce any more the scaling parameter $b > 0$ in the whole construction.

Let $\| \|_\lambda$ be the standard equivariant elliptic Quillen metric on λ, which is attached to the metrics $g^{TX\prime}, g^E$.

Now we give an extension of Theorems 8.10 and 8.11.

Theorem 10.8. *The following identity holds,*

$$\log\left(\frac{\| \|^2_{\lambda,h}}{\| \|^2_\lambda}\right)(g) = -\int_{X_g} \widetilde{\mathrm{Td}}_g\left(TX, g^{TX\prime}, \widehat{g}^{TX}\right)\mathrm{ch}_g\left(E, g^E\right) + \int_{X_g} \mathrm{Td}_g(TX)\, R_g(TX)\,\mathrm{ch}_g(E)\,. \quad (10.26)$$

Proof. The same local index theoretic as in the proof of Theorem 10.4 show that equation (10.26) is compatible with the variation of the metric $\widehat{g}^{TX}$. The anomaly formulas of [BGS88c] in the case where $g = 1$, and of [B95, Theorem 2.5] in the general case, show that equation (10.26) is also compatible with the variation of $g^{TX\prime}$. Therefore we are free to choose $g^{TX\prime} = \widehat{g}^{TX} = g^{TX}$, in which case our formula is just equation (8.47) in Theorem 8.11. The proof of our theorem is completed. □

Remark 10.9. When G is trivial, when taking $g = 1$, (10.26) is an equality of real numbers. When G is arbitrary, the considerations in Remark 8.12 are still valid.

As a by-product of Theorem 10.8, we find that the hypoelliptic metric $\| \|^2_{\lambda,h}$ does not depend on the metric g^{TX}. This can be viewed also as a consequence of Theorem 10.6. Nevertheless it is all the more striking. Indeed the hypoelliptic metric does not depend on the Kähler metric which is needed in its definition, and moreover, it is well defined for any metric $\widehat{g}^{TX}$.

10.8. Hypoelliptic torsion forms and the immersion formulas

In [BL91, B95, B97, BM04], the behaviour of the Quillen metrics and of the elliptic torsion forms under complex immersions was obtained. In these results, the genus R_g appears in exactly the same form as in Theorems 10.6 and 10.8. Here we will only discuss the case of the standard Quillen metric $\| \|_\lambda$, but the discussion can be extended to all cases.

Indeed let $i : Y \to X$ be an embedding of complex manifolds, let η be a holomorphic Hermitian vector bundle on Y, and let (ξ, v) a complex of holomorphic vector bundles on X which provides a resolution of the sheaf $i_*\mathcal{O}_Y(\eta)$. Let $\lambda(\eta), \lambda(\xi)$ be the determinant lines associated to η, ξ. Then we have a canonical isomorphism,

$$\lambda(\eta) \simeq \lambda(\xi)\,. \quad (10.27)$$

Let $\| \|_{\lambda(\eta)}, \| \|_{\lambda(\xi)}$ be Quillen metrics on $\lambda(\eta), \lambda(\xi)$. The main result of [BL91] gives a formula for $\log\left(\frac{\| \|_{\lambda(\xi)}}{\| \|_{\lambda(\eta)}}\right)^2$.

The formula in [BL91] can be written in the form,

$$\log\left(\frac{\| \|_{\lambda(\xi)}}{\| \|_{\lambda(\eta)}}\right)^2 = \int_X \operatorname{Td}(TX) R(TX) \operatorname{ch}(\xi) - \int_Y \operatorname{Td}(TY) R(TY) \operatorname{ch}(\eta) + \ldots \tag{10.28}$$

In (10.28), as explained in [BGS90b, BGS90a], ... represents the 'predictable' part of the formula from the point of view of the theorem of Riemann-Roch-Grothendieck in arithmetic geometry of Gillet and Soulé which was established in [GS92]. The terms containing the exotic genus R were also anticipated in [GS91] from this point of view.

By combining equation (10.26) in Theorem 10.8 with (10.28), we find that when $b = 1$,

$$\log\left(\frac{\| \|_{\lambda(\xi),h}}{\| \|_{\lambda(\eta),h}}\right)^2 = 2\int_X \operatorname{Td}(TX) R(TX) \operatorname{ch}(\xi) - 2\int_Y \operatorname{Td}(TY) R(TY) \operatorname{ch}(\eta) + \ldots \tag{10.29}$$

where the term ... is exactly the same one as in (10.28).

It would certainly be interesting to give a direct proof of (10.29) along the lines of [BL91]. The doubling of the genus R should appear because we would handle the resolution of η in X and also the resolution of X in TX.

However Theorems 10.6 and 10.8 can also be given another interpretation. Indeed consider the embedding $i : X \to \mathcal{X}$. Then the formulas in Theorems 10.6 and 10.8 are the exact analogues of the embedding formulas of [BL91, B95], the model of which is (10.28).

These considerations explain why the genera $R(x)$ or $R(\theta, x)$ reappear in the present context. However the analogy is misleading. First $\mathcal{X}$ is non compact. This is not serious an objection, since the harmonic oscillator compactifies $\mathcal{X}$ in a proper way. More fundamentally, the Laplacian which is considered here is hypoelliptic.

One could try to reinterpret our result from the point of view of the adiabatic limit results of [BerB94, M99, M00b]. However this would be wrong again. In particular the fact that here $\Lambda^{\cdot}(T^*X)$ is at the same time the horizontal and the vertical holomorphic exterior algebra is impossible to explain from this point of view.

11. The eta invariant

The purpose of this section is to define the reduced eta invariant mod $\mathbf{Z}$ of the hypoelliptic Dirac operator $\mathcal{D}_{Y_b}$, and to compare it to the reduced eta invariant of the classical Dirac operator D^X considered in Subsection 1.4. The reduced eta invariant for D^X was introduced by Atiyah-Patodi-Singer [APS75a, APS75b,

APS76] in their study of index problems for elliptic Dirac operators on manifolds with boundary.

The arguments given in this section being simpler than the corresponding ones for hypoelliptic holomorphic torsion, we will only sketch the proofs.

This section is organized as follows. In Subsection 11.1, we consider the case of finite-dimensional vector spaces.

In Subsection 11.2, we define the reduced eta invariant of $\mathcal{D}_{Y_b}$ as an element of $\mathbf{R}/\mathbf{Z}$.

In Subsection 11.3, we obtain a variation formula for the reduced eta invariant. In particular we show that it does not depend on $b > 0$.

Finally in Subsection 11.4, we show that when the horizontal and vertical metrics are the same, the reduced eta invariants of the elliptic and hypoelliptic Dirac operators coincide.

11.1. The finite-dimensional case

Let E be a finite-dimensional complex $\mathbf{Z}_2$-graded vector space, which is equipped with a Hermitian form ϵ. Let $A \in \operatorname{End}(E)$ be self-adjoint with respect to ϵ. Then the spectrum of A is conjugation invariant.

Consider the function of $s \in \mathbf{C}$,

$$\eta_A(s) = \operatorname{Tr_s}\left[\frac{A}{\left(A^2\right)^{(s+1)/2}}\right]. \tag{11.1}$$

In (11.1), it is implicitly assume that the zero eigenvalue is excluded from the supertrace. If $\lambda_1, \ldots, \lambda_p$ are the eigenvalues of A, we can rewrite (11.1) in the form,

$$\eta_A(s) = \sum_{\lambda \neq 0} \frac{\lambda}{\left(\lambda^2\right)^{(s+1)/2}}. \tag{11.2}$$

Observe that equations (11.1) and (11.2) are ambiguous. The ambiguity can be lifted for instance in the case where no eigenvalue λ is purely imaginary. In any case, for whatever convention we take,

$$\frac{\lambda}{\left(\lambda^2\right)^{1/2}} = \pm 1. \tag{11.3}$$

By (11.3), we find that modulo $2\mathbf{Z}$, $\eta_A(0)$ is unambiguously defined. More precisely

$$\eta_A(0) + \dim \ker A = \dim E \bmod 2\mathbf{Z}. \tag{11.4}$$

11.2. The eta invariant of the operator D_{Y_b}

We make the same assumptions as in Section 1. Here we will assume the dimension n of X to be odd. The spinor bundle S^X is then an ungraded vector bundle.

As we saw in Subsections 3.10 and 6.1, for $b > 0$, the spectrum of $\mathcal{D}^2_{Y_b}$ is discrete and conjugation invariant. Figures 3.1 and 3.2 indicate where the spectrum of $\mathcal{D}^2_{Y_b}$ is located.

Since $\mathcal{D}_{Y_b}$ commutes with $\mathcal{D}^2_{Y_b}$, corresponding properties are immediately derived for $\mathcal{D}_{Y_b}$. In particular $\mathcal{D}_{Y_b}$ acts on the characteristic spaces of $\mathcal{D}^2_{Y_b}$, which are finite dimensional. Therefore the spectrum of $\mathcal{D}_{Y_b}$ is itself discrete, and its characteristic subspaces are also finite dimensional. By Theorem 1.9, the spectrum of $\mathcal{D}_{Y_b}$ is conjugation invariant.

It follows in particular that among the eigenvalues of $\mathcal{D}_{Y_b}$ which are located on the imaginary axis, except for the zero eigenvalue, these eigenvalues come by pairs.

Definition 11.1. For $b > 0, s \in \mathbf{C}$, set

$$\eta_b(s) = \mathrm{Tr_s}\left[\frac{\mathcal{D}_{Y_b}}{\left(\mathcal{D}^2_{Y_b}\right)^{(s+1)/2}}\right]. \tag{11.5}$$

Of course the 0 eigenvalue is excluded from the supertrace in (11.5).

The function $\eta_b(s)$ will be called the eta function associated to the operator $\mathcal{D}_{Y_b}$.

Theorem 11.2. *The function $\eta_b(s)$ can be defined as a holomorphic function of $s \in \mathbf{C}, \mathrm{Re}\, s > -2$.*

Proof. By the results of Subsection 3.10, the eigenvalues of $\mathcal{D}^2_{Y_b}$ are of positive real part, except for a finite number of them.

In a first part of the proof, we will first assume that the eigenvalues of $\mathcal{D}^2_{Y_b}$ are either 0, or have positive real part. By proceeding as in [BL06, Subsection 5.3], we find that there exists $c > 0, C > 0$ such that for $t \geq 1$,

$$\left|\mathrm{Tr}\left[\mathcal{D}_{Y_b}\exp\left(-t\mathcal{D}^{X,2}_{Y_b}\right)\right]\right| \leq C\exp(-ct). \tag{11.6}$$

We will give a formula for $\eta_b(s)$ and later we will verify it does make sense under the given conditions on s. Put

$$\eta_b(s) = \frac{1}{\Gamma\left(\frac{s+1}{2}\right)}\int_0^{+\infty} t^{(s+1)/2}\mathrm{Tr_s}\left[\mathcal{D}_{Y_b}\exp\left(-t\mathcal{D}^{X,2}_{Y_b}\right)\right]\frac{dt}{t}. \tag{11.7}$$

From the above, it follows that for $\mathrm{Re}\, s$ large enough, the function $\eta_b(s)$ in (11.7) is just the one in (11.5).

To show that $\eta_b(s)$ extends to a holomorphic function of $s, \mathrm{Re}\, s > -2$, we will show that as $t \to 0$,

$$\mathrm{Tr_s}\left[\mathcal{D}_{Y_b}\exp\left(-t\mathcal{D}^2_{Y_b}\right)\right] = \mathcal{O}\left(\sqrt{t}\right). \tag{11.8}$$

Note that (11.8) is the precise analogue of the corresponding asymptotics for the operator D^X which was established by Bismut and Freed in [BF86b, Theorem 2.4], which in turn was used in [BF86b] to give a new proof of the fact that the eta function of a classical Dirac operator is holomorphic at 0.

Now the proof of [BF86b, Theorem 2.4] relies on the local families index theorem for odd-dimensional fibres, the proof of which is a simple modification of the corresponding proof in the case of even-dimensional fibres. Here we will show how to adapt the arguments of [BF86b] in the present context.

We use the notation in Subsection 6.3. Clearly

$$t^{N^H/2}\mathcal{A}^t_{Y_{b/\sqrt{t}}}t^{-N^H/2} = \psi_{1/t}\sqrt{t}\mathcal{A}_{Y_b}\psi_t. \tag{11.9}$$

We deduce from (11.9) that

$$t^{N^H/2}\mathcal{D}^t_{Y_{b/\sqrt{t}}}t^{-N^H/2} = \sqrt{t}\mathcal{D}_{Y_b}. \tag{11.10}$$

To make the correct adaptation of the arguments in [BF86b], we use the formalism of Subsection 1.3. In particular the real vector space V is assumed to be of odd dimension n. Set

$$\mathcal{B} = \operatorname{End}\left(\Lambda^{\cdot}\left(V^*\right)\right)\widehat{\otimes}c\left(\widehat{V}\right). \tag{11.11}$$

Note that $\mathcal{B}$ acts on $\Lambda^{\cdot}\left(V^*\right)\otimes\widehat{S}^V$, which is $\mathbf{Z}$-graded by the grading of $\Lambda^{\cdot}\left(V^*\right)$. If $\alpha\in\mathcal{B}$, let $\operatorname{Tr_s}[\alpha]$ denote its supertrace. If $\alpha\in\mathcal{B}$, only the even component of α in $\operatorname{End}\left(\Lambda^{\cdot}\left(V^*\right)\right)$ contributes to the supertrace. In particular if $\alpha\in\mathcal{B}^{\text{odd}}$, only the odd component of α in $c\left(\widehat{V}\right)$ contributes to the supertrace.

Let $e_1,\dots,e_n$ be an orthonormal oriented basis of V. Let us recall that by [BF86b, eq. (1.7)], among the odd monomials in the $c\left(\widehat{e}_i\right)$, up to permutation, $c\left(\widehat{e}_1\right)\dots c\left(\widehat{e}_n\right)$ is the only monomial whose trace on $\widehat{S}^V$ does not vanish.

We proceed by establishing an analogue of the local families index theorem of [BF86b, Theorem 2.10] similar to Theorem 6.1 for the superconnection $\mathbb{A}^2$ in the case where the fibres X are odd dimensional. In [BF86b], the formalism of Quillen [Q85a] of the so called odd traces was used. Let us briefly show how this can be done here. As before we may consider the Clifford variables $c\left(\widehat{e}_i\right)$ as odd, and we only consider the odd part of $(2i)^{1/2}\varphi\operatorname{Tr_s}\left[g\exp\left(-\mathbb{A}^2\right)\right]$, which is denoted $(2i)^{1/2}\varphi\operatorname{Tr_s}^{\text{odd}}\left[g\exp\left(-\mathbb{A}^2\right)\right]$. Another possibility is to consider the Clifford variables $c\left(\widehat{e}_i\right)$ as even, and to introduce an extra odd Clifford variable σ such that $\sigma^2{=}1$, while replacing $c\left(\widehat{e}_i\right)$ by $c\left(\widehat{e}_i\right)\sigma$. We would then consider the σ-supertrace $(2i)^{1/2}\varphi\operatorname{Tr_s}^{\sigma}\left[g\exp\left(-\mathbb{A}^2\right)\right]$ The same arguments as in [BF86b, Subsection 2f)] show that the two procedures are equivalent.

Using (11.9), (11.10) and proceeding as in [BF86b], we obtain this way a proof of the required result for the function $\eta_b(s)$.

Now we drop the assumption on the eigenvalues of $\mathcal{D}^2_{Y,b}$. Indeed there are only a finite family of non zero eigenvalues λ of $\mathcal{D}^2_{Y_b}$ such that $\operatorname{Re}\lambda\leq 0$. Recall that by Lidskii's theorem [RSi78, Corollary, p. 328], if A is a trace class operator, then $\operatorname{Tr}[A]$ is the sum of its eigenvalues. Once we eliminate the eigenvalues λ such that $\operatorname{Re}\lambda\leq 0$, whose contribution can be obtained by a formula similar to (11.2), we are left with a Mellin transform like in (11.7), the integrand of which still verifies the analogue of (11.6). The proof of our theorem is completed. □

From the above it follows that the definition of $\eta(0)$ is somewhat ambiguous. However $\eta(0)$ can be unambiguously defined in $\mathbf{R}/2\mathbf{Z}$. Indeed by (11.3), each of the eigenvalues contributes by a sign ± 1, which introduces an ambiguity in $2\mathbf{Z}$. The resulting formula for $\eta_b(0)$ lies in $\mathbf{R}/2\mathbf{Z}$ because the operator $\mathcal{D}_{Y_b}$ is self-adjoint with respect to the Hermitian form η in (1.50).

Let $\mathcal{S}_{0,b}$ be the characteristic subspace of $\mathcal{D}_{Y_b}^2$ which is associated to the 0 eigenvalue.

Definition 11.3. Set

$$\overline{\eta}_b(s) = \frac{1}{2}\left(\eta_b(s) + \operatorname{rk} \mathcal{S}_{0,b}\right). \tag{11.12}$$

By the above, $\overline{\eta}_b(0)$ is unambiguously defined in $\mathbf{R}/\mathbf{Z}$. This quantity will be called the reduced eta invariant of $\mathcal{D}_{Y_b}$.

11.3. The variation formula for $\overline{\eta}_b(0)$

We make the same assumptions as in Section 2, except that the fibres X are now odd dimensional. Each fibre $\mathcal{X}_s$ carries an operator $\mathcal{D}_{Y_b}$. Therefore $\overline{\eta}_b(0)$ can be viewed as a function on $S \times \mathbf{R}_+^*$.

Theorem 11.4. *The $\mathbf{R}/\mathbf{Z}$ valued function $\overline{\eta}_b(0)$ is smooth on $S \times \mathbf{R}_+^*$. Moreover,*

$$d\overline{\eta}_b(0) = \left[\int_X \widehat{A}\left(TX, \nabla^{TX}\right) \operatorname{ch}\left(E, \nabla^E\right)\right]^{(1)}. \tag{11.13}$$

In particular $\overline{\eta}_b(0) \in \mathbf{R}/\mathbf{Z}$ does not depend on b.

Proof. Each nonzero eigenvalue of $\mathcal{D}_{Y_b}$ contributes to $\overline{\eta}_b(0)$ by $\pm 1/2$. When following such an eigenvalue, it is clear that only the eigenvalue 0 could be a source of difficulty under deformation. However adding $\frac{1}{2}\operatorname{rk}\mathcal{S}_{0,b}$ exactly compensates for that. We have thus proved that $\overline{\eta}_b(0) \in \mathbf{R}/\mathbf{Z}$ depends smoothly on all parameters.

We fix $b > 0$ temporarily. By proceeding as in [APS76, p.75 and Proposition 2.11] and in [BF86b, Section2], once we know the local version of the families index theorem given in Theorem 6.1 in the case of odd-dimensional fibres, equation (11.13) is just a consequence of equation (6.8) in Theorem 6.1. The proof of our theorem is completed. □

Remark 11.5. As in Section 10, we could assume the horizontal and vertical metrics on $\mathcal{X}$ to be unrelated. The obvious analogue of Theorem 11.4 would still hold, where only the vertical metric and connection would appear.

11.4. A comparison formula for the eta invariants

Let D^X be the elliptic Dirac operator of Definition 1.1 associated to the metric g^{TX} and the connection ∇^E. Let $\overline{\eta}(0)$ be the reduced eta invariant of D^X in the sense of [APS75a]. It is the obvious analogue of $\overline{\eta}_b(0)$ in (11.12).

Theorem 11.6. *For any $b > 0$,*

$$\overline{\eta}_b(0) = \overline{\eta}(0) \text{ in } \mathbf{R}/\mathbf{Z}. \tag{11.14}$$

Proof. We briefly indicate the principle of the proof, which closely resembles the proof of Theorem 8.1.

We will use the notation of Subsection 6.4 adapted to the case where the fibres X are odd dimensional. In particular the form $(2i)^{1/2}\varphi\mathrm{Tr_s}^{\mathrm{odd}}\left[g\exp\left(-\mathbb{A}^2\right)\right]$ is closed on $S\times\mathbf{R}_+^*\times\mathbf{R}_+^*$. We define the even forms $m_{b,t}, n_{b,t}$ as in (6.7). Set

$$a = n_{b,t}\wedge dt + o_{b,t}\wedge db. \tag{11.15}$$

Let $\left(\Omega^{\cdot}(S), d^S\right)$ be the de Rham complex on S. By the obvious analogue of Theorem 6.1,

$$d^{\mathbf{R}_+^{*2}}a\in d^S\Omega^{\cdot}(S). \tag{11.16}$$

We introduce a rectangular contour Γ as in Figure 8.1. By (11.16),

$$\int_\Gamma a\in d^S\Omega^{\cdot}(S). \tag{11.17}$$

Now we take S to be a point, so that (11.17) just says that

$$\int_\Gamma a = 0. \tag{11.18}$$

Set

$$\eta_{b,\epsilon}(0) = \frac{1}{\sqrt{\pi}}\int_\epsilon^{+\infty}\mathrm{Tr_s}\left[\mathcal{D}_{Y_b}\exp\left(-t\mathcal{D}_{Y_b}^2\right)\right]\frac{dt}{\sqrt{t}}. \tag{11.19}$$

Note that there are still ambiguities in the definition of (11.19) associated with a finite number of eigenvalues, but as $\epsilon\to 0$, these ambiguities disappear mod $2\mathbf{Z}$. We will authorize ourselves to write an equality of objects depending on $\epsilon>0$ in $\mathbf{R}/\mathbf{Z}$ when their limit as $\epsilon\to 0$ is well defined in $\mathbf{R}/\mathbf{Z}$ and the corresponding equality holds.

Set

$$\overline{\eta}_{b,\epsilon}(0) = \frac{1}{2}\left(\eta_{b,\epsilon}(0) + \dim\ker\mathcal{S}_{0,b}\right). \tag{11.20}$$

By proceeding as in the proof of Theorem 11.4 and using (11.18)–(11.20), we get

$$\overline{\eta}_{b_0,\epsilon}(0) - \overline{\eta}_{\beta,\epsilon}(0) = \int_\beta^{b_0} o_{b,\epsilon}db \text{ in } \mathbf{R}/\mathbf{Z}. \tag{11.21}$$

Now we make $\epsilon\to 0$ in (11.21). By using the second identity in (6.71), and by proceeding as in the proof of Theorem 8.1, we get (11.14). The proof of our theorem is completed. □

Remark 11.7. An analogue of Theorem 11.6 similar to Theorem 10.8 can be established when the horizontal and vertical metrics are unrelated, in which the integral of a Chern-Simons class appears. Details are left to the reader.

References

[A85] M.F. Atiyah. Circular symmetry and stationary-phase approximation. *Astérisque*, (131):43–59, 1985. Colloquium in honor of Laurent Schwartz, Vol. 1 (Palaiseau, 1983).

[AB66] M.F. Atiyah and R. Bott. A Lefschetz fixed point formula for elliptic differential operators. *Bull. Amer. Math. Soc.*, 72:245–250, 1966.

[AB67] M.F. Atiyah and R. Bott. A Lefschetz fixed point formula for elliptic complexes. I. *Ann. of Math.* (2), 86:374–407, 1967.

[AB68] M.F. Atiyah and R. Bott. A Lefschetz fixed point formula for elliptic complexes. II. Applications. *Ann. of Math.* (2), 88:451–491, 1968.

[ABP73] M. Atiyah, R. Bott, and V.K. Patodi. On the heat equation and the index theorem. *Invent. Math.*, 19:279–330, 1973.

[APS75a] M.F. Atiyah, V.K. Patodi, and I.M. Singer. Spectral asymmetry and Riemannian geometry. I. *Math. Proc. Cambridge Philos. Soc.*, 77:43–69, 1975.

[APS75b] M.F. Atiyah, V.K. Patodi, and I.M. Singer. Spectral asymmetry and Riemannian geometry. II. *Math. Proc. Cambridge Philos. Soc.*, 78(3):405–432, 1975.

[APS76] M.F. Atiyah, V.K. Patodi, and I.M. Singer. Spectral asymmetry and Riemannian geometry. III. *Math. Proc. Cambridge Philos. Soc.*, 79(1):71–99, 1976.

[AS71] M.F. Atiyah and I.M. Singer. The index of elliptic operators. IV. *Ann. of Math.* (2), 93:119–138, 1971.

[BerB94] A. Berthomieu and J.-M. Bismut. Quillen metrics and higher analytic torsion forms. *J. Reine Angew. Math.*, 457:85–184, 1994.

[BeGV92] N. Berline, E. Getzler, and M. Vergne. *Heat kernels and Dirac operators.* Grundl. Math. Wiss. Band 298. Springer-Verlag, Berlin, 1992.

[B85] J.-M. Bismut. Index theorem and equivariant cohomology on the loop space. *Comm. Math. Phys.*, 98(2):213–237, 1985.

[B86] J.-M. Bismut. The Atiyah-Singer index theorem for families of Dirac operators: two heat equation proofs. *Invent. Math.*, 83(1):91–151, 1986.

[B89] J.-M. Bismut. A local index theorem for non-Kähler manifolds. *Math. Ann.*, 284(4):681–699, 1989.

[B90a] J.-M. Bismut. Equivariant Bott-Chern currents and the Ray-Singer analytic torsion. *Math. Ann.*, 287(3):495–507, 1990.

[B90b] J.-M. Bismut. Koszul complexes, harmonic oscillators, and the Todd class. *J. Amer. Math. Soc.*, 3(1):159–256, 1990. With an appendix by the author and C. Soulé.

[B94] J.-M. Bismut. Equivariant short exact sequences of vector bundles and their analytic torsion forms. *Compositio Math.*, 93(3):291–354, 1994.

[B95] J.-M. Bismut. Equivariant immersions and Quillen metrics. *J. Differential Geom.*, 41(1):53–157, 1995.

[B97] J.-M. Bismut. Holomorphic families of immersions and higher analytic torsion forms. *Astérisque*, (244):viii+275, 1997.

[B05] J.-M. Bismut. The hypoelliptic Laplacian on the cotangent bundle. *J. Amer. Math. Soc.*, 18(2):379–476 (electronic), 2005.

[B06] J.-M. Bismut. L'opérateur de Dirac hypoelliptique. *C.R. Math. Acad. Sci. Paris*, 343(10):647–651, 2006.

[BF86a] J.-M. Bismut and D.S. Freed. The analysis of elliptic families. I. Metrics and connections on determinant bundles. *Comm. Math. Phys.*, 106(1):159–176, 1986.

[BF86b] J.-M. Bismut and D.S. Freed. The analysis of elliptic families. II. Dirac operators, eta invariants, and the holonomy theorem. *Comm. Math. Phys.*, 107(1):103–163, 1986.

[BG04] J.-M. Bismut and S. Goette. Equivariant de Rham torsions. *Ann. of Math.*, 159:53–216, 2004.

[BGS88a] J.-M. Bismut, H. Gillet, and C. Soulé. Analytic torsion and holomorphic determinant bundles. I. Bott-Chern forms and analytic torsion. *Comm. Math. Phys.*, 115(1):49–78, 1988.

[BGS88b] J.-M. Bismut, H. Gillet, and C. Soulé. Analytic torsion and holomorphic determinant bundles. II. Direct images and Bott-Chern forms. *Comm. Math. Phys.*, 115(1):79–126, 1988.

[BGS88c] J.-M. Bismut, H. Gillet, and C. Soulé. Analytic torsion and holomorphic determinant bundles. III. Quillen metrics on holomorphic determinants. *Comm. Math. Phys.*, 115(2):301–351, 1988.

[BGS90a] J.-M. Bismut, H. Gillet, and C. Soulé. Complex immersions and Arakelov geometry. In *The Grothendieck Festschrift, Vol. I*, pages 249–331. Birkhäuser Boston, Boston, MA, 1990.

[BGS90b] J.-M. Bismut, H. Gillet, and C. Soulé. Bott-Chern currents and complex immersions. *Duke Math. J.*, 60(1):255–284, 1990.

[BK92] J.-M. Bismut and K. Köhler. Higher analytic torsion forms for direct images and anomaly formulas. *J. Algebraic Geom.*, 1(4):647–684, 1992.

[BL91] J.-M. Bismut and G. Lebeau. Complex immersions and Quillen metrics. *Inst. Hautes Études Sci. Publ. Math.*, (74):ii+298 pp. (1992), 1991.

[BL06] J.-M. Bismut and G. Lebeau. The hypoelliptic Laplacian and Ray-Singer metrics. *to appear in 2007*, 2006.

[BLo95] J.-M. Bismut and J. Lott. Flat vector bundles, direct images and higher real analytic torsion. *J. Amer. Math. Soc.*, 8(2):291–363, 1995.

[BM04] J.-M. Bismut and X. Ma. Holomorphic immersions and equivariant torsion forms. *J. Reine Angew. Math.*, 575:189–235, 2004.

[BoC65] R. Bott and S.S. Chern. Hermitian vector bundles and the equidistribution of the zeroes of their holomorphic sections. *Acta Math.*, 114:71–112, 1965.

[GS91] H. Gillet and C. Soulé. Analytic torsion and the arithmetic Todd genus. *Topology*, 30(1):21–54, 1991. With an appendix by D. Zagier.

[GS92] H. Gillet and C. Soulé. An arithmetic Riemann-Roch theorem. *Invent. Math.*, 110(3):473–543, 1992.

[H74] N. Hitchin. Harmonic spinors. *Advances in Math.*, 14:1–55, 1974.

[Hö67] L. Hörmander. Hypoelliptic second order differential equations. *Acta Math.*, 119:147–171, 1967.

[M99] X. Ma. Formes de torsion analytique et familles de submersions. I. *Bull. Soc. Math. France*, 127(4):541–621, 1999.

[M00a] X. Ma. Submersions and equivariant Quillen metrics. *Ann. Inst. Fourier* (*Grenoble*), 50(5):1539–1588, 2000.

[M00b] X. Ma. Formes de torsion analytique et familles de submersions. II. *Asian J. Math.*, 4(3):633–667, 2000.

[MaQ86] V. Mathai and D. Quillen. Superconnections, Thom classes, and equivariant differential forms. *Topology*, 25(1):85–110, 1986.

[MKS67] H.P. McKean, Jr. and I.M. Singer. Curvature and the eigenvalues of the Laplacian. *J. Differential Geometry*, 1(1):43–69, 1967.

[Q85a] D. Quillen. Superconnections and the Chern character. *Topology*, 24(1):89–95, 1985.

[Q85b] D. Quillen. Determinants of Cauchy-Riemann operators on Riemann surfaces. *Functional Anal. Appl.*, 19(1):31–34, 1985.

[RS71] D.B. Ray and I.M. Singer. *R*-torsion and the Laplacian on Riemannian manifolds. *Advances in Math.*, 7:145–210, 1971.

[RS73] D.B. Ray and I.M. Singer. Analytic torsion for complex manifolds. *Ann. of Math.* (2), 98:154–177, 1973.

[RSi78] M. Reed and B. Simon. *Methods of modern mathematical physics. IV. Analysis of operators.* Academic Press [Harcourt Brace Jovanovich Publishers], New York, 1978.

[W82] E. Witten. Supersymmetry and Morse theory. *J. Differential Geom.*, 17(4): 661–692 (1983), 1982.

Index

Jean-Michel Bismut
Département de Mathématique
Université Paris-Sud
Bâtiment 425
F-91405 Orsay
France
e-mail:
Jean-Michel.Bismut@math.u-psud.fr

Progress in Mathematics, Vol. 265, 247–308

Generalized Operads and Their Inner Cohomomorphisms

Dennis V. Borisov and Yuri I. Manin

Abstract. In this paper we introduce a notion of *generalized operad* containing as special cases various kinds of operad-like objects: ordinary, cyclic, modular, properads etc. We then construct inner cohomomorphism objects in their categories (and categories of algebras over them). We argue that they provide an approach to symmetry and moduli objects in non-commutative geometries based upon these "ring-like" structures. We give a unified axiomatic treatment of generalized operads as functors on categories of abstract labeled graphs. Finally, we extend inner cohomomorphism constructions to more general categorical contexts.

Mathematics Subject Classification (2000). Primary 18D50; Secondary 18D10, 20C30.

Keywords. Operads, algebras, inner cohomomorphisms, symmetry and deformations in noncommutative geometry.

0. Introduction

0.1. Inner cohomomorphisms of associative algebras

Let k be a field. Consider pairs $\mathcal{A} = (A, A_1)$ consisting of an associative k-algebra A and a finite-dimensional subspace A_1 generating A. For two such pairs $\mathcal{A} = (A, A_1)$ and $\mathcal{B} = (B, B_1)$, define the category $\mathcal{A} \Rightarrow \mathcal{B}$ by the following data.

An object of $\mathcal{A} \Rightarrow \mathcal{B}$ is a pair (F, u) where F is a k-algebra and $u : A \to F \otimes B$ is a homomorphism of algebras such that $u(A_1) \subset F \otimes B_1$ (all tensor products being taken over k).

A morphism $(F, u) \to (F', u')$ in $\mathcal{A} \Rightarrow \mathcal{B}$ is a homomorphism of algebras $v : F \to F'$ such that $u' = (v \otimes \mathrm{id}_B) \circ u$.

The following result was proved in [Ma3] (see. Prop. 2.3 in Chapter 4):

0.1.1. Theorem. *The category $\mathcal{A} \Rightarrow \mathcal{B}$ has an initial object*

$$(E,\, \delta : A \to E \otimes B)$$

defined uniquely up to unique isomorphism, together with a finite-dimensional subspace $E_1 \subset E$ generating E and satisfying $\delta(A_1) \subset E_1 \otimes B_1$.

This result can be reinterpreted as follows. Consider another category $PAlg$ whose objects are finitely generated k-algebras *together with a presentation* P, i. e. a surjection $\varphi_A : T(A_1) \to A$ where A_1 is a finite-dimensional linear space, $T(A_1)$ is its tensor algebra, and such that $\operatorname{Ker} \varphi_A \cap A_1 = \{0\}$ so that A_1 can be considered as a subspace of A (this condition is not really necessary and can be omitted as is done in Sec. 2.) Morphisms $(A, \varphi_A) \to (B, \varphi_B)$ are algebra homomorphisms $u : A \to B$ such that $u(A_1) \subset B_1$.

This category has a monoidal symmetric structure given by $(A, A_1) \bigcirc (B, B_1) := (C, C_1)$ where $C_1 = A_1 \otimes B_1$ and C is the subalgebra of $A \otimes B$ generated by C_1.

Theorem 0.1.1 establishes a functorial bijection between Hom's in this category

$$\operatorname{Hom}((A, A_1), (F, F_1) \bigcirc (B, B_1)) \cong \operatorname{Hom}((E, E_1), (F, F_1)).$$

When there is no risk of confusion, we will omit A_1, B_1 etc in notation, and denote (E, E_1) by $\underline{cohom}\,(A, B)$ so that we have the standard functorial isomorphism in $(PAlg, \bigcirc)$:

$$\operatorname{Hom}(A, F \bigcirc B) \cong \operatorname{Hom}(\underline{cohom}\,(A, B), F)$$

defining inner cohomomorphism objects.

The usual reasoning produces functorial comultiplication maps between these objects

$$\Delta_{A,B,C} : \underline{cohom}\,(A, C) \to \underline{cohom}\,(A, B) \bigcirc \underline{cohom}\,(B, C), \tag{0.1}$$

which are coassociative (compatible with the ordinary associativity constraints for $\bigcirc$).

In particular $\underline{coend}\,(A) := \underline{cohom}\,(A, A)$ has the canonical structure of a bialgebra.

0.2. Interpretation and motivation

Theorem 0.1.1 was the base of the approach to quantum groups as symmetry objects in noncommutative geometry discussed in [Ma1]–[Ma4]. Namely, consider $PAlg$ as a category of function algebras on "quantum linear spaces" so that the category of quantum linear spaces themselves will be $PAlg^{op}$. Then cohomomorphism algebras correspond to "matrix quantum spaces", and coendomorphism algeras, after passing to Hopf envelopes, become Hopf algebras of symmetries. (In fact, to obtain the conventional quantum groups, one has to add some "missing relations", cf. [Ma2], which also can be done functorially.)

In this paper we present several layers of generalizations of Theorem 0.1.1. The first step consists in extending it to operads with presentation and algebras over them, with an appropriate monoidal structure. We are motivated by the same desire to understand symmetry objects ("quantum semi-groups") in noncommutative geometry based upon operads, or algebras over an operad different from ASS. In fact, as a bonus we also get an unconventional approach to the deformation theory of operadic algebras.

Namely, let $\mathcal{P}$ be an operad, V a linear space, and $OpEnd\, V$ the operad of endomorphisms of V. The set of structures of a $\mathcal{P}$-algebra upon V is then

$$\mathrm{Hom}_{Oper}\,(\mathcal{P}, OpEnd\, V). \tag{0.2}$$

We suggest to consider the object (defined after a choice of spaces of linear generators of both operads)

$$\underline{cohom}\,(\mathcal{P}, OpEnd\, V) \tag{0.3}$$

as (an operad of functions upon) a noncommutative space. Morphisms of (0.3) to the unit object of the monoidal category of operads will then constitute its set of "classical points" (0.2).

0.3. The phantom of the operad

The next extension of Theorem 0.1.1 involves replacing operads by any of the related structures a representative list of which the reader can find, for example, in [Mar]: May and Markl operads, cyclic operads, modular operads, PROPS, properads, dioperads etc. In this paper, we use for all of them the generic name "generalized operad", or simply "operad", and call operads like May's and Markl's ones "ordinary operads", or "classical operads".

It was long recognized that one variable part of the definition of all these structures is the combinatorics and decoration of underlying graphs ("pasting schemes" of [Mar]), whereas another is the category in which components of the respective operad are supposed to lie. Operad itself for us is a functor from a category of labeled graphs to another symmetric monoidal category, as was stressed already in [KoMa], [GeKa2] and many other works. We decided to spell out the underlying formalism in Section 1 of this paper. If we appear to be too fussy, e.g., in Definition 1.3, this is because we found out that uncritical reliance on illustrative pictures can be really misleading.

One can and must approach operadic constructions from various directions and with various stocks of analogies. In this paper, we look at operads, especially those with values in abelian categories, as analogs of associative rings; collections are analogs of their generating spaces. We imagine various noncommutative geometries based upon operads, and are interested in naturally emerging symmetry and moduli objects in these noncommutative geometries.

But of course there are many more different intuitive ideas related to operads.

a) Operads provide tools for studying general algebraic structures determined by a basic set, a family of composition laws, and a family of constraints imposed upon these laws.
b) Operads embody a categorification of graph theory which can be used to study knot invariants, Feynman perturbation series etc.

c) Operads and their algebras are a formalization of computational processes and devices, in particular, tensor networks and quantum curcuits, cf. [MarkSh], [Zo] and references therein. With this in mind, we describe general endomorphism operads in 2.5 below.

It is interesting to notice that the classical theory of recursive functions must refer to a very special and in a sense universal algebra over a non-linear "computational operad", but nobody so far was able to formalize the latter. Main obstacle is this: a standard description of any partially recursive function produces a circuit that may contain cycles of an a priori unknown multiplicity and eventually infinite subprocesses producing no output at all.

0.4. Plan of the paper

In Section 1, we discuss the background topics. The centerpieces of the first part related to graphs are definitions 1.2.1 and 1.3, and the rest is devoted to collections and operads.

In Section 2, we state and prove our main theorems in two contexts: for operads in abelian categories and for algebras. However, the latter requires serious additional restrictions. We also discuss in 2.7, 2.8 some explicit descriptions of cohomomorphism objects, whenever they are known, in particular, for quadratic and more general N-homogeneous algebras and operads.

Notice that Theorem 0.1.1 was extended in [GrM] to include the case of *twisted* tensor products of algebras: see also [Ma4] where the latter appeared in the construction of the De Rham complex of quantum groups and spaces. For further developments see [GrM] and references therein. Similar generalizations might exist for operads as well.

Section 3 is dedicated to the next layer of generalizations. Namely, operads considered in Section 2 and Section 3 can be viewed as algebras over a triple whose main component is the endofunctor $\mathcal{F}$ on a category of collections, described in 1.5.5 and 1.5.6. Categories of collections in this context are abelian and endowed with a symmetric monoidal structure. In Section 3, we consider more general triples and, in particular, do not assume that the category on which the relevant endofuctor acts, is abelian.

This line of thought is continued in Section 4, where in particular operads with values in various categories of algebras are considered. This is partly motivated by the quantum cohomology operad.

Finally, in Appendix we briefly treat Markl's list ([Mar], p. 45) in terms of labeled graphs and functors on them.

Acknowledgements. Yu. M. gratefully acknowledges Bruno Vallette's comments on the preliminary drafts of this paper and discussions with him, in particular related to [Va2]. His numerous suggestions are incorporated in the text. D. B. would like to thank Ezra Getzler for his remarks relating to model categories and the theory of operads in 2-categorical setting.

1. Background

1.1. Graphs

We define objects of the category of (finite) graphs as in [KoMa], [BeMa], [GeKa2]. Geometric realizations of our graphs are not necessarily connected. This allows us to introduce a monoidal structure "disjoint union" on graphs (cf. 1.2.4), and to consider certain morphisms such as graftings and mergers which were not needed in [GeKa2] but arise naturally in more general types of operads. Moreover, our notion of a graph morphism is strictly finer than that considered in the literature: as a part of a morphism, we consider the involution j_h in the Definition 1.2.1 below. Our basic category of sets is assumed to be small.

1.1.1. Definition. *A graph τ is a family of finite sets and maps $(F_\tau, V_\tau, \partial_\tau, j_\tau)$ Elements of F_τ are called flags of τ, elements of V_τ are called vertices of τ. The map $\partial_\tau : F_\tau \to V_\tau$ associates to each flag a vertex, its boundary. The map $j_\tau : F_\tau \to F_\tau$ is an involution: $j_\tau^2 = id$.*

Marginal cases. If V_τ is empty, F_τ must be empty as well. This defines an *empty graph.* To the contrary, F_τ might be empty whereas V_τ is not. In [GeKa2] and other places, in order to treat various units, a "non-graph" with one flag and no vertices is considered. Its role in our constructions sometimes can be played by the empty graph.

Edges, tails, corollas. One vertex graphs with identical j_τ are called *corollas.* Let v be a vertex of τ, $F_\tau(v) := \partial_\tau^{-1}(v)$. Then $\tau_v := (F_\tau(v), \{v\}, \text{evident } \partial, \text{identical } j)$ is a corolla, which is called the corolla of v in τ.

Flags fixed by j_τ form the set of *tails* of τ denoted T_τ.

Two-element orbits of j_τ form the set E_τ of *edges* of τ. Elements of such an orbit are called *halves* of the respective edge.

1.1.2. Geometric realization of a graph. First, let τ be a corolla. If its set of flags is empty, its geometric realization $|\tau|$ is, by definition, a point. Otherwise construct a disjoint union of segments $[0, \frac{1}{2}]$ and identify in it all points 0. This is $|\tau|$. The image of 0 thus becomes the geometric realization of the unique vertex of τ.

Generally, to construct $|\tau|$ take a disjoint union of geometric realizations of corollas of all vertices and identify points $\frac{1}{2}$ of any two flags forming an orbit of j_τ.

A graph τ is called connected (resp. simply connected, resp. tree etc) iff its geometric realization is such. In the same vein, we can speak about connected components of a graph etc. Vertices v with empty $F_\tau(v)$ are considered as connected components.

1.2. Morphisms of graphs and monoidal category Gr

Let τ, σ be two graphs.

1.2.1. Definition. *A morphism $h : \tau \to \sigma$ is a triple (h^F, h_V, j_h), where $h^F : F_\sigma \to F_\tau$ is a contravariant map, $h_V : V_\tau \to V_\sigma$ is a covariant map, and j_h is an* *involution on the set of tails of τ contained in $F_\tau \setminus h^F(F_\sigma)$. This data must satisfy the following conditions.*

(i) *h^F is injective, h_V is surjective.*

(ii) *The image $h^F(F_\sigma)$ and its complement $F_\tau \setminus h^F(F_\sigma)$ are j_τ-invariant subsets of flags. The involution j_h fixes no tail in $F_\tau \setminus h^F(F_\sigma)$.*

It will be convenient to extend j_h to other flags in F_τ by identity.

We will say that h contracts all flags in $F_\tau \setminus h^F(F_\sigma)$. If two flags in $F_\tau \setminus h^F(F_\sigma)$ form an edge, we say that this edge is contracted by h. If two tails in $F_\tau \setminus h^F(F_\sigma)$ form an orbit of j_h, we say that it is a virtual edge contracted by h.

(iii) *If a flag f_τ is not contracted by h, that is, has the form $h^F(f_\sigma)$, then h_V sends $\partial_\tau f_\tau$ to $\partial_\sigma f_\sigma$. Two vertices of a contracted edge (actual or virtual) must have the same h_V-image.*

(iv) *The bijection $h_F^{-1} : h^F(F_\sigma) \to F_\sigma$ maps edges of τ to edges of σ.*

If it maps a pair of tails of τ to an edge of σ, we will say that h grafts these tails.

The composition of two morphisms corresponds to the set-theoretic composition of the respective maps h^F and h_V, and taking the union of two sets of virtual edges.

The resulting category is denoted Gr.

1.2.2. Geometric realization of a morphism. On geometric realizations, the action of h can be visualized as follows: we construct a subgraph of τ consisting only of flags in $h^F(F_\sigma)$, then produce its quotient, and then identify this subquotient with σ using $(h^F)^{-1}$.

The shortest description of this subquotient is this: *merge in $|\tau|$ all vertices belonging to each one fiber of h_V, then delete all flags which are contracted by h.*

It is easy to see that $(h^F)^{-1}$ identifies the geometric graph thus obtained with $|\sigma|$.

This short and intuitive description may be misleading for important concrete categories Γ of labeled graphs (see Definition 1.3). It might happen that such a category does not allow morphisms which simply delete flags, and/or does not allow morphisms that merge two vertices without contracting a path of edges (actual or virtual) connecting such a pair.

The following sequence of steps has more chances to represent a sequence of morphisms in Γ.

a) In each j_h-orbit, graft tails of $|\tau|$ belonging to $F_\tau \setminus h^F(F_\sigma)$ making an actual edge from the virtual one.

Contract to a point each connected component of the union of all actual or virtual edges whose halves belong to $F_\tau \setminus h^F(F_\sigma)$. This point becomes a new vertex.

Besides contracted halves of edges, all other flags adjacent to various vertices of the contracted component are retained and become adjacent to the vertex that is the image of this component. Thus the set of remaining flags consists exactly of the (geometric realizations of) $h^F(F_\sigma)$.

From Def. 1.2.1 (iii) it follows that all flags adjacent to the new vertex are sent by $(h^F)^{-1}$ to a subset of flags adjacent to one and the same vertex of σ.

b) Graft loose ends of each pair of remaining tails that are grafted by h.
c) Merge together those vertices of the obtained (geometric) graph whose preimages are sent by h_V to one and the same vertex of τ.

Finally, identify the resulting subquotient of $|\tau|$ with $|\sigma|$ in such a way that on flags this map becomes $(h^F)^{-1} : h^F(F_\sigma) \to F_\tau$.

Notice that all steps a)–c) could have been done in arbitrary order, for example c), b), a), with the same result. Notice also that in the geometric realization, no trace of j_h is remained: the virtual edges vanished, but we do not know how their halves were paired. This information is encoded only in the combinatorial description involving j_h.

Each step above is in fact a geometric realization of a morphism in Gr. We will now describe formally the respective classes of morphisms.

1.2.3. Contractions, graftings, mergers

a) *Virtual contractions.* A morphism $h : \tau \to \sigma$ is called *a virtual contraction*, if $F_\tau \setminus h^F(F_\sigma)$ consists only of tails, restriction of j_τ on $h^F(F_\sigma)$ coincides with the image of j_σ, and h_V is a bijection.
b) *Contractions and full contractions.* A morphism is called *a contraction*, if h^F is bijective on tails, and if for any $v \in V_\sigma$, any two different vertices in $h_V^{-1}(v)$ are connected by a path consisting of edges contracted by h.smallskip

   Let σ be a graph. Define τ as follows: $F_\tau := T_\sigma$, V_τ:={connected components of σ}. Let h^F be the identical injection $T_\sigma \to F_\sigma$. Let h_V send any vertex to the connected component in which it is contained. The resulting morphism is called *the full contraction.* Its image is a union of corollas, tails of σ are distributed among them as they are among connected components of σ. Morphisms isomorphic to such ones are also called full contractions (of their source).
c) *Grafting and total grafting.* A morphism h is called *a grafting*, if h^F and h_V are bijections.

   Let τ be a graph. Denote by $\sigma := \coprod_{v \in V_\tau} \tau_v$ the disjoint union of corollas of all its vertices. Formally, $F_\sigma = F_\tau$, $V_\sigma = V_\tau$, $\partial_\sigma = \partial_\tau$, and $j_\sigma = id$. Let $h : \sigma \to \tau$ consist of identical maps. Such a morphism is called *total grafting,* and we will reserve for it a special notation:

$$\circ_\tau : \coprod_{v \in V_\tau} \tau_v \to \tau \tag{1.1}$$

   It is defined uniquely by its target τ, up to unique isomorphism identical on τ. Any isomorphism of targets induces an isomorphism of such morphisms.

Its formal inversion cuts all edges of τ in half. Morphisms isomorphic to such one are also called total grafting (of their target).

d) *Mergers and full mergers.* A morphism h is called *a merger*, if h^F is bijective and identifies j_σ with j_τ.

A full merger projects all vertices into one. All edges become loops; tails remain tails.

Mergers play no role in the theory of ordinary operads, but are essential for treating PROPs, cf. appendix.

Each graph τ admits a morphism to a corolla bijective on tails

$$con_\tau : \tau \to con\,(\tau) \tag{1.2}$$

It can be described as the full contraction of τ followed by the full merger. It is defined uniquely by its source τ, up to unique isomorphism identical on τ. Any isomorphism of sources induces a unique isomorphism of such morphisms.

Isomorphisms constitute the intersection of all four classes.

A combinatorial argument imitating 1.2.2 shows that any morphism can be decomposed into a product of a virtual contraction, a contraction, a grafting and a merger. As soon as the order of the types of morphisms is chosen, one can define such a decomposition in a canonical way, up to a unique isomorphism.

1.2.4. Disjoint union as a monoidal structure on Gr. Disjoint union of two abstract sets having no common elements is an obvious notion. If we want to extend it to "all" sets, an appropriate formalization is that of a symmetric monoidal structure "direct sum" with empty set as the unit object. It exists, but is neither unique, nor completely obvious: what is the "disjoint union of a set with itself"? One way to introduce such a structure is described in [Bo2], Example 6.1.9.

We will focus on a small category of finite sets of all cardinalities and sketch the following method which neatly accounts for proliferation of combinatorics of symmetric groups in the standard treatments of operads.

The small category of finite sets of all cardinalities consisting of

$$\emptyset, \{1\}, \{1,2\}, \dots, \{1,2,\dots,n\}, \dots$$

admits a monoidal structure "disjoint union" $\coprod$ given by

$$\{1,2,\dots,m\} \coprod \{1,2,\dots,n\} := \{1,2,\dots,m+n\}$$

and evident commutativity and associativity constraints. For example, identification of $X \coprod Y$ with $Y \coprod X$ proceeds by putting all elements of Y before those of X and retaining the order inside groups. Empty set is the unit of this monoidal structure.

In these terms, it is clear how to extend this construction to the category of "all" totally ordered finite sets, and then to drop the orderings by passing to appropriate colimits. Thus, we can endow the category of finite sets by a monoidal structure which we keep denoting $\coprod$ and calling "disjoint union".

This monoidal structure can then be extended to Gr: $\sigma \coprod \tau$ is determined by disjoint unions of their respective flag and vertex sets, and ∂, j act on both parts as they used to.

Finally, for any finite family $\{\tau_s \mid s \in S\}$, we can define $\coprod_s \tau_s$ functorially in $\{\tau_s\}$ and S as is spelled out in [DeMi].

1.2.5. Atomization of a morphism. Let $h : \tau \to \sigma$ be a morphism of graphs. We define its *atomization* as a commutative diagram of the following form:

$$\begin{array}{ccc} \coprod_{v \in V_\sigma} \tau_v & \xrightarrow{\coprod h_v} & \coprod_{v \in V_\sigma} \sigma_v \\ {\scriptstyle k}\downarrow & & \downarrow{\scriptstyle \circ_\sigma} \\ \tau & \xrightarrow{h} & \sigma \end{array} \tag{1.3}$$

Here σ_v is the corolla of a vertex $v \in V_\sigma$, $\circ_\sigma$ is the total grafting morphism, and the remaining data are constructed as follows.

Graph τ_v. We put for $v \in V_\sigma$:

$$F_{\tau_v} := \{f \in F_\tau \mid h_V(\partial_\tau f) = v\}, \; V_{\tau_v} := \{w \in V_\tau \mid h_V(w) = v\},$$

$$\partial_{\tau_v} = \partial_\tau |_{\tau_v}, \; j_{\tau_v} = j_\tau |_{\tau_v}.$$

Morphism $h_v : \tau_v \to \sigma_v$. We put

$$h_v^F := h^F |_{F_{\sigma_v}} : F_{\sigma_v} \to F_{\tau_v}, \; h_{v,V} := h_V |_{V_{\tau_v}} : V_{\tau_v} \to V_{\sigma_v}, \; j_{h_v} := j_h |_{F_{\tau_v}}.$$

Morphism k. By definition, k^F and k_V are identical maps, hence k is a grafting.

1.2.6. Heredity. Let now $\circ_\sigma : \coprod_{v \in V_\sigma} \sigma_v \to \sigma$ be a total grafting morphism. Assume that we are given a family of morphisms $h_v : \tau_v \to \sigma_v$, $v \in V_\sigma$. Then this data can be uniquely extended to the atomization diagram (1.3) of a morphism $h : \sigma \to \tau$.

1.3. Definition

An abstract category of labelled graphs is a category Γ *endowed with a functor* $\psi : \Gamma \to Gr$ *satisfying the following conditions:*

(i) Γ *is endowed with a monoidal structure which* ψ *maps to the disjoint union in* Gr. *It will be denoted by the same sign* $\coprod$.

(ii) ψ *is faithful: if two morphisms with common source and target become equal after applying* ψ, *they are equal.*

(iii) *Call a* Γ*-corolla any object* τ *of* Γ *such that* $\psi(\tau)$ *is a corolla. Any object* $\tau \in \Gamma$ *admits a morphism to a* Γ*-corolla*

$$con_\tau : \tau \to con\,(\tau)$$

which is a lift to Γ *of the diagram of the form* (1.2) *with the source* $\psi(\tau)$. *It is defined uniquely up to unique isomorphism identical on* τ.

(iv) *Any object of* Γ *is the target of a morphism from a disjoint union of* Γ*-corollas*

$$\circ_\tau : \coprod_{v \in V_{\psi(\tau)}} \tau_v \to \tau$$

which is a lift to Γ *of the diagram of the form* (1.1) *with the target* $\psi(\tau)$*. It is defined uniquely up to unique isomorphism identical on* τ*.*

(v) *Any morphism* $h : \tau \to \sigma$ *can be embedded into a commutative diagram of the form* (1.3) *which is a lift of the atomization diagram for the morphism* $\psi(h) : \psi(\tau) \to \psi(\sigma)$*. Such a diagram is defined uniquely up to unique isomorphism.*

(vi) *Moreover, assume that we are given* σ *in* Γ*, and for each* Γ*-corolla* σ_v*,* $v \in V_{\psi(\sigma)}$*, a morphism* $h_v : \tau_v \to \sigma_v$ *where* $\{\tau_v\}$ *is a family of objects of* Γ*. Then there exists a morphism* $\tau \to \sigma$ *in* Γ *such that all this data fit into (a lift of) the atomization diagram* (1.3)*. Moreover,* τ *and* σ *are defined uniquely up to unique isomorphism.*

The last requirement formalizes what Markl calls "hereditary" property in [Mar], p. 45.

1.3.1. Comments

a) It is helpful (and usually realistic) to imagine any object $\sigma \in \Gamma$ as a pair consisting of the "underlying graph" $\psi(\sigma)$ and an additional structure on the components of $\psi(\sigma)$ such as decorating vertices by integers, a cyclic order on flags adjoining to a vertex, etc (see examples below). We will generally refer to such a structure as "labeling".

   As a rule, existence of labeling of a given type and/or additional algebraic properties required for a treatment a certain type of operadic structure put some restrictions upon underlying graphs so that ψ need not be surjective on objects. On the other hand, if these restrictions are satisfied, there might be many different compatible labeling on the same underlying graph so that ψ need not be injective on objects either.

   The functor ψ on objects simply forgets labeling.

b) In the same vein, any morphism $\sigma \to \tau$ in Γ should be imagined as a morphism $\psi(\sigma) \to \psi(\tau)$ of underlying graphs satisfying some constraints of two types: purely geometric ones which can be stated in Gr, and certain compatibility conditions with labelings (see below). This is the content of condition (ii) above.

c) Let $\sigma \in \Gamma$. Slightly abusing the language, we will call flags, vertices, edges etc. of $\psi(\sigma)$ the respective components of σ, and write, say, V_σ in place of $V_{\psi(\sigma)}$. Similar conventions will apply to morphisms in Γ. By extension, σ is called a Γ-corolla (resp. tree etc), if $\psi(\sigma)$ is a corolla (resp. tree etc).

1.3.2. Examples of labeling

a) *Oriented graphs.* Any map $F_\sigma \to \{in, out\}$ such that halves of any edge are oriented by different labels, is called *an orientation of* σ. On the geometric realization, a flag marked by *in* (resp. *out*) is oriented towards (resp. outwards) its vertex.

   Tails of σ oriented *in* (resp. *out*) are called *inputs* (resp. *outputs*) of σ. Similarly, $F_\sigma(v)$ is partitioned into inputs and outputs of the vertex v.

   Consider an orientation of σ. Its edge is called *an oriented loop*, if both its halves belong to the same vertex. Otherwise an oriented edge starts at a source vertex and ends at a different target vertex.

   More generally, a sequence of pairwise distinct edges $e_1, \dots, e_n$, is called *a simple path* of length n, if e_i and e_{i+1} have a common vertex, and the $n-1$ vertices obtained in this way are pairwise distinct. If moreover e_1 and e_n also have a common vertex distinct from the mentioned ones, this path is *a wheel* of length n. A loop is a wheel of length one. Edges in a wheel are endowed only with a cyclic order up to inversion.

   Clearly, all edges in a path (resp. a wheel) can be oriented so that the source of e_{i+1} is the target of e_i.

   If the graph is already oriented, the induced orientation on any path (resp. wheel) either has this property or not. Respectively, the wheel is called oriented or not.

   A morphism of oriented graphs h is a morphism of graphs such that h^F is compatible with orientations.

b) *Directed graphs.* An oriented graph σ is called *directed* if it satisfies the following condition:

   On each connected component of the geometric realization, one can define a continuous real valued function ("height") in such a way that moving in the direction of orientation along each flag decreases the value of this function.

   In particular, a directed graph has no oriented wheels.

   Notice that, somewhat counterintuitively, a directed graph is not necessarily oriented "from its inputs to its outputs" as is usually shown on illustrating pictures. In effect, take a corolla with only *in* flags and another corolla with only *out* flags, and graft one input to one output. The resulting graph is directed (check this) although its only edge is oriented from global outputs to global inputs.

   This is one reason why it is sometimes sensible to include in a category Γ of directed graphs only those, which have at least one input and least one output at each vertex (cf. the definition of reduced bimodules in Section 1.1 of [Va1]).

Another reason for excluding certain marginal *("unstable")* types of labeled corollas might be our desire to ensure essential finiteness of the categories denoted $\Rightarrow \sigma$ in Section 1.5.5 (cf. a description of unstable modular corollas below). In a category of (disjoint unions of) directed trees, for example, this leads to the additional requirement: corolla of any vertex has at least three flags.

This requirement might lead to some technical problems if we want to consider unital versions of our operads.

c) *Genus labeling.* A genus labeling of σ is a map $g : V_\sigma \to \mathbf{Z}_{\geq 0}, v \mapsto g_v$. The genus of a *connected* labeled graph σ is defined as

$$g(\sigma) := \sum_{v \in V_\sigma} g_v + \operatorname{rk} H_1(|\sigma|) = \sum_{v \in V_\sigma} (g_v - 1) + \operatorname{card} E_\sigma + 1.$$

Genus labeled graphs (or only connected ones) are called *modular graphs.* Corolla of any vertex of a modular graph is a modular graph.

A morphism of modular graphs is a morphism of graphs compatible with labeling in the following sense. Contraction of a looping edge raises the genus of its vertex by one. Contraction of a non-looping edge prescribes to the emerging vertex the sum of genera of two ends of the edge. Finally, grafting flags does not change genera of vertices.

Thus, a morphism between two connected modular graphs can exist only if their genera coincide.

A modular corolla with vertex of genus g and n flags is called *stable* iff $2g - 2 + n > 0$. A modular graph is called *stable,* iff corollas of all its vertices are stable.

d) *Colored graphs.* Let I be an abstract set (elements of which are called colors). *An I-colored graph* is a graph τ together with a map $F_\tau \to I$ such that two halves of each edge get the same color. Morphisms are restricted by the condition that h^F preserves color.

In [LoMa], a topological operad was studied governed by a category of colored graphs with two-element $I = \{black, white\}$. Halves of an edge in this category are always white.

e) *Cyclic labeling.* A cyclic labeling of a graph τ is a choice of cyclic order upon each set $F_\tau(v)$. Alternatively, it is a family of bijections $F_\tau(v) \to \mu_{|v|}$ where $\mu_{|v|}$ is the group of roots of unity (in $\mathbf{C}$) of degree $|v| := F_\tau(v)$; two maps which differ by a multiplication by a root of unity define the same labeling. Yet another description identifies a cyclic labeling of τ with a choice of planar structure for each corolla τ_v, that is, an isotopy class of embeddings of $|\tau_v|$ into an *oriented* plane.

In a category of cyclic labeled graphs, mergers are not allowed, whereas contractions, say, of one edge, should be compatible with cyclic labeling in an evident way: say $(0, 1, \dots, m)$ and $(0, \bar{1}, \dots, \bar{n})$ turn into $(1, \dots, m, \bar{1}, \dots, \bar{n})$, where by 0 we denoted the contracted halves of the same edge in two corollas.

An interesting variant of cyclic labeling is *unoriented cyclic labeling* (it has nothing to do with orientation of the graph itself): the cyclic orders $(0, 1, \ldots, m)$ and $(m, m-1, \ldots, 0)$ are considered equivalent. In this version, contraction of an edge leads generally to two different morphisms in Γ, with two different targets.

Combinatorics of unoriented cyclic labeled trees is very essential in the description of the topological operad of real points $\overline{M}_{0,*}(\mathbf{R})$, cf. [GoMa].

1.4. Ground categories $\mathcal{G}$

Operads of various types in this paper will be defined as certain functors from a category of labeled graphs Γ to a symmetric monoidal category $(\mathcal{G}, \otimes)$ which will be called *ground category*. The simplest example is that of finite-dimensional vector spaces over a field, or that of finite complexes of such spaces.

In order to ensure validity of various constructions we will have to postulate (locally) some additional properties of $(\mathcal{G}, \otimes)$, the most important of which are contained in the following list. At this stage, we do not assume that all of them, or some subset of them, hold simultaneously.

a) Existence of a unit object.
b) Existence of internal cohom objects $\underline{cohom}\,(X, Z)$ for any objects X, Z in $\mathcal{G}$. By definition, they fit into functorial isomorphisms
$$\mathrm{Hom}_{\mathcal{G}}(X, Y \otimes Z) = \mathrm{Hom}_{\mathcal{G}}(\underline{cohom}\,(X, Z), Y).$$
These isomorphisms are established by composition with coevaluation morphisms
$$c = c_{X,Z} : X \to \underline{cohom}\,(X, Z) \otimes Z$$
(cf. the diagram (2.3) in Section 2 below).
c) Existence of countable coproducts such that $\otimes$ is distributive with respect to these coproducts.
d) Existence of finite (and sometimes infinite) colimits.
e) $\mathcal{G}$ is an abelian category, $\otimes$ is an additive bifunctor exact in each argument.
f) $\mathcal{G}$ is a closed model category.

1.5. $\Gamma\mathcal{G}$-collections

We will denote by ΓCOR the subcategory (groupoid) of Γ consisting of Γ-corollas and *isomorphisms* between them.

1.5.1. Definition. *A $\Gamma\mathcal{G}$-collection A_1 is a functor $A_1 : \Gamma COR \to \mathcal{G}$. A morphism of $\Gamma\mathcal{G}$-collections $A_1 \to B_1$ is a functor morphism (natural transformation.)*

The category of $\Gamma\mathcal{G}$-collections will be denoted $\Gamma\mathcal{G}COLL$.

1.5.2. Examples

a) If Γ is the category of stable modular graphs, the category $\Gamma\mathcal{G}COLL$ is equivalent to the category of double sequences $A_1((g, n))$ of objects of $\mathcal{G}$ endowed with an action of $\mathbf{S}_n$ upon $A_1((g, n))$. A morphism $A_1 \to B_1$ is a sequence of morphisms in $\mathcal{G}$, $A_1((g, n)) \to B_1((g, n))$, compatible with $\mathbf{S}_n$-actions.

In effect, any Γ-collection A_1 is determined up to an isomorphism by its restriction to the category of modular corollas with flags $\{1,\dots,n\}$ and a vertex labeled by g. Their isomorphisms correspond to permutations of $\{1,\dots,n\}$.

Such collections are called stable **S**-modules in [GeKa2], (2.1).

b) Let Γ be a category of oriented graphs, containing all oriented corollas. Then the category $\Gamma\mathcal{G}$-collections is equivalent to the category of double sequences $A_1(m,n)$ of objects in $\mathcal{G}$ endowed with actions of $\mathbf{S}_m \times \mathbf{S}_n$, and equivariant componentwise isomorphisms.

This is clear: look at oriented corollas with inputs $\{1,\dots,n\}$ and outputs $\{1,\dots,m\}$: see, e.g., [Va1], 1.1.

1.5.3. White product of collections. For two collections A_1, B_1, define their white product by

$$(A_1 \bigcirc B_1)(\sigma) := A_1(\sigma) \otimes B_1(\sigma)$$

for any Γ-corolla σ, with obvious extension to morphisms.

This determines a symmetric monoidal structure $\bigcirc$ on $\Gamma\mathcal{G}COLL$.

If $(\mathcal{G}, \otimes)$ is endowed with a unit object u, then the collection U, $U(\sigma) = u$, sending each morphism of corollas to id_u is a unit object of $(\Gamma\mathcal{G}COLL, \bigcirc)$.

1.5.4. Inner cohomomorphisms for collections. If $(\mathcal{G}, \otimes)$ admits internal cohom objects, the same holds for $(\Gamma\mathcal{G}COLL, \bigcirc)$: just work componentwise.

1.5.5. Endofunctor $\mathcal{F}$ on $\Gamma\mathcal{G}COLL$. Let A_1 be a $\Gamma\mathcal{G}$-collection. Consider a Γ-corolla σ and denote by $\Rightarrow \sigma$ the category whose objects are Γ-morphisms Γ-graphs $\tau \to \sigma$, and whose morphisms are Γ-isomorphisms of morphisms identical on σ. Put

$$\mathcal{F}(A_1)(\sigma) := \mathrm{colim}_{\Rightarrow\sigma} \otimes_{v \in V_\tau} A_1(\tau_v) \tag{1.4}$$

The existence of appropriate colimits in $\mathcal{G}$ such that $\otimes$ is distributive with respect to them should be postulated at this stage.

In some important cases (e.g., stable modular graphs) any category $\Rightarrow \sigma$ is essentially finite (equivalent to a category with finitely many objects and morphisms). Therefore existence of finite colimits in $\mathcal{G}$ suffices.

Clearly, this construction is functorial with respect to isomorphisms of Γ-corollas so that we actually get a new collection $\mathcal{F}(A_1)$. Moreover, the map $A_1 \mapsto \mathcal{F}(A_1)$ extends to an endofunctor of $\Gamma\mathcal{G}COLL$.

As is well known, functor composition endows the category of endofunctors by the structure of strict monoidal category with identity.

1.5.6. Proposition. *The endofunctor $\mathcal{F}$ has a natural structure of a triple, that is, a monoid with identity in the category of endofunctors.*

Proof (sketch). The argument is essentially the same as in (2.17) of [GeKa2]. We have to construct a multiplication morphism $\mu : \mathcal{F}\cdot\mathcal{F} \to \mathcal{F}$, an identity morphism $\eta : \mathrm{Id} \to \mathcal{F}$, and to check the commutativity of several diagrams.

In other words, for a variable collection A_1, we need functorial morphisms of collections $\mu_{A_1} : \mathcal{F}^2(A_1) \to \mathcal{F}(A_1)$, $\eta_{A_1} : A_1 \to \mathcal{F}(A_1)$ fitting the relevant commutative diagrams. In turn, to define them, we have to give their values (in $\mathcal{G}$) on any Γ-corolla σ, functorially with respect to σ.

The construction of μ essentially uses the hereditary property 1.3(vi) of the category Γ. In fact, we have

$$\begin{aligned} \mathcal{F}^2(A_1)(\sigma) &= \mathrm{colim}_{\tau\to\sigma} \otimes_{v\in V_\tau} \mathcal{F}(\tau_v) \\ &= \mathrm{colim}_{\tau\to\sigma}[\mathrm{colim}_{\rho_v\to\tau_v} \otimes_{w\in V_{\rho_v}} A_1(\rho_{v,w})] \end{aligned} \tag{1.5}$$

where ρ_v are objects of Γ, and $\rho_{v,w}$ is the Γ-corolla of a vertex w of ρ_v. Using heredity, we can produce from each family $\rho_v \to \tau_v$ a morphism $\rho \to \tau$, and replace the right-hand side of (1.5) by

$$\mathrm{colim}_{\rho\to\tau\to\sigma} \otimes_{w\in V_\rho} A_1(\rho_w)$$

The latter colimit maps to $\mathcal{F}(A_1)(\sigma)$ via composition of two arrows in $\rho \to \sigma$.

As Getzler and Kapranov suggest, this construction and various similar ones needed to produce η_{A_1} and to check axioms, become more transparent if one uses the simplicial formalism.

Given σ and $k \geq 0$, define the category $\Rightarrow_k \sigma$: its objects are sequences of morphisms $(f_1, \dots, f_k)$ in Γ, $f_1 : \tau_0 \to \tau_1, \dots, f_k : \tau_{k-1} \to \tau_k$ together with an augmentation morphism $\tau_k \to \sigma$. Morphisms in $\Rightarrow_k \sigma$ are isomorphisms of such sequences compatible with augmentation.

Categories $\Rightarrow_k \sigma$ are interconnected by the standard face and degeneracy functors turning them into components of a simplicial category.

Namely, $d_i : \Rightarrow_k \sigma \to \Rightarrow_{k-1} \sigma$ skips τ_0, f_1 (resp. f_k, τ_k) for $i = 0$ (resp. $i = k$); skips τ_i and composes f_i, f_{i+1} for $1 \leq i \leq k-1$.

Similarly, $s_i : \Rightarrow_k \sigma \to \Rightarrow_{k+1} \sigma$ inserts $id : \tau_i \to \tau_i$.

An argument similar to one which we sketched above for $k = 1$ will convince the reader that one can identify the value of the $(k+1)$-th iteration $\mathcal{F}^{k+1}(A_1)$ at σ with the functor sending σ to

$$\mathrm{colim}_{\Rightarrow_k \sigma} \otimes_{v\in V_{\tau_0}} A_1(\tau_{0,v}).$$

Thus the multiplication μ is the functor induced on colimits by $d_1 : \Rightarrow_1 \sigma \to \Rightarrow_0 \sigma$. Monoidal identity maps A_1 to $\mathcal{F}(A_1)$ by sending $A_1(\sigma)$ to the diagram $\mathrm{id} : A_1(\sigma) \to A_1(\sigma)$.

Two morphisms $\mathcal{F}^3 \to \mathcal{F}^2$ corresponding to two configurations of brackets are induced by d_1 and d_2 respectively, and the associativity is expressed by the simplicial identity $d_1 d_1 = d_1 d_2$. Identity η is treated similarly. $\square$

1.6. $\Gamma\mathcal{G}$-operads as functors

Traditional approaches to operads in our context lead to three different but equivalent definitions of it. Very briefly, they can be summarized as follows.

(I) An operad is a tensor functor $(\Gamma, \coprod) \to (\mathcal{G}, \otimes)$ satisfying certain additional constraints.

(II) An operad is a collection together with a structure of an algebra over the triple $(\mathcal{F}, \mu, \eta)$.

(III) An operad is a collection together with a composition law which makes it a monoid with respect to an appropriate symmetric monoidal structure upon $\Gamma\mathcal{G}COLL$ (to be described).

We will start with the first description, and proceed to the second one in Section 1.7. As for the third description which requires first construction of a special monoidal structure on collections, it seems to be less universal. In the Appendix, we will sketch it for orientation labeling and directed graphs, following [Va1].

Consider the category whose objects are functors $A : \Gamma \to \mathcal{G}$ *compatible with the monoidal structures* $\coprod$ *and* $\otimes$ in the following sense: we are given functorial isomorphisms

$$a_{\sigma,\tau} : A(\sigma \coprod \tau) \to A(\sigma) \otimes A(\tau) \tag{1.6}$$

for all $\sigma, \tau \in \Gamma$ such that inverse isomorphisms $a^{-1}_{\sigma,\tau}$ satisfy conditions spelled out in Def. 1.8 of [DeMi]. Such functors form a category, morphisms in which are functor morphisms compatible with $a_{\sigma,\tau}$.

In [DeMi] such functors are called *tensor functors,* we will also use this terminology.

1.6.1. Definition. *The category $\Gamma\mathcal{G}OPER$ of $\Gamma\mathcal{G}$-operads is the category of those tensor functors $(A, a) : \Gamma \to \mathcal{G}$ that send any grafting morphism, in particular $\circ_\tau$, in Γ to an isomorphism.*

Morphisms are functor morphisms.

Informally, making grafting morphisms invertible means that $A(\tau)$ for any $\tau \in \Gamma$ can be canonically identified with the tensor product $\otimes_v A(\tau_v)$ where τ_v runs over Γ-corollas of all vertices of τ (or rather, of $\psi(\tau)$). To see this, one should apply (1.6) (functorially extended to disjoint unions of arbitrary finite families) to the l.h.s. of (1.1).

Moreover, the morphism (1.1) of "full contraction followed by a merger", $\circ_\tau : \tau \to con\,(\tau)$ in Γ, combined with just described tensor decomposition produces a morphism in $\mathcal{G}$

$$\otimes_{v \in V_\tau} A(\tau_v) \to A(con\,(\tau)). \tag{1.7}$$

This is our embodiment of operadic compositions.

We will omit a in the notation (A, a) for brevity, and simply treat (1.6) as identical map, as well as its extensions to arbitrary families and their disjoint unions.

1.7. From functors to algebras over the triple $(\mathcal{F}, \mu, \eta)$

Let now $A : \Gamma \to \mathcal{G}$ be an operad. Denote by A_1 its restriction to the subcategory of Γ-corollas and their isomorphisms. Let σ be a Γ-corolla. We can treat any object $\tau \to \sigma$ in $\Rightarrow \sigma$ (cf. 1.5.5) as a morphism in Γ and apply to it the functor A. We will get a morphism $\otimes_{v \in V_\tau} A_1(\tau_v) \to A_1(\sigma)$ functorial with respect to isomorphisms of τ identical on σ. Due to the universal property of colimits, these morphisms induce a morphism in $\mathcal{G}$

$$\mathcal{F}(A_1)(\sigma) \to A_1(\sigma).$$

The system of these morphisms is functorial in σ, so we get finally a morphism of collections

$$\alpha_{A_1} : \mathcal{F}(A_1) \to A_1. \tag{1.8}$$

Similarly, among the objects of $\Rightarrow \sigma$ there is the identical morphism $id : \sigma \to \sigma$. As above, it produces a morphism of collections

$$\eta_{A_1} : A_1 \to \mathcal{F}(A_1). \tag{1.9}$$

1.7.1. Proposition.

(i) *The data* $(A_1, \alpha_{A_1}, \eta_{A_1})$ *constitute an algebra over the triple* $(\mathcal{F}, \mu, \eta)$ *(see, e.g., [MarShSt], pp. 88–89).*

(ii) *The map* $A \mapsto (A_1, \alpha_{A_1}, \eta_{A_1})$ *extends to a functor coll establishing equivalence of the category of* $\Gamma\mathcal{G}$*-operads and the category of algebras over the triple* $(\mathcal{F}, \mu, \eta)$.

Proof (sketch). The proof is essentially the same as that of Proposition (2.23) in [GeKa2]. The first statement reduces to the check of commutativity of several diagrams.

The main problem in the second statement consists in extending each morphism of operads as algebras to a morphism of operads as functors. In other words, knowing the operadic compositions induced by full contractions and mergers (if the latter occur in Γ), we want to reconstruct operadic compositions induced by partial contractions such as $\tau \to \sigma$ in (a Γ-version of) a diagram (1.3). To this end, complete (1.3) by a morphism

$$f : \coprod_{w \in V_\tau} \tau_w \to \coprod_{v \in V_\sigma} \tau_v$$

which is the disjoint union of morphisms of total graftings with targets τ_v (so that τ_w are corollas whereas τ_v generally are not). Since $k \cdot f$ is a total grafting with target τ, in order to calculate the value of our functor on $\tau \to \sigma$ it suffices to know its value on the composition $\circ_\sigma(\coprod h_v)f$ which involves only graftings, full contractions and mergers of corollas. □

1.7.2. Algebra $\mathcal{F}(A_1)$. As a general formalism shows, for any A_1, the collection $\mathcal{F}(A_1)$ has the canonical structure of an operad, with $\alpha_{\mathcal{F}(A_1)} := \mu_{A_1} : \mathcal{F}^2(A_1) \to \mathcal{F}(A_1)$.

1.7.3. Proposition. *For any collection A_1 and an operad B, there exists a canonical identification*

$$\mathrm{Hom}_{\Gamma\mathcal{G}COLL}(A_1, coll(B)) = \mathrm{Hom}_{\Gamma\mathcal{G}OPER}(\mathcal{F}(A_1), B)$$

functorial in both arguments.

This means that the functor $\mathcal{F} : \Gamma\mathcal{G}COLL \to \Gamma\mathcal{G}OPER$ is a construction of the free operad freely generated by a collection, and is thus an analog of tensor algebra of a linear space.

1.8. White product of operads

For two $\Gamma\mathcal{G}$-operads A, B, define their white product by

$$(A \bigcirc B)(\sigma) := A(\sigma) \otimes B(\sigma)$$

for any object $\sigma \in \Gamma$. The extension to morphisms is evident.

This determines a symmetric monoidal structure $\bigcirc$ on $\Gamma\mathcal{G}OPER$.

As before, if $(\mathcal{G}, \otimes)$ is endowed with a unit object u, then the functor $U : \sigma \mapsto u$ sending each morphism of labeled graphs to id_u is a unit object of $(\Gamma\mathcal{G}OPER, \bigcirc)$.

1.9. Morphism j

Morphism of operads $j : \mathcal{F}(E_1 \bigcirc B_1) \to \mathcal{F}(E_1) \bigcirc \mathcal{F}(B_1)$ that is, a family of morphisms

$$\begin{aligned}&\mathrm{colim}_{\Rightarrow\sigma} \otimes_{v\in V_\tau} E_1(\tau_v) \otimes B_1(\tau_v)\\ &\to (\mathrm{colim}_{\Rightarrow\sigma} \otimes_{v\in V_{\tau'}} E_1(\tau_v{}')) \otimes (\mathrm{colim}_{\Rightarrow\sigma} \otimes_{v\in V_{\tau''}} B_1(\tau_v{}''))\end{aligned}$$

comes from the "diagonal" part of the right-hand side: $\tau = \tau' = \tau''$.

2. Inner cohomomorphism operads

2.1. Preparation

Fix a graph category Γ and a ground category $\mathcal{G}$ as above.

In this section, we will assume that $(\mathcal{G}, \otimes)$ has inner cohomomorphism objects. Moreover, for validity of the main Theorem 2.2, $\mathcal{G}$ must be abelian, with tensor product exact in each argument.

Let A be a $\Gamma\mathcal{G}$-operad, A_1 a $\Gamma\mathcal{G}$-collection, and $i_A : A_1 \to A$ a morphism of collections such that the respective morphism of operads $f_A : \mathcal{F}(A_1) \to A$ is surjective. Denote by $\mathcal{A}$ the diagram $i_A : A_1 \to A$. Such a diagram can be thought of as a presentation of A.

Let $\mathcal{B}$ be a similar data $i_B : B_1 \to B$.

As in 0.1, denote by $\mathcal{A} \Rightarrow \mathcal{B}$ the category whose objects are commutative diagrams

$$\begin{array}{ccc} A_1 & \xrightarrow{u_1} & F \bigcirc B_1 \\ {\scriptstyle i_A}\downarrow & & \downarrow{\scriptstyle id_F \bigcirc i_B} \\ A & \xrightarrow{u} & F \bigcirc B \end{array} \tag{2.1}$$

where F is an operad, u is a morphism of operads, and u_1 is a morphism of collections. In the upper row, and below in similar situations, we write F in place $coll(F)$ for brevity, identifying an operad with its underlying collection.

Notice that, unlike in 0.1, we do not assume that i_A, i_B are injective. Hence the upper row of (2.1) has to be given explicitly.

Such a diagram can be denoted (F, u, u_1) since the remaining data are determined by $\mathcal{A}$, $\mathcal{B}$.

A morphism $(F, u, u_1) \to (F', u', u'_1)$ in $\mathcal{A} \Rightarrow \mathcal{B}$ is a morphism of operads $F \to F'$ inducing a morphism of commutative diagrams (2.1) constructed for F and F' respectively.

2.2. Theorem

The category $\mathcal{A} \Rightarrow \mathcal{B}$ has an initial object

$$\begin{array}{ccc} A_1 & \xrightarrow{\delta_1} & E \bigcirc B_1 \\ {\scriptstyle i_A}\downarrow & & \downarrow{\scriptstyle id_E \bigcirc i_B} \\ A & \xrightarrow{\delta} & E \bigcirc B \end{array} \tag{2.2}$$

defined uniquely up to unique isomorphism.

Moreover, E comes together with a presentation $\mathcal{E}$, $i_E : E_1 \to E$ in which $E_1 = \underline{cohom}(A_1, B_1)$, inner cohomomorphism being taken in the category of $\Gamma\mathcal{G}$-collections.

If F is given together with its presentation $\mathcal{F}$, that is $i_F : F_1 \to F$, and u is induced by $u_1 : A_1 \to F_1 \bigcirc B_1$, then the canonical homomorphism $E \to F$ is induced by a unique morphism in the category $\mathcal{E} \Rightarrow \mathcal{F}$.

Proof. (i) *Preparation.* The morphism $u_1 : A_1 \to F \bigcirc B_1$ corresponds to a morphism $\tilde{u}_1 : E_1 \to F$ where $E_1 = \underline{cohom}(A_1, B_1)$ as above. Recall that inner cohomomorphism collections here can be constructed componentwise. Let $c : A_1 \to E_1 \bigcirc B_1$ be the coevaluation morphism. Then the diagram

$$\begin{array}{ccc} A_1 & \xrightarrow{c} & E_1 \bigcirc B_1 \\ {\scriptstyle id_{A_1}}\downarrow & & \downarrow{\scriptstyle \tilde{u}_1 \bigcirc id_{B_1}} \\ A_1 & \xrightarrow{u_1} & F \bigcirc B_1 \end{array} \tag{2.3}$$

is commutative. Composing (2.3) with (2.1), we get a commutative square of collections

$$\begin{array}{ccc} A_1 & \xrightarrow{c} & E_1 \bigcirc B_1 \\ {\scriptstyle i_A}\downarrow & & \downarrow \\ A & \xrightarrow{u} & F \bigcirc B \end{array} \tag{2.4}$$

which produces a morphism of operads with presentations

$$\begin{array}{ccc} \mathcal{F}(A_1) & \xrightarrow{\mathcal{F}(c)} & \mathcal{F}(E_1 \bigcirc B_1) \\ {\scriptstyle f_A}\downarrow & & \downarrow{\scriptstyle g} \\ A & \xrightarrow{u} & F \bigcirc B \end{array} \tag{2.5}$$

This diagram can be completed by the commutative triangle

$$\begin{array}{ccccc} \mathcal{F}(A_1) & \xrightarrow{\mathcal{F}(c)} & \mathcal{F}(E_1 \bigcirc B_1) & \xrightarrow{h} & \mathcal{F}(E_1) \bigcirc B \\ {\scriptstyle f_A}\downarrow & & \downarrow{\scriptstyle g} & \swarrow{\scriptstyle \tilde{v} \bigcirc id_B} & \\ A & \xrightarrow{u} & F \bigcirc B & & \end{array} \tag{2.6}$$

where h is the composition

$$\mathcal{F}(E_1 \bigcirc B_1) \xrightarrow{j} \mathcal{F}(E_1) \bigcirc \mathcal{F}(B_1) \xrightarrow{id \bigcirc f_B} \mathcal{F}(E_1) \bigcirc B$$

and j is described in Section 1.9.

(ii) *Construction of main objects.* Now we will construct in turn an ideal $(\tilde{R}) \subset \mathcal{F}(E_1)$, the operad $E := \mathcal{F}(E_1)/(\tilde{R})$ together with a morphism of operads $\delta : A \to E \bigcirc B$ and a morphism $v : E \to F$.

Starting with this point, we will have to use our assumption that assume that $\mathcal{G}$ is abelian, and $\otimes$ is exact.

Choose a subcollection $R \subset \operatorname{Ker} f_A$ generating $\operatorname{Ker} f_A$ as an ideal in the free operad $\mathcal{F}(A_1)$. Replace the morphism $h \circ \mathcal{F}(c) : \mathcal{F}(A_1) \to \mathcal{F}(E_1) \bigcirc B$ by the morphism canonically corresponding to it

$$\underline{cohom}(\mathcal{F}(A_1), B) \to \mathcal{F}(E_1)$$

where the inner cohomomorphisms here and below are taken in the category of collections.

Since inner cohomomorphisms are covariant functorial with respect to the first argument, we have a commutative diagram

$$\begin{array}{ccc} \underline{cohom}\,(R,B) & & \\ \downarrow & \searrow & \\ \underline{cohom}\,(\mathcal{F}(A_1),B) & \longrightarrow & \mathcal{F}(E_1) \\ \downarrow & & \downarrow \tilde{v} \\ \underline{cohom}\,(A,B) & \longrightarrow & F \end{array} \tag{2.7}$$

where $\tilde{v}$ is taken from (2.6).

Denote by $\tilde{R}$ the image of the skew arrow in (2.7). Since the composition $R \to \mathcal{F}(A_1) \to A$ is zero, the same holds for the composition of the two left vertical arrows in (2.7). This implies that the composition $\tilde{R} \to \mathcal{F}(E_1) \to F$ is zero. Since $\tilde{v}$ is a morphism of operads, its kernel contains the ideal $(\tilde{R})$ generated by $\tilde{R}$.

Rewrite the upper triangle in (2.7) as a commutative square

$$\begin{array}{ccc} \underline{cohom}\,(R,B) & \longrightarrow & \tilde{R} \\ \downarrow & & \downarrow \\ \underline{cohom}\,(\mathcal{F}(A_1),B) & \longrightarrow & \mathcal{F}(E_1) \end{array}$$

Replacing the horizontal arrows with the help of coevaluation morphisms, we get the commutative square

$$\begin{array}{ccc} R & \longrightarrow & \tilde{R} \bigcirc B \\ \downarrow & & \downarrow \\ \mathcal{F}(A_1) & \longrightarrow & \mathcal{F}(E_1) \bigcirc B \end{array} \tag{2.8}$$

which induces a morphism of operads

$$\delta : A = \mathcal{F}(A_1)/(R) \to \mathcal{F}(E_1) \bigcirc B/(\tilde{R} \bigcirc B) \cong \mathcal{F}(E_1)/(\tilde{R}) \bigcirc B = E \bigcirc B \tag{2.9}$$

(use the exactness of $\otimes$ in $\mathcal{G}$.)

(iii) *Completion of the proof.* It remains to show that $\delta : A \to E \bigcirc B$ has all the properties stated in Theorem 2.2.

From our construction, it is clear that it fits into the diagram of the form (2.2), and that it comes with the presentation $i_E : E_1 \to E$.

The morphism $\tilde{v} : \mathcal{F}(E_1) \to F$ descends to a morphism of operads $v : E \to F$, and from the commutativity of the diagram (2.6) one can infer that it induces a morphism of objects $(E, \delta, \delta_1) \to (F, u, u_1)$.

We leave the remaining checks to the reader. □

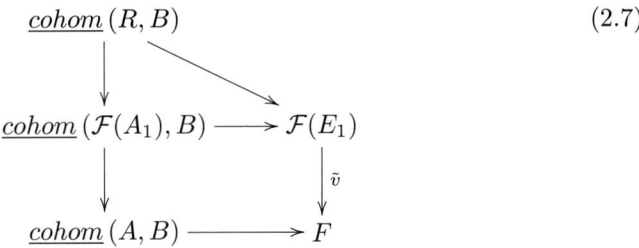

2.3. Remark

Associative algebras (without unit) can be treated as operads: functors on linear oriented trees and contractions with values in linear spaces. The Theorem 2.2 in this case reduces to the Theorem 0.1.1, or rather its extension where presentations are not supposed to be injective on components of degree 1.

2.4. Inner cohomomorphisms for operads with presentation

We can now reformulate Theorem 2.2 in the same way as it was done in the Introduction for associative algebras.

Consider the following category OP of $\Gamma\mathcal{G}$ operads given together with their presentation.

Objects of OP are pairs $\mathcal{A}$ as in 2.1. A morphism $\mathcal{A} \to \mathcal{B}$ is a pair consisting of a morphism of collections $A_1 \to B_1$ and a morphism of operads $A \to B$ compatible with i_A, i_B.

This category has a symmetric monoidal product $\bigodot$ induced by $\bigcirc$ in the following sense: $\mathcal{A} \bigodot \mathcal{B} = \mathcal{C}$ where $\mathcal{C}$ is represented by $C_1 := A_1 \bigcirc B_1$, $C :=$ the minimal suboperad containing the image $(i_A \bigcirc i_B)(C_1)$ in $A \bigcirc B$, and $i_C :=$ restriction of $i_A \bigcirc i_B$.

Theorem 2.2 can now be read as a statement that, functorially in all arguments, we have

$$\mathrm{Hom}_{OP}(\mathcal{A}, \mathcal{F} \bigodot \mathcal{B}) = \mathrm{Hom}_{OP}(\mathcal{E}, \mathcal{F}) \tag{2.10}$$

that is, $\mathcal{E}$ is an inner cohomomorphism object in OP:

$$\mathcal{E} = \underline{cohom}_{OP}(\mathcal{A}, \mathcal{B}). \tag{2.11}$$

General categorical formalism produces canonical comultiplication morphisms in OP

$$\Delta_{\mathcal{A},\mathcal{B},\mathcal{C}} : \underline{cohom}_{OP}(\mathcal{A}, \mathcal{C}) \to \underline{cohom}_{OP}(\mathcal{A}, \mathcal{B}) \bigodot \underline{cohom}_{OP}(\mathcal{B}, \mathcal{C}) \tag{2.12}$$

coassociative in an evident sense.

This is an operadic version of quantum matrices and their comultiplication.

In particular, the operad

$$\underline{coend}_{OP}\mathcal{A} := \underline{cohom}_{OP}(\mathcal{A}, \mathcal{A})$$

is endowed with a canonical coassociative comultiplication, morphism of operads

$$\Delta_{\mathcal{A}} := \Delta_{\mathcal{A},\mathcal{A},\mathcal{A}} : \underline{coend}_{OP}\mathcal{A} \to \underline{coend}_{OP}\mathcal{A} \bigodot \underline{coend}_{OP}\mathcal{A}. \tag{2.13}$$

It is generally not cocommutative, as the case of associative rings amply demonstrates.

We get thus a supply of "quantum semigroups", or Hopf algebras in the category of operads (ignoring antipode).

Notice finally that if $\mathcal{G}$ has a unit object u, we can sometimes define a unit object U in $(OP, \bigodot)$. Taking $\mathcal{F} = U$ in the adjunction formula (2.10), we see that the space of "classical homomorphisms" $\mathcal{A} \to \mathcal{B}$ in OP coincides with space of points of $\mathcal{E}$ with values in U (here we implicitly imagine operadic affine quantum spaces as objects of the dual category).

2.5. Algebras over an operad and their deformations

For certain labelings and respective graph categories Γ one can define a class of "natural" $\Gamma\mathcal{G}$-collections.

The basic example is this. Let J be an abstract set of "flavors" such that a labeling of τ consists of a map $F_\tau \to J$ (and possibly other data). (Imagine orientations, colors, or pairs (orientation, color)).

Assume furthermore that automorphisms of any Γ-corolla σ form a subgroup of permutations of its flags preserving flavors of flags. In this case any family of objects $\mathbf{V} := \{V_j \,|\, j \in J\}$ determines the following Γ-collection:

$$Coll\,(\mathbf{V})(\sigma) := \otimes_{j\in J} V_j^{\otimes F_\tau^{(j)}} \tag{2.14}$$

where $F_\tau^{(j)}$ is the subset of flags of flavour j. Automorphisms of σ act in an evident way.

Such a collection naturally extends to a functor on the groupoid of Γ-graphs τ and their isomorphisms: simply replace σ by τ in (2.14).

In order to extend it to arbitrary Γ-morphisms, consider separately three classes of morphisms.

a) *Graftings.* Graftings correspond to bijections of sets of flags preserving flavors. Hence they extend to (2.14).
b) *Mergers.* They have the same property.
c) *Contractions.* In principle, in order to accommodate contractions, we have to impose on $\mathbf{V}$ an additional structure, namely, a set of polylinear forms on (V_j) which we denote $\mathbf{v}$ and axiomatize as follows:

   Let S be a finite set and $\kappa := \{(j_s, k_s) \,|\, s \in S\}$ be a family of pairs of flavors such that in some Γ-graph there exist two vertices (perhaps coinciding) and connecting them edges which are simultaneously contracted by a Γ-morphism. The respective component v_κ of $\mathbf{v}$ is a morphism in $\mathcal{G}$

   $$v_\kappa : \otimes_{s\in S}(V_{j_s} \otimes V_{k_s}) \to u \tag{2.15}$$

   where u is the unit object of $\mathcal{G}$.

   Given $\mathbf{v}$, the prescription for extending our functor to contractions looks as follows. To map a product (2.14) for the source of a contraction to the similar product for its target, we must map identically factors corresponding to uncontracted flags, and to "kill" factors of the type $\otimes_{s\in S} V_{j_s} \otimes V_{k_s}$ with the help of (2.15).

This prescription will not necessarily describe a functor $\Gamma \to \mathcal{G}$: one should to impose upon $\mathbf{v}$ coherence conditions which we do not bother to spell out here.

In real life, this problem is avoided by specifying only bilinear forms corresponding to one-edge contractions, and then tensoring them to obtain full scale (2.15).

We can now give the main definition of this section.

2.5.1. Definition.

a) *Any functor as above* $\Gamma \to \mathcal{G}$ *with underlying collection* $Coll\,(\mathbf{V})$ *and structure forms* $\mathbf{v}$ *is called the endomorphism operad of* $(\mathbf{V}, \mathbf{v})$ *and denoted* $OpEnd\,(\mathbf{V}, \mathbf{v})$.
b) *Let* $\mathcal{P}$ *be a* $\Gamma\mathcal{G}$*-operad. Any morphism* $P \to OpEnd\,(\mathbf{V}, \mathbf{v})$ *is called a structure of* P*-algebra on* $(\mathbf{V}, \mathbf{v})$.

Examples. We will illustrate this on three types of labelling discussed in 1.3.2 above.

a) Working with a subcategory of Gr itself (as in the case of cyclic operads) we should choose one object V of $\mathcal{G}$ and a symmetric pairing $g : V \otimes V \to u$.
b) Let now Γ be a subcategory of oriented graphs. In that case one usually chooses $V_{out} = V$, $V_{in} = V^t$ (the dual object), and takes for v the canonical pairing $V^t \otimes V \to u$.
c) Finally, let Γ be a category of colored graphs, I the set of colors. In this case, one applies the full machinery of the definition above, simplifying it by caring only about one-edge contractions. So we need $V_i \otimes V_i \to u$ in the unoriented case, or else choose $V_{out,i} = V^t_{in,i}$.

Let us now return to our Definition 2.5.1. The whole space of structures of P-algebra on $(\mathbf{V}, \mathbf{v})$ is thus

$$\mathrm{Hom}_{\Gamma\mathcal{G}OPER}(P, OpEnd\,(\mathbf{V}, \mathbf{v})). \tag{2.16}$$

In the standard approach to the deformation theory of operadic algebras one chooses a space of basic operations, that is, a presentation of P (and then replaces the respective structure by a differential in an appropriate Hochschild-type complex, the step that we will not discuss here). Sometimes, one can choose a compatible presentation of $OpEnd\,(\mathbf{V}, \mathbf{v})$. For example, if connected graphs in Γ consist of oriented trees and all basic operations are binary, one can choose for generators the collection P_1 which coincides with P on corollas with three flags and is zero otherwise. Respectively, $OpEnd\,V$ is (hopefully) generated by $\underline{hom}(V^{\otimes 2}, V)$ (at least, this is the case for $\mathcal{G} = Vec_k$.)

Accepting this, we suggest to replace P by $\mathcal{P}$ which is $i : P_1 \to P$, to augment $OpEnd\,V$ accordingly, and to replace (2.16) by an appropriate set of morphisms in the category OP. After that, it is only natural to consider the operad

$$\underline{cohom}_{OP}(\mathcal{P}, \mathcal{O}p\mathcal{E}nd\,V) \tag{2.17}$$

as a non-commutative space parameterizing deformations of (a chosen collection of generators of) P-algebra structures.

2.6. Inner cohomomorphisms for algebras over an operad

P-algebras form a category $PALG$. One can try to play with P-algebras the same game as we did with associative algebras and operads of various types, and to study existence of inner cohomomorphisms in a category of P-algebras with a presentation. However, even in order to state the problem we need at least two preliminary constructions:

a) Free P-algebra $F_P(\mathbf{V}, \mathbf{v})$ generated by a family $(\mathbf{V}, \mathbf{v})$ as above. Generally, it will take values in a monoidal category larger than $\mathcal{G}$, e.g., that of inductive systems.

   This will allow us to define the notion of an algebra with presentation.

b) Symmetric white product on the category of P-algebras extending $\otimes$.

   To this end, we need a symmetric comultiplication $\Delta : P \to P \bigcirc P$ and an analog of the morphism j from Section 1.9 for free P-algebras.

Unfortunately, as our description of P-algebras above shows, we cannot do even this preliminary work in the same generality as we treated operads themselves. Therefore we step back, and for the remainder of Section 2.6 work with a version of ordinary operads.

2.6.1. Ordinary operads and their algebras. Let Γ be the category of graphs whose connected components are directed trees with exactly one output and at least one input at each vertex. Morphisms are contractions and graftings; mergers are not allowed. Let $(\mathcal{G}, \otimes, u)$ be an abelian symmetric monoidal category, such that $\otimes$ is exact in both arguments, and endowed with finite colimits and cohomomorphism objects.

Consider a $\Gamma\mathcal{G}$ operad P. Denote by $P(n)$ the value of P on the Γ-corolla with inputs $\{1, \dots, n\}$, and let $\mathbf{S}_n$ be the automorphism group of this directed corolla. Assume that $P(1) = u$ and that contracting an edge one end of which carries a corolla with one input produces canonical identifications $Q \otimes u \to Q$.

It is well known that a free P-algebra freely generated by V_1 exists in an appropriate category of inductive limits, and its underlying object is

$$F_P(V_1) = \oplus_{n=1}^{\infty} P(n) \otimes_{\mathbf{S}_n} V_1^{\otimes n}. \tag{2.18}$$

This construction is functorial in V_1. *A presentation* of an algebra V is a surjective morphism $F_P(V_1) \to V$; it can be reconstructed from its restriction $i_V : V_1 \to V$.

Assume that P is endowed with a symmetric comultiplication $\Delta : P \to P \bigcirc P$.

Then, given two objects E_1, W_1 of $\mathcal{G}$, we can define a map

$$j : F_P(E_1 \otimes W_1) \to F_P(E_1) \bigcirc F_P(W_1). \tag{2.19}$$

To construct it, first produce for each n a map

$$P(n) \otimes (E_1 \otimes W_1)^{\otimes n} \to P(n) \otimes E_1^{\otimes n} \otimes P(n) \otimes W_1^{\otimes n}$$

combining Δ with regrouping, and then the induced map of colimits

$$j(n) : P(n) \otimes_{\mathbf{S}_n} (E_1 \otimes W_1)^{\otimes n} \to (P(n) \otimes_{\mathbf{S}_n} E_1^{\otimes n}) \otimes (P(n) \otimes_{\mathbf{S}_n} W_1^{\otimes n}).$$

We put $j = \oplus_n j(n)$. One can check that this is a morphism of P-algebras.

If V, W are two P-algebras, presented by their structure morphisms $\alpha : F_P(V) \to V$, $\beta : F_P(W) \to W$, we can compose $j(n)$ with $\alpha \otimes \beta$ to define a structure of P-algebra on $V \otimes W$. This gives a symmetric monoidal structure on $PALG$ still denoted $\otimes$. We can now state

2.6.2. Proposition. *An analog of Theorem* 2.2 *holds in the category of P-algebras with presentation.*

We skip a proof which follows the same plan as that of Theorem 2.2.

2.7. Explicit constructions of cohomomorphism objects in the category of associative algebras

Existence proof of cohomomorphism objects generally is not very illuminating. In this and the following subsections, we cite several explicit constructions, valid under additional assumptions.

a) *Quadratic algebras.* This was the case first treated in [Ma1] and [Ma2]. Briefly, a quadratic algebra A (over a field k) is defined by its presentation $\alpha : F_{ASS}(A_1) \to A$ where $F_{ASS}(A_1) = T(A_1)$ is the free (tensor) algebra freely generated by a finite-dimensional vector space A_1, such that $\mathrm{Ker}\,\alpha$ is the ideal generated by the space of quadratic relations $R_A \subset A_1^{\otimes 2}$. In particular, A_1 is embedded in A, each A is naturally graded (A_1 in degree 1), and a morphism of quadratic presentations is the same as morphism of algebras which preserves grading. Moreover, each morphism is uniquely defined by its restriction to the space of generators.

This category $Qalg$ has a contravariant duality involution: $A \mapsto A^!$ where $A^! = T(A_1^*)/(R_A^{\perp})$, A_1^* denoting the dual space to A_1.

It has also two different symmetric monoidal structures, which are interchanged by !: "white product" $\circ$ and "black product" $\bullet$:

$$T(A_1)/(R_A) \circ T(B_1)/(R_B) = T(A_1 \otimes B_1)/(S_{(23)}(R_A \otimes B_1^{\otimes 2} + A_1^{\otimes 2} \otimes R_B), \quad (2.20)$$

$$T(A_1)/(R_A) \bullet T(B_1)/(R_B) = T(A_1 \otimes B_1)/(S_{(23)}(R_A \otimes R_B), \quad (2.21)$$

Here $S_{(23)} : A_1^{\otimes 2} \otimes B_1^{\otimes 2} \to (A_1 \otimes B_1)^{\otimes 2}$ interchanges two middle factors.

Both categories have unit objects: polynomials of one variable for $\circ$, and dual numbers $k[\varepsilon]/(\varepsilon^2)$ for $\bullet$ respectively. They are !-dual to each other. The generator ε combined with general categorical constructions produces differential in various versions of Koszul complex which is a base of Koszul duality.

The monoidal category $(Qalg, \bullet)$ has internal homomorphism objects. Explicitly,

$$\underline{hom}_{\bullet}(A, B) = A^! \circ B. \quad (2.22)$$

The monoidal category $(Qalg, \circ)$ has inner cohomomorphism objects. Explicitly,

$$\underline{cohom}_{\circ}(A, B) = A \bullet B^!. \quad (2.23)$$

Claim. (2.23) *is a valid description of the internal cohomomorphism object of two quadratic algebras in the total category of (algebras with) presentations.*

These formulas follow from a general functorial isomorphism (adjunction formula)

$$\mathrm{Hom}_{Qalg}(A \bullet B^!, C) = \mathrm{Hom}_{Qalg}(A, B \circ C) \tag{2.24}$$

Abstract properties of $Qalg$ expressed by (2.22)–(2.24) can be axiomatized to produce an interesting version of the notion of rigid tensor category of [DeMi]. They justify the use of the name "quantum linear spaces" for objects of $Qalg$ although the category itself is not linear or even additive.

It might be worthwhile to consider $(Qalg, \circ)$ and $(Qalg, \bullet)$ as ground categories for Γ-operads (and cooperads).

Notice that the cohomology spaces of $\overline{M}_{0,n+1}$, components of the Quantum Cohomology cooperad, are quadratic algebras (Keel's theorem).

b) N*-homogeneous algebras.* It was shown in [BerDW] that similar results hold for the category $H_N alg$ of homogeneous algebras generated in degree 1 with relations generated in degree N, for any fixed $N \geq 2$. If one continues to denote by $R_A \subset A_1^{\otimes N}$ generating relations, $A^!$ is given by relations $R_A^{\perp} \subset (A_1^*)^{\otimes N}$. Thus $! = !_N$ explicitly depends on N and gives, for example, different dual objects of a free algebra, depending on where we put its trivial relations. In the definitions (2.20) one should make straightforward modifications, replacing 2 by N. Formulas (2.22)–(2.24) still hold in the new setup. Unit object for $\bullet$ is now $k[\varepsilon]/(\varepsilon^N)$.

As a consequence, Koszul complexes become N-complexes leading to an interesting new homological effects: see [BerM] and references therein.

c) *Homogeneous algebras generated in degree one.* This case is treated in Chapter 3 of [PP] and in Section 1.3 of [GrM]; the approaches in these two papers nicely complement each other.

White product (2.21) and its $H_N alg$-version extend to this larger category as $(A \circ B)_n := A_n \otimes B_n$ (Segre product). The duality morphism ! and the black product do not survive in [PP], and internal homomorphism objects (2.22) perish. The right-hand side of the formula (2.23) is replaced by a rather long combinatorial construction which we do not reproduce here, and (2.24) becomes simply the characteristic property of cohom's:

$$\mathrm{Hom}(\underline{cohom}\,(A, B), C) = \mathrm{Hom}(A, B \circ C)$$

However, if A is quadratic, then (2.23) can be resurrected in a slightly modified form: B is the quotient of a quadratic algebra qB (leave the same generators and only quadratic relations), and we have

$$\underline{cohom}_{\circ}(A, B) = A \bullet (\mathrm{q}B)^! \tag{2.25}$$

(see [PP], Ch.3 , Proposition 4.3).

An important novelty of [PP] is a treatment of white products and cohom objects in the categories of graded modules over graded rings.

In [GrM], a version of ! and several versions of the black product are introduced. They are not as neatly packed together however as in the cases $Qalg$ and $H_N alg$. In particular, ! is not an involution. As a compensation, Theorem 1 of [GrM] establishes a nice extension of (2.25) for general homogeneous A and B involving a certain "triangle product".

A logical next step would be the introduction of these notions into "non-commutative projective algebraic geometry" where coherent sheaves appear à la Serre as quotient categories of graded modules, and cohom's of graded rings can serve as interesting non-commutative correspondences.

2.8. Explicit constructions of cohomomorphism objects in the categories of operads

One objective of the brief review above was to collect a list of patterns that could be subsequently recognized in various categories of operads. Most of the existing results which we are aware of concern quadratic operads and Koszul duality patterns. Cohomomorphism objects appear as a byproduct, although, as we have seen above, their existence is the most persistent phenomenon, even when the neat package $(\circ, \bullet, !)$ cannot be preserved.

a) *Binary quadratic operads.* The pioneering paper [GiKa] defined $(\circ, \bullet, !)$ for binary quadratic (ordinary) operads. Their construction uses a description of operads as monoids in $(COLL, \boxtimes)$, a category of collections endowed with a monoidal structure $\boxtimes$ (in 1.6 above, this is description (III)). This description was rather neglected here (see Appendix), but it makes clear the analogy between the tensor algebra of a linear space and the free operad generated by a collection V.

B. Vallette in [Va3] describes a construction of a free monoid which allows him to treat the cases when the relevant monoidal structure is not biadditive, which is the case of operads. The resulting weight grading of the free monoid can be used to define weight graded quotients, analogs of graded associative algebras: see [Fr], [Va1], [Va2].

The subcategory of ordinary operads with presentation considered in [GiKa] consists of weight graded operads generated by their binary parts (values on corollas with two inputs), with relations in weight 2. After introducing $(\circ, \bullet, !)$, Ginzburg and Kapranov prove the adjunction formula (2.24) and thus the formula (2.23) as well.

b) Developing this technique, B. Vallette in [Va1] and [Va2] defines $\circ$ for properads (see Appendix) with presentation and studies the case of (weight) quadratic relations for generators of arbitrary arity. The full adjunction formula (2.24) is established for quadratic operads with generators of one and the same arity $k \geq 2$ ([Va2], Section 4.6, Theorem 26), thus generalizing the $k = 2$ case of [GiKa].

Yet another version of this result is proved for k-ary quadratic *regular* operads, with different definitions of black and white products ([Va2], Section 5.1, Theorem 40). (A regular operad is an ordinary symmetric operad which is induced by some non-symmetric operad in a sense that will not be made precise here).

3. Non-abelian constructions

3.0. Introduction

In Section 2 it was proved that given an abelian category $\mathcal{G}$ with a symmetric monoidal structure $\otimes$ and its left adjoint <u>cohom</u>, the categories of operads in $\mathcal{G}$ and algebras over Hopf operads also possess <u>cohom</u> for the natural extension of $\otimes$. In this section we prove a version of this result in a non-abelian setting.

More precisely, we start with the description of operads as algebras over the triple $(\mathcal{F}, \mu, \eta)$, cf. Proposition 1.7.1. Here $\mathcal{F}$ is an explicit endofunctor on the category of collections $\Gamma\mathcal{G}COLL$ which inherits from $\mathcal{G}$ a monoidal structure, an abelian structure, and inner cohomomorphisms.

We replace $\Gamma\mathcal{G}COLL$ with an abstract monoidal category $(\mathcal{C}, \otimes)$ endowed with an endofunctor T which has the structure of a triple that commutes (up to a natural transformation) with the monoidal structure. In this case, with some additional assumptions, extension of <u>cohom</u> is straightforward and is given by the adjoint lifting theorem. In this formulation it is unnecessary to suppose that the monoidal product is symmetric, however, when applied to operads, the triple itself is produced using symmetric properties of the monoidal structure.

Finally we formulate the natural notion of the derived <u>cohom</u> and consider some cases when such a functor exists.

3.1. Tensor product and <u>cohom</u> for algebras over a triple

It is well known that if a category is equipped with a monoidal structure distributive with respect to direct sums, then categories of algebras over ordinary Hopf operads in this category will possess an extension of the monoidal structure. We can treat more general operads considering them in the context of triples. In this section we formulate conditions on the triples, needed to extend monoidal structure to algebras over them.

Let $(\mathcal{C}, \otimes)$ be a category with a coherently associative product (bifunctor). We do not assume $\otimes$ to be symmetric or to possess a unit object.

3.1.1. Definition. *A Hopf-like triple on $\mathcal{C}$ is a triple $T : \mathcal{C} \to \mathcal{C}$, $\mu : T \circ T \to T$, together with a natural transformation between bifunctors on $\mathcal{C}$:*

$$\tau : T \circ \otimes \ \to \ \otimes \circ (T \times T),$$

satisfying the following conditions:

a) *We have a commutative diagram of natural transformations:*

$$\begin{array}{ccc} T\circ T\circ\otimes & \xrightarrow{\tau(\mu\circ Id)} & \otimes\circ(T\times T) \\ {\scriptstyle T(\tau)}\downarrow & & \uparrow{\scriptstyle \otimes(\mu\times\mu)} \\ T\circ\otimes\circ(T\times T) & \xrightarrow{\tau} & \otimes\circ(T\times T)\circ(T\times T) \end{array}$$

where $\mu : T^2 \rightarrow T$ is the structure natural transformation of T.

b) *Let $\alpha : \otimes\circ(\otimes\times Id) \rightarrow \otimes\circ(Id\times\otimes)$ be the associativity constraint for $\otimes$. Then the following diagram of natural transformations is commutative:*

$$\begin{array}{ccccc} T\circ\otimes\circ(\otimes\times Id) & \xrightarrow{\tau} & \otimes\circ(T\times T)\circ(\otimes\times Id) & \xrightarrow{\otimes(\tau\times Id)} & \otimes\circ((\otimes\circ(T\times T))\times T) \\ {\scriptstyle T(\alpha)}\downarrow & & & & \downarrow{\scriptstyle \alpha} \\ T\circ\otimes\circ(Id\times\otimes) & \xrightarrow{\tau} & \otimes\circ(T\times T)\circ(Id\times\otimes) & \xrightarrow{\otimes(Id\times\tau)} & \otimes\circ(T\times(\otimes\circ(T\times T))) \end{array}$$

The conditions in the definition above allow us to define an associative product on the category of T-algebras, by extending it from $\mathcal{C}$. The functor morphism τ provides a definition, condition a) ensures that different ways of composing τ produce the same result, and condition b) implies coherent associativity of the resulting product on the category of T-algebras by utilizing associativity isomorphisms of $\otimes$. Here is the formal reformulation of all this.

3.1.2. Definition. *Let $(A, \alpha : T(A) \rightarrow A)$, $(B, \beta : T(B) \rightarrow B)$ be two T-algebras in $\mathcal{C}$. We define $(A,\alpha)\bigcirc(B,\beta)$ to be $(A\otimes B, \alpha\bigcirc\beta)$, where $\alpha\bigcirc\beta : T(A\otimes B) \rightarrow A\otimes B$ is the composition*

$$T(A\otimes B) \xrightarrow{\tau} T(A)\otimes T(B) \xrightarrow{\alpha\otimes\beta} A\otimes B.$$

3.1.3. Lemma. *Let $\mathcal{A}$ be the category of T-algebras in $\mathcal{C}$. Defined as above, $\bigcirc$ is a bifunctor on $\mathcal{A}$. It satisfies associativity conditions together with coherence. The associativity isomorphisms are preserved by the forgetful functor $\mathfrak{U} : \mathcal{A} \rightarrow \mathcal{C}$.*

Proof. To show that $\bigcirc$ is a bifunctor we have to show first that its value on a pair of objects in $\mathcal{A}$ is again in $\mathcal{A}$, i.e., that the following diagram is commutative

$$\begin{array}{ccc} T^2(A\otimes B) & \xrightarrow{\mu} & T(A\otimes B) \\ {\scriptstyle T(\alpha\bigcirc\beta)}\downarrow & & \downarrow{\scriptstyle \alpha\bigcirc\beta} \\ T(A\otimes B) & \xrightarrow{\alpha\bigcirc\beta} & A\otimes B \end{array}$$

This follows from condition a) in Definition 3.1.1 and the commutativity of the following diagram:

$$\begin{CD}
T(T(A)\otimes T(B)) @>{T(\alpha\otimes\beta)}>> T(A\otimes B)\\
@V{\tau}VV @VV{\tau}V\\
T^2(A)\otimes T^2(B) @>{T(\alpha)\otimes T(\beta)}>> T(A)\otimes T(B)\\
@V{\mu\otimes\mu}VV @VV{\alpha\otimes\beta}V\\
T(A)\otimes T(B) @>{\alpha\otimes\beta}>> A\otimes B
\end{CD}$$

The upper square of this diagram is commutative because T and $\otimes$ are functors and τ is a natural transformation. The lower square is commutative because it is the result of an application of $\otimes$ to two commutative squares (representing the fact that α and β are structure morphisms for T-algebras).

Thus on objects $\bigcirc$ behaves like a bifunctor on $\mathcal{A}$. We have to show that $f \bigcirc g$ is a morphism in $\mathcal{A}$ for any two morphisms $f : (A_1, \alpha_1) \to (A_2, \alpha_2)$ and $g : (B_1, \beta_1) \to (B_2, \beta_2)$ in $\mathcal{A}$, i.e., the following diagram is commutative:

$$\begin{CD}
T(A_1\otimes B_1) @>{T(f\otimes g)}>> T(A_2\otimes B_2)\\
@V{\alpha_1\bigcirc\beta_1}VV @VV{\alpha_2\bigcirc\beta_2}V\\
A_1\otimes B_1 @>{f\otimes g}>> A_2\otimes B_2
\end{CD}$$

By the definition of $\bigcirc$, this diagram can be decomposed as

$$\begin{CD}
T(A_1)\otimes T(B_1) @>{T(f\otimes g)}>> T(A_2\otimes B_2)\\
@V{\tau}VV @VV{\tau}V\\
T(A_1)\otimes T(B_1) @>{T(f)\otimes T(g)}>> T(A_2)\otimes T(B_2)\\
@V{\alpha_1\otimes\beta_1}VV @VV{\alpha_2\otimes\beta_2}V\\
A_1\otimes B_1 @>{f\otimes g}>> A_2\otimes B_2
\end{CD}$$

The upper square of this diagram is commutative because $\otimes$ and T are functors and τ is a natural transformation. The lower square is commutative because it is the result of an application of $\otimes$ to two commutative squares (expressing the fact that f and g are morphisms of T-algebras).

Clearly pairs of identities are mapped to identities by $\bigcirc$ and the composition is preserved, i.e., $\bigcirc$ is a bifunctor on $\mathcal{A}$.

It remains to show that $\bigcirc$ is coherently associative and that the forgetful functor to $\mathcal{C}$ preserves the associativity isomorphisms. The former claim actually follows from the latter, since $\otimes$ is coherently associative. So all we have to do is to show that the $\otimes$-associativity isomorphisms between the images of the forgetful functor belong to $\mathcal{A}$, i.e., the outer rectangle of the following diagram is commutative for all $A_1, A_2, A_3 \in \mathcal{A}$,

$$\begin{array}{ccccc}
T((A_1 \otimes A_2) \otimes A_3) & \xrightarrow{(\tau \otimes Id) \circ \tau} & (T(A_1) \otimes T(A_2)) \otimes T(A_3) & \xrightarrow{(\alpha_1 \otimes \alpha_2) \otimes \alpha_3} & (A_1 \otimes A_2) \otimes A_3 \\
\downarrow{\scriptstyle T(\alpha)} & & \downarrow{\scriptstyle \alpha} & & \downarrow{\scriptstyle \alpha} \\
T(A_1 \otimes (A_2 \otimes A_3)) & \xrightarrow{(Id \otimes \tau) \circ \tau} & T(A_1) \otimes (T(A_2) \otimes T(A_3)) & \xrightarrow{\alpha_1 \otimes (\alpha_2 \otimes \alpha_3)} & A_1 \otimes (A_2 \otimes A_3)
\end{array}$$

The right-hand square of this picture is commutative because $\otimes$ and T are functors and α is a natural transformation. Commutativity of the left-hand square is the contents of condition b) of Definition 3.1.1. □

Having extended the product $\otimes$ from $\mathcal{C}$ to $\mathcal{A}$, we would like to know if this extension possesses a left adjoint, given $\otimes$ does so on $\mathcal{C}$. We can infer it easily from the adjoint lifting theorem, if we assume that $\mathcal{A}$ has all coequalizers. For the question of when a category of algebras over a triple has all coequalizers see, e.g., [BarW], Section 9.3.

3.1.4. Proposition. *Suppose that $\mathcal{A}$ has all coequalizers and for every $A \in \mathcal{A}$ the functor $\mathfrak{U}(A) \otimes - : \mathcal{C} \to \mathcal{C}$ has a left adjoint. Then the functor $A \bigcirc - : \mathcal{A} \to \mathcal{A}$ has a left adjoint as well.*

Proof. It is clear that $\mathfrak{U}$ is a monadic functor ([Bo1], Definition 4.4.1) and by our construction $\mathfrak{U}(A \bigcirc -) = \mathfrak{U}(A) \otimes \mathfrak{U}(-)$, therefore (e.g., [Bo2], Theorem 4.5.6) since $\mathfrak{U}(A) \otimes -$ has a left adjoint, so does $A \bigcirc -$. This shows our assertion. □

Here we define $\underline{cohom}(A, B)$ is an object that represents $Hom(A, B \bigcirc -)$. From the last proposition we know that on $\mathcal{A}$ there is a $\underline{cohom}(-,-)$ that is functorial in the first argument. Since Yoneda embedding is full and faithful, we conclude that our construction is actually a bifunctor.

3.2. Derived $\underline{cohom}$

By now we have constructed $\underline{cohom}(-, A)$ for an algebra $A \in \mathcal{A}$, given existence of $\underline{cohom}$ on $\mathcal{C}$. Now assume that $\mathcal{C}$ carries in addition a closed model structure, such that $C \otimes -$ is a right Quillen functor for any $C \in \mathcal{C}$, i.e., it has a left adjoint and it maps fibrations and trivial fibrations to the like. Then, by a general result, $\underline{cohom}(-, C)$ is a left Quillen functor and we can define its left derived version.

Suppose that we can transport closed model structure from $\mathcal{C}$ to $\mathcal{A}$ through the adjunction of the free algebra and the forgetful functor, i.e., we can introduce a closed model structure on $\mathcal{A}$ so that a map in $\mathcal{A}$ is a weak equivalence or fibration if and only its image under the forgetful functor is such. Then for any $A \in \mathcal{A}$ we have that $A \bigcirc -$ is a right Quillen functor on $\mathcal{A}$, and hence $\underline{cohom}(-, A)$ is a left Quillen functor and we can define derived $\underline{cohom}$ as follows:

$$\mathbb{L}\underline{cohom}(-, A) := \underline{cohom}(L(-), A),$$

where L is a cofibrant replacement functor on $\mathcal{A}$.

It remains only to analyze when such a transport of model structure is possible. The general situation of a transport was considered in several papers, e.g., [Bek], [Bl], [CaGa], [Cra], [Q], [R], [S]. For our purposes it is enough to consider locally presentable categories with cofibrantly generated model structures. In this case the conditions on the forgetful functor are quite mild.

Moreover, we are mostly interested in the particular case of operads in the category of algebras over a Hopf operad in the category of dg complexes of vector spaces over a field. In the next section we will show that categories of such operads can be constructed as categories of algebras over certain triples on the category of dg complexes of vector spaces. Then the machinery of transport of model structure can be applied to these triples directly.

4. Iteration of algebraic constructions

4.0. Introduction

As it is described in Section 3, if a triple is Hopf-like with respect to a monoidal structure, it is easy to extend <u>*cohom*</u> from the ground category to the category of algebras over this triple. Hopf operads provide examples of such triples, given that monoidal structure on the ground category is distributive with respect to the direct sum. However, this is not always the case.

Consider the category $\mathfrak{A}$ of associative algebras in the category of vector spaces over a field. Since the operad of associative algebras is Hopf, $\mathfrak{A}$ has a symmetric monoidal structure, given by the tensor product of algebras. On $\mathfrak{A}$ this monoidal structure is not distributive with respect to direct sums, that is, free products of algebras. In fact, given A_1, A_2, $A_2 \in \mathfrak{A}$, in general we have no isomorphism between $A_1 \otimes (A_2 \coprod A_3)$ and $(A_1 \otimes A_2) \coprod (A_1 \otimes A_3)$. Thus we cannot represent operads in $\mathfrak{A}$ as algebras over a triple, and constructions of Section 3 concerning existence of <u>*cohom*</u> do not apply to the category of operads in $\mathfrak{A}$.

However, we can overcome these difficulties by working with the operads instead of their categories of algebras. In the example above we can represent *operads* in the category of associative algebras as *algebras over a colored operad* in the category of vector spaces.

For example, a classical operad in $\mathfrak{A}$ is a sequence $A := \{A_n\}_{n \in \mathbb{Z}_{>0}}$ of objects of $\mathfrak{A}$ and morphisms

$$\gamma_{m_1,\ldots,m_n} : A_n \otimes A_{m_1} \otimes \ldots \otimes A_{m_n} \to A_{m_1+\cdots+m_n},$$

satisfying the usual axioms. The additional structures of associative algebras on A_n's are described by a sequence of morphisms of vector spaces: $\alpha_n : A_n \otimes A_n \to A_n$, that satisfy the usual associativity axioms.

Compatibility of the operadic structure morphisms with the structures of associative algebras on individual A_n's is expressed by commutativity of the following diagrams:

$$\begin{array}{ccc}
(A_n \otimes A_n) \otimes (\otimes_{i=1}^n A_{m_i} \otimes A_{m_i}) & \xrightarrow{\alpha_n \otimes (\otimes_{i=1}^n \alpha_{m_i})} & A_n \otimes A_{m_1} \otimes \dots A_{m_n} \\
\Big\downarrow {\scriptstyle (\gamma_{m_1,\dots,m_n} \otimes \gamma_{m_1,\dots,m_n}) \circ \sigma} & & \Big\downarrow {\scriptstyle \gamma_{m_1,\dots,m_n}} \\
A_{m_1+\dots+m_n} \otimes A_{m_1+\dots+m_n} & \xrightarrow{\alpha_{m_1+\dots+m_n}} & A_{m_1+\dots+m_n}
\end{array}$$

where σ is the appropriate rearrangement of factors in the tensor products of vector spaces.

It is easy to see that we can express all of these conditions on the sequence $\{A_n\}$ as an action of a colored symmetric operad. Thus the category of classical operads in $\mathfrak{A}$ is equivalent to a category of algebras over a colored operad in vector spaces. Therefore we have the free algebra construction, results of Section 3 apply, and <u>cohom</u> extends to classical operads in $\mathfrak{A}$.

We would like to do the same in the case of general operads, as described in Section 1. To achieve the necessary degree of generality we will work with triples, which will include all operadic cases, given that the ground category has monoidal structure, which is distributive with respect to the direct sums.

The usual definition of a triple on a category $\mathcal{C}$ consists of three parts: a functor $T : \mathcal{C} \to \mathcal{C}$, and natural transformations $\zeta : T \circ T \to T$, $\eta : Id \to T$, satisfying certain associativity and unit axioms. The functor T is supposed to represent the "free algebra" construction, while ζ and η represent composition and the identity operation respectively.

Often we have more information about the triple than contained in its definition as above. We have a grading on the "free algebra" construction, given by the arity of operations involved, i.e., we can decompose $T(C)$ $(C \in \mathcal{C})$ into a direct sum of $T_n(C)$ $(n \in \mathbb{Z}_{>0})$, where $T_n(C)$ stands for applying to "generators" C all of the "n-ary operations" T_n.

Such decomposition of T is very helpful, since usually T_n's behave better with respect to monoidal structure and other triples than the whole T: see [Va3] for a closely related discussion. Using this we can mimic construction of a colored operad, as in the case of classical operads in associative algebras above, in the more general situation of decomposed triples.

In order to do so we have to formalize the notion of triples admitting such a decomposition. We should axiomatize the relationship between T_n's for different n's, so that the combined object would satisfy the associativity and the unit axioms, stated in the usual definition of a triple. The best way to do so is via representations of operads in categories.

It is often the case (in particular it is so for operads as described in Section 1) that the "n-ary operations" T_n is not just a functor on $\mathcal{C}$, but is given as a composition

$$\mathcal{C} \xrightarrow{\Delta} \mathcal{C}^{\times^n} \xrightarrow{G_n} \mathcal{C},$$

where Δ is the diagonal functor and G_n is some functor $\mathcal{C}^{\times^n} \to \mathcal{C}$. In such cases it is convenient to work with representations of operads on $\mathcal{C}$, i.e., with morphisms of operads with codomain $\{Fun(\mathcal{C}^{\times^n}, \mathcal{C})\}_{n\in\mathbb{Z}_{n>0}}$.

So we will consider strict operads in the category of categories and work with lax morphisms of such operads (i.e., with morphisms that commute with the structure functors only up to a natural transformation). A lax morphism into an endomorphism operad of a category will give us a generalized triple that under some conditions can be transformed into a usual triple. In such cases we will say that the triple is operad-like. The technique of lax morphisms was invented long ago (cf. [KS]) and applied recently to the case of operads in [Bat].

We proceed as follows: we start with the well known notion of strict pseudo-operads in categories, we organize them in a category and then extend it to include lax morphism of operads, which satisfy natural conditions of coherence (later these conditions will be shown to correspond to the associativity axioms of triples). Then we consider the notion of a strict operad in categories (i.e., with a unit) and define lax morphisms of such objects. Again we will need some coherence conditions, which later will turn out to be the unit axioms of a triple. Finally we define operad-like triples as lax morphisms into the endomorphism operad of a category and we finish by proving that iteration of algebraic constructions preserves existence of *cohom*.

4.1. Notation

We will denote functors usually by capital letters (both Greek and Latin), whereas natural transformations will be denoted by small letters.

Commutative diagrams of functors will be rarely commutative on the nose, instead we will have to endow them with natural transformations, making them commutative. When we draw a diagram as follows

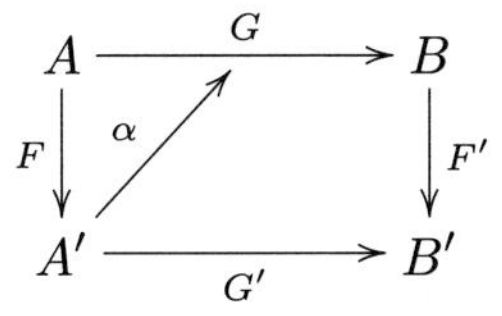

and say that α makes the diagram commutative, we mean that α is a natural transformation $G' \circ F \to F' \circ G$. Notice that α is not supposed to be an isomorphism of functors, any functor morphism is acceptable. Similarly for the diagram

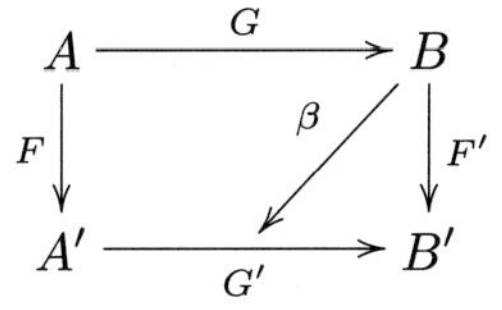

We will often encounter one of the following two situations:

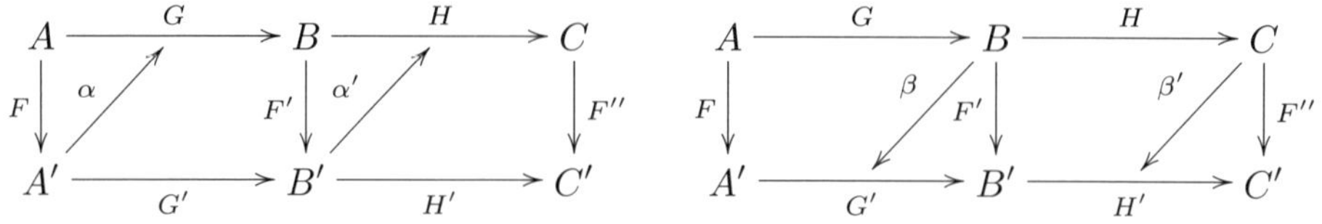

There is a well-known procedure of pasting natural transformations in such cases. In order to fix notation, we will write the relevant formulas explicitly.

4.1.1. Definition. *These compositions are defined as follows:*

$$(\alpha' * \alpha)_a := \alpha'_{G(a)} \circ H'(\alpha_a), \quad a \in A,$$
$$(\beta * \beta')_a := H'(\beta_a) \circ \beta'_{G(a)}, \quad a \in A.$$

Direct computation easily shows that both compositions are indeed natural transformations $H' \circ G' \circ F \to F'' \circ H \circ G$ and $F'' \circ H \circ G \to H' \circ G' \circ F$ respectively. Moreover, the composition $*$ is associative.

4.2. Strict pseudo-operads in categories

Let $\mathcal{CAT}$ be a small category whose objects are some small categories and morphisms are functors. We will assume that it is sufficiently rich so that all the following constructions make sense. In particular, we have a symmetric monoidal structure $\times$ given by the direct product and a choice of a category E with one object and one morphism (identity). Thus we can consider pseudo-operads in $\mathcal{CAT}$.

Although our goal is to prove existence of <u>cohom</u> for operads, these being defined in the general way as in Section 1, the categorical operads that we will use will be solely the classical ones. The reason for this is that the categories of graphs themselves, that were used in Section 1 in definition of operads, are in fact examples of classical operads in categories: see Section 4.9.3 below. Thus for our purposes there is no need to consider more general categorical operads, than the classical ones.

4.2.1. Definition. *A strict pseudo-operad in categories is a classical non-symmetric pseudo-operad in* $(\mathcal{CAT}, \times)$ *(e.g., [MarShSt], Definition* 1.18*). We will denote the category of strict pseudo-operads in* $\mathcal{CAT}$ *by* $\Psi\mathcal{OP}_{st}(\mathcal{CAT})$*. The structure functors of an object of* $\Psi\mathcal{OP}_{st}(\mathcal{CAT})$ *will be denoted by* $\{\Upsilon_{m_1,\dots,m_n}\}_{m_i \in \mathbb{Z}_{>0}}$.

We could have used of course the notion of a 2-pseudo-operad in categories and work in the more general setting of higher operads, but for our needs in this paper strict operads in categories will suffice. One of the reasons for this restriction is the following example.

4.2.2. Example. One of the most important examples of a strict pseudo-operad in categories is the endomorphism pseudo-operad $\mathfrak{E}(\mathcal{C})$ of a category $\mathcal{C}$. It is defined as follows:

$$\mathfrak{E}(\mathcal{C})_n := Fun(\mathcal{C}^{\times^n}, \mathcal{C}), \quad n \in \mathbb{Z}_{>0},$$

where Fun stands for the category of functors. The structure morphisms are given by compositions of functors.

4.3. Lax morphisms between strict pseudo-operads

As we have noted above we could have worked with 2-operads in categories instead of the usual ones, i.e., we could have incorporated natural transformations in operadic structure. However, as the example of the endomorphism operad of a category shows, it is enough for many purposes, and in particular for ours, to consider only classical operads. Yet when we start organizing these operads in categories we have to take into account the natural transformations, that we have omitted before.

We have already the category $\Psi\mathcal{OP}_{st}(\mathcal{CAT})$, whose morphisms are sequences of functors, that commute on the nose with the structure functors of pseudo-operads. This rarely happens. In most cases we have a natural transformation making these diagrams commutative. Later we will see that it is these natural transformations that define the multiplication for the triples that we will consider. So we need to enlarge $\Psi\mathcal{OP}_{st}(\mathcal{CAT})$ to include not only strict but also lax morphisms of operads. Lax morphisms were invented a long time ago (see, e.g., [KS] Section 3). For completeness we reproduce explicit definitions here. Our treatment will deviate from the classical one only when we will consider relative version of the construction (i.e., categories of lax morphisms with constant codomain) and more importantly when we introduce the notion of an equivariant operad in categories. Both of these constructions are specifically tailored for treatment of triples.

4.3.1. Definition. *Let P, P' be two strict pseudo-operads in categories. A lax morphism from P to P' is a sequence of functors $\{F_n : P_n \to P'_n\}_{n\in\mathbb{Z}_{>0}}$ and a sequence of natural transformations $\{\zeta_{m_1,\dots,m_n}\}$, making the following diagram commutative:*

$$\begin{array}{ccc}
P_n \times P_{m_1} \times \cdots \times P_{m_n} & \xrightarrow{\Upsilon_{m_1,\dots,m_n}} & P_{m_1+\cdots+m_n} \\
{\scriptstyle F^{\times n+1}}\big\downarrow & \nearrow_{\zeta_{m_1,\dots,m_n}} & \big\downarrow{\scriptstyle F} \\
P'_n \times P'_{m_1} \times \cdots \times P'_{m_n} & \xrightarrow{\Upsilon'_{m_1,\dots,m_n}} & P'_{m_1+\cdots+m_n}
\end{array}$$

Given two lax morphisms $(\{F_n\},\{\zeta_{m_1,\dots,m_n}\}) : P \to P'$ and $(\{F'_n\},\{\zeta'_{m_1,\dots,m_n}\}) : P' \to P''$ we define their composition to be $(\{F'_n \circ F_n\},\{\zeta_{m_1,\dots,m_n} * \zeta'_{m_1,\dots,m_n}\})$, where $*$ is composition of natural transformations, as defined in 4.1.1.

It is easy to see that pseudo-operads in categories and lax morphisms form a category. From the the pasting theorem of [Pow], we know that composition of ζ's is associative and therefore composition of the whole morphisms is associative. There is an identity lax morphism for every category, given by the identity functor and the trivial natural automorphism of it.

However, we are interested in a subcategory of this category, consisting of lax morphisms, that have the property of coherence. As usual coherence means that different ways of composing natural transformations are equal. Later we will see that these conditions will translate into associativity properties of the multiplication natural transformations of triples that we will construct.

4.3.2. Definition. *Let P and P' be two strict pseudo-operads in categories. Let $(F, \zeta) : P \to P'$ be a lax morphism between them. We say that (F, ζ) is coherent if the $*$-compositions of natural transformations in the following two diagrams are equal:*

$$
\begin{array}{ccccc}
P_m \times P_{\underline{m}} \times P_{\underline{n}} & \xrightarrow{\Upsilon_{\underline{m}} \times Id} & P_n \times P_{\underline{n}} & \xrightarrow{\Upsilon_{\underline{n}}} & P_{\beta_1+\dots+\beta_n} \\
\downarrow F & \nearrow \zeta_{\underline{m}} \times Id & \downarrow F & \nearrow \zeta_{\underline{n}} & \downarrow F \\
P'_m \times P'_{\underline{m}} \times P'_{\underline{n}} & \xrightarrow[\Upsilon'_{\underline{m}} \times Id]{} & P'_n \times P'_{\underline{n}} & \xrightarrow[\Upsilon'_{\underline{n}}]{} & P'_{\beta_1+\dots+\beta_n}
\end{array}
$$

$$
\begin{array}{ccccc}
P_m \times P_{\underline{m}} \times P_{\underline{n}} & \xrightarrow{Id \times \Upsilon_{\underline{n}}} & P_m \times P_{\underline{m}'} & \xrightarrow{\Upsilon_{\underline{m}'}} & P_{\beta_1+\dots+\beta_n} \\
\downarrow F & \nearrow Id \times \zeta_{\underline{n}} & \downarrow F & \nearrow \zeta_{\underline{m}'} & \downarrow F \\
P'_m \times P'_{\underline{m}} \times P'_{\underline{n}} & \xrightarrow[Id \times \Upsilon'_{\underline{n}}]{} & P'_m \times P'_{\underline{m}'} & \xrightarrow[\Upsilon'_{\underline{m}'}]{} & P'_{\beta_1+\dots+\beta_n}
\end{array} \qquad (4.1)
$$

where $\underline{m} := \{\alpha_1, \dots, \alpha_m\}$, $\underline{n} := \{\beta_1, \dots, \beta_n\}$, $\underline{m}' := \{\beta_1 + \dots + \beta_{\alpha_1}, \dots, \beta_{\alpha_1+\dots+\alpha_{m-1}+1} + \dots + \beta_n\}$, $\alpha_1 + \dots + \alpha_m = n$, α_i, $\beta_i \in \mathbb{Z}_{>0}$, *and* $P_{\underline{m}} := P_{\alpha_1} \times \dots \times P_{\alpha_m}$, *similarly for* $P_{\underline{n}}$, $P_{\underline{m}'}$ *and* P'.

4.3.3. Proposition. *Strict pseudo-operads in categories and coherent lax morphisms form a subcategory of the category of strict pseudo-operads and lax morphisms*

Proof. It is clear that the identity lax morphism for any strict pseudo-operad is coherent. We have to prove that composition of coherent lax morphisms is again coherent.

Let $P \overset{(F,\zeta)}{\to} \to P' \overset{(F',\zeta')}{\to} \to P''$ be a sequence of coherent lax morphisms between strict pseudo-operads in categories. We have to show that the $*$-product of ζ and ζ' provides a unique way of making diagrams commutative, i.e., compositions of the natural transformations as in diagrams (4.1) are equal. But these diagrams are the outer rectangles of the following diagrams:

$$
\begin{array}{ccccc}
P_m \times P_{\underline{m}} \times P_{\underline{n}} & \xrightarrow{\Upsilon_{\underline{m}} \times Id} & P_n \times P_{\underline{n}} & \xrightarrow{\Upsilon_{\underline{n}}} & P_{\beta_1+\dots+\beta_n} \\
\downarrow F & \nearrow \zeta_{\underline{m}} \times Id & \downarrow F & \nearrow \zeta_{\underline{n}} & \downarrow F \\
P'_m \times P'_{\underline{m}} \times P'_{\underline{n}} & \xrightarrow{\Upsilon'_{\underline{m}} \times Id} & P'_n \times P'_{\underline{n}} & \xrightarrow{\Upsilon'_{\underline{n}}} & P'_{\beta_1+\dots+\beta_n} \\
\downarrow F' & \nearrow \zeta'_{\underline{m}} \times Id & \downarrow F' & \nearrow \zeta'_{\underline{n}} & \downarrow F' \\
P''_m \times P''_{\underline{m}} \times P''_{\underline{n}} & \xrightarrow[\Upsilon''_{\underline{m}} \times Id]{} & P''_n \times P''_{\underline{n}} & \xrightarrow[\Upsilon''_{\underline{n}}]{} & P''_{\beta_1+\dots+\beta_n}
\end{array}
$$

$$
\begin{array}{ccccc}
P_m \times P_{\underline{m}} \times P_{\underline{n}} & \xrightarrow{Id \times \Upsilon_n} & P_m \times P_{\underline{m'}} & \xrightarrow{\Upsilon_{\underline{m'}}} & P_{\beta_1+\cdots+\beta_n} \\
\downarrow{\scriptstyle F} \quad \swarrow{\scriptstyle Id \times \zeta_{\underline{n}}} & & \downarrow{\scriptstyle F} \quad \swarrow{\scriptstyle \zeta_{\underline{m'}}} & & \downarrow{\scriptstyle F} \\
P'_m \times P'_{\underline{m}} \times P'_{\underline{n}} & \xrightarrow{Id \times \Upsilon'_n} & P'_m \times P'_{\underline{m'}} & \xrightarrow{\Upsilon'_{\underline{m'}}} & P'_{\beta_1+\cdots+\beta_n} \\
\downarrow{\scriptstyle F'} \quad \swarrow{\scriptstyle Id \times \zeta'_{\underline{n}}} & & \downarrow{\scriptstyle F'} \quad \swarrow{\scriptstyle \zeta'_{\underline{m'}}} & & \downarrow{\scriptstyle F'} \\
P''_m \times P''_{\underline{m}} \times P''_{\underline{n}} & \xrightarrow{Id \times \Upsilon''_n} & P''_m \times P''_{\underline{m'}} & \xrightarrow{\Upsilon''_{\underline{m'}}} & P''_{\beta_1+\cdots+\beta_n}
\end{array}
$$

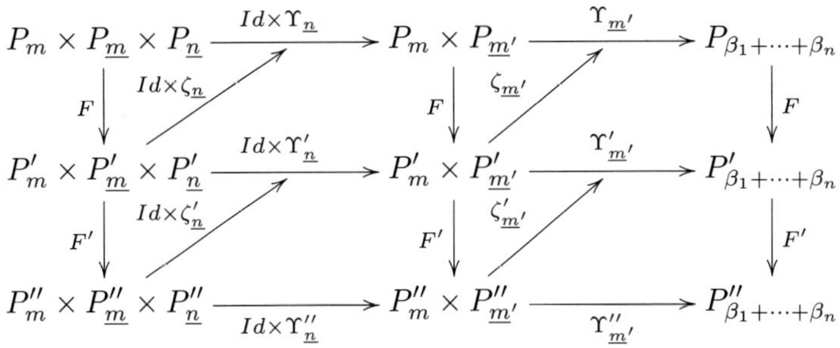

Since (F, ζ) is coherent, compositions of natural transformations in the first rows of these diagrams are equal, similarly for the second rows. Therefore compositions of first the rows and then the columns are equal. We would like to show that compositions of first the columns and then the rows are equal as well.

This follows from the pasting theorem in [Pow]. $\qquad\square$

4.3.4. Notation. We will denote the category of strict pseudo-operads in categories and coherent lax morphisms by $\Psi\mathcal{OP}(\mathcal{CAT})$.

4.4. Categories of pseudo-operads over a pseudo-operad

As with any category, it is necessary sometimes to consider an object P in $\Psi\mathcal{OP}(\mathcal{CAT})$ and all morphisms in $\Psi\mathcal{OP}(\mathcal{CAT})$ with codomain P. It will be very important for us when we will work with representations of categorical operads on a category, i.e., when P is the endomorphism operad for some category $\mathcal{C}$.

We would like of course to organize all these morphisms into a category – the category of pseudo-operads over P. But first we have to decide what shall we call a morphism between two such morphisms. In the standard way (i.e., without presence of 2-morphisms) we would define a morphism to a be a commutative (on the nose) triangle as follows

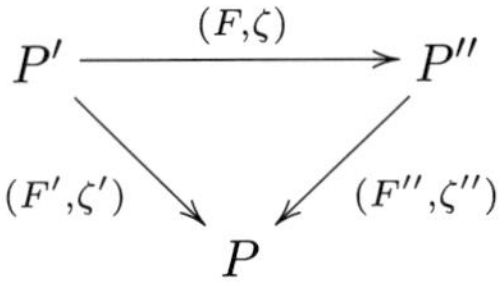

However, we have natural transformations and we should take them into account, i.e., we should define a morphism to be a diagram with a natural transformation making it commutative

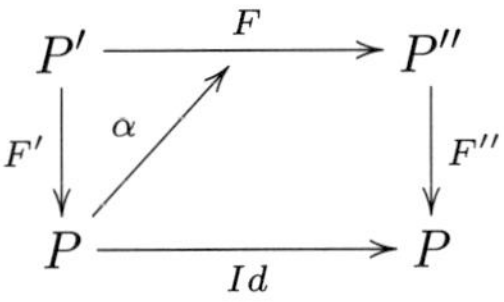

$$
\begin{array}{ccc}
P' & \xrightarrow{\;F\;} & P'' \\
{\scriptstyle F'}\downarrow \quad \nearrow{\scriptstyle \alpha} & & \downarrow{\scriptstyle F''} \\
P & \xrightarrow[Id]{} & P
\end{array}
\tag{4.2}
$$

Now we have to decide how ζ, ζ', ζ'' and α should relate to each other. If we did not have α, then the relation would have been $\zeta' = \zeta * \zeta''$. Having α we can put all of these natural transformation into one big diagram

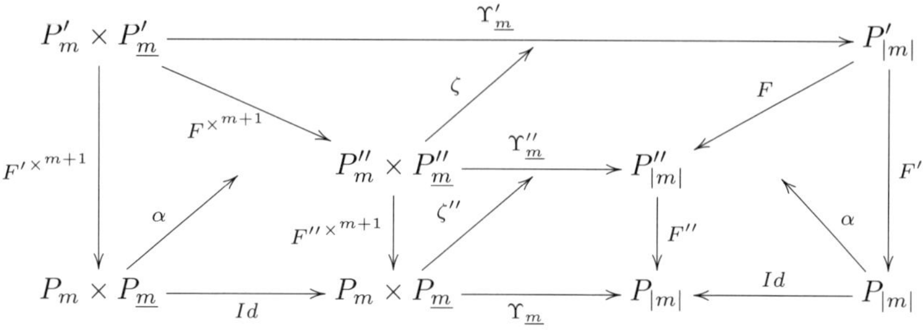

(4.3)

where for typographical reasons we have omitted ζ', that makes the outer rectangle commutative.

We see that there are two ways to construct natural transformations

$$\Upsilon_{\underline{m}} \circ F'^{\times^{m+1}} \to F'' \circ F \circ \Upsilon'_{\underline{m}}.$$

One is $\zeta{*}\zeta''{*}\alpha$ and the other is $\alpha{*}\zeta'$. Their equality is the natural condition of compatibility.

4.4.1. Definition. *The category $\Psi\mathcal{OP}(\mathcal{CAT})//P$ has coherent lax morphisms with codomain P as objects, and for any two such morphisms $(F', \zeta') : P' \to P$, $(F'', \zeta'') : P'' \to P$, a morphism from the first one to the second is a pair $((F, \zeta), \alpha)$, where $(F, \zeta) : P' \to P''$ is a coherent lax morphism of strict pseudo-operads, and α is a natural transformation, making diagram (4.2) commutative and satisfying*

$$\zeta{*}\zeta''{*}\alpha = \alpha{*}\zeta'.$$

4.4.2. Proposition. *Constructed as above $\Psi\mathcal{OP}(\mathcal{CAT})//P$ is indeed a category.*

Proof. A morphism in $\Psi\mathcal{OP}(\mathcal{CAT})//P$ consists of a coherent lax morphism and a natural transformation. From Definition 4.1.1 and Proposition 4.3.3 we know how to compose both types, so the composition in $\Psi\mathcal{OP}(\mathcal{CAT})//P$ is clear.

Since by the pasting theorem composition of natural transformations is associative ([Pow]) and we know that this is true for coherent lax morphisms (Proposition 4.3.3), all we have to do now is to show that the condition, formulated in Definition 4.4.1 is satisfied by the composition.

So let $(F', \zeta') : P' \to P$, $(F'', \zeta'') : P'' \to P$ and $(F''', \zeta''') : P''' \to P$ be three coherent lax morphisms, with codomain P. Suppose we have two morphisms $((G, \psi), \alpha) : (F', \zeta') \to (F'', \zeta'')$, $((G', \psi'), \alpha') : (F'', \zeta'') \to (F''', \zeta''')$, that satisfy conditions of Definition 4.4.1. We can organize everything into one diagram as follows:

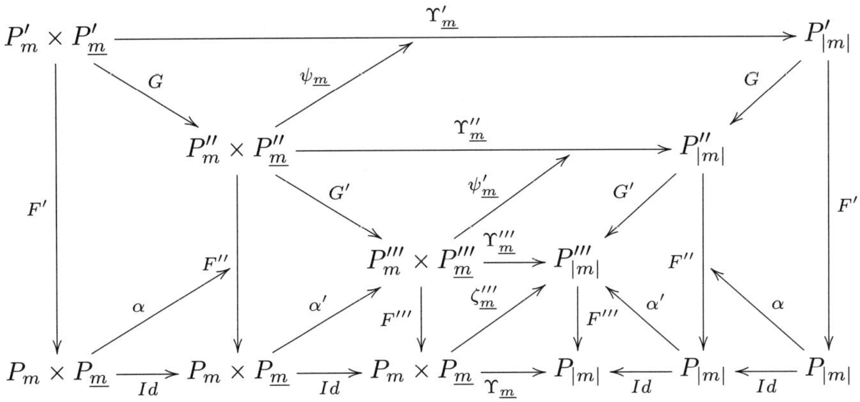

Since $((G,\psi),\alpha)$ and $((G',\psi'),\alpha')$ satisfy the conditions of Definition 4.4.1 we have the equalities

$$\psi' {*} \zeta''' {*} \alpha' = \alpha' {*} \zeta'', \qquad \psi {*} \zeta'' {*} \alpha = \alpha {*} \zeta'.$$

Using these equalities we get the following:

$$(\alpha' {*} \alpha) {*} \zeta' = \alpha' {*} \psi {*} \zeta'' {*} \alpha = \psi {*} \alpha' {*} \zeta'' {*} \alpha = (\psi {*} \psi') {*} \zeta''' {*} (\alpha' {*} \alpha),$$

where the second equality is justified by the pasting theorem ([Pow]). The combined equality is exactly the condition as in Definition 4.4.1 for the composition. □

Later we will often work with representations of categorical operads on a category, i.e., we will study lax morphisms into the endomorphism operad. We will want to construct a functor from the category of certain representations to the category of triples on that category. For that we will need a notion of the category of representations. One candidate is obviously the category of operads over the endomorphism operad, as constructed above. However, it will prove to be too relaxed. We will need a somewhat more restricted notion. Namely we will consider the subcategory with the same objects but only strict morphisms.

4.4.3. Notation. We will denote by $\Psi\mathcal{OP}_{st}(\mathcal{CAT})//P$ the subcategory of $\Psi\mathcal{OP}(\mathcal{CAT})//P$, consisting of the same objects as $\Psi\mathcal{OP}(\mathcal{CAT})//P$, but for any pair of them (P', F', ζ') and (P'', F'', ζ''), a morphism $((F,\zeta),\alpha)$ from the first to the second is in $\Psi\mathcal{OP}_{st}(\mathcal{CAT})//P$ if (F,ζ) is strict, i.e., ζ is the identity.

Since strict morphisms form a subcategory of the category of lax morphisms, we see that $\Psi\mathcal{OP}_{st}(\mathcal{CAT})//P$ is indeed a subcategory of $\Psi\mathcal{OP}(\mathcal{CAT})//P$.

4.5. Strict operads in categories

Until now we have considered pseudo-operads. Now we would like to discuss also the unital version of our constructions. Since the category $\mathcal{CAT}$ is monoidal with a unit, we have the natural notion of an operad in $\mathcal{CAT}$, as before we restrict our attention only to the classical operads. Recall that E is a choice of a category with one object and one morphism (identity).

4.5.1. Definition. *A strict operad in categories is a classical non-symmetric operad (e.g., [MarShSt], Section 1.2) in the monoidal unital category* $(\mathcal{CAT}, \times, E)$*. We will denote the category of strict operads in categories by* $\mathcal{OP}_{st}(\mathcal{CAT})$*.*

As with pseudo-operads, we would like to extend the category $\mathcal{OP}_{st}(\mathcal{CAT})$ to include lax morphism of operads. As usual that should mean making all diagrams, that before were commutative on the nose, commutative only up to natural transformations. First we list the relevant diagrams from the classical definition of operads.

As defined above, a strict operad in categories is a pseudo-operad P with a strict morphism of strict pseudo-operads $U : E \to P$, where we consider E as a pseudo-operad with $E_n := \varnothing$ for $n > 1$, and the obvious structure morphism. The strict morphism U should make the following diagrams commutative:

$$\begin{array}{ccccc} P_n & \longrightarrow & E \times P_n & \xrightarrow{U \times Id} & P_1 \times P_n \\ & \searrow^{Id} & & & \downarrow{\scriptstyle \Upsilon_n} \\ & & & & P_n \end{array} \qquad \begin{array}{ccccc} P_n & \longrightarrow & P_n \times E^{\times^n} & \xrightarrow{Id \times U^{\times^n}} & P_n \times P_1^{\times^n} \\ & \searrow^{Id} & & & \downarrow{\scriptstyle \Upsilon_{1,\dots,1}} \\ & & & & P_n \end{array}$$

A strict morphism between strict operads commutes with these U's on the nose, i.e., we have the following commutative diagram for a strict morphism of operads $F : P \to P'$:

$$\begin{array}{ccc} E & \xrightarrow{U} & P \\ {\scriptstyle Id}\downarrow & & \downarrow{\scriptstyle F} \\ E & \xrightarrow[U']{} & P' \end{array}$$

We would like our lax morphisms to do that only up to a natural transformation, that satisfies some coherence conditions. Later we will see that these conditions are translated into the usual unit axioms for triples.

4.5.2. Definition. *Let* $U : E \to P$ *and* $U' : E \to P'$ *be two strict operads in categories. A coherent lax morphism from* P *to* P' *is a coherent lax morphism of pseudo-operads* $(F, \zeta) : P \to P'$ *and a natural transformation* η*, making the following diagram commutative,*

$$\begin{array}{ccc} E & \xrightarrow{U} & P \\ {\scriptstyle Id}\downarrow & \overset{\eta}{\nearrow} & \downarrow{\scriptstyle F} \\ E & \xrightarrow[U']{} & P' \end{array}$$

such that compositions of natural transformations in the following diagrams are identities,

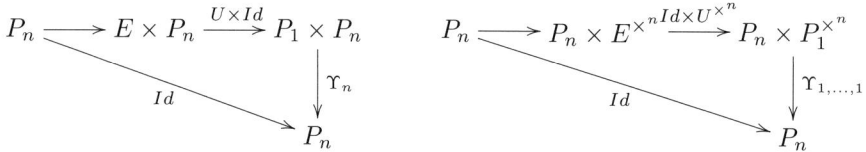

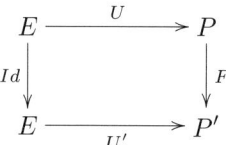

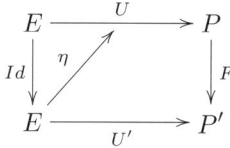

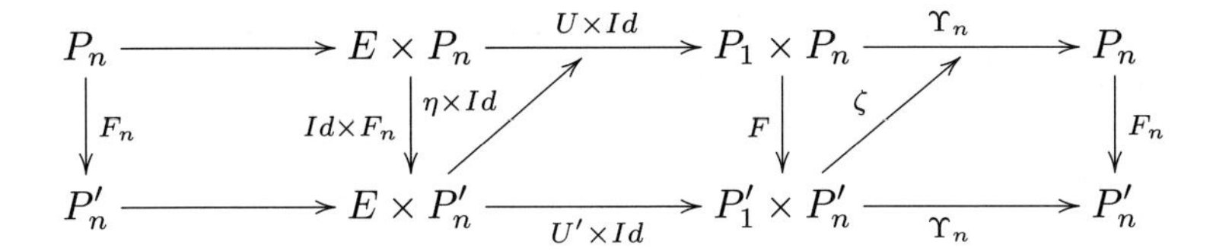

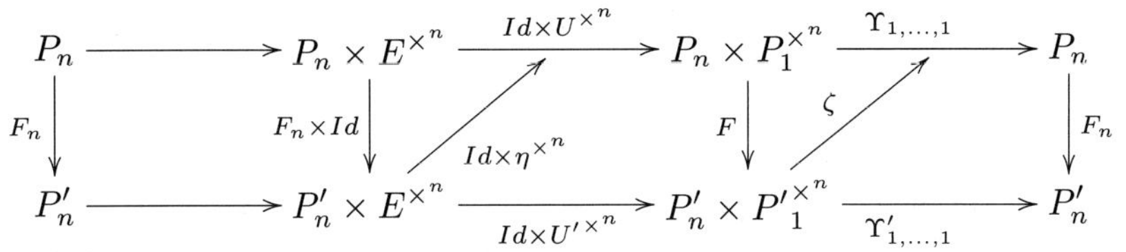

4.5.3. Proposition. *Strict operads in categories and lax morphisms between them constitute a category.*

Proof. Let $(F, \zeta, \eta) : P \to P'$ and $(F', \zeta', \eta') : P' \to P''$ be two lax morphisms between strict operads in categories. We define their composition as $(F' \circ F, \zeta * \zeta', \eta * \eta')$, where $*$ denotes composition of natural transformations as defined in 4.1.1.

It is clear that for any strict operad, sequence of identity functors and identity natural transformations in place of ζ and η constitute a lax morphism, and this morphism satisfies the conditions of identity with respect to the composition above. It remains to show that composition of lax morphisms is again a lax morphism and that this composition is associative.

To prove that composition is well defined we have to show that composition of natural transformations in the following diagrams are identities:

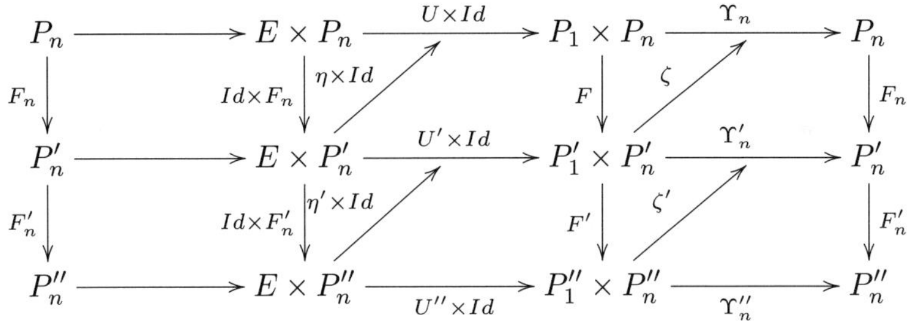

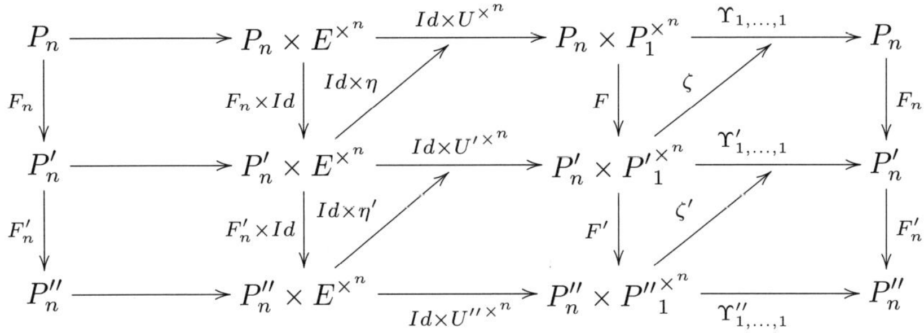

Arguing as in the proof of Proposition 4.3.3 we see that indeed compositions of these natural transformations are identities and hence composition of lax morphisms between strict operads in categories is well defined.

Since the composition of natural transformations is associative, in particular this is true for η's, hence strict operads in categories and lax morphisms between them indeed form a category. □

4.5.4. Notation. We will denote the category of strict operads in categories and lax morphisms between them by $\mathcal{OP}(\mathcal{CAT})$.

4.6. Categories of operads over an operad

Since the endomorphism pseudo-operad of a category is obviously an operad, we would like to have a notion of a category of operads over an operad, similarly to the case of pseudo-operads, that we have considered in 4.4.

Let $U : E \to P$ be a strict operad in categories. We want to organize morphisms in $\mathcal{OP}(\mathcal{CAT})$ with codomain (P, U) into a category. Given two of them $((F', \zeta'), \eta') : (P', U') \to (P, U)$ and $((F'', \zeta''), \eta'') : (P'', U'') \to (P, U)$ we would like to have the notion of a lax morphism from the first to the second. If we wanted only the ones coming from $\mathcal{OP}(\mathcal{CAT})$, we would have defined such a morphism as a lax morphism of operads (Definition 4.5.2) $((F, \zeta), \eta) : (U', P') \to (P'', U'')$, such that the following diagram is commutative,

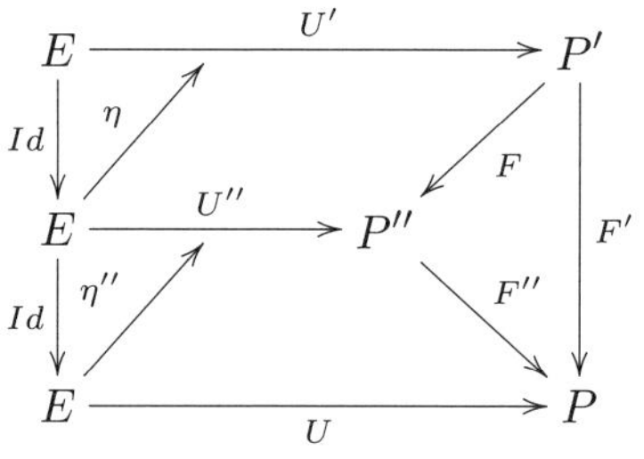

and we have the equality $\eta {*} \eta'' = \eta'$.

However, as in the case of pseudo-operads, we do not want an equality $F'' \circ F = F'$, but we usually have a natural transformation α, making the following diagram commutative,

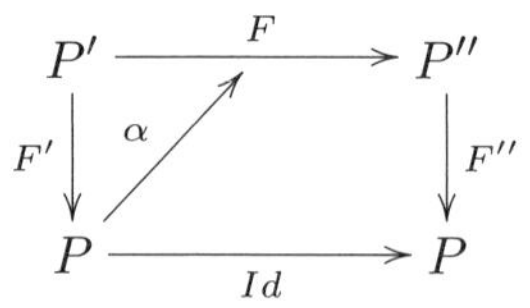

Obviously this natural transformation should satisfy the conditions of a morphism in $\Psi\mathcal{OP}(\mathcal{CAT})//P$, spelled out in Definition 4.4.1. In addition it should respect, in a sense, the unital structures η, η' and η''.

In the diagram

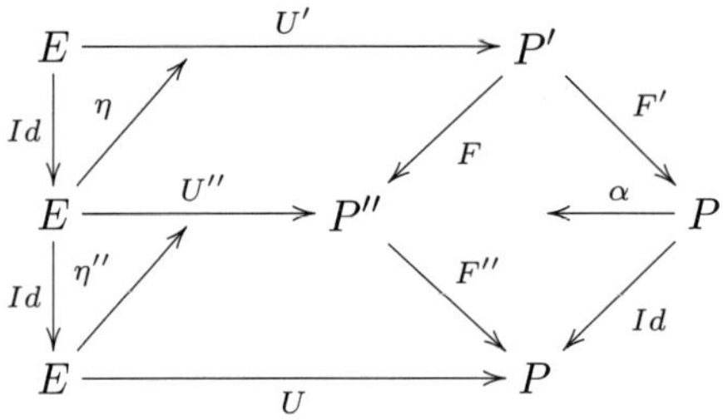

we see that there are two natural transformations $U \to F'' \circ F \circ U'$. One is $\eta * \eta''$ and the other is $\alpha * \eta'$. Their equality is the natural compatibility condition between α and the unital structures.

4.6.1. Definition. *We will denote by $\mathcal{OP}(\mathcal{CAT})//P$ the category whose objects are coherent lax morphisms of strict operads with codomain (P, U). Given two such morphisms $((F', \zeta'), \eta') : (P', U') \to (P, U)$ and $((F'', \zeta''), \eta'') : (P'', U'') \to (P, U)$ a morphism from the first one to the second is a morphism $((F, \zeta), \eta) : (P', U') \to (P'', U'')$ in $\mathcal{OP}(\mathcal{CAT})$ and a natural transformation α, such that $((F, \zeta), \alpha)$ is morphism $F' \to F''$ in $\Psi\mathcal{OP}(\mathcal{CAT})//P$, and in addition we have*

$$\alpha * \eta' = \eta * \eta''.$$

4.6.2. Proposition. *Defined as above $\mathcal{OP}(\mathcal{CAT})//P$ is indeed a category.*

Proof. Composition of two morphisms is inherited from the category of pseudo-operads over a pseudo-operad. Identities are obviously present. All we have to do is to show that the composition of two morphisms, that satisfy conditions of the above definition, also satisfies these conditions.

Let $(F', \zeta', \eta') : (P', U') \to (P, U)$, $(F'', \zeta'', \eta'') : (P'', U'') \to (P, U)$, $(F''', \zeta''', \eta''') : (P''', U''') \to (P, U)$ be three objects of $\mathcal{OP}(\mathcal{CAT})//P$. Let $((G, \psi), \phi, \alpha)$ be a morphism from the first to the second, and let $((G', \psi'), \phi', \alpha')$ be a morphism from the second to the third. We can organize everything into one diagram:

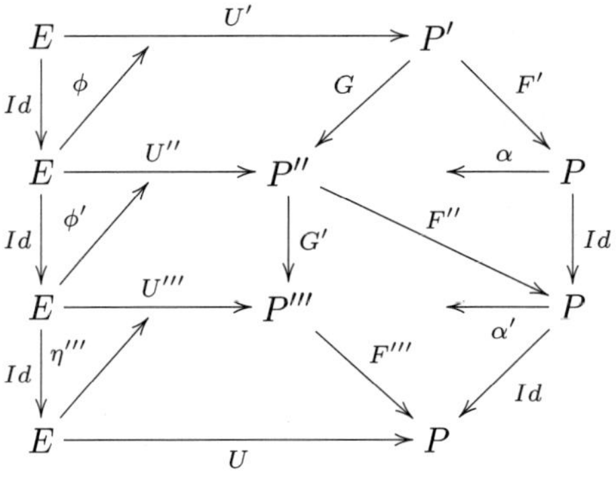

By assumption we have

$$\alpha*\eta' = \phi*\eta'' \qquad \alpha'*\eta'' = \phi'*\eta'''.$$

Using these equalities we get a sequence

$$(\alpha'*\alpha)*\eta' = \alpha'*\phi*\eta'' = \phi*\alpha'*\eta'' = (\phi*\phi')*\eta''',$$

where in the second equality we have used the pasting theorem of [Pow]. The composite equality is exactly the condition, which $\alpha'*\alpha$ should satisfy according to Definition 4.6.1. □

As it was noted in 4.4, when we will consider representations of categorical operads on a category, we would like to consider lax morphisms into the endomorphism operad of that category, and we will want to organize these representations into a category, where as morphisms we take a strict subset of morphisms in $\mathcal{OP}(\mathcal{CAT})//P$.

4.6.3. Definition. *Let (P, U) be a strict operad in categories. We define the category $\mathcal{OP}_{st}(\mathcal{CAT})//P$ as a subcategory of $\mathcal{OP}(\mathcal{CAT})//P$, consisting of the same objects, but for any morphism $((F, \zeta), \eta, \alpha)$ in $\mathcal{OP}(\mathcal{CAT})//P$, it is also a morphism in $\mathcal{OP}_{st}(\mathcal{CAT})//P$ if ζ and η are identities.*

4.7. Operad-like triples as lax representations

So far we have considered categorical operads abstractly. In this subsection we will work with specific operads, namely the endomorphism operads of categories. Example 4.2.2 shows that for any category $\mathcal{C}$, $\mathfrak{E}(\mathcal{C})$ is a pseudo-operad. Mapping E to the identity functor on $\mathcal{C}$ obviously defines a structure of an operad on $\mathfrak{E}(\mathcal{C})$.

In this section we are interested in representations of categorical operads, i.e., with lax morphisms $P \to \mathfrak{E}(\mathcal{C})$. As Proposition 4.6.2 shows, such representations form a category. We will work with them a lot, so we introduce a special term.

4.7.1. Definition. *A generalized triple on a category $\mathcal{C}$ is a coherent lax representation on it of a strict operad in categories P. The category of generalized triples on $\mathcal{C}$ will be denoted by $\mathfrak{T}(\mathcal{C})$.*

To justify the term "generalized triple" we give the following example, which is illustrative but inessential in our considerations. It was considered in [Ben] Section 5.4.

4.7.2. Example. Let P be the strict operad E. A lax representation of E on a category $\mathcal{C}$ is simply triple on $\mathcal{C}$ in the usual meaning of the term.

Indeed such representation consists first of a functor $F : E \to Fun(\mathcal{C}, \mathcal{C})$, which amounts to choosing a functor $T : \mathcal{C} \to \mathcal{C}$, secondly of a natural transformation $\zeta : T \circ T \to T$, thirdly of a natural transformation $\eta : Id_{\mathcal{C}} \to T$, such that the conditions stated in Definitions 4.5.2 and 4.3.2 are satisfied.

The condition spelled out in Definition 4.3.2 translates into associativity of ζ, i.e., into commutativity of the diagram

$$\begin{array}{ccc} T^3 & \xrightarrow{\zeta} & T^2 \\ {\scriptstyle T(\zeta)}\downarrow & & \downarrow{\scriptstyle \zeta} \\ T^2 & \xrightarrow[\zeta]{} & T \end{array}$$

The condition stated in Definition 4.5.2 means that η is a unit for the operation ζ, i.e., the following diagram is commutative

$$\begin{array}{ccccc} T & \xrightarrow{\eta\times Id} & T^2 & \xleftarrow{Id\times\eta} & T \\ & {\scriptstyle Id}\searrow & \downarrow{\scriptstyle \zeta} & \swarrow{\scriptstyle Id} & \\ & & T & & \end{array}$$

The main reason for the development of the theory of categorical operads, that we have done, is the notion of an "operad-like" triple on a category. As we have explained in introductory Section 4.0, we want to construct triples as colimits of a sequence of functors where each element of the sequence represents "operations of some arity".

However, in order for the combined object to satisfy the usual axioms of a triple, the individual elements should behave in a certain prescribed way with respect to colimits. So first we describe the conditions, which these individual functors should satisfy. Obviously we can consider colimits of any diagrams, but we will restrict our attention only to colimits of groupoids. Most of our results can be generalized to the case of arbitrary diagrams.

Consider a functor $F : \mathcal{C} \to \mathcal{C}$, and a diagram $D : \mathcal{D} \to \mathcal{C}$ in C. Since F is a functor we have a natural transformation

$$\chi_F : colim(F \circ D) \to F(colim(D)),$$

where we consider both sides as functors from $Fun(\mathcal{D}, \mathcal{C})$ to $\mathcal{C}$. We will say that a functor F commutes with colimits of groupoids if χ_F is an isomorphism, whenever $\mathcal{D}$ is a groupoid.

In dealing with generalized triples we have a more general case of functors of the type $F : \mathcal{C}^{\times^n} \to \mathcal{C}$. We would like to extend the notion of commutativity with colimits to this case too. There is an obvious way to do that, namely we will say that F commutes with colimits of groupoids if for every $1 \leq i \leq n$ and every $(n-1)$-tuple $\{C_1, \ldots, C_{i-1}, C_{i+1}, \ldots, C_n\}$ of objects of $\mathcal{C}$ the functor $F(C_1, \ldots, C_{i-1}, -, C_{i+1}, \ldots, C_n)$ commutes with colimits of groupoids.

Having a functor $F : \mathcal{C}^{\times^n} \to \mathcal{C}$ that commutes with colimits of groupoids, we will encounter situations when we have n-diagrams $\{\mathcal{D}_i\}_{1\leq i\leq n}$, $\{D_i : \mathcal{D}_i \to \mathcal{C}\}$, and we will consider $F(D_1, \ldots, D_n)$. Since F is a functor we have a morphism

$$colim(F(D_1, \ldots, D_n)) \to F(colim(D_1), \ldots, colim(D_n)).$$

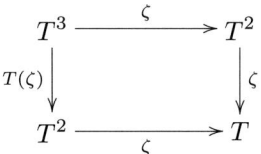

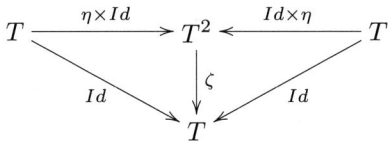

It is easy to see that this morphism factors through colimits in each variable of F, which are by assumption isomorphisms. Therefore this morphism is an isomorphism as well.

Remark. Our construction of generalized triples was specifically tailored for description of triples, i.e., monoids in the monoidal category of functors. However, since we work with operads in categories we obviously can use generalized triples to represent other objects, for example monoidal structures.

Indeed, consider a non-symmetric operad in categories, generated by one binary operation, and having isomorphisms in the category of ternary operations, connecting the two different ways of composing the binary operation with itself. If we demand that these isomorphisms satisfy the usual pentagon conditions of coherence, then a strict representation of this operad on a category is nothing else but a coherently associative product on this category. If we start with two generating operations and demand coherent associativity of both and in addition certain compatibility morphisms between their mixed compositions (these morphisms do not have to be isomorphisms), then a representation of such operad would be a 2-monoidal category as described in [Va2].

Before we proceed with the definition of operad-like triples and provide a way of constructing ordinary triples from them, we need a technical preparation. We need to prove a lemma, that allows us to combine functors, that commute with colimits of groupoids, and get a functor commuting with such colimits as well. This will be needed in the proof of associativity of the structure natural transformation of the triple, that we construct from an operad-like one.

4.7.3. Lemma. *Let F, G be two functors $\mathcal{C} \to \mathcal{C}$, that commute with colimits of groupoids. Then $F \circ G$ commutes with such colimits as well. Moreover, let $D' : \mathcal{D}' \to Fun(\mathcal{C}, \mathcal{C})$ be a diagram of functors, commuting with colimits of groupoids, and $\mathcal{D}'$ being a groupoid itself. Then for any groupoid $\mathcal{D}$ and any diagram $D : \mathcal{D} \to \mathcal{C}$ we have the following commutative diagram of natural transformations:*

$$\begin{array}{ccc} \underset{G_m \in D'(\mathcal{D}')}{colim} (colim(F \circ G_m \circ D)) & \xrightarrow{\chi_F} & F(\underset{G_m \in D'(\mathcal{D}')}{colim} (colim(G_m \circ D))) \\ \Big\downarrow{\scriptstyle \chi_{F\circ G}} & & \Big\downarrow{\scriptstyle F(\chi_G)} \\ \underset{G_m \in D'(\mathcal{D})}{colim} (F \circ G_m(colim(D))) & \xrightarrow[\chi_F]{} & F(\underset{G_m \in D'(\mathcal{D}')}{colim} (G_m(colim(D)))) \end{array}$$

Proof. Let C be in the image of $F \circ G \circ D$. There are two morphisms going out of C: one to $F(colim(G \circ D))$ and another one to $F \circ G(colim(D))$. Since F is a functor, $F(\chi_G)$ completes these morphisms to a commutative triangle. Therefore we have a factorization of the natural transformation $colim(F \circ G \circ D) \to F \circ G(colim(D))$ as χ_F applied to $G \circ D$, followed by $F(\chi_G)$. Both of these are isomorphisms, therefore so is their composition. This proves the first claim of the lemma.

The second claim is proved in a similar manner. One traces different ways to get from a object in the image of D to $F(\underset{G_m \in D'(\mathcal{D}')}{colim}(G_m(colim(D))))$ and finds that they are equal, due to functoriality of F and G_m's and the assumption that these functors commute with colimits of groupoids. □

4.7.4. Definition. *An operad-like triple on a category $\mathcal{C}$ is a generalized triple $P \to \mathfrak{E}(\mathcal{C})$, such that each component of P is a groupoid and for each object of P_n its image in $Fun(\mathcal{C}^{\times^n}, \mathcal{C})$ commutes with colimits of groupoids.*

We have defined the category $\mathfrak{T}(\mathcal{C})$ of all generalized triples on $\mathcal{C}$ by utilizing all possible morphisms of operads over an operad. With operad-like triples we want to restrict our attention to only strict subcategories as in definition 4.6.3.

We will denote by $\mathfrak{P}(\mathcal{C})$ the category whose objects are operad-like triples on $\mathcal{C}$ and whose morphisms are strict morphisms over $\mathfrak{E}(\mathcal{C})$ as defined in Definition 4.6.3.

4.7.5. Proposition. *Let $((F,\zeta),\eta) : P \to \mathfrak{E}(\mathcal{C})$ be an operad-like triple on $\mathcal{C}$. Then if we define a functor $Tot(F) : \mathcal{C} \to \mathcal{C}$ as follows,*

$$Tot(F)(C) := \coprod_{n \in \mathbb{Z}_{>0}} \underset{G_n \in F_n(P_n)}{colim} (G_n(C^{\times^n})),$$

we get a triple on $\mathcal{C}$, with the multiplication and the unit given by ζ and η respectively. In this way we get a functor $Tot : \mathfrak{P}(\mathcal{C}) \to \mathcal{T}(\mathcal{C})$ from the category of operad-like triples on $\mathcal{C}$ to the category of triples on it.

Proof. First we give the definition of the multiplication and the unit natural transformations for $Tot(F)$. The multiplication $Tot(\zeta)$ is defined as composition of the following sequence of natural transformations:

$$\begin{aligned}
&\coprod_{n \in \mathbb{Z}_{>0}} \underset{G_n \in F_n(P_n)}{colim} (G_n((\coprod_{m \in \mathbb{Z}_{>0}} \underset{G'_m \in F_m(P_m)}{colim} (G'_m(C^{\times^m})))^{\times^n})) \\
&\to \coprod_{n, m_1, \dots, m_n} \underset{G_n, G'_{m_i} \in F(P)}{colim} (G_n(G'_{m_1}(C^{\times^{m_1}}), \dots, G'_{m_n}(C^{\times^{m_n}}))) \\
&\to \coprod_{n, m_1, \dots, m_n} \underset{G_n, G'_{m_i} \in F(P)}{colim} (F(G_n \circ (G'_{m_1}, \dots, G'_{m_n}))(C^{\times^{m_1 + \dots + m_n}})),
\end{aligned}$$

where the first arrow is given by commutativity of G_n with colimits of groupoids and the second arrow is the sum of natural transformations $\zeta_{m_1,\dots,m_n}$, given by the lax representation.

The unit $Tot(\eta)$ is given as composition of the following sequence of natural transformations

$$Id_{\mathcal{C}} \to F_1(e) \to \coprod_{n \in \mathbb{Z}_{>0}} \underset{G_n \in F_n(P_n)}{colim} (G_n),$$

where e is the image under F of the identity in P, the first arrow is given by the unit η in the lax representation $((F,\zeta),\eta)$, and the second arrow is the natural inclusion of an object of a diagram into the colimit of the diagram.

We have to show that associativity and unit axioms hold. We know from definition of coherent lax morphisms of operads (Definition 4.3.2) that the natural transformation ζ satisfies associativity conditions. The multiplication for $Tot(F)$ is given as a colimit of ζ's, using commutativity with colimits of groupoids of individual functors in $F(P)$. From lemma 4.7.3 we know that different ways of composing χ's for colimits of functors produce the same result, therefore from associativity of ζ follows associativity of $Tot(\zeta)$. Similarly unit properties of η with respect to ζ imply the same for $Tot(\eta)$ with respect to $Tot(\zeta)$.

It remains to show that Tot is a functor from $\mathfrak{P}(\mathcal{C})$ to $\mathcal{T}(\mathcal{C})$. Given two operad-like triples $((F,\zeta),\eta) : P \to \mathfrak{E}(\mathcal{C})$, $((F',\zeta'),\eta) : P' \to \mathfrak{E}(\mathcal{C})$ and a morphism (F'',α) from the first to the second we have a natural transformation $Tot(\alpha) : Tot(F) \to Tot(F')$, given by α. Indeed, each $G \in F(P)$ is mapped by α to $F'(F''(G))$ in $F'(P')$. The latter is canonically included into $Tot(F')$.

Now we see that the compatibility conditions for α with ζ, ζ'' and with η, η' translate exactly to the fact that $Tot(\alpha)$ is a map between monoids in the monoidal category of endofunctors on $\mathcal{C}$. □

4.8. Symmetric operad-like triples

So far we have dealt with non-symmetric operads, and therefore with non-symmetric operad-like triples. Now we would like to introduce action of symmetric groups in our construction. Since the monoidal category $(\mathcal{CAT}, \times, E)$ is symmetric there is a standard notion of a symmetric strict operad in categories.

4.8.1. Definition. *A symmetric strict operad in categories is a classical symmetric operad (e.g., [MarShSt], Section 1.2) in the category $(\mathcal{CAT}, \times, E)$, i.e., it is a strict operad P and for each $n \in \mathbb{Z}_{>0}$ an action of the symmetric group Σ_n on the category P_n is given. We will denote the functor on P_n, that corresponds to an element $\sigma_n \in \Sigma_n$, by the same symbol σ_n. A coherent lax morphism between two symmetric strict operads is a coherent lax morphism between the operads (as defined in Definition 4.5.2), such that the functors F_n commute (on the nose) with the action of symmetric groups.*

Action of symmetric groups by means of functors provides definition of a symmetric operad in categories, but it is not useful for defining symmetric operad-like triples, since we need natural transformations for that. Therefore we introduce the notion of an equivariant symmetric operad in categories.

In order to do that we need more notation. We will denote by $\underline{\Sigma}_n$ the category whose objects are elements of the symmetric group Σ_n and the set of morphisms $Hom(\sigma_n, \sigma'_n)$ between any two of them consists of one element: $\sigma_n^{-1}\sigma'_n \in \Sigma_n$. Composition is obvious. One could call this "a regular groupoid" version of the symmetric group Σ_n.

4.8.2. Definition. *An equivariant symmetric operad in categories is a symmetric strict operad P, such that for every object $p_n \in P_n$ there is a functor $S_n : \underline{\Sigma}_n \to P_n$, such that any object $\sigma_n \in \underline{\Sigma}_n$ is mapped to $\sigma_n(p_n)$, and the following compatibility conditions are satisfied. Let $p_{m_i} \in P_{m_i}$ $1 \leq i \leq n$. Then for every $(n+1)$-tuple* *of morphisms* $\sigma_n \in \underline{\Sigma}_n$, $\sigma_{m_i} \in \underline{\Sigma}_{m_i}$, *the composition functor* $\Upsilon_{m_1,\dots,m_n} : P_n \times P_{m_1} \times \cdots \times P_{m_n} \to P_{m_1+\cdots+m_n}$ *maps* $S_n(\sigma_n) \times Id^{\times^n}$ *to* $S_{m_1+\cdots+m_n}(\sigma_{\underline{n}})$, *where the morphism* $\sigma_{\underline{n}} \in \underline{\Sigma}_{m_1+\cdots+m_n}$ *is* σ_n*-permutation of the* n *blocks. Also the composition functor maps* $Id \times S_{m_1}(\sigma_{m_1}) \times \cdots \times S_{m_n}(\sigma_{m_n})$ *to* $S_{m_1+\cdots+m_n}(\sigma_{m_1} \times \cdots \times \sigma_{m_n})$, *where the morphism* $\sigma_{m_1} \times \cdots \times \sigma_{m_n} \in \underline{\Sigma}_{m_1+\cdots+m_n}$ *corresponds to the product of permutations.*

It is clear that the compatibility conditions in the last definition are meant to reflect the standard equivariance properties of operads in symmetric categories. Indeed, when we will define symmetric operad-like triples we will see that these compatibility conditions translate into the usual equivariance.

4.8.3. Example. Let $\mathcal{C}$ be a category and consider the endomorphism operad $\mathfrak{E}(\mathcal{C})$ of $\mathcal{C}$ (Example 4.2.2.). There is one obvious symmetric structure on $\mathfrak{E}(\mathcal{C})$, namely for any $\sigma_n \in \Sigma_n$ we define a functor $\sigma_n : Fun(\mathcal{C}^{\times^n}, \mathcal{C}) \to Fun(\mathcal{C}^{\times^n}, \mathcal{C})$ as follows

$$Fun(\mathcal{C}^{\times^n}, \mathcal{C}) \ni F \mapsto F \circ \sigma_n,$$

where we consider σ_n as a functor $\mathcal{C}^{\times^n} \to \mathcal{C}^{\times^n}$, permuting the variables. It is clear that in this way we get a symmetric structure on $\mathfrak{E}(\mathcal{C})$ and we will always consider endomorphism functors with symmetric structures chosen in this way.

Note that in general an endomorphism operad is not an equivariant symmetric operad. However, we do not require equivariance in the definition of a lax morphism between two symmetric operads in categories, and consequently we can consider lax morphisms from an equivariant operad to one which is not. When the codomain is an endomorphism operad as in the last example we will have a special name for it.

4.8.4. Definition. *We will call a lax morphism from an equivariant symmetric operad* P *to an endomorphism operad* $\mathfrak{E}(\mathcal{C})$ *a symmetric generalized triple. If every component of* P *is a groupoid and every functor in the image of the generalized triple commutes with colimits of groupoids, we will call such a generalized triple a symmetric operad-like triple. We organize symmetric operad-like triples into a category, where morphisms are lax morphisms as in Definition* 4.7.4, *and in addition commuting (on the nose) with the symmetric structure (as in Definition* 4.8.1*). We will denote this category by* $\mathfrak{P}^s(\mathcal{C})$.

We have proved a proposition (Proposition 4.7.7) stating that we can get a usual triple from an operad-like one, and that this correspondence is a functor. Since equivariant symmetric operads differ from the non-symmetric ones by presence of an action of symmetric groups (functors) and representations of the "regular symmetric groupoids" (invertible morphisms) we see that the same proof applies to equivariant operad-like triples as well. So we get a functor $Tot : \mathfrak{P}^s(\mathcal{C}) \to \mathcal{T}(\mathcal{C})$.

We would like to illustrate the role of equivariance in representation of an equivariant symmetric operad on a category. So let P be an equivariant symmetric operad. Let $\mathcal{C}$ be a category and let $((F, \zeta), \eta)$ be a representation of P on $\mathcal{C}$. Let

$p_n, \{p_{m_i}\}_{1\le i\le n}$ be elements of P_n and $\{P_{m_i}\}_{1\le i\le n}$ respectively. Let $\sigma_n, \{\sigma_{m_i}\}$ be morphisms in $\underline{\Sigma}_n$ and $\{\underline{\Sigma}_{m_i}\}$. Writing explicitly the conditions for ζ to be a natural transformation we get the following commutative diagrams of natural transformations between functors on $\mathcal{C}$,

$$\begin{array}{ccc} F(p_n)\circ(F(p_{m_1})\times\cdots\times F(p_{m_n})) & \longrightarrow & F(p_n)\circ\sigma_n\circ(F(p_{m_1})\times\cdots\times F(p_{m_n})) \\ \downarrow{\scriptstyle \zeta_{m_1,\dots,m_n}} & & \downarrow{\scriptstyle \zeta_{m_1,\dots,m_n}} \\ F(\Upsilon(p_n\times p_{m_1}\times\cdots\times p_{m_n})) & \longrightarrow & F(\Upsilon(\sigma_n(p_n)\times p_{m_1}\times\cdots\times p_{m_n})) \end{array}$$

where the upper horizontal arrow is $F(S(\sigma_n))\circ(Id^{\times^n})$, and the lower horizontal arrow is $F(S(\sigma_{\underline{n}}))$.

$$\begin{array}{ccc} F(p_n)\circ(F(p_{m_1})\times\cdots\times F(p_{m_n})) & \longrightarrow & F(p_n)\circ(F(p_{m_1})\circ\sigma_{m_1}\times\cdots\times F(p_{m_n})\circ\sigma_{m_n}) \\ \downarrow{\scriptstyle \zeta_{m_1,\dots,m_n}} & & \downarrow{\scriptstyle \zeta_{m_1,\dots,m_n}} \\ F(\Upsilon(p_n\times p_{m_1}\times\cdots\times p_{m_n})) & \longrightarrow & F(\Upsilon(p_n\times\sigma_{m_1}(p_{m_1})\times\cdots\times\sigma_{m_n}(p_{m_n}))) \end{array}$$

where the upper horizontal arrow is $Id\circ(F(S(\sigma_{m_1})\times\cdots\times S(\sigma_{m_n}))$, and the lower horizontal arrow is $F(S(\sigma_{m_1}\times\cdots\times\sigma_{m_n}))$.

When we apply Tot to a symmetric operad-like triple we see that these diagrams translate to the usual equivariance diagrams for operads.

4.9. Example: operads as algebras over symmetric operad-like triples

In this subsection we would like to show that operads, as they were defined in Section 1, can be described as algebras over certain triples, that lie in the image of Tot, as described above. Until now we have considered operad-like triples as lax representations of classical operads. However, operadic constructions require working with colored operads, rather than the classical ones. The passage to the colored context is straightforward and we indicate the main steps below.

One could define colored operads in categories as colored operads in the monoidal category $(\mathcal{CAT},\times,E)$ in the usual meaning of the term. However, because we have 2-morphisms in the background, there are some minor adjustments to be made. As in the usual case, a colored operad is different from a classical one by a restriction on possible compositions. Before we define it we need a technical preparation.

Suppose we have three categories $\mathcal{C}$, $\mathcal{C}'$, $\mathcal{C}''$, and each object in all of them is given two colors from a set of colors Ω. One of the colors will be called incoming and the other outgoing. We will say that we have a colored functor $\mathcal{C}\times\mathcal{C}'\to\mathcal{C}''$ if for every two objects C, C' of $\mathcal{C}$ and $\mathcal{C}'$ respectively, such that the incoming color of C is equal to the outgoing one of C' (we will call such pairs composable), we are given an object C'' of $\mathcal{C}''$ whose incoming color is that of C' and outgoing – that of C. On morphisms such a functor should act as a usual functor, where we allow morphisms only between composable pairs in $\mathcal{C}\times\mathcal{C}'$.

4.9.1. Definition. *Let Ω be a set. A strict Ω-colored categorical operad is a sequence of categories $\{P_n\}_{n\in\mathbb{Z}_{>0}}$, for each object in P_n a set of incoming colors $(\omega_1,\ldots,\omega_n)$ and an outgoing color ω, and a set of colored functors $\{\Upsilon_{m_1,\ldots,m_n} : P_n \times P_{m_1} \times \cdots \times P_{m_n} \to P_{m_1+\cdots+P_{m_n}}\}$, satisfying the obvious associativity and unit axioms.*

Just as in case of monochrome operads we can introduce the notion of a symmetric colored operad in categories. For that we need an action of symmetric groups on the components. However, since we need to be able to permute different sets of incoming colors differently we have to use colored symmetric groups, i.e., for every set of incoming colors we have a copy of the symmetric group, that acts (by functors) on all objects with the same set of incoming colors. We omit writing explicitly the colored extension of the usual equivariance axioms (it is straightforward but long).

Similarly the equivariance structure (Definition 4.8.2) can be generalized to the colored context in a very straightforward manner. Indeed it requires connecting objects by isomorphisms with their images under permutation functors, such that the compatibility conditions from Definition 4.8.2. are satisfied. We leave writing the details explicitly to the reader.

Let $\mathcal{C}$ and $\mathcal{C}'$ be two categories. Suppose that objects in $\mathcal{C}$ have incoming and outgoing colors from a set of colors Ω and objects in $\mathcal{C}'$ - from Ω'. Suppose we have a map $f : \Omega \to \Omega'$. Then we will say that a functor $F : \mathcal{C} \to \mathcal{C}'$ preserves colors if an object $C \in \mathcal{C}$ with colors $(\omega_{in}, \omega_{out})$ is mapped by F to an object $C' \in \mathcal{C}'$ with colors $(f(\omega_{in}), f(\omega_{out}))$.

4.9.2. Definition. *A coherent lax morphism between symmetric strict colored operads in categories (colored by Ω and Ω') is a set of color preserving functors (for a choice of a map $f : \Omega \to \Omega'$) and natural transformations $((F,\zeta),\eta)$, satisfying the colored versions of coherence conditions as in Definitions* 4.3.2 *and* 4.5.2.

4.9.3. Example. The main example for us of a colored categorical operad is the operad, produced by abstract categories of labelled graphs, described in Section 1. Let Γ be a category as in Definition 1.3. Let Ω be the set of Γ-corollas. This is our set of colors.

We define a strict Ω-colored operad $\mathfrak{G}$ as follows. We set $\mathfrak{G}_n$ to be the category, whose objects are pairs of morphisms in Γ:

$$\coprod_{v\in V_\tau} \sigma_v \to \tau \to \sigma,$$

where τ is an object of Γ, σ is a Γ-corolla, and the first arrow is one of the possible atomizations of τ, provided by property (iv) of Definition 1.3. We will denote such object simply by τ. Note that corollas in the direct product are ordered.

A morphism from one such object to another is a pair of morphisms between τ's and coproducts of corollas, such that together with the identity on σ they make up a commutative ladder.

The coloring on each object of $\mathfrak{G}_n$ is obvious: as it is written above the outgoing color is σ and the incoming colors are $\{\sigma_v\}_{v \in V_\tau}$. Actions by symmetric groups are obvious as well: we just rearrange summands in the direct sum of corollas for the atomization.

To define composition functors we use property (vi) of Definition 1.3. Suppose we have $n+1$ objects of $\mathfrak{G}$: $\tau_1, \ldots, \tau_n, \tau$, such that the incoming colors of the latter are exactly the outgoing ones of the former n-tuple. According to property (vi) we have an object τ' of Γ, and a morphism $\tau' \to \tau$, fitting into diagram of the type (1.3). Here we assume that a choice of a particular τ' is made in each case. We will call this *a choice of grafting*. It is clear that we can choose atomization of τ' to be the direct sum of atomizations of τ_i's. Doing that and taking $\tau' \to \tau \to \sigma$ as the outgoing color we get a composition on $\mathfrak{G}$. Colored units are chosen in the obvious way: they are identity maps of the corollas.

Now we have to check the associativity and unit axioms. The unit ones are obvious. Associativity conditions are obvious if we take Gr itself as Γ, indeed all we do is substituting graphs in place of corollas, and this operation is associative. In case of a general Γ we make this associativity condition part of the choice of grafting.

Actions of symmetric groups obviously satisfy the conditions of Definition 4.8.1. We also have a natural equivariance structure (Definition 4.8.2) on $\mathfrak{G}$. Indeed if we take a direct sum and rearrange the summands the result is connected to the original sum by a unique isomorphism, that gives us the representation of $\underline{\Sigma}$. Here we should check that compatibility conditions from 4.8.2 are satisfied. In case Γ is Gr itself they are obvious. In general we make them part of the requirements for the choice of grafting. So we have a structure of an equivariant symmetric operad on $\mathfrak{G}$.

Now let $(\mathcal{C}, \otimes)$ be a symmetric monoidal category. We would like to define a representation of $\rho : \mathfrak{G} \to \mathfrak{E}(Fun(\Omega, \mathcal{C}))$ as follows: given an object of $\mathfrak{G}$

$$\coprod_{1 \leq i \leq n} \sigma_i \to \tau \to \sigma,$$

and an object $F : \Omega \to \mathcal{C}$, $\rho(\tau)$ acts on F by mapping it to the functor $\Omega \to \mathcal{C}$, whose value on σ is $\bigotimes_{1 \leq i \leq n} F(\sigma_i)$ and on the rest of colors the value is the initial object of $\mathcal{C}$.

If we did not have non-identity morphisms on Ω, then this definition would have been obviously correct. Indeed, then a functor from the category of colors would have been equivalent to just a choice of objects in $\mathcal{C}$, and the above choice is obviously functorial in F. However, we have to take into account the non-identity morphisms in Ω.

Let σ and σ' be two corollas. And let $\phi : \sigma \to \sigma'$ be an isomorphism. Then for any object $\coprod_{1 \leq i \leq n} \sigma_i \to \tau \to \sigma$ of $\mathfrak{G}$ we have a new object $\coprod_{1 \leq i \leq n} \sigma_i \to \tau \to \sigma'$, where the last arrow is the last arrow in the original object, followed by ϕ. We will denote this new object of $\mathfrak{G}$ by $\phi_*(\tau)$. In this way, given an object of $\mathfrak{G}$ we get a sort of Ω-diagram of such objects (it is not exactly a diagram because we have excluded morphisms from $\mathfrak{G}$, that are non-identities on the corollas).

Note that for any isomorphism ϕ as above the values of $\rho(\tau)(F)$ on σ and of $\rho(\phi_*(\tau))(F)$ on σ' are the same. Therefore, if given a F, a $\tau \in \mathfrak{G}$ we define for each corolla $\sigma' \in \Omega$, that is isomorphic to σ

$$\sigma' \mapsto \coprod \bigotimes_{1\leq i\leq n} F(\sigma_i'),$$

where the coproduct is taken over all objects from $\mathfrak{G}$, that are in the Ω-diagram corresponding to τ as described above, then we would get a new functor $\Omega \to \mathcal{C}$. Indeed, every morphism in Ω (as for example ϕ) is mapped to the identity automorphism of $\bigotimes_{1\leq i\leq n} F(\sigma_i)$.

So we get a representation of $\mathfrak{G}$ on $Fun(\Omega, \mathcal{C})$, and we will denote it by ρ. As it was noted above $\mathfrak{G}$ is an equivariant symmetric operad and hence the operad-like triple ρ is symmetric.

From Proposition 4.7.5 we conclude that there is a triple $Tot(\rho)$ on $Fun(\Omega, \mathcal{C})$. This triple is exactly the triple $\mathcal{F}$ from Section 1.5.4, and the algebras over it are the $\Gamma\mathcal{C}$-operads.

4.10. Existence of <u>cohom</u> for operads in algebras

Let T be a Hopf-like triple (Definition 3.1.1) on a category $\mathcal{C}$, that commutes with colimits of groupoids. From Lemma 3.1.3 we know that the category $\mathcal{A}$ of algebras over T has monoidal structure, and hence for any abstract category of labelled graphs Γ we can consider the category $\Gamma\mathcal{A}OPER$ of $\Gamma\mathcal{A}$-operads. Now we would like to show that if $\mathcal{C}$ possesses <u>cohom</u>, so does $\Gamma\mathcal{A}OPER$.

The key to the proof is the observation that the forgetful functor $\mathfrak{U} : \mathcal{A} \to \mathcal{C}$ maps $\Gamma\mathcal{A}$-operads to $\Gamma\mathcal{C}$-operads. Therefore an object of $\Gamma\mathcal{A}OPER$ is a sequence (parameterized by Γ-corollas) of objects in $\mathcal{C}$, such that each one of them is a T-algebra and altogether they make up a Γ-operad in $\mathcal{C}$. Of course certain compatibility conditions between these two structures should be satisfied. This situation is just a triple version of the usual instance of an action of a colored operad.

First we are going to consider sequences of objects in $\mathcal{C}$, that have both of the above structures, but with the compatibility conditions omitted. We need a very simple lemma for this, whose proof is straightforward and we leave it to the reader.

4.10.1. Lemma. *Let T and T' be two triples on $\mathcal{C}$, such that both commute with colimits of groupoids. Then the category $\mathcal{A}''$ of objects in $\mathcal{C}$, that are simultaneously algebras over T and T' is equivalent to the category of algebras over the following triple:*

$$T\coprod T' : C \mapsto \coprod T(\ldots T'(\ldots (C))),$$

where the coproduct is taken over all possible words of positive length, composed of T and T'.

In our situation we have one T (one for each Γ-corolla) but instead of T' we have a sequence (parameterized by objects of $\mathfrak{G}$) of functors $\mathcal{C}^{\times^n} \to \mathcal{C}$ (for all $n \in \mathbb{Z}_{>0}$). Each of these functors commutes with colimits of groupoids so we can form a triple out of them and T by forming all possible compositions and summing them up. This is an obvious generalization of Lemma 4.10.1. We will denote the resulting triple by $T \coprod Tot(\rho)$.

By construction $T \coprod Tot(\rho)$ is the direct product, i.e., its algebras are equivalent to operads in $\mathcal{C}$ and algebras over T, and these two structures being unrelated. Now we make this direct product into an amalgamated sum. The needed relations are provided by the Hopf-like properties of T. Recall that for T to be Hopf-like means that there is a natural transformation

$$\tau : T \circ \otimes \to \otimes \circ (T \times T),$$

satisfying certain conditions, spelled out in Definition 3.1.1.

According to property b) in Definition 3.1.1. there is a definite natural transformation

$$T \circ (\otimes \circ (Id \times \otimes)) \to \otimes \circ (Id \times \otimes) \circ (T^{\times^3}),$$

and similarly for all other possible iterations of the monoidal structure. Note that functors on both sides of the last arrow are summands in $T \coprod Tot(\rho)$ (for each Γ-corolla), therefore there are two ways to include the left side in the sum, i.e., we have a pair of parallel natural transformations (for each Γ-corolla separately):

$$\coprod_{n \geq 1} T \circ \otimes^{\circ^n} \rightrightarrows T \coprod Tot(\rho).$$

These are our relations. The left side of the two arrows is just a functor, but we have an adjunction from functors (commuting with coproducts) to triples, therefore we have a pair of morphisms between triples

$$\mathfrak{F}(\coprod_{n \geq 1} T \circ \otimes^{\circ^n}) \rightrightarrows T \coprod Tot(\rho),$$

where $\mathfrak{F}$ denotes the free triple. The coequalizer (in the category of triples) of these two morphisms is a triple whose algebras are exactly Γ-operads in $\mathcal{A}$.

Here we should discuss existence of coequalizers in the category of triples on $\mathcal{C}$. The opposite category of $Fun(\mathcal{C}, \mathcal{C})$ is equivalent to $Fun(\mathcal{C}^{op}, \mathcal{C}^{op})$, and hence the question of existence of colimits in $Fun(\mathcal{C}, \mathcal{C})$ is equivalent to the question of existence of limits in $Fun(\mathcal{C}^{op}, \mathcal{C}^{op})$. The latter can be answered by existence of colimits in $\mathcal{C}$ (e.g., [Bo1], Proposition 2.15.1.) Thus if we assume that $\mathcal{C}$ has coequalizers, so does $Fun(\mathcal{C}, \mathcal{C})$. So the question of existence of coequalizers in the category of triples on $\mathcal{C}$ is the usual question of lifting colimits from a category to a category of algebras over a triple. General conditions for their existence are very restrictive, so we will assume existence of the coequalizers above as a condition imposed on $\mathcal{C}$ itself and on T.

So far we have considered algebras over a triple, that commutes with colimits of groupoids. However, it is not always the case, as for example the triple of associative algebras in vector spaces does not commute with such colimits (in general it does not commute even with coproducts). Yet often triples that do not commute can be represented as colimits of ones, which do commute, such as the operad-like triples.

We need a reformulation of the property of a triple to be Hopf-like in the language of operad-like triples. The following definition expresses in the operad-like triple setting the property of an operad to be Hopf. This notion is well known, so we will only outline the main parts of the definition and omit the necessary coherence properties. Examples of such operad-like triples are provided by Hopf operads.

4.10.2. Definition. *Let $F : P \to \mathfrak{E}(\mathcal{C})$ be an operad-like triple on $\mathcal{C}$. It is Hopf-like if for every $G_n \in F_n(P_n)$, $n \in \mathbb{Z}_{>0}$ we have a natural transformation*

$$\tau_n : G_n \circ \otimes^{\times^n} \to \otimes \circ (G_n \times G_n) \circ \sigma_{2n},$$

where σ_{2n} is the permutation that moves all elements in the even places of a sequence to the end of it. This natural transformation should satisfy coherence conditions expressing its associativity and compatibility with the composition natural transformation on $F(P)$.

Now assume that we have a Hopf-like symmetric operad-like triple $F : P \to \mathfrak{E}(\mathcal{C})$. The family $\{\tau_n\}$ provides us with a Hopf-like structure on $Tot(F)$. Thus the category $\mathcal{A}$ of algebras over $Tot(F)$ has a symmetric monoidal structure. Let Γ be an abstract category of labelled graphs. We would like to have objects of $\Gamma\mathcal{A}OPER$ as algebras over a triple on $\mathcal{C}$.

Just as we did in case of a single T we first consider the coproduct $Tot(F)\coprod Tot(\rho)$. Here, as before, we take all possible compositions, but now we have to compose functors in several variables. Also we want to take coproduct of symmetric operad-like triples, i.e., we add up not only all compositions, but also applications to them of permutations of variables.

Finally, as before, we have a pair of parallel morphisms of triples, with codomain $Tot(F)\coprod Tot(\rho)$. Their coequalizer (if it exists) is the required triple. In total we have the following proposition.

4.10.3. Proposition. *Let $(\mathcal{C}, \otimes)$ be a symmetric monoidal category. Let $F : P \to \mathfrak{E}(\mathcal{C})$ be a symmetric operad-like triple on $\mathcal{C}$. Suppose that F is Hopf-like, and let Γ be an abstract category of labelled graphs. Then, if the category of triples on $\mathcal{C}$ has necessary coequalizers, the category of Γ-operads in the category of algebras over $Tot(F)$ is equivalent to a category of algebras over a triple on $\mathcal{C}$.*

Now using results of Section 3.1. we establish existence of <u>cohom</u> for operads in algebras over operad-like triples, given that <u>cohom</u> exists on $\mathcal{C}$.

Appendix A. Labeled graphs corresponding to various operads

This will last out a night in Russia
When nights are longest there.

W. Shakespeare, Measure for measure, 2.1.132–3

0. Operads, cyclic operads, modular operads. The graph geometry behind these structures is basically well known, and we will only briefly repeat it.

Operads. Objects: disjoint unions of directed trees with one output each. Morphisms: (generated by) contractions and graftings of an output to an input. If one considers only linear directed graphs (each vertex carries one input and one output), one gets associative algebras.

Cyclic operads. Objects: disjoint unions of (unlabeled) trees. Morphisms: contractions and graftings.

If one adds cyclic labeling, one gets the non-symmetric version of operads, resp. cyclic operads.

Modular operads. Objects: graphs of arbitrary topology with genus labeling. Morphisms: contractions and graftings compatible with labelings in the following sense.

Contraction of an edge having two distinct vertices of genera g_1, g_2, produces a new vertex of genus $g_1 + g_2$. Contraction of a loop augments the genus of its vertex by one. The effect of a general contraction is the result of the composition of contraction of edges. Grafting does not change labels.

1. PROPs. Consider first the category Γ_c whose objects are disjoint unions of oriented corollas, and morphisms are mergers (including isomorphisms). Any tensor functor $(\Gamma_c, \coprod) \to (\mathcal{G}, \otimes)$ is determined up to an isomorphism by the following data:

(i) Its values on corollas with inputs $\{1, \dots, n\}$ and outputs $\{1, \dots, m\}$ ($m = 0$ and $n = 0$ are allowed). Let such a value be denoted $P(m, n)$.
(ii) Its values upon automorphisms of such corollas. This means that each $P(m, n)$ is endowed by commuting actions of $\mathbf{S}_m$ (left) and $\mathbf{S}_n$ (right).
(iii) Its values upon merger morphisms of such corollas which are called *horizontal compositions*:

$$P(m_1, n_1) \otimes \cdots \otimes P(m_r, n_r) \to P(m_1 + \cdots + m_r, n_1 + \cdots + n_r). \qquad (A.1)$$

Consider now a larger category Γ of directed graphs without oriented wheels. This puts restrictions to morphisms compatible with orientations. In particular, if we contract an edge, we must simultaneously contract all edges connecting its ends. Mergers of two vertices connected by an oriented path also are excluded.

A tensor functor $\Gamma \to \mathcal{G}$ then produces data (i)–(iii) and moreover,

(iv) *Vertical compositions:* values of the functor upon full contractions of two-vertex directed graphs such that all inputs belong to one vertex, all outputs to another, and edges are oriented from inputs to outputs:

$$P(m,n) \otimes P(n,k) \to P(m,k), \ n \neq 0. \tag{A.2}$$

These data must satisfy some compatibility conditions which can be rephrased as existence of a monoidal category with objects $\emptyset$, $\dots$, $\{1,\dots,n\}$, $\dots$ (as in 1.2.3) enriched over $\mathcal{G}$ in such a way that its morphisms become $P(m,n)$ and their composition is given by (A.2).

Allowing mergers in PROPs, we get generally big categories $\Rightarrow \sigma$ which are main building blocks of the triple $(\mathcal{F},\mu,\eta)$ and the respective operads. In the following three operadic structures, we again exclude them.

2. Properads. Objects: all directed graphs as above. Morphisms: contractions and graftings.

3. Dioperads. Objects: all directed graphs with whose connected components are simply connected. Morphisms: contractions and graftings.

4. $\frac{1}{2}$-PROPs. Objects: directed graphs with simply connected components trees such that each edge is either unique output of its source, or unique input of its target. Morphisms: contractions and graftings.

5. Monoidal structures on the collections. Following [Va1], we will introduce the following definition, working well for the categories of directed graphs without mergers.

A directed graph τ is called *two-level* one, if there exists a partition of its vertices $V_\tau = V_\tau^1 \coprod V_\tau^2$ such that

a) *Tails at V_τ^1 are all inputs of τ, tails at V_τ^2 are all outputs of τ.*
b) *Any edge starts at V_τ^1 and ends at V_τ^2.*

Clearly, such a partition is unique, if it exists at all.

Denote by $\Rightarrow^{(2)} \sigma$ the full subcategory of $\Rightarrow \sigma$ consisting of objects whose sources are two-level graphs.

For any two collections A^1, A^2, define the third one by

$$(A^2 \boxtimes_c A^1)(\sigma) := \operatorname{colim}\,(\otimes_{v\in V_\tau^1} A^1(\tau_v)) \otimes (\otimes_{v\in V_\tau^2} A^2(\tau_v))$$

where colim is taken over $\Rightarrow^{(2)} \sigma$.

B. Vallette proves that this is a monoidal structure on collections, and that the respective operads are monoids in the resulting monoidal category.

B. Vallette treats also the case of PROPs, but here one must restrict oneself to "saturated" collections.

References

[BarW] M. Barr and Ch. Wells, *Toposes, triples and theories.* Grundlehren der mathematischen Wissenschaften 278, Springer Verlag, 1985, 358 pp.

[Bat] M. Batanin, *The Eckmann–Hilton argument, higher operads and E_n-spaces.* math.CT/0207281 Sep 2003, 58 pp.

[Bek] T. Beke, *Sheafifiable homotopy model categories. II.* J. Pure Appl. Algebra **164** (2001), no. 3, 307–324.

[Ben] J. Bénabou, *Introduction to bicategories* in 1967 *Reports of the Midwest Category Seminar*, Springer Verlag, 1–77.

[BeMa] K. Behrend and Yu. Manin, *Stacks of stable maps and Gromov–Witten invariants.* Duke Math. Journ. **85** (1996), no. 1, 1–60.

[BerDW] R. Berger, M. Dubois-Violette and M. Wambst, *Homogeneous algebras.* J. Algebra **261** (2003), 172–185. Preprint math.QA/0203035

[BerM] R. Berger and N. Marconnet, *Koszul and Gorenstein properties for homogeneous algebras.* Algebras and representation theory **9** (2006), 67–97.

[BerMo] C. Berger and I. Moerdijk, *Resolution of colored operads and rectification of homotopy algebras.* math.AT/0512576 (to appear in Cont. Math., vol. in honor of Ross Street).

[Bl] D. Blanc, *New model categories from old.* Journal of Pure and Applied Algebra **109** (1996), 37–60.

[Bo1] F. Borceux, *Handbook of categorical algebra 1. Basic category theory.* Encyclopedia of Mathematics and its applications, Cambridge University Press, 1994, 360 pp.

[Bo2] F. Borceux, *Handbook of categorical algebra 2. Categories and structures.* Encyclopedia of Mathematics and its applications, Cambridge University Press, 1994, 360 pp.

[CaGa] J.G. Cabello and A.R. Garzón, *Closed model structures for algebraic models of n-types.* Journal of Pure and Applied Algebra **103** (1995), no. 3, 287–302.

[Cra] S.E. Crans, *Quillen closed model structures for sheaves.* Journal of Pure and Applied Algebra **101** (1995), 35–57.

[DeMi] P. Deligne and J. Milne, *Tannakian categories.* In: Hodge cycles, motives and Shimura varieties, Springer Lecture Notes in Math. **900** (1982), 101–228.

[Fr] B. Fresse, *Koszul duality for operads and homology of partition posets.* Contemp. Math. **346** (2004), 115–215.

[Ga] W.L. Gan, *Koszul duality for dioperads.* Math. Res. Lett. **10** (2003), no. 1, 109–124.

[GeKa1] E. Getzler and M. Kapranov, *Cyclic operads and cyclic homology.* In: Geometry, Topology and Physics for Raoul Bott (ed. by S.-T. Yau), International Press 1995, 167–201.

[GeKa2] E. Getzler and M. Kapranov, *Modular operads.* Compositio Math. **110** (1998), no. 1, 65–126.

[GiKa] V. Ginzburg and M. Kapranov, *Koszul duality for operads.* Duke Math. J. **76** (1994), no. 1, 203–272.

[GoMa] A. Goncharov and Yu. Manin, *Multiple zeta-motives and moduli spaces* $\overline{M}_{0,n}$. Compos. Math. **140** (2004), no. 1, 1–14. Preprint math.AG/0204102.

[GrM] S. Grillo and H. Montani, *Twisted internal COHOM objects in the category of quantum spaces.* Preprint math.QA/0112233.

[Hin] V. Hinich, *Homological algebra of homotopy algebras.* Communications in Algebra **25** (1997), no. 10, 3291–3323.

[KaMa] M. Kapranov and Yu. Manin, *Modules and Morita theorem for operads.* Am. J. of Math. **123** (2001), no. 5, 811–838. Preprint math.QA/9906063.

[KoMa] M. Kontsevich and Yu. Manin, *Gromov–Witten classes, quantum cohomology, and enumerative geometry.* Comm. Math. Phys. **164** (1994), no. 3, 525–562.

[KS] G.M. Kelly and R. Street, *Review of the elements of* 2*-categories* in *Category Seminar*, Springer Lecture Notes in Mathematics **420** (1974), 75–103.

[LoMa] A. Losev and Yu. Manin, *Extended modular operad.* In: Frobenius Manifolds, ed. by C. Hertling and M. Marcolli, Vieweg & Sohn Verlag, Wiesbaden, 2004, 181–211. Preprint math.AG/0301003.

[Ma1] Yu. Manin, *Some remarks on Koszul algebras and quantum groups.* Ann. Inst. Fourier **XXXVII** (1987), no. 4, 191–205.

[Ma2] Yu. Manin, *Quantum groups and non-commutative geometry.* Publ. de CRM, Université de Montréal, 1988, 91 pp.

[Ma3] Yu. Manin, *Topics in noncommutative geometry.* Princeton University Press, 1991, 163 pp.

[Ma4] Yu. Manin, *Notes on quantum groups and quantum de Rham complexes.* Teoreticheskaya i Matematicheskaya Fizika **92** (1992), no. 3, 425–450. Reprinted in *Selected papers of Yu.I. Manin, World Scientific, Singapore 1996, 529–554.*

[Mar] M. Markl, *Operads and PROPs.* Preprint math.AT/0601129.

[MarShSt] M. Markl, St. Shnider and J. Stasheff, *Operads in Algebra, Topology and Physics.* Math. Surveys and Monographs, vol. 96, AMS 2002.

[MarkSh] I. Markov and Y. Shi, *Simulating quantum computation by contracting tensor network.* Preprint quant-ph/0511069.

[Mer] S. Merkulov, *PROP profile of deformation quantization and graph complexes with loops and wheels.* Preprint math.QA/0412257.

[PP] A. Polishchuk and L. Positselski, *Quadratic algebras.* University Lecture series, No. 37, AMS 2005.

[Pow] A.J. Power, *A 2-categorical pasting theorem.* Journ. of Algebra **129** (1990), 439–445.

[Q] D. Quillen, *Homotopical Algebra.* Springer Lecture Notes in Mathematics **43**, Berlin, 1967.

[R] C. Rezk, *Spaces of algebra structures and cohomology of operads.* Ph.D. Thesis, Massachusetts Institute of Technology, Cambridge, MA, 1996.

[S] J. Spaliński, *Strong homotopy theory of cyclic sets.* Journal of Pure and Applied Algebra **99** (1995), no. 1, 35–52.

[Val] B. Vallette, *A Koszul duality for PROPs.* Preprint math.AT/0411542 (to appear in the Transactions of the AMS).

[Va2] B. Vallette, *Manin's products, Koszul duality, Loday algebras and Deligne conjecture.* Preprint math.QA/0609002.

[Va3] B. Vallette, *Free monoid in monoidal abelian categories.* Preprint math.CT/0411543.

[Zo] P. Zograf, *Tensor networks and the enumeration of regular subgraphs.* Preprint math.CO/0605256.

Dennis V. Borisov
Department of Mathematics
Northwestern University
Evanston
USA
e-mail: dennis.borisov@gmail.com

Yuri I. Manin
Department of Mathematics
Northwestern University
Evanston
USA
and
Max-Planck-Institut für Mathematik
Bonn
Germany
e-mail: manin@math.northwestern.edu

Progress in Mathematics, Vol. 265, 309–324

Chern Character for Twisted Complexes

Paul Bressler, Alexander Gorokhovsky, Ryszard Nest and Boris Tsygan

In memory of Sasha Reznikov

Abstract. We construct the Chern character from the K-theory of twisted perfect complexes of an algebroid stack to the negative cyclic homology of the algebra of twisted matrices associated to the stack.

Mathematics Subject Classification (2000). 19E99.

Keywords. Chern character, cyclic homology, K-theory, stack.

1. Introduction

The Chern character from the algebraic K theory to the cyclic homology of associative algebras was defined by Connes and Karoubi [C], [Kar], [L]. Goodwillie and Jones [Go], [J] defined the negative cyclic homology and the Chern character with values there. In this paper we generalize this Chern character to the K-theory of twisted modules over twisted sheaves of algebras.

More precisely, we outline the construction of the Chern character of a perfect complex of twisted sheaves of modules over an algebroid stack $\mathcal{A}$ on a space M. This includes the case of a perfect complex of sheaves of modules over a sheaf of algebras $\mathcal{A}$. In the latter case, the recipient of the Chern character is the hypercohomology of M with coefficients in the sheafification of the presheaf of negative cyclic complexes. The construction of the Chern character for this case was given in [BNT1] and [K]. In the twisted case, it is not a priori clear what the recipient should be. One can construct [K2], [MC] the Chern character with values in the negative cyclic homology of the category of perfect complexes (localized by the subcategory of acyclic complexes); the question is, how to compute this cyclic homology, or perhaps how to map it into something simpler.

Ideally, the recipient of the Chern character would be the hypercohomology of M with coefficients in the negative cyclic complex of a sheaf of associative algebras. We show that this is almost the case. We construct associative algebras that form a presheaf not exactly on M but rather on a first barycentric subdivision of the nerve of a cover of M. These algebras are twisted matrix algebras. We used them in [BGNT] and [BGNT1] to classify deformations of algebroid stacks.

We construct the Chern characters

$$K_\bullet(\mathrm{Perf}(\mathcal{A})) \rightarrow \check{\mathbb{H}}^{-\bullet}(M, \mathrm{CC}_\bullet^-(\mathrm{Matr}_{\mathrm{tw}}(\mathcal{A}))) \tag{1.1}$$

$$K_\bullet(\mathrm{Perf}_Z(\mathcal{A})) \rightarrow \check{\mathbb{H}}_Z^{-\bullet}(M, \mathrm{CC}_\bullet^-(\mathrm{Matr}_{\mathrm{tw}}(\mathcal{A}))) \tag{1.2}$$

where $K_\bullet(\mathrm{Perf}(\mathcal{A}))$ is the K-theory of perfect complexes of twisted $\mathcal{A}$-modules, $K_\bullet(\mathrm{Perf}_Z(\mathcal{A}))$ is the K-theory of perfect complexes of twisted $\mathcal{A}$-modules acyclic outside a closed subset Z, and the right-hand sides are the hypercohomology of M with coefficients in the negative cyclic complex of twisted matrices, cf. Definition 3.4.2 .

Our construction of the Chern character is more along the lines of [K] than of [BNT1]. It is modified for the twisted case and for the use of twisted matrices. Another difference is a method that we use to pass from perfect to very strictly perfect complexes. This method involves a general construction of operations on cyclic complexes of algebras and categories. This general construction, in partial cases, was used before in [NT], [NT1] as a version of noncommutative calculus. We recently realized that it can be obtained in large generality by applying the functor $\mathrm{CC}_\bullet^-$ to the categories of A_∞ functors from [BLM], [K1], [Ko], [Lu], [KS], and [Ta].

The fact that these methods are applicable is due to the observation that a perfect complex, via the formalism of twisting cochains of O'Brian, Toledo, and Tong, can be naturally interpreted as an A_∞ functor from the category associated to a cover to the category of strictly perfect complexes. The fourth author is grateful to David Nadler for explaining this to him.

In the case when the stack in question is a gerbe, the recipient of the Chern character maps to the De Rham cohomology twisted by the three-cohomology class determined by this gerbe (the Dixmier-Douady class). A Chern character with values in the twisted cohomology was constructed in [MaS], [BCMMS], [AS] and generalized in [MaS1] and [TX]. The K-theory which is the source of this Chern character is rather different from the one studied here. It is called the twisted K-theory and is a generalization of the topological K-theory. Our Chern character has as its source the algebraic K-theory which probably maps to the topological one.

Acknowledgements

P.B. was supported by the Ellentuck Fund, A.G. was supported by NSF grant DMS-0400342, B.T. was supported by NSF grant DMS-0605030.

2. Gerbes and stacks

2.1.

Let M be a topological space. In this paper, by a stack on M we will mean an equivalence class of the following data:

1. an open cover $M = \cup U_i$;
2. a sheaf of rings $\mathcal{A}_i$ on every U_i;
3. an isomorphism of sheaves of rings $G_{ij} : \mathcal{A}_j|(U_i \cap U_j) \cong \mathcal{A}_i|(U_i \cap U_j)$ for every i, j;
4. an invertible element $c_{ijk} \in \mathcal{A}_i(U_i \cap U_j \cap U_k)$ for every i, j, k satisfying
$$G_{ij}G_{jk} = \mathrm{Ad}(c_{ijk})G_{ik} \tag{2.1}$$
such that, for every i, j, k, l,
$$c_{ijk}c_{ikl} = G_{ij}(c_{jkl})c_{ijl}. \tag{2.2}$$

To define equivalence, first recall the definition of a refinement. An open cover $\mathfrak{V} = \{V_j\}_{j\in J}$ is a refinement of an open cover $\mathfrak{U} = \{U_i\}_{i\in I}$ if a map $f : J \to I$ is given, such that $V_j \subset U_{f(j)}$. Open covers form a category: to say that there is a morphism from $\mathfrak{U}$ to $\mathfrak{V}$ is the same as to say that $\mathfrak{V}$ is a refinement of $\mathfrak{U}$. Composition corresponds to composition of maps f.

Our equivalence relation is by definition the weakest for which the two data $(\{U_i\}, \mathcal{A}_i, G_{ij}, c_{ijk})$ and
$$(\{V_p\}, \mathcal{A}_{f(p)}|V_p, G_{f(p)f(q)}, c_{f(p)f(q)f(r)})$$
are equivalent whenever $\{V_p\}$ is a refinement of $\{U_i\}$ (the corresponding map $\{p\} \to \{i\}$ being denoted by f).

If two data $(\{U_i'\}, \mathcal{A}_i', G_{ij}', c_{ijk}')$ and $(\{U_i''\}, \mathcal{A}_i'', G_{ij}'', c_{ijk}'')$ are given on M, define an isomorphism between them as follows. First, choose an open cover $M = \cup U_i$ refining both $\{U_i'\}$ and $\{U_i''\}$. Pass from our data to new, equivalent data corresponding to this open cover. An isomorphism is an equivalence class of a collection of isomorphisms $H_i : \mathcal{A}_i' \cong \mathcal{A}_i''$ on U_i and invertible elements b_{ij} of $\mathcal{A}_i'(U_i \cap U_j)$ such that
$$G_{ij}'' = H_i \,\mathrm{Ad}(b_{ij})G_{ij}'H_j^{-1} \tag{2.3}$$
and
$$H_i^{-1}(c_{ijk}'') = b_{ij}G_{ij}'(b_{jk})c_{ijk}'b_{ik}^{-1}. \tag{2.4}$$

If $\{V_p\}$ is a refinement of $\{U_i\}$, we pass from $(\{U_i\}, \mathcal{A}_i, G_{ij}, c_{ijk})$ to the equivalent data $(\{V_p\}, A_{f(p)f(q)}, c_{f(p)f(q)f(r)})$ as above. We define the equivalence relation to be the weakest for which, for all refinements, the data (H_i, b_{ij}) and $(H_{f(p)}, b_{f(p)f(q)})$ are equivalent.

Define composition of isomorphisms as follows. Choose a common refinement $\{U_i\}$ of the covers $\{U_i'\}$, $\{U_i''\}$, and $\{U_i'''\}$. Using the equivalence relation, identify all the stack data and all the isomorphism data with the data corresponding to the cover $\{U_i\}$. Define $H_i = H_i' \circ H_i''$ and $b_{ij} = H_i''^{-1}(b_{ij}')b_{ij}''$. It is easy to see that this composition is associative and is well defined for equivalence classes.

Now consider two isomorphisms (H'_i, b'_{ij}) and (H''_i, b''_{ij}) between the stacks $(\{U'_i\}, \mathcal{A}'_i, G'_{ij}, c'_{ijk})$ and $(\{U''_i\}, \mathcal{A}''_i, G''_{ij}, c''_{ijk})$. We can pass to a common refinement, replace our data by equivalent data, and assume that $\{U'_i\} = \{U''_i\} = \{U_i\}$. A two-morphism between the above isomorphisms is an equivalence class of a collection of invertible elements a_i of $\mathcal{A}'_i(U_i)$ such that $H''_i = H'_i \circ \mathrm{Ad}(a_i)$ and $b''_{ij} = a_i^{-1} b'_{ij} G'_{ij}(a_j)$. The equivalence relation is the weakest for which, whenever $\{V_p\}$ is a refinement of $\{U_i\}$, $\{a_i\}$ is equivalent to $\{a_{f(p)}\} : (H'_{f(p)}, b'_{f(p)f(q)}) \to (H''_{f(p)}, b''_{f(p)f(q)})$. The composition between $\{a'_i\}$ and $\{a''_i\}$ is defined by $a_i = a'_i a''_i$. This operation is well-defined at the level of equivalence classes.

With the operations thus defined, stacks form a two-groupoid.

A gerbe on a manifold M is a stack for which $\mathcal{A}_i = \mathcal{O}_{U_i}$ and $G_{ij} = 1$. Gerbes are classified up to isomorphism by cohomology classes in $H^2(M, \mathcal{O}^*_M)$.

For a stack $\mathcal{A}$ define *a twisted $\mathcal{A}$-module* over an open subset U as an equivalence class of a collection of sheaves of $\mathcal{A}_i$ -modules $\mathcal{M}_i$ on $U \cap U_i$, together with isomorphisms $g_{ij} : \mathcal{M}_j \to \mathcal{M}_i$ on $U \cap U_i \cap U_j$ such that $g_{ik} = g_{ij} G_{ij}(g_{jk}) c_{ijk}$ on $U \cap U_i \cap U_j \cap U_k$. The equivalence relation is the weakest for which, if $\{V_p\}$ is a refinement of $\{U_i\}$, the data $(\mathcal{M}_{f(p)}, g_{f(p)f(q)})$ and $(\mathcal{M}_i, g_{ij})$ are equivalent.

We leave it to the reader to define morphisms of twisted modules. A twisted module is said to be *free* if the $\mathcal{A}_i$-module $\mathcal{M}_i$ is.

2.2. Twisting cochains

Here we recall the formalism from [TT], [OTT], [OB], generalized to the case of stacks. For a stack on $M = \cup U_i$ as above, by $\mathcal{F}$ we will denote a collection $\{\mathcal{F}_i\}$ where $\mathcal{F}_i$ is a graded sheaf which is a direct summand of a free graded $\mathcal{A}_i$-module of finite rank on U_i. A p-cochain with values in $\mathcal{F}$ is a collection $a_{i_0 \ldots i_p} \in \mathcal{F}_{i_0}(U_{i_0} \cap \ldots \cap U_{i_p})$; for two collections $\mathcal{F}$ and $\mathcal{F}'$ as above, a p-cochain with values in $\mathrm{Hom}(\mathcal{F}, \mathcal{F}')$ is a collection $a_{i_0 \ldots i_p} \in \mathrm{Hom}_{\mathcal{A}_{i_0}}(\mathcal{F}_{i_p}, \mathcal{F}'_{i_0})(U_{i_0} \cap \ldots \cap U_{i_p})$ (the sheaf $\mathcal{A}_{i_0}$ acts on $\mathcal{F}_{i_p}$ via $G_{i_0 i_p}$). Define the cup product by

$$(a \smile b)_{i_0 \ldots i_{p+q}} = (-1)^{|a_{i_0 \ldots i_p}| q} a_{i_0 \ldots i_p} G_{i_p i_{p+q}}(b_{i_{p+1} \ldots i_{p+q}}) c_{i_0 i_p i_{p+q}} \tag{2.5}$$

and the differential by

$$(\check{\partial} a)_{i_0 \ldots i_{p+1}} = \sum_{k=1}^{p} (-1)^k a_{i_0 \ldots \widehat{i_k} \ldots i_{p+1}}. \tag{2.6}$$

Under these operations, $\mathrm{Hom}(\mathcal{F}, \mathcal{F})$-valued cochains form a DG algebra and $\mathcal{F}$-valued cochains a DG module.

If $\mathfrak{V}$ is a refinement of $\mathfrak{U}$ then cochains with respect to $\mathfrak{U}$ map to cochains with respect to $\mathfrak{V}$. For us, the space of cochains will be always understood as the direct limit over all the covers.

A twisting cochain is a $\mathrm{Hom}(\mathcal{F}, \mathcal{F})$-valued cochain ρ of total degree one such that

$$\check{\partial} \rho + \frac{1}{2} \rho \smile \rho = 0. \tag{2.7}$$

A morphism between twisting cochains ρ and ρ' is a cochain f of total degree zero such that $\check{\partial} f + \rho' \smile f - f \smile \rho = 0$. A homotopy between two such morphisms f and f' is a cochain θ of total degree -1 such that $f - f' = \check{\partial}\theta + \rho' \smile \theta + \theta \smile \rho$. More generally, twisting cochains form a DG category. The complex $\mathrm{Hom}(\rho, \rho')$ is the complex of $\mathrm{Hom}(\mathcal{F}, \mathcal{F}')$-valued cochains with the differential

$$f \mapsto \check{\partial} f + \rho' \smile f - (-1)^{|f|} f \smile \rho \, .$$

There is another, equivalent definition of twisting cochains. Start with a collection $\mathcal{F} = \{\mathcal{F}_i\}$ of direct summands of free graded twisted modules of finite rank on U_i (a twisted module on U_i is said to be free if the corresponding $\mathcal{A}_i$-module is). Define $\mathrm{Hom}(\mathcal{F}, \mathcal{F}')$-valued cochains as collections of morphisms of graded twisted modules $a_{i_0 \ldots i_p} : \mathcal{F}_{i_p} \to \mathcal{F}'_{i_0}$ on $U_{i_0} \cap \ldots \cap U_{i_p}$. The cup product is defined by

$$(a \smile b)_{i_0 \ldots i_{p+q}} = (-1)^{|a_{i_0 \ldots i_p}| q} a_{i_0 \ldots i_p} b_{i_{p+1} \ldots i_{p+q}} \tag{2.8}$$

and the differential by (2.6) . A twisting cochain is a cochain ρ of total degree 1 satisfying (2.7).

If one drops the requirement that the complexes $\mathcal{F}$ be direct summands of graded free modules of finite rank, we get objects that we will call *weak twisting cochains.* A morphism of (weak) twisting cochains is *a quasi-isomorphism* if f_i is for every i. Every complex $\mathcal{M}$ of twisted modules can be viewed as a weak twisting cochain, with $\mathcal{F}_i = \mathcal{M}$ for all i, $\rho_{ij} = \mathrm{id}$ for all i, j, ρ_i is the differential in $\mathcal{M}$, and $\rho_{i_0 \ldots i_p} = 0$ for $p > 2$. We denote this weak twisting cochain by $\rho_0(\mathcal{M})$. By $\boldsymbol{\rho}_0$ we denote the DG functor $\mathcal{M} \mapsto \rho_0(\mathcal{M})$ from the DG category of perfect complexes to the DG category of weak twisting cochains.

If $\{V_s\}$ is a refinement of $\{U_i\}$, we declare twisting cochains $(\mathcal{F}_i, \rho_{i_0 \ldots i_p})$ and $(\mathcal{F}_{f(s)}|V_s, \rho_{f(s_0) \ldots f(s_p)})$ equivalent. Similarly for morphisms.

A complex of twisted modules is called *perfect* (resp. *strictly perfect)* if it is locally isomorphic in the derived category (resp. isomorphic) to a direct summand of a bounded complex of finitely generated free modules. A parallel definition can of course be given for complexes of modules over associative algebras.

Lemma 2.2.1. *Let M be paracompact.*

1. *For a perfect complex $\mathcal{M}$ there exists a twisting cochain ρ together with a quasi-isomorphism of weak twisting cochains $\rho \xrightarrow{\phi} \rho_0(\mathcal{M})$.*
2. *Let $f : \mathcal{M}_1 \to \mathcal{M}_2$ be a morphism of perfect complexes. Let ρ_i, ϕ_i be twisting cochains corresponding to $\mathcal{M}_i$, $i = 1, 2$. Then there is a morphism of twisting cochains $\varphi(f)$ such that $\phi_2 \varphi(f)$ is homotopic to $f \phi_1$.*
3. *More generally, each choice $\mathcal{M} \mapsto \rho(\mathcal{M})$ extends to an A_∞ functor $\boldsymbol{\rho}$ from the DG category of perfect complexes to the DG category of twisting cochains, together with an A_∞ quasi-isomorphism $\boldsymbol{\rho} \to \boldsymbol{\rho}_0$. (We recall the definition of A_∞ functors in 3.1, and that of A_∞ morphisms of A_∞ functors in 3.2).*

Sketch of the proof. We will use the following facts about complexes of modules over associative algebras.

1) If a complex $\mathcal{F}$ is strictly perfect, for a quasi-isomorphism $\psi : \mathcal{M} \to \mathcal{F}$ there is a quasi-isomorphism $\phi : \mathcal{F} \to \mathcal{M}$ such that $\psi \circ \phi$ is homotopic to the identity.
2) If $f : \mathcal{M}_1 \to \mathcal{M}_2$ is a morphism of perfect complexes and $\phi_i : \mathcal{F}_i \to \mathcal{M}_i$, $i = 1, 2$, are quasi-isomorphisms with $\mathcal{F}_i$ strictly perfect, then there is a morphism $\varphi(f) : \mathcal{F}_1 \to \mathcal{F}_2$ such that $\phi_2\varphi$ is homotopic to $f\phi_1$.
3) If $\mathcal{F}$ is strictly perfect and $\phi : \mathcal{F} \to \mathcal{M}$ is a morphism which is zero on cohomology, then ϕ is homotopic to zero.

Let $\mathcal{M}$ be a perfect complex of twisted modules. Recall that, by our definition, locally, there is a chain of quasi-isomorphisms connecting it to a strictly perfect complex $\mathcal{F}$. Let us start by observing that one can replace that by a quasi-isomorphism from $\mathcal{F}$ to $\mathcal{M}$. In other words, locally, there is a strictly perfect complex $\mathcal{F}$ and a quasi-isomorphism $\phi : \mathcal{F} \to \mathcal{M}$. Indeed, this is true at the level of germs at every point, by virtue of 1) above. For any point, the images of generators of $\mathcal{F}$ under morphisms ϕ, resp. under homotopies s, are germs of sections of $\mathcal{M}$, resp. of $\mathcal{F}$, which are defined on some common neighborhood. Therefore quasi-isomorphisms and homotopies are themselves defined on these neighborhoods.

We get a cover $\{U_i\}$, strictly perfect complexes $\mathcal{F}_i$ with differentials ρ_i, and quasi-isomorphisms $\phi_i : \mathcal{F}_i \to \mathcal{M}$ on U_i. Now observe that, at any point of U_{ij}, the morphisms ρ_{ij} can be constructed at the level of germs because of 2). As above, we conclude that each of them can be constructed on some neighborhood of this point. Replace the cover $\{U_i\}$ by a locally finite refinement$\{U_i'\}$. Then, for every point x, find a neighborhood V_x on which all ρ_{ij} can be constructed. Cover M by such neighborhoods. Then pass to a new cover which is a common refinement of $\{U_i'\}$ and $\{V_x\}$. For this cover, the component ρ_{ij} can be defined.

Acting as above, using 2) and 3), one can construct all the components of the twisting cochain $\rho(\mathcal{M})$, of the A_∞ functor $\boldsymbol{\rho}$, and of the A_∞ morphism of A_∞ functors $\boldsymbol{\rho} \to \boldsymbol{\rho}_0$. □

Remark 2.2.2. One can assume that all components of a twisting cochain ρ lie in the space of cochains with respect to one and the same cover if the following convention is adopted: all our perfect complexes are locally quasi-isomorphic to strictly perfect complexes as *complexes of presheaves*. In other words, there is an open cover $\{U_i\}$ together with a strictly perfect complex $\mathcal{F}_i$ and a morphism $\phi_i : \mathcal{F}_i \to \mathcal{M}$ on any U_i, such that ϕ_i is a quasi-isomorphism at the level of sections on any open subset of U_i.

2.3. Twisted matrix algebras

For any p-simplex σ of the nerve of an open cover $M = \cup U_i$ which corresponds to $U_{i_0} \cap \ldots \cap U_{i_p}$, put $I_\sigma = \{i_0, \ldots, i_p\}$ and $U_\sigma = \cap_{i \in I_\sigma} U_i$. Define the algebra $\mathrm{Matr}^\sigma_{tw}(\mathcal{A})$ whose elements are finite matrices

$$\sum_{i,j \in I_\sigma} a_{ij} E_{ij}$$

such that $a_{ij} \in (\mathcal{A}_i(U_\sigma))$. The product is defined by

$$a_{ij}E_{ij} \cdot a_{lk}E_{lk} = \delta_{jl}a_{ij}G_{ij}(a_{jk})c_{ijk}E_{ik}.$$

For $\sigma \subset \tau$, the inclusion

$$i_{\sigma\tau} : \mathrm{Matr}^{\sigma}_{tw}(\mathcal{A}) \to \mathrm{Matr}^{\tau}_{tw}(\mathcal{A}),$$

$\sum a_{ij}E_{ij} \mapsto \sum(a_{ij}|U_\tau)E_{ij}$, is a morphism of algebras (not of algebras with unit). Clearly, $i_{\tau\rho}i_{\sigma\tau} = i_{\sigma\rho}$. If $\mathfrak{V}$ is a refinement of $\mathfrak{U}$ then there is a map

$$\mathrm{Matr}^{\sigma}_{tw}(\mathcal{A}) \to \mathrm{Matr}^{f(\sigma)}_{tw}(\mathcal{A})$$

which sends $\sum a_{ij}E_{ij}$ to $\sum(a_{f(i)f(j)}|V_{f(\sigma)})E_{f(i)f(j)}$.

Remark 2.3.1. For a nondecreasing map $f : I_\sigma \to I_\tau$ which is not necessarily an inclusion, we have the bimodule M_f consisting of twisted $|I_\sigma| \times |I_\tau|$ matrices. Tensoring by this bimodule defines the functor

$$f_* : \mathrm{Matr}^{\sigma}_{tw}(\mathcal{A}) - \mathrm{mod} \to \mathrm{Matr}^{\tau}_{tw}(\mathcal{A}) - \mathrm{mod}$$

such that $(fg)_* = f_*g_*$.

3. The Chern character

3.1. Hochschild and cyclic complexes

We start by recalling some facts and constructions from noncommutative geometry. Let A be an associative unital algebra over a unital algebra k. Set

$$C_p(A, A) = C_p(A) = A^{\otimes(p+1)}.$$

We denote by $b : C_p(A) \to C_{p-1}(A)$ and $B : C_p(A) \to C_{p+1}(A)$ the standard differentials from the Hochschild and cyclic homology theory (cf. [C], [L], [T]). The Hochschild chain complex is by definition $(C_\bullet(A), b)$; define

$$CC^-_\bullet(A) = (C_\bullet(A)[[u]], b + uB);$$

$$CC^{\mathrm{per}}_\bullet(A) = (C_\bullet(A)[[u, u^{-1}], b + uB);$$

$$CC_\bullet(A) = (C_\bullet(A)[[u, u^{-1}]/uC_\bullet(A)[[u]], b + uB).$$

These are, respectively, *the negative cyclic*, *the periodic cyclic*, and *the cyclic* complexes of A over k.

We can replace A by a small DG category or, more generally, by a small A_∞ category. Recall that a small A_∞ category consists of a set $\mathrm{Ob}(\mathcal{C})$ of objects and a graded k-module of $\mathcal{C}(i, j)$ of morphisms for any two objects i and j, together with compositions

$$m_n : \mathcal{C}(i_n, i_{n-1}) \otimes \ldots \otimes \mathcal{C}(i_1, i_0) \to \mathcal{C}(i_n, i_0)$$

of degree $2 - n$, $n \geq 1$, satisfying standard quadratic relations to which we refer as the A_∞ relations. In particular, m_1 is a differential on $\mathcal{C}(i, j)$. An A_∞ functor F

between two small A_∞ categories $\mathcal{C}$ and $\mathcal{D}$ consists of a map $F : \mathrm{Ob}(\mathcal{C}) \to \mathrm{Ob}(\mathcal{D})$ and k-linear maps

$$F_n : \mathcal{C}(i_n, i_{n-1}) \otimes \ldots \otimes \mathcal{C}(i_1, i_0) \to \mathcal{D}(Fi_n, Fi_0)$$

of degree $1 - n$, $n \geq 1$, satisfying another standard relation. We refer the reader to [K1] for formulas and their explanations.

For a small A_∞ category $\mathcal{C}$ one defines the Hochschild complex $C_\bullet(\mathcal{C})$ as follows:

$$C_\bullet(\mathcal{C}) = \bigoplus_{i_0,\ldots,i_n \in \mathrm{Ob}(\mathcal{C})} \mathcal{C}(i_1, i_0) \otimes \mathcal{C}(i_2, i_1) \otimes \ldots \otimes \mathcal{C}(i_n, i_{n-1}) \otimes \mathcal{C}(i_0, i_n)$$

(the total cohomological degree being the degree induced from the grading of $\mathcal{C}(i, j)$ minus n). The differential b is defined by

$$b(f_0 \otimes \ldots f_n) = \sum_{j,k} \pm m_k(f_{n-j+1}, \ldots, f_0, \ldots, f_{k-1-j}) \otimes f_{k-j} \otimes \ldots \otimes f_{n-j}$$
$$+ \sum_{j,k} \pm f_0 \otimes \ldots \otimes f_j \otimes m_k(f_{j+1}, \ldots, f_{j+k}) \otimes \ldots \otimes f_n.$$

The cyclic differential B is defined by the standard formula with appropriate signs; cf. [G].

3.2. Categories of A_∞ functors

For two DG categories $\mathcal{C}$ and $\mathcal{D}$ one can define the DG category $\mathrm{Fun}_\infty(\mathcal{C}, \mathcal{D})$. Objects of $\mathrm{Fun}_\infty(\mathcal{C}, \mathcal{D})$ are A_∞ functors $\mathcal{C} \to \mathcal{D}$. The complex $\mathrm{Fun}_\infty(\mathcal{C}, \mathcal{D})(F, G)$ of morphisms from F to G is the Hochschild cochain complex of $\mathcal{C}$ with coefficients in $\mathcal{D}$ viewed as an A_∞ bimodule over $\mathcal{C}$ via the A_∞ functors F and G, namely

$$\prod_{i_0,\ldots,i_n \in \mathrm{Ob}(\mathcal{C})} \mathrm{Hom}(\mathcal{C}(i_0, i_1) \otimes \ldots \otimes \mathcal{C}(i_{n-1}, i_n), \mathcal{D}(Fi_0, Gi_n)).$$

The DG category structure on $\mathrm{Fun}_\infty(\mathcal{C}, \mathcal{D})$ comes from the cup product. More generally, for two A_∞ categories $\mathcal{C}$ and $\mathcal{D}$, $\mathrm{Fun}_\infty(\mathcal{C}, \mathcal{D})$ is an A_∞ category. For a conceptual explanation, as well as explicit formulas for the differential and composition, cf. [Lu], [BLM], [K1], [KS].

Furthermore, for DG categories $\mathcal{C}$ and $\mathcal{D}$ there are A_∞ morphisms

$$\mathcal{C} \otimes \mathrm{Fun}_\infty(\mathcal{C}, \mathcal{D}) \to \mathcal{D} \tag{3.1}$$

(the action) and

$$\mathrm{Fun}_\infty(\mathcal{D}, \mathcal{E}) \otimes \mathrm{Fun}_\infty(\mathcal{C}, \mathcal{D}) \to \mathrm{Fun}_\infty(\mathcal{C}, \mathcal{E}) \tag{3.2}$$

(the composition). This follows from the conceptual explanation cited below; in fact these pairing were considered already in [Ko]. As a consequence, there are pairings

$$\mathrm{CC}^-_\bullet(\mathcal{C}) \otimes \mathrm{CC}^-_\bullet(\mathrm{Fun}_\infty(\mathcal{C}, \mathcal{D})) \to \mathrm{CC}^-_\bullet(\mathcal{D}) \tag{3.3}$$

and

$$\mathrm{CC}^-_\bullet(\mathrm{Fun}_\infty(\mathcal{D}, \mathcal{E})) \otimes \mathrm{CC}^-_\bullet(\mathrm{Fun}_\infty(\mathcal{C}, \mathcal{D})) \to \mathrm{CC}^-_\bullet(\mathrm{Fun}_\infty(\mathcal{C}, \mathcal{E})). \tag{3.4}$$

Indeed, recall the Getzler-Jones products [GJ]

$$\mathrm{CC}^-_\bullet(\mathcal{C}_1)\otimes\ldots\otimes\mathrm{CC}^-_\bullet(\mathcal{C}_n)\to\mathrm{CC}^-_\bullet(\mathcal{C}_1\otimes\ldots\mathcal{C}_n)[2-n]$$

which satisfy the usual A_∞ identities. To get (3.3) and (3.4), one combines these products with (3.1) and (3.2).

Example 3.2.1. Let F be an A_∞ functor from $\mathcal{C}$ to $\mathcal{D}$. Then id_F is a chain of $\mathrm{CC}^-(\mathrm{Fun}_\infty(\mathcal{C},\mathcal{D}))$ (with $n=0$). The pairing (3.3) with this chain amounts to the map of the negative cyclic complexes induced by the A_∞ functor F:

$$f_0\otimes\ldots\otimes f_n\mapsto\sum\pm F_{k_0}(\ldots f_0\ldots)\otimes F_{k_1}(\ldots)\otimes\ldots\otimes F_{k_m}(\ldots).$$

The sum is taken over all cyclic permutations of $f_0,\ldots,f_n$ and all m, k_0, $\ldots$, k_m such that f_0 is inside F_{k_0}.

Remark 3.2.2. The action (3.1) and the composition (3.1) are parts of a very nontrivial structure that was studied in [Ta].

As a consequence, this gives an A_∞ category structure $\mathrm{CC}^-(\mathrm{Fun}_\infty)$ whose objects are A_∞ categories and whose complexes of morphisms are negative cyclic complexes $\mathrm{CC}^-_\bullet(\mathrm{Fun}_\infty(\mathcal{D},\mathcal{E}))$.

From a less conceptual point of view, pairings (3.3) and (3.4) were defined, in partial cases, in [NT1] and [NT]. The A_∞ structure on $\mathrm{CC}^-(\mathrm{Fun}_\infty)$ was constructed (in the partial case when all f are identity functors) in [TT]. Cf. also [T1] for detailed proofs.

3.3. The prefibered version

We need the following modification of the above constructions. Let $\mathcal{B}$ be a category. Consider, instead of a single DG category $\mathcal{D}$, a family of DG categories $\mathcal{D}_i$, $i\in\mathrm{Ob}(\mathcal{B})$, together with a family of DG functors $f^*:\mathcal{D}_i\leftarrow\mathcal{D}_j$, $f\in\mathcal{B}(i,j)$, satisfying $(fg)^*=g^*f^*$ for any f and g. In this case we define a new DG category $\mathcal{D}$:

$$\mathrm{Ob}(\mathcal{D})=\coprod_{i\in\mathrm{Ob}(\mathcal{B})}\mathrm{Ob}(\mathcal{D}_i)$$

and, for $a\in\mathrm{Ob}(\mathcal{D}_i)$, $b\in\mathrm{Ob}(\mathcal{D}_j)$,

$$\mathcal{D}(a,b)=\oplus_{f\in\mathcal{B}(i,j)}\mathcal{D}_i(a,f^*b).$$

The composition is defined by

$$(\varphi,f)\circ(\psi,g)=(\varphi\circ f^*\psi,f\circ g)$$

for $\varphi\in\mathcal{D}_i(a,f^*b)$ and $\psi\in\mathcal{D}_j(b,g^*c)$.

We call the DG category $\mathcal{D}$ *a DG category over* $\mathcal{B}$, or, using the language of [Gil], *a prefibered DG category over* $\mathcal{B}$ *with a strict cleavage.* There is a similar construction for A_∞ categories.

Let $\mathcal{C}$, $\mathcal{D}$ be two DG categories over $\mathcal{B}$. An A_∞ functor $F : \mathcal{C} \to \mathcal{D}$ is called *an A_∞ functor over $\mathcal{B}$* if for any $a \in \mathrm{Ob}(\mathcal{C}_i)$ $Fa \in \mathrm{Ob}(\mathcal{D}_i)$, and for any $a_k \in \mathrm{Ob}(\mathcal{C}_{i_k})$, $(\varphi_k, f_k) \in \mathcal{C}(a_k, a_{k-1})$, $k = 1, \dots, n$,

$$F_n((\varphi_n, f_n), \dots, (\varphi_1, f_1)) = (\psi, f_1 \dots f_n)$$

for some $\psi \in \mathcal{D}_{i_n}$. One defines a morphism over $\mathcal{B}$ of two A_∞ functors over $\mathcal{B}$ by imposing a restriction which is identical to the one above. We get a DG category $\mathrm{Fun}_\infty^{\mathcal{B}}(\mathcal{C}, \mathcal{D})$. As in the previous section, there are A_∞ functors

$$\mathcal{C} \otimes \mathrm{Fun}_\infty^{\mathcal{B}}(\mathcal{C}, \mathcal{D}) \to \mathcal{D} \tag{3.5}$$

(the action) and

$$\mathrm{Fun}_\infty^{\mathcal{B}}(\mathcal{D}, \mathcal{E}) \otimes \mathrm{Fun}_\infty^{\mathcal{B}}(\mathcal{C}, \mathcal{D}) \to \mathrm{Fun}_\infty^{\mathcal{B}}(\mathcal{C}, \mathcal{E}) \tag{3.6}$$

(the composition), as well as

$$\mathrm{CC}_\bullet^-(\mathcal{C}) \otimes \mathrm{CC}_\bullet^-(\mathrm{Fun}_\infty^{\mathcal{B}}(\mathcal{C}, \mathcal{D})) \to \mathrm{CC}_\bullet^-(\mathcal{D}) \tag{3.7}$$

and

$$\mathrm{CC}_\bullet^-(\mathrm{Fun}_\infty^{\mathcal{B}}(\mathcal{D}, \mathcal{E})) \otimes \mathrm{CC}_\bullet^-(\mathrm{Fun}_\infty^{\mathcal{B}}(\mathcal{C}, \mathcal{D})) \to \mathrm{CC}_\bullet^-(\mathrm{Fun}_\infty^{\mathcal{B}}(\mathcal{C}, \mathcal{E})). \tag{3.8}$$

3.3.1. We need one more generalization of the above constructions. It is not necessary if one adopts the convention from Remark 2.2.2.

Suppose that instead of $\mathcal{B}$ we have a diagram of categories indexed by a category $\mathbf{U}$ (in other words, a functor from $\mathbf{U}$ to the category of categories. In our applications, $\mathbf{U}$ will be the category of open covers). Instead of a $\mathcal{B}$-category $\mathcal{D}$ we will consider a family of $\mathcal{B}_u$-categories $\mathcal{D}_u$, $u \in \mathrm{Ob}(\mathbf{U})$, together with a functor $\mathcal{D}_v \to \mathcal{D}_u$ for any morphism $u \to v$ in $\mathbf{U}$, subject to compatibility conditions that are left to the reader. The inverse limit of categories $\varprojlim_{\mathbf{U}} \mathcal{D}_u$ is then a category over the inverse limit $\varprojlim_{\mathbf{U}} \mathcal{B}_u$. We may proceed exactly as above and define the DG category of A_∞ functors over $\varprojlim \mathcal{B}_u$ from $\varprojlim \mathcal{D}_u$ to $\varprojlim \mathcal{E}_u$, etc., with the following convention: the space of maps from the inverse product, or from the tensor product of inverse products, is defined to be the inductive limit of spaces of maps from (tensor products of) individual constituents.

In this new situation, the pairings (3.6) and (3.8) still exist, while (3.7) turns into

$$\mathrm{CC}_\bullet^-(\mathrm{Fun}_\infty^{\mathcal{B}}(\mathcal{C}, \mathcal{D})) \to \varinjlim \underline{Hom}(\mathrm{CC}_\bullet^-(\mathcal{C}_u), \varprojlim \mathrm{CC}_\bullet^-(\mathcal{D}_v)). \tag{3.9}$$

3.4. The trace map for stacks

Let M be a space with a stack $\mathcal{A}$. Consider an open cover $\mathfrak{U} = \{U_i\}_{i \in I}$ such that the stack $\mathcal{A}$ can be represented by a datum $\mathcal{A}_i, G_{ij}, c_{ijk}$. Let $\mathcal{B}_\mathfrak{U}$ be the category whose set of objects is I and where for every two objects i and j there is exactly one morphism $f : i \to j$. Put $\mathcal{C}_\mathfrak{U} = k[\mathcal{B}_\mathfrak{U}]$, i.e., $(\mathcal{C}_\mathfrak{U})_i = k$ for any object i of $\mathcal{B}_\mathfrak{U}$.

There is a standard isomorphism of the stack $\mathcal{A}|U_i$ with the trivial stack associated to the sheaf of rings $\mathcal{A}_i$. Therefore one can identify twisted modules on

U_i with sheaves of A_i-modules. We will denote the twisted module corresponding to the free module $\mathcal{A}_i$ by the same letter $\mathcal{A}_i$.

Definition 3.4.1. *Define the category of* very strictly perfect complexes *on any open subset of U_i as follows. Its objects are pairs (e, d) where e is an idempotent endomorphism of degree zero of a free graded module $\sum_{a=1}^{N} \mathcal{A}_i[n_a]$ and d is a differential on* Im(e). *Morphisms between (e_1, d_1) and (e_2, d_2) are the same as morphisms between* Im(e_1) *and* Im(e_2) *in the DG category of complexes of modules. A parallel definition can be given for the category of complexes of modules over an associative algebra.*

Let $(\mathcal{D}_{\mathfrak{U}})_i$ be the category of very strictly perfect complexes of twisted $\mathcal{A}$-modules on U_i. By $\mathbf{U}$ we denote the category of open covers as above.

Strictly speaking, our situation is not exactly a partial case of what was considered in 3.3. First, $(\mathcal{D}_{\mathfrak{U}})_i$ is a presheaf of categories on U_i (in the most naive sense, i.e., it consists of a category $(\mathcal{D}_{\mathfrak{U}})_i(U)$ for any U open in U_i, and a functor $G_{UV} : (\mathcal{D}_{\mathfrak{U}})_i(V) \to (\mathcal{D}_{\mathfrak{U}})_i(U)$ for any $U \subset V$, such that $G_{UV}G_{VW} = G_{UW}$). Second, f^* are defined as functors on the subset $U_i \cap U_j$. Also, the pairing (3.7) and its generalization (3.9) are defined in a slightly restricted sense: they put in correspondence to a cyclic chain $i_0 \to i_n \to i_{n-1} \to \dots \to i_0$ a cyclic chain of the category of very strictly perfect complexes of $\mathcal{A}$-modules on $U_{i_0} \cap \dots \cap U_{i_n}$. Finally, in the notation of 3.3.1, for a morphism $f : \mathfrak{U} \to \mathfrak{V}$ in $\mathbf{U}$ and an object j of $I_{\mathfrak{V}}$, the functor $(\mathcal{D}_{\mathfrak{V}})_j \to (\mathcal{D}_{\mathfrak{U}})_{f(j)}$ induced by f is defined only on the open subset V_j.

We put $\mathcal{B} = \varprojlim \mathcal{B}_{\mathfrak{U}}$ and $\mathcal{D} = \varprojlim \mathcal{D}_{\mathfrak{U}}$.

Let Perf$(\mathcal{A})$ be the DG category of perfect complexes of twisted $\mathcal{A}$-modules on M. We denote the sheaf of categories of very strictly perfect complexes on M by $\mathrm{Perf}^{\mathrm{vstr}}(\mathcal{A})$. If Z is a closed subset of M then by $\mathrm{Perf}_Z(\mathcal{A})$ we denote the DG category of perfect complexes of twisted $\mathcal{A}$-modules on M which are acyclic outside Z.

Definition 3.4.2. *Define*

$$\check{C}^{-\bullet}(M, \mathrm{CC}^-_\bullet(\mathrm{Matr}_{\mathrm{tw}}(\mathcal{A}))) = \varinjlim_{\mathfrak{U}} \prod_{\sigma_0 \subset \sigma_1 \subset \dots \subset \sigma_p} \mathrm{CC}^-_\bullet(\mathrm{Matr}^{\sigma_p}_{\mathrm{tw}}(\mathcal{A}))$$

where σ_i run through simplices of $\mathfrak{U}$. The total differential is $b + uB + \check{\partial}$ where

$$\check{\partial} s_{\sigma_0 \dots \sigma_p} = \sum_{k=0}^{p-1} (-1)^k s_{\sigma_0 \dots \widehat{\sigma_k} \dots \sigma_p} + (-1)^p s_{\sigma_0 \dots \sigma_{p-1}} |U_{\sigma_p}.$$

For a closed subset Z of M define $\check{C}^{-\bullet}_Z(M, \mathrm{CC}^-_\bullet(\mathrm{Matr}_{\mathrm{tw}}(\mathcal{A})))$ as

$$\mathrm{Cone}(\check{C}^{-\bullet}(M, \mathrm{CC}^-_\bullet(\mathrm{Matr}_{\mathrm{tw}}(\mathcal{A}))) \to \check{C}^{-\bullet}(M \setminus Z, \mathrm{CC}^-_\bullet(\mathrm{Matr}_{\mathrm{tw}}(\mathcal{A}))))[-1] .$$

Let us construct natural morphisms

$$\mathrm{CC}^-_\bullet(\mathrm{Perf}(\mathcal{A})) \to \check{C}^{-\bullet}(M, \mathrm{CC}^-_\bullet(\mathrm{Matr}_{\mathrm{tw}}(\mathcal{A}))) \tag{3.10}$$

$$\mathrm{CC}^-_\bullet(\mathrm{Perf}_Z(\mathcal{A})) \to \check{C}^{-\bullet}_Z(M, \mathrm{CC}^-_\bullet(\mathrm{Matr}_{\mathrm{tw}}(\mathcal{A}))). \tag{3.11}$$

First, observe that the definition of a twisted cochain and Lemma 2.2.1 can be reformulated as follows.

Lemma 3.4.3.

1. *A twisting cochain is an* A_∞ *functor* $\mathcal{C} \to \mathcal{D}$ *over* $\mathcal{B}$ *in the sense of* 3.3.
2. *There is an* A_∞ *functor from the DG category of perfect complexes to the DG category* $\mathrm{Fun}_\infty(\mathcal{C}, \mathcal{D})$.

The second part of the lemma together with (3.7) give morphisms

$$\mathrm{CC}^-_\bullet(\mathrm{Perf}(\mathcal{A})) \to \mathrm{CC}^-_\bullet(\mathrm{Fun}^{\mathcal{B}}_\infty(\mathcal{C}, \mathcal{D})) \to \underline{\mathrm{Hom}}(\mathrm{CC}^-_\bullet(\mathcal{C}), \mathrm{CC}^-_\bullet(\mathcal{D}))\ .$$

As mentioned above, the image of this map is the subcomplex of those morphisms that put in correspondence to a cyclic chain $i_0 \to i_n \to i_{n-1} \to \ldots \to i_0$ a cyclic chain of the category of very strictly perfect complexes of $\mathcal{A}$-modules on $U_{i_0} \cap \ldots \cap U_{i_n}$. We therefore get a morphism

$$\mathrm{CC}^-_\bullet(\mathrm{Perf}(\mathcal{A})) \to \check{C}^{-\bullet}(M, \mathrm{CC}^-_\bullet(\mathrm{Perf}^{\mathrm{vstr}}(\mathcal{A}))).$$

Now replace the right-hand side by the quasi-isomorphic complex

$$\varinjlim_{\mathfrak{U}} \prod_{\sigma_0 \subset \sigma_1 \subset \ldots \subset \sigma_p} \mathrm{CC}^-_\bullet(\mathrm{Perf}^{\mathrm{vstr}}(\mathcal{A}(U_{\sigma_p})))$$

where σ_i run through simplices of $\mathfrak{U}$. There is a natural functor

$$\mathrm{Perf}^{\mathrm{vstr}}(\mathcal{A}(U_{\sigma_p})) \to \mathrm{Perf}^{\mathrm{vstr}}(\mathrm{Matr}^{\sigma_p}_{\mathrm{tw}}(\mathcal{A}))$$

where the right-hand side stands for the category of very strictly perfect complexes of modules over the sheaf of rings $\mathrm{Matr}^{\sigma_p}_{\mathrm{tw}}(\mathcal{A})$ on U_{σ_p}. This functor acts as follows: to a twisted module $\mathcal{M}$ it puts in correspondence the direct sum $\oplus_{i \in I_{\sigma_0}} \mathcal{M}_i$; an element $a_{ij}E_{ij}$ acts via $a_{ij}g_{ij}$.

Next, let us note that one can replace $\mathrm{CC}^-_\bullet(\mathrm{Perf}^{\mathrm{vstr}}(\mathrm{Matr}^{\sigma_p}_{\mathrm{tw}}(\mathcal{A})))$ by the complex $\mathrm{CC}^-_\bullet(\mathrm{Matr}^{\sigma_p}_{\mathrm{tw}}(\mathcal{A}))$: indeed, for any associative algebra A there is an explicit trace map

$$\mathrm{CC}^-_\bullet(\mathrm{Perf}^{\mathrm{vstr}}(A)) \to \mathrm{CC}^-_\bullet(A). \tag{3.12}$$

To construct the trace map, we use a modification of Keller's argument from [K]. We define this map as a composition

$$\mathrm{CC}^-_\bullet(\mathrm{Perf}^{\mathrm{vstr}}(A)) \to \mathrm{CC}^-_\bullet(\mathrm{Proj}(A)) \to \mathrm{CC}^-_\bullet(A);$$

the DG category in the middle is the subcategory of complexes with zero differential. The morphism on the left is $\exp(-(1 \otimes d) \times ?)$; the morphism on the right is $\mathrm{ch}(e) \times ?$ followed by the standard trace map from [L]. Let us explain the notation. The multiplication $\times$ is the binary multiplication of Getzler-Jones. The chain $\mathrm{ch}(e)$ is the Connes-Karoubi Chern character of an idempotent e, cf. [L]. To multiply $f_0 \otimes \ldots \otimes f_n$ by $\mathrm{ch}(e)$, recall that $f_k : \mathcal{F}_{i_k} \to \mathcal{F}_{i_{k-1}}$, $\mathcal{F}_{i_k}$ are free of finite rank, $e_k^2 = e_k$ in $\mathrm{Hom}(\mathcal{F}_{i_k}, \mathcal{F}_{i_k})$, $\mathcal{F}_{i_{-1}} = \mathcal{F}_{i_n}$, $e_{-1} = e_n$, and $f_k e_k = e_{k-1} f_k$. Write the usual formula for multiplication by $\mathrm{ch}(e)$, but, when a factor e stands between f_i and f_{i+1}, replace this factor by e_i. Similarly for the morphism on the left: if a factor d stands between f_i and f_{i+1}, replace this factor by d_i (the differential on the ith module). This finishes the construction of the morphism (3.10).

Next, we need to refine the map (3.12) as follows. Recall [D] that for a DG category $\mathcal{D}$ and for a full DG subcategory $\mathcal{D}_0$ the DG quotient of $\mathcal{D}$ by $\mathcal{D}_0$ is the following DG category. It has same objects as $\mathcal{D}$; its morphisms are freely generated over $\mathcal{D}$ by morphisms ϵ_i of degree -1 for any $i \in \mathrm{Ob}(\mathcal{D}_0)$, subject to $d\epsilon_i = \mathrm{id}_i$. It is easy to see that the trace map (3.12) extends to the negative cyclic complex of the Drinfeld quotient of $\mathrm{Perf}^{\mathrm{vstr}}(A)$ by the full DG subcategory of acyclic complexes. Indeed, a morphism in the DG quotient is a linear combination of monomials $f_0\epsilon_{i_0}f_1\epsilon_{i_1}\ldots\epsilon_{i_{n-1}}f_n$ where $f_k : \mathcal{F}i_k \to \mathcal{F}_{i_{k-1}}$ and $\mathcal{F}_{i_k}$ are acyclic for $k = 0, \ldots, n-1$. An acyclic very strictly perfect complex is contractible. Choose contracting homotopies s_k for $\mathcal{F}_{i_k}$. Replace all the monomials $f_0\epsilon_{i_0}f_1\epsilon_{i_1}\ldots\epsilon_{i_{n-1}}f_n$ by $f_0s_0f_1s_1\ldots s_{n-1}f_n$. Then apply the above composition to the resulting chain of $\mathrm{CC}^-_\bullet(\mathrm{Perf}^{\mathrm{vstr}}(A))$. We obtain for any associative algebra A

$$\mathrm{CC}^-_\bullet(\mathrm{Perf}^{\mathrm{vstr}}(A)_{\mathrm{Loc}}) \to \mathrm{CC}^-_\bullet(A) \tag{3.13}$$

where $_{\mathrm{Loc}}$ stands for the Drinfeld localization with respect to the full subcategory of acyclic complexes.

To construct the Chern character with supports, act as above but define $\mathcal{D}_i$ to be the Drinfeld quotient of the DG category $\mathrm{Perf}^{\mathrm{vstr}}(\mathcal{A}(U_i))$ by the full subcategory of acyclic complexes. We get a morphism

$$\mathrm{CC}^-_\bullet(\mathrm{Perf}(\mathcal{A})) \to \check{C}^{-\bullet}(M, \mathrm{CC}^-_\bullet(\mathrm{Perf}^{\mathrm{vstr}}(\mathcal{A})_{\mathrm{Loc}})) \to \check{C}^{-\bullet}(M, \mathrm{CC}^-_\bullet(\mathcal{A})).$$

From this, and from the fact that the negative cyclic complex of the localization of Perf_Z is canonically contractible outside Z, one gets easily the map (3.11).

3.5. Chern character for stacks

Now let us construct the Chern character

$$K_\bullet(\mathrm{Perf}(\mathcal{A})) \to \check{\mathbb{H}}^{-\bullet}(M, \mathrm{CC}^-_\bullet(\mathrm{Matr}_{\mathrm{tw}}(\mathcal{A}))) \tag{3.14}$$

$$K_\bullet(\mathrm{Perf}_Z(\mathcal{A})) \to \check{\mathbb{H}}^{-\bullet}_Z(M, \mathrm{CC}^-_\bullet(\mathrm{Matr}_{\mathrm{tw}}(\mathcal{A}))). \tag{3.15}$$

First, note that the K-theory in the left-hand side can be defined as in [TV]; one can easily deduce from [MC] and [K2], section 1, the Chern character from $K_\bullet(\mathrm{Perf}(\mathcal{A}))$ to the homology of the complex $\mathrm{Cone}(\mathrm{CC}^-_\bullet(\mathrm{Perf}_{\mathrm{ac}}(\mathcal{A})) \to \mathrm{CC}^-_\bullet(\mathrm{Perf}(\mathcal{A})))$. Here $\mathrm{Perf}_{\mathrm{ac}}$ stands for the category of acyclic perfect complexes.

Compose this Chern character with the trace map of 3.4. We get a Chern character from $K_\bullet(\mathrm{Perf}(\mathcal{A}))$ to

$$\check{\mathbb{H}}^{-\bullet}(M, \mathrm{Cone}(\mathrm{CC}^-_\bullet(\mathrm{Perf}^{\mathrm{vstr}}_{\mathrm{ac}}(\mathcal{A})_{\mathrm{Loc}}) \to \mathrm{CC}^-_\bullet(\mathrm{Perf}^{\mathrm{vstr}}(\mathcal{A})_{\mathrm{Loc}}))).$$

One gets the Chern characters (3.14), (3.15) easily by combining the above with (3.13).

3.6. The case of a gerbe

If $\mathcal{A}$ is a gerbe on M corresponding to a class c in $H^2(M, \mathcal{O}_M^*)$, then (in the C^∞ case) the right-hand side of (3.14) is the cohomology of M with coefficients in the complex of sheaves

$$\Omega^{-\bullet}[[u]], ud_{\mathrm{DR}} + u^2 H\wedge$$

where H is a closed three-form representing the three-class of the gerbe. In the holomorphic case, the right-hand side of (3.14) is computed by the complex $\Omega^{-\bullet,\bullet}[[u]], \overline{\partial} + \alpha\wedge + u\partial$ where α is a closed $(2,1)$ form representing the cohomology class $\partial \log c$. This can be shown along the lines of [BGNT], Theorem 7.1.2.

References

[AS] M. Atiyah and G. Segal, *Twisted K-theory and cohomology*, arXiv:math 0510674.

[BLM] Yu. Bespalov, V. Lyubashenko and O. Manzyuk, *Closed precategory of (triangulated) A_∞ categories*, in progress.

[BCMMS] P. Bouwknegt, A. Carey, V. Mathai, M. Murray and D. Stevenson, *Twisted K-theory and K-theory of bundle gerbes*, Comm. Math. Phys. **228** (2002), 1, 17–45.

[BGNT] P. Bressler, A. Gorokhovsky, R. Nest and B. Tsygan, *Deformation quantization of gerbes*, to appear in Adv. in Math., arXiv:math.QA/0512136.

[BGNT1] P. Bressler, A. Gorokhovsky, R. Nest and B. Tsygan, *Deformations of gerbes on smooth manifolds*, VASBI Conference on K-theory and Non-Commutative Geometry, Proceedings, G. Cortiñas ed., arXiv:math.QA/0701380, 2007.

[BNT] P. Bressler, R. Nest and B. Tsygan, *Riemann-Roch theorems via deformation quantization, I, II*, Adv. Math. **167** (2002), no. 1, 1–25, 26–73.

[BNT1] P. Bressler, R. Nest and B. Tsygan, *Riemann-Roch theorems via deformation quantization*, arxiv:math/9705014.

[C] A. Connes, *Noncommutative Geometry*, Academic Press, Inc., San Diego, CA, 1994.

[D] V. Drinfeld, *DG quotients of DG categories*, Journal of Algebra **272** (2004), no. 2, 643–691.

[G] E. Getzler, *Cartan homotopy formulas and the Gauss-Manin connection in cyclic homology*, Quantum deformations of algebras and their representations (Ramat-Gan, 1991/1992; Rehovot, 1991/1992), 65–78, Israel Math. Conf. Proc., 7, Bar-Ilan Univ., Ramat Gan, 1993.

[GJ] E. Getzler and J.D.S. Jones, *A_∞-algebras and the cyclic bar complex*, Illinois J. Math. **34** (1990), no. 2, 256–283.

[Gil] H. Gillet, *K-theory of twisted complexes*, Contemporary Mathematics, **55**, Part 1 (1986), 159–191.

[Gi] J. Giraud, *Cohomologie non abélienne*, Grundlehren **179**, Springer Verlag, 1971.

[Go] T. Goodwillie, *Relative algebraic K-theory and cyclic homology*, Annals of Math. **124** (1986), 347–402.

[J] J.D.S. Jones, *Cyclic homology and equivariant cohomology*, Invent. Math. **87** (1987), 403–423.

[Kar] M. Karoubi, *Homologie cyclique et K-théorie*, Astérisque **149**, Société Mathématique de France, 1987.

[K] B. Keller, *On the cyclic homology of ringed spaces and schemes*, Doc. Math. **3** (1998), 231–259.

[K1] B. Keller, *A_∞ algebras, modules and functor categories*, Trends in representation theory of algebras and related topics, 67–93, Contemporary Mathematics, **406**, AMS, 2006.

[K2] B. Keller, *On the cyclic homology of exact categories*, Journal of Pure and Applied Algebra **136** (1999), 1–56.

[Ko] M. Kontsevich, *Triangulated categories and geometry*, Course at École Normale Supérieure, Paris, 1998.

[KS] M. Kontsevich, Y. Soibelman, *Notes on A-infinity algebras, A-infinity categories and non-commutative geometry. I*, arXiv:0704.2890.

[L] J.L. Loday, *Cyclic Homology, Second Edition*, Grundlehren der Mathematischen Wissenschaften, 1997.

[Lu] V. Lyubashenko, *Category of A_∞ categories*, Homology Homotopy Appl. **5** (2003), 1, 1–48 (electronic).

[MaS] V. Mathai and D. Stevenson, *Generalized Hochschild-Kostant-Rosenberg theorem,* Advances in Mathematics **200** (2006), no. 2, 303–335.

[MaS1] V. Mathai and D. Stevenson, *Chern character in twisted K-theory: equivariant and holomorphic cases*, Comm. Math. Phys. **236** (2003), no. 1, 161–186.

[MC] R. McCarthy, *The cyclic homology of an exact category*, Journal of Pure and Applied Algebra **93** (1994), 251–296.

[NT] R. Nest and B. Tsygan, *The Fukaya type categories for associative algebras*, Deformation theory and symplectic geometry (Ascona, 1996), 285–300, Math. Phys. Stud., **20**, Kluwer Acad. Publ., Dordrecht, 1997.

[NT1] R. Nest and B. Tsygan, *On the cohomology ring of an algebra*, Advances in Geometry, 337–370, Progr. Math. **172**, Birkhäuser, Boston, Ma, 1999.

[OB] N. O'Brian, *Geometry of twisting cochains*, Compositio Math. **63** (1987), 1, 41-62.

[OTT] N. O'Brian, D. Toledo and Y.-L. Tong, *Hirzebruch–Riemann–Roch for coherent sheaves*, Amer. J. Math. **103** (1981), no. 2, 253-271.

[PS] P. Polesello and P. Schapira, *Stacks of quantization-deformation modules over complex symplectic manifolds*, International Mathematical Research Notices **49** (2004), 2637–2664.

[T] B. Tsygan, *Cyclic homology*, Cyclic homology in noncommutative geometry, 73–113, Encyclopaedia Math. Sci. **121**, Springer, Berlin, 2004.

[T1] B. Tsygan, *On the Gauss-Manin connection in cyclic homology*, math.KT/0701367.

[Ta] D. Tamarkin, *What do DG categories form?*, math.CT/0606553.

[TT] D. Tamarkin and B. Tsygan, *The ring of differential operators on forms in noncommutative calculus*, Graphs and patterns in mathematics and theoretical physics, 105–131, Proc. Sympos. Pure Math., 73, Amer. Math. Soc., Providence, RI, 2005.

[ToT] D. Toledo and Y.-L. Tong, *A parametrix for $\overline{\partial}$ and Riemann-Roch in Čech theory*, Topology **15** (1976), 273–302.

[TV] B. Toen and G. Vezzosi, *A remark on K-theory and S-categories*, *Topology* **43** (2004), no. 4, 765–791.

[TX] J.-L. Tu and P. Xu, *Chern character for twisted K-theory of orbifolds*, math.KT/0505267.

Paul Bressler
Université Claude Bernard Lyon 1
Bâtiment Doyen Jean Braconnier 43
Boulevard Du 11 Novembre 1918
F-69622 Villeurbanne CEDEX, France
e-mail: bressler@math.gmail.com

Alexander Gorokhovsky
Department of Mathematics
University of Colorado
Boulder, Colorado 80309-0395, USA
e-mail: Alexander.Gorokhovsky@colorado.edu

Ryszard Nest
University of Copenhagen
Universitetsparken 5
DK-2100 Kobenhavn 0, Denmark
e-mail: rnest@math.ku.dk

Boris Tsygan
Department of Mathematics
Northwestern University
2033 North Sherman Avenue
Evanston IL, 60201, USA
e-mail: tsygan@math.northwestern.edu

Progress in Mathematics, Vol. 265, 325–351

(C, F)-Actions in Ergodic Theory

Alexandre I. Danilenko

Abstract. This is a survey of a recent progress related to the (C, F)-construction of funny rank-one actions for locally compact groups. We exhibit a variety of examples and counterexamples produced via the (C, F)-techniques in every of the following categories: (i) probability preserving actions, (ii) infinite measure preserving actions, (iii) non-singular actions (Krieger's type III).

Mathematics Subject Classification (2000). Primary 37A15; Secondary 37A25, 37A30, 37A40.

Keywords. Ergodic transformation, mixing, rank-one action.

1. Introduction

The (C, F)-construction is a useful machinery to produce examples and counterexamples in ergodic theory. It appeared in [34] and, independently, in [19] (in a different but essentially equivalent way) as an algebraic counterpart of the classical geometric cutting-and-stacking technique. The latter has a long history but we will not discuss it here (see [46] and references therein). The (C, F)-formalism is convenient where the usual intuition related to 'stacking towers' and 'moving levels upwards' does not help much. This is the case when one constructs funny rank-one actions of general locally compact second countable (l.c.s.c.) groups without invariant ordering or groups with torsions or groups with infinitely many generators, etc. The (C, F)-construction is well suited to study Cartesian products since the product of two (C, F)-actions is again a (C, F)-action. The orbit structure of (C, F)-actions is explicit: the corresponding orbit equivalence relation is a countable inductive limit of tail equivalence relations on infinite product spaces. This is convenient to construct ergodic non-singular measures for these actions.

The outline of the paper is as follows. Section 2 is devoted to finite measure preserving (C, F)-actions. It consists of three subsections. The first one describes the (C, F)-construction. Subsection 2.2 shows how to use (C, F)-actions of auxiliary amenable groups to build the following counterexamples in the theory of probability preserving $\mathbb{Z}$-actions:

- 2-fold simple mixing transformation without prime factors [34],
- 2-fold simple mixing transformation which has square roots of all orders but no infinite square root chain [70],
- 2-fold simple mixing transformation with non-unique prime factors [28],
- mixing quasi-simple near simple transformation which is disjoint from any simple transformation [25],
- weakly mixing rigid rank-one transformations conjugate to their composition squares [27],
- weakly mixing rigid transformations with homogeneous spectra [23].

It is worthwhile to note here that the method of considering auxiliary $\mathbb{Z}^2$-actions was used earlier by Rudolph [75] to show that the second centralizer of a Bernoulli shift is just its powers and by Johnson and Park [61] to construct a simple map whose centralizer is isomorphic to $\mathbb{Z}^2$.

In Subsection 2.3 we discuss how to produce mixing rank-one actions of the following amenable groups:

- infinite direct sums of finite groups [24],
- countable locally normal groups [26],
- $\mathbb{R}^{d_1} \times \mathbb{Z}^{d_2}$ [32].

Section 3 is devoted to infinite measure preserving dynamical systems. In Subsections 3.1 and 3.2 we review various examples of (C, F)-actions from [19], [22] and [31] with 'unusual' – specific to infinite measure – properties of weak mixing and multiple recurrence respectively. Subsection 3.3 contains a 'constructive' answer to Krengel-Silva-Thieulen's question concerning possible values of the Krengel's entropy for the product of two transformations one of which is finite measure preserving. Section 4 deals with non-singular (C, F)-actions. In Subsection 4.1 we recall some concepts of non-singular ergodic theory: associated Mackey flow, AT-flow, non-singular actions of funny rank one. In Subsection 4.2 we exhibit non-singular counterparts of actions from Section 3 [19]. In Subsection 4.3 we discuss the weak mixing properties of non-singular Chacon transformations with 2-cuts (i.e., with 3 copies of the n-tower in the $(n+1)$-tower and a single spacer between the second and the third copies) [22]. In the final Section 5 we state 13 open problems which – on our opinion – may be solved with the use of the (C, F)-construction.

The definitions, motivation and brief historical remarks will be given below in the main text of the paper.

Acknowledgements. The author thanks V. Ryzhikov and the anonymous referee of this paper for the useful remarks.

2. Finite measure preserving (C, F)-actions

2.1. (C, F)-construction

We introduce here the (C, F)-construction in the most general setting (cf. [34], [19]–[32]) and review the basic properties of the (C, F)-actions.

Let G be a unimodular l.c.s.c. amenable group. Fix a (σ-finite) left Haar measure λ_G on it. Given two subsets $E, F \subset G$, by EF we mean their product, i.e., $EF = \{ef \mid e \in E, f \in F\}$. The set $\{e^{-1} \mid e \in E\}$ is denoted by E^{-1}. If E is a singleton, say $E = \{e\}$, then we will write eF for EF. For an element $g \in G$ and a subset $E \subset G$, we let $E(g) = E \cap (g^{-1}E)$.

To define a (C,F)-action of G we need two sequences $(F_n)_{n\ge 0}$ and $(C_n)_{n>0}$ of subsets in G such that the following conditions are satisfied:

$$(F_n)_{n=0}^{\infty} \text{ is a Følner sequence in } G, \tag{2.1}$$

$$C_n \text{ is finite and } \#C_n > 1, \tag{2.2}$$

$$F_n C_{n+1} \subset F_{n+1}, \tag{2.3}$$

$$F_n c \cap F_n c' = \emptyset \text{ for all } c \neq c' \in C_{n+1}. \tag{2.4}$$

We put $X_n := F_n \times \prod_{k>n} C_k$, endow X_n with the standard Borel product σ-algebra and define a Borel embedding $X_n \to X_{n+1}$ by setting

$$(f_n, c_{n+1}, c_{n+2}, \dots) \mapsto (f_n c_{n+1}, c_{n+2}, \dots). \tag{2.5}$$

It is well defined due to (2.3) and (2.4). Then we have $X_1 \subset X_2 \subset \cdots$. Hence $X := \bigcup_n X_n$ endowed with the natural Borel σ-algebra, say $\mathfrak{B}$, is a standard Borel space. Given a Borel subset $A \subset F_n$, we put

$$[A]_n := \{x \in X \mid x = (f_n, c_{n+1}, c_{n+2} \dots) \in X_n \text{ and } f_n \in A\}$$

and call this set an *n-cylinder*. It is clear that the σ-algebra $\mathfrak{B}$ is generated by the family of all cylinders.

Now we are going to define a 'canonical' measure on $(X, \mathfrak{B})$. Let κ_n stand for the equidistribution on C_n and $\nu_n := (\#C_1 \cdots \#C_n)^{-1} \lambda_G \restriction F_n$ on F_n. We define a product measure μ_n on X_n by setting $\mu_n = \nu_n \times \kappa_{n+1} \times \kappa_{n+2} \times \cdots$, $n \in \mathbb{N}$. Then the embeddings (2.5) are all measure preserving. Hence a σ-finite measure μ on X is well defined by the restrictions $\mu \restriction X_n = \mu_n$, $n \in \mathbb{N}$. To put it in another way, $(X,\mu) = \operatorname{inj\,lim}_n (X_n, \mu_n)$. Since

$$\mu_{n+1}(X_{n+1}) = \frac{\nu_{n+1}(F_{n+1})}{\nu_{n+1}(F_n C_{n+1})} \mu_n(X_n) = \frac{\lambda_G(F_{n+1})}{\lambda_G(F_n)\#C_{n+1}} \mu_n(X_n),$$

it follows that μ is finite if and only if

$$\prod_{n=0}^{\infty} \frac{\lambda_G(F_{n+1})}{\lambda_G(F_n)\#C_{n+1}} < \infty, \text{ i.e., } \sum_{n=0}^{\infty} \frac{\lambda_G(F_{n+1} \setminus (F_n C_{n+1}))}{\lambda_G(F_n)\#C_{n+1}} < \infty. \tag{2.6}$$

For the rest of Section 2 we will assume that (2.6) is satisfied. Moreover, we choose (i.e., normalize) λ_G in such a way that $\mu(X) = 1$.

To construct μ-preserving action of G on (X,μ), we fix a filtration $K_1 \subset K_2 \subset \cdots$ of G by compact subsets. Thus $\bigcup_{m=1}^\infty K_m = G$. Given $n,m \in \mathbb{N}$, we set

$$D_m^{(n)} := \Big(\bigcap_{k\in K_m} (k^{-1}F_n)\cap F_n\Big) \times \prod_{k>n} C_k \subset X_n \text{ and}$$

$$R_m^{(n)} := \Big(\bigcap_{k\in K_m} (kF_n)\cap F_n\Big) \times \prod_{k>n} C_k \subset X_n.$$

It is easy to verify that $D_{m+1}^{(n)} \subset D_m^{(n)} \subset D_m^{(n+1)}$ and $R_{m+1}^{(n)} \subset R_m^{(n)} \subset R_m^{(n+1)}$. We define a Borel mapping $K_m \times D_m^{(n)} \ni (g,x) \mapsto T_{m,g}^{(n)} x \in R_m^{(n)}$ by setting for $x = (f_n, c_{n+1}, c_{n+2}, \dots)$,

$$T_{m,g}^{(n)}(f_n, c_{n+1}, c_{n+2}\dots) := (gf_n, c_{n+1}, c_{n+2}, \dots).$$

Now let $D_m := \bigcup_{n=1}^\infty D_m^{(n)}$ and $R_m := \bigcup_{n=1}^\infty R_m^{(n)}$. Then a Borel mapping

$$T_{m,g} : K_m \times D_m \ni (g,x) \mapsto T_{m,g}x \in R_m$$

is well defined by the restrictions $T_{m,g} \restriction D_m^{(n)} = T_{m,g}^{(n)}$ for $g \in K_m$ and $n \geq 1$. It is easy to see that $D_m \supset D_{m+1}$, $R_m \supset R_{m+1}$ and $T_{m,g} \restriction D_{m+1} = T_{m+1,g}$ for all m. It follows from (2.1) that $\mu_n(D_m^{(n)}) \to 1$ and $\mu_n(R_m^{(n)}) \to 1$ as $n \to \infty$. Hence $\mu(D_m) = \mu(R_m) = 1$ for all $m \in \mathbb{N}$. Finally we set $\widehat{X} := \bigcap_{m=1}^\infty D_m \cap \bigcap_{m=1}^\infty R_m$ and define a Borel mapping $T : G \times \widehat{X} \ni (g,x) \to T_g x \in \widehat{X}$ by setting $T_g x := T_{m,g}x$ for some (and hence any) m such that $g \in K_m$. It is clear that $\mu(\widehat{X}) = 1$. Thus, we obtain that $T = (T_g)_{g\in G}$ is a free Borel measure preserving action of G on a conull subset of the standard probability space $(X, \mathfrak{B}, \mu)$. It is easy to see that T does not depend on the choice of filtration $(K_m)_{m=1}^\infty$. Throughout the paper we do not distinguish between two measurable sets (or mappings) which agree almost everywhere.

Definition 2.1. T is called the *(C,F)-action of G associated with* $(C_n, F_n)_n$.

We now list some basic properties of (X,μ,T). Given Borel subsets $A, B \subset F_n$, we have

$$[A\cap B]_n = [A]_n \cap [B]_n,\ [A\cup B]_n = [A]_n \cup [B]_n,$$

$$[A]_n = [AC_{n+1}]_{n+1} = \bigsqcup_{c\in C_{n+1}} [Ac]_{n+1},$$

$$T_g[A]_n = [gA]_n \text{ if } gA \subset F_n,$$

$$\mu([A]_n) = \#C_{n+1}\cdot \mu([Ac]_{n+1}) \text{ for every } c \in C_{n+1},$$

$$\mu([A]_n) = \frac{\lambda_G(A)}{\lambda_G(F_n)}\mu(X_n),$$

where the sign $\sqcup$ means the union of mutually disjoint sets.

In case $G = \mathbb{Z}$, it is easy to notice a similarity between the (C,F)-construction and the classical cutting-and-stacking construction of rank-one transformations

[48], [86]. Indeed, F_{n-1} (or, more precisely, the set of $(n-1)$-cylinders) corresponds to the levels of the $(n-1)$-tower and C_n corresponds to the locations of the copies of F_{n-1} inside the n-th tower F_n. (The copies $F_{n-1}c$, $c \in C_n$, are disjoint by (2.4) and they sit inside F_n by (2.3).) The remaining part of F_n, i.e., $F_n \setminus (F_{n-1}C_n)$, is the set of spacers in the n-th tower. Thus the (C, F)-construction may be regarded as a 'modified' arithmetical counterpart of the cutting-and-stacking. It is also worthwhile to note that the (C, F)-construction 'respects' Cartesian products. Namely, the product of two (C, F)-actions $(T_g^{(i)})_{g \in G_i}$ associated with $(C_n^{(i)}, F_n^{(i)})_n$, $i = 1, 2$, is the (C, F)-action of $G_1 \times G_2$ associated with $(C_n^{(1)} \times C_n^{(2)}, F_n^{(1)} \times F_n^2)_n$.

We recall now the definition of funny rank one (see [45], [46], [88]).

Definition 2.2. Let T be a measure preserving action of G on a σ-finite measure space $(X, \mathfrak{B}, \mu)$.

(i) A *Rokhlin tower or column* for T is a triple (Y, f, F), where $Y \in \mathfrak{B}$, F is a relatively compact subset of G and $f : Y \to F$ is a measurable mapping such that for any Borel subset $H \subset F$ and an element $g \in G$ with $gH \subset F$, one has $f^{-1}(gH) = T_g f^{-1}(H)$.

(ii) We say that T is of *funny rank-one* if there exists a sequence of Rokhlin towers (Y_n, f_n, F_n) such that $\lim_{n\to\infty} \mu(Y_n) = 1$ and for any subset $B \in \mathfrak{B}$, there is a sequence of Borel subsets $H_n \subset F_n$ such that

$$\lim_{n\to\infty} \mu(B \triangle f_n^{-1}(H_n)) = 0.$$

It is easy to see that any funny rank-one action is ergodic. Each (C, F)-action is of funny rank-one. Note that what we call funny rank-one is called rank-one by del Junco and Yassawi in case G is discrete and countable and $G \neq \mathbb{Z}$ [40].

2.2. Group action approach and some counterexamples in ergodic theory

Suppose we have a problem to construct an ergodic dynamical system possessing certain dynamical properties, say (D). The idea of the *group action approach* to this problem is to find an appropriate amenable group G, fix an element $h \in G$ of infinite order and to construct some special (C, F)-action $V = (V_g)_{g \in G}$ of G such that the structure of G plus the properties of V enforce the transformation V_h to possess (D). We illustrate this idea by 6 examples below. (See also [75] and [61] for earlier examples.)

2.2.1. 2-fold simple mixing transformation without prime factors. Let T be an ergodic invertible transformation of a standard probability space $(X, \mathfrak{B}, \mu)$. A T-invariant sub-σ-algebra of $\mathfrak{B}$ is called a factor of T. If $\{\emptyset, X\}$ and $\mathfrak{B}$ are the only factors of T then T is called *prime.* A 2-fold self-joining of T is a measure λ on $\mathfrak{B} \otimes \mathfrak{B}$ which is $T \times T$-invariant and has both projections equal to μ. We denote by $J_2^e(T)$ the space of all ergodic 2-fold self-joinings of T. $C(T)$ stands for the centralizer of T, i.e., the set of invertible μ-preserving maps commuting with T. We endow $C(T)$ with the weak topology [59]: $S_n \to S$ if $\mu(S_n A \triangle SA) \to 0$ for all $A \in \mathfrak{B}$. For $S \in C(T)$, the *off-diagonal* joining $\mu_S \in J_2^e(T)$ supported on the graph of S is given by

$$\mu_S(A \times B) := \mu(A \cap S^{-1}B).$$

If T is weakly mixing then $\mu \times \mu$ is also an ergodic self-joining of T. If $J_2^e(T) \subset \{\mu_S \mid S \in C(T)\} \cup \{\mu \times \mu\}$ then T is called *2-fold simple*. This definition is due to Veech [90] who proved the following result which was a starting point for the theory of simple maps (see [37], [89], [52], [80]).

Theorem 2.3. *Let T be 2-fold simple and let $\mathfrak{F}$ be a non-trivial factor. Then there is a compact subgroup K of $C(T)$ such that*

$$\mathfrak{F} = \mathrm{Fix}(K) := \{A \in \mathfrak{B} \mid \mu(A \triangle kA) = 0 \textit{ for all } k \in K\}.$$

If T is 2-fold simple then it is easy to see that either T has pure point spectrum (this case may be regarded as the trivial one) or T is weakly mixing. Many examples of weakly mixing 2-fold simple systems are constructed in [76], [37], [38], [61] [53], [28] (see also [36], [74], [89] where the simplicity of the Chacon map and some horocycle flows was established). All of these systems are either themselves prime or have prime factors. This leads naturally to the question:

(Q1) *Must every weakly mixing 2-fold simple map have a prime factor?*

Del Junco answers this question in negative in [34] via the group action approach. Let $G = \mathbb{Z} \oplus \bigoplus_{i=1}^{\infty} \mathbb{Z}/2\mathbb{Z}$. The special (C, F)-action $V = (V_g)_{g \in G}$ of G is constructed in such a way that the 'time-one' transformation $V_{(1,0)}$ is weakly mixing, 2-fold simple and $C(V_{(1,0)}) = \{V_h \mid h \in G\}$. Then by Theorem 2.3, every non-trivial factor $\mathfrak{F}$ of $V_{(1,0)}$ equals to $\mathrm{Fix}(K)$ for a finite subgroup $K \subset \bigoplus_{i=1}^{\infty} \mathbb{Z}/2\mathbb{Z}$. Hence, for n large enough, $\mathfrak{F} \supsetneq \mathrm{Fix}(\bigoplus_{i=1}^{n} \mathbb{Z}/2\mathbb{Z})$. Therefore $V_{(1,0)} \restriction \mathfrak{F}$ is not prime. To construct such a V, del Junco defines the corresponding sequence $(C_{n+1}, F_n)_{n=0}^{\infty}$ in the following inductive way. Fix a sequence $\epsilon_n \to 0$. Define recurrently two sequences $(a_n)_{n=0}^{\infty}$ and $(\widetilde{a}_n)_{n=0}^{\infty}$ by setting:

$$a_0 = \widetilde{a}_0 = 1, \ a_{n+1} := \widetilde{a}_n r_n, \ \widetilde{a}_{n+1} := a_{n+1} + (n+1)\widetilde{a}_n,$$

where r_n is an integer parameter to be specified below. Given $a > 0$, we let $I[a] := \{m \in \mathbb{Z} \mid 0 \le m < a\}$. Let $G_n := \{(g_1, g_2, \dots) \in \bigoplus_{i=1}^{\infty} \mathbb{Z}/2\mathbb{Z} \mid g_j = 0 \text{ for all } j > n\}$ and $F_n := I[a_n] \times G_n$. To define C_{n+1}, we first introduce some auxiliary notation. Let $\widetilde{F}_n := I[\widetilde{a}_n] \times G_n$ and $S_n := I[n\widetilde{a}_{n-1}] \times G_n$ and let $\phi_n : \mathbb{Z} \to G$ denote a 'tiling' homomorphism given by $\phi_n(i) := (i\widetilde{a}_n, 0)$. We now have $S_n \subset F_n$, $F_n + S_n \subset \widetilde{F}_n$,

$$F_{n+1} = \bigsqcup_{h \in I[r_n]} \widetilde{F}_n + \phi_n(h) \ \text{ and } \ S_n = \bigsqcup_{h \in I[n]} \widetilde{F}_{n-1} + \phi_{n-1}(h).$$

Given two finite sets A, B and a map $\phi : A \to B$, the probability $\frac{1}{\# A} \sum_{a \in A} \delta_{\phi(a)}$ on B will be denoted by $\mathrm{dist}_{a \in A} \phi(a)$. Given two measures κ, ρ on a finite set B, we let $\|\kappa - \rho\|_1 := \sum_{b \in B} |\kappa(b) - \rho(b)|$.

To define a so-called 'random spacer map' $s_n : I[r_n] \to S_n$ del Junco first proves the following lemma.

Lemma 2.4. *If a positive integer r_n is sufficiently large then there exists a map $s_n : I[r_n] \to S_n$ such that for any $\delta \geq n^{-2} r_n$,*

$$\|\text{dist}_{t \in I[\delta]}(s_n(h+t), s_n(h'+t)) - \lambda_{S_n} \times \lambda_{S_n}\|_1 < \epsilon_n$$

whenever $h \neq h' \in I[r_n]$ with $\{h, h'\} + I[\delta] \subset I[r_n]$, where λ_{S_n} denotes the equidistribution on S_n.

Select now r_n large so that the conclusion of Lemma 2.4 is satisfied. Take a map s_n whose existence is asserted in the lemma and finally put $C_{n+1} := \{s_n(h) + \phi_n(h) \mid h \in I[r_n]\}$. It is not difficult to see that the conditions (2.1)–(2.4) and (2.6) are satisfied for the sequence $(F_n, C_{n+1})_{n \geq 0}$. Hence the associated (C, F)-action V of G is well defined.

Note that del Junco's construction incorporates the Ornstein's idea of random spacer from [72]. Indeed, the $(n+1)$-'tower' F_{n+1} is partitioned into r_n 'windows' $\widetilde{F}_n + \phi_n(h)$, $h \in I[r_n]$. Every window contains a copy $F_n + \phi_n(h)$ of the n-tower F_n which is, in addition, 'randomly perturbed' by $s_n(h)$ inside the window.

Now we state the main result from [34].

Theorem 2.5. *The transformation $V_{(0,1)}$ is weakly mixing and 2-fold simple. Moreover, $C(V_{(0,1)}) = \{V_g \mid g \in G\}$.*

2.2.2. 2-fold simple mixing transformation which has diadic roots of all orders but no infinite square root chain. In [66] King rised a question:

(Q2) *Is there a transformation T with diadic roots of all orders but no infinite square root chain?*

In [70] Madore, a student of del Junco, answers this question in positive. He constructs a transformation T which has roots of order 2^n for any $n > 0$ and such that there is no action $(V_h)_{h \in H}$ of the group of 2-adic rationals H with $V_1 = T$. To achieve this he uses the group action approach. The auxiliary group G is a non-splitting extension of the group $L = \bigoplus_{n=1}^{\infty} \mathbb{Z}/2^n\mathbb{Z}$ via $\mathbb{Z}$. More precisely, $G = \mathbb{Z} \times L$ with the following multiplication law:

$$(i, l) + (j, m) := (i + j + \sum_{n=1}^{\infty} [(l_n + m_n)/2^n], \, l + m),$$

where [.] denotes the integer part and l_n and m_n are the n-th coordinates of l and m respectively and we assume that $0 \leq l_n, m_n < 2^n$. It is clear that the 2^n-th power of the element $(0; \underbrace{0, \ldots, 0,}_{n-1 \text{ times}} 1, 0, \ldots) \in G$ is $(1, 0)$. On the other hand, it is obvious that H does not embed into G. Slightly modifying the argument of del Junco from [34], Madore constructs a (C, F)-action V of G such that $V_{(1,0)}$ is 2-fold simple and mixing and $C(V_{(1,0)}) = \{V_g \mid g \in G\}$. It remains to note that every root of $V_{(1,0)}$ belongs to $C(V_{(1,0)})$. In fact, Madore proves a more general fact generalizing Theorem 2.5.

Theorem 2.6. *Let G be a countable Abelian group including $\mathbb{Z}^d$ as a subgroup and such that the quotient group $G/\mathbb{Z}^d$ is locally finite. Let $e = (1, 0, \dots, 0) \in \mathbb{Z}^d$. Then there exists a (C, F)-action V of G such that the transformation V_e is 2-fold simple and mixing and $C(V_e) = \{V_g \mid g \in G\}$.*

This theorem answers (Q2). Moreover, it has other applications. For instance, it provides first examples of *mixing* 2-fold simple transformations whose centralizer is $\mathbb{Z}^d$ or $\mathbb{Q}$. Non-mixing weakly mixing simple transformation T with $C(T) = \mathbb{Z}^2$ has been constructed earlier in [61].

2.2.3. 2-fold simple mixing transformation with non-unique prime factors. The first example of a 2-fold simple T with non-unique prime factors was constructed by Glasner and Weiss [53] as an inverse limit of certain horocycle flows. Some subtle results of M. Ratner on joinings of horocycle flows [74] play a crucial role in [53]. We notice also that T from [53] has many non-prime factors as well. Note that for some time it was not obvious at all whether it is possible to construct a 2-fold simple map with non-unique prime factors by means of the more elementary cutting-and-stacking technique (see [89]). To achieve this purpose del Junco and the author use the group action approach [28]. We consider an auxiliary amenable group $G = \mathbb{Z} \times (\mathbb{Z} \rtimes \mathbb{Z}/2\mathbb{Z})$ with multiplication as follows

$$(n, m, a)(n', m', a') = (n + n', m + (-1)^a m', a + a').$$

Notice that G is non-Abelian. As in 2.2.1 and 2.2.1 we construct a (C, F)-action V of G such that the transformation $V_{(1,0,0)}$ is 2-fold simple and has centralizer coinciding with the full G-action. Then apply Theorem 2.3. All non-trivial finite subgroups of G are as follows: $\{G_g \mid b \in \mathbb{Z}, b \neq 0\}$, where $G_g = \{(0, b, 1), (0, 0, 0)\}$. Note that all of them are maximal. Hence the corresponding factors $\mathrm{Fix}(G_b)$ are prime. To construct such a V we apply again Ornstein's random spacer techniques in the recurrent process of building an appropriate sequence $(C_{n+1}, F_n)_{n=0}^{\infty}$. See [28] for details. Now we state the main result of [28].

Theorem 2.7. *The transformation $V_{(1,0,0)}$ is weakly mixing and 2-fold simple. All non-trivial proper factors of $V_{(1,0,0)}$ are 2-to-1 and prime. They are as follows:* $\mathrm{Fix}(G_b)$, $b \in \mathbb{Z} \setminus \{0\}$. *Two factors* $\mathrm{Fix}(G_b)$ *and* $\mathrm{Fix}(G_{b'})$ *are isomorphic if and only if b and b' are of the same evenness.*

2.2.4. Mixing quasi-simple near simple transformation which is disjoint from any simple transformation. Let $\mathfrak{F}$ be a factor of an ergodic dynamical system $(X, \mathfrak{B}, \mu, T)$. If there exists a compact subgroup $K \subset C(T)$ such that $\mathfrak{F} = \mathrm{Fix}(K)$ then T is called a *compact extension* of $\mathfrak{F}$. Each 'intermediate' factor $\mathfrak{E}$ of T, i.e., $\mathfrak{F} \subset \mathfrak{E} \subset \mathfrak{B}$, is called an *isometric* extension of $\mathfrak{F}$. For instance, every finite-to-one extension is isometric. A 2-fold simple transformation T is called *simple* if it is 3-fold PID, i.e., $\mu \times \mu \times \mu$ is the only $T \times T \times T$-invariant measure on $X \times X \times X$ whose coordinate plane projections are all (the three of them) equal to $\mu \times \mu$ [37] (see also [89] and [52]). Two dynamical systems $(X, \mathfrak{B}_X, \mu, T)$ and $(Y, \mathfrak{B}_Y, \nu, S)$

are called *disjoint* if $\mu \times \nu$ is the only $T \times S$-invariant measure on $X \times Y$ whose marginals are μ and ν respectively [49].

An ergodic transformation T is called 2-*fold quasi-simple* if for each $\nu \in J_2^e(T)$, either $\nu = \mu \times \mu$ or the transformation $(T \times T, \nu)$ is isometric over each of the two marginals. A 2-fold simple transformation which is 3-fold PID is called *quasi-simple.* These concepts were introduced by Ryzhikov and Thouvenot in [80] and [85] (see also [35] for related, more general concepts of distal simplicity). The class of quasi-simple systems contains many natural examples. For instance, a factor of a 2-fold simple transformation is 2-fold quasi-simple. Each non-zero time transformation of any flow with a so-called R-property is quasi-simple [89]. The flows with R-property include the horocycle flows [74] and some smooth flows on 2-dimensional manifolds [47]. It was shown in [89] that each non-zero time automorphism of a horocycle flow is a factor of a simple map. In view of that Thouvenot asks (see also [35] for an analogous question):

(Q3) *Whether each quasi-simple transformation is a factor of (an isometric extension of) a simple map?*

We answer this question in negative in [25] by constructing a quasi-simple map which is disjoint from any simple transformation. Hence it is disjoint from any isometric extension of any simple map [49]. It remains to note that disjoint systems do not have isomorphic factors. To construct such a map we utilize the group action approach. Let $G = (\bigoplus_{i=1}^{\infty} \mathbb{Z}/3\mathbb{Z}) \rtimes \mathbb{Z}$ with the following multiplication law:

$$(h,n)(k,m) := (h + (-1)^n k, n+m).$$

Notice that the center of G is $\{0\} \times 2\mathbb{Z}$. We need the concept of *near simplicity* introduced in [25]. Recall that the group $\mathrm{Aut}_0(X,\mu)$ of all μ-preserving transformations is Polish when endowed with the weak topology [59]. Fix a weakly mixing transformation T. Denote by Ξ_T^e the space of probability measures on $\mathrm{Aut}_0(X,\mu)$ which are invariant and ergodic with respect to the transformation $S \mapsto T^{-1}ST$. Then every $\xi \in \Xi_T^e$ determines an ergodic self-joining μ_ξ of T by the formula

$$\mu_\xi(A \times B) := \int_{\mathrm{Aut}_0(X,\mu)} \mu(A \cap SB)\, d\xi(S).$$

If $J_2^e(T) = \{\mu_\xi \mid \xi \in \Xi_T^e\} \cup \{\mu \times \mu\}$ then T is called 2-*fold near simple.* Clearly, a 2-fold near simple transformation is 2-fold simple if and only if Ξ_T^e consists of Dirac measures only (i.e., measures supported by singletons). As was shown in [25], the theory of near simple maps is more or less parallel to the theory of simple maps. Below we will use the following fact: a factor of a 2-fold simple map is either 2-fold simple itself or non-2-fold near simple [25, Corollary 3.5, Proposition 3.6]. Utilizing the random spacer method once again we construct in [25, Section 6] a (C,F)-action V of G such that the following holds.

Theorem 2.8. (i) $J_2^e(V_{(0,1)}) = \{0.5(\mu_{V_{(g,i)}} + \mu_{V_{(-g,i)}}) \mid g \in \bigoplus_{i=1}^{\infty} \mathbb{Z}/3\mathbb{Z}\} \cup \{\mu \times \mu\}$.

(ii) $V_{(0,1)}^2$ *is simple.*

(iii) *Every non-trivial factor of* $V_{(0,1)}$ *is near simple but not 2-fold simple.*

It follows from (i) that $V_{(0,1)}$ is weakly mixing and 2-fold quasi-simple (and hence quasi-simple in view of (ii)).

Corollary 2.9. *$V_{(0,1)}$ is disjoint from any isometric (even any distal) extension of any simple transformation.*

Proof. Let S be a simple transformation. If $V_{(0,1)}$ and S are not disjoint then they have a non-trivial 'common' factor $\mathfrak{F}$ by [35, Proposition 7]. Since $\mathfrak{F}$ is a factor of S, it follows that either $S \restriction \mathfrak{F}$ is simple or $S \restriction \mathfrak{F}$ is not 2-fold near simple. However this contradicts to Theorem 2.8(iii) because $S \restriction \mathfrak{F}$ is isomorphic to $T_{(0,1)} \restriction \mathfrak{F}$. □

2.2.5. Weakly mixing rigid rank-one transformations conjugate to their composition squares. We first state a well known question (see [55], [56], [54]).

(Q4) *Whether there exist weakly mixing rank-one transformations T conjugate to T^2?*

Recently Ageev used Baire category argument to answer this question in positive [10] (see also [27, Section 1] for a short simplified proof). However these proofs are not constructive and thus no concrete example appeared in [10]. The first explicit example answering (Q4) is constructed in [27, Section 2]. For that the author used the group action approach. The auxiliary group under question is the group H of 2-adic rationals. The corresponding sequence (C_{n+1}, F_n) is constructed in such a way that

(i) $F_n = \{2^{-k_n}, 2 \cdot 2^{-k_n}, 3 \cdot 2^{-k_n}, \dots, \alpha_n\}$,
(ii) $C_{2n} = \{0, \alpha_{2n-1}, 2\alpha_{2n-1} + 1\}$,
(iii) $\#(C_{2n+1} \cap (C_{2n+1} + \alpha_{2n}))/\#C_{2n+1}$ is 'close' to 1,
(iv) for any even n, there exists an enumeration $(f_i)_{i=0}^{\#F_n - 1}$ of the elements of F_n such that $i + 2^{-k_n} \equiv f_i \pmod{\alpha_n}$, $0 \le i < \#F_n$.
(v) The cardinality of the subset $C_{2n+1}^{+} := \{c \in C_{2n+1} \mid 2 \cdot F_{2n} + 2c \subset F_{2n+1}\}$ is $0.5\#C_{2n+1}$.

The parameters α_n and k_n are chosen in an explicit way in [27]. This is in contrast with the procedures described in 2.2.1–2.2.4, where the parameters related to the 'spacer mappings' are based on 'random' choice. Denote by $T = (T_h)_{h \in H}$ the associated (C, F)-action.

Theorem 2.10 ([27]). *The transformation T_1 is weakly mixing and has rank one. Moreover, T_1 is conjugate to T_2.*

Idea of the proof. The weak mixing of T_1 can be deduced from (ii) in a rather standard way originated from [15]. The fact that T_1 has rank one follows from (iii) and (iv). Given $x = (f_{2n}, c_{2n+1}, c_{2n+2}, \dots) \in X$, we 'define' Sx by setting $Sx := (2f_{2n}, 2c_{2n+1}, 2c_{2n+2}, \dots)$. One can make this definition rigorous by using (v). Then S is measure preserving transformation of X and $ST_1S^{-1} = T_2$. □

2.2.6. Weakly mixing rigid transformations with homogeneous spectra. For each transformation $S \in \text{Aut}_0(X, \mu)$, we denote by $\mathcal{M}(S)$ the set of essential values for the multiplicity function of the unitary operator $f \mapsto f \circ S^{-1}$ on the Hilbert space $L_0^2(X, \mu) := L^2(X, \mu) \ominus \mathbb{C}$. The following Rokhlin's problem on homogeneous spectrum was open for several decades:

(Q5) *given $n > 1$, is there an ergodic transformation S with $\mathcal{M}(S) = \{n\}$?*

The affirmative answer to this problem was given for $n = 2$ in [81] and independently in [7] by showing that $\mathcal{M}(S \times S) = \{2\}$ for a generic $S \in \text{Aut}_0(X, \mu)$. Some concrete examples were shown to have homogeneous spectrum of multiplicity 2 in [82] and [8]. Some of these transformations are even mixing. For $n > 2$, Ageev developed in [9] another approach to prove the existence of ergodic transformation with homogeneous spectrum of multiplicity n. (See also [23, Section 1] for a simple proof of this fact.) This approach is based on Baire category arguments and it is not constructive.

We now present an explicit construction that appeared in [23, Section 3]. It is based on an application of the group action approach. Fix a family $e_1, \ldots, e_n$ of generators for $\mathbb{Z}^n$. Define a 'cyclic' group automorphism $A : \mathbb{Z}^n \to \mathbb{Z}^n$ by setting $Ae_1 := e_2, \ldots, Ae_{n-1} := e_n$ and $Ae_n := e_1$. Let G denote the semidirect product $\mathbb{Z}^n \rtimes_A \mathbb{Z}$ with the multiplication law as follows

$$(v, m)(w, l) := (v + A^m w, m + l), \quad v, w \in \mathbb{Z}^n, \ m, l \in \mathbb{Z}.$$

Then we have a natural embedding $v \mapsto (v, 0)$ of $\mathbb{Z}^n$ into G. We also let $e_0 := (0, 1) \in G$ and $e_{n+1} := e_0^n$. Notice that G is generated by e_1 and e_0. Moreover, $e_0 e_i e_0^{-1} = Ae_i$ for all $i = 1, \ldots, n$. Let H be the subgroup of G generated by $e_1, \ldots, e_{n+1}$. Then H is a free Abelian group with $n + 1$ generators. It is normal in G and the quotient G/H is a cyclic group of order n. Moreover, A extends naturally to H via the conjugation by e_0. We denote this extension by the same symbol A. We need the following auxiliary result.

Theorem 2.11 ([23]). *Let $(T_g)_{g \in G}$ be a measure preserving action of G. If the transformation $T_{e_{l+1} - e_1}$ is weakly mixing for each $l|n$ and T_{e_0} has a simple spectrum then $T_{e_{n+1}}$ has a homogeneous spectrum of multiplicity n.*

Thus to produce an explicit transformation with homogeneous spectrum it suffices to construct a (C, F)-action T of G satisfying the condition of Theorem 2.11. This is done in [23]. The corresponding sequence $(C_{m+1}, F_m)_{m=0}^{\infty}$ is determined via an inductive process. Since the construction procedure is rather involved, we will not reproduce it here (see [23, Section 3]) for details. The transformation T_{e_0} appears to be rigid and of rank one. Hence T_{e_0} has a simple spectrum [11].

Notice also that the explicit structure of transformations with homogeneous spectra turned out convenient to prove the following statement [23].

Theorem 2.12. *Given $n > 1$ and a subset $M \subset \mathbb{N}$ such that $M \ni 1$, there exists a weakly mixing transformation T with $\mathcal{M}(T) = n \cdot M$.*

This extends the main results of [69] and [57], where a particular case $n = 1$ was under consideration.

2.3. Mixing rank-one actions of amenable groups G

Let G be a non-compact l.c.s.c. group and $T = (T_g)_{g\in G}$ a measure preserving action of G on a standard probability space $(X, \mathfrak{B}, \mu)$.

Definition 2.13.

(i) T is called *mixing* if for all subsets $A, B \in \mathfrak{B}$ we have
$$\lim_{g\to\infty} \mu(T_g A \cap B) = \mu(A)\mu(B).$$

(ii) T is called *mixing of order l* if for any $\epsilon > 0$ and $A_0, \dots, A_l \in \mathfrak{B}$, there exists a finite subset $K \subset G$ such that
$$|\mu(T_{g_0} A_0 \cap \cdots \cap T_{g_l} A_l) - \mu(A_0) \cdots \mu(A_l)| < \epsilon$$
for each collection $g_0, \dots, g_l \in G$ with $g_i g_j^{-1} \notin K$ if $i \neq j$

In this subsection we are concerned with the following problem

(Q6) *To construct mixing (funny) rank-one actions for various amenable l.c.s.c. groups.*

We recall that mixing rank-one transformations (and actions of more general groups) have been of interest in ergodic theory since 1970 when Ornstein constructed an example of mixing transformation without square root [72]. Since then the dynamical properties of mixing rank-one transformations have been deeply investigated. Such transformations are mixing of all orders and have minimal self-joinings of all orders [63], [78], [65] (see also [84]). This implies in turn that they are prime and have trivial centralizer [76]. These results were extended partly to rank-one mixing actions of $\mathbb{R}^d$ and $\mathbb{Z}^d$ [79], [78], [83] and some other discrete countable Abelian groups [40]. Despite this progress, there are not many examples of rank-one mixing actions that are known. Most of them were obtained via stochastic cutting-and-stacking techniques using "random spacers" [72], [76], [73]. Non-random explicit rank-one constructions were shown to be mixing in [3], [18] for $\mathbb{Z}$-actions and in [6] for $\mathbb{Z}^2$-actions. Fayad constructed a smooth mixing rank-one $\mathbb{R}$-action on the 3-torus [44]. However it is not clear yet: which other amenable groups admit mixing rank-one actions and how to construct such actions?

2.3.1. G is an infinite sum of finite groups (method of 'random rotation'). Let $G = \bigoplus_{i=1}^{\infty} G_i$, where G_i is a non-trivial finite group for any i. If T is a funny rank-one action of G and every subset F_n from Definition 2.2(ii) equals $\bigoplus_{i=1}^{k_n} G_i$ for some $k_n \to \infty$ then we say that T has *rank one.*

In [24] we construct mixing rank-one (C, F)-actions of G. For this, we use a modified Ornstein's idea of random spacer [72]. However unlike the previously known examples of (C, F)-actions, the actions in [24] are constructed without adding any spacer. Instead of that on the n-th step we just cut the n-th 'tower' into 'sub-towers' and then rotate the sub-towers 'in a random way'. Thus in this context the term 'random rotation' seems to be more appropriate than the 'random spacer'.

Before passing to the construction process we state an auxiliary lemma which is analogous to Lemma 2.4. Given a finite set A, a finite group H and elements $h_1, \dots, h_l \in H$, we denote by $\pi_{h_1,\dots,h_l}$ the map $A^H \to (A^l)^H$ given by

$$(\pi_{h_1,\dots,h_l} x)(k) = (x(h_1 k), \dots, x(h_l k)).$$

Lemma 2.14. *Given $l \in \mathbb{N}$ and $\epsilon > 0$, there exists $m \in \mathbb{N}$ such that for any finite group H with $\#H > m$, one can find $s \in A^H$ such that*

$$\|\mathrm{dist}\, \pi_{h_1,\dots,h_l} s - \lambda^l\| < \epsilon \tag{2.7}$$

for all $h_1 \neq h_2 \neq \cdots \neq h_l \in H$.

Suppose now that a sequence of integers $0 < k_1 < k_2 < \cdots$ is given. Then we define $(F_n)_{n \geq 0}$ by setting $F_0 := \{1_G\}$ and $F_n := \bigoplus_{i=1}^{k_n} G_i$ for $n \geq 1$. We also set $H_0 := \bigoplus_{i=0}^{k_1} G_i$ and $H_n := \bigoplus_{i=k_n+1}^{k_{n+1}} G_i$ for $n \geq 1$. Suppose now that we are also given a sequence of maps $s_n \colon H_n \to F_n$. Then we define $c_{n+1}, \phi_n : H_n \to F_n$ by setting $\phi_n(h) := (0, h)$ and $c_{n+1}(h) := (s_n(h), h)$ and put $C_{n+1} := c_{n+1}(H_n)$. It is easy to see that the conditions (2.1)–(2.4) and (2.6) are satisfied for $(C_{n+1}, F_n)_{n \geq 0}$. To complete the definition of $(C_{n+1}, F_n)_{n \geq 0}$ it remains to choose $(k_{n+1}, s_n)_{n \geq 0}$. This will be done recurrently. Fix a sequence of reals $\epsilon_i \to 0$. Suppose we have defined k_n and s_{n-1}. Then apply Lemma 2.14 with $A := F_n$, $l := n$ and $\epsilon := \epsilon_n$ to find k_{n+1} large so that there exists $s_n \in A^{H_n}$ satisfying (2.7). Assume, in addition, that k_n grows exponentially fast. Denote by T the (C, F)-action of G associated with $(C_{n+1}, F_n)_{n=0}^{\infty}$.

Theorem 2.15 ([24]). *T is mixing of any order.*

Moreover, by perturbing s_n in an appropriate way we may obtain an uncountable collection of pairwise non-isomorphic (even pairwise disjoint in the sense of Furstenberg [49]) mixing rank-one actions of G.

2.3.2. G is a locally finite group (method of 'random rotation'). A countable discrete amenable group G is called *locally normal* if $G = \bigcup_{i=1}^{\infty} G_i$ for a nested sequence $G_1 \subset G_2 \subset \cdots$ of normal finite subgroups of G. We call such a sequence a *filtration* of G. The class of locally normal groups includes the countable direct sums of finite groups and all Abelian torsion groups. If T is a funny rank-one action of G and the corresponding sequence F_n from Definition 2.2(ii) forms a filtration of G then we say that T has *rank one*.

Developing further the stochastic 'random rotation' method from [24] we prove the following result in [26].

Theorem 2.16. *There exists an uncountable family of pairwise disjoint (and hence pairwise non-isomorphic) mixing of any orders rank-one (C, F)-actions of G.*

2.3.3. $G = \mathbb{R}^{d_1} \times \mathbb{Z}^{d_2}$ (concrete examples with explicit 'polynomial spacers'). In this subsection we present the main result of [32] – explicit construction of mixing rank-one actions of $\mathbb{R}^{d_1} \times \mathbb{Z}^{d_2}$. Unlike the stochastic constructions presented in 2.3.1 and 2.3.2, these actions are 'absolutely concrete', i.e., the parameters in the construction are all explicitly specified. The 'spacer mappings' here are polynomials with known coefficients.

We recall that a funny rank-one action T of G is said to have *rank one* (or rank-one by cubes) if there are $a_n \in \mathbb{R}$ such that the subsets F_n from Definition 2.2(ii) are as follows:

$$F_n = \{(t_1, \dots, t_{d_1+t_2}) \in G \mid 0 \le t_i < a_n \text{ for all } i = 1, \dots, d_1 + d_2\}.$$

Let $d = d_1 + d_2$. We set $H := \mathbb{Z}^d$. To define a sequence $s_n : H \ni x \mapsto s_n(x) \in G$ of 'spacer polynomials' we first introduce an auxiliary concept. Fix a family of reals $\xi_1, \dots, \xi_m$. For a nonempty subset $J \subset \{1, \dots, m\}$, we let $\xi_J := \prod_{i \in J} \xi_i$. We also let $\xi_\emptyset := 1$. If the family of reals ξ_J, J runs all the subsets of $\{1, \dots, m\}$, is independent over $\mathbb{Q}$ then we say that $\xi_1, \dots, \xi_m$ is *good*.

Case 1. If $d = d_1 > 1$, we consider a polynomial $s(x) = (s(x)_1, \dots, s(x)_d)$ given by

$$s(x)_i := (\alpha_i x_i + \gamma_i)(x_1 + \dots + x_d) + \beta x_i^2 + \delta_i x_i, \ 1 \le i \le d.$$

Here $\alpha_i, \beta_i, \gamma_i, \delta_i$ are real coefficients satisfying the following conditions: $\mathbb{Q} \ni \alpha_i > 0$, $\gamma_i \ge 0$, $\alpha_i + 2\beta_1 \ge 0$, $\alpha_i + \beta_i + \gamma_i \ge 0$ for all i and the family of reals $\alpha_1 + 2\beta_1, \dots, \alpha_d + 2\beta_d$ is good. We now set $s_n \equiv s$.

Case 2. If $d = d_1 = 1$, i.e., $G = \mathbb{R}$, we set

$$s(x) := \alpha x^2 + \beta x, \ \ \widetilde{s}(x) := \widetilde{\alpha} x^2 + \widetilde{\beta} x \ \text{ at all } x \in \mathbb{Z},$$

where α and $\widetilde{\alpha}$ are rationally independent positive reals and $\alpha + \beta \ge 0$, $\widetilde{\alpha} + \widetilde{\beta} \ge 0$.

Case 3. If $d = d_2$, we set $s_n \equiv s$, where the polynomial s is given by

$$s(x)_i = x_i(x_1 + \dots + x_d) - (x_i^2 + x_i)/2, \ i = 1, \dots, d.$$

Case 4. If $d_1 \ne 0$ and $d_2 \ne 0$, we let $\widetilde{s}_n : \mathbb{Z}^{d_1} \to \mathbb{R}^{d_1}$ and $\widehat{s}_n : \mathbb{Z}^{d_2} \to \mathbb{Z}^{d_2}$ stand for the sequences of polynomials defined in Case 1 if $d_1 > 1$ (or in Case 2 if $d_1 = 1$) and Case 3 respectively. Then we put

$$s_n(x_1, \dots, x_{d_1+d_2}) := (\widetilde{s}_n(x_1, \dots, x_{d_1}), \widehat{s}_n(x_{d_1+1}, \dots, x_{d_1+d_2}))$$

for all $(x_1, \dots, x_{d_1+d_2}) \in H$.

Next, for $(g_1, \dots, g_d) \in G$, we let $\|g\|_\infty := \max_{1 \le i \le d} |g_i|$. If $g_i \ge 0$ for all $i = 1, \dots, d$ we write $g \ge 0$. Given an increasing sequence of positive integers $(r_n)_{n=1}^\infty$, we define recurrently a real sequence $(a_n)_{n=0}^\infty$ by setting

$$a_{n+1} := \text{the integer part of } a_n r_n + \max_{h \in H_n} \|s_n(h)\|_\infty,$$

where $H_n := \{h \ge 0 \mid \|h\|_\infty < r_n\}$. We impose the following restrictions on the growth of $(r_n)_{n \ge 1}$:

$$\sum_{n=1}^\infty \frac{r_n}{a_n} < \infty \text{ and } \frac{r_n^2}{a_n} \to 0.$$

Now we are ready to determine the sequence $(C_{n+1}, F_n)_{n=0}^{\infty}$:

$$F_n := \{g \geq 0 \mid \|g\|_\infty < a_n\}, C_{n+1} := (\phi_n + s_n)(H_n),$$

where $\phi_n : H \to G$ is a 'tiling' homomorphism given by $\phi_n(h) := a_n h$. One can verify that (2.1)–(2.4) and (2.6) are all satisfied. Hence the associated rank-one (C, F)-action T of G is well defined.

Theorem 2.17 ([32]). *T is mixing.*

3. Infinite measure preserving (C, F)-actions

Recall that the canonical invariant measure for a (C, F)-action defined in 2.1 is infinite if (2.6) does not hold. This is achieved easily in the examples below. When constructing $(C_{n+1}, F_n)_n$ we follow a simple rule: after a set C_n has been already determined, just take any 'sufficiently large' Følner set for F_{n+1} (see [19], [22], [31] for details).

3.1. Weak mixing in infinite measure spaces

Let T be an ergodic transformation of a σ-finite measure space $(X, \mathfrak{B}, \mu)$. A complex number $\lambda \in \mathbb{T}$ is called an *L^∞-eigenvalue* of T if there exists a measurable function $f : X \to \mathbb{C}$ such that $f \circ T = \lambda \circ f$. It follows from the ergodicity of T that the module of f is constant a.e. The set of all L^∞-eigenvalues is called the *L^∞-spectrum* of T. The L^∞-spectrum of T is a subgroup of $\mathbb{T}$. It can be uncountable. The concept of L^∞-spectrum extends naturally to the ergodic actions of Abelian groups H (an eigenvalue of such an action is now a character of H).

We say that T has *infinite ergodic index* if $T \times T \times \cdots$ (p times) is ergodic for any p. T is *power weak mixing* if $T^{n_1} \times T^{n_2} \times \cdots \times T^{n_p}$ is ergodic for each finite sequence of non-zero integers $n_1, \dots, n_p$.

We recall that for finite measure preserving transformations the following properties are equivalent: (a) T has trivial L^∞ ($=L^2$)- spectrum, (b) $T \times T$ is ergodic, (c) T has infinite ergodic index, (d) T is power weakly mixing. For infinite measure preserving ergodic transformations we have only $(d) \Rightarrow (c) \Rightarrow (b) \Rightarrow (a)$. In [19] we prove the following theorem.

Theorem 3.1. *Let G be a countable Abelian group. Given $i \in \{1, \dots, 5\}$, there exists an infinite measure preserving free (C, F)-action $T = \{T_g\}_{g \in G}$ of G such that the property* (i) *of the following list is satisfied:*

1. *for every $g \in G$ of infinite order, the transformation T_g has infinite ergodic index,*
2. *for each finite sequence $g_1, \dots, g_n$ of G-elements of infinite order, the transformation $T_{g_1} \times \cdots \times T_{g_n}$ is ergodic,*
3. *for each $g \in G$ of infinite order, T_g has infinite ergodic index but $T_{2g} \times T_g$ is non-conservative,*
4. *the action $(T_g \times T_g)_{g \in G}$ is non-conservative,*

(5) *T has trivial L^∞-spectrum, non-ergodic Cartesian square but all k-fold Cartesian products conservative.*

We note that the first counterexamples to $(b) \Rightarrow (c)$ and $(a) \Rightarrow (b)$ for infinite measure preserving actions of $\mathbb{Z}$ were given in [62] and [1] respectively. Those transformations are infinite Markov shifts. They possess 'strong' stochastic properties and are quite different from our (C, F)-actions. Moreover, as it was noticed in [4] it is impossible to construct Markov shifts satisfying the condition (5) of Theorem 3.1. Another sort of counterexamples which are similar to our ones appeared earlier in [4], [5], [33] and [71] for $\mathbb{Z}$-actions only. Moreover, an example of $\mathbb{Z}^d$-action satisfying (2) have been constructed in [71].

In connection with Theorem 3.1(3) Silva asks:

(Q7) *Is there a non-power weakly mixing infinite measure preserving transformation T with infinite ergodic index and such that the Cartesian products $T^{n_1} \times \cdots \times T^{n_p}$ are all conservative?*

The following assertion is proved in [22]. It answers (Q7) if one takes $G = \mathbb{Z}$.

Theorem 3.2. *Let G_∞ stand for the set of G-elements of infinite order. There exists an infinite measure preserving (C, F)-action T of G such that the following properties are satisfied:*

(i) *the transformation T_g has infinite ergodic index for every $g \in G_\infty$,*
(ii) *the transformation $T_g \times T_{2g}$ is not ergodic for any $g \in G$,*
(iii) *the transformation $T_{g_1} \times \cdots \times T_{g_n}$ is conservative for every finite sequence $g_1, \dots, g_n$ of elements from G.*

3.2. Multiple recurrence of infinite measure preserving actions

Definition 3.3.

(i) Let p be a positive integer. A transformation T of a σ-finite measure space $(X, \mathfrak{B}, \mu)$ is called *p-recurrent* if for every subset $B \in \mathfrak{B}$ of positive measure there exists a positive integer k such that
$$\mu(B \cap T^{-k} B \cap \cdots \cap T^{-kp} B) > 0.$$
(ii) If T is p-recurrent for any $p > 0$, then it is called *multiply recurrent*.

By the Furnstenberg theorem [50], every finite measure preserving transformation is multiply recurrent. The situation is different in infinite measure (see [43], [2], [58]). We note that only $\mathbb{Z}$-actions are considered in those papers. In [31] Silva and the author produced new examples of infinite measure preserving actions of arbitrary countable discrete Abelian groups.

Theorem 3.4. *Given $p \in \mathbb{N} \cup \{\infty\}$, there exists a (C, F)-action $T = (T_g)_{g \in G}$ such that the transformation T_g is p-recurrent but (if $p \neq \infty$) not $(p+1)$-recurrent for every $g \in G_\infty$.*

Definition 3.3 extends naturally to actions of Abelian groups as follows.

Definition 3.5. Let G be a countable discrete infinite Abelian group and $T = (T_g)_{g\in G}$ a measure preserving action of G on a σ-finite measure space $(X, \mathfrak{B}, \mu)$.

(i) Given a positive integer $p > 0$, the action T is called *p-recurrent* if for every subset $B \in \mathfrak{B}$ of positive measure and every $g_1, \dots, g_p \in G$, there exists a positive integer k such that $\mu(B \cap T_{kg_1} B \cap \dots \cap T_{kg_p} B) > 0$.

(ii) If T is p-recurrent for any $p > 0$, then it is called *multiply recurrent.*

Clearly, T is 1-recurrent if and only if it is conservative. Every finite measure preserving G-action is multiply recurrent [51]. However in infinite measure we demonstrate the following

Theorem 3.6.

(i) *Given $p > 0$, there exists a p-recurrent (C, F)-action T such that no one transformation T_g is $(p + 1)$-recurrent, $g \in G_\infty$. (Hence T is not $(p + 1)$-recurrent.)*

(ii) *There exists a multiply recurrent (C, F)-action.*

Now let us compare Theorem 3.6(ii) with Theorem 3.4. It is easy to see that in case $G = \mathbb{Z}$, T is multiply recurrent if and only if so is T_1. Moreover, T_1 is multiply recurrent if and only if so is T_n for every $0 \neq n \in \mathbb{Z}$. Thus the multiple recurrence is equivalent to the "individual" multiply recurrence. The same holds for $G = \mathbb{Q}$ or any other group of free rank one. Hence for such groups the statements of Theorems 3.6(ii) and 3.4 are equivalent. However this is no longer true for the groups of higher free rank.

Theorem 3.7. *If the free rank of G is more than one, then there exists a (C, F)-action $T = (T_g)_{g\in G}$ which is non-2-recurrent but is individually multiply recurrent, i.e., each transformation T_g is multiply recurrent.*

Let $\mathcal{P} := \{q \in \mathbb{Q}[t] \mid q(\mathbb{Z}) \subset \mathbb{Z} \text{ and } q(0) = 0\}$. The following refinement of the concept of multiple recurrence was introduced in [14].

Definition 3.8. Let T be a measure preserving transformation of $(X, \mathfrak{B}, \mu)$.

(i) T is called *p-polynomially recurrent* if for every $q_1, \dots, q_p \in \mathcal{P}$ and $B \in \mathfrak{B}$ of positive measure there exists $n \in \mathbb{N}$ with

$$\mu(B \cap T^{q_1(n)} B \cap \dots \cap T^{q_p(n)} B) > 0.$$

(ii) If T is *p-polynomially recurrent* for every $p \in \mathbb{N}$ then it is called polynomially recurrent.

In [14] it was shown that any ergodic probability preserving transformation is polynomially recurrent. The polynomial recurrence in 'infinite measure' was investigated by Silva and the author in [31].

Theorem 3.9.

(i) *Given $p \in \mathbb{N} \cup \{\infty\}$, there exists a p-polynomially recurrent (C, F)-transformation which (if $p \neq \infty$) is not $(p + 1)$-recurrent.*

(ii) *There exists a (C, F)-transformation which is rigid (and hence multiply recurrent) but not 2-polynomially recurrent.*
(iii) *The subset of polynomially recurrent transformations is generic in the group of infinite measure preserving transformations endowed with the weak topology.*

3.3. Krengel's entropy of the product of two transformations one of which is finite measure preserving

In [67] Krengel introduced an entropy of a conservative measure preserving transformation T of an infinite σ-finite measure space $(X, \mathfrak{B}, \mu)$. If T is ergodic and A is a subset of X with $\mu(A) = 1$ then he put $h_{\mathrm{Kr}}(T) := h(T_A)$, where T_A stands for the transformation induced by T on A and $h(.)$ denotes the Kolmogorov-Sinai entropy. Notice that $h_{\mathrm{Kr}}(T)$ is well defined and it does not depend on the particular choice of A. In [67] Krengel rose a question (see also [87]):

(Q8) *Given an ergodic probability preserving transformation S with $h(S) = 0$, what is the value of $h_{\mathrm{Kr}}(T \times S)$ if $h_{\mathrm{Kr}}(T) = 0$? In particular, whether it is possible to have $h_{\mathrm{Kr}}(T \times S) = \infty$?*

In [30] Rudolph and the present author develop a conditional entropy theory for infinite measure preserving dynamical systems with respect to a σ-finite factor. Notice that T can be considered as a σ-finite factor of $T \times S$. It is not difficult to show that $h_{\mathrm{Kr}}(T \times S) = 0$ whenever S is distal (see [67] and [87] for particular cases when T has pure point spectrum). A 'converse' statement to that is one of the main results in [30]:

Theorem 3.10. *If $h(S) = 0$ but S is not distal then there is a rank-one transformation T with $h_{\mathrm{Kr}}(T) = 0$ but $h_{\mathrm{Kr}}(T \times S) = +\infty$.*

This rank-one transformation T is produced via the (C, F)-construction. The key ingredients of the proof of Theorem 3.10 are: recent results of Begun and del Junco on the existence of finite partitions with independent iterates [12], [13] and the relative orbit equivalence of the conditional entropy in the space with infinite invariant measure. The latter extends the framework of the *orbital* approach to the entropy theory initiated at [77] and developed further in [20] and [29] for finite measure preserving dynamical systems. Silva informed the present author that he with Thieullen had proved Theorem 3.10 independently of (and earlier than) [30]. As far as we know, no written version of their work is available yet.

4. Non-singular (C, F)-actions of Abelian groups

4.1. Associated flows, AT-flows and non-singular actions of funny rank one

Let G be a countable discrete amenable group and let $V = (V_g)_{g \in G}$ be an ergodic non-singular action of G on a σ-finite measure space $(X, \mathfrak{B}, \mu)$. We define a non-singular action $\widetilde{V} = (\widetilde{V}_g)_{g\in G}$ of G on the product space $(X \times \mathbb{R}, \mu \times \lambda)$ by setting

$$\widetilde{V}_g(x,t) = (V_g x, t - \log \frac{d\mu \circ V_g}{d\mu}(x)),$$

where λ is a probability measure on $\mathbb{R}$ equivalent to Lebesgue measure. Notice that $\widetilde{V}$ commutes with the $\mathbb{R}$-action on $X \times \mathbb{R}$ by translations along the second coordinate. Hence the restriction of the latter action to the σ-algebra of $\widetilde{V}$-invariant subsets is well defined. It is called the *Mackey flow associated to* V. Denote it by $W = (W_t)_{t\in\mathbb{R}}$. It is non-singular and ergodic. Hence one of the following has place:

- W is transitive and free,
- W is transitive but non-free. The stabilizer of W is $(\log\lambda)\mathbb{Z}$ for some $\lambda \in (0,1)$,
- W is trivial (on a singleton),
- W is free and non-transitive.

V is said to be of (Krieger's) type II, III_λ, III_1, III_0 respectively. We note that V is of type II if and only if there exists a V-invariant measure μ' equivalent to μ. The V-invariant measure in the class of μ is unique up to scaling. If it is finite then V is said to be *of type* II_1, otherwise V is said to be *of type* II_∞.

An action V' of another amenable group G' on $(X', \mathfrak{B}', \mu')$ is called *orbit equivalent* to V if there is a non-singular isomorphism S of X' onto X with $S\{V'_{g'}x' \mid g' \in G'\} = \{V_g Sx' \mid g \in G\}$ for a.a. $x' \in X'$. We now state a fundamental result about the orbit equivalence of amenable actions.

Theorem 4.1. *Two ergodic actions V and V' are orbit equivalent if and only if one of the following is fulfilled:*

(i) *they are both of type II_1,*
(ii) *they are both of type II_∞,*
(iii) *they are both of type III and the flows associated to them are conjugate.*

We note that originally particular cases of this theorem were established by H. Dye [42] (finite measure preserving Abelian actions) and Krieger [68] (non-singular $\mathbb{Z}$-actions). Later that was extended to general amenable actions in [16].

Definition 4.2. A non-singular flow $\{W_t\}_{t\in\mathbb{R}}$ on a standard measure space (X,μ) is called *approximately transitive* (AT) if given $\epsilon > 0$ and finitely many non-negative functions $f_1,\dots,f_n \in L^1_+(X,\mu)$ there exists a function $f \in L^1_+(X,\mu)$ and reals $t_1,\dots,t_n$ such that

$$\left\| f_i - \sum_{k=1}^m a_{ik} f \circ W_{t_k} \cdot \frac{d\mu \circ W_{t_k}}{d\mu} \right\|_1 < \epsilon, \quad i = 1,\dots,n,$$

where a_{ik}, $i = 1,\dots,n$, $j = 1,\dots,m$, are some non-negative reals.

Let $(X,\mu) = \bigotimes_{n=1}^\infty (\{0,1,\dots,m_k - 1\}, \mu_n)$, where μ_n is a non-degenerated distribution on $\{0,1,\dots,m_k - 1\}$. We assume that μ is non-atomic. Let $T : X \to X$ be the 'adding machine', i.e., $Tx := x + (1,0,0,\dots)$ coordinatewise with 'carry'.

Then T is μ-non-singular and ergodic. It is called an *odometer* (corresponding to the sequence $(\mu_n)_{n=1}^\infty$). The following statement is due to Connes and Woods [17].

Theorem 4.3. *The Mackey flow associated to an odometer is AT. Conversely, for every AT-flow W there is an odometer whose associated flow is conjugate to W.*

The following definition extends Definition 2.2 to non-singular actions.

Definition 4.4. A non-singular action S of G on a σ-finite Lebesgue space $(Y, \mathcal{A}, \nu)$ has *funny rank one* if there is a sequence $(Y_n)_{n=1}^\infty$ of measurable subsets of Y and a Følner sequence $(F_n)_{n=1}^\infty$ in G such that

(i) the subsets $S_g Y_n$, $g \in F_n$, are pairwise disjoint for each $n > 0$,
(ii) given $A \in \mathcal{A}$ of finite measure, then $\inf_{P \subset F_n} \nu(A \triangle \bigcup_{g \in P} S_g Y_n) \to 0$ as $n \to \infty$,
(iii) $\sum_{g \in F_n} \inf_{r \in \mathbb{R}} \int_{S_g Y_n} |\frac{d\nu \circ g}{d\nu} - r| d\nu \to 0$ as $n \to \infty$.

4.2. Weak mixing for non-singular (C, F)-actions

Let a sequence $(C_{n+1}, F_n)_{n=0}^\infty$ of finite subsets in G satisfy (2.1)–(2.4) and $F_0 = \{0\}$. If we assume that

$$\text{for each } g \in G, \text{ there is } N \text{ with } gF_jC_{j+1} \subset F_{j+1} \text{ for all } j > N \tag{4.1}$$

then $D_m^{(n+1)} \supset X_n$ eventually in n for any m. It follows that $D_m = R_m = \widehat{X} = X$ for all m (we utilize the notation from Subsection 2.1). This means that the associated (C, F)-action T of G is defined everywhere on X. Two points $x = (x_j)_{j \geq n}, y = (y_j)_{j \geq n} \in X_n$ belong to the same orbit of T if and only if $x_j = y_j$ for all sufficiently large j.

Given a sequence $(\kappa_n)_{n=1}^\infty$ of non-degenerated distributions on C_n such that $\bigotimes_{n=1}^\infty \kappa_n$ is non-atomic, one can construct inductively a sequence $(\tau_n)_{n=0}^\infty$ of measures on $(F_n)_{n=0}^\infty$ such that

$$\tau_0(0) = 1 \text{ and } \tau_n(fc) = \tau_{n-1}(f)\kappa_n(c) \text{ for all } f \in F_{n-1}, c \in C_n. \tag{4.2}$$

We furnish $X_n = F_n \times \prod_{k>n} C_k$ with the product measure $\mu_n := \tau_n \otimes \bigotimes_{k>n} \kappa_k$. Clearly, $\mu_n = \mu_{n+1} \restriction X_n$. Hence an inductive limit μ of $(\mu_n)_{n=1}^\infty$ is well defined. Clearly, μ is a σ-finite measure on X. We call it a *(C, F, κ)-measure*. Notice that the equivalence class of μ is determined completely by $(\kappa_n)_{n>0}$ (it does not depend on a particular choice of $(\tau_n)_{n>0}$). One can verify that T is μ-non-singular and

$$\frac{d\mu \circ T_g}{d\mu}(x) = \frac{\tau_n(y_n)}{\tau_n(x_n)} \cdot \prod_{k>n} \frac{\kappa_k(y_n)}{\kappa_k(x_n)},$$

whenever $x = (x_k)_{k \geq n}$ and $T_g x = (y_k)_{k \geq n}$ belong to X_n. Moreover, the dynamical system (X, μ, T) has funny rank one.

We now record a non-singular counterpart of Theorem 3.1. (In a similar way one can obtain non-singular counterparts of the other theorems of Section 3.)

Theorem 4.5 ([19]). *Let G be Abelian and let W be an AT-flow (see Definition 4.2). Given $i \in \{1, \dots, 5\}$, there is a non-singular funny rank-one (C, F)-action T of G such that the associated flow of T is W and the property* (i) *of Theorem* 3.1 *is satisfied.*

We note that some very particular cases of Theorem 4.5 were established earlier in [4], [5] and [71] (under assumptions that $G = \mathbb{Z}$, T is of type III_λ and $0 < \lambda \leq 1$).

4.3. Non-singular Chacon transformations

In the rest of the paper let $G = \mathbb{Z}$. We are going to distinguish a special subclass of non-singular (C, F)-actions of $\mathbb{Z}$. To this end we define a sequence $(h_n)_{n=1}^\infty$ of positive integers recurrently by setting $h_1 := 1$, $h_{n+1} := 3h_n + 1$. It is easy to see that the sequences of sets $F_{n-1} := \{0, 1, \dots, h_n - 1\}$ and $C_n := \{0, h_n, 2h_n + 1\}$ satisfy (2.1)–(2.4) but does not satisfy (4.1). Nevertheless it is not difficult to verify that the domain of the corresponding (C, F)-action is $\widehat{X} = X \setminus D$, where

$$D := \bigcup_{n \geq 0} \{x = (x_k)_{k \geq n} \in X_n \mid \text{either } x_k = 0 \text{ eventually} \\ \text{or } x_k = 2h_k + 1 \text{ eventually}\}.$$

Next, given a sequence $(\kappa'_n)_{n=1}^\infty$ of non-degenerated distributions on $\{0, 1, 2\}$ with non-atomic product $\bigotimes_{n=1}^\infty \kappa'_n$, we define measures κ_n on C_n as follows: $\kappa_n(0) := \kappa'_n(0)$, $\kappa_n(h_n) := \kappa'_n(1)$ and $\kappa_n(2h_n + 1) := \kappa'_n(2)$. Now take any sequence τ_n of measures on F_n satisfying (4.2) and denote by μ the corresponding (C, F, κ)-measure.

Definition 4.6. The corresponding dynamical system (X, μ, T), or simply T_1, is called the *non-singular* 2*-cuts Chacon transformation associated with* $(C_n, F_n, \kappa_n)_n$. We call T

- *symmetric* if $\kappa'_n(0) = \kappa'_n(2)$,
- *stationary* if $\kappa'_1 = \kappa'_2 = \cdots$,
- *κ-weakly stationary* for a distribution κ on $\{0, 1, 2\}$ if for each $n > 0$ there exists $m > n$ with $\kappa'_m = \cdots = \kappa'_{n+m} = \kappa$.

Since μ is non-atomic and D is countable, T is well defined on the μ-conull subset $\widehat{X}$.

The main theorem of [60] states that every *λ-weakly stationary symmetric* non-singular Chacon transformation has ergodic Cartesian square, where λ is the equi-distribution on $\{0, 1, 2\}$. A stronger result was obtained in [5] for *stationary symmetric* Chacon transformations. It was shown that they are power weakly mixing. However, as was noticed by the authors of [5], their methods do not work with the transformations considered in [60]. They raised the following question:

(Q9) *Whether the weakly stationary symmetric Chacon transformations have infinite ergodic index?*

We answer this question affirmatively in [22] by demonstrating

Theorem 4.7. *Each weakly stationary non-singular Chacon transformation with 2-cuts is power weakly mixing.*

5. Concluding remarks. Open problems

In these short notes we were unable to enlighten topological aspects of the (C,F)-construction. It appears that in many natural cases the (C,F)-actions appear as minimal uniquely ergodic actions on totally disconnected locally compact (or even compact) spaces. Moreover, while proving some 'measurable' statements for the (C,F)-actions we often obtain as a byproduct their topological counterparts. For this, we just need to replace the terms "ergodic index", "measurable recurrence", etc. with "index of topological transitivity", "topological recurrence", etc. See [19], [31], [22], [24], [26] for details.

We note also that the (C,F)-construction was utilized in the theory of topological orbit equivalence of locally compact minimal Cantor systems as a tool for constructing natural illustrative examples [21].

We conclude this paper by recording a list of open problems and questions which – as we believe – may be solved or answered with the use of the (C,F)-construction.

(P1) To construct a simple transformation whose centralizer is a Heizenberg group (of some order).

(P2) (Glasner, del Junco) To construct a simple $\mathbb{Z}^2$-action $T=(T_g)_{g\in\mathbb{Z}^2}$ which is isomorphic to $(T_{Ag})_{g\in\mathbb{Z}^2}$, where A is an aperiodic automorphism of $\mathbb{Z}^2$.

(P3) To construct a mixing rank-one action for each countable FC-group (i.e., a group with the finite conjugacy classes).

(P4) To construct an explicit (non-random) example of mixing rank-one action of $\bigoplus_{i=0}^{\infty}\mathbb{Z}/2\mathbb{Z}$.

(P5) To construct non-mixing weakly mixing (or even mildly mixing) counterparts of the transformations discussed in 2.2.1–2.2.4.

(P6) (Ward) Is there a rank-one non-mixing $\mathbb{Z}^2$-action $T=(T_g)_{g\in\mathbb{Z}^2}$ such that every transformation T_g, $g\neq 0$, is mixing?

(P7) (Katok) Is there an ergodic transformation T such that $\mathcal{M}(T)=\{2,5\}$?

(P8) For $n>2$, to construct a mixing transformation with homogeneous spectrum of multiplicity n.

(P9) (Ryzhikov) For $n>1$, to construct an ergodic $\mathbb{R}$-action with homogeneous spectrum of multiplicity n.

(P10) (Bergelson) Is there an infinite measure preserving transformation T with infinite ergodic index but such that $T\times T^{-1}$ is not ergodic?

(P11) (Silva) To construct a rank-one infinite measure preserving action $T=(T_g)_{g\in\mathbb{R}^2}$ of $\mathbb{R}^2$ such that the transformation T_g has infinite ergodic index for each $g\neq 0$.

(P12) To remove the restriction that W is AT from the statement of Theorem 4.5. We think that the main result of [41] (that any ergodic non-singular transformation is orbit equivalent to a non-singular Markov odometer) may help to solve this problem.

(P13) Whether the results of del Junco and Silva [39] on the centralizer and factors of Cartesian products of some special non-singular Chacon transformations extend to the weakly stationary non-singular Chacon transformations?

References

[1] J. Aaronson, M. Lin and B. Weiss, *Mixing properties of Markov operators and ergodic transformations, and ergodicity of Cartesian products.* Isr. J. Math. **33** (1979), 198–224.

[2] J. Aaronson and H. Nakada, *Multiple recurrence of Markov shifts and other infinite measure preserving transformations.* Isr. J.Math. **117** (2000), 285–310.

[3] T.M. Adams, *Smorodinsky's conjecture on rank one systems.* Proc. Amer. Math. Soc. **126** (1998), 739–744.

[4] T. Adams, N. Friedman and C.E. Silva, *Rank-one weak mixing for nonsingular transformations.* Isr. J. Math. **102** (1997), 269–281.

[5] ———, *Rank one power weak mixing nonsingular transformations.* Erg. Th. & Dyn. Sys. **21** (2001), 1321–1332.

[6] T. Adams and C.E. Silva, $\mathbb{Z}^d$*-staircase actions.* Ergod. Th. & Dynam. Sys. **19** (1999), 837–850.

[7] O.N. Ageev, *On ergodic transformations with homogeneous spectrum.* J. Dynam. Control Systems **5** (1999), 149–152.

[8] ———, *On the spectrum of Cartesian powers of classical automorphisms.* Mat. Zametki **68** (2000), 643–647.

[9] ———, *The homogeneous spectrum problem in ergodic theory.* Invent. Math. **160** (2005), 417–446.

[10] ———, *Spectral rigidity of group actions: applications to the case* $\mathrm{gr}\langle t,s;ts=st^2\rangle$. Proc. Amer. Math. Soc. **134** (2006), 1331–1338.

[11] J.R. Baxter, *A class of ergodic transformations having simple spectrum.* Proc. Amer. Math. Soc. **27** (1971), 275–279.

[12] B. Begun and A. del Junco, *Amenable groups, stationary measures and partitions with independent iterates.* Isr. J. Math., **158** (2007), 41–64.

[13] ———, *Partitions with independent iterates in random dynamical systems.* Preprint, ArXiv:0706.1607.

[14] V. Bergelson and A. Leibman, *Polynomial extensions of van der Waerden's and Semerédi's theorems.* J. Amer. Math. Soc. **9** (1996), 725–753.

[15] R.V. Chacon, *Weakly mixing transformations which are not strongly mixing.* Proc. Amer. Math. Soc. **22** (1969), 559–562.

[16] A. Connes, J. Feldman, and B. Weiss, *An amenable equivalence relation is generated by a single transformation.* Ergod. Th. & Dynam. Sys. **1** (1981), 431–450.

[17] A. Connes and E.J. Woods, *Approximately transitive flows and ITPFI factors.* Erg. Th. & Dyn. Syst. **5** (1985), 203–236.

[18] D. Creutz and C.E. Silva, *Mixing on a class of rank-one transformations.* Ergod. Th. & Dynam. Sys. **24** (2004), 407–440.

[19] A.I. Danilenko, *Funny rank-one weak mixing for nonsingular Abelian actions.* Isr. J. Math. **121** (2001), 29–54.

[20] ——, *Entropy theory from orbital point of view.* Monatsh. Math. **134** (2001), 121–141.

[21] ——, *Strong orbit equivalence of locally compact Cantor minimal systems.* Internat. J. Math. **12** (2001), 113–123.

[22] ——, *Infinite rank one actions and nonsingular Chacon transformations.* Illinois J. Math. **48** (2004), 769–786.

[23] ——, *Explicit solution of Rokhlin's problem on homogeneous spectrum and applications.* Ergod. Th. & Dyn. Syst. **26** (2006), 1467–1490.

[24] ——, *Mixing rank-one actions for infinite sums of finite groups.* Isr. J. Math. **156** (2006), 341–358.

[25] ——, *On simplicity concepts for ergodic actions.* J. d'Anal. Math. (to appear)

[26] ——, *Uncountable collection of mixing rank-one actions for locally normal groups.* Preprint.

[27] ——, *Weakly mixing rank-one transformations conjugate to their squares.* Preprint.

[28] A.I. Danilenko and A. del Junco, *Cut-and-stack simple weakly mixing map with countably many prime factors.* Proc. Amer. Math. Soc. (to appear)

[29] A.I. Danilenko and K.K. Park, *Generators and Bernoullian factors for amenable actions and cocycles on their orbits.* Erg. Th. & Dynam. Syst. **22** (2002), 1715–1745.

[30] A.I. Danilenko and D.J. Rudolph, *Conditional entropy theory in infinite measure and a question of Krengel.* Isr. J. Math. (to appear)

[31] A.I. Danilenko and C.E. Silva, *Multiple and polynomial recurrence for abelian actions in infinite measure.* J. London Math. Soc. **69** (2004), 183–200.

[32] ——, *Mixing rank-one actions of locally compact Abelian groups.* Ann. Inst. H. Poincaré, Probab. Statist. **43** (2007), 375-398.

[33] S.L. Day, B.R. Grivna, E.P. MaCartney and C.E. Silva, *Power weakly mixing infinite transformations.* New York J. Math. **5** (1999), 17–24.

[34] A. del Junco, *A simple map with no prime factors.* Isr. J. Math. **104** (1998), 301–320.

[35] A. del Junco and M. Lemańczyk, *Joinings of distally simple automorphisms.* (in preparation)

[36] A. del Junco, M. Rahe and L. Swanson, *Chacon's automorphism has minimal self-joinings.* J. Analyse Math. **37** (1980), 276–284.

[37] A. del Junco and D. Rudolph, *On ergodic actions whose self-joinings are graphs.* Erg. Th. & Dyn. Syst. **7** (1987), 531–557.

[38] ——, *A rank-one, rigid, simple, prime map.* Ergodic Theory & Dynam. Systems **7** (1987), 229–247.

[39] A. del Junco and C. Silva, *On factors of non-singular Cartesian products.* Ergodic Theory & Dynam. Systems **23** (2003), 1445–1465.

[40] A. del Junco and R. Yassawi, *Multiple mixing and rank one group actions.* Canad. J. Math. **52** (2000), 332–347.

[41] A.H. Dooley and T. Hamachi, *Nonsingular dynamical systems, Bratteli diagrams and Markov odometers.* Israel J. Math. **138** (2003), 93–123.

[42] H.A. Dye, *On groups of measure preserving transformations. I, II.* Amer. J. Math. **81** (1959), 119–159 and **85** (1963), 551–576.

[43] S. Eigen, A, Hajian and K. Halverson, *Multiple recurrence and infinite measure preserving odometers.* Isr. J. Math. **108** (1998), 37–44.

[44] B. Fayad, *Rank one and mixing differentiable flows.* Invent. Math. **160** (2005), 305–340.

[45] S. Ferenczi, *Systèmes de rang un gauche.* Ann. Inst. H. Poincaré. Probab. Statist. **21** (1985), 177–186.

[46] ———, *Systems of finite rank.* Colloq. Math. **73** (1997), 35–65.

[47] K. Frączek and M. Lemańczyk, *On mild mixing of special flows over irrational rotations under piecewise smooth maps.* Ergodic Theory & Dynam. Systems **26** (2006), 719–738.

[48] N.A. Friedman, *Introduction to ergodic theory.* Van Nostrand Reinhold Mathematical Studies, No. 29. Van Nostrand Reinhold Co., New York-Toronto, Ont.-London, 1970.

[49] H. Furstenberg, *Disjointness in ergodic theory, minimal sets and diophantine approximation.* Math. Syst. Th. **1** (1967), 1–49.

[50] ———, *Ergodic behavior of diagonal measures and a theorem of Szemerédi on arithmetic progressions.* J. d'Analyse Math. **31** (1977), 204–256.

[51] H. Furnstenberg and Y. Kazsnelson, *An ergodic Szemerédi theorem for commuting transformations.* J. d'Analyse Math. **31** (1978), 275–291.

[52] E. Glasner, *Ergodic theory via joinings.* Mathematical Surveys and Monographs, 101. American Mathematical Society Providence, RI 2003.

[53] E. Glasner and B. Weiss, *A simple weakly mixing transformation with non-unique prime factors.* Amer J. Math. **116** (1994), 361–375.

[54] G.R. Goodson, *A survey of recent results in the spectral theory of ergodic dynamical systems.* J. Dynam. Control Systems **5** (1999), 173–226.

[55] ———, *Ergodic dynamical systems conjugate to their composition squares.* Acta Math. Univ. Comenian. (N.S.) **71** (2002), 201–210.

[56] ———, *Spectral properties of ergodic dynamical systems conjugate to their composition squares.* Colloq. Math. **107** (2007), 99–118.

[57] G.R. Goodson, J. Kwiatkowski, M. Lemańczyk and P. Liardet, *On the multiplicity function of ergodic group extensions of rotations.* Studia Math. **102** (1992), 157–174.

[58] K. Gruher, F. Hines, D. Patel, C.E. Silva and R. Waelder, *Power weak mixing does not imply multiple recurrence in infinite measure and other counterexamples* New York J. of Math. **9** (2003), 1–22.

[59] P.R. Halmos, *Lectures on ergodic theory.* Publications of the Mathematical Society of Japan, no. 3. The Mathematical Society of Japan 1956.

[60] T. Hamachi and C.E. Silva, *On nonsingular Chacon transformations.* **44** (2000), 868–883.

[61] A.S.A. Johnson and K.K. Park, *A dynamical system with a Z^2 centralizer.* J. Math. Anal. Appl. **210** (1997), 337–359.

[62] S. Kakutani and W. Parry, *Infinite measure preserving transformations with "mixing"* . Bull. Amer. Math. Soc. **69** (1963), 752–756.

[63] S.A. Kalikow, *Twofold mixing implies threefold mixing for rank one transformations.* Ergodic Theory & Dynam. Systems **4** (1984), 237–259.

[64] J.L. King, *The commutant is the weak closure of the powers, for rank-*1 *transformations* Ergodic Theory & Dynam. Systems **6** (1986), 363–384.

[65] ———, *Joining-rank and the structure of finite rank mixing transformations.* J. d'Analyse Math. **51** (1988), 182–227.

[66] ———, *The generic transformation has roots of all orders.* Colloq. Math. **84/85** (2000), 521–547.

[67] U. Krengel, *Entropy of conservative transformations.* Z. Wahrscheinlichkeitstheorie verw. Geb. **7** (1967), 161–181.

[68] W. Krieger, *On ergodic flows and the isomorphism of factors.* Math. Ann. **223** (1976), 19–70.

[69] J. Kwiatkowski (jr) and M. Lemańczyk, *On the multiplicity function of ergodic group extensions. II.* Studia Math. **116** (1995), 207–215.

[70] B. Madore, *Rank-one group actions with simple mixing $\mathbb{Z}$-subactions.* New York J. Math. **10** (2004), 175–194.

[71] E.J. Muehlegger, A.S. Raich, C.E. Silva, M.P. Touloumtzis, B. Narasimhan and W. Zhao, *Infinite ergodic index $\mathbb{Z}^d$-actions in infinite measure.* Colloq. Math. **82** (1999), 167–190.

[72] D.S. Ornstein, *On the root problem in ergodic theory.* Proc. Sixth Berkley Symp. Math. Stat. Prob. (Univ. California, Berkeley, Calif., 1970/1971), Vol II: Probability Theory, pp. 347–356. Univ. of California Press Berkeley, Calif., 1972.

[73] A. Prikhodko, *Stochastic constructions of flows of rank one.* Mat. Sb. **192** (2001), 61–92.

[74] M. Ratner, *Horocycle flows, joinings and rigidity of products.* Ann. of Math. (2) **118** (1983), 277–313.

[75] D.J. Rudolph, *The second centralizer of a Bernoulli shift is just its powers.* Israel J. Math. **29** (1978), 167–178.

[76] ———, *An example of a measure preserving map with minimal self-joinings.* J. d'Analyse Math. **35** (1979), 97–122.

[77] D.J. Rudolph and B. Weiss, *Entropy and mixing for amenable group actions.* Annals of Math. **151** (2000), 1119–1150.

[78] V.V. Ryzhikov, *Mixing, rank and minimal self-joining of actions with invariant measure.* Mat. Sb. **183** (1992), 133–160.

[79] ———, *Joinings and multiple mixing of the actions of finite rank.* Funktsional. Anal. i Prilozhen. **27** (1993), 63–78.

[80] ———, *Around simple dynamical systems. Induced joinings and multiple mixing.* J. Dynam. Control Systems **3** (1997), 111–127.

[81] ———, *Transformations having homogeneous spectra.* J. Dynam. Control Systems **5** (1999), 145–148.

[82] ———, *Homogeneous spectrum, disjointness of convolutions, and mixing properties of dynamical systems.* Selected Russian Mathematics **1** (1999), 13–24.

[83] ———, *The Rokhlin problem on multiple mixing in the class of actions of positive local rank.* Funktsional. Anal. i Prilozhen **34** (2000), 90–93.

[84] ———, *Self-joinings of rank-one actions and applications.* Preprint.

[85] V.V. Ryzhikov and J.-P. Thouvenot, *Disjointness, divisibility and quasi-simplicity of measure-preserving actions.* Funktsional. Anal. i Prilozhen. **40** (2006), 85–89.

[86] P.C. Shields, *The ergodic theory of discrete sample paths.* Graduate Studies in Mathematics, 13. American Mathematical Society, Providence, RI, 1996.

[87] C.E. Silva and P. Thieullen, *A skew product entropy for nonsingular transformations.* J. London Math. Soc. **52** (1995), 497–516.

[88] A. Sokhet, *Les actions approximativement transitives dans la théory ergodique.* Thèse de doctorat, Université Paris VII, 1997.

[89] J.-P. Thouvenot, *Some properties and applications of joinings in ergodic theory.* Ergodic theory and its connections with harmonic analysis (Alexandria, 1993), pp. 207–235. London Math. Soc. Lecture Note Ser., 205. Cambridge Univ. Press, Cambridge, 1995.

[90] W.A. Veech, *A criterion for a process to be prime.* Monatsh. Math. **94** (1982), 335–341.

Alexandre I. Danilenko
Institute for Low Temperature Physics & Engineering
Ukrainian National Academy of Sciences
47 Lenin Ave.
Kharkov, 61164
Ukraine
e-mail: danilenko@ilt.kharkov.ua

Progress in Mathematics, Vol. 265, 353–375

Homomorphic Images of Branch Groups, and Serre's Property (FA)

Thomas Delzant and Rostislav Grigorchuk

Dedicated to the memory of Sasha Reznikov

Abstract. It is shown that a finitely generated branch group has Serre's property (FA) if and only if it does not surject onto the infinite cyclic group or the infinite dihedral group. An example of a finitely generated self-similar branch group surjecting onto the infinite cyclic group is constructed.

Mathematics Subject Classification (2000). 20F65.

Keywords. Property (FA), branch group, tree, hyperbolic space, indicable group.

Introduction

The study of groups acting on trees is a central subject in geometric group theory. The Bass-Serre theory establishes a dictionary between the geometric study of groups acting on trees and the algebraic study of amalgams and HNN extensions. A central topic of investigation is the fixed point property for groups acting on trees, introduced by J.-P. Serre in his book as the property (FA)[Ser80]. A fundamental result due to Tits states that a group without a free subgroup on two generators which acts on a tree by automorphisms fixes either a vertex or a point on the boundary or permutes a pair of points on the boundary; see [Tit77, PV91]. The group $SL(3,\mathbb{Z})$, and more generally, groups with Kazdhan's property (T), in particular lattices in higher rank Lie groups have the property (FA) ([dlHV89, Mar91]). A natural problem is to understand the structure of the class of (FA)-groups (the class of groups having the property (FA)). There is an algebraic characterization of enumerable (FA)-groups, due to J.-P. Serre. ([Ser80], Theorem I.6.15, page 81).

An enumerable group has the property (FA) if and only if it satisfies the following three conditions:

Partially supported by NSF grants DMS-0600975 and DMS-0456185.

(i) it is not an amalgam,
(ii) it is not indicable (i.e., admits no epimorphism onto $\mathbb{Z}$),
(iii) it is finitely generated.

But even such a nice result does not clarify the structure of the class of (FA)-groups, as the first of these properties is usually difficult to check.

The class of (FA)-groups contains the class of finite groups and is closed under quotients. As every infinite finitely generated group surjects onto a just-infinite group (i.e., an infinite group with all proper quotients finite) a natural problem is to describe just infinite (FA)-groups.

In [Gri00] (by using [Wil71]) the class (JINF) of just infinite groups is divided in three subclasses: the class (B) of branch groups, the class (HJINF) of finite extensions of finite powers of hereditary just infinite groups and the class (S) of finite extensions of finite powers of simple groups. For example, the group $SL(3,\mathbb{Z})$ belongs to the class (JINF); all infinite finitely generated simple torsion groups constructed in [Ol′79] are (FA)-groups and belong to the class (S).

A precise definition of a branch group is given in Section 1. Roughly speaking a branch group is a group which acts faithfully and level transitively on a spherically homogeneous rooted tree, and for which the structure of the lattice of subnormal subgroups mimics the structure of the tree. Branch groups may enjoy unusual properties. Among them one can find finitely generated infinite torsion groups, groups of intermediate growth, amenable but not elementary amenable groups and other surprising objects. Profinite branch groups are also related to Galois theory and other topics in Number Theory [Bos00].

In this article we discuss fixed point properties for actions of branch groups on Gromov hyperbolic spaces, in particular on $\mathbb{R}$-trees, and apply Bass-Serre theory to branch groups. Recall (see [Ser80]) that a group G is an amalgam (resp. an HNN extension) if it can be written as a free product with amalgamation $G = A *_C B$, with $C \neq A, B$ (resp. $G = A*_{tCt^{-1}=C'}$). We say that this amalgam (resp. HNN extension) is *strict* if the index of C in A is at least 3 and the index of C in B is at least 2 (resp. the indexes of C and C' in A are at least 2).

One of the corollaries of Theorem 3 is:

Theorem 1. *Let G be a finitely generated branch group. Then G is not a strict amalgam or HNN. Therefore a branch group cannot be an amalgam unless it surjects onto $\mathbb{D}_\infty$. It has Serre's property* (FA) *if and only if it is not indicable and has no epimorphism onto $\mathbb{D}_\infty$.*

We say that a group is (FL) if it has no epimorphism onto $\mathbb{Z}$ or $\mathbb{D}_\infty$. A f.g. group is (FL) if and only if it fixes a point whenever it acts isometrically on a line.

All proper quotients of branch groups are virtually abelian [Gri00]. A quotient of a branch group may be infinite: the full automorphism group of the binary rooted tree is a branch group and its abelianization is the infinite cartesian product of copies of a group of order two. It is more difficult to construct examples of finitely generated branch groups with infinite quotients (especially in the restricted setting of self-similar groups). The corresponding question was open since 1997 when the second author introduced the notion of a branch group. Perhaps the main difficulty was psychological, as he (and some other researches working in the area) was sure that all finitely generated branch groups are just infinite. Now we know that this is not correct and the second part of the paper (Section 3) is devoted to the construction of an example of an indicable finitely generated branch group (thus providing an example of a finitely generated branch group without the property (FA)). This example is the first example of a finitely generated branch group defined by a finite automaton that is not just infinite. Another example is related to Hanoi Towers group on 3 pegs H (introduced in [GŠ06] and independently in [Nek05]). Hanoi Towers group H is a 3-generated branch group [GŠ07] that has a subgroup of index 4 (the Apollonian group) which is also a branch group and is indicable (this is announced in[GNŠ06]). The group H itself is not indicable (it has finite abelianization), but it surjects onto $\mathbb{D}_\infty$, as was recently observed by Zoran Šunić. Thus H is the first example of a finitely generated branch group defined by a finite automaton that surjects onto $\mathbb{D}_\infty$.

The example of an indicable branch group presented in this paper is an elaboration of the 3-generated torsion 2-group $G = \langle a, b, c, d\rangle$ firstly constructed in [Gri80a] and later studied in [Gri89, GM93, Gri98a, Gri99] and other papers (see also the Chapter VIII of the book [dlH00] and the article [CSMS01].

Let L be the group generated by the automaton defined in Figure 1.

Theorem 2. *The group L is a branch contracting group that surjects onto $\mathbb{Z}$.*

Starting from this example, a construction of a branch group surjecting onto $\mathbb{D}_\infty$ has been proposed by Dan Segal. Let L be an indicable branch group and $l : L \to \mathbb{Z}$ an epimorphism. It is easy to see that the semi-direct product $H = \mathbb{Z}/2\mathbb{Z} \ltimes (L \times L)$ is branch. Furthermore, its surjects onto $\mathbb{D}_\infty = \mathbb{Z}/2\mathbb{Z} \ltimes \mathbb{Z}$ by the unique morphism l' whose restriction to $\mathbb{Z}/2\mathbb{Z}$ is the identity and such that $l'(g, h) = l(g) - l(h)$.

An interesting question is to understand which virtually abelian group can be realized as a quotient of a finitely generated branch group. This question is closely related to the problem of characterization of finitely generated branch groups having the Furstenberg-Tychonoff fixed ray property (FT) [Gri98b]) (existence of an invariant ray for actions on a convex cone with compact base). The problem of indicability of branch groups is also related to the recent work of D.W. Morris [Mo] who studied the action of an amenable group by homemorphisms on the line.

Acknowledgments. The authors are thankful to Zoran Šunić and Laurent Bartholdi for valuable discussions, comments, and suggestions.

1. Basic definitions and some notation

Let T be a tree, G be a group acting on T (without inversion of edges) and T^G the the set of fixed vertices of T.

Definition 1. A group G has the property (FA) if for every simplicial tree T on which G acts simplicially and without inversion, $T^G \neq \emptyset$.

The class of (FA)-groups possesses the following properties.

(i) The class of (FA)-groups is closed under taking quotients.
(ii) Let G be a group with the property (FA). If G is a subgroup of an amalgamated free product $G_1 *_A G_2$ or an HNN extension $G = G_1 *_A$, then G is contained in a conjugate of G_1 or G_2.
(iii) The class of (FA)-groups is closed under forming extensions.
(iv) If a subgroup of finite index in a group G has the property (FA), then the group G itself has the property (FA).
(v) Every finitely generated torsion group has the property (FA).

The class of (FA)-groups has certain nice structural properties and is interesting because of the strong embedding property given by (ii) and by the fact that the eigenvalues of matrices in the image of a linear representation $\rho : G \to GL_2(k)$ are integral over $\mathbb{Z}$ for any field k (Prop. 22, [Ser80]).

The property (i), the existence of just infinite quotients for finitely generated infinite groups and the trichotomy from [Gri00] mentioned in the introduction make the problem of classification of finitely generated just infinite (FA)-groups worthwhile. We are reduced to the classification of finitely generated (FA)-groups in each of the classes (B), (HJINF) and (S). Below we solve this problem, in a certain sense, for the class (B).

If a group G has a quotient isomorphic to $\mathbb{Z}$, then it acts by translations on a line and cannot be an (FA)-group. Similarly, if G surjects onto the infinite dihedral group $\mathbb{D}_\infty$, then it acts on the line via the obvious action of $\mathbb{D}_\infty$. This suggests the following definition (the first part being folklore):

Definition 2. a) A group is called *indicable* if it admits an epimorphism onto $\mathbb{Z}$.
b) A group has property (FL) (fixed point on line) if every action of G by isometry on a line fixes a point. If G is finitely generated this means that G has no epimorphism onto $\mathbb{Z}$ or $\mathbb{D}_\infty$.

In this article we will often use two other notions: the notion of a hyperbolic space and that of a branch group.

For the definition and the basic properties of Gromov hyperbolic spaces we refer the reader to [CDP90]. The theory of CAT(0)-spaces is described in [BH99]. For the definition and the study of basic properties of branch groups we refer the reader to [Gri00, BGŠ03].

Let us recall the main definition and a few important facts and notations that will be often used later.

Definition 3. A group $\mathcal{G}$ is an *algebraically branch group* if there exists a sequence of integers $\overline{k} = \{k_n\}_{n=0}^{\infty}$ and two decreasing sequences of subgroups $\{R_n\}_{n=0}^{\infty}$ and $\{V_n\}_{n=0}^{\infty}$ of $\mathcal{G}$ such that

(1) $k_n \geq 2$, for all $n > 0$, $k_0 = 1$,

(2) for all n,

$$R_n = V_n^{(1)} \times V_n^{(2)} \times \cdots \times V_n^{(k_0 k_1 \ldots k_n)}, \tag{1.1}$$

where each $V_n^{(j)}$ is an isomorphic copy of V_n,

(3) for all n, the product decomposition (1.1) of R_{n+1} is a refinement of the corresponding decomposition of R_n in the sense that the j-th factor $V_n^{(j)}$ of R_n, $j = 1, \ldots, k_0 k_1 \ldots k_n$ contains the j-th block of k_{n+1} consecutive factors

$$V_{n+1}^{((j-1)k_{n+1}+1)} \times \cdots \times V_{n+1}^{(jk_{n+1})}$$

of R_{n+1},

(4) for all n, the groups R_n are normal in $\mathcal{G}$ and

$$\bigcap_{n=0}^{\infty} R_n = 1,$$

(5) for all n, the conjugation action of $\mathcal{G}$ on R_n permutes transitively the factors in (1.1),
and

(6) for all n, the index $[\mathcal{G} : R_n]$ is finite.

A group G is a *weakly algebraically branch group* if there exists a sequence of integers $\overline{k} = \{k_n\}_{n=0}^{\infty}$ and two decreasing sequences of subgroups $\{R_n\}_{n=0}^{\infty}$ and $\{V_n\}_{n=0}^{\infty}$ of G satisfying the conditions (1)–(5).

There is a geometric counterpart of this definition.

Let $(\mathcal{T}, \emptyset)$ be a spherically homogeneous rooted tree, where $\emptyset$ is the root and $\mathcal{G}$ be a group acting on $(\mathcal{T}, \emptyset)$ by automorphisms preserving the root. Let v be a vertex, and $\mathcal{T}_v$ be the subtree consisting of the vertices w such that $v \in [w, \emptyset]$ (geodesic segment joining w with the root). The rigid stabilizer $rist_{\mathcal{G}}(v)$ of a vertex v consists of elements acting trivially on $\mathcal{T} \backslash \mathcal{T}_v$. The rigid stabilizer of the n-th level, denoted $rist_{\mathcal{G}}(n)$, is the group generated by the rigid stabilizers of the vertices on level n.

The action of $\mathcal{G}$ on $\mathcal{T}$ is called geometrically branch if it is faithfull, level transitive, and if, for any n, the rigid stabilizer $rist_{\mathcal{G}}(n)$ of $n-$th level of the tree has finite index in $\mathcal{G}$.

Observe that, in the level transitive case, the rigid stabilizers of the vertices of the same level are conjugate in $\mathcal{G}$. In this case $rist_{\mathcal{G}}(n)$ is algebraically isomorphic to the product of copies of the same group (namely the rigid stabilizer of any vertex on the given level). Hence the rigid stabilizers of the levels and vertices play the role of the subgroups R_n and V_n of the algebraic definition. A geometrically branch group is therefore algebraically branch. The algebraic definition is slightly more general than the geometric one but at the moment it is not completely clear how big the difference between the two classes of groups is. Observe that in Section 2 we will assume that the considered groups are algebraically branch, while in sections 3 and 4 we construct examples of geometrically branch groups.

When constructing these examples, we will deal only with actions on a rooted binary tree and our notation and the definition below are adapted exactly for this case. Let $\mathcal{G}$ be a branch group acting on a binary rooted tree $\mathcal{T}$. The vertices of $\mathcal{T}$ are labeled by finite sequences of 0 and 1. Let $\mathcal{T}_0, \mathcal{T}_1$ be the two subtrees consisting of the vertices starting with 0 or 1, respectively.

Notation. If $A, B, C \subset Aut(\mathcal{T})$ are three subgroups, we write $A \succeq B \times C$ if A contains the subgroup $B \times C$ of the product $Aut(\mathcal{T}_0) \times Aut(\mathcal{T}_1)$ via the canonical identification of $Aut(\mathcal{T})$ with $Aut(\mathcal{T}_i)$.

Recall that a level transitive group $\mathcal{G}$ acting on a regular rooted binary tree is called regular branch over its normal subgroup H if H has finite index in $\mathcal{G}$, $H \succeq H \times H$ and if moreover the last inclusion is of finite index.

A level transitive group $\mathcal{G}$ is called weakly regular branch over a subgroup H if H is nontrivial and $H \succeq H \times H$.

Definition 4. A group $\mathcal{G}$ acting on the rooted binary tree $(\mathcal{T}, \emptyset)$ is called *self-replicating* if, for every vertex u, the image of the stabilizer $st_{\mathcal{G}}(u)$ of u in $Aut(\mathcal{T}_u)$ (the automorphism group of the rooted tree $\mathcal{T}_u$) coincides with the group G after the canonical identification of $\mathcal{T}$ with $\mathcal{T}_u$.

Obviously a self-replicating group is level transitive if and only if it is transitive on the first level (see also Lemma A in [Gri00]).

We will use the notations $\langle R \rangle^{\mathcal{G}}$ for the normal closure in $\mathcal{G}$ of a subset $R \subset \mathcal{G}$, $x^y = y^{-1}xy$, $[x, y] = x^{-1}y^{-1}xy$. Given two subgroups, A, B in a group $\mathcal{G}$, $[A, B]$ is the subgroup of $\mathcal{G}$ generated by the commutators $[a, b]$ of elements in A and B, and $[A, B,]^{\mathcal{G}}$ its normal closure. If $\mathcal{G}$ is a group, $\gamma_3(\mathcal{G})$ denote the third member of its lower central series.

2. Fixed point properties of branch groups

Let X be a Gromov hyperbolic metric space, and ∂X its Gromov's boundary. Recall (see [Gro87] or [CDP90] Chap. 9 for instance) that a subgroup G of the group $Isom(X)$ of isometries of X is called elliptic if it has a bounded orbit (or equivalently if every orbit is bounded), parabolic if it has a unique fixed point on ∂X but is not elliptic, and loxodromic if it is not elliptic and if there exists a pair w^+, w^- of points in ∂X preserved by G. A group which is either elliptic, or parabolic or loxodromic is called elementary; this terminology is inspired by the theory of Kleinian groups. There are no constraints on the algebraic structure of elementary groups due to the following remark.

Remark. Every f.g. group G can be realized as a *parabolic* group of isometries of some proper geodesic hyperbolic space: if C is the Cayley graph of G, $C \times \mathbb{R}$ admits a G-invariant hyperbolic metric ([Gro87], 1.8.A, note that this construction is equivariant). One can also construct a finitely generated group acting on a tree with a unique fixed point at infinity. For instance the lamplighter group (semi-direct product of $\mathbb{Z}$ and $\mathbb{Z}_2[t,t^{-1}]$) fixes a unique point in the boundary of the tree of $GL_2(\mathbb{Z}_2[t^{-1},t])$. In fact, the lamplighter group can indentify with upper triangular matrices with one eigenvalue equal to 1, the other being t^n. As all these matrices have a common eigenspace, they fix one point in the boundary of the tree of GL_2 (the projective line on $\mathbb{Z}_2[t,t^{-1}]]$); but this group contains the Jordan matrix and therefore cannot fix two points in the boundary of this tree.

In what follows, X denotes a complete Gromov hyperbolic geodesic space. We will assume that either X is proper (closed balls are compact) or that X is a complete $\mathbb{R}$-tree, i.e., a complete 0-hyperbolic geodesic metric space. In the first case, $X \cup \partial X$ is a compact set (in the natural topology) and an unbounded sequence of points in X admits a subsequence which converges to a point in ∂X. Important examples of such spaces are Cayley graphs of hyperbolic groups (see [Gro87] for instance). Other examples are universal covers of compact manifolds of non positive curvature. Note that properness implies completness for a metric space, but the converse is false. Recall also that a geodesic space is proper if and only if it is complete and locally compact [Gro99]. The Gromov hyperbolicity of a geodesic space can be defined in several ways (thineness of geodesic triangles, properties of the Gromov product etc.) which are equivalent (see [CDP90] Chap. 1); we will prefer the definition in terms of the Gromov product ([CDP90] Chap. 1, Def. 1.1).

For the rest of the statements in this section we will assume that the following condition on the pair (X, G) holds:

(C) X is a complete geodesic space and X is either proper hyperbolic or is an $\mathbb{R}$-tree. G is a group and $\varphi : G \to Isom(X)$ is an isometric action of G on X.

Note that such an action extends uniquely to a continuous action on $X \cup \partial X$.

Theorem 3. *Let G be a branch group acting isometrically on a hyperbolic space X. Suppose the pair (X, G) satisfies the condition* (C)*. Then:*

a) *The image of G in $Isom(X)$ is elementary.*
b) *Suppose furthermore that G satisfies the property* (FL)*, and X is a hyperbolic graph with uniformly bounded valence of vertices. Then $\varphi(G)$ is elliptic or parabolic.*
c) *If X is $CAT(0)$ and if the group $\varphi(G)$ is elliptic, then it has a fixed point in X.*
d) *If X is $CAT(-1)$, or is an $\mathbb{R}$-tree, then $\varphi(G)$ fixes a point in X or in ∂X, or preserves a line in X.*
e) *Let X be an $\mathbb{R}$-tree. Suppose further that G is f.g.; then G cannot be parabolic.*

Corollary 1. *Let G be a f.g. branch group. G has fixed point property for actions on $\mathbb{R}$-trees if and only if it has property* (FL)*.*

Proof. A tree is $CAT(-1)$, so if G acts on a tree and does not fix a point, it must either preserve a line or a unique point on ∂X. The last possibility is excluded by e). □

From d) we also deduce:

Corollary 2. *If X is $CAT(-1)$ and G acts on $X \cup \partial X$ by isometries, then G fixes a point or contains a subgroup of index 2 which fixes two points in ∂X.*

Recall (see [Ser80]) that a group G is an amalgam (resp. an HNN extension) if it can be written as a free product with amalgamation $G = A *_C B$, with $C \neq A, B$ (resp. $G = A *_{tCt^{-1}=C'}$). We say that this amalgam (resp. HNN extension) is *strict* if the index of C in A is at least 3 and the index of C in B is at least 2 (resp. the indexes of C and C' in A are at least 2). If G splits as an amalgam or HNN extension, then G acts on a simplicial tree T without edge inversion s.t. T/G has one edge and 2 vertices in the case of an amalgam, and one edge and one vertex in the case of an HNN extension. It is easy too see that if a group is a strict amalgam or HNN extension its action on Serre's tree is not elementary. If $G = A *_C B$ with C of index 2 in A and B Serre's tree is a line, and G permutes the two ends of this line. If $G = A *_{tCt^{-1}=C'}$ and $C = C' = A$, Serre's tree is a line and G fixes the two ends of this line. If $G = A *_{tCt^{-1}=C'}$ is a strictly ascending HNN extension ($C' = A$, but $C \neq A$), the group G contains a hyperbolic element (the letter t for instance) and fixes exactly one end of the the tree. Therefore the property e) implies the following:

Corollary 3. *Let G be a f.g. branch group. Then G is neither a strict amalgam nor a strict HNN extension nor a strictly ascending HNN extension.*

Before proving Theorem 3 let us state and prove some statements that have independent interest and will be used later.

Recall that an isometry f of a hyperbolic space X is called elliptic (resp. parabolic, resp. hyperbolic) if the subgroup generated by$f$ is elliptic (resp. parabolic, resp. loxodromic). It can be proved (see [CDP90], chap. 9) that an isometry is either elliptic, or parabolic or hyperbolic, and that if X is an $\mathbb{R}-$tree an isometry cannot be parabolic. An elliptic group cannot contain a hyperbolic or a parabolic element, a loxodromic group cannot contain a parabolic element. In order to simplify the notation, if $\phi : G \to Isom(X)$ is an action of the group G, we denote by gx the image of x under the isometry $\phi(g)$.

Proposition 1. *Let the pair (X, G) satisfy* (C). *Assume that each element of G is either elliptic or parabolic. Then G is either elliptic or parabolic; if X is an $\mathbb{R}$-tree, and G is finitely generated, then G is elliptic.*

The proof of this proposition is of dynamical nature and based on the following

Lemma 1. ([CDP90], Chap. 9, Lemma 2). *Let X be a δ-hyperbolic space. Let g, h be two elliptic or parabolic isometries of X. Suppose that* $\min(d(gx, x), d(hx, x)) \geqslant 2\langle gx, hx\rangle_x + 6\delta$. *Then $g^{-1}h$ is hyperbolic.*

Recall that the Gromov product $\langle x, y\rangle_z$ is defined as $1/2(d(x,z) + d(y,z) - d(x,y))$

Proof of Proposition 1. Let us first consider the case where X is an $\mathbb{R}$-tree, which we denote by T, and G is finitely generated. Recall that the projection of a point x in a $CAT(0)$ space onto a complete convex subset Y is the unique closest point to x in Y (see [BH99], page 176). We claim that in an $\mathbb{R}$-tree T, if g is some elliptic isometry, and T^g the subtree of fixed points of g, then for every x the midpoint of the segment $[x, gx]$ is the projection of x on T^g: indeed let p be this projection, so that the image of the segment $[x, p]$ is $[gx, p]$; if the Gromov product $\langle x, gx\rangle_p = d$ is strictly positive, we can consider the point $q \in [p, x]$ s.t. $d(p, q) = d$; it is fixed by g as it belongs to $[p, x]$ and it is the unique point on this segment with $d(q, p) = d$, but q is closer than p to x, contradiction. Thus $\langle x, gx\rangle_p = 0$, and as the two segments $[x, p]$ and $[gx, p] = g[x, p]$ have the same length, p is the midpoint of $[x, gx]$. For every subset $\Sigma \subset G$, let T^Σ be the fixed subset of Σ. Let $\{g_1, \dots .g_n\}$ be a finite generating subset of G, and let us prove by induction that $T^{\{g_1,\dots g_n\}}$ is not empty. For $n = 1$ this is the hypothesis. Suppose that $T^{\{g_1,\dots g_{n-1}\}} \cap T^{\{g_n\}} = \varnothing$. The minimal distance between these two subtrees is achieved along a segment $[a, b]$, with $a \in T^{\{g_1,\dots,g_{n-1}\}}$, $b \in T^{\{g_n\}}$. Let x_0 be the midpoint of this segment: $x_0 \notin T^{\{g_n\}}$. Therefore $b \in [x_0, g_n x_0]$ is the midpoint. As $x_0 \notin T^{\{g_1,\dots g_{n-1}\}}$, we have that $x_0 \notin T^{\{g_i\}}$, for some i. The intersection $T^{\{g_i\}} \cap [a, x_0]$ is a segment $[a, c]$; the right extremity c of this segment is the projection of x_0 on $T^{\{g_i\}}$, and therefore $c \in [x_0, g_i x_0]$ is the midpoint. Thus $x_0 \in [g_i x_0, g_n x_0]$ and, in other words, $\langle g_i x_0, g_n x_0\rangle_{x_0} = 0$. Lemma 1 applies and proves that the isometry $g_i g_n$ is hyperbolic, a contradiction.

Suppose now that X is a proper geodesic hyperbolic space. Let G be as in the statement, and $x_0 \in X$ be some base-point. If the orbit Gx_0 is bounded, then it is a bounded G invariant set, and G is elliptic. Assume that Gx_0 is not bounded. We consider the set $\overline{Gx_0} \cap \partial X$.

1) Assume that this set has only one point a. It must be G invariant. Let us prove that G is parabolic. Suppose that G fixes another point b on the boundary. Then it acts on the union Y of geodesic lines between a and b. Let $L \subset Y$ be a geodesic between a and b, so that every point in Y is at distance $< 100\delta$ of L. Let $x_0 \in L$; as $\overline{Gx_0} \cap \partial X = \{a\}$, we can find two isometries g, h in G such that $d(x_0, gx_0) > 1000\delta$, $d(x_0, hx_0) > d(x_0, gx_0) + 1000\delta$ and the projections of gx_0 and hx_0 on L are on the right of x_0.

Considering these projections of gx_0 and hx_0 on L, we see that $d(x_0, hx_0) \geq d(x_0, gx_0) + d(gx_0, hx_0) - 200\delta$, thus $\langle x_0, hx_0\rangle_{g(x_0)} \leq 100\delta$. By isometry, we get $\langle g^{-1}x_0, g^{-1}hx_0\rangle_{x_0} < 100\delta < 1/2(\min(d(x_0, g^{-1}x_0), d(x_0, g^{-1}hx_0)) - 3\delta$, and h must be hyperbolic by Lemma 1.

2) Assume that $\overline{Gx_0} \cap \partial X$ has at least two points, $a, b \in \overline{Gx_0} \cap \partial X$. There exists two sequences g_n and h_n such that $g_n x_0 \to a$, and $h_n x_0 \to b$. Then $d(g_n x_0, x_0) \to \infty$ as well as $d(h_n x_0, x_0)$, but $\langle g_n x_0, h_n x_0\rangle_{x_0} \to \langle a, b\rangle_{x_0}$ and remains bounded

(by the very definition of the Gromov boundary). Lemma 1 applies and we get a contradiction. □

Corollary 4. *Let* (G, X) *satisfy* (C). *If* G *has a subgroup of finite index which is elliptic or parabolic, then* G *is also elliptic or parabolic.*

Proof. No element of G can be hyperbolic, as any power of a hyperbolic element is hyperbolic. □

Proposition 2. *Let the pair* (X, G) *satisfy* (C). *If* G *is elliptic, then it has an orbit of diameter* $\leqslant 100\delta$. *If, furthermore,* X *is* $CAT(0)$, *then* G *has a fixed point.*

Proof. In a metric space, the radius of a bounded set Y is the infimum of r s.t. there exists x with $Y \subset B(x, r)$. A center is a point c s.t. $Y \subset B(c, r')$ for every $r' > radius(Y)$. The proof of Proposition 2 is a direct consequence of the following generalization of Elie Cartan center's theorem [BH99], II.2.7. □

Proposition 3. *In a proper geodesic* δ*-hyperbolic space, the diameter of the set of centers of a bounded set is* $\leqslant 100\delta$. *In a complete* $CAT(0)$ *space, every bounded set admits a unique center.*

Proof. The second point is proved in [BH99], II.2.7. Let us prove the first assertion. Let a, b be two centers and suppose that $d(a, b) > 100\delta$. Let c be a midpoint of a, b. Let us prove that for every x in Y, $d(y, c) < r - 10\delta$ and in such way get a contradiction. By assumption $d(a, x)$ and $d(b, x)$ are less than $r + \delta$. By the 4 points definition of δ-hyperbolicity ([CDP90] Prop. 1.6) we know that $d(x, c) + d(a, b) \leq \max(d(x, a) + d(b, c), d(x, b) + d(x, c)) - 2\delta$. As $d(b, c) = d(a, c) = 1/2d(a, b) > 50\delta$ we get that $d(x, c) \leq \max(d(x, a), d(x, b)) - 48\delta \leq r - 48\delta$ and we are done. □

Proposition 4. *Let the pair* (X, G) *satisfy* (C). *If the* G*-orbit of some point of* ∂X *is finite and has at least* 3 *elements, then* G *is elliptic.*

Proof. If the orbit is finite and has at least 3 elements $w_1, \ldots, w_k$, let us construct a bounded orbit of G in X. For every triple of different points w_i, w_j, w_k in this orbit, let us consider the set C_{ijk} consisting of all points being at a distance less than 24δ from all geodesics between w_i, w_j, and w_k. By hyperbolicity this set is not empty and has diameter $\leq 100\delta$. This follows from [CDP90], Chap. 2, Prop. 2.2, p. 20. A finite union of bounded sets is bounded. Therefore, the union of the sets C_{ijk} is a bounded G invariant set. □

Proposition 5. *Let* X *be a* $\delta-$*hyperbolic graph of bounded valence. If* $G \subset Isom(X)$ *is loxodromic, then there exists an epimorphism* $m : G \to \mathbb{Z}$ *or* $m : G \to \mathbb{D}_\infty$ *such that* $\ker m$ *is elliptic.*

Proof. We will construct a combinatorial analogue of the Busemann cocycle (compare [RS95]). As G is loxodromic, the action of G fixes two points $w^{\pm}$ at infinity. It contains a subgroup of index at most two G^+ which preserves these two points, and contains some hyperbolic element h. Let U be the union of all geodesics between these two points at infinity, and choose a preferred oriented line L between this two points. If $x \in U$, there exists a point in L such that $d(x, p(x)) < 24\delta$ ([CDP90], Chap. 2, Prop. 2.2, p. 20). Choose such a point and call it a projection of x. If $x \in U$, let $R(x) = \{y \in U | d(x, y) > 1000\delta$, and the projection of y to L is on the right to that of $x\}$. Note that our hypothesis implies that for every pair x, y, $\{R(y)/R(x)\}$ is contained in the ball centered at y and of radius $d(x, y) + 2000\delta$ and is therefore finite: by definition, a point of $R(y)$ which is at distance $> d(x, y) + 2000\delta$ from y must project on L on a point which is at the distance $> 1000\delta$ of x. Note also that if h is hyperbolic, $R(h^n x)$ is strictly contained in $R(x)$ if n is $\gg 1$. Let $c(x, y) = Card\{R(y) \setminus R(x)\} - Card\{R(x) \setminus R(x)\}$. Note that $c(y, x) + c(x, y) = 0$, and that $c(x, y) + c(y, z) = c(x, z)$. Moreover, if g is in G^+, then $R(gx) = gR(x)$. Choose some point $x_0 \in U$. The formula $m(g) = c(x_0, gx_0)$ defines a nontrivial morphism $G^+ \to \mathbb{Z}$. The orbit of x_0 under the action of the kernel of m is bounded, contained in $B(x_0, 2000\delta)$, and $\ker m$ is elliptic. If G/G^+ is not trivial, and $\varepsilon \in G \setminus G^+$, then $m - m(\varepsilon g \varepsilon^{-1}) = d(g)$ extends to a nontrivial epimorphism $G \to \mathbb{D}_\infty$. □

Proof of Theorem 3. Let H_1 be the rigid stabilizer of the first level of G. It is a product of n subgroups of G, $H_1 = L_1 \times \ldots \times L_n$ conjugate in G.

i) Suppose first that L_1 contains no hyperbolic element.

Then L_1 has either (1) a bounded orbit or (2) a unique fixed point w at infinity.

(1) In the first case, let $C_1 = \{x | \forall g \in L_1, d(gx, x) < 100\delta\}$ (by Proposition 2 this set is nonempty). As L_2 commutes with L_1 it preserves C_1. Being conjugate to L_1, every orbit of L_2 is bounded. If $x_0 \in C_1$ and $D = diam(L_2 x_0)$, we see that the diameter of $(L_1 \times L_2)x_0$ is $\leqslant D + 2 \cdot 100\delta$, hence $L_1 \times L_2$ is elliptic, and the set $C_2 = \{x | \forall g \in L_1 \times L_2, d(gx, x) \leq 100\delta\}$ is not empty (Proposition 2). By induction we prove that $C_k = \{x | \forall g \in L_1 \times L_2 \times \ldots \times L_k, d(gx, x) < 100\delta\}$ is not empty; thus G admits a subgroup of finite index which is elliptic, and G is itself elliptic.

(2) In the second case, the unique fixed point w is stable under the action of the subgroup $L_2 \times \ldots \times L_n$, and G has a subgroup of finite index which is parabolic, thus G is parabolic itself.

ii) Suppose L_1 contains some hyperbolic element h. Let $w^\pm$ be the two distinct fixed points of h at infinity. As $L_2 \times \ldots \times L_n$ commutes with h this group fixes this set. Now L_2 contains a hyperbolic element h_2, conjugated to h: thus h_2 has the same fixed points at infinity as h, and H_1 must also fix the set $\{w, w^-\}$. Thus the orbit of $w^\pm$ is finite and Proposition 4 applies. The orbit of G cannot have more than 2 elements unless G is elliptic: therefore it has exactly two elements, and G is loxodromic.This proves a).

To prove b) apply Proposition 5. Proposition 2 and Proposition 3 (the unique center for bounded sets) give rise to the desired fixed point for c). Claim d) follows from the fact that between two points at infinity in a CAT(-1) space there exists a unique geodesic (visibility property). For claim e), let w be an end of a tree fixed by the group G. Let $t \to r(t)$ be a geodesic ray converging to w. Note that when the point x is fixed, the function $t \to d(x, r(t)) - t$ is constant for $t >> 1$. The value of the constant $b_w(x)$ is called the Busemann function associated to w (see [BH99], Chap. II.8. for a study of Busemann functions in CAT(0) spaces). If the point w is fixed by some isometry g, then $g.r(t)$ is another ray converging to w. But two rays converging to the same point in a tree must coincide outside a compact set. Therefore $d(gr(t), r(t)) = b(g)$ is constant for $t >> 1$, and this constant is $b_{g.w} - b_w$. By construction, $g \to b(g)$ is a homomorphism from G to $\mathbb{R}$, which is non-trivial unless every element of G is elliptic, and takes values in $\mathbb{Z}$ if X is a combinatorial tree. Suppose that the restriction of b to L_1 is trivial. Then L_1 consists of elliptic elements. Since G is finitely generated, L_1 is finitely generated as well. Thus L_1 is elliptic and i) applies. Otherwise, L_1 contains a hyperbolic element and ii) applies.

Theorem 3 is proved. □

3. An indicable branch group

Let $\mathcal{G}$ be a branch group acting on a rooted tree $\mathcal{T}$. It is proved in [Gri00] that, if $N \triangleleft \mathcal{G}$ is a nontrivial normal subgroup, then the group N contains the commutator subgroup of the rigid stabilizer $rist_{\mathcal{G}}(n)'$, for some level n. As $rist_{\mathcal{G}}(n)$ is of finite index in $\mathcal{G}$, $\mathcal{G}/rist_{\mathcal{G}}(n)$ is finite, $\mathcal{G}/rist_{\mathcal{G}}(n)'$ is virtually abelian and we have:

Proposition 6. *A proper quotient $\mathcal{G}/N$ of a branch group is a virtually abelian group.*

We construct in this section an example of a finitely generated branch group which surjects onto the infinite cyclic group. The construction starts from the finitely generated torsion 2-group firstly defined in [Gri80b] and later studied in [Gri84] and other papers (see also the Chapter VIII of the book [dlH00]).

We will list briefly some properties of G that will be used later.

Let $(\mathcal{T}, \emptyset)$ be the rooted binary tree whose vertices are the finite sequences of $0, 1$ with its natural tree structure (see [dlH00], VIII.A for details), the empty sequence $\emptyset$ being the root. If v is a vertex of $\mathcal{T}$ we denote by $\mathcal{T}_v$ the subtree consisting of the sequences starting in v. In other words, the subtree $\mathcal{T}_v$ of $\mathcal{T}$ consists of vertices w that contain v as a prefix. Deleting the first $|v|$ letters of the sequences in $\mathcal{T}_v$ yields a bijection between $\mathcal{T}_v$ and $\mathcal{T}$, called the canonical identification of these trees.

The group G (see [dlH00], VIII.B.9 for details) acts faithfully on the binary rooted tree $(\mathcal{T}, \emptyset)$ and is generated by four automorphisms a, b, c, d of the tree where a is the rooted automorphism permuting the vertices of the first level, while b, c, d are given by the recursive rules

$$b = (a, c), c = (a, d), d = (1, b).$$

This means that b acts trivially on the first level of the tree, it acts on the left subtree $\mathcal{T}_0$ as a and acts on the right subtree $\mathcal{T}_1$ as c, with similarly meaning of the relations for c and d. Here we use the canonical identifications of $\mathcal{T}$ with

$\mathcal{T}_i, i = 0, 1$. An alternative description of G is that it is the group generated by the states of the automaton drawn on the figure 1.

The group G is 3-generated as we have the relations

$$a^2 = b^2 = c^2 = d^2 = bcd = 1$$

there are many other relations and G is not finitely presented [Gri84].

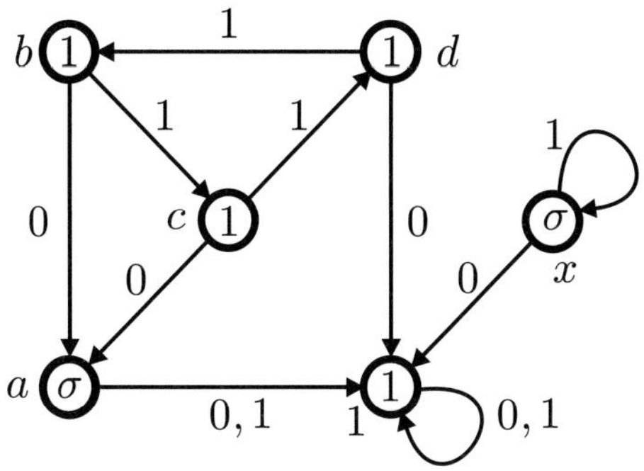

FIGURE 1. The automaton defining L

In order to study groups acting on the binary rooted tree $\mathcal{T}$, it is convenient to use the embedding

$$\psi : Aut(\mathcal{T}) \longmapsto Aut(\mathcal{T}) \wr S_2,$$
$$g \longmapsto (g_0, g_1)\alpha.$$

In this representation S_2 is a symmetric group of order 2, $\alpha \in S_2$ describes the action of g on the first level of the tree and the sections g_0, g_1 describe the action of g on the of subtrees $\mathcal{T}_0$ $\mathcal{T}_1$. We will usually identify the element g and its image $(g_0, g_1)\alpha$. Relations of this type will be often used below.

Let x be the automorphism of $\mathcal{T}$ defined by the recursive relation $x = (1, x)a$. This automorphism is called the adding machine as it imitates the adding of a unit in the ring of diadic integers [GNS00]. An important property of x is that it acts transitively on each level of $\mathcal{T}$ and therefore has infinite order.

Let $L = \langle x, G\rangle$ be the subgroup of $Aut(\mathcal{T})$ generated by G and the adding machine x.

Theorem 4. *The group L is branch, amenable, and has infinite abelization.*

The next two lemmas are the first steps towards the proof of the fact that L is a branch group.

Lemma 2. *The following formulas hold in the group L:*

$[x, a] = (x^{-1}, x)$,
$[x, d] = (x^{-1}bx, b)$
$[[x, a], d] = (1, [x, b])$,
$(1, [[x, b], c]) = [[[x, a], d], b]$.

Proof. This follows by direct computation:

$[x,a] = x^{-1}axa = a(1,x^{-1})a(1,x)aa = (x^{-1},1)(1,x) = (x^{-1},x).$

$[x,d] = x^{-1}dxd = a(1,x^{-1})(1,b)(1,x)a(1,b) = (x^{-1}bx,b)$

$[[x,a],d] = [(x^{-1},x),(1,b)] = ([x^{-1},1],[x,b]) = (1,[x,b]).$

$[[[x,a],d],b] = [(1,[x,b]),(a,c)] = (1,[[x,b],c]).$ □

Lemma 3. *The group L is self-replicating, and hence level transitive.*

Proof. Consider the elements $b = (a,c)$, $c = (a,d)$, $d = (1,b)$, $aba = (c,a)$, $xa = (1,x)$. They stabilize the two vertices of the first level of $\mathcal{T}$, and their projections on $Aut(\mathcal{T}_1) \simeq Aut(\mathcal{T})$ are c, d, b, a, x, i.e., the generators of L. Note that these elements generate L. Hence the projection of $st_L(1)$ on $Aut(\mathcal{T}_1)$ is L modulo the canonical identification of $\mathcal{T}$ and $\mathcal{T}_1$. The conjugation by a permutes the coordinates of elements in $st_L(1)$, hence the same holds for the first projection. The self-replicating property (Definition 4) follows by induction on the level. The level transitivity is an immediate consequence of the transitivity of L on the first level and the self-replicating property. □

Let

$$K = \langle [a,b] \rangle^G, S = [\langle x \rangle, G]^L,$$
$$R = \langle K, S, \gamma_3(L) \rangle^L = KS\gamma_3(L).$$

These subgroups will play an important role in our further considerations.

Lemma 4. *We have the following inclusions: $\gamma_3(G) \succeq \gamma_3(G) \times \gamma_3(G), K \succeq K \times K$, and $R \succeq S \times S$.*

Proof. The first two inclusions are known [Gri89, Gri00].

Using the commutator relations and the fact that conjugation by a permutes the coordinates we have

$$(1,[c,x]) = [(a,c),(1,x)] = [b,xa] = [b,a][b,x][[b,x],a] \in R,$$
$$(1,[x,b]) = [[x,a],d] \in R$$

by Lemma 2,

$$(1,[a,x]) = a[(a,c),(x,1)]a = a[b,(x,1)]a = ab^{-1}(x^{-1},1)b(x,1)a.$$

But $x = (1,x)a$ and $axa = (x,1)a$ which leads to

$$(1,[a,x]) = ab^{-1}x^{-1}abaxa = ab^{-1}ab[b,axa] = [a,b][b,axa].$$

Now we have

$$[b,axa] = a[aba,x]a \in S,$$

and $[a,b] \in K$ which gives $(1,[a,x]) \in R$.

Finally

$$(1,[x,d]) = (1,[x,bc]) = (1,[x,c][x,b][[x,b],c]) = (1,[x,c])(1,[x,b])(1,[[x,b],c])$$

and

$$(1,[[x,b],c]) = [[[x,a],d],b] \in R$$

by Lemma 2. Therefore the elements $(1,[a,x])$, $(1,[b,x])$, $(1,[c,x])$, $(1,[d,x])$ belong to R and, as $S = \langle [a,x],[b,x],[c,x],[d,x]\rangle^L$, the lemma is proved. □

Lemma 5. *We have the inclusion:* $\gamma_3(L) \succeq \gamma_3(L) \times \gamma_3(L)$.

Proof. Consider the subgroup $Q = \langle d, c, aca, xa\rangle \subset L$. As $d = (1,b)$, $c = (a,d)$, $aca = (d,a)$, $xa = (1,x)$, the group Q is a subdirect product in $D_4 \times L$ where $D_4 \simeq \langle a,d\rangle$ is a dihedral group of order 8. As $\gamma_3(D_4) = 1$ we get

$$\gamma_3(Q) = (1, \gamma_3(L)),$$
$$\gamma_3(aQa) = (\gamma_3(L), 1),$$

and therefore

$$\gamma_3(L) \succeq \gamma_3(L) \times \gamma_3(L).$$ □

Lemma 6. *The group L is a weakly regular branch group over R.*

Proof. We know that $K \succeq K \times K$, $\gamma_3(G) \succeq \gamma_3(G) \times \gamma_3(G)$, $\gamma_3(L) \succeq \gamma_3(L) \times \gamma_3(L)$ and $R \succeq S \times S$. But R is generated by S, $\gamma_3(L)$ and K. This implies the statement. □

In order to prove that L is a branch group, we consider its subgroup $P = \langle R, \langle x^4\rangle\rangle^L$.

Lemma 7. *The group P has finite index in L.*

Proof. Every element $g \in L$ can be written as a product $g = x^i a^j c^k d^l h f x^{4t}$, where $h \in [G,G]$, $f \in S$, $i \in \{0,1,2,3\}$, $j,k,l = 0,1$, $t \in \mathbb{Z}$. This implies that the index of P in L is $\leqslant 128$. □

Let $P_n \simeq P \times \cdots \times P \subset Aut(\mathcal{T})$ (2^n factors) be the subgroup of $Aut(\mathcal{T})$ that is the product of 2^n groups isomorphic to P that act on the corresponding 2^n subtrees rooted at the vertices on the n-th level.

Lemma 8. *The group L contains P_n for every n.*

Proof. For $n = 0$ the statement is obvious. For $n = 1$, let us consider $(xa)^4 = (1, x^4)$ which is an element of L. As L is self-replicating, for any given element $h \in L$ there exists an element k in L s.t. $k = (f,h)$. Conjugating $(1,x^4)$ by an element of L of the form (f,h), we get that $(1,(x^4)^h) \in L$. But P is generated by conjugates of x^4. This together with Lemma 6 proves the inclusion $1 \times P \succeq L$. The inclusion $P \times 1 \succeq L$ is obtained by conjugating L by a. Then we get that $P \times P = P_1 < L$.

In order to prove the lemma for $n = 2$ we observe that

$$L \ni [x,a] = (xa)^2 = (1,x^2) = (1,1,x,x)_2$$

(the index 2 indicates that we rewrite the considered while considering its action on the second level; we will use such type of notations for further levels as well). Multiplying $(1,x^2)$ (which is in L) by

$$(1,[x,a]) = (1,1,x^{-1},x)_2,$$

we get $(1,1,1,x^2)_2 \in L$. Therefore $(1,1,1,x^4)_2 \in L$ and hence $P_2 < L$ (by level transitivity and the self-replicating property of L we see that $(x^4,1,1,1)$, $(1,x^4,1,1)$ and $(1,1,x^4,1)$ also belong to L.

Let us prove the lemma by induction on $n \geq 2$. Suppose that, for every $k \leqslant n$, the inclusion $P_k < L$ holds and let us prove that $P_{n+1} < L$. Consider the element μ

$$L \ni \mu = (1,\dots,1,x^4)_{n-2} = (1,\dots,1,x^2,x^2)_{n-1},$$

As L is self-replicating, there exists an element $\rho \in St_G(u_{n-2})$, where u_{n-2} is the last vertex on the $(n-2)$-th level, whose projection at this vertex is equal to b. We have

$$\begin{aligned} L \ni [\mu,\rho] &= (1,\dots,1,[x^2,a],[x^2,c])_{n-1} \\ &= (1,\dots,1,[x,a],[x,d])_n = (1,\dots,1,x^{-1},x,x^{-1}bx,b)_{n+1}, \end{aligned} \tag{3.1}$$

As $b^2 = 1$ we get the relation

$$[\mu,\rho]^2 = (1,\dots,1,x^{-2},x^2,1,1)_{n+1}$$

Now we have

$$\begin{aligned} L \ni \eta &= (1,,,1,x^4,1)_n = (1,\dots,1,x^2,x^2,1,1)_{n+1}, \\ [\mu,\rho]^2\,\eta &= (1,\dots,1,x^4,1,1)_{n+1} \end{aligned} \tag{3.2}$$

and we come to the conclusion that $1 \times 1 \times \ldots \times 1 \times P \times 1 \times 1 \succeq L$, hence $\underbrace{P \times \ldots \times P}_{2^{n+1}} \succeq R$, and $P_{n+1} < L$, as L is level transitive. □

We can now prove that L is a branch group. This group acts transitively on each level of the rooted tree $\mathcal{T}$, and contains P_n for every $n = 1,2\ldots$. In order to prove that it is branch, as $P_n < rist_L(n)$, and L is level transitive, it is enough to check that P_n has finite index in L. We have the following diagram

$$\begin{array}{lccccccc}
L & & & & & & & \\
\uparrow & \psi_n & & & & & & \\
st_L(n) & \twoheadrightarrow & \tilde{H} & < & L & \times\ldots\times & L \\
\uparrow & & & & & & \\
rist_L(n) & & \uparrow & & \uparrow & \ldots & \uparrow \\
\uparrow & \psi_n & & & & & \\
P_n & \twoheadrightarrow & \tilde{P}_n & = & P & \times\ldots\times & P
\end{array}$$

(the vertical arrows are inclusions, $\tilde{H}$ and $\tilde{P}_n$ are ψ_n images of $st_L(n)$ and P_n respectively, where ψ_n is the n-th iteration of ψ).

As the group P has finite index in L, we get that $\tilde{P}_n$ has finite index in $\tilde{H}$ and therefore P_n has finite index in $st_L(n)$ and hence in L. This establishes the first statement of Theorem 4.

The group L is the self-similar group generated by the states of the automaton in Figure 1. The diagram of this automaton satisfies the condition of Proposition 3.9.9 of [Nek05]: it is therefore a bounded automaton in the sense of Sidki [Sid00]. This proposition states that an automaton is bounded if and only if its Moore diagram has the following property: every two nontrivial cycles are disjoint and are not connected by a directed path; a cycle is called trivial if all of its states represents the identity automorphism of the tree.

It is easy to see that automaton determining the group L satisfies this property.

By a theorem of Bondarenko and Nekrashevych (Theorem 3.9.12 in [Nek05]) every group generated by the states of a bounded automaton is contracting. Moreover, by a theorem of Bartholdi, Kaimanovich, Nekrashevych and Virag [BKNV06] such a group is amenable. This establishes the amenability of L, as well as its contracting property.

In order to compute the abelianization of L, we need to combine the contracting property of L with a rewriting process which corresponds to the embedding ψ. The combination of this rewriting process and the contraction property will produce an algorithm for solving the word problem in L: the branch algorithm. This type of algorithm appeared in [Gri84] for the first time: it is a general fact that the branch algorithm solves the word problem for contracting groups [Sav03].

The group

$$\Gamma = \langle a, b, c, d, x : a^2 = b^2 = c^2 = d^2 = bcd = 1\rangle,$$

defined by generators and relations, naturally covers L. It is isomorphic to the free product

$$\mathbb{Z}/2\mathbb{Z} * (\mathbb{Z}/2\mathbb{Z} \times \mathbb{Z}/2\mathbb{Z}) * \mathbb{Z}.$$

Therefore, the elements in Γ are uniquely represented by words $w = w(a, b, c, d, x)$ in the reduced form (for this free product structure).

Similarly the group G is naturally covered by the group

$$\langle a, b, c, d : a^2 = b^2 = c^2 = d^2 = bcd = 1\rangle \simeq \mathbb{Z}/2\mathbb{Z} * (\mathbb{Z}/2\mathbb{Z} \times \mathbb{Z}/2\mathbb{Z}).$$

The elements in G can be represented by reduced words (with respect to this free product structure).

Let w be a word representing an element of Γ, $w = u_1 x^{i_1} u_2 x^{i_2} \dots u_k x^{i_k} u_{k+1}$, where u_i are reduced words in a, b, c, d, u_i is nonempty for $i \neq 1, k+1$, and $i_j \neq 0$, for $j = 1, \dots, k$.

Let us consider the following rewriting process:

1) In each word u_i replace b, c, d by the corresponding element of the wreath product $L \wr S_2$, using the defining relations $b = (a, c), c = (a, d)$ $d = (1, b)$, $x = (1, x)a$.
2) Move all the letters a to the right using the relations $a(v_0, v_1) = (v_1, v_0)a$. Use the relation $a^2 = 1$ for simplification of words, and take the componentwise product of all involved pairs. One obtains in such a way a relation of the form $w = (w_0, w_1)a^\varepsilon$ with $\varepsilon \in \{0, 1\}$, which holds in L.
3) Reduce the words w_i in Γ, obtaining a pair (w_0, w_1) of reduced words.

Note that the length of $w_i, i = 0, 1$ is strictly shorter than of w if at least one letter a appears in the word w.

We can represent this rewriting process as a pair $\varphi = (\varphi_0, \varphi_1)$ (or a product $\varphi_0 \times \varphi_1$) of two rewritings $w \to w_0$ and $w \to w_1$. We will apply these maps to words with an even number of occurences of a, i.e., words representing the elements in $st_L(1)$: in this case $\varepsilon = 0$. We can therefore iterate this rewriting procedure φ n times for words representing elements in $st_L(n)$, and get 2^n words $w_{i_1,\dots i_n}$ with $i_j \in \{0,1\}$. (For formal definition of φ_0, φ_1 in case of the group G see [Gri98a], for L the formal description is similar. For the definition of core see [Nek05].)

Proposition 7. *The rewriting process is* 3*-step contracting with core* $\mathcal{N} = \{1, b, c, d, x, x^{-1}, bx, cx, dx, x^{-1}b, x^{-1}c, x^{-1}d, x^{-1}bx, x^{-1}cx, x^{-1}dx\}$. *Moreover, for every word* w *representing an element in* $stab_L(3)$, $\varphi^3(w)$ *consists of* 8 *words* $w_{i,j,k}, i,j,k \in \{0,1\}$ *of strictly shorter length than* w.

Proof. Let the word $w = w_1 x^{i_1} w_2 x^{i_2} \dots w_k x^{i_k} w_{k+1}$ be as above and represents an element in $stab_L(3)$. As we already have noted, if the letter a occurs in some of the w_i, then rewriting process is strictly shortening in one step. In order to study reduced words without the letter a, we will make use of the relations in Table 1.

Observe that w is a product of subwords in the form presented by the left side in the relations in Table 1, followed by an element of the set

$$\{1, b, c, d, xbx, cx, dx, xb, xc, xd, x^{-1}b, x^{-1}c, x^{-1}d, bx^{-1}, cx^{-1}, dx^{-1}\}.$$

In all relations marked by A or B the rewriting process gives shortening in one step (case A) or in two steps (case B); in the latter case note the presence of the letter a, which insures reduction of length in one more step.

If the word w is not shortened after applying twice the rewriting procedure, then either it belongs to N, or it is of the form $*x^{-1} * x \dots x^{-1} * x * t$, with $* \in \{b,c,d\}$except for the the first or last $*$ which may also represent the unit, and $t \in \{x^{-1}b, x^{-1}c, x^{-1}d, bx^{-1}, cx^{-1}, dx^{-1}\}$.

Let $x^{-1}bx = \tilde{b}$, $x^{-1}cx = \tilde{c}$, $x^{-1}dx = \tilde{d}$. These elements are of order two, and satisfy the relations $\tilde{b} = (\tilde{c}, a), \tilde{c} = (\tilde{d}, a), \tilde{d} = (\tilde{b}, 1)$. Since these relations are of the same form as the relations that hold for b, c and d, the group $\tilde{G}$ generated by $\langle a, \tilde{b}, \tilde{c}, \tilde{d}\rangle$ is isomorphic to G.

Let $A < L$ be the subgroup generated by $\langle b, c, d, \tilde{b}, \tilde{c}, \tilde{d}\rangle$. Note that A stabilizes the first level of the tree. Consider the embedding $\psi : A \to \tilde{G} \times G$ obtained by projecting the elements of A on the left and right subtrees (we use the same notation ψ for the embedding as before).

Lemma 9. *The group* $\psi(A)$ *is a subdirect product of finite index in* $\tilde{G} \times G$.

Proof. We have $\tilde{c} = (\tilde{d}, a)$, $c = (a, d)$, $\tilde{d} = (\tilde{b}, 1)$, $d = (1, b)$, $\tilde{b}\tilde{d} = \tilde{c}$ and $bd = c$. Therefore the projection of $\psi(A)$ on each of two factors is onto.

Let $B = \langle b\rangle^G$ and $\tilde{B} = \langle \tilde{b}\rangle^{\tilde{G}}$.

As $d = (1, b) \in \psi(A)$ and as for every $g \in G$ there exists some h such that $(g, h) \in \psi(A)$, we see that $(1, gbg^{-1}) \in \psi(M)$. Therefore the group $1 \times B$ is contained in $\psi(A)$ and, by a symmetric argument, $\tilde{B} \times 1$ is contained in $\psi(A)$.

$$
\begin{array}{rcll}
bx &=& (a,c)(1,x)a = (a,cx)a & \\
cx &=& (a,d)(1,x)a = (a,dx)a & \\
dx &=& (1,b)(1,x)a = (1,bx)a & \\
x^{-1}b &=& a(1,x^{-1})(a,c) = (x^{-1}c,a)a & \\
x^{-1}c &=& a(1,x^{-1})(a,d) = (x^{-1}d,a)a & \\
x^{-1}d &=& a(1,x^{-1})(1,b) = (x^{-1}b,1)a & \\
xb &=& (1,x)a(a,c) = (c,xa)a & B \\
xc &=& (1,x)a(a,d) = (d,xa)a & B \\
xd &=& (1,x)a(1,b) = (b,x)a & A \\
bx^{-1} &=& (a,c)a(1,x^{-1}) = (ax^{-1},c) & B \\
cx^{-1} &=& (a,d)a(1,x^{-1}) = (ax^{-1},d) & B \\
dx^{-1} &=& (1,b)a(1,x^{-1}) = (x^{-1},b) & A \\
xbx &=& (1,x)a(a,c)(1,x)a = (1,x)(c,a)(x,1) = (cx,xa), & A \\
x^{-1}bx^{-1} &=& a(1,x^{-1})(a,c)a(1,x^{-1}) = (x^{-1}c,ax^{-1}) & A \\
xbx^{-1} &=& (1,x)a(a,c)a(1,x^{-1}) = (1,x)(c,a)(1,x^{-1}) = (c,xax^{-1}) & B \\
x^{-1}bx &=& a(1,x^{-1})(a,c)(1,x)a = (x^{-1}cx,a), & C \\
xcx &=& (1,x)a(a,d)(1,x)a = (1,x)(d,a)(x,1) = (dx,xa) & A \\
x^{-1}cx^{-1} &=& a(1,x^{-1})(a,d)a(1,x^{-1}) = (x^{-1},1)(d,a)(1,x^{-1}) = (x^{-1}d,ax^{-1}) & A \\
xcx^{-1} &=& (1,x)a(a,d)a(1,x^{-1}) = (1,x)(d,a)(x^{-1},1) = (dx^{-1},xa) & A \\
x^{-1}cx &=& a(1,x^{-1})(a,d)(1,x)a = (x^{-1}dx,a) & C \\
xdx &=& (1,x)a(1,b)(1,x)a = (1,x)(b,1)(x,1) = (bx,x) & A \\
x^{-1}dx^{-1} &=& a(1,x^{-1})(1,b)a(1,x^{-1}) = (x^{-1},1)(b,1)(1,x^{-1}) = (x^{-1}b,x^{-1}) & A \\
xdx^{-1} &=& (1,x)a(1,b)a(1,x^{-1}) = (b,1) & A \\
x^{-1}dx &=& a(1,x^{-1})(1,b)(1,x)a = (x^{-1}bx,1) & C \\
x^2 &=& (x,x) & A \\
x^{-2} &=& (x^{-1},x^{-1}). & A
\end{array}
$$

TABLE 1. Some relations in L

Thus $\tilde{B} \times B < \psi(A) < \tilde{G} \times G$. But the groups B and $\tilde{B}$ have finite index in G and $\tilde{G}$, respectively, and the lemma is proved. □

We now finish the proof of Proposition 7. Consider a reduced word u which represents an element of L. Suppose that this element stabilizes the first level but is not shortened after applying twice the rewriting process. The word u has to be of the form $u = wb$, where w represents an element of A and

$$t \in \{x^{-1}b, x^{-1}c, x^{-1}d, bx^{-1}, cc^{-1}, dx^{-1}\}.$$

Rewrite it as a word in the letters $\langle b, c, d, \tilde{b}, \tilde{c}, \tilde{d}\rangle$. Use the relations $\tilde{b} = (\tilde{c}, a)$, $\tilde{c} = (\tilde{d}, a)$, $\tilde{d} = (\tilde{b}, 1)$, $b = (a, c), c = (a, d)$ $d = (1, b)$ to rewrite it as an element $(\tilde{w}_0, w_1)$ of $\tilde{G} \times G$.

Recall that, endowed with its natural system of generators, the group G is one step contracting with core $\mathcal{N}_0 = \{1, b, c, d\}$ (and contracting coefficient $\frac{1}{2}$ [Gri84]). In other words,applying the rewriting procedure to reduce a word v in a, b, c, d with an even number of occurrences of the letter a yields a couple a words of length $\leqslant 1/2|v|$ unless $v \in \{1, b, c, d\}$. More precisely, if $v \to (v_0, v_1)$ is obtained by rewriting in the group G, then $|v_i| \leqslant |v|/2 + 1$.

By isomorphism the same property is true for a reduced word in the alphabet $a, \tilde{b}, \tilde{c}, \tilde{d}$ determining an element in $\tilde{G}$ (and the core in this case is $\widetilde{\mathcal{N}_0} = \{1, \tilde{b}, \tilde{c}, \tilde{d}\}$).

Split the word w as a product of monads $*$ and triads $x^{-1} * x$. If there are at least two monads or at least two triads we get after rewriting shortening at each of coordinates. The remaining case is the case of a word of the form $x^{-1} * x*$ and $*x^{-1} * x$ for which one checks that reduction of length occurs in the second step.

This completes the proof of Proposition 7. □

From this proposition we get an algorithm to solve the word problem: the branch algorithm for L. Let us describe it further.

Let w be a word in the letters a, b, c, d, x. The problem is to check if $w = 1$ in L. The notation $w \equiv_L w'$ means that the two elements of L defined by the words w and w' are equal.

1) Reduce w in Γ. If w is the empty word, then in L, $w \equiv_L 1$. If it is not the empty word, compute the exponent $\exp_a w$ (that is the sum of exponents of a in w). Check if this number is even. If NO, then $w \not\equiv_L 1$. If YES go to 2).

2) Rewrite w as a pair (w_0, w_1) using the rewriting map $\varphi = (\varphi_0, \varphi_1)$. Apply 1) successively to w_0, w_1 and follow steps 1) and 2) alternatively. Either, at some step one obtains a word with odd exp_a or (after n steps) one obtains that all 2^n words represent the identity element in Γ (observe that the word problem in Γ is solvable by using the normal form for elements).

Note that $w \equiv_L 1 \Leftrightarrow (w_0 \equiv_L 1$ and $w_1 \equiv_L 1)$. Applying this procedure 3 times yields either the answer NO (the elemnt is not the identity) or a set of 8 words $w_{i,j,k}$ with $i, j, k \in \{0, 1\}$ which - by Proposition 7- are strictly shorter than w. This algorithm solves the word problem.

Lemma 10. *Let w be a word in the generators. Let $w \equiv_L (w_0, w_1)\alpha$, $\alpha = a$ or $\alpha = 1$ depending on the parity of the exponent $\exp_a w$, and the triple $(w_0, w_1), \alpha$ is obtained from w by applying once the rewriting process described above. Then*

$$\exp_x(w) = \exp_x(w_0) + \exp_x(w_1).$$

Proof. The rewriting process uses the relations $b = (a, c), c = (a, d), d = (1, b)$ and $x = (1, x)a$, $x^{-1} = (x^{-1}, 1)a$ which do not change the total exponent of x. The reduction in group Γ also doesn't change the exponent. □

Lemma 11. *The abelianization $L/[L, L]$ is infinite. The image of x in $L/[L, L]$ is of infinite order.*

Proof. Any element in the commutator group can be expressed as a product of commutators $[u, v]$. Choosing the words in a, b, c, d, x representing u and v, we get that any element in $[L, L]$ can be written as a word w with $\exp_x w = 0$. Suppose that for some $n \geqslant 1$, $x^n \in [L, L]$. We get a word $w = x^n \Pi[u_i, v_i]$ in the letters a, b, c, d, x with total exponent n for x which represents the identity element in L. Choose w of minimal length with this property. Applying the rewriting process at most 3 times to w, we get a set of 8 words $w_{ijk}, i, j, k \in 0, 1$ representing the identity element in L with the sum of exponents of the symbol x different from zero. Hence at least one of them has non zero $\exp_x$. The words w_{ijk} are shorter than w, a contradiction. □

The proof of Lemma 11 completes the proof of Theorem 4.

References

[BGŠ03] L. Bartholdi, R.I. Grigorchuk, and Z. Šunik, *Branch groups*, Handbook of algebra, Vol. 3, North-Holland, Amsterdam, 2003, pp. 989–1112. MR **2035113** **(2005f:**20046)

[BH99] M.R. Bridson and A. Haefliger, *Metric spaces of non-positive curvature*, Grundlehren der Mathematischen Wissenschaften [Fundamental Principles of Mathematical Sciences], vol. 319, Springer-Verlag, Berlin, 1999. MR **1744486** **(2000k:**53038)

[BKNV06] L. Bartholdi, V. Kaimanovich, V. Nekrashevych, and B. Virag, *Amenability of automata groups*, (preprint), 2006.

[Bos00] N. Boston, *p-adic Galois representations and pro-p Galois groups*, New horizons in pro-p groups, Progr. Math., vol. 184, Birkhäuser Boston, Boston, MA, 2000, pp. 329–348. MR **1765126** **(2001h:**11073)

[CDP90] M. Coornaert, T. Delzant, and A. Papadopoulos, *Géométrie et théorie des groupes*, Lecture Notes in Mathematics, vol. 1441, Springer-Verlag, Berlin, 1990. MR **1075994** **(92f:**57003)

[CSMS01] T. Ceccherini-Silberstein, Antonio Machì, and Fabio Scarabotti, *The Grigorchuk group of intermediate growth*, Rend. Circ. Mat. Palermo (2) **50** (2001), no. 1, 67–102. MR **1825671** **(2002a:**20044)

[dlH00] P. de la Harpe, *Topics in geometric group theory*, Chicago Lectures in Mathematics, The University of Chicago Press, 2000.

[dlHV89] P. de la Harpe and A. Valette, *La propriété (T) de Kazhdan pour les groupes localement compacts (avec un appendice de Marc Burger)*, Astérisque (1989), no. 175, 158, With an appendix by M. Burger. MR **1023471** **(90m:**22001)

[GM93] R.I. Grigorchuk and A. Maki, *On a group of intermediate growth that acts on a line by homeomorphisms*, Mat. Zametki **53** (1993), no. 2, 46–63. MR **1220809 (94c:**20008)

[GNS00] R.I. Grigorchuk, V.V. Nekrashevich, and V.I. Sushchanskiĭ, *Automata, dynamical systems, and groups*, Tr. Mat. Inst. Steklova **231** (2000), no. Din. Sist., Avtom. i Beskon. Gruppy, 134–214. MR **1841755 (2002m:**37016)

[GNŠ06] R.I. Grigorchuk, V. Nekrashevych, and Z. Šunić, *Hanoi towers groups*, Oberwolfach Reports **19** (2006), 11–14.

[Gri80a] R.I. Grigorčuk, *On Burnside's problem on periodic groups*, Funktsional. Anal. i Prilozhen. **14** (1980), no. 1, 53–54. MR **565099 (81m:**20045)

[Gri80b] ______, *On Burnside's problem on periodic groups*, Funktsional. Anal. i Prilozhen. **14** (1980), no. 1, 53–54. MR **565099 (81m:**20045)

[Gri84] R.I. Grigorchuk, *Degrees of growth of finitely generated groups and the theory of invariant means*, Izv. Akad. Nauk SSSR Ser. Mat. **48** (1984), no. 5, 939–985. MR **764305 (86h:**20041)

[Gri89] ______, *On the Hilbert-Poincaré series of graded algebras that are associated with groups*, Mat. Sb. **180** (1989), no. 2, 207–225, 304. MR **993455 (90j:**20063)

[Gri98a] ______, *An example of a finitely presented amenable group that does not belong to the class EG*, Mat. Sb. **189** (1998), no. 1, 79–100. MR **1616436 (99b:**20055)

[Gri98b] ______, *On Tychonoff groups*, Geometry and cohomology in group theory (Durham, 1994), London Math. Soc. Lecture Note Ser., vol. 252, Cambridge Univ. Press, Cambridge, 1998, pp. 170–187. MR **1709958 (2001g:**20043)

[Gri99] ______, *On the system of defining relations and the Schur multiplier of periodic groups generated by finite automata*, Groups St. Andrews 1997 in Bath, I, London Math. Soc. Lecture Note Ser., vol. 260, Cambridge Univ. Press, Cambridge, 1999, pp. 290–317. MR **1676626 (2001g:**20034)

[Gri00] ______, *Just infinite branch groups*, New horizons in pro-p groups, Progr. Math., vol. 184, Birkhäuser Boston, Boston, MA, 2000, pp. 121–179. MR **1765119 (2002f:**20044)

[Gro87] M. Gromov, *Hyperbolic groups*, Essays in group theory, Math. Sci. Res. Inst. Publ., vol. 8, Springer, New York, 1987, pp. 75–263. MR **919829 (89e:**20070)

[Gro99] M. Gromov, *Metric structures for Riemannian and non-Riemannian spaces*, Progress in Mathematics, vol. 152, Birkhäuser Boston Inc., Boston, MA, 1999, Based on the 1981 French original [MR0682063 (85e:53051)], With appendices by M. Katz, P. Pansu and S. Semmes, Translated from the French by Sean Michael Bates. MR **1699320 (2000d:**53065)

[GŠ06] R. Grigorchuk and Z. Šuniḱ, *Asymptotic aspects of Schreier graphs and Hanoi Towers groups*, C. R. Math. Acad. Sci. Paris **342** (2006), no. 8, 545–550. MR 2217913

[GŠ07] ______, *Self-similarity and branching in group theory*, Groups St. Andrews 2005, I, London Math. Soc. Lecture Note Ser., vol. 339, Cambridge Univ. Press, Cambridge, 2007, pp. 36–95.

[Mar91] G.A. Margulis, *Discrete subgroups of semisimple Lie groups*, Ergebnisse der Mathematik und ihrer Grenzgebiete (3) [Results in Mathematics and Related Areas (3)], vol. 17, Springer-Verlag, Berlin, 1991. MR **1090825 (92h:**22021)

[Mo] D.W. Morris, *Amenable groups that act on the line.* Algebr. Geom. Topol. 6 (2006), 2509–2518. MR 2286034

[Nek05] V. Nekrashevych, *Self-similar groups*, Mathematical Surveys and Monographs, vol. 117, American Mathematical Society, Providence, RI, 2005. MR **2162164 (2006e:**20047)

[Ol′79] A.Ju. Ol′šanskiĭ, *An infinite simple torsion-free Noetherian group*, Izv. Akad. Nauk SSSR Ser. Mat. (1979), no. 6, 1328–1393. MR **567039 (81i:**20033)

[PV91] I. Pays and A. Valette, *Sous-groupes libres dans les groupes d'automorphismes d'arbres*, Enseign. Math. (2) **37** (1991), no. 1-2, 151–174. MR **1115748 (92f:**20028)

[RS95] E. Rips and Z. Sela, *Canonical representatives and equations in hyperbolic groups*, Invent. Math. **120** (1995), no. 3, 489–512. MR **1334482 (96c:**20053)

[Sav03] D.M. Savchuk, *On word problem in contracting automorphism groups of rooted trees*, Vīsn. Kiïv. Unīv. Ser. Fīz.-Mat. Nauki (2003), no. 1, 51–56. MR 2018505

[Ser80] J.-P. Serre, *Trees*, Springer-Verlag, Berlin, 1980. MR **607504 (82c:**20083)

[Sid00] S. Sidki, *Automorphisms of one-rooted trees: growth, circuit structure, and acyclicity*, J. Math. Sci. (New York) **100** (2000), no. 1, 1925–1943, Algebra, 12. MR **1774362 (2002g:**05100)

[Tit77] J. Tits, *A "theorem of Lie-Kolchin" for trees*, Contributions to algebra (collection of papers dedicated to Ellis Kolchin), Academic Press, New York, 1977, pp. 377–388. MR 0578488 (58 #28205)

[Wil71] J.S. Wilson, *Groups with every proper quotient finite*, Proc. Camb. Phil. Soc. **69 (1971), 373–391.**

Thomas Delzant
Département de mathématiques
Université de Strasbourg
7 rue Descartes
F-67084 Strasbourg
e-mail: delzant@math.u-strasbg.fr

Rostislav Grigorchuk
Department of Mathematics
Texas A &M University
MS-3368, College Station, TX, 77843-3368
USA
e-mail: grigorch@math.tamu.edu

Progress in Mathematics, Vol. 265, 377–398

On Nori's Fundamental Group Scheme

Hélène Esnault, Phùng Hô Hai and Xiaotao Sun

Abstract. The aim of this note is to give a structure theorem on Nori's fundamental group scheme of a proper connected variety defined over a perfect field and endowed with a rational point.

Mathematics Subject Classification (2000). 14L17, 14L99, 14G32.

Keywords. Fundamental group scheme, Tannaka duality, finite bundles.

1. Introduction

For a proper connected reduced scheme X defined over a perfect field k Nori introduced in [8] and [9] the notion of *essentially finite bundles*. He shows that they form a k-linear abelian rigid tensor category, denoted subsequently by $\mathcal{C}^N(X)$. A k-rational point x of X endows $\mathcal{C}^N(X)$ with a fiber functor $V \mapsto V|_x$ with values in the category of finite-dimensional vector spaces over k. This makes $\mathcal{C}^N(X)$ a Tannaka category, thus by Tannaka duality ([1, 12]), the fiber functor establishes an equivalence between $\mathcal{C}^N(X)$ and the representation category $\mathrm{Rep}(\pi^N(X,x))$ of an affine group scheme $\pi^N(X,x)$, which turns out to be a pro-finite group scheme (see Section 2 for an account of Nori's construction). The purpose of this note is to study the structure of this Tannaka group scheme.

To this aim, we define two full tensor subcategories $\mathcal{C}^{\acute{e}t}(X)$ and $\mathcal{C}^F(X)$. The objects of the first one are *étale finite* bundles, that is bundles for which the corresponding representation of $\pi^N(X,x)$ factors through a finite étale group scheme, and the objects of the second one are *F-finite* bundles, that is bundles for which the corresponding representation of $\pi^N(X,x)$ factors through a finite local group scheme. As Tannaka subcategories they are the representation categories of Tannaka group schemes $\pi^{\acute{e}t}(X,x)$ and $\pi^F(X,x)$.

In fact $\pi^{\acute{e}t}(X,x)$ relates closely to the more familiar fundamental group $\pi_1(X,\bar{x})$ defined by Grothendieck ([4, Exposé V]), where $\bar{x}$ is a geometric point

Partially supported by the DFG Leibniz Preis, the DFG Heisenberg program and the grant NFSC 10025103.

above x, which is a pro-finite group. One has

$$\pi^{\acute{e}t}(X,x)(\bar{k}) \cong \pi_1(X \times_k \bar{k}, \bar{x}) \tag{1.1}$$

(see Remarks 2.10 for a detailed discussion). Thus the étale piece of Nori's group scheme takes into account only the geometric fundamental group and ignores somehow arithmetics. On the other hand, $\pi^F(X,x)$ reflects the purely inseparable covers of X. That k is perfect guarantees that inseparable covers come only from geometry, and not from the ground field.

The inclusion of $\mathcal{C}^{\acute{e}t}(X)$ (resp. $\mathcal{C}^F(X)$) in $\mathcal{C}^N(X)$ as a full tensor subcategory induces a surjective homomorphism of groups schemes $r^{\acute{e}t} : \pi^N(X,x) \to \pi^{\acute{e}t}(X)$ (resp. $r^F : \pi^N(X) \to \pi^F(X)$). Our first remark is that the natural homomorphism

$$(r^{\acute{e}t}, r^F) : \pi^N(X,x) \to \pi^{\acute{e}t}(X,x) \times \pi^F(X,x) \tag{1.2}$$

is surjective but generally not injective. We give an example which is based on Raynaud's work [11] on coverings of curves producing a new ordinary part in the Jacobian (see Corollary 3.7). In particular, it is given as a rank 1 bundle in Pic of this covering, and thus does not come from a rank 1 bundle on X. The referee observes here that the morphism induced from (1.2) on the maximal abelian quotients of $\mathcal{C}^N(X), \mathcal{C}^F(X), \mathcal{C}^{\acute{e}t}(X)$ is however an isomorphism. This provides one reason for the determination of the representation category of the kernel of (1.2): our work gives some information on the non-abelian part of Nori's category.

The central theorem of our note is the determination by its objects and morphisms of a k-linear abelian rigid tensor category $\mathcal{E}$, which is equivalent to the representation category of $\mathrm{Ker}(r^{\acute{e}t}, r^F)$ (see Definition 4.3 for the construction and Proposition 4.4 and Theorem 4.5 to see that it computes what one wishes). This is the most delicate part of the construction. If S is a finite subcategory of $\mathcal{C}^N(X)$ with an étale finite Tannaka group scheme $\pi(X,S,x)$, then the total space X_S of the $\pi(X,S,x)$-principal bundle $\pi_S : X_S \to X$ which trivializes all the objects of S has the same property as X. It is proper, reduced and connected. However, if S is finite but $\pi(X,S,x)$ is not étale, then Nori shows that X_S is still proper connected, but may not be reduced. We give a concrete example in Remark 2.3, 2), which is due to P. Deligne.

In order to describe $\mathcal{E}$, we need in some sense an extension of Nori's theory to those non-reduced covers. We define on each such X_S a full subcategory $\mathcal{F}(X_S)$ of the category of coherent sheaves, the objects of which have the property that their push down on X lies in $\mathcal{C}^N(X)$ (see Definition 2.4). We show that indeed those coherent sheaves have to be vector bundles (Proposition 2.7), so in a sense, even if the scheme X_S might be bad, objects which push down to Nori's bundles on X are still good. In particular, $\mathcal{C}^N(X_S) = \mathcal{F}(X_S)$ if $\pi(X,S,x)$ is étale (Theorem 2.9), so the definition generalizes slightly Nori's one. For given finite subcategories S and T of $\mathcal{C}^N(X)$, with $\pi(X,S,x)$ étale and $\pi(X,T,x)$ local, we introduce in Definition 4.1 a full subcategory $\mathcal{E}(X_{S\cup T}) \subset \mathcal{F}(X_{S\cup T})$ consisting of those bundles V, the push down of which on X_S is F-finite. Now the objects of $\mathcal{E}$ are pairs $(X_{S\cup T}, V)$ for V an object in $\mathcal{E}(X_{S\cup T})$. Morphisms are subtle as they do take into account the whole inductive system of such $T' \subset \mathcal{C}^F(X)$. We can formulate our main theorem (see Theorem 4.5 for a precise formulation).

Main Theorem. *The functor* $\mathcal{C}^N(X) \to \mathcal{E}$ *which assigns* $(X_S, \pi_S^*(V))$ *to* V*, where* S *is the maximal étale subcategory of the subcategory* $\langle V \rangle$ *spanned by* $V \in \mathrm{Obj}(\mathcal{C}^N(X))$*, identifies the representation category of* $\mathrm{Ker}(r^{\acute{e}t}, r^F)$ *with* $\mathcal{E}$*.*

We now describe our method of proof. We proceed in two steps. As mentioned above, the homomorphism $r^{\acute{e}t} : \pi^N(X,x) \to \pi^{\acute{e}t}(X,s)$ is surjective. We denote its kernel by $L(X,x)$ and determine its representation category in section 3. The computation is based on two results. The first one of geometric nature asserts that sections of an F-finite bundle can be computed on any principal bundle $X_S \to X$ with finite étale group scheme (see Proposition 3.2). The second one is the key to the categorial work and comes from [3, Theorem 5.8]. (For the reader's convenience, we give a short account of the categorial statement in Appendix A.) It gives a criterion for the exactness of a sequence of affine group schemes

$$1 \to L \to G \to A \to 1$$

in terms of their representation categories. Roughly speaking, assuming the exactness at L and A then the exactness at G holds if and only if the following conditions hold: (i) a representation of G becomes trivial when restricted to L if and only if it comes from a reprsentation of A; (ii) for a representation V of G considered as representation of L, its subspace of L-invariants is invariant under G; (iii) each representation of L is embedable into the restriction to L of a representation of G.

We show that the category $\mathrm{Rep}(L(X,x))$ of (finitely-dimensional) representations of $L(X,x)$ is equivalent to the category $\mathcal{D}$, whose objects are pairs (X_S, V) where $X_S \to X$ is a principal bundle under an étale finite group scheme and V is a F-finite bundle on X_S. Morphisms are defined naturally via Proposition 3.2.

By definition of $L(X,x)$, the kernel of $(r^{\acute{e}t}, r^F)$ is the kernel of restriction $r^F|_L : L(X,x) \to \pi^F(X,x)$ of r^F to $L(X,x)$. The second step consists in showing that the category $\mathcal{E}$ constructed in Section 4 is equivalent to the representation category of the kernel of $r^F|_L$. The proof is based on the strengthening of Proposition 3.2, namely Proposition 3.6 and Proposition 4.6 as well as the criterion mentioned above.

Beyond the technicalities of the proof, let us remark that any finite k-group scheme G has two natural quotients: its maximal étale quotient $G^{\acute{e}t}$ and its maximal local quotient G^F. The kernel $G^0 := \mathrm{Ker}(G \to G^{\acute{e}t})$ is the 1-component of G, in particular is local. If G is abelian, the morphism $G^0 \to G^F$ is an isomorphism, and then G is the product of $G^{\acute{e}t}$ with G^F. In general, $G^0 \to G^F$ is surjective. The article here deals in some sense with the prosystem of the kernels of $G^0 \to G^F$.

Acknowledgements. Pierre Deligne sent us his enlightening example which we reproduced in Remark 2.3, 2). It allowed us to correct the main definition of our category $\mathcal{E}$ (Section 5)) which was wrongly stated in the first version of this article. We profoundly thank him for his interest, his encouragement and his help. We also warmly thank Michel Raynaud for answering all our questions on his work on theta characteristics on curves and on their Jacobians. Finally we thank the referee for careful reading, accurate remarks and helpful suggestions.

2. Nori's category

Throughout this work we shall fix a proper reduced scheme X over a perfect field k, which is *connected* in the sense that $H^0(X, \mathcal{O}_X) = k$. We assume that $X(k) \neq \emptyset$ and fix a rational point $x \in X(k)$.

In [8], [9], Nori defines a category of "essentially finite vector bundles" which we recall now. A vector bundle on X is called *semi-stable of degree* 0 if it is semi-stable of degree 0 while restricted to each proper curve in X. This is a full subcategory of the category Qcoh(X) of quasi-coherent sheaves on X, is abelian [8, Lemma 3.6] and will be denoted by $\mathcal{S}(X)$. A vector bundle V on X is called *finite* if there are polynomials $f \neq g$ whose coefficients are non-negative integers such that $f(V)$ and $g(V)$ are isomorphic. Nori proves that finite bundles are semi-stable of degree 0 [8, Corollary 3.5] and that the full abelian subcategory of $\mathcal{S}(X)$, consisting of those bundles which are subquotients in $\mathcal{S}(X)$ of a direct sum of finite bundles, is a k-linear abelian rigid tensor category. We shall denote this category by $\mathcal{C}^N(X)$ and call its objects *Nori finite bundles.*

The fiber functor at x (where by assumption $\kappa(x) = k$)

$$|_x : \mathcal{C}^N(X) \to \mathrm{Vect}_k, \ V \mapsto V|_x := V \otimes_{\mathcal{O}_X} \kappa(x) \tag{2.1}$$

with values in the category of finite-dimensional k-linear vector spaces, implies that $\mathcal{C}^N(X)$ is a Tannaka category. We denote by $\pi^N(X, x)$ the corresponding Tannaka group scheme over k. Tannaka duality ([1, Theorem 2.11]) yields an equivalence of categories

$$\mathcal{C}^N(X) \xrightarrow{|_x \ \cong} \mathrm{Rep}(\pi^N(X, x)). \tag{2.2}$$

We denote by η the inverse functor

$$\eta : \mathrm{Rep}(\pi^N(X, x)) \to \mathcal{C}^N(X). \tag{2.3}$$

Recall that for an affine group scheme G over k, a k-morphism $j : P \to X$ is said to be *a principal* G*-bundle on* X if

(i) j is a faithfully flat affine morphism
(ii) $\phi : P \times G \to P$ defines an action of G on P such that $j \circ \phi = j \circ p_1$
(iii) $(p_1, \phi) : P \times G \to P \times_X P$ is an isomorphism.

Given such a principal G-bundle P one associates to it an exact tensor functor

$$\eta_P : \mathrm{Rep}(G) \to \mathrm{Qcoh}(X) \tag{2.4}$$

as follows. For each representation V of G, one has the diagonal action of G on the trivial bundle $\mathcal{O}_P \otimes_k V$. Using Grothendieck flat descent [4, Exposé VIII], one obtains a vector bundle $\eta_P(V)$ on X by taking the G invariants of $\mathcal{O}_P \otimes_k V$, denoted by $P \times^G V$. Conversely, consider the regular representation of G in $k[G]$ given by $(gf)(h) = f(hg)$, $g, h \in G, h \in k[G]$. Then a functor $\eta : \mathrm{Rep}(G) \to \mathrm{Qcoh}(X)$ yields a principal G-bundle on X, which is the spectrum of the $\mathcal{O}_X$-algebra $\eta(k[G])$. These two constructions are inverse to each other.

Consider the regular representation of $\pi^N(X,x)$ in $k[\pi^N(X,x)]$. Then the discussion above applied to functor η in (2.3) yields a (universal) principal $\pi^N(X,x)$-bundle $\widetilde{\pi} : \widetilde{X} \to X$ together with the identity $\widetilde{\pi}^{-1}(x) = \pi^N(X,x)$. The unit element of $\pi^N(X,x)$ yields a distinguished rational point of $\widetilde{X}$ lying above x.

The projection $\widetilde{\pi} : \widetilde{X} \to X$ is in fact pro-finite in the following sense. Let S be an abelian tensor full subcategory of $\mathcal{C}^N(X)$ generated by finitely many objects $S_i, i = 1, \dots, r$. That is, objects of S are subquotients of direct sums of tensor products of copies of S_i and $S_i^\vee$ (the dual bundle to S_i). In the sequel we shall simply call S a *finitely generated tensor subcategory* of $\mathcal{C}^N(X)$. S is a Tannaka category by means of the fibre functor at x and denote its Tannaka group by $\pi(X,S,x)$. The discussion above applied to the forgetful functor $\eta_S : S \to \mathrm{Qcoh}(X)$ yields a $\pi(X,S,x)$-principal bundle $\pi_S : X_S \to X$ and a rational point x_S lying above x. Then $\pi(X,S,x)$ is a finite group scheme, which is a quotient of $\pi^N(X,x)$ and

$$\pi^N(X,x) = \varprojlim_S \pi(X,S,x), \widetilde{X} = \varprojlim_S X_S, \ \widetilde{\pi} = \varprojlim_S \pi_S, \quad \widetilde{x} = \varprojlim_S x_S, \tag{2.5}$$

where S runs in the pro-system of finitely generated full abelian tensor subcategories of $\mathcal{C}^N(X)$. Moreover the scheme X_S is connected:

$$H^0(X_S, \mathcal{O}_{X_S}) = k \tag{2.6}$$

Furthermore, π_S is universal in the following sense:

$$V \in \mathrm{Obj}(S) \Longleftrightarrow \pi_S^*(V) \text{ trivializable.} \tag{2.7}$$

Indeed, if $\pi_S^* V$ is trivializable on X_S, then the injective map $V \hookrightarrow \pi_{S*}\pi_S^* V \cong \pi_{S*}\mathcal{O}_{X_S}^{\oplus d}$, where d is the rank of V, shows that $V \in \mathrm{Obj}(S)$. Conversely, for $V \in S$ the construction in (2.4) shows that $\pi^* V$ is trivializable on X_S.

Finally we notice that $\pi^N(X,x)$ respects base change for algebraic extensions of k, that is

$$\pi^N(X \times_k K, x \times_k K) \cong \pi^N(X,x) \times_k K \tag{2.8}$$

for any algebraic extension $K \supset k$, in particular for $K = \bar{k}$. We refer to [9, Chapters I,II] for the exposition above.

For a finite bundle V, denote by $\langle V \rangle$ the tensor subcategory generated by V.

Definition 2.1. An *étale finite* bundle is a Nori finite bundle for which $\pi(X, \langle V \rangle, x)$ is étale (equivalently is smooth). If k has characteristic $p > 0$, an *F-finite* bundle is a Nori finite bundle for which $\pi(X, \langle V \rangle, x)$ is local. We denote by $\mathcal{C}^{ét}(X)$, resp. $\mathcal{C}^F(X)$, the full tensor subcategory of $\mathcal{C}^N(X)$ of étale, resp. F-, finite bundles.

The categories $\mathcal{C}^{ét}(X)$ and $\mathcal{C}^F(X)$ are both abelian tensor full subcategories, thus via the fiber functor at x they yield Tannaka k-group schemes $\pi^{ét}(X,x)$ and

$\pi^F(X,x)$, respectively. Furthermore, one has

$$\mathcal{C}^{\text{ét}}(X) \cap \mathcal{C}^F(X) = \{\text{trivial objects}\} \tag{2.9}$$

where {trivial objects} means the full subcategory of $\mathcal{C}^N(X)$ consisting of trivializable bundles.

Following the method of [9, II, Proposition 5] we obtain the following lemma.

Lemma 2.2. *The group schemes $\pi^{\text{ét}}(X,x)$ and $\pi^F(X,x)$ respect base change for algebraic extensions of k, that is* 2.8 *holds with N replaced by ét and F.*

Remark 2.3. 1) It is shown in [7, Section 2] that a Nori finite bundle V is F-finite if and only if there is a natural number $N > 0$, such that $(F^N_{\text{abs}})^*(V)$ is trivial, where F_{abs} is the absolute Frobenius.

2) If $S \subset \mathcal{C}^N(X)$ is a finite subcategory, with $\pi(X,S,x)$ étale, then X_S is still proper, reduced and connected. However, if $\pi(X,S,x)$ is finite but not étale, X_S is still proper and connected [9, Chapter II, Proposition 3], but not necessarily reduced. Indeed there are principal bundles $Y \to X$ under a finite local group scheme, such that the total space Y is not reduced. We reproduce here an example due to P. Deligne. Let k be algebraically closed of characteristic $p > 0$, and let $X \subset \mathbb{P}^2$ be the union of a smooth conic X' and a tangent line X''. Thus $X' \cap X''$ is isomorphic to Spec $k[\epsilon]/(\epsilon^2)$ as a k-scheme. One constructs $\pi : Y \to X$ by gluing the trivial μ_p-torsors $X' \times_k \mu_p$ to $X'' \times_k \mu_p$ along a non-constant section of Spec $k[\epsilon]/(\epsilon^2) \times_k \mu_p \to$ Spec $k[\epsilon]/(\epsilon^2)$. For example, one may take the non-constant section Spec $k[\epsilon]/(\epsilon^2) \to$ Spec $k[\epsilon]/(\epsilon^2) \times_k \mu_p$ defined by $k[\xi,\epsilon]/(\xi^p - 1, \epsilon^2) \to k[\epsilon]/(\epsilon^2), \xi \mapsto 1+\epsilon$. Then Y is projective, non-reduced, and yet fulfills the condition $H^0(Y, \mathcal{O}_Y) = k$.

If X_S is not reduced, there is no good notion of semi-stable vector bundles on X_S. However, for later use in this article, we introduce a category $\mathcal{F}(X_S)$ on the principal $\pi(X,S,x)$-bundle $\pi_S : X_S \to X$ where $S \subset \mathcal{C}^N(X)$ is a finitely generated full tensor subcategory. $\mathcal{F}(X_S)$ will play on X_S the rôle $\mathcal{C}^N(X)$ plays on X.

Definition 2.4. Let $S \subset \mathcal{C}^N(X)$ be a finitely generated abelian tensor full subcategory. Define $\mathcal{F}(X_S) \subset \text{Qcoh}(X_S)$ to be the full subcategory of $\text{Qcoh}(X_S)$, the objects of which are quasi-coherent sheaves V on X_S such that $(\pi_S)_*V \in \mathcal{C}^N(X)$.

Notice that $\mathcal{F}(X_S)$ is an abelian category. In fact, for a morphism $f : V \to W$ in $\mathcal{F}(X_S)$, by the exactness of $(\pi_S)_*$, we have $(\pi_S)_*\text{Ker}f = \text{Ker}((\pi_S)_*f) \in \mathcal{C}^N(X)$, as $\mathcal{C}^N(X)$ is full in $\text{Qcoh}(X)$, and the same holds for $\text{im}f$. We will show in Proposition 2.7 below that $\mathcal{F}(X_S)$ is k-linear abelian rigid tensor category and its objects are vector bundles on X_S, and when $\pi(X,S,x)$ is reduced $\mathcal{F}(X_S)$ coincides with $\mathcal{C}^N(X_S)$.

Let $S \subset S' \subset \mathcal{C}^N(X)$ be finitely generated abelian tensor full subcategories. Then one has the following commutative diagram

$$\begin{array}{ccc} X_{S'} & \xrightarrow{\pi_{S',S}} & X_S \\ & {\scriptstyle \pi_{S'}} \searrow \quad \swarrow {\scriptstyle \pi_S} & \\ & X & \end{array} \tag{2.10}$$

Further one has a surjective (hence faithfully flat) homomorphism

$$G_{S'} := \pi(X, S', x) \to \pi(X, S, x) =: G_S \tag{2.11}$$

Lemma 2.5. *The morphism $\pi_{S',S} : X_{S'} \to X_S$ is a principal bundle under a group $G_{S',S}$ which is the kernel of the homomorphism* (2.11).

Proof. It is well known that the morphism $G_{S'} \to G_S$ is a principal bundle under the group $G_{S',S}$. In fact, the map $G_{S'} \times_{G_S} G_{S'} \to G_{S'} \times G_{S',S}$ is given by $(g, h) \mapsto (g, g^{-1}h)$ and its inverse is given by $(g, k) \mapsto (g, gk)$, where $g, h \in G_{S'}$, $k \in G_{S',S}$. Now apply the fibre functor $\eta_{S'}$ to the corresponding function algebras $k[G_{S'}]$ and $k[G_S]$ we obtain the required isomorphism

$$X_{S'} \times_{X_S} X_{S'} \cong X_{S'} \times G_{S',S}$$

(recall that $\eta_{S'}(k[G_{S'}])$ is the $\mathcal{O}_X$-algebra that determines $X_{S'}$, similarly $\eta_{S'}(k[G_S]) = \eta_S[k(G_S)]$ is the one that determines X_S). □

The principal bundle $\pi_{S',S} : X_{S'} \to X_S$ yields a tensor functor

$$\eta_{S',S} : \mathrm{Rep}(G_{S',S}) \to \mathrm{Qcoh}(X_S), \quad \eta_{S',S}(V) := X_{S'} \times^{G_{S',S}} V. \tag{2.12}$$

Lemma 2.6. *The functor $\eta_{S',S}$ in* (2.12) *is fully faithful and exact. Consequently $G_{S',S}$ is isomorphic to the Tannaka group of the category $\mathrm{im}(\eta_{S',S})$.*

Proof. It is enough to check that $\eta := \eta_{S',S}$ is full, i.e., any morphism $\eta(V) \to \eta(W)$ in $\mathrm{Qcoh}(X_S)$ is induced by a morphism $V \to W$ in $\mathrm{Rep}(G_{S',S})$. This is equivalent to showing $H^0(X_S, \eta(V)) \cong V^{G_{S',S}}$ for any $V \in \mathrm{Rep}(G_{S',S})$. Recall that $\eta(V) := X_{S'} \times^{G_{S',S}} V$. Thus

$$H^0(X_S, \eta(V)) \cong H^0(X_{S'}, \mathcal{O}_{X_{S'}} \otimes_k V)^{G_{S',S}} \cong V^{G_{S',S}} \tag{2.13}$$

since $H^0(X_{S'}, \mathcal{O}_{X_{S'}}) = k$. □

Proposition 2.7. *The category $\mathcal{F}(X_S)$ defined in Definition* 2.4 *is a Tannaka category, whose objects are vector bundles.*

Proof. We first show that for any $V \in \mathcal{F}(X_S)$, there are $W_1, W_2 \in \mathcal{C}^N(X)$ and a morphism $f : \pi_S^* W_1 \to \pi_S^* W_2$ in $\mathrm{Qcoh}(X_S)$ such that $V = \mathrm{coker}(f)$. One takes $W_2 := (\pi_S)_* V$ which by definition lies in $\mathcal{C}^N(X)$, and defines V_1 to be the kernel of the surjection $\pi_S^* W_2 \twoheadrightarrow V$. Then $W_1 := (\pi_S)_* V_1 \in \mathcal{C}^N(X)$ and $f : \pi_S^* W_1 \twoheadrightarrow V_1 \hookrightarrow \pi_S^* W_2$ satisfying $\mathrm{coker}(f) = V$.

Let $S' \subset \mathcal{C}^N(X)$ be the full tensor subcategory generated by W_1, W_2 and S. Then the pullbacks of $\pi_S^* W_1$ and $\pi_S^* W_2$ to $X_{S'}$ (under $\pi_{S',S} : X_{S'} \to X_S$) become trivial, thus $\pi_S^* W_1$ and $\pi_S^* W_2$ are in the image of

$$\eta_{S',S} : \mathrm{Rep}(\mathrm{G}_{S',S}) \to \mathrm{Qcoh}(X_S). \tag{2.14}$$

By Lemma 2.6, the functor $\eta_{S',S}$ is fully faithful, thus $V = \mathrm{coker}(f)$ is also in the image of $\eta_{S',S}$. In particular, V is a vector bundle.

It is now easy to check that $\mathcal{F}(X_S)$ is a k-linear abelian rigid tensor category. Since X_S has a rational point x_S, $\mathcal{F}(X_S)$ is a Tannaka category. □

Next we show $\mathcal{F}(X_S) = \mathcal{C}^N(X_S)$ when $\pi(X, S, x)$ is reduced. To this aim, recall that a bundle V on X is said to be *strongly semi-stable of degree 0* if for any non-singular projective curve C and any morphism $f : C \to X$, the pullback $f^* V$ is semi-stable of degree 0 on C. It is known that strongly semi-stable bundles of degree 0 on X form a k-linear Tannaka full subcategory of $\mathrm{Qcoh}(X)$ [10, Theorem 3.23]. On the other hand as Nori finiteness is preserved under pull-back by any $f : C \to X$, we see that $\mathcal{C}^N(X)$ is a full subcategory of the category of strongly semi-stable bundles of degree 0.

Lemma 2.8. *If $\pi(X, S, x)$ is a smooth finite group scheme, and if $V \in \mathcal{C}^N(X_S)$, then $\pi_S^*(\pi_{S*} V) \in \mathcal{C}^N(X_S)$ and $W := \pi_{S*} V$ is strongly semi-stable of degree 0.*

Proof. Let $G := \pi(X, S, x)$, $\pi := \pi_S$, $Y := X_S$ and $y := x_S$. Since strong semi-stability and Nori finiteness are compatible with base change by algebraic field extensions of k, we can assume that $k = \bar{k}$. Consider

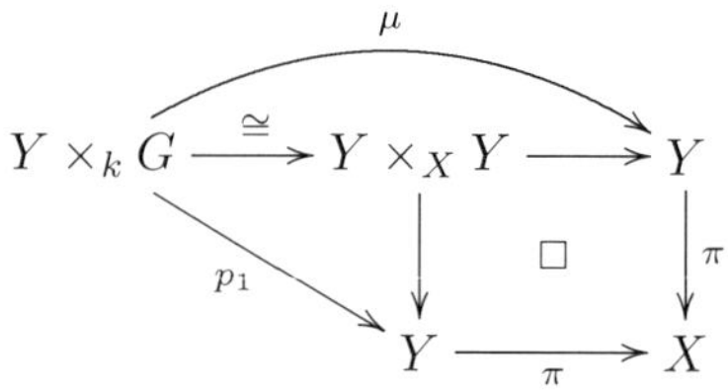

$$\tag{2.15}$$

where $\mu : Y \times_k G \to Y$ is the action of G, p_1 is the projection to the first factor and $Y \times_k G \cong Y \times_X Y$ is induced by (p_1, μ). Then

$$\pi^* \pi_* V \cong p_{1*} \mu^* V = \bigoplus_{g \in G(k)} V_g \tag{2.16}$$

where V_g is the translation of V by g. Thus $\pi^* \pi_* V \in \mathcal{C}^N(Y)$.

To show that $W := \pi_* V$ is strongly semi-stable, we consider the fiber square

$$\begin{array}{ccc} Y_C & \xrightarrow{g} & Y \\ \pi_C \downarrow & \square & \downarrow \pi \\ C & \xrightarrow{f} & X \end{array}$$

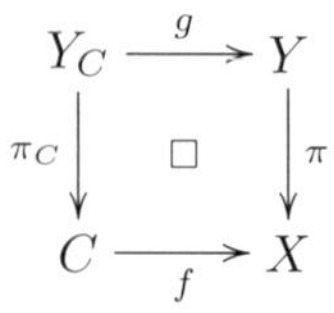

So in particular, π_C is still a principal bundle under G. Since π is finite, one has

$$f^*W = f^*(\pi_* V) = \pi_{C*}(g^*V).$$

Denote $V_C := g^*V$. Since $V_C \in \mathcal{C}^N(Y_C)$, the discussion above shows that $\pi_C^*(f^*W) = \pi_C^*(\pi_{C*}V_C) \in \mathcal{C}^N(Y_C)$. In particular, $\pi_C^*(f^*W)$ is semi-stable of degree 0, which implies that f^*W is semi-stable of degree 0. Indeed, for any subbundle $U \subset f^*W$, the bundle π_C^*U is a subbundle of $\pi_C^*(f^*W)$, hence has negative degree, consequently the degree of U is also negative. □

Theorem 2.9. *Assume that $\pi(X, S, x)$ is a smooth finite group scheme. Then $\mathcal{F}(X_S) = \mathcal{C}^N(X_S)$ and there is an exact sequence of group schemes*

$$1 \to \pi^N(X_S, x_S) \to \pi^N(X, x) \to \pi(X, S, x) \to 1 \tag{2.17}$$

Proof. By Proposition 2.7, $\mathcal{F}(X_S) \subset \mathcal{C}^N(X_S)$. We prove the inverse inclusion. Thus let $V \in \mathcal{C}^N(X_S)$. By Lemma 2.8, $W := \pi_{S_*}V$ semi-stable of degree 0 and $\pi_S^*W \in \mathcal{C}^N(X_S)$. Let $\langle W\rangle \cup S$ be the full tensor subcategory generated by W and objects of S in the Tannaka category of strongly semi-stable bundles of degree 0 and denote by G its Tannaka group with respect to the fiber functor at x. To show $W \in \mathcal{C}^N(X)$, it suffices to show that G is a finite group scheme (see construction in (2.4)).

The full subcategory of $\langle W\rangle \cup S$ whose objects become trivial when pulled-back to X_S is precisely S (see (2.7)). The functor $\pi_S^* : \langle W\rangle \cup S \to \langle \pi_S^* W\rangle$ yields a sequence of homomorphisms of group schemes

$$1 \to \pi(X_S, \langle \pi_S^* W\rangle, x_S) \to G \to \pi(X, S, x) \to 1 \tag{2.18}$$

which we claim to be exact.

The surjectivity of $G \to \pi(X, S, x)$ and the injectivity of $\pi(X_S, \langle \pi_S^* W\rangle, x_S) \to G$ follow from the definition and A.1, (i), (ii). We show the exactness at G, using Theorem A.1, (iii). Condition (a) in A.1, (iii), follows from (2.7).

We check condition (c). Let $M \in \langle \pi_S^* W\rangle$. By definition, M is a subquotient of $\pi_S^* N$, $N \in \langle W\rangle \cup S$. Thus $\pi_{S_*}M$ is a subquotient of $\pi_{S_*}\pi_S^* N = N \otimes \pi_{S_*}\pi_S^* \mathcal{O}_{X_S} \in \langle W\rangle \cup S$. Hence $\pi_{S_*}M$ lies in $\langle W\rangle \cup S$. Now we have the required surjective map $\pi_S^*(\pi_{S_*}M) \to M$.

As for (b) we use projection formula

$$H^0(X_S, \pi_S^* N) = H^0(X, \pi_{S_*}\pi_S^* N) = \mathrm{Hom}_{\mathcal{O}_X}(\pi_{S_*}\mathcal{O}_{X_S}^\vee, N) = \bigoplus_{i=1}^r k \cdot \phi_i \tag{2.19}$$

where $\phi_i : (\pi_{S_*}\mathcal{O}_{X_S})^\vee \to N$. Let $N_0 = \sum_i \mathrm{im}(\phi_i) \subset N$. Then N_0 is in S and any morphism $\phi : (\pi_{S_*}\mathcal{O}_{X_S})^\vee \to N$ has image in N_0. By comparing the ranks, we see that $\pi_S^* N_0$ is the maximal trivial subbundle in $\pi_S^* N$.

Thus the sequence in (2.18) is exact, hence G is finite. The exactness of (2.17) follows from the exactness of (2.18) by taking the projective limit on S. □

Remarks 2.10. The group scheme $\pi^{ét}(X,x)$ can be considered as the k-linearization of Grothendieck's fundamental group ([4, Exposé V]), which we recall now. Grothendieck considers the category of finite étale coverings of X with morphisms being X-morphisms. A geometric point $\bar{x} \in X(K)$ $(K = \bar{K})$ defines a fiber functor from this category to the category of finite sets: $\mathcal{G}r : (Y \xrightarrow{\pi} X) \mapsto \pi^{-1}(\bar{x})$. The fundamental group $\pi_1(X,\bar{x})$ of the connected scheme X with base point $\bar{x}$ is defined to be the automorphism group of the fiber functor. This is a pro-finite group, hence has a natural topology in which subgroups of finite index are open and form a basis of topology at the unit element. The main theorem claims an equivalence between the category of finite étale coverings and the category of finite sets with continuous action of $\mathcal{G}r$ (finite sets are endowed with discrete topology). Further there exists a pro-finitie étale covering $\hat{\pi} : \widehat{X} \to X$ which is universal in the sense that

$$\mathrm{Mor}_X(\widehat{X}, Y) \cong \mathcal{G}r(Y) \tag{2.20}$$

for any finite covering $Y \to X$ ([4, Theorem V.4.1]). One recovers the group $\pi_1(X,\bar{x})$ as the fiber $\hat{\pi}^{-1}(\bar{x})$. Notice that it suffices to check (2.20) for Galois coverings $Y \to X$.

Assume that k is moreover algebraically closed. Then the fundamental group $\pi_1(X,x)$ with base point at x is called the geometric fundamental group. Upon the algebraically closed field k, a reduced finite group scheme is uniquely determined by its k-points, which is a finite group. Therefore for any $S \subset \mathcal{C}^{ét}(X)$, $\pi_S : X_S \to X$ is a Galois covering of X under the group $\pi(X,S,x)(k)$. Conversely, any Galois covering $Y \xrightarrow{\pi} X$ under a finite group H can be considered as a principal bundle under the constant (finite) group scheme defined by H. It is easy to check that the covering $\pi^{ét} : X_{\mathcal{C}^{ét}(X)} \to X$ given in (2.5) satisfies the universal property (2.20). We conclude that the group of k-points of $\pi^{ét}(X,x)$ is isomorphic to $\pi_1(X,x)$.

If k is perfect but not algebraically closed, take $X = \mathrm{Spec}(k)$ with the rational point point $x = X \in X(k)$. Then $\mathcal{C}^{ét}(X)$ is equivalent via the fibre functor to Vect_k, and consequently $\pi^{ét}(X,x) = \{1\}$. On the other hand, Grothendieck's fundamental group is then $\mathrm{Gal}(\bar{k}/k)$, which is highly nontrivial. However, according to Lemma 2.2 we have an isomorphism of $\bar{k}$-group schemes

$$\pi^{ét}(X \times_k \bar{k}, x \times_k \bar{k}) \xrightarrow{\cong} \pi^{ét}(X,x) \times_k \bar{k}.$$

So we conclude in general

$$\pi_1(X \times_k \bar{k}, \bar{x}) = \pi^{ét}(X,x)(\bar{k}).$$

The aim of our article is to understand the relationship between the groups $\pi^N(X,x)$, $\pi^{ét}(X,x)$ and $\pi^F(X,x)$. We first notice that Theorem A.1, (i), applied to the full subcategories $\mathcal{C}^{ét}(X) \to \mathcal{C}^N(X)$, resp. $\mathcal{C}^F(X) \to \mathcal{C}^N(X)$, shows that the restriction homomorphisms $\pi^N(X,x) \xrightarrow{r^{ét}} \pi^{ét}(X,x)$, resp. $\pi^N(X,x) \xrightarrow{r^F} \pi^F(X,x)$ are faithfully flat.

Notation 2.11. We set $L(X,x) = \mathrm{Ker}(\pi^N(X,x) \to \pi^{ét}(X,x))$.

3. The representation category of the difference between Nori's fundamental group scheme and it's étale quotient

We continue to fix X/k and $x \in X(k)$ as in section 2. The purpose of this section is to determine the representation category of the kernel L of the map $\pi^N(X,x) \to \pi^{ét}(X,x)$. To this aim, we first observe the following.

Lemma 3.1. *The group scheme $\pi^{ét}(X,x)$ is the largest quotient pro-finite group scheme of $\pi^N(X,x)$ which is reduced.*

Proof. For $S \subset \mathcal{C}^N(X)$ an abelian tensor full subcategory generated by finitely many objects we set

$$S^{ét} := S \cap \mathcal{C}^{ét}(X), \tag{3.1}$$

i.e., the full subcategory consisting of objects in $\mathcal{C}^N(X)$, isomorphic both to an object in S and an object in $\mathcal{C}^{ét}(X)$. Thus $\pi(X, S^{ét}, x)$ is a reduced quotient of $\pi(X,S,x)$. We claim that this is the largest quotient of $\pi(X,S,x)$.

Tannaka duality shows that any quotient map $\pi(X,S,x) \twoheadrightarrow H$ of group schemes over k yields a fully faithful functor from $\mathrm{Rep}(H)$ to $\mathcal{C}^N(X)$ with image, say (H), lying in S, and consequently yielding an H-principal bundle $\pi_{(H)} : X_{(H)} \to X$ which is proper, connected, with a rational point $x_{(H)}$ mapping to x, so that $V \in (H)$ if and only if $\pi_{(H)}^*(V)$ is trivial. If H is reduced then (H) consists only of étale finite bundles, thus $(H) \subset S^{ét}$. Hence $\pi(X,S,x) \to H$ factors through $\pi(X,S,x) \to \pi(X,S^{ét},x) \to H$. This shows that $\pi(X,S^{ét},x)$ is the maximal reduced quotient of $\pi(X,S,x)$. Now the claim of Lemma follows by passing to the limit. □

In the rest of this section, S will denote a finitely generated tensor subcategory of $\mathcal{C}^{ét}(X)$. Thus $\pi_S : X_S \to X$ is étale and X_S is reduced.

Proposition 3.2. *Let X be a proper reduced connected scheme defined over a perfect field k. Let V be an F-finite bundle on X. Then*

$$\pi_S^* : H^0(X,V) \to H^0(X_S, \pi_S^* V)$$

is an isomorphism.

Proof. To simplify the notations, we set $Y := X_S$, $\pi := \pi_S$. Let V_0 be the maximal trivial subobject of V. Since π is étale, the bundles associated to $\pi_*\mathcal{O}_Y$, and therefore to $(\pi_*\mathcal{O}_Y)^\vee$, are étale finite. The image under a morphism of $(\pi_*\mathcal{O}_Y)^\vee$ to V is therefore at the same time étale- and F-finite, hence (see (2.9)) lies in the maximal trivial subobject V_0 of V. By projection formula we have

$$\begin{aligned} H^0(Y,\pi^*V) = H^0(X,(\pi_*\mathcal{O}_Y)\otimes V) &\cong \mathrm{Hom}_X((\pi_*\mathcal{O}_Y)^\vee, V) \\ \subset \mathrm{Hom}_X((\pi_*\mathcal{O}_Y)^\vee, V_0) &\cong H^0(Y,\pi^*V_0) = H^0(X,V_0). \end{aligned} \tag{3.2}$$

as $H^0(Y, \mathcal{O}_Y) = k$ by (2.6). Hence

$$H^0(X, V_0) \subset H^0(X, V) \subset H^0(Y, \pi^*V) \subset H^0(Y, \pi^*V_0) = H^0(X, V_0), \tag{3.3}$$

so one has everywhere equality. □

Denote by $\mathcal{S}^{ét}$ the directed system of finitely generated tensor subcategories of $\mathcal{C}^{ét}(X)$ with respect to the inclusion

$$\mathcal{S}^{ét} = \left\{S \subset \mathcal{C}^{ét}(X), \text{ finitely generated}\right\}. \tag{3.4}$$

Let $\widetilde{X}^{ét} \to X$ denote the universal pro-étale covering associated to $\pi^{ét}(X, x)$, defined similarly as in (2.5). Thus

$$\widetilde{X}^{ét} = \varprojlim_{S \in \mathcal{S}^{ét}} X_S, \quad \widetilde{\pi}_S : \widetilde{X} \twoheadrightarrow X_S. \tag{3.5}$$

By means of Proposition 3.2 we have the following isomorphism for any $V \in \mathcal{C}^F(X_S)$,

$$H^0(X_S, V) \cong H^0(\widetilde{X}, \widetilde{\pi}_S^* V). \tag{3.6}$$

Definition 3.3. The category $\mathcal{D}$ has for objects pairs (X_S, V) where $S \in \mathcal{S}^{ét}$, $V \in \mathcal{C}^F(X_S)$, and for morphisms

$$\mathrm{Hom}((X_{S_1}, V), (X_{S_2}, W)) := \mathrm{Hom}_{\widetilde{X}}(\widetilde{\pi}_{S_1}^* V, \widetilde{\pi}_{S_2}^* W).$$

For any two abelian tensor full subcategories $S_1, S_2 \in \mathcal{S}^{ét}$, denote $S_1 \cup S_2$ the abelian tensor full subcategory generated by objects of S_1 and S_2. One has $S_1 \cup S_2 \in \mathcal{S}^{ét}$. We also extend this notation for several subcategories.

Proposition 3.4. *The category $\mathcal{D}$ is an abelian, rigid k-linear tensor category, with the tensor structure defined by*

$$(X_{S_1}, V) \otimes (X_{S_2}, W) = (X_{S_1 \cup S_2}, \pi^*_{S_1 \cup S_2, S_1}(V) \otimes \pi^*_{S_1 \cup S_2, S_2}(W)) \tag{3.7}$$

The functor

$$\omega : \mathcal{D} \to \mathrm{Vect}_k, \ (X_S, V) \mapsto V|_{x_S} \tag{3.8}$$

makes $\mathcal{D}$ a Tannaka category.

Proof. We define the kernel, the image and the cokernel of a homomorphism $f : (X_{S_1}, V) \to (X_{S_2}, W)$ in $\mathcal{D}$ as follows. By means of (3.6), one has an isomorphism

$$\mathrm{Hom}_{\widetilde{X}^{ét}}(\widetilde{\pi}^*_{S_1} V, \widetilde{\pi}^*_{S_2} W) \cong \mathrm{Hom}_{X_S}(\pi^*_{S_1 \cup S_2, S_1} V, \pi^*_{S_1 \cup S_2, S_2} W), \tag{3.9}$$

under which f corresponds to f_S. Then the kernel, image and cokernel of f are defined to be the kernel, image and cokernel of f_S respectively. It is clear that $\mathcal{D}$ is an abelian category.

The unit object is $(X, \mathcal{O}_X)$, the endomorphism ring of the unit object is thus k. The dual object is given by $(X_S, V)^\vee = (X_S, V^\vee)$. □

We observe that $(X_S, \mathcal{O}_{X_S})$ is isomorphic to $(X, \mathcal{O}_X)$ in $\mathcal{D}$. More generally, for $S_1 \in \mathcal{S}^{ét}$,

$$(X_{S_1}, V) \text{ is isomorphic to } (X_{S_1 \cup S_2}, \pi^*_{S_1 \cup S_2, S_1} V) \text{ in } \mathcal{D} \text{ for all } S_2 \in \mathcal{S}^{ét}. \tag{3.10}$$

For $V \in \mathcal{C}^N(X)$, the category $\langle V \rangle^{ét}$ is defined as in (3.1). According to Lemma 2.5, for $S' = \langle V \rangle$ and $S = \langle V \rangle^{ét}$, $X_{\langle V \rangle} \to X_{\langle V \rangle^{ét}}$ is a principal bundle under the group $H = \mathrm{Ker}(G_{\langle V \rangle} \to G_{\langle V \rangle^{ét}})$, which, according to the proof of Lemma 3.1, is a local group. Therefore $\pi^*_{\langle V \rangle^{ét}}(V)$ is an F-finite bundle on $X_{\langle V \rangle^{ét}}$. Define the functor

$$q : \mathcal{C}^N(X) \to \mathcal{D}, \quad V \mapsto (X_{\langle V \rangle^{ét}}, \pi^*_{\langle V \rangle^{ét}}(V)). \tag{3.11}$$

Then q is an exact tensor functor which is compatible with the fiber functors ω and $|_x$. Thus q yields a homomorphism of group schemes

$$q^* : G(\mathcal{D}) \to \pi^N(X, x). \tag{3.12}$$

Denote by $G(\mathcal{D})$ the Tannaka group scheme over k with respect to ω. The functor q has the property that $V \in \mathcal{C}^N(X)$ is étale finite if and only if $q(V)$ is trivial in $\mathcal{D}$. Therefore the composition

$$G(\mathcal{D}) \xrightarrow{q^*} \pi^N(X, x) \to \pi^{ét}(X, x) \tag{3.13}$$

is the trivial homomorphism. That is q^* factors though a homomorphism (denoted by the same letter) to L (see Notation 2.11).

Theorem 3.5. *The representation category of the kernel $L(X, x)$ of the homomorphism $r^{ét} : \pi^N(X, x) \to \pi^{ét}(X, x)$ is equivalent to $\mathcal{D}$ by means of the functor q.*

Proof. We show that the sequence of k-group schemes (3.13) is exact. We shall use the criterion given in Theorem A.1, (iii). Condition (a) there is satisfied by (2.8).

Let (X_S, V) be an object in $\mathcal{D}$. Then, by Theorem 2.9, $W := (\pi_S)_* V$ is an object in $\mathcal{C}^N(X)$. Moreover one has a surjection $q(W) \twoheadrightarrow (X_S, V)$ in $\mathcal{D}$. Thus every object of $\mathcal{D}$ is a quotient of the image by q of an object in $\mathcal{C}^N(X)$. Condition (c) of A.1, (iii), is satisfied.

It remains to check condition (b) of A.1, (iii). For $V \in \mathcal{C}^N(X)$ set $S = \langle V \rangle^{ét}$ then $q(V) = (X_S, \pi_S^* V)$. Applying projection formula we obtain

$$H^0(X_S, \pi_S^*(V)) = \mathrm{Hom}_{\mathcal{O}_X}((\pi_{S*}\mathcal{O}_{X_S})^\vee, V) = \bigoplus_{i=1}^{r} k \cdot \varphi_i, \tag{3.14}$$

where $\varphi_i : (\pi_{S*}\mathcal{O}_{X_S})^\vee \to V$. Let $V_{ét} \subset V$ be the image of

$$\oplus_1^r \varphi_i : \quad \bigoplus_1^r (\pi_{S*}\mathcal{O}_{X_S})^\vee \to V. \tag{3.15}$$

As $(\pi_{S*}\mathcal{O}_{X_S})^\vee$ is étale finite, $V_{ét}$ is étale finite and lies in S, hence $\pi_S^* V_{ét}$ is a trivial bundle by (2.7). Thus $q(V_{ét})$ is a trivial subobject of $q(V)$. We show that it is the largest one. By the definition of $V_{ét}$, one has

$$H^0(X, (\pi_{S*}\mathcal{O}_{X_S}) \otimes V) = H^0(X, (\pi_{S*}\mathcal{O}_{X_S}) \otimes V_{ét}). \tag{3.16}$$

Applying the projection formula again, one has

$$H^0(X, (\pi_{S*}\mathcal{O}_{X_S}) \otimes V_{ét}) = H^0(X_S, \pi_S^*(V_{ét})). \tag{3.17}$$

That is, $q(V_{ét})$ is the maximal trivial subobject of $q(V)$. □

Our next aim is to study the kernel of $r^{ét} \times r^F$. For this we shall need a strengthening of Proposition 3.2.

Proposition 3.6. *Let $S \subset \mathcal{C}^{ét}(X)$ be a finitely generated tensor subcategory. Then the homomorphism of group schemes $\pi^F(X_S, x_S) \to \pi^F(X, x)$ induced by π_S^* is surjective.*

Proof. To simplify notations, we set $Y := X_S$, $\pi := \pi_S$ in the proof. According to Lemma 2.2 if suffices to consider the case where k is algebraically closed. According to Theorem A.1, (i), one has to show that for any $V \in \mathcal{C}^F(X)$ and any inclusion $\varphi : W \hookrightarrow \pi^*V$ in $\mathcal{C}^F(X_S)$, there exists an inclusion $\iota : V_0 \hookrightarrow V$ in $\mathcal{C}^F(X)$, such that $\varphi = \pi^*\iota : \pi^*V_0 \to \pi^*V$. We first assume that W is simple. The bundle V has a decomposition series $V_0 \subset V_1 \subset \ldots \subset V_N = V$ with V_i/V_{i-1} simple. Then there exists an index i such that the image of $\varphi(W)$ in $\pi^*(V_i/V_{i-1})$ is not zero. Thus we may assume that V itself is simple. It suffices now to show that φ is an isomorphism.

Using the adjointness between π_* and π^* we have

$$\begin{aligned}\mathrm{Hom}_{X_S}(W, \pi^*V) &\cong \mathrm{Hom}_{X_S}(\pi^*V^\vee, W^\vee) \cong \mathrm{Hom}_X(V^\vee, \pi_*(W^\vee)) \\ &\cong \mathrm{Hom}((\pi_*(W^\vee))^\vee, V).\end{aligned} \tag{3.18}$$

Thus φ corresponds to a non-zero morphism $\psi : \pi_*(W^\vee)^\vee \to V$. Since V is simple, ψ is surjective and hence so is $\pi^*\psi : \pi^*\pi_*(W^\vee)^\vee \to \pi^*V$. On the other hand, as in Lemma 2.8, we have

$$\pi^*\pi_*W = \bigoplus_{g \in G(k)} W_g. \tag{3.19}$$

Since W is simple, so are W_g, $g \in G(k)$. This shows that π^*V, being a quotient of a direct sum of simple objects, is semi-simple. According to Proposition 3.2, π^*V has to be simple. Therefore $W = \pi^*V$.

The general case follows by induction on the length of the decomposition of W. □

Corollary 3.7. *The natural map $\pi^N(X, x) \xrightarrow{r^{ét} \times r^F} \pi^{ét}(X, x) \times \pi^F(X, x)$ is surjective and in general not an isomorphism.*

Proof. In fact, the surjectivity of $r^{ét} \times r^F$ holds for any pro-finite group, as it holds for finite groups. In our case this can also be seen from the proof of Proposition 3.6.

The claim of the corollary is equivalent to showing that the induced homomorphism $r^F|_L : L(X,x) \to \pi^F(X,x)$ is surjective and not necessarily an isomorphism. This homomorphism is Tannaka dual to the restriction $q|_{\mathcal{C}^F(X)} : \mathcal{C}^F(X) \to \mathcal{D}$, for functor q defined in (3.11), which is the identity functor $q|_{\mathcal{C}^F(X)}(V) = (X,V)$. Now the proof of Proposition 3.6 and the injectivity criterion A.1, (ii), prove the corollary.

It remains to exhibit an example when $r^F|_L$ is not an isomorphism. According to the discussion above, this amounts to finding an F-finite bundle on X_S which does not come from X. By [11], Théorème 4.3.1, if X is a smooth projective curve of genus $g \geq 2$ over an algebraic closed field k of characteristic $p > 0$, then for $\ell \neq p$ prime with $\ell + 1 \geq (p-1)g$, there is a cyclic covering $\pi : Y \to X$ of degree ℓ (thus étale), such that $\mathrm{Pic}^0(Y)/\mathrm{Pic}^0(X)$ is ordinary. Since π is Galois cyclic of order ℓ, it is defined as $\mathrm{Spec}_X(\oplus_0^{\ell-1} L^i)$ for some L étale finite of rank 1 over X and of order ℓ, thus $\pi = \pi_{\langle L \rangle}$ and $Y = X_{\langle L \rangle}$. We conclude that there are p-power torsion rank 1 bundles on $X_{\langle L \rangle}$ which do not come from X. □

4. The representation category of the difference between Nori's fundamental group and the product of its étale and local quotients

The aim of this section is to describe the representation category of the kernel of the homomorphism $r^{\text{ét}} \times r^F$. Recall that X is a reduced proper scheme over a perfect field of characteristic $p > 0$ with a rational point $x \in X(k)$ and is connected in the sense that $H^0(X, \mathcal{O}_X) = k$.

In order to determine the representation category $\mathcal{E}$ of the kernel of $r^{\text{ét}} \times r^F$ we shall need an auxiliary category $\mathcal{E}(X_{S \cup T})$, where S is a finitely generated tensor full subcategory of $\mathcal{C}^{\text{ét}}(X)$ and T is a finitely generated tensor full subcategory of $\mathcal{C}^F(X)$.

Definition 4.1. For S a finitely generated tensor subcategory of $\mathcal{C}^{\text{ét}}(X)$ and T a finitely generated tensor full subcategory of $\mathcal{C}^F(X)$, one defines $\mathcal{E}(X_{S \cup T}) \subset \mathcal{F}(X_{S \cup T})$ (for the definition of $\mathcal{F}(XS \cup T)$, see Definition 2.4) to be the full subcategory whose objects V have the property that $(\pi_{S \cup T, S})_* V \in \mathcal{C}^F(X_S)$.

Denote by $\mathcal{T}^\ell$ the directed system of finitely generated tensor subcategories of $\mathcal{C}^F(X)$ with respect to inclusion

$$\mathcal{T}^\ell := \left\{ T \subset \mathcal{C}^F(X), \text{ finitely generated} \right\}. \tag{4.1}$$

Lemma 4.2. *Let $S \subset S' \in \mathcal{S}^{\text{ét}}$, $T \subset T' \in \mathcal{T}^\ell$ and $V \in \mathcal{E}(X_{S \cup T})$. Then:*

1) *The following commutative diagram is cartesian:*

$$\begin{array}{ccc} X_{S'\cup T} & \xrightarrow{\pi_{S'\cup T,S'}} & X_{S'} \\ \Big\downarrow{\scriptstyle \pi_{S'\cup T,S\cup T}} & & \Big\downarrow{\scriptstyle \pi_{S',S}} \\ X_{S\cup T} & \xrightarrow[\pi_{S\cup T,S}]{} & X_S \end{array}$$

2) $\mathcal{E}(X_{S\cup T})$ *is a* k*-linear abelian, rigid tensor category.*
3) $\pi^*_{S\cup T',S\cup T}V \in \mathcal{E}(X_{S\cup T'})$.
4) $\pi^*_{S'\cup T,S\cup T}V \in \mathcal{E}(X_{S'\cup T})$.
5) *The canonical homomorphism*

$$H^0(X_{S\cup T}, V) \to H^0(X_{S'\cup T}, \pi^*_{S'\cup T,S\cup T}V) \tag{4.2}$$

is an isomorphism.

Proof. 1) We first show that the following commutative diagram

$$\begin{array}{ccc} X_{S\cup T} & \xrightarrow{\pi_{S\cup T,S}} & X_S \\ \Big\downarrow{\scriptstyle \pi_{S\cup T,T}} & & \Big\downarrow{\scriptstyle \pi_S} \\ X_T & \xrightarrow[\pi_T]{} & X \end{array} \tag{4.3}$$

is cartesian. Indeed, $\pi_S : X_S \to X$ is a principal bundle under $\pi(X, S, x)$, and similarly for $\pi_T, \pi_{S\cup T}$. So the assertion is equivalent to showing that the natural homomorphism

$$\pi(X, S \cup T, x) \to \pi(X, S, x) \times \pi(X, T, x) \tag{4.4}$$

induced by the embeddings $S \subset (S \cup T), T \subset (S \cup T)$ of categories is an isomorphism. Since $\pi(X, S, x)$ (resp. $\pi(X, T, x)$) is a reduced (resp. local) quotient of $\pi(X, S \cup T, x)$, (4.4) is surjective. On the other hand, by definition, every object in $S \cup T$ is a subquotient of tensors of objects in S and objects in T, thus by A.1, (ii), (4.4) is injective. Therefore we have

$$X_{S'} \times_{X_S} X_{S\cup T} = X_{S'} \times_{X_S} (X_S \times_X X_T) = X_{S'\cup T}. \tag{4.5}$$

This shows 1).

2) Note that $\pi_{S\cup T,S} : X_{S\cup T} \to X_S$ is a principal $\pi(X, T, x)$-bundle since (4.3) is cartesian, thus $(\pi_{S\cup T,S})_*\mathcal{O}_{X_{S\cup T}}$ is F-finite on X_S. The pullback by $\pi_{S\cup T,S}$ of any F-finite bundle on X_S lies in $\mathcal{E}(X_{S\cup T})$. Then it is easy to write any $V \in \mathcal{E}(X_{S\cup T})$ as a cokernel of a morphism $\pi^*_{S\cup T,S}W_1 \to \pi^*_{S\cup T,S}W_2$, where W_1, $W_2 \in \mathcal{C}^F(X_S)$. Thus 2) follows.

3) and 4) are easy by chasing diagrams and using the projection formula, as we did already many times.

To show 5), one uses Proposition 3.2 and the projection formula

$$H^0(X_{S'\cup T}, \pi^*_{S'\cup T, S\cup T}V) = H^0(X_{S'}, (\pi_{S'\cup T,S'})_*\pi^*_{S'\cup T,S\cup T}V) \tag{4.6}$$
$$= H^0(X_{S'}, \pi^*_{S',S}(\pi_{S\cup T,S})_*V) \overset{\text{(Prop. 3.2)}}{=} H^0(X_S, (\pi_{S\cup T,S})_*V) = H^0(X_{S\cup T}, V).$$

□

Fix S in $\mathcal{S}^{\text{ét}}$ and consider the principal bundle $X_{S\cup\mathcal{C}^F(X)}$ associated to $S\cup\mathcal{C}^F(X)$, that is

$$X_{S\cup\mathcal{C}^F(X)} = \varprojlim_{T\in\mathcal{T}^\ell} X_{S\cup T}, \quad \widetilde{\pi}_{S,T} : X_{S\cup\mathcal{C}^F(X)} \twoheadrightarrow X_{S\cup T}. \tag{4.7}$$

The cartesian diagram in (4.3) implies that $X_{S\cup\mathcal{C}^F(X)}$ is the product of X_S with $\widetilde{X}^F = X_{\mathcal{C}^F(X)}$ over X. For each $V\in\mathcal{E}(X_{S\cup T})$, set

$$H^0_{\mathcal{T}^\ell}(X_{S\cup T}, V) := H^0(X_{S\cup\mathcal{C}^F(X)}, \widetilde{\pi}^*_{S,T}V). \tag{4.8}$$

Recall that (2.6) implies that $H^0(X_{S\cup\mathcal{C}^F(X)}, \mathcal{O}) = k$. Consequently, the k-vector space $H^0(X_{S\cup\mathcal{C}^F(X)}, \widetilde{\pi}^*_{S,T}V)$ is finite dimensional and one has

$$H^0(X_{S\cup\mathcal{C}^F(X)}, \widetilde{\pi}^*_{S,T}V) = \varinjlim_{T\subset T'\in\mathcal{T}^\ell} H^0(X_{S\cup T'}, \pi^*_{S\cup T', S\cup T}V). \tag{4.9}$$

So in fact,

$$H^0(X_{S\cup\mathcal{C}^F(X)}, \widetilde{\pi}^*_{S,T}V) = H^0(X_{S\cup T'}, \pi^*_{S\cup T', S\cup T}V) \text{ for some } T'\supset T, T'\in\mathcal{T}^\ell. \tag{4.10}$$

Denote $\mathcal{U} := \mathcal{C}^{\text{ét}}(X)\cup\mathcal{C}^F(X)$ and $X_\mathcal{U}$ the associated principal bundle. Thus

$$X_\mathcal{U} = \varprojlim_{\substack{S\in\mathcal{S}^{\text{ét}}\\ T\in\mathcal{T}^\ell}} X_{S\cup T}, \quad \widetilde{\pi}_{S\cup T} : X_\mathcal{U} \twoheadrightarrow X_{S\cup T}. \tag{4.11}$$

Then for any bundle $V\in\mathcal{E}(X_{S\cup T})$ we have, by means of (4.2),

$$H^0(X_\mathcal{U}, \widetilde{\pi}^*_{S\cup T}V) \cong H^0_{\mathcal{T}^\ell}(X_{S\cup T}, V). \tag{4.12}$$

Definition 4.3. The category $\mathcal{E}$ has for objects pairs $(X_{S\cup T}, V)$, where $S\in\mathcal{S}^{\text{ét}}$, $T\in\mathcal{T}^\ell$ and $V\in\mathcal{E}(X_{S\cup T})$, and for morphisms

$$\mathrm{Hom}_\mathcal{E}((X_{S_1\cup T_1}, V), (X_{S_2\cup T_2}, W)) := \mathrm{Hom}_{X_\mathcal{U}}(\widetilde{\pi}^*_{S_1\cup T_1}V, \widetilde{\pi}^*_{S_2\cup T_2}W). \tag{4.13}$$

Proposition 4.4. *The category $\mathcal{E}$ in Definition 4.3 is a Tannaka cateogry over k, with tensor product defined by ($S := S_1\cup S_1$, $T := T_1\cup T_2$)*

$$(X_{S_1\cup T_1}, V)\otimes(X_{S_2\cup T_2}, W) := (X_{S\cup T}, \pi^*_{S\cup T, S_1\cup T_1}(V)\otimes\pi^*_{S\cup T,S_2\cup T_2}(W))$$

the unit object is $(X, \mathcal{O}_X)$, and a fiber functor

$$\mathcal{E}\to\mathrm{Vect}_k, \quad (X_{S\cup T}, V)\mapsto V|_{x_{S\cup T}}. \tag{4.14}$$

Proof. We first show that $\mathcal{E}$ is an abelian category. The kernel, image and cokernel of a morphism $f : ((X_{S_1\cup T_1}, V), (X_{S_2\cup T_2}, W))$ are defined as follows. Set $S := S_1 \cup S_2$, $T := T_1 \cup T_2$. By means of (4.12), f corresponds to an element f_S of $H^0_{\mathcal{T}^\ell}(X_{S\cup T}, \pi^*_{S\cup T,S_1\cup T_1} V^\vee \otimes \pi^*_{S\cup T,S_2\cup T_2} W)$, which by means of (4.10) is represented by an element $f_{S\cup T'}$ in $\mathrm{Hom}_{X_{S\cup T'}}(\pi^*_{S\cup T',S_1\cup T_1} V, \pi^*_{S\cup T',S_2\cup T_2} W)$. Now we define the kernel, image and cokernel of f to be the kernel, image and cokernel of $f_{S\cup T'}$ respectively. It is clear that $\mathcal{E}$ is thus an abelian category. With respect to the tensor product in $\mathcal{E}$, the dual of an object is defined by $(X_{S\cup T}, V)^\vee = (X_{S\cup T}, V^\vee)$. □

One notices that the kernel (image, cokernel) of f does depend on the choice of T', in particular, for an object $(X_{S\cup T}, V)$ in $\mathcal{E}$, and for any $T' \in \mathcal{T}^\ell$, $T' \supset T$

$$(X_{S\cup T}, V) \text{ is isomorphic with } (X_{S\cup T'}, \pi^*_{S\cup T',S\cup T}(V)) \text{ in } \mathcal{E}. \tag{4.15}$$

Let $G(\mathcal{E}, x)$ be the Tannaka group scheme of $\mathcal{E}$ with respect to the given fiber functor. Consider the tautological functor $p : \mathcal{D} \to \mathcal{E}, (X_S, V) \mapsto (X_S, V)$ which is clearly compatible with the fiber functors of $\mathcal{D}$ and $\mathcal{E}$. It yields a homomorphism $p^* : G(\mathcal{E}, x) \to L(X, x)$ which is clearly injective.

Theorem 4.5. *The k-group scheme homomorphism $q^* : G(\mathcal{E}, x) \to L(X, x)$ is the kernel of the homomorphism $L(X, x) \to \pi^F(X, x)$ and consequently is the kernel of $r^{ét} \times r^F : \pi^N(X, x) \to \pi^{ét}(X, x) \times \pi^F(X, x)$. In other words the representation category of* $\mathrm{Ker}(r^{ét} \times r^F)$ *is equivalent to $\mathcal{E}$.*

Proof. We use Theorem A.1, (iii), to show that the sequence

$$G(\mathcal{E}) \to L(X, x) \to \pi^F(X, x) \tag{4.16}$$

is exact.

If $(X_{S\cup T}, V)$ is an object in $\mathcal{E}$, then

$$\pi^*_{S\cup T,S}(\pi_{S\cup T,S})_* V \twoheadrightarrow V \tag{4.17}$$

and since $(X_S, (\pi_{S\cup T,S})_* V)$ is an object of $\mathcal{D}$, every object of $\mathcal{E}$ is the quotient of an object coming from $\mathcal{D}$ via q. Thus condition (c) in A.1, (iii), is fulfilled.

The maximal trivial subobject of (X_S, V) in $\mathcal{E}$ is an object $(X_{S\cup T}, V_0)$ for some $T \in \mathcal{T}^\ell$ and V_0 is the maximal trivial subobject of $\pi^*_{S\cup T,S}(V)$ in $\mathcal{E}(X_{S\cup T})$. Proposition 4.6 below shows that there exists an F-finite bundle $W \in T$ on X and an inclusion $j : \pi^*_S W \hookrightarrow V$, such that

$$\pi^*_{S\cup T,S}(j) : \pi^*_{S\cup T} W \cong V_0.$$

Thus condition (b) in A.1, (iii), is fulfilled.

Finally recall that the homomorphism $L(X, x) \to \pi^F(X, x)$ is induced from the functor $\mathcal{C}^F(X, x) \to \mathcal{D}$, $V \mapsto (X, V)$. On the other hand the above discussion also shows that (X_S, V) is trivial if and only if the inclusion $j : \pi^*_S W \to V$ is an isomorphism, that is (X_S, V) is isomorphic to (X, W) in $\mathcal{D}$ (see 3.10). Thus condition (a) is also fulfilled. □

It just remains to prove

Proposition 4.6. *Let $V \in \mathcal{C}^F(X_S)$, and $V_0 \subset \pi^*_{S\cup T,S}(V)$ be the maximal trivial subobject of $\pi^*_{S\cup T,S}(V)$ in $\mathcal{E}(X_{S\cup T})$. Then there exists an F-finite bundle $W \in T$ on X equipped with an inclusion $j : \pi^*_S(W) \hookrightarrow V$ such that*

$$\pi^*_{S\cup T,S}(j) : \pi^*_{S\cup T}(W) = V_0 \subset \pi^*_{S\cup T,S}(V).$$

Proof. Using the Cartesian diagram (4.3), we have

$$\begin{aligned} H^0(X_{S\cup T}, \pi^*_{S\cup T,S}V) &\cong H^0(X_T, \pi_{S\cup T,T*}\pi^*_{S\cup T,S}V) \qquad (4.18)\\ &\overset{(4.3)}{\cong} H^0(X_T, \pi^*_T\pi_{S*}V) \cong H^0(X, \pi_{T*}\mathcal{O}_{X_T} \otimes \pi_{S*}V)\\ &\cong \mathrm{Hom}_X((\pi_{T*}\mathcal{O}_{X_T})^\vee, \pi_{S*}V) = \oplus_1^r k \cdot \varphi_i \end{aligned}$$

for some morphisms $\varphi_i : (\pi_{T*}\mathcal{O}_{X_T})^\vee \to \pi_{S*}V$. Let $W \subset \pi_{S*}V$ be the image of

$$\oplus_1^r\varphi_i : \oplus_1^r(\pi_{T*}\mathcal{O}_{X_T})^\vee \to \pi_{S*}V. \qquad (4.19)$$

Then W is trivializable by π_T and in particular $W \in T$ is F-finite. We show that the map $j : \pi^*_S W \to V$, induced from the inclusion $i : W \hookrightarrow \pi_{S*}V$, is injective.

Indeed, let $V' = \mathrm{im} j \subset V$ thus V' is a quotient of $\pi^*_S W$ and according to Proposition 3.6 and Theorem A.1, (i), we conclude that there is a quotient $q : W \to W'$ of W such that the quotient map $\pi^*W \to V'$ is the pull-back $\pi^*_S q : \pi^*_S W \to \pi^*_S W'$. Now the inclusion $\pi^*_S W' = V' \subset V$ corresponds to a morphism $i' : W' \to V$ which is compatible with i in the sense that $i = i' \circ q$. By assumption q is surjective and i is injective, hence i' is an isomorphism, consequently j is injective.

On the other hand, one has from (4.19)

$$\mathrm{Hom}_X((\pi_{T*}\mathcal{O}_{X_T})^\vee, \pi_{S*}V) = \mathrm{Hom}_X((\pi_{T*}\mathcal{O}_{X_T})^\vee, W) \qquad (4.20)$$

Thus

$$H^0(X_{S\cup T}, \pi^*_{S\cup T,S}V) = H^0(X_T, \pi^*_T W) = H^0(X_{S\cup T}, \pi^*_{S\cup T}W) \qquad (4.21)$$

which means that $\pi^*_{S\cup T}(W) = V_0 \subset \pi^*_{S\cup T,S}(V)$. □

Remark 4.7. Using Proposition 2.7, replacing X by X_S and X_S by $X_{S\cup T}$, one sees that the objects of $\mathcal{E}(X_{S\cup T})$ are precisely those bundles which come from a representation of a local fundamental group over k.

Appendix A. Exact sequences of Tannaka group schemes

In this appendix, we summarize the material on Tannaka categories which we used throughout the article. The statements and their proofs are taken from [3], but for the reader's convenience, we gather the information in a compact form here.

Let $L \xrightarrow{q} G \xrightarrow{p} A$ be a sequence of homomorphism of affine group scheme over a field k. It induces a sequence of functors

$$\mathrm{Rep}(A) \xrightarrow{p^*} \mathrm{Rep}(G) \xrightarrow{q^*} \mathrm{Rep}(L) \qquad (A.1)$$

where Rep denotes the category of finite-dimensional representations over k.

Theorem A.1. *With the above settings we have*

(i) *The map* $p : G \to A$ *is faithfully flat (and in particular surjective) if and only if* $p^*\mathrm{Rep}(A)$ *is a full subcategory of* $\mathrm{Rep}(G)$, *closed under taking subquotients.*

(ii) *The map* $q : L \to G$ *is a closed immersion if and only if any object of* $\mathrm{Rep}(L)$ *is a subquotient of an object of the form* $q^*(V)$ *for some* $V \in \mathrm{Rep}(G)$.

(iii) *Assume that* q *is a closed immersion and that* p *is faithfully flat. Then the sequence* $L \xrightarrow{q} G \xrightarrow{p} A$ *is exact if and only if the following conditions are fulfilled:*

  (a) *For an object* $V \in \mathrm{Rep}(G)$, $q^*(V)$ *in* $\mathrm{Rep}(L)$ *is trivial if and only if* $V \cong p^*U$ *for some* $U \in \mathrm{Rep}(A)$.

  (b) *Let* W_0 *be the maximal trivial subobject of* $q^*(V)$ *in* $\mathrm{Rep}(L)$. *Then there exists* $V_0 \subset V$ *in* $\mathrm{Rep}(G)$, *such that* $q^*(V_0) \cong W_0$.

  (c) *Any* W *in* $\mathrm{Rep}(L)$ *is embeddable in (hence, by taking duals, a quotient of)* $q^*(V)$ *for some* $V \in \mathrm{Rep}(G)$.

Proof. The statements (i) and (ii) are due to Saavedra [12]. We refer also to [1, Proposition 2.21] for a nice proof. We show (iii).

Assume that $q : L \to G$ is the kernel of $p : G \to A$. Then (a), (b) follow from the well-known properties of normal subgroups (cf. [13, Chapter 13]). It remains to show (c).

Let $\mathrm{Ind} : \mathrm{Rep}(L) \to \mathrm{Rep}(G)$ be the induced representation functor, it is the right adjoint functor to the restriction functor $\mathrm{Res} : \mathrm{Rep}(G) \to \mathrm{Rep}(L)$ that is, one has a functorial isomorphism

$$\mathrm{Hom}_G(V, \mathrm{Ind}(W)) \xrightarrow{\cong} \mathrm{Hom}_L(\mathrm{Res}(V), W). \tag{A.2}$$

It is easy to check

$$\mathrm{Ind}(W) \cong (k[G] \otimes_k W)^L \tag{A.3}$$

where L acts on $k[G]$ on the right. It is well known that $k[G]$ is faithfully flat over it subalgebra $k[A]$ ([13, Chapter 13]) and there is the following isomorphism

$$k[G] \otimes_{k[A]} k[G] \cong k[L] \otimes_k k[G] \tag{A.4}$$

which precisely means that $G \to A$ is a principal bundle under L. Consequently

$$k[G] \otimes_{k[A]} \mathrm{Ind}(W) \cong k[A] \otimes_k V \tag{A.5}$$

Thus the functor $\mathrm{Ind} : \mathrm{Rep}(L) \to \mathrm{Rep}(G)$ is exact.

Setting $V = \mathrm{Ind}(W)$ in (A.2), one obtains a canonical map $u_W : \mathrm{Ind}(W) \to W$ in $\mathrm{Rep}(L)$ which gives back the isomorphism in (A.2) as follows:

$$\mathrm{Hom}_G(V, \mathrm{Ind}(W)) \ni h \mapsto u_W \circ h \in \mathrm{Hom}_L(\mathrm{Res}(V), W).$$

The map u_W is non-zero whenever W is non-zero. Indeed, since Ind is faithfully exact, $\mathrm{Ind}(W)$ is non-zero whenever W is non-zero. Thus, if $u_W = 0$, then both sides of (A.2) were zero for any V. On the other hand, for $V = \mathrm{Ind}(W)$, the right hand side contains the identity map. This show that u_W can't vanish.

We want to show that u_W is always surjective. Let $U = \mathrm{Im}(u_W)$ and $T = W/U$. We have the following diagram

$$\begin{array}{ccccccccc} 0 & \longrightarrow & \mathrm{Ind}(U) & \longrightarrow & \mathrm{Ind}(W) & \longrightarrow & \mathrm{Ind}(T) & \longrightarrow & 0 \\ & & \downarrow{\scriptstyle u_U} & & \downarrow{\scriptstyle u_T} & & \downarrow{\scriptstyle u_W} & & \\ 0 & \longrightarrow & U & \longrightarrow & W & \longrightarrow & T & \longrightarrow & 0 \end{array} \tag{A.6}$$

By assumption, the composition $\mathrm{Ind}(W) \twoheadrightarrow \mathrm{Ind}(T) \to T$ is 0, therefore $\mathrm{Ind}(T) \to T$ is a zero map, implying $T = 0$.

Since $\mathrm{Ind}(W)$ is the union of its finite-dimensional subrepresentations, we can find a finite-dimensional G-subrepresentation $W_0(W)$ of $\mathrm{Ind}(W)$ which still maps surjectively on W. In order to obtain the statement on the embedding of W, we dualize $W_0(W^\vee) \twoheadrightarrow W^\vee$.

Conversely, assume that (a), (b), (c) are satisfied. Then it follows from (a) that for $U \in \mathrm{Rep}(A)$, $q^*p^*(U) \in \mathrm{Rep}(L)$ is trivial. Hence $pq : L \to A$ is the trivial homomorphism. Recall that by assumption, q is injective, p is surjective. Let $\bar{q} : \bar{L} \to G$ be the kernel of p. Then we have commutative diagram

$$\begin{array}{ccc} L & \xrightarrow{\ i\ } & \bar{L} \\ & {\scriptstyle q}\searrow \quad \swarrow{\scriptstyle \bar{q}} & \\ & G & \end{array} \iff \begin{array}{ccc} \mathrm{Rep}(L) & \xleftarrow{\ i^*\ } & \mathrm{Rep}(\bar{L}) \\ & {\scriptstyle q^*}\nwarrow \quad \nearrow{\scriptstyle \bar{q}^*} & \\ & \mathrm{Rep}(G) & \end{array} \tag{A.7}$$

with injective homomorphisms. It remains to show that i is surjective, which amounts to saying that the category $i^*\mathrm{Rep}(\bar{L})$ in $\mathrm{Rep}(L)$ is full and closed under taking subquotients.

We first show the fullness. Let $\bar{W}_0, \bar{W}_1$ be objects in $\mathrm{Rep}(\bar{L})$ and $\varphi : W_0 := i^*(\bar{W}_0) \to i^*(\bar{W}_1) =: W_1$ be a morphism in $\mathrm{Rep}(L)$. Since $\mathrm{Rep}(\bar{L})$ also satisfies (c), there exists V_0, V_1 in $\mathrm{Rep}(G)$ with a surjective morphism $\pi : \bar{q}^*(V_0) \to \bar{W}_0$, and an injective morphism $\iota : \bar{W}_1 \to \bar{q}^*(V_1)$. These yield a morphism

$$\psi := i^*(\iota)\varphi i^*(\pi) : q^*(V_0) \to q^*(V_1) \tag{A.8}$$

The morphism ψ induces and element in $H^0(L, q^*(V_0^\vee \otimes V_1))$. Now, by (b) and by the fact that $\mathrm{Rep}(\bar{L})$ also satisfies (b) we conclude that $\psi = i^*(\bar{\psi})$, for some $\bar{\psi} : \bar{q}^*(V_0) \to \bar{q}^*(V_1)$. Since ι is injective and π is surjective, we conclude that $\varphi = \bar{\varphi}$, for some $\bar{\varphi} : \bar{W}_0 \to \bar{W}_1$ in $\mathrm{Rep}(\bar{L})$. Thus the category $i^*\mathrm{Rep}(\bar{L})$ is full in $\mathrm{Rep}(L)$.

On the other hand, for any $W \in \mathrm{Rep}(L)$, by (c) there exist V_0, V_1 in $\mathrm{Rep}(G)$ and $\varphi : q^*(V_0) \to q^*(V_1)$ such that $W \cong \mathrm{im}\varphi$. By the fullness of $i^*\mathrm{Rep}(\bar{L})$ in $\mathrm{Rep}(L)$, $\varphi = i^*\bar{\varphi}$, hence $W \cong i^*(\mathrm{im}\bar{\varphi})$. Thus we have proved that any object in $\mathrm{Rep}(L)$ is isomorphic to the image under i^* of an object in $\mathrm{Rep}(\bar{L})$. Together with the discussion above this implies that $L \cong \bar{L}$. □

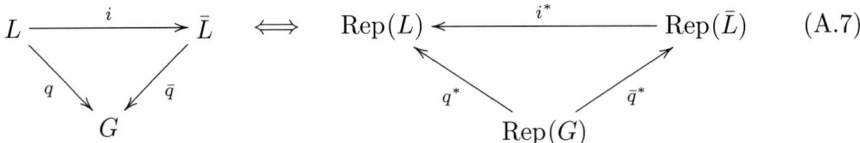

References

[1] P. Deligne and J. Milne, *Tannakian Categories*, Lectures Notes in Mathematics **900**, 101—228, Springer-Verlag (1982).

[2] P. Deligne, *Catégories tannakiennes*, The Grothendieck Festschrift, Vol. II, 111–195, Progr. Math. **87**, Birkhäuser (1990).

[3] H. Esnault and P.H. Hai, *The Gauß-Manin connection and Tannaka duality*, International Mathematics Research Notices **93878** (2006), 1–35.

[4] A. Grothendieck, *Revêtements étales et groupe fondamental*, SGA 1, Lect. Notes in Mathematics **224** (1970), Springer Verlag.

[5] N. Katz, *On the calculation of some differential Galois groups*, Invent. math. **87** (1987), 13–61.

[6] H. Langer and U. Stuhler, *Vektorbündel auf Kurven und Darstellungen der algebraischen Fundamentalgruppe*, Math. Z. **156** (1977), 73–83.

[7] V.B. Mehta and S. Subramanian, *On the fundamental group scheme*, Invent. math. **148** (2002), 143–150.

[8] M. Nori, *On the representation of the fundamental group*, Compositio math. **33** (1976), 29–41.

[9] M. Nori, *The fundamental group scheme*, Proc. Indian Acad. Sci. **91** (1982), 73–122.

[10] S. Ramanan and A. Ramanathan, *Some remarks on the instability flag*, Tohoku Math. Journal **36** (1984), 269–291.

[11] M. Raynaud, *Sections de fibrés vectoriels sur une courbe*, Bull. Soc. Math. France **110** (1982), 103–125.

[12] N. Saavedra-Rivano, *Catégories Tannakiennes*, Lectures Notes in Mathematics **265**, Springer-Verlag (1972).

[13] W.C. Waterhouse, *Introduction to affine group schemes*, Graduate Texts in Mathematics **66**, Springer-Verlag (1979).

Hélène Esnault and Phùng Hô Hai
Universität Duisburg-Essen
Mathematik
D-45117 Essen
Germany
e-mail: esnault@uni-due.de
e-mail: hai.phung@uni-due.de

Xiaotao Sun
Academy of Mathematics and Systems Science
Chinese Academy of Sciences
Beijing, 100080
P.R. of China
e-mail: xsun@math.ac.cn

Progress in Mathematics, Vol. 265, 399–414

The Reidemeister Number of Any Automorphism of a Baumslag–Solitar Group is Infinite

Alexander Fel'shtyn and Daciberg L. Gonçalves

Abstract. Let $\phi : G \to G$ be a group endomorphism where G is a finitely generated group of exponential growth, and let $R(\phi)$ denote the number of ϕ-conjugacy classes. Fel'shtyn and Hill [10] conjectured that if ϕ is injective, then $R(\phi)$ is infinite. In this paper, we show that the conjecture holds for the Baumslag-Solitar groups $B(m,n)$, where either $|m|$ or $|n|$ is greater than 1 and $|m| \neq |n|$. We also show that in the cases where $|m| = |n| > 1$ or $mn = -1$ the conjecture is true for automorphisms. In addition, we derive few results about the coincidence Reidemeister number.

Mathematics Subject Classification (2000). 20E45, 37C25, 55M20.

Keywords. Reidemeister number, twisted conjugacy classes, Baumslag-Solitar groups.

1. Introduction

J. Nielsen introduced the fixed point classes of a surface homeomorphism in [28]. Subsequently, K. Reidemeister [29] developed the algebraic foundation of the Nielsen fixed point theory for any map of any compact polyhedron. As a result of Reidemeister's work, the twisted conjugacy classes of a group homomorphism were introduced. It turns out that the fixed point classes of a map can easily be identified with the conjugacy classes of lifting of this map to the universal covering of compact polyhedron, and conjugacy classes of lifting can be identified with the twisted conjugacy classes of the homomorphism induced on the fundamental

This work was initiated during the visit of the second author to Siegen University from September 13 to September 20, 2003. The visit was partially supported by a grant of the "Projeto temático Topologia Algébrica e Geométrica-FAPESP". The second author would like to thank Professor U. Koschorke for making this visit possible and for the hospitality.

group of the polyhedron. Let G be a finitely generated group and let $\phi : G \to G$ be an endomorphism. Two elements $\alpha, \alpha' \in G$ are said to be *ϕ-conjugate* if there exists $\gamma \in G$ with $\alpha' = \gamma\alpha\phi(\gamma)^{-1}$. The number of ϕ-conjugacy classes (or twisted conjugacy classes) is called the *Reidemeister number* of an endomorphism ϕ, denoted by $R(\phi)$. If ϕ is the identity map, then the ϕ-conjugacy classes are the usual conjugacy classes in the group G. Let X be a connected compact polyhedron and $f : X \to X$ be a continuous map. The Reidemeister number $R(f)$, which is simply the cardinality of the set of ϕ-conjugacy classes where $\phi = f_\#$ is the induced homomorphism on the fundamental group, is relevant for the study of fixed points of f in the presence of the fundamental group. In fact the finiteness of Reidemeister number plays an important rôle. See for example [32], [19], [10], [13] and the introduction of [18].

A current important problem concerns obtaining a twisted analogue of the celebrated Burnside-Frobenius theorem [10, 12, 14, 11, 13]. For this purpose it is important to describe the class of groups G, such that $R(\phi) = \infty$ for any automorphism $\phi : G \to G$. A. Felshtyn and R. Hill [10] made first attempts to localize this class of groups.

Later it was proved in [7, 27] that the non-elementary Gromov hyperbolic groups belong to this class. Furthermore, using the co-Hofian property, it was shown in [7] that, if in addition G is torsion-free and freely indecomposable, then $R(\phi)$ is infinite for every injective endomorphism ϕ. This result gives supportive evidence to a conjecture of [10] which states that $R(\phi) = \infty$ if ϕ is an injective endomorphism of a finitely generated torsion-free group G with exponential growth.

This conjecture was shown to be false in general. In [18] were constructed automorphisms $\phi : G \to G$ on certain finitely generated torsion-free exponential growth groups G that are not Gromov hyperbolic with $R(\phi) < \infty$.

In the present paper we study this problem for a family of finitely generated groups which have exponential growth but are not Gromov hyperbolic. These are the Baumslag-Solitar groups, which we now define. Being indexed by pairs of integer numbers different from zero, they have the following presentation:

$$B(m,n) = \langle a, b : a^{-1}b^m a = b^n \rangle, m, n \neq 0.$$

The present work extends substantially in several directions the preliminary results obtained in [8], and simplifies some of the proofs.

The family of the Baumslag-Solitar groups has different features from the one given in [18], which is a family of groups which are metabelian having as the kernel the group $\mathbb{Z}^n$. Hence they contain a subgroup isomorphic to $\mathbb{Z} + \mathbb{Z}$. In the case of Baumslag-Solitar groups this happens if, and only if, $m = n$. For $m = n = 1$ the group $B(1,1) = \mathbb{Z} + \mathbb{Z}$ does not have exponential growth and it is also known that there are automorphisms $\phi : B(1,1) \to B(1,1)$ with $R(\phi) < \infty$. For more details about these groups $B(m,n)$ see [1, 5].

Some results in this work could be obtained by means of the classification of some of the endomorphisms of a Baumslag-Solitar group (for those, see [16] and [25]). We use only one direct consequence of the main result of [25] which concerns injective homomorphisms.

Our main results are:

Theorem. *For any injective endomorphism of* $B(m,n)$ *where* $|n| \neq |m|$ *and* $|nm| \neq 0$, *the Reidemeister number is infinite. For any automorphism of* $B(m,n)$ *where* $0 < |m| = |n|$ *and* $mn \neq 1$, *the Reidemeister number is also infinite.*

This result summarizes the results of Theorems 3.4, 4.4, , 5.4, 6.4 and Proposition 5.1 for the various values of m and n.

Theorem 7.1. *The coincidence Reidemeister number is infinite for any pair of injective endomorphisms of the group* $B(m,n)$, *where* $|n| \neq |m|$ *and* $|nm| \neq 0$.

We do not know if Theorems 5.4 and 6.4 are also true for injective homomorphisms. See Remarks 5.5 and 6.5.

We say that a group G has *property* R_∞ if any of its automorphisms ϕ has $R(\phi) = \infty$. After the preprint (in arXiv: math.GR-0405590) of this article was circulate, it was proved that the following groups have *property* R_∞ : (1) generalized Baumslag-Solitar groups, that is, finitely generated groups which act on a tree with all edge and vertex stabilizers infinite cyclic [26]; (2) lamplighter groups $Z_n \wr Z$ iff $2|n$ or $3|n$ [20]; (3) the solvable generalization Γ of $BS(1,n)$ given by the short exact sequence $1 \to Z[\frac{1}{n}] \to \Gamma \to Z^k \to 1$ as well as any group quasi-isometric to Γ [30]; (4) groups which are quasi-isometric to $BS(1,n)$ [31] (while this property is not a quasi-isometry invariant); (5) saturated weakly branch groups (including the Grigorchuk group and the Gupta-Sidki group) [15]; (6) the R. Thompson's groups [2].

We would like to complete the introduction with the following conjecture.

Conjecture 1.1. Any relatively hyperbolic group has *property* R_∞. In particular, any Kleinian group has *property* R_∞.

This paper is organized into six sections besides this one. In Section 2, we make some simple reduction of the problem to certain cases and develop some preliminaries about the Reidemeister classes of a pair of homomorphisms between short exact sequences. In Section 3, we study the case $B(\pm 1, n)$ for $|n| > 1$, with main result Theorem 3.4. In Section 4, we consider the cases $B(m,n)$ for $1 < |m| \neq |n| > 1$, with main result Theorem 4.4. In Section 5, we consider the cases $B(m,-m)$ for $|m| > 0$, with main results Propositon 5.1 and Theorem 5.4. In Section 6 we consider the cases $B(m,m)$ for $|m| > 1$, with main result Theorem 6.4. In Section 7 we derive few results about the coincidence Reidemeister number, with main result Theorem 7.1.

Acknowledgments. The authors would like to thank G. Levitt for his helpful comments improving an earlier version of this manuscript. The first author would like to thank B. Bowditch, T. Januszkiewicz, M. Kapovich and E. Troitsky for stimulating discussions and comments. The second author would like to express his thanks to D. Kochloukova for very helpful discussions.
The first author would alsom like to thank the Max-Planck-Institute für Mathematik, Bonn for its kind hospitality and support.

This article is dedicated to the memory of Sasha Reznikov.

2. Generalities and Preliminaries

In this section we first describe few elementary properties of the groups $B(m,n)$ in order to reduce our problem to certain cases. Then we recall some facts about the Reidemeister classes of a pair of homomorphisms of a short exact sequence. Recall that *a group G has* property R_∞ *if any of its automorphisms* ϕ *has* $R(\phi)=\infty$.

Recall that the Baumslag-Solitar groups are indexed by pairs of integer numbers different from zero and they have the following presentation:

$$B(m,n)=\langle a,b : a^{-1}b^m a=b^n\rangle, m,n\neq 0.$$

The first observation is that for $m=n=1$ this group is $\mathbb{Z}+\mathbb{Z}$. It is well known that this group does not have exponential growth and there are automorphisms $\phi:\mathbb{Z}+\mathbb{Z}\to\mathbb{Z}+\mathbb{Z}$ with finite Reidemeister number. So $\mathbb{Z}+\mathbb{Z}$ does not have property R_∞.

The second observation is that $B(m,n)$ is isomorphic to $B(-m,-n)$. It suffices to see that the relations $a^{-1}b^m a=b^n$ and $a^{-1}b^{-m}a=b^{-n}$ each one generates the same normal subgroup, since one relation is the inverse of the other.

The last observation is that $B(m,n)$ is isomorphic to $B(n,m)$. Suppose that $B(m,n)=\langle a,b : a^{-1}b^m a=b^n\rangle, m,n\neq 0$ and $B(n,m)=\langle c,d : c^{-1}d^n c=d^m\rangle, m,n\neq 0$. The map which sends $a\to c^{-1}$ and $b\to d$ extends to an isomorphism of the two groups.

Based on the above, we will consider only the groups $B(r,s)$, $rs\neq 0,1$ and we can show:

Proposition 2.0. *Each group* $B(r,s)$, $rs\neq 0,1$ *is isomorphic to some* $B(m,n)$, *where* m,n *satisfy* $0<m\leq|n|$ *and* $n\neq 1$.

So we will divide the problem into 4 cases. Case 1) is when $1=m<|n|$; Case 2) is when $1<m<|n|$; Case 3) when $0<m=-n$; Case 4) when $1<m=n$.

The set of the Reidemeister classes will be denoted by $R[\ ,\ ]$ and the number of such classes by $R(\ ,\ )$. When the two sequences are the same and one of the homomorphisms is the identity, then we have the usual Reidemeister classes and Reidemeister number. The main reference for this section is [17] and more details can be found there.

Let us consider a diagram of two short exact sequences of groups and maps between these two sequences:

$$\begin{array}{ccccccccc} 1 & \rightarrow & H_1 & \overset{i_1}{\rightarrow} & G_1 & \overset{p_1}{\rightarrow} & Q_1 & \rightarrow & 1 \\ & & f' \downarrow\downarrow g' & & f \downarrow\downarrow g & & \overline{f} \downarrow\downarrow \overline{g} & & \\ 1 & \rightarrow & H_2 & \overset{i_2}{\rightarrow} & G_2 & \overset{p_2}{\rightarrow} & Q_2 & \rightarrow & 1 \end{array} \tag{2.1}$$

where $f' = f|_{H_1}$, $g' = g|_{H_1}$.

We recall that the set of the Reidemeister classes $R[f_1, f_2]$ relative to homomorphisms $f_1, f_2 : K \rightarrow \pi$ is the set of the equivalence classes of elements of π given by the following relation: $\alpha \sim f_2(\tau)\alpha f_1(\tau)^{-1}$ for $\alpha \in \pi$ and $\tau \in K$.

The diagram (2.1) provides maps between sets

$$R[f', g'] \overset{\widehat{i_2}}{\rightarrow} R[f, g] \overset{\widehat{p_2}}{\rightarrow} R[\overline{f}, \overline{g}]$$

where the last map is clearly surjective. An obvious consequence of this fact will be used to solve some of the cases that we will discuss, and that will appear below as Corollary 2.2. For the remaining cases we need further information about the above sequence and we will use Corollary 2.4.

Proposition 1.2 in [17] says:

Proposition 2.1. *Given the diagram* (2.1) *we have a short sequence of sets*

$$R[f', g'] \overset{\widehat{i_2}}{\rightarrow} R[f, g] \overset{\widehat{p_2}}{\rightarrow} R[\overline{f}, \overline{g}]$$

where $\widehat{p_2}$ *is surjective and* $\widehat{p_2}^{-1}[1] = \mathrm{im}\,(\widehat{i_2})$, *where* 1 *is the identity element of* Q_2.

An immediate consequence is

Corollary 2.2. *If* $R(\overline{f}, \overline{g})$ *is infinite, then* $R(f, g)$ *is also infinite.*

Proof. Since $\widehat{p_2}$ is surjective the result follows. □

In order to study the injectivity of the map $\widehat{i_2}$, for each element $\overline{\alpha} \in Q_2$ let $H_2(\overline{\alpha}) = p_2^{-1}(\overline{\alpha})$, $C_{\overline{\alpha}} = \{\overline{\tau} \in Q_1 | \overline{g}(\overline{\tau})\overline{\alpha}\overline{f}(\overline{\tau}^{-1}) = \overline{\alpha}\}$ and let $R_{\overline{\alpha}}[f', g']$ be the set of equivalence classes of elements of $H_2(\overline{\alpha})$ given by the equivalence relation $\beta \sim g(\tau)\beta f(\tau^{-1})$, where $\beta \in H_2(\overline{\alpha})$ and $\tau \in p_1^{-1}(C_{\overline{\alpha}})$. Finally, let $R[f_{\overline{\alpha}}, g_{\overline{\alpha}}]$ be the set of equivalence classes of elements of $H_2(\overline{\alpha})$ given by the relation $\beta \sim g(\tau)\beta f(\tau^{-1})$, where $\beta \in H_2(\overline{\alpha})$ and $\tau \in G_1$.

Proposition 1.2 in [17] says:

Proposition 2.3. *Two classes of* $R(f_{\overline{\alpha}}, g_{\overline{\alpha}})$ *represent the same class of* $R(f, g)$ *if and only if they belong to the same orbit by the action of* $C_{\overline{\alpha}}$. *Further the isotropy subgroup of this action at an element* $[\beta]$ *is* $G_{[\beta]} = p_1(C_\beta) \subset C_{\overline{\alpha}}$ *where* $\beta \in [\beta]$.

An immediate consequence of this result is:

Corollary 2.4. *If $C_{\overline{\alpha}}$ is finite and $R(f_{\overline{\alpha}}, g_{\overline{\alpha}})$ is infinite for some α, then $R(f,g)$ is also infinite. In particular, this is the case if Q_2 is finite.*

Proof. The orbits of the action of $C_{\overline{\alpha}}$ on $R[f_{\overline{\alpha}}, g_{\overline{\alpha}}]$ are finite. So we have an infinite number of orbits. The last part is an easy consequence of the first part. □

3. The Cases $B(m,n)$, $1 = |m| < |n|$

From Section 2 the cases in this section reduce to Case 1), namely $B(1,n)$ for $1 < |n|$. Let $|n| > 1$ and $B(1,n) = \langle a, b : a^{-1}ba = b^n, n > 1\rangle$.
Recall from [5] that the Baumslag-Solitar groups $B(1,n)$ are finitely generated solvable groups which are not virtually nilpotent. These groups have exponential growth [23], and they are not Gromov hyperbolic. Furthermore, those groups are metabelian and torsion free.

Consider the homomorphisms $|\ |_a : B(1,n) \longrightarrow \mathbb{Z}$ which associates for each word $w \in B(1,n)$ the sum of the exponents of a in the word. It is easy to see that this is a well defined map into Z which is surjective.

Proposition 3.1. *We have a short exact sequence*

$$0 \longrightarrow K \longrightarrow B(1,n) \stackrel{|\ |_a}{\longrightarrow} \mathbb{Z} \longrightarrow 1,$$

where K, the kernel of the map $|\ |_a$, is the set of the elements which have the sum of the powers of a equal to zero. Furthermore, $B(1,n) = K \rtimes \mathbb{Z}$ (semi-direct product).

Proof. The first part is clear. The second part follows because $\mathbb{Z}$ is free, so the sequence splits. □

Proposition 3.2. *The kernel K coincides with $N\langle b\rangle$, the normalizer of $\langle b\rangle$ in $B(1,n)$.*

Proof. We have $N\langle b\rangle \subset K$. But the quotient $B/N\langle b\rangle$ has the following presentation: $\bar{a}^{-1}\bar{b}\bar{a} = \bar{b}^n, \bar{b} = 1$. Therefore this group is isomorphic to $\mathbb{Z}$ and the natural projection coincides with the map $|\ |_a$ under the obvious identification of $\mathbb{Z}$ with $B/N\langle b\rangle$. Consider the commutative diagram

$$\begin{array}{ccccccc} 0 \rightarrow N\langle b\rangle \rightarrow & & B(1,n) \rightarrow & & B/N\langle b\rangle \rightarrow & & 1 \\ \downarrow & & \downarrow & & \downarrow & & \\ 0 \rightarrow K \rightarrow & & B(1,n) \rightarrow & & \mathbb{Z} \rightarrow & & 1 \end{array}$$

where the last vertical map is an isomorphism. From the well-known five Lemma the result follows. □

The groups $B(1,n)$ are metabelian. Let ϵ be the sign of n. We recall the result that $B(1,n)$ is isomorphic to $\mathbb{Z}[1/|n|] \rtimes_\theta \mathbb{Z}$ where the action of $\mathbb{Z}$ on $\mathbb{Z}[1/|n|]$ is given by $\theta(1)(x) = x/n^\epsilon$. To see this, first observe that the map defined by $\phi(a) = (0,1)$ and $\phi(b) = (1,0)$ extends to a unique homomorphism $\phi : B \rightarrow \mathbb{Z}[1/|n|] \rtimes \mathbb{Z}$ which is clearly surjective. It suffices to show that this homomorphism is injective. Consider a word $w = a^{r_1}b^{s_1} \cdots a^{r_t}b^{s_t}$ such that

$r_1 + \cdots + r_t = 0$. Thus $w \in K$, and using the relation of the group this word is equivalent to $b^{s_1/n^{\epsilon r_1}} b^{s_2/n^{\epsilon(r_1+r_2)}} \cdots b^{s_{t-1}/n^{\epsilon(r_1+\cdots+r_{t-1})}}.b^{s_t}$. If we apply ϕ to this element, which belongs to the kernel of ϕ, we obtain that the sum of the powers $s_1/n^{r_1} + s_2/n^{\epsilon(r_1+r_2)} + \cdots + s_{t-1}/n^{\epsilon(r_1+\cdots+r_{t-1})} + s_t$ is zero. But this means that w is the trivial element, hence ϕ restricted to K is injective. Therefore the result follows.

Proposition 3.3. *Any homomorphism $\phi : B(1,n) \to B(1,n)$ is a homomorphism of the short exact sequence given in Proposition* 3.2.

Proof. Let $\bar{\phi}$ be the homomorphism induced by ϕ on the abelianization of $B(1,n)$. The abelianization of $B(1,n)$, denoted by $B(1,n)_{ab}$, is isomorphic to $Z_{|n-1|} + Z$. The torsion elements of $B(1,n)_{ab}$ form a subgroup isomorphic to $Z_{|n-1|}$ which is invariant under any homomorphism. The preimage of this subgroup under the projection to the abelianization $B(1,n) \to B(1,n)_{ab}$ is exactly the subgroup $N(b)$, i.e., the elements represented by words where the sum of the powers of a is zero. So it follows that $N(b)$ is mapped into $N(b)$. □

Theorem 3.4. *For any injective homomorphism of $B(1,n)$ the Reidemeister number is infinite.*

Proof. Let ϕ be an injective endomorphism. By Proposition 3.3 it is an endomorphism of the short exact sequence given by Proposition 3.2. The induced homomorphism on the quotient is a non-trivial endomorphism of Z. Otherwise we would have an injective homomorphism from the non-abelian group $B(1,n)$ into the abelian group K. If the induced endomorphism $\bar{\phi}$ is the identity, by Corollary 2.2 the number of Reidemeister classes is infinite and the result follows. So, assume that $\bar{\phi}$ is multiplication by $k \neq 0, 1$ and we will get a contradiction. Now we claim that there is no injective endomorphism of $B(1,n)$ such that the induced homomorphism on the quotient is multiplication by k with $k \neq 0, 1$. When we apply the homomorphism ϕ to the relation $a^{-1}ba = b^n$, using the isomorphism $B(1,n) \to Z[1/n] \rtimes Z$ we obtain: $a^{-k}\phi(b)a^k = (n^k\phi(b), 0) = (n\phi(b), 0)$, which implies that either $n^{1-k} = 1$ or $\phi(b) = 0$. Since $\phi(b) \neq 0$ and n is neither 1 or -1 we get a contradiction and the result follows. □

Remark 3.5. From the proof above we conclude that any injective homomorphism $\varphi : B(1,n) \to B(1,n)$ has the property that it induces the identity on the quotient $\mathbb{Z}$ given by the short exact sequence in Propositon 3.2. This fact will be used to study coincidence Reidemeister classes in Section 7.

4. The Case $B(m,n)$, $1 < |m| \neq |n| > 1$

From Section 2 the cases in this section reduce to Case 2), namely (m,n) for $1 < m < |n|$. The groups in this section are more complicated than the ones in the previous section. Nevertheless in order to obtain the results we will use a similar procedure to the one in the previous section. Let $1 < m < |n|$ and $B(m,n) = \langle a, b : a^{-1}b^m a = b^n \rangle$. Recall that such groups are non-virtually solvable.

Consider the homomorphism $| \ |_a : B(m,n) \longrightarrow \mathbb{Z}$ which associates to each word $w \in B(m,n)$ the sum of the powers of a in the word. It is easy to see that this is a well defined homomorphism into $\mathbb{Z}$ which is surjective.

Proposition 4.1. *We have a short exact sequence*

$$0 \longrightarrow K \longrightarrow B(m,n) \longrightarrow \mathbb{Z} \longrightarrow 1,$$

where K, the kernel of the map $| \ |_a$, is the set of the elements which have the sum of the powers of a equals to zero. Furthermore, $B(m,n) = K \rtimes \mathbb{Z}$ is a semi-direct product where the action is given with respect to some fixed section.

Proof. The first part is clear. The second part follows because $\mathbb{Z}$ is free, so the sequence splits. Since the kernel K is not abelian, the action is defined with respect to a specific section (see [3]). □

Proposition 4.2. *The kernel K coincides with $N\langle b\rangle$ which is the normalizer of $\langle b\rangle$ in $B(m,n)$.*

Proof. Similar to Proposition 3.2. □

Proposition 4.3. *Any homomorphism $\phi : B(m,n) \to B(m,n)$ is a homomorphism of the short exact sequence given in Proposition* 4.1.

Proof. Let $\bar{\phi}$ be the homomorphism induced by ϕ on the abelianized of $B(m,n)$. The abelianized of $B(m,n)$, denoted by $B(m,n)_{ab}$, is isomorphic to $\mathbb{Z}_{|n-m|} + \mathbb{Z}$. The torsion elements of $B(m,n)_{ab}$ form a subgroup isomorphic to $\mathbb{Z}_{|n-m|}$ which is invariant under any homomorphism. The preimage of this subgroup under the projection to the abelianized $B(m,n) \to B(m,n)_{ab}$ is exactly the subgroup $N(b)$, i.e., the elements represented by words where the sum of the powers of a is zero. So it follows that $N(b)$ is mapped into $N(b)$. □

In order to have a homomorphism ϕ of $B(m,n)$ which has finite Reidemeister number, the induced map on the quotient $\mathbb{Z}$ must be different from the identity by the same argument used in the proof of Theorem 3.4.

Now we will give a presentation of the group K. The group K is generated by the elements $g_i = a^{-i}ba^i$ $i \in \mathbb{Z}$ which satisfy the following relations: $\{1 = a^{-j}(a^{-1}b^m ab^{-n})a^j = g_{j+1}^m g_j^{-n}\}$ for all integers j. This presentation is a consequence of the Bass-Serre theory, see [4], Theorem 27, page 211.

Now we will prove the main result of this section. Denote by K_{ab} the abelianization of K.

Theorem 4.4. *For any injective homomorphism of $B(m,n)$ the Reidemeister number is infinite.*

Proof. Let us consider the short exact sequence, obtained from the short exact sequence given in Proposition 4.1, by taking the quotient with the commutators subgroup of K, i.e.,

$$0 \longrightarrow K_{ab} \longrightarrow B(m,n)/[K,K] \longrightarrow \mathbb{Z} \longrightarrow 1.$$

Thus we obtain a short exact sequence where the kernel K_{ab} is abelian. From the presentation of K we obtain a presentation of K_{ab} given as follows: It is generated by the elements g_i, $i \in Z$, which satisfy the following relations: $\{1 = g_{j+1}^m g_j^{-n}, g_i g_j = g_j g_i\}$ for all integers i, j. This presentation is the same as the quotient of the free abelian group generated by the elements g_i, $i \in \mathbb{Z}$ (so the direct sum of $\mathbb{Z}'s$ indexed by $\mathbb{Z}$), modulo the subgroup generated by the relations $\{1 = g_{j+1}^m g_j^{-n}\}$. Thus an element can be regarded as an equivalence class of a sequence of integers indexed by $\mathbb{Z}$, where the elements of the sequence are zero but a finite number. By abuse of notation we also denote by ϕ the induced endomorphism on $B(m,n)/[K,K]$.

Let $\phi(a) = a^k\theta$ for $\theta \in K_{ab}$ and $k \neq 1$. Recall that if $k = 1$ it follows immediately that the Reidemeister number is infinite. Since the kernel of the extension is abelian, after applying ϕ to the relation $a^{-1}b^m a = b^n$ we obtain

$$\theta^{-1}a^{-k}\phi((b)^m)a^k\theta = a^{-k}\phi(b)^m a^k = \phi(b^n) = \phi(b)^n.$$

From the main result of [25], the element $\phi(b)$ is a conjugate of a power of b, i.e., $\phi(b) = \gamma b^r \gamma^{-1}$ for some $r \neq 0$. In the abelianization the element $\gamma b^r \gamma^{-1}$ is the same as the element $a^s b a^{-s}$ for some integer s. So any power of $\phi(b)$ with exponent different from zero is a nontrivial element. Now we take both sides of the equation above to the power m^k. If $k = 0$ it follows immediately that $m = n$. Let us take $k > 1$. After applying the relation several times we obtain

$$a^{-k}\phi(b)^{mm^k}a^k = \phi(b)^{mn^k} = \phi(b)^{nm^k}.$$

Therefore it follows that $mn^k = nmk$ or $n^{k-1} = m^{k-1}$. If n is positive, since $k \neq 1$, then the only solutions are $m = n$, which is a contradiction. If n is negative then the only solutions are, $m = n$ or $m = -n$ and k even. In either case we get a contradiction. The case where $k < 0$ is similar and the result follows. □

Remark 4.5. If we apply the proof of the Theorem 4.4 above to the group $B(m,-m)$ we can conclude that any injective homomorphism $\varphi : B(m,-m) \to B(m,-m)$ has the property that it induces the multiplication by an odd number on the quotient $\mathbb{Z}$, where $\mathbb{Z}$ is given by the short exact sequence in Propositon 3.2.

5. The Case $B(m,-m)$, $1 \leq |m|$

From Section 2 the cases in this section reduce to Case 3), namely $B(m,-m)$ for $0 < m$.

We will start with the group $B(1,-1)$. This case is done in (see [9]). For sake of completeness we write another proof which is more in the spirit of the techniques used in this work. The group $B(1,-1)$ is isomorphic to the fundamental group of the Klein bottle.

Proposition 5.1. *For any automorphism ϕ of $\mathbb{Z} \rtimes \mathbb{Z}$ the Reidemeister number is infinite.*

Proof. Let us consider the short exact sequence

$$0 \to \mathbb{Z} \to \mathbb{Z} \rtimes \mathbb{Z} \to \mathbb{Z},$$

where the inclusion $\mathbb{Z} \to \mathbb{Z}\rtimes\mathbb{Z}$ sends $1 \to x$. It is well known that $\mathbb{Z}$ is characteristic in $\mathbb{Z}\rtimes\mathbb{Z}$, so any homomorphism $\varphi : \mathbb{Z}\rtimes\mathbb{Z} \to \mathbb{Z}\rtimes\mathbb{Z}$ induces a homomorphism of short exact sequence. Let φ be an automorphism. Then the induced automorhism on the quotient $\bar{\varphi} : \mathbb{Z} \to \mathbb{Z}$ is either the identity or minus the identity. In the first case we have that the Reidemeister number of φ is infinite and the result follows. So let us assume that $\bar{\varphi}$ is $-id$. The induced map on the fiber φ' is also either the identity or minus the identity. In either case, in order to compute the Reidemeister number of φ, by means of the formula given in [18], Lemma 2.1 we need to consider the homomorphism given by the composition of φ' with the conjugation by y, which is the multiplication by -1, i.e., $-\varphi'$. So either φ' or $-\varphi'$ is the identity. Again by the formula given in [18], Lemma 2.1, the result follows. □

The above result is not true for injective homomorphisms. Take for example the homomorphism defined by $\varphi(x) = x^2, \varphi(y) = y^3$. It is an injective homomorphism and $R(\varphi)$ is 4.

From now on let $1 < m$. The groups $B(m,-m)$, in contrast with others Baumslag-Solitar groups already considered, have subgroups isomorphic to $\mathbb{Z}\rtimes\mathbb{Z} = B(1,-1)$, the fundamental group of the Klein bottle.

It is straightforward to verify that Propositions 4.1, 4.2 and 4.3 are also true for $m = -n$ (this is not the case for Proposition 4.3 when $m = n$). So we have a short exact sequence

$$0 \longrightarrow K \longrightarrow B(m,-m) \longrightarrow \mathbb{Z} \longrightarrow 1,$$

where K is the kernel of the map $| \ \ |_a$. This kernel coincides with the normal subgroup generated by the element b and any homomorphism $\phi : B(m,-m) \to B(m,-m)$ is a homomorphism of the short exact sequence. Denote by $\bar{\phi}$ the induced homomorphism on the quotient $\mathbb{Z}$ and by $\phi' : K \to K$ the restriction of ϕ. Our proof have some similarities with the proof for the group $B(1,-1)$.

Proposition 5.2. *Given any automorphism $\phi : B(m,-m) \to B(m,-m)$, then the induced automorphism on the quotient $\bar{\phi}$ is either the identity or minus the identity. In the former case we have $R(\phi)$ infinite.*

Proof. Follows immediately from Corollary 2.2. □

Proposition 5.3. *Given any automorphism $\phi : B(m,-m) \to B(m,-m)$ such that the induced homomorphism on the quotient $\bar{\phi} : \mathbb{Z} \to \mathbb{Z}$ is multiplication by -1,* *either the automorphism ϕ' or the automorphism $\tau_a \circ \phi'$, where τ_a is the conjugation by the element a, have infinite Reidemeister number.*

Proof. From [25] $\phi'(b) = a^i b^\epsilon a^{-i}$ for some integer i, where ϵ is either 1 or -1 since we have an automorphism. For the other automorphism we have $\tau_a \circ \phi'(b) = a^{i+1} b^\epsilon a^{-i-1}$. We certainly have either $\epsilon = (-1)^i$ or $\epsilon = (-1)^{i+1}$. Let φ be ϕ' or $\tau_a \circ \phi'$ according to $\epsilon = (-1)^i$ or $\epsilon = (-1)^{i+1}$, respectively. A presentation of the group K was given in Section 4 before Theorem 4.4. Consider the extra relation in K given by $b = ab^{-1}a^{-1}$. Then it follows that the quotient group is $\mathbb{Z}$ and the automorphism φ induces a homomorphism on the quotient which agrees with the identity. So it follows that the Reidemeister number of φ is infinite. □

Theorem 5.4. *For any automorphism of $B(m, -m)$ the Reidemeister number is infinite.*

Proof. Let $\phi : B(m, -m) \to B(m, -m)$ an automorphism. From Proposition 5.2 we can assume that $\bar{\phi}$ is multiplication by -1. From Propositon 5.3 we know that either ϕ' or $\tau_a \circ \phi'$ has infinite Reidemeister number. By the formula given in [18], Lemma 2.1, the result follows. □

Remark 5.5. We do not know an example of an injective homomorphism on $B(m, -m)$, for $m > 1$, which has finite Reidemeister number.

6. The Case $B(m, m)$, $|m| > 1$

From Section 2 the cases in this section reduce to Case 4), namely $B(m, m)$ for $1 < m$. The proof of this case is not similar to the previous cases.

As we noted before, if $m = 1$, the group is $\mathbb{Z} + \mathbb{Z}$. Then there are automorphisms of the group which have a finite number of Reidemeister classes. For $m > 1$, in order to study its automorphisms, we describe the groups $B(m, m)$ as certain extensions. Finally we show that any automorphism of this group has infinite Reidemeister number.

These groups, in contrast with the Baumslag-Solitar groups already considered, have subgroups isomorphic to $\mathbb{Z} + \mathbb{Z}$. We remark that for $n = 2$ this is not the fundamental group of the Klein bottle. There is a surjection from $B(2, 2)$ onto the fundamental group of the Klein bottle.

We start by describing these groups. Let $|\ |_b : B(m, m) \to \mathbb{Z}$ be the homomorphism which associates to a word the sum of the powers of b which appears in the word. This is a well-defined surjective homomorphism and we have:

Proposition 6.1. *There is a splitting short exact sequence*

$$0 \to F \to B(m, m) \to \mathbb{Z} \to 1,$$

where F is the free group on m generators $x_1, \dots, x_m$ and the last map is $|\ |_b$. Further, the action of the generator $1 \in Z$ is the automorphism of F which, for $j < m$, sends x_j to x_{j+1} and x_m to x_1.

Proof. Let $F \rtimes \mathbb{Z}$ be the semi-direct product of F by $\mathbb{Z}$, where F is the free group on $x_1, ..., x_m$ and the action is given by the automorphism of F which, for $j < m$, maps x_j to x_{j+1} and x_m to x_1. We will show that $B(m, m)$ is isomorphic to $F \rtimes \mathbb{Z}$. For this consider the map $\psi : \{a, b\} \to F \rtimes \mathbb{Z}$ which sends a to x_1 and b to $1 \in Z$. This map extends to a homomorphism $B(m, m) \to F \rtimes Z$, which we also denote by ψ, since the relation which defines the group $B(m, m)$ is preserved by the map. Also ψ is a homomorphism of short exact sequences. The map restricted to the kernel of $| \ |_b$ is surjective to the free group F. Also the kernel admits a set of generators with cardinality n. So the map restricted to the kernel is an isomorphism and the result follows. □

Proposition 6.1 above shows that the groups $B(m, m)$ are policyclic. For more about the Reidemeister number of these groups see [9].

Proposition 6.2. *The center of $B(m, m)$ is the subgroup generated by b^m. Moreover, any injective homomorphism $\phi : B(m, m) \to B(m, m)$ leaves the center invariant.*

Proof. For the first part, from Proposition 6.1, we know that $B(m, m)$ is of the form $F \rtimes \mathbb{Z}$. Let $(w, b^r) \in F \rtimes \mathbb{Z}$ be in the center and $(v, 1) \in F \rtimes \mathbb{Z}$ where v is an arbitrary element of F. We have $(w, b^r)(v, 1) = (w \cdot b^r(v), b^r)$ and $(v, 1)(w, b^r) = (v \cdot w, b^r)$. We can assume that w is a word written in the reduced form which starts with $x_i^{m_i}$, for some $1 \le i \le m$. Let r_0 be the integer, $0 \le r_0 \le m - 1$, congruent to r mod m. Now we consider three cases:

(1) $r_0 = 0$. Take $v = x_{i+1}$ if $i < m$ and $v = x_1$ if $i = m$. We claim that $w.b^r(v) \neq v.w$, so the elements do not commute. To see that they do not commute observe first that $v.w$ is in the reduced form. If $w.b^r(v)$ is not reduced they cannot be equal. If it is reduced, also they can not be equal either, since they start with different letters. The argument above does not work if $w = 1$, but this is the case where the element is in the center.

(2) $r_0 \neq 0$ and $w \neq 1$. Take $v = x_i^{m_i}$. Again $v.w$ is in the reduced form starting with $x_i^{2m_i}$. If $w \cdot b^r(v)$ is not reduced they cannot be equal. If it is reduced, also they cannot be equal either, since they start with different powers of x_i, even if the word contains only one letter, since $b^r(v)$ is not a power of x_i (r is not congruent to 0 mod m).

(3) $r_0 \neq 0$ and $w = 1$. Then $r = km + r_0$, and from the relation in the group it follows that $a^{-1}b^r a = a^{-1}b^{km+r_0}ra = b^{km}a^{-1}b^{r_0}a$. But $a^{-1}b^{r_0}a = b^{r_0}$ implies $b^{r_0}ab^{-r_0} = a$, which in terms of the notation in Proposition 5.1 means $x_1 = x_{r_0}$, which is a contradiction. So the result follows.

For the second part we have to show that $\phi(b^m)$ is in the center. Since ϕ is injective, from the main result of [25], the element $\phi(b)$ is conjugated to a power of b, i.e., $\phi(b) = \gamma b^r \gamma^{-1}$ for some $r \neq 0$. Therefore $\phi(b^m) = \gamma(b^r)^m\gamma^{-1} = bmr$ and the result follows. □

Next we consider the group which is the quotient of $F \rtimes \mathbb{Z}$ by the center, where the center is the subgroup $< b^m >$. This quotient is isomorphic to $F \rtimes \mathbb{Z}_n$ and we denote the image of the generator b in $\mathbb{Z}$ by $\bar{b}$ in $\mathbb{Z}_m$.

Proposition 6.3. *Any automorphism of the group $F \rtimes \mathbb{Z}_m$ has infinite Reidemeister number.*

Proof. We know that F is the free group on the letters $x_1, ..., x_m$ and let $\theta : F \rtimes \mathbb{Z} \to \mathbb{Z}_n$ be the homomorphism defined by $\theta(x_i) = 1$ and $\theta(\bar{b}) = 0$. The kernel of this homomorphism defines a subgroup of $F \rtimes \mathbb{Z}$ of index m which is isomorphic to $F' \rtimes \mathbb{Z}_n$, where F' is the kernel of the homomorphism θ restricted to F. Now we claim that F' is invariant with respect to any homomorphism, i.e, F' is characteristic. Let $(w, \bar{1})$ be an arbitrary element of the subgroup F' with $w \neq 1$. First observe that $\theta(\phi(x_i)) = \theta(\phi(x_1))$, for all i. This follows by induction since $x_{i+1} = \bar{b}.x_i.\bar{b}^{-1}$, $\theta(\phi(x_{i+1})) = \theta(\phi(\bar{b})).\theta(\phi(x_i)).\theta(\phi(\bar{b}^{-1})) = \theta(\phi(x_i))$. Therefore $\theta(\phi(w, \bar{1})) = \theta((w, \bar{1}))\theta(\phi(x_1))$ and the subgroup F' is invariant. Therefore the automorphism ϕ provides an automorphism of the short exact sequence

$$0 \to F' \to F \rtimes \mathbb{Z}_m \to \mathbb{Z}_m + \mathbb{Z}_m \to 0$$

where the restriction to the kernel is an automorphism of a free group of finite rank. Hence, by Corollary 2.4 the result follows. □

Now we proof the main result.

Theorem 6.4. *Any automorphism ϕ of $B(m, m)$ has an infinite Reidemeister number.*

Proof. Any automorphism ϕ, from Proposition 5.2, induces an automorphism on $F \rtimes \mathbb{Z}_m$, which we denote by $\bar{\phi}$. In order to prove that ϕ has an infinite Reidemeister number, it suffices to show the same statement for $\bar{\phi}$. By Proposition 5.3 the statement is true for $\bar{\phi}$, so the theorem follows. □

Remark 6.5. Proposition 6.3 and Theorem 6.4 use only Proposition 6.2 for automorphisms, which, under this assumption, its second part of the Proposition 6.2 becomes obvious. Nevertheless, using Proposition 6.2 as stated, it is not difficult to see that Proposition 6.3 and Theorem 6.4 can be extended for injective homomorphisms if one knows the result for injective homomorphisms of a free group of finite rank. However this is still an open question.

7. Coincidence Reidemeister classes

For a pair of homomorphisms $\phi, \psi : G \to G$ one can ask similarly when a pair of homomorphisms (ϕ, ψ) has infinite coincidence Reidemeister number. If one of the homomorphisms, let us say ϕ, is an automorphism, then the problem is equivalent to the classical problem for the homomorphism $\phi^{-1} \circ \psi$. So we can apply all the results above. Theorems 3.4, 4.4, 6.2, and 6.4 can be generalized to coincidences as follows:

Theorem 7.1. *The coincidence Reidemeister number is infinite for any pair of injective endomorphisms of the group $B(m,n)$, where $|n| \neq |m|$ and $|nm| \neq 0$.*

Proof. For the cases in question, we have proved that an injective homomorphism induces a homomorphism of the short exact sequences given by Propositions 3.1 and 4.1., depending on the values of m and n, respectively. Further in any of the cases above, by the proof of the Theorems 3.4, 4.4, 6.2 and 6.4, we have that the induced homomorphisms $\bar{\phi}$ and $\bar{\psi}$ on the quotients are the identity on $\mathbb{Z}$. So the pair $(\bar{\phi}, \bar{\psi})$ has infinite coincidence Reidemeister number and the result follows from Corollary 2.2. □

The extension of Theorem 7.1 for the groups $B(m,m)$, will follow if the same result is true for a pair of injective homomophisms of a free group of finite rank. But, as pointed out in Remark 6.5, this is not known even if one of the homomorphisms is the identity.

References

[1] G. Baumslag and D. Solitar, Some two-generator one-relator non-hopfian groups, *Bull. Am. Math. Soc.* **68** (1962), 199–201.

[2] C. Bleak, A. Fel'shtyn, and D.L. Gonçalves, Twisted conjugacy classes in R. Thompson's group F, E-print arXiv:math.GR/0704.3441, 2007.

[3] K. Brown, *Cohomology of Groups*, Graduate Texts in Mathematics **87**, Springer, New York, 1982.

[4] D. Cohen, *Combinatorial group theory: a topological approach*, LMS Student Texts **14**, Cambridge University Press, 1989.

[5] B. Farb and L. Mosher, Quasi-isometric rigidity for solvable Baumslag-Solitar groups, II, *Invent. Math.* **137** no. 3 (1999), 613–649.

[6] B. Farb and L. Mosher, On the asymptotic geometry of abelian-by-cyclic groups, *Acta Math.* **184** no. 2 (2000), 145–202.

[7] A.L. Fel'shtyn, The Reidemeister number of any automorphism of a Gromov hyperbolic group is infinite, *Zapiski Nauchnych Seminarov POMI* **279** (2001), 229–241.

[8] A.L. Fel'shtyn and D.L. Gonçalves, Twisted conjugacy classes of automorphisms of Baumslag-Solitar groups, *Algebra and Discrete Mathematics*, no. 3 (2006), 36–48.

[9] A.L. Fel'shtyn, D.L. Gonçalves and P. Wong, Twisted conjugacy classes of automorphisms of Polyfree groups, (in preparation).

[10] A.L. Fel'shtyn and R. Hill, The Reidemeister zeta function with applications to Nielsen theory and a connection with Reidemeister torsion, *K-theory* **8** no. 4 (1994), 367–393.

[11] A.L. Fel'shtyn and E. Troitsky, Twisted Burnside–Frobenius theory for discrete groups, to appear in *Crelle's Journal.*

[12] A.L. Fel'shtyn and E. Troitsky, Twisted Burnside theorem for countable groups and Reidemeister numbers, *Noncommutative Geometry and Number Theory*, Vieweg, Braunschweig, 2006, 141–154.

[13] A.L. Fel'shtyn and E. Troitsky, Geometry of Reidemeister classes and twisted Burnside theorem, to appear in *K*-Theory.

[14] A.L. Fel'shtyn, E. Troitsky and A. Vershik, Twisted Burnside theorem for type II_1 groups: an example, Mathematical Research Letters **13** no. 5 (2006), 719–728.

[15] A.L. Fel'shtyn, Y. Leonov and E. Troitsky, Reidemeister numbers of saturated weakly branch groups. E-print arXiv: math.GR/0606725. Preprint MPIM 2006-79. (submitted to *Geometria Dedicata*).

[16] N.D. Gilbert, J. Howie, V. Metaftsis and E. Raptis, Tree actions of automorphism group, *J. Group Theory* **2** (2000), 213–223.

[17] D.L. Gonçalves, The coincidence Reidemeister classes on nilmanifolds and nilpotent fibrations, *Top. and its Appl.* **83** (1998), 169–186.

[18] D.L. Gonçalves and P. Wong, Twisted conjugacy classes in exponential growth groups, *Bulletin of the London Mathematical Society* **35** (2003), 261–268.

[19] D.L. Gonçalves and P. Wong, Homogeneous spaces in coincidence theory II, *Forum* **17** (2005), 297–313.

[20] D.L. Gonçalves and P. Wong, Twisted conjugacy classes in wreath products, *Internat. J. Alg. Comput.* **16** no. 5 (2006), 875–886.

[21] M. Gromov, Groups of polynomial growth and expanding maps. *Publicationes Mathematiques* **53** (1981), 53–78.

[22] M. Gromov, Hyperbolic groups, in: S. Gersten, ed., "Essays in Group Theory" *MSRI Publications* **8**, Springer-Verlag, Berlin, New York, Heidelberg, 1987, 75–263.

[23] P. de la Harpe, *Topics in Geometric Group Theory*, Chicago Lectures in Mathematics Series, The Unversity of Chicago Press, Chicago, 2000.

[24] M. Kapovich, *Hyperbolic Manifolds and Discrete Groups*, Birkhäuser, Boston, Basel, Berlin, 2000.

[25] D. Kochloukova, Injective endomorphisms of the Baumslag-Solitar group, *Algebra Colloquium* **13** no. 3 (2006), 525–534.

[26] G. Levitt, On the automorphism group of generalized Baumslag-Solitar groups, *Geometry & Topology* **11** (2007), 473–515.

[27] G. Levitt and M. Lustig, Most automorphisms of a hyperbolic group have very simple dynamics, *Ann. Scient. Éc. Norm. Sup.* **33** (2000), 507–517.

[28] J. Nielsen, Untersuchungen zur Topologie der geschlossenen zweiseitigen Flächen I, II, III, *Acta Math.* **50** (1927), 189–358; **53** (1929), 1–76; **58** (1932), 87–167.

[29] K. Reidemeister, Automorphismen von Homotopiekettenringen, *Math. Ann.* **112** (1936), 586–593.

[30] J. Taback and P. Wong, Twisted conjugacy and quasi-isometry invariance for generalized solvable Baumslag-Solitar groups, E-print arxiv:math.GR/0601271, 2006.

[31] J. Taback and P. Wong, A note on twisted conjugacy and generalized Baumslag-Solitar groups, E-print arXiv:math.GR/0606284, 2006.

[32] P. Wong, Fixed-point theory for homogeneous spaces, *Amer. J. Math.* **120** (1998), 23–42.

Alexander Fel'shtyn
Instytut Matematyki
Uniwersytet Szczecinski
ul. Wielkopolska 15
70-451 Szczecin
Poland
and
Boise State University
1910 University Drive
Boise, Idaho, 83725-155
USA
e-mail: felshtyn@diamond.boisestate.edu
felshtyn@mpim-bonn.mpg.de

Daciberg L. Gonçalves
Dept. de Matemática - IME - USP
Caixa Postal 66.281
CEP 05311-970, São Paulo - SP
Brasil
e-mail: dlgoncal@ime.usp.br

Progress in Mathematics, Vol. 265, 415–428

Pentagon Relation for the Quantum Dilogarithm and Quantized $\mathcal{M}_{0,5}^{\mathrm{cyc}}$

Alexander B. Goncharov

To the memory of Sasha Reznikov

Abstract. We give a proof of the pentagon relation for the quantum dilogarithm by using functional analysis methods. We introduce a related Schwartz space and prove that it is preserved by the intertwiner operator defined using the quantum dilogarithm. Using this we can define a representation of the quantized moduli space of configurations of 5 points on the projective line.

Mathematics Subject Classification (2000). 33E, 81T.

Keywords. The quantum dilogarithm, quantization, cluster varieties.

1. Introduction

Let $\hbar > 0$. The quantum dilogarithm function is given by the following integral:

$$\Phi^{\hbar}(z) := \exp\Big(-\frac{1}{4}\int_{\Omega}\frac{e^{-ipz}}{\mathrm{sh}(\pi p)\mathrm{sh}(\pi\hbar p)}\frac{dp}{p}\Big), \qquad \mathrm{sh}(p) = \frac{e^{p}-e^{-p}}{2}.$$

Here Ω is a path from $-\infty$ to $+\infty$ making a little half-circle going over the zero. So the integral is convergent. It goes back to Barnes [Ba], and appeared in many papers during the last 30 years: [Bax], [Sh], [Fad1], ... The function $\Phi^{\hbar}(z)$ enjoys the following properties (cf. [FG3], Section 4):

- The function $\Phi^{\hbar}(z)$ is meromorphic. Its zeros are simple zeros in the upper half-plane at the points

$$\{\pi i((2m-1)+(2n-1)\hbar)|m,n\in\mathbb{N}\}, \qquad \mathbb{N} := \{1,2,\ldots\}. \tag{1}$$

  Its poles are simple poles, located in the lower half-plane, at the points

$$\{-\pi i((2m-1)+(2n-1)\hbar)|m,n\in\mathbb{N}\}. \tag{2}$$

- The function $\Phi^\hbar(z)$ is characterized by the following difference relations. Let $q := e^{\pi i\hbar}$ and $q^\vee := e^{\pi i/\hbar}$. Then

$$\Phi^\hbar(z + 2\pi ih) = \Phi^\hbar(z)(1 + qe^z), \quad \Phi^\hbar(z + 2\pi i) = \Phi^\hbar(z)(1 + q^\vee e^{z/\hbar}), \tag{3}$$

- One has $|\Phi^\hbar(z)| = 1$ when z is on the real line.
- The function $\Phi^\hbar(z)$ is related in several ways to the dilogarithm, e.g., its asymptotic expansion when $\hbar \to 0$ is

$$\Phi^\hbar(z) \sim \exp\Big(\frac{\mathrm{L}_2(e^z)}{2\pi i\hbar}\Big), \quad \text{where } \mathrm{L}_2(x) := \int_0^x \log(1+t)\frac{dt}{t}$$

is a version of the Euler's dilogarithm function.

When $\hbar$ is a complex number with Im $\hbar > 0$, there is an infinite product expansion

$$\Phi^\hbar(z) = \frac{\Psi^q(e^z)}{\Psi^{1/q^\vee}(e^{z/\hbar})}, \qquad \text{where } \mathbf{\Psi}^q(x) := \prod_{a=1}^{\infty}(1 + q^{2a-1}x)^{-1}.$$

The function $\Phi^\hbar(z)$ provides an operator $K : L^2(\mathbb{R}) \to L^2(\mathbb{R})$, defined as a rescaled Fourier transform followed by the operator of multiplication by the quantum dilogarithm $\Phi^\hbar(x)$.

$$Kf(z) := \int_{-\infty}^{\infty} f(x)\Phi^\hbar(x)\exp(\frac{-xz}{2\pi i\hbar})dx.$$

Since $|\Phi^\hbar(x)| = 1$ on the real line, $2\pi\sqrt{\hbar}K$ is unitary.

Theorem 1.1. $(2\pi\sqrt{\hbar}K)^5 = \lambda \cdot \mathrm{Id}$, *where* $|\lambda| = 1$.

In the quasiclassical limit it gives Abel's five term relation for the dilogarithm.

The pentagon relation for the simpler version $\mathbf{\Psi}^q(x)$ of the quantum dilogarithm was discovered in [FK]. A similar pentagon relation for the function $\Phi^\hbar(z)$, which is equivalent to Theorem 1.1, was suggested in [Fad1] and proved, using different methods, in [Wo] and [FKV]. Theorem 1.1 was formulated in [CF]. However the argument presented there as a proof has a significant problem, which put on hold the program of quantization of Teichmüller spaces.

In this paper we show that the operator K is a part of a much more rigid structure, called the *quantized moduli space* $\mathcal{M}_{0,5}^{\mathrm{cyc}}$ – this easily implies Theorem 1.1.

Namely, consider the algebra generated by operators of multiplication by e^x and $e^{x/\hbar}$ and shifts by $2\pi i$ and $2\pi i\hbar$, acting as unbounded operators in $L^2(\mathbb{R})$. We use a remarkable subalgebra $\mathbf{L}$ of this $*$-algebra, and introduce a *Schwartz space* $S_{\mathbf{L}} \subset L^2(\mathbb{R})$, defined as the common domain of the operators from $\mathbf{L}$. It comes with a natural topology. Our main result, Theorem 2.6, tells that the operator K preserves the space $S_{\mathbf{L}}$, and the conjugation by K intertwines an order 5 automorphism γ of the algebra $\mathbf{L}$, see Fig. 1. This characterises the operator K up to a constant. The proof uses analytic properties of the space $S_{\mathbf{L}}$ developed in Theorem 2.3. Theorem 2.6 easily implies Theorem 1.1.

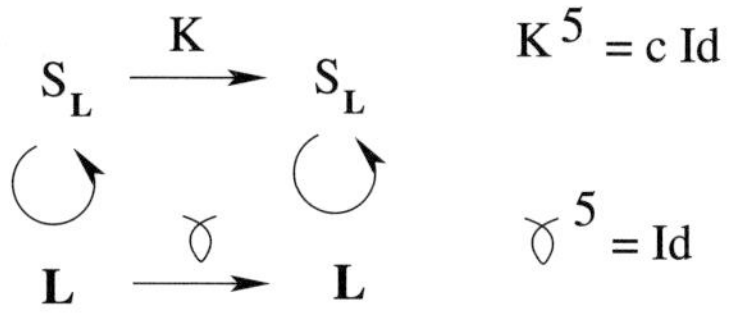

FIGURE 1. Quantized moduli space $\mathcal{M}_{0,5}^{\text{cyc}}$.

We define a space of *distributions* $S_{\mathbf{L}}^*$ as the topological dual to $S_{\mathbf{L}}$. So there is a Gelfand triple $S_{\mathbf{L}} \subset L_2(\mathbb{R}) \subset S_{\mathbf{L}}^*$. The operator K acts by its automorphisms. It would be interesting to calculate it on some distributions explicitly.

The story is similar in spirit to the Fourier transform theory developed using the algebra of polynomial differential operators:

The Fourier transform	$\longleftrightarrow$	The operator K
The algebra $\mathbf{D}$ of polynomial differential operators	$\longleftrightarrow$	The algebra $\mathbf{L}$ of difference operators
The automorphism φ of $\mathbf{D}$ given by $ix \to d/dx,\ d/dx \to -ix$	$\longleftrightarrow$	The automorphism γ of $\mathbf{L}$
The classical Schwartz space	$\longleftrightarrow$	The Schwartz space $S_{\mathbf{L}}$.

Let $\mathcal{M}_{0,5}^{\text{cyc}} \subset \overline{\mathcal{M}}_{0,5}$ be the moduli space of configurations of 5 cyclically ordered points on $\mathbb{P}^1$, where we do not allow the neighbors to collide. It carries an atlas consisting of 5 coordinate systems, providing $\mathcal{M}_{0,5}^{\text{cyc}}$ with a structure of the cluster $\mathcal{X}$-variety of type A_2. The algebra $\mathbf{L}$ is isomorphic to the algebra of regular functions on the modular double of the non-commutative q-deformation of the cluster $\mathcal{X}$-variety. The automorphism γ corresponds to a cyclic shift acting on configurations of points.

The triple $(\mathbf{L}, S_{\mathbf{L}}, \gamma)$*, see Fig.* 1*, is called the quantized moduli space* $\mathcal{M}_{0,5}^{\text{cyc}}$.

The results of this paper admit a generalization to a cluster set-up, where the role of the automorphism γ plays the cluster mapping class group. In particular this gives a definition of quantized higher Teichmüller spaces, and allows to state precisely the modular functor property of the latter.[1]

The structure of the paper. In Section 2.1 we recall the cluster $\mathcal{X}$-variety of type A_2 [FG2]. In Section 3 we identify it with $\mathcal{M}_{0,5}^{\text{cyc}}$. This clarifies formulas in Section 2.1-2.2. In Section 2.2 we recall a collection of regular functions on our cluster $\mathcal{X}$-variety. Theorem 3.2 tells that they form a basis in the space of regular functions, and in particular closed under multiplication. We introduce a q-deformed version of this basis/algebra. Its tensor product with a similar algebra for $q^\vee$ is the algebra $\mathbf{L}$. In Sections 2.3–2.5 we prove our main results.

[1] In previous versions of quantization of Teichmüller spaces/cluster $\mathcal{X}$-varieties the pair $(\mathbf{L}, S_{\mathbf{L}})$ was missing, making the resulting notion rather flabby.

Acknowledgments. I am very grateful to Joseph Bernstein for several illuminating discussions, and to Andrey Levin for reading carefully the first draft of the text and spotting some errors. I am grateful to Sergiy Kolyada for the patience and help with editing of the manuscript.

I was supported by the NSF grants DMS-0400449 and DMS-0653721. The final version of the text was written during my stay in IHES. I would like to thank IHES for hospitality and support.

2. Quantized moduli space $\mathcal{M}_{0,5}^{\mathrm{cyc}}$

2.1. Cluster varieties of type A_2

The cluster $\mathcal{X}$-variety is glued from five copies of $\mathbb{C}^* \times \mathbb{C}^*$, so that i-th copy is glued to $(i+1)$-st (indexes are modulo 5) by the map acting on the coordinate functions as follows:[2]

$$\gamma_X^* : X \longmapsto Y^{-1}, \quad Y \longmapsto (1+Y)X. \tag{4}$$

Similarly, the cluster $\mathcal{A}$-variety is glued from five copies of $\mathbb{C}^* \times \mathbb{C}^*$, so that i-th copy is glued to $(i+1)$-th by the map acting on the coordinate functions as follows:

$$\gamma_A^* : (A, B) \mapsto ((1+A)B^{-1}, A). \tag{5}$$

(Accidently, these two cluster varieties are canonically isomorphic.)

The fifth degree of each of these maps is the identity. Thus the map identifying the i-th copy of $\mathbb{C}^* \times \mathbb{C}^*$ with the $(i+1)$-st one in the standard way is an automorphism of order 5 acting on the $\mathcal{X}$- and $\mathcal{A}$-varieties. We denote it by γ.

Recall the tropical semifield $\mathbb{Z}^t$. It is the set $\mathbb{Z}$ with the operations of addition $a \oplus b := \max\{a, b\}$, and multiplication $a \otimes b := a + b$. The set $\mathcal{A}(\mathbb{Z}^t)$ of $\mathbb{Z}^t$-points of the $\mathcal{A}$-variety is defined by gluing the five copies of $\mathbb{Z}^2$ via the tropicalizations of the map (5). The map γ acts on the tropical $\mathcal{A}$-space by

$$\gamma_a : (a, b) \mapsto (\max(a, 0) - b, a), \quad \gamma^5 = \mathrm{Id}.$$

There are five cones in the tropical $\mathcal{A}$-space, shown on Fig. 2. The map γ shifts them cyclically counterclockwise. It is a piecewise linear map, whose restriction to each cone is linear.

2.2. The $*$-algebra L

The canonical basis for the cluster $\mathcal{X}$-variety of type A_2. A rational function $F(X, Y)$ is a *universally positive Laurent polynomial* on $\mathcal{X}$, if $(\gamma_X^*)^i F(X, Y)$ is a Laurent polynomial with positive integral coefficients for every i. Equivalently, it belongs to the intersection of the ring of regular functions on the scheme $\mathcal{X}$ over

[2] We use a definition which differs slightly from the standard one, but delivers the same object.

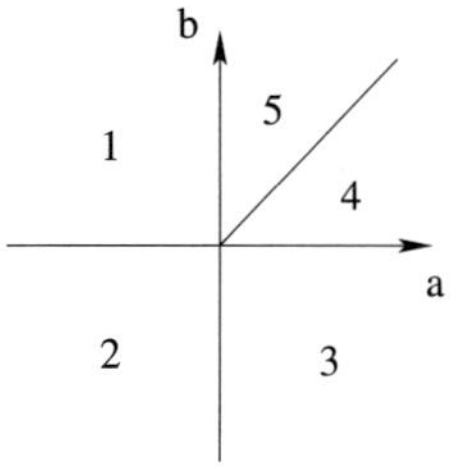

FIGURE 2. The five domains in the tropical $\mathcal{A}$-space.

$\mathbb{Z}$ with the semifield of rational functions with positive integral coefficients. There is a canonical γ-equivariant map, defined in Section 4 of [FG2]:[3]

$$\mathbb{I}_{\mathcal{A}} : \mathcal{A}(\mathbb{Z}^t) \longrightarrow \text{The space of universally positive Laurent polynomials on } \mathcal{X},$$

$$\gamma_X^* \mathbb{I}_{\mathcal{A}}(\gamma_a(a,b)) = \mathbb{I}_{\mathcal{A}}(a,b), \tag{6}$$

given by:

$$\mathbb{I}_{\mathcal{A}}(a,b) = \begin{cases} X^aY^b & \text{for} \quad a \leq 0 \text{ and } b \geq 0 \\ \left(\frac{1+X}{XY}\right)^{-b} X^a & \text{for} \quad a \leq 0 \text{ and } b \leq 0 \\ \left(\frac{1+X+XY}{Y}\right)^a \left(\frac{1+X}{XY}\right)^{-b} & \text{for} \quad a \geq 0 \text{ and } b \leq 0 \\ ((1+Y)X)^b \left(\frac{1+X+XY}{Y}\right)^{a-b} & \text{for} \quad a \geq b \geq 0 \\ Y^{b-a}((1+Y)X)^a & \text{for} \quad b \geq a \geq 0. \end{cases} \tag{7}$$

Or equivalently, showing that the leading monomial is always X^aY^b:

$$\mathbb{I}_{\mathcal{A}}(a,b) = \begin{cases} X^aY^b & \text{for} \quad a \leq 0 \text{ and } b \geq 0 \\ X^aY^b(1+X^{-1})^{-b} & \text{for} \quad a \leq 0 \text{ and } b \leq 0 \\ X^aY^b(1+X^{-1})^{-b}(1+Y^{-1}+X^{-1}Y^{-1})^a & \text{for} \quad a \geq 0 \text{ and } b \leq 0 \\ X^aY^b(1+Y^{-1})^b(1+Y^{-1}+X^{-1}Y^{-1})^{a-b} & \text{for} \quad a \geq b \geq 0 \\ X^aY^b(1+Y^{-1})^a & \text{for} \quad b \geq a \geq 0. \end{cases}$$

One can easily verifies that the formulae agree on the overlapping domains of values of a and b. The i-th row of (7) describes the restriction of the canonical map to the i-th cone.

The quantum $\mathcal{X}$-variety and the quantum canonical basis. Let T_q be the algebra generated over $\mathbb{Z}[q,q^{-1}]$ by $X^{\pm 1}, Y^{\pm 1}$, subject to the relation $q^{-1}XY - qYX = 0$. It is called the two-dimensional quantum torus algebra. It has an involutive antiautomorphism $*$ such that

$$*q = q^{-1}, \quad *X = X, \quad *Y = Y.$$

Consider the following q-deformation of the $*$-equivariant map γ:

$$\gamma_q^* : X \longmapsto Y^{-1}, \quad Y \longmapsto (1+qY)X. \tag{8}$$

[3]Observe that γ_X^* tells how the automorphism γ acts on functions, while γ_a tells the action on the tropical points.

One checks that it is an order 5 automorphism of the fraction field of T_q. The quantum $\mathcal{X}$-space $\mathcal{X}_q$ is nothing else but a pair $(\mathrm{T}_q, \gamma_q^*)$.[4]

An element $F(X,Y)$ of the fraction field of T_q is a *universally positive Laurent polynomial* on $\mathcal{X}_q$ if $(\gamma_q^*)^i F(X,Y)$ is a Laurent polynomial in X, Y, q with positive integral coefficients for every i.

Proposition 2.1. *There is a canonical γ-equivariant map*

$$\mathbb{I}^q_{\mathcal{A}} : \mathcal{A}(\mathbb{Z}^t) \longrightarrow \text{The space of universally positive Laurent polynomials on } \mathcal{X}_q.$$

Construction. It is obtained by multiplying each monomial in (7) by a (uniquely defined) power of q, making it $*$-invariant. For example, the quantum canonical map on the first cone is given by

$$\mathbb{I}^q_{\mathcal{A}}(a,b) = q^{-ab} X^a Y^b, \quad a \leq 0, b \geq 0.$$

Then we can use (6), which is valid in the q-deformed version as well.[5]

Denote by $\mathbb{L}_q$ the image of the map $\mathbb{I}^q_{\mathcal{A}}$. It is closed under multiplication. Set $\mathbf{L} := \mathbb{L}_q \otimes \mathbb{L}_{q^\vee}$.

2.3. The Schwartz space $S_{\mathbf{L}}$

Let $W \subset L^2(\mathbb{R})$ be the space of finite $\mathbb{C}$-linear combinations of the functions

$$e^{-ax^2/2+bx} P(x), \qquad \text{where } P(x) \text{ is a polynomial in } x, \text{ and } a \in \mathbb{R}_{>0}, b \in \mathbb{C}. \tag{9}$$

Set

$$\widehat{X}(f)(x) := f(x + 2\pi i\hbar), \quad \hbar \in \mathbb{R}_{>0}, \qquad \widehat{Y}(f)(x) := e^x f(x).$$
$$\widehat{X}^\vee(f)(x) := f(x + 2\pi i), \quad \hbar \in \mathbb{R}_{>0}, \qquad \widehat{Y}^\vee(f)(x) := e^{x/\hbar} f(x).$$

These are symmetric unbounded operators. They preserve W and satisfy, on W, relations

$$\widehat{X}\widehat{Y} = q^2 \widehat{Y}\widehat{X}, \quad q := e^{\pi i\hbar}.$$
$$\widehat{X}^\vee\widehat{Y}^\vee = (q^\vee)^2 \widehat{Y}^\vee\widehat{X}^\vee, \quad q^\vee := e^{\pi i/\hbar}.$$

The second pair of operators commute with the first one. Therefore these operators provide an $*$-representation of the algebra $\mathrm{T}_q \otimes \mathrm{T}_{q^\vee}$ in W.

Remark. Consider a smaller subspace $W_0 \subset W$, with $a = 1, b \in 2\pi i\hbar\mathbb{Z} + 2\pi i\mathbb{Z} + \mathbb{Z} + 1/\hbar\mathbb{Z}$ and $\deg(P) = 0$. Then acting on $e^{-x^2/2}$ we get an isomorphism of linear spaces $\mathrm{T}_q \otimes \mathrm{T}_{q^\vee} \stackrel{\sim}{=} W_0$.

In particular an element $A \in \mathbf{L}$ acts by an unbounded operator $\widehat{A}$ in W.

Definition 2.2. *The Schwartz space $S_{\mathbf{L}}$ for the $*$-algebra $\mathbf{L}$ is a subspace of $L^2(\mathbb{R})$ consisting of vectors f such that the functional $w \to (f, \widehat{A}w)$ on W is continuous for the L_2-norm.*

---

[4]Alternatively, using a geometric language, the quantum $\mathcal{X}$-space $\mathcal{X}_q$ is glued from five copies of the spectrum $\mathrm{Spec}(\mathrm{T}_q)$ of the quantum torus T_q, so that i-th copy is glued to $(i+1)$-st along the map (8)

[5]We do not use the fact that Laurent q-polynomials in the basis have positive integral coefficients.

Denote by $(*,*)$ the scalar product in L_2. The Schwartz space for the $*$-algebra $\mathbf{L}$ is the common domain of definition of operators from $\mathbf{L}$ in $L^2(\mathbb{R})$. Indeed, since W is dense in $L^2(\mathbb{R})$, the Riesz theorem implies that for any $f \in S_{\mathbf{L}}$ there exists a unique $g \in L^2(\mathbb{R})$ such that $(g, w) = (f, \widehat{A}w)$. We set $\widehat{*A}f := g$. Equivalently, let W^* be the algebraic linear dual to W. So $L^2(\mathbb{R}) \subset W^*$. Then

$$S_{\mathbf{L}} = \{v \in W^* | \widehat{A}^* v \in L^2(\mathbb{R}) \quad \text{for any } A \in \mathbf{L}\} \cap L^2(\mathbb{R}).$$

The Schwartz space $S_{\mathbf{L}}$ has a natural topology given by seminorms

$$\rho_B(f) := ||Bf||_{L_2}, \qquad B \text{ runs through a basis in } \mathbf{L}.$$

The key properties of the Schwartz space $S_{\mathbf{L}}$ which we use below are the following.

Theorem 2.3. *The space W is dense in the Schwartz space $S_{\mathbf{L}}$.*

One can interpret Theorem 2.3 by saying that the $*$*-algebra* $\mathbf{L}$ *is essentially self-adjoint in* $L^2(\mathbb{R})$.

Proof.

Lemma 2.4. *For any $w \in W, s \in S_{\mathbf{L}}$, the convolution $s * w$ lies in $S_{\mathbf{L}}$.*

Proof. Set $T_\lambda f(x) := f(x - \lambda)$. Write

$$s * w(x) = \int_{-\infty}^{\infty} w(t)(T_t s)(x)dt.$$

For any seminorm ρ_B on $S_{\mathbf{L}}$ the operator $T_\lambda : (S_{\mathbf{L}}, \rho_B) \longrightarrow (S_{\mathbf{L}}, \rho_B)$ is a bounded operator with the norm bounded by $e^{|\lambda|}$. Thus the operator $\int_{-\infty}^{\infty} w(t)T_t dt$ is a bounded operator on $(S_{\mathbf{L}}, \rho_B)$. This implies the lemma. □

Let $w_\varepsilon := (2\pi)^{-\frac{1}{2}}\varepsilon^{-1}e^{-\frac{1}{2}(x/\varepsilon)^2} \in W$ be a sequence converging as $\varepsilon \to 0$ to the δ-function at 0. Clearly one has in the topology of $S_{\mathbf{L}}$

$$\lim_{\varepsilon \to 0} w_\varepsilon * s = s(x). \tag{10}$$

Lemma 2.5. *For any $w \in W, s \in S_{\mathbf{L}}$, the Riemann sums for the integral*

$$s * w(x) = \int_{-\infty}^{\infty} s(t)\omega(x - t)dt = \int_{-\infty}^{\infty} s(t)T_t w(x)dt. \tag{11}$$

converge in the topology of $S_{\mathbf{L}}$ to the convolution $s * w$.

Proof. Let us show first that (11) is convergent in $L_2(\mathbb{R})$. The key fact is that a shift of $w \in W$ quickly becomes essentially orthogonal to w. More precisely, in the important for us case when $w = \exp(-ax^2/2 + bx)$, $a > 0$, (this includes any $w \in W_0$) we have

$$(w(x), T_\lambda w(x)) < C_w e^{-a\lambda^2/2 + (b - \bar{b})\lambda}. \tag{12}$$

Therefore in this case

$$\Big(\int_{-\infty}^{\infty} s(t)T_t w(x)dt, \int_{-\infty}^{\infty} s(t)T_t w(x)dt\Big)$$
$$\overset{(12)}{\leq} C_w \int_{-\infty}^{\infty}\int_{-\infty}^{\infty} e^{-a(t_1-t_2)^2/2+(b-\bar{b})(t_1-t_2)}|s(t_1)s(t_2)|dt_1dt_2$$
$$= C_w \int_{-\infty}^{\infty} e^{-a\lambda^2/2+(b-\bar{b})\lambda} \int_{-\infty}^{\infty} |s(t)s(t+\lambda)|dtd\lambda$$
$$\leq C_w||s||_{L_2}^2 \int_{-\infty}^{\infty} e^{-a\lambda^2/2+(b-\bar{b})\lambda}d\lambda.$$

We leave the case of an arbitrary w to the reader: it is not used in the proof of the theorem. □

The convergence with respect to the seminorm $||Bf||$ is proved by the same argument using the fact that $W_0 \subset W$ is stable under the algebra $\mathbf{L}$. The theorem follows from Lemma 2.4, (10) and Lemma 2.5. □

Remark. The same arguments show that the space W is dense in the space $S_{\mathbf{L}'}$ defined for any subalgebra $\mathbf{L}'$ of $\mathbf{L}$.

2.4. The main result

Theorem 2.6. *The operator K preserves the Schwartz space $S_{\mathbf{L}}$. It intertwines the automorphism γ of the algebra $\mathbf{L}$, i.e., for any $A \in \mathbf{L}$ and $s \in S_{\mathbf{L}}$ one has*

$$K^{-1}\widehat{A}Ks = \widehat{\gamma(A)}s. \tag{13}$$

Proof. We need the following key result.

Proposition 2.7. *For any $A \in \mathbf{L}$, $w \in W$ one has $K\widehat{\gamma(A)}w = \widehat{A}Kw$. Therefore $\widehat{A}Kw \in L^2(\mathbb{R})$.*

Proof. Let $\mathbb{L}_q'$ be the space of Laurent q-polynomials F in X, Y such that $\gamma(F)$ is again a Laurent q-polynomial. The following elements belong to $\mathbb{L}_q'$:

$$X^aY^m, \quad X^aY^{-n}(1+qX^{-1})(1+q^3X^{-1})\dots(1+q^{2n-1}X^{-1}), \quad a \in \mathbb{Z}, \ m,n \geq 0. \tag{14}$$

Indeed, $\gamma(Y^{-n}) = ((1+qY)X)^{-n} = X^{-n}\prod_{a=1}^{n}(1+q^{(2a-1)}Y)^{-1}$.

Lemma 2.8.

(i) *The monomials (14) span the space $\mathbb{L}_q'$.*

(ii) *For every $A \in \mathbb{L}_q' \otimes \mathbb{L}_{q^\vee}'$, $w \in W$ one has $K\widehat{\gamma(A)}w = \widehat{A}Kw$.*

Lemma 2.8 *implies Proposition* 2.7. Thanks to the very definition of the spaces $\mathbb{L}_q$ and $\mathbb{L}_q'$, the part (ii) of the Lemma is stronger than the commutation relation from Proposition 2.7. □

Proof of Lemma 2.8. (i) is obvious.

(ii) Let us prove first the following three basic identities:

$$K(1+q\widehat{Y})\widehat{X}w = \widehat{Y}Kw; \quad K\widehat{Y}^{-1}w = \widehat{X}Kw; \tag{15}$$

$$K\widehat{X}^{-n}w = \widehat{Y}^{-1}(1+q\widehat{X}^{-1})Kw.$$

The general case follows from this. To see this, observe that if $A_1, A_2 \in \mathbf{L}$ and $K\widehat{A}_i w = \widehat{\gamma^{-1}(A_i)}Kw$ for $i = 1, 2$, then, since $\widehat{A}_2 w \in W$, one has

$$K\widehat{A}_1\widehat{A}_2 w = \widehat{\gamma^{-1}(A_1)}K\widehat{A}_2 w = \widehat{\gamma^{-1}(A_1A_2)}Kw.$$

The first identity. Denote by C_s the line $x + is$ parallel to the x-axis. One has

$$\begin{aligned} K(1+q\widehat{Y})\widehat{X}w &= \int_{C_0}(1+qe^x)w(x+2\pi i\hbar)\Phi^\hbar(x)e^{-xz/2\pi i\hbar}dx \\ &\overset{(3)}{=} \int_{C_0} w(x+2\pi i\hbar)\Phi^\hbar(x+2\pi i\hbar)e^{-xz/2\pi ih}dx \\ &= \int_{C_{2\pi i\hbar}} w(x)\Phi^\hbar(x)e^{-(x-2\pi i\hbar)z/2\pi ih}dx \\ &\overset{(*)}{=} e^z Kw = \widehat{Y}Kw. \end{aligned}$$

To obtain the equality $(*)$ we have to move the contour $C_{2\pi i\hbar}$ down to C_0. We can justify this because: (i) the function $\Phi^\hbar(z)$ is analytic in the upper half-plane, see (2), and (ii) the function $\Phi^\hbar(z)$ growth on any horizontal line at most exponentially, while $w(x)$ decays there much faster, like e^{-x^2}.

Remark. We used here $\hbar > 0$. We would not be able to move a similar contour with negative imaginary part, since it will hit the poles of $\Phi^\hbar(z)$.

The second identity.

$$K\widehat{Y}^{-1}w(z) = Ke^{-x}w = \int_{-\infty}^{\infty} w(x)\Phi^\hbar(x)e^{-x(z+2\pi i\hbar)/2\pi i\hbar}dx = \widehat{X}Kw(z).$$

The third identity. We have

$$\begin{aligned} &\int_{C_0} w(x-2\pi i\hbar)\Phi^\hbar(x)e^{-xz/2\pi i\hbar}dx \\ &= \int_{C_{-2\pi i\hbar}} w(x)\Phi^\hbar(x+2\pi i\hbar)e^{-(x+2\pi i\hbar)z/2\pi i\hbar}dx \\ &= e^{-z}\int_{C_{-2\pi i\hbar}} w(x)\Phi^\hbar(x+2\pi i\hbar)e^{-xz/2\pi i\hbar}dx. \end{aligned}$$

We can move the contour $C_{-2\pi i\hbar}$ up towards C_0 since the function $\Phi^\hbar(x+2\pi i\hbar)$ is holomorphic above the line $C_{-2\pi i\hbar}$, and grows in horizontal directions in the area between the two contours at most exponentially, while $w(x)$ decays like e^{-x^2}. So we get

$$\begin{aligned}
&e^{-z}\int_{C_0} w(x)\Phi^{\hbar}(x+2\pi i\hbar)e^{-xz/2\pi i\hbar}dx \\
&= e^{-z}\int_{C_0} w(x)\Phi^{\hbar}(x)(1+qe^{x})e^{-xz/2\pi i\hbar}dx \\
&= Y^{-1}(1+q\widehat{X}^{-1})\int_{C_0} w(x)\Phi^{\hbar}(x)e^{-xz/2\pi i\hbar}dx.
\end{aligned}$$

Lemma 2.8 is proved. □

To show that $Ks \in S_{\mathbf{L}}$ for an $s \in S_{\mathbf{L}}$ we need to check that for any $B \in \mathbf{L}$ the functional $w \to (Ks, \widehat{B}^* w)$ is continuous. Since W is dense in $S_{\mathbf{L}}$ by Theorem 2.3, there is a sequence $v_i \in W$ converging to s in $S_{\mathbf{L}}$. This means that

$$\lim_{i\to\infty}(\widehat{B}v_i, w) = (\widehat{B}s, w) \quad \text{for any } B \in \mathbf{L},\ w \in W. \tag{16}$$

One has

$$(Ks, \widehat{B}^* w) = (s, K^{-1}\widehat{B}^* w) \overset{\text{Prop. 2.7}}{=} (s, \widehat{\gamma(B^*)}K^{-1}w) \overset{(16)}{=} \lim_{i\to\infty}(v_i, \widehat{\gamma(B^*)}K^{-1}w)$$
$$\overset{\text{def}}{=} \lim_{i\to\infty}(\widehat{\gamma(B)}v_i, K^{-1}w) \overset{(16)}{=} (\widehat{\gamma(B)}s, K^{-1}w) = (K\widehat{\gamma(B)}s, w).$$

Since the functional on the right is continuous, $Ks \in S_{\mathbf{L}}$, and we have (13). The theorem is proved. □

Since $\gamma^5 = \mathrm{Id}$, Theorem 2.6 immediately implies

Corollary 2.9. *For any $A \in \mathbf{L}$ one has $K^{-5}\widehat{A}K^5 = \widehat{A}$ on S.*

2.5. Proof of Theorem 1.1

Let $E = \{f \in L^2(\mathbb{R}) | e^{nx}f(x) \in L^2(\mathbb{R}) \quad \text{for any } n > 0\}$.

Lemma 2.10. $K^5(E) \subset E$.

Proof. Indeed, since $\widehat{Y} = e^x$ and $Y^n \in \mathbf{L}$ for any $n > 0$, one has $K^5 e^{nx} f = e^{nx}K^5 f$ for any $n > 0$, $f \in S$ by Corollary 2.9. So using the remark in the end of Section 2.3, we see that W, and hence S is dense in E, we get the claim. □

Lemma 2.11. K^5 *is the operator of multiplication by a function $F(x)$.*

Proof. We claim that the value $(K^5 f)(a)$ depends only on the value $f(a)$. Indeed, for any $f_0(x) \subset E$, $f_0(a) = f(a)$ we have $f = (e^x - e^a)\phi(x) + f_0(x)$, where $\phi(x) = (f - f_0)/(e^x - e^a) \in E$. Thus $K^5 f = (e^x - e^a)K^5\phi(x) + K^5 f_0(x)$. So $K^5 f(a) = K^5 f_0(a)$. Now define $F(a)$ from $K^5 f_0(a) = F(a)f_0(a)$. The lemma is proved. □

Proposition 2.12. *The function $F(z)$ is a constant.*

Proof. Let S_1 be the common domain of definition of the operators $\widehat{X}^a\widehat{Y}^b$, $a \leq 0, b \geq 0$.

Lemma 2.13. *The space S_1 consists of the functions $f(x)$ in L_2 which admit an analytic continuation to the upper half-plane $y > 0$, and decay faster then e^{ax} for any $a > 0$ on each line $x + iy$.*

Proof. Indeed, it is invariant under multiplication by e^{bx}, $b > 0$, and shift by $2\pi ia$, $a > 0$, which means that the Fourier transform of a function from S_1 is invariant under multiplication by e^{ax}, $a > 0$. The lemma is proved. □

Since $K^5 S \subset S$, it follows that $F(x)w \in S \subset S_1$ for $w \in W$, and hence $F(z)$ is analytic in the half-plane $y > 0$. The operator of multiplication by $F(z)$ commutes with the shifts by $2\pi i$ and $2\pi i\hbar$. Thus it commutes with the shift by $2\pi i(n + m\hbar)$, $m, n > 0$. This implies that $F(z)$ is invariant under the shifts by $2\pi i(m + n\hbar)$ where $m + n\hbar > 0$. Thus $F(z)$ is a constant when $\hbar$ is irrational. Since K^5 depends continuously on $\hbar$, we get Proposition 2.12, and hence Theorem 1.1. □

3. Algebraic geometry of $\mathcal{M}_{0,5}^{\mathrm{cyc}}$

Recall the cross-ratio $r(x_1, x_2, x_3, x_4)$ of four points on $\mathbb{P}^1$:

$$r(x_1, x_2, x_3, x_4) := \frac{(x_1 - x_2)(x_3 - x_4)}{(x_1 - x_4)(x_2 - x_3)}, \qquad r(\infty, -1, 0, z) = z.$$

It satisfies the relations

$$r(x_1, x_2, x_3, x_4) = r(x_2, x_3, x_4, x_1)^{-1} = -1 - r(x_1, x_3, x_2, x_4).$$

Let $\mathcal{M}_{0,5}$ be the moduli space of configurations of five distinct points on $\mathbb{P}^1$ considered modulo the action of PGL_2. The moduli space $\overline{\mathcal{M}}_{0,5}$ is a smooth algebraic surface compactifying $\mathcal{M}_{0,5}$. There are 10 projective lines D_{ij}, $1 \leq i < j \leq 5$, inside of $\overline{\mathcal{M}}_{0,5}$, forming "the divisor at infinity" $D = \cup D_{ij}$. The line D_{ij} parametrizes configurations of points $(x_1, x_2, x_3, x_4, x_5)$ where "x_i collides with x_j". So $\overline{\mathcal{M}}_{0,5} - D = \mathcal{M}_{0,5}$.

We picture points $x_1, \ldots, x_5$ at the vertexes of an oriented pentagon, whose orientation agrees with the cyclic order of the points. Given a triangulation T of the pentagon, let us define a pair of rational functions on the surface $\overline{\mathcal{M}}_{0,5}$, assigned to the diagonals of the triangulation. Given a diagonal E, let z_1, z_2, z_3, z_4 be the configuration of four points at the vertexes of the rectangle containing E as a diagonal, so that z_1 is a vertex of E. Then we set

$$X_E^T := r(z_1, z_2, z_3, z_4).$$

Example. Given a configuration $(\infty, -1, 0, x, y)$, and taking the triangulation related to the vertex at ∞, we get functions $X = x, Y = (y - x)/x$, see Fig. 1.1.

Definition 3.1. $\mathcal{M}_{0,5}^{\mathrm{cyc}} := \overline{\mathcal{M}}_{0,5} - \cup_{c=1}^{5} D_{c,c+1}$, *where c is modulo* 5.

The space $\mathcal{M}_{0,5}^{\mathrm{cyc}}$ is determined by a choice of cyclic order of configurations of points $(x_1, \ldots, x_5)$.

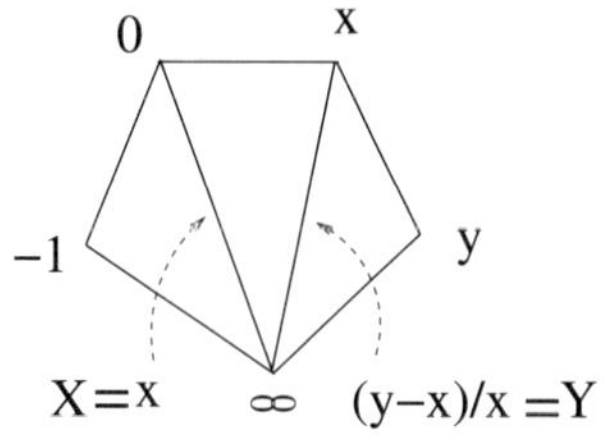

FIGURE 3

Let us define embeddings $\psi_c : \mathbb{C}^* \times \mathbb{C}^* \hookrightarrow \overline{\mathcal{M}}_{0,5}$ for $c \in \{1, \ldots, 5\}$. Set

$$\psi_1 : (X, Y) \longmapsto (\infty, -1, 0, X, X(1 + Y)).$$

One easily checks that it is an embedding. The map ψ_c is obtained from ψ_1 by the cyclic shift of the configuration of five points by $2c$. So it is also an embedding.

The following function is regular on the surface $\mathcal{M}_{0,5}^{\text{cyc}}$:

$$X_{a,b;c} := r(x_c, x_{c+1}, x_{c+2}, x_{c+3})^a r(x_c, x_{c+2}, x_{c+3}, x_{c+4})^b, \qquad a \geq 0, b \leq 0.$$

Indeed, the poles of the first factor are at the divisor $D_{c,c+1} \cup D_{c+2,c+3}$, and the poles of the second one are at the divisor $D_{c+2,c+3} \cup D_{c,c+4}$. The set of functions $\{X_{a,b;c}\}$ coincides with the one defined in Section 2.2. Indeed, one checks this for $c = 1$ using Fig. 3, and use equivariance with respect to the shifts and Fig. 4.

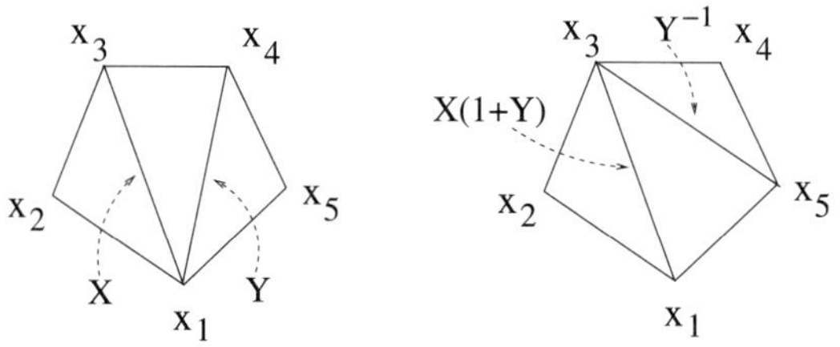

FIGURE 4. Change of the coordinates under a flip.

Theorem 3.2.

(i) *The surface $\mathcal{M}_{0,5}^{\text{cyc}}$ is the union of the five open subsets $\psi_c(\mathbb{C}^* \times \mathbb{C}^*)$ in $\overline{\mathcal{M}}_{0,5}$.*
(ii) *The functions $X_{a,b;c}$, where $a, b \in Z, a \geq 0, b \leq 0$ and c is mod 5 form a basis of the space of regular functions on the surface $\mathcal{M}_{0,5}^{\text{cyc}}$.*

Proof. (i) Straightforward.

(ii) The algebra of regular functions on $\mathcal{M}_{0,5}$ is defined as follows. Take the configuration space $\text{Conf}_5(V_2)$ of 5-tuples of vectors $(v_1, \ldots, v_5)$ in generic position in a two-dimensional symplectic vector space V_2, modulo the SL_2-action.

The group $H := (\mathbb{C}^*)^5$ acts on it by multiplying each vector v_i by a number λ_i. Let Δ_{ij} is the area in V_2 of the parallelogram $\langle v_i, v_j \rangle$. Then

$$\mathbb{Z}[\mathcal{M}_{0,5}] = \mathbb{Z}[\mathrm{Conf}_5(V_2)]^H = \mathbb{Z}[\Delta_{ij}^{\pm 1}]^H. \tag{17}$$

The subspace $\mathbb{Z}[\mathcal{M}_{0,5}^{\mathrm{cyc}}]$ is spanned by the monomials

$$\prod_{1 \le i < j \le 5} \Delta_{ij}^{a_{ij}}, \tag{18}$$

where $a_{i,i+1} \in \mathbb{Z}$, and $a_{ij} \in \mathbb{Z}_{\ge 0}$ unless $j = i \pm 1 \bmod 5$. Write the integers a_{ij} on the diagonals and sides of the pentagon. Call them the *weights* and the corresponding picture the *chord diagram.* The H-invariance means that the sum of the weights assigned to the edges and sides sharing a vertex is 0. We erase diagonals of weight 0. A monomial is *regular* if its chord diagram has no intersecting diagonals. Using the Plücker relations $\Delta_{ac}\Delta_{bd} = \Delta_{ab}\Delta_{cd} + \Delta_{ad}\Delta_{bc}$, $1 \le a < b < c < d \le 5$, and arguing by induction on the sum of the products of the weights of diagonals in the intersection points, we reduce any sum of monomials (18) to a sum of the regular ones. An easy argument with the "sum of the weights at a vertex equals zero" equations shows that for a regular monomial there exists a vertex of the pentagon such that its weights are as in Fig. 5. So the functions $X_{a,b;c}$ span the space of regular functions on $\mathcal{M}_{0,5}^{\mathrm{cyc}}$. To check that they are linearly independent, look at the monomials with maximum value of $a + b$. The theorem is proved. □

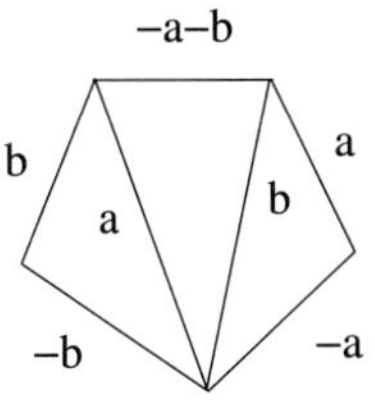

FIGURE 5. The weight diagram of a basis monomial; $a, b \ge 0$.

The quantized $\mathcal{M}_{0,5}^{\mathrm{cyc}}$ at roots of unity. Assume that q is a primitive N-th root of unity. Then the functions $x_{a,b;c} := X_{Na,Nb;c}$ generate the center of the algebra $\mathbb{L}_q$. In particular $x := X^N, y := Y^N$ are in the center. One checks ([FG2], Section 3) that the elements x, y behave under flips just like the corresponding coordinates on $\mathcal{M}_{0,5}^{\mathrm{cyc}}$. Therefore the spectrum of the center of $\mathbb{L}_q$ is identified with $\mathcal{M}_{0,5}^{\mathrm{cyc}}$. Restricting to an affine chart of $\mathcal{M}_{0,5}^{\mathrm{cyc}}$ with coordinates (α, β) we see that the localization of the algebra $\mathbb{L}_q$ at this chart is identified with the algebra generated by X, Y with the relations $X^N = \alpha, Y^N = \beta, XY = q^2 YX$. It is well know that it is a sheaf of central simple algebras over $\mathbb{C}^* \times \mathbb{C}^*$. So we get

Proposition 3.3. *Let q be a root of unity. Then the algebra $\mathbb{L}_q$ gives rise to a sheaf of Azumaya algebras on $\mathcal{M}_{0,5}^{\mathrm{cyc}}$.*

The real positive part of $\mathcal{M}_{0,5}^{\text{cyc}}$ is given by configurations of points $(\infty, -1, 0, x, y)$ with $0 < x < y$. Its closure in $\overline{\mathcal{M}}_{0,5}(\mathbb{R})$ is the pentagon. Its sides are real segments on the divisors $D_{c,c+1}$.

References

[Ba] E.W. Barnes, *The genesis of the double gamma function.* Proc. London Math. Soc. **31** (1899), 358–381.

[Bax] R. Baxter, *Exactly solved models in statistical mechanics.* Academic Press, 1982.

[CF] L. Chekhov and V. Fock, *Quantum Teichmüller spaces.* ArXive, math.QA/9908165.

[Fad1] L.D. Faddeev, *Discrete Heisenberg-Weyl group and modular group.* Lett. Math. Phys. **34** (1995), no. 3, 249–254.

[FK] L.D. Faddeev and R.M. Kashaev, *Quantum dilogarithm.* Mod.Phys.Lett. **A 9** (1994), 427. arXiv:hep-th/9310070.

[FKV] L.D. Faddeev, R.M. Kashaev and A.Yu. Volkov, *Strongly coupled quantum discrete Liouville theory. I: Algebraic approach and duality.* Commun. Math. Phys. **219** (2001), 199–219. arXiv:hep-th/0006156.

[FG2] V.V. Fock and A.B. Goncharov, *Cluster ensembles, quantization and the dilogarithm,* math.AG/0311245.

[FGII] V.V. Fock and A.B. Goncharov, *Cluster ensembles, quantization and the dilogarithm II: The intertwiner,* arXiv:math/0702398.

[FG3] V.V. Fock and A.B. Goncharov, *The quantum dilogarithm and representations of quantum cluster varieties.* arXiv:math/0702397.

[K] R. Kashaev, *Quantization of Teichmüller spaces and quantum dilogarithm.* Letters Math. Phys. **43** (1998). no. 2, 105–115. arXiv:q-alg/9705021.

[Sh] T. Shintani, *On a Kronecker limit formula for real quadratic fields.* J. Fac. Sci. Univ. Tokyo Sect 1A Math **24** (1977) 167–199.

[Wo] S.L. Woronowicz, *Quantum exponential function.* Rev. Math. Phys. **12** (2000), no. 6, 873–920.

Alexander B. Goncharov
Brown University
Dept of Mathematics
151 Thayer street
Providence RI 02912
USA
e-mail: sasha@math.brown.edu

Progress in Mathematics, Vol. 265, 429–438

Geodesic Flow on the Normal Congruence of a Minimal Surface

Brendan Guilfoyle and Wilhelm Klingenberg

Dedicated to the memory of Sasha Reznikov

Abstract. We study the geodesic flow on the normal line congruence of a minimal surface in $\mathbb{R}^3$ induced by the neutral Kähler metric on the space of oriented lines. The metric is lorentz with isolated degenerate points and the flow is shown to be completely integrable. In addition, we give a new holomorphic description of minimal surfaces in $\mathbb{R}^3$ and relate it to the classical Weierstrass representation.

Mathematics Subject Classification (2000). Primary 53B30; Secondary 53A25.

Keywords. Geodesic flow, minimal surface, oriented lines.

1. Introduction

In a recent paper [4] a neutral Kähler metric was introduced on the space $\mathbb{L}$ of oriented affine lines in $\mathbb{R}^3$. This metric is natural in the sense that it is invariant under the action induced on $\mathbb{L}$ by the Euclidean action on $\mathbb{R}^3$. Moreover, a surface in $\mathbb{L}$ is Lagrangian with respect to the associated symplectic structure iff there exist surfaces orthogonal to the associated 2-parameter family of oriented lines (or line congruence) in $\mathbb{R}^3$.

In this paper we characterise the set of oriented normals to a minimal surface in $\mathbb{R}^3$ and study the geodesic flow on the line congruence induced by this neutral Kähler metric. Along the way, we give a new holomorphic description of minimal surfaces in $\mathbb{R}^3$ and relate it to the classical Weierstrass representation.

The induced metric on a Lagrangian line congruence is either lorentz or degenerate. The null geodesics of the lorentz metric correspond to the principal foliation on the orthogonal surface and the degeneracy occurs precisely at umbilic points. We show that on the normal congruence of a minimal surface the geodesic flow is completely integrable and find the first integrals. Recently the geodesic flow on certain non-Lagrangian line congruences was investigated [5]. In that case, the metric was Riemannian with degeneracies along a curve.

The picture that emerges is this: every minimal surface carries a completely integrable dynamical system [2] [7] [9]. This is generated by geodesic motion of a lorentz metric whose null geodesics are the lines of curvatures and whose sources are the isolated umbilic points of the minimal surface. To illustrate this we compute the geodesics explicitly for the case of pure harmonic minimal surfaces. These have a unique index $-N$ umbilic point (for $N > 0$) and we show that the scattering angle for non-null geodesics is $2\pi/(N+2)$.

The next section describes the normal line congruence to a minimal surface - all background details on the geometry of the space of oriented affine lines in $\mathbb{R}^3$ can be found in [3] [4] and references therein. In [8] and papers quoted there, Reznikov studied this symplectic structure in another context. We relate the present work to the Weierstrass representation in Section 3. We then prove the result about the geodesic flow in Section 4, while we look at the case of pure harmonic minimal surfaces in the final section.

2. The Normal Line Congruence to a Minimal Surface

Let $\mathbb{L}$ be the space of oriented lines in $\mathbb{R}^3$ which we identify with the tangent bundle to the 2-sphere [6]. Let $\pi : \mathbb{L} \to \mathbb{P}^1$ be the canonical bundle and $(\mathbb{J}, \Omega, \mathbb{G})$ the neutral Kähler structure on $\mathbb{L}$ [4].

A line congruence is a 2-parameter family of oriented lines in $\mathbb{R}^3$, or equivalently, a surface $\Sigma \subset \mathbb{L}$. We are interested in characterising the line congruence formed by the oriented normal lines to a minimal surface S in $\mathbb{R}^3$:

Theorem 2.1. *A Lagrangian line congruence $\Sigma \subset \mathbb{L}$ is orthogonal to a minimal surface without flat points in $\mathbb{R}^3$ iff the congruence is the graph $\xi \mapsto (\xi, \eta = F(\xi, \bar{\xi}))$ of a local section of the canonical bundle with:*

$$\bar{\partial}\left(\frac{\partial \bar{F}}{(1+\xi\bar{\xi})^2}\right) = 0, \tag{2.1}$$

where (ξ, η) are standard holomorphic coordinates on $\mathbb{L} - \pi^{-1}\{$south pole$\}$ and ∂ represents differentiation with respect to ξ.

Proof. Let S be a minimal surface without flat points and Σ be its normal line congruence. Since the line congruence is not flat, it can be given by the graph of a local section. In terms of the canonical coordinates $(\xi, \eta = F(\xi, \bar{\xi}))$ the spin coefficients of such a line congruence are [3]:

$$\rho = \frac{\psi}{\partial F \bar{\partial} \bar{F} - \psi\bar{\psi}} \qquad \sigma = -\frac{\bar{\partial} \bar{F}}{\partial F \bar{\partial} \bar{F} - \psi\bar{\psi}},$$

with

$$\psi = \partial F + r - \frac{2\bar{\xi}F}{1+\xi\bar{\xi}}.$$

As this line congruence is orthogonal to a surface in $\mathbb{R}^3$, ρ is real, and, as the mean curvature vanishes, $\rho = 0$ on S.

Now, the graph of a Lagrangian section satisfies the following identity:

$$(1+\xi\bar{\xi})^2\bar{\partial}\left(\frac{\sigma_0}{(1+\xi\bar{\xi})^2}\right) = -\partial\psi, \tag{2.2}$$

where we have introduced $\sigma_0 = -\partial\bar{F}$. This follows from the fact that partial derivatives commute: firstly the left-hand side is

$$(1+\xi\bar{\xi})^2\bar{\partial}\left(\frac{\sigma_0}{(1+\xi\bar{\xi})^2}\right) = -\bar{\partial}\partial\bar{F} + \frac{2\xi\partial\bar{F}}{1+\xi\bar{\xi}},$$

while the right-hand side is

$$-\partial\psi = -\partial\left(\bar{\partial}\bar{F} + r - \frac{2\xi\bar{F}}{1+\xi\bar{\xi}}\right) = -\partial\bar{\partial}\bar{F} + \frac{2\xi\partial\bar{F}}{1+\xi\bar{\xi}}.$$

Here we have used the Lagrangian condition $\rho = \bar{\rho}$ and the equivalent local existence of a real function $r : \Sigma \to \mathbb{R}$ such that

$$\bar{\partial}r = \frac{2F}{(1+\xi\bar{\xi})^2}. \tag{2.3}$$

Thus, since $\rho = 0$, we have $\psi = 0$ and according to the identity (2.2), the normal congruence to a minimal surface must satisfy the holomorphic condition (2.1).

Conversely, suppose (2.1) holds for a Lagrangian line congruence Σ which is given by the graph of a local section. Then, by the identity (2.2) $\psi = C$ for some real constant C. As the orthogonal surfaces move along the line congruence in $\mathbb{R}^3$, ψ changes by $\psi \to \psi +$ constant. Thus there exists a surface S for which $\psi = 0$, and therefore $\rho = 0$, i.e., there is a minimal surface orthogonal to Σ. □

The previous theorem has two immediate consequences:

Corollary 2.2. *The normal congruence to a minimal surface is given by a local section F of the bundle $\pi : \mathbb{L} \to S^2$ with*

$$F = \sum_{n=0}^{\infty} 2\lambda_n\xi^{n+3} - \bar{\lambda}_n\bar{\xi}^{n+1}\left({\scriptstyle (n+2)(n+3)} + 2{\scriptstyle (n+1)(n+3)}\xi\bar{\xi} + {\scriptstyle (n+1)(n+2)}\xi^2\bar{\xi}^2\right),$$

for complex constants λ_n. The potential function $r : \Sigma \to \mathbb{R}$ satisfying (2.3) is:

$$r = -2\sum_{n=0}^{\infty}\frac{(3+n+(1+n)\xi\bar{\xi})(\lambda\xi^{n+2} + \bar{\lambda}\bar{\xi}^{n+2})}{1+\xi\bar{\xi}}.$$

Proof. Since the minimal surface condition is a holomorphic condition we can expand in a power series about a point:

$$\frac{\partial\bar{F}}{(1+\xi\bar{\xi})^2} = \sum_{n=0}^{\infty} a_n\xi^n.$$

This can be integrated term by term to

$$\bar{F}=\sum_{n=0}^{\infty}\beta_n\bar{\xi}^n+\alpha_n\xi^{n+1}\left(\tfrac{1}{n+1}+\tfrac{2}{n+2}\xi\bar{\xi}+\tfrac{1}{n+3}\xi^2\bar{\xi}^2\right),$$

for complex constants β_n. Now we impose the Lagrangian condition, that

$$(1+\xi\bar{\xi})\bar{\partial}\bar{F}-2\xi\bar{F}=\sum_{n=0}^{\infty}\beta_n\bar{\xi}^{n-1}(n+(n-2)\xi\bar{\xi})-2\alpha_n\xi^{n+2}\left(\tfrac{1}{(n+1)(n+2)}+\tfrac{1}{(n+2)(n+3)}\xi\bar{\xi}\right),$$

is real. This implies that $\beta_0=\beta_1=\beta_2=0$ and $(n+1)(n+2)(n+3)\beta_{n+3}=-2\bar{\alpha}_n$ for $n\geq 0$. Letting $\alpha_n=-(n+1)(n+2)(n+3)\lambda_n$ gives the stated result.

Finally it is easily checked that the expressions for r and F satisfy (2.3). □

On a minimal surface flat points are also umbilic points (and vice versa). Such points are now shown to be isolated:

Corollary 2.3. *Umbilic points on minimal surfaces are isolated and the index of the principal foliation about an umbilic point on a minimal surface is less than or equal to zero.*

Proof. An umbilic point is a point where $\partial\bar{F}=0$.

Moreover, the argument of $\bar{\partial}F$ gives the principal foliation of the surface [3]. Given that minimality implies the holomorphic condition (2.1), the zeros of $\partial\bar{F}$ are isolated and have index greater than or equal to zero. □

3. The Weierstrass Representation of a Minimal Surface

The classical Weierstrass representation constructs a minimal surface from a holomorphic curve in $\mathbb{L}$ [6]. The minimal surface in $\mathbb{R}^3$ determined by a local holomorphic section $\nu\mapsto(\nu,w(\nu))$ of the canonical bundle is given by

$$z=\frac{1}{2}w''-\frac{1}{2}\bar{\nu}^2\overline{w''}+\bar{\nu}\overline{w'}-\bar{w}$$

$$t=\frac{1}{2}\nu w''-\frac{1}{2}w'+\frac{1}{2}\bar{\nu}\overline{w''}-\frac{1}{2}\overline{w'},$$

where a prime represents differentiation with respect to the holomorphic parameter ν and $z=x^1+ix^2$, $t=x^3$ for Euclidean coordinates (x^1,x^2,x^3). The relationship between this and our approach is as follows.

Proposition 3.1. *The normal congruence of the minimal surface, in terms of the canonical coordinates ξ and η, is*

$$\xi=-\bar{\nu}\qquad\qquad \eta=\frac{1}{4}(1+\xi\bar{\xi})^3\frac{\partial^2}{\partial\bar{\xi}^2}\left(\frac{w}{1+\xi\bar{\xi}}\right)-\frac{1}{2}\bar{w}.$$

Proof. We have that

$$\frac{\partial}{\partial\nu} = \frac{1}{2}w'''\left(\frac{\partial}{\partial z} - \nu^2\frac{\partial}{\partial\bar{z}} + \nu\frac{\partial}{\partial t}\right).$$

The unit vector in $\mathbb{R}^3$ which corresponds to the point $\xi \in S^2$ is

$$e_0 = \frac{2\xi}{1+\xi\bar{\xi}}\frac{\partial}{\partial z} + \frac{2\bar{\xi}}{1+\xi\bar{\xi}}\frac{\partial}{\partial\bar{z}} + \frac{1-\xi\bar{\xi}}{1+\xi\bar{\xi}}\frac{\partial}{\partial t}.$$

The normal direction is given by the vanishing of the inner product of the preceding 2 vectors, which is easily seen to imply (for $w''' \neq 0$) $\xi = -\bar{\nu}$. At $w''' = 0$ there is an umbilic point. The remainder of the proposition follows from the incidence relation [3]:

$$\eta = \frac{1}{2}\left(z - 2t\xi - \bar{z}\xi^2\right). \qquad \square$$

The holomorphic functions of our method and that of the Weierstrass representation are related by

$$\frac{1}{(1+\xi\bar{\xi})^2}\frac{\partial F}{\partial\bar{\xi}} = \frac{1}{4}\frac{\partial^3 w}{\partial\bar{\xi}^3}.$$

4. The Geodesic Flow

We now look at the metric on Lagrangian sections:

Proposition 4.1. *The metric induced by the neutral Kähler metric on the graph of a Lagrangian section $\eta = F(\xi,\bar{\xi})$ is:*

$$ds^2 = \frac{2i}{(1+\xi\bar{\xi})^2}\left(\sigma_0 d\xi\otimes d\xi - \bar{\sigma}_0 d\bar{\xi}\otimes d\bar{\xi}\right),$$

where $\sigma_0 = -\partial\bar{F}$. Thus, for $|\sigma_0| \neq 0$ the metric is Lorentz and for $|\sigma_0| = 0$ the metric is degenerate.

Proof. The neutral Kähler metric has local expression (see equation (3.6) of [4]):

$$\mathbb{G} = \frac{2i}{(1+\xi\bar{\xi})^2}\left(d\eta\otimes d\bar{\xi} - d\bar{\eta}\otimes d\xi + \frac{2(\xi\bar{\eta}-\bar{\xi}\eta)}{1+\xi\bar{\xi}}d\xi\otimes d\bar{\xi}\right). \qquad (4.1)$$

We pull the metric back to the section:

$$\mathbb{G}|_\Sigma = \frac{2i}{(1+\xi\bar{\xi})^2}\left[\bar{\partial}F\, d\bar{\xi}\otimes d\bar{\xi} - \partial\bar{F}\, d\xi\otimes d\xi + \left(\partial F - \bar{\partial}\bar{F} + \frac{2(\xi\bar{\eta}-\bar{\xi}\eta)}{1+\xi\bar{\xi}}\right)d\xi\otimes d\bar{\xi}\right].$$

Now the Lagrangian condition says precisely that the coefficient of the $d\xi\otimes d\bar{\xi}$ term vanishes, and the result follows. $\square$

We turn now to the geodesic flow. Since the metric above is flat on the normal congruence of a minimal surface, this flow is completely integrable:

Proposition 4.2. *Consider the normal congruence to a minimal surface $\Sigma \subset \mathbb{L}$ given by $(\xi, \eta = F(\xi,\bar{\xi}))$. The geodesic flow on Σ is completely integrable with first* *integrals*

$$I_1 = \frac{2i}{(1+\xi\bar{\xi})^2}\left(\sigma_0\dot{\xi}^2 - \bar{\sigma}_0\dot{\bar{\xi}}^2\right) \qquad\qquad I_2 = \frac{\sigma_0^{\frac{1}{2}}\dot{\xi} + \bar{\sigma}_0^{\frac{1}{2}}\dot{\bar{\xi}}}{1+\xi\bar{\xi}}.$$

Proof. Consider the affinely parameterised geodesic $t \mapsto (\xi(t), \eta = F(\xi(t), \bar{\xi}(t)))$ on Σ with tangent vector

$$\mathrm{T} = \dot{\xi}\frac{\partial}{\partial\xi} + \dot{\bar{\xi}}\frac{\partial}{\partial\bar{\xi}}.$$

The geodesic equation $\mathrm{T}^j\nabla_j\mathrm{T}^k = 0$, projected onto the ξ coordinate is

$$\ddot{\xi} + \Gamma^{\xi}_{\xi\xi}\dot{\xi}^2 + 2\Gamma^{\xi}_{\xi\bar{\xi}}\dot{\xi}\dot{\bar{\xi}} + \Gamma^{\xi}_{\bar{\xi}\bar{\xi}}\dot{\bar{\xi}}^2 = 0.$$

For the induced metric (as given in Proposition 4.1) a straightforward calculation yields the Christoffel symbols:

$$\Gamma^{\xi}_{\xi\xi} = \frac{1}{2\sigma_0}\left(\partial\sigma_0 - \frac{2\sigma_0\bar{\xi}}{1+\xi\bar{\xi}}\right) \qquad \Gamma^{\bar{\xi}}_{\xi\xi} = \frac{1}{2\bar{\sigma}_0}\left(\bar{\partial}\sigma_0 - \frac{2\sigma_0\xi}{1+\xi\bar{\xi}}\right)$$

$$\Gamma^{\xi}_{\xi\bar{\xi}} = \frac{1}{2\sigma_0}\left(\bar{\partial}\sigma_0 - \frac{2\sigma_0\xi}{1+\xi\bar{\xi}}\right).$$

For the normal congruence of a minimal surface the holomorphic condition (2.1) implies that $\Gamma^{\bar{\xi}}_{\xi\xi} = 0$ and $\Gamma^{\xi}_{\xi\bar{\xi}} = 0$. Thus the geodesic equation reduces to

$$\ddot{\xi} = -\frac{1}{2}\partial\left[\ln\left(\frac{\sigma_0}{(1+\xi\bar{\xi})^2}\right)\right]\dot{\xi}^2.$$

The fact that I_1 is constant along a geodesic comes from the fact that the geodesic flow preserves the length of the tangent vector T^j. On the other hand, differentiating I_2 with respect to t:

$$\begin{aligned}
\dot{I}_2 &= \frac{1}{2}\left(\frac{\sigma_0}{(1+\xi\bar{\xi})^2}\right)^{-\frac{1}{2}}\left[\partial\left(\frac{\sigma_0}{(1+\xi\bar{\xi})^2}\right)\dot{\xi}^2 + \bar{\partial}\left(\frac{\sigma_0}{(1+\xi\bar{\xi})^2}\right)\dot{\xi}\dot{\bar{\xi}}\right] \\
&\quad + \frac{1}{2}\left(\frac{\bar{\sigma}_0}{(1+\xi\bar{\xi})^2}\right)^{-\frac{1}{2}}\left[\partial\left(\frac{\bar{\sigma}_0}{(1+\xi\bar{\xi})^2}\right)\dot{\xi}\dot{\bar{\xi}} + \bar{\partial}\left(\frac{\bar{\sigma}_0}{(1+\xi\bar{\xi})^2}\right)\dot{\bar{\xi}}^2\right] \\
&\quad + \frac{\sigma_0^{\frac{1}{2}}}{1+\xi\bar{\xi}}\ddot{\xi} + \frac{\bar{\sigma}_0^{\frac{1}{2}}}{1+\xi\bar{\xi}}\ddot{\bar{\xi}} \\
&= \frac{1}{2}\left(\frac{\sigma_0}{(1+\xi\bar{\xi})^2}\right)^{-\frac{1}{2}}\partial\left(\frac{\sigma_0}{(1+\xi\bar{\xi})^2}\right)\dot{\xi}^2 + \frac{1}{2}\left(\frac{\bar{\sigma}_0}{(1+\xi\bar{\xi})^2}\right)^{-\frac{1}{2}}\bar{\partial}\left(\frac{\bar{\sigma}_0}{(1+\xi\bar{\xi})^2}\right)\dot{\bar{\xi}}^2 \\
&\quad - \frac{1}{2}\frac{\sigma_0^{\frac{1}{2}}}{1+\xi\bar{\xi}}\partial\left[\ln\left(\frac{\sigma_0}{(1+\xi\bar{\xi})^2}\right)\right]\dot{\xi}^2 - \frac{1}{2}\frac{\bar{\sigma}_0^{\frac{1}{2}}}{1+\xi\bar{\xi}}\bar{\partial}\left[\ln\left(\frac{\bar{\sigma}_0}{(1+\xi\bar{\xi})^2}\right)\right]\dot{\bar{\xi}}^2 \\
&= 0,
\end{aligned}$$

as claimed. □

5. Examples: The Pure Harmonics

We now consider the geodesic flow for the pure harmonics, that is, the minimal surfaces with

$$\frac{\partial \bar{F}}{(1+\xi\bar{\xi})^2} = \alpha_N \xi^N,$$

for some $N \in \mathbb{N}$. These have isolated umbilic points of index $-N < 0$ at $\xi = 0$, which is also an $N+1$ –fold branch point. By a rotation we can make α_N real and rescaling the first integrals we will set it to 1.

By Proposition 4.2 above, the first integrals are

$$\mathrm{I}_1 = 2i\left(\xi^N \dot{\xi}^2 - \bar{\xi}^N \dot{\bar{\xi}}^2\right) \qquad \mathrm{I}_2 = \xi^{N/2}\dot{\xi} + \bar{\xi}^{N/2}\dot{\bar{\xi}}.$$

These can be integrated to:

$$\frac{4i\mathrm{I}_2}{N+2}\left(\xi^{\frac{N+2}{2}} - \bar{\xi}^{\frac{N+2}{2}}\right) = \mathrm{I}_1 t + c_1 \qquad \frac{2}{N+2}\left(\xi^{\frac{N+2}{2}} + \bar{\xi}^{\frac{N+2}{2}}\right) = \mathrm{I}_2 t + c_2,$$

for real constants of integration c_1 and c_2.

For null geodesics $\mathrm{I}_1 = 0$, and if we let $\xi = Re^{i\theta}$ we get two sets of null geodesics (future- and past-directed) which are given implicitly by:

$$R^{\frac{N+2}{2}} \sin\left(\frac{N+2}{2}\right)\theta = c_1 \qquad R^{\frac{N+2}{2}} \cos\left(\frac{N+2}{2}\right)\theta = c_2.$$

For $N = 0$, these form a rectangular grid, while for $N > 0$ they form the standard index $-N$ foliation about the origin. Figures 1 and 2 below shows the $N = 1$ minimal surface, and the foliation of null geodesics about the index -1 umbilic.

For non-null geodesics $\mathrm{I}_1 \neq 0$ and $\mathrm{I}_2 \neq 0$. Then the geodesics can be written parametrically:

$$R^{\frac{N+2}{2}} \sin\left(\tfrac{N+2}{2}\right)\theta = -\frac{N+2}{8\mathrm{I}_2}\left(\mathrm{I}_1 t + c_1\right) \qquad R^{\frac{N+2}{2}} \cos\left(\tfrac{N+2}{2}\right)\theta = \frac{N+2}{4}\left(\mathrm{I}_2 t + c_2\right).$$

The umbilic acts as a source of repulsion and the scattering angle can be found by noting that

$$\tan\left(\tfrac{N+2}{2}\right)\theta = -\frac{1}{2\mathrm{I}_2}\frac{\mathrm{I}_1 t + c_1}{\mathrm{I}_2 t + c_2}.$$

Thus, as $t \to \pm\infty$ we have $\tan\left(\frac{N+2}{2}\right)\theta \to -\frac{\mathrm{I}_1}{2\mathrm{I}_2^2}$. We deduce then that the scattering angle is $\frac{2\pi}{N+2}$. Figure 3 illustrates the scattering angle for a non-null geodesic about the $N = 1$ umbilic.

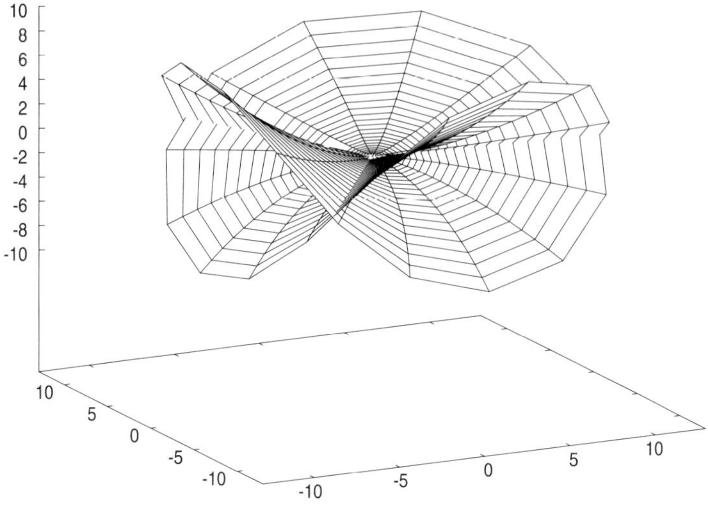

FIGURE 1

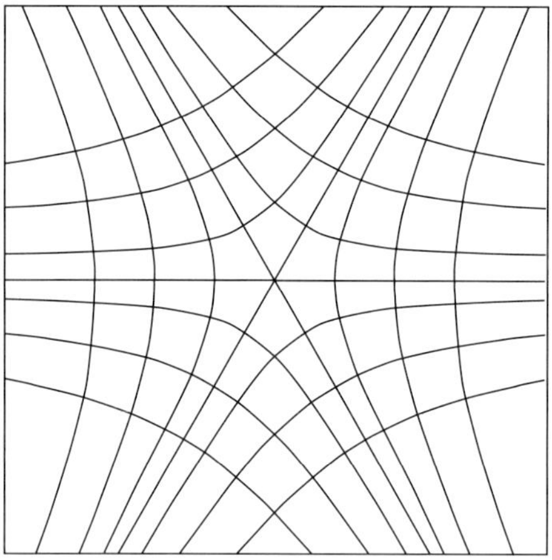

FIGURE 2

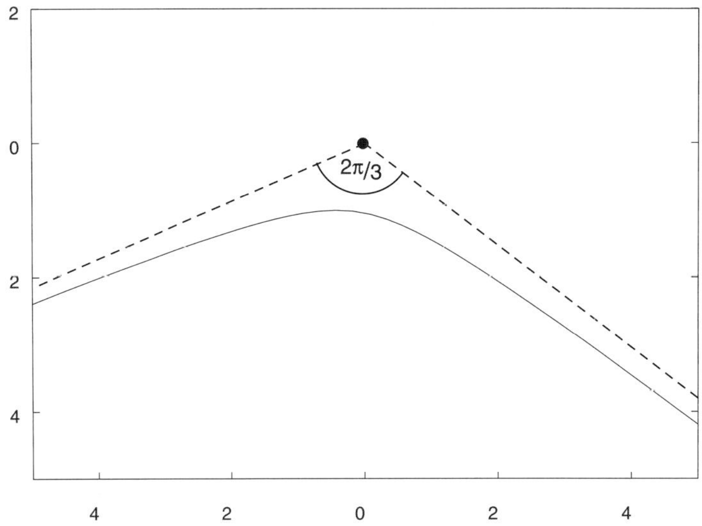

FIGURE 3

References

[1] V. Arnold and A. Givental, *Symplectic geometry*, in Encycl. of Math. Sci. **4**, Springer-Verlag, New York, (1990) 1–136.

[2] M.A. Guest, *Harmonic maps, loop groups and integrable systems*, LMS Student Texts **38**, Cambridge University Press, Cambridge (1997).

[3] B. Guilfoyle and W. Klingenberg, *Generalised surfaces in* $\mathbb{R}^3$, Math. Proc. of the R.I.A. **104A(2)** (2004) 199–209.

[4] B. Guilfoyle and W. Klingenberg, *An indefinite Kähler metric on the space of oriented lines*, J. London Math. Soc. **72**, (2005) 497–509.

[5] B. Guilfoyle and W. Klingenberg, *Geodesic flow on global holomorphic sections of* TS^2, Bull. Belg. Math. Soc. **13** (2006) 1–9.

[6] N.J. Hitchin, *Monopoles and geodesics*, Comm. Math. Phys. **83**, (1982) 579–602.

[7] J. Moser, *Various aspects of integrable Hamiltonian systems*, in Dynamical systems (C.I.M.E. Summer School, Bressanone, 1978) Progr. Math. **8**, Birkhäuser, Boston, Mass., (1980) 233–289.

[8] A. Reznikov, *Affine symplectic geometry. I: Applications to geometric inequalities*, Isr. J. Math. 80, (1992) 207–224.

[9] S. Tabachnikov, *Projectively equivalent metrics, exact transverse line fields and the geodesic flow on the ellipsoid*, Comment. Math. Helv. **74**, (1999) 306–321.

Brendan Guilfoyle
Department of Computing and Mathematics
Institute of Technology, Tralee
Clash
Tralee
Co. Kerry
Ireland
e-mail: brendan.guilfoyle@ittralee.ie

Wilhelm Klingenberg
Department of Mathematical Sciences
University of Durham
Durham DH1 3LE
United Kingdom
e-mail: wilhelm.klingenberg@durham.ac.uk

Progress in Mathematics, Vol. 265, 439–485

The Chern Character of a Parabolic Bundle, and a Parabolic Corollary of Reznikov's Theorem

Jaya N. Iyer and Carlos T. Simpson

Dedicated to Sasha Reznikov

Abstract. In this paper, we obtain an explicit formula for the Chern character of a locally abelian parabolic bundle in terms of its constituent bundles. Several features and variants of parabolic structures are discussed. Parabolic bundles arising from logarithmic connections form an important class of examples. As an application, we consider the situation when the local monodromies are semi-simple and are of finite order at infinity. In this case the parabolic Chern classes of the associated locally abelian parabolic bundle are deduced to be zero in the rational Deligne cohomology in degrees ≥ 2.

Mathematics Subject Classification (2000). Primary 14C25; Secondary 14D05, 14D20, 14D21, 14F05.

Keywords. Chow groups, logarithmic connections, parabolic bundles.

1. Introduction

Parabolic bundles were introduced by Mehta and Seshadri [Me-Se] [Se] over curves and the definition was extended over higher-dimensional varieties by Maruyama and Yokogawa [Ma-Yo], Biswas [Bi], Li [Li], Steer-Wren [Sr-Wr], Panov [Pa] and Mochizuki [Mo2]. A *parabolic bundle* F on a variety X is a collection of vector bundles F_α, indexed by a set of *weights*, i.e., α runs over a multi-indexing set $\frac{1}{n}\mathbb{Z}\times\frac{1}{n}\mathbb{Z}\times\cdots\times\frac{1}{n}\mathbb{Z}$, for some denominator n. Further, all the bundles F_α restrict on the complement $X-D$ of some normal crossing divisor $D=D_1+\cdots+D_m$ to the same bundle, the index α is an m-tuple and the F_α satisfy certain normalization/support hypothesis (see §2.1).

This work is a sequel to [Iy-Si], which in turn was motivated by Reznikov's work on characteristic classes of flat bundles [Re], [Re2]. As a long-range goal we would like to approach the Esnault conjecture [Es2] that the Chern classes of Deligne canonical extensions of motivic flat bundles vanish in the rational Chow groups. Reznikov's work shows the vanishing of an important piece of these classes, over the subset of definition of a flat bundle. We think that it should be possible to define secondary classes over a completed variety for flat connections which are quasi-unipotent at infinity, and to extend Reznikov's results to this case. At the end of this paper we treat a first and essentially easy case, when the monodromy transformations at infinity have finite order. We hope to treat the general case in the future and regain an understanding of characteristic classes such as Sasha Reznikov had.

A different method for obtaining a very partial result on the Esnault conjecture, removing a hypothesis from the GRR formula of Esnault-Viehweg [Es-Vi3], was done in [Iy-Si]. There we used a definition of the Chern character obtained from the correspondence between locally abelian parabolic bundles and usual vector bundles on a particular Deligne–Mumford stack denoted by $Z_m = X\langle \frac{D_1}{n}, \ldots, \frac{D_m}{n}\rangle$ (see [Bo], [Iy-Si, §2.3], [Cad], [Ma-Ol], [Me-Se], [Bd], [Bi]). The Chern character of F is defined to be the Chern character of the corresponding vector bundle on this stack. This was sufficient for our application in [Iy-Si], however it is clearly unsatisfactory to have only an abstract definition rather than a formula.

The aim of this note is to give an explicit formula for the Chern character in terms of the Chern character of the constituent bundles F_α and the divisor components D_i in the rational Chow groups of X. This procedure, using a DM stack to define the Chern character and then giving a computation, was first done for the parabolic degree by Borne in [Bo], however his techniques are different from ours. The parabolic aspect of the problem of extending characteristic classes for bundles from an open variety to its completion should in the future form a small part of a generalization of Reznikov's work and we hope the present paper can contribute in that direction.[1]

With our fixed denominator n, introduce the notation

$$[a_1, \ldots, a_m] := (\frac{a_1}{n}, \ldots, \frac{a_m}{n})$$

for multi-indices, so the parabolic structure is determined by the bundles $F_{[a_1,\ldots,a_m]}$ for $0 \leq a_i < n$ with a_i integers.

We prove the following statement.

Theorem 1.1. *Suppose F is a locally abelian parabolic bundle on X with respect to $D_1, \ldots, D_m$, with n as the denominator. Then we have the following formula for the Chern character of F:*

$$\mathrm{ch}(F) = \frac{\sum_{a_1=0}^{n-1} \cdots \sum_{a_m=0}^{n-1} e^{-\sum_{i=1}^{m} \frac{a_i}{n} D_i} \mathrm{ch}(F_{[a_1,\ldots,a_m]})}{\sum_{a_1=0}^{n-1} \cdots \sum_{a_m=0}^{n-1} e^{-\sum_{i=1}^{m} \frac{a_i}{n} D_i}}. \tag{1}$$

[1] Note added in July 2007: we have now been able to treat the case of unipotent monodromy along a smooth divisor, see our preprint "Regulators of canonical extensions are torsion: the smooth divisor case".

In other words, the Chern character of F is the weighted average of the Chern characters of the component bundles, with weights $e^{-\sum_{i=1}^{m}\frac{a_i}{n}D_i}$.

The proof is by showing that the parabolic bundle obtained by twisting F by a direct sum of line bundles involving D_i is *componentwise isomorphic* to a direct sum of the constituent bundles $F_{[a_1,\ldots,a_m]}$ twisted by parabolic line bundles involving D_i (see Corollary 5.7). The proof is concluded by proving the main theorem on the invariance of the Chern character under componentwise Chow isomorphism (see Theorem 2.9). It says: given locally abelian parabolic bundles F and G whose constituent bundles $F_{[a_1,\ldots,a_m]}$ and $G_{[a_1,\ldots,a_m]}$ have the same Chern character, for all a_i with $0 \leq a_i < n$, then F and G also have the same Chern character in the rational Chow groups of X.

We also give variants of the Chern character formula. One can associate a parabolic structure F to a vector bundle E on X and given filtrations on the restriction of E on the divisor components of D (see §2, §6). If X is a surface this is automatically locally abelian, but in higher dimensions it is not always the case (see Lemma 2.3). When the structure is locally abelian, we obtain a formula for $\mathrm{ch}(F)$ which involves $\mathrm{ch}(E)$ and terms under the Gysin maps on the multiple intersections of the divisor components of D (see Corollary 7.4 and Corollary 7.5). The shape of the formula depends on the way the filtrations intersect on the multiple intersections of the divisor components.

In §6 we give two easy counterexamples which show that the Chern character of a parabolic bundle cannot be obtained easily from just the Chern character of the underlying bundle and that of its filtrations taken separately, nor from the data of a filtration of subsheaves indexed by a single parameter for the whole divisor (Maruyama-Yokogawa's original definition [Ma-Yo]). These show that in order to obtain a good formula we should consider all of the bundles $F_{[a_1,\ldots,a_m]}$. This version of parabolic structure was first introduced by Li [Li], Steer-Wren [Sr-Wr] and Mochizuki [Mo2].

We treat parabolic bundles with real weights in §8. The aim is to define pull-back of a locally abelian parabolic bundle as a locally abelian parabolic bundle. This is done by approximating with the rational weights case (see Lemma 8.5). Properties like functoriality, additivity and multiplicativity of the Chern character are also discussed. In §9, on a smooth surface, parabolic structures at multiple points are discussed and a Chern character formula is obtained. Logarithmic connections were discussed by Deligne in [De]. We discuss some filtrations defined by the residue transformations of the connection at infinity. When the eigenvalues of the residues are rational and non-zero, a locally abelian parabolic bundle was associated in [Iy-Si], and this construction is considered further in §10. When the residues are nilpotent, we continue in §9 with something different: assign arbitrary weights to the pieces of the monodromy weight filtration of the nilpotent residue operators, creating a family of parabolic bundles indexed by the choices of weights. If X is a surface then these are automatically locally abelian, and as an example we make explicit the computation of the parabolic Chern character $\mathrm{ch}(F)$ in the case of a weight one unipotent Gauss-Manin system F, see Lemma 9.3.

In §10 we consider the extension of Reznikov's theory to flat bundles with finite order monodromy at infinity. Such bundles may be considered as flat bundles over a DM-stack of the form $Z_m = X\langle \frac{D_1}{n}, \ldots, \frac{D_m}{n}\rangle$, and Reznikov's theorem [Re2] applies directly (or alternatively, over a finite Kawamata covering). The only knowledge which we can add is that our formula of Theorem 1.1 gives parabolic Chern classes in terms of the parabolic structure on X deduced from the flat bundle, and Reznikov's theorem can be stated as vanishing of these classes. This might have computational content in explicit examples.

Proposition 1.2 (Parabolic corollary of Reznikov's theorem). *Suppose (E_U, ∇_U) is a flat bundle on U with rational and semisimple residues, or equivalently the monodromy transformations at infinity are of finite order. Let F denote the corresponding locally abelian parabolic bundle. Recall that $F_{[a_1,\ldots,a_m]}$ is the unique bundle on X extending E_U such that the residues of the connection over D_i have eigenvalues in the interval $[-a_i, 1-a_i)$. Using the same formula as (1) define the Deligne Chern character of F by*

$$\mathrm{ch}^{\mathcal{D}}(F) := \frac{\sum_{a_1=0}^{n-1}\cdots\sum_{a_m=0}^{n-1} e^{-\sum_{i=1}^{m}\frac{a_i}{n}D_i}\,\mathrm{ch}^{\mathcal{D}}(F_{[a_1,\ldots,a_m]})}{\sum_{a_1=0}^{n-1}\cdots\sum_{a_m=0}^{n-1} e^{-\sum_{i=1}^{m}\frac{a_i}{n}\mathrm{ch}^{\mathcal{D}}(D_i)}}$$

in the rational Deligne cohomology, and the Chern classes $\mathrm{c}_p^{\mathcal{D}}(F)$ by the usual formula. Then the classes $\mathrm{c}_p^{\mathcal{D}}(F)$ for all $p \geq 2$ vanish. This is equivalent to saying that $\mathrm{ch}_p^{\mathcal{D}}(F) = \mathrm{ch}_1^{\mathcal{D}}(F)^p/p!$.

Acknowledgements. We thank P. Deligne for having useful discussions. The first named author is supported by NSF. This material is based upon work supported by the National Science Foundation under agreement No. DMS-0111298. Any opinions, findings and conclusions or recommendations expressed in this material are those of the authors and do not necessarliy reflect the views of the National Science Foundation.

2. Parabolic bundles

Let X be a smooth projective variety over an algebraically closed field of characteristic zero, with D a normal crossing divisor on X. Write $D = D_1 + \cdots + D_m$ where D_i are the irreducible smooth components and meeting transversally. We use an approach to parabolic bundles based on multi-indices $(\alpha_1, \ldots, \alpha_m)$ of length equal to the number of components of the divisor. This approach, having its origins in the original paper of Mehta and Seshadri [Me-Se], was introduced in higher dimensions by Li [Li], Steer-Wren [Sr-Wr], Mochizuki [Mo2] and contrasts with the Maruyama-Yokogawa definition which uses a single index [Ma-Yo].

2.1. Definition

A *parabolic bundle* on (X, D) is a collection of vector bundles F_α indexed by multi-indices $\alpha = (\alpha_1, \ldots, \alpha_k)$ with $\alpha_i \in \mathbb{Q}$, together with inclusions of sheaves of $\mathcal{O}_X$-modules

$$F_\alpha \hookrightarrow F_\beta$$

whenever $\alpha_i \leq \beta_i$ (a condition which we write as $\alpha < \beta$ in what follows), subject to the following hypotheses:

- (normalization/support) let δ^i denote the multiindex $\delta^i_i = 1, \ \ \delta^i_j = 0, \ i \neq j$, then $F_{\alpha+\delta^i} = F_\alpha(D_i)$ (compatibly with the inclusion); and
- (semicontinuity) for any given α there exists $c > 0$ such that for any multi-index ε with $0 \leq \varepsilon_i < c$ we have $F_{\alpha+\varepsilon} = F_\alpha$.

It follows from the normalization/support condition that the quotient sheaves F_α/F_β for $\beta \leq \alpha$ are supported in a schematic neighborhood of the divisor D, and indeed if $\beta \leq \alpha \leq \beta + \sum n_i \delta^i$ then F_α/F_β is supported over the scheme $\sum_{i=1}^k n_i D_i$. Let $\delta := \sum_{i=1}^k \delta^i$. Then

$$F_{\alpha-\delta} = F_\alpha(-D)$$

and $F_\alpha/F_{\alpha-\delta} = F_\alpha|_D$.

The semicontinuity condition means that the structure is determined by the sheaves F_α for a finite collection of indices α with $0 \leq \alpha_i < 1$, the *weights.*

A parabolic bundle is called *locally abelian* if in a Zariski neighbourhood of any point $x \in X$ there is an isomorphism between F and a direct sum of parabolic line bundles. By Lemma 3.3 of [Iy-Si], it is equivalent to require this condition on an etale neighborhood.

The locally abelian condition first appeared in Mochizuki's paper [Mo2], in the form of his notion of *compatible filtrations.* The condition that there be a global frame splitting all of the parabolic filtrations appears as the conclusion of his Corollary 4.4 in [Mo, §4], cf Theorem 2.2 below. A somewhat similar compatibility condition appeared earlier in Li's paper [Li, Definition 2.1(a)], however his condition is considerably stronger than that of [Mo2] and some locally abelian cases such as Case B in §7.1 below will not be covered by [Li]. The notion of existence of a local frame splitting all of the filtrations, which is our definition of "locally abelian", did occur as the conclusion of [Li, Lemma 3.2].

Fix a single n which will be the denominator for all of the divisor components, to make notation easier. Let m be the number of divisor components, and introduce the notation

$$[a_1, \ldots, a_m] := (\frac{a_1}{n}, \ldots, \frac{a_m}{n})$$

for multi-indices, so the parabolic structure is determined by the bundles $F_{[a_1,\ldots,a_m]}$ for $0 \leq a_i < n$ with a_i integers.

2.2. Parabolic bundles by filtrations

Historically the first way of considering parabolic bundles was by filtrations on the restriction to divisor components [Me-Se], [Se], see also [Ma-Yo], [Bi], [IIS] [Li] [Sr-Wr] [Mo2] [Pa]. Suppose we have a vector bundle E and filtrations of $E|_{D_i}$ by saturated subbundles:

$$E|_{D_i} = F^i_0 \supset F^i_{-1} \supset \cdots \supset F^i_{-n} = 0$$

for each i, $1 \leq i \leq m$.

Consider the kernel sheaves for $-n \leq j \leq 0$,

$$0 \longrightarrow \overline{F^i_j} \longrightarrow E \longrightarrow \frac{E|_{D_i}}{F^i_j} \longrightarrow 0$$

and define

$$F_{[a_1,a_2,\ldots,a_m]} := \cap_{i=1}^m \overline{F^i_{a_i}}, \tag{2}$$

for $-n \leq a_i \leq 0$. In particular $F_{[0,\ldots,0]} = E$. This can then be extended to sheaves defined for all values of a_i using the normalization/support condition

$$F_{[a_1,\ldots,a_i+n,\ldots,a_m]} = F_{[a_1,a_2,\ldots,a_m]}(D_i). \tag{3}$$

We call this a parabolic structure *given by filtrations.*

Conversely, suppose we are given a parabolic structure $F_\cdot$ as described in (2.1) when all the component sheaves $F_{[a_1,\ldots,a_m]}$ are vector bundles. Set $E := F_{[0,\ldots,0]}$, and note that

$$E|_{D_i} = E/F_{[0,\ldots,-n,\ldots,0]}$$

where $-n$ is put in the ith place. The image of $F_{[0,\ldots,-a_i,\ldots,0]}$ in $E|_{D_i}$ is a subsheaf, and we assume that it is a saturated subbundle. This gives a parabolic structure "by filtrations". We can recover the original parabolic structure $F_\cdot$ by the intersection formula (2).

We feel that these constructions only make good sense under the locally abelian hypothesis. We note some consequences of the locally abelian property.

Lemma 2.1. *Suppose* $\{F_{[a_1,a_2,\ldots,a_m]}\}_{-n\leq a_i\leq 0}$ *define a locally abelian parabolic bundle on X with respect to* $(D_1,\ldots,D_m)$. *Let* $E := F_{[0,\ldots,0]}$, *which is a vector bundle on X. Then F comes from a construction as above using unique filtrations of* $E|_{D_i}$ *and we have the following properties:*

(a) *the* $F_{[a_1,a_2,\ldots,a_m]}$ *are locally free;*

(b) *for each k and collection of indices* $(i_1,\ldots,i_k)$, *at each point in the k-fold intersection* $P \in D_{i_1} \cap \cdots \cap D_{i_k}$ *the filtrations* $F^{i_1}_\cdot, \ldots, F^{i_k}_\cdot$ *of* E_P *admit a common splitting, hence the associated-graded*

$$Gr^{F^{i_1}}_{j_1} \cdots Gr^{F^{i_k}}_{j_k}(E_P)$$

is independent of the order in which it is taken (see [De2]*); and*

(c) *the functions*

$$P \mapsto \dim Gr^{F^{i_1}}_{j_1} \cdots Gr^{F^{i_k}}_{j_k}(E_P)$$

are locally constant functions of P on the multiple intersections $D_{i_1}\cap\cdots\cap D_{i_k}$.

Proof. Direct. □

The above conditions are essentially what Mochizuki has called "compatibility" of the filtrations [Mo, §4], and he shows that they are sufficient for obtaining a compatible local frame. Compare with [Li, Lemma 3.2] where the proof is much shorter because the compatibility condition in the hypothesis is stronger.

Theorem 2.2 (Mochizuki [Mo, Cor. 4.4]**).** *Suppose given a parabolic structure which is a collection of sheaves $F_{[a_1,a_2,\ldots,a_m]}$ obtained from filtrations on a bundle E as above. If these satisfy conditions* (a), (b) *and* (c) *of the previous lemma, then the parabolic structure is locally abelian.*

The situation is simpler in the case of surfaces which we describe here.

Lemma 2.3. *Suppose X is a surface with a normal crossings divisor $D = D_1 + \cdots + D_m \subset X$. Suppose given data of a bundle E and strict filtrations of $E|_{D_i}$ as in Lemma* 2.1. *Then this data defines a locally abelian parabolic bundle on (X, D).*

Proof. One way to prove this is to use the correspondence with bundles on the DM-stack covering $Z := X\langle \frac{D_1}{n}, \ldots, \frac{D_m}{n} \rangle$ (see [Iy-Si, Lemma 2.3]). Let Z' be the complement of the intersection points of the divisor. On Z' the given filtrations define a vector bundle, as can be seen by applying the correspondence of [Bo] [Iy-Si] in codimension 1, or more concretely just by using the filtrations to make a sequence of elementary transformations. Then, since Z is a smooth surface, this bundle extends to a unique bundle on Z, which corresponds to a locally abelian parabolic bundle on X [Iy-Si].

Another way to prove this is to note that there are only double intersections. At a point P where D_i and D_j intersect, the filtrations coming from D_i and D_j have a common splitting. This can then be extended along both D_i and D_j as a splitting of the respective filtrations, and extended in any way to the rest of X. The resulting direct sum decomposition splits the parabolic structure. This is illustrated by an example in §7.1. □

We mention here a more general notation used by Mochizuki [Mo2, §3.1] for parabolic bundles given by a filtration, starting with an origin $\mathbf{c} = (c_1, \ldots, c_m)$ which may be different from $(0, \ldots, 0)$. In this case, the underlying bundle is

$$E := F_{[c_1,\ldots,c_m]}$$

and the filtrations on $E|_{D_i}$ are denoted F^i_j indexed by $c_i - n \leq j \leq c_i$ with $F^i_{c_i} = E|_{D_i}$ and $F^i_{c_i-n} = 0$. We can go between different values of $\mathbf{c}$ by tensoring with parabolic line bundles.

2.3. Parabolic sheaves in the Maruyama-Yokogawa notation

In their original definition of parabolic structures on higher-dimensional varieties, Maruyama and Yokogawa considered the general notion of parabolic sheaf with respect to a single divisor, even if the divisor is not smooth [Ma-Yo]. Call this a *MY parabolic struture.* We can apply their definition to the full divisor

$D = D_1 + \cdots + D_m$. This is what was done for example in Biswas [Bi], Borne [Bo] and many other places. Of course for the case of curves, the two are completely equivalent because a divisor is always a disjoint union of its components; multi-indexed divisors were used by Mehta and Seshadri [Me-Se]. Some of the first places where multi-indexed divisors were used in higher dimensions were in Li [Li], Steer-Wren [Sr-Wr], Panov [Pa] and Mochizuki [Mo2]. In the MY case the parabolic structure is given by a collection of sheaves indexed by a single parameter F^{α} for $\alpha \in \mathbb{Q}$, with $F^{\alpha+1} = F^{\alpha}(D)$. We use upper indexing to distinguish this from our notation (although they would be the same in the case of a single smooth divisor). If $F_{\cdot}$ is a parabolic structure according to our notation, then we get a MY-parabolic structure by setting

$$F^{\alpha} := F_{\alpha,\ldots,\alpha}.$$

Conversely, given a MY-parabolic structure $F^{\cdot}$, if we assume that $E := F^0$ is a bundle, then the images of $F^{-\frac{a_i}{n}}$ in E_{D_i} define subsheaves at generic points of the components D_i, which we can complete to saturated subsheaves everywhere. If $F^{\cdot}$ is locally abelian (that is to say, locally a direct sum of MY-parabolic line bundles) then these saturated subsheaves are subbundles and we recover the parabolic structure via filtrations, hence the parabolic structure $F_{\cdot}$ in this way. This construction is tacitly used by Biswas in [Bi2, pp. 599, 602], although he formally sticks to the MY-parabolic notation.

In the locally abelian case, all of these different points of view permit us to represent the same objects and going between them by the various constructions we have outlined, is a commutative process in the sense that by any path we get back to the same objects in each notation. We don't attempt to identify the optimal set of hypotheses, weaker than locally abelian, on the various structures which would allow to give a more general statement of this sort of commutation of the various constructions. This doesn't seem immediately relevant since, for now, it doesn't seem clear what is the really good notion of parabolic sheaf.

2.4. Parabolic bundles on a DM-stack

Recall from [Bo] [Cad] [Ma-Ol] [Iy-Si] that given (X, D) and a denominator n, we can form a DM-stack denoted $Z := X\langle\frac{D_1}{n}, \ldots, \frac{D_m}{n}\rangle$, and there is an equivalence of categories between parabolic bundles on (X, D) with denominator n, and vector bundles on the DM-stack Z. The Chern character will be defined using this equivalence, and we would like to analyse it by an induction on the number of divisor components m. Thus, we are interested in intermediate cases of parabolic bundles on DM-stacks.

We can carry out all the above constructions in the case when X is a DM stack and D_i are smooth divisors, i.e., smooth closed substacks of codimension 1, meeting transversally on X.

Lemma 2.4. *The construction $(X, D) \mapsto Z := X\langle\frac{D}{n}\rangle$ makes sense for any smooth DM stack X and smooth divisor $D \subset X$. The stack Z is then again smooth with a morphism of stacks $Z \to X$.*

Proof. Since the construction [Cad] [Ma-Ol] [Bo] of the DM-stack $X\langle\frac{D}{n}\rangle$ when X is a variety is local for the étale topology (see [Iy-Si, §2.2]), the same construction works when X is a DM-stack. □

Let $Z_k := X\langle\frac{D_1}{n}, \ldots, \frac{D_k}{n}\rangle$. This is a DM-stack (see [Cad] [Ma-Ol] [Bo] [Iy-Si, §2.2]) and we have maps

$$\ldots \to Z_k \to Z_{k-1} \to \ldots \to Z_0 = X.$$

On Z_k we have divisors $D_j^{(k)}$ which are the pullbacks of the divisors D_j from X. When $j > k$ the divisor $D_j^{(k)}$ is smooth, whereas for $j \leq k$ the divisor $D_j^{(k)}$ has multiplicity n.

Lemma 2.5. *With the above notation, we have the inductive statement that for any* $0 \leq k < m$,

$$Z_{k+1} = Z_k\langle\frac{D_{k+1}^{(k)}}{n}\rangle.$$

Proof. Recall the definition of Z_{k+1} : if we assume D_i for $i = 1, \ldots, k+1$ is defined by equations $z_i = 0$ and on any local chart (for the étale topology) some of the components say $D_1, \ldots, D_{k'}$ occur then the local chart for Z_{k+1} with coordinates u_i is defined by the equations $z_i = u_i^n$ for $i = 1, \ldots, k'$ and $z_i = u_i$ for $i > k'$. Now $Z_k\langle\frac{D_{k+1}^{(k)}}{n}\rangle$ is obtained from Z_k by defining local chart with coordinates w_i and repeating the above construction by considering the component divisor $D_{k+1}^{(k)}$ on Z_k, by applying Lemma 2.4 and having the same denominator n. It is now clear that both the constructions define the same stack. □

Suppose X is a smooth DM stack and $D \subset X$ is a smooth divisor. Then we define the notion of *parabolic bundle* on (X, D) (with n as denominator) as follows. A parabolic structure is a collection of sheaves F_α on X (with $\alpha \in \frac{1}{n}\mathbb{Z}$) with $F_{[a]} \to F_{[a+1]}$ (remember the notation at the start here with $m = 1$ so $[a] = (\frac{a}{n})$). This is a parabolic bundle if the $F_{[a]}$ are bundles and the quotient sheaves

$$F_{[a+1]}/F_{[a]}$$

are bundles supported on D. This is equivalent to a locally abelian condition in the étale topology of X. Indeed, we can attach weights $\frac{a}{n}$ to the graded pieces $F_{[a+1]}/F_{[a]}$ whenever this is non-zero and define locally on a general point of the divisor D a direct sum L of parabolic line bundles such that if the rank of $F_{[a+1]}/F_{[a]}$ is n_a then $L = \sum_a \mathcal{O}(-\frac{a}{n}D)^{\oplus n_a}$.

Lemma 2.6. *There is an equivalence of categories between bundles on* $X\langle\frac{D}{n}\rangle$ *and parabolic bundles on* (X, D) *with* n *as denominator.*

Proof. This is proved by Borne [Bo, Theorem 5] when X is a smooth variety. In the case of a DM stack since everything is local in the étale topology it works the same way. □

Similarly if D_i are smooth divisors meeting transversally on a DM stack X then we can define a notion of locally abelian parabolic bundle on $(X; \sum_i D_i)$, as in §2.1. Here the locally abelian condition is local in the etale topology which is the only appropriate topology to work with on X.

Lemma 2.7. *With the notation of the beginning, the categories of locally abelian parabolic bundles on*

$$(Z_k; D^{(k)}_{k+1}, \ldots, D^{(k)}_m)$$

are all naturally equivalent.

Proof. When $k = m$ and for any k, so we consider Z_m and Z_k, the equivalence of vector bundles on Z_m and locally abelian parabolic bundles on Z_k is proved in [Iy-Si, Lemma 2.3] (actually it is proved when Z_k is a variety but as earlier the same proof holds for the DM-stack Z_k). This gives the equivalences of categories on any Z_k and $Z_{k'}$. □

In particular the cases $k = 0$ so $Z_k = X$ and $k = m$ where there are no further divisor components, correspond to the equivalence of categories of [Iy-Si, Lemma 2.3]:

Corollary 2.8. *The category of locally abelian parabolic bundles on X is equivalent to the category of vector bundles on $Z_m = X\langle \frac{D_1}{n}, \ldots, \frac{D_m}{n}\rangle$.*

2.5. Chern characters

We recall here the abstract definition of the Chern character of a parabolic bundle. If $F.$ is a parabolic bundle with rational weights having common denominator n, then it corresponds to a vector bundle F_{DM} on the DM-stack $Z_m = X\langle \frac{D_1}{n}, \ldots, \frac{D_m}{n}\rangle$. Let $\pi : Z_m \to X$ denote the projection. By Gillet [Gi] and Vistoli [Vi] it induces an isomorphism of rational Chow groups

$$\pi_* : CH(Z_m)_{\mathbb{Q}} \xrightarrow{\cong} CH(X)_{\mathbb{Q}}. \tag{4}$$

In [Iy-Si], following an idea of Borne [Bo], we defined the *Chern character of F* to be

$$\mathrm{ch}(F) := \pi_*(\mathrm{ch}(F_{DM})) \in CH(X)_{\mathbb{Q}}. \tag{5}$$

It is a formal consequence of this definition that Chern character is compatible additively with direct sums (or more generally extensions), multiplicatively with tensor products, and the pullback of the Chern character is the Chern character of the pullback bundle for a morphism f of varieties if the normal-crossings divisors are in standard position with respect to f.

2.6. Statement of the main theorem

Our goal is to give a formula for the Chern character defined abstractly by (5). The first main theorem is that the Chern character depends only on the Chern characters of the component bundles, and not on the inclusion morphisms between them. This is not in any way tautological, as is shown by the examples we shall consider in §5 below which show that it is not enough to consider the Chern characters of the bundle E plus the filtrations, or just the Maruyama-Yokogawa components. The full collection of component bundles $F_{[a_1,\dots,a_m]}$ is sufficient to account for the incidence data among the filtrations, and allows us to obtain the Chern character.

Theorem 2.9. *Suppose F and G are locally abelian parabolic bundles on a DM stack X with n as denominator. Suppose that for all a_i with $0 \le a_i < n$ the bundles $F_{[a_1,\dots,a_m]}$ and $G_{[a_1,\dots,a_m]}$ have the same Chern character in the rational Chow groups of X. Then the parabolic bundles F and G have the same Chern character in the rational Chow group of X.*

When we have two parabolic bundles F and G satisfying the hypothesis of the theorem, we say that F and G are *componentwise Chow equivalent.* A stronger condition is to say that F and G are *componentwise isomorphic*, meaning that the $F_{[a_1,\dots,a_n]}$ and $G_{[a_1,\dots,a_n]}$ are isomorphic bundles on X. This obviously implies that they are componentwise Chow equivalent, and so the theorem will imply that they have the same Chern character.

Once we have Theorem 2.9, it is relatively straightforward to give an explicit calculation of the Chern character by exhibiting a componentwise isomorphism of parabolic bundles. The componentwise isomorphism which will come into play, will not, however, come from an isomorphism of parabolic structures because the individual isomorphisms on component bundles will not respect the inclusion maps in the parabolic structure. The resulting formula is a weighted average as stated in Theorem 1.1, proven as Theorem 5.8 below.

3. Reduction to the case of one divisor

In this section and the next, we prove Theorem 2.9. In this section we will use the intermediate stacks Z_k in order to reduce to the case of only one smooth divisor component; then in the next section we prove the formula for that case. To see how the reduction works we have to note what happens to the component bundles in the equivalence of Lemma 2.7.

Fix $0 < k \le m$ and consider the equivalence of Lemma 2.7 which we denote (a) in what follows: suppose E is a locally abelian parabolic bundle on

$$(Z_{k-1}; D_k^{(k-1)}, \dots, D_m^{(k-1)}),$$

then it corresponds to F which is a locally abelian parabolic bundle on

$$(Z_k; D_{k+1}^{(k)}, \dots, D_m^{(k)}).$$

Recall that $Z_k = Z_{k-1}\langle\frac{D_k^{(k-1)}}{n}\rangle$ and that we have an equivalence (b) between bundles on Z_k, and parabolic bundles on Z_{k-1} with respect to the divisor $D_k^{(k-1)}$. For any $b_{k+1}, \dots, b_m$ we can let a_k vary, and using E we obtain a parabolic bundle

$$H^{[b_{k+1},\dots,b_m]} := a_k \mapsto E_{[a_k, b_{k+1},\dots,b_m]}$$

on Z_{k-1} with respect to the divisor $D_k^{(k-1)}$.

Lemma 3.1. *Suppose that E and F correspond via the equivalence* (a) *as in the above notation, and define the parabolic bundle $H^{[b_{k+1},\dots,b_m]}$ as above, which for any $b_{k+1},\dots,b_m$ is a parabolic bundle on Z_{k-1} with respect to the divisor $D_k^{(k-1)}$. Then this parabolic bundle $H^{[b_{k+1},\dots,b_m]}$ is the one which corresponds via the equivalence* (b) *to the component vector bundle $F_{[b_{k+1},\dots,b_m]}$ of the parabolic bundle F.*

Proof. We use the definition of the pushforward ([Iy-Si, §2.2]) which provides the explicit equivalence in Lemma 2.7. For simplicity, we assume that $k = m-1$ so we are looking at the case

$$Z_m \xrightarrow{p} Z_{m-1} \xrightarrow{q} Z_{m-2}.$$

Let G be the vector bundle on Z_m corresponding to E or F, using the equivalence in Lemma 2.7. Consider the vector bundle $F_{[b_m]}$ on Z_{m-1}. We want to check that the associated parabolic bundle $q_*F_{[b_m]}$ is H^{b_m}. The following equalities prove this claim.

$$\begin{aligned}
(q_*F_{[b_m]})_{a_{m-1}} &= q_*(F_{[b_m]}(a_{m-1}R_{m-1}^{(m-1)})) \\
&= q_*((p_*G)_{[b_m]}(a_{m-1}R_{m-1}^{(m-1)})) \\
&= q_*(p_*(G(b_mR_m^{(m)}))(a_{m-1}R_{m-1}^{(m-1)})) \\
&= (q\circ p)_*(G(b_mR_m^{(m)} + a_{m-1}R_{m-1}^{(m)})) \\
&= E_{[a_{m-1},b_m]} \\
&= (H^{[b_m]})_{[a_{m-1}]}.
\end{aligned}$$

Here $R_{m-1}^{(m-1)}$, $R_{m-1}^{(m)}$ and $R_m^{(m)}$ are the n-th roots of D_{m-1}, D_{m-1}, and D_m respectively, over Z_{m-1}, Z_m and Z_m respectively. □

A corollary of this observation is that we can reduce for Theorem 2.9 to the case of a single divisor.

Corollary 3.2. *Suppose that Theorem* 2.9 *is known for $m = 1$, that is, for a single smooth divisor. Then it holds in general.*

Proof. Fix X with $D_1,\dots,D_m$ and define the sequence of intermediate stacks Z_k as above. Suppose F and G are locally abelian parabolic bundles on $X = Z_0$ which are componentwise Chow equivalent. For any k let $F^{(k)}$ and $G^{(k)}$ denote the corresponding locally abelian parabolic bundles on Z_k with respect to the remaining divisors $D_{k+1}^{(k)},\dots,D_m^{(k)}$. We claim by induction on $0 \le k \le m$ that the $F^{(k)}$ and $G^{(k)}$ are componentwise Chow equivalent. This is tautologically true for $k = 0$. Fix $0 < k \le m$ and suppose it is true for $k-1$. Then $F^{(k-1)}$ and $G^{(k-1)}$ induce for any $b_{k+1},\dots,b_m$ parabolic bundles which we can denote by $H_F^{[b_{k+1},\dots,b_m]}$ and $H_G^{[b_{k+1},\dots,b_m]}$, as in Lemma 3.1. These are parabolic bundles on Z_{k-1} with respect to the single smooth divisor $D_k^{(k-1)}$. The components of these parabolic bundles are Chow equivalent, since they come from the components of $F^{(k-1)}$ and $G^{(k-1)}$ which by the induction hypothesis are componentwise Chow equivalent. Therefore, considered as parabolic bundles with respect to a single divisor, $H_F^{[b_{k+1},\ldots,b_m]}$ and $H_G^{[b_{k+1},\ldots,b_m]}$ are componentwise Chow equivalent. In the present corollary we are assuming that Theorem 2.9 is known for the case $m = 1$ of a single divisor. Applying this case of Theorem 2.9 we get that the bundles on Z_k associated to $H_F^{[b_{k+1},\ldots,b_m]}$ and $H_G^{[b_{k+1},\ldots,b_m]}$ are Chow equivalent. However, by Lemma 3.1 applied to the comparison between $F^{(k-1)}$ and $F^{(k)}$, the bundle on Z_k corresponding to the parabolic bundle $H_F^{[b_{k+1},\ldots,b_m]}$ is exactly the component

$$F^{(k)}_{[b_{k+1},\ldots,b_m]}.$$

Similarly, applying Lemma 3.1 to the comparison between $G^{(k-1)}$ and $G^{(k)}$, the bundle on Z_k corresponding to the parabolic bundle $H_G^{[b_{k+1},\ldots,b_m]}$ is exactly the component

$$G^{(k)}_{[b_{k+1},\ldots,b_m]}.$$

Thus the result of our application of the single divisor case of Theorem 2.9 is that the bundles $F^{(k)}_{[b_{k+1},\ldots,b_m]}$ and $G^{(k)}_{[b_{k+1},\ldots,b_m]}$ are Chow equivalent. This exactly says that the parabolic bundles $F^{(k)}$ and $G^{(k)}$ are componentwise Chow equivalent, which completes our induction step.

When $k = m$ at the end of the induction, $F^{(m)}$ and $G^{(m)}$ are componentwise Chow equivalent. But these are usual bundles on Z_m, so their Chern characters coincide. The Chern characters of F and G are defined as the pushforwards of those of $F^{(m)}$ and $G^{(m)}$, so these are the same too, giving the statement of Theorem 2.9. □

4. The single divisor case

By Corollary 3.2, it now suffices to prove Theorem 2.9 in the case $m = 1$. Simplify notation. Suppose we have a smooth DM stack X and a smooth divisor D, and suppose we have a parabolic bundle F on X with respect to D. It is a collection of bundles denoted $F_{[a]}$ with $a \in \mathbb{Z}$ (as usual without saying so we assume that the denominator is n). Let $Z := X\langle \frac{D}{n} \rangle$, so F corresponds to a vector bundle E on Z. According to the definition (5) we would like to show that the Chern character of E in the rational Chow group of Z depends only on the Chern characters of the $F_{[a]}$ in the rational Chow group of X, noting the identification (4).

Let $p : Z \to X$ denote the map of DM stacks. The inverse image $p^*(D)$ is a divisor in Z which has multiplicity n, because p is totally ramified of degree n over D. In particular, there is a divisor $R \subset Z$ such that

$$p^*(D) = n \cdot R.$$

This R is well defined as a smooth closed substack of codimension 1 in Z. However, R is a gerb over D. More precisely, we have a map $R \to D$ and there is a covering of D in the etale topology by maps $U \to D$ such that there is a lifting $U \to R$. If we are given such a lifting then this gives a trivialization

$$U \times_D R \cong U \times B(\mathbb{Z}/n),$$

where $B(\mathbb{Z}/n)$ is the one-point stack with group $\mathbb{Z}/n$. This can be summed up by saying that R is a gerb over D with group $\mathbb{Z}/n$. It is in general not trivial. (We conjecture that the obstruction is the same as the obstruction to the normal bundle $N_{D/X}$ having an n-th root as line bundle on D.) On the other hand, the character theory for R over D is trivialized in the following sense. There is a line bundle $N := \mathcal{O}_X(R)|_R$ on R with the property that on any fiber of the form $B(\mathbb{Z}/n)$, N is the primitive character of $\mathbb{Z}/n$.

Using N, we get a canonical decomposition of bundles on R. Suppose E is a bundle on R. Then $p_{R,*}E$ is a bundle on D which corresponds in each fiber to the trivial character. Here p_R is the map p restricted to R. For any i we have a map

$$p_R^*(p_{R,*}(E \otimes N^{\otimes -i})) \otimes N^{\otimes i} \to E.$$

Lemma 4.1. *If E is a bundle on R then the above maps put together for $0 \leq i < n$ give a direct sum decomposition*

$$\bigoplus_{i=0}^{n-1} p_R^*(p_{R,*}(E \otimes N^{\otimes -i})) \otimes N^{\otimes i} \xrightarrow{\cong} E.$$

Proof. The maps exist globally. To check that the map is an isomorphism it suffices to do it locally over D in the etale topology (since the map p_R is involved). As noted above, locally over D the gerb R is a product of the form $U \times B(\mathbb{Z}/n)$. A bundle E on the product is the same thing as a bundle on U together with an action of the group $\mathbb{Z}/n$. In turn this is the same thing as a bundle with action of the group algebra $\mathcal{O}_U[\mathbb{Z}/n]$ but relative *Spec* of this algebra over U is a disjoint union of n copies of U, so E decomposes as a direct sum of pieces corresponding to these sections. This decomposition may be written as $E = \bigoplus_\chi E_\chi$ where the χ are characters of $\mathbb{Z}/n$ and $\mathbb{Z}/n$ acts on E_χ via the character χ. In terms of the DM stack this means that E decomposes as a direct sum of bundles on U tensored with characters of $\mathbb{Z}/n$ considered as line bundles on $B(\mathbb{Z}/n)$. Using this decomposition we can check that the above map is an isomorphism (actually it gives back the same decomposition). □

Now suppose E is a bundle on Z. Then its restriction to R, noted E_R, decomposes according to the above lemma. Define two pieces as follows: $E_{R,\text{fix}}$ is the piece corresponding to $i = 0$ in the decomposition. Thus

$$E_{R,\text{fix}} = p_R^*(p_{R,*}E_R).$$

On the other hand, let $E_{R,\text{var}}$ denote the direct sum of the other pieces in the decomposition. The decomposition of Lemma 4.1 thus gives a direct sum decomposition

$$E_R = E_{R,\text{fix}} \oplus E_{R,\text{var}}.$$

Define the *standard elementary transformation* $\mathbf{e}(E)$ of a bundle E over Z, as the kernel

$$0 \to \mathbf{e}(E) \to E \to E_{R,\mathrm{var}} \to 0. \tag{6}$$

Lemma 4.2. *Suppose E is a bundle on Z. Then we have the following exact sequence for the restriction of the standard elementary transformation of E:*

$$0 \to E_{R,\mathrm{var}} \otimes N^* \to (\mathbf{e}(E))_R \to E_{R,\mathrm{fix}} \to 0.$$

Proof. Consider the exact sequence :

$$0 \longrightarrow E \otimes \mathcal{O}(-R) \longrightarrow E \longrightarrow E_R \longrightarrow 0.$$

Since $E_R = E_{R,\mathrm{fix}} \oplus E_{R,\mathrm{var}}$, and $\mathbf{e}(E)$ is the kernel of the composed map

$$E \longrightarrow E_R \longrightarrow E_{R,\mathrm{var}}$$

there is an induced injective map

$$E \otimes \mathcal{O}(-R) \longrightarrow \mathbf{e}(E)$$

inducing the restriction map on R

$$(E_{R,\mathrm{fix}} \oplus E_{R,\mathrm{var}}) \otimes \mathcal{O}(-R)_{|R} \longrightarrow \mathbf{e}(E)_{|R}$$

The kernel of the restriction

$$(\mathbf{e}(E))_R \longrightarrow E_R \longrightarrow E_{R,\mathrm{fix}}$$

is clearly $E_{R,\mathrm{var}} \otimes \mathcal{O}(-R)_{|R} = E_{R,\mathrm{var}} \otimes N^*$. □

Suppose E is a bundle on Z. Define $\rho(E)$ to be the largest integer k with $0 \le k < n$ such that the piece

$$p_R^*(p_{R,*}(E_R \otimes N^{\otimes -k})) \otimes N^{\otimes k}$$

in the decomposition of Lemma 4.1 is nonzero.

Actually we may consider this definition for any vector bundle on R.

Corollary 4.3. *The invariant ρ decreases under the standard elementary transformation: if $\rho(E) > 0$ then*

$$\rho(\mathbf{e}(E)) < \rho(E).$$

Proof. Consider the exact sequence from Lemma 4.2 :

$$0 \to E_{R,\mathrm{var}} \otimes N^* \to (\mathbf{e}(E))_R \to E_{R,\mathrm{fix}} \to 0.$$

Using the pushforward and pullback operations on this exact sequence, after twisting by powers of N, we notice that it suffices to check that $\rho(E_{R,\mathrm{var}} \otimes N^*) < \rho(E)$ and $\rho(E_{R,\mathrm{fix}}) = 0$.

Now

$$\begin{aligned} p_R^* p_{R*}(E_{R,\mathrm{fix}} \otimes N^{-k}) \otimes N^k &= p_R^* p_{R*}(p_R^* p_{R*} E \otimes N^{-k}) \otimes N^k \\ &= p_R^*(p_{R*} E \otimes p_{R*} N^{-k}) \otimes N^k \\ &= 0 \text{ if } \mathrm{k} \neq 0. \end{aligned}$$

Also,

$$\begin{aligned} & p_R^* p_{R*}(E_{R,\mathrm{var}} \otimes N^{-1} \otimes N^{-k}) \otimes N^k \\ = \; & p_R^* p_{R*}\left((\textstyle\sum_{i=1}^{\rho(E)} p_R^* p_{R*}(E_R \otimes N^{-i}) \otimes N^i) \otimes N^{-1} \otimes N^{-k} \right) \otimes N^k \\ = \; & p_R^* \left(\textstyle\sum_{i=1}^{\rho(E)} p_{R*}(E_R \otimes N^{-i}) \otimes p_{R*} N^{i-1-k} \right) \otimes N^k. \end{aligned}$$

The summands in the above term corresponding to $i-1-k \neq 0$ are zero. In other words, the only term left is for $i = k+1$, but if $k \geq \rho(E)$ then this doesn't occur and the whole is zero. Hence $\rho(E_{R,\mathrm{var}} \otimes N^*) < \rho(E)$. □

We now describe the pieces in the decomposition of Lemma 4.1 for E_R in terms of the parabolic structure on X. Introduce the following notation: if F is a parabolic bundle on X along the divisor D, then for any $a \in \mathbb{Z}$ set $\mathbf{gr}_{[a]}(F) := F_{[a]}/F_{[a-1]}$. It is a vector bundle on the divisor D.

Lemma 4.4. *Suppose E is a bundle on Z corresponding to a parabolic bundle F over X. Then for any $a \in \mathbb{Z}$ we have*

$$p_{R,*}(E_R \otimes N^{\otimes a}) \cong \mathbf{gr}_{[a]}(F).$$

Proof. We have

$$F_{[a]} = p_*(E(aR)).$$

Note that $R^1 p_*$ vanishes on coherent sheaves, since p is a finite map in the etale topology. Thus p_* is exact. This gives

$$\mathbf{gr}_{[a]}(F) = p_*(E \otimes (\mathcal{O}_Z(aR)/\mathcal{O}_Z((a-1)R))).$$

However, $(\mathcal{O}_Z(aR)/\mathcal{O}_Z((a-1)R))$ is a bundle on R which is equal to $N^{\otimes a}$. This gives the statement. □

We say that two bundles on R are *Chow equivalent relative to Z* if their Chern characters map to the same thing in the rational Chow group of Z. Caution: this is different from their being Chow equivalent on R, because the map $CH(R)_{\mathbb{Q}} \to CH(Z)_{\mathbb{Q}}$ might not be injective.

Lemma 4.5. *Suppose $p : Z = X\langle \frac{1}{n} \rangle \longrightarrow X$ is a morphism of DM-stacks as in the beginning of this section. Then the following diagram commutes:*

$$\begin{array}{ccc} CH^{\cdot}(R)_{\mathbb{Q}} & \longrightarrow & CH^{\cdot}(Z)_{\mathbb{Q}} \\ \downarrow\cong & & \downarrow\cong \\ CH^{\cdot}(D)_{\mathbb{Q}} & \longrightarrow & CH^{\cdot}(X)_{\mathbb{Q}} \end{array}$$

Proof. Use composition of proper pushforwards [Vo]. The vertical isomorphisms come from the fact that $Z \to X$ and $R \to D$ induce isomorphisms of coarse moduli schemes, and [Vi] [Gi]. □

Corollary 4.6. *Suppose E and G are vector bundles on Z corresponding to parabolic bundles F and H respectively on X. If F and H are componentwise Chow equivalent then each of the components in the decompositions of Lemma 4.1 for E_R and G_R are Chow equivalent relative to Z.*

Proof. Since F and H are componentwise Chow equivalent the graded pieces $\mathbf{gr}_{[a]}(F)$ and $\mathbf{gr}_{[a]}(H)$ are Chow equivalent on X. Hence by Lemma 4.4, $p_{R*}(E \otimes N^a)$ and $p_{R*}(G \otimes N^a)$ are Chow equivalent on X, in other words they are vector bundles on D which are Chow equivalent relative to X. The pullback of Chow equivalent objects on D relative to X are Chow equivalent objects on R relative to Z, by Lemma 4.5. Thus, in the sum decomposition of E_R and G_R as in Lemma 4.1, we conclude that the component sheaves are Chow equivalent relative to Z. □

Corollary 4.7. *Suppose E and G are vector bundles on Z corresponding to parabolic bundles F and H respectively on X. Suppose that F and H are componentwise Chow equivalent. Then the sheaves $E_{R,\mathrm{fix}}$ and $G_{R,\mathrm{fix}}$ are Chow equivalent on Z. Similarly, the sheaves $E_{R,\mathrm{var}}$ and $G_{R,\mathrm{var}}$ are Chow equivalent on Z.*

Proof. These sheaves come from the components of the decomposition for E_R and G_R. □

Lemma 4.8. *Suppose E and G are vector bundles on Z corresponding to parabolic bundles F and H respectively on X. Suppose that F and H are componentwise Chow equivalent. As a matter of notation, let $\mathbf{e}_X F$ and $\mathbf{e}_X H$ denote the parabolic bundles on X corresponding to the vector bundles $\mathbf{e}E$ and $\mathbf{e}G$. Then $\mathbf{e}_X F$ and $\mathbf{e}_X H$ are componentwise Chow equivalent.*

Proof. Firstly, we claim that

$$(\mathbf{e}_X F)_{[0]} = F_{[0]}. \tag{7}$$

To prove the claim, note that $F_{[0]} = p_*(E)$. On the other hand, since $E_{R,\mathrm{var}}$ has only components which have trivial direct images, we have $p_*(E_{R,\mathrm{var}}) = 0$, so the left exact sequence for the direct image of (6), shows that

$$p_*(\mathbf{e}E) = p_*(E).$$

This gives the claim.

The same claim holds for H.

Now twist the exact sequence in Lemma 4.2 by N^a, and take the pushforward (which is exact). Do this for both bundles E and G, yielding the exact sequences

$$0 \longrightarrow p_{R*}(E_{R,\mathrm{var}} \otimes N^{-1+a}) \longrightarrow p_*(\mathbf{e}(E)_R \otimes N^a) \longrightarrow p_*(E_{R,\mathrm{fix}} \otimes N^a) \longrightarrow 0$$

and

$$0 \longrightarrow p_{R*}(G_{R,\mathrm{var}} \otimes N^{-1+a}) \longrightarrow p_*(\mathbf{e}(G)_R \otimes N^a) \longrightarrow p_*(G_{R,\mathrm{fix}} \otimes N^a) \longrightarrow 0.$$

By the hypothesis, Corollary 4.6 applies to say that the various components in the decomposition of Lemma 4.1 for $E_{R,\mathrm{var}}$ and $E_{R,\mathrm{fix}}$ are Chow equivalent relative Z to the corresponding components of $G_{R,\mathrm{var}}$ and $G_{R,\mathrm{fix}}$. Thus the left and right terms of both exact sequences are Chow equivalent relative to X, so $p_*(\mathbf{e}(E)_R \otimes N^a)$ and $p_*(\mathbf{e}(G)_R \otimes N^a)$ are Chow equivalent relative to X.

Hence by Lemma 4.4, $\mathbf{gr}_{[a]}(\mathbf{e}_X F)$ and $\mathbf{gr}_{[a]}(\mathbf{e}_X H)$ are Chow equivalent relative to X. Together with the above claim (7), we deduce that the constituent bundles of $\mathbf{e}_X F$ and $\mathbf{e}_X H$ are Chow equivalent on X. □

We can iterate the operation of doing the elementary transform, denoted $E \mapsto \mathbf{e}^p E$. This corresponds to a parabolic bundle on X denoted by $F \mapsto \mathbf{e}^p_X F$. Note that this is indeed the iteration of the notation $\mathbf{e}_X$

Exercise 4.9. *Give an explicit description of $\mathbf{e}_X$ in terms of parabolic bundles.*

Because the invariant $\rho(E)$ decreases under the operation of doing the standard elementary transform (until we get to $\rho = 0$) it follows that $\rho(\mathbf{e}^p E) = 0$ for some $p \leq n$.

Lemma 4.10. *Suppose E is a bundle on E with $\rho(E) = 0$. Then E is the pullback of a bundle on X.*

Proof. In this case, we have $E_R \simeq p_R^* p_{R*} E_R$. Hence by Lemma 4.4, it follows that $\mathbf{gr}_{[a]}(F) = 0$ for $a > 0$. This implies that F has only one constituent bundle $F_{[0]}$ and is a usual bundle on X. Hence E is the pullback of $F_{[0]}$. □

The next lemma gives the induction step for the proof of the theorem.

Lemma 4.11. *Suppose E and G are vector bundles on Z corresponding to parabolic bundles F and H respectively on X. Suppose that F and H are componentwise Chow equivalent. Suppose also that $\mathbf{e}E$ and $\mathbf{e}G$ are Chow equivalent on Z. Then E and G are Chow equivalent on Z.*

Proof. The componentwise Chow equivalence gives from Corollary 4.7 that $E_{R,\mathrm{var}}$ and $G_{R,\mathrm{var}}$ are Chow equivalent relative to Z. The exact sequence of Lemma 4.2 gives that E and G are Chow equivalent on Z. □

Finally we can prove Theorem 2.9 in the single divisor case.

Theorem 4.12. *Suppose E and G are vector bundles on Z corresponding to parabolic bundles F and H respectively on X. Suppose that F and H are componentwise Chow equivalent. Then E and G are Chow equivalent on Z.*

Proof. Do this by descending induction with respect to the number p given above Lemma 4.10. There is some p_0 such that $\rho(\mathbf{e}^{p_0} E) = 0$ and $\rho(\mathbf{e}^{p_0} G) = 0$. These come from bundles on X. By Lemma 4.8, these bundles (which are the zero components of the corresponding parabolic bundles) are Chow equivalent. Thus $\mathbf{e}^{p_0} E$ and $\mathbf{e}^{p_0} G$ are Chow equivalent. On the other hand, by Lemma 4.8, all of the $\mathbf{e}^p_X F$ and $\mathbf{e}^p_X H$ are componentwise Chow equivalent. It follows from Lemma 4.11, if we know that $\mathbf{e}^{p+1} E$ and $\mathbf{e}^{p+1} G$ are Chow equivalent then we get that $\mathbf{e}^p E$ and $\mathbf{e}^p G$ are Chow equivalent. By descending induction on p we get that E and G are Chow equivalent. □

Using Corollary 3.2, we have now completed the proof of Theorem 2.9.

5. A formula for the parabolic Chern character

Now we would like to use Theorem 2.9 to help get a formula for the Chern classes. Go back to the general situation of a smooth variety X with smooth divisors $D_1, \ldots, D_m$ intersecting transversally. Once we know the formula for the Chern character of a line bundle, we will no longer need to use the stack $Z = X[\frac{D_1}{n}, \ldots, \frac{D_m}{n}]$.

Lemma 5.1. *Let F be a parabolic bundle on X with respect to $D_1, \ldots, D_m$. Then we can form the twisted parabolic bundle $F \otimes \mathcal{O}(\sum_{i=0}^m \frac{b_i}{n} D_i)$. We have the formulae*

$$\left(F \otimes \mathcal{O}(\sum_{i=0}^m \frac{b_i}{n} D_i)\right)_{[a_1,\ldots,a_m]} = F_{[a_1+b_1,\ldots,a_m+b_m]}$$

and

$$\mathrm{ch}\left(F \otimes \mathcal{O}(\sum_{i=0}^m \frac{b_i}{n} D_i)\right) = e^{\sum_{i=0}^m \frac{b_i}{n} D_i} \mathrm{ch}(F).$$

Proof. Consider the projection $p : Z = X\langle \frac{D_1}{n}, \ldots, \frac{D_m}{n} \rangle \longrightarrow X$. Let E be the vector bundle on Z corresponding to F on X and $\mathcal{O}(\sum_i b_i R_i)$ be the line bundle on Z corresponding to $\mathcal{O}(\sum_i \frac{b_i}{n} D_i)$ on X. Here R_i denotes the divisor on Z such that $p^* D_i = n.R_i$.

Notice that

$$\begin{aligned}\left(F \otimes \mathcal{O}(\sum_i \frac{b_i}{n} D_i)\right)_{[a_1,\ldots,a_m]} &= p_*(E \otimes \mathcal{O}(\sum_i b_i R_i) \otimes \mathcal{O}(a_i R_i)) \\ &= p_*(E \otimes \mathcal{O}(\sum_i (a_i + b_i) R_i)) \\ &= F_{[a_1+b_1,\ldots,a_m+b_m]}.\end{aligned}$$

The formula for the Chern character is due to the fact that the Chern character defined as we are doing through DM-stacks is multiplicative for tensor products, and coincides with the exponential for rational divisors, see [Iy-Si]. □

Lemma 5.2. *We have the formula for the trivial line bundle $\mathcal{O}$ considered as a parabolic bundle:*

$$\mathcal{O}_{[a_1,\ldots,a_m]} = \mathcal{O}(\sum_{i=0}^m [\frac{a_i}{n}] D_i)$$

where the square brackets on the right signify the greatest integer function (on the left they are the notation we introduced at the beginning).

Proof. This follows from the definition as in Lemma 5.1. □

Corollary 5.3. *Suppose E is a vector bundle on X considered as a parabolic bundle with its trivial structure. Then*

$$\left(E \otimes \mathcal{O}(\sum_{i=0}^{m} \frac{b_i}{n} D_i)\right)_{[a_1,\dots,a_m]} = E(\sum_{i=0}^{m} [\frac{a_i + b_i}{n}] D_i).$$

Proof. Use the definition of associated parabolic bundle as in Lemma 5.1. □

Suppose F is a parabolic bundle on X with respect to $D_1, \dots, D_m$. We will now show by calculation that the two parabolic bundles

$$\left(\bigoplus_{k_1=0}^{n-1} \cdots \bigoplus_{k_m=0}^{n-1} \mathcal{O}(-\sum_{i=1}^{m} \frac{k_i}{n} D_i)\right) \otimes F$$

and

$$\left(\bigoplus_{u_1=0}^{n-1} \cdots \bigoplus_{u_m=0}^{n-1} F_{[u_1,\dots,u_m]} \otimes \mathcal{O}(-\sum_{i=1}^{m} \frac{u_i}{n} D_i)\right)$$

are componentwise isomorphic (and hence, componentwise Chow equivalent). Notice that the second bundle is a direct sum of vector bundles on X, the component bundles of F, tensored with parabolic line bundles, whereas the first is F tensored with a bundle of positive rank. This will then allow us to get a formula for $\mathrm{ch}(F)$.

Lemma 5.4. *For any $0 \le a_i < n$ we have*

$$\left(\left(\bigoplus_{k_1=0}^{n-1} \cdots \bigoplus_{k_m=0}^{n-1} \mathcal{O}(-\sum_{i=1}^{m} \frac{k_i}{n} D_i)\right) \otimes F\right)_{[a_1,\dots,a_m]} \cong \bigoplus_{k_1=0}^{n-1} \cdots \bigoplus_{k_m=0}^{n-1} F_{[a_1-k_1,\dots,a_m-k_m]}.$$

Proof. Indeed, we have

$$\left(\mathcal{O}(-\sum_{i=1}^{m} \frac{k_i}{n} D_i) \otimes F\right)_{[a_1,\dots,a_m]} \cong F_{[a_1-k_1,\dots,a_m-k_m]}$$

by Lemma 5.1 above. □

Lemma 5.5.

$$\left(\bigoplus_{u_1=0}^{n-1} \cdots \bigoplus_{u_m=0}^{n-1} F_{[u_1,\dots,u_m]} \otimes \mathcal{O}(-\sum_{i=1}^{m} \frac{u_i}{n} D_i)\right)_{[a_1,\dots,a_m]}$$

$$\cong \bigoplus_{u_1=0}^{n-1} \cdots \bigoplus_{u_m=0}^{n-1} F_{[u_1,\dots,u_m]} \otimes \mathcal{O}(\sum_{i=1}^{m} [\frac{a_i - u_i}{n}] D_i).$$

Proof. We have

$$\left(\mathcal{O}(-\sum_{i=1}^{m} \frac{u_i}{n} D_i)\right)_{[a_1,\dots,a_m]} = \mathcal{O}(\sum_{i=1}^{m} [\frac{a_i - u_i}{n}] D_i)$$

and hence, since $F_{[u_1,\ldots,u_m]}$ is just a vector bundle on X,

$$\left(F_{[u_1,\ldots,u_m]} \otimes \mathcal{O}(-\sum_{i=1}^{m} \frac{u_i}{n} D_i)\right)_{[a_1,\ldots,a_m]} \cong F_{[u_1,\ldots,u_m]} \otimes \mathcal{O}(\sum_{i=1}^{m} [\frac{a_i - u_i}{n}] D_i). \quad \square$$

We put these two together with the following.

Lemma 5.6.

$$\bigoplus_{u_1=0}^{n-1} \cdots \bigoplus_{u_m=0}^{n-1} F_{[u_1,\ldots,u_m]} \otimes \mathcal{O}(\sum_{i=1}^{m} [\frac{a_i - u_i}{n}] D_i)$$

$$\cong \bigoplus_{k_1=0}^{n-1} \cdots \bigoplus_{k_m=0}^{n-1} F_{[a_1-k_1,\ldots,a_m-k_m]}.$$

Proof. For given integers $0 \le a_i < n$ and $0 \le u_i < n$, set

$$k_i := a_i - u_i - n \cdot [\frac{a_i - u_i}{n}],$$

so that

$$a_i - k_i = u_i + n \cdot [\frac{a_i - u_i}{n}].$$

With this definition of k_i we have

$$F_{[u_1,\ldots,u_m]} \otimes \mathcal{O}(\sum_{i=1}^{m} [\frac{a_i - u_i}{n}] D_i) \cong F_{[a_1-k_1,\ldots,a_m-k_m]},$$

due to the periodicity of the parabolic structure.

Note that $0 \le k_i < n$, because $a_i - u - i < 0$ if and only if the greatest integer piece in the definition of k_i is equal to -1 (otherwise it is 0).

For a fixed $(a_1, \ldots, a_m)$, as $(u_1, \ldots, u_m)$ ranges over all possible choices with $0 \le u_i < n$ the resulting $(k_1, \ldots, k_m)$ also ranges over all possible choices with $0 \le k_i < n$. Thus we get the isomorphism which is claimed. $\square$

Corollary 5.7. *If F is a parabolic bundle on X with respect to $D_1, \ldots, D_m$ then the parabolic bundles*

$$\left(\bigoplus_{k_1=0}^{n-1} \cdots \bigoplus_{k_m=0}^{n-1} \mathcal{O}(-\sum_{i=1}^{m} \frac{k_i}{n} D_i)\right) \otimes F$$

and

$$\bigoplus_{u_1=0}^{n-1} \cdots \bigoplus_{u_m=0}^{n-1} F_{[u_1,\ldots,u_m]} \otimes \mathcal{O}(-\sum_{i=1}^{m} \frac{u_i}{n} D_i)$$

are componentwise isomorphic, hence componentwise Chow equivalent.

Proof. Putting together Lemmas 5.4, 5.5 and 5.6 gives, for any $0 \leq a_i < n$

$$\left(\left(\bigoplus_{k_1=0}^{n-1}\cdots\bigoplus_{k_m=0}^{n-1}\mathcal{O}(-\sum_{i=1}^{m}\frac{k_i}{n}D_i)\right)\otimes F\right)_{[a_1,\ldots,a_m]}$$

$$\cong\left(\bigoplus_{u_1=0}^{n-1}\cdots\bigoplus_{u_m=0}^{n-1}F_{[u_1,\ldots,u_m]}\otimes\mathcal{O}(-\sum_{i=1}^{m}\frac{u_i}{n}D_i)\right)_{[a_1,\ldots,a_m]}. \qquad \square$$

We can now calculate with the previous corollary.

Theorem 5.8. *Suppose F is a parabolic bundle on X with respect to $D_1,\ldots,D_m$, with n as denominator. Then we have the following formula for the Chern character of F:*

$$\mathrm{ch}(F)=\frac{\sum_{a_1=0}^{n-1}\cdots\sum_{a_m=0}^{n-1}e^{-\sum_{i=1}^{m}\frac{a_i}{n}D_i}\mathrm{ch}(F_{[a_1,\ldots,a_m]})}{\sum_{a_1=0}^{n-1}\cdots\sum_{a_m=0}^{n-1}e^{-\sum_{i=1}^{m}\frac{a_i}{n}D_i}}.$$

In other words, the Chern character of F is the weighted average of the Chern characters of the component bundles in the range $0 \leq a_i < n$, with weights $e^{-\sum_{i=1}^{m}\frac{a_i}{n}D_i}$.

Proof. By Theorem 2.9, two componentwise Chow equivalent parabolic bundles have the same Chern character. From the general theory over a DM stack we know that Chern character of parabolic bundles is additive and multiplicative, and Lemma 5.1 says that it behaves as usual on line bundles. Therefore the Chern characters of both parabolic bundles appearing in the statement of Corollary 5.7 are the same. This gives the formula

$$(\sum_{a_1=0}^{n-1}\cdots\sum_{a_m=0}^{n-1}e^{-\sum_{i=1}^{m}\frac{a_i}{n}D_i})\cdot\mathrm{ch}(F)=\sum_{a_1=0}^{n-1}\cdots\sum_{a_m=0}^{n-1}e^{-\sum_{i=1}^{m}\frac{a_i}{n}D_i}\mathrm{ch}(F_{[a_1,\ldots,a_m]}).$$

The term multiplying $\mathrm{ch}(F)$ on the left side is an element of the Chow group which has nonzero term in degree zero. Therefore, in the rational Chow group it can be inverted and we get the formula stated in the theorem. $\square$

Remark 5.9. *The function*

$$(a_1,\ldots,a_m)\mapsto e^{-\sum_{i=1}^{m}\frac{a_i}{n}D_i}\mathrm{ch}(F_{[a_1,\ldots,a_m]})$$

is periodic in the variables a_i, that is, the value for a_i+n is the same as the value for a_i.

Remark 5.10. *Also the formula is clearly additive.*

6. Examples with parabolic line bundles

We verify the formula of Theorem 5.8 for parabolic line bundles, and then give some examples which are direct sums of line bundles which show why it is necessary to include all of the terms $F_{[a_1,\ldots,a_m]}$ in the formula.

6.1. Verification for line bundles

Suppose $L = \mathcal{O}(\alpha.D)$ is a parabolic line bundle on (X, D) where D is an irreducible and smooth divisor and $\alpha = \frac{h}{n} \in \mathbb{Q}$. The formula of Theorem 5.8 is obviously invariant if we tensor the parabolic bundle by a vector bundle on X, in particular we can always tensor with an integer power of $\mathcal{O}(D)$ so it suffices to check when $0 \le h < n$.

Notice that the constituent bundles are

$$\begin{aligned} L_{[a_i]} &= \mathcal{O} && \text{if } 0 \le a_i \le n-h-1 \\ &= \mathcal{O}(D) && \text{if } n-h \le a_i < n. \end{aligned}$$

We have to check that the Chern character of L is

$$\mathrm{ch}(L) = e^{\alpha.D}.$$

The formula in Theorem 5.8 gives

$$\begin{aligned} \mathrm{ch}(L) &= \frac{1 + e^{-\frac{1}{n}.D} + \cdots + e^{-\frac{(n-h-1)}{n}.D} + e^{-\frac{(n-h)}{n}.D}.e^D + \cdots + e^{-\frac{(n-1)}{n}.D}.e^D}{1 + e^{-\frac{1}{n}.D} + \cdots + e^{-\frac{(n-1)}{n}.D}} \\ &= \frac{1 + e^{-\frac{1}{n}.D} + \cdots + e^{-\frac{(n-h-1)}{n}.D} + e^{-\frac{(n-h)}{n}.D}.e^D + \cdots + e^{-\frac{(n-1)}{n}.D}.e^D}{(\frac{1}{e^{\frac{h}{n}.D}})(e^{\frac{h}{n}.D} + e^{\frac{h-1}{n}.D} + \cdots + 1 + \cdots + e^{-\frac{n-1-h}{n}.D})} \\ &= e^{\frac{h}{n}.D}. \end{aligned}$$

Suppose $D_1, D_2, \ldots, D_m$ are distinct smooth divisors which have normal crossings on X. Let $L_i = \mathcal{O}(\alpha_i.D_i)$ be parabolic line bundles with $\alpha_i \in \mathbb{Q}$, for $1 \le i \le m$. Then the constituent bundles of the tensor product $L := L_1 \otimes L_2 \otimes \cdots \otimes L_m$ are

$$(L_1 \otimes L_2 \otimes \cdots \otimes L_m)_{[a_1,a_2,\ldots,a_m]} = (L_1)_{[a_1]} \otimes (L_2)_{[a_2]} \otimes \cdots \otimes (L_m)_{[a_m]}$$

and

$$\begin{aligned} &\mathrm{ch}\left((L_1 \otimes L_2 \otimes \cdots \otimes L_m)_{[a_1,a_2,\ldots,a_m]}\right) \\ &= \mathrm{ch}\left((L_1)_{[a_1]}\right).\mathrm{ch}\left((L_2)_{[a_2]}\right) \cdots \mathrm{ch}\left((L_m)_{[a_m]}\right). \end{aligned}$$

The formula in Theorem 5.8 is now easily verified for the case when L is a parabolic line bundle as above, once it is verified for the parabolic line bundles L_i. Indeed, the formula in this case is essentially the product of the Chern characters of L_i, for $1 \le i \le m$.

6.2. The case of two divisors and $n = 2$

Suppose we have two divisor components D_1 and D_2, and suppose the denominator is $n = 2$. Then a parabolic bundle may be written as a 2×2 matrix

$$F = \begin{pmatrix} F_{[0,0]} & F_{[0,1]} \\ F_{[1,0]} & F_{[1,1]} \end{pmatrix}.$$

In particular by Lemma 5.2 we have

$$\mathcal{O}\left(\frac{D_1}{2}\right) = \begin{pmatrix} \mathcal{O} & \mathcal{O} \\ \mathcal{O}(D_1) & \mathcal{O}(D_1) \end{pmatrix}, \quad \mathcal{O}\left(\frac{D_2}{2}\right) = \begin{pmatrix} \mathcal{O} & \mathcal{O}(D_2) \\ \mathcal{O} & \mathcal{O}(D_2) \end{pmatrix},$$

$$\mathcal{O}(D_1) = \begin{pmatrix} \mathcal{O}(D_1) & \mathcal{O}(D_1) \\ \mathcal{O}(D_1) & \mathcal{O}(D_1) \end{pmatrix}, \quad \mathcal{O}\left(D_1 + \frac{D_2}{2}\right) = \begin{pmatrix} \mathcal{O}(D_1) & \mathcal{O}(D_1+D_2) \\ \mathcal{O}(D_1) & \mathcal{O}(D_1+D_2) \end{pmatrix},$$

and

$$\mathcal{O}\left(\frac{D_1}{2} + \frac{D_2}{2}\right) = \begin{pmatrix} \mathcal{O} & \mathcal{O}(D_2) \\ \mathcal{O}(D_1) & \mathcal{O}(D_1+D_2) \end{pmatrix}.$$

6.3. Counterexample for filtrations

Giving a parabolic bundle by filtrations amounts essentially to considering the bundle $E = F_{[0,0]}$ together with its subsheaves $F_{[-1,0]}$ and $F_{[0,-1]}$. By the formula (3) these subsheaves are determined by $F_{[1,0]}$ and $F_{[0,1]}$, that is, the upper right and lower left places in the matrix. The lower right place doesn't intervene in the filtration notation. This lets us construct an example: if

$$F := \mathcal{O}\left(\frac{D_1}{2}\right) \oplus \mathcal{O}\left(\frac{D_2}{2}\right), \quad G := \mathcal{O} \oplus \mathcal{O}\left(\frac{D_1}{2} + \frac{D_2}{2}\right)$$

then F and G have the same underlying bundle $E = \mathcal{O} \oplus \mathcal{O}$, and the Chern data for their filtrations are the same, however their Chern characters are different. For example if $X = \mathbb{P}^2$ and D_1 and D_2 are two distinct lines whose class is denoted H then

$$\mathrm{ch}(F) = \mathrm{ch}\left(\mathcal{O}_X\left(\frac{1}{2}D_1 + \frac{1}{2}D_2\right) \oplus \mathcal{O}_X\right) = 1 + e^{\frac{1}{2}D_1 + \frac{1}{2}D_2} = 2 + H + \frac{H^2}{2} \tag{8}$$

and

$$\mathrm{ch}(G) = \mathrm{ch}\left(\mathcal{O}_X\left(\frac{1}{2}D_1\right) \oplus \mathcal{O}_X\left(\frac{1}{2}D_2\right)\right) = e^{\frac{1}{2}D_1} + e^{\frac{1}{2}D_2} = 2e^{H/2} = 2 + H + \frac{H^2}{4}. \tag{9}$$

6.4. Counterexample for MY structure

Similarly, the MY-parabolic structure consists of $F_{[0,0]}$ and $F_{[1,1]}$, that is, the diagonal terms in the matrix, and the off-diagonal terms don't intervene. A different example serves to show that there is no easy formula for the Chern character in terms of these pieces only. Put

$$F := \mathcal{O}\left(\frac{D_1}{2} + \frac{D_2}{2}\right) \oplus \mathcal{O}(D_1), \quad G := \mathcal{O}\left(D_1 + \frac{D_2}{2}\right) \oplus \mathcal{O}\left(\frac{D_1}{2}\right).$$

Then

$$F_{[0,0]} = \mathcal{O} \oplus \mathcal{O}(D_1) = G_{[0,0]}$$

and

$$F_{[1,1]} = \mathcal{O}(D_1 + D_2) \oplus \mathcal{O}(D_1) = G_{[1,1]}.$$

On the other hand, again in the example $X = \mathbb{P}^2$ and D_1 and D_2 are lines whose class is denoted H we have

$$\mathrm{ch}(F) = e^{\frac{1}{2}D_1+\frac{1}{2}D_2} + e^{D_1} = 2e^H = 2 + 2H + H^2$$

whereas

$$\mathrm{ch}(G) = e^{D_1+\frac{D_2}{2}} + e^{\frac{D_1}{2}} = e^{\frac{3H}{2}} + e^{\frac{H}{2}} = 2 + 2H + \frac{5}{4}H^2.$$

In both of these examples, of course the structure with filtrations or the MY-parabolic structure permits to obtain back the full multi-indexed structure and therefore to get the Chern character, however these examples show that the Chern character cannot be written down easily just in terms of the Chern characters of the component pieces.

7. A formula involving intersection of filtrations

In this section we will give another expression for the parabolic Chern character formula, when the parabolic structure is viewed as coming from filtrations on the divisor components. This formula will involve terms on the multiple intersections of the divisor components, of intersections of the various filtrations. To see how these terms show up in the formula, we first illustrate it by an example below.

7.1. Example on surfaces

We consider more closely how the intersection of the filtrations on D_1 and D_2 comes into play for determining the Chern character. Panov [Pa] and Mochizuki [Mo2] considered this situation and obtained formulas for the second parabolic Chern class involving intersections of the filtrations.

For this example we keep the hypothesis that X is a surface and the denominator is $n = 2$, also assuming that there are only two divisor components D_1 and D_2 intersecting at a point P. The typical example is $X = \mathbb{P}^2$ and the D_i are distinct lines meeting at P.

Let $E = F_{[0,0]}$ be a rank two bundle. Consider rank one strict subbundles $B_i \subset E|_{D_i}$. Note that

$$E|_{D_1} = F_{[0,0]}/F_{[-2,0]}, \quad E|_{D_2} = F_{[0,0]}/F_{[0,-2]}.$$

There is a unique parabolic structure with

$$B_1 = F_{[-1,0]}/F_{[-2,0]},$$

and

$$B_2 = F_{[0,-1]}/F_{[0,-2]}.$$

The quotient $(E|_{D_1})/B_1$ is a line bundle on D_1 and similarly for D_2, and if the parabolic structure corresponds to the B_i as above then

$$(E|_{D_1})/B_1 = F_{[0,0]}/F_{[-1,0]}$$

and similarly on D_2.

In particular, $F_{[-1,0]}$ is defined by the exact sequence

$$0 \to F_{[-1,0]} \to E \to (E|_{D_1})/B_1 \to 0.$$

Similarly, $F_{[0,-1]}$ is defined by the exact sequence

$$0 \to F_{[0,-1]} \to E \to (E|_{D_2})/B_2 \to 0.$$

Note that the Chern characters of $F_{[-1,0]}$ and $F_{[0,-1]}$ don't depend on the intersection of the B_i over $D_1 \cap D_2$. On the other hand, $F_{[-1,-1]}$ is a vector bundle, by the locally abelian condition. Furthermore, as a subsheaf of E it is equal to $F_{[-1,0]}$ along D_1 and $F_{[0,-1]}$ along D_2. Thus, in fact $F_{[-1,-1]}$ is the subsheaf of E which is the intersection of these two subsheaves. To prove this note that the intersection of two reflexive subsheaves of a reflexive sheaf is again reflexive because it has the Hartogs exension property. In dimension two, reflexive sheaves are vector bundles, and they are determined by what they are in codimension one.

We have a left exact sequence

$$0 \to F_{[-1,-1]} \to E \to (E|_{D_1})/B_1 \oplus (E|_{D_2})/B_2.$$

Here is where the intersection of the filtrations comes in: in our example $D_1 \cap D_2$ is a single point, denote it by P. We have one-dimensional subspaces of the two-dimensional fiber of E over P:

$$B_{1,P}, B_{2,P} \subset E_P.$$

There are two cases: either they coincide, or they don't.

Case A: they coincide – In this case we can choose a local frame for E in which B_1 and B_2 are both generated by the first basis vector. We are basically in the direct sum of two rank one bundles, one of which containing the two subspaces and the other not. In this case there is an exact sequence

$$0 \to F_{[-1,-1]} \to E \to (E|_{D_1})/B_1 \oplus (E|_{D_2})/B_2 \to Q \to 0$$

where Q is a rank one skyscraper sheaf at P. This is because the fibers of $(E|_{D_1})/B_1$ and $(E|_{D_2})/B_2$ coincide at P, and Q is by definition this fiber with the map being the difference of the two elements. Things coming from E go to the same in both fibers so they map to zero in Q.

An example of this situation would be the parabolic bundle $\mathcal{O}_X(\frac{1}{2}D_1 + \frac{1}{2}D_2) \oplus \mathcal{O}_X$.

Case B: they differ – In this case we can choose a local frame for E in which B_1 and B_2 are generated by the two basis vectors respectively. In this case the map in question is surjective so we get a short exact sequence

$$0 \to F_{[-1,-1]} \to E \to (E|_{D_1})/B_1 \oplus (E|_{D_2})/B_2 \to 0.$$

An example of this situation would be the parabolic bundle $\mathcal{O}_X(\frac{1}{2}D_1) \oplus \mathcal{O}_X(\frac{1}{2}D_2)$.

The formula for the Chern character will involve the Chern character of E, the Chern characters of the bundles B_i, and a correction term for the intersection.

All other things being equal, the formulas in the two cases will differ by $\mathrm{ch}(Q)$ at the place $F_{[1,1]}$ (this is the same as for $F_{[-1,-1]}$). When the weighted average is taken, this comes in with a coefficient of $(\frac{1}{4}+\cdots)$, but the higher-order terms multiplied by the codimension 2 class $\mathrm{ch}(Q)$ come out to zero because we are on a surface. Therefore, the formulae in case A and case B will differ by $\frac{1}{4}\mathrm{ch}(Q)$. Fortunately enough this is what actually happens in the examples of the previous section!

7.2. Changing the indexing

When describing a parabolic bundle by filtrations, we most naturally get to the bundles $F_{[a_1,\ldots,a_m]}$ with $-n \le a_i \le 0$. On the other hand, the weighted average in Theorem 5.8 is over a_i in the positive interval $[0, n-1]$. It is convenient to have a formula which brings into play the bundles in a general product of intervals. The need for such was seen in the example of the previous subsection.

We have the following result which meets up with Mochizuki's notation and discussion in [Mo2, §3.1].

Proposition 7.1. *Let* $\mathbf{b} = (b_1, \ldots, b_m)$ *be any multi-index of integers. Then the Chern character of the parabolic bundle* $\mathrm{ch}(F)$ *is obtained by taking the weighted average of the* $\mathrm{ch}(F_{[a_1,\ldots,a_m]})$ *with weights* $e^{-\sum_{i=1}^m \frac{a_i}{n} D_i}$, *over the product of intervals* $b_i \le a_i < b_i + n$, *and then multiplying by* $e^{-\sum_{i=1}^m \frac{b_i}{n} D_i}$ *(that is the weight for the smallest multi-index in the range). This formula may also be written as:*

$$\mathrm{ch}(F) = \frac{\sum_{a_1=b_1}^{b_1+n-1} \cdots \sum_{a_m=b_m}^{b_m+n-1} e^{-\sum_{i=1}^m \frac{a_i}{n} D_i} \mathrm{ch}(F_{[a_1,\ldots,a_m]})}{\sum_{a_1=0}^{n-1} \cdots \sum_{a_m=0}^{n-1} e^{-\sum_{i=1}^m \frac{a_i}{n} D_i}}. \tag{10}$$

Proof. If a_i and a_i' differ by integer multiples of n then by using condition (3) of §2.2, we have

$$e^{-\sum_{i=1}^m \frac{a_i'}{n} D_i} \mathrm{ch}(F_{[a_1',\ldots,a_m']}) = e^{-\sum_{i=1}^m \frac{a_i}{n} D_i} \mathrm{ch}(F_{[a_1,\ldots,a_m]}).$$

Thus, the numerator in the formula (10) is equal to the numerator of the formula in Theorem 5.8. The denominators are the same. On the other hand, if we form the weighted average as described in the first sentence of the proposition, then the numerator will be the same as in (10). The denominator of the weighted average is

$$\sum_{a_1=b_1}^{b_1+n-1} \cdots \sum_{a_m=b_m}^{b_m+n-1} e^{-\sum_{i=1}^m \frac{a_i}{n} D_i} = e^{-\sum_{i=1}^m \frac{b_i}{n} D_i} \sum_{a_1=0}^{n-1} \cdots \sum_{a_m=0}^{n-1} e^{-\sum_{i=1}^m \frac{a_i}{n} D_i}.$$

Hence, when we multiply the weighted average by $e^{-\sum_{i=1}^m \frac{b_i}{n} D_i}$ we get (10). □

Remark 7.2. *If we replace the denominator* n *by a new one* np *then the formulae of Theorem 5.8 or the previous proposition, give the same answers.*

Indeed, the parabolic structure $\widetilde{F}$ for denominator np contains the same sheaves, but each one is copied p^m times:

$$\widetilde{F}_{[pa_1+q_1,\ldots,pa_m+q_m]} = F_{[a_1,\ldots,a_m]}$$

for $0 \leq q_i \leq p-1$. Therefore, both the numerator and the denominator in our formulae are multiplied by

$$\sum_{q_1=0}^{p-1} \cdots \sum_{q_m=0}^{p-1} e^{-\sum_{i=1}^{m} \frac{q_i}{np} D_i},$$

and the quotient stays the same.

7.3. A general formula involving intersection of filtrations

We can generalize the example of surfaces in §7.1, to get a formula which generalizes the codimension 2 formulae of Panov [Pa] and Mochizuki [Mo2].

In this section we suppose we are working with the notation of a locally abelian parabolic structure F given by filtrations, on a vector bundle $E := F_{[0,\ldots,0]}$ with filtrations

$$E|_{D_i} = F^i_0 \supset F^i_{-1} \supset \cdots \supset F^i_{-n} = 0.$$

Then for $-n \leq a_i \leq 0$ define the quotient sheaves supported on D_i

$$Q^i_{[a_i]} := \frac{E|_{D_i}}{F^i_{a_i}}$$

and the parabolic structure $F.$ is given by

$$F_{[a_1,\ldots,a_m]} = \ker\left(E \to \oplus_{i=0}^m Q^i_{[a_i]}\right). \tag{11}$$

More generally define a family of multi-indexed quotient sheaves by

$$\begin{aligned} Q^i_{[a_i]} &:= \frac{E|_{D_i}}{F^i_{a_i}} \text{ on } D_i \\ Q^{i,j}_{[a_i,a_j]} &:= \frac{E|_{D_i\cap D_j}}{F^i_{a_i} + F^j_{a_j}} \text{ on } D_i \cap D_j \\ &\vdots \\ Q_{[a_1,a_2,\ldots,a_m]} &:= \frac{E|_{D_1\cap\ldots\cap D_m}}{F^1_{a_1} + \cdots + F^m_{a_m}} \text{ on } D_1 \cap D_2 \cap \ldots \cap D_m. \end{aligned}$$

In these notations we have $-n \leq a_i \leq 0$.

If we consider quotient sheaves as corresponding to linear subspaces of the Grothendieck projective bundle associated to E, then the multiple quotients above are multiple intersections of the $Q^i_{[a_i]}$. The formula (11) extends to a Koszul-style resolution of the component sheaves of the parabolic structure.

Lemma 7.3. *Suppose that the filtrations give a locally abelian parabolic structure, in particular they satisfy the conditions of Lemma 2.1. Then for any $-n \le a_i \le 0$ the following sequence is well defined and exact:*

$$0 \to F_{[a_1,a_2,\ldots,a_m]} \to E \to \bigoplus_{i=1}^{m} Q^i_{[a_i]} \to \bigoplus_{i<j} Q^{i,j}_{[a_i,a_j]} \to \cdots \to Q_{[a_1,a_2,\ldots,a_m]} \to 0.$$

Proof. The maps in the exact sequence are obtained from the quotient structures of the terms with alternating signs like in the Čech complex. We just have to prove exactness. This is a local question. By the locally abelian condition, we may assume that E with its filtrations is a direct sum of rank one pieces. The formation of the sequence, and its exactness, are compatible with direct sums. Therefore we may assume that E has rank one, and in fact $E \cong \mathcal{O}_X$.

In the case where E is the rank one trivial bundle, the filtration steps are either 0 or all of $\mathcal{O}_{D_i}$. In particular, there is $-n < b_i \le 0$ such that $F^i_j = \mathcal{O}_{D_i}$ for $j \ge b_i$ and $F^i_j = 0$ for $j < b_i$. Then

$$Q^{i_1,\ldots,i_k}_{[a_{i_1},\ldots,a_{i_k}]} = \mathcal{O}_{D_{i_1}\cap\cdots\cap D_{i_k}}$$

if $a_{i_j} < b_{i_j}$ for all $j = 1, \ldots, k$, and the quotient is zero otherwise.

The sequence is defined for each multiindex $a_1, \ldots, a_m$. Up to reordering the coordinates which doesn't affect the proof, we may assume that there is $p \in [0, m]$ such that $a_i < b_i$ for $i \le p$, but $a_i \ge b_i$ for $i > p$. In this case, the quotient is nonzero only when $i_1, \ldots, i_k \le p$. Furthermore,

$$F_{[a_1,\ldots,a_m]} = \mathcal{O}(-D_1 - \cdots - D_p).$$

In local coordinates, the divisors $D_1, \ldots, D_p$ are coordinate divisors. Everything is constant in the other coordinate directions which we may ignore. The complex in question becomes

$$\mathcal{O}(-D_1 - \cdots - D_p) \to \mathcal{O} \to \oplus_{1\le i\le p}\mathcal{O}_{D_i} \to \oplus_{1\le i<j\le p}\mathcal{O}_{D_i\cap D_j} \to \ldots \to \mathcal{O}_{D_1\cap\cdots\cap D_p}.$$

Etale locally, this is exactly the same as the exterior tensor product of p copies of the resolution of $\mathcal{O}_{\mathbb{A}^1}(-D)$ on the affine line $\mathbb{A}^1$ with divisor D corresponding to the origin,

$$\mathcal{O}_{\mathbb{A}^1}(-D) \longrightarrow \mathcal{O}_{\mathbb{A}^1} \longrightarrow \mathcal{O}_D \longrightarrow 0.$$

In particular, the exterior tensor product complex is exact except at the beginning, where it resolves $\mathcal{O}(-D_1 - \cdots - D_p)$ as required. □

Using the resolution of Lemma 7.3 we can compute the Chern character of $F_{[a_1,a_2,\ldots,a_m]}$ in terms of the Chern character of sheaves supported on intersection of the divisors $D_{i_1} \cap \cdots \cap D_{i_r}$. This gives us

$$\mathrm{ch}(F_{[a_1,a_2,\ldots,a_m]}) = \mathrm{ch}(E) + \sum_{k=1}^{m}(-1)^k \sum_{i_1<i_2<\cdots<i_k} \mathrm{ch}(Q^{i_1,\ldots,i_k}_{[a_{i_1},\ldots,a_{i_k}]}).$$

Substituting this formula for $\mathrm{ch}(F_{[a_1,a_2,\ldots,a_m]})$ into Theorem 5.8, or rather into (10) of Proposition 7.1 with $b_i = -n$, we obtain the following formula for the associated parabolic bundle. Note that the limits of the sums are different in the numerator and denominator, as in (10). Also the term $\mathrm{ch}(E)$ occurs with a different factor in the numerator and denominator; the ratio of these factors is $e^{\sum_{i=1}^m D_i} = e^D$.

Corollary 7.4. *If F is a locally abelian parabolic bundle then*

$$\mathrm{ch}(F) = e^D \mathrm{ch}(E) +$$

$$\frac{\sum_{a_1=-n}^{-1} \cdots \sum_{a_m=-n}^{-1} e^{-\sum_{i=1}^m \frac{a_i}{n} D_i} \sum_{k=1}^m (-1)^k \sum_{i_1<i_2<\cdots<i_k} \mathrm{ch}(Q^{i_1,\ldots,i_k}_{[a_{i_1},\ldots,a_{i_k}]})}{\sum_{a_1=0}^{n-1} \cdots \sum_{a_m=0}^{n-1} e^{-\sum_{i=1}^m \frac{a_i}{n} D_i}}.$$

In fact, we can also write the formula in terms of an associated graded. For this, fix $1 \le i_1 < \cdots < i_k \le m$ and analyze the quotient $Q^{i_1,\ldots,i_k}_{[a_{i_1}-1,\ldots,a_{i_k}-1]}$ along the multiple intersection $D_{i_1\cdots i_k}$. There, the bundle $E|_{D_{i_1\cdots i_k}}$ has k filtrations $F^{i_j}_{a_{i_j}}|_{D_{i_1\cdots i_k}}$ indexed by $-n \le a_{i_j} \le 0$, leading to a multiple-associated-graded defined as follows. For $-n \le a_{i_j} \le 0$ put

$$F^{i_1,\ldots,i_k}_{[a_{i_1},\ldots,a_{i_k}]} := \bigcap_{j=1}^{k} F^{i_j}_{a_{i_j}}|_{D_{i_1\cdots i_k}}.$$

Then define

$$Gr^{i_1,\ldots,i_k}_{[a_{i_1},\ldots,a_{i_k}]} := \frac{F^{i_1,\ldots,i_k}_{[a_{i_1},\ldots,a_{i_k}]}}{\sum_{q=1}^k F^{i_1,\ldots,i_k}_{[a_{i_1},\ldots,a_{i_q}-1,\ldots a_{i_k}]}} \tag{12}$$

where the indices in the denominator are almost all a_{i_j} but one $a_{i_q} - 1$. A good way to picture this when $k = 2$ is to draw a square divided into a grid whose sides are the intervals $[-n, 0]$. The filtrations correspond to horizontal and vertical half-planes intersected with the square. Pieces of the associated-graded are indexed by grid squares, indexed by their upper right points. Thus the pieces are defined for $1 - n \le a_{i_j} \le 0$.

If the parabolic structure is locally abelian then the filtrations admit a common splitting and we have

$$Gr^{i_1,\ldots,i_k}_{[a_{i_1},\ldots,a_{i_k}]} = Gr^{F^{i_1}}_{a_{i_1}} Gr^{F^{i_2}}_{a_{i_2}} \cdots Gr^{F^{i_k}}_{a_{i_k}} (E|_{D_{i_1\cdots i_k}}),$$

or more generally the same thing in any order. Without the common splitting hypothesis, the multi-graded defined previously would not even have dimensions which add up.

The multi-quotient has an induced multiple filtration whose associated-graded is a sum of pieces of the multi-graded defined above. In the $k = 2$ picture, the multi-quotient corresponds to a rectangle in the upper right corner of the square. For example, we have

$$Gr^{i_1,\ldots,i_k}_{[a_{i_1},\ldots,a_{i_k}]} \cong \ker\left(Q^{i_1,\ldots,i_k}_{[a_{i_1}-1,\ldots,a_{i_k}-1]} \to \bigoplus_{j=1}^{k} Q^{i_1,\ldots,i_k}_{[a_{i_1}-1,\ldots,a_{i_j},\ldots,a_{i_k}-1]}\right)$$

where in the direct sum, the indices are all $a_{i_l}-1$ except for one which is a_{i_j}.

Thus in the Grothendieck group of sheaves on $D_{i_1}\cap\cdots\cap D_{i_k}$, we have an equivalence

$$Q^{i_1,\ldots,i_k}_{[a_{i_1},\ldots,a_{i_k}]} \sim \bigoplus_{\underline{c},\, a_{i_j}<c_{i_j}\leq 0} Gr^{i_1,\ldots,i_k}_{[c_{i_1},\ldots,c_{i_k}]}.$$

This gives us the following formula, based on Corollary 7.4 which in turn comes from (10) of Proposition 7.1 (thus as before the limits of the sum in the numerator and denominator are different).

Corollary 7.5. *Suppose F is a locally abelian parabolic structure. Define the multi-associated-graded by* (12) *above. Then we have the formula*

$$\mathrm{ch}(F) = e^D\mathrm{ch}(E) +$$

$$\frac{\sum_{-n\leq a_1,\ldots,a_m<0} e^{-\sum_{i=1}^m \frac{a_i}{n}D_i}\sum_{k=1}^m(-1)^k\sum_{i_1<i_2<\cdots<i_k}\sum_{a_{i_j}<c_{i_j}\leq 0}\mathrm{ch}(Gr^{i_1,\ldots,i_k}_{[c_{i_1},\ldots,c_{i_k}]})}{\sum_{a_1=0}^{n-1}\cdots\sum_{a_m=0}^{n-1} e^{-\sum_{i=1}^m\frac{a_i}{n}D_i}}.$$

7.4. The case of a single smooth divisor

In the case when there is only one smooth divisor component D this formula becomes

$$\mathrm{ch}(F) = e^D\mathrm{ch}(E) - \frac{\sum_{-n<c\leq 0}\left(\sum_{-n\leq a<c} e^{-\frac{a}{n}D}\right)\mathrm{ch}(Gr_{[c]})}{\sum_{0\leq a<n} e^{-\frac{a}{n}D}}. \tag{13}$$

This can be simplified using the identity $(1+x+\cdots+x^{n-1}) = (1-x)^{-1}(1-x^n)$ applied to $x = e^{-\frac{1}{n}D}$, which gives

$$\mathrm{ch}(F) = e^D\mathrm{ch}(E) - \sum_{-n<c\leq 0}\frac{e^D - e^{-\frac{c}{n}D}}{1-e^{-D}}\mathrm{ch}(Gr_{[c]}).$$

We can again rewrite this in terms of the rational indexing in the interval $(-1,0]$, denoting by Gr_α the graded $Gr_{[n\alpha]}$. The formula becomes

$$\mathrm{ch}(F) = e^D\mathrm{ch}(E) - \sum_{-1<\alpha\leq 0}\frac{e^D - e^{-\alpha D}}{1-e^{-D}}\mathrm{ch}(Gr_\alpha). \tag{14}$$

The expression on the right should be interpreted formally, in the sense that the exponentials are written as power series, then the division is done formally, and finally the resulting power series is applied to $D\in CH^1(X)_{\mathbb{Q}}$. The result is a polynomial in D because of the nilpotence of the product structure on $CH^{>0}(X)_{\mathbb{Q}}$.

Our formula still is not in optimal form. One checks that it gives the right formula for a line bundle $F = \mathcal{O}(\frac{b}{n}D)$. We leave it to the reader to make the analogous transformations of the formula in the case of several divisors, possibly meeting only pairwise as a start, and to compare the result with the codimension 2 formulae of Panov [Pa] and Mochizuki [Mo2].

A.J. de Jong pointed out that one would also like to compare this with the formula given by Esnault and Viehweg [Es-Vi, Corollary (B.3), p. 186] for the global Newton class of a flat bundle in terms of local contributions. Given a flat bundle on $X - D$, one associates a parabolic bundle in a natural way and we would expect the formula of [Es-Vi] to be a simple consequence of the fact that the parabolic Chern classes of the resulting bundle are zero at least in rational cohomology. Indeed, the overall shape of the formula in [Es-Vi] is very similar to the ones we are considering here, namely the global contribution from the bundle on X is balanced out by local contributions from the graded pieces of the parabolic structure. However, it seems that the comparison with [Es-Vi] is not immediate: one would need to make use of some additional special identities which must be satisfied by the $\mathrm{ch}(Gr_\alpha(E))$ due to the fact that the parabolic structure comes from a flat bundle. All in all, it seems clear that there is much room for further progress in understanding this question.

8. Parabolic bundles with real weights

In this section we consider parabolic bundles with real weights and define their Chern character and pullback bundles.

Let X be a smooth variety and D be a normal crossing divisor on X. Write $D = D_1 + \cdots + D_m$ where D_i are the irreducible smooth components and meeting transversally.

A *parabolic bundle* on (X, D) is a collection of vector bundles F_α indexed by multi-indices $\alpha = (\alpha_1, \ldots, \alpha_k)$ with $\alpha_i \in \mathbb{R}$, satisfying the same conditions as recalled in §2. The structure is determined by the sheaves F_α for a finite collection of indices α with $0 \leq \alpha_i < 1$, the *weights*.

Remark 8.1. *A parabolic bundle with rational weights and denominator n can be considered as a parabolic bundle with real weights by setting*

$$F_{(t_1, t_2, \ldots, t_m)} := F_{[[nt_1],[nt_2],\ldots,[nt_m]]} = F_{(\frac{[nt_1]}{n}, \frac{[nt_2]}{n}, \ldots, \frac{[nt_m]}{n})}$$

where $[nt_i]$ is the greatest integer less than or equal to nt_i, for any $t_i \in [0, 1) \subset \mathbb{R}$.

We say that F is *locally abelian* if in a Zariski neighbourhood of any point $x \in X$, F is isomorphic to a direct sum of parabolic line bundles with real coefficients.

8.1. Perturbation of parabolic bundles with real weights

The following construction is a simplified version of the one Mochizuki [Mo2, §3.3] considered, and which suffices for our purpose. Variations of parabolic weights were considered earlier in [Me-Se], [Bd-Hu], [Th].

Suppose F is a parabolic bundle with real weights on a smooth variety (X, D). Consider the real weights

$$\{\alpha = (\alpha_1, \alpha_2, \ldots, \alpha_m) : 0 \leq \alpha_i \leq 1\}.$$

By definition

$$F_\alpha|_{D_i} = \frac{F_\alpha}{F_{\alpha-\delta_i}}$$

and denote the image

$$\overline{F_{\alpha;D_i,\gamma_i}} := \mathrm{Im}\left(\mathrm{F}_{(\alpha_1,\dots,\gamma_i,\dots,\alpha_m)} \longrightarrow \mathrm{F}_\alpha|_{\mathrm{D_i}}\right)$$

whenever $\alpha_i - 1 < \gamma_i \leq \alpha_i$.

Note that if γ is a multiindex with $\alpha_i - 1 < \gamma_i \leq \alpha_i$ then we have an exact sequence

$$0 \to F_\gamma \to F_\alpha \to \bigoplus_i \frac{F_\alpha|_{D_i}}{\overline{F_{\alpha;D_i,\gamma_i}}}.$$

Consider the graded sheaves

$$\mathbf{gr}^i_{\alpha;\gamma_i} F := \frac{F_\alpha|_{D_i}}{\overline{F_{\alpha;D_i,\gamma_i}}}.$$

By the semicontinuity condition there are finitely many indices and γ_i such that the graded sheaves $\mathbf{gr}^i_{\alpha_i-\gamma_i} F$ are non-zero.

Let

$$r_{\alpha_i} = \min\{|\alpha_\mathrm{i} - \gamma_\mathrm{i}| \;:\; \mathbf{gr}^\mathrm{i}_{\alpha_\mathrm{i}/\gamma_\mathrm{i}} \mathrm{F} \neq 0\}$$

Choose ϵ_{α_i} such that $\epsilon_{\alpha_i} < r_{\alpha_i}$ and $\alpha_i + \epsilon_i$ is a rational number, for each i.

The following construction was used by Mochizuki in [Mo2, §3.4].

Definition 8.2. *A parabolic bundle F^ϵ with rational weights $a_i = \alpha_i + \epsilon_{\alpha_i}$ is defined by setting:*

$$F^\epsilon_{[a_1,a_2,\dots,a_m]} := F_{\alpha_1+\epsilon_{\alpha_1},\dots,\alpha_m+\epsilon_{\alpha_m}}.$$

We call F^ϵ an ϵ–perturbation of F on X.

For any rational weights $t = [t_1, \dots, t_m]$, we have the inclusion of sheaves

$$F_t \hookrightarrow F^\epsilon_t$$

In other words, we can write

$$F \hookrightarrow F^\epsilon.$$

Write $\epsilon = \{\epsilon_{\alpha_i}\}$, where α_i runs over the finite set of real weights which determine F.

Suppose $\{F^i\}_{i\in I}$ is a projective system of parabolic bundles indexed by an ordered set I with inclusions $F^i \hookrightarrow F^j$ for $i \leq j$. Define the *intersection* by the formula

$$\left(\bigcap_{i\in I} F^i\right)_\alpha := \bigcap_{i\in I} F^i_\alpha.$$

This defines a parabolic sheaf. We say that the collection $\{F^i\}_{i\in I}$ is *simultaneously locally abelian* if there is an etale covering of X such that on the pullback to this etale covering, each of the F^i admits a direct sum decomposition as a sum of parabolic line bundles, and the inclusion maps are compatible with these direct sum decompositions. Inclusions of parabolic line bundles are just inequalities of real divisors, and the intersection of a family of parabolic line bundles just corresponds to taking the inf of the family of real coefficients. Thus we have the following useful fact.

Lemma 8.3. *If $\{F^i\}_{i\in I}$ is a simultaneously locally abelian projective system of inclusions of parabolic bundles, then the intersection $\bigcap_{i\in I} F^i$ is a locally abelian parabolic bundle.*

Lemma 8.4. *Suppose F is a locally abelian parabolic bundle with real weights $\alpha = (\alpha_1, \ldots, \alpha_m)$ on (X, D). Then any ϵ–perturbation F^ϵ of F is also locally abelian with the same decomposition. Thus the family of F^ϵ is a simultaneously locally abelian projective system of inclusions. Taking the intersection we have*

$$F = \bigcap_{\epsilon \to 0} F^\epsilon.$$

Proof. Since this is a local question, we assume that

$$F = \oplus_j \mathcal{O}(\sum \gamma_j^i.D_i)^{n_j}$$

for some $\gamma_j^i \in \mathbb{R}$. Any ϵ-perturbation of F is

$$F^\epsilon = \oplus_j \mathcal{O}(\sum a_j^i.D_i)^{n_j}$$

where $a_j^i = \gamma_j^i + \epsilon_j^i$ are rational numbers and ϵ_j^i are small. Hence F^ϵ is locally abelian. □

8.2. Pullback of parabolic bundles with real weights

Consider a morphism

$$f : (Y, D') \longrightarrow (X, D)$$

such that $f^{-1}(D) \subset D'$. Here X, Y are smooth varieties and D, D' are normal crossing divisors on X and Y respectively.

In [Iy-Si, Lemma 2.6], the pullback of a locally abelian parabolic bundle with rational weights was defined, using its correspondence with usual vector bundles on a DM-stack. Our aim here is to define the pullback f^*F on (Y, D') of a locally abelian parabolic bundle F with real weights on (X, D).

Lemma 8.5. *Suppose F is a locally abelian parabolic bundle with real weights on (X, D). For any morphism $f : (Y, D') \longrightarrow (X, D)$ such that $f^{-1}D \subset D'$, we can define the pullback f^*F on (Y, D) as a locally abelian parabolic bundle with real weights.*

Proof. By Lemma 8.4, we can write

$$F = \bigcap_{\epsilon \to 0} F^\epsilon.$$

By [Iy-Si, Lemma 2.6], f^*F^ϵ is a locally abelian parabolic bundle with rational weights.

Locally, by Lemma 8.4, each F^ϵ is locally abelian, and the decompositions are compatible for different ϵ. Thus we can write locally

$$f^*F^\epsilon = \oplus_j \mathcal{O}(\sum a^i_j(\epsilon).D'_i)^{n_j}$$

where $a^i_j(\epsilon)$ are rational numbers depending on ϵ. In other words, the pullbacks form a simultaneously locally abelian projective system. By Lemma 8.4, we can define the pullback of F as the intersection

$$f^*F := \bigcap_{\epsilon \to 0} f^*F^\epsilon,$$

and it is a locally abelian parabolic bundle. In fact, locally let $\alpha^i_j = \lim_{\epsilon\to 0} \mathrm{a}^i_j(\epsilon)$ (which converges and is a real number), then

$$f^*F = \oplus_j \mathcal{O}(\sum a^i_j.D'_i)^{n_j}.$$

□

8.3. Tensor products of parabolic bundles with real weights

Suppose F and G are two locally abelian parabolic bundles with real weights. We would like to define their tensor product. Recall that by [Iy-Si, Lemma 2.3], the tensor product of locally abelian parabolic bundles with rational weights can be defined using the correspondence with usual vector bundles on a DM–stack.

Lemma 8.6. *Suppose F and G are locally abelian parabolic bundles with real weights on (X, D). Then we can define $F \otimes G$ as a locally abelian parabolic bundle with real weights.*

Proof. By Lemma 8.4, we can write

$$F = \bigcap_{\epsilon\to 0} F^\epsilon, \quad G = \bigcap_{\epsilon\to 0} G^\epsilon$$

The families $\{F^\epsilon\}_{\epsilon\to 0}$ and $\{G^{\epsilon'}\}_{\epsilon'\to 0}$ are simultaneously locally abelian, and we can take a common refinement of the two coverings so that they are locally abelian with respect to the same covering. Then the family of tensor products $\{F^\epsilon \otimes G^{\epsilon'}\}_{\epsilon,\epsilon'\to 0}$ is again simultaneously locally abelian with respect to the same decomposition and we can define

$$F \otimes G := \bigcap_{\epsilon,\epsilon'\to 0} F^\epsilon \otimes G^{\epsilon'}.$$

□

One can also consider duals and internal $\underline{Hom}$.

8.4. Description by filtrations on a linear constructible decomposition of the space of weights

For both of the operations defined above, the description in terms of filtrations can jump when the parabolic weights cross "walls". Fix a vector bundle E and filtrations of E_{D_i}. These filtrations determine an open subset of possible assignments of weights α^j_i to the filtrations F^j_i with $\alpha^{j-1}_i < \alpha^j_i$. This defines an open subset $W(E, \{F_i^j\}) \subset \mathbb{R}^N$. Note that the locally abelian condition doesn't depend on the choice of weights but is just a statement about the filtrations. However, when we apply the pullback operation for a map $(Y, D') \to (X, D)$ the filtrations on the pullback bundle might depend on the choice of weights $\underline{\alpha} \in W(E, \{F_i^j\})$.

A subset of $\mathbb{R}^N$ is *linear-constructible* if it is defined by a finite number of linear equalities and inequalities. It is $\mathbb{Q}$*-linear-constructible* if the equalities and inequalities have coefficients in $\mathbb{Q}$.

The filtrations for the pullback parabolic bundle are fixed over a $\mathbb{Q}$-linear constructible stratification of the space of weights. This phenomenon is somewhat similar to what was observed by Budur in [Bu].

Proposition 8.7. *Suppose $f : (Y, D') \to (X, D)$ is a morphism of smooth varieties with normal crossings divisors in good position. Suppose $(E, \{F_i^j\})$ is a locally abelian datum of filtrations for a parabolic structure on (X, D). There is a stratification of $W(E, \{F_i^j\})$ into a finite disjoint union of $\mathbb{Q}$-linear constructible sets $W(p)$ such that over each stratum, there is a fixed collection of filtrations $\widetilde{F}_i^j(p)$ for the pullback bundle $\widetilde{E} := f^*E$ and a $\mathbb{Q}$-linear function of weights $f^*(p) : W(p) \to W(\widetilde{E}, \{\widetilde{F}_i^j(p)\})$ such that for $\alpha \in W(p)$ the pullback of the parabolic bundle $(E, \{F_i^j\}, \alpha)$ is equal to $(f^*E, \{\widetilde{F}_i^j(p)\}, f^*(p)(\alpha))$.*

We leave the proof to the reader.

A similar statement holds for tensor product, which is again left to the reader.

8.5. Chern character of parabolic bundles with real weights

Suppose $K \subset \mathbb{R}$ is a subfield, and suppose V is a K-vector space. If $f \in V \otimes K[x]$ then we can define in a formal way $\int_0^1 f \in V$. The same is true if f is a formal piecewise polynomial function whose intervals of different definitions are defined over K. A similar remark holds for multiple integrals – in the case we shall consider the domains of piecewise definition will be products of intervals defined over K but this could also extend to K-linear constructible regions.

Using this meaning, the formula of Theorem 5.8 may be rewritten replacing sums by integrals:

$$\mathrm{ch}(F) = \frac{\int_{\alpha_1=0}^1 \cdots \int_{\alpha_m=0}^1 e^{-\sum_{i=1}^m \alpha_i D_i} \mathrm{ch}(F_\alpha)}{\int_{\alpha_1=0}^1 \cdots \int_{\alpha_m=0}^1 e^{-\sum_{i=1}^m \alpha_i D_i}}. \tag{15}$$

In this formula note that the exponentials of real combinations of divisors are interpreted as formal polynomials. The power series for the exponential terminates because the product structure of $CH^{>0}(X)$ is nilpotent.

If F is a parabolic bundle with rational weights, then this still takes values in $CH^{\cdot}(X)_{\mathbb{Q}}$.

If F is a parabolic bundle with real weights, then the formula (15) may be taken as the *definition* of $\mathrm{ch}(F) \in CH^{\cdot}(X)_{\mathbb{R}} := CH^{\cdot}(X) \otimes_{\mathbb{Z}} \mathbb{R}$. No topology or metric structure is needed on $CH^{\cdot}(X)_{\mathbb{R}}$ because the integrals involved are piecewise polynomials.

Theorem 8.8. *The Chern character of locally abelian parabolic bundles with real weights, is additive for exact sequences, multiplicative for tensor products, and functorial for pullbacks along good morphisms of varieties with normal crossings divisors.*

Proof. Additivity for exact sequences follows from the shape of the formula. Suppose $f : (Y, D') \to (X, D)$ is a good morphism of varieties with normal crossings divisors. Fix a bundle and collection of filtrations $(E, \{F_i^j\})$ on (X, D). The Chern character may then be viewed as a function

$$\mathrm{ch} : W(E, \{F_i^j\}) \to CH^{\cdot}(X)_{\mathbb{R}}.$$

This function is obtained as a polynomial with coefficients which are rational linear combinations of the various Chern classes of the intersections of the filtrations, see §7.3. The same may be said of the Chern character of parabolic bundles over (Y, D') once filtrations are fixed. Use Proposition 8.7 to decompose the space $W(E, \{F_i^j\})$ into a finite union of $\mathbb{Q}$-linear constructible subsets on which the filtrations of the pullback parabolic structure will be invariant. Over these subsets the Chern character of the pullback parabolic structures are again polynomials with coefficients in $CH^{\cdot}(X)_{\mathbb{Q}}$. On the other hand, by [Iy-Si, Lemma 2.8], whenever the weights are rational we have that the Chern character of the pullback is the pullback of the Chern character. We therefore have two polynomials with $CH^{\cdot}(X)_{\mathbb{Q}}$ coefficients which agree on the rational points of a certain $\mathbb{Q}$-linear constructible set. It follows that the polynomial functions into $CH^{\cdot}(X)_{\mathbb{R}}$ agree on the real points of the $\mathbb{Q}$-linear constructible set. This proves compatibility of the Chern character for pullbacks of real parabolic bundles.

The proof for tensor products is similar, using the analogue of Proposition 8.7. □

9. Variants

In this section we consider a variant of the notion of parabolic structures for the case of a divisor with multiple points, and also a variant of the construction of parabolic bundle associated to a logarithmic connection, concerning the case of unipotent monodromy at infinity. In both cases, we will restrict to the case when X is a smooth projective surface.

9.1. Parabolic structures at multiple points

Let X be a nonsingular projective surface. Let $D \subset X$ be a divisor such that $D = \cup_{i=1}^m D_i$ and D_i are smooth and irreducible curves. Let $P = \{P_1, \ldots, P_r\}$ be a set of points. Assume that the points P_j are crossing points of D_i, and that they are general multiple points, that is, through a crossing point P_j we have divisors $D_1, \ldots, D_k$ which are pairwise transverse. Assume that D has normal crossings outside of the set of points P.

Let $\pi : X' \longrightarrow X$ be the blow–up of X at P and E be the exceptional divisor on X'; note that E is a sum of disjoint exceptional components E_j over the points P_j respectively. The pullback divisor $D' = \sum_{i=1}^{m} D'_i + E$ is a normal crossing divisor, where D'_i is the strict transform of D_i, for $1 \leq i \leq m$.

We define a notion of *exceptionally constant parabolic structure on* (X, D, P). The term "exceptionally constant" means that the parabolic structure pulls back to one which is constant along the exceptional divisors. Following notation of Mochizuki [Mo2] we fix an origin for the filtrations which is a multi-index $\mathbf{c}$. This may be important in the present case since the structures might differ for different values of $\mathbf{c}$.

Definition 9.1. *Fix a positive integer n for the denominator, and an uplet of integers* $\mathbf{c} = (c_{D,1}, \ldots, c_{D,m}, c_{P,1}, \ldots, c_{P,r})$. *An* exceptionally constant parabolic structure on (X, D, P) (denoted by $(H, F^{\cdot}_{\cdot}, G^{\cdot}_{\cdot})$) with origin $\mathbf{c}$ *consists of a vector bundle H on X together with filtrations F^i on the restrictions H_{D_i} of H on D_i, and furthermore filtrations G^j of the vector spaces H_{P_j}. The indexing of these filtrations is F^i_j for $c_{D,i} - n \leq j \leq c_{D,i}$ with $F^i_{c_{D,i}} = H|_{D_i}$ and $F^i_{c_{D,i}-n} = 0$, and G^j_k for $c_{P,j} - n \leq k \leq c_{P,j}$ with analogous end conditions.*

Let $H' = \pi^* H$ be the pullback of the vector bundle H. The filtrations F^j_i along the D'_i and G^j_k along the exceptional divisors E_j determine a parabolic structure denoted $\Phi(H, F^{\cdot}_{\cdot}, G^{\cdot}_{\cdot})$ over $(X', D'+E)$. By Lemma 2.3, it is automatically locally abelian.

We can use the formula of Theorem 5.8 to obtain a formula for the Chern character of $\Phi(H, F^{\cdot}_{\cdot}, G^{\cdot}_{\cdot})$

Consider the push–forward map

$$\pi_* : CH_{\cdot}(X') \otimes \mathbb{Q} \longrightarrow CH_{\cdot}(X) \otimes \mathbb{Q}$$

We define the Chern character of the exceptionally constant parabolic structure on X, $(H, F^{\cdot}_{\cdot}, G^{\cdot}_{\cdot})$, to be

$$\mathrm{ch}(H, F^{\cdot}_{\cdot}, G^{\cdot}_{\cdot}) := \pi_* \mathrm{ch}\ \Phi(H, F^{\cdot}_{\cdot}, G^{\cdot}_{\cdot}).$$

9.2. Parabolic bundles associated to unipotent monodromy at infinity

Recall that one can associate a parabolic bundle to a logarithmic connection with rational residues, in a canonical way, such that the weights correspond to the eigenvalues of the residues (see [Iy-Si] or §10 below). In this section, we point out that one can do something substantially different, in the case of nilpotent residues. Suppose (E, ∇) is a logarithmic connection on X, with singularities along a normal-crossings divisor $D = D_1 + \cdots + D_m$, such that the residue η_i of ∇ are nilpotent, for each $i = 1, \ldots, m$. In other words, (E, ∇) is the Deligne extension of a flat bundle with unipotent monodromy at infinity.

In this case, we still have some different natural filtrations along divisor components, but the eigenvalues of the residue are zero so there is no canonical choice of weights. Instead, define some characteristic numbers by arbitrarily assigning weights to these filtrations. Assume that X is a surface here, so that the resulting parabolic structures will automatically be locally abelian. It seems to be an interesting question to determine when the locally abelian condition holds for these kinds of filtrations in the case of dimension ≥ 3.

Consider the *Image filtration* on the restriction E_{D_i} of E to a divisor component:

$$E_{D_i} = F_0^i \supset F_1^i \supset \cdots \supset F_{l_i-1}^i \supset F_{l_i+1}^i = 0$$

where

$$F_j^i := \text{image } (\eta_i^j : E_{D_i} \longrightarrow E_{D_i}),$$

$\eta_i^j := \eta_i \circ \eta_i \circ \cdots \circ \eta_i$ (j-times) and $l_i + 1$ is the order of η_i.

Alternatively, we can consider the *Kernel filtration* induced by the kernels of the operator η_i: write

$$F_j^i := \text{kernel } (\eta_i^{l_i+1-j} : E_{D_i} \longrightarrow E_{D_i}).$$

Mixing these two filtrations gives rise to the *monodromy weight filtration* $\{W_l\}$ defined by Deligne [De3]. This is an increasing filtration

$$\{0\} \subset W_0 \subset W_1 \subset \cdots \subset W_{2l_i} = E_{D_i}$$

uniquely determined by the conditions:

- $\eta_i(W_l) \subset W_{l-2}$
- the induced map $\eta_i^l : \mathrm{Gr}_{k+l}(W_*) \to \mathrm{Gr}_{k-l}(W_*)$ is an isomorphism for each l.

Here $\mathrm{Gr}_l(W_*) := W_l/W_{l-1}$.

Explicitly, the filtration is defined by induction as follows: let

$$W_0 = \text{image}(\eta_i^{l_i}) \text{ and } W_{2l_i-1} = \ker(\eta_i^{l_i}).$$

Now fix some $l < l_i + 1$; if

$$0 \subset W_{l-1} \subset W_{2l_i-l} \subset W_{2l_i} = E_{D_i}$$

has already been defined in such a way that

$$\eta_i^{l_i-l+1}(W_{2l_i-l}) \subset W_{l-1}$$

then we define

$$W_l/W_{l-1} = \text{image}(\eta_i^{l_i-l} : W_{2l_i-l}/W_{l-1} \longrightarrow W_{2l_i-l}/W_{l-1})$$

and W_l, W_{2l_i-l-1} to be the corresponding inverse images. Notice that

$$W_l/W_{l-1} \subset W_{2l_i-l-1}/W_{l-1}$$

so that $W_l \subset W_{2l_i-l-1}$. Clearly, $\eta_i^{l_i-1}(W_{2l_i-l-1}) \subset W_l$, so that the induction hypothesis is satisfied.

Lemma 9.2. *Suppose X is a surface. Consider the* Image *or the* Kernel *or the* monodromy weight filtrations *considered above, on the restrictions E_{D_i} of E to the divisor components. We can associate a locally abelian parabolic bundle on (X, D) with respect to (E_U, ∇_U) together with either of these filtrations by assigning aribitrary weights.*

Proof. By Lemma 2.3, the parabolic structure defined by the filtrations is automatically locally abelian. □

9.3. Examples arising from families

Suppose $\pi : X \longrightarrow S$ is a semi-stable family of projective varieties such that $\pi_U : X_U \longrightarrow U$ is a smooth morphism, for some open subvariety $U \subset S$ and $D := S - U$ is a normal crossing divisor. Let d be the relative dimension of $X \longrightarrow S$.

In this situation, the Gauss–Manin bundles $\mathcal{H}^l := R^l\pi_*(\Omega^{\bullet}_{X/S}(\pi^{-1}D))$ for $0 \leq l \leq 2d$, are equipped with a logarithmic flat connection. Furthermore, the local monodromies are unipotent and $\mathcal{H}^l$ is the Deligne extension of the restriction $\mathcal{H}^l_U$ (see [St]). Let η_i be the residue transformations along the divisor components D_i. Unipotency of the monodromy operators implies nilpotency of η_i and the order of nilpotency is at most $l+1$ (see [La]). In particular, the length of the Image and the Kernel filtrations in the previous subsection is at most $l+1$ and the monodromy weight filtration is of length at most $2l+1$. We make an explicit computation of the Chern character of the associated locally abelian parabolic bundle in the following case:

Suppose S is a surface and $X \longrightarrow S$ is a semi-stable family of abelian varieties. We consider the Gauss-Manin system $\mathcal{H}^1$ of weight one on S. For simplicity assume that D is a smooth irreducible divisor. Then the residue transformation η has order of nilpotency two and in this case the *monodromy weight filtration* is written as

$$\mathcal{H}^1{}_{|D} = W_2 \supset W_1 \supset W_0 \supset W_{-1} = 0.$$

Here $W_1 = \text{kernel}(\eta)$ and $W_0 = \text{image}(\eta)$. The graded pieces

$$\mathbf{gr}_m := \frac{W_m}{W_{m-1}}$$

carry a polarized pure Hodge structure of weight m (see [Sc]). Also, the graded piece of weight two is isomorphic to the piece of weight zero, by the monodromy operator N (in [Sc], N polarizes the mixed Hodge structures).

By Lemma 9.2, we can associate a locally abelian parabolic bundle F on S corresponding to $\{W_{\cdot}\}$, with arbitrary weights $(\alpha_0, \alpha_1, \alpha_2)$ with $-1 < \alpha_0 < \alpha_1 < \alpha_2 \leq 0$.

Lemma 9.3. *Suppose $X \longrightarrow S$ is a semi-stable family of abelian varieties of genus g. Let g_i denote the rank of $\mathbf{gr}_i$ for $i = 0, 1, 2$, thus $g = g_0 + g_1 + g_2$ and $g_0 = g_2$. With notation as above, assigning weights $(\alpha_0, \alpha_1, \alpha_2)$, the Chern character of the locally abelian parabolic bundle F is given by the formula*

$$\text{ch}(F) = \sum_{i=0}^{2} g_i e^{-\alpha_i D} \in CH^{\cdot}(S)_{\mathbb{Q}}.$$

In other words it is Chow-equivalent to a direct sum of parabolic line bundles.

Proof. Let $k : D \hookrightarrow X$ denote the inclusion. Suppose A is a rank r bundle along D whose Chern character is $r \in CH^0(D)_{\mathbb{Q}}$. Then, the sheaf $k_*(A)$ on X has Chern character given by a Riemann-Roch formula. This formula depends only on the Chern character of A on D, in particular it is r times the value for the case $A = \mathcal{O}_D$. In that case we can use the exact sequence

$$0 \to \mathcal{O}(-D) \to \mathcal{O} \to \mathcal{O}_D \to 0$$

to conclude that the Chern character of $k_*(A)$ is $r(1 - e^{-D})$.

Turn now to the situation of the lemma. By [vdG] or [Es-Vi3], we have

$$\mathrm{ch}(F) = g \in CH^0(S)_{\mathbb{Q}}$$

and similarly for $\mathrm{ch}(\mathbf{gr}_1)$ which corresponds to a family of abelian varieties along D, we get

$$\mathrm{ch}(\mathbf{gr}_1) = g_1 \in CH^0(D)_{\mathbb{Q}}.$$

Clearly, $\mathrm{ch}(\mathbf{gr}_0) = g_0 \in CH^0(D)_{\mathbb{Q}}$, thus $\mathrm{ch}(\mathbf{gr}_2) = g_2 \in CH^0(D)_{\mathbb{Q}}$ by the isomorphism between the weight two and weight zero piece given by the monodromy operator. Plugging these into the formula (14) of §7.4 and using the previous paragraph for the Chern characters of $k_*(A)$ we get the formula

$$\mathrm{ch}(F) = e^D g - \sum_{i=0}^{2} \frac{e^D - e^{-\alpha_i D}}{(1 - e^{-D})} g_i (1 - e^{-D}) \in CH^{\cdot}(S)_{\mathbb{Q}}.$$

Simplifying with $g = g_0 + g_1 + g_2$ gives the stated formula. □

10. Extended Reznikov theory for finite order monodromy at infinity

Suppose U is a nonsingular variety defined over the complex numbers. Consider a nonsingular compactification X of U such that $D := X - U$ is a normal crossing divisor. Suppose (E_U, ∇_U) is a bundle with a flat connection on U. Consider the canonical extension (E, ∇) of (E_U, ∇_U) on X (see [De]). Here ∇ is a logarithmic connection on E, i.e.,

$$\nabla : E \longrightarrow E \otimes \Omega_X(\log D)$$

is a $\mathbb{C}$-linear map and satisfies the Leibniz rule. Flatness implies that $\nabla \circ \nabla = 0$.

Consider the sequence induced by the Poincaré residue map

$$E \longrightarrow E \otimes \Omega_X(\log D) \xrightarrow{res} E \otimes \mathcal{O}_D.$$

This induces an operator

$$\eta_i : E_{D_i} \longrightarrow (E \otimes \Omega_X(\log D))_{|D_i} \xrightarrow{res} E_{D_i}$$

called the residue transformation along the component D_i and $\eta_i \in \mathrm{End}(E_{D_i})$.

Definition 10.1. *We say that* (E, ∇) has rational residues *if the eigenvalues of the residue transformations* η_i *above are rational numbers.*

This is equivalent to saying that the local monodromy transformations around the divisor components D_i of D are quasi-unipotent.

If α_i are the rational residues then [De]

$$e^{2\pi i \alpha_i} = \text{eigenvalues of the local monodromy.}$$

Suppose the residues of (E, ∇) are non-zero and rational. In [Iy-Si, Lemma 3.3], a locally abelian parabolic bundle $\mathcal{E}$ on (X, D) was associated to (E, ∇). In fact, $\mathcal{E}$ was associated to the flat connection (E_U, ∇_U) on U and the constituent bundles were defined, using a construction due to Deligne-Manin. If we choose the extension (E, ∇) on X such that the rational residues lie in the interval $[0, 1)$ then the weights are precisely the negatives of the rational residues. In other words, if $0 \leq -\alpha_i^1 < -\alpha_i^2 < \cdots < -\alpha_i^{n_i} < 1$ are the rational residues along D_i then the weights are $\alpha_i^1 > \alpha_i^2 > \cdots > \alpha_i^{n_i}$ along D_i.

10.1. Residues are rational and semisimple

Suppose that the residues are rational and furthermore on the associated-graded of the parabolic structure, the residue of the connection induces a semisimple operator. In this case, the monodromy transformations of the corresponding local system are semisimple with eigenvalues which are roots of unity, thus they are of finite order. If n denotes a common denominator for the rational residues of the connection (and hence for the corresponding parabolic weights) then the monodromy transformations have order n. This implies that the connection extends to a flat connection on the DM-stack $Z := X\langle \frac{D_1}{n}, \ldots, \frac{D_m}{n} \rangle$. Conversely any flat connection on the DM-stack Z gives rise to a connection on U with semisimple and rational residues.

The locally abelian parabolic bundle on X corresponds to the vector bundle on Z underlying the flat bundle as extended over Z. Indeed, when the monodromy transformations have order n, the monodromy around the divisor at infinity in Z is trivial, and in this case the Deligne canonical extension is the vector bundle underlying the extended flat bundle. By [Iy-Si] the Deligne canonical extension over Z is the vector bundle corresponding to the parabolic bundle on X.

10.2. Reznikov's theory in the case of rational semisimple residues

The theory of secondary characteristic classes works equally well on the DM-stack Z. In particular, we can define the rational Deligne cohomology

$$H^{2p}_{\mathcal{D}}(Z, \mathbb{Q}(p)) := \mathbb{H}^{2p}(Z^{\text{an}}; \mathbb{Q}(p) \to \Omega^0_{Z^{\text{an}}} \to \ldots \to \Omega^p_{Z^{\text{an}}}),$$

and also the cohomology

$$H^{2p-1}(Z, \mathbb{C}/\mathbb{Q}) = \mathbb{H}^{2p}(Z^{\text{an}}; \mathbb{Q} \to \Omega^{\cdot}_{Z^{\text{an}}}).$$

Dividing by the Hodge filtration provides a map

$$H^{2p-1}(Z, \mathbb{C}/\mathbb{Q}) \to H^{2p}_{\mathcal{D}}(Z, \mathbb{Q}(p)). \tag{16}$$

On the other hand, the Deligne cycle class map from Chow groups to Deligne cohomology is a map

$$CH^p(Z)_{\mathbb{Q}} \to H^{2p}_{\mathcal{D}}(Z, \mathbb{Q}(p)). \tag{17}$$

If E is a vector bundle on Z then its Chern character in $CH^{\cdot}(Z)_{\mathbb{Q}}$ maps to its Chern character in $\oplus_p H^{2p}_{\mathcal{D}}(Z, \mathbb{Q}(p))$.

Lemma 10.2. *Pullback for the map $Z \to X$ gives an isomorphism of Deligne cohomology groups*

$$H^{2p}_{\mathcal{D}}(X, \mathbb{Q}(p)). \xrightarrow{\cong} H^{2p}_{\mathcal{D}}(Z, \mathbb{Q}(p))$$

compatible with the isomorphism of rational Chow groups and the map (17). *It also induces an isomorphism*

$$H^{2p-1}(X, \mathbb{C}/\mathbb{Q}) \xrightarrow{\cong} H^{2p-1}(Z, \mathbb{C}/\mathbb{Q})$$

and this is compatible with the projection (16).

Suppose F is a locally abelian parabolic bundle on X. Define the Chern character of F in Deligne cohomology of X by using the formula of Theorem 5.8 and taking the Chern characters of the pieces $F_{[a_1,\ldots,a_m]}$ in the Deligne cohomology of X. Thus

$$\mathrm{ch}^{\mathcal{D}}(F) := \frac{\sum_{a_1=0}^{n-1} \cdots \sum_{a_m=0}^{n-1} e^{-\sum_{i=1}^{m} \frac{a_i}{n} c_1^{\mathcal{D}}(D_i)} \mathrm{ch}^{\mathcal{D}}(F_{[a_1,\ldots,a_m]})}{\sum_{a_1=0}^{n-1} \cdots \sum_{a_m=0}^{n-1} e^{-\sum_{i=1}^{m} \frac{a_i}{n} c_1^{\mathcal{D}}(D_i)}}. \tag{18}$$

The products are taken with the product structure of Deligne cohomology which is compatible with the intersection product in Chow groups [Es-Vi2].

Corollary 10.3. *Suppose F is a locally abelian parabolic bundle on X with n as common denominator for the rational weights, corresponding to a vector bundle E on Z. Then $\mathrm{ch}^{\mathcal{D}}(F)$ as given by the above formula* (18), *pulls back to $\mathrm{ch}^{\mathcal{D}}(E)$ on Z via the isomorphism of Lemma* 10.2.

Proof. By Theorem 5.8 this is the case for the Chern character in Chow groups, and we have the compatibility of the isomorphism of Lemma 10.2 with the projection (17). □

Now, go back to the situation where (E_U, ∇_U) is a flat bundle on U with rational and semisimple residues. It extends to a flat bundle (E, ∇) on Z and also the local system L_U on U extends to a local system L on Z.

Consider a Kawamata cover (see [Kaw])

$$f : Y \longrightarrow X$$

so that Y is a smooth projective variety. Then there is a factorization

$$Y \xrightarrow{h} Z \xrightarrow{\pi} X$$

such that $f = \pi \circ h$. The flat connection on Z pulls back to a flat connection (E_Y, ∇_Y) on Y. Thus, Esnault's theory of secondary classes for flat bundles [Es] gives a class $\widehat{c}_p(L) \in H^{2p-1}(Z, \mathbb{C}/\mathbb{Q})$. By [Es], this class projects under the map (16) to the Deligne Chern class $c_p^{\mathcal{D}}(E)$ for the vector bundle E on Z.

Proposition 10.4. *Reznikov's result on the vanishing of the rational secondary classes works equally well over a smooth projective DM-stack. Thus, with the above notation* $\widehat{c}_p(L) = 0$ *in* $H^{2p-1}(Z, \mathbb{C}/\mathbb{Q})$*, for* $p \geq 2$.

Proof. Either of Reznikov's proofs of [Re] work equally well over the DM-stack Z. Alternatively, we can reduce to the utilisation of [Re] on the finite cover Y as follows: by Reznikov's theorem the secondary classes of (E_Y, ∇_Y) are trivial in the $\mathbb{C}/\mathbb{Q}$-cohomology in degrees ≥ 3 of Y. The map $Y \to Z$ induces an injection $H^i(Z, V) \hookrightarrow H^i(Y, V)$ for any $\mathbb{Q}$-vector space V, in particular $V = \mathbb{C}/\mathbb{Q}$. This implies that the secondary classes vanish on Z

$$\widehat{c}_p(L) = 0 \in H^{2p-1}(Z, \mathbb{C}/\mathbb{Q})$$

for $p \geq 2$. □

Combining with our formula of Theorem 5.8 we obtain a formula for an element of the Deligne cohomology over the compactification X of U which vanishes by Reznikov's theorem.

Corollary 10.5. *Suppose* (E_U, ∇_U) *is a flat bundle on* U *with rational and semisimple residues, or equivalently the monodromy transformations at infinity are of finite order. Let* F *denote the corresponding locally abelian parabolic bundle. Define the Deligne Chern character* $\mathrm{ch}^{\mathcal{D}}(F)$ *on* X *by the formula* (18). *Then the rational Deligne Chern classes* $\mathrm{c}_p^{\mathcal{D}}(F)$ *in all degrees* ≥ 2 *vanish.*

Proof. This follows from Corollary 10.3 and Proposition 10.4. □

References

[Bi] I. Biswas, *Parabolic bundles as orbifold bundles*, Duke Math. J. **88** (1997), no. 2, 305–325.

[Bi2] I. Biswas, *Chern classes for parabolic bundles*, J. Math. Kyoto Univ. **37** (1997), no. 4, 597–613.

[Bl-Es] S. Bloch and H. Esnault, *Algebraic Chern-Simons theory*, Amer. J. Math. **119** (1997), no. 4, 903–952.

[Bd] H.U. Boden, *Representations of orbifold groups and parabolic bundles*, Comment. Math. Helv. **66** (1991), no. 3, 389–447.

[Bd-Hu] H.U. Boden and Y. Hu, *Variations of moduli of parabolic bundles*, Math. Ann. **301** (1995), no. 3, 539–559.

[Bo] N. Borne, *Fibrés paraboliques et champ des racines*, preprint 2005, `math.AG/0604458`.

[Bu] N. Budur, *Unitary local systems, multiplier ideals, and polynomial periodicity of Hodge numbers*, preprint 2006, `math.AG/0610382`.

[Cad] C. Cadman, *Using stacks to impose tangency conditions on curves*, Preprint `math.AG/0312349`.

[Ch-Sm] J. Cheeger and J. Simons, *Differential characters and geometric invariants*, Geometry and topology (College Park, Md., 1983/84), 50–80, Lecture Notes in Math., **1167**, Springer, Berlin, 1985.

[Cn-Sm] S.S. Chern and J. Simons, *Characteristic forms and geometric invariants*, Ann. of Math. (2) **99** (1974), 48–69.

[De] P. Deligne, *Equations différentielles a points singuliers reguliers.* Lect. Notes in Math. **163**, 1970.

[De2] P. Deligne, *Théorie de Hodge II.*, Inst. Hautes Etudes Sci. Publ. Math. **40** (1971), 5–57.

[De3] P. Deligne, *La conjecture de Weil. II.* (French) [Weil's conjecture. II] Inst. Hautes Études Sci. Publ. Math. No. **52** (1980), 137–252.

[Es] H. Esnault, *Characteristic classes of flat bundles*, Topology **27** (1988), no. 3, 323–352.

[Es2] H. Esnault, *Recent developments on characteristic classes of flat bundles on complex algebraic manifolds*, Jahresber. Deutsch. Math.-Verein. **98** (1996), no. 4, 182–191.

[Es-Vi] H. Esnault and E. Viehweg, *Logarithmic De Rham complexes and vanishing theorems*, Invent.Math., **86** (1986), 161–194.

[Es-Vi2] H. Esnault and E. Viehweg, *Deligne-Beilinson cohomology,* in: Beilinson's conjectures on special values of L-functions, 43–91, Perspect. Math., **4**, Academic Press, Boston, MA, 1988.

[Es-Vi3] H. Esnault and E. Viehweg, *Chern classes of Gauss-Manin bundles of weight 1 vanish*, K-Theory **26** (2002), no. 3, 287–305.

[Gi] H. Gillet, *Intersection theory on algebraic stacks and Q-varieties*, Proceedings of the Luminy conference on algebraic K-theory (Luminy, 1983). J. Pure Appl. Algebra **34** (1984), 193–240.

[IIS] M. Inaba, K. Iwasaki and M.-H. Saito, *Moduli of Stable Parabolic Connections, Riemann-Hilbert correspondence and Geometry of Painlevé equation of type VI, Part I*, Preprint `math.AG/0309342`.

[IKN] L. Illusie, K. Kato and C. Nakayama, *Quasi-unipotent logarithmic Riemann-Hilbert correspondences*, J. Math. Sci. Univ. Tokyo **12** (2005), no. 1, 1–66.

[Iy-Si] J.N. Iyer and C.T. Simpson, *A relation between the parabolic Chern characters of the de Rham bundles*, arXiv `math.AG/0604196`, to appear in Math. Annalen.

[KN] K. Kato and C. Nakayama, *Log Betti cohomology, log étale cohomology, and log de Rham cohomology of log schemes over C*, Kodai Math. J. **22** (1999), no. 2, 161–186.

[Kaw] Y. Kawamata, *Characterization of abelian varieties*, Compositio Math. **43** (1981), no. 2, 253–276.

[Kr] A. Kresch, *Cycle groups for Artin stacks*, Invent. Math. **138** (1999), no. 3, 495–536.

[La] A. Landman, *On the Picard-Lefschetz transformation for algebraic manifolds acquiring general singularities*, Trans. Amer. Math. Soc. **181** (1973), 89–126.

[Li] J. Li, *Hermitian-Einstein metrics and Chern number inequalities on parabolic stable bundles over Kähler manifolds.* Comm. Anal. Geom. **8** (2000), no. 3, 445–475.

[Ma-Yo] M. Maruyama and K. Yokogawa, *Moduli of parabolic stable sheaves*, Math. Ann. **293** (1992), no. 1, 77–99.

[Ma-Ol] K. Matsuki and M. Olsson, *Kawamata-Viehweg vanishing and Kodaira vanishing for stacks*, Math. Res. Lett. **12** (2005), 207–217.

[Me-Se] V.B. Mehta and C.S. Seshadri, *Moduli of vector bundles on curves with parabolic structures*, Math. Ann. **248** (1980), no. 3, 205–239.

[Mo] T. Mochizuki, *Asymptotic behaviour of tame harmonic bundles and an application to pure twistor D-modules*, Preprint `math.DG/0312230`.

[Mo2] T. Mochizuki, *Kobayashi-Hitchin correspondence for tame harmonic bundles and an application*, Preprint `math.DG/0411300`, preprint Kyoto-Math 2005-15.

[Mu] D. Mumford, *Towards an Enumerative Geometry of the Moduli Space of Curves*, Arithmetic and geometry, Vol. II, 271–328, Progr. Math., **36**, Birkhäuser Boston, Boston, MA, 1983.

[Oh] M. Ohtsuki, *A residue formula for Chern classes associated with logarithmic connections*, Tokyo J. Math. **5** (1982), no. 1, 13–21.

[Pa] D. Panov, *Doctoral thesis*, 2005.

[Re] A. Reznikov, *Rationality of secondary classes*, J. Differential Geom. **43** (1996), no. 3, 674–692.

[Re2] A. Reznikov, *All regulators of flat bundles are torsion*, Ann. of Math. (2) **141** (1995), no. 2, 373–386.

[Sc] W. Schmid, *Variation of Hodge structure: the singularities of the period mapping*, Invent. Math. **22** (1973), 211–319.

[Se] C.S. Seshadri, *Moduli of vector bundles on curves with parabolic structures.* Bull. Amer. Math. Soc. **83** (1977), 124–126.

[St] J. Steenbrink, *Limits of Hodge structures*, Invent. Math. **31** (1975/76), 229–257.

[Sr-Wr] B. Steer and A. Wren, *The Donaldson-Hitchin-Kobayashi correspondence for parabolic bundles over orbifold surfaces.* Canad. J. Math. **53** (2001), no. 6, 1309–1339.

[Th] M. Thaddeus, *Variation of moduli of parabolic Higgs bundles*, J. Reine Angew. Math. **547** (2002), 1–14.

[vdG] G. van der Geer, *Cycles on the moduli space of abelian varieties*, Moduli of curves and abelian varieties, 65–89, Aspects Math. E33, Vieweg, Braunschweig 1999.

[Vi] A. Vistoli, *Intersection theory on algebraic stacks and on their moduli spaces*, Invent. Math. **97** (1989), 613–670.

[Vo] C. Voisin, *Théorie de Hodge et géométrie algébrique complexe*, Cours Spécialisés **10**. S.M.F., Paris, 2002.

Jaya N. Iyer
School of Mathematics
Institute for Advanced Study
1 Einstein Drive
Princeton NJ, 08540
USA.
e-mail: jniyer@ias.edu

The Institute of Mathematical Sciences
CIT Campus
Taramani, Chennai 600113
India
e-mail: jniyer@imsc.res.in

Carlos T. Simpson
CNRS, Laboratoire J.-A.Dieudonné
Université de Nice – Sophia Antipolis
Parc Valrose
06108 Nice Cedex 02
France
e-mail: carlos@math.unice.fr

Progress in Mathematics, Vol. 265, 487–564

Kleinian Groups in Higher Dimensions

Michael Kapovich

To the memory of Sasha Reznikov

Abstract. This is a survey of higher-dimensional Kleinian groups, i.e., discrete isometry groups of the hyperbolic n-space $\mathbb{H}^n$ for $n \geq 4$. Our main emphasis is on the topological and geometric aspects of higher-dimensional Kleinian groups and their contrast with the discrete groups of isometry of $\mathbb{H}^3$.

Mathematics Subject Classification (2000). Primary 30F40; Secondary 20F67.

Keywords. Kleinian groups, hyperbolic manifolds.

1. Introduction

The goal of this survey is to give an overview (mainly from the topological perspective) of the theory of Kleinian groups in higher dimensions. The survey grew out of a series of lectures I gave in the University of Maryland in the Fall of 1991. An early (much shorter) version of this paper appeared as the preprint [110]. In this survey I collect well-known facts as well as less-known and new results. Hopefully, this will make the survey interesting to both non-experts and experts. We also refer the reader to Tukia's short survey [219] of higher-dimensional Kleinian groups.

There is a vast variety of Kleinian groups in higher dimensions: It appears that there is no hope for a comprehensive structure theory similar to the theory of discrete groups of isometries of $\mathbb{H}^3$. I do not know a good guiding principle for the taxonomy of higher-dimensional Kleinian groups. In this paper the higher-dimensional Kleinian groups are organized according to the topological complexity of their limit sets. In this setting one of the key questions that I will address is the interaction between the geometry and topology of the limit set and the algebraic and topological properties of the Kleinian group.

During this work the I was partially supported by various NSF grants, especially DMS-8902619 at the University of Maryland and DMS-04-05180 at UC Davis. Most of this work was done when I was visiting the Max Plank Institute for Mathematics in Bonn.

This paper is organized as follows. In Section 2 we consider the most basic concepts of the theory of Kleinian groups, e.g., domain of discontinuity, limit set, geometric finiteness, etc. In Section 3 we discuss various ways to construct Kleinian groups and list the tools of the theory of Kleinian groups in higher dimensions. In Section 4 we review the homological algebra used in the paper. In Section 5 we state topological rigidity results of Farrell and Jones and the coarse compact core theorem for higher-dimensional Kleinian groups. In Section 6 we discuss various notions of equivalence between Kleinian groups: From the weakest (isomorphism) to the strongest (conjugacy). In Section 7 we consider groups with zero-dimensional limit sets; such groups are relatively well understood. Convex-cocompact groups with 1-dimensional limit sets are discussed in Section 8. Although the topology of the limit sets of such groups is well understood, their group-theoretic structure is a mystery. We know very little about Kleinian groups with higher-dimensional limit sets, thus we restrict the discussion to Kleinian groups whose limit sets are topological spheres (Section 9). We then discuss Ahlfors finiteness theorem and its failure in higher dimensions (Section 10). We then consider the representation varieties of Kleinian groups (Section 11). Lastly we discuss algebraic and topological constraints on Kleinian groups in higher dimensions (Section 12).

Acknowledgments. I am grateful to C. McMullen, T. Delzant, A. Nabutovsky and J. Souto for several suggestions, and to L. Potyagailo for a number of comments, suggestions and corrections. I am also grateful to the referee for numerous corrections.

Contents

2. Basic definitions

Möbius transformations. For the lack of space, our discussion of the basics of Kleinian groups below is somewhat sketchy. For the detailed treatment we refer the reader to [18, 39, 116, 138, 190]. We let $\mathbb{B}^{n+1}$ denote the closed ball $\mathbb{H}^{n+1} \cup \mathbb{S}^n$; its boundary $\mathbb{S}^n$ is identified via the stereographic projection with $\overline{\mathbb{R}^n} = \mathbb{R}^n \cup \{\infty\}$. A *horoball* B in $\mathbb{H}^{n+1}$ is a round ball in $\mathbb{H}^{n+1}$ which is tangent to the boundary sphere $\mathbb{S}^n$. The point of tangency is called the (hyperbolic) *center* of B.

Let $\mathbf{Mob}(\mathbb{S}^n)$ denote the group of all *Möbius* transformations of the n-sphere $\mathbb{S}^n$, i.e., compositions of inversions in $\mathbb{S}^n$. The group $\mathbf{Mob}(\mathbb{S}^n)$ admits an extension to the hyperbolic space $\mathbb{H}^{n+1}$, so that $\mathbf{Mob}(\mathbb{S}^n) = \mathrm{Isom}(\mathbb{H}^{n+1})$, the isometry group of $\mathbb{H}^{n+1}$.

For elements $\gamma \in \mathbf{Mob}(\mathbb{S}^n)$ define the *displacement function*

$$d_\gamma(x) := d(x, \gamma(x)), \quad x \in \mathbb{H}^{n+1}.$$

The elements γ of $\mathbf{Mob}(\mathbb{S}^n)$ are classified as:

1. *Hyperbolic:* The function d_γ is bounded away from zero. Its minimum is attained on a geodesic $A_\gamma \subset \mathbb{H}^{n+1}$ invariant under γ. The ideal end-points of A_γ are the fixed points of γ in $\mathbb{S}^n$.
2. *Parabolic*: The function d_γ is positive but has zero infimum on $\mathbb{H}^{n+1}$; such elements have precisely one fixed point in $\mathbb{S}^n$.
3. *Elliptic*: γ fixes a point in $\mathbb{H}^{n+1}$.

The group $\mathbf{Mob}(\mathbb{S}^n)$ is isomorphic to an index 2 subgroup in the Lorentz group $O(n+1,1)$, see, e.g., [190]. In particular, $\mathbf{Mob}(\mathbb{S}^n)$ is a matrix group. Selberg's lemma [203] implies that every finitely generated group of matrices contains a finite index subgroup which is torsion-free. A group Γ is said to *virtually* satisfy a property X if it contains a finite index subgroup $\Gamma' \subset \Gamma$, such that Γ' satisfies X. Therefore, every finitely generated group of matrices is *virtually torsion-free.* Moreover, every finitely-generated matrix group is *residually finite*, i.e., the intersection of all its finite-index subgroups is trivial, see [151, 203]. This, of course, applies to the finitely generated subgroups of $\mathbf{Mob}(\mathbb{S}^n)$ as well.

Definition 2.1. A discrete subgroup $\Gamma \subset \mathbf{Mob}(\mathbb{S}^n)$ is called a *Kleinian group.*

Dynamical notions. The *discontinuity set* $\Omega(\Gamma)$ of a group $\Gamma \subset \mathbf{Mob}(\mathbb{S}^n)$, is the largest open subset in $\mathbb{S}^n$ where Γ acts properly discontinuously. Its complement $\mathbb{S}^n \setminus \Omega(\Gamma)$ is the *limit set* $\Lambda(\Gamma)$ of the group Γ. Equivalently, the limit set of a Kleinian group can be described as the accumulation set in the sphere $\mathbb{S}^n$ of an orbit $\Gamma \cdot o$. Here o is an arbitrary point in $\mathbb{H}^{n+1}$. A Kleinian group is called *elementary* if its limit set is finite, i.e., is either empty, or consists of one or of two points.

We will use the notation $M^n(\Gamma)$ for the n-dimensional quotient $\Omega(\Gamma)/\Gamma$ and $\bar{M}^{n+1}(\Gamma)$ for the $n+1$-dimensional quotient $(\mathbb{H}^{n+1} \cup \Omega(\Gamma))/\Gamma$.

For a closed subset $\Lambda \subset \mathbb{S}^n$, let $Hull(\Lambda)$ denote its *convex hull* in $\mathbb{H}^{n+1}$, i.e., the smallest closed convex subset H of $\mathbb{H}^{n+1}$ such that

$$cl_{\mathbb{B}^{n+1}}(H) \cap \mathbb{S}^n = \Lambda.$$

Clearly, if Λ is a point, then $Hull(\Lambda)$ does not exist. Otherwise, $Hull(\Lambda)$ exists and is unique. We declare $Hull(\Lambda)$ to be empty in the case when Λ is a single point.

One way to visualize the convex hull $Hull(\Lambda)$ is to consider the *projective model* of the hyperbolic space, where the geodesic lines are straight line segments contained in the interior of $\mathbb{B}^{n+1}$. Therefore, the Euclidean notion of convexity coincides with the hyperbolic notion. This implies that the convex hull in this model can be described as follows: $Hull(\Lambda)$ is the intersection of the Euclidean convex hull of Λ with the interior of $\mathbb{B}^{n+1}$.

Suppose that $\Lambda = \Lambda(\Gamma)$ is the limit set of a Kleinian group $\Gamma \subset \mathbf{Mob}(\mathbb{S}^n)$. The quotient $Hull(\Lambda)/\Gamma$ is called the *convex core* of the orbifold $N = \mathbb{H}^{n+1}/\Gamma$. It is characterized by the property that it is the smallest closed convex subset in N, whose inclusion to N is a homotopy-equivalence. For $\epsilon > 0$ consider the open ϵ-neighborhood $Hull_\epsilon(\Lambda)$ of $Hull(\Lambda)$ in $\mathbb{H}^{n+1}$. Since $Hull_\epsilon(\Lambda)$ is Γ-invariant, we can form the quotient $Hull_\epsilon(\Lambda)/\Gamma$. Then $Hull_\epsilon(\Lambda)/\Gamma$ is the ϵ-neighborhood of the convex core.

Geometric finiteness. We now arrive to one of the key notions in the theory of Kleinian groups:

Definition 2.2. A Kleinian group $\Gamma \subset \mathbf{Mob}(\mathbb{S}^n)$ is called *geometrically finite* if:

(1) Γ is finitely generated, and
(2) $vol(Hull_\epsilon(\Lambda(\Gamma))/\Gamma) < \infty$.

In a number of important special cases, e.g., when Γ is torsion-free, or $n = 2$, or when $\Lambda(\Gamma) = \mathbb{S}^n$, the assumption (1) follows from (2), see [39]. However, E. Hamilton [93] constructed an example of a Kleinian group $\Gamma \subset \mathbf{Mob}(\mathbb{S}^3)$ for which (2) holds but (1) fails. This group contains finite order elements of arbitrarily high order; by Selberg's lemma such groups cannot be finitely-generated.

A Kleinian group $\Gamma \subset \mathbf{Mob}(\mathbb{S}^n)$ is called a *lattice* if $\mathbb{H}^{n+1}/\Gamma$ has finite volume. Equivalently, $\Lambda(\Gamma) = \mathbb{S}^n$ and Γ is geometrically finite. A lattice is *cocompact* (or *uniform*) if $\mathbb{H}^{n+1}/\Gamma$ is compact.

One can characterize geometrically finite groups in terms of their limit sets. Before stating this theorem we need two more definitions.

Definition 2.3. A limit point $\xi \in \Lambda(\Gamma)$ is called a *conical limit point* if there exists a geodesic $\alpha \subset \mathbb{H}^{n+1}$ asymptotic to ξ, a point $o \in \mathbb{H}^{n+1}$, a number $r < \infty$, and a sequence $\gamma_i \in \Gamma$ so that

1. $\lim_i \gamma_i(o) = \xi$.
2. $d(\gamma_i(o), \alpha) \leq r$.

The reason for this name comes from the shape of the r-neighborhood of the vertical geodesic α in the upper half-space model of $\mathbb{H}^{n+1}$: It is a Euclidean cone with the axis α. Equivalently, one can describe the conical limit points of nonelementary groups as follows (see [14, 39]):

$\xi \in \Lambda(\Gamma)$ is a conical limit point if and only if for every $\eta \in \Lambda(\Gamma) \setminus \{\xi\}$ there exists a point ψ and a sequence $\gamma_i \in \Gamma$ such that:

1. $\lim_i \gamma_i(\zeta) = \xi$ for every $\zeta \in \Lambda(\Gamma) \setminus \{\psi\}$.
2. $\lim_i \gamma_i^{-1}(\xi) \neq \lim_i \gamma_i^{-1}(\eta)$.

The set of conical limit points of a Kleinian group Γ is denoted $\Lambda_c(\Gamma)$.

Definition 2.4. A point $\xi \in \Lambda(\Gamma)$ is called a *bounded parabolic point* if it is the fixed point of a parabolic subgroup $\Pi \subset \Gamma$ and $(\Lambda(\Gamma) - \{\xi\})/\Pi$ is compact.

Below is a *dynamical* characterizations of geometrically finite groups:

Theorem 2.5. (*A. Beardon and B. Maskit* [14], *B. Bowditch* [39]) *A Kleinian group Γ is geometrically finite if and only if each limit point $\xi \in \Lambda(\Gamma)$ is either a* conical limit point *or a* bounded parabolic point.

C. Bishop [32] proved that one can drop the word *bounded* in the above theorem. We refer the reader to Bowditch's paper [39] for the proof of other criteria of geometric finiteness collected in Theorems 2.6, 2.7, 2.8 below. (The case $n = 2$ is treated in [152] and [213].)

Theorem 2.6.

1. *If a Kleinian subgroup* $\Gamma \subset \mathbf{Mob}(\mathbb{S}^n)$ *admits a convex fundamental polyhedron with finitely many faces then it is geometrically finite.*
2. *Let* $\Gamma \subset \mathbf{Mob}(\mathbb{S}^n)$ *be a geometrically finite Kleinian group so that either* (a) $n \leq 2$, *or* (b) Γ *contains no parabolic elements, or* (c) Γ *is a lattice.*

   Then Γ *admits a convex fundamental polyhedron with finitely many faces.*

On the other hand, there are geometrically finite subgroups of $\mathbf{Mob}(\mathbb{S}^3)$ which do not admit a convex fundamental polyhedron with finitely many faces, see [10].

Theorem 2.7. *Let* $\Gamma \subset \mathbf{Mob}(\mathbb{S}^n)$ *be a Kleinian subgroup containing no parabolic elements. Then the following are equivalent:*

(a) Γ *is geometrically finite.*
(b) $Hull(\Lambda(\Gamma))/\Gamma$ *is compact.*
(c) $\bar{M}^{n+1}(\Gamma)$ *is compact.*

If Γ is geometrically finite and contains no parabolic elements, it is called *convex-cocompact.* We will frequently use the fact that every convex-cocompact Kleinian group is *Gromov-hyperbolic*, see, e.g., [44].

The criterion given in Theorem 2.7 generalizes to the case of groups with parabolic elements, although the statement becomes more complicated:

Theorem 2.8. *The following are equivalent:*

(a) Γ *is geometrically finite.*
(b) *There exists a pairwise disjoint* Γ*-invariant collection of open horoballs* $B_i \subset \mathbb{H}^{n+1}, i \in I$, *which are centered at fixed points of parabolic subgroups of* Γ, *such that the quotient*
$$\left(Hull(\Lambda(\Gamma)) \setminus \bigcup_{i \in I} B_i \right) /\Gamma$$
is compact.
(c) *Let* $\Pi_i, i \in I$ *be the collection of maximal (virtually) parabolic subgroups of* Γ. *For each* i *there exists a* Π_i*-invariant convex subset* $C_i \subset \mathbb{B}^{n+1}$, *so that the quotient*
$$\left(\mathbb{H}^{n+1} \cup \Omega(\Gamma) \setminus \bigcup_{i \in I} C_i \right) /\Gamma$$
is compact. If $\Omega(\Gamma) = \emptyset$, *then one can take* $C_i = B_i$, *a horoball in* $\mathbb{H}^{n+1}$.

If $n = 1$, then every finitely generated Kleinian group is geometrically finite. The proof is rather elementary, see, e.g., [53]. For $n \geq 2$ this implication is no longer true. The first (implicit) examples were given by L. Bers, they are *singly-degenerate groups*:

Definition 2.9. A finitely generated nonelementary Kleinian subgroup of $\mathbf{Mob}(\mathbb{S}^2)$ is *singly degenerate* if its domain of discontinuity is simply-connected, i.e., homeomorphic to the 2-disk.

L. Bers [19] proved that singly degenerate Kleinian groups exist and are never geometrically finite. The first *explicit* examples of finitely generated geometrically infinite Kleinian subgroups Γ of $\mathbf{Mob}(\mathbb{S}^2)$ were given by T. Jørgensen [104]. In Jørgensen's examples, Γ appears as a normal subgroup of a lattice $\hat{\Gamma} \subset \mathbf{Mob}(\mathbb{S}^2)$ with $\hat{\Gamma}/\Gamma \cong \mathbb{Z}$. Remarkably, all known examples of finitely-generated geometrically infinite Kleinian subgroups of $\mathbf{Mob}(\mathbb{S}^n)$ can be traced to the 2-dimensional examples. More precisely, every *known* finitely-generated geometrically infinite Kleinian subgroup $\Gamma \subset \mathbf{Mob}(\mathbb{S}^n)$ admits a decomposition as the graph of groups

$$(\mathcal{G}, \Gamma_v, \Gamma_e),$$

where at least one of the vertex groups Γ_v is either a geometrically infinite subgroup contained in $\mathbf{Mob}(\mathbb{S}^2)$, or is a quasiconformal deformations of such.

Problem 2.10. *Construct examples of finitely-generated geometrically infinite subgroups of* $\mathbf{Mob}(\mathbb{S}^n)$, $n \geq 3$, *which do not have the* 2*-dimensional origin as above.*

Assumption 2.11. *From now on we will assume that all Kleinian groups are finitely generated and torsion-free, unless stated otherwise.*

Note that the second part of this assumption is not very restrictive because of Selberg's lemma.

Cusps and tubes. The Γ-conjugacy classes $[\Pi]$ of maximal parabolic subgroups Π of a Kleinian group Γ are called *cusps* of Γ. More geometrically, cusps of Γ can be described using the *thick-thin decomposition* of the quotient manifold $M = \mathbb{H}^{n+1}/\Gamma$. Given a positive number $\epsilon > 0$, let $M_{(0,\epsilon]}$ denote the collection of points x in M such that there exists a homotopically nontrivial loop α based at x, so that the length of α is at most ϵ. Then $M_{(\epsilon,\infty)}$ is the complement of $M_{(0,\epsilon]}$ in M. According to Kazhdan-Margulis lemma [130], there exists a number $\mu = \mu_{n+1} > 0$ such that for every Kleinian group Γ and every $0 < \epsilon \leq \mu$, every component of $M_{(0,\epsilon]}$ has a *virtually abelian* fundamental group. The submanifold $M_{(0,\epsilon]}$ is called the *thin part* of M and its complement the *thick part* of M. The compact components of $M_{(0,\epsilon]}$ are called *tubes* and the noncompact components are called *cusps.*

Then the cusps of Γ are in bijective correspondence with the cusps in $M_{(0,\epsilon]}$:

For every cusp $[\Pi]$ in Γ, there exists a noncompact component $C \subset M_{(0,\epsilon]}$, so that $\Pi = \pi_1(C)$. Conversely, for each cusp $C \subset M$, there exists a maximal parabolic subgroup $\Pi \subset \Gamma$ such that $\Pi = \pi_1(C)$.

Taking the Γ-conjugacy class of Π reflects the ambiguity in the choice of the base-point needed to identify $\pi_1(M)$ and Γ.

If $n \leq 2$ and the manifold M is oriented, then the components C_i of $M_{(0,\epsilon]}$ are convex: The cusps in M are quotients of horoballs in $\mathbb{H}^{n+1}$, while the compact components T_i of $M_{(0,\epsilon]}$ are metric R_i-neighborhoods of closed geodesics $\gamma_i \subset M$. In higher dimensions ($n \geq 3$) convexity (in general) fails. However every tube T_i in $M_{(0,\epsilon]}$ is a finite union of convex sets containing a certain closed geodesic $\gamma_i \subset T_i$. In particular, every tube T_i is homeomorphic to a disk bundle over $\mathbb{S}^1$. A similar, although more complicated description, holds for the cusps, where one has to consider (in general) a union of infinitely many convex subsets. See, e.g., [121].

Möbius structures. In this paper we shall also discuss the subject closely related to the theory of Kleinian groups, namely *Möbius structures.* When M is a smooth manifold of dimension ≥ 3, a *Möbius* (or *flat conformal*) structure K on a M is the conformal class of a conformally-Euclidean Riemannian metric on M. Topologically, K is a maximal *Möbius atlas* on M, i.e., an atlas with Möbius transition maps. Thus, for each Kleinian group Γ and Γ-invariant subset $\Omega \subset \Omega(\Gamma)$, the standard Möbius structure on $\Omega \subset \mathbb{S}^n$ projects to a Möbius structure K_Γ on the manifold Ω/Γ. The Möbius structures of this type are called *uniformizable.*

Complex-hyperbolic Kleinian groups. Instead of considering the isometry group of the hyperbolic space, one can consider other negatively curved symmetric spaces, for instance, the *complex-hyperbolic n-space* $\mathbb{CH}^n$ and its group of automorphisms $PU(n,1)$. From the analytical viewpoint, $\mathbb{CH}^n$ is the unit ball in $\mathbb{C}^n$ and $PU(n,1)$ is the group of biholomorphic automorphisms of this ball. The Bergman metric on $\mathbb{CH}^n$ is a Kähler metric of negative sectional curvature. The discrete subgroups of $PU(n,1)$ are *complex-hyperbolic Kleinian groups.* They share many properties with Kleinian groups. In fact, *nearly all positive results* stated in this survey for Kleinian subgroups of $\mathbf{Mob}(\mathbb{S}^n)$ ($n \geq 3$) are also valid for the complex-hyperbolic Kleinian groups! (One has to replace *virtually abelian* with *virtually nilpotent* in the discussion of cusps.) There exists an isometric embedding $\mathbb{H}^n \to \mathbb{CH}^n$ which induces an embedding of the isometry groups and therefore complex-hyperbolic Kleinian groups ($n \geq 4$) also inherit the pathologies of the higher-dimensional Kleinian groups. We refer the reader to [80, 81, 199] for detailed discussion.

3. Ways and means of Kleinian groups

3.1. Ways: Sources of Kleinian groups

The following is a list of ways to construct Kleinian groups.

(a) **Poincaré fundamental polyhedron theorem** (see, e.g., [190] for a very detailed discussion, as well as [157]). This source is, in principle, the most general. The Poincaré fundamental polyhedron theorem asserts that given a polyhedron Φ in $\mathbb{H}^{n+1}$ and a collection of elements $\gamma_1, \gamma_2, \ldots, \gamma_k, \ldots$ of $\mathbf{Mob}(\mathbb{S}^n)$, pairing the faces of Φ, under certain conditions on this data, the group Γ generated by $\gamma_1, \gamma_2, \ldots, \gamma_k, \ldots$ is Kleinian and Φ is a *fundamental domain* for the action of the group Γ on $\mathbb{H}^{n+1}$.

Every Kleinian group has a convex fundamental polyhedron (for example, the *Dirichlet fundamental domain*). However, in practice, the Poincaré fundamental polyhedron theorem is not always easy to use, especially if Φ has many faces and n is large. This theorem was used, for instance, to construct non-arithmetic lattices in $\mathbf{Mob}(\mathbb{S}^n)$ (see [149, 150, 221]), as well as other interesting Kleinian groups, see, e.g., [60, 98, 111, 139, 191].

(b) **Klein–Maskit Combination Theorems** (see, e.g., [138] and [157]). Suppose that we are given two Kleinian groups $\Gamma_1, \Gamma_2 \subset \mathbf{Mob}(\mathbb{S}^n)$ which share a common subgroup Γ_3, or a single Kleinian group Γ_1 and a Möbius transformation $\tau \in \mathbf{Mob}(\mathbb{S}^n)$ which conjugates subgroups $\Gamma_3, \Gamma_3' \subset \Gamma_1$. The Combination Theorems provide conditions which guarantee that the group $\Gamma \subset \mathbf{Mob}(\mathbb{S}^n)$ generated by Γ_1 and Γ_2 (or by Γ_1 and τ) is again Kleinian and is isomorphic to the amalgam

$$\Gamma \cong \Gamma_1 *_{\Gamma_3} \Gamma_2,$$

or to the HNN extension

$$\Gamma \cong \Gamma_1 *_{\Gamma_3} = HNN(\Gamma_1, \tau).$$

The proofs of the Combination Theorems generalize the classical "ping-pong" argument due to Schottky and Klein. The Combination Theorems also show that the quotient manifold $M^n(\Gamma)$ of the group Γ is obtained from $M^n(\Gamma_1)$, $M^n(\Gamma_2)$ (or $M(\Gamma_1)$) via some "cut-and-paste" operation. Moreover, Combination Theorems generalize to graph of groups. There should be a generalization of Combination Theorems to *complexes of groups* (see, e.g., [44] for the definition); however, to the best of my knowledge, nobody worked out the general result, see [118] for a special case.

(c) **Arithmetic groups and their subgroups** (see, e.g., [148] and [222]). A subgroup $\Gamma \subset O(n,1)$ is called *arithmetic* if there exists an embedding

$$\iota : O(n,1) \hookrightarrow GL(N, \mathbb{R}),$$

such that the image $\iota(\Gamma)$ is *commensurable* with the intersection

$$\iota(O(n,1)) \cap GL(N, \mathbb{Z}).$$

Recall that two subgroups $\Gamma_1, \Gamma_2 \subset G$ are called *commensurable* if $\Gamma_1 \cap \Gamma_2$ has finite index in both Γ_1 and Γ_2.

Below is a specific construction of arithmetic groups. Let f be a quadratic form of signature $(n,1)$ in $n+1$ variables with coefficients in a totally real algebraic number field $K \subset \mathbb{R}$ satisfying the following condition:

(*) For every nontrivial (i.e., different from the identity) embedding $\sigma : K \to \mathbb{R}$, the quadratic form f^σ is positive definite.

Without loss of generality one may assume that this quadratic form is diagonal. For instance, take

$$f(x) = -\sqrt{2}x_0^2 + x_1^2 + \cdots + x_n^2.$$

We now define discrete subgroups of $\mathrm{Isom}(\mathbb{H}^n)$ using the form f. Let A denote the ring of integers of K. We define the group $\Gamma := O(f, A)$ consisting of matrices with entries in A preserving the form f. Then Γ is a discrete subgroup of $O(f, \mathbb{R})$. Moreover, it is a lattice: its index 2 subgroup

$$\Gamma' = O'(f, A) := O(f, A) \cap O'(f, \mathbb{R})$$

acts on $\mathbb{H}^n$ so that $\mathbb{H}^n/\Gamma'$ has finite volume. Such groups Γ (and subgroups of $\mathrm{Isom}(\mathbb{H}^n)$ commensurable to them) are called *arithmetic subgroups of the simplest type* in $O(n, 1)$, see [222].

Remark 3.1. If $\Gamma \subset O(n, 1)$ is an arithmetic lattice so that either Γ is non-cocompact or n is even, then it follows from the classification of rational structures on $O(n, 1)$ that Γ is commensurable to an arithmetic lattice of the simplest type. For odd n there is another family of arithmetic lattices given as the groups of units of appropriate skew-Hermitian forms over quaternionic algebras. Yet other families of arithmetic lattices exist for $n = 3$ and $n = 7$. See, e.g., [222].

We refer the reader to [148] for the detailed treatment of geometry and topology of arithmetic subgroups of $\mathbf{Mob}(\mathbb{S}^2)$.

(d) **Small deformations of a given Kleinian group**. We discuss this construction in detail in Section 11.1. The idea is to take a Kleinian group $\Gamma \subset \mathbf{Mob}(\mathbb{S}^n)$ and to "perturb it a little bit", by modifying the generators slightly (within $\mathbf{Mob}(\mathbb{S}^n)$) and preserving the relators. The result is a new group Γ' which may or may not be Kleinian and even if it is, Γ' is not necessarily isomorphic to Γ. However if Γ is convex-cocompact, Γ' is again a convex-cocompact group isomorphic to Γ, see Theorem 11.12.

(e) **Limits of sequences of Kleinian groups**, see Section 11.3. Take a sequence Γ_i of Kleinian subgroups of $\mathbf{Mob}(\mathbb{S}^n)$ and assume that it has a limit Γ: It turns out that there are two ways to make sense of this procedure (*algebraic* and *geometric* limit). In any case, Γ is again a Kleinian group. Even if the (algebraic) limit does not exist as a subgroup of $\mathbf{Mob}(\mathbb{S}^n)$, there is a way to make sense of the limiting group as a group of isometries of a metric tree. This logic turns out to be useful for proving compactness theorems for sequences of Kleinian groups.

(f) **Differential-geometric constructions of hyperbolic metrics.** The only (but spectacular) example where it has been used is Perelman's work on Ricci flow and proof of Thurston's geometrization conjecture. See [134, 173, 183, 184]. However applicability of this tool at the moment appears to be limited to 3-manifolds.

A beautiful example of application of (b) and (c) is the construction of M. Gromov and I. Piatetski-Shapiro [88] of *non-arithmetic* lattices in $\mathbf{Mob}(\mathbb{S}^n)$. Starting with two arithmetic groups Γ_j $(j = 1, 2)$ they first "cut these groups in half", take "one half" $\Delta_j \subset \Gamma_j$ of each, and then combine Δ_1 and Δ_2 via Maskit Combination. The construction of Kleinian groups in [89] (see also Section 9.1) is an application of (b), (c) and (d). Thurston's hyperbolic Dehn surgery theorem is an example of (e). One of the most sophisticated constructions of Kleinian groups is given by Thurston's hyperbolization theorem (see, e.g., [116], [180], [181]); still, it is essentially a combination (a very complicated one!) of (b), (d) and (e).

Remark 3.2. There is potentially the sixth source of Kleinian groups in higher dimensions: monodromy of linear ordinary differential equations. However, to the best of my knowledge, the only example of its application relevant to Kleinian groups, is the construction of lattices in $PU(n,1)$ (i.e., the isometry group of the complex-hyperbolic n-space) by Deligne and Mostow, see [62].

3.2. Means: Tools of the theory of Kleinian groups in higher dimensions

Several key tools of the "classical" theory of Kleinian subgroups of $\mathbf{Mob}(\mathbb{S}^2)$ (mainly, the Beltrami equation and pleated hypersurfaces) are missing in higher dimensions. Below is the list of main tools that are currently available.

(a) **Dynamics, more specifically, the convergence property.** Namely, every sequence of Möbius transformations $\gamma_i \in \mathbf{Mob}(\mathbb{S}^n)$ either contains a convergent subsequence or contains a subsequence which converges to a constant map away from a point in $\mathbb{S}^n$. See, e.g., [116].

(b) **Kazhdan-Margulis lemma** and its corollaries.

It turns out that the lion share of the general results about higher-dimensional Kleinian groups is a combination of (a) and (b), together with some hyperbolic geometry.

(c) **Group actions on trees and Rips theory**. This is a very potent tool for proving compactness results for families of representations of Kleinian groups, see for instance Theorem 11.16.

(d) **Barycentric maps**. These maps were originally introduced by A. Douady and C. Earle [65] as a tool of the Teichmüller theory of Riemann surfaces. In the hands of G. Besson, G. Courtois and S. Gallot these maps became a powerful analytic tool of the theory of Kleinian groups in higher dimensions, see, e.g., [21, 22], as well as Theorems 10.21 and 11.24 in this survey. In contrast, equivariant harmonic maps which proved so useful in the study of, say, Kähler groups, seem at the moment to be only of a very limited use in the theory of Kleinian groups in higher dimensions.

(e) **Ergodic theory of the actions of Γ on its limit set and Patterson–Sullivan measures.** See for instance [177, 208] and the survey of P. Tukia [219].

(f) **Conformal geometric analysis.** This is a branch of (conformal) differential geometry concerned with the analysis of the conformally-flat Riemannian metrics on $M^n(\Gamma) = \Omega(\Gamma)/\Gamma$. This tool tends to work rather well in the case when $M^n(\Gamma)$ is compact. The most interesting examples of this technique are due to R. Schoen and S-T. Yau [198], S. Nayatani [176], A. Chang, J. Qing, J. and P. Yang, [54], and H. Izeki [100, 101, 102].

(g) **Infinite-dimensional representation theory** of the group $\mathbf{Mob}(\mathbb{S}^n)$. The only (but rather striking) example of its application is Y. Shalom's work [204].

(h) **Topological rigidity theorems of Farrell and Jones**: See Section 5.

4. A bit of homological algebra

Why does one need homological algebra in order to study higher-dimensional Kleinian groups?

Essentially the only time one encounters group cohomology with twisted coefficients in the study of Kleinian subgroups of $\mathbf{Mob}(\mathbb{S}^2)$, is in the proof of Ahlfors' finiteness theorem, see [136]. Another (minor) encounter appears in the proof of the smoothness theorem for deformation spaces of Kleinian groups, Section 8.8 in [116]. Otherwise, homological algebra is hardly ever needed. The main reason for this, I believe, is 3-fold:

1. Solvability of the 2-dimensional Beltrami equation, which implies smoothness of the deformation spaces of Kleinian groups in the most interesting situations.

2. Scott compact core theorem [201, 202] ensures that every finitely-generated Kleinian group $\Gamma \subset \mathbf{Mob}(\mathbb{S}^2)$ satisfies a very strong finiteness property: Not only it is finitely-presented, it is also (canonically) isomorphic to the fundamental group of a compact aspherical 3-manifold with boundary (Scott compact core).

3. The separation between Kleinian groups of the cohomological dimension 1, 2 and 3 comes rather easily: Free groups, "generic" Kleinian groups, and lattices. Moreover, every Kleinian group $\Gamma \subset \mathbf{Mob}(\mathbb{S}^2)$ which is not a lattice, splits as

$$\Gamma \cong \Gamma_0 * \Gamma_1 * \cdots * \Gamma_k, \tag{4.1}$$

where Γ_0 is free and each Γ_i, $i \geq 1$, is freely indecomposable, 2-dimensional group. In the language of homological algebra, the group Γ_0 has cohomological dimension 1, while the groups Γ_i, $i \geq 1$, are *two-dimensional duality groups.*

All this changes rather dramatically in higher dimensions:

1. Solvability of the Beltrami equation fails, which, in particular, leads to non-smoothness of the deformation spaces of Kleinian groups, Theorem 11.4. In order to study the local structure of character varieties one then needs the first and the second group cohomology with (finite-dimensional) twisted coefficients.

2. Scott compact core theorem fails for Kleinian subgroups of $\mathbf{Mob}(\mathbb{S}^3)$, for instance, they do not have to be finitely-presented, see Section 10. Therefore, it appears that one has to reconsider the assumption that Kleinian groups are finitely-generated. It is quite likely, that in higher dimensions, in order to get good structural results, one has to restrict to Kleinian groups of *finite type*, i.e., type FP, defined below. This definition requires homological algebra.

3. One has to learn how to separate k-dimensional from m-dimensional in the algebraic structure of Kleinian groups. For the subgroups of $\mathbf{Mob}(\mathbb{S}^2)$ this separation comes in the form of the free product decomposition (4.1). It appears at the moment that "truly" m-dimensional groups are the *m-dimensional duality groups.* For instance, for Kleinian subgroups Γ of $\mathbf{Mob}(\mathbb{S}^n)$ which are n-dimensional duality groups, one can prove a coarse form of the Scott compact core theorem, [124]. In particular, every such group admits the structure of an $n+1$-dimensional *Poincaré duality pair* (Γ, Δ). The latter is a homological analogue of the fundamental group of a compact aspherical $n+1$-manifold with boundary (where the

boundary corresponds to the collection of subgroups Δ in Γ). See Section 5.3 for more details.

4. The (co)homological dimension appears to be an integral part of the discussion of the critical exponent of higher-dimensional Kleinian groups, see Section 10.2 and Izeki's papers [100, 101].

(Co)homology of groups. Some of the above discussion was rather speculative; we now return to the firm ground of homological algebra. We refer the reader to [29, 47] for the comprehensive treatment of (co)homologies of groups.

Throughout this section we let R be a commutative ring with a unit. The examples that the reader should have in mind are $R = \mathbb{Z}$, $\mathbb{Z}/p\mathbb{Z}$ and $R = \mathbb{R}$. The *group ring* $R\Gamma$ of a group Γ consists of finite linear combinations of the form

$$\sum_{\gamma\in\Gamma} r_\gamma \gamma,$$

with $r_\gamma \in R$ equal to zero for all but finitely many $\gamma \in \Gamma$. Let V be a (left) $R\Gamma$-module. Basic examples include $V = R$ (with the trivial $R\Gamma$-module structure) and $V = R\Gamma$. If R is a field, then V is nothing but a vector space over R equipped with a linear action of the group Γ. The very useful (for the theory of Kleinian groups) example is the following:

Let $G = \mathbf{Mob}(\mathbb{S}^n)$, $\mathfrak{g}$ be the Lie algebra of G. Then G acts on $\mathfrak{g}$ via the *adjoint representation* $Ad = Ad_G$. Therefore $\mathfrak{g}$ becomes an $\mathbb{R}G$-module. For every abstract group Γ and a representation $\rho : \Gamma \to G$ we obtain the $\mathbb{R}\Gamma$-module

$$V = \mathfrak{g}_{Ad(\rho)},$$

where the action of Γ is given by the composition $Ad \circ \rho$. We will abbreviate this module to $Ad(\rho)$. From the theory of Kleinian groups viewpoint, the most important example of this module is when Γ is a Kleinian subgroup of G and ρ is the identity embedding.

A *projective* $R\Gamma$-module, is a module P, such that every exact sequence of $R\Gamma$-modules

$$Q \to P \to 0$$

splits. For instance, every free $R\Gamma$-module is projective.

Assume now that V be an $R\Gamma$-module. A *resolution of* V is an exact sequence of $R\Gamma$-modules:

$$\cdots \to P_n \to \cdots \to P_0 \to V \to 0.$$

Every $R\Gamma$-module has a unique projective resolution up to a chain homotopy equivalence.

Example. Let $V = \mathbb{Z}$, be the trivial $\mathbb{Z}\Gamma$-module. Let K be a cell complex which is $K(\Gamma, 1)$, i.e., K is connected, $\pi_1(K) \cong \Gamma$ and $\pi_i(K) = 0$ for $i \geq 2$. Let X denote the universal cover of K. Lift the cell complex structure from K to X. The group action $\Gamma \curvearrowright X$, determines a natural structure of a $\mathbb{Z}\Gamma$-module on the cellular chain complex $C_*(X)$. Since the latter is acyclic, we obtain a resolution of $\mathbb{Z}$ with

$$P_i = C_i(X),$$

and the homomorphism $P_0 \to \mathbb{Z}$ given by the augmentation. Moreover, as the group Γ acts freely on X, each module P_i is a *free* $\mathbb{Z}\Gamma$-module:

$$P_i \cong \oplus_{j \in C_i} \mathbb{Z}\Gamma,$$

where C_i is the set of i-cells in K.

A group Γ is said to be of *finite type*, or FP (over R), if there exists a resolution by finitely generated projective $R\Gamma$-modules

$$0 \to P_k \to P_{k-1} \to \cdots \to P_0 \to R \to 0.$$

For example, if there exists a finite cell complex $K = K(\Gamma, 1)$, then Γ has finite type for an arbitrary ring R. Every group of finite type is finitely generated, although it does not have to be finitely-presented, see [24].

The *cohomology* of Γ with coefficients in an $R\Gamma$-module V, $H^*(\Gamma, V)$, is defined as the homology of chain complex

$$\mathrm{Hom}_{R\Gamma}(P_*, M),$$

where P_* is a projective resolution of the trivial $R\Gamma$-module R. The *homology* of Γ with coefficients in V, $H_*(\Gamma, V)$, is the homology of the chain complex

$$P_* \otimes_{R\Gamma} V.$$

An example to keep in mind is the following. Suppose that K is a manifold, or, more generally, a cell complex, which is an Eilenberg-MacLane space $K(\Gamma, 1)$. Then one can use the chain complex $C_*(X, R)$ as the resolution P_*. Therefore, for the trivial Γ-module R we have

$$H^*(\Gamma, R) \cong H^*(K, R), \quad H_*(\Gamma, R) \cong H_*(K, R).$$

For the more general modules V, in order to compute $H^*(\Gamma, V)$ and $H_*(\Gamma, V)$, one uses the (co)homology of K with coefficients in an appropriate bundle over K.

Similarly, given a collection Π of subgroups of Γ, one defines the relative (co)homology groups $H^*(\Gamma, \Pi; V)$ and $H_*(\Gamma, \Pi; V)$. Whenever discussing (co)homology with $R = \mathbb{Z}$, trivial $\mathbb{Z}\Gamma$-module, we will use the notation $H^*(\Gamma)$, $H_*(\Gamma)$.

The (co)homology of groups behaves in a manner similar to the more familiar (co)homology of cell complexes. For instance, if Γ admits an n-dimensional $K(\Gamma, 1)$, then $H^i(\Gamma, V) = H_i(\Gamma, V) = 0$ for all $i > n$ and all $R\Gamma$-modules.

(Co)homological dimension. For a group Γ, let $cd_R(\Gamma)$ and $hd_R(\Gamma)$ denote the *cohomological* and *homological* dimensions of Γ (over R):

$$cd_R(\Gamma) = \sup\{n : \exists \text{ an } R\Gamma\text{-module } V \text{ so that } H^n(\Gamma, V) \neq 0\},$$

$$hd_R(\Gamma) = \sup\{n : \exists \text{ an } R\Gamma\text{-module } V \text{ so that } H_n(\Gamma, V) \neq 0\}.$$

We will omit the subscript $\mathbb{Z}$ whenever $R = \mathbb{Z}$.

Using the relative (co)homology one defines the relative (co)homological dimension of Γ with respect to a collection Π of its subgroups, $cd_R(\Gamma, \Pi)$ and $hd_R(\Gamma, \Pi)$.

We will use this definition in the case when Γ is a Kleinian group as follows. Let $\mathcal{P}$ denote the set of all maximal (elementary) subgroups of Γ which contain $\mathbb{Z}^2$. For every Γ-conjugacy class $[\Pi_i]$ in $\mathcal{P}$, choose a representative $\Pi_i \subset \Gamma$. Then Π will denote the set of all these representatives Π_i. By abusing the notation, we will refer to the set Π as the set of cusps of virtual rank ≥ 2 in Γ.

If Γ is of type FP, then

$$hd_R(\Gamma) = cd_R(\Gamma), \quad \forall \text{ rings } R,$$

see for instance [29]. In general,

$$hd_R(\Gamma) \leq cd_R(\Gamma) \leq hd_R(\Gamma) + 1.$$

Example. Let Γ be a free group of finite rank $k > 0$. Then $hd_R(\Gamma) = cd_R(\Gamma)$ for all rings R. Indeed, Γ admits a finite $K(\Gamma, 1)$ which is the bouquet B of k circles. Since B is 1-dimensional,

$$hd_R(\Gamma) = cd_R(\Gamma) \leq 1.$$

On the other hand, by taking the trivial $R\Gamma$-module $V = R$ we obtain

$$H_1(B, R) = R^k,$$

the direct sum of k copies of R, and hence is nontrivial.

It turns out that the converse to this example is also true, which is an application of the famous theorem of J. Stallings on the ends of groups:

Theorem 4.1. (*J. Stallings* [205].) *If Γ is a finitely generated group with $cd(\Gamma) = 1$, then Γ is free.*

This result was generalized by M. Dunwoody:

Theorem 4.2. (*M. Dunwoody* [66].) *Let R be an commutative ring with a unit.*

1. *If Γ is a finitely generated torsion-free group with $cd_R(\Gamma) = 1$, then Γ is free.*
2. *If Γ is finitely-presented and $cd_R(\Gamma) = 1$ then Γ is a free product of finite and cyclic groups with amalgamation over finite subgroups. In particular, Γ is virtually free.*

Duality groups. A group Γ is said to be an m-dimensional *duality group*, if Γ has type FP and

$$H^i(\Gamma, R\Gamma) \neq 0, \text{ for } i = m \text{ and } H^i(\Gamma, R\Gamma) = 0, \text{ for } i \neq m.$$

For instance, a finitely-presented group Γ is a 2-dimensional duality group (over $\mathbb{Z}$) if and only if $cd(\Gamma) = 2$ and Γ does not split as a nontrivial free product.

Poincaré duality groups. Poincaré duality groups are homological generalizations of the fundamental groups of closed aspherical manifolds.

Definition 4.3. A group Γ is an (oriented) m-dimensional Poincaré duality group over R (a $PD(m)$-group for short) if Γ is of type FP and

$$H^i(\Gamma, R\Gamma) \cong R, \text{ for } i = m \text{ and } H^i(\Gamma, R\Gamma) = 0, \text{ for } i \neq m.$$

The basic examples are the fundamental groups of closed oriented aspherical n-manifolds.

This definition generalizes to (possibly non-oriented) $PD(n)$-groups, where we have to twist the module $V = R\Gamma$ by an appropriate orientation character $\chi : R\Gamma \to R$. The basic examples are the fundamental groups of closed aspherical n-manifolds M. The character χ in this case corresponds to the orientation character $\pi_1(M) \to R$.

We will need (in Section 5.3) the following relative version of the $PD(n)$ groups.

Definition 4.4. Let Γ be an $(n-1)$-dimensional group of type FP, and let

$$\Delta_1, \dots, \Delta_k \subset \Gamma$$

be $PD(n-1)$ subgroups of Γ. Set $\Delta := \{\Delta_1, \dots, \Delta_k\}$. Then, the group pair (Γ, Δ) is an *n-dimensional Poincaré duality pair, or a $PD(n)$ pair*, if the double of Γ over the Δ_i's is a $PD(n)$ group.

We recall that the double of Γ over the Δ_i's is the fundamental group of the graph of groups $\mathcal{G}$, where $\mathcal{G}$ has two vertices labelled by Γ, k edges with the i-th edge labelled by Δ_i, and edge monomorphisms are the inclusions $\Delta_i \to \Gamma$.

An alternate homological definition of $PD(n)$ pairs is the following: A group pair (Γ, Δ) is a $PD(n)$ pair if Γ and each Δ_i has type FP, and

$$H^*(\Gamma, \Delta; \mathbb{Z}\Gamma) \simeq H^*_c(\mathbb{R}^n).$$

If (Γ, Δ) is a $PD(n)$ pair, where Γ and each Δ_i admit a finite Eilenberg–MacLane space X and Y_i respectively, then the inclusions $\Delta_i \to \Gamma$ induce a map

$$\sqcup_i Y_i \to X$$

whose mapping cylinder C gives a *Poincaré pair* $(C, \sqcup_i Y_i)$. The latter is a pair which satisfies Poincaré duality for manifolds with boundary with local coefficients, where $\sqcup_i Y_i$ serves as the boundary of C. The most important example of a $PD(n)$ pair is the following. Let M be a compact manifold which is $K(\Gamma, 1)$. We suppose that the boundary of M is the disjoint union

$$\partial M = N_1 \cup \dots \cup N_k,$$

of π_1-injective components, each of which is a $K(\Delta_i, 1)$, $i = 1, \dots, k$. Then the pair

$$(\Gamma, \{\Delta_1, \dots, \Delta_k\})$$

is a $PD(n)$ pair. See [30] for the details.

The following is one of the major problems in higher-dimensional topology:

Conjecture 4.5. (*C.T.C. Wall, see a very detailed discussion in* [131].) *Suppose that* Γ *is an finitely-presented* n*-dimensional Poincaré duality group over* $\mathbb{Z}$. *Then there exits a closed* n*-dimensional manifold* M *which is* $K(\Gamma, 1)$.

This problem is open for all $n \geq 3$. The case $n = 1$ is an easy corollary of the Stallings-Dunwoody theorem. In the case $n = 2$, the positive solution is due to Eckmann, Linnel and Muller, see [67, 68]. This result was extended to the case of fields R by B. Bowditch [41] and for general rings R by M. Kapovich and B. Kleiner [123] and B. Kleiner [133]. Cannon's conjecture below is a special case (after Perelman's work) of Wall's problem for $n = 3$:

Conjecture 4.6 (J. Cannon). *Suppose that* Γ *is a Gromov-hyperbolic group whose ideal boundary is homeomorphic to* $\mathbb{S}^2$. *Then* Γ *admits a cocompact properly discontinuous isometric action on* $\mathbb{H}^3$.

5. Topological rigidity and coarse compact core theorem

First, few historical remarks. After the work of B. Maskit [156] and A. Marden [152], it became clear that the major developments in the 3-dimensional topology occurring at that time (in the 1960s and the early 1970s) were of extreme importance to the theory of Kleinian groups. The key topological results were:

1. Topological rigidity theorems of Stallings and Waldhausen. Under appropriate assumptions they proved that homotopy equivalence of Haken manifolds implies homeomorphism, see [95] for the detailed discussion. In the context of Kleinian groups, it meant that the (properly understood) algebraic structure of a geometrically finite Kleinian group $\Gamma \subset \mathbf{Mob}(\mathbb{S}^2)$ determines the topology of the associated hyperbolic 3-manifold $\mathbb{H}^3/\Gamma$.

2. Dehn Lemma, Loop Theorem and their consequences. The most important (for Kleinian groups) of these consequences was the *Scott compact core theorem* [201, 202]. This theorem meant for (possibly geometrically infinite) Kleinian groups, that the hyperbolic 3-manifold $M = \mathbb{H}^3/\Gamma$ admits a deformation retraction to an (essentially canonical) compact submanifold $M_c \subset M$ (the *compact core* of M).

Remark 5.1. Of course, after W. Thurston entered the area of Kleinian groups, the theory experienced yet another radical change and became the theory of hyperbolic 3-manifolds. However, this is another story.

We now turn to the higher dimensions.

5.1. Results of Farrell and Jones

The following conjecture is a natural generalization of the topological rigidity of 3-manifolds:

Conjecture 5.2 (A. Borel). *Let* M, N *be closed aspherical* n*-manifolds and* $f : M \to N$ *is a homotopy-equivalence. Then* f *is homotopic to a homeomorphism. (There is also a relative version of this conjecture.)*

We refer the reader to [131] for a detailed discussion of Borel's Conjecture and its relation to Wall's Conjecture 4.5. Although, in full generality, Conjecture 5.2 is expected to be false, in the last 20 years there has been a remarkable progress in proving this conjecture in the context to Kleinian groups. Most of these results appear in the works of T. Farrell and L. Jones. We collect some of them below.

Theorem 5.3. (*T. Farrell and L. Jones* [71].) *Suppose that* $\Gamma \subset \mathbf{Mob}(\mathbb{S}^n)$ *is a convex-cocompact Kleinian group,* $n \geq 4$, N *is a compact aspherical manifold (possibly with nonempty boundary* ∂N*) and* $f : (\bar{M}^{n+1}(\Gamma), M^n(\Gamma)) \to (N, \partial N)$ *is a homotopy-equivalence which is a homeomorphism on the boundary. Then* f *is homotopic to a homeomorphism (rel.* $M^n(\Gamma)$*).*

Theorem 5.4. (*T. Farrell and L. Jones,* [72, Theorem 0.1]) *Suppose that* X *is a nonpositively curved closed Riemannian manifold,* Y *is a closed aspherical manifold of dimension* ≥ 5 *and* $f : X \to Y$ *is a homotopy-equivalence. Then* f *is homotopic to a homeomorphism.*

Theorem 5.5. (*T. Farrell and L. Jones,* [73, Proposition 0.10]) *For each Kleinian group* Γ *the Whitehead group* $Wh(\Gamma)$ *is trivial.*

By combining Theorem 5.5 with the s-cobordism theorem (see, e.g., [137, 186, 196]), one gets:

Corollary 5.6. *Suppose that* W^{n+1} *is a topological (resp. PL, smooth)* h*-cobordism so that* $n \geq 4$ *and* $\pi_1(W^{n+1})$ *is isomorphic to a Kleinian group. Then* W *is trivial in the topological (resp. PL, smooth) category.*

5.2. Limit sets and homological algebra

Let $\Gamma \subset \mathbf{Mob}(\mathbb{S}^n)$ be a convex-cocompact subgroup with the limit set Λ and R be a ring. (One can also deal with geometrically finite groups by using relative cohomology, see [121].) The following theorem establishes a link between topology of the limit sets and the cohomology of Γ:

Theorem 5.7. (*M. Bestina, G. Mess* [28].)

$$H^*(\Gamma, R\Gamma) \cong H^*_c(Hull(\Lambda), R) \cong \tilde{H}^{*-1}(\Lambda, R).$$

Here we are using the Chech cohomology of the limit set.

In particular, Γ is an $m+1$-dimensional duality group over $\mathbb{Z}$ (see Section 4) iff $\tilde{H}^{*-1}(\Lambda)$ vanishes except in dimension m. In this case Λ, homologically, looks like an infinite bouquet of m-spheres. Moreover

$$cd(\Gamma) = \dim(\Lambda),$$

which gives a geometric interpretation of the cohomological dimension of Γ.

5.3. Coarse compact core

In this section we state the best (presently) available higher-dimensional generalization of the Scott compact core theorem. The main drawback of this result is that it applies only to Kleinian groups $\Gamma \subset \mathbf{Mob}(\mathbb{S}^n)$ which are n-dimensional duality groups.

We first need some definitions. We recall that every end e of a manifold M admits a *basic system of neighborhoods*, which is a decreasing sequence (E_j) of nested connected subsets of M with compact frontier, so that

$$\bigcap_{j \in \mathbb{N}} E_j = \emptyset.$$

Given such (E_j), we obtain the inverse system of the fundamental groups $(\pi_1(E_j))$. Consider the image Γ_j of $\pi_1(E_j)$ in $\pi_1(M) = \Gamma$. Then

$$\Gamma_1 \supset \Gamma_2 \supset \cdots$$

The *fundamental group* of e (or, rather, its image in Γ) is defined as

$$\pi_1(e) = \bigcap_{j \in \mathbb{N}} \Gamma_j.$$

An end e is called *almost stable* if every sequence (Γ_j) as above is eventually constant, in which case

$$\pi_1(e) = \Gamma_j$$

for all sufficiently large j. (This notion is weaker than the notion of *semistabe* ends, which the reader might be familiar with.) For instance, if M is an open handlebody of finite genus, then M has unique end e, which is almost stable, whose fundamental group $\pi_1(e)$ is the free group $\pi_1(M)$. On the other hand, if S is a surface of infinite genus with a unique end e, then e is not almost stable. If M is the complement to a Cantor set in $\mathbb{S}^2$, then no end of M is almost stable.

The following is the *Coarse Compact Core Theorem* proved in [124] in the general context of *coarse Poincaré duality spaces.*

Theorem 5.8. (*M. Kapovich, B. Kleiner* [124].) *Let $\Gamma \subset \mathbf{Mob}(\mathbb{S}^n)$ be a Kleinian subgroup which is an n-dimensional duality group. Then the manifold $M = \mathbb{H}^{n+1}/\Gamma$ contains a compact submanifold M_c (the* coarse compact core*) satisfying the following:*

1. *Γ contains a finite collection Δ of $PD(n-1)$ subgroups $\Delta_1, \ldots, \Delta_k$.*
2. *The pair (Γ, Δ) is an n-dimensional Poincaré duality pair.*
3. *The group $\pi_1(M_c)$ maps onto $\pi_1(M) = \Gamma$.*
4. *The manifold M has exactly k ends $e_1, \ldots, e_k$, each of which is almost stable; the components $E_1, \ldots, E_k$ of $M \setminus M_c$ are basic neighborhoods of $e_1, \ldots, e_k$.*
5. *For every $i = 1, \ldots, k$, $\pi_1(e_i) = \Delta_i$ is the image of $\pi_1(E_i)$ in Γ.*

See Figure 1. In the case when $n = 3$, this theorem, of course, is a special casc of thc *Scott compact corc thcorcm* [201, 202]. Morc prccisely, it covers the case when Scott compact core has incompressible boundary, for otherwise Γ splits as a

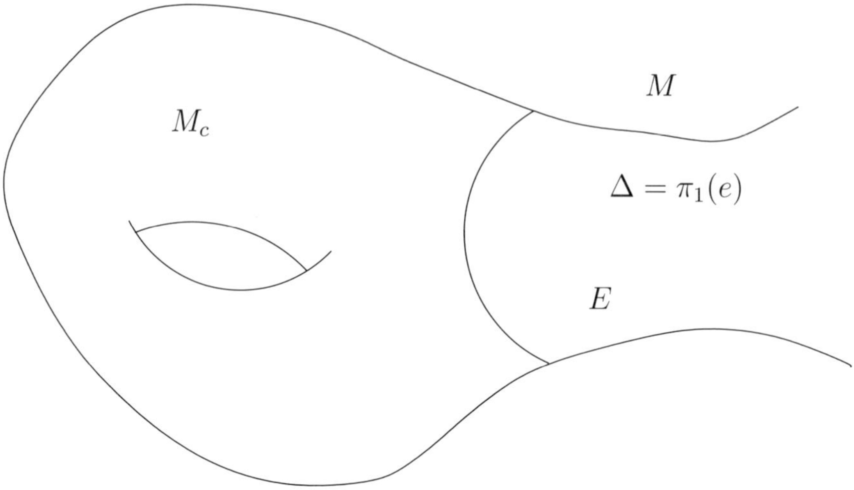

FIGURE 1. *A coarse compact core.*

free product and is not a 2-dimensional duality group. If M is a tame manifold, e.g., Γ is geometrically finite, this theorem is also obvious. At the moment, all known examples of Kleinian groups in $\mathbf{Mob}(\mathbb{S}^n)$, $n \geq 3$, which are n-dimensional duality groups, are geometrically finite.

Problem 5.9. *Generalize Theorem* 5.8 *to groups of type FP which are not n-dimensional duality groups. (Of course, the conclusion of Part* 1 *of Theorem* 5.8 *would have to be suitably modified.)*

6. Notions of equivalence for Kleinian groups

In this section we discuss various equivalence relations for Kleinian subgroups Γ_1, Γ_2 of $\mathbf{Mob}(\mathbb{S}^n)$. We start with the weakest one and end with the strongest.

(0) **Algebraic**: Γ_1 is isomorphic to Γ_2 as an abstract group.

(1) **Dynamical**: there exists a homeomorphism $f : \Lambda(\Gamma_1) \to \Lambda(\Gamma_2)$ such that $f\Gamma_1 f^{-1} = \Gamma_2$; i.e., the groups Γ_1 and Γ_2 have the same topological dynamics on their limit sets. Thus, Γ_1 is geometrically finite iff Γ_2 is, since geometric finiteness can be stated in terms of topological dynamics of a group on its limit set (Theorem 2.5).

(2) **Topological conjugation**: there exists a homeomorphism $f : \mathbb{S}^n \to \mathbb{S}^n$ such that $f\Gamma_1 f^{-1} = \Gamma_2$. (One can relax this by assuming that f is defined only on the domain of discontinuity of Γ_1.)

(3) **Quasiconformal conjugation**: in (2) one can find a quasiconformal homeomorphism. In the case $n = 1$ one should replace *quasiconformal* with *quasisymmetric.*

(4) **Topological isotopy**: in (2) there exists a continuous family of homeomorphisms $h_t : \mathbb{S}^n \to \mathbb{S}^n$ such that: $h_0 = id$, $\forall t$, $h_t\Gamma_1 h_t^{-1} \subset \mathbf{Mob}(\mathbb{S}^n)$ and $h_1\Gamma_1 h_1^{-1} = \Gamma_2$.
(5) **Quasiconformal isotopy**: in (4) all homeomorphisms are quasiconformal (quasisymmetric).
(6) **Möbius conjugation:** there is $f \in \mathbf{Mob}(\mathbb{S}^n)$ such that $f\Gamma_1 f^{-1} = \Gamma_2$.

We refer the reader to [94, 99] for the definitions of quasisymmetric and quasiconformal homeomorphisms.

Below is a collection of facts about the relation between different notions of equivalence of Kleinian groups.

Suppose that both groups Γ_j are *geometrically finite* and $\varphi : \Gamma_1 \to \Gamma_2$ is an isomorphism which preserves the *type* of elements, i.e., for $\gamma \in \Gamma_1$, $\varphi(\gamma)$ is hyperbolic if and only if γ is hyperbolic. It is clear that the above assumptions are necessary for getting the equivalence (1). The following theorem shows that these assumptions are also sufficient.

Theorem 6.1. (*P. Tukia* [217].) *Under the above assumptions, the isomorphism φ can be realized by the equivalence (1), i.e., there exists a (quasisymmetric) homeomorphism f of the limit sets, so that $f\gamma f^{-1} = \varphi(\gamma)$ for all $\gamma \in \Gamma_1$. Moreover, if $f : \Omega(\Gamma_1) \to \Omega(\Gamma_2)$ is a φ-equivariant quasiconformal (quasisymmetric) homeomorphism, then f admits a φ-equivariant quasiconformal (quasisymmetric) extension to the entire sphere.*

Question 6.2. (Quasiconformal vs. topological.) *Suppose that two Kleinian groups $\Gamma_1, \Gamma_2 \subset \mathbf{Mob}(\mathbb{S}^n)$ are topologically conjugate by a homeomorphism f (defined either on $\Omega(\Gamma_1)$, or on $\Lambda(\Gamma_1)$, or on the entire $\mathbb{S}^n$), which induces a type-preserving isomorphism $\varphi : \Gamma_1 \to \Gamma_2$. Does it imply that φ is induced by a quasiconformal (quasisymmetric) homeomorphism with the same domain as f?*

Note that, for every n, the above question actually consists of 3 subquestions, depending on the domain of f. Here is what is currently known about these questions:

1. If $n = 1$ then all three questions have the affirmative answer and the proof is rather elementary. It also follows for instance from Theorem 6.1.

2. If $n = 2$ then the answer to all three questions is again positive, but the proof is highly nontrivial. The easiest case is when the homeomorphism f is defined on $\Omega(\Gamma_1)$. Then we get the induced homeomorphism $\bar{f}$ of the quotient surfaces $S_1 \to S_2$, where $S_i = \Omega(\Gamma_i)/\Gamma_i$. The existence of a diffeomorphism $S_1 \to S_2$ homotopic to $\bar{f}$ follows from the uniqueness of the smooth structure on surfaces. If S_1 is compact, then this diffeomorphism lifts to an equivariant quasiconformal homeomorphism $\Omega(\Gamma_1) \to \Omega(\Gamma_2)$. Two noncompact surfaces can be diffeomorphic but not quasiconformally homeomorphic: For instance, the open disk is not quasiconformally equivalent to the complex plane. However, since φ is type-preserving, Ahlfors Finiteness Theorem [3] in conjunction with a lemma of Bers and Maskit

(see, e.g., [116, Corollary 4.85]), implies the existence of a quasiconformal homeomorphism $S_1 \to S_2$.

If f is defined on the limit set and Γ_1, Γ_2 are geometrically finite, then the positive answer is a special case of Tukia's theorem 6.1. However, if the groups Γ_i are not geometrically finite, the proof becomes very difficult and is a corollary of the solution of the Ending Lamination Conjecture in the work of J. Brock, R. Canary and Y. Minsky in [46, 166, 167], and M. Rees [193].

Combination of the Ahlfors Finiteness Theorem with the Ending Lamination Conjecture also gives the positive answer in the case when f is defined on $\mathbb{S}^2$.

3. If f is defined on $\Omega(\Gamma_1)$, then the answer is positive provided that $n \neq 4$ and $M^n(\Gamma_1)$ is compact. This is a consequence of the theorem of D. Sullivan [207], who proved uniqueness of the quasiconformal structure on compact n-manifolds ($n \neq 4$): Apply Sullivan's theorem to the manifolds $M^n(\Gamma_i)$, $i = 1, 2$, and lift the quasiconformal homeomorphism to the domain of discontinuity.

Remark 6.3. An alternative proof of Sullivan's theorem and its generalization was given by J. Luukkainen in [147], see also [220].

If $n = 4$, f is defined on $\Omega(\Gamma_1)$, and $M^4(\Gamma_1)$ is compact, then the situation is unclear but one probably should expect the negative answer since the uniqueness of quasiconformal structures in dimension 4 was disproved by S. Donaldson and D. Sullivan [64].

Question 6.4. *Is there a pair of Kleinian groups $\Gamma_1, \Gamma_2 \subset \mathbf{Mob}(\mathbb{S}^n)$ so that the manifolds $M^n(\Gamma_1), M^n(\Gamma_2)$ are homeomorphic but not diffeomorphic?*

Note that in view of the examples in [70], the positive answer to the above question would not be too surprising.

If f is defined on $\Omega(\Gamma_1)$ and we do not assume compactness of $M^n(\Gamma_1)$, then the answer to Question 6.2 is negative in a variety of ways.

(a) For instance, take *singly degenerate groups* $\Gamma_1, \Gamma_2 \subset \mathbf{Mob}(\mathbb{S}^2)$, which are both isomorphic to the fundamental group of a closed oriented surface S, contain no parabolic elements and have distinct ending laminations. Then $\Omega(\Gamma_i) \subset \mathbb{S}^2$ are open disks D_i for both i. There exists an equivariant homeomorphism $h : D_1 \to D_2$, which induces an isomorphism $\varphi : \Gamma_1 \to \Gamma_2$. However, since the ending laminations are different, there is no equivariant homeomorphism $\Lambda(\Gamma_1) \to \Lambda(\Gamma_2)$.

Now extend both groups to the 3-sphere so that $\Gamma_i \subset \mathbf{Mob}(\mathbb{S}^3)$, $i = 1, 2$. Then the 3-dimensional domains of discontinuity B_i of both groups are diffeomorphic to the open 3-ball, $i = 1, 2$; the quotient manifolds are

$$M^3(\Gamma_i) = B_i/\Gamma_i \cong S \times \mathbb{R}, \quad i = 1, 2.$$

Therefore there exists an equivariant diffeomorphism $f : B_1 \to B_2$. We claim that this map cannot be quasiconformal. Indeed, otherwise it would extend to an equivariant homeomorphism of the limit sets (which are planar subsets of $\mathbb{R}^3$). This is a contradiction.

(b) One can construct geometrically finite examples as well. The reason is that even though all (orientation-preserving) parabolic elements of $\mathbf{Mob}(\mathbb{S}^2)$ are quasiconformally conjugate, the analogous assertion is false for the parabolic elements of $\mathbf{Mob}(\mathbb{S}^3)$. Suppose that τ is the translation in $\mathbb{R}^3$ by a nonzero vector v. Let R_{θ_i}, $i = 1, 2$, denote the rotations around v by the angles $\theta_1, \theta_2 \in [0, \pi]$. Then the skew motions

$$\gamma_i = R_{\theta_i} \circ \tau_i, \quad i = 1, 2$$

are parabolic elements of $\mathbf{Mob}(\mathbb{S}^3)$. One can show that

Proposition 6.5. *The Möbius transformations γ_1 and γ_2 are quasiconformally conjugate in $\mathbb{S}^3$ if and only if $\theta_1 = \theta_2$.*

The proof is based on a calculation of the extremal length of a certain family of curves in $\mathbb{R}^3$ and we will not present it here.

Note that the cyclic groups $\Gamma_i = \langle \gamma_i \rangle$ are geometrically finite, the isomorphism $\varphi : \Gamma_1 \to \Gamma_2$ sending γ_1 to γ_2 is type-preserving. The quotient manifolds $M^3(\Gamma_i)$ are both diffeomorphic to $\mathbb{R}^2 \times \mathbb{S}^1$, therefore there exists a φ-equivariant diffeomorphism $f : \Omega(\Gamma_1) \to \Omega(\Gamma_2)$ which, of course, extends to a homeomorphism $\mathbb{S}^3 \to \mathbb{S}^3$. However, according to Proposition 6.5, this homeomorphism cannot be made quasiconformal.

These examples do not resolve the following:

Question 6.6. *Suppose that $\Gamma_1, \Gamma_2 \subset \mathbf{Mob}(\mathbb{S}^n)$, $n \geq 3$, are Kleinian groups and $f : \Lambda(\Gamma_1) \to \Lambda(\Gamma_2)$ is a homeomorphism which induces an isomorphism $\Gamma_1 \to \Gamma_2$. Does it follow that f is quasisymmetric?*

If $n \leq 2$, then (in the list of equivalences between Kleinian groups) we have the implication

$$(3) \Rightarrow (5).$$

Indeed, consider a quasiconformal homeomorphism f conjugating Kleinian groups Γ_1 and Γ_2 and let μ denote the Beltrami differential of f. Then for $t \in [0, 1]$ the solutions of the Beltrami equation

$$\frac{\partial f_t}{\partial \bar{z}} = t\mu \frac{\partial f}{\partial z}$$

also conjugate Γ_1 to Kleinian subgroups of $\mathbf{Mob}(\mathbb{S}^2)$, see, e.g., [20]. This gives the required quasiconformal isotopy. Since (2) is equivalent to (3) for $n \leq 2$, it follows that for $n \leq 2$ we have

$$(2) \iff (3) \iff (4) \iff (5)$$

This argument however fails completely in higher dimensions, since the Beltrami equation in $\mathbb{R}^n$ for $n \geq 3$ is overdetermined.

Question 6.7. *In the list of equivalences between Kleinian groups:*

(a) *Does* (2) $\Rightarrow$ (4) *?*
(b) *Does* (3) $\Rightarrow$ (5) *?*

One can show (using quasiconformal stability, see Section 11.2, cf. [154, Theorem 7.2]) that for convex-cocompact groups parts (a) and (b) of the above question are equivalent. In Theorem 11.11 we give examples of convex-cocompact Kleinian groups in $\mathbf{Mob}(\mathbb{S}^n)$, $n \geq 5$, for which the answer to Question 6.7 is negative. The situation in dimensions 3 and 4 at the moment is unclear, but we expect in these dimensions the answer to be negative as well.

The implications (i)⇒(6) for i$\leq$ 5 are, of course, extremely rare. The most celebrated example is provided by the Mostow rigidity theorem:

Theorem 6.8. *Suppose that* $\Gamma_1, \Gamma_2 \subset \mathbf{Mob}(\mathbb{S}^n)$ *are lattices and* $n \geq 2$. *Then* (0) $\Rightarrow$ (6) *for these groups.*

See [175] for G. Mostow's original proof or [120] for a more elementary argument along the same lines which uses only the analytical properties of quasiconformal mappings. A completely different argument due to M. Gromov can be found in [18]. Yet another proof is an application of the barycentric maps [21]. Note that, presently, there are no proofs using equivariant harmonic maps.

Mostow's ergodic arguments were greatly generalized by D. Sullivan in [208], see also [5]:

Theorem 6.9. (*D. Sullivan* [208].) *Suppose that* $\Gamma_1, \Gamma_2 \subset \mathbf{Mob}(\mathbb{S}^n)$ *are Kleinian groups whose limit set is the entire* $\mathbb{S}^n$ *and so that the action of* Γ_1 *on* $\mathbb{S}^n$ *is recurrent. Then* (3) $\Rightarrow$ (6) *for these groups.*

The action of $\Gamma \subset \mathbf{Mob}(\mathbb{S}^n)$ on $\mathbb{S}^n$ is called *recurrent* if for every measurable subset $E \subset \mathbb{S}^n$ of positive Lebesgue measure, the measure of the intersection $\gamma(E) \cap E$ is positive for some $\gamma \in \Gamma \setminus \{1\}$.

7. Groups with zero-dimensional limit sets

In what follows, we let dim denote the covering dimension of topological spaces, see for instance [96]. Suppose that $\Gamma \subset \mathbf{Mob}(\mathbb{S}^n)$ is a non-elementary Kleinian subgroup of $\mathbf{Mob}(\mathbb{S}^n)$ and $\dim(\Lambda(\Gamma)) = 0$; hence $\Lambda(\Gamma)$ is totally disconnected (its only connected components are points). Recall that a *discontinuum* is a nonempty perfect totally disconnected compact topological space, see, e.g., [8]. Hence $\Lambda(\Gamma)$ is a discontinuum. It follows (see, e.g., [8]) that $\Lambda(\Gamma)$ is homeomorphic to the standard Cantor set $K \subset [0, 1]$. Below is a couple of examples of Kleinian groups whose limit sets are totally disconnected.

Example. (A **Schottky group**, see, e.g., [138, 157].) Let $n, k \geq 1$. Suppose that we are given a collection of disjoint closed topological n-balls

$$B_1^+, \dots, B_k^+, B_1^-, \dots, B_k^- \subset \mathbb{S}^n$$

and Möbius transformations $\gamma_j \in \mathbf{Mob}(\mathbb{S}^n)$ so that $\gamma_j(\mathrm{int}(B_j^+)) = \mathrm{ext}(B_j^-)$. We assume that for each pair B_j^+, B_j^- there exists a diffeomorphism of $\mathbb{S}^n$ which carries these balls to the round balls.[1] Then

$$\Phi := \mathbb{S}^n - \bigcup_{j=1}^{k} (B_j^+ \cup \mathrm{int}(B_j^-))$$

is a fundamental domain for the group Γ generated by $\gamma_1, \ldots, \gamma_k$. The group Γ is called a *Schottky group.* It is isomorphic to a free group of rank k, and the limit set of Γ is a discontinuum provided that $k \geq 2$. Every nontrivial element of Γ is hyperbolic.

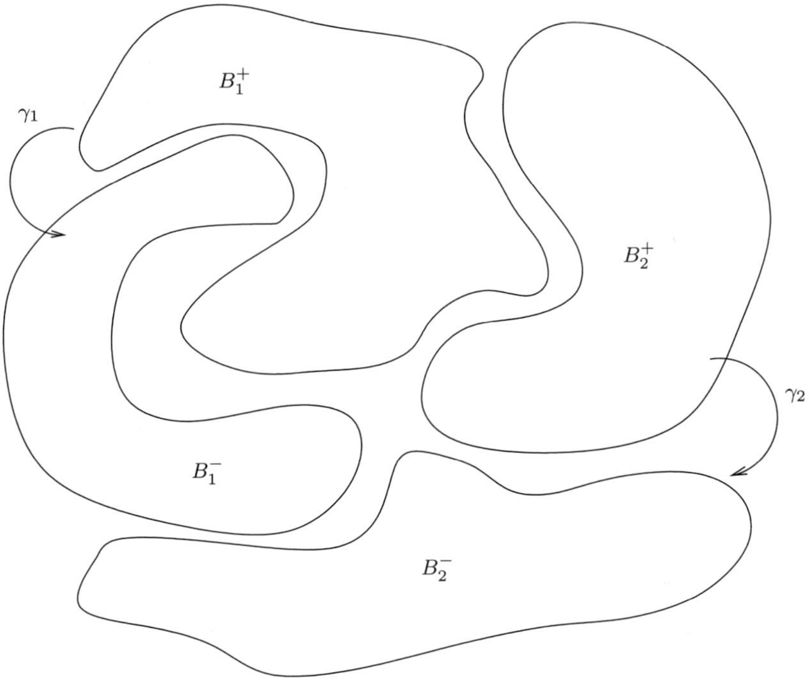

FIGURE 2. *A Schottky group.*

Before giving the next example we need a definition. Suppose that $\Gamma \subset$ **Mob**$(\mathbb{S}^n)$ is a nontrivial elementary subgroup. Then, after conjugating Γ if necessary, we can assume that either:

1. Γ fixes $0, \infty \in \overline{\mathbb{R}^n} = \mathbb{S}^n$ and therefore is generated by $\gamma(x) = Ax$, where A is the product of a scalar $c > 1$ by an orthogonal matrix.
2. Or Γ acts on $\mathbb{R}^n \subset \mathbb{S}^n$ by Euclidean isometries.

[1] By the smooth Schoenflies theorem, for $n \neq 4$ it suffices to assume that the balls B_j^+ have smooth boundary.

In the first case we take the fundamental domain Φ for the action of Γ on $\mathbb{S}^n$ to be an annulus bounded by two disjoint round spheres. In the second case we take a Dirichlet fundamental domain $\Phi \subset \mathbb{R}^n$ for Γ.

We refer to the fundamental domains Φ as *standard* fundamental domains. A fundamental domain for Γ is *topologically standard* if it is the image of a standard fundamental domain of Γ under a diffeomorphism of $\Omega(\Gamma)$ commuting with Γ. For instance, the fundamental domain for a rank 1 Schottky group is topologically standard. Therefore, the fundamental domain Φ for the Schottky group satisfies the property that it is the intersection of topologically standard fundamental domains

$$\Phi_j = \mathbb{S}^n - (B_j^+ \cup \mathrm{int}(B_j^-))$$

for the groups $\Gamma_j = \langle \gamma_j \rangle$.

Given a domain $\Phi \subset \mathbb{S}^n$, we let $\Phi^c \subset \mathbb{S}^n$ denote the closure of the complement of Φ. We are now ready for the second example which is a generalization of the first.

Example. (**Schottky-type groups, see, e.g.,** [138, 157]**.**) Start with a collection of elementary Kleinian groups $\Gamma_i \subset \mathbf{Mob}(\mathbb{S}^n)$, $i = 1, \ldots, k$. Let $\Phi_i \subset \mathbb{S}^n$ be topologically standard fundamental domains for these groups. Assume that

$$\Phi_i^c \cap \Phi_j^c = \emptyset$$

for all $i \neq j$. Let $\Gamma \subset \mathbf{Mob}(\mathbb{S}^n)$ be the group generated by $\Gamma_1, \ldots, \Gamma_k$. Then:

1. As an abstract group, Γ is isomorphic to the free product $\Gamma_1 * \cdots * \Gamma_k$.
2. $\Phi := \Phi_1 \cap \cdots \cap \Phi_k$ is a fundamental domain for the group Γ.
3. The limit set of Γ is totally disconnected.

The groups Γ obtained in this fashion are called *Schottky-type groups.*

A Schottky-type group is called *classical* if it admits a fundamental domain $\Phi := \Phi_1 \cap \ldots \cap \Phi_k$, so that each Φ_i is geometrically standard. It is not difficult to see that Schottky-type groups are geometrically finite. For instance, consider the case of a Schottky group Γ of rank k, for $n \geq 2$. We have the map

$$j : \mathbb{Z} = H_n(M^n(\Gamma)) \to H_n(\bar{M}^{n+1}(\Gamma))$$

induced by the inclusion of manifolds. Since the manifold $\bar{M} = \bar{M}^{n+1}(\Gamma)$ is $K(\Gamma, 1)$, it follows that

$$H^n(\bar{M}) = H^n(\Gamma) = H^n(B) = 0,$$

where B is the bouquet of k circles. Therefore $j = 0$. Hence $\bar{M}$ is compact and hence Γ is convex-cocompact. A similar argument works for Schottky-type groups, provided that one uses cohomology relative to the cusps.

The quotient manifolds of the Schottky-type groups Γ have a rather simple topology, as it follows from the explicit description of their fundamental domains. Namely, let $M_i = M^n(\Gamma_i)$, $i = 1, \ldots, k$. Then we get the smooth connected sum decomposition

$$M^n(\Gamma) = M_1 \# \cdots \# M_k.$$

By combining this with Theorem 6.1 we obtain

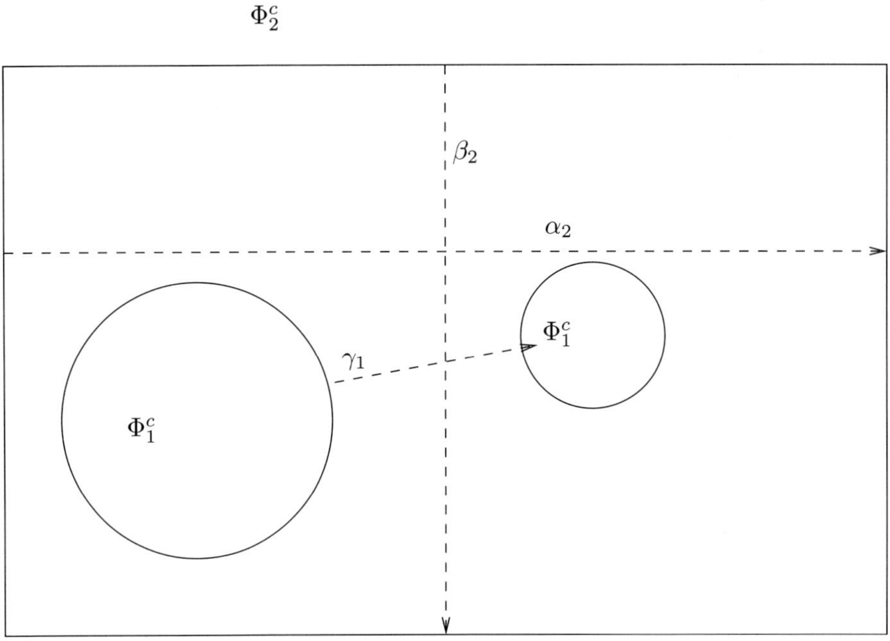

FIGURE 3. *A classical Schottky-type group isomorphic to* $\mathbb{Z} * \mathbb{Z}^2$ *with* $\mathbb{Z} = \langle \gamma_1 \rangle$, $\mathbb{Z}^2 = \langle \alpha_2, \beta_2 \rangle$.

Proposition 7.1. 1. *Suppose that* Γ, Γ' *are Schottky groups of the same rank. Then there exists a quasiconformal homeomorphism* $f : \mathbb{S}^n \to \mathbb{S}^n$ *which conjugates* Γ *to* Γ', *i.e.,* $f \Gamma f^{-1} = \Gamma'$.

2. *Suppose that* Γ, Γ' *are Schottky-type groups and* $\varphi : \Gamma \to \Gamma'$ *is a type-preserving isomorphism, so that for every free factor* Γ_i *in* Γ, *the restriction* $\varphi : \Gamma_i \to \Gamma'_i \subset \Gamma'$ *is induced by a quasiconformal homeomorphism of* $\mathbb{S}^n$. *Then there exists a quasiconformal homeomorphism* $f : \mathbb{S}^n \to \mathbb{S}^n$ *which induces the isomorphism* φ.

Question 7.2. *Let* $n \geq 3$. *Is there a quasiconformal isotopy between* Γ *and* Γ' *in the above theorem (either part* 1 *or part* 2*)?*

In the case when Γ and Γ' are both classical, the positive answer follows rather easily. In the non-classical case the above question is open even if $n = 3$ and Γ is a Schottky group.

Schottky subgroups of $\mathbf{Mob}(\mathbb{S}^2)$ can be characterized as follows:

Theorem 7.3. (*B. Maskit* [155].) *A Kleinian subgroup* $\Gamma \subset \mathbf{Mob}(\mathbb{S}^2)$ *is a Schottky group if and only if* Γ *is free, has nonempty domain of discontinuity in* $\mathbb{S}^2$ *and consists only of hyperbolic elements.*

This result was generalized by N. Gusevskii and N. Zindinova [92]:

Theorem 7.4. *Let* $\Gamma \subset \mathbf{Mob}(\mathbb{S}^2)$ *be a Kleinian subgroup, which has nonempty domain of discontinuity in* $\mathbb{S}^2$ *and is isomorphic to a Schottky-type group* Γ' *via a type-preserving isomorphism* $\Gamma \to \Gamma'$. *Then* Γ *is a Schottky-type group.*

Both theorems are easy under the assumption that Γ is geometrically finite, the key point here is that (in dimension 2) one can prove geometric finiteness under the above mild assumptions.

If Γ is a Kleinian subgroup of $\mathbf{Mob}(\mathbb{S}^3)$, then the above results are not longer true, moreover, Γ can be geometrically infinite. For instance, take a free finitely generated purely hyperbolic discrete subgroup of $PSL(2,\mathbb{C})$, whose limit set is the 2-sphere (the existence of such groups was first established by V. Chuckrow [56]). The Möbius extension of this group to $\mathbb{S}^3$ has nonempty domain of discontinuity, but is not geometrically finite.

Tameness of limit sets. Below we address the following:

Question 7.5. *Suppose that* $\Gamma \subset \mathbf{Mob}(\mathbb{S}^n)$ *is a Kleinian group, whose limit set is a discontinuum. What can be said about the embedding* $\Lambda(\Gamma) \subset \mathbb{S}^n$*?*

A discontinuum $D \subset \mathbb{S}^n$ is called *tame* if there exists a homeomorphism $f : \mathbb{S}^n \to \mathbb{S}^n$ which carries D to the Cantor set $K \subset [0,1]$ and is called *wild* otherwise. It is a classical (and easy) result that every discontinuum in $\mathbb{S}^2$ is tame, see, e.g., [31]. The (historically) first example of a wild discontinuum was the *Antoine's necklace* $A \subset \mathbb{S}^3$:

$$\pi_1(\mathbb{S}^3 \setminus A) \neq \{1\},$$

which explains why A is wild, see [31]. D. DeGryse and R. Osborne [61] constructed for every $n \geq 3$ examples of wild discontinua $D_n \subset \mathbb{S}^n$, such that

$$\pi_1(\mathbb{S}^n \setminus D_n) = \{1\}.$$

See also [77] for infinitely many inequivalent 3-dimensional examples of this type.

The algebraic structure of Kleinian groups with totally disconnected limit sets is given by

Theorem 7.6. (*R. Kulkarni* [141].) *Suppose that a Kleinian group* $\Gamma \subset \mathbf{Mob}(\mathbb{S}^n)$ *has a totally disconnected limit set. Then* Γ *is isomorphic to a Schottky-type group.*

One can even describe (to some extent) fundamental domains of such groups:

Theorem 7.7. (*N. Gusevskii* [90].) *Suppose that the limit set of* $\Gamma \subset \mathbf{Mob}(\mathbb{S}^n)$ *is totally disconnected. Then* Γ *admits a fundamental domain* Φ *of the same shape as in Example* 7, *only the fundamental domains* Φ_i *for* Γ_i*'s are not required to be topologically standard.*

The proof of Theorem 7.7 is based on the following

Theorem 7.8. (*M. Brin* [45].) *Let $\tilde{M}$ be a smooth oriented n-manifold of dimension > 2, so that $H^1(\tilde{M}) = 0$. Let $\Gamma \curvearrowright \tilde{M}$ is a smooth properly discontinuous free action.*

Then, for every smooth oriented compact hypersurface Σ in $\tilde{M}$ and an open neighborhood U of $\Gamma \cdot \Sigma$, there exists a smooth compact connected oriented hypersurface $\Sigma^* \subset U$ such that for every $\gamma \in \Gamma$ either $\gamma\Sigma^* \cap \Sigma^* = \emptyset$ or $\gamma\Sigma^* = \Sigma^*$.

This theorem allows one to split (inductively) the Kleinian group Γ as a free product in a "geometric fashion": Start with a compact hypersurface in $\Omega(\Gamma)$ which separates components of $\Lambda(\Gamma)$. Find Σ^* as in Brin's theorem which still separates. Then cut open the manifold $M^n(\Gamma)$ along the projection of Σ^*. This decomposition yields a free product decomposition $\Gamma = \Gamma' * \Gamma''$ so that Γ is a Klein combination of the groups Γ', Γ''. Continue inductively. Finite generation of Γ implies that the decomposition process will terminate and the terminal groups must be elementary. Note that if all Σ^* were spheres, then this decomposition would be of Schottky-type.

Corollary 7.9. *Every Kleinian group with a totally disconnected limit set is geometrically finite.*

Proof. Repeat the arguments which we used to establish geometric finiteness of Schottky groups. □

Problem 7.10. *Suppose that $\Gamma \subset \mathbf{Mob}(\mathbb{S}^n)$ is such that $\Lambda(\Gamma)$ is a tame discontinuum. Does is follow that Γ is a Schottky-type group?*

If $n = 2$ then the affirmative answer to this question follows for instance from Maskit's theorem. If $n = 3$ then the answer is again positive; moreover,

Proposition 7.11. *Suppose that $\Gamma \subset \mathbf{Mob}(\mathbb{S}^3)$ is such that $\Lambda(\Gamma)$ is totally disconnected and $\pi_1(\Omega(\Gamma)) = 1$. Then Γ is a Schottky-type group*

Proof. Under the above assumptions, $\pi_2(\Omega(\Gamma)) \neq \{0\}$; hence, by the Sphere Theorem (see, e.g., [95]), we can find a smooth hypersurface Σ^* as in Brin's theorem, so that Σ^* is diffeomorphic to $\mathbb{S}^2$. Therefore, as we saw above, it follows that Γ is a Schottky-type group. □

This argument however fails for $n \geq 4$, where Problem 7.10 is still open. On the other hand, there are Kleinian subgroups of $\mathbf{Mob}(\mathbb{S}^3)$ with wild discontinua as limit sets. The first such example was given by M. Bestvina and D. Cooper:

Theorem 7.12. (*M. Bestvina, D. Cooper* [25].) *There exists a Kleinian group $\Gamma \subset \mathbf{Mob}(\mathbb{S}^3)$ which contains parabolic elements, so that $\Lambda(\Gamma)$ is a wild discontinuum.*

The proof that $\pi_1(\Omega(\Gamma)) \neq 1$ presented in [25] was incomplete; however the gap was filled several years later by S. Matsumoto:

Theorem 7.13. (*S. Matsumoto* [158, 159], *see also* [91].) *There are Kleinian groups Γ in $\mathbf{Mob}(\mathbb{S}^3)$ without parabolic elements whose limit sets are wild discontinua.*

8. Groups with one-dimensional limit sets

The simplest examples of 1-dimensional limit sets of Kleinian groups are topological circles. For instance, the limit set of a lattice $\Gamma \subset \mathrm{Isom}(\mathbb{H}^2)$ is the round circle. Of course, even if the limit set of $\Gamma \subset \mathbf{Mob}(\mathbb{S}^n)$ is a topological circle, its embedding in $\mathbb{S}^n$ can be complicated. We will discuss this issue later on. For now, we are only interested in the topology of the limit set itself.

Given convex-cocompact Kleinian groups $\Gamma_1, \Gamma_2 \subset \mathbf{Mob}(\mathbb{S}^n)$ with 1-dimensional limit sets, one can use Klein–Maskit Combination theorems in order to get convex-cocompact Kleinian groups $\Gamma \subset \mathbf{Mob}(\mathbb{S}^n)$ isomorphic to

$$\Gamma_1 *_\Delta \Gamma_2,$$

where Δ is either trivial or infinite cyclic. The limit sets of the resulting groups are again 1-dimensional. For instance, if $\Lambda(\Gamma_i)$ is a topological circle for $i = 1, 2$ then the limit set of $\Gamma = \Gamma_1 * \Gamma_2$ will be disconnected: The connected components of $\Lambda(\Gamma)$ are topological circles and points. Similarly, if $\Delta = \mathbb{Z}$, then the limit set of $\Gamma = \Gamma_1 *_\Delta \Gamma_2$ will have *cut pairs*: The complement to the 2-point set $\Lambda(\Delta)$ in $\Lambda(\Gamma)$ is disconnected.

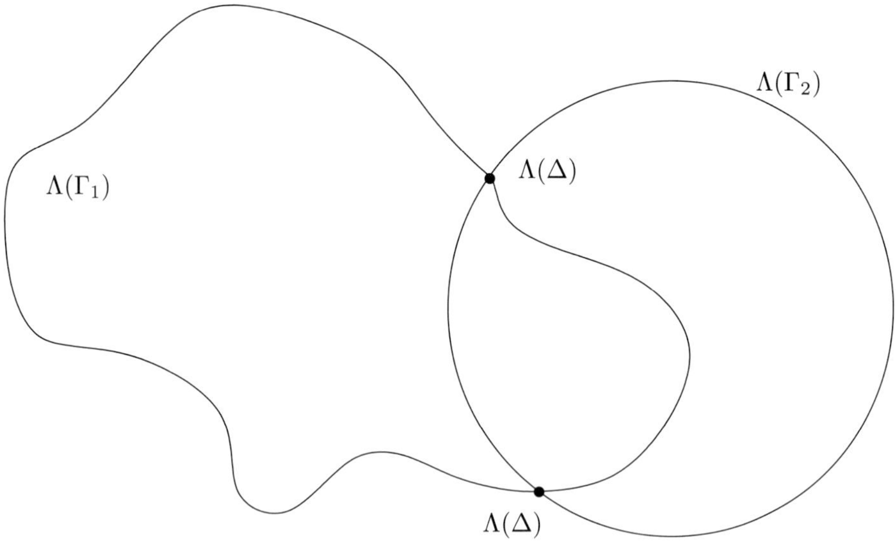

FIGURE 4. *Combination of two 1-quasifuchsian groups:* $\Gamma = \Gamma_1 *_{\mathbb{Z}} \Gamma_2$.

These constructions are, of course, not very illuminating. Therefore we are interested in examples of 1-dimensional limit sets which are connected and which do not contain cut-pairs. It turns out that there are only two such examples:

1. **The Sierpinski carpet $\mathcal{S}$.** Start with the unit square $S = I \times I$. Subdivide this square into 9 squares of the size $\frac{1}{3} \times \frac{1}{3}$ and then remove from S the open middle square $(\frac{1}{3}, \frac{2}{3}) \times (\frac{1}{3}, \frac{2}{3})$. Repeat this for each of the remaining $\frac{1}{3} \times \frac{1}{3}$ sub-squares in S and continue inductively. After removing a countable collection of open squares we are left with a compact subset $\mathcal{S} \subset \mathbb{R}^2$, called the *Sierpinski carpet.*
2. **The Menger curve $\mathcal{M}$.** Start with the unit cube $Q = I \times I \times I$. Each face F_i of Q contains a copy of the Sierpinski carpet $\mathcal{S}_i$. Let $p_i : Q \to F_i$ denote the orthogonal projection. Then
$$\mathcal{M} := \bigcap_i p_i^{-1}(\mathcal{S}_i)$$
is called the *Menger curve.*

Example. There exists a convex-cocompact subgroup $G \subset \mathbf{Mob}(\mathbb{S}^2)$ whose limit set is homeomorphic to the Sierpinski carpet $\mathcal{S}$.

To construct such an example start with a compact hyperbolic manifold M^3 with nonempty totally-geodesic boundary. Thus we get an embedding of $\Gamma = \pi_1(M^3)$ into $\mathbf{Mob}(\mathbb{S}^2)$ as a convex-cocompact Kleinian subgroup. The limit set of Γ is homeomorphic to the Sierpinski carpet. To see this note that the convex hull $Hull(\Lambda(\Gamma))$ in $\mathbb{H}^3$ is obtained by removing from $\mathbb{H}^3$ a countable collection of disjoint open half-spaces $H_j \subset \mathbb{H}^3$. The ideal boundary of each H_j is the open round disk $D_j \subset \mathbb{S}^2$. Thus
$$\Lambda(\Gamma) = \mathbb{S}^2 \setminus \bigcup_j \operatorname{int}(D_j).$$
Clearly, $D_j \cap D_i = \emptyset$, unless $i = j$; since $\Lambda(\Gamma)$ has empty interior. See Figure 5. According to Claytor's theorem [57], it follows that $\Lambda(\Gamma)$ is homeomorphic to $\mathcal{S}$. Moreover, it is easy to see that this homeomorphism extends to the 2-sphere, since it sends the boundary circles of $\Lambda(\Gamma)$ to the boundary squares of $\mathcal{S}$.

The construction of Kleinian groups whose limit sets are homeomorphic to $\mathcal{M}$ is more complicated:

Example. (M. Bourdon [37]; see also [118].) There exists a convex-cocompact subgroup $\Gamma \subset \mathbf{Mob}(\mathbb{S}^3)$ whose limit set is homeomorphic to the Menger curve $\mathcal{M}$.

The following theorem is proved in [122] in the more general context of Gromov-hyperbolic groups:

Theorem 8.1. (*M. Kapovich, B. Kleiner* [122].) *Suppose that* $\Gamma \subset \mathbf{Mob}(\mathbb{S}^n)$ *is a (torsion-free) nonelementary convex-cocompact subgroup such that:*
(a) Γ *does not split as a free product,*
(b) Γ *does not split as an amalgam over* $\mathbb{Z}$*,*
(c) $\dim(\Lambda(\Gamma)) = 1$.
Then $\Lambda(\Gamma)$ *is either homeomorphic to the Sierpinski carpet or to the Menger curve.*

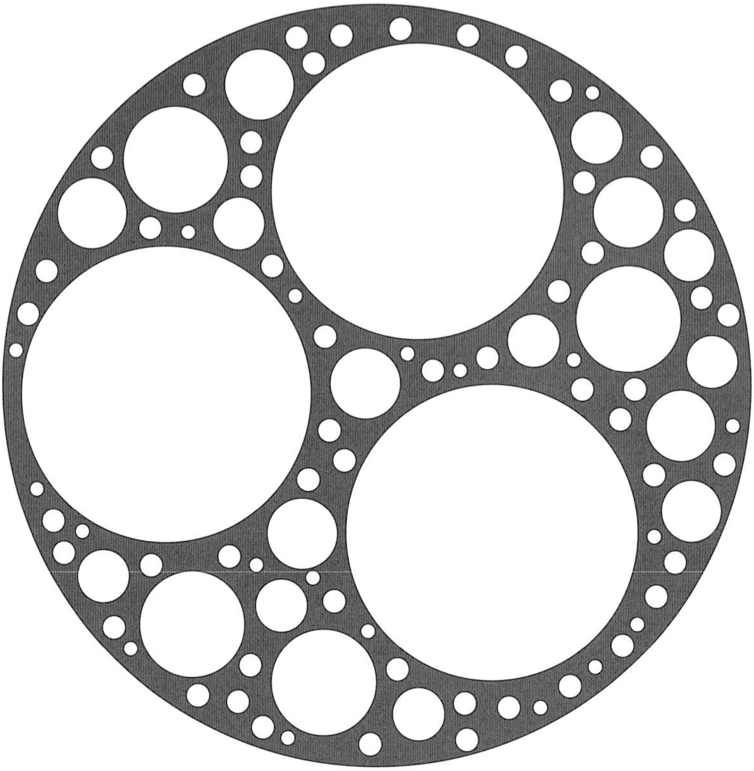

FIGURE 5. *A limit set homeomorphic to the Sierpinski carpet.*

Conjecture 8.2. (*M. Kapovich, B. Kleiner* [122].) *If* $\Gamma \subset \mathbf{Mob}(\mathbb{S}^n)$ *is a (torsion-free) convex-cocompact Kleinian group whose limit set is homeomorphic to the Sierpinski carpet, then* Γ *is isomorphic to a convex-cocompact subgroup in* $\mathbf{Mob}(\mathbb{S}^2)$.

It was proved in [122] that this conjecture would follow either from the positive solution of the 3-dimensional Wall's problem (Problem 4.5) or from Cannon's conjecture (Conjecture 4.6).

Topology of the limit sets of geometrically infinite Kleinian groups can be more complicated. A *dendroid* is a compact locally connected simply-connected 1-dimensional topological space.

Theorem 8.3. (*J. Cannon and W. Thurston* [51], *see also* [1] *and* [58].) *There exist singly-degenerate Kleinian groups whose limit sets are* dendroids.

Conjecturally, limit sets of all singly-degenerate Kleinian groups are dendroids and the following problem is open even for $n = 2$:

Problem 8.4. *Suppose that $\Gamma \subset \mathbf{Mob}(\mathbb{S}^n)$ is a Kleinian group whose limit set is connected and 1-dimensional. Is it true that $\Lambda(\Gamma)$ is locally connected?*

See [163], [165] and [168, 169] for partial results in dimension 2.

9. Groups whose limit sets are topological spheres

Definition 9.1. A Kleinian group $\Gamma \subset \mathbf{Mob}(\mathbb{S}^n)$ is called *i-fuchsian*[2] if $\Lambda(\Gamma)$ is a round i-dimensional sphere in $\mathbb{S}^n$.

To construct examples of i-fuchsian groups start with a lattice $\Gamma \subset \mathbf{Mob}(\mathbb{S}^i)$. The limit set of Γ is the round sphere $\mathbb{S}^i$. Define the *canonical embedding*

$$\iota : \mathbf{Mob}(\mathbb{S}^i) \hookrightarrow \mathbf{Mob}(\mathbb{S}^n)$$

induced by the embedding of the Lorentz groups

$$O(i+1,1) \hookrightarrow O(n+1,1)$$

$$A \mapsto \begin{bmatrix} A & 0 \\ 0 & I \end{bmatrix},$$

where I is the identity matrix. Therefore we get the *canonical* embedding

$$\iota : \Gamma \hookrightarrow \mathbf{Mob}(\mathbb{S}^n).$$

One can modify this construction as follows. Note that the stabilizer of $\mathbb{S}^i$ in $\mathbf{Mob}(\mathbb{S}^n)$ contains $\mathbf{Mob}(\mathbb{S}^i) \times SO(n-i)$. Choose a homomorphism $\phi : \Gamma \to SO(n-i)$. For instance, being residually finite, the group Γ will have many epimorphisms to finite groups, which then can be embedded in $SO(n-i)$ if $n-i$ is sufficiently large. Alternatively, in many cases the group Γ will have infinite abelianization Γ^{ab}. The abelian group Γ^{ab} admits many embeddings into $SO(n-i)$ provided that $n-i \geq 2$. Then the image of

$$\rho = \iota \times \phi : \Gamma \to \mathbf{Mob}(\mathbb{S}^i) \times SO(n-i) \subset \mathbf{Mob}(\mathbb{S}^n)$$

is also an i-fuchsian group, since $\rho(\Gamma)$ preserves $\mathbb{S}^i$ and the action of $\rho(\Gamma)$ on $\mathbb{S}^i$ is the same as the action of Γ.

Definition 9.2. A Kleinian group $\Gamma \subset \mathbf{Mob}(\mathbb{S}^n)$ is called *i-quasifuchsian* if its limit set is a topological i-dimensional sphere.

We will refer to the number $n-i$ as the *codimension* of a (quasi)fuchsian group Γ.

Example. Suppose that Γ is an i-fuchsian subgroup of $\mathbf{Mob}(\mathbb{S}^n)$ and $\Gamma' \subset \mathbf{Mob}(\mathbb{S}^n)$ is another group which is topologically conjugate to Γ (with a homeomorphism f defined on the entire n-sphere). Then Γ' is i-quasifuchsian. However, as we will see, there are i-quasifuchsian groups (for $n \geq 3$) which cannot be obtained in this fashion.

---

[2] Our definition is somewhat different from the classical: fuchsian subgroups of $PSL(2,\mathbb{C})$ are usually required to preserves a round disk in $\mathbb{S}^2$.

The (quasiconformal) homeomorphisms f as in the previous example exist in abundance if $i = 1, n = 2$, due to the solvability of the Beltrami equation. If $i \geq 2$, the situation is very different and it is not so easy to construct nontrivial examples of i-quasifuchsian groups which are not fuchsian. Some of these examples will be discussed below.

The following result was proved by M. Bestvina and G. Mess [28] in the context of Gromov-hyperbolic groups:

Theorem 9.3. *Each convex-cocompact i-quasifuchsian group is a* Poincaré duality group *(over $\mathbb{Z}$) of dimension $i+1$. Conversely, if $\Gamma \subset \mathbf{Mob}(\mathbb{S}^n)$ is a convex-cocompact Poincaré duality group, then $\Lambda(\Gamma)$ is a homology manifold which is a homology sphere.*

Question 9.4. *Is it true that each convex-cocompact quasifuchsian group is isomorphic to the fundamental group of a closed aspherical manifold?*

This is, of course, a special case of Wall's problem (Problem 4.5).

Question 9.5. *Is there a convex-cocompact group $\Gamma \subset \mathbf{Mob}(\mathbb{S}^n)$ whose limit set is a homology manifold which is homology sphere, so that $\Lambda(\Gamma)$ is not homeomorphic to a sphere?*

9.1. Quasifuchsian groups of codimension 1

The situation in the case of $n = 2$ is completely understood due to the following:

Theorem 9.6. (*B. Maskit* [156], *see also* [152].) *Let $\Gamma \subset \mathbf{Mob}(\mathbb{S}^2)$ be a Kleinian group whose domain of discontinuity $\Omega(\Gamma)$ consists of precisely two components. Then:*

1. *Γ is 1-quasifuchsian and geometrically finite.*
2. *Γ is quasiconformally conjugate to a 1-fuchsian group.*
3. *$\bar{M}^3(\Gamma) = (\mathbb{H}^3 \cup \Omega(\Gamma))/\Gamma$ is homeomorphic to an interval bundle over a surface S, which is 2-fold covered by $\Omega(\Gamma)/\Gamma$.*

Our goal is to compare the higher-dimensional situation with this theorem. Suppose that $\Gamma \subset \mathbf{Mob}(\mathbb{S}^n)$ is a codimension 1 quasifuchsian group. Then $\Omega(\Gamma)$ consists of two components, Ω_1, Ω_2. After replacing Γ by an appropriate index 2 subgroup, we can assume that each Ω_i is Γ-invariant; hence $M^n(\Gamma) = M_1 \cup M_2$, where $M_i := \Omega_i/\Gamma$. Then, by the Alexander duality, $H_*(\Omega_i) \cong H_*(point)$, $i = 1, 2$. Therefore, if Ω_i is simply-connected, then Ω_i is contractible. Below we discuss what is currently known about such quasifuchsian groups for $n \neq 4$.

Theorem 9.7. *Suppose that both M_i are compact and both Ω_i are simply-connected. Then $\bar{M}^{n+1}(\Gamma)$ is diffeomorphic to $M_1 \times [0,1]$ provided that $n \geq 5$.*

Proof. Note that, for homological reasons, $W = \bar{M}^{n+1}(\Gamma)$ is compact, hence Γ is convex-cocompact, see Theorem 2.7. Since both $\Omega_1, \Omega_2, \mathbb{H}^{n+1}$ are contractible, the inclusions

$$M_i \hookrightarrow W, \quad i = 1, 2$$

are homotopy-equivalences. Therefore W defines a smooth h-cobordism between the aspherical manifolds M_1 and M_2. According to Corollary 5.6, this h-cobordism is smoothly trivial. □

Suppose that $n = 3$ and both Ω_1, Ω_2 are contractible. Then

$$\pi_1(M_1) \cong \pi_1(M_2) \cong \Gamma,$$

the manifolds M_1 and M_2 are both irreducible and have infinite fundamental groups. If Γ were to contain a subgroup Π isomorphic to $\mathbb{Z}^2$, the subgroup Π would be parabolic. This would contradict compactness of W. Therefore, according to Perelman's solution of Thurston's hyperbolization conjecture, there exists a closed hyperbolic 3-manifold M_0 which is homeomorphic to M_1 and M_2. Since M_0 is hyperbolic, its fundamental group Γ_0 acts as a 2-fuchsian group on $\mathbb{S}^3$. Therefore, according to our discussion in Section 6, the group Γ is quasiconformally conjugate to Γ_0. It is not difficult to see that passage to the index 2 subgroup which we used above does no harm and we obtain:

Proposition 9.8. *Suppose that* $\Gamma \subset \mathbf{Mob}(\mathbb{S}^3)$ *is a codimension* 1 *quasifuchsian subgroup, so that both components of* $\Omega(\Gamma)$ *are simply-connected and* $M^3(\Gamma)$ *is compact. Then* Γ *is quasiconformally conjugate to a* 2*-fuchsian group* $\Gamma_0 \subset \mathbf{Mob}(\mathbb{S}^3)$.

On the other hand, we do not know if the 4-dimensional manifold $\bar{M}(\Gamma)$ is homeomorphic (or diffeomorphic) to an interval bundle over a 3-manifold.

Proposition 9.8 fails for $n \geq 4$:

Theorem 9.9. *For every* $n \geq 4$ *there are codimension* 1 *convex-cocompact quasifuchsian subgroups* $\Gamma \subset \mathbf{Mob}(\mathbb{S}^n)$*, so that both components of* $\Omega(\Gamma)$ *are simply-connected, but* Γ *is not isomorphic to a fuchsian group.*

Proof. We give only a sketch of the proof, the details will appear elsewhere. Fix $n \geq 4$. M. Gromov and W. Thurston in [89] construct examples of negatively curved compact conformally-flat n-manifolds M^n, so that M^n is not homotopy-equivalent to any closed hyperbolic n-manifold N^n. (See also [119] for a review of the Gromov–Thurston examples and for a construction of a convex projective structure on M^n.)

By choosing parameters in the construction of [89] more carefully, one can construct an example of a *uniformizable* flat conformal manifold M^n with the same properties. Moreover, $M^n = \Omega_1/\Gamma$, $\Gamma \subset \mathbf{Mob}(\mathbb{S}^n)$ is convex-cocompact, and $\Omega(\Gamma) = \Omega_1 \cup \Omega_2$ is the union of two simply-connected components. Then $\Lambda(\Gamma)$ is homeomorphic to $\mathbb{S}^{n-1}$, since the limit set of Γ is homeomorphic to the ideal boundary of the universal cover of the negatively curved manifold M^n. If Γ were isomorphic to an $n-1$-fuchsian group Γ', then Γ' would be isomorphic to the fundamental group of a closed hyperbolic n-manifold N^n, which is a contradiction. □

The above examples have another interesting property. Let Ω^{n+1} denote the domain of discontinuity of the group Γ (regarded as a subgroup of $\mathbf{Mob}(\mathbb{S}^{n+1})$).

Note that Ω^{n+1} is connected and $\pi_1(\Omega^{n+1}) \cong \mathbb{Z}$. Since both Ω_1, Ω_2 are contractible, it follows that

$$\pi_i(\Omega^{n+1}) = 0, \quad i \geq 2.$$

Set $M^{n+1} = \Omega^{n+1}/\Gamma$. We have the short exact sequence

$$1 \to \mathbb{Z} = \pi_1(\Omega^{n+1}) \to \pi_1(M^{n+1}) \to \Gamma \to 1.$$

The embedding $M_1 \to M^{n+1}$ determines a splitting of this sequence. Hence the manifolds M^{n+1} and $\mathbb{S}^1 \times M^n$ are homotopy-equivalent. Given the existence of a metric of negative curvature on M^n, we obtain a metric of nonpositive curvature on $\mathbb{S}^1 \times M^n$. Therefore, by Theorem 5.4, the manifolds M^{n+1} and $\mathbb{S}^1 \times M^n$ are homeomorphic.

Let $k\mathbb{Z} \subset \mathbb{Z} \subset \pi_1(M^{n+1})$ be the index k subgroup in the center of $\pi_1(M^{n+1})$. Then we obtain the k-fold covering $X_k \to X_1 = M^{n+1}$, where

$$\pi_1(X_k) = k\mathbb{Z} \times \Gamma \subset \pi_1(X_1).$$

Since the manifolds X_k have isomorphic fundamental groups and $\pi_i(X_k) = 0$ for all $i \geq 2, k \in \mathbb{N}$, these manifolds are all homeomorphic to the smooth manifold X_1 by Theorem 5.4. By [132, Essay IV], there only finitely many smooth structures on the manifold X_1. Therefore we obtain an infinite family of diffeomorphic manifolds

$$M_j^{n+1} := X_{k_j}, j \in \mathbb{N}$$

and smooth covering maps $p_j : M_j^{n+1} \to M^{n+1}$.

The $(n+1)$-manifold $M^{n+1} = \Omega^{n+1}/\Gamma$ has the flat conformal structure K_1 uniformized by the group Γ. Let K_j denote the flat conformal structure on M^{n+1}, which is the lift of K_1 via p_j.[3] We thus obtain an infinite family of diffeomorphic flat conformal manifolds

$$(M_j^{n+1}, K_j), j = 1, 2, \ldots$$

Question 9.10. *Suppose that M is a closed hyperbolic n-manifold. Is there a finite cover $f : M' \to M$ such that the pull-back map $f^* : H^3(M, \mathbb{Z}/2) \to H^3(M', \mathbb{Z}/2)$ is trivial? (Recall* [132] *that the group $H^3(M, \mathbb{Z}/2)$ classifies PL structures on M if $n \geq 5$.)*

We regard the structures K_j as elements of $\mathfrak{M}(X)$, the moduli space of the flat conformal structures on a fixed smooth manifold X. The proof of the following claim is similar to [108], where it was proved in the context of 3-manifolds:

Proposition 9.11. *For different i, j the structures K_i, K_j lie in different connected components of the moduli space $\mathfrak{M}(X)$. Thus $\mathfrak{M}(X)$ consists of infinitely many connected components.*

We note that K. Scannell in [197] constructed examples of hyperbolic 3-manifolds X for which $\mathfrak{M}(X)$ consists of infinitely many connected components.

[3] The structures K_j are obtained via *grafting* of (M^{n+1}, K_1) along the hypersurface M^n.

To get the same phenomenon in dimension 4 consider one of the hyperbolic manifolds M^3 obtained by Dehn surgery on a 2-bridge knot, so that the natural embedding

$$\Gamma = \pi_1(M) \hookrightarrow \mathbf{Mob}(\mathbb{S}^3)$$

is (locally) rigid (see Example 11.1.2). Then the natural embedding of $\Gamma \to \mathbf{Mob}(\mathbb{S}^4)$ is also rigid and hence the manifold $M^4 = M^4(\Gamma) \cong M \times \mathbb{S}^1$ has rigid flat conformal structure. Taking k-fold covers of this manifold we obtain infinitely many rigid flat conformal structures on M^4. By combining these results we obtain

Theorem 9.12. *For every $n \geq 3$ there exists a smooth compact n-manifold X^n such that $\mathfrak{M}(X^n)$ consists of infinitely many connected components.*

We now return to our discussion of Kleinian groups, restricting to $n = 3$. Suppose that Γ is a convex-cocompact 2-quasifuchsian group, such that both components of $\Omega(\Gamma)$ are simply-connected. Then, by proposition 9.8, the limit set of Γ is *tame*, i.e., there is a homeomorphism of $\mathbb{S}^3$ which maps $\Lambda(\Gamma)$ to the round sphere.

Theorem 9.13. (*B. Apanasov and A. Tetenov* [11].) *There exists a convex-cocompact 2-quasifuchsian group $\Gamma \subset \mathbf{Mob}(\mathbb{S}^3)$ whose limit set is a* wild *2-sphere, i.e., there is no homeomorphism of $\mathbb{S}^3$ which maps $\Lambda(\Gamma)$ to the round sphere. Moreover, one component of $\Omega(\Gamma)$ is simply-connected.*

9.2. 1-quasifuchsian subgroups of $\mathbf{Mob}(\mathbb{S}^3)$

Given a Kleinian subgroup $\Gamma \subset \mathbf{Mob}(\mathbb{S}^3)$ whose limit set is a topological circle C, we would like to analyze the embedding $C \hookrightarrow \mathbb{S}^3$. It is clear that C could be an unknot in $\mathbb{S}^3$ (i.e., there exists a homeomorphism of $\mathbb{S}^3$ which maps C to a round circle), take for instance any 1-fuchsian subgroup of $\mathbf{Mob}(\mathbb{S}^3)$.

A topological circle C in $\mathbb{S}^3$ is called *tame* if it is isotopic to a polygonal knot in $\mathbb{S}^3$; if C is not tame, it is called *wild*.

Proposition 9.14. 1. *If Γ is a 1-quasifuchsian subgroup of $\mathbf{Mob}(\mathbb{S}^3)$, then either $\Lambda(\Gamma)$ is an unknot or it is a wild knot K such that $\pi_1(\mathbb{S}^3 \setminus K)$ is infinitely generated.*
2. *Each 1-quasifuchsian group is geometrically finite.*

Proof. Since Γ is a 1-quasifuchsian subgroup of $\mathbf{Mob}(\mathbb{S}^3)$, this group is nonelementary. The fundamental group of $M = M^3(\Gamma)$ is finitely generated (since Γ is) and we have the exact sequence:

$$1 \to \pi_1(\Omega(\Gamma)) \to \pi_1(M) \to \Gamma \to 1.$$

Suppose that $\pi_1(\Omega(\Gamma))$ is finitely generated. Then, according to Jaco-Hempel's Theorem [95], $\pi_1(\Omega(\Gamma)) \cong \mathbb{Z}$. This immediately excludes tame nontrivial knots (the result proved by R. Kulkarni [140]). It remains to exclude wild knots with

$$\Delta := \pi_1(\Omega(\Gamma)) \cong \mathbb{Z}.$$

Without loss of generality (after passing to an index 2 subgroup in Γ), we can assume that Δ is contained in the center of $\pi_1(M)$. Note that M is a Seifert manifold since its fundamental group has infinite center. Hence M admits an $\mathbb{S}^1$-action. Lift this action to $\Omega(\Gamma)$ and then extend it to the entire 3-sphere (so that the fixed point set is the limit set). Raymond's classification [192] of topological $\mathbb{S}^1$-actions on $\mathbb{S}^3$ implies that this $\mathbb{S}^1$-action is topologically conjugate to the orthogonal action, hence $\Lambda(\Gamma)$ is an unknot. This proves (1).

To prove (2) note that the group Γ acts as a *convergence group* on $\mathbb{S}^1 = \Lambda(\Gamma)$ (see [218]). Hence, according to [218], there exists a homeomorphism $f : \Lambda(\Gamma) \to \mathbb{S}^1$ such that

$$f\Gamma f^{-1} = \Gamma' \subset PSL(2,\mathbb{R}).$$

Since finitely generated discrete subgroups of $PSL(2,\mathbb{R})$ are geometrically finite, it follows that Γ' is geometrically finite. As geometric finiteness is an invariant of the topological dynamics on the limit set (see Theorem 2.5), the group Γ is geometrically finite as well. □

On the other hand, even if $\Lambda(\Gamma)$ is an unknot, the 3-manifold $\Omega(\Gamma)/\Gamma$ is not necessarily a product:

Theorem 9.15. (*M. Gromov, B. Lawson, W. Thurston* [87], *N. Kuiper* [139], *and M. Kapovich* [107, 112].) *There are* 1*-quasifuchsian groups* $\Gamma \subset \mathbf{Mob}(\mathbb{S}^3)$ *such that* $\Lambda(\Gamma)$ *are unknotted but* Γ *are not topologically conjugate to* 1*-fuchsian groups.*

In the examples constructed in this theorem, the manifolds $M^3(\Gamma)$ are nontrivial oriented circle bundles over orientable surfaces. On the other hand, for every 1-fuchsian group $\Gamma_0 \subset \mathbf{Mob}(\mathbb{S}^3)$, the manifold $M^3(\Gamma_0)$ is a 3-dimensional Seifert manifold with the zero Euler number, since it admits a natural $\mathbb{H}^2 \times \mathbb{R}$-structure. Hence, in this case, $M^3(\Gamma_0)$ admits a finite cover which is homeomorphic to the product of $\mathbb{S}^1$ and a surface.

Theorem 9.16. (*B. Apanasov* [10], *B. Maskit* [157], *see also* [87].) *There are* 1*-quasifuchsian groups* $\Gamma \subset \mathbf{Mob}(\mathbb{S}^3)$ *such that* $\Lambda(\Gamma)$ *are wild knots.*

Proof. (Sketch.) Start with a necklace of round balls

$$B_0, B_1, \ldots, B_{m-1} \subset \mathbb{S}^3,$$

so that B_i is tangent to B_j, if $j = i+1 \in \mathbb{Z}/m\mathbb{Z}$ and is disjoint otherwise. Assume that this necklace is *knotted*, i.e., the polygonal knot obtained by connecting the consecutive points of tangency is a nontrivial knot $K \subset \mathbb{S}^3$. See Figure 6.

Let $\gamma_i \in \mathbf{Mob}(\mathbb{S}^3)$ denote the inversion in the sphere ∂B_i, $i = 0, 1, \ldots, m-1$. Let $\Gamma \subset \mathbf{Mob}(\mathbb{S}^n)$ be the group generated by these inversions. By the Poincaré fundamental polyhedron theorem,

$$\Phi = \mathbb{S}^3 \setminus \bigcup_{i=0}^{m-1} B_i$$

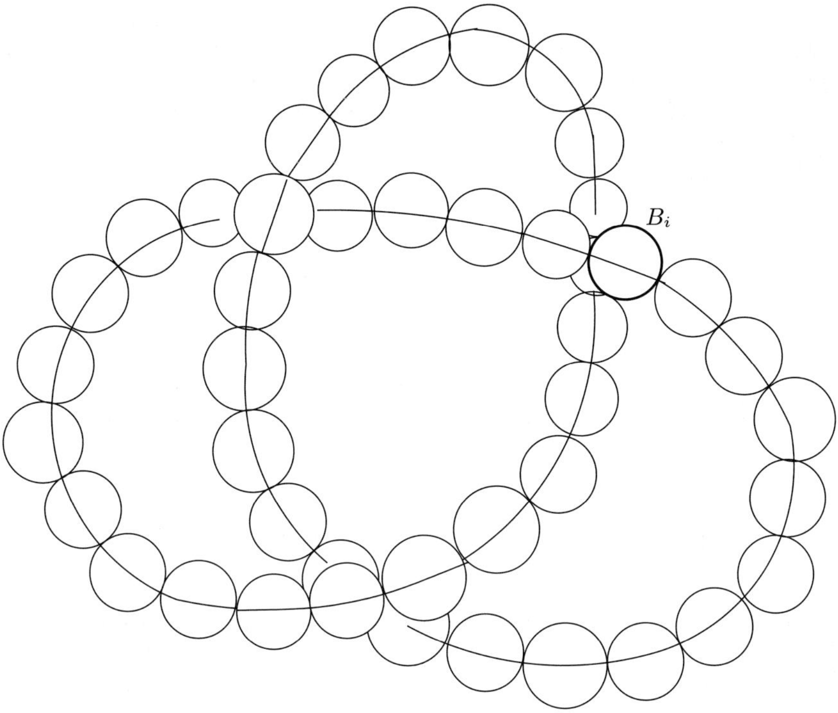

FIGURE 6

is a fundamental domain for Γ, the group Γ is geometrically finite and is isomorphic to a 1-fuchsian group Γ'. Therefore, Tukia's theorem 6.1 applied to the isomorphism $\Gamma \to \Gamma'$, implies that Γ is 1-quasifuchsian. By Seifert–van Kampen Theorem, $\pi_1(\mathbb{S}^3 \setminus K)$ embeds in $\pi_1(\Omega(\Gamma))$. Therefore $\pi_1(\Omega(\Gamma))$ is not isomorphic to $\mathbb{Z}$ and, hence, the limit set of Γ is a wild knot. $\square$

By modifying the above construction, S. Hwang proved

Theorem 9.17. (*S. Hwang* [97].) *Let L be a polygonal link in $\mathbb{S}^3$. Then there exists a (torsion-free) convex-cocompact Kleinian group $\Gamma \subset$ **Mob**$(\mathbb{S}^3)$ with a fundamental domain $\Phi \subset \mathbb{S}^3$ such that the complement $\mathbb{S}^3 \setminus \Phi$ is isotopic to a regular neighborhood of L.*

The above theorem is the key for proving

Theorem 9.18. (*S. Hwang* [97].) *Let M^3 be a closed oriented 3-manifold. Then there exists a closed oriented 3-manifold N^3 such that the connected sum $M^3\#N^3$ admits a Möbius structure.*

Results similar to Theorems 9.17 and 9.18 hold in dimension 4, see [117], although one has to assume that M^4 is a Spin-manifold. One of the key ingredients in [117] is the following:

Theorem 9.19. *Let $Q \subset \mathbb{R}^4 \subset \mathbb{S}^4$ be a finite 2-dimensional subcomplex in the standard cubulation of $\mathbb{R}^4$. Then there exists a convex-cocompact Kleinian subgroup $\Gamma \subset \mathbf{Mob}(\mathbb{S}^4)$ (generated by reflections) with a fundamental domain $\Phi \subset \mathbb{S}^4$, such that the complement $\mathbb{S}^4 \setminus \Phi$ is isotopic to a regular neighborhood of Q.*

Very little is known about quasifuchsian groups in $\mathbf{Mob}(\mathbb{S}^n)$ whose limit sets have dimension between 2 and $n-2$. Perhaps the most interesting result here is obtained by I. Belegradek [15] who used the construction from [87] to get

Theorem 9.20. *There exist convex-cocompact 2-quasifuchsian subgroups $\Gamma_1, \Gamma_2 \subset \mathbf{Mob}(\mathbb{S}^4)$ so that:*

1) *$\Lambda(\Gamma_1)$ is a wild 2-sphere in $\mathbb{S}^4$.*
2) *$\Lambda(\Gamma_2)$ is tame but the group Γ_2 is not topologically conjugate to a 2-fuchsian group: $M^4(\Gamma_2)$ is a nontrivial circle bundle over a hyperbolic 3-manifold.*

Similar results probably hold for codimension 2 quasifuchsian subgroups in $\mathbf{Mob}(\mathbb{S}^n)$, $n \geq 5$.

10. Ahlfors finiteness theorem in higher dimensions: Quest for the holy grail

10.1. The holy grail

One of the most fundamental results of the theory of Kleinian subgroups of $\mathbf{Mob}(\mathbb{S}^2)$ is the *Ahlfors Finiteness Theorem* (the "Holy Grail"), which we state here together with its companions:

Theorem 10.1. *Suppose that $\Gamma \subset PSL(2, \mathbb{C})$ is a Kleinian group*[4] *which may have torsion. Then the following hold:*

1. *(L. Ahlfors [3], L. Greenberg [83].) The group Γ is analytically finite, i.e., the quotient $O := \Omega(\Gamma)/\Gamma$ is a complex orbifold of finite conformal type*[5]*. In particular, O is homotopy-equivalent to a finite CW complex.*
2. *(D. Sullivan [209].) Γ has only finitely many cusps.*
3. *(M. Feighn and G. Mess [75].) Γ has only finitely many Γ-conjugacy classes of finite order elements.*
4. *(P. Scott [201, 202].) Γ is finitely presentable and the orbifold $\mathbb{H}^3/\Gamma$ is finitely covered by a manifold $\mathbb{H}^3/\Gamma'$, which is homotopy-equivalent to a compact 3-manifold.*

---

[4] Recall that all Kleinian groups are assumed to be finitely generated.

[5] I.e., as a Riemann surface it biholomorphic to a compact Riemann surface with a finite subset removed; as an orbifold it has only finitely many singular cone-points.

5. (*L. Ahlfors* [4].) *The action of* Γ *on* $\Lambda(\Gamma)$ *is* recurrent *with respect to the Lebesgue measure* μ.

Alternative analytical proofs of Part 1 (i.e., the original Ahlfors' finiteness theorem) are given for instance in [116, Section 8.14] and [153]. A geometric proof (valid even in the context of manifolds of negative curvature) follows from the solution of Tameness Conjecture, see [2].

Corollary 10.2. *If* Γ *is as above, then:*

(a) *For each component* Ω_0 *of* $\Omega(\Gamma)$, *the limit set of the stabilizer of* Ω_0 *in* Γ *equals* $\partial\Omega_0$ *(follows directly from Part* 1 *of Theorem* 10.1*). In particular, no component of* $\Omega(\Gamma)$ *has trivial stabilizer.*
(b) *Kleinian subgroups* Γ *of* $\mathbf{Mob}(\mathbb{S}^2)$ *are* coherent, *i.e., each finitely generated subgroup of* Γ *is also finitely presented (follows from Part* 4 *of Theorem* 10.1*).*
(c) (*W. Thurston, see* [170].) *If* $\Gamma \subset \mathbf{Mob}(\mathbb{S}^2)$ *is geometrically finite with* $\Omega(\Gamma) \neq \emptyset$, *then each finitely generated subgroup* $\Delta \subset \Gamma$ *is geometrically finite as well.*

We also now fully understand the topology of the manifold (orbifold) $\mathbb{H}^3/\Gamma$:

Theorem 10.3 (Former tameness conjecture). *The quotient* $\mathbb{H}^3/\Gamma$ *is* tame, *i.e., it is homeomorphic to the interior of a compact manifold (orbifold) with boundary.*

The above theorem was proved for freely indecomposable groups Γ by F. Bonahon [34] and by I. Agol [2], D. Calegari and D. Gabai [49] in the general case.

The next theorem is a combination of a result by Thurston [213], who proved ergodicity for tame Kleinian subgroups of $\mathbf{Mob}(\mathbb{S}^2)$, and the proof of the tameness conjecture:

Theorem 10.4. *If* Γ *is as above, then the action of* Γ *on* $\Lambda(\Gamma)$ *is ergodic with respect to the Lebesgue measure: each measurable* Γ*-invariant function on* $\Lambda(\Gamma)$ *is constant a.e..*

Note, that the conglomerate of assertions presented above contains statements of different nature: algebraic, topological, dynamical. For a while it was hoped that a theorem analogous to Theorem 10.1 can be proved for Kleinian groups in higher dimensions; an attempt to develop analytical technique to achieve this was made by Ahlfors in [6] (see also [179]).

Nearly all algebraic and topological assertions of Theorem 10.1 and the Corollary 10.2 have been disproved in the case of Kleinian groups acting in higher dimensions (M. Kapovich and L. Potyagailo [114], [127], [128], [187], [188]):

Theorem 10.5. *There exist Kleinian subgroups* $\Gamma_1, \ldots, \Gamma_5 \subset \mathbf{Mob}(\mathbb{S}^3)$ *so that:*

1. *The group* Γ_1 *is not finitely presentable.*
2. *For each* i, *the manifold* $M(\Gamma_i) = \Omega(\Gamma_i)/\Gamma_i$ *contains a component with infinitely generated fundamental group.*
3. Γ_2 *is free and has infinitely many cusps (of rank* 1*).*

4. Γ_3 *is not torsion-free and has infinitely many conjugacy classes of finite order elements.*
5. Γ_4 *is a normal subgroup of a convex-cocompact group* $\widehat{\Gamma}_4 \subset \mathbf{Mob}(\mathbb{S}^3)$ *and satisfies* (1), (2) *and* (4).
6. (*B. Bowditch, G. Mess* [42].) *The group* Γ_5 *satisfies* (1) *and* (2) *and is contained in a cocompact lattice* $\widehat{\Gamma}_5 \subset \mathbf{Mob}(\mathbb{S}^3)$.
7. *Groups* Γ_i, $i = 1, \ldots, 4$ *are normal subgroups of geometrically finite groups* $\widehat{\Gamma}_i$ *so that* $\widehat{\Gamma}_i/\Gamma_i \cong \mathbb{Z}$.

Remark 10.6. By modifying Γ_3 one can also construct an example $\Gamma_6 \subset \mathbf{Mob}(\mathbb{S}^3)$ such that $\Omega(\Gamma_6)/\Gamma_6$ has infinitely many connected components.

At the time when the above examples were constructed, they were regarded as a "rare pathology". It appears however that such examples are rather common:

Conjecture 10.7. (*M. Kapovich, L. Potyagailo, E. Vinberg* [129].) *Suppose that* $\Gamma \subset \mathbf{Mob}(\mathbb{S}^n)$ *is an arithmetic lattice, where* $n \geq 3$. *Then* Γ *is* noncoherent, *i.e., it contains a finitely generated subgroup* Δ *which is not finitely presentable.*

This conjecture was proved in [129] in a number of special cases, e.g., for all non-cocompact arithmetic lattices provided that $n \geq 5$.

All the examples Γ_i in the above theorem are based upon the existence of hyperbolic 3-manifolds M^3 of finite volume which fiber over the circle: the groups Γ_i are obtained by manipulating with the normal surface subgroups in $\pi_1(M^3)$.

Problem 10.8. *Find examples similar to* Γ_i*'s without using hyperbolic* 3*-manifolds fibering over the circle.*

Problem 10.9. *Construct a finitely generated Kleinian subgroup* $\Gamma \subset \mathbf{Mob}(\mathbb{S}^n)$ *such that Part* (a) *of Corollary* 10.2 *fails for* Γ.

Problem 10.10. (*G. Mess.*) *Construct a finitely-presented Kleinian subgroup of* $\mathbf{Mob}(\mathbb{S}^n)$ ($n \geq 3$) *which contains no parabolic elements and for which any of the assertions of Theorem* 10.1 *fail. (In Part* (a) *one would need to replace analytical finiteness with finiteness of the homotopy type.)*

Problem 10.11. *Construct a finitely generated Kleinian subgroup* $\Gamma \subset \mathbf{Mob}(\mathbb{S}^3)$ *such that* $\Omega(\Gamma)$ *contains a contractible component* Ω_0 *so that:*

The stabilizer Γ_0 *of* Ω_0 *in* Γ *is finitely-generated, but the manifold* Ω_0/Γ_0 *is not tame.*

Note however that although algebra and topology fail, the assertions of dynamical nature (part 5 of Theorem 10.1, part (a) of Corollary 10.2, and Theorem 10.4) remain open in higher dimensions. Moreover, an attempt to construct a higher-dimensional counter-example to Theorem 10.1 (part 5) along the lines of the examples Γ_i, is doomed to failure:

Theorem 10.12. (*K. Matsuzaki* [160].) *Let* $\widehat{\Gamma}$ *be a geometrically finite subgroup*[6] *in* $\mathbf{Mob}(\mathbb{S}^n)$. *Suppose that* $\Gamma \subset \widehat{\Gamma}$ *is a normal subgroup (which does not have to be finitely generated). Then the action of* Γ *on its limit set is recurrent.*

Ergodicity fails however for discrete subgroups of $PU(2,1)$ (it probably also fails for Kleinian groups in higher dimensions but an example would be difficult to construct):

Theorem 10.13. *There exists a finitely generated (but not finitely presentable!) discrete group* Γ *of isometries of complex-hyperbolic* 2*-plane* $\mathbb{CH}^2$ *so that the limit set of* Γ *is the* 3*-sphere and the action of* Γ *on* $\mathbb{S}^3$ *is not ergodic.*

Proof. There are examples (the first was constructed by R. Livne in his thesis [145], see also [62]) of cocompact torsion-free discrete subgroups $\widehat{\Gamma} \subset PU(2,1)$ such that the complex 2-manifold $M = \mathbb{CH}^2/\widehat{\Gamma}$ admits a nonconstant holomorphic map $f : M \to S$ to a Riemann surface S of genus ≥ 2. The fundamental group of the generic fiber of f maps onto a normal subgroup Γ in $\widehat{\Gamma}$, so that Γ is finitely generated but is not finitely presentable [115]. By lifting f to the universal covers we get a nonconstant holomorphic Γ-invariant function

$$\tilde{f} : \mathbb{CH}^2 \to \mathbb{H}^2.$$

Then the bounded harmonic function $Re(\tilde{f})$ is also Γ-invariant and nonconstant. This harmonic function admits a measurable extension h to $\mathbb{S}^3$, the boundary of the complex ball $\mathbb{CH}^2$, so that h is Γ-invariant and not a.e. constant. □

10.2. Groups with small limit sets

So far, our quest for the holy grail mostly resembled Monty Python's: We are not sure what to look for in higher dimensions. Nevertheless, there is a glimmer of hope.

Recall that the Hausdorff dimension $\dim_H$ of a subset $E \subset \mathbb{R}^n$ is defined as follows. For each $\alpha > 0$ consider the α-Hausdorff measure of E:

$$mes_\alpha(E) = \lim_{\rho \to 0} \inf\{\sum_i r_i^\alpha : \quad r_i \leq \rho, E \text{ is contained in the union of } r_i\text{-balls}\}.$$

The Hausdorff dimension of E is

$$\dim_H(E) = \inf_\alpha \{\alpha : mes_\alpha(E) = 0\}.$$

According to [96], for every bounded subset $E \subset \mathbb{R}^n$ one has the inequality

$$\dim(E) \leq \dim_H(E)$$

between topological and Hausdorff dimensions. In particular, if Γ is a Kleinian group with $\dim_H(\Lambda(\Gamma)) < 1$, then Γ is geometrically finite and is isomorphic to a Schottky-type group, see Theorem 7.7.

Conjecture 10.14. *If* $\dim_H(\Lambda(\Gamma)) < 1$, *then* Γ *is a Schottky-type group. Moreover,* Γ *is classical.*

[6] Actually, the proof also works for subgroups of any rank 1 Lie group.

The *critical exponent* of a Kleinian group $\Gamma \subset \mathbf{Mob}(\mathbb{S}^n)$ is

$$\delta(\Gamma) := \inf\{s > 0 : \sum_{\gamma \in \Gamma} e^{-sd(x,\gamma x)} < \infty\},$$

where d is the hyperbolic metric in $\mathbb{H}^{n+1}$. The following theorem is the result of combined efforts of a large number of mathematicians, including P. Tukia, D. Sullivan and P. Nicholls, we refer to [177], [33] for the proofs:

Theorem 10.15. *For every Kleinian subgroup* $\Gamma \subset \mathbf{Mob}(\mathbb{S}^n)$,

$$\delta(\Gamma) = \dim_H(\Lambda_c(\Gamma)).$$

Recall that $\Lambda_c(\Gamma)$ is the conical limit set of Γ.

The critical exponent relates to λ_0, the bottom of the spectrum of Laplacian on the hyperbolic manifold $\mathbb{H}^{n+1}/\Gamma$, by the following

Theorem 10.16. (*D. Sullivan* [211])

$$\lambda_0 = \left(\frac{n}{2}\right)^2, \quad \text{if} \quad 0 \le \delta(\Gamma) \le \frac{n}{2},$$
$$\lambda_0 = \delta(\Gamma)(n - \delta(\Gamma)), \quad \text{if} \quad \frac{n}{2} < \delta(\Gamma) \le n.$$

The expectation is that Kleinian groups in $\mathbf{Mob}(\mathbb{S}^n)$ with *small* limit sets behave analogously to the Kleinian subgroups of $\mathbf{Mob}(\mathbb{S}^2)$.

Conjecture 10.17. *Suppose that* Γ *is a (finitely generated) subgroup of* $\mathbf{Mob}(\mathbb{S}^n)$ *so that* $\Lambda(\Gamma)$ *has Hausdorff dimension* < 2. *Then* Γ *is geometrically finite.*

For $n = 2$, this conjecture is a theorem of C. Bishop and P. Jones [33]. A partial generalization of [33] was proved by A. Chang, J. Qing, J. and P. Yang [54]:

Theorem 10.18. *Suppose that* Γ *is a (finitely generated)* conformally finite[7] *subgroup of* $\mathbf{Mob}(\mathbb{S}^n)$ *such that* $\dim_H(\Lambda(\Gamma)) < n$. *Then* Γ *is geometrically finite.*

The converse to the above theorem was proved earlier by P. Tukia [216].

Theorem 10.19. (*Y. Shalom* [204].) *Suppose that* Γ *is a geometrically finite subgroup of* $\mathbf{Mob}(\mathbb{S}^n)$ *such that* $\dim_H(\Lambda(\Gamma)) < 2$ *and* $\Delta \subset \Gamma$ *is a finitely generated normal subgroup. Then* Δ *has finite index in* Γ. *In particular,* Δ *is geometrically finite as well.*

Thus, attempts to construct geometrically infinite groups using normal subgrooups in geometrically finite Kleinian groups with *small* limit sets, are doomed to failure. On the other hand, the assumption that $\delta(\Gamma)$ is small should impose strong restrictions on the algebraic properties of the group Γ.

Conjecture 10.20. *Suppose that* Γ *is a Kleinian group in* $\mathbf{Mob}(\mathbb{S}^n)$ *which does not contain parabolic elements. Then:*

[7] I.e., $M^n(\Gamma) = \Omega(\Gamma)/\Gamma$ is compact modulo cusps.

1. $cd(\Gamma) - 1 \leq \delta(\Gamma)$.
2. *In the case of equality,* Γ *is an* i*-fuchsian convex-cocompact group,* $i = \delta(\Gamma)$.

Recall that *cd* and *hd* stand for the cohomological and homological dimensions of a group. A partial confirmation of Part 1 of this conjecture is obtained in

Theorem 10.21. (*M. Kapovich* [121].) *Suppose that* $\Gamma \subset \mathbf{Mob}(\mathbb{S}^n)$ *is a Kleinian group. Then for every ring* R,

$$hd_R(\Gamma, \Pi) - 1 \leq \delta(\Gamma),$$

where $\Pi \subset \Gamma$ *is the set of virtually abelian subgroups of* Γ *of (virtual) rank* ≥ 2 *and* $hd_R(\Gamma, \Pi)$ *is the relative homological dimension.*

We refer the reader to the series of papers by H. Izeki [100, 101, 102] for the related results.

Corollary 10.22. (*M. Kapovich* [121].) *Suppose that the group* Γ *is finitely-presented and* $\delta(\Gamma) < 1$. *Then* Γ *is free.*

Proof. Since $\delta(\Gamma) < 1$, it follows that Γ contains no rank 2 abelian subgroups. Then we have the inequalities

$$cd(\Gamma) \leq 1 + hd(\Gamma) \leq \delta(\Gamma) + 2 < 3.$$

Combined with finite presentability of Γ, the inequality $cd(\Gamma) \leq 2$ implies that Γ has *finite type*; therefore

$$cd(\Gamma) = hd(\Gamma) \leq \delta(\Gamma) + 1 < 2.$$

Hence Γ is free by Stallings' Theorem 4.1. □

An inequality similar to Conjecture 10.20 was proved by A. Reznikov: For a (finitely-generated) group Γ define

$$\alpha(\Gamma) := \inf\{p \in [1, \infty] : \ell_p H^1(\Gamma) \neq 0\}.$$

Here $\ell_p H^1$ is the first ℓ_p-cohomology of the group Γ, see [38] for the precise definition. Then

Theorem 10.23. (*A. Reznikov* [194], *see also* [38] *for the detailed proof in the case of isometries of* $CAT(-1)$ *spaces.*) *For every Kleinian group* $\Gamma \subset \mathbf{Mob}(\mathbb{S}^n)$,

$$\alpha(\Gamma) \leq \max(\delta(\Gamma), 1).$$

Question 10.24. *What can be said about* Γ *in the case of equality in Reznikov's theorem?*

In the case of geometrically finite groups, Part 2 of Conjecture 10.20 holds:

Theorem 10.25. 1. (*Chenbo Yue* [228], *see also* [22] *and* [35].) *Suppose that* Γ *is convex-cocompact and* $i = \delta(\Gamma) = cd(\Gamma) - 1$. *Then* Γ *is* i*-fuchsian.*

2. (*M. Kapovich* [121].) *Suppose that* Γ *is geometrically finite. Then the following three conditions are equivalent:*
   $i = \delta(\Gamma) = cd(\Gamma, \Pi) - 1,$
   $i = \dim(\Lambda(\Gamma)) = \dim_H(\Lambda(\Gamma)),$
   Γ *is* i*-fuchsian.*

Conjecture 10.26. *Suppose that* Γ *is a Kleinian group in* $\mathbf{Mob}(\mathbb{S}^n)$ *whose limit set is not totally disconnected and has Hausdorff dimension* 1*. Then* Γ *is* 1*-fuchsian.*

This conjecture is known to be true for $n = 2$, see [50].

Problem 10.27. (*The* gap *problem, L. Bowen, cf.* [206].)

1. *Is there a number* $d_n < n$ *such that for every Schottky subgroup* $\Gamma \subset \mathbf{Mob}(\mathbb{S}^n)$, $n \geq 3$, *we have:*
$$\delta(\Gamma) < d_n.$$
2. *More generally, consider a sequence* $\Gamma_j \subset \mathbf{Mob}(\mathbb{S}^n)$ *of convex-cocompact groups isomorphic to a fixed group* Γ *so that:* $\Lambda(\Gamma_j) \neq \mathbb{S}^n$ *for each* j*. Is it true that*
$$\limsup_{j\to\infty} \delta(\Gamma_j) < n\ ?$$

By the work of R. Phillips and P. Sarnak [185], the answer to the Part 1 of this question is positive in the class of classical Schottky groups.

11. Representation varieties of Kleinian groups

For a finitely-generated Γ consider the *representation variety* of Γ:
$$R_n(\Gamma) := Hom(\Gamma, \mathbf{Mob}(\mathbb{S}^n)).$$
If Γ has the presentation
$$\Gamma = \langle x_1, \ldots, x_m | r_1, \ldots, r_k, \ldots \rangle,$$
the representation variety is given by
$$\{(g_1, \ldots, g_m) \in (\mathbf{Mob}(\mathbb{S}^n))^m : r_1(g_1, \ldots, g_m) = 1, \ldots r_k(g_1, \ldots, g_m) = 1, \ldots .\}.$$
The group $\mathbf{Mob}(\mathbb{S}^n)$ acts on $R_n(\Gamma)$ via conjugation:
$$\theta \cdot \rho(\gamma) = \theta\rho(\gamma)\theta^{-1}, \quad \theta \in \mathbf{Mob}(\mathbb{S}^n).$$
Given this action, one can form the quotient variety
$$X_n(\Gamma) := R_n(\Gamma) // \mathbf{Mob}(\mathbb{S}^n),$$
called the *character variety*. Roughly speaking, the elements of $X_n(\Gamma)$ are represented by conjugacy classes of representations $\rho : \Gamma \to \mathbf{Mob}(\mathbb{S}^n)$. This is literally true for "most" representations, the ones for which $\rho(\Gamma)$ does not contain a normal parabolic subgroup, see [103]. In general, the representations ρ_1, ρ_2 project to the same point in $X_n(\Gamma)$ iff the closures of their $\mathbf{Mob}(\mathbb{S}^n)$-orbits have nonempty intersection. We let $[\rho]$ denote the projection of $\rho \in R_n(\Gamma)$ to $X_n(\Gamma)$.

A *trivial deformation* of a representation $\rho_0 \in R_n(\Gamma)$ is a connected curve $\rho_t \in R_n(\Gamma)$ which projects to a point in $X_n(\Gamma)$. A representation ρ_0 is called *rigid* if it admits no nontrivial deformations.

We will be mostly interested in representations $\rho \in R_n(\Gamma)$ which have discrete, nonelementary image, however much of our discussion is more general.

In this section we address the following issues related to the character varieties:

(i) Local structure of $X_n(\Gamma)$ and existence of small deformations of a given Kleinian group (rigidity vs. flexibility).
(ii) Connectedness of the subspace $\mathcal{D}_n(\Gamma)$ of discrete and faithful representations in $X_n(\Gamma)$.
(iii) Structural stability: What happens to a Kleinian group in $\mathbf{Mob}(\mathbb{S}^n)$ if we deform it a little bit? Does it stay Kleinian?
(iv) Compactness of $\mathcal{D}_n(\Gamma)$ and estimates on various natural continuous functionals on $\mathcal{D}_n(\Gamma)$.
(v) Difficulties in constructing "truly higher-dimensional" geometrically infinite Kleinian groups.

11.1. Local theory

We start by considering the *local* structure of $X_n(\Gamma)$. Given an abstract group Γ and a representation $\rho \in R_n(\Gamma)$, we have the adjoint action of $\rho(\Gamma)$ on the Lie algebra $\mathfrak{g}$ of $\mathbf{Mob}(\mathbb{S}^n)$ and the associated first cohomology group

$$H^1(\Gamma, Ad(\rho)) = Z^1(\Gamma, Ad(\rho))/B^1(\Gamma, Ad(\rho)),$$

see Section 4. It was first observed by A. Weil [226] (in the general context of representations to Lie groups) that if $X_n(\Gamma)$ is smooth at $[\rho] \in X_n(\Gamma)$ then $H^1(\Gamma, Ad(\rho))$ is isomorphic to the tangent space to $X_n(\Gamma)$ at $[\rho]$. Moreover, Weil proved that if $H^1(\Gamma, Ad(\rho)) = 0$ then $[\rho]$ is an isolated point on $X_n(\Gamma)$, i.e., ρ is rigid.

Therefore, the elements of $H^1(\Gamma, Ad(\rho))$ can be regarded as *infinitesimal deformations* of the representation ρ. An infinitesimal deformation $[\xi] \in H^1(\Gamma, Ad(\rho))$ is called *integrable* if it is tangent to a smooth curve in $X_n(\Gamma)$. The obstructions to integrability are cohomological in nature, they are certain elements of $H^2(\Gamma, Ad(\rho))$, called *Massey products*. However, in practice, these cohomology classes are very difficult to compute. The first such obstruction is the *cup-product*:

$$\phi([\xi]) = [\xi] \cup [\xi] \in H^2(\Gamma, Ad(\rho)),$$

see for instance [82]. Here $\phi([\xi])$ is represented by the 2-cocycle

$$\tau(x, y) = [\xi(x), Ad \circ \rho(x)\xi(y)],$$

where $[,]$ is the Lie bracket on the Lie algebra $\mathfrak{g}$. If the first obstruction vanishes and Γ is the fundamental group of a surface, then $X_n(\Gamma)$ is smooth at ρ, see [82], where a much more general result is proved.

We will be mostly interested in the case where $\rho : \Gamma \hookrightarrow \mathbf{Mob}(\mathbb{S}^n)$ is a discrete embedding, whose image we will identify with Γ. Then, by abusing the terminology, we will talk of small deformations of ρ in $R_n(\Gamma)$ as *small deformations of* Γ itself.

11.1.1. Small deformations of 1-quasifuchsian groups. Recall that $\mathbf{Mob}(\mathbb{S}^n)$ has dimension $d = (n+2)(n+1)/2$. Suppose that $\Gamma \subset \mathbf{Mob}(\mathbb{S}^n)$ is 1-quasifuchsian; in this subsection we allow Γ to have nontrivial finite order elements. We assume however that Γ contains no elements fixing the circle $C = \Lambda(\Gamma)$ pointwise. Therefore we obtain the injective map

$$\Gamma \to \mathrm{Isom}(\mathbb{H}^2)$$

given by the restriction of the elements of Γ to the round circle C. To simplify the discussion we assume that Γ preserves the orientation on C. Then Γ embeds as a lattice in $PSL(2,\mathbb{R})$.

If Γ contains no parabolic elements then it has the presentation:

$$\langle a_1, b_1, \ldots, a_q, b_q, c_1, \ldots, c_k | [a_1, b_1] \cdots [a_q, b_q] \cdot c_1 \cdot \cdots \cdot c_k = 1, c_j^{r_j} = 1, j = 1, \ldots, k \rangle.$$

For a representation $\rho : \Gamma \to \mathbf{Mob}(\mathbb{S}^n)$ we let

$$e_j := d - \dim\{\xi \in \mathfrak{g} : Ad \circ \rho(c_j)(\xi) = \xi\};$$

in other words, e_j is the codimension of the centralizer of $\rho(c_j)$ in $\mathbf{Mob}(\mathbb{S}^n)$. Let s denote the dimension of the centralizer of $\rho(\Gamma)$ in $\mathbf{Mob}(\mathbb{S}^n)$.

Theorem 11.1. (*A. Weil* [226].)

$$h = \dim H^1(\Gamma, Ad(\rho)) = (2q-2)d + 2s + e_1 + \cdots + e_k. \tag{11.1}$$

Moreover, if $s = 0$, then $X_n(\Gamma)$ near $[\rho]$ is a smooth h-dimensional manifold.

For instance, if $n = 1$, $\Gamma \subset PSL(2,\mathbb{R}) \subset \mathbf{Mob}(\mathbb{S}^1)$; therefore $d = 3$, we get $e_i = 1$ for each $i = 1, \ldots, k$, $s = 0$. Hence

$$h = 6q - 6 + k,$$

which is the familiar formula for the dimension of the Teichmüller space of the orbifold $O = \mathbb{H}^2/\Gamma$. If $n = 2$, we, of course, obtain $h = 2(6q - 6 + k)$ which is the (real) dimension of the space of the quasifuchsian deformations of Γ in $PSL(2,\mathbb{C})$.

To better understand the difficulties which one encounters in the case of i-fuchsian groups for $i \geq 2$, we consider the *hyperbolic triangle groups* Γ. The reason for considering these groups is that they are rigid in $PSL(2,\mathbb{R})$ (similarly to rigidity of lattices in $\mathbf{Mob}(\mathbb{S}^n)$, $n \geq 2$).

The triangle groups are the 1-fuchsian groups with $q = 0$, $k = 3$; every such Γ has the presentation

$$\langle c_1, c_2, c_3 | c_1 \cdot c_2 \cdot c_3 = 1, c_j^{r_j} = 1, j = 1, 2, 3 \rangle,$$

where $r_1^{-1} + r_2^{-1} + r_3^{-1} < 1$. Such group embeds discretely into $PSL(2,\mathbb{R})$ and we will denote the image of this embedding by $\Delta = \Delta(r_1, r_2, r_3)$

As a subgroup of $\mathbf{Mob}(\mathbb{S}^2)$, the group Δ is rigid (which follows from vanishing of H^1). Moreover, triangle groups are "strongly rigid" in $\mathbf{Mob}(\mathbb{S}^2)$, i.e., every discrete embedding of Γ into $\mathbf{Mob}(\mathbb{S}^2)$ is induced by conjugation of the identity embedding, see [84] for the complete description of $X_2(\Gamma)$.

The situation changes somewhat if we consider representations into $\mathbf{Mob}(\mathbb{S}^3)$. First, let $\rho_0 : \Delta \to \Gamma' \subset \mathbf{Mob}(\mathbb{S}^3)$ be the embedding obtained as the composition

$$\Delta \subset \mathbf{Mob}(\mathbb{S}^1) \hookrightarrow \mathbf{Mob}(\mathbb{S}^2) \hookrightarrow \mathbf{Mob}(\mathbb{S}^3).$$

of natural embeddings. Then $\dim H^1(\Delta, Ad(\rho_0)) = 0$ and hence ρ_0 is still rigid in $\mathbf{Mob}(\mathbb{S}^3)$. The easiest way to see this is to use Weil's formula (11.1):

$$d = 10, s = 1, e_j = 6, \qquad \text{for } j = 1, 2, 3$$

and hence

$$h = -20 + 2 + 6 + 6 + 6 = 0.$$

However, instead of ρ_0 we can take a *twisted* extension. Suppose that we can find numbers $m_j \in \mathbb{Z}, 1 < |m_j| < r_j - 1$ $(j = 1, 2, 3)$ such that:

$$m_1^{-1} + m_2^{-1} + m_3^{-1} = 0, \text{ and } \forall j,\ m_j \text{ divides } r_j.$$

(This is satisfied for instance by $m_1 = m_2 = 4, m_3 = -2$ and $r_j = 8$ for all j.)

Define a homomorphism $\theta : \Delta \to SO(2)$ by sending c_i to the rotation by $2\pi/m_i$. Then define $\rho : \Delta \to \mathbf{Mob}(\mathbb{S}^1) \times SO(2) \subset \mathbf{Mob}(\mathbb{S}^3)$ by twisting ρ_0 via θ:

$$\rho(\gamma) = \rho_0(\gamma) \times \theta(\gamma), \quad \gamma \in \Delta.$$

It is clear that $\rho(\Delta)$ is again a 1-fuchsian subgroup in $\mathbf{Mob}(\mathbb{S}^3)$. If $r_j > 3$ for each j, then $e_j = 8$, $s = 1$ and the formula (11.1) gives the dimension $h = 6$ for $H^1(\Delta, Ad(\rho))$. I do not know if any of these *infinitesimal* deformations is integrable. To decide this one has to analyze the quadratic form

$$\phi : H^1(\Delta, Ad(\rho)) \to H^2(\Delta, Ad(\rho)) = \mathbb{R},$$

given by the cup-product. According to [82], the quadratic cone $\{\phi = 0\}$ is analytically isomorphic to a neighborhood of $[\rho]$ in $X_3(\Delta)$; hence it suffices to find a nontrivial 1-cocycle ξ for which $\phi([\xi]) = 0$ to get nontrivial deformations of the representation ρ.

On the other hand, one can use (11.1) to show that every representation ρ of the group $\Delta = \Delta(2, 3, r_3)$ into $\mathbf{Mob}(\mathbb{S}^3)$ has zero cohomology $H^1(\Delta, Ad(\rho))$. Therefore $X_3(\Delta)$ is a zero-dimensional algebraic variety and, hence, is a finite set.

This situation is somewhat typical for representations of lattices in $\mathbf{Mob}(\mathbb{S}^n)$ $(n \geq 2)$ into $\mathbf{Mob}(\mathbb{S}^{n+1})$: In a number of cases we can prove rigidity by making cohomological computations; in some cases we can only conclude that H^1 is nonzero, without being able to make a definitive conclusion about existence of nontrivial deformations.

11.1.2. Small deformations of i-quasifuchsian groups for $i \geq 2$. In the case of $(n-1)$-quasifuchsian groups Γ $(n \geq 2)$, the existence of nontrivial deformations of Γ in $\mathbf{Mob}(\mathbb{S}^n)$ is not at all clear. Suppose that $\Gamma \subset \mathrm{Isom}(\mathbb{H}^n)$ is a cocompact lattice. Then the identity embedding

$$\iota : \Gamma \hookrightarrow \mathrm{Isom}(\mathbb{H}^n),$$

is rigid by Mostow's theorem.

Remark 11.2. Actually, (local) rigidity of ι was known prior to the work of Mostow; it was first established by E. Calabi [48], whose proof was later generalized by A. Weil [224, 225]. These arguments were based on proving that $H^1(\Gamma, Ad(\iota)) = 0$.

Consider now the composition of ι with the natural embedding:

$$\rho : \Gamma \to \mathrm{Isom}(\mathbb{H}^n) \hookrightarrow \mathrm{Isom}(\mathbb{H}^{n+1}).$$

Then

$$H^1(\Gamma, Ad(\iota)) \cong H^1(\Gamma, V_n),$$

where $V_n = \mathbb{R}^{n,1}$ and Γ acts on the Lorentz space $\mathbb{R}^{n,1}$ via the usual embedding $\Gamma \hookrightarrow O(n, 1)$.

It turns out that ρ may or may not be rigid, even if Γ is torsion-free: Rigidity depends on the lattice Γ. One has the following list ((a) through (e)) of constructions of deformations and infinitesimal deformations of $[\rho]$ in $X_n(\Gamma)$:

(a) *Bending*, see [103], [135]. Given a connected properly embedded totally-geodesic hypersurface $S \subset M = \mathbb{H}^n/\Gamma$, one associates with S a smooth curve through $[\rho]$ in $X_n(\Gamma)$, called the *bending deformation* of $[\rho]$. More generally, given a disjoint collections of such hypersurfaces $S_1, \dots, S_k$, one obtains a k-dimensional smooth submanifold in $X_n(\Gamma)$ containing $[\rho]$. This construction is completely analogous to bending deformations of 1-fuchsian subgroups in $\mathbf{Mob}(\mathbb{S}^2)$ defined by W. Thurston in [213]. We let $[\xi_S]$ denote the element of $H^1(\Gamma, Ad(\rho))$ corresponding to a connected totally-geodesic hypersurface $S \subset M$.

There are numerous groups Γ satisfying assumptions of the bending construction. Namely, start with an arithmetic group $O'(f, A)$ of the simplest type (see Section 2), where

$$f = a_0 x_0^2 + a_1 x_1^2 + \cdots + a_n x_n^2,$$

and $a_0 < 0, a_i > 0, i = 1, \dots, n$. Identify $\mathbb{H}^n$ with a component of the hyperboloid $\{f(x) = -1\}$. Then the stabilizer of the hyperplane $P = \{x_n = 0\}$ in $O'(f, A)$ is an arithmetic lattice in $\mathrm{Isom}(\mathbb{H}^{n-1})$. The intersection $H = P \cap \mathbb{H}^n$ is a hyperplane in $\mathbb{H}^n$. After passing to an appropriate finite-index subgroup Γ in $O'(f, A)$, one obtains a totally-geodesic embedding of the hypersurface $S = H/\Gamma'$ into H/Γ, where $\Gamma' = \Gamma \cap O'(f, A)$. We refer the reader to [164] for the details.

Problem 11.3 (I. Rivin). *Construct examples of hyperbolic n-manifolds M of finite volume $(n \geq 4)$ such that M contains a* separating *properly embedded totally-geodesic hypersurface $S \subset M$. Note that the main objective of* [164] *was to construct* nonseparating *hypersurfaces.*

The idea of bending deformations of representations is quite simple and has nothing to do with the hyperbolic space. Below is a general description of bending as defined by D. Johnson and J. Millson in [103]. Suppose that we are given a graph of groups $\mathcal{G}$ with the vertex groups Γ_v and the edge groups Γ_e. Let $\Gamma = \pi_1(\mathcal{G})$ be the fundamental group of $\mathcal{G}$. For instance, the amalgam

$$\Gamma = \Gamma_{v_1} *_{\Gamma_e} \Gamma_{v_2} \tag{11.2}$$

is the fundamental group of a graph of groups which is a single edge e with two vertices v_1, v_2. Let $\rho_0 : \Gamma \to G$ be a representation of Γ to a Lie group G. A *bending deformation* of ρ_0 is a curve of representations $\rho_t : \Gamma \to G, t \in [-1, 1]$, such that for each vertex group Γ_v, we have

$$\rho_t|\Gamma_v = g_{v,t}(\rho_0|\Gamma_v)g_{v,t}^{-1},$$

for some curve $g_{v,t}$ of elements of G.

Therefore, the restriction of ρ_t to each vertex group determines a trivial deformation of the representation of this group. The trick is that the deformation of the representation of the entire group Γ may be still nontrivial. For instance, in the case of the amalgam (11.2), let $g_t \in G, g_0 = 1$, be a curve of elements centralizing $\rho(\Gamma_e)$, but not $\rho(\Gamma_{v_1}), \rho(\Gamma_{v_2})$. Define the family of representations

$$\rho_t : \Gamma \to G, \quad \rho_t|\Gamma_{v_1} = \rho_0|\Gamma_{v_1}, \quad \rho_t|\Gamma_{v_2} = g_t(\rho_0|\Gamma_{v_1})g_t^{-1}.$$

In the case of the HNN extension

$$\Gamma = \Gamma_{v_1} *_{\Gamma_e}$$

generated by Γ_{v_1} and $\tau \in \Gamma$ such $\tau\Gamma_e\tau^{-1} = \Gamma'_e \subset \Gamma_{v_1}$, we take

$$\rho_t|\Gamma_{v_1} = \rho_0|\Gamma_{v_1}, \quad \rho_t(\tau) = \rho_0(\tau)g_t.$$

This is a nontrivial deformation of the representation ρ_0.

We now return to the case when $\Gamma = \pi_1(M)$, M is a hyperbolic n-manifold containing pairwise disjoint totally geodesic hypersurfaces $S_i, i = 1, \dots, k$. Then the group Γ splits as the graph of groups $\mathcal{G}$, so that the vertex subgroups Γ_{v_j} are the fundamental groups of the components of $M \setminus (S_1 \cup \cdots \cup S_k)$ and the edge groups Γ_{e_i} are the fundamental groups $\pi_1(S_i)$. Therefore

1. The centralizer of each $\Gamma_{e_i} = \pi_1(S_i)$ in $\mathbf{Mob}(\mathbb{S}^n)$ is 1-dimensional (the group of elliptic rotations around the limit set of Γ_{e_i}).
2. The centralizer in $\mathbf{Mob}(\mathbb{S}^n)$ of the fundamental group of each Γ_{v_j} is zero-dimensional (i.e., $\mathbb{Z}_2$).

Hence one obtains nontrivial bending deformations ρ_t of the identity embedding of $\rho : \Gamma \hookrightarrow \mathbf{Mob}(\mathbb{S}^n)$. The set of bending parameters $t = (t_1, \dots, t_k)$ can be identified with $(\mathbb{S}^1)^k$, as the centralizer of each Γ_{e_i} in $\mathbf{Mob}(\mathbb{S}^n)$ is the circle $SO(2)$.

Theorem 11.4. (*D. Johnson and J. Millson* [103].) *For every $n \geq 4$ there exists a uniform lattice $\Gamma \subset \mathrm{Isom}(\mathbb{H}^n)$ and intersecting hypersurfaces $S_1, S_2 \subset \mathbb{H}^n/\Gamma$, so that*

$$[\xi_{S_1}] \cup [\xi_{S_2}] \in H^2(\Gamma, Ad(\rho)) \neq 0.$$

In particular, $X_n(\Gamma)$ is not smooth at $[\rho]$.

In contrast with this result, the character varieties $X_2(\Gamma)$ tend to be smooth:

Theorem 11.5. (*M. Kapovich* [116, Theorem 8.44].) *Let* $\Gamma \subset \mathbf{Mob}(\mathbb{S}^2)$ *be a discrete subgroup. Then the identity embedding* $\iota : \Gamma \to \mathbf{Mob}(\mathbb{S}^2)$ *determines a smooth point on* $X_2(\Gamma)$.

On the other hand, there are cocompact lattices $\Gamma \subset \mathbf{Mob}(\mathbb{S}^2)$ and (nondiscrete) representations $\rho : \Gamma \to \mathbf{Mob}(\mathbb{S}^2)$ for which $X_2(\Gamma)$ has nonquadratic singularity at $[\rho]$, see [126].

(b) *Generalized bending* associated with a collection of compact totally-geodesic submanifolds with boundary in M^n, see [12][8], [125], [161], [13].

The idea of the generalized bending is that instead of considering fundamental groups of graphs of groups, one looks at the more general *complexes of groups.* The only examples which had been worked out are 2-dimensional complexes of groups. Let $\mathcal{X}$ be such a complex with the vertex groups Γ_v. Let $\pi_1(\mathcal{X}) = \Gamma$ and $\rho_0 : \Gamma \to G$ be a representation to a Lie group. Then, as in the definition of bending, a *generalized bending* of ρ_0 is a curve of representations $\rho_t : \Gamma \to G, t \in [-1, 1]$, whose restrictions to each vertex subgroup Γ_v are trivial deformations of $\rho_0|\Gamma_v$.

(c) Suppose that a lattice $\Gamma \subset \mathrm{Isom}(\mathbb{H}^n)$ is a *reflection group*, i.e., it is generated by reflections in the faces of a convex acute polyhedron $\Phi \subset \mathbb{H}^n$ of finite volume (the fundamental domain of Γ). If f is the number of facets of Φ, then one can show that

$$\dim H^1(\Gamma, Ad(\rho)) = f - n - 1,$$

see [113]. The facets of Φ correspond to vectors spanning H^1. If $n \geq 4$, it is unclear which (if any) of these infinitesimal deformations are integrable. Of course, in *some* examples *some* of these infinitesimal deformations are integrable, since they appear as infinitesimal bending deformations. If $n = 3$, then $X_3(\Gamma)$ is smooth near $[\rho]$ and has dimension $f - 4$, see [113].

And that's all for $n \geq 3$.

Problem 11.6 (P. Storm). *Let* M *be a compact hyperbolic* $(n+1)$*-dimensional manifold with nonempty totally-geodesic boundary,* $n \geq 3$. *Let* $\Gamma := \pi_1(M) \subset \mathbf{Mob}(\mathbb{S}^n)$. *Is it true that* Γ *is rigid in* $\mathbf{Mob}(\mathbb{S}^n)$*?*

Note that (by Mostow rigidity) rigidity of Γ would follow if we knew that for each component S of ∂M, the fundamental group $\Gamma_S := \pi_1(S)$ is rigid in $\mathbf{Mob}(\mathbb{S}^n)$. At the moment, we do not have results in either direction of this problem:

1. It is unclear if any of the rigid hyperbolic 3-manifolds, or their disjoint union (see Example 11.1.2), bounds a compact hyperbolic 4-manifold.
2. Even if some Γ_S is not rigid, it is unclear if its deformations extend to deformations of Γ.

[8]Some of the theorems stated in this paper are probably incorrect since they do not take into account the restrictions on the angles between the totally-geodesic submanifolds.

The only known example of a rigid group Γ (as in Problem 11.6) is the fundamental group of a 4-dimensional hyperbolic *orbifold*. Moreover, in this example the group Γ_S is *not* rigid:

Consider the 120-cell $D^4 \subset \mathbb{H}^4$ which appears in [60]. Pick a facet $F \subset D^4$. Let $\Gamma \subset \mathbf{Mob}(\mathbb{S}^3)$ be the Kleinian group generated by reflections in all facets of D^4 except for F. Then Γ is the fundamental group of a right-angled 4-dimensional reflection orbifold $\mathcal{O}$ with boundary (the boundary $S = \partial\mathcal{O}$ corresponds to the facet F). The subgroup $\Gamma_S := \pi_1(S)$ is the Coxeter group generated by reflections in the facets of the regular right-angled hyperbolic dodecahedron. In particular, $X_3(\Gamma)$ is a smooth 8-dimensional manifold near $[\iota]$, where $\iota : \Gamma_S \hookrightarrow \mathbf{Mob}(\mathbb{S}^2) \hookrightarrow \mathbf{Mob}(\mathbb{S}^3)$ is the identity embedding.

Theorem 11.7 (M. Kapovich). *Γ is rigid in $\mathbf{Mob}(\mathbb{S}^3)$.*

Assume now that $n = 3$.

(d) The first obstruction to integrability of infinitesimal deformations is always zero, see [126].

Question 11.8. *Suppose that $\Gamma \subset \mathbf{Mob}(\mathbb{S}^2)$ is a cocompact lattice. It it true that the character variety $X_3(\Gamma)$ is smooth at the point $[\rho]$?*

(e) Finally, there are several constructions which work for specific examples of lattices $\Gamma \subset \mathbf{Mob}(\mathbb{S}^2)$, e.g., *stumping deformations* [9], generalized in [212].

We recall the following

Conjecture 11.9. *Suppose that $\Gamma \subset \mathbf{Mob}(\mathbb{S}^n)$ is a lattice. Then Γ contains a finite-index subgroup Γ' such that Γ' has infinite abelianization, i.e., $H^1(\Gamma', \mathbb{R}) \neq 0$.*

We refer the reader to [144, 146, 164, 189, 200] for various results towards this conjecture in the case of arithmetic lattices in $\mathrm{Isom}(\mathbb{H}^n)$. The methods used in these papers for proving virtual nonvanishing of the first cohomology group usually also apply to the cohomology groups $H^1(\Gamma, Ad(\rho))$, where $\rho : \Gamma \to \mathrm{Isom}(\mathbb{H}^{n+1})$ is the natural embedding. On the other hand, the proofs of special cases of Conjecture 11.9 for hyperbolic 3-manifolds which use the methods of 3-dimensional topology (see, e.g., [142]), usually provide no information about rigidity of Γ in $\mathrm{Isom}(\mathbb{H}^4)$.

Conjecture 11.10. *Suppose that $\Gamma \subset \mathrm{Isom}(\mathbb{H}^n)$ is a lattice. Then there exists a finite-index subgroup $\Gamma' \subset \Gamma$ so that $H^1(\Gamma', Ad(\rho)) \neq 0$.*

On the other hand, some uniform torsion-free lattices in $\mathbf{Mob}(\mathbb{S}^2)$ are rigid in $\mathbf{Mob}(\mathbb{S}^3)$:

Example. In [113] we constructed examples of (torsion-free) cocompact lattices Γ in $\mathbf{Mob}(\mathbb{S}^2)$ for which $H^1(\Gamma, Ad(\rho)) = 0$, where $\rho : \Gamma \to \mathbf{Mob}(\mathbb{S}^3)$ is the natural embedding. The quotient manifolds $\mathbb{H}^3/\Gamma$ in these examples are non-Haken. K. Scannell [197] constructed analogous examples with Haken quotients $\mathbb{H}^3/\Gamma$.

More specifically, it was proved in [113] that for every hyperbolic 2-bridge knot $K \subset \mathbb{S}^3$, there are infinitely many (hyperbolic) Dehn surgeries on K, so that for the resulting manifolds M_j, $j \in \mathbb{N}$, we have

$$H^1(\Gamma_j, Ad(\rho)) = 0, \text{ where } \Gamma_j = \pi_1(M_j).$$

11.1.3. Failure of quasiconformal isotopy. The goal of this section is to construct examples of Kleinian groups which are quasiconformally conjugate, but cannot be deformed to each other. As the reader will see, the tools for constructing such examples were available 12 years ago. I realized that such examples exist only recently, while working on this survey.

Theorem 11.11. *There exists a pair of convex-cocompact Kleinian groups $\Delta_1, \Delta_2 \subset$ $\mathbf{Mob}(\mathbb{S}^5)$ and a quasiconformal homeomorphism $f : \mathbb{S}^5 \to \mathbb{S}^5$ conjugating Δ_1 to Δ_2, which is not isotopic to the identity through homeomorphisms $h_t : \mathbb{S}^5 \to \mathbb{S}^5$ such that*

$$h_t \Delta_1 h_t^{-1} \subset \mathbf{Mob}(\mathbb{S}^5)\,.$$

Proof. We begin with a lattice $\Gamma = \pi_1(N)$, where $N = M_j$ is as in the discussion of Example 11.1.2 and $K \subset \mathbb{S}^3$ is the figure 8 knot. Consider the representation

$$\rho_1 : \Gamma \hookrightarrow \mathbf{Mob}(\mathbb{S}^2) \hookrightarrow \mathbf{Mob}(\mathbb{S}^5)$$

obtained by the composition of natural embeddings. Then

$$H^1(\Gamma, Ad(\rho_1)) = H^1(\Gamma, V_3 \oplus V_3 \oplus V_3 \oplus \mathbb{R}^3),$$

where $V_3 = \mathbb{R}^{3,1}$ and $\mathbb{R}^3$ is the trivial 3-dimensional $\mathbb{R}\Gamma$-module. Since $H^1(\Gamma, V_3) =$ 0 by [113] and $H^1(\Gamma, \mathbb{R}^3) = 0$ since N is a rational homology sphere, we obtain

$$H^1(\Gamma, Ad(\rho_1)) = 0.$$

Therefore ρ_1 is rigid. If N is an integer homology 3-sphere, then nonvanishing of the Casson invariant of K implies that Γ admits a nontrivial homomorphism

$$\theta : \Gamma \to SO(3),$$

which lifts to $SU(2)$, see [7]. If M_j is not an integer homology sphere, then Γ has nontrivial abelianization and hence we also obtain a nontrivial homomorphism $\theta : \Gamma \to SO(3)$ with cyclic image. In any case, we twist the representation ρ_1 by θ:

$$\rho_2 = \rho_1 \times \theta : \Gamma \to \mathbf{Mob}(\mathbb{S}^2) \times SO(3) \subset \mathbf{Mob}(\mathbb{S}^5)\,.$$

It is clear that $[\rho_1], [\rho_2]$ are distinct points of $X_5(\Gamma)$. The images of ρ_1 and ρ_2 are 2-fuchsian, convex-cocompact groups $\Delta_1, \Delta_2 \subset \mathbf{Mob}(\mathbb{S}^5)$. We obtain the isomorphism

$$\rho := \rho_2 \circ \rho_1^{-1} : \Delta_1 \to \Delta_2$$

Clearly,

$$M^5(\Delta_1) = N \times \mathbb{S}^2,$$

while $M^5(\Delta_2)$ is the 2-sphere bundle over N associated with θ. It is easy to see that the latter bundle is (smoothly) trivial. Therefore we obtain a diffeomorphism

$$h : M^5(\Delta_1) \to M^5(\Delta_2)$$

which lifts to a ρ-equivariant diffeomorphism $f : \Omega(\Delta_1) \to \Omega(\Delta_2)$. The latter extends to a quasiconformal homeomorphism $f : \mathbb{S}^5 \to \mathbb{S}^5$ by Theorem 6.1. If there was a continuous family of homomorphisms ρ_t connecting ρ to the identity embedding $\Delta_1 \to \Delta_2$, then the representation ρ_1 would not be rigid in $X_5(\Gamma)$. Contradiction. □

By embedding naturally the groups Δ_1, Δ_2 to $\mathbf{Mob}(\mathbb{S}^n)$ for $n \geq 6$ one obtains higher-dimensional examples.

11.2. Stability theorem

Let $\Gamma \subset \mathbf{Mob}(\mathbb{S}^n)$ be a geometrically finite Kleinian group. Consider the set of cusps in Γ:

$$[\Pi_1], \ldots, [\Pi_m],$$

where Π_i are maximal parabolic subgroups of Γ. We define the *(topologically) relative* representation variety

$$R_n^{top}(\Gamma) = \{\rho : \Gamma \to \mathbf{Mob}(\mathbb{S}^n) : \rho(\Pi_i) \text{ is topologically conjugate to } \Pi_i \text{ in } \mathbb{S}^n, \forall i\}$$

and the *(quasiconformally) relative* representation variety

$$R_n^{qc}(\Gamma) = \{\rho : \Gamma \to \mathbf{Mob}(\mathbb{S}^n) : \rho(\Pi_i) \text{ is quasiconformally conjugate to } \Pi_i \text{ in } \mathbb{S}^n\}.$$

Let $Homeo(\mathbb{S}^n)$ and $QC(\mathbb{S}^n)$ be the groups of homeomorphisms and quasiconformal homeomorphisms of $\mathbb{S}^n$ with the topology of uniform convergence. Let $X_n^{top}(\Gamma), X_n^{qc}(\Gamma)$ be the projections of $R_n^{top}(\Gamma), R_n^{qc}(\Gamma)$ to $X_n(\Gamma)$. Let $\iota : \Gamma \to \mathbf{Mob}(\mathbb{S}^n)$ be the identity embedding. Then the *Stability Theorem* for geometrically finite groups states that every homomorphism ρ of Γ sufficiently close to ι is induced by a (quasiconformal) homeomorphism h_ρ close to the identity and depending continuously on ρ. More precisely:

Theorem 11.12. (*Stability theorem, see* [40, 69, 79, 108, 152, 210].) *There exist neighborhoods U^{top}, U^{qc} of ι in $R_n^{top}(\Gamma), R_n^{qc}(\Gamma)$ respectively, and continuous maps*

$$L^{top} : U^{top} \to Homeo(\mathbb{S}^n), \quad L^{qc} : U^{qc} \to QC(\mathbb{S}^n)$$

so that

$$L^{top}(\iota) = L^{qc}(\iota) = Id,$$

and for every $\rho \in U^{top}$, resp. $\rho \in U^{qc}$, the homeomorphism $L^{top}(\rho)$, resp. $L^{qc}(\rho)$ is ρ-equivariant.

This theorem was first proved by A. Marden in [152] in the case $n = 3$. Marden was using convex finitely-sided fundamental domains with simplicial links of vertices: Such polyhedra are generic among the Dirichlet fundamental domains, see [106]. Marden then argued that a small perturbation of such fundamental domain is again a fundamental domain (by the Poincaré fundamental polyhedron theorem). Moreover, the *simplicial* assumption implies that the combinatorics of the fundamental domain does not change under a small perturbation. This allowed Marden to construct an equivariant quasiconformal homeomorphism close to the identity. This argument does not readily generalize to higher dimensions, mainly because finiteness of the number of faces is not equivalent to geometric finiteness. (Otherwise, the same argument goes through.)

D. Sullivan [210] considered the case of general n, but assumed that Γ is convex-cocompact. Then he proved the existence of a homeomorphism h_ρ defined on the limit set of Γ and the fact that it depends continuously on ρ. The fact that

$$h_\rho : \Lambda(\Gamma) \to \Lambda(\rho(\Gamma))$$

is necessarily quasi-symmetric, then follows from Tukia's theorem 6.1. One then has to show existence of a ρ-equivariant diffeomorphism of the domains of discontinuity

$$f_\rho : \Omega(\Gamma) \to \Omega(\rho(\Gamma))$$

smoothly depending on ρ. This is achieved by appealing to Thurston's holonomy theorem (see [69, 79]) for the Möbius structures on the manifold $M^n(\Gamma)$, as it is done in [101, 108]. The homeomorphisms h_ρ and f_ρ yield a ρ-equivariant quasi-conformal homeomorphism of the n-sphere by Theorem 6.1.

The proof in [69] is a good alternative to the above argument; it is also sufficiently flexible to handle the case of geometrically finite Kleinian groups with cusps. Namely, instead of working with the n-dimensional manifold $M^n(\Gamma)$ one works with the convex hyperbolic $(n+1)$-manifold

$$H(\Gamma) := Hull_\epsilon(\Lambda(\Gamma))/\Gamma.$$

An analogue of Thurston's holonomy theorem for manifolds with boundary applies in this case. Thus, for $\rho \in U^{top}$, there exists a hyperbolic structure $s(\rho)$ (with the holonomy ρ) on the thick part

$$H(\Gamma)_{[\mu,\infty)}$$

of the manifold $H(\Gamma)$. Moreover, convexity of the boundary for the new hyperbolic structures (away from the cusps) persists under small perturbations of the hyperbolic structure. Therefore, if Γ is convex-cocompact, $\Gamma' := \rho(\Gamma)$ is again convex-cocompact and $\rho : \Gamma \to \Gamma'$ is an isomorphism. If Γ is merely geometrically finite, because ρ belongs to the relative representation variety, it follows that the hyperbolic structure $s(\rho)$ extends to a convex complete hyperbolic structure on the cusps. This argument also yields a ρ-equivariant diffeomorphism

$$Hull_\epsilon(\Lambda(\Gamma)) \to Hull_\epsilon(\Lambda(\Gamma'))$$

depending continuously on ρ. To get from the convex hulls to the domain of discontinuity one uses the existence of the canonical equivariant diffeomorphisms ("the nearest-point projections")

$$\Omega(\Gamma) \to \partial Hull_\epsilon(\Lambda(\Gamma)), \quad \Omega(\Gamma') \to \partial Hull_\epsilon(\Lambda(\Gamma')).$$

We refer the reader to [69] for the details.

Sullivan also had a converse to the Stability Theorem for (finitely-generated) subgroups on $\mathbf{Mob}(\mathbb{S}^2)$:

Theorem 11.13. (*D. Sullivan,* [210, Theorem A′]) *If a (finitely-generated) Kleinian subgroup of* $\mathbf{Mob}(\mathbb{S}^2)$ *is stable in the sense of Theorem* 11.12, *then it is geometrically finite or its identity embedding in* $\mathbf{Mob}(\mathbb{S}^2)$ *is rigid in* $X_2(\Gamma)$.

It was proved in [116] that every rigid Γ in the above theorem has to be geometrically finite. Now it, of course, follows from the positive solution of the Bers–Thurston density conjecture (geometrically finite groups are dense among Kleinian subgroups of $\mathbf{Mob}(\mathbb{S}^2)$).

Question 11.14. *Does Theorem* 11.13 *hold for subgroups of* $\mathbf{Mob}(\mathbb{S}^n)$, $n \geq 3$*?*

We expect the answer to be negative.

11.3. Space of discrete and faithful representations

Let $\mathcal{D}_n(\Gamma) \subset X_n(\Gamma)$ denote the subset corresponding to discrete, injective and nonelementary representations of Γ.

Theorem 11.15 (Chuckrow–Jørgensen–Wielenberg). $\mathcal{D}_n(\Gamma) \subset X_n(\Gamma)$ *is closed. See for instance* [227, 154].

It turns out that there exists another way to construct limits of sequences of Kleinian groups, by regarding them as *closed subsets* of $\mathbf{Mob}(\mathbb{S}^n)$. This leads to the topology of *geometric convergence* of Kleinian groups. With few exceptions, the space of Kleinian groups is again closed in this topology (see, e.g., [215, 116]). In general, $\mathcal{D}_n(\Gamma)$ is not compact. Nevertheless, this space can be compactified by *projective classes* of nontrivial Γ-actions on real trees. This compactification generalizes Thurston's compactification of the Teichmüller space. The compactification by actions on trees was first defined by J. Morgan and P. Shalen [172] and J. Morgan [171] using algebraic geometry. More flexible, geometric, definitions of this compactification were introduced by M. Bestvina [23] and F. Paulin [182]. See also [116] for the construction of this compactification using ultralimits of metric spaces.

This viewpoint provides a powerful tool for proving compactness of $\mathcal{D}_n(\Gamma)$ for certain classes of groups: If $\mathcal{D}_n(\Gamma)$ is non-compact then Γ admits nontrivial action on a certain $\mathbb{R}$-tree. One then proves that such action cannot exist. The tools for proving such non-existence theorems are originally due to Morgan and Shalen (but limited to the fundamental groups of 3-manifolds, see [172]); a much more general method is due to E. Rips (Rips theory), see [27]. One then obtains the following (see, e.g., [116]):

Theorem 11.16 (Rips–Thurston Compactness theorem). *Suppose that* Γ *is a finitely-presented group which does not split as an amalgam over a virtually abelian group. Then* $\mathcal{D}_n(\Gamma)$ *is compact.*

Remark 11.17. W. Thurston [214] proved this theorem for a certain class of 3-manifold groups in the case $n = 2$.

Unfortunately, none of the known proofs of Theorem 11.16 gives an explicit bound on the "size" of $\mathcal{D}_n(\Gamma)$.

Problem 11.18. *Find a "constructive" proof of Theorem* 11.16. *More precisely, consider a group* Γ *with a finite presentation* $\langle g_1, .., g_k | R_1, .., R_m \rangle$. *Given* $[\rho] \in \mathcal{D}_{n-1}(\Gamma)$ *define*

$$B_{n-1}([\rho]) := \inf_{x \in \mathbb{H}^n} \max_{i=1,\ldots,k} d(x, \rho(g_i)(x)).$$

Find an explicit constant C, *which depends on* n, k, m *and the lengths of the words* R_i, *so that the function* $B_{n-1} : \mathcal{D}_{n-1}(\Gamma) \to \mathbb{R}$ *is bounded from above by* C.

Remark 11.19. In the case of Coxeter groups Γ, such explicit bound was obtained by Y. Lai [143]: The constant C depends only on the rank of the Coxeter group and n.

Theorem 11.16 suggests that one should also look for *geometric bounds* on $[\rho] \in \mathcal{D}_n(\Gamma)$: Even if $\mathcal{D}_n(\Gamma)$ is noncompact (or its "size" is unknown), one can still try to find some natural functionals on $\mathcal{D}_n(\Gamma)$ and obtain explicit bounds (from below and from above) on these functionals.

Definition 11.20 (Diameter of a representation). Given a discrete embedding $\rho : \Gamma \to \Gamma' = \rho(\Gamma) \subset \mathbf{Mob}(\mathbb{S}^n)$, consider the set S of connected subgraphs $\sigma \subset \mathbb{H}^{n+1}/\Gamma'$ with the property: The map $\pi_1(\sigma) \to \pi_1(M)$ is surjective.

Then the *diameter* of ρ is

$$\mathrm{diam}(\rho) := \inf\{\mathrm{length}(\sigma) : \sigma \in S\}.$$

Problem 11.21. *Given a group* Γ *as in Theorem* 11.16, *find explicit bounds on* $\mathrm{diam}(\rho)$ *(in terms of the presentation of* Γ*) for representations* $[\rho] \in \mathcal{D}_n(\Gamma)$.

Note that the positive *lower bound* on $\mathrm{diam}(\rho)$ is an easy corollary of the Kazhdan-Margulis lemma.

Definition 11.22. (Volumes of a representation) Fix a homology class $[\zeta] \in Z_p(\Gamma)$, $2 \le p \le cd(\Gamma)$. For a representation $\rho \in \mathcal{D}_n(\Gamma)$ consider the quotient manifold $M = \mathbb{H}^n/\rho(\Gamma)$. Define the set $E(\zeta)$ of singular p-cycles $\zeta' \in Z_p(M)$ which represent the homology class $[\zeta]$ under the isomorphism

$$H_p(\Gamma) \to H_p(M)$$

induced by the isomorphism $\rho : \Gamma \to \pi_1(M)$. Lastly, define the *$\rho$-volume* of the class $[\zeta]$ by

$$Vol_\rho(\zeta) := \inf\{Vol(\zeta') : \zeta' \in E(\zeta)\}.$$

Let $||\zeta||$ denote the Gromov-norm of the class $[\zeta]$ and let c_p denote the volume of the regular ideal geodesic p-simplex in $\mathbb{H}^p$. Then an easy application of Thurston's "chain-straightening" is the inequality

$$Vol_\rho(\zeta) \le c_p ||\zeta||$$

for all ρ, $[\zeta]$ and $p \ge 2$. However good lower bounds on the volume are considerably more difficult to get.

Given a hyperbolic manifold M define $H_p^{par}(M)$ to be the image in $H_p(M)$ of the p-th homology group of the union of all cusps of M. Then for every *parabolic* class $[\zeta] \in H_p^{par}(M)$ and every ρ, we clearly have

$$Vol_\rho(\zeta) = 0.$$

However there exists a positive constant $\epsilon = \epsilon(p, n)$ such that for every $p > 0$, every non-cuspidal class $[\zeta]$ and every ρ, we obtain

$$Vol_\rho(\zeta) \geq \epsilon,$$

see [121]. Below are some more interesting lower bounds on the volume:

Theorem 11.23. (*Follows directly from* [86, Theorem 5.38][9]). *Let* Γ *be isomorphic to the fundamental group of a compact aspherical k-manifold* N *and* $[\zeta] = [N]$ *be the fundamental class of* M. *Then there exists a universal (explicit) constant* $c(p, n) > 0$ *depending only on* p *and* n, *such that*

$$Vol_\rho(\zeta) \geq c(p, n)||N||.$$

One gets better estimates using the work of Besson, Courtois and Gallot [21] [10]:

Theorem 11.24. *Fix a representation* $[\phi] \in \mathcal{D}_n(\Gamma)$. *Then for every* $[\rho] \in \mathcal{D}_n(\Gamma)$ *and* $p \geq 3$ *we obtain*

$$\left(\frac{p}{n}\right)^{p+1} Vol_\phi(\zeta) \leq Vol_\rho(\zeta).$$

For instance, if $\Gamma' := \phi(\Gamma)$ happens to be a uniform lattice in $\mathrm{Isom}(\mathbb{H}^p)$, we obtain

Corollary 11.25. *For every* $[\rho] \in \mathcal{D}_n(\Gamma)$ *and* $p \geq 3$ *we have*

$$\left(\frac{p}{n}\right)^{p+1} Vol(M') \leq Vol_\rho(\zeta),$$

where $M' = \mathbb{H}^p/\Gamma'$ *and* $[\zeta]$ *is the fundamental class.*

11.4. Why is it so difficult to construct higher-dimensional geometrically infinite Kleinian groups?

(i). The oldest trick for proving existence of geometrically infinite groups is due to L. Bers [19]:

Start with (say) a convex-cocompact subgroup $\Gamma \subset \mathbf{Mob}(\mathbb{S}^n)$. Let $Q(\Gamma) \subset \mathcal{D}_n(\Gamma)$ be the (open) subset of representations induced by quasiconformal conjugation. Let $Q_0(\Gamma)$ denote the component of $Q(\Gamma)$ containing the (conjugacy class of) identity representation $[\rho_0]$. We assume that the closure of $Q_0(\Gamma)$ is not open in $X_n(\Gamma)$. Then there exists a curve $[\rho_t] \in X_n(\Gamma), t \in [0, 1]$, so that ρ_1 is either nondiscrete or non-injective. Since $\mathcal{D}_n(\Gamma)$ is closed, it follows that there exists $s \in (0, 1)$ such that $[\rho_s(\Gamma)]$ belongs to $\mathcal{D}_n(\Gamma)$ but $\rho_s(\Gamma)$ is not convex-cocompact.

[9] I am grateful to A. Nabutovsky for this reference.

[10] I am grateful to J. Souto for pointing this out.

If $\Gamma' = \rho_s(\Gamma)$ contains no parabolic elements, it would follow that Γ' is isomorphic to Γ and is not geometrically finite. However, it could happen that the frontier of $Q_0(\Gamma)$ consists entirely of the classes $[\rho]$ for which $\rho(\Gamma)$ contains parabolic elements.

The latter cannot occur if $n = 2$ for dimension reasons: The set of parabolic elements of $PSL(2, \mathbb{C})$ has real codimension 2 and, hence, does not separate. However for all $n \neq 2$, the set of parabolic elements has real codimension 1 and this argument is inconclusive.

One can try to apply the above argument in the case of a codimension 1 fuchsian group $\Gamma \subset \mathbf{Mob}(\mathbb{S}^n)$ which acts as a cocompact lattice on $\mathbb{H}^n \subset \mathbb{H}^{n+1}$. Suppose that $M = \mathbb{H}^n/\Gamma$ contains a totally-geodesic compact hypersurface S. Then we have the circle $\mathbb{S}^1$ worth of bending deformations ρ_t along S. As $t = \pi$, the image of ρ_t is again contained in $\mathbf{Mob}(\mathbb{S}^{n-1})$. Therefore ρ_π is either nondiscrete or non-injective. However, conceivably, in all such cases, for $[\rho_s] \in \partial Q_0(\Gamma)$ the representation ρ_s is geometrically finite (because its image may contain parabolic elements). It happens, for instance, if Γ is a reflection group.

Note that even when $n = 2$ and we are bending a 1-fuchsian group Γ, it is hard to predict which simple closed geodesics $\alpha \subset \mathbb{H}^2/\Gamma$ yield geometrically infinite groups (via bending along α).

(ii). One can try to construct explicit examples of fundamental domains, following, say, T. Jørgensen [104] or A. Marden and T. Jørgensen [105].

The trouble is that constructing fundamental polyhedra with infinitely many faces in $\mathbb{H}^4$ is quite a bit harder than in $\mathbb{H}^3$. One can try to find a lattice $\widehat{\Gamma} \subset \mathbf{Mob}(\mathbb{S}^3)$ which contains a nontrivial finitely-generated normal subgroup Γ of infinite index.[11] This is, probably, the most promising approach, since it works for complex-hyperbolic lattices in $PU(2,1)$, cf. [115]. One can try to imitate Livne's examples, by constructing $\Gamma \subset \widehat{\Gamma}$ such that $\widehat{\Gamma}/\Gamma$ is isomorphic to a surface group. This would require coming up with a specific compact convex polyhedron in $\mathbb{H}^4$ such that the associated 4-manifold appears as a (singular) fibration over a surface.

(iii). One can try to use combinatorial group theory. Note that there are plenty of examples of (mostly 2-dimensional) Gromov-hyperbolic groups $\widehat{\Gamma}$ which contain nontrivial finitely-generated normal subgroups Γ of infinite index. See, e.g., [26, 43, 174, 195] for the examples which are not 3-manifold groups. However embedding a given hyperbolic group $\widehat{\Gamma}$ in $\mathbf{Mob}(\mathbb{S}^n)$ is a nontrivial task, cf. [37, 118] and discussion in Section 12. The groups considered in [118], probably provide the best opportunity here, since most of them do not pass the *perimeter test* of J. McCammond and D. Wise [162]. (If a geometrically finite group $\widehat{\Gamma}$ satisfies the perimeter test, then every finitely-generated subgroup of $\widehat{\Gamma}$ is geometrically finite.)

(iv). *What would geometrically infinite examples look like?* Let $\Gamma \subset \mathbf{Mob}(\mathbb{S}^2)$ be a singly-degenerate group; assume for simplicity that the injectivity radius of

[11] If $\widehat{\Gamma}$ is a Kleinian group containing a nontrivial normal subgroup Γ of infinite index, then Γ is necessarily geometrically infinite.

$\mathbb{H}^3/\Gamma$ is bounded away from zero. Let S denote the boundary of

$$Hull(\Lambda(\Gamma))/\Gamma$$

and $\lambda \subset S$ be the ending lamination of Γ. Then every leaf of λ lifts to an exponentially distorted curve κ in $\mathbb{H}^3$: Given points $x, y \in \kappa$, their extrinsic distance $d(x, y)$ in $\mathbb{H}^3$ is roughly the logarithm of their intrinsic distance along κ.

One would like to imitate this behavior in dimension 4. Let M be a closed hyperbolic 3-manifold containing an embedded compact totally-geodesic surface $S \subset M$. Let $\lambda \subset S$ be an ending lamination from the above example. One would like to construct a complete hyperbolic 4-manifold N homotopy-equivalent to M, so that under the (smooth) homotopy-equivalence $f : M \to N$ we have:

For every leaf L of λ, $f(L)$ lifts to an exponentially distorted curve in $\mathbb{H}^4$.

Then $\pi_1(N)$ will necessarily be a geometrically infinite subgroup Γ of $\mathbf{Mob}(\mathbb{S}^4)$. At the moment it is not even clear how to make this work with a hyperbolic metric on N replaced by a complete Riemannian metric of negatively pinched sectional curvature, although constructing a Gromov-hyperbolic metric with this behavior is not that difficult. (Recall that a Riemannian metric is said to be *negatively pinched* if its sectional curvature varies between two negative numbers.) An example Γ of this type is likely to have two components of $\Omega(\Gamma)$: One contractible and one not.

More ambitiously, one can try to get a singly degenerate group $\Gamma \subset \mathbf{Mob}(\mathbb{S}^3)$ (so that $\Omega(\Gamma)$ is contractible and $M^3(\Gamma)$ is compact). How would such an example look like? One can imagine taking a 1-dimensional quasi-geodesic foliation λ of the 3-manifold M as above and then requiring that for every leaf $L \subset \lambda$, the curve $f(L)$ lifts to an exponentially distorted curve in $\mathbb{H}^4$. At the moment I do not see even a Gromov-hyperbolic model of this behavior. Another option would be to work with 2-dimensional laminations ν (with simply-connected leaves) in M and require every leaf $L \subset \nu$ to correspond to an exponentially distorted surface in $\mathbb{H}^4$ (or a negatively-curved simply-connected 4-manifold), which limits to a single point in $\mathbb{S}^3$.

Problem 11.26. *Construct a complete negatively pinched* 4*-dimensional Riemannian manifold N homotopy-equivalent to a hyperbolic* 3*-manifold M, so that the convex core of N either has exactly one boundary component or equals N itself.*

Question 11.27. *Is there a geometrically infinite Kleinian subgroup of* $\mathbf{Mob}(\mathbb{S}^n)$ *whose limit is homeomorphic to the Menger curve? Is there a geometrically infinite Kleinian subgroup of* $\mathbf{Mob}(\mathbb{S}^n)$ *which is isomorphic to the fundamental group of a closed aspherical manifold of dimension* ≥ 3*? Are there examples of such groups acting isometrically on complete negatively pinched manifolds? Are there examples of hyperbolic (or even negatively curved)* 4*-manifolds M such that $\pi_1(M) = \Gamma$ fits into a short exact sequence*

$$1 \to \pi_1(S) \to \Gamma \to \pi_1(F) \to 1,$$

where S, F are closed hyperbolic surfaces? Note that there are no complex-hyperbolic examples of this type, see [115].

12. Algebraic and topological constraints on Kleinian groups

Sadly, there are only few known algebraic and topological restrictions on Kleinian subgroups in $\mathbf{Mob}(\mathbb{S}^n)$ that do not follow from the *elementary* restrictions, which come from the restrictions on geometry of complete negatively curved Riemannian manifolds. Examples of the elementary restrictions on a Kleinian group Γ are:

1. Every solvable subgroup of a Kleinian group is virtually abelian.
2. The normalizer (in Γ) of an infinite cyclic subgroup of Γ is virtually abelian.
3. Every elementary (i.e., virtually abelian) subgroup $\Delta \subset \Gamma$ is contained in a unique maximal elementary subgroup $\tilde{\Delta} \subset \Gamma$.
4. Every Kleinian group has finite (virtual) cohomological dimension.

In this section we review known *nonelementary* algebraic and topological restrictions on Kleinian groups.

12.1. Algebraic constraints

Definition 12.1. An *abstract Kleinian group* is a group Γ which admits a discrete embedding in $\mathbf{Mob}(\mathbb{S}^n)$ for some n. Such a group is called *elementary* if it is virtually abelian.

In order to eliminate trivial restrictions on abstract Kleinian groups one can restrict attention to Gromov-hyperbolic Kleinian groups. Below is the list of known algebraic constraints on Kleinian groups under this extra assumption:

1. Kleinian groups are residually finite and virtually torsion-free.[12] (This, of course, holds for all finitely generated matrix groups.)
2. Kleinian groups satisfy the *Haagerup property*, in particular, infinite Kleinian groups do not satisfy property (T), see [55].
3. If a Kleinian group Γ is *Kähler*, then Γ is virtually isomorphic to the fundamental group of a compact Riemann surface. This is a deep theorem of J. Carlson and D. Toledo [52], who proved that every homomorphism of a Kähler group to $\mathbf{Mob}(\mathbb{S}^n)$ either factors through a virtually surface group, or its image fixes a point in $\mathbb{B}^{n+1}$.

Recall that a topological group G is said to satisfy the *Haagerup property* if it admits a (metrically) proper continuous isometric action on a Hilbert space H. An action of a metrizable topological group G on H is *metrically proper* if for every bounded subset $B \subset H$, the set

$$\{g \in G : g(B) \cap B \neq \emptyset\}$$

is a bounded subset of G. Since $\mathbf{Mob}(\mathbb{S}^n)$ satisfies the Haagerup property for every n (see, e.g., [55]), all Kleinian groups also do.

A group π is called *Kähler* if it is isomorphic to the fundamental group of a compact Kähler manifold. For instance, every uniform lattice in $\mathbb{CH}^n$ is Kähler; therefore it cannot be an abstract Kleinian group unless $n = 1$.

[12] It is widely believed that there are Gromov-hyperbolic groups which are not residually finite.

Remark 12.2. A (finitely-generated) group satisfies the Haagerup property if and only if it admits an isometric (metrically) properly discontinuous action on the infinite dimensional hyperbolic space $\mathbb{H}^\infty$, see [85, 7.A.III]. The result of Carlson and Toledo shows that (for Gromov-hyperbolic groups) there are nontrivial obstructions to replacing these infinite-dimensional actions with finite-dimensional ones.

Observation 12.3. *All currently known nontrivial restrictions on abstract Kleinian groups can be traced to* 1, 2 *or* 3.

Problem 12.4. *Find other restrictions on abstract Kleinian groups.*

Potentially, some new restrictions would follow from the Rips-Thurston compactness theorem. The difficulty comes from the following. Let Γ be a Gromov-hyperbolic group which admits no nontrivial isometric actions on $\mathbb{R}$-trees. Then (see [116]) there exists $C < \infty$, such that for every sequence $[\rho_j] \in \mathcal{D}_n(\Gamma)$, we obtain a uniform bound

$$B_n([\rho_j]) \le C, \tag{12.1}$$

where $B_n : \mathcal{D}_n(\Gamma) \to \mathbb{R}$ is the minimax function defined in Problem 11.18. If n were fixed, then the sequence (ρ_j) would subconverge to a representation to $\mathbf{Mob}(\mathbb{S}^n)$ (for some choice of representations ρ_j in the classes $[\rho_j]$). However, since we are not fixing the dimension of the hyperbolic space on which our Γ is supposed to act, the inequality (12.1) does not seem to yield any useful information. By taking an ultralimit of ρ_j's we will get an action of Γ on an infinite-dimensional hyperbolic space. This action, however, may have a fixed point, since

$$\lim_{n\to\infty} \mu_n = 0,$$

where μ_n is the Margulis constant for $\mathbb{H}^{n+1}$. See also Remark 12.2.

Example. Let M^3 be a closed non-Haken hyperbolic 3-manifold, so that $\Pi := \pi_1(M)$ contains a maximal 1-fuchsian subgroup F. For each automorphism $\phi : F \to F$ we define the HNN extension

$$\Gamma_\phi := \Pi *_{F \cong_\phi F} = \langle \Pi, t | tgt^{-1} = \phi(g), \forall g \in F \rangle.$$

Then Γ_ϕ is Gromov-hyperbolic for all pseudo-Anosov automorphisms ϕ, see [27]. It is a direct corollary of Theorem 11.16 that for every n, only finitely many of the groups Γ_ϕ embed in $\mathbf{Mob}(\mathbb{S}^n)$ as Kleinian subgroups. Is it true that there exists a pseudo-Anosov automorphism ϕ such that Γ_ϕ is not an abstract Kleinian group?

Infinite finitely-generated Gromov-hyperbolic Coxeter groups are all linear, satisfy the Haagerup property and are not Kähler (except for the virtually surface groups).

Problem 12.5. *Is it true that every finitely-generated Gromov-hyperbolic Coxeter group* Γ *is an abstract Kleinian group?*

Note that there are Gromov-hyperbolic Coxeter groups Γ which do not admit discrete embeddings $\rho : \Gamma \to \mathbf{Mob}(\mathbb{S}^n)$ (for any n), so that the Coxeter generators of Γ act as reflections in the faces of a fundamental domain of $\rho(\Gamma)$, see [76].

The answer to the next question is probably negative, but the examples would be tricky to construct:

Question 12.6. *Is it true that a group weakly commensurable to a Kleinian group is also a Kleinian group? Even more ambitiously: Is the property of being Kleinian a quasi-isometry invariant of a group?*

Recall that two groups Γ and Γ' are called *weakly commensurable* if there exists a chain of groups and homomorphisms

$$\Gamma = \Gamma_0 \to \Gamma_1 \leftarrow \Gamma_2 \to \Gamma_3 \cdots . \leftarrow \Gamma_{k-1} \to \Gamma_k = \Gamma',$$

where each arrow $\Gamma_i \to \Gamma_{i\pm1}$ is a homomorphism whose kernel and cokernel are finite.

There are few more known algebraic restrictions on geometrically finite Kleinian groups. All such groups are relatively hyperbolic.

We recall that a group Γ is called *cohopfian* if every injective endomorphism $\Gamma \to \Gamma$ is also surjective.

Remark 12.7. A group Γ is called *hopfian* if every epimorphism $\Gamma \to \Gamma$ is injective. Every residually finite group Γ is hopfian, see [151]. In particular, every Kleinian group is hopfian.

For instance, free groups and free abelian groups are not cohopfian. More generally, if Γ splits as a nontrivial free product,

$$\Gamma \cong \Gamma_1 * \Gamma_2,$$

then Γ is not cohopfian: Indeed, for nontrivial elements $\gamma_1 \in \Gamma_1, \gamma_2 \in \Gamma_2$, set $\alpha := \gamma_1\gamma_2$, and

$$\Gamma_1' := \alpha\Gamma_1\alpha^{-1}.$$

Then

$$\Gamma \cong \Gamma_1' * \Gamma_2$$

is a proper subgroup of Γ. On the other hand, lattices in $\mathrm{Isom}(\mathbb{H}^n)$, $n \geq 3$, and uniform lattices in $\mathrm{Isom}(\mathbb{H}^2)$ are cohopfian. Indeed, Mostow rigidity theorem implies that if M_1, M_2 are hyperbolic n-manifolds of finite volume (and $n \geq 3$) or compact hyperbolic surfaces, and $M_1 \to M_2$ is a d-fold covering, then

$$Vol(M_1) = dVol(M_2).$$

On the other hand, if M is a hyperbolic manifold of finite volume (or a compact hyperbolic surfaces), then every proper embedding

$$\pi_1(M) \to \pi_1(M)$$

induces a d-fold covering $M \to M$, with $d \in \{2, 3, \dots, \infty\}$. Hence $\pi_1(M)$ is cohopfian.

If a Kleinian group $\Gamma \subset \mathbf{Mob}(\mathbb{S}^n)$ fails to be cohopfian, we can iterate a proper embedding $\phi : \Gamma \to \Gamma$, thereby obtaining a sequence of discrete and faithful representations

$$\rho_i = \underbrace{\phi \circ \cdots \circ \phi}_{i \text{ times}}$$

of Γ into $\mathbf{Mob}(\mathbb{S}^n)$. By analyzing such sequences, T. Delzant and L. Potyagailo [63] obtained a characterization of geometrically finite Kleinian groups which are *cohopfian.* We will need two definitions in order to describe their result.

Definition 12.8. If Γ is a Kleinian group and $\Delta \subset \Gamma$ is an elementary subgroup, let $\tilde{\Delta}$ denote the maximal elementary subgroup of Γ containing Δ.

Definition 12.9. Suppose a group Γ splits as a graph of groups

$$\Gamma \cong \pi_1(\mathcal{G}, \Gamma_e, \Gamma_v), \tag{12.2}$$

and suppose that edge groups Γ_e of this graph are elementary. We say that the edge group Γ_e is *essentially non-maximal* if the subgroup $\tilde{\Gamma}_e \subset \Gamma$, is not conjugate into any of the vertex subgroups of the graph of groups $\mathcal{G}$. The splitting is *essentially non-maximal* if there exists at least one such an edge group. Otherwise we say that the splitting is essentially maximal.

For instance, if every edge subgroup is a maximal elementary subgroup of Γ, then the splitting is essentially maximal.

Theorem 12.10. (*T. Delzant and L. Potyagailo* [63].) *Let Γ be a non-elementary, geometrically finite, one-ended Kleinian group without 2-torsion. Then Γ is cohopfian if and only if the following two conditions are satisfied:*

1) *Γ has no essentially non-maximal splittings.*
2) *Γ does not split as an amalgamated free product*

$$\Gamma = \Gamma_1 *_{\Gamma_3} \tilde{\Gamma}_3,$$

with $\tilde{\Gamma}_3$ maximal elementary, such that the normal closure of the subgroup Γ_3 in $\tilde{\Gamma}_3$ is of infinite index in $\tilde{\Gamma}_3$.

One of the ingredients in the proof of this theorem was the fact that nonelementary geometrically finite groups Γ do not contain subgroups Γ', which are conjugate to Γ in $\mathbf{Mob}(\mathbb{S}^n)$, see [223].

Question 12.11 (L. Potyagailo). *Let $\Gamma \subset \mathbf{Mob}(\mathbb{S}^n)$ be a finitely generated non-elementary Kleinian group. Suppose that $\alpha \in \mathbf{Mob}(\mathbb{S}^n)$ is such that*

$$\Gamma' = \alpha \Gamma \alpha^{-1} \subset \Gamma.$$

Does it follow that $\Gamma' = \Gamma$?

The affirmative answer to this question for $n = 2$ was given in a paper of L. Potyagailo and K.-I. Ohshika [178] (modulo Tameness Conjecture, Theorem 10.3).

Question 12.12. *Is the isomorphism problem solvable within the class of all finitely-presented Kleinian groups? Note that the work of F. Dahmani and D. Groves* [59] *implies solvability of the isomorphism problem in the category of geometrically finite Kleinian groups.*

It was proved by M. Bonk and O. Schramm [36] that every Gromov-hyperbolic group Γ embeds quasi-isometrically in the usual hyperbolic space $\mathbb{H}^n$ for some $n = n(\Gamma)$. A natural question is if one can prove an *equivariant* version of this result. Note that there are many Gromov-hyperbolic groups which are not Kleinian, e.g., groups with property (T) and Gromov-hyperbolic Kähler groups. Therefore one has to relax the *isometric* assumption. The natural category for this is the *uniformly quasiconformal* actions. Such an action is a monomorphism

$$\rho : \Gamma \hookrightarrow QC(\mathbb{S}^n)$$

whose image consists of K-quasiconformal homeomorphisms with K depending only on ρ.

Problem 12.13. *Let Γ be a Gromov-hyperbolic group. Does Γ admit a uniformly quasiconformal discrete action on $\mathbb{S}^n$ for some n? For instance, is there such an action if Γ is a uniform lattice in $PU(n,1)$ or satisfies the property* (T)*?*

T. Farrell and J. Lafont [74] proved that the topological counterpart of this problem has positive solution. A corollary of their work is that every Gromov-hyperbolic group Γ admits a *convergence* action ρ on the closed n-ball, so that the limit set of $\Gamma' = \rho(\Gamma)$ is equivariantly homeomorphic to the ideal boundary of Γ and $\Omega(\Gamma')/\Gamma'$ is compact and connected. We refer the reader to [78] for the definition of a convergence action.

12.2. Topological constraints

The basic problem here is to find topological restrictions on the hyperbolic manifold $\mathbb{H}^{n+1}/\Gamma$ and on the conformally-flat manifold $\Omega(\Gamma)/\Gamma$, which do not follow from the algebraic restrictions on the group Γ and from the general algebraic topology restrictions (e.g., vanishing of the characteristic classes). There are only few nontrivial results in this direction. For $n = 3$ we have:

Theorem 12.14. (*M. Kapovich* [109].) *There exists a function $c(\chi)$ with the following property. Let S be a closed hyperbolic surface. Suppose that M^4 is a complete hyperbolic 4-manifold which is homeomorphic to the total space of an $\mathbb{R}^2$-bundle $\xi : E \to S$ with the Euler number $e(\xi)$. Then*

$$|e(\xi)| \leq c(\chi(S)).$$

More generally,

Theorem 12.15. (*M. Kapovich* [109].) *There exists a function $C(\chi_1, \chi_2)$ with the following property. Suppose that M^4 is a complete oriented hyperbolic 4-manifold. Let $\sigma_j : \Sigma_j \to M^4$ $(j = 1, 2)$ be π_1-injective maps of closed oriented surfaces Σ_j. Then*

$$|\langle \sigma_1, \sigma_2 \rangle| \leq C(\chi(\Sigma_1), \chi(\Sigma_2)).$$

Here $\langle , \rangle$ is the intersection pairing on $H_2(M^4)$. The bounds appearing in these theorems are explicit but astronomically high. The expected bounds are linear in $\chi(S)$ and $\chi(S_i)$, $i = 1, 2$, cf. [87].

Other known restrictions are applications of the compactness theorem 11.16 and therefore explicit bounds in the following theorems are unknown.

Theorem 12.16. (*M. Kapovich* [111].) *Given a closed hyperbolic n-manifold B ($n \geq 3$) there exists a number $c(B)$ so that the following is true. Suppose that M^{2n} a complete hyperbolic $2n$-manifold which is homeomorphic to the total space of an $\mathbb{R}^n$-bundle $\xi: E \to B$ with the Euler number $e(\xi)$. Then*

$$|e(\xi)| \leq c(B).$$

I. Belegradek greatly improved this result:

Theorem 12.17. (*I. Belegradek* [16].) *Given a closed hyperbolic n-manifold B ($n \geq 3$) there exists a number $C(B, k)$ so that the number of inequivalent $\mathbb{R}^k$-bundles $\xi : E \to B$ whose total space admits a complete hyperbolic metric, is at most $C(B, k)$.*

Given a group π, let $\mathcal{M}_{\pi,n}$ denote the set of n-manifolds, whose fundamental group is isomorphic to π and which admit complete hyperbolic metrics.

Theorem 12.18. (*I. Belegradek* [17].) *Suppose that π is a finitely-presented group with finite Betti numbers. Assume that π does not split as an amalgam over a virtually abelian subgroup. The set $\mathcal{M}_{\pi,n}$ breaks into finitely many intersection preserving homotopy types.*

References

[1] W. Abikoff, *Kleinian groups – geometrically finite and geometrically perverse*, in "Geometry of group representations", vol. 74 of Contemp. Math., Amer. Math. Soc., 1988, pp. 1–50.

[2] I. Agol, *Tameness of hyperbolic 3-manifolds.* Preprint math.GT/0405568, 2004.

[3] L. Ahlfors, *Finitely generated Kleinian groups*, Amer. J. of Math., 86 (1964), pp. 413–429.

[4] ——, *Remarks on the limit point set of a finitely generated Kleinian group.*, in "Advances in the Theory of Riemann Surfaces" (Proc. Conf., Stony Brook, 1969), Ann. of Math. Studies, No. 66, Princeton Univ. Press, Princeton, N.J., 1971, pp. 19–26.

[5] ——, *Ergodic properties of Moebius transformations*, in "Analytic functions", vol. 798 of Springer Lecture Notes in Mathematics, Springer, 1980, pp. 19–25.

[6] ——, *Möbius transformations in several dimensions*, Ordway Professorship Lectures in Mathematics, University of Minnesota School of Mathematics, Minneapolis, Minn., 1981.

[7] S. Akbulut and J. McCarthy, *Casson's invariant for oriented homology 3-spheres, an exposition*, Mathematical Notes 36, Princeton University Press, Princeton, 1990.

[8] P. S. Alexandroff, *Einführung in die Mengenlehre und in die allgemeine Topologie*, Hochschulbücher für Mathematik, 85, 1984.

[9] B. Apanasov, *Bending and stamping deformations of hyperbolic manifolds*, Ann. Global Anal. Geom., 8 (1990), pp. 3–12.

[10] ——, *Discrete groups in space and uniformization problems*, Kluwer Academic Publishers Group, Dordrecht, 1991.

[11] B. Apanasov and A. Tetenov, *Nontrivial cobordisms with geometrically finite hyperbolic structures*, J. Differential Geom., 28 (1988), pp. 407–422.

[12] ——, *Deformations of hyperbolic structures along surfaces with boundary and pleated surfaces*, in "Low-dimensional topology" (Knoxville, 1992), Internat. Press, 1994, pp. 1–14.

[13] A. Bart and K. Scannell, *The generalized cuspidal cohomology problem*, Canad. J. Math., 58 (2006), pp. 673–690.

[14] A. Beardon and B. Maskit, *Limit points of Kleinian groups and finite sided fundamental polyhedra*, Acta Math., 132 (1974), pp. 1–12.

[15] I. Belegradek, *Some curious Kleinian groups and hyperbolic 5-manifolds*, Transform. Groups, 2 (1997), pp. 3–29.

[16] ——, *Intersections in hyperbolic manifolds*, Geometry and Topology, 2 (1998), pp. 117–144.

[17] ——, *Pinching, Pontrjagin classes, and negatively curved vector bundles*, Invent. Math., 144 (2001), pp. 353–379.

[18] R. Benedetti and C. Pertonio, *Lectures on Hyperbolic Geometry*, Springer, 1992.

[19] L. Bers, *On the boundaries of Teichmüller spaces and on Kleinian groups, I*, Annals of Math., 91 (1970), pp. 570–600.

[20] ——, *Uniformization, moduli and Kleinian groups*, Bull. London Math. Society, 4 (1972), pp. 257–300.

[21] G. Besson, G. Courtois, and S. Gallot, *Lemme de Schwarz réel et applications géométriques*, Acta Math., 183 (1999), pp. 145–169.

[22] ——, *Hyperbolic manifolds, amalgamated products and critical exponents*, C. R. Math. Acad. Sci. Paris, 336 (2003), pp. 257–261.

[23] M. Bestvina, *Degenerations of hyperbolic space*, Duke Math. Journal, 56 (1988), pp. 143–161.

[24] M. Bestvina and N. Brady, *Morse theory and finiteness properties of groups*, Inventiones Math., 129 (1997), pp. 445–470.

[25] M. Bestvina and D. Cooper, *A wild Cantor set as the limit set of a conformal group action on* $\mathbf{S}^3$, Proc. of AMS, 99 (1987), pp. 623–627.

[26] M. Bestvina and M. Feighn, *A combination theorem for negatively curved groups*, J. Differential Geom., 35 (1992), pp. 85–101.

[27] ——, *Stable actions of groups on real trees*, Invent. Math., 121 (1995), pp. 287–321.

[28] M. Bestvina and G. Mess, *The boundary of negatively curved groups*, Journal of the AMS, 4 (1991), pp. 469–481.

[29] R. Bieri, *Homological dimension of discrete groups*, Mathematical Notes, Queen Mary College, 1976.

[30] R. Bieri and B. Eckmann, *Relative homology and Poincaré duality for group pairs*, J. Pure Appl. Algebra, 13 (1978), pp. 277–319.

[31] R.H. Bing, *The geometric topology of 3-manifolds*, vol. 40 of American Mathematical Society Colloquium Publications, American Mathematical Society, Providence, RI, 1983.

[32] C. Bishop, *On a theorem of Beardon and Maskit*, Ann. Acad. Sci. Fenn. Math., 21 (1996), pp. 383–388.

[33] C. Bishop and P. Jones, *Hausdorff dimension and Kleinian groups*, Acta Math., 179 (1997), pp. 1–39.

[34] F. Bonahon, *Boutes des varietes hyperboliques de dimension trois*, Ann. of Math, 124 (1986), pp. 71–158.

[35] M. Bonk and B. Kleiner, *Rigidity for quasi-Fuchsian actions on negatively curved spaces*, Int. Math. Res. Not., (2004), pp. 3309–3316.

[36] M. Bonk and O. Schramm, *Embeddings of Gromov hyperbolic spaces*, Geom. Funct. Anal., 10 (2000), pp. 266–306.

[37] M. Bourdon, *Sur la dimension de Hausdorff de l'ensemble limite d'une famille de sous-groupes convexes co-compacts*, C. R. Acad. Sci. Paris Sér. I Math., 325 (1997), pp. 1097–1100.

[38] M. Bourdon, F. Martin, and A. Valette, *Vanishing and non-vanishing for the first L^p-cohomology of groups*, Comment. Math. Helv., 80 (2005), pp. 377–389.

[39] B.H. Bowditch, *Geometrical finiteness for hyperbolic groups*, Journal of Functional Anal., 113 (1993), pp. 245–317.

[40] B.H. Bowditch, *Spaces of geometrically finite representations*, Ann. Acad. Sci. Fenn. Math., 23 (1998), pp. 389–414.

[41] B.H. Bowditch, *Planar groups and the Seifert conjecture*, J. Reine Angew. Math., 576 (2004), pp. 11–62.

[42] B.H. Bowditch and G. Mess, *A 4-dimensional Kleinian group*, Transactions of AMS, 14 (1994), pp. 391–405.

[43] N. Brady, *Branched coverings of cubical complexes and subgroups of hyperbolic groups*, J. London Math. Soc. (2), 60 (1999), pp. 461–480.

[44] M. Bridson and A. Haefliger, *Metric spaces of non-positive curvature*, Springer-Verlag, Berlin, 1999.

[45] M. Brin, *Torsion-free action on 1-acyclic manifolds and the loop theorem*, Topology, 20 (1981), pp. 353–364.

[46] J. Brock, R. Canary, and Y. Minsky, *The classification of Kleinian surface groups, II: The Ending Lamination Conjecture.* Preprint, 2004.

[47] K. Brown, *Cohomology of Groups*, vol. 87 of Graduate Texts in Mathematics, Springer, 1982.

[48] E. Calabi, *On compact Riemannian manifolds with constant curvature*, in Proc. Sympos. Pure Math., Vol. III, American Mathematical Society, Providence, R.I., 1961, pp. 155–180.

[49] D. Calegari and D. Gabai, *Shrinkwrapping and the taming of hyperbolic 3-manifolds*, J. Amer. Math. Soc., 19 (2006), pp. 385–446.

[50] R. Canary and E. Taylor, *Hausdorff dimension and limits of Kleinian groups*, Geom. Funct. Anal., 9 (1999), pp. 283–297.

[51] J. Cannon and W.P. Thurston, *Group invariant Peano curves*. Preprint.

[52] J. Carlson and D. Toledo, *Harmonic mappings of Kähler manifolds to locally symmetric spaces*, Inst. Hautes Études Sci. Publ. Math., (1989), pp. 173–201.

[53] J. Casson and S. Bleiler, *Automorphisms of Surfaces After Nielsen and Thurston*, vol. 9 of LMS Student Texts, Cambridge University Press, 1988.

[54] A. Chang, J. Qing, and P. Yang, *On finiteness of Kleinian groups in gcncral dimension*, J. Reine Angew. Math., 571 (2004), pp. 1–17.

[55] P.-A. Cherix, M. Cowling, P. Jolissaint, P. Julg, and A. Valette, *Groups with the Haagerup property*, vol. 197 of Progress in Mathematics, Birkhäuser, 2001.

[56] V. Chuckrow, *Schottky groups with applications to Kleinian groups*, Ann. of Math., 88 (1968), pp. 47–61.

[57] S. Claytor, *Topological immersion of Peanian continua in the spherical surface*, Ann. of Math., 35 (1924), pp. 809–835.

[58] D. Cooper, D. Long, and A. Reid, *Bundles and finite foliations*, Inventiones Mathematicae, 118 (1994), pp. 255–283.

[59] F. Dahmani and D. Groves, *The isomorphism problem for toral relatively hyperbolic groups*. Preprint, 2005.

[60] M.W. Davis, *A hyperbolic 4-manifold*, Proc. Amer. Math. Soc., 93 (1985), pp. 325–328.

[61] D.G. DeGryse and R.P. Osborne, *A wild Cantor set in E^n with simply connected complement*, Fund. Math., 86 (1974), pp. 9–27.

[62] P. Deligne and G.D. Mostow, *Commensurabilities among lattices in* $\mathrm{PU}(1,n)$, vol. 132 of Annals of Mathematics Studies, Princeton University Press, 1993.

[63] T. Delzant and L. Potyagailo, *Endomorphisms of Kleinian groups*, Geom. Funct. Anal., 13 (2003), pp. 396–436.

[64] S. Donaldson and D. Sullivan, *Quasiconformal 4-manifolds*, Acta Math., (1989), pp. 181–252.

[65] A. Douady and C. Earle, *Conformally natural extension of homeomorphisms of the circle*, Acta Mathematica, 157 (1986), pp. 23–48.

[66] M.J. Dunwoody, *Accessibility and groups of cohomological dimension one*, Proc. London Math. Soc. (3), 38 (1979), pp. 193–215.

[67] B. Eckmann and P. Linnel, *Poincaré dualtiy groups of dimension 2, II*, Comm. Math. Helv., 58 (1983), pp. 111–114.

[68] B. Eckmann and H. Müller, *Poincaré duality groups of dimension two*, Comm. Math. Helv., 55 (1980), pp. 510–520.

[69] D. Epstein, R. Canary, and P. Green, *Notes on notes of Thurston*, in "Analytical and geometric aspects of hyperbolic space", vol. 111 of London Math. Soc. Lecture Notes, Cambridge Univ. Press, 1987, pp. 3–92.

[70] F. Farrell and L. Jones, *Negatively curved manifolds with exotic smooth structures*, J. Amer. Math. Soc., 2 (1989), pp. 899–908.

[71] ——, *A topological analogue of Mostow's rigidity theorem*, Journal of the AMS, 2 (1989), pp. 257–370.

[72] ——, *Topological rigidity for compact non-positively curved manifolds*, in "Differential geometry"., vol. 54 of Proc. of Symp. in Pure Math., 1993, pp. 229–274.

[73] ——, *Rigidity for aspherical manifolds with* $\pi_1 \subset GL_m(R)$, Asian J. Math., 2 (1998), pp. 215–262.

[74] F.T. Farrell and J.-F. Lafont, *EZ-structures and topological applications*, Comment. Math. Helv., 80 (2005), pp. 103–121.

[75] M. Feighn and G. Mess, *Conjugacy classes of finite subgroups in Kleinian groups*, Amer. J. of Math., 113 (1991), pp. 179–188.

[76] A. Felikson and P. Tumarkin, *A series of word-hyperbolic Coxeter groups.* Preprint, 2005.

[77] D. Garity, D. Repovš, and M. Željko, *Rigid Cantor sets in* $\mathbf{R}^3$ *with simply connected complement*, Proc. Amer. Math. Soc., 134 (2006), pp. 2447–2456.

[78] F.W. Gehring and G.J. Martin, *Discrete convergence groups*, in "Complex analysis, I" (College Park, Md., 1985–86), Springer, Berlin, 1987, pp. 158–167.

[79] W. Goldman, *Geometric structures on manifolds and varieties of representations*, in "Geometry of group representations", vol. 74 of Contemporary Mathematics, 1987, pp. 169–198.

[80] ——, *Complex hyperbolic Kleinian groups*, in "Complex Geometry". Proceedings of the Osaka International Conference, vol. 143 of Lecture notes in pure and applied mathematics, 1992, pp. 31–52.

[81] ——, *Complex hyperbolic geometry*, Oxford Mathematical Monographs, Oxford University Press, 1999.

[82] W. Goldman and J. J. Millson, *The deformation theory of representations of fundamental groups of compact Kähler manifolds*, Math. Publications of IHES, 67 (1988), pp. 43–96.

[83] L. Greenberg, *On a theorem of Ahlfors and conjugate subgroups of Kleinian groups*, Amer. J. of Math., 89 (1967), pp. 56–68.

[84] ——, *Homomorphisms of triangle groups into* PSL(2, **C**), in Riemann surfaces and related topics: Proceedings of the 1978 Stony Brook Conference, vol. 97 of Ann. of Math. Stud., Princeton Univ. Press, 1981, pp. 167–181.

[85] M. Gromov, *Asymptotic invariants of infinite groups*, in "Geometric groups theory", volume 2, Proc. of the Symp. in Sussex 1991, G. Niblo and M. Roller, eds., vol. 182 of Lecture Notes series, Cambridge University Press, 1993.

[86] ——, *Metric structures for Riemannian and non-Riemannian spaces*, Birkhäuser Boston, 1999.

[87] M. Gromov, H. Lawson, and W. Thurston, *Hyperbolic* 4*-manifolds and conformally flat* 3*-manifolds*, Publ. Math. of IHES, 68 (1988), pp. 27–45.

[88] M. Gromov and I. Piatetski-Shapiro, *Non-arithmetic groups in Lobachevsky spaces*, Publ. Math. of IHES, 66 (1988), pp. 93–103.

[89] M. Gromov and W. Thurston, *Pinching constants for hyperbolic manifolds*, Inventiones Mathematicae, 89 (1987), pp. 1–12.

[90] N. Gusevskiĭ, *Geometric decomposition of spatial Kleinian groups*, Soviet Math. Dokl., 38 (1989), pp. 98–101.

[91] ———, *Strange actions of free Kleinian groups*, Sibirsk. Mat. Zh., 37 (1996), pp. 90–107.

[92] N. Gusevskiĭ and N. Zindinova, *Characterization of extended Schottky groups*, Soviet Math. Doklady, 33 (1986), pp. 239–241.

[93] E. Hamilton, *Geometric finiteness for hyperbolic orbifolds*, Topology, 37 (1998), pp. 635–657.

[94] J. Heinonen, *Lectures on analysis on metric spaces*, Universitext, Springer-Verlag, New York, 2001.

[95] J. Hempel, *3-manifolds*, vol. 86 of Annals of Mathematics Studies, Princeton University Press, 1976.

[96] W. Hurewicz and H. Wallman, *Dimension Theory*, Princeton University Press, 1941.

[97] S. Hwang, *Moebius structures on 3-manifolds*. Ph. D. Thesis, University of Utah, 2001.

[98] D. Ivanšić, J.G. Ratcliffe, and S.T. Tschantz, *Complements of tori and Klein bottles in the 4-sphere that have hyperbolic structure*, Algebr. Geom. Topol., 5 (2005), pp. 999–1026.

[99] T. Iwaniec and G. Martin, *Geometric function theory and non-linear analysis*, Oxford Mathematical Monographs, Oxford University Press, New York, 2001.

[100] H. Izeki, *Limit sets of Kleinian groups and conformally flat Riemannian manifolds*, Invent. Math., 122 (1995), pp. 603–625.

[101] ———, *Quasiconformal stability of Kleinian groups and an embedding of a space of flat conformal structures*, Conform. Geom. Dyn., 4 (2000), pp. 108–119.

[102] ———, *Convex-cocompactness of Kleinian groups and conformally flat manifolds with positive scalar curvature*, Proc. Amer. Math. Soc., 130 (2002), pp. 3731–3740.

[103] D. Johnson and J.J. Millson, *Deformation spaces associated to compact hyperbolic manifolds*, in "Discrete Groups in Geometry and Analysis", Papers in honor of G.D. Mostow on his 60th birthsday, vol. 67 of Progress in Mathematics, Birkhäuser, 1987, pp. 48–106.

[104] T. Jørgensen, *Compact 3-manifolds of constant negative curvature fibering over the circle*, Ann. of Math., 106 (1977), pp. 61–72.

[105] T. Jørgensen and A. Marden, *Two doubly degenerate groups*, Quart. J. Math. Oxford Ser. (2), 30 (1979), pp. 143–156.

[106] ———, *Generic fundamental polyhedra for Kleinian groups*, in "Holomorphic functions and moduli", Vol. II (Berkeley, 1986), vol. 11 of Math. Sci. Res. Inst. Publ., Springer, New York, 1988, pp. 69–85.

[107] M. Kapovich, *Flat conformal structures on three-dimensional manifolds: the existence problem. I*, Sibirsk. Mat. Zh., 30 (1989), pp. 60–73.

[108] ——, *Deformation spaces of flat conformal structures*, in "Proceedings of the Second Soviet-Japan Symposium of Topology" (Khabarovsk, 1989). Questions and Answers in General Topology, 8 (1990), pp. 253–264.

[109] ——, *Intersection pairing on hyperbolic 4-manifolds*. Preprint, MSRI, 1992.

[110] ——, *Topological aspects of Kleinian groups in several dimensions*. Preprint, MSRI, 1992.

[111] ——, *Flat conformal structures on 3-manifolds. I. Uniformization of closed Seifert manifolds*, J. Differential Geom., 38 (1993), pp. 191–215.

[112] ——, *Hyperbolic 4-manifolds fibering over surfaces*. Preprint, 1993.

[113] ——, *Deformations of representations of discrete subgroups of* $SO(3,1)$, Math. Ann., 299 (1994), pp. 341–354.

[114] ——, *On the absence of Sullivan's cusp finiteness theorem in higher dimensions*, in "Algebra and analysis" (Irkutsk, 1989), Amer. Math. Soc., Providence, RI, 1995, pp. 77–89.

[115] ——, *On normal subgroups in the fundamental groups of complex surfaces*. Preprint, 1998.

[116] ——, *Hyperbolic manifolds and discrete groups*, Birkhäuser Boston Inc., Boston, MA, 2001.

[117] ——, *Conformally flat metrics on 4-manifolds*, J. Differential Geom., 66 (2004), pp. 289–301.

[118] ——, *Representations of polygons of finite groups*, Geom. Topol., 9 (2005), pp. 1915–1951.

[119] ——, *Convex projective Gromov-Thurston examples*. Preprint, 2006.

[120] ——, *Lectures on the geometric group theory*. Preprint, 2006.

[121] ——, *Homological dimension and critical exponent of Kleinian groups*. Preprint, 2007.

[122] M. Kapovich and B. Kleiner, *Hyperbolic groups with low-dimensional boundary*, Ann. Sci. École Norm. Sup. (4), 33 (2000), pp. 647–669.

[123] ——, *Geometry of quasi-planes*. Preprint, 2004.

[124] ——, *Coarse Alexander duality and duality groups*, Journal of Diff. Geometry, 69 (2005), pp. 279–352.

[125] M. Kapovich and J. J. Millson, *Bending deformations of representations of fundamental groups of complexes of groups*. Preprint, 1996.

[126] ——, *On the deformation theory of representations of fundamental groups of closed hyperbolic 3-manifolds*, Topology, 35 (1996), pp. 1085–1106.

[127] M. Kapovich and L. Potyagailo, *On absence of Ahlfors' and Sullivan's finiteness theorems for Kleinian groups in higher dimensions*, Siberian Math. Journ., 32 (1991), pp. 227–237.

[128] ——, *On absence of Ahlfors' finiteness theorem for Kleinian groups in dimension 3*, Topology and its Applications, 40 (1991), pp. 83–91.

[129] M. Kapovich, L. Potyagailo, and E.B. Vinberg, *Non-coherence of some non-uniform lattices in* $\mathrm{Isom}(\mathbb{H}^n)$, Geometry and Topology, (to appear).

[130] D. Kazhdan and G. Margulis, *A proof of Selberg's hypothesis*, Math. Sbornik, 75 (1968), pp. 162–168.

[131] R. Kirby, *Problems in low-dimensional topology*, in "Geometric topology" (Athens, GA, 1993), Providence, RI, 1997, Amer. Math. Soc., pp. 35–473.

[132] R. Kirby and L. Siebenmann, *Foundational essays on topological manifolds, smoothings and triangulations*, vol. 88 of Annals of Mathematics Studies, Princeton University Press, 1977.

[133] B. Kleiner. In preparation.

[134] B. Kleiner and J. Lott, *Notes of Perelman's papers.* Preprint, math.DG/0605667, 2006.

[135] C. Kourouniotis, *Deformations of hyperbolic structures*, Math. Proc. Cambridge Philos. Soc., 98 (1985), pp. 247–261.

[136] I. Kra, *Automorphic Forms and Kleinian Groups*, Benjamin Reading, Massachusetts, 1972.

[137] M. Kreck and W. Lück, *The Novikov conjecture*, vol. 33 of Oberwolfach Seminars, Birkhäuser Verlag, Basel, 2005. Geometry and algebra.

[138] S.L. Krushkal', B.N. Apanasov, and N.A. Gusevskiĭ, *Kleinian groups and uniformization in examples and problems*, vol. 62 of Translations of Mathematical Monographs, American Mathematical Society, 1986.

[139] N. Kuiper, *Hyperbolic 4-manifolds and tesselations*, Math. Publ. of IHES, 68 (1988), pp. 47–76.

[140] R. Kulkarni, *Groups with domains of discontinuity. Some topological aspects of Kleinian groups*, Amer. J. of Math., 100 (1978), pp. 897–911.

[141] ——, *Infinite regular coverings*, Duke Math. Journal, 45 (1978), pp. 781–796.

[142] M. Lackenby, *Heegaard splittings, the virtually Haken conjecture and property* (τ), Invent. Math., 164 (2006), pp. 317–359.

[143] Y. Lai. In preparation.

[144] J.-S. Li and J.J. Millson, *On the first Betti number of a hyperbolic manifold with an arithmetic fundamental group*, Duke Math. J., 71 (1993), pp. 365–401.

[145] R. Livne, *On certain covers of the universal elliptic curve.* Ph. D. Thesis, Harvard University, 1981.

[146] A. Lubotzky, *Eigenvalues of the Laplacian, the first Betti number and the congruence subgroup problem*, Ann. of Math., 144 (1996), pp. 441–452.

[147] J. Luukkainen, *Lipschitz and quasiconformal approximation of homeomorphism pairs*, Topology Appl., 109 (2001), pp. 1–40.

[148] C. Maclachlan and A. Reid, *The arithmetic of hyperbolic 3-manifolds*, vol. 219 of Graduate Texts in Mathematics, Springer-Verlag, New York, 2003.

[149] V.S. Makarov, *On a certain class of discrete groups of Lobačevskiĭspace having an infinite fundamental region of finite measure*, Dokl. Akad. Nauk SSSR, 167 (1966), pp. 30–33.

[150] ——, *The Fedorov groups of four-dimensional and five-dimensional Lobačevskiĭ space*, in Studies in General Algebra, No. 1 (Russian), Kišinev. Gos. Univ., Kishinev, 1968, pp. 120–129.

[151] A. Malćev, *On isomorphic matrix representations of infinite groups*, Mat. Sb., 8 (1940), pp. 405–422.

[152] A. Marden, *Geometry of finitely generated Kleinian groups*, Ann. of Math., 99 (1974), pp. 383–496.

[153] ——, *A proof of the Ahlfors finiteness theorem*, in "Spaces of Kleinian groups", vol. 329 of London Math. Soc. Lecture Note Ser., Cambridge Univ. Press, 2006, pp. 247–257.

[154] G. Martin, *On discrete Möbius groups in all dimensions: A generalization of Jorgensen's inequality*, Acta Math., 163 (1989), pp. 253–289.

[155] B. Maskit, *A characterization of Schottky groups*, J. d' Anal. Math., 19 (1967), pp. 227–230.

[156] ——, *On boundaries of Teichmüller spaces and on Kleinian groups, II*, Annals of Math., 91 (1970), pp. 607-663.

[157] ——, *Kleinian groups*, vol. 287 of Grundlehren der Math. Wissenschaften, Springer, 1987.

[158] S. Matsumoto, *Topological aspects of conformally flat manifolds*, Proc. Jap. Acad. of Sci., 65 (1989), pp. 231–234.

[159] ——, *Foundations of flat conformal structure*, in "Aspects of low-dimensional manifolds", Kinokuniya, Tokyo, 1992, pp. 167–261.

[160] K. Matsuzaki, *Ergodic properties of discrete groups; inheritance to normal subgroups and invariance under quasiconformal deformations*, J. Math. Kyoto Univ., 33 (1993), pp. 205–226.

[161] J. Maubon, *Variations d'entropies et déformations de structures conformes plates sur les variétés hyperboliques compactes*, Ergodic Theory Dynam. Systems, 20 (2000), pp. 1735–1748.

[162] J.P. McCammond and D.T. Wise, *Coherence, local quasiconvexity, and the perimeter of 2-complexes*, Geom. Funct. Anal., 15 (2005), pp. 859–927.

[163] C. McMullen, *Local connectivity, Kleinian groups and geodesics on the blowup of the torus*, Invent. Math., 146 (2001), pp. 35–91.

[164] J.J. Millson, *On the first Betti number of a constant negatively curved manifold*, Ann. of Math., 104 (1976), pp. 235–247.

[165] Y. Minsky, *On rigidity, limit sets and invariants of hyperbolic 3-manifolds*, Journal of the AMS, 7 (1994), pp. 539–588.

[166] ——, *The classification of Kleinian surface groups, I: Models and bounds.* Preprint, 2003.

[167] ——, *Combinatorial and geometrical aspects of hyperbolic 3-manifolds*, in "Kleinian groups and hyperbolic 3-manifolds" (Warwick, 2001), vol. 299 of London Math. Soc. Lecture Note Ser., Cambridge Univ. Press, Cambridge, 2003, pp. 3–40.

[168] M. Mj, *Cannon-Thurston maps, I: Bounded geometry and a theorem of McMullen.* Preprint, 2005.

[169] ——, *Cannon-Thurston maps for surface groups, II: Split geometry and the Minsky model.* Preprint, 2006.

[170] J. Morgan, *On Thurston's uniformization theorem for three-dimensional manifolds*, in "The Smith conjecture", J. Morgan and H. Bass, eds., Academic Press, 1984, pp. 37–125.

[171] ——, *Group actions on trees and the compactification of the space of classes of $SO(n,1)$ representations*, Topology, 25 (1986), pp. 1–33.

[172] J. Morgan and P. Shalen, *Degenerations of hyperbolic structures, III: Actions of 3-manifold groups on trees and Thurston's compactness theorem*, Ann. of Math., 127 (1988), pp. 457–519.

[173] J. Morgan and G. Tian, *Ricci flow and the Poincaré conjecture*. Preprint, 2006.

[174] L. Mosher, *Hyperbolic-by-hyperbolic hyperbolic groups*, Proceedings of AMS, 125 (1997), pp. 3447–3455.

[175] G.D. Mostow, *Strong rigidity of locally symmetric spaces*, vol. 78 of Annals of mathematical studies, Princeton University Press, Princeton, 1973.

[176] S. Nayatani, *Patterson-Sullivan measure and conformally flat metrics*, Math. Z., 225 (1997), pp. 115–131.

[177] P.J. Nicholls, *The ergodic theory of discrete groups*, Cambridge University Press, 1989.

[178] K.-I. Ohshika and L. Potyagailo, *Self-embeddings of Kleinian groups*, Annales de l'École Normale Supérieure, 31 (1998), pp. 329–343.

[179] H. Ohtake, *On Ahlfors' weak finiteness theorem*, J. Math. Kyoto Univ., 24 (1984), pp. 725–740.

[180] J.-P. Otal, *Le théorème d'hyperbolisation pour les variétès fibrées de dimension* 3, Astérisque 235, Société mathématique de France, 1996.

[181] ——, *Thurston's hyperbolization of Haken manifolds*, in "Surveys in differential geometry", Vol. III (Cambridge, MA, 1996), Int. Press, Boston, MA, 1998, pp. 77–194.

[182] F. Paulin, *Topologie de Gromov equivariant, structures hyperboliques et arbres reels*, Inventiones Mathematicae, 94 (1988), pp. 53–80.

[183] G. Perelman, *Entropy formula for the Ricci flow and its geometric applications*. Preprint, math.DG/0211159, 2002.

[184] ——, *Ricci flow with surgery on 3-manifolds*. Preprint, math.DG/0303109, 2003.

[185] R.S. Phillips and P. Sarnak, *The Laplacian for domains in hyperbolic space and limit sets of Kleinian groups*, Acta Math., 155 (1985), pp. 173–241.

[186] V. Poénaru, *Le théorème de s-cobordisme*, in Séminaire Bourbaki, 23ème année (1970/1971), Exp. No. 392, Springer, Berlin, 1971, pp. 197–219. Lecture Notes in Math., Vol. 244.

[187] L. Potyagailo, *The problem of finiteness for Kleinian groups in* 3*-space*, in "Knots 90" (Osaka, 1990), de Gruyter, Berlin, 1992, pp. 619–623.

[188] ——, *Finitely generated Kleinian groups in* 3*-space and* 3*-manifolds of infinite homotopy type*, Transactions of the AMS, 344 (1994), pp. 57–77.

[189] M.S. Raghunathan and T.N. Venkataramana, *The first Betti number of arithmetic groups and the congruence subgroup problem*, in "Linear algebraic groups and their representations" (Los Angeles, 1992), vol. 153 of Contemp. Math., Amer. Math. Soc., 1993, pp. 95–107.

[190] J. Ratcliffe, *Foundations of hyperbolic manifolds*, Springer, 1994.

[191] J. Ratcliffe and S. Tschantz, *Some examples of aspherical* 4*-manifolds that are homology* 4*-spheres*, Topology, 44 (2005), pp. 341–350.

[192] F. Raymond, *Classification of the actions of the circle on 3-manifolds*, Trans. Amer. Math. Soc., 131 (1968), pp. 51–78.

[193] M. Rees, *The Ending Laminations Theorem direct from Teichmuller geodesics.* Preprint, math.GT/0404007, 2004.

[194] A. Reznikov, *Analytic topology of groups, actions, strings and varietes.* Preprint, 1999.

[195] E. Rips, *Subgroups of small cancellation groups*, Bull. London Math. Soc., 14 (1982), pp. 45–47.

[196] C. Rourke and B. Sanderson, *Introduction to piecewise-linear topology*, Springer study edition, Springer, 1982.

[197] K. Scannell, *Local rigidity of hyperbolic 3-manifolds after Dehn surgery*, Duke Math. J., 114 (2002), pp. 1–14.

[198] R. Schoen and S.-T. Yau, *Conformally flat manifolds, Kleinian groups and scalar curvature*, Invent. Math., 92 (1988), pp. 47–71.

[199] R. Schwartz, *Complex hyperbolic triangle groups*, in Proceedings of the International Congress of Mathematicians, Vol. II (Beijing, 2002), Beijing, 2002, Higher Ed. Press, pp. 339–349.

[200] J. Schwermer, *Special cycles and automorphic forms on arithmetically defined hyperbolic 3-manifolds*, Asian J. Math., 8 (2004), pp. 837–859.

[201] P. Scott, *Compact submanifolds of 3-manifolds*, Journ. of the LMS, 6 (1973), pp. 437–448.

[202] ——, *Compact submanifolds of 3-manifolds*, Journ. of the LMS, 7 (1973), pp. 246–250.

[203] A. Selberg, *On discontinuous groups in higher-dimensional symmetric spaces*, in "Contributions to function theory", Bombay, Tata Institute, 1960, pp. 147–164.

[204] Y. Shalom, *Rigidity, unitary representations of semisimple groups, and fundamental groups of manifolds with rank one transformation group*, Ann. of Math. (2), 152 (2000), pp. 113–182.

[205] J. Stallings, *On torsion-free groups with infinitely many ends*, Ann. of Math., 88 (1968), pp. 312–334.

[206] B. Stratmann, *A note on geometric upper bounds for the exponent of convergence of convex cocompact Kleinian groups*, Indag. Math. (N.S.), 13 (2002), pp. 269–280.

[207] D. Sullivan, *Hyperbolic geometry and homeomorphisms*, in "Proceedings of Georgia Conference on Geometric Topology", Academic Press, 1977, pp. 543–555.

[208] ——, *On ergodic theorey at infinity of arbitrary discrete groups of hyperbolic motions*, Ann. Math. Stud., 97 (1981), pp. 465–496.

[209] ——, *On finiteness theorem for cusps*, Acta Math., 147 (1981), pp. 289–299.

[210] ——, *Quasiconformal homeomorphisms and dynamics. Structural stablity implies hyperbolicity*, Acta Math., 155 (1985), pp. 243–260.

[211] ——, *Related aspects of positivity in Riemannian geometry*, Journal of Differential Geometry, 25 (1987), pp. 327–351.

[212] S. Tan, *Deformations of flat conformal structures on a hyperbolic 3-manifold*, J. Differential Geom., 37 (1993), pp. 161–176.

[213] W. Thurston, *Geometry and topology of 3-manifolds*. Princeton University Lecture Notes, 1978–1981.

[214] ——, *Hyperbolic structures on 3-manifolds, I*, Ann. of Math., 124 (1986), pp. 203–246.

[215] ——, *Three-Dimensional Geometry and Topology, I*, vol. 35 of Princeton Mathematical Series, Princeton University Press, 1997.

[216] P. Tukia, *The Hausdorff dimension of the limit set of a geometrically finite Kleinian group*, Acta Math., 152 (1984), pp. 127–140.

[217] ——, *On isomorphisms of geometrically finite Moebius groups*, Mathematical Publications of IHES, 61 (1985), pp. 171–214.

[218] ——, *Homeomorphic conjugates of Fuchsian groups*, J. Reine Angew. Math., 391 (1988), pp. 1–54.

[219] ——, *A survey of Möbius groups*, in Proceedings of the International Congress of Mathematicians, Vol. 1, 2 (Zürich, 1994), Basel, 1995, Birkhäuser, pp. 907–916.

[220] P. Tukia and J. Väisälä, *Lipschitz and quasiconformal approximation and extension*, Ann. Acad. Sci. Fenn. Ser. A I Math., 6 (1981), pp. 303–342.

[221] È.B. Vinberg, *Discrete groups generated by reflections in Lobačevskiĭ spaces*, Mat. Sb. (N.S.), 72 (114) (1967), pp. 471–488; correction, ibid. 73 (115) (1967), 303.

[222] È.B. Vinberg and O.V. Shvartsman, *Discrete groups of motions of spaces of constant curvature*, in Geometry, II, vol. 29 of Encyclopaedia Math. Sci., Springer, 1993, pp. 139–248.

[223] S. Wang and Q. Zhou, *On the proper conjugation of Kleinian groups*, Geom. Dedicata, 56 (1995), pp. 145–154.

[224] A. Weil, *Discrete subgroups of Lie groups, I*, Ann. of Math., 72 (1960), pp. 69–384.

[225] ——, *Discrete subgroups of Lie groups, II*, Ann. of Math., 75 (1962), pp. 578–602.

[226] ——, *Remarks on the cohomology of groups*, Ann. of Math., 80 (1964), pp. 149–157.

[227] N. Wielenberg, *Discrete Moebius groups: fundamental polyhedra and convergence*, Amer. Journ. Math., 99 (1977), pp. 861–878.

[228] C. Yue, *Dimension and rigidity of quasifuchsian representations*, Ann. of Math. (2), 143 (1996), pp. 331–355.

Michael Kapovich
Department of Mathematics,
University of California,
Davis, CA 95616
USA
e-mail: kapovich@math.ucdavis.edu

Progress in Mathematics, Vol. 265, 565–645

A_∞-bimodules and Serre A_∞-functors

Volodymyr Lyubashenko and Oleksandr Manzyuk

To the memory of Oleksandr Reznikov

Abstract. We define A_∞-bimodules similarly to Tradler and show that this notion is equivalent to an A_∞-functor with two arguments which takes values in the differential graded category of complexes of $\Bbbk$-modules, where $\Bbbk$ is a ground commutative ring. Serre A_∞-functors are defined via A_∞-bimodules likewise Kontsevich and Soibelman. We prove that a unital closed under shifts A_∞-category $\mathcal{A}$ over a field $\Bbbk$ admits a Serre A_∞-functor if and only if its homotopy category $H^0\mathcal{A}$ admits a Serre $\Bbbk$-linear functor. The proof uses categories enriched in $\mathcal{K}$, the homotopy category of complexes of $\Bbbk$-modules, and Serre $\mathcal{K}$-functors. Also we use a new A_∞-version of the Yoneda Lemma generalizing the previously obtained result.

Mathematics Subject Classification (2000). Primary 18D20, 18G55, 55U15; Secondary 18D05.

Keywords. A_∞-categories, A_∞-modules, A_∞-bimodules, Serre A_∞-functors, Yoneda Lemma.

Serre–Grothendieck duality for coherent sheaves on a smooth projective variety was reformulated by Bondal and Kapranov in terms of Serre functors [BK89]. Being an abstract category theory notion Serre functors were discovered in other contexts as well, for instance, in Kapranov's studies of constructible sheaves on stratified spaces [Kap90]. Reiten and van den Bergh showed that Serre functors in categories of modules are related to Auslander–Reiten sequences and triangles [RvdB02].

Often Serre functors are considered in triangulated categories and it is reasonable to lift them to their origin – pretriangulated **dg**-categories or A_∞-categories. Soibelman defines Serre A_∞-functors in [Soi04], based on Kontsevich and Soibelman work which is a sequel to [KS06]. In the present article we consider Serre A_∞-functors in detail. We define them via A_∞-bimodules in Section 6 and use enriched categories to draw conclusions about existence of Serre A_∞-functors.

A_∞-modules over A_∞-algebras are introduced by Keller [Kel01]. A_∞-bimodules over A_∞-algebras are defined by Tradler [Tra01, Tra02]. A_∞-modules and

A_∞-bimodules over A_∞-categories over a field were first considered by Lefèvre-Hasegawa [LH03] under the name of polydules and bipolydules. A_∞-modules over A_∞-categories were developed further by Keller [Kel06]. We study A_∞-bimodules over A_∞-categories over a ground commutative ring $\Bbbk$ in Section 5 and show that this notion is equivalent to an A_∞-functor with two arguments which takes values in the **dg**-category $\underline{\mathsf{C}}_\Bbbk$ of complexes of $\Bbbk$-modules. A similar notion of A_∞-modules over an A_∞-category $\mathcal{C}$ from Section 4 is equivalent to an A_∞-functor $\mathcal{C} \to \underline{\mathsf{C}}_\Bbbk$. The latter point of view taken by Seidel [Sei06] proved useful for ordinary and differential graded categories as well, see Drinfeld's article [Dri04, Appendix III].

Any unital A_∞-category $\mathcal{A}$ determines a $\mathcal{K}$-category $\mathsf{k}\mathcal{A}$ [Lyu03, BLM06], where $\mathcal{K}$ is the homotopy category of complexes of $\Bbbk$-modules. Respectively, an A_∞-functor f determines a $\mathcal{K}$-functor $\mathsf{k}f$. In particular, a Serre A_∞-functor $S : \mathcal{A} \to \mathcal{A}$ determines a Serre $\mathcal{K}$-functor $\mathsf{k}S : \mathsf{k}\mathcal{A} \to \mathsf{k}\mathcal{A}$. We prove also the converse: if $\mathsf{k}\mathcal{A}$ admits a Serre $\mathcal{K}$-functor, then $\mathcal{A}$ admits a Serre A_∞-functor (Corollary 6.3). This shows the importance of enriched categories in the subject.

Besides enrichment in $\mathcal{K}$ we consider in Section 2 also categories enriched in the category **gr** of graded $\Bbbk$-modules. When $\Bbbk$ is a field, we prove that a Serre $\mathcal{K}$-functor exists in $\mathsf{k}\mathcal{A}$ if and only if the cohomology **gr**-category $H^\bullet\mathcal{A} \overset{\text{def}}{=} H^\bullet(\mathsf{k}\mathcal{A})$ admits a Serre **gr**-functor (Corollary 2.16). If the **gr**-category $H^\bullet\mathcal{A}$ is closed under shifts, then it admits a Serre **gr**-functor if and only if the $\Bbbk$-linear category $H^0\mathcal{A}$ admits a Serre $\Bbbk$-linear functor (Corollary 2.18, Proposition 2.21). Summing up, a unital closed under shifts A_∞-category $\mathcal{A}$ over a field $\Bbbk$ admits a Serre A_∞-functor if and only if its homotopy category $H^0\mathcal{A}$ admits a Serre $\Bbbk$-linear functor (Theorem 6.5). This applies, in particular, to a pretriangulated A_∞-enhancement $\mathcal{A}$ of a triangulated category $H^0\mathcal{A}$ over a field $\Bbbk$.

In the proofs we use a new A_∞-version of the Yoneda Lemma (Theorem A.1). It generalizes the previous result that the Yoneda A_∞-functor is homotopy full and faithful [Fuk02, Theorem 9.1], [LM04, Theorem A.11], as well as a result of Seidel [Sei06, Lemma 2.12] which was proven over a ground field $\Bbbk$. The proof of the Yoneda Lemma occupies Appendix A. It is based on the theory of A_∞-bimodules developed in Section 5.

Acknowledgment. The second author would like to thank the Mathematisches Forschungsinstitut Oberwolfach and its director Prof. Dr. Gert-Martin Greuel personally for financial support. Commutative diagrams were typeset with the package `diagrams.sty` by Paul Taylor.

0.1. Notation and conventions

Notation follows closely the usage of the book [BLM06]. In particular, $\mathscr{U}$ is a ground universe containing an element which is an infinite set, and $\Bbbk$ denotes a $\mathscr{U}$-small commutative associative ring with unity. A *graded quiver* $\mathcal{C}$ typically means a $\mathscr{U}$-small set of objects $\operatorname{Ob}\mathcal{C}$ together with $\mathscr{U}$-small $\mathbb{Z}$-graded $\Bbbk$-modules of morphisms $\mathcal{C}(X,Y)$, given for each pair $X, Y \in \operatorname{Ob}\mathcal{C}$. For any graded $\Bbbk$-module M there is another graded $\Bbbk$-module $sM = M[1]$, its *suspension*, with the shifted grading $(sM)^k = M[1]^k = M^{k+1}$. The mapping $s : M \to sM$ given by the identity maps $M^k \Longrightarrow M[1]^{k-1}$ has degree -1. The composition of maps, morphisms, functors, etc. is denoted $fg = f \cdot g = \xrightarrow{f} \xrightarrow{g} = g \circ f$. A function (or a functor) $f : X \to Y$ applied to an element is denoted $f(x) = xf = x.f = x{\scriptstyle\bullet}f$ and occasionally fx.

Objects of the (large) Abelian $\mathscr{U}$-category $\mathsf{C}_\Bbbk$ of complexes of $\Bbbk$-modules are $\mathscr{U}$-small differential graded $\Bbbk$-modules. Morphisms of $\mathsf{C}_\Bbbk$ are chain maps. The category $\mathsf{C}_\Bbbk$ is symmetric closed monoidal: the inner hom-complexes $\underline{\mathsf{C}}_\Bbbk(X,Y)$ are $\mathscr{U}$-small, therefore, objects of $\mathsf{C}_\Bbbk$. This determines a (large) differential graded $\mathscr{U}$-category $\underline{\mathsf{C}}_\Bbbk$. In particular, $\underline{\mathsf{C}}_\Bbbk$ is a non-small graded $\mathscr{U}$-quiver.

Speaking about a symmetric monoidal category $(\mathcal{C}, \otimes, \mathbb{1}, c)$ we actually mean the equivalent notion of a symmetric Monoidal category $(\mathcal{C}, \otimes^I, \lambda^f)$ [Lyu99, Definitions 1.2.2, 1.2.14], [BLM06, Chapter 3]. It is equipped with tensor product functors $\otimes^I : (X_i)_{i\in I} \mapsto \otimes^{i\in I} X_i$, where I are finite linearly ordered (index) sets. The isomorphisms $\lambda^f : \otimes^{i\in I} X_i \to \otimes^{j\in J} \otimes^{i\in f^{-1}j} X_i$ given for any map $f : I \to J$ can be thought of as constructed from the associativity isomorphisms a and commutativity isomorphisms c. When f is non-decreasing, the isomorphisms λ^f can be ignored similarly to associativity isomorphisms. The coherence principle of [BLM06, Section 3.25] allows to write down canonical isomorphisms ω_c (products of λ^f's and their inverses) between iterated tensor products, indicating only the permutation of arguments ω. One of them, $\sigma_{(12)} : \otimes^{i\in I} \otimes^{j\in J} X_{ij} \to \otimes^{j\in J} \otimes^{i\in I} X_{ij}$ is defined explicitly in [BLM06, (3.28.1)]. Sometimes, when the permutation of arguments reads clearly, we write simply perm for the corresponding canonical isomorphism.

A symmetric multicategory $\widehat{\mathcal{C}}$ is associated with a lax symmetric Monoidal category $(\mathcal{C}, \otimes^I, \lambda^f)$, where natural transformations λ^f are not necessarily invertible, see [BLM06, Section 4.20].

The category of graded $\Bbbk$-linear quivers has a natural symmetric Monoidal structure. For given quivers $\mathcal{Q}_i$ the tensor product quiver $\boxtimes^{i\in I}\mathcal{Q}_i$ has the set of objects $\prod_{i\in I} \operatorname{Ob}\mathcal{Q}_i$ and the graded $\Bbbk$-modules of morphisms

$$(\boxtimes^{i\in I}\mathcal{Q}_i)((X_i)_{i\in I}, (Y_i)_{i\in I}) = \otimes^{i\in I}\mathcal{Q}_i(X_i, Y_i).$$

For any graded quiver $\mathcal{C}$ and a sequence of objects $(X_0, \dots, X_n)$ of $\mathcal{C}$ we use in this article the notation

$$\bar{T}^n s\mathcal{C}(X_0, \dots, X_n) = s\mathcal{C}(X_0, X_1) \otimes \cdots \otimes s\mathcal{C}(X_{n-1}, X_n),$$
$$T^n s\mathcal{C}(X_0, X_n) = \oplus_{X_1,\dots,X_{n-1}\in \operatorname{Ob}\mathcal{C}} \bar{T}^n s\mathcal{C}(X_0, \dots, X_n).$$

When the list of arguments is obvious we abbreviate the notation $\bar{T}^n s\mathcal{C}(X_0, \dots, X_n)$ to $\bar{T}^n s\mathcal{C}(X_0, X_n)$. For $n = 0$ we set $T^0 s\mathcal{C}(X, Y) = \Bbbk$ if $X = Y$, and 0 otherwise. The tensor quiver is $Ts\mathcal{C} = \oplus_{n\geqslant 0} T^n s\mathcal{C} = \oplus_{n\geqslant 0} (s\mathcal{C})^{\otimes n}$.

An *A_∞-category* means a graded quiver $\mathcal{C}$ with n-ary compositions $b_n : T^n s\mathcal{C}(X_0, X_n) \to s\mathcal{C}(X_0, X_n)$ of degree 1 given for all $n \geqslant 1$ (we assume for simplicity that $b_0 = 0$) such that $b^2 = 0$ for the $\Bbbk$-linear map $b : Ts\mathcal{C} \to Ts\mathcal{C}$ of degree 1

$$b = \sum_{\substack{r+n+t=k \\ r+1+t=l}} 1^{\otimes r} \otimes b_n \otimes 1^{\otimes t} : T^k s\mathcal{C} \to T^l s\mathcal{C}. \tag{0.1}$$

The composition $b \cdot \mathrm{pr}_1 = (0, b_1, b_2, \dots) : Ts\mathcal{C} \to s\mathcal{C}$ is denoted $\check{b}$.

The tensor quiver $T\mathcal{C}$ becomes a counital coalgebra when equipped with the cut comultiplication $\Delta_0 : T\mathcal{C}(X,Y) \to \oplus_{Z \in \mathrm{Ob}\,\mathcal{C}} T\mathcal{C}(X,Z) \bigotimes_{\Bbbk} T\mathcal{C}(Z,Y)$, $h_1 \otimes h_2 \otimes \dots \otimes h_n \mapsto \sum_{k=0}^n h_1 \otimes \dots \otimes h_k \bigotimes h_{k+1} \otimes \dots \otimes h_n$. The map b given by (0.1) is a coderivation with respect to this comultiplication. Thus b is a *codifferential.*

We denote by $\mathbf{n}$ the linearly ordered index set $\{1 < 2 < \dots < n\}$.

An A_∞*-functor* $f : \mathcal{A} \to \mathcal{B}$ is a map of objects $f = \mathrm{Ob}\, f : \mathrm{Ob}\,\mathcal{A} \to \mathrm{Ob}\,\mathcal{B}$, $X \mapsto Xf$ and $\Bbbk$-linear maps $f : Ts\mathcal{A}(X,Y) \to Ts\mathcal{B}(Xf, Yf)$ of degree 0 which agree with the cut comultiplication and commute with the codifferentials b. Such f is determined in a unique way by its components $f_k = f\,\mathrm{pr}_1 : T^k s\mathcal{A}(X,Y) \to s\mathcal{B}(Xf, Yf)$, $k \geqslant 1$ (we require that $f_0 = 0$). This generalizes to the case of A_∞-functors with many arguments $f : (\mathcal{A}_i)_{i \in \mathbf{n}} \to \mathcal{B}$. Such A_∞*-functor* is a quiver map $f : \boxtimes^{i \in \mathbf{n}} Ts\mathcal{A}_i \to Ts\mathcal{B}$ of degree 0 which agrees with the cut comultiplication and commutes with the differentials. Denote $\check{f} = f \cdot \mathrm{pr}_1 : \boxtimes^{i \in \mathbf{n}} Ts\mathcal{A}_i \to s\mathcal{B}$. The restrictions $f_{(k_i)_{i \in \mathbf{n}}}$ of $\check{f}$ to $\boxtimes^{i \in \mathbf{n}} T^{k_i} s\mathcal{A}_i$ are called the components of f. It is required that the restriction $f_{00\dots0}$ of $\check{f}$ to $\boxtimes^{i \in \mathbf{n}} T^0 s\mathcal{A}_i$ vanishes. The components determine coalgebra homomorphism f in a unique way. Commutation with the differentials means that the following compositions are equal

$$\bigl(\boxtimes^{i \in \mathbf{n}} Ts\mathcal{A}_i \xrightarrow{f} Ts\mathcal{B} \xrightarrow{\check{b}} s\mathcal{B}\bigr) = \bigl(\boxtimes^{i \in \mathbf{n}} Ts\mathcal{A}_i \xrightarrow{\sum_{i=1}^n 1^{\boxtimes(i-1)} \boxtimes b \boxtimes 1^{\boxtimes(n-i)}} \boxtimes^{i \in \mathbf{n}} Ts\mathcal{A}_i \xrightarrow{\check{f}} s\mathcal{B}\bigr).$$

The set of A_∞-functors $(\mathcal{A}_i)_{i \in \mathbf{n}} \to \mathcal{B}$ is denoted by $\mathsf{A}_\infty((\mathcal{A}_i)_{i \in \mathbf{n}}; \mathcal{B})$. There is a natural way to compose A_∞-functors, the composition is associative, and for an arbitrary A_∞-category $\mathcal{A}$, the identity A_∞-functor $\mathrm{id}_\mathcal{A} : \mathcal{A} \to \mathcal{A}$ is the unit with respect to the composition. Thus, A_∞-categories and A_∞-functors constitute a symmetric multicategory A_∞ [BLM06, Chapter 12].

With a family $(\mathcal{A}_i)_{i=1}^n, \mathcal{B}$ of A_∞-categories we associate a graded quiver $\underline{\mathsf{A}_\infty}((\mathcal{A}_i)_{i=1}^n; \mathcal{B})$. Its objects are A_∞-functors with n entries. Morphisms are A_∞-transformations between such A_∞-functors f and g, that is, (f,g)-coderivations. Such coderivation r can be identified with the collection of its components $\check{r} = r \cdot \mathrm{pr}_1$, thus,

$$\begin{aligned} s\underline{\mathsf{A}_\infty}((\mathcal{A}_i)_{i=1}^n; \mathcal{B})(f,g) &\simeq \prod_{X,Y \in \prod_{i=1}^n \mathrm{Ob}\,\mathcal{A}_i} \underline{\mathsf{C}}_\Bbbk\bigl((\boxtimes^{i \in \mathbf{n}} Ts\mathcal{A}_i)(X,Y), s\mathcal{B}(Xf, Yg)\bigr) \\ &= \prod_{(X_i)_{i \in \mathbf{n}}, (Y_i)_{i \in \mathbf{n}} \in \prod_{i=1}^n \mathrm{Ob}\,\mathcal{A}_i} \underline{\mathsf{C}}_\Bbbk\bigl(\otimes^{i \in \mathbf{n}}[Ts\mathcal{A}_i(X_i, Y_i)], s\mathcal{B}((X_i)_{i \in \mathbf{n}} f, (Y_i)_{i \in \mathbf{n}} g)\bigr). \end{aligned}$$

Moreover, $\underline{\mathsf{A}_\infty}((\mathcal{A}_i)_{i=1}^n; \mathcal{B})$ has a distinguished A_∞-category structure which together with the evaluation A_∞-functor

$$\mathrm{ev}^{\mathsf{A}_\infty} : (\mathcal{A}_i)_{i=1}^n, \underline{\mathsf{A}_\infty}((\mathcal{A}_i)_{i=1}^n; \mathcal{B}) \to \mathcal{B}, \qquad (X_1, \dots, X_n, f) \mapsto (X_1, \dots, X_n)f$$

turns the symmetric multicategory A_∞ into a closed multicategory [BLM06]. Thus, for arbitrary A_∞-categories $(\mathcal{A}_i)_{i\in\mathbf{n}}$, $(\mathcal{B}_j)_{j\in\mathbf{m}}$, $\mathcal{C}$, the map

$$\varphi^{\mathsf{A}_\infty}: \mathsf{A}_\infty((\mathcal{B}_j)_{j\in\mathbf{m}};\underline{\mathsf{A}_\infty}((\mathcal{A}_i)_{i\in\mathbf{n}};\mathcal{C})) \longrightarrow \mathsf{A}_\infty((\mathcal{A}_i)_{i\in\mathbf{n}},(\mathcal{B}_j)_{j\in\mathbf{m}};\mathcal{C}),$$
$$f \longmapsto ((\mathrm{id}_{\mathcal{A}_i})_{i\in\mathbf{n}}, f)\,\mathrm{ev}^{\mathsf{A}_\infty}$$

is bijective. It follows from the general properties of closed multicategories that the bijection $\varphi^{\mathsf{A}_\infty}$ extends uniquely to an isomorphism of A_∞-categories

$$\underline{\varphi}^{\mathsf{A}_\infty}: \underline{\mathsf{A}_\infty}((\mathcal{B}_j)_{j\in\mathbf{m}};\underline{\mathsf{A}_\infty}((\mathcal{A}_i)_{i\in\mathbf{n}};\mathcal{C})) \to \underline{\mathsf{A}_\infty}((\mathcal{A}_i)_{i\in\mathbf{n}},(\mathcal{B}_j)_{j\in\mathbf{m}};\mathcal{C})$$

with $\mathrm{Ob}\,\underline{\varphi}^{\mathsf{A}_\infty} = \varphi^{\mathsf{A}_\infty}$. In particular, if $\mathcal{C}$ is a unital A_∞-category, $\varphi^{\mathsf{A}_\infty}$ maps isomorphic A_∞-functors to isomorphic.

The components $\mathrm{ev}^{\mathsf{A}_\infty}_{k_1,\dots,k_n;m}$ of the evaluation A_∞-functor vanish if $m>1$ by formula [BLM06, (12.25.4)]. For $m=0,1$ they are

$$\mathrm{ev}^{\mathsf{A}_\infty}_{k_1,\dots,k_n;0}: [\otimes^{i\in\mathbf{n}}T^{k_i}s\mathcal{A}_i(X_i,Y_i)]\otimes T^0 s\underline{\mathsf{A}_\infty}((\mathcal{A}_i)_{i=1}^n;\mathcal{B})(f,f) \xrightarrow{f_{k_1,\dots,k_n}} s\mathcal{B}((X_i)_{i\in\mathbf{n}}f,(Y_i)_{i\in\mathbf{n}}f),$$

$$\mathrm{ev}^{\mathsf{A}_\infty}_{k_1,\dots,k_n;1} = \big[(\otimes^{i\in\mathbf{n}}T^{k_i}s\mathcal{A}_i(X_i,Y_i))\otimes s\underline{\mathsf{A}_\infty}((\mathcal{A}_i)_{i=1}^n;\mathcal{B})(f,g) \xrightarrow{1\otimes\mathrm{pr}_{k_1,\dots,k_n}} [\otimes^{i\in\mathbf{n}}T^{k_i}s\mathcal{A}_i(X_i,Y_i)]\otimes\underline{\mathsf{C}}_{\Bbbk}(\otimes^{i\in\mathbf{n}}[T^{k_i}s\mathcal{A}_i(X_i,Y_i)], s\mathcal{B}((X_i)_{i\in\mathbf{n}}f,(Y_i)_{i\in\mathbf{n}}g)) \xrightarrow{\mathrm{ev}^{\mathsf{C}_\Bbbk}} s\mathcal{B}((X_i)_{i\in\mathbf{n}}f,(Y_i)_{i\in\mathbf{n}}g)\big]. \tag{0.2}$$

When $(\mathcal{A}_i)_{i=1}^n,\mathcal{B}$ are unital A_∞-categories, we define $\underline{\mathsf{A}^{\mathrm{u}}_\infty}((\mathcal{A}_i)_{i=1}^n;\mathcal{B})$ as a full A_∞-subcategory of $\underline{\mathsf{A}_\infty}((\mathcal{A}_i)_{i=1}^n;\mathcal{B})$, whose objects are unital A_∞-functors. Equipped with a similar evaluation $\mathrm{ev}^{\mathsf{A}^{\mathrm{u}}_\infty}$ the collection $\mathsf{A}^{\mathrm{u}}_\infty$ of unital A_∞-categories and unital A_∞-functors also becomes a closed multicategory. Similarly to the case of A_∞, there is a natural bijection

$$\varphi^{\mathsf{A}^{\mathrm{u}}_\infty}: \mathsf{A}^{\mathrm{u}}_\infty((\mathcal{B}_j)_{j\in\mathbf{m}};\underline{\mathsf{A}^{\mathrm{u}}_\infty}((\mathcal{A}_i)_{i\in\mathbf{n}};\mathcal{C})) \longrightarrow \mathsf{A}^{\mathrm{u}}_\infty((\mathcal{A}_i)_{i\in\mathbf{n}},(\mathcal{B}_j)_{j\in\mathbf{m}};\mathcal{C}),$$
$$f \longmapsto ((\mathrm{id}_{\mathcal{A}_i})_{i\in\mathbf{n}}, f)\,\mathrm{ev}^{\mathsf{A}^{\mathrm{u}}_\infty}$$

for arbitrary unital A_∞-categories $(\mathcal{A}_i)_{i\in\mathbf{n}}$, $(\mathcal{B}_j)_{j\in\mathbf{m}}$, $\mathcal{C}$.

In the simplest version graded spans $\mathcal{P}$ consist of a $\mathscr{U}$-small set $\mathrm{Ob}_s\,\mathcal{P}$ of source objects, a $\mathscr{U}$-small set $\mathrm{Ob}_t\,\mathcal{P}$ of target objects, and $\mathscr{U}$-small graded $\Bbbk$-modules $\mathcal{P}(X,Y)$ for all $X\in\mathrm{Ob}_s\,\mathcal{P}$, $Y\in\mathrm{Ob}_t\,\mathcal{P}$. Graded quivers $\mathcal{A}$ are particular cases of spans, distinguished by the condition $\mathrm{Ob}_s\,\mathcal{A}=\mathrm{Ob}_t\,\mathcal{A}$. The tensor product $\mathcal{P}\otimes\mathcal{Q}$ of two spans $\mathcal{P}$, $\mathcal{Q}$ exists if $\mathrm{Ob}_t\,\mathcal{P}=\mathrm{Ob}_s\,\mathcal{Q}$ and equals

$$(\mathcal{P}\otimes\mathcal{Q})(X,Z) = \bigoplus_{Y\in\mathrm{Ob}_t\,\mathcal{P}} \mathcal{P}(X,Y)\otimes_\Bbbk\mathcal{Q}(Y,Z).$$

Details can be found in [BLM06].

Next we explain our notation for closed symmetric monoidal categories which differs slightly from [Kel82, Chapter 1].

Let $(\mathcal{V}, \otimes, \mathbb{1}, c)$ be a closed symmetric monoidal $\mathscr{U}$-category. For each pair of objects $X, Y \in \operatorname{Ob}\mathcal{V}$, let $\underline{\mathcal{V}}(X,Y)$ denote the inner hom-object. Denote by $\mathrm{ev}^{\mathcal{V}} : X \otimes \underline{\mathcal{V}}(X,Y) \to Y$ and $\mathrm{coev}^{\mathcal{V}} : Y \to \underline{\mathcal{V}}(X, X\otimes Y)$ the evaluation and coevaluation morphisms, respectively. Then the mutually inverse adjunction isomorphisms are explicitly given as follows:

$$\begin{aligned} \mathcal{V}(Y, \underline{\mathcal{V}}(X,Z)) &\longleftrightarrow \mathcal{V}(X\otimes Y, Z), \\ f &\longmapsto (1_X \otimes f)\,\mathrm{ev}_{X,Z}, \\ \mathrm{coev}_{X,Y}\, \underline{\mathcal{V}}(X,g) &\longleftarrow\!\shortmid\; g. \end{aligned}$$

There is a $\mathcal{V}$-category $\underline{\mathcal{V}}$ whose objects are those of $\mathcal{V}$, and for each pair of objects X and Y, the object $\underline{\mathcal{V}}(X,Y) \in \operatorname{Ob}\mathcal{V}$ is the inner hom-object of $\mathcal{V}$. The composition is found from the following equation:

$$\begin{aligned} &\big[X\otimes \underline{\mathcal{V}}(X,Y)\otimes \underline{\mathcal{V}}(Y,Z) \xrightarrow{1\otimes \mu_{\underline{\mathcal{V}}}} X\otimes \underline{\mathcal{V}}(X,Z) \xrightarrow{\mathrm{ev}^{\mathcal{V}}} Z\big] \\ &\qquad = \big[X\otimes \underline{\mathcal{V}}(X,Y)\otimes \underline{\mathcal{V}}(Y,Z) \xrightarrow{\mathrm{ev}^{\mathcal{V}}\otimes 1} Y\otimes \underline{\mathcal{V}}(Y,Z) \xrightarrow{\mathrm{ev}^{\mathcal{V}}} Z\big]. \qquad (0.3) \end{aligned}$$

The identity morphism $1^{\underline{\mathcal{V}}}_X : \mathbb{1} \to \underline{\mathcal{V}}(X,X)$ is found from the following equation:

$$\big[X \xrightarrow{\sim} X\otimes \mathbb{1} \xrightarrow{1\otimes 1^{\underline{\mathcal{V}}}_X} X\otimes \underline{\mathcal{V}}(X,X) \xrightarrow{\mathrm{ev}^{\mathcal{V}}} X\big] = \mathrm{id}_X .$$

For our applications we need several categories $\mathcal{V}$, for instance the Abelian category $\mathsf{C}_{\Bbbk}$ of complexes of $\Bbbk$-modules and its quotient category $\mathcal{K}$, the homotopy category of complexes of $\Bbbk$-modules. The tensor product is the tensor product of complexes, the unit object is $\Bbbk$, viewed as a complex concentrated in degree 0, and the symmetry is the standard symmetry $c : X\otimes Y \to Y\otimes X$, $x\otimes y \mapsto (-)^{xy} y\otimes x$. We shorten up the usual notation $(-1)^{\deg x\cdot\deg y}$ to $(-)^{xy}$. Similarly, $(-)^x$ means $(-1)^{\deg x}$, $(-)^{x+y}$ means $(-1)^{\deg x+\deg y}$, etc. For each pair of complexes X and Y, the inner hom-object $\underline{\mathcal{K}}(X,Y)$ is the same as the inner hom-complex $\underline{\mathsf{C}}_{\Bbbk}(X,Y)$ in the symmetric closed monoidal Abelian category $\mathsf{C}_{\Bbbk}$. The evaluation morphism $\mathrm{ev}^{\mathcal{K}} : X\otimes\underline{\mathcal{K}}(X,Y) \to Y$ and the coevaluation morphism $\mathrm{coev}^{\mathcal{K}} : Y \to \underline{\mathcal{K}}(X, X\otimes Y)$ in $\mathcal{K}$ are the homotopy classes of the evaluation morphism $\mathrm{ev}^{\mathsf{C}_{\Bbbk}} : X\otimes\underline{\mathsf{C}}_{\Bbbk}(X,Y) \to Y$ and the coevaluation morphism $\mathrm{coev}^{\mathsf{C}_{\Bbbk}} : Y \to \underline{\mathsf{C}}_{\Bbbk}(X, X\otimes Y)$ in $\mathsf{C}_{\Bbbk}$, respectively.

It is easy to see that $\mu_{\underline{\mathcal{K}}} = m_2^{\underline{\mathsf{C}}_{\Bbbk}}$ and $1^{\underline{\mathcal{K}}}_X = 1^{\underline{\mathsf{C}}_{\Bbbk}}_X$, therefore $\underline{\mathcal{K}} = \mathsf{k}\underline{\mathsf{C}}_{\Bbbk}$.

Also we use as $\mathcal{V}$ the category $\mathbf{gr} = \mathbf{gr}(\Bbbk\text{-}\mathbf{Mod})$ of graded $\Bbbk$-modules, and the familiar category $\Bbbk$-$\mathbf{Mod}$ of $\Bbbk$-modules.

The following identity holds for any homogeneous $\Bbbk$-linear map $a : X \to A$ of arbitrary degree by the properties of the closed monoidal category $\underline{\mathsf{C}}_{\Bbbk}$:

$$\begin{aligned} &\big(\underline{\mathsf{C}}_{\Bbbk}(A,B)\otimes\underline{\mathsf{C}}_{\Bbbk}(B,C) \xrightarrow{m_2} \underline{\mathsf{C}}_{\Bbbk}(A,C) \xrightarrow{\underline{\mathsf{C}}_{\Bbbk}(a,C)} \underline{\mathsf{C}}_{\Bbbk}(X,C)\big) \\ &= \big(\underline{\mathsf{C}}_{\Bbbk}(A,B)\otimes\underline{\mathsf{C}}_{\Bbbk}(B,C) \xrightarrow{\underline{\mathsf{C}}_{\Bbbk}(a,B)\otimes 1} \underline{\mathsf{C}}_{\Bbbk}(X,B)\otimes\underline{\mathsf{C}}_{\Bbbk}(B,C) \xrightarrow{m_2} \underline{\mathsf{C}}_{\Bbbk}(X,C)\big). \qquad (0.4) \end{aligned}$$

Let $f : A\otimes X \to B$, $g : B\otimes Y \to C$ be two homogeneous $\Bbbk$-linear maps of arbitrary degrees, that is, $f \in \underline{\mathsf{C}}_{\Bbbk}(A\otimes X, B)^{\bullet}$, $g \in \underline{\mathsf{C}}_{\Bbbk}(B\otimes Y, C)^{\bullet}$. Then the following identity is proven in [LM04] as equation (A.1.2):

$$\begin{aligned}
\big(X \otimes Y \xrightarrow{\mathrm{coev}_{A,X} \otimes \mathrm{coev}_{B,Y}} \underline{\mathsf{C}}_{\Bbbk}(A, A \otimes X) \otimes \underline{\mathsf{C}}_{\Bbbk}(B, B \otimes Y) \\
\xrightarrow{\underline{\mathsf{C}}_{\Bbbk}(A,f)\otimes\underline{\mathsf{C}}_{\Bbbk}(B,g)} \underline{\mathsf{C}}_{\Bbbk}(A,B) \otimes \underline{\mathsf{C}}_{\Bbbk}(B,C) \xrightarrow{m_2} \underline{\mathsf{C}}_{\Bbbk}(A,C)\big) \\
= \big(X\otimes Y \xrightarrow{\mathrm{coev}_{A,X\otimes Y}} \underline{\mathsf{C}}_{\Bbbk}(A, A\otimes X\otimes Y) \xrightarrow{\underline{\mathsf{C}}_{\Bbbk}(A,f\otimes 1)} \underline{\mathsf{C}}_{\Bbbk}(A, B\otimes Y) \xrightarrow{\underline{\mathsf{C}}_{\Bbbk}(A,g)} \underline{\mathsf{C}}_{\Bbbk}(A,C)\big).
\end{aligned} \tag{0.5}$$

1. $\mathcal{V}$-categories

We refer the reader to [Kel82, Chapter 1] for the basic theory of enriched categories. The category of unital (resp. non-unital) $\mathcal{V}$-categories (where $\mathcal{V}$ is a closed symmetric monoidal category) is denoted $\mathcal{V}\text{-}\mathcal{C}at$ (resp. $\mathcal{V}\text{-}\mathcal{C}at^{nu}$).

1.1. Opposite $\mathcal{V}$-categories

Let $\mathcal{A}$ be a $\mathcal{V}$-category, not necessarily unital. Its opposite $\mathcal{A}^{\mathrm{op}}$ is defined in the standard way. Namely, $\mathrm{Ob}\,\mathcal{A}^{\mathrm{op}} = \mathrm{Ob}\,\mathcal{A}$, and for each pair of objects $X, Y \in \mathrm{Ob}\,\mathcal{A}$, $\mathcal{A}^{\mathrm{op}}(X,Y) = \mathcal{A}(Y,X)$. The composition in $\mathcal{A}^{\mathrm{op}}$ is given by

$$\mu_{\mathcal{A}^{\mathrm{op}}} = \big[\mathcal{A}^{\mathrm{op}}(X,Y) \otimes \mathcal{A}^{\mathrm{op}}(Y,Z) = \mathcal{A}(Y,X) \otimes \mathcal{A}(Z,Y) \xrightarrow{c} \mathcal{A}(Z,Y) \otimes \mathcal{A}(Y,X) \xrightarrow{\mu_{\mathcal{A}}} \mathcal{A}(Z,X) = \mathcal{A}^{\mathrm{op}}(X,Z)\big].$$

More generally, for each $n \geqslant 1$, the iterated n-ary composition in $\mathcal{A}^{\mathrm{op}}$ is

$$\mu^{\mathbf{n}}_{\mathcal{A}^{\mathrm{op}}} = \big[\otimes^{i\in\mathbf{n}}\mathcal{A}^{\mathrm{op}}(X_{i-1}, X_i) = \otimes^{i\in\mathbf{n}}\mathcal{A}(X_i, X_{i-1}) \xrightarrow{\omega^0_c} \otimes^{i\in\mathbf{n}}\mathcal{A}(X_{n-i+1}, X_{n-i}) \xrightarrow{\mu^{\mathbf{n}}_{\mathcal{A}}} \mathcal{A}(X_n, X_0) = \mathcal{A}^{\mathrm{op}}(X_0, X_n)\big], \tag{1.1}$$

where the permutation $\omega^0 = \left(\begin{smallmatrix} 1 & 2 & \dots & n-1 & n \\ n & n-1 & \dots & 2 & 1 \end{smallmatrix}\right)$ is the longest element of $\mathfrak{S}_n$, and ω^0_c is the corresponding signed permutation, the action of ω^0 in tensor products via standard symmetry. Note that if $\mathcal{A}$ is unital, then so is $\mathcal{A}^{\mathrm{op}}$, with the same identity morphisms.

Let $f : \mathcal{A} \to \mathcal{B}$ be a $\mathcal{V}$-functor, not necessarily unital. It gives rise to a $\mathcal{V}$-functor $f^{\mathrm{op}} : \mathcal{A}^{\mathrm{op}} \to \mathcal{B}^{\mathrm{op}}$ with $\mathrm{Ob}\, f^{\mathrm{op}} = \mathrm{Ob}\, f : \mathrm{Ob}\,\mathcal{A} \to \mathrm{Ob}\,\mathcal{B}$, and

$$f^{\mathrm{op}} = \big[\mathcal{A}^{\mathrm{op}}(X,Y) = \mathcal{A}(Y,X) \xrightarrow{f} \mathcal{B}(Yf, Xf) = \mathcal{B}^{\mathrm{op}}(Xf, Yf)\big], \quad X, Y \in \mathrm{Ob}\,\mathcal{A}.$$

Note that f^{op} is a unital $\mathcal{V}$-functor if so is f. Clearly, the correspondences $\mathcal{A} \mapsto \mathcal{A}^{\mathrm{op}}$, $f \mapsto f^{\mathrm{op}}$ define a functor $-^{\mathrm{op}} : \mathcal{V}\text{-}\mathcal{C}at^{nu} \to \mathcal{V}\text{-}\mathcal{C}at^{nu}$ which restricts to a functor $-^{\mathrm{op}} : \mathcal{V}\text{-}\mathcal{C}at \to \mathcal{V}\text{-}\mathcal{C}at$. The functor $-^{\mathrm{op}}$ is symmetric monoidal. More precisely, for arbitrary $\mathcal{V}$-categories $\mathcal{A}_i$, $i \in \mathbf{n}$, the equation $\boxtimes^{i\in\mathbf{n}}\mathcal{A}_i^{\mathrm{op}} = (\boxtimes^{i\in\mathbf{n}}\mathcal{A}_i)^{\mathrm{op}}$ holds. Indeed, the underlying $\mathcal{V}$-quivers of both categories coincide, and so do the identity morphisms if the categories $\mathcal{A}_i$, $i \in \mathbf{n}$, are unital. The composition in $\boxtimes^{i\in\mathbf{n}}\mathcal{A}_i^{\mathrm{op}}$ is given by

$$\mu_{\boxtimes^{i\in\mathbf{n}}\mathcal{A}_i^{\mathrm{op}}} = \bigl[\bigl(\otimes^{i\in\mathbf{n}}\mathcal{A}_i^{\mathrm{op}}(X_i,Y_i)\bigr)\otimes\bigl(\otimes^{i\in\mathbf{n}}\mathcal{A}_i^{\mathrm{op}}(Y_i,Z_i)\bigr)$$
$$\xrightarrow{\sigma_{(12)}} \otimes^{i\in\mathbf{n}}\bigl(\mathcal{A}_i(Y_i,X_i)\otimes\mathcal{A}_i(Z_i,Y_i)\bigr) \xrightarrow{\otimes^{i\in\mathbf{n}}c} \otimes^{i\in\mathbf{n}}\bigl(\mathcal{A}_i(Z_i,Y_i)\otimes\mathcal{A}_i(Y_i,X_i)\bigr)$$
$$\xrightarrow{\otimes^{i\in\mathbf{n}}\mu_{\mathcal{A}_i}} \otimes^{i\in\mathbf{n}}\mathcal{A}_i(Z_i,X_i) = \otimes^{i\in\mathbf{n}}\mathcal{A}_i^{\mathrm{op}}(X_i,Z_i)\bigr].$$

The composition in $(\boxtimes^{i\in\mathbf{n}}\mathcal{A}_i)^{\mathrm{op}}$ is given by

$$\mu_{(\boxtimes^{i\in\mathbf{n}}\mathcal{A}_i)^{\mathrm{op}}} = \bigl[(\boxtimes^{i\in\mathbf{n}}\mathcal{A}_i)^{\mathrm{op}}((X_i)_{i\in\mathbf{n}},(Y_i)_{i\in\mathbf{n}})\otimes(\boxtimes^{i\in\mathbf{n}}\mathcal{A}_i)^{\mathrm{op}}((Y_i)_{i\in\mathbf{n}},(Z_i)_{i\in\mathbf{n}})$$
$$= \bigl(\otimes^{i\in\mathbf{n}}\mathcal{A}_i(Y_i,X_i)\bigr)\otimes\bigl(\otimes^{i\in\mathbf{n}}\mathcal{A}_i(Z_i,Y_i)\bigr) \xrightarrow{c} \bigl(\otimes^{i\in\mathbf{n}}\mathcal{A}_i(Z_i,Y_i)\bigr)\otimes\bigl(\otimes^{i\in\mathbf{n}}\mathcal{A}_i(Y_i,X_i)\bigr)$$
$$\xrightarrow{\sigma_{(12)}} \otimes^{i\in\mathbf{n}}\bigl(\mathcal{A}_i(Z_i,Y_i)\otimes\mathcal{A}_i(Y_i,X_i)\bigr) \xrightarrow{\otimes^{i\in\mathbf{n}}\mu_{\mathcal{A}_i}} \otimes^{i\in\mathbf{n}}\mathcal{A}_i(Z_i,X_i) = \otimes^{i\in\mathbf{n}}\mathcal{A}_i^{\mathrm{op}}(X_i,Z_i)\bigr].$$

The equation $\mu_{\boxtimes^{i\in\mathbf{n}}\mathcal{A}_i^{\mathrm{op}}} = \mu_{(\boxtimes^{i\in\mathbf{n}}\mathcal{A}_i)^{\mathrm{op}}}$ follows from the following equation in $\mathcal{V}$:

$$\bigl[\bigl(\otimes^{i\in\mathbf{n}}\mathcal{A}_i(Y_i,X_i)\bigr)\otimes\bigl(\otimes^{i\in\mathbf{n}}\mathcal{A}_i(Z_i,Y_i)\bigr) \xrightarrow{\sigma_{(12)}} \otimes^{i\in\mathbf{n}}\bigl(\mathcal{A}_i(Y_i,X_i)\otimes\mathcal{A}_i(Z_i,Y_i)\bigr)$$
$$\xrightarrow{\otimes^{i\in\mathbf{n}}c} \otimes^{i\in\mathbf{n}}\bigl(\mathcal{A}_i(Z_i,Y_i)\otimes\mathcal{A}_i(Y_i,X_i)\bigr)\bigr]$$
$$= \bigl[\bigl(\otimes^{i\in\mathbf{n}}\mathcal{A}_i(Y_i,X_i)\bigr)\otimes\bigl(\otimes^{i\in\mathbf{n}}\mathcal{A}_i(Z_i,Y_i)\bigr) \xrightarrow{c} \bigl(\otimes^{i\in\mathbf{n}}\mathcal{A}_i(Z_i,Y_i)\bigr)\otimes\bigl(\otimes^{i\in\mathbf{n}}\mathcal{A}_i(Y_i,X_i)\bigr)$$
$$\xrightarrow{\sigma_{(12)}} \otimes^{i\in\mathbf{n}}\bigl(\mathcal{A}_i(Z_i,Y_i)\otimes\mathcal{A}_i(Y_i,X_i)\bigr)\bigr],$$

which is a consequence of coherence principle of [BLM06, Lemma 3.26, Remark 3.27]. Therefore, $-^{\mathrm{op}}$ induces a symmetric multifunctor $-^{\mathrm{op}} : \widehat{\mathcal{V}\text{-}\mathcal{C}at^{nu}} \to \widehat{\mathcal{V}\text{-}\mathcal{C}at^{nu}}$ which restricts to a symmetric multifunctor $-^{\mathrm{op}} : \widehat{\mathcal{V}\text{-}\mathcal{C}at} \to \widehat{\mathcal{V}\text{-}\mathcal{C}at}$.

1.2. Hom-functor

A $\mathcal{V}$-category $\mathcal{A}$ gives rise to a $\mathcal{V}$-functor $\mathrm{Hom}_{\mathcal{A}} : \mathcal{A}^{\mathrm{op}}\boxtimes\mathcal{A} \to \underline{\mathcal{V}}$ which maps a pair of objects $(X,Y)\in\mathrm{Ob}\,\mathcal{A}\times\mathrm{Ob}\,\mathcal{A}$ to $\mathcal{A}(X,Y)\in\mathrm{Ob}\,\underline{\mathcal{V}}$, and whose action on morphisms is given by

$$\mathrm{Hom}_{\mathcal{A}} = \bigl[\mathcal{A}^{\mathrm{op}}(X,Y)\otimes\mathcal{A}(U,V) = \mathcal{A}(Y,X)\otimes\mathcal{A}(U,V) \xrightarrow{\mathrm{coev}^{\mathcal{V}}}$$
$$\underline{\mathcal{V}}(\mathcal{A}(X,U),\mathcal{A}(X,U)\otimes\mathcal{A}(Y,X)\otimes\mathcal{A}(U,V)) \xrightarrow{\underline{\mathcal{V}}(1,(c\otimes 1)\mu^{\mathbf{3}}_{\mathcal{A}})} \mathcal{V}(\mathcal{A}(X,U),\mathcal{A}(Y,V))\bigr].$$

Equivalently, $\mathrm{Hom}_{\mathcal{A}}$ is found by closedness of $\mathcal{V}$ from the diagram

$$\begin{array}{ccc}
\mathcal{A}(X,U)\otimes\mathcal{A}(Y,X)\otimes\mathcal{A}(U,V) & \xrightarrow{1\otimes\mathrm{Hom}_{\mathcal{A}}} & \mathcal{A}(X,U)\otimes\underline{\mathcal{V}}(\mathcal{A}(X,U),\mathcal{A}(Y,V)) \\
{\scriptstyle c\otimes 1}\downarrow & & \downarrow{\scriptstyle \mathrm{ev}^{\mathcal{V}}} \\
\mathcal{A}(Y,X)\otimes\mathcal{A}(X,U)\otimes\mathcal{A}(U,V) & \xrightarrow{\mu^{\mathbf{3}}_{\mathcal{A}}} & \mathcal{A}(Y,V)
\end{array} \tag{1.2}$$

Lemma 1.3. *Let $\mathcal{A}$ be a $\mathcal{V}$-category. Then*

$$\mathrm{Hom}_{\mathcal{A}^{\mathrm{op}}} = \bigl[\mathcal{A}\boxtimes\mathcal{A}^{\mathrm{op}} \xrightarrow{c} \mathcal{A}^{\mathrm{op}}\boxtimes\mathcal{A} \xrightarrow{\mathrm{Hom}_{\mathcal{A}}} \underline{\mathcal{V}}\bigr].$$

Proof. Using (1.1), we obtain:

$$\begin{aligned}
\mathrm{Hom}_{\mathcal{A}^{\mathrm{op}}} &= \big[\mathcal{A}(X,Y)\otimes\mathcal{A}^{\mathrm{op}}(U,V) = \mathcal{A}^{\mathrm{op}}(Y,X)\otimes\mathcal{A}^{\mathrm{op}}(U,V) \\
&\xrightarrow{\mathrm{coev}^{\mathcal{V}}} \underline{\mathcal{V}}(\mathcal{A}^{\mathrm{op}}(X,U),\mathcal{A}^{\mathrm{op}}(X,U)\otimes\mathcal{A}^{\mathrm{op}}(Y,X)\otimes\mathcal{A}^{\mathrm{op}}(U,V)) \\
&\qquad\xrightarrow{\underline{\mathcal{V}}(1,(c\otimes 1)\mu^{\mathbf{3}}_{\mathcal{A}^{\mathrm{op}}})} \underline{\mathcal{V}}(\mathcal{A}^{\mathrm{op}}(X,U),\mathcal{A}^{\mathrm{op}}(Y,V))\big] \\
&= \big[\mathcal{A}(X,Y)\otimes\mathcal{A}(V,U) \xrightarrow{\mathrm{coev}^{\mathcal{V}}} \underline{\mathcal{V}}(\mathcal{A}(U,X),\mathcal{A}(U,X)\otimes\mathcal{A}(X,Y)\otimes\mathcal{A}(V,U)) \\
&\qquad\xrightarrow{\underline{\mathcal{V}}(1,(c\otimes 1)\omega^0_c\mu^{\mathbf{3}}_{\mathcal{A}})} \underline{\mathcal{V}}(\mathcal{A}(U,X),\mathcal{A}(V,Y))\big],
\end{aligned}$$

where $\omega^0 = (13) \in \mathfrak{S}_3$. Clearly, $(c\otimes 1)\omega^0_c = (1\otimes c)(c\otimes 1)$, therefore

$$\begin{aligned}
\mathrm{Hom}_{\mathcal{A}^{\mathrm{op}}} &= \big[\mathcal{A}(X,Y)\otimes\mathcal{A}(V,U) \xrightarrow{\mathrm{coev}^{\mathcal{V}}} \underline{\mathcal{V}}(\mathcal{A}(U,X),\mathcal{A}(U,X)\otimes\mathcal{A}(X,Y)\otimes\mathcal{A}(V,U)) \\
&\qquad\xrightarrow{\underline{\mathcal{V}}(1,(1\otimes c)(c\otimes 1)\mu^{\mathbf{3}}_{\mathcal{A}})} \underline{\mathcal{V}}(\mathcal{A}(U,X),\mathcal{A}(V,Y))\big] \\
&= \big[\mathcal{A}(X,Y)\otimes\mathcal{A}(V,U) \xrightarrow{c} \mathcal{A}(V,U)\otimes\mathcal{A}(X,Y) \xrightarrow{\mathrm{coev}^{\mathcal{V}}} \\
&\quad \underline{\mathcal{V}}(\mathcal{A}(U,X),\mathcal{A}(U,X)\otimes\mathcal{A}(V,U)\otimes\mathcal{A}(X,Y)) \xrightarrow{\underline{\mathcal{V}}(1,(c\otimes 1)\mu^{\mathbf{3}}_{\mathcal{A}})} \underline{\mathcal{V}}(\mathcal{A}(U,X),\mathcal{A}(V,Y))\big] \\
&= \big[\mathcal{A}(X,Y)\otimes\mathcal{A}(V,U) \xrightarrow{c} \mathcal{A}(V,U)\otimes\mathcal{A}(X,Y) \xrightarrow{\mathrm{Hom}_{\mathcal{A}}} \underline{\mathcal{V}}(\mathcal{A}(U,X),\mathcal{A}(V,Y))\big].
\end{aligned}$$

The lemma is proven. □

An object X of $\mathcal{A}$ defines a $\mathcal{V}$-functor $X : \mathbb{1} \to \mathcal{A}$, $* \mapsto X$, $\mathbb{1}(*,*) = \mathbb{1} \xrightarrow{1^{\mathcal{A}}_X} \mathcal{A}(X,X)$, whose source $\mathbb{1}$ is a $\mathcal{V}$-category with one object $*$. This $\mathcal{V}$-category is a unit of tensor multiplication $\boxtimes$. The $\mathcal{V}$-functors $\mathcal{A}(\_,Y) = \mathrm{Hom}_{\mathcal{A}}(\_,Y) : \mathcal{A}^{\mathrm{op}} \to \underline{\mathcal{V}}$ and $\mathcal{A}(X,\_) = \mathrm{Hom}_{\mathcal{A}}(X,\_) : \mathcal{A} \to \underline{\mathcal{V}}$ are defined as follows:

$$\begin{aligned}
\mathcal{A}(_,Y) &= \big[\mathcal{A}^{\mathrm{op}} \xrightarrow{\sim} \mathcal{A}^{\mathrm{op}}\boxtimes\mathbb{1} \xrightarrow{1\boxtimes Y} \mathcal{A}^{\mathrm{op}}\boxtimes\mathcal{A} \xrightarrow{\mathrm{Hom}_{\mathcal{A}}} \underline{\mathcal{V}}\big], \\
\mathcal{A}(X,_) &= \big[\mathcal{A} \xrightarrow{\sim} \mathbb{1}\boxtimes\mathcal{A} \xrightarrow{X\boxtimes 1} \mathcal{A}^{\mathrm{op}}\boxtimes\mathcal{A} \xrightarrow{\mathrm{Hom}_{\mathcal{A}}} \underline{\mathcal{V}}\big].
\end{aligned}$$

Thus, the $\mathcal{V}$-functor $\mathcal{A}(\_,Y)$ maps an object X to $\mathcal{A}(X,Y)$, and its action on morphisms is given by

$$\begin{aligned}
\mathcal{A}(_,Y) = \big[\mathcal{A}^{\mathrm{op}}(W,X) = \mathcal{A}(X,W) &\xrightarrow{\mathrm{coev}^{\mathcal{V}}} \underline{\mathcal{V}}(\mathcal{A}(W,Y),\mathcal{A}(W,Y)\otimes\mathcal{A}(X,W)) \\
&\xrightarrow{\underline{\mathcal{V}}(1,c\mu_{\mathcal{A}})} \underline{\mathcal{V}}(\mathcal{A}(W,Y),\mathcal{A}(X,Y))\big]. \qquad (1.3)
\end{aligned}$$

Similarly, the $\mathcal{V}$-functor $\mathcal{A}(X,\_)$ maps an object Y to $\mathcal{A}(X,Y)$, and its action on morphisms is given by

$$\begin{aligned}
\mathcal{A}(X,_) = \big[\mathcal{A}(Y,Z) &\xrightarrow{\mathrm{coev}^{\mathcal{V}}} \underline{\mathcal{V}}(\mathcal{A}(X,Y),\mathcal{A}(X,Y)\otimes\mathcal{A}(Y,Z)) \\
&\xrightarrow{\underline{\mathcal{V}}(1,\mu_{\mathcal{A}})} \underline{\mathcal{V}}(\mathcal{A}(X,Y),\mathcal{A}(X,Z))\big]. \qquad (1.4)
\end{aligned}$$

1.4. Duality functor

The unit object $\mathbb{1}$ of $\mathcal{V}$ defines the duality $\mathcal{V}$-functor $\underline{\mathcal{V}}(\_,\mathbb{1}) = \mathrm{Hom}_{\mathcal{V}}(\_,\mathbb{1}) : \underline{\mathcal{V}}^{\mathrm{op}} \to \underline{\mathcal{V}}$. The functor $\underline{\mathcal{V}}(\_,\mathbb{1})$ maps an object M to its dual $\underline{\mathcal{V}}(M,\mathbb{1})$, and its action on morphisms is given by

$$\underline{\mathcal{V}}(\_,\mathbb{1}) = \big[\underline{\mathcal{V}}^{\mathrm{op}}(M,N) = \underline{\mathcal{V}}(N,M) \xrightarrow{\mathrm{coev}^{\mathcal{V}}} \underline{\mathcal{V}}(\underline{\mathcal{V}}(M,\mathbb{1}), \underline{\mathcal{V}}(M,\mathbb{1}) \otimes \underline{\mathcal{V}}(N,M)) \xrightarrow{\underline{\mathcal{V}}(1,c)}$$
$$\underline{\mathcal{V}}(\underline{\mathcal{V}}(M,\mathbb{1}), \underline{\mathcal{V}}(N,M) \otimes \underline{\mathcal{V}}(M,\mathbb{1})) \xrightarrow{\underline{\mathcal{V}}(1,\mu_{\underline{\mathcal{V}}})} \underline{\mathcal{V}}(\underline{\mathcal{V}}(M,\mathbb{1}), \underline{\mathcal{V}}(N,\mathbb{1}))\big]. \quad (1.5)$$

For each object M there is a natural morphism $e : M \to \underline{\mathcal{V}}(\underline{\mathcal{V}}(M,\mathbb{1}),\mathbb{1})$ which is a unique solution of the following equation in $\mathcal{V}$:

$$\begin{array}{ccc}
\underline{\mathcal{V}}(M,\mathbb{1}) \otimes M & \xrightarrow{\quad c \quad} & M \otimes \underline{\mathcal{V}}(M,\mathbb{1}) \\
{\scriptstyle 1\otimes e}\big\downarrow & & \big\downarrow{\scriptstyle \mathrm{ev}^{\mathcal{V}}} \\
\underline{\mathcal{V}}(M,\mathbb{1}) \otimes \underline{\mathcal{V}}(\underline{\mathcal{V}}(M,\mathbb{1}),\mathbb{1}) & \xrightarrow{\quad \mathrm{ev}^{\mathcal{V}} \quad} & \mathbb{1}
\end{array}$$

Explicitly,

$$e = \big[M \xrightarrow{\mathrm{coev}^{\mathcal{V}}} \underline{\mathcal{V}}(\underline{\mathcal{V}}(M,\mathbb{1}), \underline{\mathcal{V}}(M,\mathbb{1}) \otimes M) \xrightarrow{\underline{\mathcal{V}}(1,c)}$$
$$\underline{\mathcal{V}}(\underline{\mathcal{V}}(M,\mathbb{1}), M \otimes \underline{\mathcal{V}}(M,\mathbb{1})) \xrightarrow{\underline{\mathcal{V}}(1,\mathrm{ev}^{\mathcal{V}})} \underline{\mathcal{V}}(\underline{\mathcal{V}}(M,\mathbb{1}),\mathbb{1})\big].$$

An object M is *reflexive* if e is an isomorphism in $\mathcal{V}$.

1.5. Representability

Let us state for the record the following

Proposition 1.6 (Weak Yoneda Lemma). *Let $F : \mathcal{A} \to \underline{\mathcal{V}}$ be a $\mathcal{V}$-functor, X an object of $\mathcal{A}$. There is a bijection between elements of $F(X)$,* i.e., *morphisms $t : \mathbb{1} \to F(X)$, and natural transformations $\mathcal{A}(X,\_) \to F : \mathcal{A} \to \underline{\mathcal{V}}$ defined as follows: with an element $t : \mathbb{1} \to F(X)$ a natural transformation is associated whose components are given by*

$$\mathcal{A}(X,Z) \xrightarrow{t\otimes F_{X,Z}} F(X) \otimes \underline{\mathcal{V}}(F(X),F(Z)) \xrightarrow{\mathrm{ev}^{\mathcal{V}}} F(Z), \quad Z \in \mathrm{Ob}\,\mathcal{A}.$$

In particular, F is representable if and only if there is an object $X \in \mathrm{Ob}\,\mathcal{A}$ and an element $t : \mathbb{1} \to F(X)$ such that for each object $Z \in \mathrm{Ob}\,\mathcal{A}$ the above composite is invertible.

Proof. Standard, see [Kel82, Section 1.9]. □

2. Serre functors for $\mathcal{V}$-categories

Serre functors for enriched categories are for us a bridge between ordinary Serre functors and Serre A_∞-functors.

2.1. Serre $\mathcal{V}$-functors

Let $\mathcal{C}$ be a $\mathcal{V}$-category, $S : \mathcal{C} \to \mathcal{C}$ a $\mathcal{V}$-functor. Consider a natural transformation ψ as in the diagram below:

$$\begin{array}{ccc} \mathcal{C}^{\mathrm{op}} \boxtimes \mathcal{C} & \xrightarrow{1 \boxtimes S} & \mathcal{C}^{\mathrm{op}} \boxtimes \mathcal{C} \\ {\scriptstyle \mathrm{Hom}^{\mathrm{op}}_{\mathcal{C}^{\mathrm{op}}}} \Big\downarrow & \overset{\psi}{\swarrow} & \Big\downarrow {\scriptstyle \mathrm{Hom}_{\mathcal{C}}} \\ \underline{\mathcal{V}}^{\mathrm{op}} & \xrightarrow[\underline{\mathcal{V}}(\_,\mathbb{1})]{} & \underline{\mathcal{V}} \end{array} \tag{2.1}$$

The natural transformation ψ is a collection of morphisms of $\mathcal{V}$

$$\psi_{X,Y} : \mathbb{1} \to \underline{\mathcal{V}}(\mathcal{C}(X, YS), \underline{\mathcal{V}}(\mathcal{C}(Y, X), \mathbb{1})), \quad X, Y \in \operatorname{Ob} \mathcal{C}.$$

Equivalently, ψ is given by a collection of morphisms $\psi_{X,Y} : \mathcal{C}(X, YS) \to \underline{\mathcal{V}}(\mathcal{C}(Y, X), \mathbb{1})$ of $\mathcal{V}$, for $X, Y \in \operatorname{Ob} \mathcal{C}$. Naturality of ψ may be verified variable-by-variable.

Definition 2.2. Let $\mathcal{C}$ be a $\mathcal{V}$-category. A $\mathcal{V}$-functor $S : \mathcal{C} \to \mathcal{C}$ is called a *right Serre functor* if there exists a natural isomorphism ψ as in (2.1). If moreover S is a self–equivalence, it is called a *Serre functor.*

This terms agree with the conventions of Mazorchuk and Stroppel [MS05] and up to taking dual spaces with the terminology of Reiten and van den Bergh [RvdB02].

Lemma 2.3. *Let $S : \mathcal{C} \to \mathcal{C}$ be a $\mathcal{V}$-functor. Fix an object Y of $\mathcal{C}$. A collection of morphisms $(\psi_{X,Y} : \mathcal{C}(X, YS) \to \underline{\mathcal{V}}(\mathcal{C}(Y, X), \mathbb{1}))_{X \in \operatorname{Ob} \mathcal{C}}$ of $\mathcal{V}$ is natural in X if and only if*

$$\psi_{X,Y} = \bigl[\mathcal{C}(X, YS) \xrightarrow{\mathrm{coev}^{\mathcal{V}}} \underline{\mathcal{V}}(\mathcal{C}(Y, X), \mathcal{C}(Y, X) \otimes \mathcal{C}(X, YS)) \xrightarrow{\underline{\mathcal{V}}(1,\mu_{\mathcal{C}})} \underline{\mathcal{V}}(\mathcal{C}(Y, X), \mathcal{C}(Y, YS)) \xrightarrow{\underline{\mathcal{V}}(1,\tau_Y)} \underline{\mathcal{V}}(\mathcal{C}(Y, X), \mathbb{1})\bigr], \tag{2.2}$$

where

$$\tau_Y = \bigl[\mathcal{C}(Y, YS) \xrightarrow{1^{\mathcal{C}}_Y \otimes 1} \mathcal{C}(Y, Y) \otimes \mathcal{C}(Y, YS) \xrightarrow{1 \otimes \psi_{Y,Y}} \mathcal{C}(Y, Y) \otimes \underline{\mathcal{V}}(\mathcal{C}(Y, Y), \mathbb{1}) \xrightarrow{\mathrm{ev}^{\mathcal{V}}} \mathbb{1}\bigr]. \tag{2.3}$$

Proof. The collection $(\psi_{X,Y})_{X \in \operatorname{Ob} \mathcal{C}}$ is a natural $\mathcal{V}$-transformation

$$\begin{array}{ccc} \mathcal{C}^{\mathrm{op}} & \xrightarrow{\quad \mathcal{C}(\_, YS) \quad} & \underline{\mathcal{V}} \\ & {\scriptstyle \mathcal{C}(Y,\_)^{\mathrm{op}}} \searrow \quad \Big\Downarrow {\scriptstyle \psi_{-,Y}} \quad \nearrow {\scriptstyle \underline{\mathcal{V}}(\_,\mathbb{1})} & \\ & \underline{\mathcal{V}}^{\mathrm{op}} & \end{array}$$

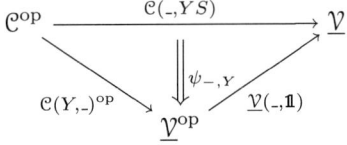

if the following diagram commutes:

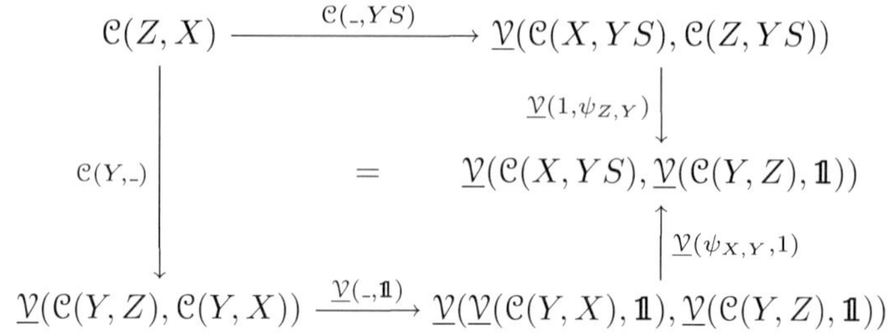

By closedness, this is equivalent to commutativity of the exterior of the following diagram:

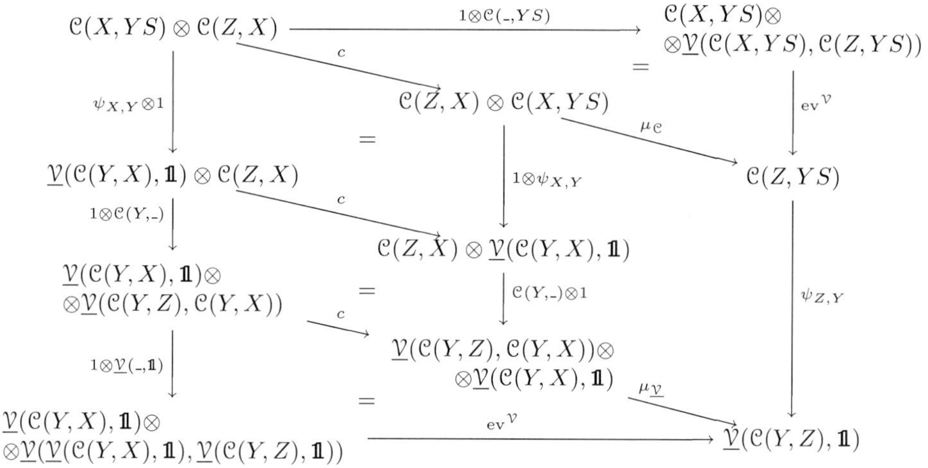

The right upper quadrilateral and the left lower quadrilateral commute by definition of $\mathcal{C}(\_, YS)$ and $\underline{\mathcal{V}}(\_, \mathbb{1})$ respectively, see (1.3) and its particular case (1.5). Since c is an isomorphism, commutativity of the exterior is equivalent to commutativity of the pentagon. Again, by closedness, this is equivalent to commutativity of the exterior of the following diagram:

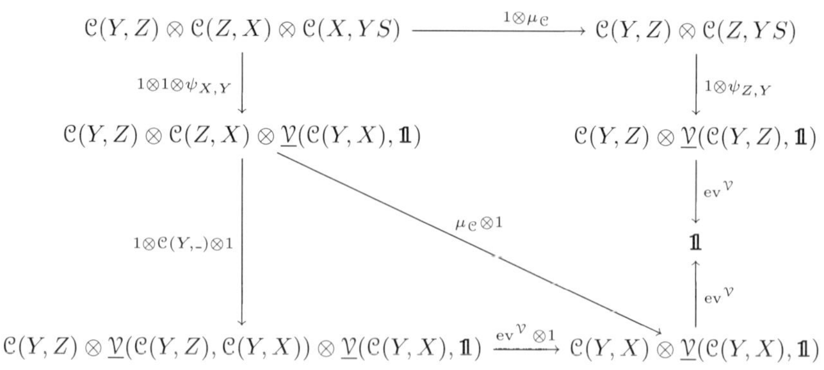

The triangle commutes by definition of $\mathcal{C}(Y,\_)$, see (1.4). It follows that naturality of $\psi_{-,Y}$ is equivalent to commutativity of the hexagon:

$$\begin{array}{ccc} \mathcal{C}(Y,Z)\otimes\mathcal{C}(Z,X)\otimes\mathcal{C}(X,YS) & \xrightarrow{1\otimes\mu_{\mathcal{C}}\psi_{Z,Y}} & \mathcal{C}(Y,Z)\otimes\underline{\mathcal{V}}(\mathcal{C}(Y,Z),\mathbb{1}) \\ {\scriptstyle \mu_{\mathcal{C}}\otimes\psi_{X,Y}}\downarrow & & \downarrow{\scriptstyle \mathrm{ev}^{\mathcal{V}}} \\ \mathcal{C}(Y,X)\otimes\underline{\mathcal{V}}(\mathcal{C}(Y,X),\mathbb{1}) & \xrightarrow{\mathrm{ev}^{\mathcal{V}}} & \mathbb{1} \end{array} \tag{2.4}$$

Assume that $\psi_{-,Y}$ is natural, so the above diagram commutes, and consider a particular case, $Z=Y$. Composing both paths of the diagram with the morphism $1^{\mathcal{C}}_Y\otimes1\otimes1:\mathcal{C}(Y,X)\otimes\mathcal{C}(X,YS)\to\mathcal{C}(Y,Y)\otimes\mathcal{C}(Y,X)\otimes\mathcal{C}(X,YS)$, we obtain:

$$\begin{array}{ccc} \mathcal{C}(Y,X)\otimes\mathcal{C}(X,YS) & \xrightarrow{\mu_{\mathcal{C}}} & \mathcal{C}(Y,YS) \\ {\scriptstyle 1\otimes\psi_{X,Y}}\downarrow & & \downarrow{\scriptstyle \tau_Y} \\ \mathcal{C}(Y,X)\otimes\underline{\mathcal{V}}(\mathcal{C}(Y,X),\mathbb{1}) & \xrightarrow{\mathrm{ev}^{\mathcal{V}}} & \mathbb{1} \end{array} \tag{2.5}$$

where τ_Y is given by expression (2.3). By closedness, the above equation admits a unique solution $\psi_{X,Y}$, namely, (2.2).

Assume now that $\psi_{X,Y}$ is given by (2.2). Then (2.5) holds true. Plugging it into (2.4), whose commutativity has to be proven, we obtain the equation

$$\begin{array}{ccccc} \mathcal{C}(Y,Z)\otimes\mathcal{C}(Z,X)\otimes\mathcal{C}(X,YS) & \xrightarrow{1\otimes\mu_{\mathcal{C}}} & \mathcal{C}(Y,Z)\otimes\mathcal{C}(Z,YS) & \xrightarrow{\mu_{\mathcal{C}}} & \mathcal{C}(Y,YS) \\ {\scriptstyle \mu_{\mathcal{C}}\otimes1}\downarrow & & & & \downarrow{\scriptstyle \tau_Y} \\ \mathcal{C}(Y,X)\otimes\mathcal{C}(X,YS) & \xrightarrow{\mu_{\mathcal{C}}} & \mathcal{C}(Y,YS) & \xrightarrow{\tau_Y} & \mathbb{1} \end{array}$$

which holds true by associativity of composition. □

Lemma 2.4. *Let $S:\mathcal{C}\to\mathcal{C}$ be a $\mathcal{V}$-functor. Fix an object X of $\mathcal{C}$. A collection of morphisms $(\psi_{X,Y}:\mathcal{C}(X,YS)\to\underline{\mathcal{V}}(\mathcal{C}(Y,X),\mathbb{1}))_{Y\in\mathrm{Ob}\,\mathcal{C}}$ of $\mathcal{V}$ is natural in Y if and only if for each $Y\in\mathrm{Ob}\,\mathcal{C}$*

$$\begin{aligned} \psi_{X,Y} = \big[\mathcal{C}(X,YS) &\xrightarrow{\mathrm{coev}^{\mathcal{V}}} \underline{\mathcal{V}}(\mathcal{C}(Y,X),\mathcal{C}(Y,X)\otimes\mathcal{C}(X,YS)) \xrightarrow{\underline{\mathcal{V}}(1,S\otimes1)} \\ \underline{\mathcal{V}}(\mathcal{C}(Y,X),\mathcal{C}(YS,XS)\otimes\mathcal{C}(X,YS)) &\xrightarrow{\underline{\mathcal{V}}(1,c\mu_{\mathcal{C}})} \underline{\mathcal{V}}(\mathcal{C}(Y,X),\mathcal{C}(X,XS)) \\ &\xrightarrow{\underline{\mathcal{V}}(1,\tau_X)} \underline{\mathcal{V}}(\mathcal{C}(Y,X),\mathbb{1})\big], \end{aligned} \tag{2.6}$$

where τ_X is given by (2.3).

Proof. Naturality of $\psi_{X,-}$ presented by the square

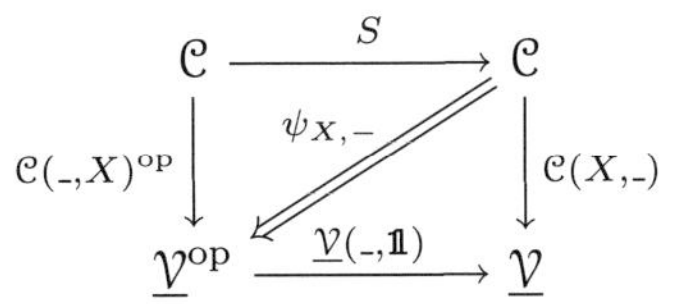

is expressed by commutativity in $\mathcal{V}$ of the following diagram:

$$\begin{array}{ccc}
\mathcal{C}(Y,Z) & \xrightarrow{\quad S \quad} & \mathcal{C}(YS,ZS) \\
\Big\downarrow{\scriptstyle \mathcal{C}(_,X)} & & \Big\downarrow{\scriptstyle \mathcal{C}(X,_)} \\
\underline{\mathcal{V}}(\mathcal{C}(Z,X),\mathcal{C}(Y,X)) & & \underline{\mathcal{V}}(\mathcal{C}(X,YS),\mathcal{C}(X,ZS)) \\
\Big\downarrow{\scriptstyle \underline{\mathcal{V}}(_,\mathbb{1})} & & \Big\downarrow{\scriptstyle \underline{\mathcal{V}}(1,\psi_{X,Z})} \\
\underline{\mathcal{V}}(\underline{\mathcal{V}}(\mathcal{C}(Y,X),\mathbb{1}),\underline{\mathcal{V}}(\mathcal{C}(Z,X),\mathbb{1})) & \xrightarrow{\underline{\mathcal{V}}(\psi_{X,Y},1)} & \underline{\mathcal{V}}(\mathcal{C}(X,YS),\underline{\mathcal{V}}(\mathcal{C}(Z,X),\mathbb{1}))
\end{array}$$

By closedness, the latter is equivalent to commutativity of the exterior of the diagram displayed on the facing page. Since c is an isomorphism, it follows that the polygon marked by $\boxed{*}$ is commutative. By closedness, this is equivalent to commutativity of the exterior of the following diagram:

$$\begin{array}{ccc}
\mathcal{C}(Z,X)\otimes\mathcal{C}(X,YS)\otimes\mathcal{C}(Y,Z) & \xrightarrow{1\otimes1\otimes S} & \mathcal{C}(Z,X)\otimes\mathcal{C}(X,YS)\otimes\mathcal{C}(YS,ZS) \\
\Big\downarrow{\scriptstyle 1\otimes c} & & \Big\downarrow{\scriptstyle 1\otimes\mu_{\mathcal{C}}} \\
\mathcal{C}(Z,X)\otimes\mathcal{C}(Y,Z)\otimes\mathcal{C}(X,YS) & & \mathcal{C}(Z,X)\otimes\mathcal{C}(X,ZS) \\
\Big\downarrow{\scriptstyle 1\otimes1\otimes\psi_{X,Y}} & & \Big\downarrow{\scriptstyle 1\otimes\psi_{X,Z}} \\
\mathcal{C}(Z,X)\otimes\mathcal{C}(Y,Z)\otimes\underline{\mathcal{V}}(\mathcal{C}(Y,X),\mathbb{1}) & & \mathcal{C}(Z,X)\otimes\underline{\mathcal{V}}(\mathcal{C}(Z,X),\mathbb{1}) \\
 & \searrow{\scriptstyle c\mu_{\mathcal{C}}\otimes1} & \Big\downarrow{\scriptstyle \mathrm{ev}^{\mathcal{V}}} \\
\Big\downarrow{\scriptstyle 1\otimes\mathcal{C}(_,X)\otimes1} & & \mathbb{1} \\
 & & \Big\uparrow{\scriptstyle \mathrm{ev}^{\mathcal{V}}} \\
\mathcal{C}(Z,X)\otimes\underline{\mathcal{V}}(\mathcal{C}(Z,X),\mathcal{C}(Y,X))\otimes\underline{\mathcal{V}}(\mathcal{C}(Y,X),\mathbb{1}) & \xrightarrow{\mathrm{ev}^{\mathcal{V}}\otimes1} & \mathcal{C}(Y,X)\otimes\underline{\mathcal{V}}(\mathcal{C}(Y,X),\mathbb{1})
\end{array}$$

The triangle commutes by (1.3). Therefore, the remaining polygon is commutative as well:

$$\begin{array}{ccc}
\mathcal{C}(Z,X)\otimes\mathcal{C}(X,YS)\otimes\mathcal{C}(Y,Z) & \xrightarrow{1\otimes1\otimes S} & \mathcal{C}(Z,X)\otimes\mathcal{C}(X,YS)\otimes\mathcal{C}(YS,ZS) \\
\Big\downarrow{\scriptstyle (123)^{\sim}} & & \Big\downarrow{\scriptstyle 1\otimes\mu_{\mathcal{C}}} \\
\mathcal{C}(Y,Z)\otimes\mathcal{C}(Z,X)\otimes\mathcal{C}(X,YS) & & \mathcal{C}(Z,X)\otimes\mathcal{C}(X,ZS) \\
\Big\downarrow{\scriptstyle \mu_{\mathcal{C}}\otimes1} & & \Big\downarrow{\scriptstyle 1\otimes\psi_{X,Z}} \\
\mathcal{C}(Y,X)\otimes\mathcal{C}(X,YS) & & \mathcal{C}(Z,X)\otimes\underline{\mathcal{V}}(\mathcal{C}(Z,X),\mathbb{1}) \\
\Big\downarrow{\scriptstyle 1\otimes\psi_{X,Y}} & & \Big\downarrow{\scriptstyle \mathrm{ev}^{\mathcal{V}}} \\
\mathcal{C}(Y,X)\otimes\underline{\mathcal{V}}(\mathcal{C}(Y,X),\mathbb{1}) & \xrightarrow{\quad \mathrm{ev}^{\mathcal{V}} \quad} & \mathbb{1}
\end{array} \tag{2.7}$$

Suppose that the collection of morphisms $(\psi_{X,Y}\colon \mathcal{C}(X,YS)\to\underline{\mathcal{V}}(\mathcal{C}(Y,X),\mathbb{1}))_{Y\in\mathrm{Ob}\,\mathcal{C}}$ is natural in Y. Consider diagram (2.7) with $Z = X$. Composing both paths with

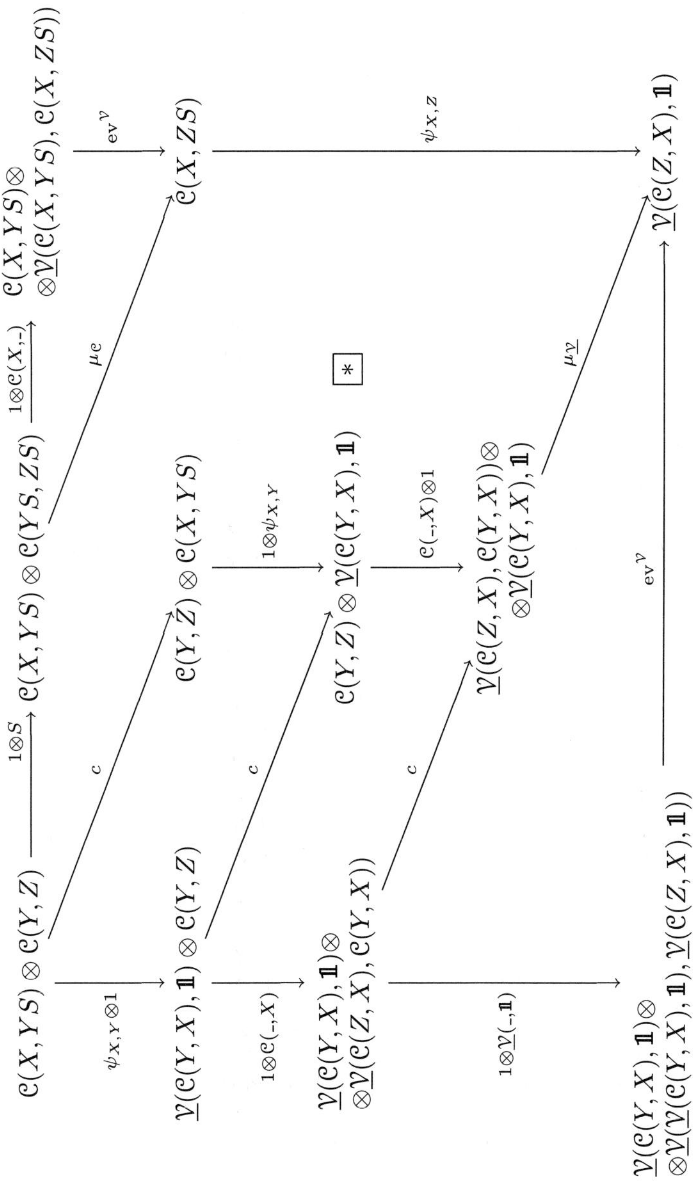
$\mathcal{C}(X,YS)\otimes\mathcal{C}(Y,Z)$
$1\otimes S$
$\mathcal{C}(X,YS)\otimes\mathcal{C}(YS,ZS)$
$1\otimes\mathcal{C}(X,\_)$
$\mathcal{C}(X,YS)\otimes$
$\otimes\underline{\mathcal{V}}(\mathcal{C}(X,YS),\mathcal{C}(X,ZS))$
$\mathrm{ev}^{\mathcal{V}}$
$\mu_{\mathcal{C}}$
c
$\psi_{X,Y}\otimes 1$
$\underline{\mathcal{V}}(\mathcal{C}(Y,X),\mathbb{1})\otimes\mathcal{C}(Y,Z)$
$\mathcal{C}(Y,Z)\otimes\mathcal{C}(X,YS)$
$\mathcal{C}(X,ZS)$
$1\otimes\psi_{X,Y}$
$\psi_{X,Z}$
$1\otimes\mathcal{C}(\_,X)$
$\underline{\mathcal{V}}(\mathcal{C}(Y,X),\mathbb{1})\otimes$
$\otimes\underline{\mathcal{V}}(\mathcal{C}(Z,X),\mathcal{C}(Y,X))$
$\mathcal{C}(Y,Z)\otimes\underline{\mathcal{V}}(\mathcal{C}(Y,X),\mathbb{1})$
$*$
$\mathcal{C}(\_,X)\otimes 1$
$\underline{\mathcal{V}}(\mathcal{C}(Z,X),\mathcal{C}(Y,X))\otimes$
$\otimes\underline{\mathcal{V}}(\mathcal{C}(Y,X),\mathbb{1})$
$\mu_{\underline{\mathcal{V}}}$
$1\otimes\underline{\mathcal{V}}(\_,\mathbb{1})$
$\underline{\mathcal{V}}(\mathcal{C}(Y,X),\mathbb{1})\otimes$
$\otimes\underline{\mathcal{V}}(\underline{\mathcal{V}}(\mathcal{C}(Y,X),\mathbb{1}),\underline{\mathcal{V}}(\mathcal{C}(Z,X),\mathbb{1}))$
$\mathrm{ev}^{\mathcal{V}}$
$\underline{\mathcal{V}}(\mathcal{C}(Z,X),\mathbb{1})$

FIGURE 1.

the morphism $1^{\mathcal{C}}_X \otimes 1 \otimes 1 : \mathcal{C}(X,YS) \otimes \mathcal{C}(Y,X) \to \mathcal{C}(X,X) \otimes \mathcal{C}(X,YS) \otimes \mathcal{C}(Y,X)$ gives an equation:

$$
\begin{array}{ccccc}
\mathcal{C}(Y,X)\otimes\mathcal{C}(X,YS) & \xrightarrow{S\otimes 1} & \mathcal{C}(YS,XS)\otimes\mathcal{C}(X,YS) & \xrightarrow{c\mu_{\mathcal{C}}} & \mathcal{C}(X,XS) \\
{\scriptstyle 1\otimes\psi_{X,Y}}\downarrow & & = & & \downarrow{\scriptstyle \tau_X} \\
\mathcal{C}(Y,X)\otimes\underline{\mathcal{V}}(\mathcal{C}(Y,X),\mathbb{1}) & & \xrightarrow{\qquad\mathrm{ev}^{\mathcal{V}}\qquad} & & \mathbb{1}
\end{array}
\tag{2.8}
$$

The only solution to the above equation is given by (2.6).

Conversely, suppose equation (2.8) holds. It suffices to prove that diagram (2.7) is commutative. Plugging in the expressions for $(1 \otimes \psi_{X,Y})\,\mathrm{ev}^{\mathcal{V}}$ and $(1 \otimes \psi_{X,Z})\,\mathrm{ev}^{\mathcal{V}}$ into (2.7), we obtain (cancelling a common permutation of the factors of the source object):

$$
\begin{array}{ccc}
\mathcal{C}(X,YS)\otimes\mathcal{C}(Y,Z)\otimes\mathcal{C}(Z,X) & \xrightarrow{1\otimes S\otimes S} & \mathcal{C}(X,YS)\otimes\mathcal{C}(YS,ZS)\otimes\mathcal{C}(ZS,XS) \\
{\scriptstyle 1\otimes\mu_{\mathcal{C}}}\downarrow & & \downarrow{\scriptstyle \mu_{\mathcal{C}}\otimes 1} \\
\mathcal{C}(X,YS)\otimes\mathcal{C}(Y,X) & & \mathcal{C}(X,ZS)\otimes\mathcal{C}(ZS,XS) \\
{\scriptstyle 1\otimes S}\downarrow & & \downarrow{\scriptstyle \mu_{\mathcal{C}}} \\
\mathcal{C}(X,YS)\otimes\mathcal{C}(YS,XS) & & \mathcal{C}(X,XS) \\
{\scriptstyle \mu_{\mathcal{C}}}\downarrow & & \downarrow{\scriptstyle \tau_X} \\
\mathcal{C}(X,XS) & \xrightarrow{\qquad\tau_X\qquad} & \mathbb{1}
\end{array}
$$

Commutativity of the diagram follows from associativity of $\mu_{\mathcal{C}}$ and the fact that S is a $\mathcal{V}$-functor. The lemma is proven. □

Proposition 2.5. *Assume that $S : \mathcal{C} \to \mathcal{C}$ is a $\mathcal{V}$-functor, and ψ is a natural transformation as in (2.1). Then the following diagram commutes (in $\mathcal{V}$):*

$$
\begin{array}{ccc}
\mathcal{C}(Y,X) & \xrightarrow{e} & \underline{\mathcal{V}}(\underline{\mathcal{V}}(\mathcal{C}(Y,X),\mathbb{1}),\mathbb{1}) \\
{\scriptstyle S}\downarrow & & \downarrow{\scriptstyle \underline{\mathcal{V}}(\psi_{X,Y},\mathbb{1})} \\
\mathcal{C}(YS,XS) & \xrightarrow{\psi_{YS,X}} & \underline{\mathcal{V}}(\mathcal{C}(X,YS),\mathbb{1})
\end{array}
$$

In particular, if for each pair of objects $X, Y \in \mathrm{Ob}\,\mathcal{C}$ the object $\mathcal{C}(Y,X)$ is reflexive, and ψ is an isomorphism, then S is fully faithful.

Proof. By closedness, it suffices to prove commutativity of the following diagram:

$$\begin{array}{ccc}
\mathcal{C}(X,YS)\otimes\mathcal{C}(Y,X) & \xrightarrow{1\otimes e} & \mathcal{C}(X,YS)\otimes\underline{\mathcal{V}}(\underline{\mathcal{V}}(\mathcal{C}(Y,X),\mathbb{1}),\mathbb{1}) \\
\downarrow{\scriptstyle 1\otimes S} & & \downarrow{\scriptstyle 1\otimes\underline{\mathcal{V}}(\psi_{X,Y},\mathbb{1})} \\
\mathcal{C}(X,YS)\otimes\mathcal{C}(YS,XS) & & \mathcal{C}(X,YS)\otimes\underline{\mathcal{V}}(\mathcal{C}(X,YS),\mathbb{1}) \\
\downarrow{\scriptstyle 1\otimes\psi_{YS,X}} & & \downarrow{\scriptstyle \mathrm{ev}^{\mathcal{V}}} \\
\mathcal{C}(X,YS)\otimes\underline{\mathcal{V}}(\mathcal{C}(X,YS),\mathbb{1}) & \xrightarrow{\mathrm{ev}^{\mathcal{V}}} & \mathbb{1}
\end{array}$$

Using (2.5) and the definition of e, the above diagram can be transformed as follows:

$$\begin{array}{ccc}
\mathcal{C}(X,YS)\otimes\mathcal{C}(Y,X) & \xrightarrow{c} & \mathcal{C}(Y,X)\otimes\mathcal{C}(X,YS) \\
\downarrow{\scriptstyle 1\otimes S} & & \downarrow{\scriptstyle 1\otimes\psi_{X,Y}} \\
\mathcal{C}(X,YS)\otimes\mathcal{C}(YS,XS) & & \mathcal{C}(Y,X)\otimes\underline{\mathcal{V}}(\mathcal{C}(Y,X),\mathbb{1}) \\
\downarrow{\scriptstyle \mu_{\mathcal{C}}} & & \downarrow{\scriptstyle \mathrm{ev}^{\mathcal{V}}} \\
\mathcal{C}(X,XS) & \xrightarrow{\tau_X} & \mathbb{1}
\end{array}$$

It is commutative by (2.8). □

Proposition 2.5 implies that a right Serre functor is fully faithful if and only if $\mathcal{C}$ is hom-reflexive, *i.e.*, if $\mathcal{C}(X,Y)$ is a reflexive object of $\mathcal{V}$ for each pair of objects $X,Y\in\operatorname{Ob}\mathcal{C}$. If this is the case, a right Serre functor will be a Serre functor if and only if it is essentially surjective on objects. The most natural reason for hom-reflexivity is, of course, $\Bbbk$ being a field. When $\Bbbk$ is a field, an object C of $\mathbf{gr}(\Bbbk\text{-vect})$ is reflexive iff all spaces C^n are finite-dimensional. The ring $\Bbbk$ being a field, the homology functor $H^\bullet:\mathcal{K}\to\mathbf{gr}(\Bbbk\text{-vect})$ is an equivalence (see e.g. [GM03, Chapter III, § 2, Proposition 4]). Hence, an object C of $\mathcal{K}$ is reflexive iff all homology spaces H^nC are finite-dimensional. A projective module of finite rank over an arbitrary commutative ring $\Bbbk$ is reflexive as an object of a rigid monoidal category [DM82, Example 1.23]. Thus, an object C of $\mathbf{gr}(\Bbbk\text{-}\mathbf{Mod})$ whose components C^n are projective $\Bbbk$-modules of finite rank is reflexive.

Proposition 2.6. *Let $\mathcal{C}$ be a $\mathcal{V}$-category. There exists a right Serre $\mathcal{V}$-functor $S:\mathcal{C}\to\mathcal{C}$ if and only if for each object $Y\in\operatorname{Ob}\mathcal{C}$ the $\mathcal{V}$-functor*

$$\operatorname{Hom}_{\mathcal{C}}(Y,\_)^{\mathrm{op}}\cdot\underline{\mathcal{V}}(\_,\mathbb{1})=\underline{\mathcal{V}}(\mathcal{C}(Y,\_)^{\mathrm{op}},\mathbb{1}):\mathcal{C}^{\mathrm{op}}\to\underline{\mathcal{V}}$$

is representable.

Proof. Standard, see [Kel82, Section 1.10]. □

2.7. Commutation with equivalences

Let $\mathcal{C}$ and $\mathcal{C}'$ be $\mathcal{V}$-categories with right Serre functors $S : \mathcal{C} \to \mathcal{C}$ and $S' : \mathcal{C}' \to \mathcal{C}'$, respectively. Let ψ and ψ' be isomorphisms as in (2.1). For objects $Y \in \operatorname{Ob}\mathcal{C}$, $Z \in \operatorname{Ob}\mathcal{C}'$, define τ_Y, τ'_Z by (2.3). Let $T : \mathcal{C} \to \mathcal{C}'$ be a $\mathcal{V}$-functor, and suppose that T is fully faithful. Then there is a natural transformation $\varkappa : ST \to TS'$ such that, for each object $Y \in \operatorname{Ob}\mathcal{C}$, the following equation holds:

$$\big[\mathcal{C}(Y,YS) \xrightarrow{T} \mathcal{C}'(YT,YST) \xrightarrow{\mathcal{C}'(YT,\varkappa)} \mathcal{C}'(YT,YTS') \xrightarrow{\tau'_{YT}} \mathbb{1}\big] = \tau_Y. \tag{2.9}$$

Indeed, the left-hand side of equation (2.9) equals

$$\big[\mathcal{C}(Y,YS) \xrightarrow{T} \mathcal{C}'(YT,YST) \xrightarrow{1\otimes\varkappa_Y} \mathcal{C}'(YT,YST)\otimes\mathcal{C}'(YST,YTS') \xrightarrow{\mu_{\mathcal{C}'}} \mathcal{C}'(YT,YTS') \xrightarrow{\tau'_{YT}} \mathbb{1}\big].$$

Using relation (2.5) between τ'_{YT} and $\psi'_{YST,YT}$, we get:

$$\big[\mathcal{C}(Y,YS) \xrightarrow{T} \mathcal{C}'(YT,YST) \xrightarrow{1\otimes\varkappa_Y} \mathcal{C}'(YT,YST)\otimes\mathcal{C}'(YST,YTS') \xrightarrow{1\otimes\psi'_{YST,YT}} \mathcal{C}'(YT,YST)\otimes\underline{\mathcal{V}}(\mathcal{C}'(YT,YST),\mathbb{1}) \xrightarrow{\mathrm{ev}^{\mathcal{V}}} \mathbb{1}\big].$$

Therefore, equation (2.9) is equivalent to the following equation:

$$\big[\mathcal{C}(Y,YS) \xrightarrow{1\otimes\varkappa_Y} \mathcal{C}(Y,YS)\otimes\mathcal{C}'(YST,YTS') \xrightarrow{1\otimes\psi'_{YST,YT}} \mathcal{C}(Y,YS)\otimes\underline{\mathcal{V}}(\mathcal{C}'(YT,YST),\mathbb{1}) \xrightarrow{1\otimes\underline{\mathcal{V}}(T,\mathbb{1})} \mathcal{C}(Y,YS)\otimes\underline{\mathcal{V}}(\mathcal{C}(Y,YS),\mathbb{1}) \xrightarrow{\mathrm{ev}^{\mathcal{V}}} \mathbb{1}\big] = \tau_Y.$$

It implies that the composite

$$\mathbb{1} \xrightarrow{\varkappa_Y} \mathcal{C}'(YST,YTS') \xrightarrow{\psi'_{YST,YT}} \underline{\mathcal{V}}(\mathcal{C}'(YT,YST),\mathbb{1}) \xrightarrow{\underline{\mathcal{V}}(T,\mathbb{1})} \underline{\mathcal{V}}(\mathcal{C}(Y,YS),\mathbb{1})$$

is equal to $\underline{\tau_Y} : \mathbb{1} \to \underline{\mathcal{V}}(\mathcal{C}(Y,YS),\mathbb{1})$, the morphism that corresponds to τ_Y by closedness of the category $\mathcal{V}$. Since the morphisms $\psi'_{YST,YT}$ and $\underline{\mathcal{V}}(T,\mathbb{1})$ are invertible, the morphism $\varkappa_Y : \mathbb{1} \to \mathcal{C}'(YST,YTS')$ is uniquely determined.

Lemma 2.8. *The transformation $\varkappa$ satisfies the following equation:*

$$\psi_{X,Y} = \big[\mathcal{C}(X,YS) \xrightarrow{T} \mathcal{C}'(XT,YST) \xrightarrow{\mathcal{C}'(XT,\varkappa)} \mathcal{C}(XT,YTS') \xrightarrow{\psi'_{XT,YT}} \underline{\mathcal{V}}(\mathcal{C}'(YT,XT),\mathbb{1}) \xrightarrow{\underline{\mathcal{V}}(T,\mathbb{1})} \underline{\mathcal{V}}(\mathcal{C}(Y,X),\mathbb{1})\big],$$

for each pair of objects $X,Y \in \operatorname{Ob}\mathcal{C}$.

Proof. The exterior of the following diagram commutes:

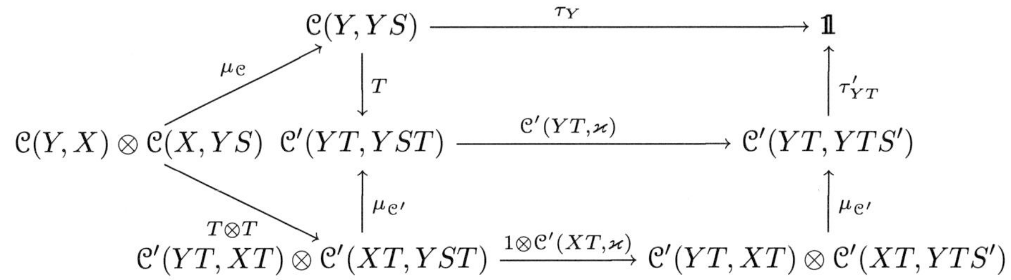
$\mathcal{C}(Y,YS)$
τ_Y
$\mathbb{1}$
$\mu_{\mathcal{C}}$
T
τ'_{YT}
$\mathcal{C}(Y,X)\otimes\mathcal{C}(X,YS)$
$\mathcal{C}'(YT,YST)$
$\mathcal{C}'(YT,\varkappa)$
$\mathcal{C}'(YT,YTS')$
$\mu_{\mathcal{C}'}$
$\mu_{\mathcal{C}'}$
$T\otimes T$
$\mathcal{C}'(YT,XT)\otimes\mathcal{C}'(XT,YST)$
$1\otimes\mathcal{C}'(XT,\varkappa)$
$\mathcal{C}'(YT,XT)\otimes\mathcal{C}'(XT,YTS')$

The right upper square commutes by the definition of $\varkappa$, commutativity of the lower square is a consequence of associativity of $\mu_{\mathcal{C}'}$. The left quadrilateral is commutative since T is a $\mathcal{V}$-functor. Transforming both paths with the help of equation (2.5) yields the following equation:

$$\big[\mathcal{C}(Y,X)\otimes\mathcal{C}(X,YS)\xrightarrow{1\otimes\psi_{X,Y}}\mathcal{C}(Y,X)\otimes\underline{\mathcal{V}}(\mathcal{C}(Y,X),\mathbb{1})\xrightarrow{\mathrm{ev}^{\mathcal{V}}}\mathbb{1}\big]$$
$$=\big[\mathcal{C}(Y,X)\otimes\mathcal{C}(X,YS)\xrightarrow{T\otimes T}\mathcal{C}'(YT,XT)\otimes\mathcal{C}'(XT,YST)\xrightarrow{1\otimes\mathcal{C}'(XT,\varkappa)}$$
$$\mathcal{C}'(YT,XT)\otimes\mathcal{C}'(XT,YTS')\xrightarrow{1\otimes\psi'_{XT,YT}}\mathcal{C}'(YT,XT)\otimes\underline{\mathcal{V}}(\mathcal{C}'(YT,XT),\mathbb{1})\xrightarrow{\mathrm{ev}^{\mathcal{V}}}\mathbb{1}\big]$$
$$=\big[\mathcal{C}(Y,X)\otimes\mathcal{C}(X,YS)\xrightarrow{1\otimes T}\mathcal{C}(Y,X)\otimes\mathcal{C}'(XT,YST)$$
$$\xrightarrow{1\otimes\mathcal{C}'(XT,\varkappa)}\mathcal{C}(Y,X)\otimes\mathcal{C}'(XT,YTS')\xrightarrow{1\otimes\psi'_{XT,YT}}\mathcal{C}(Y,X)\otimes\underline{\mathcal{V}}(\mathcal{C}'(YT,XT),\mathbb{1})$$
$$\xrightarrow{1\otimes\underline{\mathcal{V}}(T,\mathbb{1})}\mathcal{C}(Y,X)\otimes\underline{\mathcal{V}}(\mathcal{C}(Y,X),\mathbb{1})\xrightarrow{\mathrm{ev}^{\mathcal{V}}}\mathbb{1}\big].$$

The required equation follows by closedness of $\mathcal{V}$. □

Corollary 2.9. *If T is an equivalence, then the natural transformation $\varkappa : ST \to TS'$ is an isomorphism.*

Proof. Lemma 2.8 implies that $\mathcal{C}'(XT,\varkappa) : \mathcal{C}'(XT,YST) \to \mathcal{C}'(XT,XTS')$ is an isomorphism, for each $X \in \operatorname{Ob}\mathcal{C}$. Since T is essentially surjective, it follows that the morphism $\mathcal{C}'(Z,\varkappa) : \mathcal{C}'(Z,YST) \to \mathcal{C}'(Z,YTS')$ is invertible, for each $Z \in \operatorname{Ob}\mathcal{C}'$, thus $\varkappa$ is an isomorphism. □

Corollary 2.10. *A right Serre $\mathcal{V}$-functor is unique up to an isomorphism.*

Proof. Suppose $S, S' : \mathcal{C} \to \mathcal{C}$ are right Serre functors. Applying Corollary 2.9 to the functor $T = \mathrm{Id}_{\mathcal{C}} : \mathcal{C} \to \mathcal{C}$ yields a natural isomorphism $\varkappa : S \to S'$. □

2.11. Trace functionals determine the Serre functor

Combining for a natural transformation ψ diagrams (2.5) and (2.8) we get the equation

$$\begin{array}{ccccc} \mathcal{C}(Y,X)\otimes\mathcal{C}(X,YS) & \xrightarrow{\mu_{\mathcal{C}}} & \mathcal{C}(Y,YS) & \xrightarrow{\tau_Y} & \mathbb{1} \\ \downarrow{\scriptstyle S\otimes 1} & & = & & \uparrow{\scriptstyle \tau_X} \\ \mathcal{C}(YS,XS)\otimes\mathcal{C}(X,YS) & \xrightarrow{c} & \mathcal{C}(X,YS)\otimes\mathcal{C}(YS,XS) & \xrightarrow{\mu_{\mathcal{C}}} & \mathcal{C}(X,XS) \end{array} \tag{2.10}$$

The above diagram can be written as the equation

$$\begin{array}{ccc} \mathcal{C}(X,YS)\otimes\mathcal{C}(Y,X) & \xrightarrow{1\otimes S} & \mathcal{C}(X,YS)\otimes\mathcal{C}(YS,XS) \\ \downarrow{\scriptstyle c} & = & \downarrow{\scriptstyle \phi_{YS,X}} \\ \mathcal{C}(Y,X)\otimes\mathcal{C}(X,YS) & \xrightarrow{\phi_{X,Y}} & \mathbb{1} \end{array} \tag{2.11}$$

When S is a fully faithful right Serre functor, the pairing

$$\phi_{X,Y} = \big[\mathcal{C}(Y,X)\otimes\mathcal{C}(X,YS) \xrightarrow{\mu_{\mathcal{C}}} \mathcal{C}(Y,YS) \xrightarrow{\tau_Y} \mathbb{1}\big] \tag{2.12}$$

is perfect. Namely, the induced by it morphism $\psi_{X,Y} : \mathcal{C}(X,YS) \to \underline{\mathcal{V}}(\mathcal{C}(Y,X),\mathbb{1})$ is invertible, and induced by the pairing

$$c\cdot\phi_{X,Y} = \big[\mathcal{C}(X,YS)\otimes\mathcal{C}(Y,X) \xrightarrow{c} \mathcal{C}(Y,X)\otimes\mathcal{C}(X,YS) \xrightarrow{\phi_{X,Y}} \mathbb{1}\big]$$

the morphism $\psi' : \mathcal{C}(Y,X) \to \underline{\mathcal{V}}(\mathcal{C}(X,YS),\mathbb{1})$ is invertible. In fact, diagram (2.11) implies that

$$\psi' = \big[\mathcal{C}(Y,X) \xrightarrow{S} \mathcal{C}(YS,XS) \xrightarrow{\psi_{YS,X}} \underline{\mathcal{V}}(\mathcal{C}(X,YS),\mathbb{1})\big].$$

Diagram (2.11) allows to restore the morphisms $S : \mathcal{C}(Y,X) \to \mathcal{C}(YS,XS)$ unambiguously from $\operatorname{Ob} S$ and the trace functionals τ, due to $\psi_{YS,X}$ being isomorphisms.

Proposition 2.12. *A map $\operatorname{Ob} S$ and trace functionals τ_X, $X \in \operatorname{Ob}\mathcal{C}$, such that the induced $\psi_{X,Y}$ from (2.2) are invertible, define a unique right Serre $\mathcal{V}$-functor $(S,\psi_{X,Y})$.*

Proof. Let us show that the obtained morphisms $S : \mathcal{C}(Y,X) \to \mathcal{C}(YS,XS)$ preserve the composition in $\mathcal{C}$. In fact, due to associativity of composition we have

$$\begin{aligned}
&\big[\mathcal{C}(X,ZS)\otimes\mathcal{C}(Z,Y)\otimes\mathcal{C}(Y,X) \xrightarrow{1\otimes S\otimes S} \mathcal{C}(X,ZS)\otimes\mathcal{C}(ZS,YS)\otimes\mathcal{C}(YS,XS)\\
&\qquad \xrightarrow{1\otimes\mu_{\mathcal{C}}} \mathcal{C}(X,ZS)\otimes\mathcal{C}(ZS,XS) \xrightarrow{\phi_{ZS,X}} \mathbb{1}\big]\\
&= \big[\mathcal{C}(X,ZS)\otimes\mathcal{C}(Z,Y)\otimes\mathcal{C}(Y,X) \xrightarrow{1\otimes S\otimes S} \mathcal{C}(X,ZS)\otimes\mathcal{C}(ZS,YS)\otimes\mathcal{C}(YS,XS)\\
&\qquad \xrightarrow{\mu_{\mathcal{C}}\otimes 1} \mathcal{C}(X,YS)\otimes\mathcal{C}(YS,XS) \xrightarrow{\mu_{\mathcal{C}}} \mathcal{C}(X,XS) \xrightarrow{\tau_X} \mathbb{1}\big]\\
&= \big[\mathcal{C}(X,ZS)\otimes\mathcal{C}(Z,Y)\otimes\mathcal{C}(Y,X) \xrightarrow{(1\otimes S\otimes 1)(\mu_{\mathcal{C}}\otimes 1)} \mathcal{C}(X,YS)\otimes\mathcal{C}(Y,X)\\
&\qquad \xrightarrow{1\otimes S} \mathcal{C}(X,YS)\otimes\mathcal{C}(YS,XS) \xrightarrow{\mu_{\mathcal{C}}} \mathcal{C}(X,XS) \xrightarrow{\tau_X} \mathbb{1}\big]\\
&= \big[\mathcal{C}(X,ZS)\otimes\mathcal{C}(Z,Y)\otimes\mathcal{C}(Y,X) \xrightarrow{(1\otimes S\otimes 1)(\mu_{\mathcal{C}}\otimes 1)} \mathcal{C}(X,YS)\otimes\mathcal{C}(Y,X)\\
&\qquad \xrightarrow{c} \mathcal{C}(Y,X)\otimes\mathcal{C}(X,YS) \xrightarrow{\mu_{\mathcal{C}}} \mathcal{C}(Y,YS) \xrightarrow{\tau_Y} \mathbb{1}\big]\\
&= \big[\mathcal{C}(X,ZS)\otimes\mathcal{C}(Z,Y)\otimes\mathcal{C}(Y,X) \xrightarrow{(123)_c(1\otimes 1\otimes S)} \mathcal{C}(Y,X)\otimes\mathcal{C}(X,ZS)\otimes\mathcal{C}(ZS,YS)\\
&\qquad \xrightarrow{1\otimes\mu_{\mathcal{C}}} \mathcal{C}(Y,X)\otimes\mathcal{C}(X,YS) \xrightarrow{\mu_{\mathcal{C}}} \mathcal{C}(Y,YS) \xrightarrow{\tau_Y} \mathbb{1}\big]\\
&= \big[\mathcal{C}(X,ZS)\otimes\mathcal{C}(Z,Y)\otimes\mathcal{C}(Y,X) \xrightarrow{(123)_c(1\otimes 1\otimes S)} \mathcal{C}(Y,X)\otimes\mathcal{C}(X,ZS)\otimes\mathcal{C}(ZS,YS)\\
&\qquad \xrightarrow{\mu_{\mathcal{C}}\otimes 1} \mathcal{C}(Y,ZS)\otimes\mathcal{C}(ZS,YS) \xrightarrow{\mu_{\mathcal{C}}} \mathcal{C}(Y,YS) \xrightarrow{\tau_Y} \mathbb{1}\big]\\
&= \big[\mathcal{C}(X,ZS)\otimes\mathcal{C}(Z,Y)\otimes\mathcal{C}(Y,X) \xrightarrow{(123)_c(\mu_{\mathcal{C}}\otimes 1)} \mathcal{C}(Y,ZS)\otimes\mathcal{C}(Z,Y)\\
&\qquad \xrightarrow{1\otimes S} \mathcal{C}(Y,ZS)\otimes\mathcal{C}(ZS,YS) \xrightarrow{\mu_{\mathcal{C}}} \mathcal{C}(Y,YS) \xrightarrow{\tau_Y} \mathbb{1}\big]\\
&= \big[\mathcal{C}(X,ZS)\otimes\mathcal{C}(Z,Y)\otimes\mathcal{C}(Y,X) \xrightarrow{(123)_c(\mu_{\mathcal{C}}\otimes 1)} \mathcal{C}(Y,ZS)\otimes\mathcal{C}(Z,Y)\\
&\qquad \xrightarrow{c} \mathcal{C}(Z,Y)\otimes\mathcal{C}(Y,ZS) \xrightarrow{\mu_{\mathcal{C}}} \mathcal{C}(Z,ZS) \xrightarrow{\tau_Z} \mathbb{1}\big]\\
&= \big[\mathcal{C}(X,ZS)\otimes\mathcal{C}(Z,Y)\otimes\mathcal{C}(Y,X) \xrightarrow{(321)_c} \mathcal{C}(Z,Y)\otimes\mathcal{C}(Y,X)\otimes\mathcal{C}(X,ZS)\\
&\qquad \xrightarrow{1\otimes\mu_{\mathcal{C}}} \mathcal{C}(Z,Y)\otimes\mathcal{C}(Y,ZS) \xrightarrow{\mu_{\mathcal{C}}} \mathcal{C}(Z,ZS) \xrightarrow{\tau_Z} \mathbb{1}\big].
\end{aligned}$$

On the other hand

$$\begin{aligned}
&\big[\mathcal{C}(X,ZS)\otimes\mathcal{C}(Z,Y)\otimes\mathcal{C}(Y,X) \xrightarrow{1\otimes\mu_{\mathcal{C}}} \mathcal{C}(X,ZS)\otimes\mathcal{C}(Z,X)\\
&\qquad \xrightarrow{1\otimes S} \mathcal{C}(X,ZS)\otimes\mathcal{C}(ZS,XS) \xrightarrow{\phi_{ZS,X}} \mathbb{1}\big]\\
&= \big[\mathcal{C}(X,ZS)\otimes\mathcal{C}(Z,Y)\otimes\mathcal{C}(Y,X) \xrightarrow{1\otimes\mu_{\mathcal{C}}} \mathcal{C}(X,ZS)\otimes\mathcal{C}(Z,X)\\
&\qquad \xrightarrow{c} \mathcal{C}(Z,X)\otimes\mathcal{C}(X,ZS) \xrightarrow{\phi_{X,Z}} \mathbb{1}\big]\\
&= \big[\mathcal{C}(X,ZS)\otimes\mathcal{C}(Z,Y)\otimes\mathcal{C}(Y,X) \xrightarrow{(321)_c} \mathcal{C}(Z,Y)\otimes\mathcal{C}(Y,X)\otimes\mathcal{C}(X,ZS)\\
&\qquad \xrightarrow{\mu_{\mathcal{C}}\otimes 1} \mathcal{C}(Z,X)\otimes\mathcal{C}(X,ZS) \xrightarrow{\mu_{\mathcal{C}}} \mathcal{C}(Z,ZS) \xrightarrow{\tau_Z} \mathbb{1}\big]
\end{aligned}$$

$$= \big[\mathcal{C}(X,ZS)\otimes\mathcal{C}(Z,Y)\otimes\mathcal{C}(Y,X) \xrightarrow{(321)_c} \mathcal{C}(Z,Y)\otimes\mathcal{C}(Y,X)\otimes\mathcal{C}(X,ZS)$$
$$\xrightarrow{1\otimes\mu_{\mathcal{C}}} \mathcal{C}(Z,Y)\otimes\mathcal{C}(Y,ZS) \xrightarrow{\mu_{\mathcal{C}}} \mathcal{C}(Z,ZS) \xrightarrow{\tau_Z} \mathbb{1}\big].$$

The last lines of both expressions coincide, hence $(S\otimes S)\mu_{\mathcal{C}} = \mu_{\mathcal{C}} S$.

Let us prove that the morphisms $S : \mathcal{C}(X,X) \to \mathcal{C}(XS,XS)$ of $\mathcal{V}$ preserve units. Indeed, the exterior of the following diagram commutes:

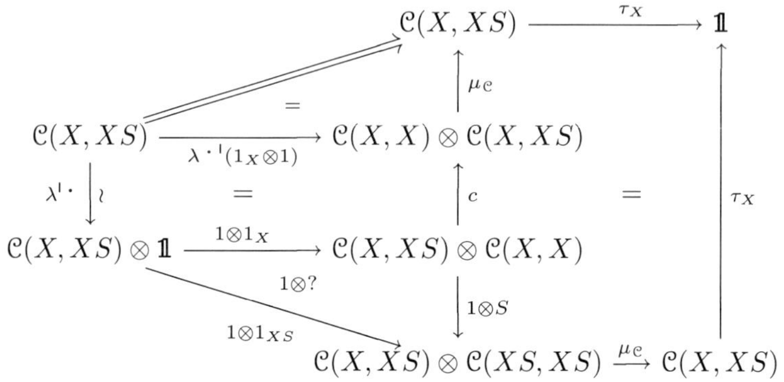

Therefore, both paths from $\mathcal{C}(X,XS)$ to $\mathbb{1}$, going through the isomorphism $\lambda^{I\cdot}$, sides of triangle marked '$1\otimes ?$', $\mu_{\mathcal{C}}$ and τ_X, compose to the same morphism τ_X. Invertibility of $\psi_{X,X}$ implies that the origin '?' of the mentioned triangle commutes, that is,

$$1_{XS} = \big[\mathbb{1} \xrightarrow{1_X} \mathcal{C}(X,X) \xrightarrow{S} \mathcal{C}(XS,XS)\big].$$

Summing up, the constructed $S : \mathcal{C} \to \mathcal{C}$ is a $\mathcal{V}$-functor. Applying Lemma 2.3 we deduce that $\psi_{-,Y}$ is natural in the first argument for all objects Y of $\mathcal{C}$. Recall that $\psi_{X,Y}$ is a unique morphism which makes diagram (2.5) commutative. Due to equation (2.10) $\psi_{X,Y}$ makes commutative also diagram (2.8). This means that $\psi_{X,Y}$ can be presented also in the form (2.6). Applying Lemma 2.4 we deduce that $\psi_{X,-}$ is natural in the second argument for all objects X of $\mathcal{C}$. Being natural in each variable ψ is natural as a whole [Kel82, Section 1.4]. □

2.13. Base change

Let $\mathcal{V} = (\mathcal{V}, \otimes^I_{\mathcal{V}}, \lambda^f_{\mathcal{V}})$, $\mathcal{W} = (\mathcal{W}, \otimes^I_{\mathcal{W}}, \lambda^f_{\mathcal{W}})$ be closed symmetric Monoidal $\mathscr{U}$-categories. Let $(B, \beta^I) : (\mathcal{V}, \otimes^I_{\mathcal{V}}, \lambda^f_{\mathcal{V}}) \to (\mathcal{W}, \otimes^I_{\mathcal{W}}, \lambda^f_{\mathcal{W}})$ be a lax symmetric Monoidal functor. Denote by $\widehat{B} : \widehat{\mathcal{V}} \to \widehat{\mathcal{W}}$ the corresponding multifunctor. According to [Man07], (B, β^I) gives rise to a lax symmetric Monoidal $\mathcal{C}at$-functor $(B_*, \beta^I_*) : \mathcal{V}\text{-}\mathcal{C}at \to \mathcal{W}\text{-}\mathcal{C}at$. Since the multicategories $\widehat{\mathcal{V}}$ and $\widehat{\mathcal{W}}$ are closed, the multifunctor $\widehat{B}$ determines the closing transformation $\underline{\widehat{B}}$. In particular, we have a $\mathcal{W}$-functor $B_*\underline{\mathcal{V}} \to \underline{\mathcal{W}}$, $X \mapsto BX$, which is denoted by $\underline{\widehat{B}}$ by abuse of notation, whose action on morphisms is found from the following equation in $\mathcal{W}$:

$$\big[BX \otimes B(\underline{\mathcal{V}}(X,Y)) \xrightarrow{1\otimes\underline{\widehat{B}}} BX \otimes \underline{\mathcal{W}}(BX,BY) \xrightarrow{\mathrm{ev}^{\mathcal{W}}} BY\big] = \widehat{B}(\mathrm{ev}^{\mathcal{V}}). \qquad (2.13)$$

Let $\widehat{B_*} : \widehat{\mathcal{V}\text{-}\mathcal{C}at} \to \widehat{\mathcal{W}\text{-}\mathcal{C}at}$ denote the symmetric $\mathcal{C}at$-multifunctor that corresponds to the lax symmetric Monoidal $\mathcal{C}at$-functor (B_*, β_*^I). Clearly, $\widehat{B_*}$ commutes with taking opposite.

In the sequel, the tensor product in the categories $\mathcal{V}$ and $\mathcal{W}$ is denoted by $\otimes$, the unit objects in both categories are denoted by $\mathbb{1}$.

Let $\mathcal{A}$ be a $\mathcal{V}$-category. We claim that the $\mathcal{W}$-functor

$$\widehat{B_*}\,\mathrm{Hom}_{\mathcal{A}} \cdot \underline{\widehat{B}} = \big[B_*(\mathcal{A})^{\mathrm{op}} \boxtimes B_*(\mathcal{A}) \xrightarrow{\widehat{B_*}\,\mathrm{Hom}_{\mathcal{A}}} B_*\underline{\mathcal{V}} \xrightarrow{\underline{\widehat{B}}} \underline{\mathcal{W}}\big]$$

coincides with $\mathrm{Hom}_{B_*\mathcal{A}}$. Indeed, both functors send a pair of objects $(X,Y) \in \mathrm{Ob}\,\mathcal{A} \times \mathrm{Ob}\,\mathcal{A}$ to the object $B(\mathcal{A}(X,Y)) = (B_*\mathcal{A})(X,Y)$ of $\mathcal{W}$. Applying $\widehat{B}$ to equation (1.2) yields a commutative diagram

$$\begin{array}{ccc}
B(\mathcal{A}(X,U))\otimes B(\mathcal{A}(Y,X))\otimes B(\mathcal{A}(U,V)) & \xrightarrow{1\otimes \widehat{B}\,\mathrm{Hom}_{\mathcal{A}}} & B(\mathcal{A}(X,U))\otimes B\underline{\mathcal{V}}(\mathcal{A}(X,U),\mathcal{A}(Y,V)) \\
\downarrow{\scriptstyle c\otimes 1} & & \downarrow{\scriptstyle \widehat{B}(\mathrm{ev}^{\mathcal{V}})} \\
B(\mathcal{A}(Y,X))\otimes B(\mathcal{A}(X,U))\otimes B(\mathcal{A}(U,V)) & \xrightarrow[\mu^{\mathbf{3}}_{B_*\mathcal{A}}]{\widehat{B}(\mu^{\mathbf{3}}_{\mathcal{A}})\;\Vert} & B(\mathcal{A}(Y,V))
\end{array}$$

Expanding $\widehat{B}(\mathrm{ev}^{\mathcal{V}})$ according to (2.13) we transform the above diagram as follows:

$$\begin{array}{ccc}
B(\mathcal{A}(X,U))\otimes B(\mathcal{A}(Y,X))\otimes B(\mathcal{A}(U,V)) & \xrightarrow{1\otimes \widehat{B}\,\mathrm{Hom}_{\mathcal{A}}\cdot\underline{\widehat{B}}} & B(\mathcal{A}(X,U))\otimes \underline{\mathcal{W}}(B(\mathcal{A}(X,U)),B(\mathcal{A}(Y,V))) \\
\downarrow{\scriptstyle c\otimes 1} & & \downarrow{\scriptstyle \mathrm{ev}^{\mathcal{W}}} \\
B(\mathcal{A}(Y,X))\otimes B(\mathcal{A}(X,U))\otimes B(\mathcal{A}(U,V)) & \xrightarrow{\mu^{\mathbf{3}}_{B_*\mathcal{A}}} & B(\mathcal{A}(Y,V))
\end{array}$$

It follows that the functors $\widehat{B_*}\,\mathrm{Hom}_{\mathcal{A}} \cdot \underline{\widehat{B}}$ and $\mathrm{Hom}_{B_*\mathcal{A}}$ satisfy the same equation, therefore they must coincide by closedness of $\mathcal{W}$.

There is a natural transformation of $\mathcal{W}$-functors ζ' as in the diagram below:

$$\begin{array}{ccc}
B_*\underline{\mathcal{V}}^{\mathrm{op}} & \xrightarrow{B_*\underline{\mathcal{V}}(_,\mathbb{1})} & B_*\underline{\mathcal{V}} \\
\downarrow{\scriptstyle (\underline{\widehat{B}})^{\mathrm{op}}} & \swarrow\!\!\swarrow{\scriptstyle \zeta'} & \downarrow{\scriptstyle \underline{\widehat{B}}} \\
\underline{\mathcal{W}}^{\mathrm{op}} & \xrightarrow{\underline{\mathcal{W}}(_,B\mathbb{1})} & \underline{\mathcal{W}}
\end{array}$$

For each object X, the morphism $\zeta'_X : B(\underline{\mathcal{V}}(X,\mathbb{1})) \to \underline{\mathcal{W}}(BX, B\mathbb{1})$ in $\mathcal{W}$ comes from the map $\widehat{B}(\mathrm{ev}^{\mathcal{V}}) : BX \otimes B\underline{\mathcal{V}}(X,\mathbb{1}) \to B\mathbb{1}$ by closedness of $\mathcal{W}$. In other words, $\zeta'_X = \underline{\widehat{B}}_{X,\mathbb{1}}$. Naturality of ζ' is expressed by the following equation in $\mathcal{W}$:

$$\begin{array}{ccc}
B\underline{\mathcal{V}}(Y,X) & \xrightarrow{B\underline{\mathcal{V}}(_,\mathbb{1})} & B\underline{\mathcal{V}}(\underline{\mathcal{V}}(X,\mathbb{1}),\underline{\mathcal{V}}(Y,\mathbb{1})) \\
\downarrow{\scriptstyle \underline{\widehat{B}}} & & \downarrow{\scriptstyle \underline{\widehat{B}}} \\
\underline{\mathcal{W}}(BY,BX) & & \underline{\mathcal{W}}(B\underline{\mathcal{V}}(X,\mathbb{1}),B\underline{\mathcal{V}}(Y,\mathbb{1})) \\
\downarrow{\scriptstyle \underline{\mathcal{W}}(_,B\mathbb{1})} & & \downarrow{\scriptstyle \underline{\mathcal{W}}(1,\zeta'_Y)} \\
\underline{\mathcal{W}}(\underline{\mathcal{W}}(BX,B\mathbb{1}),\underline{\mathcal{W}}(BY,B\mathbb{1})) & \xrightarrow{\underline{\mathcal{W}}(\zeta'_X,1)} & \underline{\mathcal{W}}(B\underline{\mathcal{V}}(X,\mathbb{1}),\underline{\mathcal{W}}(BY,B\mathbb{1}))
\end{array}$$

By closedness of $\mathcal{W}$, it is equivalent to the following equation:

$$\begin{array}{ccc}
B\underline{\mathcal{V}}(X,\mathbb{1})\otimes B\underline{\mathcal{V}}(Y,X) & \xrightarrow{1\otimes B\underline{\mathcal{V}}(_,\mathbb{1})} & B\underline{\mathcal{V}}(X,\mathbb{1})\otimes B\underline{\mathcal{V}}(\underline{\mathcal{V}}(X,\mathbb{1}),\underline{\mathcal{V}}(Y,\mathbb{1})) \\
{\scriptstyle \zeta'_X\otimes\underline{\widehat{B}}}\downarrow & & \downarrow{\scriptstyle \widehat{B}(\mathrm{ev}^{\mathcal{V}})} \\
\underline{\mathcal{W}}(BX,B\mathbb{1})\otimes\underline{\mathcal{W}}(BY,BX) & & B\underline{\mathcal{V}}(Y,\mathbb{1}) \\
{\scriptstyle 1\otimes\underline{\mathcal{W}}(_,B\mathbb{1})}\downarrow & & \downarrow{\scriptstyle \zeta'_Y} \\
\underline{\mathcal{W}}(BX,B\mathbb{1})\otimes\underline{\mathcal{W}}(\underline{\mathcal{W}}(BX,B\mathbb{1}),\underline{\mathcal{W}}(BY,B\mathbb{1})) & \xrightarrow{\mathrm{ev}^{\mathcal{W}}} & \underline{\mathcal{W}}(BY,B\mathbb{1})
\end{array}$$

By (1.3), the above equation reduces to the equation

$$\begin{array}{ccc}
B\underline{\mathcal{V}}(Y,X)\otimes B\underline{\mathcal{V}}(X,\mathbb{1}) & \xrightarrow[\mu_{B_*\underline{\mathcal{V}}}]{\widehat{B}(\mu_{\underline{\mathcal{V}}})\ \Vert} & B\underline{\mathcal{V}}(Y,\mathbb{1}) \\
{\scriptstyle \underline{\widehat{B}}\otimes\zeta'_X\ =\ \underline{\widehat{B}}_{Y,X}\otimes\underline{\widehat{B}}_{X,\mathbb{1}}}\downarrow & & \downarrow{\scriptstyle \underline{\widehat{B}}_{Y,\mathbb{1}}\ =\ \zeta'_Y} \\
\underline{\mathcal{W}}(BY,BX)\otimes\underline{\mathcal{W}}(BX,B\mathbb{1}) & \xrightarrow{\mu_{\underline{\mathcal{W}}}} & \underline{\mathcal{W}}(BY,B\mathbb{1})
\end{array}$$

which expresses the fact that $\underline{\widehat{B}} : B_*\underline{\mathcal{V}} \to \underline{\mathcal{W}}$ is a $\mathcal{W}$-functor.

Suppose that $\beta^{\varnothing} : \mathbb{1} \to B\mathbb{1}$ is an isomorphism. Then there is a natural isomorphism of functors

$$\underline{\mathcal{W}}(1,(\beta^{\varnothing})^{-1}) : \underline{\mathcal{W}}(\_,B\mathbb{1}) \to \underline{\mathcal{W}}(\_,\mathbb{1}) : \underline{\mathcal{W}}^{\mathrm{op}} \to \underline{\mathcal{W}}.$$

Pasting it with ζ' gives a natural transformation ζ as in the diagram below:

$$\begin{array}{ccc}
B_*\underline{\mathcal{V}}^{\mathrm{op}} & \xrightarrow{B_*\underline{\mathcal{V}}(_,\mathbb{1})} & B_*\underline{\mathcal{V}} \\
{\scriptstyle (\underline{\widehat{B}})^{\mathrm{op}}}\downarrow & \overset{\zeta}{\swarrow} & \downarrow{\scriptstyle \underline{\widehat{B}}} \\
\underline{\mathcal{W}}^{\mathrm{op}} & \xrightarrow{\underline{\mathcal{W}}(_,\mathbb{1})} & \underline{\mathcal{W}}
\end{array} \qquad (2.14)$$

Proposition 2.14. *Suppose ζ is an isomorphism. Let $\mathcal{C}$ be a $\mathcal{V}$-category, and suppose $S : \mathcal{C} \to \mathcal{C}$ is a right Serre $\mathcal{V}$-functor. Then $B_*(S) : B_*(\mathcal{C}) \to B_*(\mathcal{C})$ is a right Serre $\mathcal{W}$-functor.*

Proof. Let ψ be a natural isomorphism as in (2.1). Applying the $\mathcal{C}at$-multifunctor $\widehat{B_*}$ and patching the result with diagram (2.14) yields the following diagram:

$$\begin{array}{ccc}
B_*(\mathcal{C})^{\mathrm{op}}\boxtimes B_*(\mathcal{C}) & \xrightarrow{1\boxtimes B_*(S)} & B_*(\mathcal{C})^{\mathrm{op}}\boxtimes B_*(\mathcal{C}) \\
{\scriptstyle \widehat{B_*}(\mathrm{Hom}_{\mathcal{C}^{\mathrm{op}}})^{\mathrm{op}}}\downarrow & \overset{\widehat{B_*}(\psi)}{\swarrow} & \downarrow{\scriptstyle \widehat{B_*}\,\mathrm{Hom}_{\mathcal{C}}} \\
B_*\underline{\mathcal{V}}^{\mathrm{op}} & \xrightarrow{B_*\underline{\mathcal{V}}(_,\mathbb{1})} & B_*\underline{\mathcal{V}} \\
{\scriptstyle (\underline{\widehat{B}})^{\mathrm{op}}}\downarrow & \overset{\zeta}{\swarrow} & \downarrow{\scriptstyle \underline{\widehat{B}}} \\
\underline{\mathcal{W}}^{\mathrm{op}} & \xrightarrow{\underline{\mathcal{W}}(_,\mathbb{1})} & \underline{\mathcal{W}}
\end{array} \qquad (2.15)$$

Since $\widehat{B_*}\,\mathrm{Hom}_{\mathcal{C}} \cdot \underline{\widehat{B}} = \mathrm{Hom}_{B_*(\mathcal{C})}$ and $\widehat{B_*}\,\mathrm{Hom}_{\mathcal{C}^{\mathrm{op}}} \cdot \underline{\widehat{B}} = \mathrm{Hom}_{B_*(\mathcal{C})^{\mathrm{op}}}$, we obtain a natural transformation

$$\begin{array}{ccc} B_*(\mathcal{C})^{\mathrm{op}} \boxtimes B_*(\mathcal{C}) & \xrightarrow{1\boxtimes B_*(S)} & B_*(\mathcal{C})^{\mathrm{op}} \boxtimes B_*(\mathcal{C}) \\ {\scriptstyle \mathrm{Hom}^{\mathrm{op}}_{B_*(\mathcal{C})^{\mathrm{op}}}}\downarrow & \Swarrow & \downarrow{\scriptstyle \mathrm{Hom}_{B_*(\mathcal{C})}} \\ \underline{\mathcal{W}}^{\mathrm{op}} & \xrightarrow{\underline{\mathcal{W}}(\_,\mathbb{1})} & \underline{\mathcal{W}} \end{array}$$

It is invertible since so are ψ and ζ. It follows that a right Serre $\mathcal{V}$-functor $S : \mathcal{C} \to \mathcal{C}$ induces a right Serre $\mathcal{W}$-functor $B_*(S) : B_*(\mathcal{C}) \to B_*(\mathcal{C})$. □

2.15. From $\mathcal{K}$-categories to gr-categories

Consider the lax symmetric Monoidal base change functor $(H^\bullet, \kappa^I) : \mathcal{K} \to \mathbf{gr}$, $X \mapsto H^\bullet X = (H^n X)_{n\in\mathbb{Z}}$, where for each $I \in \mathrm{Ob}\,\mathcal{S}$ the morphism $\kappa^I : \otimes^{i\in I} H^\bullet X_i \to H^\bullet \otimes^{i\in I} X_i$ is the Künneth map. There is a $\mathbf{gr}$-functor $\underline{\widehat{H^\bullet}} : H^\bullet_* \underline{\mathcal{K}} \to \underline{\mathbf{gr}}$, $X \mapsto H^\bullet X$, that acts on morphisms via the map

$$\mathcal{K}(X[-n], Y) = H^n \underline{\mathcal{K}}(X,Y) \to \underline{\mathbf{gr}}(H^\bullet X, H^\bullet Y)^n = \prod_{d\in\mathbb{Z}} \underline{\Bbbk\text{-}\mathbf{Mod}}(H^{d-n}X, H^d Y)$$

which sends the homotopy class of a chain map $f : X[-n] \to Y$ to $(H^d(f))_{d\in\mathbb{Z}}$. Note that $H^\bullet$ preserves the unit object, therefore there is a natural transformation

$$\begin{array}{ccc} H^\bullet_* \underline{\mathcal{K}}^{\mathrm{op}} & \xrightarrow{H^\bullet_* \underline{\mathcal{K}}(\_,\Bbbk)} & H^\bullet_* \underline{\mathcal{K}} \\ {\scriptstyle (\underline{\widehat{H^\bullet}})^{\mathrm{op}}}\downarrow & \overset{\zeta}{\Swarrow} & \downarrow{\scriptstyle \underline{\widehat{H^\bullet}}} \\ \underline{\mathbf{gr}}^{\mathrm{op}} & \xrightarrow{\underline{\mathbf{gr}}(\_,\Bbbk)} & \underline{\mathbf{gr}} \end{array}$$

Explicitly, the map $\zeta_X = \underline{\widehat{H^\bullet}}_{X,\Bbbk} : H^\bullet(\underline{\mathcal{K}}(X,\Bbbk)) \to \underline{\mathbf{gr}}(H^\bullet X, H^\bullet \Bbbk) = \underline{\mathbf{gr}}(H^\bullet X, \Bbbk)$ is given by its components

$$\mathcal{K}(X[-n], \Bbbk) = H^n \underline{\mathcal{K}}(X,\Bbbk) \to \underline{\mathbf{gr}}(H^\bullet X, \Bbbk)^n = \underline{\Bbbk\text{-}\mathbf{Mod}}(H^{-n}X, \Bbbk), \quad f \mapsto H^0(f).$$

In general, ζ is not invertible. However, if $\Bbbk$ is a field, ζ is an isomorphism. In fact, in this case $H^\bullet : \mathcal{K} \to \mathbf{gr}$ is an equivalence. A quasi-inverse is given by the functor $F : \mathbf{gr} \to \mathcal{K}$ which equips a graded $\Bbbk$-module with the trivial differential.

Corollary 2.16. *Suppose $\Bbbk$ is a field. Let $S : \mathcal{C} \to \mathcal{C}$ be a (right) Serre $\mathcal{K}$-functor. Then $H^\bullet_*(S) : H^\bullet_*(\mathcal{C}) \to H^\bullet_*(\mathcal{C})$ is a (right) Serre* **gr**-*functor. Moreover, $H^\bullet_*$ reflects (right) Serre functors: if $H^\bullet_*(\mathcal{C})$ admits a (right) Serre* **gr**-*functor, then $\mathcal{C}$ admits a (right) Serre $\mathcal{K}$-functor.*

Proof. The first assertion follows from Proposition 2.14. For the proof of the second, note that the symmetric Monoidal functor $F : \mathbf{gr} \to \mathcal{K}$ induces a symmetric Monoidal $\mathcal{C}at$-functor $F_* : \widehat{\mathbf{gr}\text{-}\mathcal{C}at} \to \widehat{\mathcal{K}\text{-}\mathcal{C}at}$. The corresponding $\mathcal{K}$-functor $\underline{\widehat{F}} : F_*\underline{\mathbf{gr}} \to \underline{\mathcal{K}}$ acts as the identity on morphisms (the complex $\underline{\mathcal{K}}(FX, FY)$ carries the trivial differential and coincides with $\underline{\mathbf{gr}}(X,Y)$ as a graded $\Bbbk$-module). Furthermore, F preserves the unit object, therefore Proposition 2.14 applies. It follows that if $\bar{S} : H^\bullet_*(\mathcal{C}) \to H^\bullet_*(\mathcal{C})$ is a right Serre **gr**-functor, then

$F_*(\bar{S}) : F_*H^\bullet_*(\mathcal{C}) \to F_*H^\bullet_*(\mathcal{C})$ is a right Serre $\mathcal{K}$-functor. Since the $\mathcal{K}$-category $F_*H^\bullet_*(\mathcal{C})$ is isomorphic to $\mathcal{C}$, the right Serre $\mathcal{K}$-functor $F_*(\bar{S})$ translates to a right Serre $\mathcal{K}$-functor on $\mathcal{C}$. □

2.17. From gr-categories to $\Bbbk$-categories

Consider a lax symmetric Monoidal base change functor $(N, \nu^I) : \mathbf{gr} \to \Bbbk\text{-}\mathbf{Mod}$, $X = (X^n)_{n\in\mathbb{Z}} \mapsto X^0$, where for each $I \in \operatorname{Ob}\mathcal{S}$ the map $\nu^I : \otimes^{i\in I} NX_i = \otimes^{i\in I} X_i^0 \to N \otimes^{i\in I} X_i = \oplus_{\sum_{i\in I} n_i = 0} X_i^{n_i}$ is the natural embedding. The $\Bbbk$-functor $\widehat{\underline{N}} : N_*\underline{\mathbf{gr}} \to \underline{\Bbbk\text{-}\mathbf{Mod}}$, $X \mapsto NX = X^0$, acts on morphisms via the projection

$$N\underline{\mathbf{gr}}(X,Y) = \underline{\mathbf{gr}}(X,Y)^0 = \prod_{d\in\mathbb{Z}} \underline{\Bbbk\text{-}\mathbf{Mod}}(X^d, Y^d) \\ \to \underline{\Bbbk\text{-}\mathbf{Mod}}(X^0, Y^0) = \underline{\Bbbk\text{-}\mathbf{Mod}}(NX, NY).$$

The functor N preserves the unit object, therefore there exists a natural transformation

$$\begin{array}{ccc} N_*\underline{\mathbf{gr}}^{\mathrm{op}} & \xrightarrow{N_*\underline{\mathbf{gr}}(\_,\Bbbk)} & N_*\underline{\mathbf{gr}} \\ {\scriptstyle (\widehat{\underline{N}})^{\mathrm{op}}}\big\downarrow & \overset{\zeta}{\Swarrow} & \big\downarrow{\scriptstyle \widehat{\underline{N}}} \\ \underline{\Bbbk\text{-}\mathbf{Mod}}^{\mathrm{op}} & \xrightarrow{\underline{\Bbbk\text{-}\mathbf{Mod}}(\_,\Bbbk)} & \underline{\Bbbk\text{-}\mathbf{Mod}} \end{array}$$

Explicitly, the map $\zeta_X = \widehat{\underline{N}}_{X,\Bbbk}$ is the identity map

$$N\underline{\mathbf{gr}}(X,\Bbbk) = \underline{\mathbf{gr}}(X,\Bbbk)^0 \to \underline{\Bbbk\text{-}\mathbf{Mod}}(X^0,\Bbbk) = \underline{\Bbbk\text{-}\mathbf{Mod}}(NX,\Bbbk).$$

Corollary 2.18 (to Proposition 2.14). *Suppose $S : \mathcal{C} \to \mathcal{C}$ is a right Serre* **gr***-functor. Then $N_*(S) : N_*(\mathcal{C}) \to N_*(\mathcal{C})$ is a right Serre $\Bbbk$-functor.*

If $N_*(\mathcal{C})$ possess a right Serre $\Bbbk$-functor, it does not imply, in general, that $\mathcal{C}$ has a right Serre **gr**-functor. However, this will be the case if $\mathcal{C}$ is closed under shifts, as explained in the next section.

2.19. Categories closed under shifts

As in [BLM06, Chapter 10] denote by $\mathcal{Z}$ the following algebra (strict monoidal category) in the symmetric monoidal category of **dg**-categories, $\mathcal{K}$-categories or **gr**-categories. As a graded quiver $\mathcal{Z}$ has $\operatorname{Ob}\mathcal{Z} = \mathbb{Z}$ and $\mathcal{Z}(m,n) = \Bbbk[n-m]$. In the first two cases $\mathcal{Z}$ is supplied with zero differential. Composition in the category $\mathcal{Z}$ comes from the multiplication in $\Bbbk$:

$$\mu_{\mathcal{Z}} : \mathcal{Z}(l,m) \otimes_\Bbbk \mathcal{Z}(m,n) = \Bbbk[m-l] \otimes_\Bbbk \Bbbk[n-m] \to \Bbbk[n-l] = \mathcal{Z}(l,n), \\ 1s^{m-l} \otimes 1s^{n-m} \mapsto 1s^{n-l}.$$

The elements $1 \in \Bbbk = \mathcal{Z}(n,n)$ are identity morphisms of $\mathcal{Z}$.

The object $\mathcal{Z}$ of $(\mathcal{V}\text{-}\mathcal{C}at, \boxtimes)$ (where $\mathcal{V}$ is **dg**, $\mathcal{K}$ or **gr**) is equipped with an algebra (a strict monoidal category) structure, given by multiplication – the $\mathcal{V}$-functor

$$\otimes_\psi : \mathcal{Z} \boxtimes \mathcal{Z} \to \mathcal{Z}, \qquad m \times n \mapsto m + n,$$

$$\otimes_\psi : (\mathcal{Z} \boxtimes \mathcal{Z})(n \times m, k \times l) = \mathcal{Z}(n,k) \otimes \mathcal{Z}(m,l) \to \mathcal{Z}(n+m, k+l),$$

$$1s^{k-n} \otimes 1s^{l-m} \mapsto (-1)^{k(m-l)} s^{k+l-n-m}.$$

Therefore, for the three mentioned $\mathcal{V}$ the functor $- \boxtimes \mathcal{Z} : \mathcal{V}\text{-}\mathcal{C}at \to \mathcal{V}\text{-}\mathcal{C}at$, $\mathcal{C} \mapsto \mathcal{C} \boxtimes \mathcal{Z}$, is a monad. It takes a $\mathcal{V}$-category $\mathcal{C}$ to the $\mathcal{V}$-category $\mathcal{C} \boxtimes \mathcal{Z}$ with the set of objects $\operatorname{Ob} \mathcal{C} \boxtimes \mathcal{Z} = \operatorname{Ob} \mathcal{C} \times \mathbb{Z}$ and with the graded modules of morphisms $(\mathcal{C} \boxtimes \mathcal{Z})((X,n),(Y,m)) = \mathcal{C}(X,Y) \otimes \Bbbk[m-n]$. The composition is given by the following morphism in $\mathcal{V}$:

$$\begin{aligned} \mu_{\mathcal{C}\boxtimes\mathcal{Z}} = \big[&(\mathcal{C} \boxtimes \mathcal{Z})((X,n),(Y,m)) \otimes (\mathcal{C} \boxtimes \mathcal{Z})((Y,m),(Z,p)) = \\ &\mathcal{C}(X,Y)\otimes\Bbbk[m-n]\otimes\mathcal{C}(Y,Z)\otimes\Bbbk[p-m] \xrightarrow{1\otimes c\otimes 1} \mathcal{C}(X,Y)\otimes\mathcal{C}(Y,Z)\otimes\Bbbk[m-n]\otimes\Bbbk[p-m] \\ &\xrightarrow{\mu_\mathcal{C}\otimes\mu_\mathcal{Z}} \mathcal{C}(X,Z) \otimes \Bbbk[p-n] = (\mathcal{C} \boxtimes \mathcal{Z})((X,n),(Z,p))\big]. \end{aligned} \tag{2.16}$$

The unit $u_{[]} = \mathrm{id} \boxtimes \mathbb{1}_\mathcal{Z} : \mathcal{C} \to \mathcal{C} \boxtimes \mathcal{Z}$ of the monad $- \boxtimes \mathcal{Z} : \mathcal{V}\text{-}\mathcal{C}at \to \mathcal{V}\text{-}\mathcal{C}at$ is the natural embedding $X \mapsto (X,0)$ bijective on morphisms. Here $\mathbb{1}_\mathcal{Z} : \Bbbk \to \mathcal{Z}$, $* \mapsto 0$ is the unit of the algebra $\mathcal{Z}$, whose source is the graded category $\Bbbk$ with one object.

The $\mathcal{V}$-category $\mathcal{C} \boxtimes \mathcal{Z}$ admits an isomorphic form $\mathcal{C}^{[]}$ whose set of objects is $\operatorname{Ob} \mathcal{C}^{[]} = \operatorname{Ob} \mathcal{C} \times \mathbb{Z}$, likewise $\mathcal{C} \boxtimes \mathcal{Z}$. The graded $\Bbbk$-modules of morphisms are $\mathcal{C}^{[]}((X,n),(Y,m)) = \mathcal{C}(X,Y)[m-n]$. This graded quiver is identified with $\mathcal{C} \boxtimes \mathcal{Z}$ via the isomorphism

$$\text{ß} = \big[\mathcal{C}(X,Y)[m-n] \xrightarrow{s^{n-m}} \mathcal{C}(X,Y) \xrightarrow{\lambda^{\mathsf{I}\cdot}(1\otimes s^{m-n})} \mathcal{C}(X,Y) \otimes \mathcal{Z}(n,m)\big]$$

in [BLM06, Chapter 10]. Therefore, in the cases of $\mathcal{V} = \mathsf{C}_\Bbbk$ or $\mathcal{V} = \mathcal{K}$ the graded $\Bbbk$-module $\mathcal{C}^{[]}((X,n),(Y,m))$ is equipped with the differential $(-1)^{m-n} s^{n-m} d_\mathcal{C} s^{m-n}$. Multiplication in $\mathcal{C}^{[]}$ is found from (2.16) as

$$\begin{aligned} \mu_{\mathcal{C}^{[]}} &= \big[\mathcal{C}(X,Y)[m-n]\otimes\mathcal{C}(Y,Z)[p-m] \xrightarrow{\text{ß}\otimes\text{ß}} \mathcal{C}(X,Y)\otimes\Bbbk[m-n]\otimes\mathcal{C}(Y,Z)\otimes\Bbbk[p-m] \\ &\qquad \xrightarrow{\mu_{\mathcal{C}\boxtimes\mathcal{Z}}} \mathcal{C}(X,Z) \otimes \Bbbk[p-n] \xrightarrow{\text{ß}^{-1}} \mathcal{C}(X,Z)[p-n]\big] \\ &= \big[\mathcal{C}(X,Y)[m-n] \otimes \mathcal{C}(Y,Z)[p-m] \xrightarrow{(s^{m-n}\otimes s^{p-m})^{-1}} \mathcal{C}(X,Y) \otimes \mathcal{C}(Y,Z) \\ &\qquad \xrightarrow{\mu_\mathcal{C}} \mathcal{C}(X,Z) \xrightarrow{s^{p-n}} \mathcal{C}(X,Z)[p-n]\big]. \end{aligned} \tag{2.17}$$

Definition 2.20. We say that a $\mathcal{V}$-category $\mathcal{C}$ is *closed under shifts* if every object (X,n) of $\mathcal{C}^{[]}$ is isomorphic in $\mathcal{C}^{[]}$ to some object $(Y,0)$, $Y = X[n] \in \operatorname{Ob} \mathcal{C}$.

Clearly, $-^{[]} : \mathcal{V}\text{-}\mathcal{C}at \to \mathcal{V}\text{-}\mathcal{C}at$ is also a monad, whose unit $u_{[]} : \mathcal{C} \to \mathcal{C}^{[]}$ is the natural embedding $X \mapsto (X, 0)$ identity on morphisms. Immediately one finds that a $\mathcal{V}$-category $\mathcal{C}$ is closed under shifts if and only if the functor $u_{[]} : \mathcal{C} \to \mathcal{C}^{[]}$ is an equivalence.

The lax symmetric Monoidal base change functor $(H^\bullet, \kappa^I) : \mathcal{K} \to \mathbf{gr}$ gives, in particular, the Künneth functor $\kappa : H^\bullet\mathcal{C} \boxtimes H^\bullet\mathcal{Z} \to H^\bullet(\mathcal{C} \boxtimes \mathcal{Z})$, identity on objects. It is an isomorphism of $\mathbf{gr}$-categories because $\mathcal{Z}(m, n) = \Bbbk[n - m]$ are flat graded $\Bbbk$-modules. Clearly, $\mathcal{Z}$ coincides with $H^\bullet\mathcal{Z}$ as a graded $\Bbbk$-quiver, hence we have the isomorphism $\kappa : (H^\bullet\mathcal{C}) \boxtimes \mathcal{Z} \to H^\bullet(\mathcal{C} \boxtimes \mathcal{Z})$. Equivalently we may write the isomorphism $(H^\bullet\mathcal{C})^{[]} \simeq H^\bullet(\mathcal{C}^{[]})$. From the lax monoidality of $(H^\bullet, \kappa^I)$ we deduce the following equation:

$$H^\bullet(u_{[]}) = \big[H^\bullet\mathcal{C} \xrightarrow{u_{[]}} (H^\bullet\mathcal{C}) \boxtimes \mathcal{Z} \xrightarrow[\sim]{\kappa} H^\bullet(\mathcal{C} \boxtimes \mathcal{Z})\big].$$

Therefore, if $\mathcal{C}$ is a $\mathcal{K}$-category closed under shifts, then $H^\bullet\mathcal{C}$ is a $\mathbf{gr}$-category closed under shifts.

For a $\mathbf{gr}$-category $\mathcal{C}$, the components of the graded $\Bbbk$-module $\mathcal{C}(X, Y)$ are denoted by $\mathcal{C}(X, Y)^n = \mathcal{C}^n(X, Y)$, $X, Y \in \operatorname{Ob}\mathcal{C}$, $n \in \mathbb{Z}$. The $\Bbbk$-category $N_*(\mathcal{C})$ is denoted by $\mathcal{C}^0$.

Proposition 2.21. *Let $\mathcal{C}$ be a $\mathbf{gr}$-category closed under shifts. Suppose $S^0 : \mathcal{C}^0 \to \mathcal{C}^0$ is a right Serre $\Bbbk$-functor. Then there exists a right Serre $\mathbf{gr}$-functor $S : \mathcal{C} \to \mathcal{C}$ such that $N_*(S) = S^0$.*

Proof. Let $\psi^0 = (\psi^0_{X,Y} : \mathcal{C}^0(X, YS^0) \to \underline{\Bbbk\text{-}\mathbf{Mod}}(\mathcal{C}^0(Y, X), \Bbbk))_{X,Y \in \operatorname{Ob}\mathcal{C}}$ be a natural isomorphism. Let $\phi^0_{X,Y} : \mathcal{C}^0(Y, X) \otimes \mathcal{C}^0(X, YS) \to \Bbbk$, $X, Y \in \operatorname{Ob}\mathcal{C}$, denote the corresponding pairings from (2.12). Define trace functionals $\tau^0_X : \mathcal{C}^0(X, XS) \to \Bbbk$, $X \in \operatorname{Ob}\mathcal{C}$, by formula (2.3). We are going to apply Proposition 2.12. For this we need to specify a map $\operatorname{Ob} S : \operatorname{Ob}\mathcal{C} \to \operatorname{Ob}\mathcal{C}$ and trace functionals $\tau_X : \mathcal{C}(X, XS) \to \Bbbk$, $X \in \operatorname{Ob}\mathcal{C}$. Set $\operatorname{Ob} S = \operatorname{Ob} S^0$. Let the 0-th component of τ_X be equal to the map τ^0_X, the other components necessarily vanish since $\Bbbk$ is concentrated in degree 0. Let us prove that the pairings $\phi_{X,Y}$ given by (2.12) are perfect. For $n \in \mathbb{Z}$, the restriction of $\phi_{X,Y}$ to the summand $\mathcal{C}^n(Y, X) \otimes \mathcal{C}^{-n}(X, YS)$ is given by

$$\phi_{X,Y} = \big[\mathcal{C}^n(Y, X) \otimes \mathcal{C}^{-n}(X, YS) \xrightarrow{\mu_\mathcal{C}} \mathcal{C}^0(Y, YS) \xrightarrow{\tau^0_Y} \Bbbk\big].$$

It can be written as follows:

$$\phi_{X,Y} = \big[\mathcal{C}^n(Y, X) \otimes \mathcal{C}^{-n}(X, YS) = \mathcal{C}^{[]}((Y, 0), (X, n))^0 \otimes \mathcal{C}^{[]}((X, n), (YS, 0))^0 \\ \xrightarrow{(-)^n \mu_{\mathcal{C}^{[]}}} \mathcal{C}^{[]}((Y, 0), (YS, 0))^0 = \mathcal{C}^0(Y, YS) \xrightarrow{\tau^0_Y} \Bbbk\big].$$

Since $\mathcal{C}$ is closed under shifts, there exist an object $X[n] \in \operatorname{Ob}\mathcal{C}$ and an isomorphism $\alpha : (X,n) \to (X[n],0)$ in $\mathcal{C}^{[]}$. Using associativity of $\mu_{\mathcal{C}^{[]}}$, we obtain:

$$\begin{aligned}
\phi_{X,Y} = \big[&\mathcal{C}^n(Y,X) \otimes \mathcal{C}^{-n}(X,YS) = \mathcal{C}^{[]}((Y,0),(X,n))^0 \otimes \mathcal{C}^{[]}((X,n),(YS,0))^0 \\
&\xrightarrow{\mathcal{C}^{[]}(1,\alpha)^0 \otimes \mathcal{C}^{[]}(\alpha^{-1},1)^0} \mathcal{C}^{[]}((Y,0),(X[n],0))^0 \otimes \mathcal{C}^{[]}((X[n],0),(YS,0))^0 \\
&\xrightarrow{(-)^n \mu_{\mathcal{C}^{[]}}} \mathcal{C}^{[]}((Y,0),(YS,0))^0 = \mathcal{C}^0(Y,YS) \xrightarrow{\tau^0_Y} \Bbbk\big] \\
= \big[&\mathcal{C}^n(Y,X) \otimes \mathcal{C}^{-n}(X,YS) = \mathcal{C}^{[]}((Y,0),(X,n))^0 \otimes \mathcal{C}^{[]}((X,n),(YS,0))^0 \\
&\xrightarrow{\mathcal{C}^{[]}(1,\alpha)^0 \otimes \mathcal{C}^{[]}(\alpha^{-1},1)^0} \mathcal{C}^{[]}((Y,0),(X[n],0))^0 \otimes \mathcal{C}^{[]}((X[n],0),(YS,0))^0 \\
&= \mathcal{C}^0(Y,X[n]) \otimes \mathcal{C}^0(X[n],YS) \xrightarrow{(-)^n \mu_{\mathcal{C}^0}} \mathcal{C}^0(Y,YS) \xrightarrow{\tau^0_Y} \Bbbk\big] \\
= \big[&\mathcal{C}^n(Y,X) \otimes \mathcal{C}^{-n}(X,YS) = \mathcal{C}^{[]}((Y,0),(X,n))^0 \otimes \mathcal{C}^{[]}((X,n),(YS,0))^0 \\
&\xrightarrow{\mathcal{C}^{[]}(1,\alpha)^0 \otimes \mathcal{C}^{[]}(\alpha^{-1},1)^0} \mathcal{C}^{[]}((Y,0),(X[n],0))^0 \otimes \mathcal{C}^{[]}((X[n],0),(YS,0))^0 \\
&= \mathcal{C}^0(Y,X[n]) \otimes \mathcal{C}^0(X[n],YS) \xrightarrow{(-)^n \phi^0_{X[n],Y}} \Bbbk\big].
\end{aligned}$$

Since $\phi^0_{X[n],Y}$ is a perfect pairing and the maps $\mathcal{C}^{[]}(1,\alpha)^0$ and $\mathcal{C}^{[]}(\alpha^{-1},1)^0$ are invertible, the pairing $\phi_{X,Y}$ is perfect as well. Indeed, it is easy to see that the corresponding maps $\psi^{-n}_{X,Y}$ and $\psi^0_{X[n],Y}$ are related as follows:

$$\begin{aligned}
\psi^{-n}_{X,Y} = \big[&\mathcal{C}^{-n}(X,YS) \xrightarrow{\mathcal{C}^{[]}(\alpha^{-1},1)^0} \mathcal{C}^0(X[n],YS) \xrightarrow{(-)^n \psi^0_{X[n],Y}} \\
&\operatorname{Hom}_\Bbbk(\mathcal{C}^0(Y,X[n]),\Bbbk) \xrightarrow{\operatorname{Hom}_\Bbbk(\mathcal{C}^{[]}(1,\alpha))^0,1)} \operatorname{Hom}_\Bbbk(\mathcal{C}^n(Y,X),\Bbbk)\big].
\end{aligned}$$

Proposition 2.12 implies that there is a right Serre **gr**-functor $S : \mathcal{C} \to \mathcal{C}$. Its components are determined unambiguously by equation (2.10). Applying the multifunctor $\widehat{N}$ to it we find that the functor $N_*(S) : \mathcal{C}^0 \to \mathcal{C}^0$ satisfies the same equation the functor $S^0 : \mathcal{C}^0 \to \mathcal{C}^0$ does. By uniqueness of the solution, $N_*(S) = S^0$. □

3. A_∞-categories and $\mathcal{K}$-categories

In this section we recall and deepen the relationship between A_∞-categories and $\mathcal{K}$-categories. It is implemented by a multifunctor $\mathsf{k} : \mathsf{A}^{\mathrm{u}}_\infty \to \widehat{\mathcal{K}\text{-}\mathcal{C}at}$ from [BLM06, Chapter 13], where $\mathcal{K}\text{-}\mathcal{C}at$ is the symmetric Monoidal category of $\mathcal{K}$-categories and $\mathcal{K}$-functors, and $\widehat{\mathcal{K}\text{-}\mathcal{C}at}$ is the corresponding symmetric multicategory. This multifunctor extends to non-unital A_∞-categories as a sort of multifunctor $\mathsf{k} : \mathsf{A}_\infty \to \widehat{\mathcal{K}\text{-}\mathcal{C}at^{nu}}$, where $\mathcal{K}\text{-}\mathcal{C}at^{nu}$ is the symmetric Monoidal category of non-unital $\mathcal{K}$-categories and $\mathcal{K}$-functors, and $\widehat{\mathcal{K}\text{-}\mathcal{C}at^{nu}}$ is the corresponding symmetric multicategory [loc. cit.].

3.1. Opposite A_∞-categories

Recall the following definitions from [LM04, Appendix A]. Let $\mathcal{A}$ be a graded $\Bbbk$-quiver. Then its *opposite quiver* $\mathcal{A}^{\mathrm{op}}$ is defined as the quiver with the same class of objects $\operatorname{Ob}\mathcal{A}^{\mathrm{op}} = \operatorname{Ob}\mathcal{A}$, and with graded $\Bbbk$-modules of morphisms $\mathcal{A}^{\mathrm{op}}(X,Y) = \mathcal{A}(Y,X)$.

Let $\gamma : Ts\mathcal{A}^{\mathrm{op}} \to Ts\mathcal{A}$ denote the following anti-isomorphism of coalgebras and algebras (free categories):

$$\gamma = (-1)^k \omega^0_c : s\mathcal{A}^{\mathrm{op}}(X_0,X_1) \otimes \dots \otimes s\mathcal{A}^{\mathrm{op}}(X_{k-1},X_k) \\ \to s\mathcal{A}(X_k,X_{k-1}) \otimes \dots \otimes s\mathcal{A}(X_1,X_0), \quad (3.1)$$

where $\omega^0 = \left(\begin{smallmatrix} 1 & 2 & \dots & k-1 & k \\ k & k-1 & \dots & 2 & 1 \end{smallmatrix}\right) \in \mathfrak{S}_k$. Clearly, $\gamma\Delta_0 = \Delta_0(\gamma\otimes\gamma)c = \Delta_0 c(\gamma\otimes\gamma)$, which is the anti-isomorphism property. Notice also that $(\mathcal{A}^{\mathrm{op}})^{\mathrm{op}} = \mathcal{A}$ and $\gamma^2 = \mathrm{id}$.

When $\mathcal{A}$ is an A_∞-category with the codifferential $b : Ts\mathcal{A} \to Ts\mathcal{A}$, then $\gamma b\gamma : Ts\mathcal{A}^{\mathrm{op}} \to Ts\mathcal{A}^{\mathrm{op}}$ is also a codifferential. Indeed,

$$\gamma b\gamma\Delta_0 = \gamma b\Delta_0 c(\gamma\otimes\gamma) = \gamma\Delta_0(1\otimes b + b\otimes 1)c(\gamma\otimes\gamma) \\ = \Delta_0(\gamma\otimes\gamma)c(1\otimes b + b\otimes 1)c(\gamma\otimes\gamma) = \Delta_0(\gamma b\gamma\otimes 1 + 1\otimes\gamma b\gamma).$$

The *opposite A_∞-category* $\mathcal{A}^{\mathrm{op}}$ to an A_∞-category $\mathcal{A}$ is the opposite quiver, equipped with the codifferential $b^{\mathrm{op}} = \gamma b\gamma : Ts\mathcal{A}^{\mathrm{op}} \to Ts\mathcal{A}^{\mathrm{op}}$. The components of b^{op} are computed as follows:

$$b^{\mathrm{op}}_k = (-)^{k+1}\bigl[s\mathcal{A}^{\mathrm{op}}(X_0,X_1) \otimes \dots \otimes s\mathcal{A}^{\mathrm{op}}(X_{k-1},X_k) \xrightarrow{\omega^0_c} \\ s\mathcal{A}(X_k,X_{k-1}) \otimes \dots \otimes s\mathcal{A}(X_1,X_0) \xrightarrow{b_k} s\mathcal{A}(X_k,X_0) = s\mathcal{A}^{\mathrm{op}}(X_0,X_k)\bigr]. \quad (3.2)$$

The sign $(-1)^k$ in (3.1) ensures that the above definition agrees with the definition of the opposite usual category, meaning that, for an arbitrary A_∞-category $\mathcal{A}$, $\mathsf{k}(\mathcal{A}^{\mathrm{op}}) = (\mathsf{k}\mathcal{A})^{\mathrm{op}}$. Indeed, clearly, both categories have $\operatorname{Ob}\mathcal{A}$ as the set of objects. Furthermore, for each pair of objects $X, Y \in \operatorname{Ob}\mathcal{A}$,

$$m^{\mathrm{op}}_1 = sb^{\mathrm{op}}_1 s^{-1} = sb_1 s^{-1} = m_1 : \mathcal{A}^{\mathrm{op}}(X,Y) = \mathcal{A}(Y,X) \to \mathcal{A}(Y,X) = \mathcal{A}^{\mathrm{op}}(X,Y),$$

therefore $(\mathsf{k}\mathcal{A}^{\mathrm{op}})(X,Y) = (\mathcal{A}^{\mathrm{op}}(X,Y), m^{\mathrm{op}}_1) = (\mathcal{A}(Y,X), m_1) = (\mathsf{k}\mathcal{A})^{\mathrm{op}}(X,Y)$. Finally, the compositions in both categories coincide:

$$\mu_{\mathsf{k}\mathcal{A}^{\mathrm{op}}} = m^{\mathrm{op}}_2 = (s\otimes s)b^{\mathrm{op}}_2 s^{-1} = -(s\otimes s)cb_2 s^{-1} = c(s\otimes s)b_2 s^{-1} = cm_2 = \mu_{(\mathsf{k}\mathcal{A})^{\mathrm{op}}}.$$

In particular, it follows that $\mathcal{A}^{\mathrm{op}}$ is unital if so is $\mathcal{A}$, with the same unit elements.

For an arbitrary A_∞-functor $f : \boxtimes^{i\in\mathbf{n}} Ts\mathcal{A}_i \to Ts\mathcal{B}$ there is another A_∞-functor f^{op} defined by the commutative square

$$\begin{array}{ccc} \boxtimes^{i\in\mathbf{n}} Ts\mathcal{A}_i & \xrightarrow{f} & Ts\mathcal{B} \\ {\scriptstyle\boxtimes^{\mathbf{n}}\gamma}\downarrow & & \downarrow{\scriptstyle\gamma} \\ \boxtimes^{i\in\mathbf{n}} Ts\mathcal{A}^{\mathrm{op}}_i & \xrightarrow{f^{\mathrm{op}}} & Ts\mathcal{B}^{\mathrm{op}} \end{array}$$

Since $\gamma^2 = \mathrm{id}$, the A_∞-functor f^{op} is found as the composite

$$f^{\mathrm{op}} = \bigl[\boxtimes^{i\in\mathbf{n}} Ts\mathcal{A}^{\mathrm{op}} \xrightarrow{\boxtimes^{\mathbf{n}}\gamma} \boxtimes^{i\in\mathbf{n}} Ts\mathcal{A}_i \xrightarrow{f} Ts\mathcal{B} \xrightarrow{\gamma} Ts\mathcal{B}^{\mathrm{op}}\bigr].$$

A non-unital $\mathcal{K}$-functor $\mathsf{k}f : \boxtimes^{i\in\mathbf{n}}\mathsf{k}\mathcal{A}_i \to \mathsf{k}\mathcal{B}$ is associated with f in [BLM06, Chapter 13]. This $\mathcal{K}$-functor acts on objects in the same way as f. It is determined by the components f_{e_j}, $e_j = (0,\dots,0,1,0,\dots,0)$:

$$\mathsf{k}f = \bigl[\otimes^{j\in\mathbf{n}}\mathsf{k}\mathcal{A}_j(X_j,Y_j) \xrightarrow{\otimes^{j\in\mathbf{n}}(sf_{e_j}s^{-1})}$$
$$\otimes^{j\in\mathbf{n}}\mathsf{k}\mathcal{B}\bigl(((Y_i)_{i<j},(X_i)_{i\geqslant j})f,((Y_i)_{i\leqslant j},(X_i)_{i>j})f\bigr) \xrightarrow{\mu^{\mathbf{n}}_{\mathsf{k}\mathcal{B}}} \mathsf{k}\mathcal{B}\bigl((X_i)_{i\in\mathbf{n}}f,(Y_i)_{i\in\mathbf{n}}f\bigr)\bigr],$$

The case of $n = 1$ was considered in [Lyu03, Proposition 8.6]. According to [BLM06] an A_∞-functor f is called *unital*, if the $\mathcal{K}$-functor $\mathsf{k}f$ is unital. The set of unital A_∞-functors $\boxtimes^{i\in\mathbf{n}}Ts\mathcal{A}_i \to Ts\mathcal{B}$ is denoted $\mathsf{A}^{\mathrm{u}}_\infty((\mathcal{A}_i)_{i\in\mathbf{n}};\mathcal{B})$. The assignment $\mathcal{A}\mapsto\mathsf{k}\mathcal{A}$, $f\mapsto\mathsf{k}f$ gives a multifunctor $\mathsf{k}:\mathsf{A}^{\mathrm{u}}_\infty \to \widehat{\mathcal{K}\text{-}\mathcal{C}at}$, see [BLM06, Chapter 13].

Lemma 3.2. *For an arbitrary A_∞-functor $f : \boxtimes^{i\in\mathbf{n}}Ts\mathcal{A}_i \to Ts\mathcal{B}$, the $\mathcal{K}$-functors $\mathsf{k}f^{\mathrm{op}}, (\mathsf{k}f)^{\mathrm{op}} : \boxtimes^{i\in\mathbf{n}}\mathsf{k}\mathcal{A}_i^{\mathrm{op}} \to \mathsf{k}\mathcal{B}^{\mathrm{op}}$ coincide.*

Proof. The case $n = 1$ is straightforward. We provide a proof in the case $n = 2$, which we are going to use later.

Let $f : Ts\mathcal{A}\boxtimes Ts\mathcal{B} \to Ts\mathcal{C}$ be an A_∞-functor. The components of f^{op} are given by

$$\begin{aligned} f^{\mathrm{op}}_{kn} &= (-)^{k+n-1}\bigl[s\mathcal{A}^{\mathrm{op}}(X_0,X_1)\otimes\cdots\otimes s\mathcal{A}^{\mathrm{op}}(X_{k-1},X_k)\otimes \\ &\qquad\qquad \otimes s\mathcal{B}^{\mathrm{op}}(U_0,U_1)\otimes\cdots\otimes s\mathcal{B}^{\mathrm{op}}(U_{n-1},U_n) \\ &= s\mathcal{A}(X_1,X_0)\otimes\cdots\otimes s\mathcal{A}(X_k,X_{k-1})\otimes s\mathcal{B}(U_1,U_0)\otimes\cdots\otimes s\mathcal{B}(U_n,U_{n-1}) \\ &\xrightarrow{\pi^{kn}_c} s\mathcal{A}(X_k,X_{k-1})\otimes\cdots\otimes s\mathcal{A}(X_1,X_0)\otimes s\mathcal{B}(U_n,U_{n-1})\otimes\cdots\otimes s\mathcal{B}(U_1,U_0) \\ &\qquad \xrightarrow{f_{kn}} s\mathcal{C}((X_k,U_n)f,(X_0,U_0)f) = s\mathcal{C}^{\mathrm{op}}((X_0,U_0)f,(X_k,U_n)f)\bigr], \end{aligned} \tag{3.3}$$

where $\pi^{kn} = \left(\begin{smallmatrix} 1 & 2 & \dots & k & k+1 & k+2 & \dots & k+n \\ k & k-1 & \dots & 1 & k+n & k+n-1 & \dots & k+1 \end{smallmatrix}\right) \in \mathfrak{S}_{k+n}$, and π^{kn}_c is the corresponding signed permutation.

Clearly, both $\mathsf{k}f^{\mathrm{op}}$ and $(\mathsf{k}f)^{\mathrm{op}}$ act as $\operatorname{Ob} f$ on objects. Let $X,Y\in\operatorname{Ob}\mathcal{A}$, $U,V\in\operatorname{Ob}\mathcal{B}$. Then

$$\mathsf{k}f^{\mathrm{op}} = \bigl[\mathcal{A}^{\mathrm{op}}(X,Y)\otimes\mathcal{B}^{\mathrm{op}}(U,V) \xrightarrow{sf^{\mathrm{op}}_{10}s^{-1}\otimes sf^{\mathrm{op}}_{01}s^{-1}}$$
$$\mathcal{C}^{\mathrm{op}}((X,U)f,(Y,U)f)\otimes\mathcal{C}^{\mathrm{op}}((Y,U)f,(Y,V)f) \xrightarrow{\mu_{\mathsf{k}\mathcal{C}^{\mathrm{op}}}} \mathcal{C}^{\mathrm{op}}((X,U)f,(Y,V)f)\bigr].$$

By (3.3),

$$f^{\mathrm{op}}_{10} = f_{10} : s\mathcal{A}^{\mathrm{op}}(X,Y) \to s\mathcal{C}((Y,U)f,(X,U)f) = s\mathcal{C}^{\mathrm{op}}((X,U)f,(Y,U)f),$$
$$f^{\mathrm{op}}_{01} = f_{01} : s\mathcal{B}^{\mathrm{op}}(U,V) \to s\mathcal{C}((Y,V)f,(Y,U)f) = s\mathcal{C}^{\mathrm{op}}((Y,U)f,(Y,V)f),$$

therefore

$$\begin{aligned}\mathsf{k}f^{\mathrm{op}} = \bigl[\mathcal{A}(Y,X)\otimes\mathcal{B}(V,U) &\xrightarrow{sf_{10}s^{-1}\otimes sf_{01}s^{-1}} \\ &\mathcal{C}((Y,U)f,(X,U)f)\otimes\mathcal{C}((Y,V)f,(Y,U)f) \xrightarrow{c} \\ &\mathcal{C}((Y,V)f,(Y,U)f)\otimes\mathcal{C}((Y,U)f,(X,U)f) \xrightarrow{\mu_{\mathsf{k}\mathcal{C}}} \mathcal{C}((Y,V)f,(X,U)f)\bigr] \\ = \bigl[\mathcal{A}(Y,X)\otimes\mathcal{B}(V,U) &\xrightarrow{c} \mathcal{B}(V,U)\otimes\mathcal{A}(Y,X) \xrightarrow{sf_{01}s^{-1}\otimes sf_{10}s^{-1}} \\ &\mathcal{C}((Y,V)f,(Y,U)f)\otimes\mathcal{C}((Y,U)f,(X,U)f) \xrightarrow{\mu_{\mathsf{k}\mathcal{C}}} \mathcal{C}((Y,V)f,(X,U)f)\bigr].\end{aligned}$$

Further,

$$\begin{aligned}(\mathsf{k}f)^{\mathrm{op}} = \bigl[\mathcal{A}(Y,X)\otimes\mathcal{B}(V,U) &\xrightarrow{sf_{10}s^{-1}\otimes sf_{01}s^{-1}} \\ &\mathcal{C}((Y,V)f,(X,V)f)\otimes\mathcal{C}((X,V)f,(X,U)f) \xrightarrow{\mu_{\mathsf{k}\mathcal{C}}} \mathcal{C}((Y,V)f,(X,U)f)\bigr].\end{aligned}$$

We must therefore prove the following equation in $\mathcal{K}$:

$$\begin{aligned}\bigl[\mathcal{A}(Y,X)\otimes\mathcal{B}(V,U) &\xrightarrow{c} \mathcal{B}(V,U)\otimes\mathcal{A}(Y,X) \xrightarrow{sf_{01}s^{-1}\otimes sf_{10}s^{-1}} \\ &\mathcal{C}((Y,V)f,(Y,U)f)\otimes\mathcal{C}((Y,U)f,(X,U)f) \xrightarrow{\mu_{\mathsf{k}\mathcal{C}}} \mathcal{C}((Y,V)f,(X,U)f)\bigr] \\ = \bigl[\mathcal{A}(Y,X)\otimes\mathcal{B}(V,U) &\xrightarrow{sf_{10}s^{-1}\otimes sf_{01}s^{-1}} \mathcal{C}((Y,V)f,(X,V)f)\otimes\mathcal{C}((X,V)f,(X,U)f) \\ &\xrightarrow{\mu_{\mathsf{k}\mathcal{C}}} \mathcal{C}((Y,V)f,(X,U)f)\bigr]. \qquad (3.4)\end{aligned}$$

By definition of an A_∞-functor the equation $fb = (b\boxtimes 1 + 1\boxtimes b)f : Ts\mathcal{A}\boxtimes Ts\mathcal{B} \to Ts\mathcal{C}$ holds. Restricting it to $s\mathcal{A}\boxtimes s\mathcal{B}$ and composing with $\mathrm{pr}_1 : Ts\mathcal{C} \to s\mathcal{C}$, we obtain

$$\begin{aligned}&(f_{10}\otimes f_{01})b_2 + c(f_{01}\otimes f_{10})b_2 + f_{11}b_1 \\ &\qquad = (1\otimes b_1 + b_1\otimes 1)f_{11} : s\mathcal{A}(Y,X)\otimes s\mathcal{B}(V,U) \to s\mathcal{C}((Y,V)f,(X,U)f).\end{aligned}$$

Thus, $(f_{10}\otimes f_{01})b_2 + c(f_{01}\otimes f_{10})b_2$ is a boundary. Therefore,

$$(s\otimes s)(f_{10}\otimes f_{01})b_2 = c(s\otimes s)(f_{01}\otimes f_{10})b_2$$

in $\mathcal{K}$. This implies equation (3.4). □

In particular, f^{op} is a unital A_∞-functor if f is unital.

Proposition 3.3. *The correspondences $\mathcal{A} \mapsto \mathcal{A}^{\mathrm{op}}$, $f \mapsto f^{\mathrm{op}}$ define a symmetric multifunctor $-^{\mathrm{op}} : \mathsf{A}_\infty \to \mathsf{A}_\infty$ which restricts to a symmetric multifunctor $-^{\mathrm{op}} : \mathsf{A}^{\mathrm{u}}_\infty \to \mathsf{A}^{\mathrm{u}}_\infty$.*

Proof. Straightforward. □

As an arbitrary symmetric multifunctor between closed multicategories, $-^{\mathrm{op}}$ possesses a closing transformation $\underline{\mathrm{op}} : \underline{\mathsf{A}_\infty}((\mathcal{A}_i)_{i\in I};\mathcal{B})^{\mathrm{op}} \to \underline{\mathsf{A}_\infty}((\mathcal{A}_i^{\mathrm{op}})_{i\in I};\mathcal{B}^{\mathrm{op}})$ uniquely determined by the following equation in A_∞:

$$\big[(\mathcal{A}_i^{\mathrm{op}})_{i\in I}, \underline{\mathsf{A}_\infty}((\mathcal{A}_i)_{i\in I};\mathcal{B})^{\mathrm{op}} \xrightarrow{1,\underline{\mathrm{op}}} (\mathcal{A}_i^{\mathrm{op}})_{i\in I}, \underline{\mathsf{A}_\infty}((\mathcal{A}_i^{\mathrm{op}})_{i\in I};\mathcal{B}^{\mathrm{op}}) \xrightarrow{\mathrm{ev}^{\mathsf{A}_\infty}} \mathcal{B}^{\mathrm{op}}\big] = (\mathrm{ev}^{\mathsf{A}_\infty})^{\mathrm{op}}. \quad (3.5)$$

The A_∞-functor $(\mathrm{ev}^{\mathsf{A}_\infty})^{\mathrm{op}}$ acts on objects in the same way as $\mathrm{ev}^{\mathsf{A}_\infty}$. It follows that $(X_i)_{i\in I}(f)\underline{\mathrm{op}} = (X_i)_{i\in I}f$ for an arbitrary A_∞-functor $f : (\mathcal{A}_i)_{i\in I} \to \mathcal{B}$ and a family of objects $X_i \in \operatorname{Ob}\mathcal{A}_i$, $i \in I$. The components

$$(\mathrm{ev}^{\mathsf{A}_\infty})^{\mathrm{op}}_{(m_i),m} = -\big[\boxtimes^{i\in I} T^{m_i} s\mathcal{A}_i^{\mathrm{op}} \boxtimes T^m s\underline{\mathsf{A}_\infty}((\mathcal{A}_i)_{i\in I};\mathcal{B})^{\mathrm{op}} \xrightarrow{\boxtimes^I(\gamma)_I \boxtimes \gamma} \boxtimes^{i\in I} T^{m_i} s\mathcal{A}_i \boxtimes T^m s\underline{\mathsf{A}_\infty}((\mathcal{A}_i)_{i\in I};\mathcal{B}) \xrightarrow{\mathrm{ev}^{\mathsf{A}_\infty}_{(m_i),m}} s\mathcal{B}\big] \quad (3.6)$$

vanish unless $m = 0$ or $m = 1$ since the same holds for $\mathrm{ev}^{\mathsf{A}_\infty}_{(m_i),m}$. From equations (3.5) and (3.6) we infer that

$$\begin{aligned}
&(\otimes^{i\in I} 1^{\otimes m_i} \otimes \operatorname{Ob}\underline{\mathrm{op}})\, \mathrm{ev}^{\mathsf{A}_\infty}_{(m_i),0} \\
&= -\big[\otimes^{i\in I} \otimes^{p_i\in\mathbf{m_i}} s\mathcal{A}_i^{\mathrm{op}}(X^i_{p_i-1}, X^i_{p_i}) \otimes T^0 s\underline{\mathsf{A}_\infty}((\mathcal{A}_i)_{i\in I};\mathcal{B})^{\mathrm{op}}(f,f) \\
&\qquad \simeq \otimes^{i\in I} \otimes^{p_i\in\mathbf{m_i}} s\mathcal{A}_i^{\mathrm{op}}(X^i_{p_i-1}, X^i_{p_i}) \xrightarrow{\otimes^{i\in I}(-)^{m_i}\omega_c^0} \\
&\qquad \otimes^{i\in I} \otimes^{p_i\in\mathbf{m_i}} s\mathcal{A}_i(X^i_{m_i-p_i}, X^i_{m_i-p_i+1}) \xrightarrow{f_{(m_i)}} s\mathcal{B}((X^i_0)_{i\in I}f, (X^i_{m_i})_{i\in I}f)\big],
\end{aligned}$$

therefore $(f)\underline{\mathrm{op}} = f^{\mathrm{op}} : (\mathcal{A}_i^{\mathrm{op}})_{i\in I} \to \mathcal{B}^{\mathrm{op}}$. Similarly,

$$\begin{aligned}
&(\otimes^{i\in I} 1^{\otimes m_i} \otimes \underline{\mathrm{op}}_1)\, \mathrm{ev}^{\mathsf{A}_\infty}_{(m_i),1} \\
&= \big[\otimes^{i\in I} \otimes^{p_i\in\mathbf{m_i}} s\mathcal{A}_i^{\mathrm{op}}(X^i_{p_i-1}, X^i_{p_i}) \otimes s\underline{\mathsf{A}_\infty}((\mathcal{A}_i)_{i\in I};\mathcal{B})^{\mathrm{op}}(f,g) \\
&\xrightarrow{\otimes^{i\in I}(-)^{m_i}\omega_c^0\otimes 1} \otimes^{i\in I} \otimes^{p_i\in\mathbf{m_i}} s\mathcal{A}_i(X^i_{m_i-p_i}, X^i_{m_i-p_i+1}) \otimes s\underline{\mathsf{A}_\infty}((\mathcal{A}_i)_{i\in I};\mathcal{B})(g,f) \\
&\xrightarrow{\otimes^{i\in I}\otimes^{p_i\in\mathbf{m_i}} 1\otimes\mathrm{pr}} \otimes^{i\in I} \otimes^{p_i\in\mathbf{m_i}} s\mathcal{A}_i(X^i_{m_i-p_i}, X^i_{m_i-p_i+1}) \\
&\qquad \otimes \underline{\mathsf{C}}_{\Bbbk}(\otimes^{i\in I} \otimes^{p_i\in\mathbf{m_i}} s\mathcal{A}_i(X^i_{m_i-p_i}, X^i_{m_i-p_i+1}), s\mathcal{B}((X^i_{m_i})_{i\in I}g, (X^i_0)_{i\in I}f)) \\
&\qquad \xrightarrow{\mathrm{ev}^{\mathsf{C}_\Bbbk}} s\mathcal{B}((X^i_{m_i})_{i\in I}g, (X^i_0)_{i\in I}f) = s\mathcal{B}^{\mathrm{op}}((X^i_0)_{i\in I}f^{\mathrm{op}}, (X^i_{m_i})_{i\in I}g^{\mathrm{op}})\big].
\end{aligned}$$

It follows that the map

$$\underline{\mathrm{op}}_1 : s\underline{\mathsf{A}_\infty}((\mathcal{A}_i)_{i\in I};\mathcal{B})^{\mathrm{op}}(f,g) \to s\underline{\mathsf{A}_\infty}((\mathcal{A}_i^{\mathrm{op}})_{i\in I};\mathcal{B}^{\mathrm{op}})(f^{\mathrm{op}},g^{\mathrm{op}})$$

takes an A_∞-transformation $r : g \to f : (\mathcal{A}_i)_{i\in I} \to \mathcal{B}$ to the opposite A_∞-transformation $r^{\mathrm{op}} \overset{\mathrm{def}}{=} (r)\underline{\mathrm{op}}_1 : f^{\mathrm{op}} \to g^{\mathrm{op}} : (\mathcal{A}_i^{\mathrm{op}})_{i\in I} \to \mathcal{B}^{\mathrm{op}}$ with the components

$$[(r)\underline{\mathrm{op}}_1]_{(m_i)} = (-)^{m_1+\dots+m_n}\big[\otimes^{i\in I}\otimes^{p_i\in\mathbf{m_i}} s\mathcal{A}_i^{\mathrm{op}}(X^i_{p_i-1},X^i_{p_i}) \xrightarrow{\otimes^{i\in I}\omega^0_c}$$
$$\otimes^{i\in I}\otimes^{p_i\in\mathbf{m_i}} s\mathcal{A}_i(X^i_{m_i-p_i},X^i_{m_i-p_i+1}) \xrightarrow{r_{(m_i)}} s\mathcal{B}((X^i_{m_i})_{i\in I}g,(X^i_0)_{i\in I}f)\big].$$

The higher components of $\underline{\mathrm{op}}$ vanish. Similar computations can be performed in the multicategory $\mathsf{A}^{\mathrm{u}}_\infty$. They lead to the same formulas for $\underline{\mathrm{op}}$, which means that the A_∞-functor $\underline{\mathrm{op}}$ restricts to a unital A_∞-functor $\underline{\mathrm{op}} : \underline{\mathsf{A}^{\mathrm{u}}_\infty((\mathcal{A}_i)_{i\in I};\mathcal{B})^{\mathrm{op}}} \to \underline{\mathsf{A}^{\mathrm{u}}_\infty}((\mathcal{A}_i^{\mathrm{op}})_{i\in I};\mathcal{B}^{\mathrm{op}})$ if the A_∞-categories $\mathcal{A}_i$, $i\in I$, $\mathcal{B}$ are unital.

As an easy application of the above considerations note that if $r : f \to g : (\mathcal{A}_i)_{i\in I} \to \mathcal{B}$ is an isomorphism of A_∞-functors, then $r^{\mathrm{op}} : g^{\mathrm{op}} \to f^{\mathrm{op}} : (\mathcal{A}_i^{\mathrm{op}})_{i\in I} \to \mathcal{B}^{\mathrm{op}}$ is an isomorphism as well.

3.4. The $\Bbbk$-$\mathcal{C}at$-multifunctor k

The multifunctor $\mathsf{k} : \mathsf{A}^{\mathrm{u}}_\infty \to \widehat{\mathcal{K}\text{-}\mathcal{C}at}$ is defined in [BLM06, Chapter 13]. Here we construct its extension to natural A_∞-transformations as follows. Let $f, g : (\mathcal{A}_i)_{i\in I} \to \mathcal{B}$ be unital A_∞-functors, $r : f \to g : (\mathcal{A}_i)_{i\in I} \to \mathcal{B}$ a natural A_∞-transformation. It gives rise to a natural transformation of $\mathcal{K}$-functors $\mathsf{k}r : \mathsf{k}f \to \mathsf{k}g : \boxtimes^{i\in I}\mathsf{k}\mathcal{A}_i \to \mathsf{k}\mathcal{B}$. Components of $\mathsf{k}r$ are given by

$${}_{(X_i)_{i\in I}}\mathsf{k}r = {}_{(X_i)_{i\in I}}r_0 s^{-1} : \Bbbk \to \mathcal{B}((X_i)_{i\in I}f,(X_i)_{i\in I}g), \quad X_i \in \operatorname{Ob}\mathcal{A}_i, \quad i\in I.$$

Since $r_0 b_1 = 0$, $\mathsf{k}r$ is a chain map. Naturality is expressed by the following equation in $\mathcal{K}$:

$$\begin{array}{ccc}
\boxtimes^{i\in I}\mathsf{k}\mathcal{A}_i(X_i,Y_i) & \xrightarrow{\mathsf{k}f} & \mathsf{k}\mathcal{B}((X_i)_{i\in I}f,(Y_i)_{i\in I}f) \\
\Big\downarrow{\scriptstyle \mathsf{k}g} & = & \Big\downarrow{\scriptstyle (1\otimes{}_{(Y_i)_{i\in I}}\mathsf{k}r)\mu_{\mathsf{k}\mathcal{B}}} \\
\mathsf{k}\mathcal{B}((X_i)_{i\in I}g,(Y_i)_{i\in I}g) & \xrightarrow{({}_{(X_i)_{i\in I}}\mathsf{k}r\otimes 1)\mu_{\mathsf{k}\mathcal{B}}} & \mathsf{k}\mathcal{B}((X_i)_{i\in I}f,(Y_i)_{i\in I}g).
\end{array}$$

Associativity of $\mu_{\mathsf{k}\mathcal{B}}$ allows to write it as follows:

$$\big[\otimes^{i\in I}\mathsf{k}\mathcal{A}_i(X_i,Y_i) \xrightarrow{\otimes^{i\in I}sf_{e_i}s^{-1}\otimes{}_{(Y_i)_{i\in I}}r_0s^{-1}}$$
$$\otimes^{i\in I}\mathsf{k}\mathcal{B}(((Y_j)_{j<i},(X_j)_{j\geqslant i})f,((Y_j)_{j\leqslant i},(X_j)_{j>i})f)\otimes\mathsf{k}\mathcal{B}((Y_i)_{i\in I}f,(Y_i)_{i\in I}g)$$
$$\xrightarrow{\mu^{I\sqcup\mathbf{1}}_{\mathsf{k}\mathcal{B}}} \mathsf{k}\mathcal{B}((X_i)_{i\in I}f,(Y_i)_{i\in I}g)\big]$$
$$= \big[\otimes^{i\in I}\mathsf{k}\mathcal{A}_i(X_i,Y_i) \xrightarrow{{}_{(X_i)_{i\in I}}r_0s^{-1}\otimes\otimes^{i\in I}sg_{e_i}s^{-1}}$$
$$\mathsf{k}\mathcal{B}((X_i)_{i\in I}f,(X_i)_{i\in I}g)\otimes\otimes^{i\in I}\mathsf{k}\mathcal{B}(((Y_j)_{j<i},(X_j)_{j\geqslant i})g,((Y_j)_{j\leqslant i},(X_j)_{j>i})g)$$
$$\xrightarrow{\mu^{\mathbf{1}\sqcup I}_{\mathsf{k}\mathcal{B}}} \mathsf{k}\mathcal{B}((X_i)_{i\in I}f,(Y_i)_{i\in I}g)\big].$$

This equation is a consequence of the following equation in $\mathcal{K}$:

$$\begin{aligned}(s(f|_i^{(Y_j)_{j<i},(X_j)_{j>i}})_1 s^{-1} \otimes {}_{(Y_j)_{j\leqslant i},(X_j)_{j>i}} r_0 s^{-1})\mu_{\Bbbk\mathcal{B}} \\ = ({}_{(Y_j)_{j<i},(X_j)_{j\geqslant i}} r_0 s^{-1} \otimes s(g|_i^{(Y_j)_{j<i},(X_j)_{j>i}})_1 s^{-1})\mu_{\Bbbk\mathcal{B}} : \\ \mathcal{A}(X_i,Y_i) \to \mathcal{B}(((Y_j)_{j<i},(X_j)_{j\geqslant i})f, ((Y_j)_{j\leqslant i},(X_j)_{j>i})g),\end{aligned}$$

which in turn follows from the equation $(rB_1)_{e_i} = 0$:

$$\begin{aligned}(sf_{e_i}s^{-1} \otimes r_0 s^{-1})m_2 - (r_0 s^{-1} \otimes sg_{e_i}s^{-1})m_2 + sr_{e_i}s^{-1}m_1 + m_1 sr_{e_i}s^{-1} \\ = s[(f_{e_i} \otimes r_0)b_2 + (r_0 \otimes g_{e_i})b_2 + r_{e_i}b_1 + b_1 r_{e_i}]s^{-1} = 0 : \\ \mathcal{A}(X_i,Y_i) \to \mathcal{B}(((Y_j)_{j<i},(X_j)_{j\geqslant i})f, ((Y_j)_{j\leqslant i},(X_j)_{j>i})g).\end{aligned}$$

The 2-category $\mathcal{K}$-$\mathcal{C}at$ is naturally a symmetric Monoidal $\Bbbk$-$\mathcal{C}at$-category, therefore $\widehat{\mathcal{K}\text{-}\mathcal{C}at}$ is a symmetric $\Bbbk$-$\mathcal{C}at$-multicategory. According to the general recipe, for each map $\phi : I \to J$, the composition in $\widehat{\mathcal{K}\text{-}\mathcal{C}at}$ is given by the $\Bbbk$-linear functor

$$\begin{aligned}\mu_\phi^{\widehat{\mathcal{K}\text{-}\mathcal{C}at}} = \big[\boxtimes^{J\sqcup\mathbf{1}}[(\mathcal{K}\text{-}\mathcal{C}at(\boxtimes^{i\in\phi^{-1}j}\mathcal{A}_i,\mathcal{B}_j))_{j\in J}, \mathcal{K}\text{-}\mathcal{C}at(\boxtimes^{j\in J}\mathcal{B}_j,\mathcal{C})] \xrightarrow[\sim]{\Lambda^\phi_{\Bbbk\text{-}\mathcal{C}at}} \\ \boxtimes^{j\in J}\mathcal{K}\text{-}\mathcal{C}at(\boxtimes^{i\in\phi^{-1}j}\mathcal{A}_i,\mathcal{B}_j) \boxtimes \mathcal{K}\text{-}\mathcal{C}at(\boxtimes^{j\in J}\mathcal{B}_j,\mathcal{C}) \xrightarrow{\boxtimes^J\boxtimes 1} \\ \mathcal{K}\text{-}\mathcal{C}at(\boxtimes^{j\in J}\boxtimes^{i\in\phi^{-1}j}\mathcal{A}_i, \boxtimes^{j\in J}\mathcal{B}_j)\boxtimes\mathcal{K}\text{-}\mathcal{C}at(\boxtimes^{j\in J}\mathcal{B}_j,\mathcal{C}) \xrightarrow{\lambda^\phi\cdot - \cdot -} \mathcal{K}\text{-}\mathcal{C}at(\boxtimes^{i\in I}\mathcal{A}_i,\mathcal{C})\big].\end{aligned}$$

In particular, the action on natural transformations is given by the map

$$\begin{aligned}\otimes^{j\in J}\mathcal{K}\text{-}\mathcal{C}at(\boxtimes^{i\in\phi^{-1}j}\mathcal{A}_i,\mathcal{B}_j)(f_j,g_j) \otimes \mathcal{K}\text{-}\mathcal{C}at(\boxtimes^{j\in J}\mathcal{B}_j,\mathcal{C})(h,k) \\ \to \mathcal{K}\text{-}\mathcal{C}at(\boxtimes^{i\in I}\mathcal{A}_i,\mathcal{C})((f_j)_{j\in J}\cdot h, (g_j)_{j\in J}\cdot k), \quad \otimes^{j\in J} r^j \otimes p \mapsto (r^j)_{j\in J}\cdot p,\end{aligned}$$

where for each collection of objects $X_i \in \mathcal{A}_i$, $i \in I$,

$$\begin{aligned}{}_{(X_i)_{i\in I}}[(r^j)_{j\in J}\cdot p] = \big[\Bbbk \xrightarrow[\sim]{\lambda^{\varnothing\to J\sqcup\mathbf{1}}} \otimes^{J\sqcup\mathbf{1}}\Bbbk \xrightarrow{\otimes^{j\in J}{}_{(X_i)_{i\in\phi^{-1}j}}r^j \otimes {}_{((X_i)_{i\in\phi^{-1}j}g_j)_{j\in J}}p} \\ \otimes^{j\in J}\mathcal{B}((X_i)_{i\in\phi^{-1}j}f_j, (X_i)_{i\in\phi^{-1}j}g_j) \otimes \mathcal{C}(((X_i)_{i\in\phi^{-1}j}g_j)_{j\in J}h, ((X_i)_{i\in\phi^{-1}j}g_j)_{j\in J}k) \\ \xrightarrow{h\otimes 1} \mathcal{C}((X_i)_{i\in I}(f_j)_{j\in J}h, (X_i)_{i\in I}(g_j)_{j\in J}h) \otimes \mathcal{C}((X_i)_{i\in I}(g_j)_{j\in J}h, (X_i)_{i\in I}(g_j)_{j\in J}k) \\ \xrightarrow{\mu_{\mathcal{C}}} \mathcal{C}((X_i)_{i\in I}(f_j)_{j\in J}h, (X_i)_{i\in I}(g_j)_{j\in J}k)\big].\end{aligned}$$

The base change functor $H^0 : \mathsf{A}^{\mathrm{u}}_\infty \to \Bbbk$-$\mathcal{C}at$ turns the symmetric $\mathsf{A}^{\mathrm{u}}_\infty$-multicategory $\underline{\mathsf{A}}^{\mathrm{u}}_\infty$ into a symmetric $\Bbbk$-$\mathcal{C}at$-multicategory, which we denote by A^u_∞. That is, the objects of A^u_∞ are unital A_∞-categories, and for each collection $(\mathcal{A}_i)_{i\in I}$, $\mathcal{B}$ of unital A_∞-categories, there is a $\Bbbk$-linear category $A^u_\infty((\mathcal{A}_i)_{i\in I};\mathcal{B}) = H^0\underline{\mathsf{A}}^{\mathrm{u}}_\infty((\mathcal{A}_i)_{i\in I};\mathcal{B})$, whose objects are unital A_∞-functors, and whose morphisms are equivalence classes of natural A_∞-transformations. The composition in A^u_∞ is given by the $\Bbbk$-linear functor $\mu_\phi^{A^u_\infty} = H^0(\mu_\phi^{\mathsf{A}^u_\infty}) = H^0(\mathsf{k}\mu_\phi^{\mathsf{A}^u_\infty})$, where

$$\mathsf{k}\mu_\phi^{\mathsf{A}^u_\infty} = \Big[\otimes^{j\in J}\underline{\mathsf{A}^u_\infty}((\mathcal{A}_i)_{i\in\phi^{-1}j};\mathcal{B}_j)(f_j,g_j)\otimes\underline{\mathsf{A}^u_\infty}((\mathcal{B}_j)_{j\in J};\mathcal{C})(h,k) \xrightarrow{\otimes^{j\in J}sM_{e_j0}s^{-1}\otimes sM_{0\dots01}s^{-1}}$$

$$\otimes^{j\in J}\underline{\mathsf{A}^u_\infty}((\mathcal{A}_i)_{i\in I};\mathcal{C})(((g_l)_{l<j},(f_l)_{l\geqslant j})h,((g_l)_{l\leqslant j},(f_l)_{l>j})h)$$
$$\otimes\underline{\mathsf{A}^u_\infty}((\mathcal{A}_i)_{i\in I};\mathcal{C})((g_j)_{j\in J}h,(g_j)_{j\in J}k)$$
$$\xrightarrow{\mu^{J\sqcup\mathbf{1}}_{\mathsf{k}\underline{\mathsf{A}^u_\infty}((\mathcal{A}_i)_{i\in I};\mathcal{C})}}\underline{\mathsf{A}^u_\infty}((\mathcal{A}_i)_{i\in I};\mathcal{C})((f_j)_{j\in J}h,(g_j)_{j\in J}k)\Big].$$

Proposition 3.5. *There is a symmetric $\Bbbk$-$\mathcal{C}at$-multifunctor $\mathsf{k}: A^u_\infty \to \widehat{\mathcal{K}\text{-}\mathcal{C}at}$.*

Proof. It remains to prove compatibility of k with $\mu_\phi^{A^u_\infty}$ on the level of transformations. Let $r^j \in s\underline{\mathsf{A}^u_\infty}((\mathcal{A}_i)_{i\in\phi^{-1}j};\mathcal{B}_j)(f_j,g_j)$, $j\in J$, $p\in s\underline{\mathsf{A}^u_\infty}((\mathcal{B}_j)_{j\in J};\mathcal{C})$ be natural A_∞-transformations. Then $((r^js^{-1})_{j\in J},ps^{-1})\mu_\phi^{A^u_\infty}$ is the equivalence class of the following A_∞-transformation:

$$[\otimes^{j\in J}((g_l)_{l<j},r^j,(f_l)_{l>j},h)M_{e_j0}s^{-1}\otimes((g_j)_{j\in J},p)M_{0\dots01}s^{-1}]\mu^{J\sqcup\mathbf{1}}_{\mathsf{k}\underline{\mathsf{A}^u_\infty}((\mathcal{A}_i)_{i\in I};\mathcal{C})}.$$

In order to find $\mathsf{k}[((r^js^{-1})_{j\in J},ps^{-1})\mu_\phi^{A^u_\infty}]$ we need the 0-th components of the above expression. Since $[(t\otimes q)B_2]_0 = (t_0\otimes q_0)b_2$, for arbitrary composable A_∞-transformations t and q, it follows that

$$\begin{aligned}
&{}_{(X_i)_{i\in I}}\mathsf{k}[((r^js^{-1})_{j\in J},ps^{-1})\mu_\phi^{A^u_\infty}]\\
&=(\otimes^{j\in J}{}_{(X_i)_{i\in\phi^{-1}j}}[((g_l)_{l<j},r^j,(f_l)_{l>j},h)M_{e_j0}]_0s^{-1}\otimes\\
&\qquad\qquad\otimes{}_{(X_i)_{i\in I}}[((g_j)_{j\in J},p)M_{0\dots01}]_0s^{-1})\mu^{J\sqcup\mathbf{1}}_{\mathsf{k}\mathcal{C}}\\
&=(\otimes^{j\in J}{}_{(X_i)_{i\in\phi^{-1}j}}r^j_0s^{-1}s(h\big|_j^{((X_i)_{i\in\phi^{-1}l}g_l)_{l<j},((X_i)_{i\in\phi^{-1}l}f_l)_{l>j}})_1s^{-1}\otimes\\
&\qquad\qquad\otimes{}_{((X_i)_{i\in\phi^{-1}j}g_j)_{j\in J}}p_0s^{-1})\mu^{J\sqcup\mathbf{1}}_{\mathsf{k}\mathcal{C}}.
\end{aligned}$$

By associativity of $\mu_{\mathsf{k}\mathcal{C}}$, this equals

$$\begin{aligned}
&(\otimes^{j\in J}{}_{(X_i)_{i\in\phi^{-1}j}}r^j_0s^{-1}\cdot\otimes^{j\in J}s(h\big|_j^{((X_i)_{i\in\phi^{-1}l}g_l)_{l<j},((X_i)_{i\in\phi^{-1}l}f_l)_{l>j}})_1s^{-1}\mu^J_{\mathsf{k}\mathcal{C}}\\
&\qquad\qquad\otimes{}_{((X_i)_{i\in\phi^{-1}j}g_j)_{j\in J}}p_0s^{-1})\mu_{\mathsf{k}\mathcal{C}}\\
&=(\otimes^{j\in J}{}_{(X_i)_{i\in\phi^{-1}j}}\mathsf{k}r^j\cdot\mathsf{k}h\otimes{}_{((X_i)_{i\in\phi^{-1}j}g_j)_{j\in J}}\mathsf{k}p)\mu_{\mathsf{k}\mathcal{C}} = {}_{(X_i)_{i\in I}}[(\mathsf{k}r^j)_{j\in J}\cdot\mathsf{k}p].
\end{aligned}$$

Therefore, $\mathsf{k}[((r^js^{-1})_{j\in J},ps^{-1})\mu_\phi^{A^u_\infty}] = ((\mathsf{k}r^j)_{j\in J},\mathsf{k}p)\mu_\phi^{\widehat{\mathcal{K}\text{-}\mathcal{C}at}}$, hence k is a multifunctor. □

The quotient functor $Q:\mathbf{dg}=\mathsf{C}_\Bbbk\to\mathcal{K}$ equipped with the identity transformation $\otimes^{i\in I}QX_i\to Q\otimes^{i\in I}X_i$ is a symmetric Monoidal functor. It gives rise to a symmetric Monoidal $\mathcal{C}at$-functor $Q_*:\mathbf{dg}\text{-}\mathcal{C}at\to\mathcal{K}\text{-}\mathcal{C}at$. Let $\widehat{Q_*}:\widehat{\mathbf{dg}\text{-}\mathcal{C}at}\to\widehat{\mathcal{K}\text{-}\mathcal{C}at}$ denote the corresponding symmetric $\mathcal{C}at$-multifunctor.

Proposition 3.6. *There is a multinatural isomorphism*

$$\begin{array}{ccc} \mathsf{A}^{\mathrm{u}}_\infty \times \widehat{\mathbf{dg}\text{-}\mathcal{C}at} & \xrightarrow{\boxdot} & \mathsf{A}^{\mathrm{u}}_\infty \\ {\scriptstyle \mathsf{k}\times\widehat{Q_*}}\downarrow & \overset{\xi}{\swarrow\!\!\!\!\!\!=} & \downarrow{\scriptstyle \mathsf{k}} \\ \widehat{\mathcal{K}\text{-}\mathcal{C}at} \times \widehat{\mathcal{K}\text{-}\mathcal{C}at} & \xrightarrow{\boxtimes} & \widehat{\mathcal{K}\text{-}\mathcal{C}at} \end{array}$$

where $\boxdot : \mathsf{A}^{\mathrm{u}}_\infty \boxtimes \widehat{\mathbf{dg}\text{-}\mathcal{C}at} \to \mathsf{A}^{\mathrm{u}}_\infty$ *is the action of differential graded categories on unital* A_∞*-categories constructed in* [BLM06, Appendices C.10–C.13].

Proof. Given a unital A_∞-category $\mathcal{A}$ and a differential graded category $\mathcal{C}$, define an isomorphism of $\mathcal{K}$-quivers $\xi : \mathsf{k}(\mathcal{A} \boxdot \mathcal{C}) \to \mathsf{k}\mathcal{A} \boxtimes Q_*\mathcal{C}$, identity on objects, as follows. For $X, Y \in \operatorname{Ob}\mathcal{A}$, $U, V \in \operatorname{Ob}\mathcal{C}$, we have

$$\begin{aligned} \mathsf{k}(\mathcal{A} \boxdot \mathcal{C})((X,U),(Y,V)) &= ((\mathcal{A} \boxdot \mathcal{C})((X,U),(Y,V)), m_1^{\mathcal{A}\boxdot\mathcal{C}}) \\ &= ((s\mathcal{A}(X,Y) \otimes \mathcal{C}(U,V))[-1], sb^{\mathcal{A}\boxdot\mathcal{C}}s^{-1}), \\ (\mathsf{k}\mathcal{A} \boxtimes Q_*\mathcal{C})((X,U),(Y,V)) &= \mathsf{k}\mathcal{A}(X,Y) \otimes Q_*\mathcal{C}(U,V) \\ &= (\mathcal{A}(X,Y) \otimes \mathcal{C}(U,V), m_1 \otimes 1 + 1 \otimes d). \end{aligned}$$

Define ξ by

$$\xi = s(s^{-1} \otimes 1) : (s\mathcal{A}(X,Y) \otimes \mathcal{C}(U,V))[-1] \to \mathcal{A}(X,Y) \otimes \mathcal{C}(U,V).$$

The morphism ξ commutes with the differential since

$$\begin{aligned} m_1^{\mathcal{A}\boxdot\mathcal{C}} \cdot \xi = sb^{\mathcal{A}\boxdot\mathcal{C}}s^{-1} \cdot s(s^{-1} \otimes 1) &= s(b_1 \otimes 1 - 1 \otimes d)(s^{-1} \otimes 1) \\ &= s(s^{-1} \otimes 1)(sb_1s^{-1} \otimes 1 + 1 \otimes d) = \xi \cdot (m_1 \otimes 1 + 1 \otimes d), \end{aligned}$$

therefore it is an isomorphism of $\mathcal{K}$-quivers. We claim that it also respects the composition. Indeed, suppose $X, Y, Z \in \operatorname{Ob}\mathcal{A}$, $U, V, W \in \operatorname{Ob}\mathcal{C}$. From [BLM06, (C.10.1)] we find that

$$\begin{aligned} \mu_{\mathsf{k}(\mathcal{A}\boxdot\mathcal{C})} = m_2^{\mathcal{A}\boxdot\mathcal{C}} = \big[&(\mathcal{A} \boxdot \mathcal{C})((X,U),(Y,V)) \otimes (\mathcal{A} \boxdot \mathcal{C})((Y,V),(Z,W)) \xrightarrow{s\otimes s} \\ &(s\mathcal{A}(X,Y) \otimes \mathcal{C}(U,V)) \otimes (s\mathcal{A}(Y,Z) \otimes \mathcal{C}(V,W)) \xrightarrow{\sigma_{(12)}} \\ &(s\mathcal{A}(X,Y) \otimes s\mathcal{A}(Y,Z)) \otimes (\mathcal{C}(U,V) \otimes \mathcal{C}(V,W)) \xrightarrow{b_2\otimes\mu_{\mathcal{C}}} \\ &s\mathcal{A}(X,Z) \otimes \mathcal{C}(U,W) \xrightarrow{s^{-1}} (\mathcal{A} \boxdot \mathcal{C})((X,U),(Z,W))\big], \\ \mu_{\mathsf{k}\mathcal{A}\boxtimes Q_*\mathcal{C}} = \big[&(\mathcal{A}(X,Y) \otimes \mathcal{C}(U,V)) \otimes (\mathcal{A}(Y,Z) \otimes \mathcal{C}(V,W)) \xrightarrow{\sigma_{(12)}} \\ &(\mathcal{A}(X,Y) \otimes \mathcal{A}(Y,Z)) \otimes (\mathcal{C}(U,V) \otimes \mathcal{C}(V,W)) \xrightarrow{m_2\otimes\mu_{\mathcal{C}}} \mathcal{A}(X,Z) \otimes \mathcal{C}(U,W)\big]. \end{aligned}$$

It follows that

$$\mu_{\mathsf{k}(\mathcal{A}\boxdot\mathcal{C})}\cdot\xi = \big[(\mathcal{A}\boxdot\mathcal{C})((X,U),(Y,V))\otimes(\mathcal{A}\boxdot\mathcal{C})((Y,V),(Z,W)) \xrightarrow{s\otimes s}$$
$$(s\mathcal{A}(X,Y)\otimes\mathcal{C}(U,V))\otimes(s\mathcal{A}(Y,Z)\otimes\mathcal{C}(V,W)) \xrightarrow{\sigma_{(12)}}$$
$$(s\mathcal{A}(X,Y)\otimes s\mathcal{A}(Y,Z))\otimes(\mathcal{C}(U,V)\otimes\mathcal{C}(V,W)) \xrightarrow{b_2s^{-1}\otimes\mu_{\mathcal{C}}} \mathcal{A}(X,Z)\otimes\mathcal{C}(U,W)\big],$$
$$(\xi\otimes\xi)\cdot\mu_{\mathsf{k}\mathcal{A}\boxtimes Q_*\mathcal{C}} = \big[(\mathcal{A}\boxdot\mathcal{C})((X,U),(Y,V))\otimes(\mathcal{A}\boxdot\mathcal{C})((Y,V),(Z,W))$$
$$\xrightarrow{s(s^{-1}\otimes 1)\otimes s(s^{-1}\otimes 1)} (\mathcal{A}(X,Y)\otimes\mathcal{C}(U,V))\otimes(\mathcal{A}(Y,Z)\otimes\mathcal{C}(V,W)) \xrightarrow{\sigma_{(12)}}$$
$$(\mathcal{A}(X,Y)\otimes\mathcal{A}(Y,Z))\otimes(\mathcal{C}(U,V)\otimes\mathcal{C}(V,W)) \xrightarrow{m_2\otimes\mu_{\mathcal{C}}} \mathcal{A}(X,Z)\otimes\mathcal{C}(U,W)\big]$$
$$= \big[(\mathcal{A}\boxdot\mathcal{C})((X,U),(Y,V))\otimes(\mathcal{A}\boxdot\mathcal{C})((Y,V),(Z,W)) \xrightarrow{s\otimes s}$$
$$(s\mathcal{A}(X,Y)\otimes\mathcal{C}(U,V))\otimes(s\mathcal{A}(Y,Z)\otimes\mathcal{C}(V,W)) \xrightarrow{\sigma_{(12)}}$$
$$(s\mathcal{A}(X,Y)\otimes s\mathcal{A}(Y,Z))\otimes(\mathcal{C}(U,V)\otimes\mathcal{C}(V,W)) \xrightarrow{-(s^{-1}\otimes s^{-1})m_2\otimes\mu_{\mathcal{C}}} \mathcal{A}(X,Z)\otimes\mathcal{C}(U,W)\big],$$

therefore $\mu_{\mathsf{k}(\mathcal{A}\boxdot\mathcal{C})}\cdot\xi = (\xi\otimes\xi)\cdot\mu_{\mathsf{k}\mathcal{A}\boxtimes Q_*\mathcal{C}}$, as $b_2s^{-1} = -(s^{-1}\otimes s^{-1})m_2$. The morphism ξ also respects the identity morphisms since

$$1^{\mathsf{k}(\mathcal{A}\boxdot\mathcal{C})}_{(X,U)}\xi = ({}_X\mathbf{i}^{\mathcal{A}}_0\otimes 1^{\mathcal{C}}_U)s^{-1}\cdot s(s^{-1}\otimes 1) = ({}_X\mathbf{i}^{\mathcal{A}}_0 s^{-1}\otimes 1^{\mathcal{C}}_U) = (1^{\mathsf{k}\mathcal{A}}_X\otimes 1^{\mathcal{C}}_U) = 1^{\mathsf{k}\mathcal{A}\boxtimes Q_*\mathcal{C}}_{(X,U)}.$$

Thus, ξ is an isomorphism of $\mathcal{K}$-categories.

Multinaturality of ξ reduces to the following problem. Let $f:\boxtimes^{i\in I}Ts\mathcal{A}_i\to Ts\mathcal{B}$ be an A_∞-functor, $g:\boxtimes^{i\in I}\mathcal{C}_i\to\mathcal{D}$ a differential graded functor. Then the diagram

$$\begin{array}{ccc}
\boxtimes^{i\in I}\mathsf{k}(\mathcal{A}_i\boxdot\mathcal{C}_i) & \xrightarrow{\boxtimes^I\xi} & \boxtimes^{i\in I}(\mathsf{k}\mathcal{A}_i\boxtimes Q_*\mathcal{C}_i) \\
{\scriptstyle \mathsf{k}(f\boxdot g)}\big\downarrow & & \big\downarrow{\scriptstyle \sigma_{(12)}\cdot(\mathsf{k}f\boxtimes\widehat{Q_*}g)} \\
\mathsf{k}(\mathcal{B}\boxdot\mathcal{D}) & \xrightarrow{\quad\xi\quad} & \mathsf{k}\mathcal{B}\boxtimes Q_*\mathcal{D}
\end{array}$$

must commute; let us prove this. Let $X_i,Y_i\in\operatorname{Ob}\mathcal{A}_i$, $U_i,V_i\in\operatorname{Ob}\mathcal{C}_i$, $i\in I$, be families of objects. Then

$$\mathsf{k}(f\boxdot g) = \big[\otimes^{i\in I}(\mathcal{A}_i\boxdot\mathcal{C}_i)((X_i,U_i),(Y_i,V_i)) \xrightarrow{\otimes^{i\in I}s(f\boxdot g)_{e_i}s^{-1}}$$
$$\otimes^{i\in I}(\mathcal{B}\boxdot\mathcal{D})\big((((Y_j)_{j<i},(X_j)_{j\geqslant i})f,((V_j)_{j<i},(U_j)_{j\geqslant i})g),$$
$$(((Y_j)_{j\leqslant i},(X_j)_{j>i})f,((V_j)_{j\leqslant i},(U_j)_{j>i})g)\big)$$
$$\xrightarrow{\mu^I_{\mathsf{k}(\mathcal{B}\boxdot\mathcal{D})}} (\mathcal{B}\boxdot\mathcal{D})\big(((X_i)_{i\in I}f,(U_i)_{i\in I}g),((Y_i)_{i\in I}f,(V_i)_{i\in I}g)\big)\big].$$

From [BLM06, (C.5.1)] we infer that

$$(f \boxdot g)_{e_i} = \big[sA_i(X_i,Y_i) \otimes \mathcal{C}(U_i,V_i) \xrightarrow{f_{e_i}\otimes g_{e_i}} s\mathcal{B}(((Y_j)_{j<i},(X_j)_{j\geqslant i})f,((Y_j)_{j\leqslant i},(X_j)_{j>i})f)\otimes \otimes \mathcal{D}(((V_j)_{j<i},(U_j)_{j\geqslant i})g,((V_j)_{j\leqslant i},(U_j)_{j>i})g)\big],$$

where

$$g_{e_i} = \big[\mathcal{C}_i(U_i,V_i) \xrightarrow{\lambda^{i:\mathbf{1}\hookrightarrow I}\cdot\otimes^{j\in I}[(1_{V_j})_{j<i},\mathrm{id},(1_{U_j})_{j>i}]} \otimes^{j\in I}[(\mathcal{C}_j(V_j,V_j))_{j<i},\mathcal{C}_i(U_i,V_i),(\mathcal{C}_j(U_j,U_j))_{j>i}] \xrightarrow{g} \mathcal{D}(((V_j)_{j<i},(U_j)_{j\geqslant i})g,((V_j)_{j\leqslant i},(U_j)_{j>i})g)\big].$$

Therefore

$$\mathsf{k}(f\boxdot g)\cdot\xi = \big[\otimes^{i\in I}(\mathcal{A}_i\boxdot\mathcal{C}_i)((X_i,U_i),(Y_i,V_i)) \xrightarrow{\otimes^{i\in I}s(f_{e_i}\otimes g_{e_i})s^{-1}} \otimes^{i\in I}(\mathcal{B}\boxdot\mathcal{D})((((Y_i)_{j<i},(X_j)_{j\geqslant i})f,((V_i)_{j<i},(U_j)_{j\geqslant i})g), (((Y_j)_{j\leqslant i},(X_j)_{j>i})f,((V_j)_{j\leqslant i},(U_j)_{j>i})g))$$
$$\xrightarrow{\otimes^{i\in I}\xi} \otimes^{i\in I}[\mathcal{B}(((Y_i)_{j<i},(X_j)_{j\geqslant i})f,((Y_j)_{j\leqslant i},(X_j)_{j>i})f) \otimes \mathcal{D}(((V_i)_{j<i},(U_j)_{j\geqslant i})g,((V_j)_{j\leqslant i},(U_j)_{j>i})g)]$$
$$\xrightarrow{\mu^I_{\mathsf{k}\mathcal{B}\boxtimes Q_*\mathcal{D}}} \mathcal{B}((X_i)_{i\in I}f,(Y_i)_{i\in I}f)\otimes\mathcal{D}((U_i)_{i\in I}g,(V_i)_{i\in I}g)\big]$$

since ξ respects the composition. The above expression can be transformed as follows:

$$\mathsf{k}(f\boxdot g)\cdot\xi = \big[\otimes^{i\in I}(\mathcal{A}_i\boxdot\mathcal{C}_i)((X_i,U_i),(Y_i,V_i)) \xrightarrow{\otimes^{i\in I}s(s^{-1}\otimes 1)} \otimes^{i\in I}(\mathcal{A}_i(X_i,Y_i)\otimes\mathcal{C}_i(U_i,V_i)) \xrightarrow{\otimes^{i\in I}(sf_{e_i}s^{-1}\otimes g_{e_i})}$$
$$\otimes^{i\in I}[\mathcal{B}(((Y_i)_{j<i},(X_j)_{j\geqslant i})f,((Y_j)_{j\leqslant i},(X_j)_{j>i})f) \otimes \mathcal{D}(((V_i)_{j<i},(U_j)_{j\geqslant i})g,((V_j)_{j\leqslant i},(U_j)_{j>i})g)]$$
$$\xrightarrow{\sigma_{(12)}} (\otimes^{i\in I}\mathcal{B}(((Y_j)_{j<i},(X_j)_{j\geqslant i})f,((Y_j)_{j\leqslant i},(X_j)_{j>i})f)) \otimes (\otimes^{i\in I}\mathcal{D}(((V_j)_{j<i},(U_j)_{j\geqslant i})g,((V_j)_{j\leqslant i},(U_j)_{j>i})g))$$
$$\xrightarrow{\mu^I_{\mathsf{k}\mathcal{B}}\otimes\mu^I_{\mathcal{D}}} \mathcal{B}((X_i)_{i\in I}f,(Y_i)_{i\in I}f)\otimes\mathcal{D}((U_i)_{i\in I}g,(V_i)_{i\in I}g)\big]$$
$$= \big[\otimes^{i\in I}(\mathcal{A}_i\boxdot\mathcal{C}_i)((X_i,U_i),(Y_i,V_i)) \xrightarrow{\otimes^{i\in I}\xi} \otimes^{i\in I}(\mathcal{A}_i(X_i,Y_i)\otimes\mathcal{C}_i(U_i,V_i)) \xrightarrow{\sigma_{(12)}} (\otimes^{i\in I}\mathcal{A}_i(X_i,Y_i))\otimes(\otimes^{i\in I}\mathcal{C}_i(U_i,V_i))$$
$$\xrightarrow{(\otimes^{i\in I}sf_{e_i}s^{-1})\mu^I_{\mathsf{k}\mathcal{B}}\otimes(\otimes^{i\in I}g_{e_i})\mu_{\mathcal{D}}} \mathcal{B}((X_i)_{i\in I}f,(Y_i)_{i\in I}f)\otimes\mathcal{D}((U_i)_{i\in I}g,(V_i)_{i\in I}g)\big]$$
$$= \otimes^{i\in I}\xi\cdot\sigma_{(12)}\cdot(\mathsf{k}f\otimes g),$$

due to the definition of $\mathsf{k}f$ and the identity

$$\bigl[\otimes^{i\in I}\mathcal{C}_i(U_i,V_i) \xrightarrow{\otimes^{i\in I} g_{e_i}} \otimes^{i\in I}\mathcal{D}(((V_j)_{j<i},(U_j)_{j\geqslant i})g,((V_j)_{j\leqslant i},(U_j)_{j>i})g) \xrightarrow{\mu^I_{\mathcal{D}}} \mathcal{D}((U_i)_{i\in I}g,(V_i)_{i\in I}g)\bigr] = g,$$

which is a consequence of g being a functor and associativity of $\mu_{\mathcal{D}}$. The proposition is proven. □

3.7. A_∞-categories closed under shifts

Unital A_∞-categories closed under shifts are defined in [BLM06, Chapter 15] similarly to Definition 2.20. A unital A_∞-category $\mathcal{C}$ is closed under shifts if and only if the A_∞-functor $u_{[]} : \mathcal{C} \to \mathcal{C}^{[]} \simeq \mathcal{C} \boxdot \mathcal{Z} = \mathcal{C} \boxtimes \mathcal{Z}$ is an equivalence.

For an arbitrary A_∞-category $\mathcal{C}$ the operations in $\mathcal{C}^{[]}$ are described explicitly in [BLM06, (15.2.2)]. In particular,

$$\begin{aligned} b_2^{\mathcal{C}^{[]}} &= (-)^{p-n}\bigl[s\mathcal{C}^{[]}((X,n),(Y,m)) \otimes s\mathcal{C}^{[]}((Y,m),(Z,p)) \\ &= s\mathcal{C}(X,Y)[m-n] \otimes s\mathcal{C}(Y,Z)[p-m] \xrightarrow{(s^{m-n}\otimes s^{p-m})^{-1}} s\mathcal{C}(X,Y)\otimes s\mathcal{C}(Y,Z) \\ &\xrightarrow{b_2^{\mathcal{C}}} s\mathcal{C}(X,Z) \xrightarrow{s^{p-n}} s\mathcal{C}(X,Z)[p-n] = s\mathcal{C}^{[]}((X,n),(Z,p))\bigr]. \end{aligned}$$

The above proposition implies that the binary operation $m_2^{\mathcal{C}^{[]}} = (s\otimes s)b_2^{\mathcal{C}^{[]}}s^{-1}$ in $\mathcal{C}^{[]}$ is homotopic to multiplication in $\mathsf{k}\mathcal{C}^{[]}$ given by formula (2.17). Actually, $m_2^{\mathcal{C}^{[]}}$ is given precisely by chain map (2.17), as one easily deduces from the above expression for $b_2^{\mathcal{C}^{[]}}$.

We have denoted the algebra $\mathcal{Z}$ in **dg**-$\mathcal{C}at$, "the same" algebra $Q_*\mathcal{Z}$ in $\mathcal{K}$-$\mathcal{C}at$ and "the same" algebra $H^\bullet(Q_*\mathcal{Z})$ in **gr**-$\mathcal{C}at$ all by the same letter $\mathcal{Z}$ by abuse of notation. Since units of the monads $-\boxdot\mathcal{Z}$ and $-\boxtimes\mathcal{Z}$ reduce essentially to the unit of the algebra $\mathcal{Z}$, Proposition 3.6 implies the following relation between them:

$$\bigl[\mathsf{k}\mathcal{A} \xrightarrow{\mathsf{k}u_{[]}} \mathsf{k}(\mathcal{A}\boxdot\mathcal{Z}) \xrightarrow[\sim]{\xi} \mathsf{k}\mathcal{A}\boxtimes Q_*\mathcal{Z}\bigr] = \bigl[\mathsf{k}\mathcal{A} \xrightarrow{u_{[]}} \mathsf{k}\mathcal{A}\boxtimes\mathcal{Z} = \mathsf{k}\mathcal{A}\boxtimes Q_*\mathcal{Z}\bigr].$$

Thus, if one of the $\mathcal{K}$-functors $\mathsf{k}u_{[]} : \mathsf{k}\mathcal{A} \to \mathsf{k}(\mathcal{A}\boxdot\mathcal{Z})$ and $u_{[]} : \mathsf{k}\mathcal{A} \to \mathsf{k}\mathcal{A}\boxtimes\mathcal{Z}$ is an equivalence, then so is the other. The former is a $\mathcal{K}$-equivalence if and only if the A_∞-functor $u_{[]} : \mathcal{A} \to \mathcal{A}\boxdot\mathcal{Z}$ is an equivalence. Therefore, the A_∞-category $\mathcal{A}$ is closed under shifts if and only if the $\mathcal{K}$-category $\mathsf{k}\mathcal{A}$ is closed under shifts.

3.8. Shifts as differential graded functors

Let $f = (f^i : C^i \to D^{i+\deg f})_{i\in\mathbb{Z}} \in \underline{\mathsf{C}}_{\Bbbk}(C,D)$ be a homogeneous element (a $\Bbbk$-linear map $f : C \to D$ of certain degree $d = \deg f$). Define $f^{[n]} = (-)^{fn}s^{-n}fs^n = (-)^{dn}\bigl(C[n]^i = C^{i+n} \xrightarrow{f^{i+n}} D^{i+n+d} = D[n]^{i+d}\bigr)$, which is an element of $\underline{\mathsf{C}}_{\Bbbk}(C[n],D[n])$ of the same degree $\deg f$.

Define the shift differential graded functor $[n] : \underline{\mathsf{C}}_{\Bbbk} \to \underline{\mathsf{C}}_{\Bbbk}$ as follows. It takes a complex (C,d) to the complex $(C[n],d^{[n]})$, $d^{[n]} = (-)^n s^{-n}ds^n$. On morphisms it acts via $\underline{\mathsf{C}}_\Bbbk(s^{-n},1)\cdot\underline{\mathsf{C}}_\Bbbk(1,s^n):\underline{\mathsf{C}}_\Bbbk(C,D)\to\underline{\mathsf{C}}_\Bbbk(C[n],D[n])$, $f\mapsto f^{[n]}$. Clearly, $[n]\cdot[m]=[n+m]$.

4. A_∞-modules

Consider the monoidal category $(\mathscr{Q}/S,\otimes)$ of graded $\Bbbk$-quivers. When $S=\mathbf{1}$ it reduces to the category of graded $\Bbbk$-modules used by Keller [Kel01] in his definition of A_∞-modules over A_∞-algebras. Let C, D be coassociative counital coalgebras; let $\psi:C\to D$ be a homomorphism; let $\delta:M\to M\otimes C$ and $\delta:N\to N\otimes D$ be counital comodules; let $f:M\to N$ be a ψ-comodule homomorphism, $f\delta=\delta(f\otimes\psi)$; let $\xi:C\to D$ be a (ψ,ψ)-coderivation, $\xi\Delta_0=\Delta_0(\psi\otimes\xi+\xi\otimes\psi)$. Define a (ψ,f,ξ)-*connection* as a morphism $r:M\to N$ of certain degree such that

$$\begin{array}{ccc} M & \xrightarrow{\delta} & M\otimes C \\ {\scriptstyle r}\downarrow & = & \downarrow{\scriptstyle f\otimes\xi+r\otimes\psi} \\ N & \xrightarrow{\delta} & N\otimes D \end{array}$$

compare with Tradler [Tra01]. Let (C,b^C) be a differential graded coalgebra. Let a counital comodule M have a $(1,1,b^C)$-connection $b^M:M\to M$ of degree 1, that is, $b^M\delta=\delta(1\otimes b^C+b^M\otimes1)$. Its *curvature* $(b^M)^2:M\to M$ is always a C-comodule homomorphism of degree 2. If it vanishes, b^M is called a flat connection (a differential) on M.

Equivalently, we consider the category $({}^d\mathscr{Q}/S,\otimes)$ of differential graded quivers, and coalgebras and comodules therein. For A_∞-applications it suffices to consider coalgebras (resp. comodules) whose underlying graded coalgebra (resp. comodule) has the form $Ts\mathcal{A}$ (resp. $s\mathcal{M}\otimes Ts\mathcal{C}$).

Let $\mathcal{M}\in\operatorname{Ob}\mathscr{Q}/S$ be graded quiver such that $\mathcal{M}(X,Y)=\mathcal{M}(Y)$ depends only on $Y\in S$. For any quiver $\mathcal{C}\in\operatorname{Ob}\mathscr{Q}/S$ the tensor quiver $C=(Ts\mathcal{C},\Delta_0)$ is a coalgebra. The comodule $\delta=1\otimes\Delta_0:M=s\mathcal{M}\otimes Ts\mathcal{C}\to s\mathcal{M}\otimes Ts\mathcal{C}\otimes Ts\mathcal{C}$ is counital. Let $(\mathcal{C},b^{\mathcal{C}})$ be an A_∞-category. Equivalently, we consider augmented coalgebras in $({}^d\mathscr{Q}/S,\otimes)$ of the form $(Ts\mathcal{C},\Delta_0,b^{\mathcal{C}})$. Let $b^{\mathcal{M}}:s\mathcal{M}\otimes Ts\mathcal{C}\to s\mathcal{M}\otimes Ts\mathcal{C}$ be a $(1,1,b^{\mathcal{C}})$-connection. Define the matrix coefficients of $b^{\mathcal{M}}$ to be

$$b^{\mathcal{M}}_{mn}=(1\otimes\operatorname{in}_m)\cdot b^{\mathcal{M}}\cdot(1\otimes\operatorname{pr}_n):s\mathcal{M}\otimes T^ms\mathcal{C}\to s\mathcal{M}\otimes T^ns\mathcal{C},\quad m,n\geqslant0.$$

The coefficients $b^{\mathcal{M}}_{m0}:s\mathcal{M}\otimes T^ms\mathcal{C}\to s\mathcal{M}$ are abbreviated to $b^{\mathcal{M}}_m$ and called components of $b^{\mathcal{M}}$.

A version of the following statement occurs in [LH03, Lemme 2.1.2.1].

Lemma 4.1. *Any $(1,1,b^{\mathcal{C}})$-connection $b^{\mathcal{M}}:s\mathcal{M}\otimes Ts\mathcal{C}\to s\mathcal{M}\otimes Ts\mathcal{C}$ is determined in a unique way by its components $b^{\mathcal{M}}_n:s\mathcal{M}\otimes T^ns\mathcal{C}\to s\mathcal{M}$, $n\geqslant0$. The matrix coefficients of $b^{\mathcal{M}}$ are expressed via components of $b^{\mathcal{M}}$ and components of the* *codifferential* $b^{\mathcal{C}}$ *as follows:*

$$b^{\mathcal{M}}_{mn} = b^{\mathcal{M}}_{m-n} \otimes 1^{\otimes n} + \sum_{\substack{p+k+q=m \\ p+1+q=n}} 1^{\otimes 1+p} \otimes b^{\mathcal{C}}_k \otimes 1^{\otimes q} : s\mathcal{M} \otimes T^m s\mathcal{C} \to s\mathcal{M} \otimes T^n s\mathcal{C}$$

for $m \geqslant n$. *If* $m < n$, *the matrix coefficient* $b^{\mathcal{M}}_{mn}$ *vanishes.*

Such comodules are particular cases of bimodules discussed below. That is why statements about comodules are only formulated. We prove more general results in the next section.

The morphism $(b^{\mathcal{M}})^2 : s\mathcal{M} \otimes Ts\mathcal{C} \to s\mathcal{M} \otimes Ts\mathcal{C}$ is a $(1,1,0)$-connection of degree 2, therefore equation $(b^{\mathcal{M}})^2 = 0$ is equivalent to its particular case $(b^{\mathcal{M}})^2(1 \otimes \mathrm{pr}_0) = 0 : s\mathcal{M} \otimes Ts\mathcal{C} \to s\mathcal{M}$. Thus $b^{\mathcal{M}}$ is a flat connection if for each $m \geqslant 0$ the following equation holds:

$$\sum_{n=0}^{m} (b^{\mathcal{M}}_{m-n} \otimes 1^{\otimes n}) b^{\mathcal{M}}_n + \sum_{p+k+q=m} (1^{\otimes 1+p} \otimes b^{\mathcal{C}}_k \otimes 1^{\otimes q}) b^{\mathcal{M}}_{p+1+q} = 0 : \\ s\mathcal{M} \otimes T^m s\mathcal{C} \to s\mathcal{M}. \quad (4.1)$$

Equivalently, such a $Ts\mathcal{C}$-comodule with a flat connection is the $Ts\mathcal{C}$-comodule $(s\mathcal{M} \otimes Ts\mathcal{C}, b^{\mathcal{M}})$ in the category $({}^{d}\mathcal{Q}/S, \otimes)$. It consists of the following data: a graded $\Bbbk$-module $\mathcal{M}(X)$ for each object X of $\mathcal{C}$; a family of $\Bbbk$-linear maps of degree 1

$$b^{\mathcal{M}}_n : s\mathcal{M}(X_0) \otimes s\mathcal{C}(X_0, X_1) \otimes \cdots \otimes s\mathcal{C}(X_{n-1}, X_n) \to s\mathcal{M}(X_n), \quad n \geqslant 0,$$

subject to equations (4.1). Equation (4.1) for $m = 0$ implies $(b^{\mathcal{M}}_0)^2 = 0$, that is, $(s\mathcal{M}(X), b^{\mathcal{M}}_0)$ is a chain complex, for each object $X \in \operatorname{Ob}\mathcal{C}$. We call a $Ts\mathcal{C}$-comodule with a flat connection $(s\mathcal{M} \otimes Ts\mathcal{C}, b^{\mathcal{M}})$, $\mathcal{M}(*, Y) = \mathcal{M}(Y)$, a $\mathcal{C}$-*module* (an A_∞-*module over* $\mathcal{C}$). $\mathcal{C}$-modules form a differential graded category $\mathcal{C}$-mod. The notion of a module over some kind of A_∞-category was introduced by Lefèvre-Hasegawa under the name of polydule [LH03].

Proposition 4.2. *An arbitrary* A_∞*-functor* $\phi : \mathcal{C} \to \underline{\mathsf{C}}_\Bbbk$ *determines a* $Ts\mathcal{C}$*-comodule* $s\mathcal{M} \otimes Ts\mathcal{C}$ *with a flat connection* $b^{\mathcal{M}}$ *by the formulae:* $\mathcal{M}(X) = X\phi$, *for each object* X *of* $\mathcal{C}$, $b^{\mathcal{M}}_0 = s^{-1}ds : s\mathcal{M}(X) \to s\mathcal{M}(X)$, *where* d *is the differential in the complex* $X\phi$, *and for* $n > 0$

$$b^{\mathcal{M}}_n = \big[s\mathcal{M}(X_0) \otimes s\mathcal{C}(X_0, X_1) \otimes \cdots \otimes s\mathcal{C}(X_{n-1}, X_n) \\ \xrightarrow{1 \otimes \phi_n} s\mathcal{M}(X_0) \otimes s\underline{\mathsf{C}}_\Bbbk(\mathcal{M}(X_0), \mathcal{M}(X_n)) \xrightarrow{(s \otimes s)^{-1}} \mathcal{M}(X_0) \otimes \underline{\mathsf{C}}_\Bbbk(\mathcal{M}(X_0), \mathcal{M}(X_n)) \\ \xrightarrow{\mathrm{ev}^{\mathsf{C}_\Bbbk}} \mathcal{M}(X_n) \xrightarrow{s} s\mathcal{M}(X_n)\big]. \quad (4.2)$$

This mapping from A_∞*-functors to* $\mathcal{C}$*-modules is bijective. Moreover, the differential graded categories* $\underline{\mathsf{A}_\infty}(\mathcal{C}; \underline{\mathsf{C}}_\Bbbk)$ *and* $\mathcal{C}$-mod *are isomorphic.*

A $\mathcal{C}$-*module* (an A_∞-*module over* $\mathcal{C}$) is defined as an A_∞-functor $\phi : \mathcal{C} \to \underline{\mathsf{C}}_\Bbbk$ by Seidel [Sei06, Section 1j]. The above proposition shows that the both definitions of $\mathcal{C}$-modules are equivalent. In the differential graded case $\mathcal{C}$-modules are actively used by Drinfeld [Dri04].

Definition 4.3. Let $\mathcal{C}$ be a unital A_∞-category. A $\mathcal{C}$-module $\mathcal{M}$ determined by an A_∞-functor $\phi : \mathcal{C} \to \underline{\mathsf{C}}_\Bbbk$ is called *unital* if ϕ is unital.

Proposition 4.4. *A $\mathcal{C}$-module $\mathcal{M}$ is unital if and only if for each $X \in \operatorname{Ob}\mathcal{C}$ the composition*

$$\big[s\mathcal{M}(X) \simeq s\mathcal{M}(X) \otimes \Bbbk \xrightarrow{1\otimes {}_X\mathbf{i}_0^{\mathcal{C}}} s\mathcal{M}(X) \otimes s\mathcal{C}(X,X) \xrightarrow{b_1^{\mathcal{M}}} s\mathcal{M}(X)\big]$$

is homotopic to identity map.

Proof. The second statement expands to the property that

$$\big[s\mathcal{M}(X) \simeq s\mathcal{M}(X) \otimes \Bbbk \xrightarrow{s^{-1}\otimes {}_X\mathbf{i}_0^{\mathcal{C}}} \mathcal{M}(X) \otimes s\mathcal{C}(X,X) \\ \xrightarrow{1\otimes\phi_1 s^{-1}} \mathcal{M}(X) \otimes \underline{\mathsf{C}}_\Bbbk(\mathcal{M}(X),\mathcal{M}(X)) \xrightarrow{\mathrm{ev}^{\mathsf{C}_\Bbbk}} \mathcal{M}(X) \xrightarrow{s} s\mathcal{M}(X)\big]$$

is homotopic to identity. That is,

$${}_X\mathbf{i}_0^{\mathcal{C}}\phi_1 s^{-1} = 1_{s\mathcal{M}(X)} + v m_1^{\underline{\mathsf{C}}_\Bbbk}, \quad \text{or,} \quad {}_X\mathbf{i}_0^{\mathcal{C}}\phi_1 = 1_{\mathcal{M}(X)}s + vsb_1^{\underline{\mathsf{C}}_\Bbbk}.$$

In other words, A_∞-functor ϕ is unital. □

5. A_∞-bimodules

Consider monoidal category $(\mathscr{Q}/S, \otimes)$ of graded $\Bbbk$-quivers. When $S = \mathbf{1}$ it reduces to the category of graded $\Bbbk$-modules used by Tradler [Tra01, Tra02] in his definition of A_∞-bimodules over A_∞-algebras. We extend his definitions of A_∞-bimodules improved in [TT06] from graded $\Bbbk$-modules to graded $\Bbbk$-quivers. The notion of a bimodule over some kind of A_∞-categories was introduced by Lefèvre-Hasegawa under the name of bipolydule [LH03].

Definition 5.1. Let A, C be coassociative counital coalgebras in $(\mathscr{Q}/R, \otimes)$ resp. $(\mathscr{Q}/S, \otimes)$. A *counital (A, C)-bicomodule* (P, δ^P) consists of a graded $\Bbbk$-span ($\mathbf{gr}$-span) P with $\operatorname{Ob}_s P = R$, $\operatorname{Ob}_t P = S$, $\operatorname{Par} P = \operatorname{Ob}_s P \times \operatorname{Ob}_t P$, $\mathrm{src} = \mathrm{pr}_1$, $\mathrm{tgt} = \mathrm{pr}_2$ and a coaction $\delta^P = (\delta', \delta'') : P \to (A \otimes_R P) \oplus (P \otimes_S C)$ of degree 0 such that the following diagram commutes

$$\begin{array}{ccc} P & \xrightarrow{\quad\delta\quad} & (A \otimes_R P) \oplus (P \otimes_S C) \\ {\scriptstyle\delta}\downarrow & & \downarrow{\scriptstyle(\Delta\otimes1)\oplus(\delta\otimes1)} \\ (A \otimes_R P) \oplus (P \otimes_S C) & \xrightarrow{(1\otimes\delta)\oplus(1\otimes\Delta)} & (A \otimes_R A \otimes_R P) \oplus (A \otimes_R P \otimes_S C) \\ & & \oplus (P \otimes_S C \otimes_S C) \end{array}$$

and $\delta' \cdot (\varepsilon \otimes 1) = 1 = \delta'' \cdot (1 \otimes \varepsilon) : P \to P$.

The equation presented on the diagram consists in fact of three equations claiming that P is a left A-comodule, a right C-comodule and the coactions commute.

Let A, B, C, D be coassociative counital coalgebras; let $\phi : A \to B$, $\psi : C \to D$ be homomorphisms; let $\chi : A \to B$ be a (ϕ, ϕ)-coderivation and let $\xi : C \to D$ be a (ψ, ψ)-coderivation of certain degree, that is, $\chi\Delta = \Delta(\phi \otimes \chi + \chi \otimes \phi)$, $\xi\Delta = \Delta(\psi\otimes\xi+\xi\otimes\psi)$. Let $\delta : P \to (A\otimes P)\oplus(P\otimes C)$ be a counital (A, C)-bicomodule and let $\delta : Q \to (B\otimes Q)\oplus(Q\otimes D)$ be a counital (B, D)-bicomodule. A $\Bbbk$-span morphism $f : P \to Q$ of degree 0 with $\operatorname{Ob}_s f = \operatorname{Ob}\phi$, $\operatorname{Ob}_t f = \operatorname{Ob}\psi$ is a (ϕ,ψ)-*bicomodule homomorphism* if $f\delta' = \delta'(\phi\otimes f) : P \to B\otimes Q$ and $f\delta'' = \delta''(f\otimes\psi) : P \to Q\otimes D$. Define a $(\phi, \psi, f, \chi, \xi)$-*connection* as a $\Bbbk$-span morphism $r : P \to Q$ of certain degree with $\operatorname{Ob}_s r = \operatorname{Ob}\phi$, $\operatorname{Ob}_t r = \operatorname{Ob}\psi$ such that

$$\begin{array}{ccc}
P & \xrightarrow{\ \delta\ } & (A\otimes P)\oplus(P\otimes C) \\
{\scriptstyle r}\big\downarrow & & \big\downarrow{\scriptstyle (\phi\otimes r+\chi\otimes f)\oplus(f\otimes\xi+r\otimes\psi)} \\
Q & \xrightarrow{\ \delta\ } & (B\otimes Q)\oplus(Q\otimes D)
\end{array}$$

Let (A, b^A), (C, b^C) be differential graded coalgebras and let P be an (A, C)-bicomodule with an $(\mathrm{id}_A, \mathrm{id}_C, \mathrm{id}_P, b^A, b^C)$-connection $b^P : P \to P$ of degree 1, that is, $b^P\delta' = \delta'(1 \otimes b^P + b^A \otimes 1)$ and $b^P\delta'' = \delta''(1 \otimes b^C + b^P \otimes 1)$. Its *curvature* $(b^P)^2 : P \to P$ is always an (A, C)-bicomodule homomorphism of degree 2. If it vanishes, b^P is called a *flat connection* (a differential) on P.

Taking for (A, b^A) the trivial differential graded coalgebra $\Bbbk$ with the trivial coactions we recover the notions introduced in Section 4. Namely, an (A, C)-bicomodule P with an $(\mathrm{id}_A, \mathrm{id}_C, \mathrm{id}_P, b^A, b^C)$-connection $b^P : P \to P$ of degree 1 is the same as a C-comodule with a $(1, 1, b^C)$-connection, both flatness conditions coincide, etc.

Equivalently, bicomodules with flat connections are bicomodules which live in the category of differential graded spans. The set of A-C-bicomodules becomes the set of objects of a differential graded category A-C-bicomod. For differential graded bicomodules P, Q, the k-th component of the graded $\Bbbk$-module A-C-bicomod(P, Q) consists of $(\mathrm{id}_A, \mathrm{id}_C)$-bicomodule homomorphisms $t : P \to Q$ of degree k. The differential of t is the commutator $tm_1 = tb^Q - (-)^t b^P t : P \to Q$, which is again a homomorphism of bicomodules, naturally of degree $k + 1$. Composition of homomorphisms of bicomodules is the ordinary composition of $\Bbbk$-span morphisms.

The main example of a bicomodule is the following. Let $\mathcal{A}$, $\mathcal{B}$, $\mathcal{C}$, $\mathcal{D}$ be graded $\Bbbk$-quivers. Let $\mathcal{P}$, $\mathcal{Q}$ be **gr**-spans with $\operatorname{Ob}_s \mathcal{P} = \operatorname{Ob}\mathcal{A}$, $\operatorname{Ob}_t \mathcal{P} = \operatorname{Ob}\mathcal{C}$, $\operatorname{Ob}_s \mathcal{Q} = \operatorname{Ob}\mathcal{B}$, $\operatorname{Ob}_t \mathcal{Q} = \operatorname{Ob}\mathcal{D}$, $\operatorname{Par}\mathcal{P} = \operatorname{Ob}_s \mathcal{P} \times \operatorname{Ob}_t \mathcal{P}$, $\operatorname{Par}\mathcal{Q} = \operatorname{Ob}_s \mathcal{Q} \times \operatorname{Ob}_t \mathcal{Q}$, $\mathrm{src} = \mathrm{pr}_1$, $\mathrm{tgt} = \mathrm{pr}_2$. Take coalgebras $A = Ts\mathcal{A}$, $B = Ts\mathcal{B}$, $C = Ts\mathcal{C}$, $D = Ts\mathcal{D}$ and bicomodules $P = Ts\mathcal{A}\otimes s\mathcal{P}\otimes Ts\mathcal{C}$, $Q = Ts\mathcal{B}\otimes s\mathcal{Q}\otimes Ts\mathcal{D}$ equipped with the cut comultiplications

(coactions)

$$\Delta_0(a_1,\dots,a_n) = \sum_{i=0}^{n}(a_1,\dots,a_i)\otimes(a_{i+1},\dots,a_n),$$

$$\delta(a_1,\dots,a_k,p,c_{k+1},\dots,c_{k+l}) = \sum_{i=0}^{k}(a_1,\dots,a_i)\otimes(a_{i+1},\dots,p,\dots,c_{k+l}) + \sum_{i=k}^{k+l}(a_1,\dots,p,\dots,c_i)\otimes(c_{i+1},\dots,c_{k+l}).$$

Notice that a graded quiver $\mathcal{M} \in \operatorname{Ob}\mathscr{Q}/S$ such that $\mathcal{M}(X,Y) = \mathcal{M}(Y)$ depends only on $Y \in S$ is nothing else but a **gr**-span $\mathcal{M}$ with $\operatorname{Ob}_s\mathcal{M} = \{*\}$, $\operatorname{Ob}_t\mathcal{M} = S$. Thus, $Ts\mathcal{C}$-comodules of the form $s\mathcal{M}\otimes Ts\mathcal{C}$ from Section 4 are nothing else but $Ts\mathcal{A}$-$Ts\mathcal{C}$-bicomodules $Ts\mathcal{A}\otimes s\mathcal{M}\otimes Ts\mathcal{C}$ for the graded quiver $\mathcal{A} = \mathbb{1}_u$ with one object $*$ and with $\mathbb{1}_u(*,*) = 0$. Furthermore, A_∞-modules $\mathcal{M}$ over an A_∞-category $\mathcal{C}$ are the same as $\mathbb{1}_u$-$\mathcal{C}$-bimodules, as defined before Proposition 5.3.

Let $\phi : Ts\mathcal{A} \to Ts\mathcal{B}$, $\psi : Ts\mathcal{C} \to Ts\mathcal{D}$ be augmented coalgebra morphisms. Let $g : P \to Q$ be a $\Bbbk$-span morphism of certain degree with $\operatorname{Ob}_s g = \operatorname{Ob}\phi$, $\operatorname{Ob}_t g = \operatorname{Ob}\psi$. Define the matrix coefficients of g to be

$$g_{kl;mn} = (\mathrm{in}_k\otimes 1\otimes \mathrm{in}_l)\cdot g\cdot(\mathrm{pr}_m\otimes 1\otimes \mathrm{pr}_n) : T^ks\mathcal{A}\otimes s\mathcal{P}\otimes T^ls\mathcal{C} \to T^ms\mathcal{B}\otimes s\mathcal{Q}\otimes T^ns\mathcal{D}, \quad k,l,m,n \geqslant 0.$$

The coefficients $g_{kl;00} : T^ks\mathcal{A}\otimes s\mathcal{P}\otimes T^ls\mathcal{C} \to s\mathcal{Q}$ are abbreviated to g_{kl} and called components of g. Denote by $\breve{g}$ the composite $g\cdot(\mathrm{pr}_0\otimes 1\otimes \mathrm{pr}_0) : Ts\mathcal{A}\otimes s\mathcal{P}\otimes Ts\mathcal{C} \to s\mathcal{Q}$. The restriction of $\breve{g}$ to the summand $T^ks\mathcal{A}\otimes s\mathcal{P}\otimes T^ls\mathcal{C}$ is precisely the component g_{kl}.

Let $f : P \to Q$ be a (ϕ,ψ)-bicomodule homomorphism. It is uniquely recovered from its components similarly to Tradler [Tra01, Lemma 4.2]. Let us supply the details. The coaction δ^P has two components,

$$\delta' = \Delta_0\otimes 1\otimes 1 : Ts\mathcal{A}\otimes s\mathcal{P}\otimes Ts\mathcal{C} \to Ts\mathcal{A}\otimes Ts\mathcal{A}\otimes s\mathcal{P}\otimes Ts\mathcal{C},$$

$$\delta'' = 1\otimes 1\otimes \Delta_0 : Ts\mathcal{A}\otimes s\mathcal{P}\otimes Ts\mathcal{C} \to Ts\mathcal{A}\otimes s\mathcal{P}\otimes Ts\mathcal{C}\otimes Ts\mathcal{C},$$

and similarly for δ^Q. As f is a (ϕ,ψ)-bicomodule homomorphism, it satisfies the equations

$$f(\Delta_0\otimes 1\otimes 1) = (\Delta_0\otimes 1\otimes 1)(\phi\otimes f) : Ts\mathcal{A}\otimes s\mathcal{P}\otimes Ts\mathcal{C} \to Ts\mathcal{B}\otimes Ts\mathcal{B}\otimes s\mathcal{Q}\otimes Ts\mathcal{D},$$

$$f(1\otimes 1\otimes \Delta_0) = (1\otimes 1\otimes \Delta_0)(f\otimes \psi) : Ts\mathcal{A}\otimes s\mathcal{P}\otimes Ts\mathcal{C} \to Ts\mathcal{B}\otimes s\mathcal{Q}\otimes Ts\mathcal{D}\otimes Ts\mathcal{D}.$$

It follows that

$$f(\Delta_0\otimes 1\otimes \Delta_0) = (\Delta_0\otimes 1\otimes \Delta_0)(\phi\otimes f\otimes \psi) : Ts\mathcal{A}\otimes s\mathcal{P}\otimes Ts\mathcal{C} \to Ts\mathcal{B}\otimes Ts\mathcal{B}\otimes s\mathcal{Q}\otimes Ts\mathcal{D}\otimes Ts\mathcal{D}.$$

Composing both sides with the morphism

$$1\otimes \mathrm{pr}_0\otimes 1\otimes \mathrm{pr}_0\otimes 1 : Ts\mathcal{B}\otimes Ts\mathcal{B}\otimes s\mathcal{Q}\otimes Ts\mathcal{D}\otimes Ts\mathcal{D} \to Ts\mathcal{B}\otimes s\mathcal{Q}\otimes Ts\mathcal{D}, \quad (5.1)$$

and taking into account the identities $\Delta_0(1\otimes\mathrm{pr}_0) = 1$, $\Delta_0(\mathrm{pr}_0\otimes 1) = 1$, we obtain

$$f = (\Delta_0 \otimes 1 \otimes \Delta_0)(\phi \otimes \check{f} \otimes \psi). \tag{5.2}$$

This equation implies the following formulas for the matrix coefficients of f:

$$f_{kl;mn} = \sum_{\substack{i_1+\dots+i_m+p=k\\ j_1+\dots+j_n+q=l}} (\phi_{i_1}\otimes\dots\otimes\phi_{i_m}\otimes f_{pq}\otimes\psi_{j_1}\otimes\dots\otimes\psi_{j_n}) :$$
$$T^k s\mathcal{A}\otimes s\mathcal{P}\otimes T^l s\mathcal{C} \to T^m s\mathcal{B}\otimes s\mathcal{Q}\otimes T^n s\mathcal{D}, \quad k,l,m,n \geqslant 0. \tag{5.3}$$

In particular, if $k < m$ or $l < n$, the matrix coefficient $f_{kl;mn}$ vanishes.

Let $r : P \to Q$ be a (ϕ,ψ,f,χ,ξ)-connection. It satisfies the following equations:

$$r(\Delta_0\otimes 1\otimes 1) = (\Delta_0\otimes 1\otimes 1)(\phi\otimes r + \chi\otimes f) :$$
$$Ts\mathcal{A}\otimes s\mathcal{P}\otimes Ts\mathcal{C} \to Ts\mathcal{B}\otimes Ts\mathcal{B}\otimes s\mathcal{Q}\otimes Ts\mathcal{D},$$
$$r(1\otimes 1\otimes \Delta_0) = (1\otimes 1\otimes\Delta_0)(f\otimes\xi + r\otimes\psi) :$$
$$Ts\mathcal{A}\otimes s\mathcal{P}\otimes Ts\mathcal{C} \to Ts\mathcal{B}\otimes s\mathcal{Q}\otimes Ts\mathcal{D}\otimes Ts\mathcal{D}.$$

They imply that

$$r(\Delta_0\otimes 1\otimes\Delta_0) = (\Delta_0\otimes 1\otimes\Delta_0)(\phi\otimes f\otimes\xi + \phi\otimes r\otimes\psi + \chi\otimes f\otimes\psi) :$$
$$Ts\mathcal{A}\otimes s\mathcal{P}\otimes Ts\mathcal{C} \to Ts\mathcal{B}\otimes Ts\mathcal{B}\otimes s\mathcal{Q}\otimes Ts\mathcal{D}\otimes Ts\mathcal{D}.$$

Composing both side with the morphism (5.1) we obtain

$$r = (\Delta_0\otimes 1\otimes\Delta_0)(\phi\otimes\check{f}\otimes\xi + \phi\otimes\check{r}\otimes\psi + \chi\otimes\check{f}\otimes\psi). \tag{5.4}$$

From this equation we find the following expression for the matrix coefficient $r_{kl;mn}$:

$$\sum_{\substack{i_1+\dots+i_m+i=k\\ j+j_1+\dots+j_p+t+j_{p+1}+\dots+j_{p+q}=l}}^{p+1+q=n} \phi_{i_1}\otimes\dots\otimes\phi_{i_m}\otimes f_{ij}\otimes\psi_{j_1}\otimes\dots\otimes\psi_{j_p}\otimes\xi_t\otimes\psi_{j_{p+1}}\otimes\dots\otimes\psi_{j_{p+q}}$$
$$+ \sum_{\substack{i_1+\dots+i_m+i=k\\ j+j_1+\dots+j_n=l}} \phi_{i_1}\otimes\dots\otimes\phi_{i_m}\otimes r_{ij}\otimes\psi_{j_1}\otimes\dots\otimes\psi_{j_n}$$
$$+ \sum_{\substack{i_1+\dots+i_a+u+i_{a+1}+\dots+i_{a+c}+i=k\\ j+j_1+\dots+j_n=l}}^{a+1+c=m} \phi_{i_1}\otimes\dots\otimes\phi_{i_a}\otimes\chi_u\otimes\phi_{i_{a+1}}\otimes\dots\otimes\phi_{i_{a+c}}\otimes f_{ij}\otimes\psi_{j_1}\otimes\dots\otimes\psi_{j_n} :$$
$$T^k s\mathcal{A}\otimes s\mathcal{P}\otimes T^l s\mathcal{C} \to T^m s\mathcal{B}\otimes s\mathcal{Q}\otimes T^n s\mathcal{D}, \quad k,l,m,n \geqslant 0. \tag{5.5}$$

Let $\mathcal{A}$, $\mathcal{C}$ be A_∞-categories and let $\mathcal{P}$ be a **gr**-span with $\mathrm{Ob}_s\mathcal{P} = \mathrm{Ob}\mathcal{A}$, $\mathrm{Ob}_t\mathcal{P} = \mathrm{Ob}\mathcal{C}$, $\mathrm{Par}\mathcal{P} = \mathrm{Ob}_s\mathcal{P}\times\mathrm{Ob}_t\mathcal{P}$, $\mathrm{src} = \mathrm{pr}_1$, $\mathrm{tgt} = \mathrm{pr}_2$. Let $A = Ts\mathcal{A}$, $C = Ts\mathcal{C}$, and consider the bicomodule $P = Ts\mathcal{A}\otimes s\mathcal{P}\otimes Ts\mathcal{C}$. The set of $(1,1,1,b^{\mathcal{A}},b^{\mathcal{C}})$-connections $b^{\mathcal{P}} : P \to P$ of degree 1 with $(b^{\mathcal{P}}_{00})^2 = 0$ is in bijection with the set of augmented coalgebra homomorphisms $\phi^{\mathcal{P}} : Ts\mathcal{A}^{\mathrm{op}}\boxtimes Ts\mathcal{C} \to Ts\underline{\mathsf{C}}_{\Bbbk}$. Indeed, collections of complexes $(\phi^{\mathcal{P}}(X,Y),d)^{X\in\operatorname{Ob}\mathcal{A}}_{Y\in\operatorname{Ob}\mathcal{C}}$ are identified with the **dg**-spans $(\mathcal{P}, sb^{\mathcal{P}}_{00}s^{-1})$. In particular, for each pair of objects $X\in\operatorname{Ob}\mathcal{A}$, $Y\in\operatorname{Ob}\mathcal{C}$ holds $(\phi^{\mathcal{P}}(X,Y))[1]=(s\mathcal{P}(X,Y),-b^{\mathcal{P}}_{00})$. The components $b^{\mathcal{P}}_{kn}$ and $\phi^{\mathcal{P}}_{kn}$ are related for $(k,n)\neq(0,0)$ by the formula

$$
\begin{aligned}
b^{\mathcal{P}}_{kn} = \big[&s\mathcal{A}(X_k,X_{k-1})\otimes\cdots\otimes s\mathcal{A}(X_1,X_0)\otimes s\mathcal{P}(X_0,Y_0)\otimes s\mathcal{C}(Y_0,Y_1)\otimes\cdots \\
&\otimes s\mathcal{C}(Y_{n-1},Y_n)\xrightarrow{\tilde\gamma\otimes1^{\otimes n}} \\
&s\mathcal{P}(X_0,Y_0)\otimes s\mathcal{A}^{\mathrm{op}}(X_0,X_1)\otimes\cdots\otimes s\mathcal{A}^{\mathrm{op}}(X_{k-1},X_k)\otimes s\mathcal{C}(Y_0,Y_1)\otimes\cdots\otimes s\mathcal{C}(Y_{n-1},Y_n) \\
&\xrightarrow{1\otimes\phi^{\mathcal{P}}_{kn}} s\mathcal{P}(X_0,Y_0)\otimes s\underline{\mathsf{C}}_{\Bbbk}(\mathcal{P}(X_0,Y_0),\mathcal{P}(X_k,Y_n)) \\
&\xrightarrow{1\otimes s^{-1}} s\mathcal{P}(X_0,Y_0)\otimes \underline{\mathsf{C}}_{\Bbbk}(\mathcal{P}(X_0,Y_0),\mathcal{P}(X_k,Y_n)) \\
&\xrightarrow{1\otimes[1]} s\mathcal{P}(X_0,Y_0)\otimes \underline{\mathsf{C}}_{\Bbbk}(s\mathcal{P}(X_0,Y_0),s\mathcal{P}(X_k,Y_n)) \xrightarrow{\mathrm{ev}^{\mathsf{C}_{\Bbbk}}} s\mathcal{P}(X_k,Y_n)\big],
\end{aligned}
$$

where $\tilde\gamma=(12\dots k+1)\cdot\gamma$, and anti-isomorphism γ is defined by (3.1).

The components of $b^{\mathcal{P}}$ can be written in a more concise form. Given objects $X,Y\in\operatorname{Ob}\mathcal{A}$, $Z,W\in\operatorname{Ob}\mathcal{C}$, define

$$
\begin{aligned}
\check b^{\mathcal{P}}_{+} = \big[&Ts\mathcal{A}(Y,X)\otimes s\mathcal{P}(X,Z)\otimes Ts\mathcal{C}(Z,W) \\
&\xrightarrow{c\otimes1} s\mathcal{P}(X,Z)\otimes Ts\mathcal{A}(Y,X)\otimes Ts\mathcal{C}(Z,W) \\
&\xrightarrow{1\otimes\gamma\otimes1} s\mathcal{P}(X,Z)\otimes Ts\mathcal{A}^{\mathrm{op}}(X,Y)\otimes Ts\mathcal{C}(Z,W) \\
&\xrightarrow{1\otimes\check\phi^{\mathcal{P}}} s\mathcal{P}(X,Z)\otimes s\underline{\mathsf{C}}_{\Bbbk}(\mathcal{P}(X,Z),\mathcal{P}(Y,W)) \\
&\xrightarrow{(s\otimes s)^{-1}} \mathcal{P}(X,Z)\otimes \underline{\mathsf{C}}_{\Bbbk}(\mathcal{P}(X,Z),\mathcal{P}(Y,W)) \xrightarrow{\mathrm{ev}^{\mathsf{C}_{\Bbbk}}} \mathcal{P}(Y,W) \xrightarrow{s} s\mathcal{P}(Y,W)\big] \\
= \big[&Ts\mathcal{A}(Y,X)\otimes s\mathcal{P}(X,Z)\otimes Ts\mathcal{C}(Z,W) \xrightarrow{c\otimes1} s\mathcal{P}(X,Z)\otimes Ts\mathcal{A}(Y,X)\otimes Ts\mathcal{C}(Z,W) \\
&\xrightarrow{1\otimes\gamma\otimes1} s\mathcal{P}(X,Z)\otimes Ts\mathcal{A}^{\mathrm{op}}(X,Y)\otimes Ts\mathcal{C}(Z,W) \\
&\xrightarrow{1\otimes\check\phi^{\mathcal{P}}} s\mathcal{P}(X,Z)\otimes s\underline{\mathsf{C}}_{\Bbbk}(\mathcal{P}(X,Z),\mathcal{P}(Y,W)) \\
&\xrightarrow{1\otimes s^{-1}[1]} s\mathcal{P}(X,Z)\otimes \underline{\mathsf{C}}_{\Bbbk}(s\mathcal{P}(X,Z),s\mathcal{P}(Y,W)) \xrightarrow{\mathrm{ev}^{\mathsf{C}_{\Bbbk}}} s\mathcal{P}(Y,W)\big],
\end{aligned}
\tag{5.6}
$$

where $\gamma: Ts\mathcal{A}\to Ts\mathcal{A}^{\mathrm{op}}$ is the coalgebra anti-isomorphism (3.1), and $\check\phi^{\mathcal{P}}=\phi^{\mathcal{P}}\,\mathrm{pr}_1 : Ts\mathcal{A}^{\mathrm{op}}\boxtimes Ts\mathcal{C}\to s\underline{\mathsf{C}}_{\Bbbk}$. Conversely, components of the A_∞-functor $\phi^{\mathcal{P}}$ can be found as

$$
\begin{aligned}
\check\phi^{\mathcal{P}} = \big[&Ts\mathcal{A}^{\mathrm{op}}(X,Y)\otimes Ts\mathcal{C}(Z,W)\xrightarrow{\gamma\otimes1} Ts\mathcal{A}(Y,X)\otimes Ts\mathcal{C}(Z,W) \\
&\xrightarrow{\mathrm{coev}^{\mathsf{C}_{\Bbbk}}} \underline{\mathsf{C}}_{\Bbbk}(s\mathcal{P}(X,Z), s\mathcal{P}(X,Z)\otimes Ts\mathcal{A}(Y,X)\otimes Ts\mathcal{C}(Z,W)) \\
&\xrightarrow{\underline{\mathsf{C}}_{\Bbbk}(1,(c\otimes1)\check b^{\mathcal{P}}_{+})} \underline{\mathsf{C}}_{\Bbbk}(s\mathcal{P}(X,Z), s\mathcal{P}(Y,W)) \xrightarrow{[-1]s} s\underline{\mathsf{C}}_{\Bbbk}(\mathcal{P}(X,Z),\mathcal{P}(Y,W))\big].
\end{aligned}
\tag{5.7}
$$

Define also

$$\check{b}^{\mathcal{P}}_0 = \bigl[Ts\mathcal{A}(Y,X)\otimes s\mathcal{P}(X,Z)\otimes Ts\mathcal{C}(Z,W) \xrightarrow{\mathrm{pr}_0\otimes 1\otimes \mathrm{pr}_0} s\mathcal{P}(X,Z) \xrightarrow{b^{\mathcal{P}}_{00}} s\mathcal{P}(X,Z)\bigr]. \tag{5.8}$$

Note that $\check{b}^{\mathcal{P}}_+$ vanishes on $T^0s\mathcal{A}(Y,X)\otimes s\mathcal{P}(X,Z)\otimes T^0s\mathcal{C}(Z,W)$ since $\check{\phi}^{\mathcal{P}}$ vanishes on $T^0s\mathcal{A}^{\mathrm{op}}(X,Y)\otimes T^0s\mathcal{C}(Z,W)$. It follows that $\check{b}^{\mathcal{P}} = \check{b}^{\mathcal{P}}_+ + \check{b}^{\mathcal{P}}_0$.

The following statement was proven by Lefèvre-Hasegawa in assumption that the ground ring is a field [LH03, Lemme 5.3.0.1].

Proposition 5.2. *$b^{\mathcal{P}}$ is a flat connection, that is, $(Ts\mathcal{A}\otimes s\mathcal{P}\otimes Ts\mathcal{C}, b^{\mathcal{P}})$ is a bicomodule in ${}^d\mathcal{Q}$, if and only if the corresponding augmented coalgebra homomorphism $\phi^{\mathcal{P}} : Ts\mathcal{A}^{\mathrm{op}}\boxtimes Ts\mathcal{C}\to Ts\underline{\mathsf{C}}_{\Bbbk}$ is an A_∞-functor.*

This is proven by a straightforward computation. The full proof is given in archive version [LM07] of this article.

Let $\mathcal{A}$, $\mathcal{C}$ be A_∞-categories. The full subcategory of the differential graded category $Ts\mathcal{A}$-$Ts\mathcal{C}$-bicomod consisting of **dg**-bicomodules whose underlying graded bicomodule has the form $Ts\mathcal{A}\otimes s\mathcal{P}\otimes Ts\mathcal{C}$ is denoted by $\mathcal{A}$-$\mathcal{C}$-bimod. Its objects are called *A_∞-bimodules*, extending the terminology of Tradler [Tra01].

Proposition 5.3. *The differential graded categories* $\mathcal{A}$-$\mathcal{C}$-bimod *and* $\underline{\mathsf{A}_\infty}(\mathcal{A}^{\mathrm{op}},\mathcal{C};\underline{\mathsf{C}}_{\Bbbk})$ *are isomorphic.*

Proposition 5.2 establishes a bijection between the sets of objects of the differential graded categories $\underline{\mathsf{A}_\infty}(\mathcal{A}^{\mathrm{op}},\mathcal{C};\underline{\mathsf{C}}_{\Bbbk})$ and $\mathcal{A}$-$\mathcal{C}$-bimod. It extends to an isomorphism of differential graded categories. The full proof is given in archive version [LM07] of this article.

Let us write explicitly the inverse map

$$\Phi^{-1} : \mathcal{A}\text{-}\mathcal{C}\text{-bimod}(\mathcal{P},\mathcal{Q}) \to \underline{\mathsf{A}_\infty}(\mathcal{A}^{\mathrm{op}},\mathcal{C};\underline{\mathsf{C}}_{\Bbbk})(\phi,\psi).$$

It takes a bicomodule homomorphism $t : Ts\mathcal{A}\otimes s\mathcal{P}\otimes Ts\mathcal{C}\to Ts\mathcal{A}\otimes s\mathcal{Q}\otimes Ts\mathcal{C}$ to an A_∞-transformation $rs^{-1}\in\underline{\mathsf{A}_\infty}(\mathcal{A}^{\mathrm{op}},\mathcal{C};\underline{\mathsf{C}}_{\Bbbk})(\phi,\psi)$ given by its components

$$\begin{aligned}\check{r} = (-)^t\bigl[&Ts\mathcal{A}^{\mathrm{op}}(Y,X)\otimes Ts\mathcal{C}(Z,W) \xrightarrow{\gamma\otimes 1} Ts\mathcal{A}(X,Y)\otimes Ts\mathcal{C}(Z,W) \xrightarrow{\mathrm{coev}^{\underline{\mathsf{C}}_{\Bbbk}}} \\ &\underline{\mathsf{C}}_{\Bbbk}(s\mathcal{P}(Y,Z), s\mathcal{P}(Y,Z)\otimes Ts\mathcal{A}(X,Y)\otimes Ts\mathcal{C}(Z,W)) \xrightarrow{\underline{\mathsf{C}}_{\Bbbk}(1,(c\otimes 1)\check{t})} \\ &\underline{\mathsf{C}}_{\Bbbk}(s\mathcal{P}(Y,Z), s\mathcal{Q}(X,W)) \xrightarrow{[-1]s} s\underline{\mathsf{C}}_{\Bbbk}(\mathcal{P}(Y,Z),\mathcal{Q}(X,W))\bigr].\end{aligned} \tag{5.9}$$

5.4. Regular A_∞-bimodule

Let $\mathcal{A}$ be an A_∞-category. Extending the notion of regular A_∞-bimodule given by Tradler [Tra01, Lemma 5.1(a)] from the case of A_∞-algebras to A_∞-categories, define the *regular $\mathcal{A}$-$\mathcal{A}$-bimodule* $\mathcal{R} = \mathcal{R}_{\mathcal{A}}$ as follows. Its underlying quiver coincides with $\mathcal{A}$. Components of the codifferential $b^{\mathcal{R}}$ are given by

$$\check{b}^{\mathcal{R}} = \bigl[Ts\mathcal{A}\otimes s\mathcal{A}\otimes Ts\mathcal{A} \xrightarrow{\mu_{Ts\mathcal{A}}} Ts\mathcal{A} \xrightarrow{\check{b}^{\mathcal{A}}} s\mathcal{A}\bigr],$$

where $\mu_{Ts\mathcal{A}}$ is the multiplication in the tensor quiver $Ts\mathcal{A}$. Equivalently, $b^{\mathcal{R}}_{kn} = b^{\mathcal{A}}_{k+1+n}$, $k, n \geqslant 0$. Flatness of $b^{\mathcal{R}}$ in the form

$$(b^{\mathcal{A}} \otimes 1 \otimes 1 + 1 \otimes 1 \otimes b^{\mathcal{C}})\check{b}^{\mathcal{P}} + (\Delta_0 \otimes 1 \otimes \Delta_0)(1 \otimes \check{b}^{\mathcal{P}} \otimes 1)\check{b}^{\mathcal{P}} = 0. \tag{5.10}$$

is equivalent to the A_∞-identity $b^{\mathcal{A}} \cdot \check{b}^{\mathcal{A}} = 0$. Indeed, the three summands of the left-hand side of (5.10) correspond to three kinds of subintervals of the interval $[1, k+1+n] \cap \mathbb{Z}$. Subintervals of the first two types miss the point $k+1$ and those of the third type contain it.

Definition 5.5. Define an A_∞-functor $\operatorname{Hom}_{\mathcal{A}} : \mathcal{A}^{\mathrm{op}}, \mathcal{A} \to \underline{\mathsf{C}}_{\Bbbk}$ as the A_∞-functor $\phi^{\mathcal{R}}$ that corresponds to the regular $\mathcal{A}$-$\mathcal{A}$-bimodule $\mathcal{R} = \mathcal{R}_{\mathcal{A}}$.

The A_∞-functor $\operatorname{Hom}_{\mathcal{A}}$ takes a pair of objects $X, Z \in \operatorname{Ob}\mathcal{A}$ to the chain complex $(\mathcal{A}(X,Z), m_1)$. The components of $\operatorname{Hom}_{\mathcal{A}}$ are found from equation (5.7):

$$\begin{aligned}(\operatorname{Hom}_{\mathcal{A}})_{kn} = \bigl[T^k s\mathcal{A}^{\mathrm{op}}(X,Y) \otimes T^n s\mathcal{A}(Z,W) \xrightarrow{\gamma\otimes 1} T^k s\mathcal{A}(Y,X) \otimes T^n s\mathcal{A}(Z,W) \\ \xrightarrow{\operatorname{coev}^{\mathsf{C}_{\Bbbk}}} \underline{\mathsf{C}}_{\Bbbk}(s\mathcal{A}(X,Z), s\mathcal{A}(X,Z) \otimes T^k s\mathcal{A}(Y,X) \otimes T^n s\mathcal{A}(Z,W)) \\ \xrightarrow{\underline{\mathsf{C}}_{\Bbbk}(1,(c\otimes 1)b^{\mathcal{A}}_{k+1+n})} \underline{\mathsf{C}}_{\Bbbk}(s\mathcal{A}(X,Z), s\mathcal{A}(Y,W)) \xrightarrow{[-1]^s} s\underline{\mathsf{C}}_{\Bbbk}(\mathcal{A}(X,Z), \mathcal{A}(Y,W))\bigr].\end{aligned} \tag{5.11}$$

Closedness of the multicategory A_∞ [BLM06, Theorem 12.19] implies that there exists a unique A_∞-functor $\mathscr{Y} : \mathcal{A} \to \underline{\mathsf{A}_\infty}(\mathcal{A}^{\mathrm{op}}; \underline{\mathsf{C}}_{\Bbbk})$ (called the *Yoneda A_∞-functor*) such that

$$\operatorname{Hom}_{\mathcal{A}} = \bigl[\mathcal{A}^{\mathrm{op}}, \mathcal{A} \xrightarrow{1,\mathscr{Y}} \mathcal{A}^{\mathrm{op}}, \underline{\mathsf{A}_\infty}(\mathcal{A}^{\mathrm{op}}; \underline{\mathsf{C}}_{\Bbbk}) \xrightarrow{\operatorname{ev}^{\mathsf{A}_\infty}} \underline{\mathsf{C}}_{\Bbbk}\bigr].$$

Explicit formula [BLM06, (12.25.4)] for evaluation component $\operatorname{ev}^{\mathsf{A}_\infty}_{k0}$ shows that the value of $\mathscr{Y}$ on an object Z of $\mathcal{A}$ is given by the restriction A_∞-functor

$$Z\mathscr{Y} = H^Z = H^Z_{\mathcal{A}} = \operatorname{Hom}_{\mathcal{A}}\bigr|^Z : \mathcal{A}^{\mathrm{op}} \to \underline{\mathsf{C}}_{\Bbbk}, \quad X \mapsto (\mathcal{A}(X,Z), m_1) = \operatorname{Hom}_{\mathcal{A}}(X,Z)$$

with the components

$$\begin{aligned}H^Z_k = (\operatorname{Hom}_{\mathcal{A}})_{k0} = (-1)^k \bigl[T^k s\mathcal{A}^{\mathrm{op}}(X,Y) \xrightarrow{\operatorname{coev}^{\mathsf{C}_{\Bbbk}}} \\ \underline{\mathsf{C}}_{\Bbbk}(s\mathcal{A}(X,Z), s\mathcal{A}(X,Z) \otimes T^k s\mathcal{A}^{\mathrm{op}}(X,Y)) \xrightarrow{\underline{\mathsf{C}}_{\Bbbk}(1,\omega^0_c b^{\mathcal{A}}_{k+1})} \\ \underline{\mathsf{C}}_{\Bbbk}(s\mathcal{A}(X,Z), s\mathcal{A}(Y,Z)) \xrightarrow{[-1]^s} s\underline{\mathsf{C}}_{\Bbbk}(\mathcal{A}(X,Z), \mathcal{A}(Y,Z))\bigr],\end{aligned} \tag{5.12}$$

where $\omega^0 = \left(\begin{smallmatrix} 0 & 1 & \dots & k-1 & k \\ k & k-1 & \dots & 1 & 0 \end{smallmatrix}\right) \in \mathfrak{S}_{k+1}$, and ω^0_c is the corresponding signed permutation. Restrictions of A_∞-functors in general are defined in [BLM06, Section 12.18], in particular, the k-th component of $\operatorname{Hom}_{\mathcal{A}}\bigr|^Z_1$ described by [*loc. cit.*, (12.18.2)] equals $(1, \operatorname{Ob}\mathscr{Y})\operatorname{ev}^{\mathsf{A}_\infty}_{k0}$. Equivalently, components of the A_∞-functor $H^Z : \mathcal{A}^{\mathrm{op}} \to \underline{\mathsf{C}}_{\Bbbk}$ are determined by the equation

$$\begin{aligned}s^{\otimes k} H^Z_k s^{-1} = (-1)^{k(k+1)/2+1} \bigl[T^k \mathcal{A}^{\mathrm{op}}(X,Y) \xrightarrow{\operatorname{coev}^{\mathsf{C}_{\Bbbk}}} \\ \underline{\mathsf{C}}_{\Bbbk}(\mathcal{A}(X,Z), \mathcal{A}(X,Z) \otimes T^k \mathcal{A}^{\mathrm{op}}(X,Y)) \xrightarrow{\underline{\mathsf{C}}_{\Bbbk}(1,\omega^0_c m^{\mathcal{A}}_{k+1})} \underline{\mathsf{C}}_{\Bbbk}(\mathcal{A}(X,Z), \mathcal{A}(Y,Z))\bigr].\end{aligned}$$

Notice that $\mathrm{ev}^{\mathsf{A}_\infty}_{km}$ vanishes unless $m \leqslant 1$. Formula [BLM06, (12.25.4)] for the component $\mathrm{ev}^{\mathsf{A}_\infty}_{k1}$ implies that the component $(\mathrm{Hom}_{\mathcal{A}})_{kn}$ is determined for $n \geqslant 1$, $k \geqslant 0$ by $\mathscr{Y}_{nk}$ which is the composition of $\mathscr{Y}_n$ with

$$\mathrm{pr}_k : s\underline{\mathsf{A}_\infty}(\mathcal{A}^{\mathrm{op}}; \underline{\mathsf{C}}_\Bbbk)(H^Z, H^W) \to \underline{\mathsf{C}}_\Bbbk(T^k s\mathcal{A}^{\mathrm{op}}(X,Y), s\underline{\mathsf{C}}_\Bbbk(XH^Z, YH^W))$$

as follows:

$$(\mathrm{Hom}_{\mathcal{A}})_{kn} = \bigl[T^k s\mathcal{A}^{\mathrm{op}}(X,Y) \otimes T^n s\mathcal{A}(Z,W) \xrightarrow{1\otimes\mathscr{Y}_{nk}} T^k s\mathcal{A}^{\mathrm{op}}(X,Y) \otimes \underline{\mathsf{C}}_\Bbbk\bigl(T^k s\mathcal{A}^{\mathrm{op}}(X,Y), s\underline{\mathsf{C}}_\Bbbk(\mathcal{A}(X,Z), \mathcal{A}(Y,W))\bigr) \xrightarrow{\mathrm{ev}^{\mathsf{C}_\Bbbk}} s\underline{\mathsf{C}}_\Bbbk(\mathcal{A}(X,Z), \mathcal{A}(Y,W))\bigr].$$

Conversely, the component $\mathscr{Y}_n$ is determined by the components $(\mathrm{Hom}_{\mathcal{A}})_{kn}$ for all $k \geqslant 0$ via the formula

$$\mathscr{Y}_{nk} = \bigl[T^n s\mathcal{A}(Z,W) \xrightarrow{\mathrm{coev}^{\mathsf{C}_\Bbbk}} \underline{\mathsf{C}}_\Bbbk(T^k s\mathcal{A}^{\mathrm{op}}(X,Y), T^k s\mathcal{A}^{\mathrm{op}}(X,Y) \otimes T^n s\mathcal{A}(Z,W)) \xrightarrow{\underline{\mathsf{C}}_\Bbbk(1,(\mathrm{Hom}_{\mathcal{A}})_{kn})} \underline{\mathsf{C}}_\Bbbk\bigl(T^k s\mathcal{A}^{\mathrm{op}}(X,Y), s\underline{\mathsf{C}}_\Bbbk(\mathcal{A}(X,Z), \mathcal{A}(Y,W))\bigr)\bigr].$$

Plugging in expression (5.11) we get

$$\mathscr{Y}_{nk} = \bigl[T^n s\mathcal{A}(Z,W) \xrightarrow{\mathrm{coev}^{\mathsf{C}_\Bbbk}} \underline{\mathsf{C}}_\Bbbk(T^k s\mathcal{A}^{\mathrm{op}}(X,Y), T^k s\mathcal{A}^{\mathrm{op}}(X,Y) \otimes T^n s\mathcal{A}(Z,W)) \xrightarrow{\underline{\mathsf{C}}_\Bbbk(1,\mathrm{coev}^{\mathsf{C}_\Bbbk})}$$
$$\underline{\mathsf{C}}_\Bbbk\bigl(T^k s\mathcal{A}^{\mathrm{op}}(X,Y), \underline{\mathsf{C}}_\Bbbk(s\mathcal{A}(X,Z), s\mathcal{A}(X,Z) \otimes T^k s\mathcal{A}^{\mathrm{op}}(X,Y) \otimes T^n s\mathcal{A}(Z,W))\bigr)$$
$$\xrightarrow{\underline{\mathsf{C}}_\Bbbk(1,\underline{\mathsf{C}}_\Bbbk(1,(1\otimes\gamma\otimes1)(c\otimes1)b^{\mathcal{A}}_{k+1+n}))} \underline{\mathsf{C}}_\Bbbk\bigl(T^k s\mathcal{A}^{\mathrm{op}}(X,Y), \underline{\mathsf{C}}_\Bbbk(s\mathcal{A}(X,Z), s\mathcal{A}(Y,W))\bigr)$$
$$\xrightarrow{\underline{\mathsf{C}}_\Bbbk(1,[-1]s)} \underline{\mathsf{C}}_\Bbbk\bigl(T^k s\mathcal{A}^{\mathrm{op}}(X,Y), s\underline{\mathsf{C}}_\Bbbk(\mathcal{A}(X,Z), \mathcal{A}(Y,W))\bigr)\bigr].$$

Another kind of the Yoneda A_∞-functor $Y : \mathcal{A} \to \underline{\mathsf{A}_\infty}(\mathcal{A}^{\mathrm{op}}; \underline{\mathsf{C}}_\Bbbk)$ was introduced in [LM04, Appendix A]. Actually, it was defined there as an A_∞-functor from $\mathcal{A}^{\mathrm{op}}$ to $\underline{\mathsf{A}_\infty}(\mathcal{A}; \underline{\mathsf{C}}_\Bbbk)$. It turns out that Y which we shall call the *shifted Yoneda A_∞-functor* differs from $\mathscr{Y}$ by a shift:

$$Y = \mathscr{Y} \cdot \underline{\mathsf{A}_\infty}(1;[1]) : \mathcal{A} \to \underline{\mathsf{A}_\infty}(\mathcal{A}^{\mathrm{op}}; \underline{\mathsf{C}}_\Bbbk). \tag{5.13}$$

Indeed, an object Z of $\mathcal{A}$ is taken by Y to the A_∞-functor $ZY = h^Z : \mathcal{A}^{\mathrm{op}} \to \underline{\mathsf{C}}_\Bbbk$, $X \mapsto (s\mathcal{A}(X,Z), -b_1) = (\mathcal{A}(X,Z), m_1)[1] = (XH^Z)[1]$. The components of $H^Z \cdot [1]$

$$H^Z_k s^{-1}[1]s = (-1)^k \bigl[T^k s\mathcal{A}^{\mathrm{op}}(X,Y) \xrightarrow{\mathrm{coev}^{\mathsf{C}_\Bbbk}} \underline{\mathsf{C}}_\Bbbk(s\mathcal{A}(X,Z), s\mathcal{A}(X,Z) \otimes T^k s\mathcal{A}^{\mathrm{op}}(X,Y)) \xrightarrow{\underline{\mathsf{C}}_\Bbbk(1,\omega^0_c b^{\mathcal{A}}_{k+1})} \underline{\mathsf{C}}_\Bbbk(s\mathcal{A}(X,Z), s\mathcal{A}(Y,Z)) \xrightarrow{s} s\underline{\mathsf{C}}_\Bbbk(s\mathcal{A}(X,Z), s\mathcal{A}(Y,Z))\bigr]$$

coincide with the components h^Z_k by [LM04, Appendix A]. Therefore, $h^Z = H^Z{\cdot}[1]$. Furthermore, the components Y_n are determined by $Y_{nk} = Y_n \cdot \mathrm{pr}_k$, which turn out [*loc. cit.*] to coincide with

$$\begin{aligned}
\mathscr{Y}_{nk} \cdot \underline{\mathsf{C}}_{\Bbbk}(1, s^{-1}[1]s) = \big[T^n s\mathcal{A}(Z,W) \xrightarrow{\mathrm{coev}^{\mathsf{C}_{\Bbbk}}} \\
\underline{\mathsf{C}}_{\Bbbk}(T^k s\mathcal{A}^{\mathrm{op}}(X,Y), T^k s\mathcal{A}^{\mathrm{op}}(X,Y) \otimes T^n s\mathcal{A}(Z,W)) \xrightarrow{\underline{\mathsf{C}}_{\Bbbk}(1,\mathrm{coev}^{\mathsf{C}_{\Bbbk}})} \\
\underline{\mathsf{C}}_{\Bbbk}\big(T^k s\mathcal{A}^{\mathrm{op}}(X,Y), \underline{\mathsf{C}}_{\Bbbk}(s\mathcal{A}(X,Z), s\mathcal{A}(X,Z) \otimes T^k s\mathcal{A}^{\mathrm{op}}(X,Y) \otimes T^n s\mathcal{A}(Z,W))\big) \\
\xrightarrow{\underline{\mathsf{C}}_{\Bbbk}(1,\underline{\mathsf{C}}_{\Bbbk}(1,(1\otimes\gamma\otimes1)(c\otimes1)b^{\mathcal{A}}_{k+1+n}))} \underline{\mathsf{C}}_{\Bbbk}\big(T^k s\mathcal{A}^{\mathrm{op}}(X,Y), \underline{\mathsf{C}}_{\Bbbk}(s\mathcal{A}(X,Z), s\mathcal{A}(Y,W))\big) \\
\xrightarrow{\underline{\mathsf{C}}_{\Bbbk}(1,s)} \underline{\mathsf{C}}_{\Bbbk}\big(T^k s\mathcal{A}^{\mathrm{op}}(X,Y), s\underline{\mathsf{C}}_{\Bbbk}(s\mathcal{A}(X,Z), s\mathcal{A}(Y,W))\big)\big] \\
= \big[T^n s\mathcal{A}(Z,W) \xrightarrow{\mathrm{coev}^{\mathsf{C}_{\Bbbk}}} \\
\underline{\mathsf{C}}_{\Bbbk}(s\mathcal{A}(X,Z) \otimes T^k s\mathcal{A}^{\mathrm{op}}(X,Y), s\mathcal{A}(X,Z) \otimes T^k s\mathcal{A}^{\mathrm{op}}(X,Y) \otimes T^n s\mathcal{A}(Z,W)) \\
\xrightarrow{\underline{\mathsf{C}}_{\Bbbk}(1,(1\otimes\gamma\otimes1)(c\otimes1)b^{\mathcal{A}}_{k+1+n})} \underline{\mathsf{C}}_{\Bbbk}(s\mathcal{A}(X,Z) \otimes T^k s\mathcal{A}^{\mathrm{op}}(X,Y), s\mathcal{A}(Y,W)) \\
\xrightarrow[\sim]{\underline{\varphi}^{-1}} \underline{\mathsf{C}}_{\Bbbk}\big(T^k s\mathcal{A}^{\mathrm{op}}(X,Y), \underline{\mathsf{C}}_{\Bbbk}(s\mathcal{A}(X,Z), s\mathcal{A}(Y,W))\big) \\
\xrightarrow{\underline{\mathsf{C}}_{\Bbbk}(1,s)} \underline{\mathsf{C}}_{\Bbbk}\big(T^k s\mathcal{A}^{\mathrm{op}}(X,Y), s\underline{\mathsf{C}}_{\Bbbk}(s\mathcal{A}(X,Z), s\mathcal{A}(Y,W))\big)\big].
\end{aligned}$$

The natural isomorphism $\underline{\varphi}^{\mathsf{C}_{\Bbbk}} : \underline{\mathsf{C}}_{\Bbbk}(A, \underline{\mathsf{C}}_{\Bbbk}(B,C)) \to \underline{\mathsf{C}}_{\Bbbk}(B \otimes A, C)$ is found from the equation

$$\begin{aligned}
\big[B \otimes A \otimes \underline{\mathsf{C}}_{\Bbbk}(A, \underline{\mathsf{C}}_{\Bbbk}(B,C)) \xrightarrow{1\otimes\mathrm{ev}^{\mathsf{C}_{\Bbbk}}} B \otimes \underline{\mathsf{C}}_{\Bbbk}(B,C) \xrightarrow{\mathrm{ev}^{\mathsf{C}_{\Bbbk}}} C\big] \\
= \big[B \otimes A \otimes \underline{\mathsf{C}}_{\Bbbk}(A, \underline{\mathsf{C}}_{\Bbbk}(B,C)) \xrightarrow{\underline{\varphi}^{\mathsf{C}_{\Bbbk}}} B \otimes A \otimes \underline{\mathsf{C}}_{\Bbbk}(B \otimes A, C) \xrightarrow{\mathrm{ev}^{\mathsf{C}_{\Bbbk}}} C\big].
\end{aligned}$$

Its solvability is implied by closedness of $\mathsf{C}_{\Bbbk}$. Summing up, (5.13) holds and the two Yoneda A_∞-functors agree.

5.6. Restriction of scalars

Let $f : \mathcal{A} \to \mathcal{B}$, $g : \mathcal{C} \to \mathcal{D}$ be A_∞-functors. Let $\mathcal{P}$ be a $\mathcal{B}$-$\mathcal{D}$-bimodule, $\phi : \mathcal{B}^{\mathrm{op}}, \mathcal{D} \to \underline{\mathsf{C}}_{\Bbbk}$ the corresponding A_∞-functor. Define an $\mathcal{A}$-$\mathcal{C}$-bimodule ${}_f\mathcal{P}_g$ as the bimodule corresponding to the composite

$$\mathcal{A}^{\mathrm{op}}, \mathcal{C} \xrightarrow{f^{\mathrm{op}},g} \mathcal{B}^{\mathrm{op}}, \mathcal{D} \xrightarrow{\phi} \underline{\mathsf{C}}_{\Bbbk}.$$

Its underlying **gr**-span is given by ${}_f\mathcal{P}_g(X,Y) = \mathcal{P}(Xf, Yg)$, $X \in \mathrm{Ob}\,\mathcal{A}$, $Y \in \mathrm{Ob}\,\mathcal{C}$. Components of the codifferential $b^{{}_f\mathcal{P}_g}$ are found using formulas (5.6) and (5.8):

$$\begin{aligned}
\check{b}_+^{f\mathcal{P}_g} = \bigl[&Ts\mathcal{A}(X,Y) \otimes s\mathcal{P}(Yf,Zg) \otimes Ts\mathcal{C}(Z,W) \\
&\xrightarrow{c\otimes 1} s\mathcal{P}(Yf,Zg) \otimes Ts\mathcal{A}(X,Y) \otimes Ts\mathcal{C}(Z,W) \\
&\xrightarrow{1\otimes\gamma\otimes 1} s\mathcal{P}(Yf,Zg) \otimes Ts\mathcal{A}^{\mathrm{op}}(Y,X) \otimes Ts\mathcal{C}(Z,W) \\
&\xrightarrow{1\otimes[(f^{\mathrm{op}},g)\phi]^\vee} s\mathcal{P}(Yf,Zg) \otimes s\underline{\mathsf{C}}_{\Bbbk}(\mathcal{P}(Yf,Zg),\mathcal{P}(Xf,Wg)) \\
&\xrightarrow{1\otimes s^{-1}[1]} s\mathcal{P}(Yf,Zg) \otimes \underline{\mathsf{C}}_{\Bbbk}(s\mathcal{P}(Yf,Zg),s\mathcal{P}(Xf,Wg)) \xrightarrow{\mathrm{ev}^{\mathsf{C}_{\Bbbk}}} s\mathcal{P}(Xf,Wg)\bigr] \\
= \bigl[&Ts\mathcal{A}(X,Y) \otimes s\mathcal{P}(Yf,Zg) \otimes Ts\mathcal{C}(Z,W) \xrightarrow{f\otimes 1\otimes g} \\
&Ts\mathcal{B}(Xf,Yf) \otimes s\mathcal{P}(Yf,Zg) \otimes Ts\mathcal{D}(Zg,Wg) \xrightarrow{c\otimes 1} \\
&s\mathcal{P}(Yf,Zg) \otimes Ts\mathcal{B}(Xf,Yf) \otimes Ts\mathcal{D}(Zg,Wg) \xrightarrow{1\otimes\gamma\otimes 1} \\
&s\mathcal{P}(Yf,Zg) \otimes Ts\mathcal{B}^{\mathrm{op}}(Yf,Xf) \otimes Ts\mathcal{D}(Zg,Wg) \xrightarrow{1\otimes\check{\phi}} \\
&s\mathcal{P}(Yf,Zg) \otimes s\underline{\mathsf{C}}_{\Bbbk}(\mathcal{P}(Yf,Zg),\mathcal{P}(Xf,Wg)) \xrightarrow{1\otimes s^{-1}[1]} \\
&s\mathcal{P}(Yf,Zg) \otimes \underline{\mathsf{C}}_{\Bbbk}(s\mathcal{P}(Yf,Zg),s\mathcal{P}(Xf,Wg)) \xrightarrow{\mathrm{ev}^{\mathsf{C}_{\Bbbk}}} s\mathcal{P}(Xf,Wg)\bigr], \\
\check{b}_0^{f\mathcal{P}_g} = \bigl[&Ts\mathcal{A}(X,Y) \otimes s\mathcal{P}(Yf,Zg) \otimes Ts\mathcal{C}(Z,W) \xrightarrow{\mathrm{pr}_0\otimes 1\otimes\mathrm{pr}_0} s\mathcal{P}(Yf,Zg) \\
&\xrightarrow{b_{00}^{\mathcal{P}}} s\mathcal{P}(Yf,Zg)\bigr].
\end{aligned}$$

These equations can be combined into a single formula

$$\begin{aligned}
\check{b}^{f\mathcal{P}_g} = \bigl[&Ts\mathcal{A}(X,Y) \otimes s\mathcal{P}(Yf,Zg) \otimes Ts\mathcal{C}(Z,W) \xrightarrow{f\otimes 1\otimes g} \\
&Ts\mathcal{B}(Xf,Yf) \otimes s\mathcal{P}(Yf,Zg) \otimes Ts\mathcal{D}(Zg,Wg) \xrightarrow{\check{b}^{\mathcal{P}}} s\mathcal{P}(Xf,Wg)\bigr]. \qquad (5.14)
\end{aligned}$$

Let $f : \mathcal{A} \to \mathcal{B}$ be an A_∞-functor. Define an $(\mathrm{id}_{Ts\mathcal{A}}, \mathrm{id}_{Ts\mathcal{A}})$-bicomodule homomorphism $t^f : \mathcal{R}_{\mathcal{A}} = \mathcal{A} \to {}_f\mathcal{B}_f = {}_f(\mathcal{R}_{\mathcal{B}})_f$ of degree 0 by its components

$$\check{t}^f = \bigl[Ts\mathcal{A}(X,Y) \otimes s\mathcal{A}(Y,Z) \otimes Ts\mathcal{A}(Z,W) \xrightarrow{\mu_{Ts\mathcal{A}}} Ts\mathcal{A}(X,W) \xrightarrow{\check{f}} s\mathcal{B}(Xf,Wf)\bigr],$$

or in extended form,

$$\begin{aligned}
t^f_{kn} = \bigl[&s\mathcal{A}(X_k,X_{k-1}) \otimes \dots \otimes s\mathcal{A}(X_1,X_0) \otimes s\mathcal{A}(X_0,Z_0) \otimes \\
&\otimes s\mathcal{A}(Z_0,Z_1) \otimes \dots \otimes s\mathcal{A}(Z_{n-1},Z_n) \xrightarrow{f_{k+1+n}} s\mathcal{B}(X_kf,Z_nf)\bigr]. \qquad (5.15)
\end{aligned}$$

We claim that $t^f d = 0$. As usual, it suffices to show that $(t^f d)^\vee = 0$. From the identity

$$\begin{aligned}
(t^f d)^\vee = t^f \cdot \check{b}^{f(\mathcal{R}_{\mathcal{B}})_f} - b^{\mathcal{R}_{\mathcal{A}}} \cdot \check{t}^f &= (\Delta_0 \otimes 1 \otimes \Delta_0)(1 \otimes \check{t}^f \otimes 1)\check{b}^{f(\mathcal{R}_{\mathcal{B}})_f} \\
&- (b^{\mathcal{A}} \otimes 1 \otimes 1 + 1 \otimes 1 \otimes b^{\mathcal{A}})\check{t}^f - (\Delta_0 \otimes 1 \otimes \Delta_0)(1 \otimes \check{b}^{\mathcal{R}_{\mathcal{A}}} \otimes 1)\check{t}^f
\end{aligned}$$

it follows that

$$(t^f d)^\vee = \bigl[Ts\mathcal{A}(X,Y) \otimes s\mathcal{A}(Y,Z) \otimes Ts\mathcal{A}(Z,W) \xrightarrow{\Delta_0 \otimes 1 \otimes \Delta_0}$$

$$\bigoplus_{U,V \in \operatorname{Ob}\mathcal{A}} Ts\mathcal{A}(X,U) \otimes Ts\mathcal{A}(U,Y) \otimes s\mathcal{A}(Y,Z) \otimes Ts\mathcal{A}(Z,V) \otimes Ts\mathcal{A}(V,W) \xrightarrow{\sum 1 \otimes \mu_{Ts\mathcal{A}} \otimes 1}$$

$$\bigoplus_{U,V \in \operatorname{Ob}\mathcal{A}} Ts\mathcal{A}(X,U) \otimes Ts\mathcal{A}(U,V) \otimes Ts\mathcal{A}(V,W) \xrightarrow{\sum 1 \otimes \check{f} \otimes 1}$$

$$\bigoplus_{U,V \in \operatorname{Ob}\mathcal{A}} Ts\mathcal{A}(X,U) \otimes s\mathcal{B}(Uf,Vf) \otimes Ts\mathcal{A}(V,W) \xrightarrow{\sum f \otimes 1 \otimes f}$$

$$\bigoplus_{U,V \in \operatorname{Ob}\mathcal{A}} Ts\mathcal{B}(Xf,Uf) \otimes s\mathcal{B}(Uf,Vf) \otimes Ts\mathcal{B}(Vf,Wf) \xrightarrow{\mu_{Ts\mathcal{B}}}$$

$$Ts\mathcal{B}(Xf,Wf) \xrightarrow{\check{b}^{\mathcal{B}}} s\mathcal{B}(Xf,Wf)\bigr]$$

$$- \bigl[Ts\mathcal{A}(X,Y) \otimes s\mathcal{A}(Y,Z) \otimes Ts\mathcal{A}(Z,W) \xrightarrow{b^{\mathcal{A}} \otimes 1 \otimes 1 + 1 \otimes 1 \otimes b^{\mathcal{A}}}$$

$$Ts\mathcal{A}(X,Y) \otimes s\mathcal{A}(Y,Z) \otimes Ts\mathcal{A}(Z,W) \xrightarrow{\mu_{Ts\mathcal{A}}} Ts\mathcal{A}(X,W) \xrightarrow{\check{f}} s\mathcal{B}(Xf,Wf)\bigr]$$

$$- \bigl[Ts\mathcal{A}(X,Y) \otimes s\mathcal{A}(Y,Z) \otimes Ts\mathcal{A}(Z,W) \xrightarrow{\Delta_0 \otimes 1 \otimes \Delta_0}$$

$$\bigoplus_{U,V \in \operatorname{Ob}\mathcal{A}} Ts\mathcal{A}(X,U) \otimes Ts\mathcal{A}(U,Y) \otimes s\mathcal{A}(Y,Z) \otimes Ts\mathcal{A}(Z,V) \otimes Ts\mathcal{A}(V,W)$$

$$\xrightarrow{\sum 1 \otimes \mu_{Ts\mathcal{A}} \otimes 1} \bigoplus_{U,V \in \operatorname{Ob}\mathcal{A}} Ts\mathcal{A}(X,U) \otimes Ts\mathcal{A}(U,V) \otimes Ts\mathcal{A}(V,W) \xrightarrow{\sum 1 \otimes \check{b}^{\mathcal{A}} \otimes 1}$$

$$\bigoplus_{U,V \in \operatorname{Ob}\mathcal{A}} Ts\mathcal{A}(X,U) \otimes s\mathcal{A}(U,V) \otimes Ts\mathcal{A}(V,W) \xrightarrow{\sum \mu_{Ts\mathcal{A}}} Ts\mathcal{A}(X,W)$$

$$\xrightarrow{\check{f}} s\mathcal{B}(Xf,Wf)\bigr].$$

Likewise Section 5.4 we see that the equation $(t^f d)^\vee = 0$ is equivalent to $f \cdot \check{b}^{\mathcal{B}} = b^{\mathcal{A}} \cdot \check{f}$.

Corollary 5.7. *Let $f : \mathcal{A} \to \mathcal{B}$ be an A_∞-functor. There is a natural A_∞-transformation $r^f : \operatorname{Hom}_{\mathcal{A}} \to (f^{\mathrm{op}}, f) \cdot \operatorname{Hom}_{\mathcal{B}} : \mathcal{A}^{\mathrm{op}}, \mathcal{A} \to \underline{\mathsf{C}}_{\Bbbk}$ depicted as follows:*

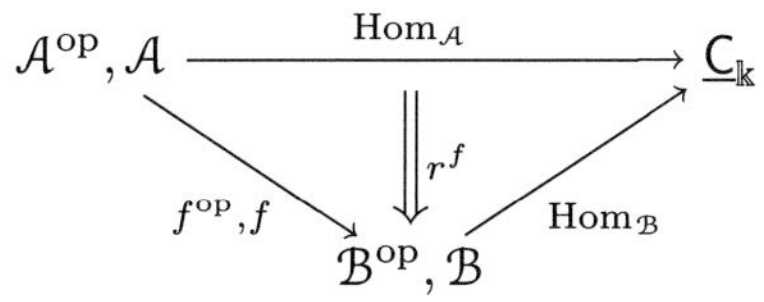

It is invertible if f is homotopy full and faithful.

Proof. Define $r^f = (t^f)\Phi^{-1}s \in s\underline{\mathsf{A}_\infty}(\mathcal{A}^{\mathrm{op}}, \mathcal{A}; \underline{\mathsf{C}}_{\Bbbk})(\operatorname{Hom}_{\mathcal{A}}, (f^{\mathrm{op}}, f)\operatorname{Hom}_{\mathcal{B}})$, where $t^f : \mathcal{A} \to {}_f\mathcal{B}_f$ is the closed bicomodule homomorphism defined above. Since Φ is an invertible chain map, it follows that r^f is a natural A_∞-transformation. Suppose f

is homotopy full and faithful. That is, its first component f_1 is homotopy invertible. This implies that the $(0,0)$-component

$$ {}_{X,Z}r^f_{00} = \bigl[\Bbbk \xrightarrow{\mathrm{coev}} \underline{\mathsf{C}}_\Bbbk(s\mathcal{A}(X,Z), s\mathcal{A}(X,Z)) \xrightarrow{\underline{\mathsf{C}}_\Bbbk(1,f_1)} \\ \underline{\mathsf{C}}_\Bbbk(s\mathcal{A}(X,Z), s\mathcal{B}(Xf,Zf)) \xrightarrow{[-1]s} s\underline{\mathsf{C}}_\Bbbk(\mathcal{A}(X,Z), \mathcal{B}(Xf,Zf))\bigr], $$

found from (5.9) and (5.15), is invertible modulo boundaries in

$$ s\underline{\mathsf{C}}_\Bbbk(\mathcal{A}(X,Z), \mathcal{B}(Xf,Zf)), $$

thus r^f is invertible by [BLM06, Lemma 13.9]. The corollary is proven. □

5.8. Opposite bimodule

Let $\mathcal{P}$ be an $\mathcal{A}$-$\mathcal{C}$-bimodule, $\phi : \mathcal{A}^{\mathrm{op}}, \mathcal{C} \to \underline{\mathsf{C}}_\Bbbk$ the corresponding A_∞-functor. Define an *opposite bimodule* $\mathcal{P}^{\mathrm{op}}$ as the $\mathcal{C}^{\mathrm{op}}$-$\mathcal{A}^{\mathrm{op}}$-bimodule corresponding to the A_∞-functor

$$ (\mathrm{id}_\mathcal{C}, \mathrm{id}_{\mathcal{A}^{\mathrm{op}}}, \phi)\mu^{\mathsf{A}_\infty}_{\mathsf{X}:\mathbf{2}\to\mathbf{2}} = \bigl[Ts\mathcal{C} \boxtimes Ts\mathcal{A}^{\mathrm{op}} \xrightarrow{c} Ts\mathcal{A}^{\mathrm{op}} \boxtimes Ts\mathcal{C} \xrightarrow{\phi} Ts\underline{\mathsf{C}}_\Bbbk\bigr]. $$

Its underlying $\mathbf{gr}$-span is given by $\mathcal{P}^{\mathrm{op}}(Y,X) = \mathcal{P}(X,Y)$, $X \in \operatorname{Ob}\mathcal{A}$, $Y \in \operatorname{Ob}\mathcal{C}$. Components of the differential $b^{\mathcal{P}^{\mathrm{op}}}$ are found from equations (5.6) and (5.8):

$$ \check{b}^{\mathcal{P}^{\mathrm{op}}}_{+} = \bigl[Ts\mathcal{C}^{\mathrm{op}}(W,Z) \otimes s\mathcal{P}^{\mathrm{op}}(Z,Y) \otimes Ts\mathcal{A}^{\mathrm{op}}(Y,X) \xrightarrow{c\otimes 1} \\ s\mathcal{P}(Y,Z) \otimes Ts\mathcal{C}^{\mathrm{op}}(W,Z) \otimes Ts\mathcal{A}^{\mathrm{op}}(Y,X) \xrightarrow{1\otimes\gamma\otimes 1} s\mathcal{P}(Y,Z) \otimes Ts\mathcal{C}(Z,W) \otimes Ts\mathcal{A}^{\mathrm{op}}(Y,X) \\ \xrightarrow{1\otimes c} s\mathcal{P}(Y,Z) \otimes Ts\mathcal{A}^{\mathrm{op}}(Y,X) \otimes Ts\mathcal{C}(Z,W) \xrightarrow{1\otimes\check{\phi}} s\mathcal{P}(Y,Z) \otimes s\underline{\mathsf{C}}_\Bbbk(\mathcal{P}(Y,Z), \mathcal{P}(X,W)) \\ \xrightarrow{1\otimes s^{-1}[1]} s\mathcal{P}(Y,Z) \otimes \underline{\mathsf{C}}_\Bbbk(s\mathcal{P}(Y,Z), s\mathcal{P}(X,W)) \xrightarrow{\mathrm{ev}^{\underline{\mathsf{C}}_\Bbbk}} s\mathcal{P}(X,W) = s\mathcal{P}^{\mathrm{op}}(W,X)\bigr] \\ = \bigl[Ts\mathcal{C}^{\mathrm{op}}(W,Z) \otimes s\mathcal{P}^{\mathrm{op}}(Z,Y) \otimes Ts\mathcal{A}^{\mathrm{op}}(Y,X) \xrightarrow{(13)^\sim} \\ Ts\mathcal{A}^{\mathrm{op}}(Y,X) \otimes s\mathcal{P}(Y,Z) \otimes Ts\mathcal{C}^{\mathrm{op}}(W,Z) \xrightarrow{\gamma\otimes 1\otimes\gamma} \\ Ts\mathcal{A}(X,Y) \otimes s\mathcal{P}(Y,Z) \otimes Ts\mathcal{C}(Z,W) \xrightarrow{\check{b}^{\mathcal{P}}_{+}} s\mathcal{P}(X,W) = s\mathcal{P}^{\mathrm{op}}(W,X)\bigr], $$

$$ \check{b}^{\mathcal{P}^{\mathrm{op}}}_{0} = \bigl[Ts\mathcal{C}^{\mathrm{op}}(W,X) \otimes s\mathcal{P}^{\mathrm{op}}(Z,Y) \otimes Ts\mathcal{A}^{\mathrm{op}}(Y,X) \xrightarrow{\mathrm{pr}_0\otimes 1\otimes \mathrm{pr}_0} s\mathcal{P}(Y,Z) \\ \xrightarrow{b^{\mathcal{P}}_{00}} s\mathcal{P}(Y,Z)\bigr]. $$

These equations are particular cases of a single formula:

$$\begin{aligned}\check{b}^{\mathcal{P}^{\mathrm{op}}} &= \bigl[Ts\mathcal{C}^{\mathrm{op}}(W,Z)\otimes s\mathcal{P}^{\mathrm{op}}(Z,Y)\otimes Ts\mathcal{A}^{\mathrm{op}}(Y,X) \xrightarrow{(13)^\sim} \\ &\qquad Ts\mathcal{A}^{\mathrm{op}}(Y,X)\otimes s\mathcal{P}(Y,Z)\otimes Ts\mathcal{C}^{\mathrm{op}}(W,Z) \xrightarrow{\gamma\otimes 1\otimes\gamma} \\ &\qquad Ts\mathcal{A}(X,Y)\otimes s\mathcal{P}(Y,Z)\otimes Ts\mathcal{C}(Z,W) \xrightarrow{\check{b}^{\mathcal{P}}} s\mathcal{P}(X,W) = s\mathcal{P}^{\mathrm{op}}(W,X)\bigr] \\ &= -\bigl[Ts\mathcal{C}^{\mathrm{op}}(W,Z)\otimes s\mathcal{P}^{\mathrm{op}}(Z,Y)\otimes Ts\mathcal{A}^{\mathrm{op}}(Y,X) \xrightarrow{(13)^\sim} \\ &\qquad Ts\mathcal{A}^{\mathrm{op}}(Y,X)\otimes s\mathcal{P}^{\mathrm{op}}(Z,Y)\otimes Ts\mathcal{C}^{\mathrm{op}}(W,Z) \xrightarrow{\gamma\otimes\gamma\otimes\gamma} \\ &\qquad Ts\mathcal{A}(X,Y)\otimes s\mathcal{P}(Y,Z)\otimes Ts\mathcal{C}(Z,W) \xrightarrow{\check{b}^{\mathcal{P}}} s\mathcal{P}(X,W) = s\mathcal{P}^{\mathrm{op}}(W,X)\bigr]. \qquad (5.16)\end{aligned}$$

Proposition 5.9. *Let $\mathcal{A}$ be an A_∞-category. Then $\mathcal{R}^{\mathrm{op}}_{\mathcal{A}} = \mathcal{R}_{\mathcal{A}^{\mathrm{op}}}$ as $\mathcal{A}^{\mathrm{op}}$-$\mathcal{A}^{\mathrm{op}}$-bimodules.*

Proof. Clearly, the underlying **gr**-spans of the both bimodules coincide. Computing $\check{b}^{\mathcal{R}^{\mathrm{op}}_{\mathcal{A}}}$ by formula (5.16) yields

$$\begin{aligned}\check{b}^{\mathcal{R}^{\mathrm{op}}_{\mathcal{A}}} &= -\bigl[Ts\mathcal{A}^{\mathrm{op}}(W,Z)\otimes s\mathcal{A}^{\mathrm{op}}(Z,Y)\otimes Ts\mathcal{A}^{\mathrm{op}}(Y,X) \xrightarrow{(13)^\sim} \\ &\quad Ts\mathcal{A}^{\mathrm{op}}(Y,X)\otimes s\mathcal{A}^{\mathrm{op}}(Z,Y)\otimes Ts\mathcal{A}^{\mathrm{op}}(W,Z) \xrightarrow{\gamma\otimes\gamma\otimes\gamma} \\ &\qquad Ts\mathcal{A}(X,Y)\otimes s\mathcal{A}(Y,Z)\otimes Ts\mathcal{A}(Z,W) \xrightarrow{\mu_{Ts\mathcal{A}}} Ts\mathcal{A}(X,W) \xrightarrow{\check{b}^{\mathcal{A}}} s\mathcal{A}(X,W)\bigr] \\ &= -\bigl[Ts\mathcal{A}^{\mathrm{op}}(W,Z)\otimes s\mathcal{A}^{\mathrm{op}}(Z,Y)\otimes Ts\mathcal{A}^{\mathrm{op}}(Y,X) \xrightarrow{\mu_{Ts\mathcal{A}^{\mathrm{op}}}} Ts\mathcal{A}^{\mathrm{op}}(W,X) \\ &\qquad\qquad \xrightarrow{\gamma} Ts\mathcal{A}(X,W) \xrightarrow{\check{b}^{\mathcal{A}}} s\mathcal{A}(X,W)\bigr]\end{aligned}$$

since $\gamma : Ts\mathcal{A}^{\mathrm{op}} \to Ts\mathcal{A}$ is a category anti-isomorphism. Since $b^{\mathcal{A}^{\mathrm{op}}} = \gamma b^{\mathcal{A}}\gamma$, it follows that $\check{b}^{\mathcal{A}^{\mathrm{op}}} = -\gamma\check{b}^{\mathcal{A}} : Ts\mathcal{A}^{\mathrm{op}}(W,X) \to s\mathcal{A}(W,X)$, therefore

$$\begin{aligned}\check{b}^{\mathcal{R}^{\mathrm{op}}_{\mathcal{A}}} &= \bigl[Ts\mathcal{A}^{\mathrm{op}}(W,Z)\otimes s\mathcal{A}^{\mathrm{op}}(Z,Y)\otimes Ts\mathcal{A}^{\mathrm{op}}(Y,X) \xrightarrow{\mu_{Ts\mathcal{A}^{\mathrm{op}}}} \\ &\qquad Ts\mathcal{A}^{\mathrm{op}}(W,X) \xrightarrow{\check{b}^{\mathcal{A}^{\mathrm{op}}}} s\mathcal{A}^{\mathrm{op}}(W,X)\bigr] = \check{b}^{\mathcal{R}_{\mathcal{A}^{\mathrm{op}}}}.\end{aligned}$$

The proposition is proven. □

Corollary 5.10. *Let $\mathcal{A}$ be an A_∞-category. Then*

$$\mathrm{Hom}_{\mathcal{A}^{\mathrm{op}}} = \bigl[Ts\mathcal{A}\boxtimes Ts\mathcal{A}^{\mathrm{op}} \xrightarrow{c} Ts\mathcal{A}^{\mathrm{op}}\boxtimes Ts\mathcal{A} \xrightarrow{\mathrm{Hom}_{\mathcal{A}}} Ts\underline{\mathsf{C}}_{\Bbbk}\bigr].$$

Proposition 5.11. *For an arbitrary A_∞-category $\mathcal{A}$, $\mathsf{k}\,\mathrm{Hom}_{\mathcal{A}} = \mathrm{Hom}_{\mathsf{k}\mathcal{A}} : \mathsf{k}\mathcal{A}^{\mathrm{op}}\boxtimes \mathsf{k}\mathcal{A} \to \underline{\mathcal{K}}$.*

Proof. Let $X,Y,U,V \in \mathrm{Ob}\,\mathcal{A}$. Then

$$\begin{aligned}\mathsf{k}\,\mathrm{Hom}_{\mathcal{A}} &= \bigl[\mathcal{A}^{\mathrm{op}}(X,Y)\otimes\mathcal{A}(U,V) \xrightarrow{s(\mathrm{Hom}_{\mathcal{A}})_{10}s^{-1}\otimes s(\mathrm{Hom}_{\mathcal{A}})_{01}s^{-1}} \\ &\quad \underline{\mathsf{C}}_{\Bbbk}(\mathcal{A}(X,U),\mathcal{A}(Y,U))\otimes\underline{\mathsf{C}}_{\Bbbk}(\mathcal{A}(Y,U),\mathcal{A}(Y,V)) \xrightarrow{m_2^{\underline{\mathsf{C}}_{\Bbbk}}} \underline{\mathsf{C}}_{\Bbbk}(\mathcal{A}(X,U),\mathcal{A}(Y,V))\bigr].\end{aligned}$$

According to (5.12),

$$
\begin{aligned}
s(\mathrm{Hom}_{\mathcal{A}})_{10}s^{-1} &= -\big[\mathcal{A}(Y,X) \xrightarrow{s} s\mathcal{A}(Y,X) \\
&\qquad \xrightarrow{\mathrm{coev}^{\underline{\mathsf{C}}_\Bbbk}} \underline{\mathsf{C}}_\Bbbk(s\mathcal{A}(X,U), s\mathcal{A}(X,U)\otimes s\mathcal{A}(Y,X)) \\
&\qquad \xrightarrow{\underline{\mathsf{C}}_\Bbbk(1,cb_2)} \underline{\mathsf{C}}_\Bbbk(s\mathcal{A}(X,U), s\mathcal{A}(Y,U)) \xrightarrow[\underline{\mathsf{C}}_\Bbbk(s,1)\cdot\underline{\mathsf{C}}_\Bbbk(1,s^{-1})]{[-1]} \underline{\mathsf{C}}_\Bbbk(\mathcal{A}(X,U),\mathcal{A}(Y,U))\big] \\
&= -\big[\mathcal{A}(Y,X) \xrightarrow{\mathrm{coev}^{\underline{\mathsf{C}}_\Bbbk}} \underline{\mathsf{C}}_\Bbbk(\mathcal{A}(X,U),\mathcal{A}(X,U)\otimes\mathcal{A}(Y,X)) \xrightarrow{\underline{\mathsf{C}}_\Bbbk(1,s\otimes s)} \\
&\qquad \underline{\mathsf{C}}_\Bbbk(\mathcal{A}(X,U), s\mathcal{A}(X,U)\otimes s\mathcal{A}(Y,X)) \xrightarrow{\underline{\mathsf{C}}_\Bbbk(1,cb_2s^{-1})} \underline{\mathsf{C}}_\Bbbk(\mathcal{A}(X,U),\mathcal{A}(Y,U))\big] \\
&= \big[\mathcal{A}(Y,X) \xrightarrow{\mathrm{coev}^{\underline{\mathsf{C}}_\Bbbk}} \underline{\mathsf{C}}_\Bbbk(\mathcal{A}(X,U),\mathcal{A}(X,U)\otimes\mathcal{A}(Y,X)) \xrightarrow{\underline{\mathsf{C}}_\Bbbk(1,c)} \\
&\qquad \underline{\mathsf{C}}_\Bbbk(\mathcal{A}(X,U),\mathcal{A}(Y,X)\otimes\mathcal{A}(X,U)) \xrightarrow{\underline{\mathsf{C}}_\Bbbk(1,m_2)} \underline{\mathsf{C}}_\Bbbk(\mathcal{A}(X,U),\mathcal{A}(Y,U))\big]. \qquad (5.17)
\end{aligned}
$$

Similarly we obtain from equation (5.11)

$$
\begin{aligned}
s(\mathrm{Hom}_{\mathcal{A}})_{01}s^{-1} &= \big[\mathcal{A}(U,V) \xrightarrow{s} s\mathcal{A}(U,V) \\
&\qquad \xrightarrow{\mathrm{coev}^{\underline{\mathsf{C}}_\Bbbk}} \underline{\mathsf{C}}_\Bbbk(s\mathcal{A}(Y,U), s\mathcal{A}(Y,U)\otimes s\mathcal{A}(U,V)) \\
&\qquad \xrightarrow{\underline{\mathsf{C}}_\Bbbk(1,b_2)} \underline{\mathsf{C}}_\Bbbk(s\mathcal{A}(Y,U), s\mathcal{A}(Y,V)) \xrightarrow{[-1]} \underline{\mathsf{C}}_\Bbbk(\mathcal{A}(Y,U),\mathcal{A}(Y,V))\big] \\
&= \big[\mathcal{A}(U,V) \xrightarrow{\mathrm{coev}^{\underline{\mathsf{C}}_\Bbbk}} \underline{\mathsf{C}}_\Bbbk(\mathcal{A}(Y,U),\mathcal{A}(Y,U)\otimes\mathcal{A}(U,V)) \xrightarrow{\underline{\mathsf{C}}_\Bbbk(1,m_2)} \underline{\mathsf{C}}_\Bbbk(\mathcal{A}(Y,U),\mathcal{A}(Y,V))\big].
\end{aligned}
$$

It follows that

$$
\begin{aligned}
&\mathsf{k}\,\mathrm{Hom}_{\mathcal{A}} = \big[\mathcal{A}(Y,X)\otimes\mathcal{A}(U,V) \xrightarrow{\mathrm{coev}^{\underline{\mathsf{C}}_\Bbbk}\otimes\mathrm{coev}^{\underline{\mathsf{C}}_\Bbbk}} \\
&\underline{\mathsf{C}}_\Bbbk(\mathcal{A}(X,U),\mathcal{A}(X,U)\otimes\mathcal{A}(Y,X))\otimes\underline{\mathsf{C}}_\Bbbk(\mathcal{A}(Y,U),\mathcal{A}(Y,U)\otimes\mathcal{A}(U,V)) \xrightarrow{\underline{\mathsf{C}}_\Bbbk(1,cm_2)\otimes\underline{\mathsf{C}}_\Bbbk(1,m_2)} \\
&\qquad \underline{\mathsf{C}}_\Bbbk(\mathcal{A}(X,U),\mathcal{A}(Y,U))\otimes\underline{\mathsf{C}}_\Bbbk(\mathcal{A}(Y,U),\mathcal{A}(Y,V)) \xrightarrow{m_2^{\underline{\mathsf{C}}_\Bbbk}} \underline{\mathsf{C}}_\Bbbk(\mathcal{A}(X,U),\mathcal{A}(Y,V))\big].
\end{aligned}
$$

Equation (A.1.2) of [LM04] allows to write the above expression as follows:

$$
\begin{aligned}
&\mathsf{k}\,\mathrm{Hom}_{\mathcal{A}} = \big[\mathcal{A}(Y,X)\otimes\mathcal{A}(U,V) \xrightarrow{\mathrm{coev}^{\underline{\mathsf{C}}_\Bbbk}} \underline{\mathsf{C}}_\Bbbk(\mathcal{A}(X,U),\mathcal{A}(X,U)\otimes\mathcal{A}(Y,X)\otimes\mathcal{A}(U,V)) \\
&\quad \xrightarrow{\underline{\mathsf{C}}_\Bbbk(1,cm_2\otimes1)} \underline{\mathsf{C}}_\Bbbk(\mathcal{A}(X,U),\mathcal{A}(Y,U)\otimes\mathcal{A}(U,V)) \xrightarrow{\underline{\mathsf{C}}_\Bbbk(1,m_2)} \underline{\mathsf{C}}_\Bbbk(\mathcal{A}(X,U),\mathcal{A}(Y,V))\big] \\
&= \big[\mathcal{A}(Y,X)\otimes\mathcal{A}(U,V) \xrightarrow{\mathrm{coev}^{\underline{\mathsf{C}}_\Bbbk}} \underline{\mathsf{C}}_\Bbbk(\mathcal{A}(X,U),\mathcal{A}(X,U)\otimes\mathcal{A}(Y,X)\otimes\mathcal{A}(U,V)) \xrightarrow{\underline{\mathsf{C}}_\Bbbk(1,c\otimes1)} \\
&\underline{\mathsf{C}}_\Bbbk(\mathcal{A}(X,U),\mathcal{A}(Y,X)\otimes\mathcal{A}(X,U)\otimes\mathcal{A}(U,V)) \xrightarrow{\underline{\mathsf{C}}_\Bbbk(1,(m_2\otimes1)m_2)} \underline{\mathsf{C}}_\Bbbk(\mathcal{A}(X,U),\mathcal{A}(Y,V))\big] \\
&\qquad = \mathrm{Hom}_{\mathsf{k}\mathcal{A}}\,.
\end{aligned}
$$

The proposition is proven. □

5.12. Duality A_∞-functor

The regular module $\Bbbk$, viewed as a complex concentrated in degree 0, determines the duality A_∞-functor $D = H^\Bbbk = h^\Bbbk\cdot[-1] : \underline{\mathsf{C}}_\Bbbk^{\mathrm{op}} \to \underline{\mathsf{C}}_\Bbbk$. It maps a complex M to its dual $(\underline{\mathsf{C}}_\Bbbk(M,\Bbbk), m_1) = (\underline{\mathsf{C}}_\Bbbk(M,\Bbbk), -\underline{\mathsf{C}}_\Bbbk(d,1))$. Since $\underline{\mathsf{C}}_\Bbbk$ is a differential graded category, the components D_k vanish if $k > 1$, due to (5.12). The component D_1 is given by

$$D_1 = -\bigl[s\underline{\mathsf{C}}_{\Bbbk}{}^{\mathrm{op}}(M,N) = s\underline{\mathsf{C}}_{\Bbbk}(N,M) \xrightarrow{\mathrm{coev}^{\mathsf{C}_{\Bbbk}}} \underline{\mathsf{C}}_{\Bbbk}(s\underline{\mathsf{C}}_{\Bbbk}(M,\Bbbk), s\underline{\mathsf{C}}_{\Bbbk}(M,\Bbbk) \otimes s\underline{\mathsf{C}}_{\Bbbk}(N,M)) \xrightarrow{\underline{\mathsf{C}}_{\Bbbk}(1,cb_2^{\underline{\mathsf{C}}_{\Bbbk}})} \underline{\mathsf{C}}_{\Bbbk}(s\underline{\mathsf{C}}_{\Bbbk}(M,\Bbbk), s\underline{\mathsf{C}}_{\Bbbk}(N,\Bbbk)) \xrightarrow{[-1]} \underline{\mathsf{C}}_{\Bbbk}(\underline{\mathsf{C}}_{\Bbbk}(M,\Bbbk), \underline{\mathsf{C}}_{\Bbbk}(N,\Bbbk)) \xrightarrow{s} s\underline{\mathsf{C}}_{\Bbbk}(\underline{\mathsf{C}}_{\Bbbk}(M,\Bbbk), \underline{\mathsf{C}}_{\Bbbk}(N,\Bbbk))\bigr].$$

It follows that

$$sD_1s^{-1} = \bigl[\underline{\mathsf{C}}_{\Bbbk}{}^{\mathrm{op}}(M,N) = \underline{\mathsf{C}}_{\Bbbk}(N,M) \xrightarrow{\mathrm{coev}^{\mathsf{C}_{\Bbbk}}} \underline{\mathsf{C}}_{\Bbbk}(\underline{\mathsf{C}}_{\Bbbk}(M,\Bbbk), \underline{\mathsf{C}}_{\Bbbk}(M,\Bbbk) \otimes \underline{\mathsf{C}}_{\Bbbk}(N,M)) \xrightarrow{\underline{\mathsf{C}}_{\Bbbk}(1,cm_2^{\underline{\mathsf{C}}_{\Bbbk}})} \underline{\mathsf{C}}_{\Bbbk}(\underline{\mathsf{C}}_{\Bbbk}(M,\Bbbk), \underline{\mathsf{C}}_{\Bbbk}(N,\Bbbk))\bigr],$$

cf. (5.17). It follows from (1.5) that $\mathsf{k}D = \underline{\mathcal{K}}(\_,\Bbbk) : \mathsf{k}\underline{\mathsf{C}}_{\Bbbk}{}^{\mathrm{op}} = \underline{\mathcal{K}}^{\mathrm{op}} \to \mathsf{k}\underline{\mathsf{C}}_{\Bbbk} = \underline{\mathcal{K}}$.

5.13. Dual A_∞-bimodule

Let $\mathcal{A}$, $\mathcal{C}$ be A_∞-categories, and let $\mathcal{P}$ be an $\mathcal{A}$-$\mathcal{C}$-bimodule with a flat $(1,1,1,b^{\mathcal{A}},b^{\mathcal{C}})$-connection $b^{\mathcal{P}} : Ts\mathcal{A} \otimes s\mathcal{P} \otimes Ts\mathcal{C} \to Ts\mathcal{A} \otimes s\mathcal{P} \otimes Ts\mathcal{C}$, and let $\phi^{\mathcal{P}} : Ts\mathcal{A}^{\mathrm{op}} \boxtimes Ts\mathcal{C} \to Ts\underline{\mathsf{C}}_{\Bbbk}$ be the corresponding A_∞-functor. Define the dual $\mathcal{C}$-$\mathcal{A}$-bimodule $\mathcal{P}^*$ as the bimodule that corresponds to the following A_∞-functor:

$$\phi^{\mathcal{P}^*} = ((\phi^{\mathcal{P}})^{\mathrm{op}}, D)\mu^{\mathsf{A}_\infty}_{\mathbf{2}\to\mathbf{1}} \centerdot \mathsf{A}_\infty(\mathsf{X};\underline{\mathsf{C}}_{\Bbbk}) : \mathcal{C}^{\mathrm{op}}, \mathcal{A} \to \underline{\mathsf{C}}_{\Bbbk},$$

where $\mathsf{X} : \mathbf{2} \to \mathbf{2}$, $1 \mapsto 2$, $2 \mapsto 1$, and the map $\mathsf{A}_\infty(\mathsf{X};\underline{\mathsf{C}}_{\Bbbk})$ is given by the composite

$$\mathsf{A}_\infty(\mathcal{A},\mathcal{C}^{\mathrm{op}};\underline{\mathsf{C}}_{\Bbbk}) \xrightarrow{\mathrm{id}_{\mathcal{C}^{\mathrm{op}}} \times \mathrm{id}_{\mathcal{A}} \times 1} \mathsf{A}_\infty(\mathcal{C}^{\mathrm{op}};\mathcal{C}^{\mathrm{op}}) \times \mathsf{A}_\infty(\mathcal{A};\mathcal{A}) \times \mathsf{A}_\infty(\mathcal{A},\mathcal{C}^{\mathrm{op}};\underline{\mathsf{C}}_{\Bbbk}) \xrightarrow{\mu^{\mathsf{A}_\infty}_{\mathsf{X}:\mathbf{2}\to\mathbf{2}}} \mathsf{A}_\infty(\mathcal{C}^{\mathrm{op}},\mathcal{A};\underline{\mathsf{C}}_{\Bbbk}).$$

Equivalently,

$$\phi^{\mathcal{P}^*} = \bigl(Ts\mathcal{C}^{\mathrm{op}} \boxtimes Ts\mathcal{A} \xrightarrow{c} Ts\mathcal{A} \boxtimes Ts\mathcal{C}^{\mathrm{op}} \xrightarrow{\gamma\boxtimes\gamma} Ts\mathcal{A}^{\mathrm{op}} \boxtimes Ts\mathcal{C} \xrightarrow{\phi^{\mathcal{P}}} Ts\underline{\mathsf{C}}_{\Bbbk} \xrightarrow{\gamma} Ts\underline{\mathsf{C}}_{\Bbbk}{}^{\mathrm{op}} \xrightarrow{D} Ts\underline{\mathsf{C}}_{\Bbbk}\bigr). \tag{5.18}$$

The underlying **gr**-span of $\mathcal{P}^*$ is given by $\mathrm{Ob}_s\,\mathcal{P}^* = \mathrm{Ob}_t\,\mathcal{P} = \mathrm{Ob}\,\mathcal{C}$, $\mathrm{Ob}_t\,\mathcal{P}^* = \mathrm{Ob}_s\,\mathcal{P} = \mathrm{Ob}\,\mathcal{A}$, $\mathrm{Par}\,\mathcal{P}^* = \mathrm{Ob}_s\,\mathcal{P}^* \times \mathrm{Ob}_t\,\mathcal{P}^*$, $\mathrm{src} = \mathrm{pr}_1$, $\mathrm{tgt} = \mathrm{pr}_2$, $\mathcal{P}^*(X,Y) = \underline{\mathsf{C}}_{\Bbbk}(\mathcal{P}(Y,X),\Bbbk)$, $X \in \mathrm{Ob}\,\mathcal{C}$, $Y \in \mathrm{Ob}\,\mathcal{A}$. Moreover,

$$\check{\phi}^{\mathcal{P}^*} = \phi^{\mathcal{P}^*}\mathrm{pr}_1 = \bigl[Ts\mathcal{C}^{\mathrm{op}}(Y,Z) \otimes Ts\mathcal{A}(X,W) \xrightarrow{c(\gamma\otimes\gamma)} Ts\mathcal{A}^{\mathrm{op}}(W,X) \otimes Ts\mathcal{C}(Z,Y) \xrightarrow{\check{\phi}^{\mathcal{P}}} s\underline{\mathsf{C}}_{\Bbbk}(\mathcal{P}(W,Z),\mathcal{P}(X,Y)) \xrightarrow{\mathrm{coev}^{\mathsf{C}_{\Bbbk}}} \underline{\mathsf{C}}_{\Bbbk}(s\underline{\mathsf{C}}_{\Bbbk}(\mathcal{P}(X,Y),\Bbbk), s\underline{\mathsf{C}}_{\Bbbk}(\mathcal{P}(X,Y),\Bbbk) \otimes s\underline{\mathsf{C}}_{\Bbbk}(\mathcal{P}(W,Z),\mathcal{P}(X,Y))) \xrightarrow{\underline{\mathsf{C}}_{\Bbbk}(1,cb_2^{\underline{\mathsf{C}}_{\Bbbk}})} \underline{\mathsf{C}}_{\Bbbk}(s\underline{\mathsf{C}}_{\Bbbk}(\mathcal{P}(X,Y),\Bbbk), s\underline{\mathsf{C}}_{\Bbbk}(\mathcal{P}(W,Z),\Bbbk)) \xrightarrow{[-1]} \underline{\mathsf{C}}_{\Bbbk}(\underline{\mathsf{C}}_{\Bbbk}(\mathcal{P}(X,Y),\Bbbk), \underline{\mathsf{C}}_{\Bbbk}(\mathcal{P}(W,Z),\Bbbk)) \xrightarrow{s} s\underline{\mathsf{C}}_{\Bbbk}(\underline{\mathsf{C}}_{\Bbbk}(\mathcal{P}(X,Y),\Bbbk), \underline{\mathsf{C}}_{\Bbbk}(\mathcal{P}(W,Z),\Bbbk))\bigr]$$

(the minus sign present in $\gamma : s\underline{\mathsf{C}}_{\Bbbk}{}^{\mathrm{op}} \to s\underline{\mathsf{C}}_{\Bbbk}$ cancels that present in D_1). According to (5.6),

$$\begin{aligned}
\check{b}_+^{\mathcal{P}^*} &= \bigl[Ts\mathcal{C}(X,Y)\otimes s\mathcal{P}^*(Y,Z)\otimes Ts\mathcal{A}(Z,W) \xrightarrow{c\otimes 1} \\
& s\mathcal{P}^*(Y,Z)\otimes Ts\mathcal{C}(X,Y)\otimes Ts\mathcal{A}(Z,W) \xrightarrow{1\otimes\gamma\otimes 1} s\mathcal{P}^*(Y,Z)\otimes Ts\mathcal{C}^{\mathrm{op}}(Y,X)\otimes Ts\mathcal{A}(Z,W) \\
& \xrightarrow{1\otimes c} s\mathcal{P}^*(Y,Z)\otimes Ts\mathcal{A}(Z,W)\otimes Ts\mathcal{C}^{\mathrm{op}}(Y,X) \xrightarrow{1\otimes\gamma\otimes\gamma} \\
& s\mathcal{P}^*(Y,Z)\otimes Ts\mathcal{A}^{\mathrm{op}}(W,Z)\otimes Ts\mathcal{C}(X,Y) \xrightarrow{1\otimes\check{\phi}^{\mathcal{P}}} s\mathcal{P}^*(Y,Z)\otimes s\underline{\mathsf{C}}_{\Bbbk}(\mathcal{P}(W,X),\mathcal{P}(Z,Y)) \\
& \xrightarrow{1\otimes\mathrm{coev}^{\mathsf{C}_{\Bbbk}}} s\mathcal{P}^*(Y,Z)\otimes\underline{\mathsf{C}}_{\Bbbk}(s\underline{\mathsf{C}}_{\Bbbk}(\mathcal{P}(Z,Y),\Bbbk), s\underline{\mathsf{C}}_{\Bbbk}(\mathcal{P}(Z,Y),\Bbbk)\otimes s\underline{\mathsf{C}}_{\Bbbk}(\mathcal{P}(W,X),\mathcal{P}(Z,Y))) \\
& \xrightarrow{1\otimes\underline{\mathsf{C}}_{\Bbbk}(1,cb_2^{\mathsf{C}_{\Bbbk}})} s\mathcal{P}^*(Y,Z)\otimes\underline{\mathsf{C}}_{\Bbbk}(s\mathcal{P}^*(Y,Z),s\mathcal{P}^*(X,W)) \xrightarrow{\mathrm{ev}^{\mathsf{C}_{\Bbbk}}} s\mathcal{P}^*(X,W)\bigr] \\
&= \bigl[Ts\mathcal{C}(X,Y)\otimes s\mathcal{P}^*(Y,Z)\otimes Ts\mathcal{A}(Z,W) \xrightarrow{(123)} Ts\mathcal{A}(Z,W)\otimes Ts\mathcal{C}(X,Y)\otimes s\mathcal{P}^*(Y,Z) \\
& \xrightarrow{\gamma\otimes 1\otimes 1} Ts\mathcal{A}^{\mathrm{op}}(W,Z)\otimes Ts\mathcal{C}(X,Y)\otimes s\mathcal{P}^*(Y,Z) \xrightarrow{\check{\phi}^{\mathcal{P}}\otimes 1} \\
& s\underline{\mathsf{C}}_{\Bbbk}(\mathcal{P}(W,X),\mathcal{P}(Z,Y))\otimes s\underline{\mathsf{C}}_{\Bbbk}(\mathcal{P}(Z,Y),\Bbbk) \xrightarrow{b_2^{\mathsf{C}_{\Bbbk}}} s\underline{\mathsf{C}}_{\Bbbk}(\mathcal{P}(W,X),\Bbbk) = s\mathcal{P}^*(X,W)\bigr],
\end{aligned}$$

by properties of closed monoidal categories. Similarly, by (5.8)

$$\begin{aligned}
\check{b}_0^{\mathcal{P}^*} = \bigl[Ts\mathcal{C}(X,Y)\otimes s\mathcal{P}^*(Y,Z)\otimes Ts\mathcal{A}(Z,W) &\xrightarrow{\mathrm{pr}_0\otimes 1\otimes\mathrm{pr}_0} s\mathcal{P}^*(Y,Z) \\
&\xrightarrow{-s^{-1}\underline{\mathsf{C}}_{\Bbbk}(d,1)s} s\mathcal{P}^*(Y,Z)\bigr],
\end{aligned}$$

where d is the differential in the complex $\phi^{\mathcal{P}}(Z,Y) = \mathcal{P}(Z,Y)$.

Proposition 5.14. *The map $\check{b}^{\mathcal{P}^*} : Ts\mathcal{C}\otimes s\mathcal{P}^*\otimes Ts\mathcal{A} \to s\mathcal{P}^*$ satisfies the following equation:*

$$\begin{aligned}
&\bigl[s\mathcal{P}(W,X)\otimes Ts\mathcal{C}(X,Y)\otimes s\mathcal{P}^*(Y,Z)\otimes Ts\mathcal{A}(Z,W) \xrightarrow{1\otimes\check{b}^{\mathcal{P}^*}} s\mathcal{P}(W,X)\otimes s\mathcal{P}^*(X,W) \\
&\qquad \xrightarrow{s^{-1}\otimes s^{-1}} \mathcal{P}(W,X)\otimes\underline{\mathsf{C}}_{\Bbbk}(\mathcal{P}(W,X),\Bbbk) \xrightarrow{\mathrm{ev}^{\mathsf{C}_{\Bbbk}}} \Bbbk\bigr] \\
&= -\bigl[s\mathcal{P}(W,X)\otimes Ts\mathcal{C}(X,Y)\otimes s\mathcal{P}^*(Y,Z)\otimes Ts\mathcal{A}(Z,W) \xrightarrow{(1234)} \\
&Ts\mathcal{A}(Z,W)\otimes s\mathcal{P}(W,X)\otimes Ts\mathcal{C}(X,Y)\otimes s\mathcal{P}^*(Y,Z) \xrightarrow{\check{b}^{\mathcal{P}}\otimes 1} s\mathcal{P}(Z,Y)\otimes s\mathcal{P}^*(Y,Z) \\
&\qquad \xrightarrow{s^{-1}\otimes s^{-1}} \mathcal{P}(Z,Y)\otimes\underline{\mathsf{C}}_{\Bbbk}(\mathcal{P}(Z,Y),\Bbbk) \xrightarrow{\mathrm{ev}^{\mathsf{C}_{\Bbbk}}} \Bbbk\bigr].
\end{aligned}$$

Proof. It suffices to prove that the pairs of morphisms $\check{b}^{\mathcal{P}}_{+}$, $\check{b}^{\mathcal{P}^*}_{+}$ and $\check{b}^{\mathcal{P}}_{0}$, $\check{b}^{\mathcal{P}^*}_{0}$ are related by similar equations. The corresponding equation for $\check{b}^{\mathcal{P}}_{+}$, $\check{b}^{\mathcal{P}^*}_{+}$ is given below:

$$\begin{aligned}
&\bigl[s\mathcal{P}(W,X)\otimes Ts\mathcal{C}(X,Y)\otimes s\mathcal{P}^*(Y,Z)\otimes Ts\mathcal{A}(Z,W) \xrightarrow{1\otimes\check{b}^{\mathcal{P}^*}_{+}} s\mathcal{P}(W,X)\otimes s\mathcal{P}^*(X,W)\\
&\qquad \xrightarrow{s^{-1}\otimes s^{-1}} \mathcal{P}(W,X)\otimes \underline{\mathsf{C}}_{\Bbbk}(\mathcal{P}(W,X),\Bbbk) \xrightarrow{\mathrm{ev}^{\mathsf{C}_\Bbbk}} \Bbbk\bigr]\\
&= -\bigl[s\mathcal{P}(W,X)\otimes Ts\mathcal{C}(X,Y)\otimes s\mathcal{P}^*(Y,Z)\otimes Ts\mathcal{A}(Z,W) \xrightarrow{(1234)}\\
&Ts\mathcal{A}(Z,W)\otimes s\mathcal{P}(W,X)\otimes Ts\mathcal{C}(X,Y)\otimes s\mathcal{P}^*(Y,Z) \xrightarrow{\check{b}^{\mathcal{P}}_{+}\otimes 1} s\mathcal{P}(Z,Y)\otimes s\mathcal{P}^*(Y,Z)\\
&\qquad \xrightarrow{s^{-1}\otimes s^{-1}} \mathcal{P}(Z,Y)\otimes \underline{\mathsf{C}}_{\Bbbk}(\mathcal{P}(Z,Y),\Bbbk) \xrightarrow{\mathrm{ev}^{\mathsf{C}_\Bbbk}} \Bbbk\bigr]. \qquad (5.19)
\end{aligned}$$

Its left-hand side equals

$$\begin{aligned}
&\bigl[s\mathcal{P}(W,X)\otimes Ts\mathcal{C}(X,Y)\otimes s\mathcal{P}^*(Y,Z)\otimes Ts\mathcal{A}(Z,W) \xrightarrow{1\otimes(123)}\\
&\quad s\mathcal{P}(W,X)\otimes Ts\mathcal{A}(Z,W)\otimes Ts\mathcal{C}(X,Y)\otimes s\mathcal{P}^*(Y,Z) \xrightarrow{1\otimes\gamma\otimes1\otimes1}\\
&\quad s\mathcal{P}(W,X)\otimes Ts\mathcal{A}^{\mathrm{op}}(W,Z)\otimes Ts\mathcal{C}(X,Y)\otimes s\mathcal{P}^*(Y,Z) \xrightarrow{1\otimes\check{\phi}^{\mathcal{P}}\otimes1}\\
&\quad s\mathcal{P}(W,X)\otimes s\underline{\mathsf{C}}_{\Bbbk}(\mathcal{P}(W,X),\mathcal{P}(Z,Y))\otimes s\underline{\mathsf{C}}_{\Bbbk}(\mathcal{P}(Z,Y),\Bbbk) \xrightarrow{1\otimes b_2^{\mathsf{C}_\Bbbk}}\\
&s\mathcal{P}(W,X)\otimes s\underline{\mathsf{C}}_{\Bbbk}(\mathcal{P}(W,X),\Bbbk) \xrightarrow{s^{-1}\otimes s^{-1}} \mathcal{P}(W,X)\otimes \underline{\mathsf{C}}_{\Bbbk}(\mathcal{P}(W,X),\Bbbk) \xrightarrow{\mathrm{ev}^{\mathsf{C}_\Bbbk}} \Bbbk\bigr].
\end{aligned}$$

Note that $(1234)(c\otimes1\otimes1) = 1\otimes(123)$. Using (5.6), the right-hand side can be written as follows:

$$\begin{aligned}
&-\bigl[s\mathcal{P}(W,X)\otimes Ts\mathcal{C}(X,Y)\otimes s\mathcal{P}^*(Y,Z)\otimes Ts\mathcal{A}(Z,W) \xrightarrow{1\otimes(123)}\\
&\quad s\mathcal{P}(W,X)\otimes Ts\mathcal{A}(Z,W)\otimes Ts\mathcal{C}(X,Y)\otimes s\mathcal{P}^*(Y,Z) \xrightarrow{1\otimes\gamma\otimes1\otimes1}\\
&\quad s\mathcal{P}(W,X)\otimes Ts\mathcal{A}^{\mathrm{op}}(W,Z)\otimes Ts\mathcal{C}(X,Y)\otimes s\mathcal{P}^*(Y,Z) \xrightarrow{1\otimes\check{\phi}^{\mathcal{P}}\otimes1}\\
&\quad s\mathcal{P}(W,X)\otimes s\underline{\mathsf{C}}_{\Bbbk}(\mathcal{P}(W,X),\mathcal{P}(Z,Y))\otimes s\underline{\mathsf{C}}_{\Bbbk}(\mathcal{P}(Z,Y),\Bbbk) \xrightarrow{1\otimes s^{-1}\otimes1}\\
&\quad s\mathcal{P}(W,X)\otimes \underline{\mathsf{C}}_{\Bbbk}(\mathcal{P}(W,X),\mathcal{P}(Z,Y))\otimes s\underline{\mathsf{C}}_{\Bbbk}(\mathcal{P}(Z,Y),\Bbbk) \xrightarrow{1\otimes[1]\otimes1}\\
&\quad s\mathcal{P}(W,X)\otimes \underline{\mathsf{C}}_{\Bbbk}(s\mathcal{P}(W,X),s\mathcal{P}(Z,Y))\otimes s\underline{\mathsf{C}}_{\Bbbk}(\mathcal{P}(Z,Y),\Bbbk) \xrightarrow{\mathrm{ev}^{\mathsf{C}_\Bbbk}\otimes1}\\
&\quad s\mathcal{P}(Z,Y)\otimes s\underline{\mathsf{C}}_{\Bbbk}(\mathcal{P}(Z,Y),\Bbbk) \xrightarrow{s^{-1}\otimes s^{-1}} \mathcal{P}(Z,Y)\otimes \underline{\mathsf{C}}_{\Bbbk}(\mathcal{P}(Z,Y),\Bbbk) \xrightarrow{\mathrm{ev}^{\mathsf{C}_\Bbbk}} \Bbbk\bigr].
\end{aligned}$$

Equation (5.19) follows from the following equation, which we are going to prove:

$$\begin{aligned}
&\bigl[s\mathcal{P}(W,X)\otimes s\underline{\mathsf{C}}_{\Bbbk}(\mathcal{P}(W,X),\mathcal{P}(Z,Y))\otimes s\underline{\mathsf{C}}_{\Bbbk}(\mathcal{P}(Z,Y),\Bbbk) \xrightarrow{1\otimes b_2^{\mathsf{C}_\Bbbk}}\\
&\quad s\mathcal{P}(W,X)\otimes s\underline{\mathsf{C}}_{\Bbbk}(\mathcal{P}(W,X),\Bbbk) \xrightarrow{s^{-1}\otimes s^{-1}} \mathcal{P}(W,X)\otimes \underline{\mathsf{C}}_{\Bbbk}(\mathcal{P}(W,X),\Bbbk) \xrightarrow{\mathrm{ev}^{\mathsf{C}_\Bbbk}} \Bbbk\bigr]\\
&= -\bigl[s\mathcal{P}(W,X)\otimes s\underline{\mathsf{C}}_{\Bbbk}(\mathcal{P}(W,X),\mathcal{P}(Z,Y))\otimes s\underline{\mathsf{C}}_{\Bbbk}(\mathcal{P}(Z,Y),\Bbbk) \xrightarrow{1\otimes s^{-1}\otimes1}
\end{aligned}$$

$$s\mathcal{P}(W,X)\otimes\underline{\mathsf{C}}_{\Bbbk}(\mathcal{P}(W,X),\mathcal{P}(Z,Y))\otimes s\underline{\mathsf{C}}_{\Bbbk}(\mathcal{P}(Z,Y),\Bbbk)\xrightarrow{1\otimes[1]\otimes1}$$
$$s\mathcal{P}(W,X)\otimes\underline{\mathsf{C}}_{\Bbbk}(s\mathcal{P}(W,X),s\mathcal{P}(Z,Y))\otimes s\underline{\mathsf{C}}_{\Bbbk}(\mathcal{P}(Z,Y),\Bbbk)\xrightarrow{\mathrm{ev}^{\mathsf{C}_{\Bbbk}}\otimes1}$$
$$s\mathcal{P}(Z,Y)\otimes s\underline{\mathsf{C}}_{\Bbbk}(\mathcal{P}(Z,Y),\Bbbk)\xrightarrow{s^{-1}\otimes s^{-1}}\mathcal{P}(Z,Y)\otimes\underline{\mathsf{C}}_{\Bbbk}(\mathcal{P}(Z,Y),\Bbbk)\xrightarrow{\mathrm{ev}^{\mathsf{C}_{\Bbbk}}}\Bbbk\big].$$

Composing both sides with the morphism $-s\otimes s\otimes s$ we obtain an equivalent equation:

$$\big[\mathcal{P}(W,X)\otimes\underline{\mathsf{C}}_{\Bbbk}(\mathcal{P}(W,X),\mathcal{P}(Z,Y))\otimes\underline{\mathsf{C}}_{\Bbbk}(\mathcal{P}(Z,Y),\Bbbk)\xrightarrow{1\otimes m_2^{\mathsf{C}_{\Bbbk}}}$$
$$\mathcal{P}(W,X)\otimes\underline{\mathsf{C}}_{\Bbbk}(\mathcal{P}(W,X),\Bbbk)\xrightarrow{\mathrm{ev}^{\mathsf{C}_{\Bbbk}}}\Bbbk\big]$$
$$=-\big[\mathcal{P}(W,X)\otimes\underline{\mathsf{C}}_{\Bbbk}(\mathcal{P}(W,X),\mathcal{P}(Z,Y))\otimes\underline{\mathsf{C}}_{\Bbbk}(\mathcal{P}(Z,Y),\Bbbk)\xrightarrow{s\otimes1\otimes s}$$
$$s\mathcal{P}(W,X)\otimes\underline{\mathsf{C}}_{\Bbbk}(\mathcal{P}(W,X),\mathcal{P}(Z,Y))\otimes s\underline{\mathsf{C}}_{\Bbbk}(\mathcal{P}(Z,Y),\Bbbk)\xrightarrow{1\otimes[1]\otimes1}$$
$$s\mathcal{P}(W,X)\otimes\underline{\mathsf{C}}_{\Bbbk}(s\mathcal{P}(W,X),s\mathcal{P}(Z,Y))\otimes s\underline{\mathsf{C}}_{\Bbbk}(\mathcal{P}(Z,Y),\Bbbk)\xrightarrow{\mathrm{ev}^{\mathsf{C}_{\Bbbk}}\otimes1}$$
$$s\mathcal{P}(Z,Y)\otimes s\underline{\mathsf{C}}_{\Bbbk}(\mathcal{P}(Z,Y),\Bbbk)\xrightarrow{s^{-1}\otimes s^{-1}}\mathcal{P}(Z,Y)\otimes\underline{\mathsf{C}}_{\Bbbk}(\mathcal{P}(Z,Y),\Bbbk)\xrightarrow{\mathrm{ev}^{\mathsf{C}_{\Bbbk}}}\Bbbk\big].$$

It follows from the definition of **dg**-functor [1] that

$$(s\otimes1)(1\otimes[1])\,\mathrm{ev}^{\mathsf{C}_{\Bbbk}}s^{-1}=(s\otimes1)(s^{-1}\otimes1)\,\mathrm{ev}^{\mathsf{C}_{\Bbbk}}ss^{-1}=\mathrm{ev}^{\mathsf{C}_{\Bbbk}},$$

therefore the right-hand side of the above equation is equal to $(\mathrm{ev}^{\mathsf{C}_{\Bbbk}}\otimes1)\,\mathrm{ev}^{\mathsf{C}_{\Bbbk}}$, and it equals the left-hand side by the definition of $m_2^{\mathsf{C}_{\Bbbk}}$.

The corresponding equation for $\check{b}_0^{\mathcal{P}}$ and $\check{b}_0^{\mathcal{P}^*}$ reads as follows:

$$\big[s\mathcal{P}(W,X)\otimes T^0s\mathcal{C}(X,Y)\otimes s\mathcal{P}^*(Y,Z)\otimes T^0s\mathcal{A}(Z,W)\xrightarrow{1\otimes\check{b}_0^{\mathcal{P}^*}}$$
$$s\mathcal{P}(W,X)\otimes s\mathcal{P}^*(X,W)\xrightarrow{s^{-1}\otimes s^{-1}}\mathcal{P}(W,X)\otimes\underline{\mathsf{C}}_{\Bbbk}(\mathcal{P}(W,X),\Bbbk)\xrightarrow{\mathrm{ev}^{\mathsf{C}_{\Bbbk}}}\Bbbk\big]$$
$$=-\big[s\mathcal{P}(W,X)\otimes T^0s\mathcal{C}(X,Y)\otimes s\mathcal{P}^*(Y,Z)\otimes T^0s\mathcal{A}(Z,W)\xrightarrow{(1234)}$$
$$T^0s\mathcal{A}(Z,W)\otimes s\mathcal{P}(W,X)\otimes T^0s\mathcal{C}(X,Y)\otimes s\mathcal{P}^*(Y,Z)\xrightarrow{\check{b}_0^{\mathcal{P}}\otimes1}$$
$$s\mathcal{P}(Z,Y)\otimes s\mathcal{P}^*(Y,Z)\xrightarrow{s^{-1}\otimes s^{-1}}\mathcal{P}(Z,Y)\otimes\underline{\mathsf{C}}_{\Bbbk}(\mathcal{P}(Z,Y),\Bbbk)\xrightarrow{\mathrm{ev}^{\mathsf{C}_{\Bbbk}}}\Bbbk\big].$$

It is non-trivial only if $X=Y$ and $Z=W$. In this case, up to obvious isomorphism the left-hand side equals

$$\big[s\mathcal{P}(Z,X)\otimes s\mathcal{P}^*(X,Z)\xrightarrow{s^{-1}\otimes s^{-1}}\mathcal{P}(Z,X)\otimes\underline{\mathsf{C}}_{\Bbbk}(\mathcal{P}(Z,X),\Bbbk)$$
$$\xrightarrow{1\otimes\underline{\mathsf{C}}_{\Bbbk}(d,1)}\mathcal{P}(Z,X)\otimes\underline{\mathsf{C}}_{\Bbbk}(\mathcal{P}(Z,X),\Bbbk)\xrightarrow{\mathrm{ev}^{\mathsf{C}_{\Bbbk}}}\Bbbk\big]$$
$$=\big[s\mathcal{P}(Z,X)\otimes s\mathcal{P}^*(X,Z)\xrightarrow{s^{-1}\otimes s^{-1}}\mathcal{P}(Z,X)\otimes\underline{\mathsf{C}}_{\Bbbk}(\mathcal{P}(Z,X),\Bbbk)$$
$$\xrightarrow{d\otimes1}\mathcal{P}(Z,X)\otimes\underline{\mathsf{C}}_{\Bbbk}(\mathcal{P}(Z,X),\Bbbk)\xrightarrow{\mathrm{ev}^{\mathsf{C}_{\Bbbk}}}\Bbbk\big]$$
$$=-\big[s\mathcal{P}(Z,X)\otimes s\mathcal{P}^*(X,Z)\xrightarrow{s^{-1}ds\otimes1}s\mathcal{P}(Z,X)\otimes s\underline{\mathsf{C}}_{\Bbbk}(\mathcal{P}(Z,X),\Bbbk)$$

$$\xrightarrow{s^{-1}\otimes s^{-1}} \mathcal{P}(Z,X)\otimes \underline{\mathsf{C}}_{\Bbbk}(\mathcal{P}(Z,X),\Bbbk) \xrightarrow{\mathrm{ev}^{\mathsf{C}_{\Bbbk}}} \Bbbk\big],$$

which is the right-hand side (up to obvious isomorphism). □

Proposition 5.15. *Let $\mathcal{A}$ be an A_∞-category. Denote by $\mathcal{R}$ the regular $\mathcal{A}$-bimodule. Then*

$$\phi^{\mathcal{R}^*} = (\mathrm{Hom}^{\mathrm{op}}_{\mathcal{A}^{\mathrm{op}}}, D)\mu^{\mathsf{A}_\infty}_{\mathbf{2}\to\mathbf{1}} = \mathrm{Hom}^{\mathrm{op}}_{\mathcal{A}^{\mathrm{op}}}\cdot D : \mathcal{A}^{\mathrm{op}},\mathcal{A} \to \underline{\mathsf{C}}_{\Bbbk}.$$

Proof. Formula (5.18) implies that

$$\begin{aligned}\phi^{\mathcal{R}^*} &= \big[Ts\mathcal{A}^{\mathrm{op}}\boxtimes Ts\mathcal{A} \xrightarrow{c} Ts\mathcal{A}\boxtimes Ts\mathcal{A}^{\mathrm{op}} \xrightarrow{\gamma\boxtimes\gamma} Ts\mathcal{A}^{\mathrm{op}}\boxtimes Ts\mathcal{A} \\ &\qquad \xrightarrow{\mathrm{Hom}_{\mathcal{A}}} Ts\underline{\mathsf{C}}_{\Bbbk} \xrightarrow{\gamma} Ts\underline{\mathsf{C}}_{\Bbbk}{}^{\mathrm{op}} \xrightarrow{D} Ts\underline{\mathsf{C}}_{\Bbbk}\big] \\ &= \big[Ts\mathcal{A}^{\mathrm{op}}\boxtimes Ts\mathcal{A} \xrightarrow{\gamma\boxtimes\gamma} Ts\mathcal{A}\boxtimes Ts\mathcal{A}^{\mathrm{op}} \xrightarrow{c} Ts\mathcal{A}^{\mathrm{op}}\boxtimes Ts\mathcal{A} \\ &\qquad \xrightarrow{\mathrm{Hom}_{\mathcal{A}}} Ts\underline{\mathsf{C}}_{\Bbbk} \xrightarrow{\gamma} Ts\underline{\mathsf{C}}_{\Bbbk}{}^{\mathrm{op}} \xrightarrow{D} Ts\underline{\mathsf{C}}_{\Bbbk}\big].\end{aligned}$$

By Corollary 5.10, $\mathrm{Hom}_{\mathcal{A}^{\mathrm{op}}} = [Ts\mathcal{A}\boxtimes Ts\mathcal{A}^{\mathrm{op}} \xrightarrow{c} Ts\mathcal{A}^{\mathrm{op}}\boxtimes Ts\mathcal{A} \xrightarrow{\mathrm{Hom}_{\mathcal{A}}} Ts\underline{\mathsf{C}}_{\Bbbk}]$, therefore

$$\begin{aligned}\phi^{\mathcal{R}^*} &= \big[Ts\mathcal{A}^{\mathrm{op}}\boxtimes Ts\mathcal{A} \xrightarrow{\gamma\boxtimes\gamma} Ts\mathcal{A}\boxtimes Ts\mathcal{A}^{\mathrm{op}} \xrightarrow{\mathrm{Hom}_{\mathcal{A}^{\mathrm{op}}}} Ts\underline{\mathsf{C}}_{\Bbbk} \xrightarrow{\gamma} Ts\underline{\mathsf{C}}_{\Bbbk}{}^{\mathrm{op}} \xrightarrow{D} Ts\underline{\mathsf{C}}_{\Bbbk}\big] \\ &= \big[Ts\mathcal{A}^{\mathrm{op}}\boxtimes Ts\mathcal{A} \xrightarrow{\mathrm{Hom}^{\mathrm{op}}_{\mathcal{A}^{\mathrm{op}}}} Ts\underline{\mathsf{C}}_{\Bbbk}{}^{\mathrm{op}} \xrightarrow{D} Ts\underline{\mathsf{C}}_{\Bbbk}\big].\end{aligned}$$

The proposition is proven. □

5.16. Unital A_∞-bimodules

Definition 5.17. An $\mathcal{A}$-$\mathcal{C}$-bimodule $\mathcal{P}$ corresponding to an A_∞-functor $\phi : \mathcal{A}^{\mathrm{op}},\mathcal{C} \to \underline{\mathsf{C}}_{\Bbbk}$ is called *unital* if the A_∞-functor ϕ is unital.

Proposition 5.18. *An $\mathcal{A}$-$\mathcal{C}$-bimodule $\mathcal{P}$ is unital if and only if for all $X\in\mathrm{Ob}\,\mathcal{A}$, $Y\in\mathrm{Ob}\,\mathcal{C}$ the compositions*

$$\begin{aligned}&\big[s\mathcal{P}(X,Y) = s\mathcal{P}(X,Y)\otimes\Bbbk \xrightarrow{1\otimes{}_Y\mathbf{i}^{\mathcal{C}}_0} s\mathcal{P}(X,Y)\otimes s\mathcal{C}(Y,Y) \xrightarrow{b^{\mathcal{P}}_{01}} s\mathcal{P}(X,Y)\big], \\ -&\big[s\mathcal{P}(X,Y) = \Bbbk\otimes s\mathcal{P}(X,Y) \xrightarrow{{}_X\mathbf{i}^{\mathcal{A}}_0\otimes 1} s\mathcal{A}(X,X)\otimes s\mathcal{P}(X,Y) \xrightarrow{b^{\mathcal{P}}_{10}} s\mathcal{P}(X,Y)\big]\end{aligned}$$

are homotopic to the identity map.

Proof. The second statement expands to the property that

$$\begin{aligned}&\big[s\mathcal{P}(X,Y) \xrightarrow{s^{-1}\otimes{}_Y\mathbf{i}^{\mathcal{C}}_0\phi_{01}s^{-1}} \mathcal{P}(X,Y)\otimes\underline{\mathsf{C}}_{\Bbbk}(\mathcal{P}(X,Y),\mathcal{P}(X,Y)) \xrightarrow{\mathrm{ev}^{\mathsf{C}_{\Bbbk}}s} s\mathcal{P}(X,Y)\big], \\ &\big[s\mathcal{P}(X,Y) \xrightarrow{s^{-1}\otimes{}_X\mathbf{i}^{\mathcal{A}}_0\phi_{10}s^{-1}} \mathcal{P}(X,Y)\otimes\underline{\mathsf{C}}_{\Bbbk}(\mathcal{P}(X,Y),\mathcal{P}(X,Y)) \xrightarrow{\mathrm{ev}^{\mathsf{C}_{\Bbbk}}s} s\mathcal{P}(X,Y)\big]\end{aligned}$$

are homotopic to identity. That is,

$${}_Y\mathbf{i}^{\mathcal{C}}_0\big(\phi_{01}|^X_{\mathcal{C}}\big) - 1_{s\mathcal{P}(X,Y)}s \in \mathrm{Im}\,b_1, \qquad {}_X\mathbf{i}^{\mathcal{A}}_0\big(\phi_{10}|^Y_{\mathcal{A}^{\mathrm{op}}}\big) - 1_{s\mathcal{P}(X,Y)}s \in \mathrm{Im}\,b_1$$

for all $X\in\mathrm{Ob}\,\mathcal{A}$, $Y\in\mathrm{Ob}\,\mathcal{C}$. By [BLM06, Proposition 13.6] the A_∞-functor ϕ is unital. □

Remark 5.19. Suppose $\mathcal{A}$ is a unital A_∞-category. By the above criterion, the regular $\mathcal{A}$-bimodule $\mathcal{R}_\mathcal{A}$ is unital, and therefore the A_∞-functor $\operatorname{Hom}_\mathcal{A} : \mathcal{A}^{\mathrm{op}}, \mathcal{A} \to \underline{\mathsf{C}}_\Bbbk$ is unital. In particular, for each object Z of $\mathcal{A}$, the representable A_∞-functor

$$H^Z = \operatorname{Hom}_\mathcal{A}|_1^Z : \mathcal{A}^{\mathrm{op}} \to \underline{\mathsf{C}}_\Bbbk$$

is unital, by [BLM06, Proposition 13.6]. Thus, the Yoneda A_∞-functor $\mathscr{Y} : \mathcal{A} \to \underline{\mathsf{A}_\infty}(\mathcal{A}^{\mathrm{op}}; \underline{\mathsf{C}}_\Bbbk)$ takes values in the full A_∞-subcategory $\underline{\mathsf{A}^{\mathrm{u}}_\infty}(\mathcal{A}^{\mathrm{op}}; \underline{\mathsf{C}}_\Bbbk)$ of $\underline{\mathsf{A}_\infty}(\mathcal{A}^{\mathrm{op}}; \underline{\mathsf{C}}_\Bbbk)$. Furthermore, by the closedness of the multicategory $\mathsf{A}^{\mathrm{u}}_\infty$, the A_∞-functor $\mathscr{Y}$ is unital.

Let $\mathcal{A}$, $\mathcal{B}$ be unital A_∞-categories. A unital A_∞-functor $f : \mathcal{A} \to \mathcal{B}$ is called *homotopy fully faithful* if the corresponding $\mathcal{K}$-functor $\mathsf{k}f : \mathsf{k}\mathcal{A} \to \mathsf{k}\mathcal{B}$ is fully faithful. That is, f is homotopy fully faithful if and only if its first component is homotopy invertible. Equivalently, f is homotopy fully faithful if and only if it admits a factorization

$$\mathcal{A} \xrightarrow{g} \mathcal{I} \xhookrightarrow{e} \mathcal{B}, \tag{5.20}$$

where $\mathcal{I}$ is a full A_∞-subcategory of $\mathcal{B}$, $e : \mathcal{I} \to \mathcal{B}$ is the embedding, and $g : \mathcal{A} \to \mathcal{I}$ is an A_∞-equivalence.

Lemma 5.20. *Suppose $f : \mathcal{A} \to \mathcal{B}$ is a homotopy fully faithful A_∞-functor. Then for an arbitrary A_∞-category $\mathcal{C}$ the A_∞-functor $\underline{\mathsf{A}_\infty}(1; f) : \underline{\mathsf{A}_\infty}(\mathcal{C}; \mathcal{A}) \to \underline{\mathsf{A}_\infty}(\mathcal{C}; \mathcal{B})$ is homotopy fully faithful.*

Proof. Factorize f as in (5.20). Then the A_∞-functor $\underline{\mathsf{A}_\infty}(1; f)$ factorizes as

$$\underline{\mathsf{A}_\infty}(\mathcal{C}; \mathcal{A}) \xrightarrow{\underline{\mathsf{A}_\infty}(1;g)} \underline{\mathsf{A}_\infty}(\mathcal{C}; \mathcal{I}) \xrightarrow{\underline{\mathsf{A}_\infty}(1;e)} \underline{\mathsf{A}_\infty}(\mathcal{C}; \mathcal{B}).$$

The A_∞-functor $\underline{\mathsf{A}_\infty}(1; g)$ is an A_∞-equivalence since $\underline{\mathsf{A}_\infty}(\mathcal{C}; -)$ is a A^u_∞-2-functor, so it suffices to show that $\underline{\mathsf{A}_\infty}(1; e)$ is a full embedding. Since e is a strict A_∞-functor, so is $\underline{\mathsf{A}_\infty}(1; e)$. Its first component is given by

$$s\underline{\mathsf{A}_\infty}(\mathcal{C}; \mathcal{I})(\phi, \psi) = \prod_{n\geqslant 0}^{X,Y\in \operatorname{Ob}\mathcal{C}} \underline{\mathsf{C}}_\Bbbk(T^n s\mathcal{C}(X, Y), s\mathcal{I}(X\phi, Y\psi)) \xrightarrow{\underline{\mathsf{C}}_\Bbbk(1,e_1)}$$

$$s\underline{\mathsf{A}_\infty}(\mathcal{C}; \mathcal{B})(\phi e, \psi e) = \prod_{n\geqslant 0}^{X,Y\in \operatorname{Ob}\mathcal{C}} \underline{\mathsf{C}}_\Bbbk(T^n s\mathcal{C}(X, Y), s\mathcal{B}(X\phi, Y\psi)),$$

that is, $r = (r_n) \mapsto re = (r_n e_1)$. Since $s\mathcal{I}(X\phi, Y\psi) = s\mathcal{B}(X\phi, Y\psi)$ and $e_1 : s\mathcal{I}(X\phi, Y\psi) \to s\mathcal{B}(X\phi, Y\psi)$ is the identity morphism, the above map is the identity morphism, and the proof is complete. □

Example 5.21. Let $g : \mathcal{C} \to \mathcal{A}$ be an A_∞-functor. Then an $\mathcal{A}$-$\mathcal{C}$-bimodule $\mathcal{A}^g$ is associated to it via the A_∞-functor

$$\begin{aligned} \mathcal{A}^g &= \big[\mathcal{A}^{\mathrm{op}}, \mathcal{C} \xrightarrow{1,g} \mathcal{A}^{\mathrm{op}}, \mathcal{A} \xrightarrow{1,\mathscr{Y}} \mathcal{A}^{\mathrm{op}}, \underline{\mathsf{A}_\infty}(\mathcal{A}^{\mathrm{op}}, \underline{\mathsf{C}}_\Bbbk) \xrightarrow{\mathrm{ev}^{\mathsf{A}_\infty}} \underline{\mathsf{C}}_\Bbbk\big] \\ &= \big[\mathcal{A}^{\mathrm{op}}, \mathcal{C} \xrightarrow{1,g} \mathcal{A}^{\mathrm{op}}, \mathcal{A} \xrightarrow{\operatorname{Hom}_\mathcal{A}} \underline{\mathsf{C}}_\Bbbk\big]. \end{aligned} \tag{5.21}$$

As we already noticed, $\mathcal{A}$-$\mathcal{C}$-bimodules are objects of the differential graded category $\underline{\mathsf{A}_\infty}(\mathcal{A}^{\mathrm{op}},\mathcal{C};\underline{\mathsf{C}}_\Bbbk) \simeq \underline{\mathsf{A}_\infty}(\mathcal{C};\underline{\mathsf{A}_\infty}(\mathcal{A}^{\mathrm{op}};\underline{\mathsf{C}}_\Bbbk))$. Thus it makes sense to speak about their isomorphisms in the homotopy category $H^0(\underline{\mathsf{A}_\infty}(\mathcal{A}^{\mathrm{op}},\mathcal{C};\underline{\mathsf{C}}_\Bbbk))$.

Proposition 5.22. *There is an A_∞-functor $\underline{\mathsf{A}_\infty}(\mathcal{C};\mathcal{A}) \to \underline{\mathsf{A}_\infty}(\mathcal{A}^{\mathrm{op}},\mathcal{C};\underline{\mathsf{C}}_\Bbbk)$, $g \mapsto \mathcal{A}^g$, homotopy fully faithful if $\mathcal{A}$ is a unital A_∞-category. In that case, A_∞-functors $g,h:\mathcal{C}\to\mathcal{A}$ are isomorphic if and only if the bimodules $\mathcal{A}^g$ and $\mathcal{A}^h$ are isomorphic. If both $\mathcal{C}$ and $\mathcal{A}$ are unital, then the above A_∞-functor restricts to $\underline{\mathsf{A}^{\mathrm{u}}_\infty}(\mathcal{C};\mathcal{A}) \to \underline{\mathsf{A}^{\mathrm{u}}_\infty}(\mathcal{A}^{\mathrm{op}},\mathcal{C};\underline{\mathsf{C}}_\Bbbk)$. Moreover, g is unital if and only if $\mathcal{A}^g$ is unital.*

Proof. The functor in question is the composite

$$\underline{\mathsf{A}_\infty}(\mathcal{C};\mathcal{A}) \xrightarrow{\mathsf{A}_\infty(1;\mathscr{Y})} \underline{\mathsf{A}_\infty}(\mathcal{C};\underline{\mathsf{A}_\infty}(\mathcal{A}^{\mathrm{op}};\underline{\mathsf{C}}_\Bbbk)) \xrightarrow[\sim]{\varphi^{\mathsf{A}_\infty}} \underline{\mathsf{A}_\infty}(\mathcal{A}^{\mathrm{op}},\mathcal{C};\underline{\mathsf{C}}_\Bbbk),$$

where $\mathscr{Y}$ is the Yoneda A_∞-functor. When $\mathcal{A}$ is unital, $\mathscr{Y}:\mathcal{A}\to\underline{\mathsf{A}^{\mathrm{u}}_\infty}(\mathcal{A}^{\mathrm{op}};\underline{\mathsf{C}}_\Bbbk)$ is homotopy fully faithful by Corollary A.9, see also [LM04, Theorem A.11]. Thus, the first claim follows from Lemma 5.20.

Assume that A_∞-categories $\mathcal{C}$ and $\mathcal{A}$ are unital, and A_∞-functor (5.21) is unital. Let us prove that $g:\mathcal{C}\to\mathcal{A}$ is unital. Denote $f = g\cdot\mathscr{Y}:\mathcal{C}\to\underline{\mathsf{A}_\infty}(\mathcal{A}^{\mathrm{op}};\underline{\mathsf{C}}_\Bbbk)$. The A_∞-functor

$$f' = \bigl[\mathcal{A}^{\mathrm{op}},\mathcal{C} \xrightarrow{1,f} \mathcal{A}^{\mathrm{op}},\underline{\mathsf{A}^{\mathrm{u}}_\infty}(\mathcal{A}^{\mathrm{op}},\underline{\mathsf{C}}_\Bbbk) \xrightarrow{\mathrm{ev}} \underline{\mathsf{C}}_\Bbbk\bigr]$$

is unital by assumption. The bijection

$$\varphi^{\mathsf{A}_\infty}:\mathsf{A}_\infty(\mathcal{C};\underline{\mathsf{A}_\infty}(\mathcal{A}^{\mathrm{op}};\underline{\mathsf{C}}_\Bbbk)) \to \mathsf{A}_\infty(\mathcal{A}^{\mathrm{op}},\mathcal{C};\underline{\mathsf{C}}_\Bbbk)$$

shows that given f' can be obtained from a unique f. The bijection

$$\varphi^{\mathsf{A}^{\mathrm{u}}_\infty}:\mathsf{A}^{\mathrm{u}}_\infty(\mathcal{C};\underline{\mathsf{A}^{\mathrm{u}}_\infty}(\mathcal{A}^{\mathrm{op}};\underline{\mathsf{C}}_\Bbbk)) \to \mathsf{A}^{\mathrm{u}}_\infty(\mathcal{A}^{\mathrm{op}},\mathcal{C};\underline{\mathsf{C}}_\Bbbk)$$

shows that such A_∞-functor f is unital.

Thus, the composition of $g:\mathcal{C}\to\mathcal{A}$ with the unital homotopy fully faithful A_∞-functor $\mathscr{Y}:\mathcal{A}\to\underline{\mathsf{A}^{\mathrm{u}}_\infty}(\mathcal{A}^{\mathrm{op}};\underline{\mathsf{C}}_\Bbbk)$ is unital. Denote by $\mathrm{Rep}(\mathcal{A}^{\mathrm{op}},\underline{\mathsf{C}}_\Bbbk)$ the essential image of $\mathscr{Y}$, the full differential graded subcategory of $\underline{\mathsf{A}^{\mathrm{u}}_\infty}(\mathcal{A}^{\mathrm{op}};\underline{\mathsf{C}}_\Bbbk)$ whose objects are representable A_∞-functors $(X)\mathscr{Y} = H^X$. The composition of g with the A_∞-equivalence $\mathscr{Y}:\mathcal{A}\to\mathrm{Rep}(\mathcal{A}^{\mathrm{op}},\underline{\mathsf{C}}_\Bbbk)$ is unital. Denote by $\Psi:\mathrm{Rep}(\mathcal{A}^{\mathrm{op}},\underline{\mathsf{C}}_\Bbbk)\to\mathcal{A}$ a quasi-inverse to $\mathscr{Y}$. We find that g is isomorphic to a unital A_∞-functor $g\cdot\mathscr{Y}\cdot\Psi:\mathcal{C}\to\mathcal{A}$. Thus, g is unital itself by [Lyu03, (8.2.4)]. □

Proposition 5.23. *Let $\mathcal{A},\mathcal{C}$ be A_∞-categories, and suppose $\mathcal{A}$ is unital. Let $\mathcal{P}$ be an $\mathcal{A}$-$\mathcal{C}$-bimodule, $\phi^{\mathcal{P}}:\mathcal{A}^{\mathrm{op}},\mathcal{C}\to\underline{\mathsf{C}}_\Bbbk$ the corresponding A_∞-functor. The $\mathcal{A}$-$\mathcal{C}$-bimodule $\mathcal{P}$ is isomorphic to $\mathcal{A}^g$ for some A_∞-functor $g:\mathcal{C}\to\mathcal{A}$ if and only if for each object $Y\in\mathrm{Ob}\,\mathcal{C}$ the A_∞-functor $\phi^{\mathcal{P}}|_1^Y:\mathcal{A}^{\mathrm{op}}\to\underline{\mathsf{C}}_\Bbbk$ is representable.*

Proof. The "only if" part is obvious. For the proof of "if", consider the A_∞-functor $f = (\varphi^{\mathsf{A}_\infty})^{-1}(\phi^{\mathcal{P}}):\mathcal{C}\to\underline{\mathsf{A}_\infty}(\mathcal{A}^{\mathrm{op}};\underline{\mathsf{C}}_\Bbbk)$. It acts on objects by $Y\mapsto\phi^{\mathcal{P}}|_1^Y$, $Y\in\mathrm{Ob}\,\mathcal{C}$, therefore it takes values in the A_∞-subcategory $\mathrm{Rep}(\mathcal{A}^{\mathrm{op}},\underline{\mathsf{C}}_\Bbbk)$ of representable A_∞-functors. Denote by $\Psi:\mathrm{Rep}(\mathcal{A}^{\mathrm{op}};\underline{\mathsf{C}}_\Bbbk)\to\mathcal{A}$ a quasi-inverse to $\mathscr{Y}$. Let

g denote the A_∞-functor $f \cdot \Psi : \mathcal{C} \to \mathcal{A}$. Then the composite $g \cdot \mathscr{Y} = f \cdot \Psi \cdot \mathscr{Y} : \mathcal{C} \to \underline{A_\infty}(\mathcal{A}^{\mathrm{op}}; \underline{\mathsf{C}_{\Bbbk}})$ is isomorphic to f, therefore the A_∞-functor

$$\varphi^{A_\infty}(g \cdot \mathscr{Y}) = \left[\mathcal{A}^{\mathrm{op}}, \mathcal{C} \xrightarrow{1, g \cdot \mathscr{Y}} \mathcal{A}^{\mathrm{op}}, \underline{A_\infty}(\mathcal{A}^{\mathrm{op}}; \underline{\mathsf{C}_{\Bbbk}}) \xrightarrow{\mathrm{ev}^{A_\infty}} \underline{\mathsf{C}_{\Bbbk}} \right],$$

corresponding to the bimodule $\mathcal{A}^g$, is isomorphic to $\varphi^{A_\infty}(f) = \phi^{\mathcal{P}}$. Thus, $\mathcal{A}^g$ is isomorphic to $\mathcal{P}$. $\square$

Lemma 5.24. *If $\mathcal{A}$-$\mathcal{C}$-bimodule $\mathcal{P}$ is unital, then the dual $\mathcal{C}$-$\mathcal{A}$-bimodule $\mathcal{P}^*$ is unital as well.*

Proof. The A_∞-functor $\phi^{\mathcal{P}^*}$ is the composite of two A_∞-functors, $(\phi^{\mathcal{P}})^{\mathrm{op}} : \mathcal{A}, \mathcal{C}^{\mathrm{op}} \to \underline{\mathsf{C}_{\Bbbk}}^{\mathrm{op}}$ and $D = H^{\Bbbk} : \underline{\mathsf{C}_{\Bbbk}}^{\mathrm{op}} \to \underline{\mathsf{C}_{\Bbbk}}$. The latter is unital by Remark 5.19. The former is unital if and only if $\phi^{\mathcal{P}}$ is unital. $\square$

6. Serre A_∞-functors

Here we present Serre A_∞-functors as an application of A_∞-bimodules.

Definition 6.1 (cf. Soibelman [Soi04], Kontsevich and Soibelman, sequel to [KS06]). A *right Serre A_∞-functor $S : \mathcal{A} \to \mathcal{A}$* in a unital A_∞-category $\mathcal{A}$ is an A_∞-functor for which the $\mathcal{A}$-bimodules $\mathcal{A}^S = \left[\mathcal{A}^{\mathrm{op}}, \mathcal{A} \xrightarrow{1, S} \mathcal{A}^{\mathrm{op}}, \mathcal{A} \xrightarrow{\mathrm{Hom}_{\mathcal{A}}} \underline{\mathsf{C}_{\Bbbk}} \right]$ and $\mathcal{A}^*$ are isomorphic. If, moreover, S is an A_∞-equivalence, it is called a *Serre A_∞-functor*.

By Lemma 5.24 and Proposition 5.22, if a right Serre A_∞-functor exists, then it is unital and unique up to an isomorphism.

Proposition 6.2. *If $S : \mathcal{A} \to \mathcal{A}$ is a (right) Serre A_∞-functor, then $\Bbbk S : \Bbbk \mathcal{A} \to \Bbbk \mathcal{A}$ is a (right) Serre $\mathcal{K}$-functor.*

Proof. Let $p : \mathcal{A}^S \to \mathcal{A}^*$ be an isomorphism. More precisely, p is an isomorphism

$$(1, S, \mathrm{Hom}_{\mathcal{A}})\mu_{\mathrm{id}:2 \to 2}^{A_\infty^u} \to (\mathrm{Hom}_{\mathcal{A}^{\mathrm{op}}}^{\mathrm{op}}, D)\mu_{2 \to 1}^{A_\infty^u} : \mathcal{A}^{\mathrm{op}}, \mathcal{A} \to \underline{\mathsf{C}_{\Bbbk}}.$$

We visualize this by the following diagram:

$$
\begin{array}{ccc}
\mathcal{A}^{\mathrm{op}}, \mathcal{A} & \xrightarrow{1, S} & \mathcal{A}^{\mathrm{op}}, \mathcal{A} \\
{\scriptstyle \mathrm{Hom}_{\mathcal{A}^{\mathrm{op}}}^{\mathrm{op}}} \downarrow & \overset{p}{\diagup} & \downarrow {\scriptstyle \mathrm{Hom}_{\mathcal{A}}} \\
\underline{\mathsf{C}_{\Bbbk}}^{\mathrm{op}} & \xrightarrow{D} & \underline{\mathsf{C}_{\Bbbk}}
\end{array}
$$

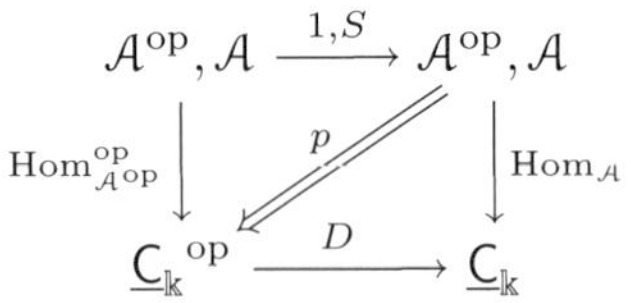

Applying the $\Bbbk$-$\mathcal{C}at$-multifunctor $\Bbbk$, and using Lemma 3.2, Proposition 5.11, and results of Section 5.12, we get a similar diagram in $\mathcal{K}$-$\mathcal{C}at$:

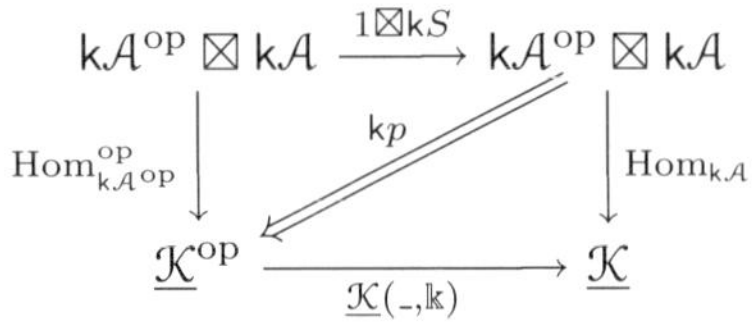

Since $\Bbbk p$ is an isomorphism, it follows that $\Bbbk S$ is a right Serre $\mathcal{K}$-functor.

The A_∞-functor S is an equivalence if and only if $\mathsf{k}S$ is a $\mathcal{K}$-equivalence. $\square$

When $\mathcal{A}$ is an A_∞-algebra and S is its identity endomorphism, the natural transformation $p : \mathcal{A} \to \mathcal{A}^*$ identifies with an ∞-inner-product on $\mathcal{A}$, as defined by Tradler [Tra01, Definition 5.3].

Corollary 6.3. *Let $\mathcal{A}$ be a unital A_∞-category. Then $\mathcal{A}$ admits a (right) Serre A_∞-functor if and only if $\mathsf{k}\mathcal{A}$ admits a (right) Serre $\mathcal{K}$-functor.*

Proof. The "only if" part is proven above. Suppose $\mathsf{k}\mathcal{A}$ admits a Serre $\mathcal{K}$-functor. By Proposition 2.6 this implies representability of the $\mathcal{K}$-functor

$$\mathrm{Hom}_{\mathsf{k}\mathcal{A}}(Y,\_)^{\mathrm{op}} \cdot \underline{\mathcal{K}}(\_,\Bbbk) = \mathsf{k}[\mathrm{Hom}_{\mathcal{A}}(Y,\_)^{\mathrm{op}} \cdot D] : \mathsf{k}\mathcal{A}^{\mathrm{op}} \to \underline{\mathcal{K}} = \mathsf{k}\underline{\mathsf{C}}_{\Bbbk},$$

for each object $Y \in \mathrm{Ob}\,\mathcal{A}$. By Corollary A.6 the A_∞-functor

$$\mathrm{Hom}_{\mathcal{A}}(Y,\_)^{\mathrm{op}} \cdot D = (\mathrm{Hom}^{\mathrm{op}}_{\mathcal{A}^{\mathrm{op}}}, D)\mu^{\mathsf{A}_\infty}_{\mathbf{2}\to\mathbf{1}}\big|^{Y}_{1} : \mathcal{A}^{\mathrm{op}} \to \underline{\mathsf{C}}_{\Bbbk}$$

is representable for each $Y \in \mathrm{Ob}\,\mathcal{A}$. By Proposition 5.23 the bimodule $\mathcal{A}^*$ corresponding to the A_∞-functor $(\mathrm{Hom}^{\mathrm{op}}_{\mathcal{A}^{\mathrm{op}}}, D)\mu^{\mathsf{A}_\infty}_{\mathbf{2}\to\mathbf{1}}$ is isomorphic to $\mathcal{A}^S$ for some A_∞-functor $S : \mathcal{A} \to \mathcal{A}$. $\square$

Corollary 6.4. *Suppose $\mathcal{A}$ is a Hom-reflexive A_∞-category,* i.e., *the complex $\mathcal{A}(X,Y)$ is reflexive in $\mathcal{K}$ for each pair of objects $X,Y \in \mathrm{Ob}\,\mathcal{A}$. If $S : \mathcal{A} \to \mathcal{A}$ is a right Serre A_∞-functor, then S is homotopy fully faithful.*

Proof. The $\mathcal{K}$-functor $\mathsf{k}S$ is fully faithful by Proposition 2.5. $\square$

In particular, the above corollary applies if $\Bbbk$ is a field and all homology spaces $H^n(\mathcal{A}(X,Y))$ are finite dimensional. If $\mathcal{A}$ is closed under shifts, the latter condition is equivalent to requiring that $H^0(\mathcal{A}(X,Y))$ be finite dimensional for each pair $X,Y \in \mathrm{Ob}\,\mathcal{A}$. Indeed, $H^n(\mathcal{A}(X,Y)) = H^n(\mathsf{k}\mathcal{A}(X,Y)) = H^0(\mathsf{k}\mathcal{A}(X,Y)[n]) = H^0((\mathsf{k}\mathcal{A})^{[]}((X,0),(Y,n)))$. The $\mathcal{K}$-category $\mathsf{k}\mathcal{A}$ is closed under shifts by results of Section 3.7, therefore there exists an isomorphism $\alpha : (Y,n) \to (Z,0)$ in $(\mathsf{k}\mathcal{A})^{[]}$, for some $Z \in \mathrm{Ob}\,\mathcal{A}$. It induces an isomorphism $(\mathsf{k}\mathcal{A})^{[]}(1,\alpha) : (\mathsf{k}\mathcal{A})^{[]}((X,0),(Y,n)) \to (\mathsf{k}\mathcal{A})^{[]}((X,0),(Z,0)) = \mathsf{k}\mathcal{A}(X,Z)$ in $\mathcal{K}$, thus an isomorphism in homology

$$H^n(\mathcal{A}(X,Y)) = H^0((\mathsf{k}\mathcal{A})^{[]}((X,0),(Y,n))) \simeq H^0(\mathsf{k}\mathcal{A}(X,Z)) = H^0(\mathcal{A}(X,Z)).$$

The latter space is finite dimensional by assumption.

Theorem 6.5. *Suppose $\Bbbk$ is a field, $\mathcal{A}$ is a unital A_∞-category closed under shifts. Then the following conditions are equivalent:*

(a) *$\mathcal{A}$ admits a (right) Serre A_∞-functor;*
(b) *$\mathsf{k}\mathcal{A}$ admits a (right) Serre $\mathcal{K}$-functor;*
(c) *$H^\bullet\mathcal{A} \overset{\mathrm{def}}{=} H^\bullet(\mathsf{k}\mathcal{A})$ admits a (right) Serre* **gr***-functor;*
(d) *$H^0(\mathcal{A})$ admits (right) Serre $\Bbbk$-linear functor.*

Proof. Equivalence of (a) and (b) is proven in Corollary 6.3. Conditions (b) and (c) are equivalent due to Corollary 2.16, because $H^\bullet : \mathcal{K} \to \mathbf{gr}$ is an equivalence of symmetric monoidal categories. Condition (c) implies (d) for arbitrary $\mathbf{gr}$-category by Corollary 2.18, in particular, for $H^\bullet\mathcal{A}$. Note that $H^\bullet\mathcal{A}$ is closed under shifts by Section 3.7 and the discussion preceding Proposition 2.21. Therefore, (d) implies (c) due to Proposition 2.21. □

An application of this theorem is the following. Let $\Bbbk$ be a field. Drinfeld's construction of quotients of pretriangulated $\mathbf{dg}$-categories [Dri04] allows to find a pretriangulated $\mathbf{dg}$-category $\mathcal{A}$ such that $H^0(\mathcal{A})$ is some familiar derived category (e.g. category $D^b_{\text{coh}}(X)$ of complexes of coherent sheaves on a projective variety X). If a right Serre functor exists for $H^0(\mathcal{A})$, then $\mathcal{A}$ admits a right Serre A_∞-functor S by the above theorem. That is the case of $H^0(\mathcal{A}) \simeq D^b_{\text{coh}}(X)$, where X is a smooth projective variety [BK89, Example 3.2]. Notice that $S : \mathcal{A} \to \mathcal{A}$ does not have to be a $\mathbf{dg}$-functor.

Proposition 6.6. *Let $S : \mathcal{A} \to \mathcal{A}$, $S' : \mathcal{B} \to \mathcal{B}$ be right Serre A_∞-functors. Let $g : \mathcal{B} \to \mathcal{A}$ be an A_∞-equivalence. Then the A_∞-functors $S'g : \mathcal{B} \to \mathcal{A}$ and $gS : \mathcal{B} \to \mathcal{A}$ are isomorphic.*

Proof. Consider the following diagram:

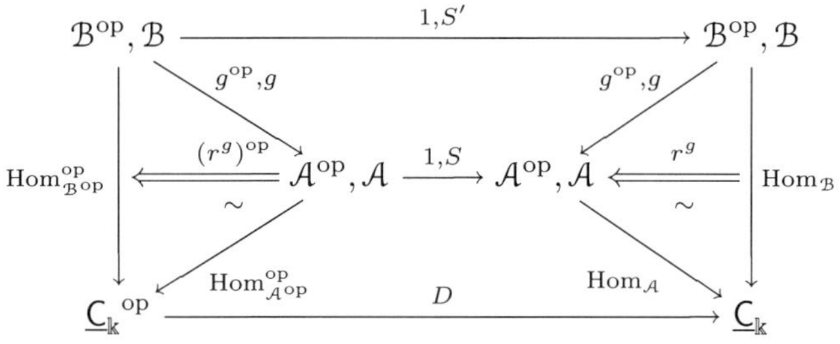

Here the natural A_∞-isomorphism r^g is that constructed in Corollary 5.7. The exterior and the lower trapezoid commute up to natural A_∞-isomorphisms by definition of right Serre functor. It follows that the A_∞-functors $(g^{\text{op}}, S'g)\operatorname{Hom}_\mathcal{A} : \mathcal{B}^{\text{op}}, \mathcal{B} \to \underline{\mathsf{C}}_\Bbbk$ and $(g^{\text{op}}, gS)\operatorname{Hom}_\mathcal{A} : \mathcal{B}^{\text{op}}, \mathcal{B} \to \underline{\mathsf{C}}_\Bbbk$ are isomorphic. Consider the A_∞-functors

$$\mathcal{B} \xrightarrow{S'g} \mathcal{A} \xrightarrow{\mathscr{Y}} \underline{\mathsf{A}}^{\text{u}}_\infty(\mathcal{A}^{\text{op}}; \underline{\mathsf{C}}_\Bbbk) \xrightarrow{\underline{\mathsf{A}}^{\text{u}}_\infty(g^{\text{op}};1)} \underline{\mathsf{A}}^{\text{u}}_\infty(\mathcal{B}^{\text{op}}; \underline{\mathsf{C}}_\Bbbk)$$

and

$$\mathcal{B} \xrightarrow{gS} \mathcal{A} \xrightarrow{\mathscr{Y}} \underline{\mathsf{A}}^{\text{u}}_\infty(\mathcal{A}^{\text{op}}; \underline{\mathsf{C}}_\Bbbk) \xrightarrow{\underline{\mathsf{A}}^{\text{u}}_\infty(g^{\text{op}};1)} \underline{\mathsf{A}}^{\text{u}}_\infty(\mathcal{B}^{\text{op}}; \underline{\mathsf{C}}_\Bbbk)$$

that correspond to $(g^{\text{op}}, S'g)\operatorname{Hom}_\mathcal{A} : \mathcal{B}^{\text{op}}, \mathcal{B} \to \underline{\mathsf{C}}_\Bbbk$ and $(g^{\text{op}}, gS)\operatorname{Hom}_\mathcal{A} : \mathcal{B}^{\text{op}}, \mathcal{B} \to \underline{\mathsf{C}}_\Bbbk$ by closedness. More precisely, the upper line is equal to

$$(\underline{\varphi}^{\mathsf{A}^{\text{u}}_\infty})^{-1}((g^{\text{op}}, S'g)\operatorname{Hom}_\mathcal{A})$$

and the bottom line is equal to

$$(\underline{\varphi}^{\mathsf{A}^{\text{u}}_\infty})^{-1}((g^{\text{op}}, gS)\operatorname{Hom}_\mathcal{A}),$$

where

$$\underline{\varphi}^{\mathsf{A}^{\mathrm{u}}_\infty} : \underline{\mathsf{A}^{\mathrm{u}}_\infty}(\mathcal{B}; \underline{\mathsf{A}^{\mathrm{u}}_\infty}(\mathcal{B}^{\mathrm{op}}; \underline{\mathsf{C}}_{\Bbbk})) \to \underline{\mathsf{A}^{\mathrm{u}}_\infty}(\mathcal{B}^{\mathrm{op}}, \mathcal{B}; \underline{\mathsf{C}}_{\Bbbk})$$

is the natural isomorphism of A_∞-categories coming from the closed structure. It follows that the A_∞-functors $S' \cdot g \cdot \mathscr{Y} \cdot \underline{\mathsf{A}^{\mathrm{u}}_\infty}(g^{\mathrm{op}}; 1)$ and $g \cdot S \cdot \mathscr{Y} \cdot \underline{\mathsf{A}^{\mathrm{u}}_\infty}(g^{\mathrm{op}}; 1)$ are isomorphic. Obviously, the A_∞-functor $\overline{g^{\mathrm{op}}}$ is an equivalence, therefore so is the A_∞-functor $\underline{\mathsf{A}^{\mathrm{u}}_\infty}(g^{\mathrm{op}}; 1)$ since $\underline{\mathsf{A}^{\mathrm{u}}_\infty}(\_; \underline{\mathsf{C}}_{\Bbbk})$ is an A^u_∞-2-functor. Therefore, the A_∞-functors $S' \cdot g \cdot \mathscr{Y}$ and $g \cdot S \cdot \mathscr{Y}$ are isomorphic. However, this implies that the A_∞-functors $(1, S'g) \operatorname{Hom}_{\mathcal{A}} = \underline{\varphi}^{\mathsf{A}^{\mathrm{u}}_\infty}(S' \cdot g \cdot \mathscr{Y})$ and $(1, gS) \operatorname{Hom}_{\mathcal{A}} = \underline{\varphi}^{\mathsf{A}^{\mathrm{u}}_\infty}(g \cdot S \cdot \mathscr{Y})$ are isomorphic as well. These A_∞-functors correspond to $(\mathcal{A}, \mathcal{B})$-$A_\infty$-bimodules $\mathcal{A}^{S'g}$ and $\mathcal{A}^{gS}$ respectively. Proposition 5.22 implies an isomorphism between the A_∞-functors $S'g$ and gS. □

Remark 6.7. Let $\mathcal{A}$ be an A_∞-category, let $S : \mathcal{A} \to \mathcal{A}$ be an A_∞-functor. The $(0,0)$-component of a cycle $p \in \underline{\mathsf{A}_\infty}(\mathcal{A}^{\mathrm{op}}, \mathcal{A}; \underline{\mathsf{C}}_{\Bbbk})(\mathcal{A}^S, \mathcal{A}^*)[1]^{-1}$ determines for all objects X, Y of $\mathcal{A}$ a degree 0 map

$$\Bbbk \simeq T^0 s\mathcal{A}^{\mathrm{op}}(X, X) \otimes T^0 s\mathcal{A}(Y, Y) \xrightarrow{p_{00}} s\underline{\mathsf{C}}_{\Bbbk}(\mathcal{A}(X, YS), \mathcal{A}^*(X, Y)) \\ \xrightarrow{s^{-1}} \underline{\mathsf{C}}_{\Bbbk}(\mathcal{A}(X, YS), \underline{\mathsf{C}}_{\Bbbk}(\mathcal{A}(Y, X), \Bbbk)).$$

The obtained mapping $\mathcal{A}(X, YS) \to \underline{\mathsf{C}}_{\Bbbk}(\mathcal{A}(Y, X), \Bbbk)$ is a chain map, since $p_{00} s^{-1} m_1 = 0$. Its homotopy class gives $\psi_{X,Y}$ from (2.1) when the pair (S, p) is projected to $(\mathsf{k}S, \psi = \mathsf{k}p)$ via the multifunctor k.

6.8. A strict case of a Serre A_∞-functor

Let us consider a particularly simple case of an A_∞-category $\mathcal{A}$ with a right Serre functor $S : \mathcal{A} \to \mathcal{A}$ which is a strict A_∞-functor (only the first component does not vanish) and with the invertible natural A_∞-transformation $p : \mathcal{A}^S \to \mathcal{A}^* : \mathcal{A}^{\mathrm{op}}, \mathcal{A} \to \underline{\mathsf{C}}_{\Bbbk}$ whose only non-vanishing component is $p_{00} : T^0 s\mathcal{A}^{\mathrm{op}}(X, X) \otimes T^0 s\mathcal{A}(Y, Y) \to s\underline{\mathsf{C}}_{\Bbbk}(\mathcal{A}(X, YS), \underline{\mathsf{C}}_{\Bbbk}(\mathcal{A}(Y, X), \Bbbk))$. Invertibility of p, equivalent to invertibility of p_{00}, means that the induced chain maps $r_{00} : \mathcal{A}(X, YS) \to \underline{\mathsf{C}}_{\Bbbk}(\mathcal{A}(Y, X), \Bbbk)$ are homotopy invertible for all objects X, Y of $\mathcal{A}$. General formulae for pB_1 give the components $(pB_1)_{00} = p_{00} b_1$ and

$$(pB_1)_{kn} = ([(1, S) \operatorname{Hom}_{\mathcal{A}}]_{kn} \otimes p_{00}) b_2^{\underline{\mathsf{C}}_{\Bbbk}} + (p_{00} \otimes [\operatorname{Hom}^{\mathrm{op}}_{\mathcal{A}^{\mathrm{op}}} \cdot D]_{kn}) b_2^{\underline{\mathsf{C}}_{\Bbbk}} \tag{6.1}$$

for $k + n > 0$. Since p is natural, $pB_1 = 0$, thus the right-hand side of (6.1) has to vanish. Expanding the first summand we get

$$(-)^k \big[T^k s\mathcal{A}^{\mathrm{op}}(X_0, X_k) \otimes T^n s\mathcal{A}(Y_0, Y_n) \xrightarrow{\mathrm{coev}} \\ \underline{\mathsf{C}}_{\Bbbk}\big(s\mathcal{A}(X_0, Y_0 S), s\mathcal{A}(X_0, Y_0 S) \otimes T^k s\mathcal{A}^{\mathrm{op}}(X_0, X_k) \otimes T^n s\mathcal{A}(Y_0, Y_n)\big) \\ \xrightarrow{\underline{\mathsf{C}}_{\Bbbk}(1, \rho_c(1^{\otimes k+1} \otimes S_1^{\otimes n}) b_{k+1+n} r_{00})} \underline{\mathsf{C}}_{\Bbbk}\big(s\mathcal{A}(X_0, Y_0 S), s\underline{\mathsf{C}}_{\Bbbk}(\mathcal{A}(Y_n, X_k), \Bbbk)\big) \\ \xrightarrow{[-1]s} s\underline{\mathsf{C}}_{\Bbbk}\big(\mathcal{A}(X_0, Y_0 S), \underline{\mathsf{C}}_{\Bbbk}(\mathcal{A}(Y_n, X_k), \Bbbk)\big)\big].$$

Expanding the second summand we obtain

$$-(-)^k\bigl[T^k s\mathcal{A}^{\mathrm{op}}(X_0,X_k)\otimes T^n s\mathcal{A}(Y_0,Y_n)$$
$$\xrightarrow{\mathrm{coev}} \underline{\mathsf{C}}_{\Bbbk}\bigl(s\mathcal{A}(Y_n,X_k), s\mathcal{A}(Y_n,X_k)\otimes T^k s\mathcal{A}^{\mathrm{op}}(X_0,X_k)\otimes T^n s\mathcal{A}(Y_0,Y_n)\bigr)$$
$$\xrightarrow{\underline{\mathsf{C}}_{\Bbbk}(1,(123)_c(1\otimes 1\otimes\omega_c^0))} \underline{\mathsf{C}}_{\Bbbk}\bigl(s\mathcal{A}(Y_n,X_k), T^n s\mathcal{A}(Y_0,Y_n)\otimes s\mathcal{A}(Y_n,X_k)\otimes T^k s\mathcal{A}(X_k,X_0)\bigr)$$
$$\xrightarrow{\underline{\mathsf{C}}_{\Bbbk}(1,b_{n+1+k})} \underline{\mathsf{C}}_{\Bbbk}\bigl(s\mathcal{A}(Y_n,X_k), s\mathcal{A}(Y_0,X_0)\bigr) \xrightarrow{[-1]s} s\underline{\mathsf{C}}_{\Bbbk}\bigl(\mathcal{A}(Y_n,X_k),\mathcal{A}(Y_0,X_0)\bigr)$$
$$\xrightarrow{\mathrm{coev}} \underline{\mathsf{C}}_{\Bbbk}\bigl(s\mathcal{A}(X_0,Y_0S), s\mathcal{A}(X_0,Y_0S)\otimes s\underline{\mathsf{C}}_{\Bbbk}(\mathcal{A}(Y_n,X_k),\mathcal{A}(Y_0,X_0))\bigr) \xrightarrow{\underline{\mathsf{C}}_{\Bbbk}(1,(r_{00}\otimes 1)cb_2)}$$
$$\underline{\mathsf{C}}_{\Bbbk}\bigl(s\mathcal{A}(X_0,Y_0S), s\underline{\mathsf{C}}_{\Bbbk}(\mathcal{A}(Y_n,X_k),\Bbbk)\bigr) \xrightarrow{[-1]s} s\underline{\mathsf{C}}_{\Bbbk}\bigl(\mathcal{A}(X_0,Y_0S),\underline{\mathsf{C}}_{\Bbbk}(\mathcal{A}(Y_n,X_k),\Bbbk)\bigr)\bigr].$$

The sum of the two above expressions must vanish. The obtained equation can be simplified further by closedness of $\mathsf{C}_{\Bbbk}$. The homotopy isomorphism r_{00} induces the pairing

$$q_{00} = \bigl[\mathcal{A}(Y,X)\otimes\mathcal{A}(X,YS) \xrightarrow{1\otimes r_{00}} \mathcal{A}(Y,X)\otimes\underline{\mathsf{C}}_{\Bbbk}(\mathcal{A}(Y,X),\Bbbk) \xrightarrow{\mathrm{ev}} \Bbbk\bigr].$$

Using it we write down the naturality condition for p as follows: for all $k\geqslant 0$, $n\geqslant 0$

$$\bigl[\mathcal{A}(Y_n,X_k)\otimes T^k\mathcal{A}(X_k,X_0)\otimes\mathcal{A}(X_0,Y_0S)\otimes T^n\mathcal{A}(Y_0,Y_n)$$
$$\xrightarrow{(1^{\otimes 3}\otimes(sS_1s^{-1})^{\otimes n})(1\otimes m_{k+1+n})} \mathcal{A}(Y_n,X_k)\otimes\mathcal{A}(X_k,Y_nS) \xrightarrow{q_{00}} \Bbbk\bigr]$$
$$= (-)^{(k+1)(n+1)}\bigl[\mathcal{A}(Y_n,X_k)\otimes T^k\mathcal{A}(X_k,X_0)\otimes\mathcal{A}(X_0,Y_0S)\otimes T^n\mathcal{A}(Y_0,Y_n)$$
$$\xrightarrow{(1234)_c} T^n\mathcal{A}(Y_0,Y_n)\otimes\mathcal{A}(Y_n,X_k)\otimes T^k\mathcal{A}(X_k,X_0)\otimes\mathcal{A}(X_0,Y_0S)$$
$$\xrightarrow{m_{n+1+k}\otimes 1} \mathcal{A}(Y_0,X_0)\otimes\mathcal{A}(X_0,Y_0S) \xrightarrow{q_{00}} \Bbbk\bigr]. \tag{6.2}$$

Let us give a sufficient condition for this equation to hold true.

Proposition 6.9. *Let $\mathcal{A}$ be an A_∞-category, and let $S:\mathcal{A}\to\mathcal{A}$ be a strict A_∞-functor. Suppose given a pairing $q_{00}:\mathcal{A}(Y,X)\otimes\mathcal{A}(X,YS)\to\Bbbk$ for all objects X, Y of $\mathcal{A}$. Assume that for all $X,Y\in\operatorname{Ob}\mathcal{A}$*

(a) *q_{00} is a chain map;*
(b) *the induced chain map*

$$r_{00} = \bigl[\mathcal{A}(X,YS) \xrightarrow{\mathrm{coev}} \underline{\mathsf{C}}_{\Bbbk}(\mathcal{A}(Y,X),\mathcal{A}(Y,X)\otimes\mathcal{A}(X,YS)) \xrightarrow{\underline{\mathsf{C}}_{\Bbbk}(1,q_{00})} \underline{\mathsf{C}}_{\Bbbk}(\mathcal{A}(Y,X),\Bbbk)\bigr]$$

is homotopy invertible;

(c) *the pairing q_{00} is symmetric in a sense similar to diagram (2.11), namely, the following diagram of chain maps commutes:*

$$\begin{array}{ccc}
\mathcal{A}(X,YS)\otimes\mathcal{A}(Y,X) & \xrightarrow{1\otimes sS_1s^{-1}} & \mathcal{A}(X,YS)\otimes\mathcal{A}(YS,XS) \\
{\scriptstyle c}\downarrow & = & \downarrow{\scriptstyle q_{00}} \\
\mathcal{A}(Y,X)\otimes\mathcal{A}(X,YS) & \xrightarrow{\quad q_{00}\quad} & \Bbbk
\end{array} \tag{6.3}$$

(d) *the following equation holds for all $k \geqslant 0$ and all objects $X_0, \dots, X_k, Y$*

$$\big[\mathcal{A}(Y,X_k)\otimes T^k\mathcal{A}(X_k,X_0)\otimes\mathcal{A}(X_0,YS) \xrightarrow{1\otimes m_{k+1}} \mathcal{A}(Y,X_k)\otimes\mathcal{A}(X_k,YS) \xrightarrow{q_{00}} \Bbbk\big]$$
$$= (-)^{k+1}\big[\mathcal{A}(Y,X_k)\otimes T^k\mathcal{A}(X_k,X_0)\otimes\mathcal{A}(X_0,YS)$$
$$\xrightarrow{m_{1+k}\otimes 1} \mathcal{A}(Y,X_0)\otimes\mathcal{A}(X_0,YS) \xrightarrow{q_{00}} \Bbbk\big]. \quad (6.4)$$

Then the natural A_∞-transformation $p : \mathcal{A}^S \to \mathcal{A}^* : \mathcal{A}^{\mathrm{op}},\mathcal{A} \to \underline{\mathsf{C}}_\Bbbk$ with the only non-vanishing component $p_{00} : 1 \mapsto r_{00}$ is invertible and $S : \mathcal{A} \to \mathcal{A}$ is a Serre A_∞-functor.

Notice that (6.4) is precisely the case of (6.2) with $n = 0$. On the other hand, diagram (2.11) written for $\mathcal{K}$-category $\mathcal{C} = \mathsf{k}\mathcal{A}$ and the pairing $\phi = [q_{00}]$ says that (6.3) has to commute only up to homotopy. Thus, condition (c) is sufficient but not necessary.

Proof. We have to prove equation (6.2) for all $k \geqslant 0$, $n \geqslant 0$. The case of $n = 0$ holds by condition (d). Let us proceed by induction on n. Assume that (6.2) holds true for all $k \geqslant 0$, $0 \leqslant n < N$. Let us prove equation (6.2) for $k \geqslant 0$, $n = N$. We have

$$\begin{aligned}
&(-)^{(k+1)(n+1)}(13524)_c\cdot(m_{n+1+k}\otimes 1)\cdot q_{00}\\
&\overset{(d)}{=} (-)^{(k+1)(n+1)+k+n+1}(13524)_c\cdot(1\otimes m_{n+k+1})\cdot q_{00}\\
&= (-)^{kn}(12345)_c\cdot(m_{n+k+1}\otimes 1)\cdot c\cdot q_{00}\\
&\overset{(c)}{=} (-)^{kn}(12345)_c\cdot(m_{n+k+1}\otimes sS_1s^{-1})\cdot q_{00}\\
&= (-)^{(k+2)n}(1^{\otimes 3}\otimes sS_1s^{-1}\otimes 1)\cdot(12345)_c\cdot(m_{n+k+1}\otimes 1)\cdot q_{00}\\
&\underset{\text{for } k+1,n-1}{\overset{\text{by } (6.2)}{=}} (1^{\otimes 3}\otimes sS_1s^{-1}\otimes 1)\cdot(1^{\otimes 4}\otimes T^{n-1}(sS_1s^{-1}))\cdot(1\otimes m_{k+1+n})\cdot q_{00} :\\
&\mathcal{A}(Y_n,X_k)\otimes T^k\mathcal{A}(X_k,X_0)\otimes\mathcal{A}(X_0,Y_0S)\otimes\mathcal{A}(Y_0,Y_1)\otimes T^{n-1}\mathcal{A}(Y_1,Y_n)\to\Bbbk.
\end{aligned}$$

This is just equation (6.2) for k, n. □

Some authors like to consider a special case of the above in which $S = [d]$ is the shift functor (when it makes sense), the paring q_{00} is symmetric and cyclically symmetric with respect to n-ary compositions, cf. [Cos04, Section 6.2]. Then $\mathcal{A}$ is called a Calabi–Yau A_∞-category. General Serre A_∞-functors cover wider scope, although they require more data to work with.

Appendix A. The Yoneda Lemma

A version of the classical Yoneda Lemma is presented in Mac Lane's book [Mac88, Section III.2] as the following statement. For any category $\mathcal{C}$ there is an isomorphism of functors

$$\mathrm{ev}^{\mathcal{C}at} \simeq \bigl[\mathcal{C} \times \underline{\mathcal{C}at}(\mathcal{C}, \mathcal{S}et) \xrightarrow{\mathscr{Y}^{\mathrm{op}}\times 1} \underline{\mathcal{C}at}(\mathcal{C}, \mathcal{S}et)^{\mathrm{op}} \times \underline{\mathcal{C}at}(\mathcal{C}, \mathcal{S}et) \xrightarrow{\mathrm{Hom}_{\underline{\mathcal{C}at}(\mathcal{C},\mathcal{S}et)}} \mathcal{S}et\bigr],$$

where $\mathscr{Y} : \mathcal{C}^{\mathrm{op}} \to \underline{\mathcal{C}at}(\mathcal{C}, \mathcal{S}et)$, $X \mapsto \mathcal{C}(X, \_)$, is the Yoneda embedding. Here we generalize this to A_∞-setting.

Theorem A.1 (The Yoneda Lemma). *For any A_∞-category $\mathcal{A}$ there is a natural A_∞-transformation*

$$\Omega : \mathrm{ev}^{\mathsf{A}_\infty} \to \bigl[\mathcal{A}, \underline{\mathsf{A}_\infty}(\mathcal{A}; \underline{\mathsf{C}}_\Bbbk) \xrightarrow{\mathscr{Y}^{\mathrm{op}},1} \underline{\mathsf{A}_\infty}(\mathcal{A}; \underline{\mathsf{C}}_\Bbbk)^{\mathrm{op}}, \underline{\mathsf{A}_\infty}(\mathcal{A}; \underline{\mathsf{C}}_\Bbbk) \xrightarrow{\mathrm{Hom}_{\underline{\mathsf{A}_\infty}(\mathcal{A};\underline{\mathsf{C}}_\Bbbk)}} \underline{\mathsf{C}}_\Bbbk\bigr].$$

If the A_∞-category $\mathcal{A}$ is unital, Ω restricts to an invertible *natural A_∞-transformation*

$$\begin{array}{ccc} \mathcal{A}, \underline{\mathsf{A}^{\mathrm{u}}_\infty}(\mathcal{A}; \underline{\mathsf{C}}_\Bbbk) & \xrightarrow{\qquad\mathrm{ev}^{\mathsf{A}^{\mathrm{u}}_\infty}\qquad} & \underline{\mathsf{C}}_\Bbbk \\ & \Omega \Downarrow & \\ {\scriptstyle \mathscr{Y}^{\mathrm{op}},1} \searrow & \underline{\mathsf{A}^{\mathrm{u}}_\infty}(\mathcal{A}; \underline{\mathsf{C}}_\Bbbk)^{\mathrm{op}}, \underline{\mathsf{A}^{\mathrm{u}}_\infty}(\mathcal{A}; \underline{\mathsf{C}}_\Bbbk) & \nearrow {\scriptstyle \mathrm{Hom}_{\underline{\mathsf{A}^{\mathrm{u}}_\infty}(\mathcal{A};\underline{\mathsf{C}}_\Bbbk)}} \end{array}$$

Previously published A_∞-versions of Yoneda Lemma contented with the statement that for unital A_∞-category $\mathcal{A}$, the Yoneda A_∞-functor $\mathscr{Y} : \mathcal{A}^{\mathrm{op}} \to \underline{\mathsf{A}^{\mathrm{u}}_\infty}(\mathcal{A}; \underline{\mathsf{C}}_\Bbbk)$ is homotopy full and faithful [Fuk02, Theorem 9.1], [LM04, Theorem A.11]. A more general form of the Yoneda Lemma is considered by Seidel [Sei06, Lemma 2.12] over a ground field $\Bbbk$. We shall see that these are corollaries of the above theorem.

Proof. First of all we describe the A_∞-transformation Ω for an arbitrary A_∞-category $\mathcal{A}$. The discussion of Section 5.6 applied to the A_∞-functor

$$\psi = \bigl[\mathcal{A}, \underline{\mathsf{A}_\infty}(\mathcal{A}; \underline{\mathsf{C}}_\Bbbk) \xrightarrow{\mathscr{Y}^{\mathrm{op}},1} \underline{\mathsf{A}_\infty}(\mathcal{A}; \underline{\mathsf{C}}_\Bbbk)^{\mathrm{op}}, \underline{\mathsf{A}_\infty}(\mathcal{A}; \underline{\mathsf{C}}_\Bbbk) \xrightarrow{\mathrm{Hom}_{\underline{\mathsf{A}_\infty}(\mathcal{A};\underline{\mathsf{C}}_\Bbbk)}} \underline{\mathsf{C}}_\Bbbk\bigr]$$

presents the corresponding $\mathcal{A}^{\mathrm{op}}$-$\underline{\mathsf{A}_\infty}(\mathcal{A}; \underline{\mathsf{C}}_\Bbbk)$-bimodule $\mathcal{Q} = {}_{\mathscr{Y}}\underline{\mathsf{A}_\infty}(\mathcal{A}; \underline{\mathsf{C}}_\Bbbk)_1$ via the regular A_∞-bimodule. Thus,

$$(\mathcal{Q}(X, f), sb^{\mathcal{Q}}_{00}s^{-1}) = (\underline{\mathsf{A}_\infty}(\mathcal{A}; \underline{\mathsf{C}}_\Bbbk)(H^X, f), sB_1 s^{-1}).$$

According to (5.12) $H^X = \mathcal{A}^{\mathrm{op}}(\_, X) = \mathcal{A}(X, \_)$ has the components

$$\begin{aligned} H^X_k = (\mathrm{Hom}_{\mathcal{A}^{\mathrm{op}}})_{k0} = \bigl[T^k s\mathcal{A}(Y, Z) &\xrightarrow{\mathrm{coev}^{\underline{\mathsf{C}}_\Bbbk}} \underline{\mathsf{C}}_\Bbbk(s\mathcal{A}(X, Y), s\mathcal{A}(X, Y) \otimes T^k s\mathcal{A}(Y, Z)) \\ &\xrightarrow{\underline{\mathsf{C}}_\Bbbk(1, b^{\mathcal{A}}_{1+k})} \underline{\mathsf{C}}_\Bbbk(s\mathcal{A}(X, Y), s\mathcal{A}(X, Z)) \xrightarrow{[-1]s} s\underline{\mathsf{C}}_\Bbbk(\mathcal{A}(X, Y), \mathcal{A}(X, Z))\bigr]. \quad \text{(A.1)} \end{aligned}$$

We have $b^{\mathcal{Q}}_{00} = B_1$ and, moreover, by (5.14)

$$\check{b}^{\mathcal{Q}} = \bigl[Ts\mathcal{A}^{\mathrm{op}}(Y,X) \otimes s\mathcal{Q}(X,f) \otimes Ts\underline{\mathsf{A}_\infty}(\mathcal{A};\underline{\mathsf{C}}_\Bbbk)(f,g)$$
$$\xrightarrow{\mathscr{Y}\otimes 1\otimes 1} Ts\underline{\mathsf{A}_\infty}(\mathcal{A};\underline{\mathsf{C}}_\Bbbk)(H^Y,H^X) \otimes s\underline{\mathsf{A}_\infty}(\mathcal{A};\underline{\mathsf{C}}_\Bbbk)(H^X,f) \otimes Ts\underline{\mathsf{A}_\infty}(\mathcal{A};\underline{\mathsf{C}}_\Bbbk)(f,g)$$
$$\xrightarrow{\check{B}} s\underline{\mathsf{A}_\infty}(\mathcal{A};\underline{\mathsf{C}}_\Bbbk)(H^Y,g) = s\mathcal{Q}(Y,g)\bigr].$$

Since $\underline{\mathsf{A}_\infty}(\mathcal{A};\underline{\mathsf{C}}_\Bbbk)$ is a differential graded category, $B_p = 0$ for $p > 2$. Therefore, $b^{\mathcal{Q}}_{kn} = 0$ if $n > 1$, and $b^{\mathcal{Q}}_{k1} = 0$ if $k > 0$. The non-trivial components are (for $k > 0$)

$$b^{\mathcal{Q}}_{k0} = \bigl[T^k s\mathcal{A}^{\mathrm{op}}(Y,X) \otimes s\mathcal{Q}(X,f) \otimes T^0 s\underline{\mathsf{A}_\infty}(\mathcal{A};\underline{\mathsf{C}}_\Bbbk)(f,f) \xrightarrow{\mathscr{Y}_k\otimes 1\otimes 1}$$
$$s\underline{\mathsf{A}_\infty}(\mathcal{A};\underline{\mathsf{C}}_\Bbbk)(H^Y,H^X)\otimes s\underline{\mathsf{A}_\infty}(\mathcal{A};\underline{\mathsf{C}}_\Bbbk)(H^X,f) \xrightarrow{B_2} s\underline{\mathsf{A}_\infty}(\mathcal{A};\underline{\mathsf{C}}_\Bbbk)(H^Y,f) = s\mathcal{Q}(Y,f)\bigr],$$
$$b^{\mathcal{Q}}_{01} = \bigl[T^0 s\mathcal{A}^{\mathrm{op}}(X,X) \otimes s\mathcal{Q}(X,f) \otimes s\underline{\mathsf{A}_\infty}(\mathcal{A};\underline{\mathsf{C}}_\Bbbk)(f,g)$$
$$\xrightarrow{B_2} s\underline{\mathsf{A}_\infty}(\mathcal{A};\underline{\mathsf{C}}_\Bbbk)(H^X,g) = s\mathcal{Q}(X,g)\bigr]. \quad \text{(A.2)}$$

Denote by $\mathcal{E}$ the $\mathcal{A}^{\mathrm{op}}$-$\underline{\mathsf{A}_\infty}(\mathcal{A};\underline{\mathsf{C}}_\Bbbk)$-bimodule corresponding to the A_∞-functor $\mathrm{ev}^{\mathsf{A}_\infty} : \mathcal{A}, \underline{\mathsf{A}_\infty}(\mathcal{A};\underline{\mathsf{C}}_\Bbbk) \to \underline{\mathsf{C}}_\Bbbk$. For any object X of $\mathcal{A}$ and any A_∞-functor $f : \mathcal{A} \to \underline{\mathsf{C}}_\Bbbk$ the complex $(\mathcal{E}(X,f), sb^{\mathcal{E}}_{00}s^{-1})$ is (Xf,d). According to (5.6)

$$\check{b}^{\mathcal{E}}_{+} = \bigl[Ts\mathcal{A}^{\mathrm{op}}(Y,X) \otimes s\mathcal{E}(X,f) \otimes Ts\underline{\mathsf{A}_\infty}(\mathcal{A};\underline{\mathsf{C}}_\Bbbk)(f,g)$$
$$\xrightarrow{c\otimes 1} s\mathcal{E}(X,f) \otimes Ts\mathcal{A}^{\mathrm{op}}(Y,X) \otimes Ts\underline{\mathsf{A}_\infty}(\mathcal{A};\underline{\mathsf{C}}_\Bbbk)(f,g) \xrightarrow{1\otimes\gamma\otimes 1}$$
$$s\mathcal{E}(X,f) \otimes Ts\mathcal{A}(X,Y) \otimes Ts\underline{\mathsf{A}_\infty}(\mathcal{A};\underline{\mathsf{C}}_\Bbbk)(f,g) \xrightarrow{1\otimes\check{\mathrm{ev}}^{\mathsf{A}_\infty}} s\mathcal{E}(X,f) \otimes s\underline{\mathsf{C}}_\Bbbk(Xf,Yg)$$
$$\xrightarrow{1\otimes s^{-1}[1]} Xf[1] \otimes \underline{\mathsf{C}}_\Bbbk(Xf[1],Yg[1]) \xrightarrow{\mathrm{ev}^{\mathsf{C}_\Bbbk}} Yg[1] = s\mathcal{E}(Y,g)\bigr].$$

Explicit formula (0.2) for $\mathrm{ev}^{\mathsf{A}_\infty}$ shows that $b^{\mathcal{E}}_{kn} = 0$ if $n > 1$. The remaining components are described as

$$b^{\mathcal{E}}_{k0} = \bigl[T^k s\mathcal{A}^{\mathrm{op}}(Y,X) \otimes s\mathcal{E}(X,f) \otimes T^0 s\underline{\mathsf{A}_\infty}(\mathcal{A};\underline{\mathsf{C}}_\Bbbk)(f,f)$$
$$\xrightarrow{c\otimes 1} s\mathcal{E}(X,f) \otimes T^k s\mathcal{A}^{\mathrm{op}}(Y,X) \xrightarrow{1\otimes\gamma} s\mathcal{E}(X,f) \otimes T^k s\mathcal{A}(X,Y) \xrightarrow{1\otimes f_k}$$
$$Xf[1]\otimes s\underline{\mathsf{C}}_\Bbbk(Xf,Yf) \xrightarrow{1\otimes s^{-1}[1]} Xf[1]\otimes\underline{\mathsf{C}}_\Bbbk(Xf[1],Yf[1]) \xrightarrow{\mathrm{ev}^{\mathsf{C}_\Bbbk}} Yf[1] = s\mathcal{E}(Y,f)\bigr]$$

for $k > 0$, and if $k \geqslant 0$ there is

$$b^{\mathcal{E}}_{k1} = \bigl[T^k s\mathcal{A}^{\mathrm{op}}(Y,X) \otimes s\mathcal{E}(X,f) \otimes s\underline{\mathsf{A}_\infty}(\mathcal{A};\underline{\mathsf{C}}_\Bbbk)(f,g)$$
$$\xrightarrow{c\otimes 1} s\mathcal{E}(X,f) \otimes T^k s\mathcal{A}^{\mathrm{op}}(Y,X) \otimes s\underline{\mathsf{A}_\infty}(\mathcal{A};\underline{\mathsf{C}}_\Bbbk)(f,g)$$
$$\xrightarrow{1\otimes\gamma\otimes 1} s\mathcal{E}(X,f) \otimes T^k s\mathcal{A}(X,Y) \otimes s\underline{\mathsf{A}_\infty}(\mathcal{A};\underline{\mathsf{C}}_\Bbbk)(f,g)$$
$$\xrightarrow{1\otimes 1\otimes \mathrm{pr}_k} s\mathcal{E}(X,f) \otimes T^k s\mathcal{A}(X,Y) \otimes \underline{\mathsf{C}}_\Bbbk(T^k s\mathcal{A}(X,Y), s\underline{\mathsf{C}}_\Bbbk(Xf,Yg)) \xrightarrow{1\otimes\mathrm{ev}^{\underline{\mathsf{C}}_\Bbbk}}$$
$$Xf[1]\otimes s\underline{\mathsf{C}}_\Bbbk(Xf,Yg) \xrightarrow{1\otimes s^{-1}[1]} Xf[1]\otimes\underline{\mathsf{C}}_\Bbbk(Xf[1],Yg[1]) \xrightarrow{\mathrm{ev}^{\mathsf{C}_\Bbbk}} Yg[1] = s\mathcal{E}(Y,g)\bigr].$$

The A_∞-transformation Ω in question is constructed via a homomorphism

$$\mho = (\Omega s^{-1})\Phi : Ts\mathcal{A}^{\mathrm{op}} \otimes s\mathcal{E} \otimes T s\underline{\mathsf{A}_\infty}(\mathcal{A};\underline{\mathsf{C}}_\Bbbk) \to Ts\mathcal{A}^{\mathrm{op}} \otimes s\mathcal{Q} \otimes T s\underline{\mathsf{A}_\infty}(\mathcal{A};\underline{\mathsf{C}}_\Bbbk)$$

of $Ts\mathcal{A}^{\mathrm{op}}$-$Ts\underline{\mathsf{A}_\infty}(\mathcal{A};\underline{\mathsf{C}}_\Bbbk)$-bicomodules thanks to Proposition 5.3. Its matrix coefficients are recovered from the components via formula (5.3) as

$$\mho_{kl;mn} = \sum_{\substack{m+p=k\\ q+n=l}} (1^{\otimes m} \otimes \mho_{pq} \otimes 1^{\otimes n}) :$$
$$T^k s\mathcal{A}^{\mathrm{op}} \otimes s\mathcal{E} \otimes T^l s\underline{\mathsf{A}_\infty}(\mathcal{A};\underline{\mathsf{C}}_\Bbbk) \to T^m s\mathcal{A}^{\mathrm{op}} \otimes s\mathcal{Q} \otimes T^n s\underline{\mathsf{A}_\infty}(\mathcal{A};\underline{\mathsf{C}}_\Bbbk).$$

The composition of the morphism

$$\mho_{pq} : T^p s\mathcal{A}^{\mathrm{op}}(X_0,X_p) \otimes X_p f_0[1] \otimes T^q s\underline{\mathsf{A}_\infty}(\mathcal{A};\underline{\mathsf{C}}_\Bbbk)(f_0,f_q) \to s\underline{\mathsf{A}_\infty}(\mathcal{A};\underline{\mathsf{C}}_\Bbbk)(H^{X_0},f_q)$$

with the projection

$$\mathrm{pr}_n : s\underline{\mathsf{A}_\infty}(\mathcal{A};\underline{\mathsf{C}}_\Bbbk)(H^{X_0},f_q) \to \underline{\mathsf{C}}_\Bbbk(T^n s\mathcal{A}(Z_0,Z_n), s\underline{\mathsf{C}}_\Bbbk(\mathcal{A}(X_0,Z_0),Z_n f_q)) \tag{A.3}$$

is given by the composite

$$\mho_{pq;n} \overset{\mathrm{def}}{=} \mho_{pq}\cdot \mathrm{pr}_n = (-)^{p+1}\big[T^p s\mathcal{A}^{\mathrm{op}}(X_0,X_p) \otimes X_p f_0[1] \otimes T^q s\underline{\mathsf{A}_\infty}(\mathcal{A};\underline{\mathsf{C}}_\Bbbk)(f_0,f_q)$$
$$\xrightarrow{\mathrm{coev}^{\underline{\mathsf{C}}_\Bbbk}} \underline{\mathsf{C}}_\Bbbk(s\mathcal{A}(X_0,Z_0)\otimes T^n s\mathcal{A}(Z_0,Z_n),$$
$$s\mathcal{A}(X_0,Z_0)\otimes T^n s\mathcal{A}(Z_0,Z_n)\otimes T^p s\mathcal{A}^{\mathrm{op}}(X_0,X_p)\otimes X_p f_0[1]\otimes T^q s\underline{\mathsf{A}_\infty}(\mathcal{A};\underline{\mathsf{C}}_\Bbbk)(f_0,f_q))$$
$$\xrightarrow{\underline{\mathsf{C}}_\Bbbk(1,\mathrm{perm})} \underline{\mathsf{C}}_\Bbbk(s\mathcal{A}(X_0,Z_0)\otimes T^n s\mathcal{A}(Z_0,Z_n),$$
$$X_p f_0[1]\otimes T^p s\mathcal{A}(X_p,X_0)\otimes s\mathcal{A}(X_0,Z_0)\otimes T^n s\mathcal{A}(Z_0,Z_n)\otimes T^q s\underline{\mathsf{A}_\infty}(\mathcal{A};\underline{\mathsf{C}}_\Bbbk)(f_0,f_q))$$
$$\xrightarrow{\underline{\mathsf{C}}_\Bbbk(1,1\otimes \mathrm{ev}^{\mathsf{A}_\infty}_{p+1+n,q})} \underline{\mathsf{C}}_\Bbbk(s\mathcal{A}(X_0,Z_0)\otimes T^n s\mathcal{A}(Z_0,Z_n), X_p f_0[1]\otimes s\underline{\mathsf{C}}_\Bbbk(X_p f_0, Z_n f_q))$$
$$\xrightarrow{\underline{\mathsf{C}}_\Bbbk(1,1\otimes s^{-1}[1])} \underline{\mathsf{C}}_\Bbbk(s\mathcal{A}(X_0,Z_0)\otimes T^n s\mathcal{A}(Z_0,Z_n), X_p f_0[1]\otimes \underline{\mathsf{C}}_\Bbbk(X_p f_0[1], Z_n f_q[1]))$$
$$\xrightarrow{\underline{\mathsf{C}}_\Bbbk(1,\mathrm{ev}^{\underline{\mathsf{C}}_\Bbbk})} \underline{\mathsf{C}}_\Bbbk(s\mathcal{A}(X_0,Z_0)\otimes T^n s\mathcal{A}(Z_0,Z_n), Z_n f_q[1])$$
$$\xrightarrow{(\varphi^{\underline{\mathsf{C}}_\Bbbk})^{-1}} \underline{\mathsf{C}}_\Bbbk(T^n s\mathcal{A}(Z_0,Z_n), \underline{\mathsf{C}}_\Bbbk(s\mathcal{A}(X_0,Z_0), Z_n f_q[1]))$$
$$\xrightarrow{\underline{\mathsf{C}}_\Bbbk(1,[-1]s)} \underline{\mathsf{C}}_\Bbbk(T^n s\mathcal{A}(Z_0,Z_n), s\underline{\mathsf{C}}_\Bbbk(\mathcal{A}(X_0,Z_0), Z_n f_q))\big]. \tag{A.4}$$

Thus, an element $x_1\otimes\cdots\otimes x_p\otimes y\otimes r_1\otimes\cdots\otimes r_q \in T^p s\mathcal{A}^{\mathrm{op}}(X_0,X_p)\otimes X_p f_0[1]\otimes T^q s\underline{\mathsf{A}_\infty}(\mathcal{A};\underline{\mathsf{C}}_\Bbbk)(f_0,f_q)$ is mapped to an A_∞-transformation $(x_1\otimes\cdots\otimes x_p\otimes y\otimes r_1\otimes\cdots\otimes r_q)\mho_{pq} \in s\underline{\mathsf{A}_\infty}(\mathcal{A};\underline{\mathsf{C}}_\Bbbk)(H^{X_0},f_q)$ with components

$$[(x_1\otimes\cdots\otimes x_p\otimes y\otimes r_1\otimes\cdots\otimes r_q)\mho_{pq}]_n : T^n s\mathcal{A}(Z_0,Z_n) \to s\underline{\mathsf{C}}_\Bbbk(\mathcal{A}(X_0,Z_0),Z_n f_q),$$
$$z_1\otimes\cdots\otimes z_n \mapsto (z_1\otimes\cdots\otimes z_n\otimes x_1\otimes\cdots\otimes x_p\otimes y\otimes r_1\otimes\cdots\otimes r_q)\mho'_{pq;n},$$

where $\mho'_{pq;n} \overset{\text{def}}{=} (1^{\otimes n} \otimes \mho_{pq;n}) \operatorname{ev}^{\mathsf{C}_\Bbbk} = (1^{\otimes n} \otimes \mho_{pq} \cdot \operatorname{pr}_n) \operatorname{ev}^{\mathsf{C}_\Bbbk} = (1^{\otimes n} \otimes \mho_{pq}) \operatorname{ev}^{\mathsf{A}_\infty}_{n1}$ is given by

$$\mho'_{pq;n} = (-)^{p+1} \big[T^n s\mathcal{A}(Z_0, Z_n) \otimes T^p s\mathcal{A}^{\mathrm{op}}(X_0, X_p) \otimes X_p f_0[1] \otimes T^q s\underline{\mathsf{A}_\infty}(\mathcal{A}; \underline{\mathsf{C}}_\Bbbk)(f_0, f_q)$$
$$\xrightarrow{\operatorname{coev}^{\mathsf{C}_\Bbbk}} \underline{\mathsf{C}}_\Bbbk(s\mathcal{A}(X_0, Z_0), s\mathcal{A}(X_0, Z_0) \otimes T^n s\mathcal{A}(Z_0, Z_n)$$
$$\otimes T^p s\mathcal{A}^{\mathrm{op}}(X_0, X_p) \otimes X_p f_0[1] \otimes T^q s\underline{\mathsf{A}_\infty}(\mathcal{A}; \underline{\mathsf{C}}_\Bbbk)(f_0, f_q))$$
$$\xrightarrow{\underline{\mathsf{C}}_\Bbbk(1,\mathrm{perm})} \underline{\mathsf{C}}_\Bbbk(s\mathcal{A}(X_0, Z_0), X_p f_0[1] \otimes T^p s\mathcal{A}(X_p, X_0)$$
$$\otimes s\mathcal{A}(X_0, Z_0) \otimes T^n s\mathcal{A}(Z_0, Z_n) \otimes T^q s\underline{\mathsf{A}_\infty}(\mathcal{A}; \underline{\mathsf{C}}_\Bbbk)(f_0, f_q))$$
$$\xrightarrow{\underline{\mathsf{C}}_\Bbbk(1,1\otimes \operatorname{ev}^{\mathsf{A}_\infty}_{p+1+n,q})} \underline{\mathsf{C}}_\Bbbk(s\mathcal{A}(X_0, Z_0), X_p f_0[1] \otimes s\underline{\mathsf{C}}_\Bbbk(X_p f_0, Z_n f_q))$$
$$\xrightarrow{\underline{\mathsf{C}}_\Bbbk(1,1\otimes s^{-1}[1])} \underline{\mathsf{C}}_\Bbbk(s\mathcal{A}(X_0, Z_0), X_p f_0[1] \otimes \underline{\mathsf{C}}_\Bbbk(X_p f_0[1], Z_n f_q[1]))$$
$$\xrightarrow{\underline{\mathsf{C}}_\Bbbk(1,\operatorname{ev}^{\mathsf{C}_\Bbbk})} \underline{\mathsf{C}}_\Bbbk(s\mathcal{A}(X_0, Z_0), Z_n f_q[1]) \xrightarrow{[-1]s} s\underline{\mathsf{C}}_\Bbbk(\mathcal{A}(X_0, Z_0), Z_n f_q) \big].$$

It follows that $\mho_{pq} : T^p s\mathcal{A}^{\mathrm{op}} \otimes s\mathcal{E} \otimes T^q s\underline{\mathsf{A}_\infty}(\mathcal{A}; \underline{\mathsf{C}}_\Bbbk) \to s\mathcal{Q}$ vanishes if $q > 1$. The other components are given by

$$\mho'_{p0;n} = (-)^{p+1} \big[T^n s\mathcal{A}(Z_0, Z_n) \otimes T^p s\mathcal{A}^{\mathrm{op}}(X_0, X_p) \otimes X_p f[1]$$
$$\xrightarrow{\operatorname{coev}^{\mathsf{C}_\Bbbk}} \underline{\mathsf{C}}_\Bbbk(s\mathcal{A}(X_0, Z_0), s\mathcal{A}(X_0, Z_0) \otimes T^n s\mathcal{A}(Z_0, Z_n) \otimes T^p s\mathcal{A}^{\mathrm{op}}(X_0, X_p) \otimes X_p f[1])$$
$$\xrightarrow{\underline{\mathsf{C}}_\Bbbk(1,\mathrm{perm})} \underline{\mathsf{C}}_\Bbbk(s\mathcal{A}(X_0, Z_0), X_p f[1] \otimes T^p s\mathcal{A}(X_p, X_0) \otimes s\mathcal{A}(X_0, Z_0) \otimes T^n s\mathcal{A}(Z_0, Z_n))$$
$$\xrightarrow{\underline{\mathsf{C}}_\Bbbk(1,1\otimes f_{p+1+n})} \underline{\mathsf{C}}_\Bbbk(s\mathcal{A}(X_0, Z_0), X_p f[1] \otimes s\underline{\mathsf{C}}_\Bbbk(X_p f, Z_n f))$$
$$\xrightarrow{\underline{\mathsf{C}}_\Bbbk(1,1\otimes s^{-1}[1])} \underline{\mathsf{C}}_\Bbbk(s\mathcal{A}(X_0, Z_0), X_p f[1] \otimes \underline{\mathsf{C}}_\Bbbk(X_p f[1], Z_n f[1]))$$
$$\xrightarrow{\underline{\mathsf{C}}_\Bbbk(1,\operatorname{ev}^{\mathsf{C}_\Bbbk})} \underline{\mathsf{C}}_\Bbbk(s\mathcal{A}(X_0, Z_0), Z_n f[1]) \xrightarrow{[-1]s} s\underline{\mathsf{C}}_\Bbbk(\mathcal{A}(X_0, Z_0), Z_n f) \big] \quad \text{(A.5)}$$

and

$$\mho'_{p1;n} = (-)^{p+1} \big[T^n s\mathcal{A}(Z_0, Z_n) \otimes T^p s\mathcal{A}^{\mathrm{op}}(X_0, X_p) \otimes X_p f[1] \otimes s\underline{\mathsf{A}_\infty}(\mathcal{A}; \underline{\mathsf{C}}_\Bbbk)(f, g)$$
$$\xrightarrow{\operatorname{coev}^{\mathsf{C}_\Bbbk}} \underline{\mathsf{C}}_\Bbbk(s\mathcal{A}(X_0, Z_0), s\mathcal{A}(X_0, Z_0) \otimes T^n s\mathcal{A}(Z_0, Z_n)$$
$$\otimes T^p s\mathcal{A}^{\mathrm{op}}(X_0, X_p) \otimes X_p f[1] \otimes s\underline{\mathsf{A}_\infty}(\mathcal{A}; \underline{\mathsf{C}}_\Bbbk)(f, g))$$
$$\xrightarrow{\underline{\mathsf{C}}_\Bbbk(1,\mathrm{perm})} \underline{\mathsf{C}}_\Bbbk(s\mathcal{A}(X_0, Z_0), X_p f[1] \otimes T^p s\mathcal{A}(X_p, X_0)$$
$$\otimes s\mathcal{A}(X_0, Z_0) \otimes T^n s\mathcal{A}(Z_0, Z_n) \otimes s\underline{\mathsf{A}_\infty}(\mathcal{A}; \underline{\mathsf{C}}_\Bbbk)(f, g))$$
$$\xrightarrow{\underline{\mathsf{C}}_\Bbbk(1,1\otimes 1^{\otimes p+1+n} \otimes \operatorname{pr}_{p+1+n})} \underline{\mathsf{C}}_\Bbbk(s\mathcal{A}(X_0, Z_0), X_p f[1] \otimes T^p s\mathcal{A}(X_p, X_0) \otimes s\mathcal{A}(X_0, Z_0)$$
$$\otimes T^n s\mathcal{A}(Z_0, Z_n) \otimes \underline{\mathsf{C}}_\Bbbk(T^p s\mathcal{A}(X_p, X_0) \otimes s\mathcal{A}(X_0, Z_0) \otimes T^n s\mathcal{A}(Z_0, Z_n), s\underline{\mathsf{C}}_\Bbbk(X_p f, Z_n g)))$$
$$\xrightarrow{\underline{\mathsf{C}}_\Bbbk(1,1\otimes \operatorname{ev}^{\mathsf{C}_\Bbbk})} \underline{\mathsf{C}}_\Bbbk(s\mathcal{A}(X_0, Z_0), X_p f[1] \otimes s\underline{\mathsf{C}}_\Bbbk(X_p f, Z_n g))$$
$$\xrightarrow{\underline{\mathsf{C}}_\Bbbk(1,1\otimes s^{-1}[1])} \underline{\mathsf{C}}_\Bbbk(s\mathcal{A}(X_0, Z_0), X_p f[1] \otimes \underline{\mathsf{C}}_\Bbbk(X_p f[1], Z_n g[1]))$$

$$\xrightarrow{\underline{\mathsf{C}}_{\Bbbk}(1,\mathrm{ev}^{\underline{\mathsf{C}}_{\Bbbk}})} \underline{\mathsf{C}}_{\Bbbk}(s\mathcal{A}(X_0,Z_0),Z_ng[1]) \xrightarrow{[-1]s} s\underline{\mathsf{C}}_{\Bbbk}(\mathcal{A}(X_0,Z_0),Z_ng)\big].$$

Naturality of the A_∞-transformation Ω is implied by the following

Lemma A.2. *The bicomodule homomorphism $\mho$ is a chain map.*

This is proven by a straightforward computation. The full proof is given in archive version [LM07] of this article.

Let $\mathcal{A}$ be an A_∞-category, and let $f:\mathcal{A}\to\underline{\mathsf{C}}_{\Bbbk}$ be an A_∞-functor. Denote by $\mathcal{M}$ the $\mathcal{A}$-module determined by f in Proposition 4.2. Denote

$$\Upsilon = \mho_{00} : s\mathcal{M}(X) = s\mathcal{E}(X,f) = Xf[1] \to s\underline{\mathsf{A}_\infty}(\mathcal{A};\underline{\mathsf{C}}_{\Bbbk})(H^X,f) \tag{A.6}$$

for the sake of brevity. The composition of Υ with the projection pr_n from (A.3) is given by the particular case $p=q=0$ of (A.4):

$$\begin{aligned}\Upsilon_n = -\big[& s\mathcal{M}(X) \xrightarrow{\mathrm{coev}^{\underline{\mathsf{C}}_{\Bbbk}}} \\ & \underline{\mathsf{C}}_{\Bbbk}(s\mathcal{A}(X,Z_0)\otimes T^ns\mathcal{A}(Z_0,Z_n), s\mathcal{A}(X,Z_0)\otimes T^ns\mathcal{A}(Z_0,Z_n)\otimes s\mathcal{M}(X)) \\ & \xrightarrow{\underline{\mathsf{C}}_{\Bbbk}(1,\tau_c)} \underline{\mathsf{C}}_{\Bbbk}(s\mathcal{A}(X,Z_0)\otimes T^ns\mathcal{A}(Z_0,Z_n), s\mathcal{M}(X)\otimes s\mathcal{A}(X,Z_0)\otimes T^ns\mathcal{A}(Z_0,Z_n)) \\ & \xrightarrow{\underline{\mathsf{C}}_{\Bbbk}(1,b^{\mathcal{M}}_{n+1})} \underline{\mathsf{C}}_{\Bbbk}(s\mathcal{A}(X,Z_0)\otimes T^ns\mathcal{A}(Z_0,Z_n), s\mathcal{M}(Z_n)) \\ & \xrightarrow{(\varphi^{\underline{\mathsf{C}}_{\Bbbk}})^{-1}} \underline{\mathsf{C}}_{\Bbbk}(T^ns\mathcal{A}(Z_0,Z_n), \underline{\mathsf{C}}_{\Bbbk}(s\mathcal{A}(X,Z_0), s\mathcal{M}(Z_n))) \\ & \xrightarrow{\underline{\mathsf{C}}_{\Bbbk}(1,[-1])} \underline{\mathsf{C}}_{\Bbbk}(T^ns\mathcal{A}(Z_0,Z_n), \underline{\mathsf{C}}_{\Bbbk}(\mathcal{A}(X,Z_0), \mathcal{M}(Z_n))) \\ & \xrightarrow{\underline{\mathsf{C}}_{\Bbbk}(1,s)} \underline{\mathsf{C}}_{\Bbbk}(T^ns\mathcal{A}(Z_0,Z_n), s\underline{\mathsf{C}}_{\Bbbk}(\mathcal{A}(X,Z_0), \mathcal{M}(Z_n)))\big],\end{aligned}$$

where $n\geqslant 0$, $\tau = \left(\begin{smallmatrix} 0 & 1 & \dots & n & n+1 \\ 1 & 2 & \dots & n+1 & 0\end{smallmatrix}\right)\in\mathfrak{S}_{n+2}$. An element $r\in s\mathcal{M}(X)$ is mapped to an A_∞-transformation $(r)\Upsilon$ with the components

$$\begin{gathered}(r)\Upsilon_n : T^ns\mathcal{A}(Z_0,Z_n) \to s\underline{\mathsf{C}}_{\Bbbk}(\mathcal{A}(X,Z_0),\mathcal{M}(Z_n)), \quad n\geqslant 0, \\ z_1\otimes\dots\otimes z_n \mapsto (z_1\otimes\dots\otimes z_n\otimes r)\Upsilon'_n,\end{gathered}$$

where

$$\begin{aligned}\Upsilon'_n = -\big[& T^ns\mathcal{A}(Z_0,Z_n)\otimes s\mathcal{M}(X) \xrightarrow{\mathrm{coev}^{\underline{\mathsf{C}}_{\Bbbk}}} \\ & \underline{\mathsf{C}}_{\Bbbk}(s\mathcal{A}(X,Z_0), s\mathcal{A}(X,Z_0)\otimes T^ns\mathcal{A}(Z_0,Z_n)\otimes s\mathcal{M}(X)) \xrightarrow{\underline{\mathsf{C}}_{\Bbbk}(1,\tau_c)} \\ & \underline{\mathsf{C}}_{\Bbbk}(s\mathcal{A}(X,Z_0), s\mathcal{M}(X)\otimes s\mathcal{A}(X,Z_0)\otimes T^ns\mathcal{A}(Z_0,Z_n)) \xrightarrow{\underline{\mathsf{C}}_{\Bbbk}(1,b^{\mathcal{M}}_{n+1})} \\ & \underline{\mathsf{C}}_{\Bbbk}(s\mathcal{A}(X,Z_0), s\mathcal{M}(Z_n)) \xrightarrow{[-1]} \underline{\mathsf{C}}_{\Bbbk}(\mathcal{A}(X,Z_0),\mathcal{M}(Z_n)) \xrightarrow{s} s\underline{\mathsf{C}}_{\Bbbk}(\mathcal{A}(X,Z_0),\mathcal{M}(Z_n))\big].\end{aligned}$$

Since $\mho$ is a chain map by Lemma A.2, the map

$$\mho_{00} = \Upsilon : (s\mathcal{M}(X), b^{\mathcal{E}}_{00} = s^{-1}d^{Xf}s = b^{\mathcal{M}}_0) \to (s\underline{\mathsf{A}_\infty}(\mathcal{A};\underline{\mathsf{C}}_{\Bbbk})(H^X,f), b^{\mathcal{Q}}_{00} = B_1)$$

is a chain map as well. The following result generalizes previously known A_∞-version of the Yoneda Lemma [Fuk02, Theorem 9.1], [LM04, Proposition A.9], and gives the latter if $f=H^W$ for $W\in\operatorname{Ob}\mathcal{A}$. It can also be found in Seidel's book

[Sei06, Lemma 2.12], where it is proven assuming that the ground ring $\Bbbk$ is a field, and the proof is based on a spectral sequence argument. The proof presented in archive version [LM07] of this article is considerably longer than that of Seidel, however it works in the case of an arbitrary commutative ground ring.

Proposition A.3. *Let $\mathcal{A}$ be a unital A_∞-category, let X be an object of $\mathcal{A}$, and let $f : \mathcal{A} \to \underline{\mathsf{C}}_\Bbbk$ be a unital A_∞-functor. Then the map Υ is homotopy invertible.*

Proof. The A_∞-module $\mathcal{M}$ corresponding to f is unital by Proposition 4.4. The components of f are expressed via the components of $b^\mathcal{M}$ as follows ($k \geqslant 1$):

$$f_k = \big[T^k s\mathcal{A}(Z_0, Z_k) \xrightarrow{\mathrm{coev}^{\underline{\mathsf{C}}_\Bbbk}} \underline{\mathsf{C}}_\Bbbk(s\mathcal{M}(Z_0), s\mathcal{M}(Z_0) \otimes T^k s\mathcal{A}(Z_0, Z_k))$$
$$\xrightarrow{\underline{\mathsf{C}}_\Bbbk(1, b^\mathcal{M}_k)} \underline{\mathsf{C}}_\Bbbk(s\mathcal{M}(Z_0), s\mathcal{M}(Z_k)) \xrightarrow{[-1]} \underline{\mathsf{C}}_\Bbbk(\mathcal{M}(Z_0), \mathcal{M}(Z_k))$$
$$\xrightarrow{s} s\underline{\mathsf{C}}_\Bbbk(\mathcal{M}(Z_0), \mathcal{M}(Z_k))\big]. \quad (A.7)$$

Define a map $\alpha : s\underline{\mathsf{A}_\infty}(\mathcal{A}; \underline{\mathsf{C}}_\Bbbk)(H^X, f) \to s\mathcal{M}(X)$ as follows:

$$\alpha = \big[s\underline{\mathsf{A}_\infty}(\mathcal{A}; \underline{\mathsf{C}}_\Bbbk)(H^X, f) \xrightarrow{\mathrm{pr}_0} s\underline{\mathsf{C}}_\Bbbk(\mathcal{A}(X, X), \mathcal{M}(X)) \xrightarrow{s^{-1}} \underline{\mathsf{C}}_\Bbbk(\mathcal{A}(X, X), \mathcal{M}(X))$$
$$\xrightarrow{[1]} \underline{\mathsf{C}}_\Bbbk(s\mathcal{A}(X, X), s\mathcal{M}(X)) \xrightarrow{\underline{\mathsf{C}}_\Bbbk({}_X\mathbf{i}^\mathcal{A}_0, 1)} \underline{\mathsf{C}}_\Bbbk(\Bbbk, s\mathcal{M}(X)) = s\mathcal{M}(X)\big]. \quad (A.8)$$

The map α is a chain map. Indeed, pr_0 is a chain map, and

$$s^{-1}[1]\underline{\mathsf{C}}_\Bbbk({}_X\mathbf{i}^\mathcal{A}_0, 1) b^\mathcal{M}_0 = s^{-1}[1]\underline{\mathsf{C}}_\Bbbk({}_X\mathbf{i}^\mathcal{A}_0, 1)\underline{\mathsf{C}}_\Bbbk(1, b^\mathcal{M}_0)$$
$$= s^{-1}[1](-\underline{\mathsf{C}}_\Bbbk(1, b^\mathcal{M}_0) + \underline{\mathsf{C}}_\Bbbk(b_1, 1))\underline{\mathsf{C}}_\Bbbk({}_X\mathbf{i}^\mathcal{A}_0, 1) = s^{-1}[1] m_1^{\underline{\mathsf{C}}_\Bbbk}\underline{\mathsf{C}}_\Bbbk({}_X\mathbf{i}^\mathcal{A}_0, 1)$$
$$= s^{-1} m_1^{\underline{\mathsf{C}}_\Bbbk}[1]\underline{\mathsf{C}}_\Bbbk({}_X\mathbf{i}^\mathcal{A}_0, 1) = b_1^{\underline{\mathsf{C}}_\Bbbk} s^{-1}[1]\underline{\mathsf{C}}_\Bbbk({}_X\mathbf{i}^\mathcal{A}_0, 1),$$

since ${}_X\mathbf{i}^\mathcal{A}_0$ is a chain map, and $[1]$ is a differential graded functor. Let us compute $\Upsilon\alpha$:

$$\Upsilon\alpha = \Upsilon_0 s^{-1}[1]\underline{\mathsf{C}}_\Bbbk({}_X\mathbf{i}^\mathcal{A}_0, 1) = -\big[s\mathcal{M}(X) \xrightarrow{\mathrm{coev}^{\underline{\mathsf{C}}_\Bbbk}} \underline{\mathsf{C}}_\Bbbk(s\mathcal{A}(X, X), s\mathcal{A}(X, X) \otimes s\mathcal{M}(X))$$
$$\xrightarrow{\underline{\mathsf{C}}_\Bbbk(1, c)} \underline{\mathsf{C}}_\Bbbk(s\mathcal{A}(X, X), s\mathcal{M}(X) \otimes s\mathcal{A}(X, X)) \xrightarrow{\underline{\mathsf{C}}_\Bbbk(1, b^\mathcal{M}_1)} \underline{\mathsf{C}}_\Bbbk(s\mathcal{A}(X, X), s\mathcal{M}(X))$$
$$\xrightarrow{\underline{\mathsf{C}}_\Bbbk({}_X\mathbf{i}^\mathcal{A}_0, 1)} \underline{\mathsf{C}}_\Bbbk(\Bbbk, s\mathcal{M}(X)) = s\mathcal{M}(X)\big]$$
$$= \big[s\mathcal{M}(X) \xrightarrow{\mathrm{coev}^{\underline{\mathsf{C}}_\Bbbk}} \underline{\mathsf{C}}_\Bbbk(s\mathcal{A}(X, X), s\mathcal{A}(X, X) \otimes s\mathcal{M}(X)) \xrightarrow{\underline{\mathsf{C}}_\Bbbk({}_X\mathbf{i}^\mathcal{A}_0, 1)}$$
$$\underline{\mathsf{C}}_\Bbbk(\Bbbk, s\mathcal{A}(X, X) \otimes s\mathcal{M}(X)) \xrightarrow{\underline{\mathsf{C}}_\Bbbk(1, cb^\mathcal{M}_1)} \underline{\mathsf{C}}_\Bbbk(\Bbbk, s\mathcal{M}(X)) = s\mathcal{M}(X)\big]$$
$$= \big[s\mathcal{M}(X) \xrightarrow{\mathrm{coev}^{\underline{\mathsf{C}}_\Bbbk}} \underline{\mathsf{C}}_\Bbbk(\Bbbk, \Bbbk \otimes s\mathcal{M}(X)) \xrightarrow{\underline{\mathsf{C}}_\Bbbk(1, {}_X\mathbf{i}^\mathcal{A}_0 \otimes 1)} \underline{\mathsf{C}}_\Bbbk(\Bbbk, s\mathcal{A}(X, X) \otimes s\mathcal{M}(X))$$
$$\xrightarrow{\underline{\mathsf{C}}_\Bbbk(1, cb^\mathcal{M}_1)} \underline{\mathsf{C}}_\Bbbk(\Bbbk, s\mathcal{M}(X)) = s\mathcal{M}(X)\big]$$
$$= \big[s\mathcal{M}(X) \xrightarrow{1 \otimes {}_X\mathbf{i}^\mathcal{A}_0} s\mathcal{M}(X) \otimes s\mathcal{A}(X, X) \xrightarrow{b^\mathcal{M}_1} s\mathcal{M}(X)\big]. \quad (A.9)$$

Since $\mathcal{M}$ is a unital A_∞-module by Proposition 4.4, it follows that $\Upsilon\alpha$ is homotopic to identity. Let us prove that $\alpha\Upsilon$ is homotopy invertible.

The graded $\Bbbk$-module $s\underline{\mathsf{A}_\infty}(\mathcal{A};\underline{\mathsf{C}}_\Bbbk)(H^X,f)$ is $V=\prod_{n=0}^\infty V_n$, where

$$V_n = \prod_{Z_0,\dots,Z_n\in\operatorname{Ob}\mathcal{A}} \underline{\mathsf{C}}_\Bbbk(\bar{T}^n s\mathcal{A}(Z_0,Z_n), s\underline{\mathsf{C}}_\Bbbk(\mathcal{A}(X,Z_0),\mathcal{M}(Z_n)))$$

and all products are taken in the category of graded $\Bbbk$-modules. In other terms, for $d\in\mathbb{Z}$, $V^d=\prod_{n=0}^\infty V_n^d$, where

$$V_n^d = \prod_{Z_0,\dots,Z_n\in\operatorname{Ob}\mathcal{A}} \underline{\mathsf{C}}_\Bbbk(\bar{T}^n s\mathcal{A}(Z_0,Z_n), s\underline{\mathsf{C}}_\Bbbk(\mathcal{A}(X,Z_0),\mathcal{M}(Z_n)))^d.$$

We consider V_n^d as Abelian groups with discrete topology. The Abelian group V^d is equipped with the topology of the product. Thus, its basis of neighborhoods of 0 is given by $\Bbbk$-submodules $\Phi_m^d = 0^{m-1}\times\prod_{n=m}^\infty V_n^d$. They form a filtration $V^d=\Phi_0^d\supset\Phi_1^d\supset\Phi_2^d\supset\dots$. We call a $\Bbbk$-linear map $a:V\to V$ of degree p continuous if the induced maps $a^{d,d+p}=a\big|_{V^d}:V^d\to V^{d+p}$ are continuous for all $d\in\mathbb{Z}$. This holds if and only if for any $d\in\mathbb{Z}$ and $m\in\mathbb{N}\overset{\text{def}}{=}\mathbb{Z}_{\geqslant 0}$ there exists an integer $\kappa=\kappa(d,m)\in\mathbb{N}$ such that $(\Phi_\kappa^d)a\subset\Phi_m^{d+p}$. We may assume that

$$m'<m'' \quad\text{implies}\quad \kappa(d,m')\leqslant\kappa(d,m''). \tag{A.10}$$

Indeed, a given function $m\mapsto\kappa(d,m)$ can be replaced with the function $m\mapsto\kappa'(d,m)=\min_{n\geqslant m}\kappa(d,n)$ and κ' satisfies condition (A.10). Continuous linear maps $a:V\to V$ of degree p are in bijection with families of $\mathbb{N}\times\mathbb{N}$-matrices $(A^{d,d+p})_{d\in\mathbb{Z}}$ of linear maps $A_{nm}^{d,d+p}:V_n^d\to V_m^{d+p}$ with finite number of non-vanishing elements in each column of $A^{d,d+p}$. Indeed, to each continuous map $a^{d,d+p}:V^d\to V^{d+p}$ corresponds the inductive limit over m of $\kappa(d,m)\times m$-matrices of maps $V^d/\Phi^d_{\kappa(d,m)}\to V^{d+p}/\Phi_m^{d+p}$. On the other hand, to each family $(A^{d,d+p})_{d\in\mathbb{Z}}$ of $\mathbb{N}\times\mathbb{N}$-matrices with finite number of non-vanishing elements in each column correspond obvious maps $a^{d,d+p}:V^d\to V^{d+p}$, and they are continuous. Thus, $a=(a^{d,d+p})_{d\in\mathbb{Z}}$ is continuous. A continuous map $a:V\to V$ can be completely recovered from one $\mathbb{N}\times\mathbb{N}$-matrix $(a_{nm})_{n,m\in\mathbb{N}}$ of maps $a_{nm}=(A_{nm}^{d,d+p})_{d\in\mathbb{Z}}:V_n\to V_m$ of degree p. Naturally, not any such matrix determines a continuous map, however, if the number of non-vanishing elements in each column of (a_{nm}) is finite, then this matrix does determine a continuous map.

The differential $D\overset{\text{def}}{=}B_1:V\to V$, $r\mapsto(r)B_1=rb-(-)^r br$ is continuous and the function κ for it is simply $\kappa(d,m)=m$. Its matrix is given by

$$D=B_1=\begin{bmatrix} D_{0,0} & D_{0,1} & D_{0,2} & \dots\\ 0 & D_{1,1} & D_{1,2} & \dots\\ 0 & 0 & D_{2,2} & \dots\\ \vdots & \vdots & \vdots & \ddots \end{bmatrix},$$

where

$$D_{k,k} = \underline{\mathsf{C}}_\Bbbk(1, b_1^{\underline{\mathsf{C}}_\Bbbk}) - \underline{\mathsf{C}}_\Bbbk\big(\sum_{p+1+q=k} 1^{\otimes p} \otimes b_1 \otimes 1^{\otimes q}, 1\big) : V_k \to V_k,$$

$$r_k D_{k,k} = r_k b_1^{\underline{\mathsf{C}}_\Bbbk} - (-)^r \sum_{p+1+q=k} (1^{\otimes p} \otimes b_1 \otimes 1^{\otimes q}) r_k,$$

(one easily recognizes the differential in the complex V_k),

$$r_k D_{k,k+1} = (r_k \otimes f_1) b_2^{\underline{\mathsf{C}}_\Bbbk} + (H_1^X \otimes r_k) b_2^{\underline{\mathsf{C}}_\Bbbk} - (-)^r \sum_{p+q=k-1} (1^{\otimes p} \otimes b_2 \otimes 1^{\otimes q}) r_k.$$

Further we shall see that we do not need to compute other components.

Composition of $\alpha\Upsilon$ with pr_n equals

$$\begin{aligned}
\alpha\Upsilon_n = -\big[& s\underline{\mathsf{A}_\infty}(\mathcal{A}; \underline{\mathsf{C}}_\Bbbk)(H^X, f) \xrightarrow{\mathrm{pr}_0} s\underline{\mathsf{C}}_\Bbbk(\mathcal{A}(X,X), \mathcal{M}(X)) \\
& \xrightarrow{s^{-1}[1]} \underline{\mathsf{C}}_\Bbbk(s\mathcal{A}(X,X), s\mathcal{M}(X)) \xrightarrow{\underline{\mathsf{C}}_\Bbbk(x\mathbf{i}_0^{\mathcal{A}},1)} \underline{\mathsf{C}}_\Bbbk(\Bbbk, s\mathcal{M}(X)) = s\mathcal{M}(X) \\
& \xrightarrow{\mathrm{coev}^{\underline{\mathsf{C}}_\Bbbk}} \underline{\mathsf{C}}_\Bbbk(s\mathcal{A}(X, Z_0) \otimes T^n s\mathcal{A}(Z_0, Z_n), s\mathcal{A}(X, Z_0) \otimes T^n s\mathcal{A}(Z_0, Z_n) \otimes s\mathcal{M}(X)) \\
& \xrightarrow{\underline{\mathsf{C}}_\Bbbk(1,\tau_c)} \underline{\mathsf{C}}_\Bbbk(s\mathcal{A}(X, Z_0) \otimes T^n s\mathcal{A}(Z_0, Z_n), s\mathcal{M}(X) \otimes s\mathcal{A}(X, Z_0) \otimes T^n s\mathcal{A}(Z_0, Z_n)) \\
& \xrightarrow{\underline{\mathsf{C}}_\Bbbk(1,b_{n+1}^{\mathcal{M}})} \underline{\mathsf{C}}_\Bbbk(s\mathcal{A}(X, Z_0) \otimes T^n s\mathcal{A}(Z_0, Z_n), s\mathcal{M}(Z_n)) \\
& \xrightarrow{(\varphi^{\underline{\mathsf{C}}_\Bbbk})^{-1}} \underline{\mathsf{C}}_\Bbbk(T^n s\mathcal{A}(Z_0, Z_n), \underline{\mathsf{C}}_\Bbbk(s\mathcal{A}(X, Z_0), s\mathcal{M}(Z_n))) \\
& \xrightarrow{\underline{\mathsf{C}}_\Bbbk(1,[-1]s)} \underline{\mathsf{C}}_\Bbbk(T^n s\mathcal{A}(Z_0, Z_n), s\underline{\mathsf{C}}_\Bbbk(\mathcal{A}(X, Z_0), \mathcal{M}(Z_n)))\big].
\end{aligned}$$

Clearly, $\alpha\Upsilon$ is continuous (take $\kappa(d,m) = 1$). Its $\mathbb{N} \times \mathbb{N}$-matrix has the form

$$\alpha\Upsilon = \begin{bmatrix} * & * & * & \dots \\ 0 & 0 & 0 & \dots \\ 0 & 0 & 0 & \dots \\ \vdots & \vdots & \vdots & \ddots \end{bmatrix}.$$

Lemma A.4. *The map $\alpha\Upsilon : V \to V$ is homotopic to a continuous map $V \to V$, whose $\mathbb{N} \times \mathbb{N}$-matrix is upper-triangular with the identity maps $\mathrm{id} : V_k \to V_k$ on the diagonal.*

The proof is analogous to proof of [LM04, Lemma A.10]. The full proof is given in archive version [LM07] of this article.

The continuous map of degree 0

$$\alpha\Upsilon + B_1(K - K') + (K - K')B_1 = 1 + N : V \to V,$$

obtained in Lemma A.4, is invertible (its inverse is determined by the upper-triangular matrix $\sum_{i=0}^\infty (-N)^i$, which is well-defined). Therefore, $\alpha\Upsilon$ is homotopy invertible. We have proved earlier that $\Upsilon\alpha$ is homotopic to identity. Viewing α, Υ as morphisms of the homotopy category $H^0(\underline{\mathsf{C}}_\Bbbk)$, we see that both of them are homotopy invertible. Hence, they are homotopy inverse to each other. Proposition A.3 is proven. □

Homotopy invertibility of $\Upsilon = \mho_{00}$ implies invertibility of the cycle Ω_{00} up to boundaries. Hence the natural A_∞-transformation Ω is invertible and Theorem A.1 is proven. □

Corollary A.5. *There is a bijection between elements of $H^0(\mathcal{M}(X), d)$ and equivalence classes of natural A_∞-transformations $H^X \to f : \mathcal{A} \to \underline{\mathsf{C}}_\Bbbk$.*

The following representability criterion has been proven independently by Seidel [Sei06, Lemma 3.1] in the case when the ground ring $\Bbbk$ is a field.

Corollary A.6. *A unital A_∞-functor $f : \mathcal{A} \to \underline{\mathsf{C}}_\Bbbk$ is isomorphic to H^X for an object $X \in \operatorname{Ob}\mathcal{A}$ if and only if the $\mathcal{K}$-functor $\mathsf{k}f : \mathsf{k}\mathcal{A} \to \underline{\mathcal{K}} = \mathsf{k}\underline{\mathsf{C}}_\Bbbk$ is representable by X.*

Proof. The A_∞-functor f is isomorphic to H^X if and only if there is an invertible natural A_∞-transformation $H^X \to f : \mathcal{A} \to \underline{\mathsf{C}}_\Bbbk$. By Proposition A.3, this is the case if and only if there is a cycle $t \in \mathcal{M}(X)$ of degree 0 such that the natural A_∞-transformation $(ts)\Upsilon$ is invertible. By [Lyu03, Proposition 7.15], invertibility of $(ts)\Upsilon$ is equivalent to invertibility modulo boundaries of the 0-th component $(ts)\Upsilon_0$ of $(ts)\Upsilon$. For each $Z \in \operatorname{Ob}\mathcal{A}$, the element ${}_Z(ts)\Upsilon_0$ of $\underline{\mathsf{C}}_\Bbbk(\mathcal{A}(X,Z), \mathcal{M}(Z))$ is given by

$$
\begin{aligned}
{}_Z(ts)\Upsilon_0 &= -\big[\mathcal{A}(X,Z) \xrightarrow{s} s\mathcal{A}(X,Z) \xrightarrow{(ts\otimes 1)b_1^{\mathcal{M}}} s\mathcal{M}(Z) \xrightarrow{s^{-1}} \mathcal{M}(Z)\big] \\
&= -\big[\mathcal{A}(X,Z) \xrightarrow{s} s\mathcal{A}(X,Z) \xrightarrow{ts\otimes f_1 s^{-1}} s\mathcal{M}(X) \otimes \underline{\mathsf{C}}_\Bbbk(\mathcal{M}(X), \mathcal{M}(Z)) \xrightarrow{1\otimes[1]} \\
&\qquad s\mathcal{M}(X) \otimes \underline{\mathsf{C}}_\Bbbk(s\mathcal{M}(X), s\mathcal{M}(Z)) \xrightarrow{\mathrm{ev}} s\mathcal{M}(Z) \xrightarrow{s^{-1}} \mathcal{M}(Z)\big] \\
&= -\big[\mathcal{A}(X,Z) \xrightarrow{s} s\mathcal{A}(X,Z) \xrightarrow{ts\otimes f_1 s^{-1}} s\mathcal{M}(X) \otimes \underline{\mathsf{C}}_\Bbbk(\mathcal{M}(X), \mathcal{M}(Z)) \xrightarrow{s^{-1}\otimes 1} \\
&\qquad \mathcal{M}(X) \otimes \underline{\mathsf{C}}_\Bbbk(\mathcal{M}(X), \mathcal{M}(Z)) \xrightarrow{\mathrm{ev}} \mathcal{M}(Z)\big] \\
&= \big[\mathcal{A}(X,Z) \xrightarrow{t\otimes s f_1 s^{-1}} \mathcal{M}(X) \otimes \underline{\mathsf{C}}_\Bbbk(\mathcal{M}(X), \mathcal{M}(Z)) \xrightarrow{\mathrm{ev}} \mathcal{M}(Z)\big].
\end{aligned}
$$

By Proposition 1.6, the above composite is invertible in $\underline{\mathsf{C}}_\Bbbk(\mathcal{A}(X,Z), \mathcal{M}(Z))$ modulo boundaries, i.e., homotopy invertible, if and only if $\mathsf{k}f$ is representable by the object X. □

Proposition A.7. *So defined Ω turns the pasting*

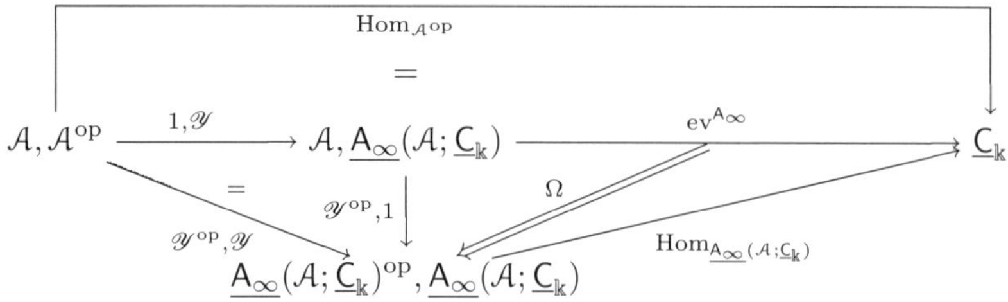

into the natural A_∞-transformation $r^{\mathscr{Y}}$ defined in Corollary 5.7. Equivalently, the homomorphism of $Ts\mathcal{A}^{\mathrm{op}}$-$Ts\mathcal{A}^{\mathrm{op}}$-bicomodules $t^{\mathscr{Y}} : \mathcal{R}_{\mathcal{A}^{\mathrm{op}}} \to {}_{\mathscr{Y}}\underline{\mathsf{A}_\infty}(\mathcal{A};\underline{\mathsf{C}}_{\Bbbk})_{\mathscr{Y}}$ coincides with $(1\otimes 1\otimes \mathscr{Y})\cdot \mho : {}_1\mathcal{E}_{\mathscr{Y}} \to {}_{\mathscr{Y}}\underline{\mathsf{A}_\infty}(\mathcal{A};\underline{\mathsf{C}}_{\Bbbk})_{\mathscr{Y}}$. In other terms,

$$\begin{aligned}&\bigl[Ts\mathcal{A}^{\mathrm{op}}\otimes s\mathcal{A}^{\mathrm{op}}\otimes Ts\mathcal{A}^{\mathrm{op}} \xrightarrow{\mu_{Ts\mathcal{A}^{\mathrm{op}}}} Ts\mathcal{A}^{\mathrm{op}} \xrightarrow{\check{\mathscr{Y}}} s\underline{\mathsf{A}_\infty}(\mathcal{A};\underline{\mathsf{C}}_{\Bbbk})\bigr]\\ &= \bigl[Ts\mathcal{A}^{\mathrm{op}}\otimes s\mathcal{A}^{\mathrm{op}}\otimes Ts\mathcal{A}^{\mathrm{op}} \xrightarrow{1\otimes 1\otimes \mathscr{Y}} Ts\mathcal{A}^{\mathrm{op}}\otimes s\mathcal{E}_{\mathscr{Y}}\otimes Ts\underline{\mathsf{A}_\infty}(\mathcal{A};\underline{\mathsf{C}}_{\Bbbk}) \xrightarrow{\check{\mho}} s\underline{\mathsf{A}_\infty}(\mathcal{A};\underline{\mathsf{C}}_{\Bbbk})\bigr].\end{aligned}$$

This is proven by a straightforward computation. The full proof is given in archive version [LM07] of this article.

Corollary A.8 (Fukaya [Fuk02] **Theorem 9.1, Lyubashenko and Manzyuk** [LM04] **Proposition A.9).** *The A_∞-functor $\mathscr{Y} : \mathcal{A}^{\mathrm{op}} \to \underline{\mathsf{A}^{\mathrm{u}}_\infty}(\mathcal{A};\underline{\mathsf{C}}_{\Bbbk})$ is homotopy fully faithful.*

Proof. We have

$$\mathscr{Y}_1 = \mho_{00} : s\mathcal{A}^{\mathrm{op}}(X,Y) \to s\underline{\mathsf{A}^{\mathrm{u}}_\infty}(\mathcal{A};\underline{\mathsf{C}}_{\Bbbk})(H^X,H^Y),$$

for each pair $X, Y \in \operatorname{Ob}\mathcal{A}$. By Proposition A.3, the component $\mho_{00}$ is homotopy invertible, hence so is $\mathscr{Y}_1$. □

Let $\operatorname{Rep}(\mathcal{A},\underline{\mathsf{C}}_{\Bbbk})$ denote the essential image of $\mathscr{Y} : \mathcal{A}^{\mathrm{op}} \to \underline{\mathsf{A}^{\mathrm{u}}_\infty}(\mathcal{A};\underline{\mathsf{C}}_{\Bbbk})$, i.e., the full differential graded subcategory of $\underline{\mathsf{A}^{\mathrm{u}}_\infty}(\mathcal{A};\underline{\mathsf{C}}_{\Bbbk})$ whose objects are representable A_∞-functors $(X)\mathscr{Y} = H^X : \mathcal{A} \to \underline{\mathsf{C}}_{\Bbbk}$, for $X \in \operatorname{Ob}\mathcal{A}$, which are unital by Remark 5.19. The differential graded category $\operatorname{Rep}A^u_\infty(\mathcal{A},\underline{\mathsf{C}}_{\Bbbk})$ is $\mathscr{U}$-small. Thus, the Yoneda A_∞-functor $\mathscr{Y} : \mathcal{A}^{\mathrm{op}} \to \underline{\mathsf{A}^{\mathrm{u}}_\infty}(\mathcal{A};\underline{\mathsf{C}}_{\Bbbk})$ takes values in the $\mathscr{U}$-small subcategory $\operatorname{Rep}(\mathcal{A},\underline{\mathsf{C}}_{\Bbbk})$.

Corollary A.9 (Fukaya [Fuk02] **Theorem 9.1, Lyubashenko and Manzyuk** [LM04] **Theorem A.11).** *Let $\mathcal{A}$ be a unital A_∞-category. Then the restricted Yoneda A_∞-functor $\mathscr{Y} : \mathcal{A}^{\mathrm{op}} \to \operatorname{Rep}(\mathcal{A},\underline{\mathsf{C}}_{\Bbbk})$ is an equivalence.*

In particular, each $\mathscr{U}$-small unital A_∞-category is A_∞-equivalent to a $\mathscr{U}$-small differential graded category.

References

[BK89] A.I. Bondal and M.M. Kapranov, *Representable functors, Serre functors, and reconstructions*, Izv. Akad. Nauk SSSR Ser. Mat. **53** (1989), no. 6, 1183–1205, 1337.

[BLM06] Yu. Bespalov, V.V. Lyubashenko, and O. Manzyuk, *Closed multicategory of pretriangulated A_∞-categories*, book in progress, 2006, `http://www.math.ksu.edu/~lub/papers.html`.

[Cos04] K.J. Costello, *Topological conformal field theories and Calabi-Yau categories*, 2004, `math.QA/0412149`.

[DM82] P. Deligne and J.S. Milne, *Tannakian categories*, Hodge cycles, motives, and Shimura varieties, Lecture Notes in Math., no. 900, Springer-Verlag, Berlin, Heidelberg, New York, 1982, pp. 101–228.

[Dri04] V.G. Drinfeld, *DG quotients of DG categories*, J. Algebra **272** (2004), no. 2, 643–691, math.KT/0210114.

[Fuk02] K. Fukaya, *Floer homology and mirror symmetry. II*, Minimal surfaces, geometric analysis and symplectic geometry (Baltimore, MD, 1999), Adv. Stud. Pure Math., vol. 34, Math. Soc. Japan, Tokyo, 2002, pp. 31–127.

[GM03] S.I. Gelfand and Y.I. Manin, *Methods of homological algebra*, Springer Monographs in Mathematics, Springer-Verlag, Berlin, 1996, 2003.

[Kap90] M.M. Kapranov, *Surgeries and Serre functors in constructive sheaves*, Funktsional. Anal. i Prilozhen. **24** (1990), no. 2, 85–86.

[Kel01] B. Keller, *Introduction to A-infinity algebras and modules*, Homology, Homotopy and Applications **3** (2001), no. 1, 1–35, math.RA/9910179 http://intlpress.com/HHA/v3/n1/a1/.

[Kel06] B. Keller, *A-infinity algebras, modules and functor categories*, Trends in representation theory of algebras and related topics, Contemp. Math., vol. 406, Amer. Math. Soc., Providence, RI, 2006, math.RT/0510508, pp. 67–93.

[Kel82] G.M. Kelly, *Basic concepts of enriched category theory*, London Mathematical Society Lecture Notes, vol. 64, Cambridge University Press, 1982, http://www.tac.mta.ca/tac/reprints/articles/10/tr10abs.html.

[KS06] M. Kontsevich and Y.S. Soibelman, *Notes on A-infinity algebras, A-infinity categories and non-commutative geometry. I*, 2006, math.RA/0606241.

[LH03] K. Lefèvre-Hasegawa, *Sur les A_∞-catégories*, Ph.D. thesis, Université Paris 7, U.F.R. de Mathématiques, 2003, math.CT/0310337.

[Lyu99] V.V. Lyubashenko, *Squared Hopf algebras*, Mem. Amer. Math. Soc. **142** (1999), no. 677, 184 pp.

[Lyu03] V.V. Lyubashenko, *Category of A_∞-categories*, Homology, Homotopy and Applications **5** (2003), no. 1, 1–48, math.CT/0210047 http://intlpress.com/HHA/v5/n1/a1/.

[LM04] V.V. Lyubashenko and O. Manzyuk, *Quotients of unital A_∞-categories*, Max-Planck-Institut fur Mathematik preprint, MPI 04-19, 2004, math.CT/0306018.

[LM07] V.V. Lyubashenko and O. Manzyuk, *A_∞-bimodules and Serre A_∞-functors*, 2007, math.CT/0701165.

[Mac88] S. Mac Lane, *Categories for the working mathematician*, GTM, vol. 5, Springer-Verlag, New York, 1971, 1988.

[Man07] O. Manzyuk, *A_∞-bimodules and Serre A_∞-functors*, Ph.D. thesis, Technische Universität Kaiserslautern, Fachbereich Mathematik, 2007.

[MS05] V. Mazorchuk and C. Stroppel, *Projective-injective modules, Serre functors and symmetric algebras*, 2005, math.RT/0508119.

[RvdB02] I. Reiten and M. van den Bergh, *Noetherian hereditary abelian categories satisfying Serre duality*, J. Amer. Math. Soc. **15** (2002), no. 2, 295–366 (electronic), math.RT/9911242.

[Sei06] P. Seidel, *Fukaya categories and Picard–Lefschetz theory*, http://www.math.uchicago.edu/~seidel book in progress, 2006.

[Soi04] Y.S. Soibelman, *Mirror symmetry and noncommutative geometry of A_∞-categories*, J. Math. Phys. **45** (2004), no. 10, 3742–3757.

[Tra01] T. Tradler, *Infinity-inner-products on A-infinity-algebras*, 2001, math.AT/0108027.

[Tra02] T. Tradler, *The BV algebra on Hochschild cohomology induced by infinity inner products*, 2002, math.QA/0210150.

[TT06] J. Terilla and T. Tradler, *Deformations of associative algebras with inner products*, Homology, Homotopy Appl. **8** (2006), no. 2, 115–131 (electronic), math.QA/0305052.

Volodymyr Lyubashenko
Institute of Mathematics NASU
3 Tereshchenkivska st.
Kyiv-4, 01601 MSP
Ukraine
e-mail: lub@imath.kiev.ua

Oleksandr Manzyuk
Fachbereich Mathematik
Postfach 3049
67653 Kaiserslautern
Germany
e-mail: manzyuk@mathematik.uni-kl.de

Progress in Mathematics, Vol. 265, 647–667

Geometrization of Probability

Vitali D. Milman

Dedicated to the memory of Alexander (Sasha) Reznikov, a remarkable mathematician with tragic fate, and who called me his advisor, of which I was always proud.

Abstract. It was recently observed that asymptotic theory of high-dimensional convexity is extended in a very broad sense to the category of log-concave measures, and moreover, this extension is needed to understand and to solve some problems of asymptotic theory of high-dimensional convexity proper. Many important geometric inequalities were interpreted and extended to such category. On the other hand, some typical probabilisitic results are interpreted and proved in a geometric framework. Even more importantly, such extension to the log-concave category was needed to solve some central problems of a purely geometric nature. The goal of this article is to outline this development and to demonstrate examples of results which confirm this picture.

Mathematics Subject Classification (2000). 60D, 52A.

Keywords. Asymptotic geometric analysis; functional unequalities; log-concave measures; isotropicity.

1. Introduction

1.a. A few historical remarks

The framework of the subject we will discuss in this survey involves very high dimensional spaces (normed spaces, convex bodies) and accompanying asymptotic (by increasing dimension) phenomena.

The starting point of this direction was the open problems of Geometric Functional Analysis (in the '60s and '70s). This development naturally led to the Asymptotic Theory of Finite Dimensional spaces (in '80s and '90s). See the books [MS86], [Pi89] and the survey [LM93] where this point of view still prevails.

Supported in part by an Israel Science Foundation Grant and by the Minkowski Minerva Centre for Geometry.

During this period, the problems and methods of Classical Convexity were absorbed by the Asymptotic Theory (including geometric inequalites and many geometric, i.e., "isometric" as opposed to "isomorphic" problems).

As an outcome, we derived a new theory: Asymptotic Geometric Analysis. (Two surveys, [GM01] and [GM04] give a proper picture of this theory at this stage.)

One of the most important points of already the first stage of this development is a change in intuition about the behavior of high-dimensional spaces. Instead of the diversity expected in high dimensions and chaotic behavior, we observe a unified behavior with very little diversity. We analyze this change of intuition in [M98] and [M00]. We refer the reader to [M00] for some examples which illustrate this. Also in [M04], we attempt to describe the main principles and phenonema governing the asymptotic behavior of high-dimensional convex bodies and normed spaces.

Acknowledgement. I would like to thank B. Klartag for very useful advice.

1.b. "Convergence" of geometric functional analysis and classical convexity, creating asymptotic geometric analysis

In this introduction, we will give only one result from the past, but will present it in two different forms: one which corresponds to the spirit of Functional Analysis, and the other in the spirit of Convexity Theory. We will meet this result in our main text later. I mean the result which is often called the "Quotient of a Subspace Theorem".

Theorem [M85]. *There is a universal constant $c > 0$ such that for any λ, $1/2 \leq \lambda < 1$, and any n-dimensional normed space X, there exist subspaces $F \hookrightarrow E \hookrightarrow X$ with*

$$k = \dim E/F \geq \lambda n\,,$$

and

$$\operatorname{dist}(E/F, \ell_2^k) \leq c\frac{|\log(1-\lambda)|}{1-\lambda}\,.$$

Here $\operatorname{dist}(X, Y)$ is the (multiplicative) distance between two normed spaces X and Y which is called the Banach–Mazur distance, and which is formally defined by

$$\operatorname{dist}(X,Y) = \inf\big\{\|T\| \cdot \|T^{-1}\| \;\big|\; T : X \to Y \text{ is an invertible linear operator between spaces } X \text{ and } Y\big\}\,.$$

This distance is defined as infinity if such an invertible operator does not exist.

Some additional remarks: Of course, we may consider the proportion $\lambda > 0$ to be below $1/2$. In this case (i.e., for $0 < \lambda < 1/2$) there is another universal constant $C > 0$ such that

$$\operatorname{dist}\left(E/F, \ell_2^k\right) \leq 1 + C\sqrt{\lambda}\,.$$

However, this is already an automatic consequence of the well-known and old results of the Asymptotic Theory (see [MS86]).

But the case of λ to be close to 1 is of very special importance. This is already a structural fact. One may start to feel how we can approach and deal with an arbitrary convex body and normed space.

We now present the above theorem in a geometric form. We often call it the global version of the QS-Theorem.

Theorem [M91]. *Let $K \subset \mathbb{R}^n$ be a convex compact body and 0 be its barycenter. There are two linear operators $u_1, u_2 \in SL_n$, such that if $T = K \cap u_1 K$ then $Q = \mathrm{Conv}(T \cup u_2 T)$ is c-isomorphic to an ellipsoid $\mathcal{E}$ (for a universal constant $c > 0$), i.e., $\frac{1}{c}\mathcal{E} \subset Q \subset c\mathcal{E}$. Also, the volume of $\mathcal{E}$ remains the same as the volume of the original body K.*

Note, that constant c doesn't depend on the dimension n or the body K. It is universal, and to feel the meaning of the theorem, one should think of n being very large. In this sense, both theorems above are *asymptotic* and their meaning and strength are revealed when dimension n increases to infinity.

2. Extension of the category of convex bodies to the category of log-concave measures

Let us first define the class of log-concave measures and functions.

Definitions. A Borel measure μ on $\mathbb{R}^n$ is log-concave iff for any $0 < \lambda < 1$ and any $A, B \subset \mathbb{R}^n$ such that all involved sets $(A, B, \lambda A + (1-\lambda)B)$ are measurable

$$\mu(\lambda A + (1-\lambda)B) \geq \mu(A)^\lambda \mu(B)^{1-\lambda}.$$

Here λA is a homothety and $+$ is the Minkowski sum, i.e., $\lambda A + (1-\lambda)B = \{\lambda x + (1-\lambda)y \mid x \in A \text{ and } y \in B\}$.

A few very important examples of log-concave measures:

(i) The standard volume on $\mathbb{R}^n$, $\mu(K) = \mathrm{Vol}\, K$ (by Brunn–Minkowski inequality).
(ii) The restriction of volume on a convex set K: $\mu_K(A) = \mathrm{Vol}(K \cap A)$, K-convex.
(iii) Marginals of volume restricted to a convex set.

Let μ be a measure on $\mathbb{R}^n$ with the density function $f(x)$, i.e., $d\mu = f(x)dx$. Let E be a subspace of $\mathbb{R}^n$. Then we define marginal $\mathrm{Proj}_E\, \mu$ of μ on E the measure on E with density

$$(\mathrm{Proj}_E\, f) = \int_{x+E^\perp} f(y)dy\,,$$

where $E^\perp$ is the orthogonal subspace of E.

Obviously, marginals of log-concave measures are log-concave measures. In particular, for a convex set K, we consider the measure $\mu_K = 1_K dx$ (where 1_K is the characteristic function of K) and the marginals of this measure are log-concave measures.

Function $f(x) \geq 0$ is called log-concave if $\log f$ is concave, i.e., $f(x) = e^{-\varphi(x)}$ and φ is convex.

The connection between log-concavity of measures and functions was established by C. Borell [Bo74]: Let the support of a measure μ, $\operatorname{Supp}\mu$, not belong to any affine hyperplane. Then μ is log-concave iff μ is absolutely continuous on $\operatorname{Supp}\mu$ and the density f is a log-concave function.

Now we have many more examples of log-concave densities: Let $|x|$ define the standard euclidean norm on $\mathbb{R}^n$ and $\|x\|$ be any norm on $\mathbb{R}^n$. Then any of the following functions is the density of a log-concave measure:

(i) $e^{-|x|}$ (exponential distribution);
(ii) $\frac{1}{(\sqrt{2\pi})^n} e^{-|x|^2/2}$ (the gaussian distribution);
(iii) $e^{-\|x\|^p/p}$, for any norm and $1 \leq p < \infty$.

Log-concavity was used in Convexity Theory already from the 1950s (Henstock–MacBeath) and later, say, Prékopa–Leindler extension of Brunn–Minkowsky inequality (see [Pi89]), or the use of log-concave functions to study volume of sections of ℓ_p^n by Meyer–Pajor [MP88]. But a purely geometric study of log-concavity waited until the end of the 1980s, and was initiated by K. Ball [Ba86], who extended the study of some geometric problems of convexity to a larger category of log-concave measures. In particular, he studied isotropicity of such measures and connected it with isotropicity of convex bodies. He also considered some important geometric inequalities in the extended framework of log-concave measures ("functional versions" of geometric inequalities). However, just recently it was observed that such an extension is much broader than we thought, and is needed to understand and to solve some problems of asymptotic theory of high-dimensional convexity proper.

Three features characterize this extension.

(i) On the one hand, important geometric inequalities (and other kinds of geometric statements) are interpreted, extended and proved for log-concave measures.
(ii) On the other hand, some typical probabilistic results (and thinking) are interpreted and proved in a geometric framework.
(iii) And most importantly, an extension of the geometric approach to the log-concave category is needed to solve some central problems of a purely geometric nature.

The goal of this article is to demonstrate examples of results to confirm this picture.

We consider only finite measures, and only normalization distinguishes them from probability measures. This is the reason I call this extension "Geometrization of Probability". In this extension we identify K with the measure

$$\mu_K := \mathrm{Vol}_{|K} \quad (\text{i.e., } \mu(A) = \mathrm{Vol}(A \cap K)) .$$

3. Functional form of some geometric inequalities

3.a. Prékopa–Leindler inequality (functional version of Brunn–Minkowski inequality)

We introduce first sup-convolution which we call, following [AKM04], the Asplund product:

$$(f \star g)(x) = \mathrm{Sup}_{x_1+x_2=x}\, f(x_1)g(x_2).$$

EXAMPLE. $1_K \star 1_T = 1_{K+T}$.
Also λ-homothety for function is defined by

$$(\lambda \cdot f)(x) := f^{\lambda}\left(\frac{x}{\lambda}\right), \quad \lambda > 0$$

(So $f \star f = 2 \cdot f$)

In this language, the Prékopa–Leindler inequality stated that, for $f, g : \mathbb{R}^n \to [0, \infty)$, $0 < \lambda < 1$,

$$\int (\lambda \cdot f) \star ((1-\lambda) \cdot g) \geq \left(\int f\right)^{\lambda} \cdot \left(\int g\right)^{1-\lambda}.$$

In this formulation, Prékopa-Leindler is a functional analogue of the multiplicative, dimensional free, form of Brunn–Minkowski inequality:

$$\mathrm{Vol}\left(\lambda A + (1-\lambda)B\right) \geq (\mathrm{Vol}\, A)^{\lambda} \cdot (\mathrm{Vol}\, B)^{1-\lambda}$$

(for any subsets A and B of $\mathbb{R}^n$ and $0 < \lambda < 1$ such that all sets involved are measurable). Also "isomorphic" inequalities have their functional form. E.g. geometric statement:

Reverse Brunn–Minkowski inequality (Milman [M86]):
$\exists C$ such that for any convex, symmetric $K, P \subset \mathbb{R}^n$, there are linear transforms $T_K, T_P \in SL_n$ (where T_K depends solely on K, and T_P depends solely on P), such that if $\tilde{K} = T_K(K)$, $\tilde{P} = T_P(P)$, then

$$\mathrm{Vol}(\tilde{K} + \tilde{P})^{1/n} < C\left[\mathrm{Vol}(\tilde{K})^{1/n} + \mathrm{Vol}(\tilde{P})^{1/n}\right].$$

Its *functional analogue* is the following statement (Klartag–Milman, [KM05]): For any even log-concave $f, g : \mathbb{R}^n \to (0, \infty)$ there are $T_f, T_g \in SL_n$, such that $\tilde{f} = f \circ T_f$, $\tilde{g} = g \circ T_g$ satisfy

$$\left[\int \tilde{f} \star \tilde{g}\right]^{1/n} < C\left[\left(\int \tilde{f}\right)^{1/n} + \left(\int \tilde{g}\right)^{1/n}\right]$$

where T_f depends solely on f and T_g solely on g (and C is, as before, a universal constant).

3.b. Notion of polarity for log-concave measures; functional version of the Santaló inequality

Let $K \subset \mathbb{R}^n$, convex, $0 \in K$. The polar set K° is defined by

$$K^\circ := \left\{ x \in \mathbb{R}^n : (x, y) \le 1 \ \forall y \in K \right\}.$$

[Functional Analysis interpretation: If $K = -K$, $\|x\|_K$ – Minkowski functional of K, i.e., K is the unit ball of $X = (\mathbb{R}^n, \|\cdot\|_K)$. Then $X^* = (\mathbb{R}^n, \|\cdot\|_K^*)$ has K° its unit ball.]

Let D be the unit euclidean ball. The following well-known geometric fact is called *Blaschke–Santaló inequality*:

Let $K = -K$, then

$$|K| \cdot |K^\circ| \le |D|^2$$

(i.e., maximum is achieved on $K := D$).

Let us recall a well-known problem: What is $\min |K| \cdot |K^\circ|$ (Mahler, ∼'39)? The asymptotic answer to this problem is given in Bourgain–Milman [BM85;87]: $\exists c > 0$ universal such that

$$c \le \left(\frac{|K| \cdot |K^\circ|}{|D|^2} \right)^{1/n}.$$

Very recently, G. Kuperberg [Ku07] gave a different proof of this inequality which does not use the standard technique of the Asymptotic Theory.

For a general not necessarily centrally-symmetric convex body K, the Blaschke–Santaló inequality is also correct for a suitable shift of K: There exists x_0 such that, for $\widehat{K} = K - x_0$,

$$|\widehat{K}| \cdot |\widehat{K}^\circ| \le |D|^2$$

($\min_x |K| \cdot |(K - x)^\circ|$ is achieved for x_0 called the Santaló point of K; then 0 is the barycenter of $(K - x_0)^\circ$.)

Now the functional version of these inequalities:

We start with *Legendre transform*

$$\mathcal{L}\varphi(x) = \sup_{y \in \mathbb{R}^n} \left[(x, y) - \varphi(y) \right].$$

If φ is convex and low semi-continuous, then $\mathcal{L}\mathcal{L}\varphi = \varphi$.

We define polarity for non-negative functions by [AKM04]

$$f^\circ = e^{-\mathcal{L}(-\log f)}, \quad \text{i.e.,} -\log f^\circ = \mathcal{L}(-\log f),$$

or

$$f^\circ(x) = \inf_{y \in \mathbb{R}^n} \frac{e^{-(x,y)}}{f(y)}.$$

If f is log-concave upper semi-continuous then $(f^\circ)^\circ = f$.

EXAMPLES. For any convex body K, such that $0 \in \overset{\circ}{K}$,

$$1_K^\circ = e^{-\|x\|_{K^\circ}}, \quad \left(e^{-\|x\|_K^2/2}\right)^\circ = e^{-\|x\|_{K^\circ}^2/2}.$$

So, the following triple is associated with K:

$$\left(1_K; e^{-\|x\|_K^2/2}; e^{-\|x\|_K}\right)$$

and its polar

$$\left(1_{K^\circ}; e^{-\|x\|_{K^\circ}^2/2}; e^{-\|x\|_{K^\circ}}\right).$$

The only f such that $f^\circ = f$ is the standard Gaussian density, which plays the role of Euclidean ball D, in the "functional" extension of convexity theory we are discussing.

Some elementary properties of *polarity*:

$$(f \star g)^\circ = f^\circ \cdot g^\circ$$

(and therefore, for log-concave functions $(f \cdot g)^\circ = f^\circ \star g^\circ$);

$$(\lambda \cdot f)^\circ = (f^\circ)^\lambda$$

(note, that the dot-product $\lambda \cdot f$ here is the λ-homothety defined in 3a).

Theorem (Artstein, Klartag, Milman [AKM04]). *Let $f : \mathbb{R}^n \to \mathbb{R}^+$, $\int f < \infty$. Then:*

(i) *For some x_0 and $\widetilde{f}(x) = f(x - x_0)$,*

$$\int \widetilde{f} \cdot \int \widetilde{f}^\circ \le (2\pi)^n . \tag{1}$$

For log-concave f, we may take $x_0 = \int xf / \int f$.

In the case of f-even, obviously $x_0 = 0$, and the inequality (1) was proved by K. Ball in his thesis [Ba86].

(ii) $\min_{x_0} \int \widetilde{f} \cdot \int \widetilde{f}^\circ = (2\pi)^n$ *iff f is a.e. a gaussian density.*

The standard geometric Santaló inquality for convex bodies follows from (1): apply (1) to $f = e^{-\|x\|_K^2/2}$. Then $\int_{\mathbb{R}^n} f\, dx = c_n|K|$ where $c_n = (2\pi)^{n/2}/|D|$, and similarly for f° which implies Blaschke–Santaló's inequality.

Let us repeat the statement without using the polarity notion:

Theorem [AKM04]. *Let $f : \mathbb{R}^n \to \mathbb{R}^+$, $\int f < \infty$. Then, for some x_0,*

$$\int f \cdot \int \left[\inf_{y \in \mathbb{R}^n} \frac{e^{-(x,y)}}{f(y)}\right] e^{-(x,x_0)} dx \le (2\pi)^n. \tag{2}$$

For log-concave f, we may take $x_o = \int xf / \int f$. Also, $\min_{x_0}$ of that expression is equal to $(2\pi)^n$ iff f is a.e. a gaussian density function $f(x) = \exp[(Ax, x) + (x, z) + a]$ for some vector z, and $a \in \mathbb{R}$ and an operator $A \ge 0$.

Also the reverse inequality is true in the functional form.

Theorem (Klartag–Milman [KM05]). *$\exists c > 0$, such that for every log-concave $f : \mathbb{R}^n \to \mathbb{R}^+$, $\int f < \infty$, we have*

$$c < \left(\int_{\mathbb{R}^n} f \cdot \int_{\mathbb{R}^n} f^\circ \right)^{1/n} .$$

We call a function $f \geq 0$ α-concave ($0 < \alpha < \infty$) if $f^{1/\alpha}$ is concave on $\operatorname{Supp} f$.

Important Example: Let $K \subset \mathbb{R}^{n+\alpha}$ be a convex set and E be a subspace, $\dim E = n$. Then, $f := \operatorname{Proj}_E \mathbb{1}_K$ is α-concave. Obviously, an α-concave function is log-concave.

FACT. *Any log-concave function $f : \mathbb{R}^n \to [0, \infty)$ is locally uniform on $\mathbb{R}^n$ approximated by α-concave functions f_α, $f_\alpha(x) \to f(x)$ ($\alpha \to \infty$), for*

$$f_\alpha(x) = \left(1 + \frac{\log f(x)}{\alpha}\right)_+^\alpha \leq f(x) ,$$

Here, $\varphi(x)_+ = \max\{\varphi(x); 0\}$.

Define "α-duality" by

$$\mathcal{L}_\alpha f(x) = \inf_{y; f(y)>0} \frac{\left(1 - \frac{(x,y)}{\alpha}\right)_+^\alpha}{f(y)} \leq f^\circ(x) ,$$

for $\alpha \geq 1$, and

$$\mathcal{L}_\alpha f(x) = \inf_{y; f(y)>0} \frac{(1 - (x,y))_+^\alpha}{f(y)} ,$$

for $0 < \alpha < 1$. Clearly, $\mathcal{L}_\alpha f$ is α-concave. Note also that $\mathcal{L}_\alpha f \to f^\circ$ for $\alpha \to \infty$ and $\mathcal{L}_\alpha 1_K \to 1_{K^\circ}$ for $\alpha \to 0$ and 1_T is the characteristic function of the set T.

FACT. *If f is upper semicontinuous and α-concave, $f(0) > 0$, then $\mathcal{L}_\alpha \mathcal{L}_\alpha f = f$.*

Theorem [AKM04]. *Let f be α-concave on $\mathbb{R}^n$, α is an integer, $\mathbb{E} f < \infty$ and $\int x f(x) = 0$. Then*

$$\int_{\mathbb{R}^n} f \cdot \int_{\mathbb{R}^n} \mathcal{L}_\alpha(f) \leq \frac{\alpha^n \kappa_{n+\alpha}^2}{\kappa_\alpha^2} \quad \left(\underset{\alpha \to \infty}{\longrightarrow} (2\pi)^n \right) , \tag{3}$$

where $\kappa_k = \operatorname{Vol} D_k$, and the inequality is exact.

Historical remark: The origin of the transform $\mathcal{L}_\alpha$ is from the 1960s. I searched for duality for new moduli, I worked with. Today they are called "asymptotic moduli".

The necessary transform was [M71a]

$$K\varphi = \operatorname{Sup}_y \frac{(x,y) + 1}{\varphi} .$$

To deal with this transform we consider the following substitutions. We consider the function $f = \varphi - 1$ and the transform $L_1 f = K\varphi - 1$ to come to

$$L_1(f) = \operatorname{Sup}_y \frac{(x,y) - f(y)}{1 + f(y)} .$$

Consider it as a part of the family L_μ:

$$L_\mu(f) = \mathrm{Sup}_{y\in\mathbb{R}^{n-1}} \frac{(x,y)-f(y)}{1+\mu f(y)}\,,$$

where f is convex on $\mathbb{R}^{n-1}$. Of course, $\mu = 0$ gives the Legendre transform.

To understand the meaning and inversion formula introduce a norm on $\mathbb{R}^n$:

$$\left\|\left(y;\frac{1}{\sqrt{\mu}}\right)\right\| = \frac{1+\mu f(y)}{\sqrt{\mu}}\,.$$

Then

$$\left\|\left(x;\frac{1}{\sqrt{\mu}}\right)\right\|^* = \frac{1+\mu L_\mu f(y)}{\sqrt{\mu}}\,.$$

Reflexivity of finite-dimensional space implies

$$L_\mu L_\mu f = f\,.$$

Interestingly, only $\mu = 0$, i.e., the case of the Legendre transform proper, lacks this geometric interpretation.

The inequality (3) was written in [AKM04] only for integer value of α. We take later $\alpha \to \infty$ to derive the inequality (2). However, a natural tensoration argument provides a similar inequality for any rational $\alpha > 0$ and, taking the limit, also any $\alpha > 0$. Such tensoration arguments were used by Klartag for proving Theorem 2.1 in [K07a]. At the same time, it is also a particular case of the result by Fradelizi–Meyer [FM07]. They prove the following fact.

Theorem [FM07]. *Let $\rho : \mathbb{R}^+ \to \mathbb{R}^+$ be a log-concave nonincreasing function and let φ be a convex function such that $0 < \int_{\mathbb{R}^n} \rho(\varphi(x))dx < \infty$. Define a shifted Legendre transform $\mathcal{L}^z$ by*

$$\mathcal{L}^z\varphi(y) = \mathrm{Sup}_x\left((x-z,y-z)-\varphi(x)\right)$$

for any $y \in \mathbb{R}^n$. Then, for some $z \in \mathbb{R}^n$,

$$\int_{\mathbb{R}^n} \rho(\varphi(x))dx \int \rho\big(\mathcal{L}^z(\varphi(y))\big)dy \le \left(\int_{\mathbb{R}^n} \rho\left(\frac{|x|^2}{2}\right)dx\right)^2.$$

([FM07] *also provides equality conditions under the condition that ρ is a decreasing function). The particular cases corresponding to functions $\rho(t) = e^{-t}$ and $\rho(t) = (1-t)_+^\alpha$ lead to the previous results from* [AKM04].

There are many inequalities in the spirit of the above theorems. Some of them may be developed by the original approach of Ball [Ba86], and also by the method of [AKM04] or using the correspondence between log-concave functions and convex bodies as was put forward by Ball in [Ba86], [Ba88] and used in [KM05]. For other inequalities in this style, see [FM07]. However, we will concentrate our attention on some surprizing extensions which appeared in attempts to answer a question raised by D. Cordero-Erausquin.

He conjectured the following (very unusual) inequality:

Let K and T be any convex centrally symmetric bodies and D be the Euclidean ball. Is it true that

$$Vol(K \cap T) \cdot (K^\circ \cap T) \leq \mathrm{Vol}(D \cap T)^2? \tag{4}$$

He proved this conjecture [C02] for the case where K and $T \subset \mathbb{R}^{2n}$ could be realized as unit balls of complex Banach norms and, in addition T is invariant under complex conjugation. One may see (4) as a " localization" of the standard Blaschke–Santaló inequality.

The surprizing fact is that the functional version of (4) has been proved by Klartag [K07a] and Barthe–Cordero-Erausquin (unpublished) but the geometric conjecture (4) does not follow from it (or, at least, we can't see how it may follow). So, the proved theorem is

Theorem (Klartag [K07a]; Barthe–Cordero-Erausquin). *Let $f : \mathbb{R}^n \to (-\infty, \infty]$ be an even measurable function, and assume that μ is an even log-concave measure on $\mathbb{R}^n$. Then,*

$$\int_{\mathbb{R}^n} e^{-f} d\mu \int_{\mathbb{R}^n} e^{-\mathcal{L}f} d\mu \leq \left(\int_{\mathbb{R}^n} e^{-\frac{|x|^2}{2}} d\mu \right)^2,$$

whenever at least one of the integrals on the left-hand side is both finite and non-zero.

To describe one geometric consequence, we need the following:

Definition. If A is the unit ball of the norm $\|\cdot\|_A$ and B is the unit ball of the norm $\|\cdot\|_B$, then $A \cap_2 B$ is defined as the unit ball of the norm $\|x\|_{A\cap_2 B} = \sqrt{\|x\|_A^2 + \|x\|_B^2}$.

Corollary (Klartag [K07a]). *Let $K, T \subset \mathbb{R}^n$ be centrally-symmetric, convex bodies. Then,*

$$\mathrm{Vol}_n(K \cap_2 T)\, \mathrm{Vol}_n(K^\circ \cap_2 T) \leq \mathrm{Vol}_n(D \cap_2 T)^2.$$

Note that $A \cap B \subset A \cap_2 B \subset \sqrt{2}(A \cap B)$ for any centrally-symmetric convex sets $A, B \subset \mathbb{R}^n$. Thus, the theorem immediately implies that

$$\mathrm{Vol}_n(K \cap T)\, \mathrm{Vol}_n(K^\circ \cap T) \leq 2^n\, \mathrm{Vol}_n(D \cap T)^2.$$

Let us show one more fact in this spirit from [K07a].

Let $\psi : \mathbb{R}^n \to (-\infty, \infty]$ be a convex, even function, and let $\alpha > 0$ be a parameter. Let μ be a measure on $\mathbb{R}^n$ whose density $F = d\mu/dx$ is

$$F(x) = \int_0^\infty t^{n+1} e^{-\alpha t^2} e^{-\psi(tx)} dt\,.$$

Then, for any centrally-symmetric, convex body $K \subset \mathbb{R}^n$,

$$\mu(K)\mu(K^\circ) \leq \mu(D)^2.$$

An example of a measure which is covered by this theorem is, e.g. the measure with density $\frac{1}{(1+\|x\|^2)^{n+2}}$ where $\|\cdot\|$ is a norm on $\mathbb{R}^n$. So, such measures may have "heavy tails" and not be log-concave.

3.c. Functional form of Urysohn inequality
(Urysohn inequality for log-concave functions)

Recall the classical Urysohn inequality:

$$\left(\frac{\operatorname{Vol} K}{\operatorname{Vol} D}\right)^{1/n} \le M^\star(K) := \int_{S^{n-1}} \sup_{y\in K}(x,y)d\sigma(x)$$

and, by Steiner formula,

$$\operatorname{Vol}(D+\varepsilon K) = \operatorname{Vol} D + \varepsilon n M^\star(K)\operatorname{Vol} D + O(\varepsilon^2)\,.$$

So, we may define the analogous quantity. Let $G(x) = e^{-|x|^2/2}$. Then define

$$V_G(f) = \lim_{\varepsilon\to O+} \frac{\int G\star[\varepsilon\cdot f] - \int G}{\varepsilon}$$

(one may show that lim exists).

Denote $M^\star(f) = 2\frac{V_G(f)}{n\int G} = \frac{V_G(f)}{\frac{n}{2}(2\pi)^{n/2}}$. Then $M^\star(G) = 1$.

If $f = 1_K$ then (calculation)

$$V_G(1_K) = \frac{(2\pi)^{\frac{n-1}{2}} n\kappa_n}{\kappa_{n-1}} M^\star(K)$$

($\kappa_n = \operatorname{Vol} D_n$).

So $M^\star(K) = c_n M^\star(1_K)$ for $c_n \sim \sqrt{n}$.

The quantity $M^*(f)$ has the following properties:

(i) Linearity:: $M^\star(f\star g) = M^\star(f) + M^\star(g)$;
(ii) Homogenuity:: $M^\star(\lambda\cdot f) = \lambda M^\star(f)$, $\lambda > 0$.

Theorem [KM05]. *Let $f:\mathbb{R}^n\to[0,\infty]$ be an even log-concave function such that $\int f = \int G$ $(=(2\pi)^{n/2})$. Then*

$$M^\star(f) \ge M^\star(G) = 1\,.$$

3.d. Mixed measures – what are they?

Introducing $M^*(f)$ in the previous section creates a feeling that there is a natural and clear notion of mixed measures which extends the notion of mixed volumes. However, the situation is not so, and what mixed measures are is absolutely not yet clear to me. This stage of "geometrization of probability" is still ahead of us.

We see only some examples, mostly on the level of "experiments", which demonstrate, however, the high interest the theory should generate. I will describe below a couple of examples (from Klartag [K07a]).

For $f:\mathbb{R}^n\to[0,\infty)$, concave on $\operatorname{Supp} f$, define a variant of the Legendre transform

$$\mathcal{L}'f = \sup_{y;f(y)>0}\left[f(y) - (x,y)\right]$$

(note $\mathcal{L}'f$ is convex).

For $f_i : \mathbb{R}^n \to [0,\infty)$, $i = 0,1,\dots,n$, compactly supported, concave on their Supp, denote

$$V(f_0,\dots,f_n) = \int_{\mathbb{R}^n} [\mathcal{L}' f_o](x) D\big(\mathrm{Hess}[\mathcal{L}' f_1](x),\dots,\mathrm{Hess}[\mathcal{L}' f_n]\big)\,dx\,.$$

(See the Appendix for a definition and a few properties of mixed discriminants $D(A_1,\dots,A_n)$ of matrices $A_i \geq 0$.)

Then the following is true: The multilinear form V is

(i) fully symmetric with respect to permutations of $\{0,1,\dots,n\}$;
(ii) monotone; i.e., if f_i and g_i as above and $f_i \leq g_i$ then $V(f_0,\dots,f_n) \leq V(g_0,\dots,g_n)$.
(iii) satisfies "hyperbolic" Alexandrov–Fenchel type inequality

$$V(f_0,f_1,\dots,f_n)^2 \geq V(f_0,f_0,f_2\dots,f_n)\cdot V(f_1,f_1,f_2,\dots,f_n).$$

And now "the dual" statement:

Let $K \subset \mathbb{R}^n$ be convex compact. Let $f_i : \mathbb{R}^n \to [0,\infty)$, $i = 0,1,\dots,n$, be concave, vanishing on ∂K, with bounded second derivatives in $\overset{\circ}{K}$. Denote:

$$I(f_0,\dots,f_n) = \int_K f_0(x) D(-\mathrm{Hess}\, f_1,\dots,-\mathrm{Hess}\, f_n)\,dx\,.$$

Then, the multilinear form I is:

(i) fully symmetric with respect to permutations;
(ii) monotone (in the above class of functions);
(iii) the following "elliptic-type" inequality is satisfied:

$$I(f_0,f_1,\dots,f_n)^2 \leq I(f_0,f_0,f_2\dots,f_n)\cdot I(f_1,f_1,f_2,\dots,f_n)\,.$$

So, the Legendre transform "transforms" elliptic type inequalities into hyperbolic type! Why? We could not observe this kind of phenomenon in the category of convex sets because the functional duality is not closed in this category.

4. A Central Limit Theorem (CLT) for convex sets and log-concave measures

In the classical geometric approach, we study a geometric shape of projections (or sections) of convex body K, and we know that they are, with high probability, close to euclidean balls for small enough rank of projections.

The exact old estimate stated [M71b] that, with high probability, a random projection P_E of a convex body K in $\mathbb{R}^n$ of rank $k^* < cn\big(\frac{M^*(K)}{\mathrm{diam}\, K}\big)^2$ is isomorphic upto a constant 2 to a euclidean k^*-dimensional ball. Here c is a universal constant, $M^*(K)$ was defined in 3.c and $\mathrm{diam}\, K$ is the diameter of K.

But what about measure projections (marginals) of convex bodies in place of geometric projections? This question was first asked by Gromov [Gr88]. He made some initial observations, but recently the structure of random marginals was understood completely. To describe the results we need some notions.

Normalize the convex body $K \subset \mathbb{R}^n$ such that

$$\operatorname{Vol} K = 1, \quad \int_K \vec{x}dx = 0, \quad \int_K \langle x, \theta \rangle^2 dx = |\theta|^2 L_K^2,$$

for any $\theta \in \mathbb{R}^n$. We say that K is in "isotropic" position and the constant L_K is called the isotropic constant of K.

Theorem (Klartag [K07b], [K07c]). *Suppose $K \subset \mathbb{R}^n$ is convex and isotropic, and X is distributed uniformly in K. Then $\exists \Theta \subset S^{n-1}$ with $\sigma_{n-1}(\Theta) \geq 1 - \delta_n$, such that for $\theta \in \Theta$,*

$$\sup_{A \subset \mathbb{R}} \left| \operatorname{Prob}\left\{ \langle X, \theta \rangle \in A \right\} - \frac{1}{L_K \sqrt{2\pi}} \int_A e^{-t^2/2L_K^2} dt \right| \leq \epsilon_n .$$

Here, say, $\delta_n < \exp(-cn^{0.9})$, $\epsilon_n < Cn^{-1/100}$.

Progress towards this goal was obtained earlier by Brehm–Voigt [BV00] and Anttila–Ball–Perissinaki [ABP03]. There is an analogue multi-dimensional version

Theorem (Klartag [K07b]). *Let $K \subset \mathbb{R}^n$ be convex and isotropic. The r.v. X is distributed uniformly in K. Suppose $\epsilon > 0$ and $k < c\epsilon^2 \frac{\log n}{\log \log n}$.*

Then $\exists \mathcal{E} \subset G_{n,k}$ with $\sigma_{n,k}(\mathcal{E}) \geq 1 - \exp(-cn^{0.9})$, such that for $E \in \mathcal{E}$,

$$\sup_{A \subset E} \left| \operatorname{Prob}\left\{ \operatorname{Proj}_E(X) \in A \right\} - \frac{1}{L_K^k} \int_A \frac{e^{-|x|^2/2L_K^2}}{(2\pi)^{k/2}} dx \right| \leq \epsilon .$$

Very recently, Klartag [K07c] improved all estimates in the two previous results: instead of log-type estimates in the previous result, he proved a polynomial type estimate. This means that there is a principle difference between the dimension k^* such that geometric shape of projections on subspaces of this dimension can be approximately euclidean and the dimension of marginals which are approximately gussian. In the first case, in some examples, say a cross-polytope – the unit ball of ℓ_1^n space, k^* cannot be above $\sim \log n$, but in the second case we have $\sim$ gaussian marginals in dimensions of the order of say $n^{1/20}$.

5. Isotropic position and isotropic constant

We again recall that a convex body $K \subset \mathbb{R}^n$, with the barycenter of K at 0, is in isotropic position iff $\operatorname{Vol} K = 1$ and, $\forall i, j = 1, \dots, n$,

$$\int_K x_i x_i ds = \delta_{ij} L_K^2$$

($x = (x_1, \dots, x_n)$). We call L_K the isotropic constant of K. It is an old and famous problem of Bourgain if isotropic constants $\{L_K\}$ are uniformly bounded (by dim. n and convex bodies in $\mathbb{R}^n$). A well-known 20-year-old estimate of Bourgain's states that $L_K \leq Cn^{1/4} \log n$. However, recently Klartag proved

Theorem (Klartag [K06]). *For any convex body $K \subset \mathbb{R}^n$ and $\epsilon > 0$ there exists a convex body $T \subset \mathbb{R}^n$, such that*

$$(1-\epsilon)T \subset K - x_0 \subset (1+\epsilon)T$$

and $L_T < c/\sqrt{\epsilon}$.

Corollary (Klartag [K06], relying on Paouris' recent theorem [P06]).

$$L_K < Cn^{1/4} \quad \text{when } K \subset \mathbb{R}^n .$$

It is important to note that the proof of the last theorem requires the extension of Asymptotic Theory of Convexity to the category of log-concave measures.

In a very rough sketch of his proof, Klartag considered the 'momentum' map

$$F(x) = \log \int_K e^{\langle x,y\rangle} dy$$

(K is a convex body in the isotropic position) which produces (by considering gradient) the transportation of measure from $\mathbb{R}^n$ to K. This creates the family $\{f_x(y) = e^{\langle x,y\rangle} 1_K(y)\}_{x \in nK^\circ}$ of log-concave densities.

The boundedness of the isotropic constant for any of these measures (the isotropic constant of a measure should be defined) would imply the theorem (it would construct an approximation T). In the next step, this fact is proved in the average (which means the existence of one such measure). The proof uses the reverse Santaló inequality [BM87].

To give some details of the proof of the theorem, we need to establish a connection between log-concavity and convex bodies.

For any even log-concave $f : \mathbb{R}^n \to \mathbb{R}^+$ we associate a norm (K. Ball [Ba86])

$$\|x\|_f = \left(\int_0^\infty nf(rx) r^{n-1} dr \right)^{-1/n} .$$

Denote K_f be the unit ball of $\| \cdot \|_f$.

Let us note a few properties of this correspondence:

1. $\operatorname{Vol} K_f = \int f$
2. Define $\overline{\overline{K}}_f = \{x \in \mathbb{R}^n : f(x) > e^{-n}\}$. Then, for a universal $c > 0$,

$$K_f \subset \overline{\overline{K}}_f \subset cK_f .$$

3. Let f and g be log-concave functions and $f(0) = g(0) = 1$. Then, for some universal constants c_1 and c_2

$$c_1 K_{f\star g} \subset K_f + K_g \subset c_2 K_{f\star g}$$

and

$$c_1 n K_f^\circ \subset K_{f^\circ} \subset c_2 n K_f^\circ .$$

Let us now define the isotropic constant of a log-concave measure. We say that f is in the *isotropic position* if

$$\operatorname{Sup}_{x\in\mathbb{R}^n} f(x) = 1 = \int f(x)dx \quad \text{and}$$

$$\int_{x\in\mathbb{R}^n} x_i x_j\, f dx = \delta_{ij} L_f^2$$

and the constant L_f is called the *isotropic constant* of the measure fdx.

One may write a formula for L_f without "putting" fdx in the isotropic position,

$$L_f = \left(\frac{\operatorname{Sup}_{x\in\mathbb{R}^n} f(x)}{\int_{\mathbb{R}^n} fdx}\right)^{1/n} (\det \operatorname{Cov} f)^{1/2n}$$

where covariance matrix

$$\operatorname{Cov} f = \big(\operatorname{Cov}_f(x_i, x_j)\big)\,,$$

$$\operatorname{Cov}_f(x_i, x_j) = \frac{\int_{\mathbb{R}^n} x_i x_j f dx}{\int f dx} - \frac{\int x_i f}{\int f} \cdot \frac{\int x_j f}{\int f}\,.$$

Then, for any K convex, $L_K = L_{1_K}$.

A sketch of Klartag's proof of a solution of the "isomorphic" slicing problem. Let K be convex compact, $O \in K$, $\operatorname{Vol} K = 1$. We will divide the proof into a few steps, and we will refer to [K06] for the proofs which will not be presented.

1. Let $f : K \to [0, \infty)$ be a log-concave function. Assume

$$\left(\frac{\operatorname{Sup}_{x\in K} f}{\inf_{x\in K} f}\right)^{1/n} < C\,.$$

Then K_f isomorphic to K, i.e., $\exists\, c_1 := c_1(C)$ such that

$$\tfrac{1}{c_1} K_f \subset K \subset c_1 K_f$$

(here, as before,

$$K_f = \{x \in \mathbb{R}^n;\ \int_0^\infty n f(rx) r^{n-1} dr \geq 1\}$$

and is a convex set by K. Ball).

2. (K. Ball [Ba86]) $L_f \simeq L_{K_f}$. So, our goal is to find such an f that $L_f <$ const. (which implies that $L_{K_f} <$ const.).
3. Consider a (convex) function $F_K(x) := F(x)$

$$F(x) = \log \int_K e^{\langle x,y\rangle} dy\,.$$

   (a) This function produces a transportation of measure

$$\nabla F := \psi : \mathbb{R}^n \longrightarrow \overset{\circ}{K}$$

(similar to the so called 'momentum' map). Recall the notation of transportation. Let μ_1 and μ_2 be two Borel measures in $\mathbb{R}^n$ and $T : \mathbb{R}^n \to \mathbb{R}^n$ such that, for any measurable set $A \subset \mathbb{R}^n$,

$$\mu_2(A) = \mu_1(T^{-1}A)\,.$$

Then we say that T transports μ_1 to μ_2. Equivalently, $\forall \varphi \in C^+(\mathbb{R}^n)$

$$\int_{\mathbb{R}^n} \varphi(x) d\mu_2(x) = \int_{\mathbb{R}^n} \varphi(Tx) d\mu_1(x)\,.$$

The following fact is straightforward.

Fact. *Let $F : \mathbb{R}^n \to \mathbb{R}$ be C^2-smooth strictly convex and $K = \mathrm{Im}(\nabla F)$. Let measure μ have density $\frac{d\mu}{dx} = \det \mathrm{Hess}\, F(x)$. Then $\nabla F : \mathbb{R}^n \to \mathbb{R}^n$ transports μ to $\mathrm{Vol}_{|K}$.*

Applying this to our situation, we see that ∇F transports the measure μ to the uniform measure on K.
Using this, we see that, for any measurable set $A \subset \mathbb{R}^n$,

$$\int_A \det \mathrm{Hess}\, F = \mathrm{Vol}((\nabla F)A) \leq 1\,.$$

(b) Note that $\nabla F(x) = \int y\, d\mu_{K,x}(y)$ and the density of $\mu_{K,x}$ is

$$\frac{e^{\langle x,y\rangle} 1_K(y)}{\int_K e^{\langle x,z\rangle} dz}\,.$$

Also $\mathrm{Hess}(F)(x) = \mathrm{Cov}(\mu_{K,x})$. Therefore

$$\det \mathrm{Hess}\, F(x) = \left(\int f_x / \operatorname{Sup} f_x \right)^2 \cdot L_{f_x}^{2n}$$

where $f_x(y) = e^{\langle x,y\rangle} 1_K(y)$.
So, we consider the family of log-concave functions and we search for a function as in 1. and 2. inside this family.

4. Let $x \in nK^\circ$. (Note that the volume $|K| = 1$ implies $|nK^\circ|^{1/n} \sim 1$ by the Bourgain–Milman reverse Santaló inequality [BM87].)
   (a) Then

$$\left(\frac{\operatorname{Sup}_{y\in K} f_x(y)}{\inf_{y\in K} f_x(y)} \right)^{1/n} < C.$$

   Indeed, $\operatorname{Sup}_{y\in K} f_x(y) = \operatorname{Sup}_{y\in K} e^{\langle x,y\rangle} \leq e^{\|x\|^*} \leq e^n$. (Similarly for $\inf \geq e^{-n}$.) So we know that $K_{f_x} \sim K$ for any $x \in nK^\circ$.
   (b) We want to find $x \in nK^\circ$ such that $L_{f_x} <$ Const., i.e., to estimate from above by some constant

$$\big(\det \mathrm{Hess}\, F(x) \big)^{1/2n} \left(\frac{\operatorname{Sup} f_x}{\int f_x} \right)^{1/n}.$$

Actually, it is enough to find $x \in nK^\circ$ such that

$$\det \operatorname{Hess} F(x) < \text{Const.}^n.$$

We prove this "on average":

$$\frac{1}{|nK^\circ|}\int_{nK^\circ} \det \operatorname{Hess} F(x) \le \frac{1}{|nK^\circ|} \operatorname{Vol}(\operatorname{Im}(\nabla F)) \le \frac{1}{|nK^\circ|} \le C^n$$

(this is the reverse Santaló inequality we already mentioned).

6. Is further extension possible?

Does the family of log-concave measures (we discussed in Sections 2 and 3) represent the largest class of probability measures where Geometry is extended so naturally?

This is not clear. But let us consider a much larger class of "convex measures" (I also like the terminology "hyperbolic measures").

In Section 3b, we introduced the class of α-concave functions for $0 < \alpha < \infty$. We used there the terminology from [GrM87]. We now extend this class to negative α but also we will change the notation and follow C. Borell's approach. The new "s-concavity", for positive s, will correspond to $1/\alpha$-concavity above, i.e., $s = 1/\alpha$.

Definition (C. Borell, '74). Fix $-\infty \le s \le 1$; a measure μ on $\mathbb{R}^n$ is s-concave iff $\forall\, A, B \subset \mathbb{R}^n$ non-empty and measurable, $t \in (0,1)$,

$$\mu\big(tA + (1-t)B\big) \ge \big(t\mu(A)^s + (1-t)\mu(B)^s\big)^{1/s}.$$

Note, for $s = 0$, it is exactly the log-concavity condition:

$$\mu\big(tA + (1-t)B\big) \ge \mu(A)^t \mu(B)^{1-t},$$

and, for $s = -\infty$,

$$\mu\big(tA + (1-t)B\big) \ge \min\big(\mu(A), \mu(B)\big)\,. \tag{5}$$

Denote $\mathcal{M}(s)$ the class of all finite s-concave measures. Clearly $\mathcal{M}(s_1) \supseteq \mathcal{M}(s_2)$ for $s_1 < s_2$, and so for any s an s-concave measure satisfies (5).

New example: Cauchy distribution with density

$$p(x) = \frac{c_n}{(1+|x|^2)^{\frac{n+1}{2}}};$$

in this case $s = -1$ (the so-called "heavy tails" distributions).

C. Borell:

(i) $\forall\,\mu \in \mathcal{M}(-\infty)$ has a convex $\operatorname{supp} K \subset \mathbb{R}^n$ and μ is absolutely continuous (w.r.t Lebesgue measure) on K;

(ii) If μ is s-concave, then $s \le 1/\dim K$;

(iii) If $\dim K = n$, the density p of μ satisfies $\forall\, x, y \in K$

$$p\big(tx + (1-t)y\big) \geq \big(tp(x)^{s_n} + (1-t)p(y)^{s_n}\big)^{1/s_n}$$

for $s_n = \frac{s}{1-ns}$. (So, if μ is log-concave then also its density is a log-concave function; however, if $s = -\infty$ then its density is $(-1/n)$-concave.)

Also, levels of densities of convex measures are boundaries of convex sets. Recently, interest in s-concave measures for negative s has been revived; see [B06].

Connection with the classical convexity. The definition of convex measures corresponds to the unified principle behind most (or, perhaps, all) geometric inequalites, a principle of minimization:

$$f(A; B) \geq \min\big\{f(A; A), f(B; B)\big\}$$

["the minimum is achieved on equal objects"].

EXAMPLES.

(i) Alexandrov–Fenchel inequality is equivalent to the above miminization principle

$$V(A; B; C_1, \dots) \geq \min\big(V(A; A; C_1, \dots); V(B; B; C_1, \dots)\big).$$

(ii) Brunn–Minkowski inequality: $\forall\, t, \tau > 0$ and A, B convex:

$$|tA + \tau B|^{1/n} \geq t|A|^{1/n} + \tau|B|^{1/n}$$

is again equivalent to the minimization principle:

$$|tA + \tau B| \geq \min\big(|(t+\tau)A|, |(t+\tau)B|\big).$$

And so on (see, [GM04] for more examples and a discussion on this subject).

Is this an incidental similarity? Or does a deeper meaning lie behind it?

Appendix: Mixed discriminants

Consider the space S_n of real symmetric $n \times n$ matrices. We polarize the function $A \to \det A$ to obtain the symmetric multilinear form

$$D(A_1, \dots, A_n) = \frac{1}{n!} \sum_{\varepsilon \in \{0,1\}^n} (-1)^{n + \sum \varepsilon_i} \det\Big(\sum \epsilon_i A_i\Big),$$

where $A_i \in S_n$. Then, if $t_1, \dots, t_m > 0$ and $A_1, \dots, A_m \in S_n$, the determinant of $t_1 A_1 + \dots + t_m A_m$ is a homogeneous polynomial of degree n in t_i, which we write in the form

$$\det(t_1 A_1 + \dots + t_m A_m) = \sum_{1 \leq i_1 \leq \dots \leq i_n \leq m} n! D(A_{i_1}, \dots, A_{i_n}) t_{i_n} \cdots t_{i_n},$$

where the coefficients $D(A_{i_1}, \dots, A_{i_n})$ are independent of permutations of variables A_i. The coefficient $D(A_1, \dots, A_n)$ is called the mixed discriminant of $A_1, \dots, A_n$. Note that $D(A, \dots, A) = \det A$. The fact that the polynomial $P(t) = \det(A +$

tI) has only real roots for any $A \in S_n$ plays the central role in the proof of a number of very interesting inequalities connecting mixed discriminants, which are quite similar to the classical Newton inequalities. They were first discovered by Alexandrov [Al38] in one of his approaches to what is now called Alexandrov–Fenchel inequalities. Today, they are part of a more general theory (see, e.g., [H94] or the Appendix in [K07a]). For example, if all matrices involved are positive, Alexandrov proved,

$$D(A_1, \ldots, A_{n-2}, B, C)^2 \geq D(A_1, \ldots, A_{n-2}, B, B) \cdot D(A_1, \ldots, A_{n-2}, C, C).$$

There are many interesting inequalities for matrices which are corollaries of this remarkable inequality. For example,

$$D(A_1, A_2, \ldots, A_n) \geq \prod_{i=1}^{n} [\det A_i]^{1/n}.$$

References

[Al38] ALEXANDROV, A.D, On the theory of mixed volumes of convex bodies IV : Mixed discriminants and mixed volumes (in Russian), Math. Sb. N.S. 3 (1938), 227–251.

[ABP03] ANTTILA, M., BALL, K., PERISSINAKI, I., The central limit problem for convex bodies, Trans. Amer. Math. Soc. 355:12 (2003), 4723–4735.

[AKM04] ARTSTEIN-AVIDAN, S., KLARTAG, B., MILMAN, V., The Santaló point of a function, and a functional form of the Santalo inequality, Mathematika 51:1-2 (2004), 33–48.

[Ba86] BALL, K., Isometric problems in ℓ_p and sections of convex sets, PhD dissertation, Cambridge (1986).

[Ba88] BALL, K., Logarithmically concave functions and sections of convex sets in R^n, Studia Math. 88:1 (1988), 69–84.

[B06] BOBKOV, S.G., Large deviations and isoperimetry over convex probability measures, preprint (2006).

[Bo74] BORELL, C., Convex measures on locally convex spaces, Ark. Mat. 12 (1974), 239–252.

[BM85] BOURGAIN, J., MILMAN, V.D., Sections euclidiennes et volume des corps symétriques convexes dans $\mathbb{R}^n$ (in French), C.R. Acad. Sci. Paris Ser. I Math. 300:13 (1985), 435–438.

[BM87] BOURGAIN, J., MILMAN, V.D., New volume ratio properties for convex symmetric bodies in $\mathbb{R}^n$, Invent. Math. 88:2 (1987), 319–340.

[BV00] BREHM, U., VOIGT, J., Asymptotics of cross sections for convex bodies, Beiträge Algebra Geom. 41:2 (2000), 437–454.

[C02] CORDERO-ERAUSQUIN, D., Santaló's inequality on $\mathbb{C}^n$ by complex interpolation. C.R. Math. Acad. Sci. Paris 334:9 (2002), 767–772.

[FM07] FRADELIZI, M., MEYER, M., Some functional forms of Blaschke–Santaló inequality, Math. Z. 256 (2007), no. 2, 379–395.

[GM01] Giannopoulos, A.A., Milman, V.D., Euclidean structure in finite dimensional normed spaces, Handbook of the geometry of Banach spaces, I, North-Holland, Amsterdam (2001), 707–779.

[GM04] Giannopoulos, A.A., Milman, V.D., Asymptotic convex geometry: short overview, Different faces of geometry, Int. Math. Ser. (N.Y.) 3, Kluwer/Plenum, New York (2004), 87–162.

[Gr88] Gromov, M., Dimension, nonlinear spectra and width, Geometric Aspects of Functional Analysis (1986/87), 132–184, Lecture Notes in Math., 1317, Springer, Berlin, 1988.

[GrM87] Gromov, M., Milman, V.D., Generalization of the spherical isoperimetric inequality to uniformly convex Banach spaces, Compositio Math. 62:3 (1987), 263–282.

[H94] Hörmander, L., Notions of convexity, Progress in Mathematics 127, Birkhäuser Boston Inc., Boston, MA, 1994.

[K06] Klartag, B., On convex perturbations with a bounded isotropic constant, GAFA, Geom. funct. anal. 16:6 (2006), 1274–1290.

[K07a] Klartag, B., Marginals of geometric inequalities, GAFA Seminar, Springer Lect. Notes in Math. 1910 (2007), 133–166.

[K07b] Klartag, B., A central limit theorem for convex sets, Invent. Math. 168:1 (2007), 91–131.

[K07c] Klartag, B., Power law estimates in the central limit theorem for convex sets, J. Funct. Anal. 245:1 (2007), 284–310.

[KM05] Klartag, B., Milman, V.D., Geometry of log-concave functions and measures, Geom. Dedicata 112 (2005), 169–182.

[Ku07] Kuperberg, G., From the Mahler conjecture to Gauss linking intergrals, GAFA, Geom. funct. anal., to appear.

[LM93] Lindenstrauss, J., Milman, V.D., The local theory of normed spaces and its applications to convexity, Handbook of Convex geometry A,B, North-Holland, Amsterdam (1993), 1149–1220.

[MP88] Meyer, M, Pajor, A., Sections of the unit ball of L_p^n. J. Funct. Anal. 80:1 (1988), 109–123.

[M71a] Milman, V.D., Geometric theory of Banach spaces. II Geometry of the unit ball (in Russian), Uspehi Mat. Nauk 26 (1971), no. 6(162), 73–149.

[M71b] Milman, V.D., A new proof of A. Dvoretzky's theorem on cross-sections of convex bodies (in Russian), Funkcional. Anal. i Priložen. 5:4 (1971), 28–37.

[M85] Milman, V.D., Almost Euclidean quotient spaces of subspaces of a finite-dimensional normed space, Proc. Amer. Math. Soc. 94:3 (1985), 445–449.

[M86] Milman, V.D., Inégalité de Brunn–Minkowski inverse et applications à la theorie locale des espaces normés [An inverse form of the Brunn-Minkowski inequality, with applications to the local theory of normed spaces], C.R. Acad. Sci. Paris Ser. I Math. 302:1 (1986), 25–28.

[M91] Milman, V., Some applications of duality relations, Geometric Aspects of Functional Analysis (1989–90), 13–40, Lecture Notes in Math., 1469, Springer, Berlin, 1991.

[M98] MILMAN, V., Surprising geometric phenomena in high-dimensional convexity theory, European Congress of Mathematics, Vol. II (Budapest, 1996), 73–91, Progr. Math. 169, Birkhäuser, Basel, 1998.

[M00] MILMAN, V., Topics in asymptotic geometric analysis, GAFA 2000 (Tel Aviv, 1999). Geom. Funct. Anal. 2000, Special Volume, Part II, 792–815.

[M04] MILMAN, V.D., Phenomena that occur in high dimensions (in Russian), Uspekhi Mat. Nauk 59:1 (2004), (355), 157–168; English transl. in Russian Math. Surveys 59:1 (2004), 159–169.

[MS86] MILMAN, V.D.; SCHECHTMAN, G., Asymptotic Theory of Finite-Dimensional Normed Spaces, with an appendix by M. GROMOV, Lecture Notes in Mathematics, 1200. Springer-Verlag, Berlin, 1986, viii+156 pp.

[P06] Paouris, G., Concentration of mass on convex bodies GAFA, Geom. funct. anal. 16:5 (2006), 1021–1049.

[Pi89] PISIER, G., The Volume of Convex Bodies and Banach Space Geometry. Cambridge Tracts in Mathematics, 94. Cambridge University Press, Cambridge, 1989. xvi+250 pp.

Vitali D. Milman
School of Mathematical Sciences
Sackler Faculty of Exact Sciences
Tel Aviv University
Ramat-Aviv, Tel-Aviv 69978
Israel

Progress in Mathematics, Vol. 265, 669–683

Milnor Invariants and l-Class Groups

Masanori Morishita

To the memory of Professor Alexander Reznikov

Abstract. Following the analogies between knots and primes, we introduce arithmetic analogues of higher linking matrices for prime numbers by using the arithmetic Milnor numbers. As an application, we describe the Galois module structure of the l-class group of a cyclic extension of $\mathbb{Q}$ of degree l (l being a prime number) in terms of the arithmetic higher linking matrices. In particular, our formula generalizes the classical formula of Rédei on the 4 and 8 ranks of the 2-class group of a quadratic field.

Mathematics Subject Classification (2000). Primary 11R; 57M.

Keywords. Arithmetic topology, Milnor invariant, class group.

Introduction

Let k be a quadratic field $\mathbb{Q}(\sqrt{d})$, $d = p_1 \cdots p_n$, where p_i's are distinct prime numbers and assumed to be congruent to 1 mod 4 for simplicity. A classical theorem by C. F. Gauss [G] asserts that the 2-rank of the narrow ideal class group H_k of k is $n-1$ so that the 2-primary part $H_k(2)$ of H_k has the form

$$H_k(2) = \bigoplus_{i=1}^{n-1} \mathbb{Z}/2^{a_i}\mathbb{Z} \quad (a_i \geq 1)$$

as abelian group. Since Gauss' time, it has been a problem to determine the whole structure of $H_k(2)$, namely to describe the 2^q-rank for an integer $q \geq 1$

$$e_q := \#\{i \mid a_i \geq q\}.$$

Among many works on this problem, L. Rédei [R1] showed the following formula

$$e_2 = n - 1 - \mathrm{rank}_{\mathbb{F}_2}(L_2), \quad L_2 = \begin{pmatrix} \left(\frac{d/p_1}{p_1}\right) & \left(\frac{p_2}{p_1}\right) & \cdots & \left(\frac{p_n}{p_1}\right) \\ \left(\frac{p_1}{p_2}\right) & \left(\frac{d/p_2}{p_2}\right) & \cdots & \left(\frac{p_n}{p_2}\right) \\ \vdots & \vdots & \ddots & \vdots \\ \left(\frac{p_1}{p_n}\right) & \left(\frac{p_2}{p_n}\right) & \cdots & \left(\frac{d/p_n}{p_n}\right) \end{pmatrix} \tag{0.1}$$

where $\left(\frac{p}{q}\right) \in \{\pm 1\}$ denotes the Legendre symbol identified with an element of the field $\mathbb{F}_2$ with 2 elements, and further he gave similar formulas for e_3 in some cases using the triple symbol introduced by himself [R2]. Though many authors have studied this problem, in particular the case of $n = 2$, by using the power residue symbols and arithmetical consideration such as Pell's equation (see for example [B], [BS], [Ha], [Y] etc), it still remains a problem to obtain general formulas extending the above-mentioned one, due to Rédei, for higher e_q's.

As one can easily see, this problem has an immediate generalization for a cyclic extension k over $\mathbb{Q}$ of arbitrary prime degree l and is formulated as a problem on the Galois module structure of the l-primary part $H_k(l)$ of the narrow ideal class group H_k of k. Let $p_1, \ldots, p_n$ be the distinct prime numbers ramified in $k/\mathbb{Q}$ which are congruent to 1 mod l, and let γ be a generator of the Galois group of k over $\mathbb{Q}$. The narrow l-class group $H_k(l)$ is then regarded as a module over the complete discrete valuation ring $\mathcal{O} = \mathbb{Z}_l[\langle\gamma\rangle]/(\gamma^{l-1} + \cdots + \gamma + 1) = \mathbb{Z}_l[\zeta]$ where $\mathbb{Z}_l$ denotes the ring of l-adic integers and $\zeta = \gamma \bmod (\gamma^{l-1} + \cdots + \gamma + 1)$. Then the genus theory by Iyanaga-Tamagawa [IT] tells us that the $\mathcal{O}$-module $H_k(l)$ has the form

$$H_k(l) = \bigoplus_{i=1}^{n-1} \mathcal{O}/\mathfrak{m}^{a_i} \quad (a_i \geq 1)$$

where $\mathfrak{m}$ is the maximal ideal of $\mathcal{O}$ generated by $\zeta - 1$. Hence the determination of the $\mathcal{O}$-module structure of $H_k(l)$ is again equivalent to the determinination of

$$\text{the } \mathfrak{m}^q\text{-rank } e_q := \#\{i \mid a_i \geq q\} \ (q \geq 1).$$

The results on this generalized problem have not been obtained along the line of the case of $l = 2$ (cf.[R3]).

The purpose of this paper is to study this problem (for general l) in light of the analogy with knot theory. To explain our underlying idea more precisely, let us recall a part of the basic analogies between knots and primes ([Ka], [Mo4], [Re]):

$$\begin{array}{lcl} \text{knot } K : S^1 = K(\mathbb{Z}, 1) \hookrightarrow S^3 & \longleftrightarrow & \text{prime} : \mathrm{Spec}(\mathbb{F}_p) = K(\hat{\mathbb{Z}}, 1) \hookrightarrow \mathrm{Spec}(\mathbb{Z}) \cup \{\infty\} \\ \text{link } L = K_1 \cup \cdots \cup K_n & & \text{primes } S = \{(p_1), \ldots, (p_n)\} \end{array}$$

tubular n.b.d V_K boundary ∂V_K	$\longleftrightarrow$	p-adic integers $\mathrm{Spec}(\mathbb{Z}_p)$ p-adic field $\mathrm{Spec}(\mathbb{Q}_p)$
$\pi_1(V_K) = \langle \beta \rangle$ $\pi_1(\partial V_K) = \langle \alpha, \beta \mid [\alpha, \beta] = 1 \rangle$ α : meridian β : longitude	$\longleftrightarrow$	$\pi_1^{\text{ét}}(\mathrm{Spec}(\mathbb{Z}_p)) = \langle \sigma \rangle$ $\pi_1^{\text{tame}}(\mathrm{Spec}(\mathbb{Q}_p)) = \langle \tau, \sigma \mid \tau^{p-1}[\tau, \sigma] = 1 \rangle$ τ : monodromy σ : Frobenius automorphism
$\pi_1(S^3 \setminus L)$	$\longleftrightarrow$	$\pi_1^{\text{ét}}(\mathrm{Spec}(\mathbb{Z}) \setminus S)$
3-manifold $M \to S^3$	$\longleftrightarrow$	number ring $\mathrm{Spec}(\mathcal{O}_k) \to \mathrm{Spec}(\mathbb{Z})$
$C_2(M, \mathbb{Z}) \overset{\partial}{\to} C_1(M, \mathbb{Z})$ $\Sigma \mapsto \partial \Sigma$	$\longleftrightarrow$	$k^\times \to \bigoplus_{\mathfrak{p}:\text{primes}} \mathbb{Z}$ $a \mapsto a\mathcal{O}_k$
$H_1(M, \mathbb{Z})$	$\longleftrightarrow$	H_k
linking number	$\longleftrightarrow$	power residue symbol
Milnor invariant	$\longleftrightarrow$	arithmetic Milnor invariant

where π_1^{tame} denotes the tame fundamental group. The analogy between linking numbers and power residue symbols was studied and arithmetic analogues for prime numbers of the Milnor link invariants [Mi] (higher linking numbers) were introduced in our previous papers [Mo2,3] (see also [V]). For a systematic account on analogies between 3-dimensional topology and algebraic number theory, we consult [Mo4].

Based on the analogies above, we can consider a link-theoretic counterpart of our problem as follows. Let $M \to S^3$ be an l-fold cyclic covering branched along a link $L = K_1 \cup \cdots \cup L_n$ with Galois group $\langle \gamma \rangle \simeq \mathbb{Z}/l\mathbb{Z}$ (l being a prime number). Assume that M is a rational homology 3-sphere so that the homology group $H_1(M, \mathbb{Z})$ is finite. Then, as in the case of the l-class group $H_k(l)$, the l-primary part $H_1(M, \mathbb{Z})(l)$ of $H_1(M, \mathbb{Z})$ is regarded as a module over the the complete discrete valuation ring $\mathcal{O}$ and a result in [Mo1] shows that the $\mathfrak{m}$-rank of $H_1(M, \mathbb{Z})(l)$ is $n-1$ so that $H_1(M, \mathbb{Z})$ has the form

$$H_1(M, \mathbb{Z})(l) = \bigoplus_{i=1}^{n-1} \mathcal{O}/\mathfrak{m}^{a_i} \quad (a_i \geq 1).$$

In [HMM], we described the $\mathfrak{m}^q$-rank e_q of $H_1(M, \mathbb{Z})(l)$ by introducing the *higher linking matrices* obtained by the truncating the l-adic *Traldi matrix*. This is seen as an l-adic strengthening of the method by W. Massey [Ma] and L. Traldi [Tr].

We note that for the $\mathfrak{m}^2$-rank e_2, our formula reads

$$e_2 = n - 1 - \mathrm{rank}_{\mathbb{F}_l}(C \bmod l) \tag{0.2}$$

where $C = (C_{ij})$ is the *linking matrix* defined by $C_{ij} := \mathrm{lk}(K_i, K_j) :=$ the linking number of K_i and K_j for $i \neq j$ and $C_{ii} = -\sum_{j \neq i} \mathrm{lk}(K_i, K_j)$. Note that (0.2) is exactly analogous to (0.1) in view of the analogy between the linking number and the power residue symbol.

Now the main idea of this paper is to regard Rédei's matrix L_2 as the mod 2 *arithmetic linking matrix* for prime numbers and describe the e_q's by introducing the *arithmetic higher linking matrices* in terms of the arithmetic Milnor invariants (Theorem 4.5). Our argument is the adaptation of the method of [HMM] in our arithmetic situation. The analogies are so close that we can translate the whole argument of topology side into arithmetic.

The contents of this paper are as follows. In Section 1, we recall the Milnor invariants for prime numbers which play a central role as analogues of higher linking numbers in our approach. In Section 2, we introduce the *Alexander module* for prime numbers and give its presentation matrix by the *universal higher linking matrix*, called the *Traldi matrix*, in terms of the Milnor invariants. In Section 3, we give the relation between the Alexander module and the l-class group. This is seen as an analog of the connection between the Alexander module of a link and the homology of a cyclic branched cover of the link. Combining these, we present formulas for e_q's above in terms of the higher linking matrices.

Acknowledgement. I would like to thank M. Kapranov for suggesting an application of my Milnor invariants to the 2-class groups of quadratic fields. This work is partly supported by the Grants-in-Aid for Scientific Research, Ministry of Education, Culture, Sports, Science and Technology, Japan.

1. Milnor invariants for prime numbers

In this section, we recall the Milnor invariants for prime numbers introduced in [Mo2,3] and review their basic properties. Throughout this paper, we fix a prime number l.

Let S be a finite set of n distinct prime numbers $p_1, \ldots, p_n$ such that $p_i \equiv 1 \bmod l$, $1 \leq i \leq n$. We write $p_i - 1 = l^{f_i} q_i$, $(l, q_i) = 1$, $1 \leq i \leq n$, and set $e = \min\{f_i \mid 1 \leq i \leq n\}$ and $m_S = l^e$. In the following, we fix a power m of l with $1 < m \leq m_S$. Let G_S be the maximal pro-l quotient of the étale fundamental group of the complement of S in $\mathrm{Spec}(\mathbb{Z})$ which is the Galois group of the maximal pro-l extension $\mathbb{Q}_S$ of the rational number field $\mathbb{Q}$ unramified outside $S \cup \{\infty\}$ where ∞ denotes the infinite prime of $\mathbb{Q}$. Choose a prime $\mathfrak{P}_i$ of $\mathbb{Q}_S$ lying over p_i for $1 \leq i \leq n$ and let D_i be the decomposition group of $\mathfrak{P}_i$ which is identified with the Galois group G_i of the maximal pro-l extension $\mathbb{Q}_{p_i}(l)$ of the p_i-adic field $\mathbb{Q}_{p_i}$. Let ζ_j be a primitive l^j-th root of 1 so that $\zeta_{j+1}^l = \zeta_j$ for $j \geq 1$ and we fix an embedding of $\mathbb{Q}(\zeta_j)$ into $\mathbb{Q}_{p_i}(l)$. (Note $\mathbb{Q}(\zeta_e) \subset \mathbb{Q}_{p_i}$.) The field $\mathbb{Q}_{p_i}(l)$ is generated over $\mathbb{Q}_{p_i}$ by ζ_j and $\sqrt[l^j]{p_i}$ for $j \geq 1$ and the local Galois group G_i is topologically generated by 2 elements τ_i and σ_i defined by

$$\tau_i(\zeta_j) = \zeta_j, \qquad \tau_i(\sqrt[l^j]{p_i}) = \zeta_j \sqrt[l^j]{p_i},$$
$$\sigma_i(\zeta_j) = \zeta_j^{p_i}, \qquad \sigma_i(\sqrt[l^j]{p_i}) = \sqrt[l^j]{p_i}$$

subject to the relation $\tau_i^{p_i-1}[\tau_i, \sigma_i] = 1$. We denote by the same τ_i and σ_i the corresponding elements in D_i so that τ_i is a topological generator of the inertia group I_i of $\mathfrak{P}_i$ and σ_i is an extension of the Frobenius automorphism over p_i of the subfield of $\mathbb{Q}_S$ fixed by I_i. We call τ_i a *monodromy over* p_i and σ_i a *Frobenius automorphism over* p_i. Then the Galois group G_S has the following presentation as a pro-l group [Ko]. Let F be the free pro-l group on n generators $x_1, \ldots, x_n$ and let $\pi : F \to G_S$ be the continuous homomorphism defined by $\pi(x_i) = \tau_i$ for $1 \leq i \leq n$. Then π is surjective and the kernel of π is the closed subgroup of F generated normally by $x_1^{p_1-1}[x_1, y_1], \ldots, x_n^{p_n-1}[x_n, y_n]$ where $y_i \in F$ represents σ_i in G_S and $[x_i, y_i] = x_i y_i x_i^{-1} y_i^{-1}$:

$$G_S = \langle x_1, \ldots, x_n \mid x_1^{p_1-1}[x_1, y_1] = \cdots = x_n^{p_n-1}[x_n, y_n] = 1 \rangle. \tag{1.1}$$

On the other hand, for a link L consisting of n knots $K_1, \ldots, K_n$ in the 3-sphere S^3, the pro-l completion $\widehat{G_L}$ of the topological fundamental group $G_L = \pi_1(S^3 \setminus L)$ of the complement of L in S^3 is shown to have the following presentation [HMM]

$$\widehat{G_L} = \langle x_1, \ldots, x_n \mid [x_1, y_1] = \cdots = [x_n, y_n] = 1 \rangle \tag{1.2}$$

where x_i and y_i represent a meridian α_i and a longitude β_i around K_i respectively. Our idea is to regard (1.1) as an arithmetic analogy of (1.2). Note that the pair (τ_i, σ_i) of a monodromy and a Frobenius automorphism over p_i corresponds to the pair (α_i, β_i) of a meridian and a longitude around K_i. In view of this analogy, we introduce an arithmetic analogue for prime numbers S of the Milnor link invariants (higher linking numbers) [Mi],[Tu].

Let $\mathbb{Z}_l$ be the ring of l-adic integers. For a pro-l group G, we denote by $\mathbb{Z}_l[[G]]$ the completed group ring of G over $\mathbb{Z}_l$. Let $\partial_i = \frac{\partial}{\partial x_i} : \mathbb{Z}_l[[F]] \to \mathbb{Z}_l[[F]]$ be the Fox derivative on the free pro-l group F for $1 \leq i \leq n$ ([F],[Ih]), and let $\epsilon : \mathbb{Z}_l[[F]] \to \mathbb{Z}_l$ be the augmentation map. For a multi-index $I = (i_1 \cdots i_r)$, we set

$$\epsilon_I(\alpha) = \epsilon(\partial_{i_1} \cdots \partial_{i_r}(\alpha)), \quad \alpha \in \mathbb{Z}_l[[F]]. \tag{1.3}$$

Let $\mathbb{Z}_l\langle\langle X_1, \ldots, X_n \rangle\rangle$ be the formal power series ring over the ring $\mathbb{Z}_l$ in non-commuting variables $X_1, \ldots, X_n$ which is compact in the topology taking the ideals $I(r)$ of power series with homogeneous components of degree $\geq r$ as the system of neighborhood of 0. The pro-l Magnus embedding $M : F \to \mathbb{Z}_l\langle\langle X_1, \ldots, X_n \rangle\rangle^\times$ defined by

$$M(x_i) = 1 + X_i, \quad M(x_i^{-1}) = 1 - X_i + X_i^2 - \cdots \quad (1 \leq i \leq n),$$

gives the isomorphism $\mathbb{Z}_l$-algebra isomorphism $\mathbb{Z}_l[[F]] \simeq \mathbb{Z}_l\langle\langle X_1, \ldots, X_n\rangle\rangle$ of compact $\mathbb{Z}_l$-algebras. The resulting expansion of $\alpha \in \mathbb{Z}_l[[F]]$ is then given by the form

$$M(\alpha) = \epsilon(\alpha) + \sum_I \epsilon_I(\alpha) X_I \tag{1.4}$$

where I ranges over multi-indices I of length $|I| \geq 1$ and we set $X_I = X_{i_1} \cdots X_{i_r}$ for $I = (i_1 \cdots i_r)$. We write $\epsilon_I(\alpha)_m = \epsilon(\partial_{i_1} \cdots \partial_{i_r}(\alpha))_m$ for the image of $\epsilon_I(\alpha)$ under the natural projection $\mathbb{Z}_l \to \mathbb{Z}/m\mathbb{Z}$. We then define, for $I = (i_1 \cdots i_r)$,

$$\mu(I) = \epsilon_{I'}(y_{i_r}), \quad \text{and} \quad \mu_m(I) = \epsilon_{I'}(y_{i_r})_m \tag{1.5}$$

where $I' = (i_1 \cdots i_{r-1})$. By convention, we set $\mu(I) = 0$ for $|I| = 1$. We call $\mu(I)$ (resp. $\mu_m(I)$) the (resp. *mod m*) *Milnor number*. For a multi-index I, $1 \leq |I| < m_S$, we define the indeterminacy $\Delta(I)$ to be the ideal of $\mathbb{Z}/m\mathbb{Z}$ generated by the binomial coefficients $\binom{m_S}{t}$ and $\mu_m(J)$ where $1 \leq t < |I|$ and J ranges over all cyclic permutations of proper subsequences of I. We set

$$\overline{\mu}_m(I) = \mu_m(I) \bmod \Delta(I) \tag{1.6}$$

and we call them the Milnor $\overline{\mu}_m$ invariant for prime numbers S.

The Milnor invariants are interpreted as arithmetic symbols describing the prime decomposition law in Heisenberg extensions. Let $N_r(R)$ be the upper Heisenberg group of degree r over a commutative ring R, namely the group of all upper triangular $r \times r$ matrices with 1 along the diagonal entries over R. For a multi-index $I = (i_1 \cdots i_r)$, $2 \leq r < m_S$ such that $\Delta(I) \neq \mathbb{Z}/m\mathbb{Z}$, we define a representation

$$\rho_I \;:\; F \longrightarrow N_r((\mathbb{Z}/m\mathbb{Z})/\Delta(I))$$

by

$$\rho_I(f) = \begin{bmatrix} 1 & \epsilon(\partial_{i_1}(f))_m & \epsilon(\partial_{i_1}\partial_{i_2}(f))_m & \cdots & \epsilon(\partial_{i_1} \cdots \partial_{i_{r-1}}(f))_m \\ & 1 & \epsilon(\partial_{i_2}(f))_m & \cdots & \epsilon(\partial_{i_2} \cdots \partial_{i_{r-1}}(f))_m \\ & & \ddots & \ddots & \vdots \\ & 0 & & \ddots & \epsilon(\partial_{i_{r-1}}(f))_m \\ & & & & 1 \end{bmatrix} \bmod \Delta(I).$$

Theorem 1.7. ([Mo3]). *Notation being as above,*

(1) *the representation ρ_I factors through G_S, and it gives a surjective representation of G_S onto $N_r((\mathbb{Z}/m\mathbb{Z})/\Delta(I))$ if $i_1, \ldots, i_{r-1}$ are distinct each other;*
(2) *suppose $i_1, \ldots, i_{r-1}$ are distinct each other; if k_r denotes the extension of $\mathbb{Q}$ corresponding to* $\mathrm{Ker}(\rho_I)$, *$k_r/\mathbb{Q}$ is a Galois extension ramified over $p_{i_1}, \ldots, p_{i_{r-1}}$ with Galois group $N_r((\mathbb{Z}/m\mathbb{Z})/\Delta(I))$ and we have*

$$\rho_I(\sigma_{i_r}) = \begin{bmatrix} 1 & 0 & \cdots & 0 & \mu_m(I) \\ & 1 & 0 & \cdots & 0 \\ & & \ddots & \ddots & \vdots \\ & 0 & & 1 & 0 \\ & & & & 1 \end{bmatrix} \bmod \Delta(I)$$

The following theorem tells us that the power residue symbol and the Rédei triple symbol [R2] are arithmetic analogues of the linking number and the triple linking number respectively.

Theorem 1.8. ([Mo2], [V]).

(1) *For $i \neq j$, we have* $\left(\frac{p_j}{p_i}\right)_m = \zeta_m^{\mu_m(ij)}$ *where* $\left(\frac{p_j}{p_i}\right)_m$ *denotes the m-th power residue symbol in* $\mathbb{Q}_{p_i}$.

(2) *Assume that $p_i \equiv 1 \bmod 4$ and the Legendre symbols* $\left(\frac{p_j}{p_i}\right) = 1$ *for* $1 \leq i \neq j \leq 3$, *and let $[p_i, p_j, p_k]$ be the Rédei triple symbol for any permutation ijk of 123. Then we have*

$$[p_i, p_j, p_k] = (-1)^{\mu_2(ijk)}.$$

2. The Alexander module of the Galois group G_S

In this section, we recall the Alexander module of the Galois group G_S introduced in [Mo2] and give its presentation matrix explicitly in terms of the Milnor numbers. We keep the same notations as in Section 1.

Let H_S be the abelianization of G_S and let $\psi : \mathbb{Z}_l[[G_S]] \to \mathbb{Z}_l[[H_S]]$ be the $\mathbb{Z}_l$-algebra homomorphism of the completed group rings induced by the natural map $G_S \to H_S$. Since $H_S \simeq \mathbb{Z}/m_1\mathbb{Z} \times \cdots \times \mathbb{Z}/m_n\mathbb{Z}$, $\mathbb{Z}_l[[H_S]]$ is isomorphic to $\mathbb{Z}_l[t_1, \dots, t_n]/(t_1^{m_1} - 1, \dots t_n^{m_n} - 1)$ which is identified with $\Lambda_S = \mathbb{Z}_l[[X_1, \dots, X_n]]/((1+X_1)^{m_1} - 1, \dots, (1+X_n)^{m_n} - 1)$ by sending t_i to $1 + X_i$. Let $\pi : \mathbb{Z}_l[[F]] \to \mathbb{Z}_l[[G_S]]$ be the $\mathbb{Z}_l$-algebra homomorphism of the completed group rings induced by π so that the composite $\psi \circ \pi$ is regarded as a $\mathbb{Z}_l$-algebra homomorphism $\mathbb{Z}_l[[F]] \to \Lambda_S$. The *Alexander module* A_S of the Galois group G_S is defined to be the ψ-derived module over Λ_S and is also given by using the pro-l Fox free differential calculus as follows [Mo2]. By virtue of the presentation (1.1) of G_S, we define the *Alexander matrix* $P_S = (P_S(i,j))$ of G_S by

$$P_S(i,j) = \psi \circ \pi \left(\partial_j (x_i^{p_i - 1}[x_i, y_i])\right) \tag{2.1}$$

and then the Alexander module A_S of G_S is given as the Λ_S-module presented by P_S:

$$A_S = \operatorname{Coker}({\Lambda_S}^n \stackrel{P_S}{\longrightarrow} {\Lambda_S}^n). \tag{2.2}$$

Let H be the cyclic group $\langle t \mid t^m = 1 \rangle$ of order m and let $\lambda : H_S \to H$ be the homomorphism defined by $\lambda(t_i) = t$ for $1 \leq i \leq n$. The group ring $\mathbb{Z}_l[H]$ is identified with $\mathbb{Z}_l[t]/(t^m - 1) \simeq \mathbb{Z}_l[[X]]/((1+X)^m - 1)$ by which we denote Λ. We use the same λ for the $\mathbb{Z}_l$-algebra homomorphism $\Lambda_S \to \Lambda$ induced by λ. The *reduced Alexander matrix* $\overline{P}_S$ is then defined by $\lambda(P_S)$ and the *reduced Alexander module* $\overline{A}_S$ of G_S by the Λ-module presented by $\overline{P}_S$:

$$\overline{A}_S = \operatorname{Coker}(\Lambda^n \stackrel{\overline{P}_S}{\longrightarrow} \Lambda^n) = A_S \otimes_{\Lambda_S} \Lambda. \tag{2.3}$$

Next, we introduce the arithmetic analog of the Traldi matrix [Tr] as follows.

Definition 2.4. The *Traldi matrix* $T_S = (T_S(i,j))$ of G_S over Λ_S is defined by

$$T_S(i,j) = \begin{cases} X_i^{-1}((1+X_i)^{p_i-1}-1) - \sum_{r\geq 1}\sum_{\substack{1\leq i_1,\ldots,i_r\leq n \\ i_r\neq i}} \mu(i_1\cdots i_r i)X_{i_1}\cdots X_{i_r}, & i=j \\ \mu(ji)X_i + \sum_{r\geq 1}\sum_{1\leq i_1\cdots i_r\leq n} \mu(i_1\cdots i_r ji)X_iX_{i_1}\cdots X_{i_r}, & i\neq j. \end{cases}$$

where $T_S(i,j)$ is regarded as an element of Λ_S and we also define the *reduced Traldi matrix* $\overline{T}_S$ of G_S over Λ by

$$\overline{T}_S = \lambda(T_S) = T_S(X,\ldots,X).$$

Our theorem is then stated as follows.

Theorem 2.5. *The Traldi matrix T_S (resp. reduced Traldi matrix $\overline{T}_S$) gives a presentation matrix of the Alexander module A_S (resp. reduced Alexander module $\overline{A}_S$) over Λ_S (resp. Λ).*

Proof. By (2.1), (2.2) and (2.3), it suffices to show $\psi\circ\pi\left(\partial_j(x_i^{p_i-1}[x_i,y_i])\right) = T_S(i,j)$. We compute $\partial_j(x_i^{p_i-1}[x_i,y_i])$ by the Fox differential calculus. Since $\partial_j(x_i^{p_i-1}[x_i,y_i]) = \partial_j(x_i^{p_i-1}) + x_i^{p_i-1}\partial_j([x_i,y_i])$ and $\psi\circ\pi(x_i^{p_i-1}) = (1+X_i)^{p_i-1} = 1$ in Λ_S, we have

$$\psi\circ\pi\left(\partial_j(x_i^{p_i-1}[x_i,y_i])\right) = \psi\circ\pi\left(\partial_j(x_i^{p_i-1})\right) + \psi\circ\pi\left(\partial_j([x_i,y_i])\right). \quad (1)$$

For the 1st term, $\partial_j(x_i^{p_i-1}) = \frac{x_i^{p_i-1}-1}{x_i-1}\partial_j(x_i)$ yields

$$\psi\circ\pi\left(\partial_j(x_i^{p_i-1})\right) = X_i^{-1}((1+X_i)^{p_i-1}-1)\delta_{i,j}. \quad (2)$$

For the 2nd term, $\partial_j([x_i,y_i]) = (1-x_iy_ix_i^{-1})\partial_j(x_i) + x_i(1-y_ix_i^{-1}y_i^{-1})\partial_j(y_i)$ together with $y_i = 1+\sum_{r\geq 1}\sum_{1\leq i_1,\ldots,i_r\leq n}\mu(i_1\cdots i_r i)(x_{i_1}-1)\cdots(x_{i_r}-1)$ ((1.3)–(1.5)) yields

$$\begin{aligned}\psi\circ\pi\left(\partial_j([x_i,y_i])\right) &= \delta_{i,j} - \sum_{r\geq 1}\sum_{1\leq i_1,\ldots,i_r\leq n}\mu(i_1\cdots i_r i)X_{i_1}\cdots X_{i_r} \\ &\quad + \mu(ji)X_i + \sum_{r\geq 1}\sum_{1\leq i_1,\ldots,i_r\leq n}\mu(i_1\cdots i_r ji)X_iX_{i_1}\cdots X_{i_r}.\end{aligned} \quad (3)$$

Putting (1), (2) and (3) together, we get the assertion. □

Finally, we introduce the *truncated Traldi matrices* as follows.

Definition 2.6. For $q \geq 2$, the *q-th truncated Traldi matrix* ${T_S}^{(q)} = ({T_S}^{(q)}(i,j))$ is defined by

$$
{T_S}^{(q)}(i,j) = \begin{cases} X_i^{-1}((1+X_i)^{p_i-1}-1) - \sum_{r=1}^{q-1} \sum_{\substack{1 \leq i_1,\ldots,i_r \leq n \\ i_r \neq i}} \mu(i_1 \cdots i_r i) X_{i_1} \cdots X_{i_r}, & i = j \\ \mu(ji)X_i + \sum_{r=1}^{q-2} \sum_{1 \leq i_1,\ldots,i_r \leq n} \mu(i_1 \cdots i_r ji) X_i X_{i_1} \cdots X_{i_r}, & i \neq j \end{cases}
$$

and we also define the *q-th truncated reduced Traldi matrix* ${\overline{T}_S}^{(k)}$ by

$$
{\overline{T}_S}^{(q)} := \lambda({T_S}^{(q)}) = {T_S}^{(q)}(X,\ldots,X).
$$

Remark 2.7. We note ${\overline{T}_S}^{(2)} = X \cdot L^{(2)}$ where $L^{(2)} = (L^{(2)}(i,j))$ is given by $L^{(2)}(i,i) = -\sum_{j \neq i} \mu(ji), L^{(2)}(i,j) = \mu(ji)$ for $i \neq j$. By (1) of Theorem 1.8, Rédei's matrix L_2 [R1] mentioned in the introduction is essentially same as the *linking matrix* $L^{(2)}$. Thus our Traldi matrix T_S is regarded as a *universal higher linking matrix.*

3. Relation between the Alexander module and the l-class group

In this section, we give the relation between the Alexander module A_S and the l-class group of a cyclic subextension of $\mathbb{Q}_S/\mathbb{Q}$ of degree l. We keep the same notations as in Sections 1, 2.

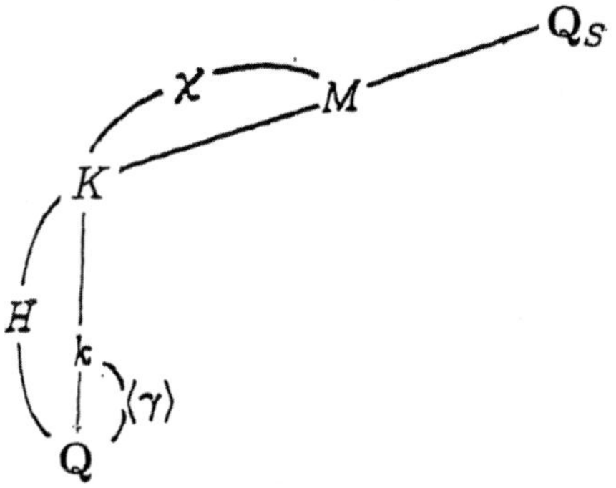

Let K be the subextension of $\mathbb{Q}_S/\mathbb{Q}$ corresponding to the kernel $\mathcal{Y}$ of $\lambda \circ \psi : G_S \to H$ so that K is a cyclic extension of degree m over $\mathbb{Q}$ with Galois group $\mathrm{Gal}(K/\mathbb{Q}) = H = \langle t \rangle$. Let k be the (unique) subfield of K of degree l over $\mathbb{Q}$ so that k is a cyclic extension of $\mathbb{Q}$ with Galois group $\mathrm{Gal}(k/\mathbb{Q}) = H/H^l = \langle \gamma \rangle$, $\gamma = t \bmod H^l$. Let M be the maximal abelian subextension of $\mathbb{Q}_S/K$ and let $\mathcal{X}$ denote the Galois group of M over K, $\mathcal{X} = \mathcal{Y}/[\mathcal{Y},\mathcal{Y}]$, on which H acts via inner automorphism so that $\mathcal{X}$ is regarded as a Λ-module. Let $\tilde{k}$ be the narrow Hilbert l-class field of k (the maximal abelian extension of k which is unramified at all finite primes) so that the Galois group $\mathrm{Gal}(\tilde{k}/k)$ is isomorphic to the narrow l-class group $H_k(l)$ of k by the Artin reciprocity map.

Firstly, we recall an arithmetic analog of the Crowell exact sequence ([Mo2]) in our context, which gives the relation between the reduced Alexander module $\overline{A}_S$ and the Galois group $\mathcal{X}$.

Theorem 3.1 ([Mo2,Theorem 2.2.9]). *There is a split exact sequence of Λ-modules*

$$0 \longrightarrow \mathcal{X} \xrightarrow{\iota} \overline{A}_S \xrightarrow{\kappa} I_\Lambda \longrightarrow 0$$

where I_Λ is the augmentation ideal of Λ and ι and κ are given as follows:

$$\iota(g \bmod [\mathcal{Y},\mathcal{Y}]) = (\lambda \circ \psi \circ \pi(\partial_i(f))) \text{ for } \pi(f) = g,$$

$$\kappa((\alpha_i) \bmod \mathrm{Im}(\overline{P}_S)) = (t-1)\sum_{i=1}^{n} \alpha_i.$$

Next, we give the connection between $\mathcal{X}$ and the narrow l-class group $H_k(l)$. We set $\nu_l(t) = 1 + t + \cdots + t^{l-1}$. Since the norm $\nu_l(\gamma)$ acts trivially on $H_k(l)$, $H_k(l)$ is regarded as a module over the complete discrete valuation ring $\mathcal{O} := \mathbb{Z}_l[H]/(\nu_l(t)) = \Lambda/(\nu_l(1+X)) = \mathbb{Z}_l[\zeta]$ where $\zeta = t \bmod (\nu_l(t))$. Then we have the following.

Theorem 3.2. *Notation being as above, we have an isomorphism of $\mathcal{O}$-modules*

$$\mathcal{X}/\nu_l(t)\mathcal{X} \simeq H_k(l).$$

The proof of this theorem is similar to the standard argument in Iwasawa theory (e.g, [W,§13.3]).

Let $\mathfrak{p}_i$ be the prime of M lying below $\mathfrak{P}_i$ and let $I_i(M/k)$ be the inertia group of $\mathfrak{p}_i$ over k, which is generated by s_i^l where $s_i = \tau_i|_M$. Since $\tilde{k}$ is the maximal abelian subextension of M/k which is unramified over S, we have

$$H_k(l) \simeq \mathrm{Gal}(\tilde{k}/k) \simeq \mathrm{Gal}(M/k)/\langle \mathrm{Gal}(M/k)', I_i(M/k)\ (1 \le i \le n)\rangle \tag{3.3}$$

where $\mathrm{Gal}(M/k)'$ denotes the topological commutator subgroup of $\mathrm{Gal}(M/k)$ and $\langle A\rangle$ stands for the closed subgroup generated by A. Since $s_i|_L = s_1|_L = t$ for $1 \le i \le n$, we have $\mathcal{X}s_i = \mathcal{X}s_1$ and so $s_i = u_i s_1$ for some $u_i \in \mathcal{X}$ $(1 \le i \le n)$. In the following, we write x^α for the action $\alpha x \alpha^{-1}$ of $\alpha \in H$ on $x \in \mathcal{X}$ if the notation $\alpha(x)$ may cause confusion. Two lemmas are in order.

Lemma 3.4. $\mathrm{Gal}(M/k) = \mathcal{X} I_i(M/k)$ *and one has* $s_i^l = u_i^{\nu_l(t)} \cdot s_1^l$ *for* $1 \le i \le n$.

Proof. Since $s_i|_L = t$, the inertia group $I_i(M/k) = \langle s_i^l\rangle$ is mapped onto $\langle t^l\rangle = \mathrm{Gal}(L/k) = \mathrm{Gal}(M/k)/\mathcal{X}$. Hence the first assertion follows. The second assertion is shown by

$$\begin{aligned} s_i^l &= (u_i s_1)^l \\ &= u_i s_1 u_i s_1^{-1} s_1^2 \cdots s_1^{l-1} u_i s_1^{-(l-1)} s_1^l \\ &= u_i u_i^{s_1} \cdots u_i^{s_1^{l-1}} s_1^l \\ &= u_i^{\nu_l(t)} s_1^l. \end{aligned}$$

□

Lemma 3.5. $\mathrm{Gal}(M/k)' = (t^l - 1)\mathcal{X}$.

Proof. Take $a, b \in \mathrm{Gal}(M/k)$ and write $a = \alpha x$ and $b = \beta y$ where $x, y \in \mathcal{X}$ and $\alpha = \tilde{t}^{li}, \beta = \tilde{t}^{lj}$, $\tilde{t}$ being a lift of t to $\mathrm{Gal}(M/k)$. Then one has

$$\begin{aligned}[a,b] &= \alpha x \beta y x^{-1} \alpha^{-1} y^{-1} \beta^{-1} \\ &= x^{\alpha}(yx^{-1})^{\alpha\beta} \alpha\beta\alpha^{-1} y^{-1} \beta^{-1} \\ &= (x^{\alpha})^{1-\beta}(y^{\beta})^{\alpha-1} \quad (\langle t^l \rangle \text{ is abelian}).\end{aligned}$$

Letting $\beta = 1$ and $\alpha = \tilde{t}^l$ in the above equation, we have $y^{\tilde{t}^l - 1} = [a, b]$ for any $y \in \mathcal{X}$. Hence we have $(t^l - 1)\mathcal{X} \subset \mathrm{Gal}(M/k)'$. On the other hand, since $1 - \beta = 1 - \tilde{t}^{lj} = (1 - \tilde{t}^l)\nu_j(\tilde{t}^l)$, $(x^{\alpha})^{1-\beta} \in (t^l - 1)\mathcal{X}$. Similarly, we have $(y^{\beta})^{\alpha - 1} \in (t^l - 1)\mathcal{X}$. Hence we have $[a, b] \in \mathcal{X}$ for any $a, b \in \mathrm{Gal}(M/k)$, which yields the converse inclusion $\mathrm{Gal}(M/k)' \subset (t^l - 1)\mathcal{X}$. □

Proof of Theorem 3.2. By (3.3) and Lemmas 3.4 and 3.5, we have $\mathcal{O}$-isomorphisms

$$\begin{aligned}\mathrm{Gal}(\tilde{k}/k) &\simeq \mathcal{X} I_1(M/k)/\langle (t^l - 1)\mathcal{X}, \nu_l(t)u_i \cdot s_1^l \ (1 \le i \le n)\rangle \\ &\simeq \mathcal{X}/\nu_l(t)\langle (t-1)\mathcal{X}, u_i \ (1 \le i \le n)\rangle.\end{aligned} \tag{3.6}$$

Replacing k by $\mathbb{Q}$, the same argument as above yields

$$1 = \mathcal{X}/\langle (t-1)\mathcal{X}, u_i \ (1 \le i \le n)\rangle \tag{3.7}$$

since the narrow Hilbert l-class field of $\mathbb{Q}$ is $\mathbb{Q}$ itself.

Putting (3.6) and (3.7) together, we get the assertion. □

By Theorems 3.1 and 3.2, we obtain the following relation between the reduced Alexander module $\overline{A}_S$ and the l-class group $H_k(l)$, which is analogous to the relation between the reduced Alexander module of a link and the homology of a cyclic branched cover (cf. [Hi, 5.4, 5.7]).

Theorem 3.8. *We have an isomorphism of $\mathcal{O}$-modules*

$$\overline{A}_S \otimes_{\Lambda} \mathcal{O} \simeq H_k(l) \oplus \mathcal{O}.$$

Proof. This follows from tensoring $\overline{A}_S \simeq \mathcal{X} \oplus \Lambda$ with $\mathcal{O} = \Lambda/(\nu_l(1+X))$ over Λ. □

4. Galois module structure of the l-class group

In this section, combining the results in Sections 2 and 3, we determine the Galois module structure of the narrow l-class group $H_k(l)$ in terms of the higher linking matrices. We keep the same notations as in the previous sections.

We first recall the genus theory for the number field k ([IT]). Let $\mathfrak{m}$ be the maximal ideal of $\mathcal{O}$ generated by $\varpi = \zeta - 1$ with residue field $\mathcal{O}/\mathfrak{m} = \mathbb{F}_l$ of l elements.

Lemma 4.1. *The dimension of $H_k(l) \otimes_{\mathcal{O}} \mathbb{F}_l$ over $\mathbb{F}_l$ is $n - 1$.*

Proof. Let $\langle\zeta_1\rangle$ be the group generated by the primitive l-th root ζ_1 of 1 chosen in Section 1. The genus theory [IT] tells us that the map $\chi : H_k(l) \to \langle\zeta_1\rangle^n$ defined by

$$\chi([\mathfrak{a}]) = \left(\left(\frac{N\mathfrak{a}}{p_1}\right)_l, \dots, \left(\frac{N\mathfrak{a}}{p_n}\right)_l \right)$$

gives rise to the isomorphism

$$H_k/(\gamma-1)H_k \simeq \{(z_1,\dots,z_n) \in \langle\zeta_1\rangle^n \mid \prod_{i=1}^{n} z_i = 1\} \simeq \mathbb{F}_l^{n-1}$$

where $N\mathfrak{a}$ is the norm of an integral ideal $\mathfrak{a}$ and $\left(\frac{*}{p_i}\right)_l$ is the l-th power residue symbol in $\mathbb{Q}_{p_i}$ (Note that $\langle\zeta_1\rangle \subset \mathbb{Q}_{p_i}$). Since $H_k/(\gamma-1)H_k = H_k(l) \otimes_{\mathcal{O}} \mathcal{O}/(\varpi) = H_k(l) \otimes_{\mathcal{O}} \mathbb{F}_l$, the assertion follows. □

By Lemma 4.1, we have the isomorphism

$$H_k(l) \simeq \bigoplus_{i=1}^{n-1} \mathcal{O}/\mathfrak{m}^{a_i} \quad (a_i \geq 1) \tag{4.2}$$

of $\mathcal{O}$-modules. Hence the determination of the $\mathcal{O}$-module structure of $H_k(l)$ is equivalent to that of the $\mathfrak{m}^q$-rank

$$e_q = \#\{i \mid a_i \geq q\} = \dim_{\mathbb{F}_l} H_k(l) \otimes_{\mathcal{O}} \mathfrak{m}^{q-1}/\mathfrak{m}^q \quad (q \geq 1).$$

We describe the $\mathfrak{m}^q$-rank e_q in terms of the *higher linking matrices* obtained from the truncated reduced Traldi matrices (2.6) evaluated at $X = \varpi$.

Definition 4.3. The *higher linking matrix* $L_S = (L_S(i,j))$ over $\mathcal{O}$ is defined by

$$L_S(i,j) = \overline{T}_S(i,j)(\varpi) = \begin{cases} -\sum\limits_{r\geq 1} \sum\limits_{\substack{1\leq i_1,\dots,i_r\leq n \\ i_r\neq i}} \mu(i_1\cdots i_r i)\varpi^r, & i=j \\ \mu(ji)\varpi + \sum\limits_{r\geq 1} \sum\limits_{1\leq i_1,\dots,i_r\leq n} \mu(i_1\cdots i_r ji)\varpi^{r+1}, & i\neq j. \end{cases}$$

For $q \geq 2$, the *q-th truncated higher linking matrix* $L_S^{(q)}$ is defined by

$$L_S^{(q)}(i,j) = \overline{T}_S^{(q)}(i,j)(\varpi) = \begin{cases} -\sum\limits_{r=1}^{q-1} \sum\limits_{\substack{1\leq i_1,\dots,i_r\leq n \\ i_r\neq i}} \mu(i_1\cdots i_r i)\varpi^r, & i=j \\ \mu(ji)\varpi + \sum\limits_{r=1}^{q-2} \sum\limits_{1\leq i_1,\dots,i_r\leq n} \mu(i_1\cdots i_r ji)\varpi^{r+1}, & i\neq j. \end{cases}$$

By Theorems 2,5 and 3.8, we have

Theorem 4.4. *The higher linking matrix L_S gives a presentation matrix for the $\mathcal{O}$-module $H_k(l) \oplus \mathcal{O}$. For $q \geq 2$, the q-th truncated higher linking matrix $L_S^{(q)}$ gives a presentation matrix for the $\mathcal{O}/\mathfrak{m}^q$-module $(H_k(l) \otimes_{\mathcal{O}} \mathcal{O}/\mathfrak{m}^q) \oplus \mathcal{O}/\mathfrak{m}^q$.*

Restating Theorem 4.4 in terms of e_q, we obatin our main formula.

Theorem 4.5. *For $q \geq 2$, let $\varepsilon_1^{(q)}, \dots, \varepsilon_{n-1}^{(q)}, \varepsilon_n^{(q)} = 0$ be the elementary divisors of $L_S^{(q)}$, where $\varepsilon_i^{(d)} | \varepsilon_{i+1}^{(d)}$ $(1 \leq i \leq n-1)$. Then we have*

$$e_q = \#\{i \,|\, \varepsilon_i^{(q)} \equiv 0 \bmod \mathfrak{m}^q\} - 1.$$

For the initial term of $k = 2$, we recover Rédei's formula for arbitrary l.

Corollary 4.6. *We have*

$$e_2 = n - 1 - \mathrm{rank}_{\mathbb{F}_l}(L \bmod l)$$

where $L = (L_{ij})$ is the linking matrix defined by $L_{i,i} = -\sum_{j \neq i} \mu(ji), L_{ij} = \mu(ji)$ for $i \neq j$.

Proof. In fact, $L_S^{(2)} = \varpi L$ and so $e_2 = n-1-\mathrm{rank}_{\mathbb{F}_l}(L \bmod l)$ by Theorem 4.5. □

Let us see the case of $n = 2$. By Lemma 4.1, $H_k(l)$ has $\mathfrak{m}$-rank 1

$$H_k(l) = \mathcal{O}/\mathfrak{m}^{a_i} \quad (a_i \geq 1)$$

and $e_q = 0$ or 1. By Theorem 4.5 we have

$$e_q = 1 \Longleftrightarrow L_S^{(q)} \equiv O_2 \bmod \varpi^q.$$

where O_2 is the 2 by 2 zero matrix. Since $L_S^{(q)}(1,2) = -L_S^{(q)}(1,1)$, $L_S^{(q)}(2,2) = -L_S^{(q)}(2,1)$, we have the following.

Corollary 4.7. *Suppose $n = 2$. For each $q \geq 1$, assuming $e_q = 1$, we have*

$$e_{q+1} = 1 \Longleftrightarrow \begin{cases} \displaystyle\sum_{r=1}^{q} \sum_{i_1,\dots,i_{r-1}=1,2} \mu(i_1 \cdots i_{r-1} 21)\varpi^r \equiv 0 \bmod \varpi^{q+1}, \\ \displaystyle\sum_{r=1}^{q} \sum_{i_1,\dots,i_{r-1}=1,2} \mu(i_1 \cdots i_{r-1} 12)\varpi^r \equiv 0 \bmod \varpi^{q+1}. \end{cases}$$

Example 4.8. (Borromean primes) D. Vogel [V] computed many Milnor invariants by making the computer program. He finds that the triple $S = \{13, 61, 937\}$ is really a mod 2 arithmetical analog of the Borromean ring in the sense that $\mu_2(ij) = 0$ for all $1 \leq i,j \leq 3$ and $\mu_2(ijk) = 1$ for any permutation ijk of 123 and $\mu_2(ijk) = 0$ for other ijk.

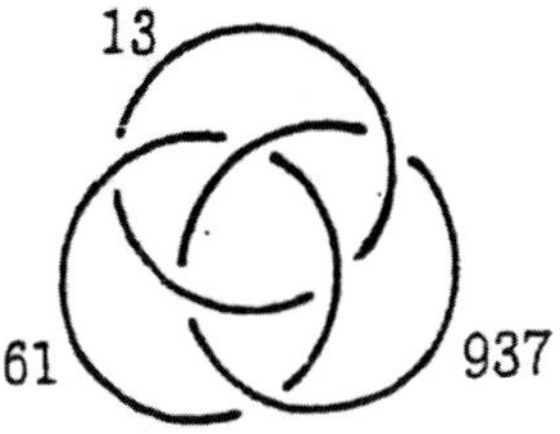

Further, we have all $\mu_4(ij) = 0$ (I owe this computation to K. Yamamura). We then find that $L_S^{(2)} \equiv O_3 \bmod 4$ and

$$L_S^{(3)} \equiv \begin{pmatrix} 0 & 4 & 4 \\ 4 & 0 & 4 \\ 4 & 4 & 0 \end{pmatrix} \sim \begin{pmatrix} 4 & 0 & 0 \\ 0 & 4 & 0 \\ 0 & 0 & 0 \end{pmatrix} \bmod 8$$

and so $e_2 = 2$ and $e_3 = 0$ by Theorem 4.4. Hence $H_k(2) = \mathbb{Z}/4\mathbb{Z} \oplus \mathbb{Z}/4\mathbb{Z}$ for $k = \mathbb{Q}(\sqrt{13 \cdot 61 \cdot 937})$.

On the other hand, for the triple $\{5, 101, 8081\}$, all Milnor number $\mu_2(I)$ vanishes if $|I| \leq 3$ ([V]). Further all $\mu_4(ij) = 0$ (Yamamura) and so $L_S^{(3)} \equiv O_3 \bmod 8$. Hence $e_3 = 2$ and $H_k \otimes \mathbb{Z}/8\mathbb{Z} = \mathbb{Z}/8\mathbb{Z} \oplus \mathbb{Z}/8\mathbb{Z}$ for $k = \mathbb{Q}(\sqrt{5 \cdot 101 \cdot 8081})$.

References

[B] H. Bauer, Zur Berechnung der 2-Klassenzahl der quadratischen Zahlköper mit genau zwei verschiedenen Diskriminantenprimteilern, *J. Reine Angew. Math.* **248** (1971), 42–46.

[BS] W. Bosma and P. Stevenhagen, On the computation of quadratic 2-class groups, *J. Théor. Nombres Bordeaux* (2) **8** (1996), 283–313; Erratum: **9** (1997), 249.

[F] R.H. Fox, Free differential calculus. I: Derivation in the free group ring, *Ann. of Math.* **57**, (1953), 547–560.

[G] C.F. Gauss, *Disquisitiones arithmeticae*, Yale Univ. 1966.

[Ha] H. Hasse, An algorithm for determining the structure of the 2-Sylow-subgroups of the divisor class group of a quadratic number field, *Symposia Mathematica*, Vol. XV (Convegno di Strutture in Corpi Algebrici, INDAM, Rome, 1973), Academic Press, London, (1975), 341–352.

[Hi] J.A. Hillman, *Algebraic invariants of links*, Series on Knots and Everything, **32**, World Scientific Publishing Co. 2002.

[HMM] J. Hillman, D. Matei and M. Morishita, Pro-p link groups and p-homology groups, in: *Primes and Knots*, Contemporary Math., AMS. **416** (2006), 121–136.

[Ih] Y. Ihara, On Galois representations arising from towers of coverings of $\mathbb{P}^1 \backslash \{0, 1, \infty\}$, *Invent. Math.* **86**, (1986), 427–459.

[IT] S. Iyanaga and T. Tamagawa, Sur la theorie du corps de classes sur le corps des nombres rationnels, *J. Math. Soc. Japan* **3** (1951), 220–227.

[Ka] M. Kapranov, Analogies between number fields and 3-manifolds, unpublished note, MPI, 1996.

[Ko] H. Koch, *Galoissche Theorie der p-Erweiterungen*, Springer, Berlin-New York, VEB Deutscher Verlag der Wissenschaften, Berlin, 1970.

[Ma] W. Massey, Completion of link modules, *Duke Math. J.* **47**, (1980), 399–420.

[Mi] J. Milnor, Isotopy of links, in: *Algebraic Geometry and Topology, A symposium in honour of S. Lefschetz* (edited by R.H. Fox, D.S. Spencer and W. Tucker), Princeton Univ. Press, Princeton, (1957), 280–306.

[Mo1] M. Morishita, A theory of genera for cyclic coverings of links, *Proc. Japan. Academy* **77**, no. 7 (2001), 115–118.

[Mo2] ——, On certain analogies between knots and primes, *J. Reine Angew. Math.* **550**, (2002), 141–167.

[Mo3] ——, Milnor invariants and Massey products for prime numbers, *Compositio Math.* **140** (2004), 69–83.

[Mo4] ——, Analogies between knots and primes, 3-manifolds and number rings, to appear in *Sugaku Expositions*, AMS. (Analogies between prime numbers and knots, (Japanese) *Sugaku* **58** no. 1 (2006), 40–63.)

[R1] L. Rédei, Arithmetischer Beweis des Satzes über die Anzahl der durch vier teilbaren Invarianten der absoluten Klassengruppe im quadratischen Zahlkörper, *J. Reine Angew. Math.* **171** (1934), 55–60.

[R2] ——, Ein neues zahlentheoretisches Symbol mit Anwendungen auf die Theorie der quadratischen Zahlkörper, I, *J. Reine Angew. Math.* **180** (1938), 1–43.

[R3] ——, Über die Klassengruppen und Klassenköper algebraischer Zahlköper, *J. Reine Angew. Math.* **186** (1944), 80–90.

[Re] A. Reznikov, Embedded incompressible surfaces and homology of ramified coverings of three-manifolds, *Sel. math. New ser.* **6** (2000), 1–39.

[Tr] L. Traldi, Milnor's invariants and the completions of link modules, *Trans. Amer. Math. Soc.* **284** (1984), 401–424.

[Tu] V. Turaev, Milnor's invariants and Massey products, English transl. *J. Soviet Math.* **12** (1979), 128–137.

[V] D. Vogel, On the Galois group of 2-extensions with restricted ramification, *J. Reine Angew. Math.* **581** (2005), 117–150.

[Y] Y. Yamamoto, Divisibility by 16 of class number of quadratic fields whose 2-class groups are cyclic, *Osaka J. Math.* **21** (1984), 1–22.

[W] L. Washington, *Introduction to cyclotomic fields*, Graduate Texts in Mathematics, **83**, Springer-Verlag, New York, 1982.

Masanori Morishita
Graduate School of Mathematics
Kyushu University
Fukuoka, 812-8581
Japan
e-mail: `morisita@math.kyushu-u.ac.jp`

Progress in Mathematics, Vol. 265, 685–695

Three Topological Properties of Small Eigenfunctions on Hyperbolic Surfaces

Jean-Pierre Otal

To the memory of Sasha Reznikov

Abstract. We apply topological methods for studying eigenfunctions on finite volume hyperbolic surfaces. From the Lemma saying that any non-zero eigenfunction on an hyperbolic surface with eigenvalue $\leq \frac{1}{4}$ has an incompressible nodal set, we deduce the following propositions: 1) the non-existence of cuspidal eigenfunctions with eigenvalue $\leq \frac{1}{4}$ on surfaces of genus 0 or 1 (a result already obtained by Huxley); 2) the dimension of a cuspidal eigenspace with eigenvalue $\leq \frac{1}{4}$ is not more than $2g-3$ (a generalization of 1)); 3) a dichotomy for functions in an eigenspace with eigenvalue $\leq \frac{1}{4}$: either an eigenfunction exists which has a smooth nodal set / or all functions in the eigenspace have common zeroes.

Mathematics Subject Classification (2000). Primary 35P05, 58G20, 22E30, 43A85, 58G25, 32M15; Secondary 58J5.

Keywords. Hyperbolic surfaces, eigenfunctions, small eigenvalue.

1. Introduction

Let S be a hyperbolic surface, identified with the quotient $\mathbb{H}/\Gamma$ of the Poincaré upper half-plane $\mathbb{H}$ by a *Fuchsian group* Γ; i.e., a discrete and torsion-free subgroup of $\mathrm{PSL}(2,\mathbb{R})$. The Laplacian on $\mathbb{H}$ is the differential operator which associates to a C^2-function f the function $\Delta f(z) = y^2(\frac{\partial^2 f}{\partial x^2} + \frac{\partial^2 f}{\partial y^2})$. It induces a differential operator on S which extends to a self-adjoint operator Δ_S densely defined on $L^2(S)$. Its domain is the Sobolev space $\mathcal{H}$ consisting of the functions $\phi \in L^2(S)$ whose gradient in the sense of distributions is a measurable vector field which satisfies $\int_S \|\nabla\phi\|^2 dv < \infty$. The Laplacian is a non-positive operator whose spectrum is contained in a smallest interval $]-\infty, -\lambda_0(S)] \subset \mathbb{R}^- \cup \{0\}$. The Rayleigh quotients allow to characterize (*the bottom of the spectrum*) $\lambda_0(S)$:

$$\lambda_0(S) = \inf \frac{\int_S \|\nabla\phi\|^2 dv}{\int_S \phi^2 dv},$$

the infimum being taken over all the non-zero functions $\phi \in \mathcal{H}$. By density, the infimum is the same if one restricts to C^∞-functions with compact support. The bottom of the spectrum of the Laplacian on $\mathbb{H}$ is $\lambda_0(\mathbb{H}) = \frac{1}{4}$ (cf. [Cha]).

Definition. Let $\lambda > 0$. A function $f : S \to \mathbb{R}$ is a λ-*eigenfunction* if $f \in L^2(S)$ and satisfies $\Delta f + \lambda f = 0$. When $0 < \lambda \leq \frac{1}{4}$, f is called a *small eigenfunction.* We denote by $\mathcal{E}_\lambda$ the space of λ-eigenfunctions.

Recall that if S is a finite area hyperbolic surface, then S is conformally equivalent to a compact Riemann surface $\overline{S}$ from which some points have been removed ; these points are the *punctures.* We say that f is *cuspidal* if $f(z)$ tends to 0 when z tends to the punctures. One denotes by $\mathcal{E}^c_\lambda$ the space of the cuspidal λ-eigenfunctions (when S is compact, $\mathcal{E}_\lambda = \mathcal{E}^c_\lambda$).

Definition. Let f be a non-zero λ-eigenfunction. *The nodal set of f* is the set of points where f vanishes; we denote this by $\mathcal{Z}(f)$.

Let us recall now a theorem of S.-Y. Cheng [Che] which describes locally the nodal set. In a neighborhood of a regular point $p \in \mathcal{Z}(f)$ (such that $\nabla f(p) \neq 0$), the implicit function theorem says that $\mathcal{Z}(f)$ is a submanifold. In a neighborhood of a critical point $p \in \mathcal{Z}(f)$, a theorem due to Bers [Be] implies that in a geodesic chart around p, f is asymptotic to an harmonic polynomial (in our setting, that is in constant curvature, it is just a consequence of the analyticity of f which implies that f vanishes up to a finite order). Hence there is some integer $k \geq 2$ such that in an exponential chart around p, f is asymptotic to some polynomial $c\Re(z^k)$. Cheng has shown that after composition with a diffeomorphism which is defined in a neighborhood of p and which is tangent to the identity map at p, f becomes equal to this polynomial. In particular, the intersection of $\mathcal{Z}(f)$ with a small neighborhood of p is diffeomorphic to the set of zeroes of $\Re(z^k)$. When S is compact, $\mathcal{Z}(f)$ is therefore a finite graph; its vertices are the critical points which are contained in $\mathcal{Z}(f)$, the valency at a vertex is twice k, $k \geq 2$ being the vanishing order of f at this vertex.

The starting point of this paper is a simple topological property of the nodal set.

Definition. A graph in a surface Σ is *incompressible* if the fundamental group of any of its connected components maps injectively into $\pi_1(\Sigma)$.

Lemma 1. *Let S be a hyperbolic surface. Let $0 < \lambda \leq \frac{1}{4}$ and let $f : S \to \mathbb{R}$ be a λ-eigenfunction. Then, the graph $\mathcal{Z}(f)$ is incompressible and the Euler characteristic of each connected component of $S - \mathcal{Z}(f)$ is negative.*

We deduce from this lemma the following propositions.

Proposition 2. *Let S be a finite area hyperbolic surface with genus 0 or 1. Then S does not carry any non-zero cuspidal eigenfunction with $0 < \lambda \leq \frac{1}{4}$.*

This result had also been obtained by Huxley using another method [Hu].

Proposition 3. *Let S be a finite area hyperbolic surface and let λ be any non-zero eigenvalue of the Laplacian on S. Then, if $\lambda \leq \frac{1}{4}$, the dimension of the eigenspace $\mathcal{E}_\lambda$ is at most $-\chi(S) - 1$ and the dimension of the cuspidal eigenspace $\mathcal{E}_\lambda^c$ is at most $-\chi(\overline{S}) - 1 = 2g - 3$.*

The proof of Proposition 3 is an adaptation of the proof by Bruno Sevennec of the existence of an upperbound for the multiplicity of the second eigenvalue of Schrödinger operators on compact surfaces [Se]. In our case, the incompressibility of the components of $S - \mathcal{Z}(f)$ allows one to simplify certain arguments and to obtain a better bound.

Proposition 4. *Let S be a finite area hyperbolic surface. Let $E \subset \mathcal{E}_\lambda$ be a vector space of cuspidal λ-eigenfunctions with $\lambda \leq \frac{1}{4}$. Then one of the following holds:*

1. *E contains a non-zero function whose nodal set is a smooth submanifold;*
2. *there is a finite set $\{p_1, \cdots, p_n\} \subset S$ and integers ≥ 2, $\{k_1, \cdots, k_n\}$ such that any function in E vanishes at the points p_i with an order $\geq k_i$.*

Remark on surfaces with variable curvature. The results above all follow from topological arguments. They hold more generally when S is a surface of negative Euler characteristic carrying a finite area complete Riemannian metric, after replacing in the statements the constant $\dfrac{1}{4}$ by $\lambda_0(\tilde{S})$, the bottom of the spectrum of the Laplacian on the universal cover of S. Such eigenvalues and their eigenfunctions are called *small.* (More generally, an eigenfunction is said to be *parabolic* if its nodal set enters each puncture arbitrarily far.)

2. Incompressibility of the nodal set. Proof of Proposition 2

Proof of Lemma 1. In order to prove the incompressibility of the graph, we argue by contradiction. Let us consider a closed curve contained in $\mathcal{Z}(f)$ which is essential on this graph but which is homotopic to 0 on S. This curve can be lifted to $\mathbb{H}$, and we can also suppose that its lift is a Jordan curve, bounding a disc $\tilde{D}$. Denote by π the covering map $\mathbb{H} \to S$ and by $\tilde{f}$ the function $f \circ \pi$. Denote by $\phi : \mathbb{H} \to \mathbb{R}$ the function which equals $\tilde{f}$ on $\tilde{D}$ and which vanishes identically on $\mathbb{H} - \tilde{D}$. One has

$$\int_{\mathbb{H}} \phi^2 dv = \int_{\tilde{D}} \tilde{f}^2 dv \leq \int_S f^2 dv < \infty.$$

Since ϕ vanishes on $\partial\tilde{D}$, its gradient (in the distribution sense) is the vector field on $\mathbb{H}$ which is 0 outside from $\tilde{D}$ and which equals ∇f on $\tilde{D}$. One has

$$\int_{\mathbb{H}} \|\nabla\phi\|^2 dv = \int_{\tilde{D}} \|\nabla\tilde{f}\|^2 dv \leq \int_S \|\nabla f\|^2 dv < \infty.$$

Therefore, ϕ belongs to the Sobolev space $\mathcal{H}$, which is the domain of the Laplacian on $\mathbb{H}$. Using the Green's formula and the fact that f vanishes on $\partial\tilde{D}$, one sees that the Rayleigh quotient $\dfrac{\int \|\nabla\phi\|^2 dv}{\int \phi^2 dv}$ is equal to λ. This contradicts the fact that the bottom of the spectrum of the Laplacian on $\mathbb{H}$ is $\frac{1}{4}$. Indeed, if the Rayleigh quotient of a function in the Sobolev space $\mathcal{H}$ equals $\frac{1}{4}$, then this function has to be a $\frac{1}{4}$-eigenfunction; in particular, it does not vanish on any open sets.

Now let C be a connected component of $S - \mathcal{Z}(f)$. Saying that its Euler characteristic is negative is equivalent to saying that C is not homeomorphic to a disc or to an annulus. Let us show by contradiction that both situations are impossible. We suppose first that C is homeomorphic to a disc. Let $\tilde{C}$ be a connected component of the preimage of C in the universal covering. Observe that the difference with the foregoing situation is that $\tilde{C}$ is not necessarily compact. However, the covering map π, when restricted to $\tilde{C}$, is a homeomorphism onto its image. Let us denote again by ϕ the function which is equal to $\tilde{f}$ on $\tilde{C}$ and which vanishes in the complement $\mathbb{H} - \tilde{C}$. One shows as before that ϕ belongs to $\mathcal{H}$. Since its Rayleigh quotient is λ we get a contradiction.

We suppose now that C is homeomorphic to an annulus. Let us consider the hyperbolic annulus $\tilde{S}$, which is the covering of S with fundamental group the image of $\pi_1(C)$ in $\pi_1(S)$. The annulus C can be lifted to an annulus $\tilde{C}$ in $\tilde{S}$. The function ϕ which equals $\tilde{f}$ on $\tilde{C}$, and which is zero on $\tilde{S} - \tilde{C}$ is in the domain of the Laplacian on $\tilde{S}$ and its Rayleigh equals λ. Since $\tilde{S}$ is the quotient of $\mathbb{H}$ by an amenable group (or after an explicit computation), we know that the bottom of the spectrum on $\tilde{S}$ equals $\frac{1}{4}$. Once again, one obtains a contradiction. □

Lemma 1 is a particular case of the next result which can be proven in the same way.

Lemma 5. *Let S be a hyperbolic surface and $f \in S \to \mathbb{R}$ be λ-eigenfunction with $\lambda \leq \frac{1}{4}$. Let C be a connected component of $S - \mathcal{Z}(f)$, $\overline{C}$ be its closure in S and $\tilde{S}_{\overline{C}}$ be the covering of S with fundamental group $\pi_1(\overline{C})$. Then the bottom of the spectrum of the Laplacian on $\tilde{S}_{\overline{C}}$ satisfies $\lambda_0(S_{\overline{C}}) < \lambda$.*

Remark on the Hausdorff dimension of subgroups. When S has finite area, it is well-known that the fundamental group of $\overline{C}$ is geometrically finite (a Schottky group in our case). The last Lemma implies in particular : $\lambda_0(S_{\overline{C}}) < \frac{1}{4}$. Hence, by Patterson and Sullivan, the Hausdorff dimension $\delta_{\overline{C}}$ of the limit set of $\pi_1(\overline{C})$ is bigger than $\frac{1}{2}$ [Pa, Su] ; the bottom of the spectrum is reached by the positive eigenfunction which is the Poisson-Helgason transform of the Hausdorff measure on the limit set ([He], [O]). If one writes the eigenvalue λ as $\lambda = s(1-s)$ choosing $s \geq \frac{1}{2}$, one thus has : $s < \delta$.

The Theorem of Cheng says that the nodal set of an eigenfunction is a locally finite graph but nothing does guarantee, even for a finite area surface that the number of critical points of f which are contained in $\mathcal{Z}(f)$ is finite. However, for the eigenvalues $\lambda \leq \frac{1}{4}$, Cheng's result generalizes as follow. Let us recall first a definition of a cuspidal eigenfunction which is equivalent to the one that was given in the introduction.

Definition. Let S be a hyperbolic surface with finite area. A neighborhood of a puncture in S is isometric to an horoball V_A, the quotient of a slit $z = x + iy \in \mathbb{H}, y \geq A$ by the translation $z \to z+1$. This neighborhood is foliated by the images of the horizontal lines, called *horocycles*. Let $f : S \to \mathbb{R}$ be a λ-eigenfunction. One says that *f is cuspidal at the puncture p* when the average of f on each horocycle centered at p is 0 (this definition agrees with the one given in the introduction). One says that *f is cuspidal* when it is cuspidal at any puncture.

Let us identify a neighborhood of p in S with the horoball V_A: when f is cuspidal at p, one has: $f(z) = o(e^{-2\pi y})$ [I, p. 64].

Lemma 6. *Let S be a hyperbolic surface with finite area and let $f : S \to \mathbb{R}$ be a λ-eigenfunction with $\lambda \leq \frac{1}{4}$.*

1. *The closure of $\mathcal{Z}(f)$ in $\overline{S}$ is a graph;*
2. *when f is cuspidal, each puncture in $\overline{S} - S$ is a vertex of this graph which has even valency.*

It may be that this lemma holds also for all the eigenvalues λ, however the proof that will follow uses Lemma 1 and depends therefore strongly on the hypothesis $\lambda \leq \frac{1}{4}$. When S has variable curvature, this lemma also holds when the eigenvalue $\lambda > 0$ is less than the bottom of the spectrum of the Laplacian on the universal covering.

Proof of Lemma 6. Each connected component of $S - \mathcal{Z}(f)$ has Euler characteristic ≤ -1. Since S has finite area, its topological type is finite also; in particular S can only contain a finite number of incompressible surfaces which are disjoint and have negative Euler characteristic. Therefore $S - \mathcal{Z}(f)$ has only finitely many connected components. Let C^+ be the reunion of the connected components where f is positive. Denote by s_i the vertices of the graph $\mathcal{Z}(f)$ (a priori, there could be infinitely many such vertices). All these vertices are in the closure of C^+. If, in a neighborhood of s_i, f is conjugate to the polynomial $c_i\Re(z^{k_i})$, then s_i appears with multiplicity k_i in the frontier of C^+. The Euler characteristic $\chi(\overline{C^+})$ equals $\chi(C^+) - \sum(k_i - 1)$. As the frontier of $\overline{C^+}$ is contained in the incompressible graph $\mathcal{Z}(f)$, the fundamental group of $\overline{C^+}$ maps injectively to $\pi_1(S)$ and therefore : $\chi(\overline{C^+}) \geq \chi(S)$. It follows that the set of vertices of $\mathcal{Z}(f)$ is finite.

Furthermore, only finitely many connected components of $\mathcal{Z}(f)$ are homeomorphic to a circle. Indeed there exists an integer $n(S)$ which depend only on the topology of S and which has the following property: if $n(S)$ disjoint curves are drawn on S, then two of them bound an annulus. If the graph $\mathcal{Z}(f)$ did contain more than $n(S)$ connected components, the complement of $\mathcal{Z}(f)$ would contain a disc or an annulus, contradicting Lemma 1.

Let p be a puncture of S; let us choose as neighborhoods of p in S the quotients V_A of the half-planes $z = x + iy \in \mathbb{H}, y > A$ by the translation $z \to z + 1$. Let us write λ as $\lambda = s(1-s)$ with $s \geq \frac{1}{2}$. Since $f \in L^2(S)$, there exists (cf. [I]) $\alpha \in \mathbb{R}$ and a function $g : S \to \mathbb{R}$ such that one can write $f(z) = \alpha y^{1-s} + g(z)$ with

1. $g(z) = o(e^{-2\pi y})$ and
2. $\int_0^1 g(x+iy)dx = 0$ for all y.

If (the Fourier coefficient) α is not 0, then $|f(z)| \to \infty$ when $z \to p$. Thus there is a neighborhood of p in S which is disjoint from $\mathcal{Z}(f)$; the level lines of f in this neighborhood are closed curves around p.

If $\alpha = 0$, property (2) implies that $\mathcal{Z}(f)$ accumulates on p. Let us choose A sufficiently large so that the horoball V_A does not contain any vertex of the graph $\mathcal{Z}(f)$: the intersection of this graph with V_A is therefore a 1-dimensional submanifold Z_A of V_A. Each connected component of Z_A is homeomorphic to one of the following :

1. an arc which connects the boundary of the annulus V_A to itself ;
2. an arc properly embedded in V_A which connects ∂V_A to the puncture p ;
3. an arc which connects the puncture p to itself.

Let us show that Z_A does not contain any arc of type (3). Otherwise, the connected component of the complement of this arc which is contained in the annulus V_A would be simply connected. This component would contain a component of $S - \mathcal{Z}(f)$ which would also be simply connected in contradiction with the Lemma 1.

It follows that Z_A is the reunion of a finite number of arcs connecting ∂V_A to itself and a finite number of arcs connecting ∂V_A to the puncture p. Therefore, if we complete V_A in a disc D_A by adding the puncture p, the closure of $\mathcal{Z}(f)$ is a graph in D_A. It is not hard to see that the valency of p is even (maybe 2).

When f is cuspidal, the constant in the Fourier expansion is 0 at each puncture. The closure of $\mathcal{Z}(f)$ in $\overline{S}$ is then a graph which contains all the punctures. □

Proof of Proposition 2. Let S be a hyperbolic surface with finite area. Let f be a non-zero λ-eigenfunction. Let $\overline{S}$ be the closed surface obtained from S by filling in the punctures. One applies the Euler-Poincaré formula to the folllowing cell-decomposition of $\overline{S}$: the 2 skeleton is the complement in S of the nodal set $\overline{\mathcal{Z}}(f)$, the 1-skeleton is the nodal set and the 0 skeleton is the set of the punctures of $\overline{S}$ which are not in $\overline{\mathcal{Z}}(f)$. If k is the number of those punctures the Euler-Poincaré formula states :

$$\chi(\overline{S}) - k = \chi(C^+) + \chi(C^-) + \chi(\overline{\mathcal{Z}}(f)).$$

Suppose that f is *cuspidal.* Then each cusp of $\overline{S}$ is contained in $\overline{\mathcal{Z}}(f)$ so that $k = 0$. Since the Euler characteristic of each connected component of $C^{\pm}$ is negative and since $\chi(\overline{\mathcal{Z}}(f)) \leq 0$, one has : $\chi(\overline{S}) < 0$. □

The same method also gives:

Proposition 7. *Let S be a hyperbolic surface with finite area, which is homeomorphic to a once punctured torus or to a thrice punctured sphere. Then S does not possess any non-zero λ-eigenfunction with $0 < \lambda \leq \frac{1}{4}$.*

Proof. Let us suppose first that S is an once punctured torus and that there exists a non-zero λ-eigenfunction f defined on S. By Proposition 2, we can suppose that f is not cuspidal ; it keeps therefore a constant sign near the puncture, say it is positive. Thus the connected components of C^- are contained in the complement of the closure of the component of C^+ which contains this neighborhood. Since this complement is an annulus or a reunion of discs, we find a contradiction with Lemma 1.

Let S be homeomorphic to a thrice-punctured sphere. If f is a λ-eigenfunction on S, with $\lambda \leq \frac{1}{4}$, we can suppose that f is not cuspidal. Then one of the components of $S - \mathcal{Z}(f)$ is an annulus, which is impossible again. □

Remark on groups with torsion. Lemma 1 can be generalized to cofinite Fuchsian groups Γ which contain torsion elements. The quotient $S = \mathbb{H}/\Gamma$ is then a hyperbolic orbifold, locally isometric to the quotient of $\mathbb{H}$ be a finite elliptic subgroup. It is homeomorphic to a compact surface with finitely many punctures and with finitely many marked points : these points are in 1-1 correspondance with the conjugacy classes of non-trivial elliptic subgroups, and are marked by an integer invariant, the order of the isotropy group. In this generality, Lemma 1 states the same but in the language of orbifolds : *the components of C^+ and C^- are orbifolds with negative Euler characteristic.* This excludes that a component of $C^{\pm}$ be a disc or an annulus, but also a disc with one marked point with order $n \geq 2$ or a disc with two marked points of order 2. One deduces that when S has genus 0 or 1, then S does not admit any non-zero λ-eigenfunction for $\lambda \leq \frac{1}{4}$ which is cuspidal and which vanishes at all singular points on S except at at most 3 points when the genus is 0, or at at most 1 when the genus is 1 (cf. [Hu, p. 352]). (When all singular points have order 2, one can say little more.)

Remark. Proposition 2 can be compared with a theorem of Peter Zograf, which says roughly that when the number of punctures on a hyperbolic orbifold with finite area is large with respect to the genus, then this orbifold admits a non-zero small eigenfunction. The proof given in [Z] consists in constructing, under this hypothesis, a function with 0 average and whose Rayleigh quotient is less than $\frac{1}{4}$. The existence of a non-zero λ-eigenfunction for some $\lambda < \frac{1}{4}$ follows from the variational characterization of eigenvalues ; however, it can not be guaranteed in general that this eigenfunction is cuspidal. In the same paper, Zograf constructs a family of hyperbolic orbifolds which admit cuspidal λ-eigenfunctions for arbitrarily small values of λ : these are genus 1 orbifolds with a single puncture and several singular points of order 2.

Remark. Proposition 2 applies to certain congruence groups $\Gamma_0(N)$, $\Gamma_1(N)$, $\Gamma(N)$, those such that the associated hyperbolic surface has genus 0 or 1 (cf. [Hu, p. 353]; see also [CP], [CL], [Cum] for tabulations of the subgroups of $\mathrm{PSL}(2,\mathbb{Z})$ with small genus).

3. The multiplicity of the small eigenvalues. Proof of Proposition 3

We study now some consequences of Lemma 1 on the dimension of the eigenspaces $\mathcal{E}_\lambda$ when $\lambda \leq \frac{1}{4}$.

We follow closely the proof of [Se, Theorem 4] which rests on a beautiful application of the Borsuk-Ulam Theorem. As we said in the introduction, the proof gets simplified in our setting, thanks to the incompressibility of the nodal components. In particular, we will consider a stratification analoguous to the one introduced in [Se] ; but it will be a stratification of the eigenspace itself and not of the space of pairs of disjoint open sets in S like in [Se]. However, besides this, the proof is the same.

Proof of Proposition 3. Let E be one of the vector spaces $\mathcal{E}_\lambda$ or $\mathcal{E}^c_\lambda$. Let m be the dimension of E. Denote by $\mathbb{S}^{m-1}$ the unit sphere of this space (for an arbitrary norm), by τ the antipodal involution and by $\mathbb{P}^{m-1}$ the quotient projective space. The covering $\mathbb{S}^{m-1} \to \mathbb{P}^{m-1}$ is described by a cohomology class $H^1(\mathbb{P}^{m-1}, \mathbb{Z}_2)$ denoted by α.

For $i \in \mathbb{Z}$, one defines $\tilde{\mathcal{P}}(i) \subset \mathbb{S}^{m-1}$ as the set of the functions f such that

$$\mathcal{X}(f) = \chi(C^+(f)) + \chi(C^-(f)) = i.$$

By Lemma 1, one has : $\mathbb{S}^{m-1} = \bigcup_{i=\chi(S)}^{i=-2} \tilde{\mathcal{P}}(i)$ if $E = \mathcal{E}_\lambda$ and $\mathbb{S}^{m-1} = \bigcup_{i=\chi(\overline{S})}^{i=-2} \tilde{\mathcal{P}}(i)$ if $E = \mathcal{E}^c_\lambda$. We observe that (although that won't be used in the proof) for any j with $\chi(\overline{S}) \leq j \leq -2$, the sets $\cup_{i=j}^{i=-2}\tilde{\mathcal{P}}(i)$ are closed in $\mathbb{S}^{m-1}$ (see also the proof of Proposition 4). Therefore each atom $\tilde{\mathcal{P}}(i)$ is locally closed in $\mathbb{S}^{m-1}$ and the $\tilde{\mathcal{P}}(i)$'s can be viewed as the strata of a stratification $\tilde{\mathcal{P}}$ of $\mathbb{S}^{m-1}$. Each stratum is invariant under the involution τ and one obtains therefore a stratification $\mathcal{P}$ of $\mathbb{P}^{m-1}$ by taking the quotient spaces $\mathcal{P}(i) = \tilde{\mathcal{P}}(i)/\tau$. The covering $\tilde{\mathcal{P}}(i) \to \mathcal{P}(i)$ is described by the Čech-cohomology class $\alpha|_{\mathcal{P}(i)}$.

Let us show that for each i, the covering $\tilde{\mathcal{P}}(i) \to \mathcal{P}(i)$ is trivial. The reasonning in [Se, Proposition 13] gets simplified thanks to the incompressibility of the connected components of $C^\pm(f)$. Indeed, for g sufficiently near to f, there is an isotopy of S which sends $C^+(g)$ (resp. $C^-(g)$)) onto $C^+(f)$ (resp. $C^-(f)$)). To see this, let us consider *compact cores* C^+, C^- of $C^+(f)$, $C^-(f)$, i.e., compact surfaces C^+, C^- which are respectively contained in $C^+(f)$, $C^-(f)$ and are such that the inclusion into $C^+(f)$, $C^-(f)$ is an homotopy equivalence. Then for g in a neighborhood $\mathcal{V}(f)$ of f in $\mathbb{S}^{m-1}$, C^+ et C^- are incompressible surfaces which are contained in $C^+(g)$ and $C^-(g)$ respectively: the Euler characteristic of C^+ (resp. of C^-) is thus not more than that of $C^+(g)$ (resp. of $C^-(g)$) with equality if and only if C^+ (resp. C^-) is a compact core of $C^+(g)$ (resp. of $C^-(g)$). But, if $\tilde{g} \in \tilde{\mathcal{P}}(i)$, then $\mathcal{X}(g) = \mathcal{X}(f) = \chi(C^-) + \chi(C^+)$: hence C^+ (resp. C^-) is a compact core of $C^+(g)$ (resp. of $C^-(g)$). One deduces, for $g \in \mathcal{V}(f) \cap \tilde{\mathcal{P}}(i)$, an isotopy of S which sends $C^{\pm}(f)$ to $C^{\pm}(g)$. An isotopy with the same property exists therefore when f and g are in the same connected component of $\tilde{\mathcal{P}}(i)$. One has also that $C^+(f)$ is not isotopic to $C^-(f)$ since both are incompressible subsurfaces of S disjoint and with negative Euler characteristic. Thus, each connected component of $\tilde{\mathcal{P}}(i)$ meets each fiber of the covering $\tilde{\mathcal{P}}(i) \to \mathcal{P}(i)$ in at most one point : this covering is therefore trivial.

Thus, the Čech-cohomology class $\alpha|_{\mathcal{P}(i)}$ vanishes. It follows that the cup-product α^k vanishes also, where k is the number of strata in the stratification $\mathcal{P}$ (cf. [Se, Lemma 8]). It is classical that α has order m in the $\mathbb{Z}_2$-cohomology ring of $\mathbb{P}^{m-1}$. Therefore $m \leq k$. Since k is not more than $-\chi(S) - 1$ when $E = \mathcal{E}_\lambda$ and $-\chi(\overline{S}) - 1 = 2g - 3$ when $E = \mathcal{E}^c_\lambda$, this proves Proposition 3. □

Remark. Note that this upperbound for the dimension of *any* eigenspace with eigenvalue below $\frac{1}{4}$ goes (slowly) in the direction of the conjectural bound $\lambda_{2g-2} \geq \frac{1}{4}$ (cf. [Bu, p. 223], [Sc]). In fact, if Lemma 1 were true for functions f in the vector space which is the direct sum of the eigenspaces $\mathcal{E}_{\lambda_i}$, for $0 < \lambda_i \leq \dfrac{1}{4}$, then the method for proving Proposition 3 would give the bound $2g - 3$ for the dimension of this space. As a consequence, we would have $\lambda_{2g-2} > \frac{1}{4}$.

4. Common zeros to eigenfunctions. Proof of Proposition 4

Let $\lambda \leq \frac{1}{4}$ and consider a vector space $E \subset \mathcal{E}_\lambda$. Let f be a function in E which minimizes the functional $g \to \mathcal{X}(g) = \chi(C^+(g)) + \chi(C^-(g))$.

If $\mathcal{X}(f) = -2\chi(\overline{S})$, then, the nodal set of f is a submanifold of $\overline{S}$, by the Euler-Poincaré formula.

Let us suppose that $\mathcal{X}(f) > -2\chi(\overline{S})$. We are going to show that any function in E vanishes at all vertices of the graph $\mathcal{Z}(f)$. Let us choose a chart V_j around each finite vertex p_j of this graph in which f can be written $f(z) = c_j \Re(z^{k_j})$. Let $\mathcal{V}$ be a neighborhood of the set of vertices, which is the union of the V_i's and of a neighborhood of the punctures of $\overline{S}$. The norm of the gradient of f on $\mathcal{Z}(f) - \mathcal{V}$ is bounded from below. For any function $h \in E$, there exists thus some $\epsilon_0 > 0$ so that for all ϵ, $|\epsilon| \leq \epsilon_0$, the intersections of the nodal sets of $f + \epsilon h$ and f with $\overline{S} - \mathcal{V}$ are isotopic relatively to $\partial\mathcal{V}$. Let $h \in E$ and suppose that $h(p_i) \neq 0$. After eventually replacing V_j by a smaller open set, h keeps a constant sign, say positive, on V_j. Then, for all $\epsilon > 0$ sufficiently small, the nodal set of $f + \epsilon h$ intersects V_j

as a union of k_j arcs properly embedded (each of those arcs is contained in one of the k_j sectors where f is negative). In particulier, one has

$$\chi(V_j \cap C^+(f+\epsilon h)) < \chi(V_j \cap C^+(f)), \quad \chi(V_j \cap C^-(f+\epsilon h)) = \chi(V_j \cap C^-(f)),$$

whereas the other components V of $\mathcal{V}$ satisfy $\chi(V \cap C^+(f+\epsilon h)) \leq \chi(V \cap C^+(f))$ and also $\chi(V \cap C^-(f+\epsilon h)) \leq \chi(V \cap C^-(f))$. Using the additivity of the Euler characteristic, we find $\mathcal{X}(f+\epsilon h) < \mathcal{X}(f)$. This contradicts that f minimizes the functional $\mathcal{X}$.

Proof of Proposition 4. Let us choose, among the functions which minimize the functional $\mathcal{X}$ a function f which maximizes the number of finite vertices on the nodal set.

Let $p_1, \cdots, p_n$ be the vertices of the graph $\mathcal{Z}(f)$ which are contained in S and suppose that f vanishes up to the order k_j ($k_j \geq 2$) at the point p_j. Let $h \in E$. Choose a chart around each vertex p_j in which f can be written as an harmonic polynomial $c_j \Re(z^{k_j})$. As before, there exists some $\epsilon_0 > 0$ such that if $|\epsilon| \leq \epsilon_0$, the intersections $\mathcal{Z}(f) \cap (S - \mathcal{V})$ et $\mathcal{Z}(f+\epsilon h) \cap (S - \mathcal{V})$ are isotopic relatively to $\partial \mathcal{V}$. We know already that h vanishes at all points p_i : let κ_i be its vanishing order at p_i. Then, in the chart V_j around p_j, $f + \epsilon h$ vanishes at p_j up to the order exactly κ_j if $\kappa_j < k_j$ and up to the order k_j at least if $\kappa_j \geq k_j$. For $\epsilon \leq \epsilon_0$, the nodal set $\mathcal{Z}(f+\epsilon h)$ intersects V_j as a graph denoted by $\mathcal{Z}_j(f+\epsilon h)$; this graph is properly embedded in V_j and it has exactly $2k_j$ vertices on ∂V_j. Since f minimizes $\mathcal{X}$, one has, for all ϵ :

$$\chi(\mathcal{Z}(f+\epsilon h)) \leq \chi(\mathcal{Z}(f)).$$

For $\epsilon \leq \epsilon_0$, the additivity of the Euler characteristic gives :

$$\sum_j \chi(\mathcal{Z}_j(f+\epsilon h)) \leq \sum_j \chi(\mathcal{Z}_j(f)).$$

Thus, for all j, $\mathcal{Z}_j(f+\epsilon h)$ is connected (in fact a tree). Let us suppose that there is some index ι such that $\kappa_\iota < k_\iota$. Then $\mathcal{Z}_\iota(f+\epsilon h)$ is a graph with Euler characteristic $+1$, which has one vertex of valency $2\kappa_\iota$ and $2k_\iota$ free vertices (of valency 1). There must be therefore another vertex of valency ≥ 3. This contradicts the property of f that it maximizes the number of vertices on its nodal set.

□

References

[Be] L. Bers, *Local behaviour of solutions of general linear elliptic equations*, Comm. Pure Appl. Math. **8** (1955), 475–504.

[Bu] P. Buser, *Geometry and Spectra of Compact Riemann Surface*, Birkhäuser, Boston, 1992.

[Cha] I. Chavel, *Eigenvalues In Riemannian Geometry*, Academic Press, Orlando, 1984.

[Che] S.-Y. Cheng, *Eigenfunctions and nodal sets*, Comment. Math. Helvetici **51** (1976) 43–55.

[Cum] C.J. Cummins, *Congruence subgroups of groups commensurable with* $PSL(2, \mathbb{Z})$, Exper. Mathematics **13** (2004), 361–382.

[CL] K.S. Chua and M.L. Lang, *Congruence subgroups associated to the Monster*, Exper. Math. **13** (2004), 343–360.

[CP] C.J. Cummins et S. Pauli, *Congruence subgroups of* $PSL(2, \mathbb{Z})$ *of genus less than or equal to* 24, Exper. Mathematics **12** no. 2 (2004), 243–255.

[He] S. Helgason, *Topics in Harmonic Analysis on Homogeneous Spaces*, Progress in Mathematics, vol. 13, Birkhäuser, Boston-Basel-Stuttgart (1981).

[Hu] M.N. Huxley, *Cheeger's Inequality with a boundary term*, Commentarii Mathematici Helvetici **58** (1983).

[I] H. Iwaniec, *Introduction to the spectral theory of automorphic forms*, Biblioteca de la Revista Matemática Iberoamericana Madrid, 1995.

[O] J.-P. Otal, *Sur les fonctions propres du Laplacien du disque hyperbolique*, Comptes Rendus de l'Académie des Sciences, Paris **327** (1998), 161–166.

[Pa] S.J. Patterson, *The limit set of a Fuchsian group*, Acta Math. **136** no. 3–4 (1976), 161–166.

[Sc] P. Schmutz, *Small eigenvalues on Riemann surfaces of genus 2*, Invent. Math. **106** (1991), 121–138.

[Se] B. Sevennec, *Multiplicity of the second Schrödinger eigenvalue on closed surfaces*, Math. Ann. **324** (2002), 195–211.

[Su] D. Sullivan, *Related aspects of positivity in Riemannian geometry*, J. Differential Geometry **25** no. 3 (1987), 327–351.

[Z] P. Zograf, *Small eigenvalues of automorphic Laplacians in spaces of cusp forms*, Soviet Math. Dokl. **27** no. 2 (1983), 420–422.

Jean-Pierre Otal
Laboratoire Émile Picard
Université Paul Sabatier
118, route de Narbonne
F-31062 Toulouse cedex 9
France

Progress in Mathematics, Vol. 265, 697–719

Quantum p-adic Spaces and Quantum p-adic Groups

Yan Soibelman

To the memory of Sasha Reznikov

Abstract. We discuss examples of non-commutative spaces over non-archimedean fields. Those include non-commutative and quantum affinoid algebras, quantized K3 surfaces and quantized locally analytic p-adic groups.

Mathematics Subject Classification (2000). 12J25.

Keywords. Berkovich spectrum, deformation quantization, mirror symmetry, quantum groups.

1. Introduction

The paper is devoted to examples of quantum spaces over non-archimedean fields and is, in a sense, a continuation of [So1] (part of the material is borrowed from the loc. cit). There are three classes of examples which I discuss in this paper: quantum affinoid spaces, quantum non-archimedean Calabi-Yau varieties and quantum p-adic groups. Let us recall the definitions and discuss the contents of the paper.

Quantum affinoid algebras are defined similarly to the "classical case" $q = 1$. It is a special case of a more general notion of non-commutative affinoid algebra introduced in [So1]. Let k be a Banach field and $k\langle\langle T_1, \dots, T_n\rangle\rangle$ be the algebra of formal series in free variables $T_1, \dots, T_n$. For each $r = (r_1, \dots, r_n), r_i \geq 0, 1 \leq i \leq n$ we define a subspace $k\langle\langle T_1, \dots, T_n\rangle\rangle_r$ consisting of series $f = \sum_{i_1,\dots,i_m} a_{i_1,\dots,i_m} T_{i_1} \cdots T_{i_m}$ such that $\sum_{i_1,\dots,i_m} |a_{i_1,\dots,i_m}| r_{i_1} \cdots r_{i_m} < +\infty$. Here the summation is taken over all sequences $(i_1, \dots, i_m), m \geq 0$ and $|\bullet|$ denotes the norm in k. In this paper we consider the case when k is a valuation field (i.e., a Banach field with respect to a multiplicative non-archimedean norm). In the non-archimedean case the convergency condition is replaced by the following one: $max\, |a_{i_1,\dots,i_m}| r_{i_1} \cdots r_{i_m} \to 0$ as $i_1 + \cdots + i_m \to \infty$. Clearly each $k\langle\langle x_1, \dots, x_n\rangle\rangle_r$ is a Banach algebra called the algebra of analytic functions on a non-commutative k-polydisc $E_{NC}(0, r)$ centered at zero and having the (multi)radius $r = (r_1, \dots, r_n)$.

The norm is given by $max\,|a_{i_1,\dots,i_m}|r_{i_1}\cdots r_{i_m}$. A *non-commutative k-affinoid algebra* is an admissible noetherian quotient of this algebra (cf. [Be1], Definition 2.1.1). Let us fix $q \in k^*$ such that $|q| = 1$, and $r = (r_1,\dots,r_n), r_i \geq 0$. A *quantum k-affinoid algebra* is a special case of the previous definition. It is defined as an admissible quotient of the algebra $k\{T\}_{q,r} := k\{T_1,\dots,T_n\}_{q,r}$ of the series $f = \sum_{l=(l_1,\dots,l_n)\in \mathbf{Z}_+^n} a_l T_1^{l_1}\cdots T_n^{l_n}$ such that $a_l \in k$, $T_iT_j = qT_jT_i, j < i$, and $max|a_l|r^{|l|} \to 0$ as $|l| := l_1+\cdots+l_n \to \infty$. The latter is also called the algebra of analytic functions on the quantum polydisc $E_q(0,r)$. It is less useful notion than the one of non-commutative affinoid algebra since there are few two-sided closed ideals in the algebra $k\{T_1,\dots,T_n\}_{q,r}$. Nevertheless quantum affinoid algebras appear in practice (e.g., in the case of quantum Calabi-Yau manifolds considered below). In the case when all $r_i = 1$ we speak about *strictly k-affinoid* non-commutative (resp. quantum) algebras, similarly to [Be1]. Any non-archimedean extension K of k gives rise to a non-commutative (resp. quantum) affinoid k-algebra, cf. loc.cit. There is a generalization of quantum affinoid algebras which we will also call quantum affinoid algebras. Namely, let $Q = ((q_{ij}))$ be an $n \times n$ matrix with entries from k such that $q_{ij}q_{ji} = 1, |q_{ij}| = 1$ for all i,j. Then we define the quantum affinoid algebra as an admissible quotient of the algebra $k\{T_1,\dots,T_n\}_{Q,r}$. The latter defined similarly to $k\{T_1,\dots,T_n\}_{q,r}$, but now we use polynomials in variables $T_i, 1 \leq i \leq n$ such that $T_iT_j = q_{ij}T_jT_i$. One can think of $k\{T_1,\dots,T_n\}_{Q,r}$ as of the quotient of $k\langle\langle T_i, t_{ij}\rangle\rangle_{r,\mathbf{1}_{ij}}$, where $1 \leq i,j \leq n$ and $\mathbf{1}_{ij}$ is the unit $n \times n$ matrix, by the two-sided ideal generated by the relations

$$t_{ij}t_{ji} = 1, T_iT_j = t_{ij}T_jT_i, t_{ij}a = at_{ij},$$

for all indices i,j and all $a \in k\langle\langle T_i, t_{ij}\rangle\rangle_{r,\mathbf{1}_{ij}}$. In other words, we treat q_{ij} as variables which belong to the center of our algebra and have the norms equal to one. Having the above-discussed generalizations of affinoid algebras we can consider their Berkovich spectra (sets of multiplicative seminorms). Differently from the commutative case, the theory of non-commutative and quantum analytic spaces is not developed yet (see discussion in [So1]).

Quantum Calabi-Yau manifolds provide examples of topological spaces equipped with rings of non-commutative affinoid algebras, which are quantum affinoid outside of a "small" subspace. More precisely, let $k = \mathbf{C}((t))$ be the field of Laurent series, equipped with its standard valuation (order of the pole) and the corresponding non-archimedean norm. Quantum Calabi-Yau manifold of dimension n over $\mathbf{C}((t))$, is defined as a ringed space $(X, \mathcal{O}_{q,X})$ which consists of an analytic Calabi-Yau manifold X of dimension n over $\mathbf{C}((t))$ and a sheaf of $\mathbf{C}((t))$-algebras $\mathcal{O}_{q,X}$ on X such that $\mathcal{O}_{q,X}(U)$ is a non-commutative affinoid algebra for any affinoid $U \subset X$ and the following two conditions are satisfied:

1) X is a $\mathbf{C}((t))$-analytic manifold corresponding to a maximally degenerate algebraic Calabi-Yau manifold X^{alg} of dimension n (see [KoSo2] for the definitions).

2) Let $Sk(X)$ be the skeleton of X defined in [KoSo1], and let us choose a projection $\pi : X \to Sk(X)$ described in the loc.cit. Then the direct image $\pi_*(\mathcal{O}_{q,X})$ is locally isomorphic (outside of a topological subvariety of the codimension at least two) to the sheaf of $\mathbf{C}((t))$-algebras $\mathcal{O}^{can}_{q,\mathbf{R}^n}$ on $\mathbf{R}^n$ which is characterized by the property that for any open connected subset $U \subset \mathbf{R}^n$ we have $\mathcal{O}^{can}_{q,\mathbf{R}^n}(U) = \{\sum_{l\in\mathbf{Z}^n} a_l z^l\}$ such that $a_l \in \mathbf{C}((t))$ and $\sup_{l\in\mathbf{Z}^n}(\log|a_l| + \langle l, x\rangle) < \infty$ for any $x \in U$. Here $\langle (l_1, \ldots, l_n), (x_1, \ldots, x_n)\rangle = \sum_{1\le i\le n} l_i x_i$.

For a motivation of this definition see [KoSo1-2] in the "commutative" case $q = 1$. Roughly speaking, in that case the above definition requires the Calabi-Yau manifold X to be locally isomorphic (outside of a "small" subspace) to an analytic torus fibration $\pi_{can} : (\mathbf{G}^{an}_m)^n \to \mathbf{R}^n$, where on $\mathbf{C}((t))$-points the canonical projection π_{can} is the "tropical" map $(z_1, \ldots, z_n) \mapsto (\log|z_1|, \ldots, \log|z_n|)$. This is a "rigid-analytic" implementation of the Strominger-Yau-Zaslow conjecture in Mirror Symmetry (see [KoSo1-2] for more on this topic). In present paper we discuss the case $n = 2$, essentially following [KoSo1], [So1]. Perhaps the higher-dimensional case can be studied by the technique developed in a recent paper [GroSie1] (which in some sense generalizes to the higher-dimensional case ideas of [KoSo1]). We plan to return to this problem in the future.

Finally, we discuss the notion of p-adic quantum group. Quantum groups over p-adic fields and their representations will be discussed in more detail in the forthcoming paper [So2]. We have borrowed some material from there. Recall that quantum groups are considered in the literature either in the framework of algebraic groups or in some special examples of locally compact groups over $\mathbf{R}$. In the case of groups over $\mathbf{R}$ or $\mathbf{C}$ there is the following problem: how to describe, say, smooth or analytic (or rapidly descreasing) functions on a complex or real Lie group in terms of the representation theory of its enveloping algebra? Finite-dimensional representations give rise to the algebra of regular functions (via Peter-Weyl theorem), but more general classes of functions are not so easy to handle. The case of p-adic fields is different for two reasons. First, choosing a good basis in the enveloping algebra, we can consider series with certain restrictions on the growth of norms of their coefficients. This allows us to describe a basis of compact open neighbourhoods of the unit of the corresponding p-adic group. Furthermore, combining the ideas of [ShT1] with the approach of [So3] one can define the algebra of *locally analytic* functions on a compact p-adic group as a certain completion of the coordinate ring of the group. Dualizing, one obtains the algebra of locally-analytic distributions. According to [ShT1] modules over the latter provides an interesting class of p-adic representations, which contains, e.g., principal series representations. The above considerations can be "quantized", giving rise to quantum locally-analytic groups.

The present paper contains a discussion of the above-mentioned three classes of examples of non-commutative spaces. The proofs are omitted and will appear in separate publications. We should warn the reader that the paper does not present a piece of developed theory. This explains its sketchy character. My aim is to show interesting classes of non-archimedean non-commutative spaces which can be obtained as analytic non-commutative deformations of the corresponding classical spaces. They deserve further study (for the quantum groups case see [So2]).

When talking about rigid analytic spaces we use the approach of Berkovich, which seems to be more suitable in the non-commutative framework. For this reason our terminology is consistent with [Be1].

Acknowledgements. I am grateful to many people who shared with me their ideas and insights, especially to Vladimir Berkovich, Joseph Bernstein, Matthew Emerton and Maxim Kontsevich. Excellent papers [SchT1-6] by Peter Schneider and Jeremy Teitelbaum played a crucial role in convincing me that the theory of quantum p-adic groups should exist. I thank IHES for the hospitality and excellent research conditions. This work was partially supported by an NSF grant.

2. Quantum affinoid algebras

Let k be a valuation field.

We recall here the definition already given in the Introduction. Let us fix $r = (r_1, \dots, r_n) \in \mathbf{R}^n_{>0}$. We start with the algebra $k\langle T\rangle := k\langle T_1, \dots, T_n\rangle$ of polynomials in n free variables and consider its completion $k\langle\langle T\rangle\rangle_r$ with respect to the norm $|\sum_{\lambda\in P(\mathbf{Z}^n_+)} a_\lambda T^\lambda| = \max_\lambda |a_\lambda| r^\lambda$. Here $P(\mathbf{Z}^n_+)$ is the set of finite paths in $\mathbf{Z}^n_+$ starting at the origin, and $T^\lambda = T_1^{\lambda_1} T_2^{\lambda_2} \cdots$ for the path which moves λ_1 steps in the direction $(1, 0, 0, \dots)$ then λ_2 steps in the direction $(0, 1, 0, 0, \dots)$, and so on (repetitions are allowed, so we can have a monomial like $T_1^{\lambda_1} T_2^{\lambda_2} T_1^{\lambda_3}$).

Definition 2.0.1. *We say that a noetherian Banach unital algebra A is non-commutative affinoid k-algebra if there is an admissible surjective homomorphism $k\langle\langle T\rangle\rangle_r \to A$ (admissibility means that the norm on the image is the quotient norm).*

In particular, affinoid algebras in the sense of [Be1] belong to this class (unfortunately the terminology is confusing in this case: commutative affinoid algebras give examples of non-commutative affinoid algebras!). Another class of examples is formed by quantum affinoid algebras defined in the Introduction.

Let us now recall the following definition (see [Be1]).

Definition 2.0.2. *Berkovich spectrum $M(A)$ of a unital Banach ring A consists of bounded multiplicative seminorms on A.*

If A is a k-algebra, we require that seminorms extend the norm on k. It is well-known (see [Be1], Th. 1.2.1) that if A is commutative then $M(A)$ is a non-empty compact Hausdorff topological space (in the weak topology). If $\nu \in M(A)$ then $Ker\, \nu$ is a two-sided closed prime ideal in A. Therefore it is not clear whether $M(A)$ is non-empty in the non-commutative case.

Algebras of analytic functions on the non-commutative and quantum polydiscs carry multiplicative "Gauss norms" (see Introduction), hence the Berkovich spectrum is non-empty in each of those cases. The following example can be found in [So3], [SoVo].

Let L be a free abelian group of finite rank d, $\varphi : L \times L \to \mathbf{Z}$ be a skew-symmetric bilinear form, $q \in K^*$ satisfies the condition $|q| = 1$. Then $|q^{\varphi(\lambda,\mu)}| = 1$ for any $\lambda, \mu \in L$. We denote by $A_q(T(L, \varphi))$ the *algebra of regular functions on the quantum torus* $T_q(L, \varphi)$. By definition, it is a k-algebra with generators $e(\lambda), \lambda \in L$, subject to the relation

$$e(\lambda)e(\mu) = q^{\varphi(\lambda,\mu)}e(\lambda + \mu).$$

The algebra of analytic functions on the analytic quantum torus $T_q^{an}(L, \varphi)$ consists by definition of series $\sum_{\lambda \in L} a(\lambda)e(\lambda), a(\lambda) \in k$ such that for all $r = (r_1, \ldots, r_d), r_i > 0$ one has: $|a(\lambda)|r^\lambda \to 0$ as $|\lambda| \to \infty$ (here $|(\lambda_1, \ldots, \lambda_d)| = \sum_i |\lambda_i|$).

Quantum affinoid algebra $k\{T\}_{q,r}$ discussed in the Introduction is the algebra of analytic functions on quantum polydisc of the (multi)radius $r = (r_1, \ldots, r_n)$. It was shown in [So1] that $M(k\{T\}_{q,r})$ can be quite big as long as $|q - 1| < 1$. In particular, it contains "quantum" analogs of the norms $|f|_{E(a,\rho)}$ which is the "maximum norm" of an analytic function f on the polydisc centered at $a = (a_1, \ldots, a_n)$ of the radius $\rho = (\rho_1, \ldots, \rho_n)$, with the condition $a_i \leq \rho_i < r_i, 1 \leq i \leq n$. Similar result holds for the quantum analytic torus. This observation demonstrates an interesting phenomenon: differently from the formal deformation quantization, the non-archimedean analytic quantization "preserves" some part of the spectrum of the "classical" object.

The conventional definition of the quatization can be carried out to the analytic case with obvious changes. Indeed, the notion of Poisson algebra admits a straightforward generalization to the analytic case (Poisson bracket is required to be a bi-analytic map). Furthermore, for any commutative affinoid algebra A there is a notion of non-commutative A-affinoid algebra, which is a natural generalization of the notion of k-affinoid algebra (we use $A\langle\langle T_1, \ldots, T_n\rangle\rangle_r$ instead of $k\langle\langle T_1, \ldots, T_n\rangle\rangle_r$).

Let now $\mathcal{O}(E(0, r))$ be the algebra of analytic functions on a 1-dimensional polydisc $E(0, r) = M(k\{r^{-1}T\})$ of the radius r (the notation is from [Be1], Chapter 2). We say that a non-commutative $\mathcal{O}(E(0, r))$-affinoid algebra A is an *analytic quantization of a k-affinoid commutative Poisson algebra A_0 over the polydisc $E(0, r)$* if the following two conditions are satisfied:

1) A is a topological $\mathcal{O}(E(0, r))$-algebra, free as a topological $\mathcal{O}(E(0, r))$-module.
2) The quotient A/TA is isomorphic to A_0 as a k-affinoid Poisson algebra.

Then a quantization of a k-analytic space $(X, \mathcal{O}_X)$ iss a ringed space $(X, \mathcal{O}_{q,X})$ such that for any affinoid $U \subset X$ the algebra $\mathcal{O}_{q,X}(U)$ is an analytic quantization of $\mathcal{O}_X(U)$ over some polydisc $E(0, r)$.

Notice that the projection $A \to A_0$ induces an embedding of Berkovich spectra $M(A_0) \to M(A)$. Every element of A can be thought of as analytic function on $E(0,r)$ with values in a non-commutative k-affinoid algebra. Suppose that $A \simeq A_0\{r^{-1}T\}$ as a $k\{r^{-1}T\}$-module. Then the topological vector space A is isomorphic to the space of analytic functions on $E(0,r)$ with values in A_0 (but the product is not a pointwise product of functions). Assume that $r \leq 1$ and consider the subspace A_1 of analytic functions $a(x)$ as above such that $|a(0)|_{A_0} \leq 1, |a(x) - a(0)|_{A_0} \leq |T(x)|, x \in E(0,r)$, where $\bullet|_{A_0}$ denotes the norm on A_0. Here x is interpreted as a seminorm on the Banach k-algebra $k\{r^{-1}T\}$, hence $|T(x)|$ is the norm of the generator T in the completition of the residue field $k\{r^{-1}T\}/Ker\, x$. It is clear that A_1 is in fact a Banach k-algebra. Hence the natural projection $a(x) \mapsto a(0)$ defines an embedding $M(A_0) \to M(A_1)$.

Suppose that X is an analytic spaces for which there is a notion of a skeleton $Sk(X)$ either in the sense of [KoSo1] (then X is assumed to be Calabi-Yau) or in the sense of [Be2,Be3]. Then in either of these cases there is a continuous retraction $\pi : X \to Sk(X)$. Suppose that the there is a quantization $(X, \mathcal{O}_{q,X})$ of $(X, \mathcal{O}_X)$ in the above sense.

Conjecture 2.0.3. *For any closed $V \subset X$ there is a natural embedding $i_V : V \subset M(\mathcal{O}_{q,X}(\pi^{-1}(V)))$ such that $\pi \circ i_V = id_V$. Moreover if $V_1 \subset V_2$ then the restriction of i_{V_1} to V_2 is equal to V_2.*

In other words, the skeleton survives an analytic quantization. The above conjecture is not very precise, because there is no general definition of a skeleton. The definition given in [KoSo1] is different from the one in [Be2,3] even for Calabi-Yau manifolds. Hence the conjecture is an "experimental fact" at this time.

3. Quantum Calabi-Yau varieties

Let X^{alg} be a maximally degenerate (in the sense of [KoSo1-2]) algebraic Calabi-Yau manifold over $\mathbf{C}((t))$ of dimension n and X be the corresponding $\mathbf{C}((t))$-analytic space. Then one can associate with X a PL-manifold $Sk(X)$ of real dimension n, called the skeleton of X (see [KoSo1-2]). A choice of Kähler structure on X^{alg} defines (conjecturally) a continuous retraction $\pi : X \to Sk(X)$. This map satisfies the condition 2) from the Introduction. In other words, it defines a (singular) analytic torus fibration over $Sk(X)$ with the generic fiber, isomorphic to the analytic space $M(k[T_1^{\pm 1}, \dots, T_n^{\pm 1}]/(|T_i| = c_i, 1 \leq i \leq n))$, where $c_i > 0, 1 \leq i \leq n$ are some numbers. Since the projection is Stein (see loc. cit), one can reconstruct X (as a ringed space) from the knowledge of $(Sk(X), \pi_*(\mathcal{O}_X))$, where $\mathcal{O}_X$ is the sheaf of analytic functions on X.

Let $B = Sk(X)$ and B^{sing} be the "singular subvariety" of real codimension two (see Introduction). It was observed in [KoSo1] that the norms of elements of the direct image sheaf $\pi_*(\mathcal{O}_X^{\times})$ define an integral affine structure on $B^0 := B \setminus B^{\mathrm{sing}}$. Hence we would like to reconstruct the analytic space starting with a PL-manifold equipped with a (singular) integral affine structure. As we will explain in the next subsection the same data give rise to a sheaf of quantum affinoid algebras on B^0.

3.1. Integral affine structures and quantized canonical sheaf

Here we explain following [KoSo1] and [So1] how a manifold with integral affine structure defines a sheaf of (quantum) affinoid algebras.

Recall that an integral affine structure ($\mathbf{Z}$-affine structure for short) on an n-dimensional topological manifold Y is given by a maximal atlas of charts such that the change of coordinates between any two charts is described by the formula

$$x'_i = \sum_{1\le j\le n} a_{ij}x_j + b_i,$$

where $(a_{ij}) \in GL(n,\mathbf{Z}), (b_i) \in \mathbf{R}^n$. In this case one can speak about the sheaf of $\mathbf{Z}$-affine functions, i.e., those which can be locally expressed in affine coordinates by the formula $f = \sum_{1\le i\le n} a_i x_i + b, a_i \in \mathbf{Z}, b \in \mathbf{R}$. An equivalent description: $\mathbf{Z}$-affine structure is given by a covariant lattice $T^{\mathbf{Z}} \subset TY$ in the tangent bundle (recall that an affine structure on Y is the same as a torsion free flat connection on the tangent bundle TY).

Let Y be a manifold with $\mathbf{Z}$-affine structure. The sheaf of $\mathbf{Z}$-affine functions $Aff_{\mathbf{Z}} := Aff_{\mathbf{Z},Y}$ gives rise to an exact sequence of sheaves of abelian groups

$$0 \to \mathbf{R} \to Aff_{\mathbf{Z}} \to (T^*)^{\mathbf{Z}} \to 0,$$

where $(T^*)^{\mathbf{Z}}$ is the sheaf associated with the dual to the covariant lattice $T^{\mathbf{Z}} \subset TY$.

Let us recall the following notion introduced in [KoSo1], Section 7.1. Let k be a valuation field.

Definition 3.1.1. *A k-affine structure on Y compatible with the given $\mathbf{Z}$-affine structure is a sheaf Aff_k of abelian groups on Y, an exact sequence of sheaves*

$$1 \to k^\times \to Aff_k \to (T^*)^{\mathbf{Z}} \to 1,$$

together with a homomorphism Φ of this exact sequence to the exact sequence of sheaves of abelian groups

$$0 \to \mathbf{R} \to Aff_{\mathbf{Z}} \to (T^*)^{\mathbf{Z}} \to 0,$$

such that $\Phi = id$ on $(T^*)^{\mathbf{Z}}$ and $\Phi = \mathrm{val}$ on $k^\times$, where val is the valuation map.

Since Y carries a $\mathbf{Z}$-affine structure, we have the corresponding $GL(n,\mathbf{Z}) \ltimes \mathbf{R}^n$-torsor on Y, whose fiber over a point x consists of all $\mathbf{Z}$-affine coordinate systems at x.

Then one has the following equivalent description of the notion of k-affine structure.

Definition 3.1.2. *A k-affine structure on Y compatible with the given $\mathbf{Z}$-affine structure is a $GL(n,\mathbf{Z}) \ltimes (k^\times)^n$-torsor on Y such that the application of $val^{\times n}$ to $(k^\times)^n$ gives the initial $GL(n,\mathbf{Z}) \ltimes \mathbf{R}^n$-torsor.*

Assume that Y is oriented and carries a k-affine structure compatible with a given $\mathbf{Z}$-affine structure. Orientation allows us to reduce the structure group of the torsor defining the k-affine structure to $SL(n,\mathbf{Z}) \ltimes (k^\times)^n$.

Let $q \in k, |q| = 1$, and $z_1, \ldots, z_n$ be invertible variables such that $z_i z_j = q z_j z_i$, for all $1 \le i < j \le n$. We define the sheaf of k-algebras $\mathcal{O}_q^{can}$ on $\mathbf{R}^n, n \ge 2$ by the formulas:

$$\mathcal{O}_q^{can}(U) = \left\{ \sum_{I=(I_1,\ldots,I_n)\in \mathbf{Z}^n} c_I z^I \;\middle|\; \forall (x_1,\ldots,x_n) \in U \ \sup_I \left(\log(|c_I|) + \sum_{1\le m\le n} I_m x_m \right) < \infty \right\},$$

where $z^I = z_1^{I_1} \ldots z_n^{I_n}$. Since $|q| = 1$ the convergency condition does not depend on the order of variables.

The sheaf $\mathcal{O}_q^{can}$ can be lifted to Y (we keep the same notation for the lifting). In order to do that it suffices to define the action of the group $SL(n,\mathbf{Z}) \ltimes (k^\times)^n$ on the canonical sheaf on $\mathbf{R}^n$. Namely, the inverse to an element $(A, \lambda_1, \ldots, \lambda_n) \in SL(n,\mathbf{Z}) \ltimes (k^\times)^n$ acts on monomials as

$$z^I = z_1^{I_1} \ldots z_n^{I_n} \mapsto \left(\prod_{i=1}^n \lambda_i^{I_i} \right) z^{A(I)} \ .$$

The action of the same element on $\mathbf{R}^n$ is given by a similar formula:

$$x = (x_1, \ldots, x_n) \mapsto A(x) - (val(\lambda_1), \ldots, val(\lambda_n)) \ .$$

Any n-dimensional manifold Y with integral affine structure admits a covering by charts with transition functions being integral affine transformations. This allows to define the sheaf $\mathcal{O}_{q,Y}^{can}$ as the one which is locally isomorphic to $\mathcal{O}_q^{can} = \mathcal{O}_{q,\mathbf{R}^n}^{can}$.

It is explained in [KoSo1] (see also [So1], Section 7.2) that for any open $U \subset \mathbf{R}^n$ the topological space $M(\mathcal{O}_q^{can}(U))$ for $q = 1$ is an analytic torus fibration in the sense of Introduction. Recall that an analytic torus fibration is a fiber bundle (X, Y, π) consisting of a commutative k-analytic space, a topological manifold Y and a continuous map $\pi : X \to Y$ such that it is locally isomorphic to the torus fibration $\mathbf{G}_m^n \to \mathbf{R}^n$ from Introduction. In that case π is a Stein map, and we have: $\pi^{-1}(U) = M(\mathcal{O}_{q=1}^{can}(U))$. Therefore we can think of the ringed space $(Y, \mathcal{O}_{q,Y}^{can})$ as of quantization of this torus fibration.

3.2. Model sheaf near a singular point

In the case of maximally degenerate K3 surfaces the skeleton is homeomorphic to $B = S^2$ (the two-dimensional sphere) equipped with an integral affine structure outside of the subset B^{sing} consisting of 24 points (see [KoSo1], Section 6.4, where the affine structure is described). The construction of the previous subsection gives rise to a sheaf of quantum $\mathbf{C}((t))$-affinoid algebras over $B^0 = B \setminus B^{\text{sing}}$. In order to complete the quantization procedure we need to extend the sheaf $\mathcal{O}_{q,B^0}^{can}$ to a neighbourhood of B^{sing}. It is explained in [KoSo1] (case $q = 1$) and in [So1] (case $|q| = 1$) that one has to modify this sheaf in order to extend it to singular points.

Summarizing, the quantization is achieved in two steps. First, we define a sheaf of non-commutative $\mathbf{C}((t))$-affinoid algebras in a neighbourhood of B^{sing} such that it is locally isomorphic to the canonical sheaf $\mathcal{O}^{can}_{q,B^0}$ outside of B^{sing} and gives a "local model" for the future sheaf $\pi_*(\mathcal{O}_{q,X})$ at the singularities. Second, we modify the sheaf $\mathcal{O}^{can}_{q,B^0}$ by applying (infinitely many times) automorphisms associated with edges of an infinite tree embedded in B, such that its external vertices belong to B^{sing}. Those modifications ensure that the resulting sheaf can be glued with the model sheaf at the singularities, and that it is indeed the direct image of the sheaf of analytic functions on a compact $\mathbf{C}((t))$-analytic K3 surface. More precisely, we do the following.

We start with an open covering of $\mathbf{R}^2$ by the following sets $U_i, 1 \leq i \leq 3$. Let us fix a number $0 < \varepsilon < 1$ and define

$$\begin{aligned} U_1 &= \{(x,y) \in \mathbf{R}^2 | x < \varepsilon|y|\} \\ U_2 &= \{(x,y) \in \mathbf{R}^2 | x > 0, y < \varepsilon x\} \\ U_3 &= \{(x,y) \in \mathbf{R}^2 | x > 0, y > 0\} \end{aligned}$$

Clearly $\mathbf{R}^2 \setminus \{(0,0)\} = U_1 \cup U_2 \cup U_3$. We will also need a slightly modified domain $U_2' \subset U_2$ defined as $\{(x,y) \in \mathbf{R}^2 | x > 0, y < \frac{\varepsilon}{1+\varepsilon} x\}$.

Let $\pi_{can} : (\mathbf{G}_m^{an})^2 \to \mathbf{R}^2$ be the canonical map defined in the Introduction (see also [KoSo1]). We define the following three open subsets of the two-dimensional analytic torus: $T_i := \pi_{can}^{-1}(U_i), i = 1,3$ and $T_2 := \pi_{can}^{-1}(U_2')$. There are natural projections $\pi_i : T_i \to U_i$ given by the formulas

$$\pi_i(|\bullet|) = \pi_{can}(|\bullet|) = (\log|\xi_i|, \log|\eta_i|), \quad i = 1,3$$

$$\pi_2(|\bullet|) = \begin{cases} (\log|\xi_2|, \log|\eta_2|) & \text{if } |\eta_2| < 1 \\ (\log|\xi_2| - \log|\eta_2|, \log|\eta_2|) & \text{if } |\eta_2| \geq 1 \end{cases}$$

To each T_i we assign the algebra $\mathcal{O}_q(T_i)$ of series $\sum_{m,n} c_{mn}\xi_i^m\eta_i^n$ such that $\xi_i\eta_i = q\eta_i\xi_i$, $c_{mn} \in \mathbf{C}((t))$, and for the seminorm $|\bullet|$ corresponding to a point of T_i (which means that $(log|\xi_i|, log|\eta_i|) \in U_i$) one has: $\sup_{m,n}(m\, log|\xi_i| + n\, log|\eta_i|) < +\infty$. Similarly, we can define $\mathcal{O}_q(U)$ for any $U \subset U_i$. In this way we obtain a sheaf of quantum $\mathbf{C}((t))$-affinoid algebras on the set U_i. We will denote this sheaf by $\pi_{i*}(\mathcal{O}_{q,T_i})$.

We define the sheaf $\mathcal{O}_q^{can}$ on $\mathbf{R}^2 \setminus \{(0,0)\}$ as $\pi_{i*}(\mathcal{O}_{q,T_i})$ on each domain U_i, with identifications

$$\begin{aligned} (\xi_1,\eta_1) &= (\xi_2,\eta_2) && \text{on } U_1 \cap U_2 \\ (\xi_1,\eta_1) &= (\xi_3,\eta_3) && \text{on } U_1 \cap U_3 \\ (\xi_2,\eta_2) &= (\xi_3\eta_3,\eta_3) && \text{on } U_2 \cap U_3 \end{aligned}$$

The notation for the sheaf is consistent with the previous subsection since $\mathcal{O}_q^{can}$ is locally isomorphic to the canonical sheaf associated with the standard integral affine structure.

Let us modify the canonical sheaf $\mathcal{O}_q^{can}$ in the following way. On the sets U_1 and $U_2 \cup U_3$ the new sheaf $\mathcal{O}_q^{mod}$ is isomorphic to $\mathcal{O}_q^{can}$ (by identifying of coordinates (ξ_1,η_1) and glued coordinates (ξ_2,η_2) and (ξ_3,η_3) respectively). On the intersection $U_1 \cap (U_2 \cup U_3)$ we identify two copies of the canonical sheaf by an automorphism φ of $\mathcal{O}_q^{can}$ given (we skip the index of the coordinates) by

$$\varphi(\xi,\eta) = \begin{cases} (\xi(1+\eta),\eta) & \text{on} \quad U_1 \cap U_2 \\ (\xi(1+\eta^{-1}),\eta) & \text{on} \quad U_1 \cap U_3 \end{cases}$$

Finally we are going to introduce a sheaf of $\mathbf{C}((t))$-algebras $\mathcal{O}_q^{\text{sing}}$ on a small open disc $W \subset \mathbf{R}^2, \{(0,0)\} \in W$ such that $\mathcal{O}_q^{\text{sing}}|_{W\setminus\{(0,0)\}}$ is isomorphic to $\mathcal{O}_q^{\text{mod}}|_{W\setminus\{(0,0)\}}$. The sheaf $\mathcal{O}_q^{\text{sing}}$ provides a non-commutative deformation of the "local model sheaf" near a singular point (see [KoSo1], Section 8 about the latter).

Let us consider a non-commutative $\mathbf{C}((t))$-algebra $A_q(S)$ generated by α,β,γ subject to the following relations:

$$\alpha\gamma = q\gamma\alpha, q\beta\gamma = \gamma\beta,$$
$$\beta\alpha - q\alpha\beta = 1 - q,$$
$$(\alpha\beta - 1)\gamma = 1.$$

For $q = 1$ this algebra coincides with the algebra of regular functions on the surface $S \subset \mathbf{A}^3_{\mathbf{C}((t))}$ given by the equation $(\alpha\beta-1)\gamma = 1$ and moreover, it is a flat deformation of the latter with respect to the parameter $q-1$. It is explained in [KoSo1], Section 8, that there is a natural map $p : S^{an} \to \mathbf{R}^2$ of the corresponding analytic surface such that $p_*(\mathcal{O}_{S^{an}})$ is a local model near a singularity of the sheaf $\pi_*(\mathcal{O}_X)$, where X is the maximally degenerate K3 surface and π is the projection to the skeleton $Sk(X)$.

Let us denote by $\mathcal{O}_{q,r_1,r_2,r_3}(S^{an})$ the non-commutative affinoid algebra which is the quotient of $\mathbf{C}((t))\langle\langle\alpha,\beta,\gamma\rangle\rangle_{r_1,r_2,r_3}$ by the closed two-sided ideal generated by the above three relations for $A_q(S)$. Here $r_i, i = 1,2,3$ are arbitrary non-negative numbers. We denote by $\mathcal{O}_q(S^{an})$ the intersection of all algebras $\mathcal{O}_{q,r_1,r_2,r_3}(S^{an})$.

We define homomorphisms of non-commutative algebras $g_i\colon A_q(S)\to\mathcal{O}_q(T_i)$, $1 \le i \le 3$ by the following formulas (the notation is obvious):

$$\begin{aligned} g_1(\alpha,\beta,\gamma) &= ({\xi_1}^{-1}, \xi_1(1+\eta_1), {\eta_1}^{-1}) \\ g_2(\alpha,\beta,\gamma) &= ((1+\eta_2){\xi_2}^{-1}, \xi_2, {\eta_2}^{-1}) \\ g_3(\alpha,\beta,\gamma) &= ((1+\eta_3)(\xi_3\eta_3)^{-1}, \xi_3\eta_3, (\eta_3)^{-1}) \end{aligned}$$

These homomorphisms correspond to the natural embeddings $T_i \hookrightarrow S^{an}, i = 1,2,3$. One can use these homomorphisms in order to show an existence of non-trivial multiplicative seminorms on $A_q(S)$ and construct explicitly some of the corresponding representations of $A_q(S)$ in a k-Banach vector space.

For example, let us consider a Banach vector space V_r consisting of series $\sum_{i\in\mathbf{Z}} a_iT^i, a_i \in k$ such that $|a_i|r^i \to 0$ as $|i| \to \infty$, where $r > 0$ is some number. Let $\tau : V_r \to V_r$ be the shift operator: $\tau(f)(T) = f(qT)$. Define $\alpha = T$ (operator of multiplication by T), $\gamma = -\tau^{-1}$ and $\beta = T^{-1}\circ(1-\tau)$. One checks that all the relations of $A_q(S)$ are satisfied, and moreover, the seminorm on $A_q(S)$ induced by the operator norm is multiplicative. (Similar considerations apply to the analytic quantum torus derived from $\xi\eta = q\eta\xi$. Then the element $\sum_{n,m\in\mathbf{Z}} a_{nm}\xi^n\eta^m$ transforms the series $f = \sum_{i\in\mathbf{Z}} c_i T^i$ into $\sum_{n,m\in\mathbf{Z}} a_{nm} q^{nm} T^m f(q^n T)$.) Rescaling the action of α and γ by arbitrary non-zero numbers one can adjust the action of β in such a way that the norms of operators α, β, γ "cover" an open neighborhood of the point $(1,1,1)$. More precisely, let us consider the map $f : M(\mathcal{O}_q(S^{an})) \to \mathbf{R}^3$ defined by the formula $f(x) = (a,b,c)$ where $a = \max(0, \log|\alpha|_x), b = \max(0, \log|\beta|_x), c = \log|\gamma|_x = -\log|\alpha\beta - 1|_x$. Here $|\cdot|_x = \exp(-val_x(\cdot))$ denotes the mulitplicative seminorm corresponding to the point $x \in M(\mathcal{O}_q(S^{an}))$. Then the image of f is homeomorphic to $\mathbf{R}^2$, similarly to the case $q = 1$ considered in [KoSo1]. More precisely, let us decompose $M(\mathcal{O}_q(S^{an})) = S_- \cup S_0 \cup S_+$ according to the sign of $\log|\gamma|_x$ where $x \in M(\mathcal{O}_q(S^{an}))$. Then

$$\begin{aligned} f(S_-) &= \{(a,b,c) \in \mathbf{R}^3 \,|\, c < 0, a \geq 0, b \geq 0,\ ab(a+b+c) = 0\} \\ f(S_0) &= \{(a,b,c) \in \mathbf{R}^3 \,|\, c = 0, a \geq 0, b \geq 0,\ ab = 0\} \\ f(S_+) &= \{(a,b,c) \in \mathbf{R}^3 \,|\, c > 0, a \geq 0, b \geq 0,\ ab = 0\} \end{aligned}$$

In fact the image of the map f coincides with the image of the embedding $j : \mathbf{R}^2 \to \mathbf{R}^3$ given by formula

$$j(x,y) = \begin{cases} (-x\,,\, \max(x+y,0)\,,\, -y) & \text{if} \quad x \leq 0 \\ (0\,,\, x + \max(y,0)\,,\, -y) & \text{if} \quad x \geq 0 \end{cases}$$

Proofs of the above observations are different from the case $q = 1$. Indeed, there are no one-dimensional modules over $A_q(S)$ corresponding to the points of the surface S. Therefore it is not obvious that there are multiplicative seminorms x on $A_q(S)$ with the prescribed value of $f(x)$. Seminorms on $A_q(S)$ arise from representations of this algebra in k-Banach vector spaces: if $\rho : A_q(S) \to \mathrm{End}_k(V)$ is such a representaion then we can define $|a|_\rho = ||\rho(a)||$, where $||\rho(a)||$ is the operator norm in the Banach algebra $\mathrm{End}_k(V)$ of bounded operators on V. Such seminorms are, in general, submultiplicative: $|ab|_\rho \leq |a|_\rho |b|_\rho$. We are interested in those which are multiplicative. This can be achieved, e.g., by mapping of $A_q(S)$ into an algebra which admits multiplicative seminorms. We discussed above the homomorphisms g_i of $A_q(S)$ into analytic quantum tori. Let us consider a different example of such homomorphism. Let $\delta = (\alpha\beta - 1)\gamma$. One checks that δ is a central element in the quantum affinoid algebra $B_q(S)$ generated by the first three relations for $A_q(S)$ (i.e., we drop the relation $\delta = 1$). Let us consider the quantum affinoid algebra B generated by $\beta^{\pm 1}, \gamma^{\pm 1}, \delta$ subject to the relations:

$$\beta\delta = \delta\beta, \gamma\delta = \delta\gamma, \gamma\beta = q\beta\gamma,$$

and such that β^{-1} is inverse to β and γ^{-1} is inverse to γ. There is an embedding of algebras $A_q(S) \to B/(\delta - 1)B$ induced by the linear map $A_q(S) \to B$ such that β and γ are mapped into the corresponding elements of B and $\alpha \mapsto (1 + \delta\gamma^{-1})\beta^{-1}$. Notice that for any $r_1 > 0$, $r_2 > 0$, $r_3 \geq 0$ one can define a multiplicative norm $|\bullet|_{r_1,r_2,r_3}$ on B such that $|\sum_{n\in\mathbf{Z},m\in\mathbf{Z},l\in\mathbf{Z}_+} c_{nml}\beta^n\gamma^m\delta^l|_{r_1,r_2,r_3} = \max_{n,m,l} |c_{nml}| r_1^n r_2^m r_3^l$. Moreover, we can complete B with respect to this multiplicative norm and obtain a quantum affinoid algebra B_{r_1,r_2,r_3}. We can also invert

δ and do the same construction. In this way we obtain the quantum affinoid algebra denoted by $B^{(1)}_{r_1,r_2,r_3}$. Since $A_q(S)$ is embedded into the quotient of any of these algebras by the central ideal, we obtain plenty multiplicative norms on $A_q(S)$ and on its completions.

We denote by p the composition $j^{-1}\circ f$. In the case $q=1$ it is an analytic torus fibration over the set $\mathbf{R}^2\setminus\{(0,0)\}$. Let now W be a small disc in $\mathbf{R}^2$ centered at the origin. We need to define the non-commutative affinoid algebra $\mathcal{O}_q^{\mathrm{sing}}(W)$. In the commutative case $q=1$ it is defined as $\mathcal{O}_{S^{an}}^{\mathrm{sing}}(p_*^{-1}(W)) = p_*(\mathcal{O}_{S^{an}})(W)$.

For each $i=1,2,3$ we define the $\mathbf{C}((t))$-affinoid algebras $\mathcal{O}_q^{\mathrm{sing}}(p^{-1}(U_i))$ such that for every $x\in M(\mathcal{O}_q^{\mathrm{sing}}(p^{-1}(U_i))$ one has $p(x)\in U_i$ (it coincides with the intersection of all completions of $A_q(S)$ with respect to multiplicative seminorms x such that $p(x)\in U_i, i=1,2,3$). Similarly we define algebras ${\mathcal{O}_q}^{\mathrm{sing}}(p^{-1}(W))$ and ${\mathcal{O}_q}^{\mathrm{sing}}(p^{-1}(W^0))$, where $W^0=W\setminus\{(0,0)\}$ or, more generally, any $\mathcal{O}_q^{\mathrm{sing}}(p^{-1}(U))$ for U being an open subset of W. Using homomorphisms $g_i, i=1,2,3$ one proves that if $U\subset U_i$ then $\mathcal{O}_q^{\mathrm{sing}}(p^{-1}(U))$ is isomorphic to $\mathcal{O}_q^{\mathrm{mod}}(\pi_i^{-1}(U))$. The latter is defined as the set of series $\sum_{m,n\in\mathbf{Z}} c_{mn}\xi_i^m\eta_i^n$ such that $\pi_i(|\bullet|)\in U$ for any multiplicative seminorm $|\bullet|$ such that $\sup_{m,n}(\log|c_{mn}|+m\log|\xi_i|+n\log|\eta_i|<\infty)$, if $(x,y)\in U$. The isomorphism of sheaves $\mathcal{O}_q^{\mathrm{sing}}|_{W\setminus\{(0,0)\}}\simeq {\mathcal{O}_q}^{\mathrm{mod}}|_{W\setminus\{(0,0)\}}$ follows. Details of this construction will be explained elsewhere.

3.3. Trees, automorphisms and gluing

As was explained in [KoSo1] in the case of $q=1$ and in [So1] in the case $|q|=1$, one has to modify the canonical sheaf in order to glue it with $\mathcal{O}_q^{\mathrm{sing}}$. Here we explain the construction following [KoSo1], [So1], leaving the details to a separate publication. The starting point for the construction is a subset $\mathcal{L}\subset B$ which is an infinite tree. We called it *lines* in [KoSo1].

The definition is quite general. Here we discuss the 2-dimensional case, while a much more complicated higher-dimensional case was considered in the recent paper [GroSie1]. For a manifold Y which carries a $\mathbf{Z}$-affine structure a *line* l is defined by a continuous map $f_l:(0,+\infty)\to Y$ and a covariantly constant (with respect to the connection which gives the affine structure) nowhere vanishing integer-valued 1-form $\alpha_l\in\Gamma((0,+\infty),f_l^*((T^*)^{\mathbf{Z}})$. A set $\mathcal{L}$ of lines is required to be decomposed into a disjoint union $\mathcal{L}=\mathcal{L}_{in}\cup\mathcal{L}_{com}$ of *initial* and *composite* lines. Each composite line is obtained as a result of a finite number of "collisions" of initial lines. A collision is described by a Y-shape figure, where the bottom leg of Y is a composite line, while two other segments are "parents" of the leg, so that the leg is obtained as a result of the collision. A construction of the set $\mathcal{L}$ satisfying the axioms from [KoSo1] was proposed in [KoSo1], Section 9.3. Generalization to the higher-dimensional case can be found in [GroSie1]. In the two-dimensional case the lines form an infinite tree embedded into B. The edges have rational slopes with respect to the integral affine structure. The tree is dense in B^0.

With each line l (i.e., edge of the tree) we associate a continuous family of automorphisms of stalks of sheaves of algebras $\varphi_l(t):(\mathcal{O}_q^{can})_{Y,f_l(t)}\to(\mathcal{O}_q^{can})_{Y,f_l(t)}$.

Automorphisms φ_l can be defined in the following way (see [KoSo1], Section 10.4).

First we choose affine coordinates in a neighborhhod of a point $b \in B \setminus B^{\mathrm{sing}}$, identifyin b with the point $(0,0) \in \mathbf{R}^2$. Let $l = l_+ \in \mathcal{L}_{in}$ be (in the standard affine coordinates) a line in the half-plane $y > 0$ emerging from $(0,0)$ (there is another such line l_- in the half-plane $y < 0$, see [KoSo1] for the details). Assume that t is sufficiently small. Then we define $\varphi_l(t)$ on topological generators ξ, η by the formula

$$\varphi_l(t)(\xi, \eta) = (\xi(1 + \eta^{-1}), \eta).$$

In order to extend $\varphi_l(t)$ to the interval $(0, t_0)$, where t_0 is not small, we cover the corresponding segment of l by open charts. Then a change of affine coordinates transforms η into a monomial multiplied by a constant from $(\mathbf{C}((t)))^{\times}$. Moreover, one can choose the change of coordinates in such a way that $\eta \mapsto C\eta$ where $C \in (\mathbf{C}((t)))^{\times}, |C| < 1$ (such change of coordinates preserve the 1-form dy. Constant C is equal to $exp(-L)$, where L is the length of the segment of l between two points in different coordinate charts). Therefore η extends analytically in a unique way to an element of $\Gamma((0, +\infty), f_l^*((\mathcal{O}_q^{can})^{\times}))$. Moreover the norm $|\eta|$ strictly decreases as t increases, and remains strictly smaller than 1. Similarly to [KoSo1], Section 10.4 one deduces that $\varphi_l(t)$ can be extended for all $t > 0$. This defines $\varphi_l(t)$ for $l \in \mathcal{L}_{in}$.

Next step is to extend $\varphi_l(t)$ to the case when $l \in \mathcal{L}_{com}$, i.e., to the case when the line is obtained as a result of a collision of two lines belonging to $\mathcal{L}_{in}$. Following [KoSo1], Section 10, we introduce a group G which contains all the automorphisms $\varphi_l(t)$, and then prove the factorization theorem (see [KoSo1], Theorem 6) which allows us to define $\varphi_l(0)$ in the case when l is obtained as a result of a collision of two lines l_1 and l_2. Then we extend $\varphi_l(t)$ analytically for all $t > 0$ similarly to the case $l \in \mathcal{L}_{in}$.

More precisely, the construction of G goes such as follows. Let $(x_0, y_0) \in \mathbf{R}^2$ be a point, $\alpha_1, \alpha_2 \in (\mathbf{Z}^2)^*$ be 1-covectors such that $\alpha_1 \wedge \alpha_2 > 0$. Denote by $V = V_{(x_0,y_0),\alpha_1,\alpha_2}$ the closed angle

$$\{(x,y) \in \mathbf{R}^2 | \langle \alpha_i, (x,y) - (x_0,y_0) \rangle \geq 0, i = 1,2\}$$

Let $\mathcal{O}_q(V)$ be a $\mathbf{C}((t))$-algebra consisting of series $f = \sum_{n,m \in \mathbf{Z}} c_{n,m} \xi^n \eta^m$, such that $\xi\eta = q\eta\xi$ and $c_{n,m} \in \mathbf{C}((t))$ satisfy the condition that for all $(x,y) \in V$ we have:

1. if $c_{n,m} \neq 0$ then $\langle (n,m), (x,y) - (x_0,y_0) \rangle \leq 0$, where we identified $(n,m) \in \mathbf{Z}^2$ with a covector in $(T_p^*Y)^{\mathbf{Z}}$;
2. $\log |c_{n,m}| + nx + my \to -\infty$ as long as $|n| + |m| \to +\infty$.

For an integer covector $\mu = adx + bdy \in (\mathbf{Z}^2)^*$ we denote by $R_\mu := R_{(a,b)}$ the monomial $\xi^a \eta^b$. Then $R_{(a,b)} R_{(c,d)} = q^{ad-bc} R_{(c,d)} R_{(a,b)} = q^{-bc} R_{(a+c,b+d)}$. We define a prounipotent group $G := G(q, \alpha_1, \alpha_2, V)$ which consists of automorphisms of $\mathcal{O}_q(V)$ having the form $f \mapsto e^g f e^{-g}$ where

$$g = \sum_{n_1,n_2 \geq 0, n_1+n_2>0} c_{n_1,n_2} R_{\alpha_1}^{-n_1} R_{\alpha_2}^{-n_2}$$

where $c_{n_1,n_2} \in \mathbf{C}((t))$ and

$$\log|c_{n,m}| - n_1\langle \alpha_1, (x,y)\rangle - n_2\langle \alpha_2, (x,y)\rangle \leq 0 \;\; \forall\, (x,y) \in V.$$

The latter condition is equivalent to $\log|c_{n,m}| - \langle n_1\alpha_1 + n_2\alpha_2, (x_0, y_0)\rangle \leq 0$. The assumption $|q| = 1$ ensures that the product is well defined.

Let us consider automorphisms as above such that in the series for g the ratio $\lambda = n_2/n_1 \in [0, +\infty]_{\mathbf{Q}} := \mathbf{Q}_{\geq 0} \cup \infty$ is fixed. Such automorphism form a commutative subgroup $G_\lambda := G_\lambda(q, \alpha_1, \alpha_2, V) \subset G$. There is a natural map $\prod_\lambda G_\lambda \to G$, defined as in [KoSo1], Section 10.2. The factorization theorem proved in the loc. cit states that this map is a bijection of sets.

Example 3.3.1. *Let us consider the automorphism, discussed above:*

$$\varphi(\xi, \eta) = (\xi(1 + \eta^{-1}), \eta).$$

One can check that the transformation $\xi \mapsto \xi(1 + \eta^{-1})$ has the form

$$exp(Li_{2,q}(\eta^{-1})/(q-1))\xi exp(-Li_{2,q}(\eta^{-1})/(q-1)),$$

where $Li_{2,q}(x)$ is the quantum dilogarithm function (see, e.g., [BR]). It satisfies the property $(x;q)_\infty = exp(Li_{2,q}(-x)/(q-1))$, where $(a;q)_N = \prod_{0 \leq n \leq N}(1 - aq^n)$ for $1 \leq N \leq \infty$. Using the formula $(x;q)_\infty = \sum_{n\geq 0} \frac{(-1)^n q^{n(n-1)/2} x^n}{(q;q)_n}$ one can show that $lim_{q\to 1} Li_{2,q}(x) = Li_2(x) = \sum_{n \geq 1}(-1)^n x^n/n^2$, which is the ordinary dilogarithm function (the latter appeared in [KoSo1], *Section* 10.4 *in the reconstruction problem of rigid analytic* K3 *surfaces).*

Let us now assume that lines l_1 and l_2 collide at $p = f_{l_1}(t_1) = f_{l_2}(t_2)$, generating the line $l \in \mathcal{L}_{com}$. Then $\varphi_l(0)$ is defined with the help of factorization theorem. More precisely, we set $\alpha_i := \alpha_{l_i}(t_i)$, $i = 1, 2$ and the angle V is the intersection of certain half-planes $P_{l_1,t_1} \cap P_{l_2,t_2}$ defined in [KoSo1], Section 10.3. The half-plane $P_{l,t}$ is contained in the region of convergence of $\varphi_l(t)$. By construction, the elements $g_0 := \varphi_{l_1}(t_1)$ and $g_{+\infty} := \varphi_{l_2}(t_2)$ belong respectively to G_0 and $G_{+\infty}$. The we have:

$$g_{+\infty} g_0 = \prod\nolimits_{\to} \left((g_\lambda)_{\lambda \in [0,+\infty]_{\mathbf{Q}}}\right) = g_0 \cdots g_{1/2} \cdots g_1 \cdots g_{+\infty}.$$

Each term g_λ with $0 < \lambda = n_1/n_2 < +\infty$ corresponds to the newborn line l with the direction covector $n_1\alpha_{l_1}(t_1) + n_2\alpha_{l_2}(t_2)$. Then we set $\varphi_l(0) := g_\lambda$. This transformation is defined by a series which is convergent in a neighborhood of p, and using the analytic continuation we obtain $\varphi_l(t)$ for $t > 0$, as we said above. Recall that every line carries an integer 1-form $\alpha_l = adx + bdy$. By construction, $\varphi_l(t) \in G_\lambda$, where λ is the slope of α_l.

Having automorphisms φ_l assigned to lines $l \in \mathcal{L}$ we proceed as in [KoSo1], Section 11, modifying the sheaf $\mathcal{O}_q^{can}$ along each line. We denote the resulting sheaf by $\mathcal{O}_q^{\mathrm{mod}}$. It is isomorphic to the previously constructed sheaf $\mathcal{O}_q^{\mathrm{mod}}$ in a neighborhood of the point $(0,0)$.

Remark 3.3.2. *The appearance of the dilogarithm function in the above example can be illustrated in the picture of collision of two lines (say,* $(x,0)$ *and* $(0,y)$ $x,y \geq 0$*) which leads to the appearance of the new line, which is the diagonal* $(x,x), x \geq 0$*. Then the factorization theorem gives rise to the five-term identity* $g_\infty g_0 = g_0 g_1 g_\infty$*, which is the quantum version of the famous five-term identity for the dilogarithm function.*

4. p-adic quantum groups

4.1. How to quantize p-adic groups

Let L be a finite algebraic extension of the field $\mathbf{Q}_p$ of p-adic numbers, and K be a discretely valued subfield of the field of complex p-adic numbers $\mathbf{C}_p$ containing L. All the fields carry non-archimedean norms, which we will denote simply by $|\bullet|$ (sometimes we will be more specific, using the notation like $|x|_K$ in order to specify which field we consider). We denote by $\mathcal{O}_L$ the ring of integers of L and by m_L the maximal ideal of $\mathcal{O}_L$. Let G be a locally L-analytic group, which is the group of L-points of a split reductive algebraic group $\mathbf{G}$ over L. Let $H \subset G$ be an open maximal compact subgroup. We would like to define quantum analogs of the algebras $C^{la}(G,K), C^{la}(H,K)$ of locally analytic functions on G and H, as well as their strong duals $D^{la}(G,K)$, $D^{la}(H,K)$, which are the algebras of locally analytic distributions on G and H respectively (see [SchT1], [Em1]). Modules over the algebras of locally analytic distributions were used in [SchT1-5], [Em1] for a description of locally analytic admissible representations of locally L-analytic groups. Our aim is to derive "quantum" analogs of those results. In this paper we will discuss definitions of the algebras only.

First of all we are going to define locally analytic functions and locally analytic distributions on the quantized compact group H. Let us explain our approach in the case of the "classical" (i.e., non-quantum) group H. We are going to present definitions of $C^{la}(H,K)$ and $D^{la}(H,K)$ in such a way that they can be generalized to the case of quantum groups. The difficulty which one needs to overcome is to define everything using only two Hopf algebras: the universal enveloping algebra $U(g)$, $g = \mathrm{Lie}(G)$ and the algebra $K[G]$ of *regular* functions on the algebraic group G. Our construction consists of three steps.

1) For a "sufficiently small" open compact subgroup $H_r \subset H$ we define the algebra $C^{an}(H_r,K)$ of *analytic* functions on H_r. Here $0 < r \leq 1$ is a parameter, such that $H_1 = H$, and if $r_1 < r_2$ then $H_{r_1} \subset H_{r_2}$. The strong dual to $C^{an}(H_r,K)$ is denoted by $D^{an}(H_r,K)$. It is, by definition, the algebra of *analytic* distributions on H_r. It can be described (see [Em1], Section 5.2) as a certain completion of the universal enveloping algebra $U(g)$ of the Lie algebra $g = \mathrm{Lie}(G)$.

2) For any $r \leq 1$ we define a norm on the algebra of regular functions $K[G]$ such that the completion with respect to this norm is the algebra of *continuous* functions $C(H_r, K)$ on the group H_r.
3) In order to define *locally analytic* functions on H we consider a family of seminorms $|f|_l$ on $K[G]$, where $f \in K[G]$ and l runs through the set $D^{an}(H_r, K)$. More precisely, for every $l \in D^{an}(H_r, K)$ we define a seminorm $|f|_l = ||(id \otimes l)(\delta(f))||$, where δ is the coproduct on the Hopf algebra $K[G]$ and $|| \bullet ||$ is the norm defined above on the step 2). The completion of $K[G]$ with respect to the topology defined by the family of seminorms $|\bullet|_l$ is the algebra $C^{la}(H, K)$ of locally analytic functions on H defined in [SchT1]. The strong dual $D^{la}(H, K) := C^{la}(H, K)'_b$ is the algebra of locally analytic distributions introduced in [SchT1].

We recall that the locally analytic representation theory of G developed in [SchT1-6] is based on the notion of coadmissible module over the algebra $D^{la}(H, K)$, where $H \subset G$ is an open compact subgroup. Therefore, from the point of view of representation theory, it suffices to quantize $D^{la}(H, K)$.

4.2. Quantization of "small" compact subgroups

We would like to quantize $D^{la}(H, K)$ following the above considerations. We will do that for the class of algebraic quantum groups introduced by Lusztig (see [Lu1], [Lu2]). Let us fix $q \in L^{\times}$ such that $|q| = 1$ (this restriction is not necessary for algebraic quantization, but it will be important when we discuss convergent series). We will assume that there is $h \in \mathcal{O}_L$ such that $|h| < 1, exp(h) = q$. Let $\mathbf{G}$ be a semisimple simply-connected algebraic group over $\mathbf{Z}$, associated with a Cartan matrix $((a_{ij}))$ (more precisely, in order to be consistent with the terminology of [Lu1] we start with a *root datum* of finite type associated with a Cartan datum, see [Lu1], Chapter 2. These data give rise to the Cartan matrix in the ordinary sense). The algebraic group $\mathbf{G}(\mathbf{C})$ of $\mathbf{C}$-points of $\mathbf{G}$ was quantized by Drinfeld (see, e.g., [KorSo] Chapters 1,2). We will need a $\mathbf{Z}$-form of the quantized algebraic group $\mathbf{G}$ introduced by Lusztig (see [Lu1]). It allows us to define the quantized group over an arbitrary field. We need to be more specific when speaking about "quantized" group. More precisely, following [Lu1] one can define Hopf L-algebras $U_q(g_L)$ and $L[G]_q$, which are the quantized enveloping algebra of the Lie algebra g_L of $\mathbf{G}(L)$ and the algebra of *regular* functions on the algebraic quantum group $\mathbf{G}(L)$ respectively. Extending scalars to K we obtain Hopf K-algebras $U_q(g_K) = K \otimes_L U_q(g_L)$ and $K[G]_q = K \otimes_L L[G]_q$. We will also need $\mathbf{Z}$-forms of the above Hopf algebras, which will be denoted by $U := U_A$ and $A[G]_v$ respectively. The latter are Hopf algebras over the ring $A = \mathbf{Z}[v, v^{-1}]$, where v is a variable. The algebras U_L and $L[G]_q$ are obtained by tensoring of U and $A[G]_v$ respectively with L in such a way that v acts on L by multiplication by q.

As an A-module the algebra U is isomorphic to the tensor product $U \simeq U^+ \otimes U^0 \otimes U^-$ where $U^{\pm}$ are the quantized Borel subalgebras and U^0 is the quantized Cartan subalgebra (see [Lu2]). Recall that U^+ (resp. U^-) is an A-algebra generated by the divided powers $E_i^{(N)}$ (resp $F_i^{(N)}$) of the Chevalley generators E_i (resp. F_i) of the quantized enveloping $\mathbf{Q}(v)$-algebra $\mathbf{U}$, where $1 \le i \le n := \operatorname{rank} g_L$ (see [Lu2]). The algebra U^0 is generated over A by the generators $K_i^{s_i}\binom{K_i}{n_i}_v$, where $\binom{K_i}{n_i}_v = \prod_{1\le m\le n_i} \frac{K_i v^{d_i(-m+1)} - K_i^{-1} v^{d_i(m-1)}}{v^{d_i m} - v^{-d_i m}}$. Here $n_i \in \mathbf{Z}_+$, $s_i \in \{0,1\}$, and $K_i, 1 \le i \le n$ are the standard Chevalley generators of $\mathbf{U}$. The integer numbers $d_i \in \{1,2,3\}$ satisfy the condition that $((d_i a_{ij}))$ is a symmetric positive definite matrix with $a_{ii} = 2$, and $a_{ij} \le 0$ if $i \ne j$. Recall that there is a canonical reduction of U_A at $v = 1$, which is a Hopf $\mathbf{Z}$-algebra $U_{\mathbf{Z}}$. It is the universal enveloping of the integer Lie algebra $g_{\mathbf{Z}}$ of the corresponding Chevalley group. We will denote by $t_i \in g_L$ the generators at $v = 1$ corresponding to K_i, keeping the same notation $E_i^{(N)}, F_i^{(N)}$ for the rest of the generators of $g_{\mathbf{Z}}$. Thus we have the standard decompostion $g_{\mathbf{Z}} = g_{\mathbf{Z}}^+ \oplus h_{\mathbf{Z}} \oplus g_{\mathbf{Z}}^-$, where the Lie algebra $g_{\mathbf{Z}}^+$ is generated (as a Lie algebra over $\mathbf{Z}$) by $E_i^{(N)}$, the Lie algebra $g_{\mathbf{Z}}^-$ is generated by $F_i^{(N)}$ and the commutative Lie algebra $h_{\mathbf{Z}}$ is generated by $\binom{t_i}{N} := t_i(t_i-1)\cdots(t_i-N+1)/N!$, $1 \le i \le n$ (see [St], Theorem 2). Lie algebra g_L is generated by the standard Chevalley generators $E_i = E_i^{(1)}, F_i = F_i^{(1)}, t_i, 1 \le i \le n$. The Hopf algebra $U/(v-1)U$ is the universal enveloping algebra $U(g_L)$ of g_L.

In what follows, while keeping the above notation, we will assume for simplicity that $L = \mathbf{Q}_p$.

We will need the following extension of $U_q(g_L)$. Let us fix a basis $\{\alpha_i\}_{1\le i\le n}$ of simple roots of g_L, as well as invariant bilinear form on this Lie algebra such that $(\alpha_i, \alpha_j) = d_i a_{ij}$.

Let $h_{\mathcal{O}_L} = \oplus_{1\le i\le n} \mathbf{Z}_p t_i$. We fix a global chart $\psi : h_{\mathcal{O}_L} \to T^0$, where $T^0 = \mathbf{T}(\mathcal{O}_L)$ is the maximal compact torus. Then any element $a \in T^0$ can be written as an analytic function $t = \psi(\sum_{1\le i\le n} x_i t_i) := t(x_1, \dots, x_n)$, where $x_i \in \mathbf{Z}_p$. Let us introduce a unital topological Hopf L-algebra $U_q^{an}(g_L)$ which is a Hopf L-algebra generated by $E_i^{(N)}, F_i^{(N)}, 1 \le i \le n, N \ge 1$ and the elements $t(x) = t(x_1, \dots, x_n) \in T^0$ as above, such that the relations between $E_i^{(N)}, F_i^{(N)}$ are the same as in $U_q(g_L)$, and $t(x)E_i = E_i t(x+v_i), t(x)F_i = F_i t(x - v_i)$, where $v_i = (a_{1i}, a_{2i}, \dots, a_{ni})$. The elements $K_i^{\pm 1} = \exp(\pm d_i h)$ (recall that $\exp(h) = q$) belong to this algebra and together with $E_i^{(N)}, F_i^{(N)}, 1 \le i \le n, N \ge 1$ generate the Hopf algebra isomorphic to $U_q(g_L)$.

There is a natural non-degenerate pairing $U_q^{an}(g_L) \otimes L[G]_q \to L$ which extends the natural non-degenerate pairing $U_A \otimes A[G]_v \to A$ defined in [Lu2]. Extending scalars we obtain the algebra $U_q^{an}(g_K)$ and the pairing $U_q^{an}(g_K) \otimes K[G]_q \to K$.

For the rest of this subsection we will assume that $d_i = 1$, i.e., $((a_{ij}))$ is symmetric, and $L = \mathbf{Q}_p$. These conditions can be relaxed. We make them in order to simplify formulas.

There is a natural action $U_q^{an}(g_K) \otimes K[G]_q \to K[G]_q$ (right action) given by the formula $l(f) = (id \otimes l)(\delta(f))$, where $l \in U_q^{an}(g_K), f \in K[G]_q$ and $\delta : K[G]_q \to K[G]_q \otimes K[G]_q$ is the coproduct.

Recall that $K[G]_q \simeq \oplus_\Lambda m_\Lambda V(\Lambda)$ which is the sum of irreducible finite-dimensional highest weight $U_q(g_K)$-modules $V(\Lambda)$ with multiplicities m_Λ. This is also an isomorphism of $U_q^{an}(g_K)$-modules. Each element $E_i^{(N)}, F_i^{(N)}$ acts locally nilpotently on $K[G]_q$, while each $t(x)$ acts as a semi-simple linear map.

Let $R = \oplus_{1\le i\le n}\mathbf{Z}\alpha_i$ be the set of roots of $g_{\mathbf{Z}}$. We denote by R^+ (resp R^-) the set of positive (resp. negative) roots. We will often write $\alpha > 0$ (resp. $\alpha < 0$) if $\alpha \in R^+$ (resp. $\alpha \in R^-$). Following [KorSo], Chapter 4, or [Lu1], Chapter 3, 41, one can construct quantum root vectors $E_\alpha^{(N)}, F_\alpha^{(N)} \in U_q(g_K), \alpha > 0, N \ge 1$, such that $E_{\alpha_i}^{(N)} = E_i^{(N)}, F_{\alpha_i}^{(N)} = F_i^{(N)}$ (in order to keep track of integrality of the coefficients we are going to use the formulas from [Lu1]). Let us fix a convex linear order on the set of roots, such that all negative roots preceed all positive roots (convexity means that $\alpha < \alpha+\beta < \beta$ for positive roots and the oppoiste inequalities for negative roots) .

For every $0 < r \le 1$ we define $U_q(g_K)(r)$ as a K-vector space consisting of series

$$\xi = \sum_{m\in\mathbf{Z}^n,\alpha>0,s_\alpha,p_\alpha\ge 0} c_{m,s,p}t^m/m! \prod_{\alpha>0} F_\alpha^{(p_\alpha)}E_\alpha^{(s_\alpha)},$$

such that $c_{m,s,p} \in K, t^m/m! = t_1^{m_1}/m_1!\cdots t_n^{m_n}/m_n!, |c_{m,s,p}|r^{-(|s|+|p|+|m|)} \to 0$ as $|m|+|s|+|p| \to \infty$. Here and below m, s, p denote multi-indices. We define $|\xi|_r = \sup_{m,s,p}|c_{m,s,p}|r^{-(|s|+|p|+|m|)}$. Let $t_{\alpha_i}(x), 1 \le i \le n$ be an ordered basis of T^0 (see [SchT1], Section 4). Then, as a topological K-vector space (with the topology defined by the norm $|\bullet|_r$) the space $U_q(g_K)(r)$ is isomorphic to the K-vector space of infinite series

$$\eta = \sum_{N_i,M_i\ge 0,l_i\in\mathbf{Z}} b_{M,N} \prod_{1\le i\le n} (t_{\alpha_i}(x)-1)^{l_i}F_i^{(M_i)}E_i^{(N_i)},$$

such that $M = (M_i), N = (N_i)$, and $|b_{M,N,l}||t_{\alpha_i}(x)-1|_r r^{-(|M|+|N|)} \to 0$ as $|M| + |N| + |l| \to \infty$, equipped with the norm defined by

$$|\eta|'_r = \sup_{M,N,l} |b_{M,N,l}||t_{\alpha_i}(x)-1|_r r^{-(|M|+|N|)}.$$

It is easy to see that $U_q(g_K)(r)$ is a K-Banach vector space. It contains Banach vector subspaces $U_q^+(g_K)(r)$ (resp. $U_q^-(g_K)(r)$) which are closures of vector subspaces generated by all the elements $E_\alpha^{(N)}$ (resp. $F_\alpha^{(N)}$). It also contains an analytic neighborhood of $1 \in T^0$, which is an analytic group isomorphic to the ball of radius r in the Lie algebra $h_{\mathcal{O}_L}$. The latter is an analytic Lie group via Campbell-Hausdorff formula. We can always assume that r belongs to the algebraic closure of L, thus the corresponding analytic groups are in fact affinoid.

Proposition 4.2.1. *The norm $|\xi|_r$ (equivalently the norm $|\eta|'_r$) gives rise to a Banach K-algebra structure on $U_q(g_K)(r)$.*

Similarly to the case $q = 1$ (see [Em1], Section 5.2) one can ask whether the algebra $U_q(g_K)(r)$ corresponds to a "good" analytic group. Let us consider the completion of the tensor product $U_q(g_K)(r) \otimes U_q(g_K)(r)$ with respect to the minimal Banach norm. Then we have the following result.

Proposition 4.2.2. *The Hopf algebra structure on $U_q^{an}(g_K)$ admits a continuous extension to $U_q(g_K)(r)$, making it into a topological Hopf algebra.*

Let us consider the topological K-algebra $U_q^{(1)}(g_K)$ which is the projective limit of $U_q(g_K)(r)$ for all $0 < r < 1$. Then we have the following result, which is a corollary of the previous proposition.

Proposition 4.2.3. *The Hopf algebra structure on $U_q^{an}(g_K)$ admits a continuous extension to $U_q^{(1)}(g_K)$, making it into a topological Hopf algebra.*

Since the elements E_α, F_α act locally nilpotently on $K[G]_q$, there is a well-defined action of $U_q(g_K)(r)$ on $K[G]_q$, which extends to the action of $U_q^{(1)}(g_K)$ on $K[G]_q$. Notice that the pairing $U_q(g_K)(r) \otimes K[G]_q \to K, (l, f) \mapsto l(f)$ is non-degenerate. In particular we can define the norm on $K[G]_q$ by the formula $||f||_r = \sup_{l \neq 0} \frac{|l(f)|}{|l|_r}, l \in U_q(g_K)(r)$.

Let now $H_r, r = p^{-N}$ be a "small" compact open subgroup of G. This means that the exponential map $exp : \mathbf{Z}^d = h_{\mathbf{Z}} \oplus g_{\mathbf{Z}}^+ \oplus g_{\mathbf{Z}}^- \to G$ defines an analytic isomorphism $B(0, r) \to H_r$, where $B(0, r) \subset \mathbf{Z}_p^d$ is the ball consisting of points $(x_i, x_\alpha, y_\alpha) \in \mathbf{Z}_p^d, 1 \le i \le n, \alpha > 0$ such that $x_\alpha, y_\alpha \in p^N \mathbf{Z}_p, x_i \in p^N \mathbf{Z}_p$, for all $\alpha > 0, 1 \le i \le n$.

Definition 4.2.4. *The space of analytic functions on the quantum group H_r (notation $C^{an}(H_r, K)_q$) is the completion of $K[G]_q$ with respect to the norm $|| \bullet ||_r$.*

Proposition 4.2.5. *The space $C^{an}(H_r, K)_q$ is a Banach Hopf K-algebra.*

Definition 4.2.6. *The algebra of analytic distributions on the quantum group H_r (notation $D^{an}(H_r, K)_q$) is the strong dual to $C^{an}(H_r, K)_q$.*

One can define a norm $|| \bullet ||$ on $K[G]_q$ such that the completion with respect to this norm is by definition the algebra $C(H, K)_q$ of continuous functions on the open maximal compact subgroup $H = H_1$ (in the case of $q = 1$ this is a theorem, not a definition). Then we proceed as follows.

Any linear functional $l \in D^{an}(H_r, K)_q$ defines a seminorm $| \bullet |_l$ on $K[G]_q$ such that

$$|f|_l = ||(id \otimes l)\delta(f)||.$$

The collection of seminorms $| \bullet |_l, l \in D^{an}(H_r, K)_q$ gives rise to a locally convex topology on $K[G]_q$.

Definition 4.2.7. *The space $C^{an}(H, H_r, K)_q$ of functions on the quantum group H which are locally analytic with respect to the quantum group H_r is the completion of $K[G]_q$ in the topology defined by the collection of seminorms $| \bullet |_l$.*

Definition 4.2.8.

a) *The space* $C^{la}(H,K)_q$ *of locally analytic functions on the quantum group* H *is the inductive limit* $\varinjlim_{r\leq 1} \mathcal{O}^{an}(H,H_r,K)$ *(i.e., it consists of functions on quantum group* H *which are locally analytic with respect to some* H_r, $r<1$*).*

b) *The space* $D^{la}(H,K)_q$ *of locally analytic distributions on the quantum group* H *is the strong dual to* $C^{la}(H,K)_q$*.*

Since some details related to the proof of the following results are not finished, I formulate it as a conjecture.

Conjecture 4.2.9. *Both spaces* $C^{la}(H,K)_q$ *and* $D^{la}(H,K)_q$ *are topological Hopf* K*-algebras. Furthermore,* $D^{la}(H,K)_q$ *is a Frechét-Stein algebra in the sense of* [SchT1].

In the next subsection we will explain the definition of the norm $||\bullet||$ in the case of the group $SL_2(\mathbf{Z}_p)$. The general case is similar, but requires more details. It will be considered in [So2].

4.3. The $GL_2(\mathbf{Z}_p)$-case

We will use the notation $K[GL_2(\mathbf{Q}_p)]_q$ for the algebra of regular K-valued functions on the algebraic quantum group $GL_2(\mathbf{Q}_p)$. It is known (see [KorSo], Chapter 3) that $K[GL_2(\mathbf{Q}_p)]_q$ is generated by generators $t_{ij}, 1\leq i,j\leq 2$ subject to the relations

$$t_{11}t_{12}=q^{-1}t_{12}t_{11}, \qquad t_{11}t_{21}=q^{-1}t_{21}t_{11}, \tag{1}$$

$$t_{12}t_{22}=q^{-1}t_{22}t_{12}, \qquad t_{21}t_{22}=q^{-1}t_{22}t_{21}, \tag{2}$$

$$t_{12}t_{21}=t_{21}t_{12}, \quad t_{11}t_{22}-t_{22}t_{11}=\left(q^{-1}-q\right)t_{12}t_{21}, \tag{3}$$

The element $det_q = t_{11}t_{22}-q^{-1}t_{12}t_{21}$ generates the center of the above algebra. As a result, the algebra $K[SL_2(\mathbf{Q}_p)]_q$ of regular functions on quantum group $SL_2(\mathbf{Q}_p)$ is obtained from the above algebra by adding one more equation

$$t_{11}t_{22}-q^{-1}t_{12}t_{21}=1. \tag{4}$$

We are going to use the ideas of the representation theory of quantized algebras of functions (see [KorSo]).

Let V be a separable K-Banach vector space. This means that V contains a dense K-vector subspace spanned by the orthonormal basis $e_m, m\geq 0$ (orthonormal means that $||e_m||=1$ for all m). Let us consider the following representations $V_c, c\in K$ of $K[GL_2(\mathbf{Q}_p)]_q$ in V (cf. [KorSo], Chapter 4, Section 4.1):

$$t_{11}(e_m)=a_{11}(m)e_{m-1}, t_{21}(e_m)=a_{21}(m)e_m,$$

$$t_{12}(e_m)=a_{12}(m)e_m, t_{22}(e_m)=a_{22}(m)e_{m+1},$$

$$det_q=c.$$

Here $a_{ij}(m)\in K$ and $e_m=0$ for $m<0$. In particular the line Ke_0 is invariant with respect to the subalgebra A_+ generated by t_{11},t_{21}. Let us assume that not all $a_{11}(m)$ are equal to zero. Then it is easy to see from the commutation relations between t_{ij} that

$$a_{21}(m) = a_{21}(0)q^{-m}, a_{12}(m) = a_{12}(0)q^{-m}, m \geq 0.$$

Moreover

$$a_{11}(m+1)a_{22}(m) - a_{11}(m)a_{22}(m-1) = (q^{-1} - q)q^{-2m}h_0,$$

where $h_0 = a_{21}(0)a_{12}(0)$. Let $s(m) = a_{11}(m)a_{22}(m-1), m \geq 1$. Then we have

$$s(m+1) - s(m) = (q^{-1} - q)q^{-2m}h_0, s(1) = (q^{-1} - q)h_0.$$

It follows that

$$s(m) = (q^{-1} - q)(1 + q^{-2} + \cdots + q^{-2(m-1)})h_0 = q(q^{-2m} - 1)h_0,$$

for all $m \geq 1$. Since the quantum determinant is equal to c, we have

$$s(m+1) = a_{11}(m+1)a_{22}(m) = c + q^{-2m-1}h_0.$$

Comparing two formulas for $s(m)$ we see that

$$h_0 = a_{21}(0)a_{12}(0) = -cq^{-1}.$$

From now on we will assume that $|1 - q| < 1$.

Then the operators t_{12} and t_{21} are bounded. We also have $|s(m)| = |c(q^{-m} - 1)| = |c|, m \geq 1$.

Assume that $a_{21}(0) \neq 0$. Then the above representations (which are algebraically irreducible as long as q is not a root of 1) depend on the parameters $a_{21}(0), a_{11}(m), a_{22}(m), m \geq 0$ subject to the relations $a_{11}(m)a_{22}(m-1) = c(1 - q^{-2m})$. We will further specify restrictions on these parameters. The idea is the same as in [KorSo], Chapter 3, where in order to define continuous functions on the quantum group $SU(2)$ we singled out irreducible representations of $\mathbf{C}[SL_2(\mathbf{C})]_q$ corresponding to the intersection of the group $SU(2)$ with the big Bruhat cell for $SL_2(\mathbf{C})$. This intersection is the union of symplectic leaves of the Poisson-Lie group $SU(2)$. Kernel of an irreducible representation defines a symplectic leaf ("orbit method") which explains the relationship of representation theory and symplectic geometry. Notice that in the case $q = 1$ one can define the algebra of continuous function $C(SU(2))$ in the following way. For any function $f \in \mathbf{C}[SL_2(\mathbf{C})]$ one takes its restriction to the above-mentioned union of symplectic leaves. Since the latter is dense in $SU(2)$, the completion of the algebra $C[SL_2(\mathbf{C})]$ with respect to the sup-norm taken over all irreducible representations corresponding to the symplectic leaves is exactly $C(SU(2))$. Now we observe that symplectic leaves in $SL_2(\mathbf{C})$ are algebraic subvarieties, therefore they exist over any field. We will use the same formulas in the case of any p-adic field L (in this section we take $L = \mathbf{Q}_p$). In order to specify a symplectic leaf in $GL_2(\mathbf{C})$ we need in addition to fix the value of the determinant (it belongs to the center of the Poisson algebra $\mathbf{C}[GL_2(\mathbf{C})]$).

Let us recall (see [KorSo], Chapter 3) that to every element $t, c \in K^{\times}$ one can assign a 1-dimensional representation $W_{c,t} = \mathbf{Q}_p e_0$ of $K[GL_2(L)]_q$ such that

$t_{11}(e_0) = te_0, t_{22}(e_0) = ct^{-1}e_0$, and the rest of generators act on e_0 trivially. Recall (see [KorSo], Chapter 1) that complex 2-dimensional symplectic leaves of $GL_2(\mathbf{C})$ are algebraic subvarieties $S_{c,t}$ given by the equations:

$$t_{11}t_{22} - t_{21}t_{12} = c, t_{12} = t^2 t_{21},$$

where c, t are non-zero complex numbers. We define symplectic leaves over $\mathbf{Q}_p$ by the same formulas, taking $c, t \in \mathbf{Q}_p$. In order to define the norm of the restriction of a regular function $f \in \mathbf{C}[GL_2(\mathbf{Q}_p)]$ on $GL_2(\mathbf{Z}_p)$ we can choose a subset in the set of symplectic leaves $S_{c,t}$ such that the union of their intersection with $GL_2(\mathbf{Z}_p)$ is dense in the latter group. It suffices to take those leaves $S_{c,t}$ for which $|c| \leq 1, t \in \mathbf{Z}_p^{\times}$, and both t_{12} and t_{21} are non-zero.

Let us consider infinite-dimensional representation $V_{c,t}$ as above for which $a_{12}(0) = t^2 a_{21}(0), |c| \leq 1$ for a fixed $i \geq 0$, and $a_{21}(0) \neq 0$. We will also assume that the norm of the operators corresponding to t_{11} and t_{22} is less or equal than 1. It follows from the equality $t^2 a_{21}^2(0) = -cq^{-1}$ that $|a_{21}(0)| = |c|$, hence the norm of the operators corresponding to t_{12} and t_{21} is less or equal than $|c| \leq 1$. It follows that the norm of the operator $\pi_{c,t}(f)$ corresponding to an element $f \in K[GL_2(\mathbf{Q}_p)]_q$ acting in $V_{c,t}$ is bounded from above as $V_{c,t}$ run through the set of irreducible representations with the above restrictions on c, t. In addition, we are going to consider only those $c \in K^{\times}$ for which $-cq^{-1}$ is a square in K. We define the norm $||f||_{GL_2(\mathbf{Z}_p),q}$, $f \in K[GL_2(\mathbf{Q}_p)]_q$ as the supremum of norms of the operators $\pi_{c,t}(f)$ corresponding an element f in all representations $V_{c,t}$ as above. This is the desired sup-norm which we used in our definition of the algebra of locally analytic functions.

References

[BR] V. Bazhanov and N. Reshetikhin, Remarks on the quantum dilogarithm, *J. Phys. A: Math. Gen.* **28** (1995), 2217–2226.

[Be1] V. Berkovich, *Spectral theory and analytic geometry over non-archimedean fields*, Amer. Math. Soc., 1990.

[Be2] V. Berkovich, Smooth p-adic analytic spaces are locally contractible, *Invent. Math.* **137** (1999), 1–84.

[Be3] V. Berkovich, Smooth p-adic analytic spaces are locally contractible II, In: *Geometric aspects of Dwork theory*, Berlin, 293–370.

[Fro] H. Frommer, *The locally analytic principal series of split reductive groups*, preprint 265, University of Münster, 2003.

[GroSie1] M. Gross and B. Siebert, *From real affine geometry to complex geometry*, math.AG/0703822.

[Em1] M. Emerton, *Locally analytic vectors in representations of locally p-adic analytic groups*, http://www.math.northwestern.edu/ emerton/preprints.html, to appear in Memoirs of the AMS.

[Em2] M. Emerton, Locally analytic representation theory of p-adic reductive groups: A summary of some recent developments, to appear in the Proceedings of the LMS symposium on *L-functions and Galois representations.*

[KoSo1] M. Kontsevich and Y. Soibelman, *Affine structures and non-archimedean analytic spaces*, math.AG/0406564.

[KoSo2] M. Kontsevich and Y. Soibelman, *Homological mirror symmetry and torus fibrations*, math.SG/0011041.

[KorSo] L. Korogodsky and Y. Soibelman, *Algebras of functions on quantum groups I.* AMS, 1998.

[Lu1] G. Lusztig, *Introduction to quantum groups*, Birkhäuser, 1993.

[Lu2] G. Lusztig, Quantum groups at roots of 1, *Geometriae Dedicata 35* (1989), 89–114.

[Lu3] G. Lusztig, Finite-dimensional Hopf algebras arising from quantized universal enveloping algebras, *J. Amer. Math. Soc.* **3** no. 1 (1990), 257–296.

[Sch-NFA] P. Schneider, *Non-archimedean functional analysis*, Springer.

[SchT1] P. Schneider and J.Teitelbaum, *Algebras of p-adic distributions and admissible representations*, math.NT/0206056.

[SchT2] P. Schneider and J. Teitelbaum, *p-adic Fourier Theory*, math.NT/0102012.

[SchT3] P. Schneider and J. Teitelbaum, *$U(g)$-finite locally analytic representations*, math.NT/0005072.

[SchT4] P. Schneider and J. Teitelbaum, *Banach space representations and Iwasawa theory*, math.NT/0005066.

[SchT5] P. Schneider and J. Teitelbaum, *Locally analytic distributions and p-adic representation theory, with applications to GL_2*, math.NT/9912073.

[SchT6] P. Schneider and J. Teitelbaum, *p-adic boundary values*, math.NT/9901159.

[So1] Y. Soibelman, *On non-commutative spaces over non-archimedean fields*, math.QA/0606001.

[So2] Y. Soibelman, *Quantum p-adic groups and their representations*, in preparation.

[So3] Y. Soibelman, Algebras of functions on a compact quantum group and its representations, *Leningrad Math. Journ.* **2** (1991), 193–225.

[SoVo] Y. Soibelman and V. Vologodsky, Non-commutative compactifications and elliptic curves, math.AG/0205117, published in *Int. Math. Res. Notes* **28** (2003).

Yan Soibelman
Department of Mathematics
Kansas State University
Manhattan, KS 66506
USA
e-mail: soibel@math.ksu.edu

Progress in Mathematics, Vol. 265, 721–742

Convolution Equations on Lattices: Periodic Solutions with Values in a Prime Characteristic Field

Mikhail Zaidenberg

Dedicated to the memory of Sasha Reznikov

Abstract. These notes are inspired by the theory of cellular automata. The latter aims, in particular, to provide a model for inter-cellular or inter-molecular interactions. A linear cellular automaton on a lattice Λ is a discrete dynamical system generated by a convolution operator $\Delta_a : f \longmapsto f * a$ with kernel a concentrated in the nearest neighborhood ω of 0 in Λ. In [Za$_1$] we gave a survey (limited essentially to the characteristic 2 case) on the σ^+-cellular automaton with kernel the constant function 1 in ω. In the present paper we deal with general convolution operators over a field of characteristic $p > 0$. Our approach is based on the harmonic analysis. We address the problem of determining the spectrum of a convolution operator in the spaces of pluri-periodic functions on Λ. This is equivalent to the problem of counting points on the associate algebraic hypersurface in an algebraic torus according to their torsion multi-orders. These problems lead to a version of the Chebyshev-Dickson polynomials parameterized this time by the set of all finite index sublattices of Λ and not by the naturals as in the classical case. It happens that the divisibility property of the classical Chebyshev-Dickson polynomials holds in this more general setting.

Mathematics Subject Classification (2000). 11B39, 11T06, 11T99, 31C05, 37B15, 43A99.

Keywords. Cellular automaton, Chebyshev-Dickson polynomial, convolution operator, lattice, sublattice, finite field, discrete Fourier transform, discrete harmonic function, pluri-periodic function.

MP: *Do you yourself perceive a fundamental difference between pure and applied mathematics?*
Stanislaw Ulam: *I really don't. I think it's a question of language, and perhaps habits.*

Acknowledgements. This work partially was done during the author's visit to the MPIM at Bonn. The author thanks this institution for a generous support and excellent working conditions.

Introduction

These notes are inspired by the theory of linear cellular automata. Such an automaton on the integer lattice $\mathbb{Z}^s$ can be viewed as a discrete dynamical system generated by a convolution operator $f \longmapsto \Delta_a f = f * a$, acting on functions $f : \mathbb{Z}^s \to K$ with values in a Galois field $K = \mathrm{GF}(p)$. Usually the kernel a of Δ_a is concentrated in the nearest neighborhood of $0 \in \mathbb{Z}^s$. We are interested more generally in systems of convolution equations

$$\Delta_{a_j} f = 0, \qquad j = 1, \dots, t \tag{1}$$

with kernel $\bar{a} = (a_1, \dots, a_t)$ of bounded (and so finite) support. We address the following questions.

Problem. *Describe the set of all possible pluri-periods of the pluri-periodic solutions of* (1)*. More precisely, given a pluri-period $\bar{n} \in \mathbb{N}^s$ compute the spectral multiplicities of* (1) *on the space of all $\bar{n}$-periodic functions on $\mathbb{Z}^s$ and the dimension of the corresponding kernel* $\ker \Delta_{\bar{a}}$.

At present these problems seem to be out of reach. We provide however different interpretations that could be useful in future approaches.

Counting pluri-periods amounts to counting points on the associated affine algebraic variety $\Sigma_{\bar{a}}$ (called symbolic variety) according to their multi-orders. The symbolic variety $\Sigma_{\bar{a}}$ is a subvariety in the algebraic torus $(\bar{K}^\times)^s$, where $\bar{K}$ stands for the algebraic closure of K. The multiplicative group $\bar{K}^\times$ being a torsion group, the torus is covered by the finite subgroups. We are interested in the distribution of points on $\Sigma_{\bar{a}}$ according to the filtration of $(\bar{K}^\times)^s$ by finite subgroups.

The spectral multiplicities which appear in the above problem can be described via Chebyshev-Dickson polynomial systems. Such a system associates a degree d polynomial in $K[\lambda]$ to any sublattice of $\mathbb{Z}^s$ of index d, namely the characteristic polynomial of (1) in the corresponding function space. The classical Chebyshev-Dickson polynomials appear in the simplest case where $s = 1$. In Theorem 0.1 we establish the divisibility property for Chebyshev-Dickson systems. Besides, we give in Proposition 2.11 a description of these systems via iterated resultants.

Using the harmonic analysis we interpret the points of $\Sigma_{\bar{a}}$ as $\bar{a}$-harmonic lattice characters; see Theorem 0.2 below. Here '$\bar{a}$-harmonic' simply means 'satisfying (1)'. In the classical case, for a solution of (1) the value in a lattice point is a sum of its values over the neighbor points[1]; this explains our terminology.

Resuming, in these notes we explore interplay between periods of solutions of a system of convolution equations on a lattice, on one hand, torsion orders of points on the associate symbolic variety, on the other hand, and harmonic characters. Let us develop along these lines in more detail.

[1]In positive characteristic, one has to replace averaging by summation.

1. σ^+*-automaton.* In [Za$_1$] we gave a survey on the σ^+-automata on rectangular and toric grids. Let us recall the setup. On the integral lattice $\Lambda = \mathbb{Z}^s$ we consider the following function a^+ with values in the binary Galois field GF(2):

$$a^+ = \delta_0 + \sum_{i=1}^{s} (\delta_{e_i} + \delta_{-e_i}), \tag{2}$$

where $e_1, \ldots, e_s$ stands for the canonical lattice basis. We let Δ_{a^+} denote the convolution operator $f \longmapsto f * a^+$ acting on binary functions $f : \Lambda \to \mathrm{GF}(2)$. It generates a discrete dynamical system on Λ called a σ^+-automaton studied e.g., in [MOW], [Su], [GKW], [BR], [SB], [HMP]; see further references in [Za$_1$].

2. σ^+*-game.* The σ^+-automaton on the plane lattice $\Lambda = \mathbb{Z}^2$ is related to the solitaire game 'Lights Out', also called a σ^+-game. Let us describe the game. Suppose that the offices in a department, which will be our table of game, correspond to the vertices of a grid $P_{m,n} = L_m \times L_n$, where L_m stands for the linear graph with m vertices. Suppose also that the interrupters are synchronized in such an uncommon way that turning off or on in one room changes automatically to the opposite the states in all rooms neighbors through a wall. The question arises whether the last person leaving the department can always manage to turn all the lights off.

It is possible to reduce this problem to an analogous one for the toric grid $\mathbb{T}_{m',n'} = C_{m'} \times C_{n'}$, where $m' = m+1$, $n' = n+1$ and C_m stands for the circular graph with m vertices. We let $\mathcal{F} = \mathcal{F}(\mathbb{T}_{m,n}, \mathrm{GF}(2))$ be the function space on the torus equipped with the standard bilinear form $\langle \cdot, \cdot \rangle$. The move at a vertex v of $\mathbb{T}_{m,n}$ in the σ^+-game, applied to a function (a 'pattern') $f \in \mathcal{F}(\mathbb{T}_{m,n}, \mathrm{GF}(2))$, consists in the addition

$$f \longmapsto f + a_v^+ \mod 2\,,$$

where $a_v^+(u) = a^+(u+v)$ is the shifted star function (2) centered at $v \in \mathbb{T}_{m,n}$. Thus the σ^+-game on the torus $\mathbb{T}_{m,n}$ is winning starting with the initial pattern f_0 if and only if f_0 can be decomposed into a sum of shifts of the star function a^+.

The linear invariants of the σ^+-game form a subspace $\mathcal{H} \subseteq \mathcal{F}$ orthogonal to all shifts a_v^+, $v \in \mathbb{T}_{m,n}$. Indeed

$$h \in \mathcal{H} \iff \langle h, f + a_v^+ \rangle \equiv \langle h, f \rangle \mod 2 \qquad \forall v \in \mathbb{T}_{m,n},\ \forall f \in \mathcal{F}\,.$$

Moreover the initial pattern f_0 is winning if and only if $f_0 \in \mathcal{H}^\perp$. The functions $h \in \mathcal{H}$ are called harmonic [Za$_1$]. This is justified by the following property: for any vertex v of the grid $\mathbb{T}_{m,n}$, the value $h(v)$ is the sum modulo 2 of the values of h over the neighbors of v in $\mathbb{T}_{m,n}$. Actually $\mathcal{H} = \ker(\Delta_{a^+})$. Thus the σ^+-game on a toric grid $\mathbb{T}_{m,n}$ is winning for any initial pattern if and only if $0 \notin \mathrm{spec}(\Delta_{a^+})$. The latter is known to be equivalent to the condition $\gcd(T_m, T_n^+) = 1$, where T_m stands for the classical mth Chebyshev-Dickson polynomial over GF(2) and $T_n^+(x) = T_n(x+1)$, see [Za$_1$, 2.35] and the references therein.

3. This discussion leads to the following problems.

- *Determine the set of all winning toric grids $\mathbb{T}_{m,n}$. Equivalently, determine the set of all pairs (m,n) such that the polynomials T_m and T_n^+ are coprime. Or, which is complementary, determine the set of all toric grids $\mathbb{T}_{m,n}$ admitting a nonzero binary harmonic function.*
- *Given $(m,n) \in \mathbb{N}^2$ compute the dimension $d(m,n)$ of the subspace $\mathcal{H}$ of all harmonic functions on $\mathbb{T}_{m,n}$ or, equivalently, the dimension of the subspace $\mathcal{H}^\perp$ of all winning patterns.*

For m,n odd we provide several different interpretations of $d(m,n)$. In particular we will show that, over the algebraic closure $\bar{K}$ of the base field $K = \mathrm{GF}(2)$, there is an orthonormal basis of $\mathcal{H} \otimes \bar{K}$ consisting of harmonic characters on $\mathbb{T}_{m,n}$ with values in the multiplicative group $\bar{K}^\times$. An initial pattern f_0 on $\mathbb{T}_{m,n}$ is winning if and only if f_0 is orthogonal to all harmonic characters on $\mathbb{T}_{m,n}$. The latter ones are in one-to-one correspondence with the (m,n)-bi-torsion points on the symbolic hypersurface. In our case the symbolic hypersurface is the elliptic cubic in the torus $(\bar{K}^\times)^2 = (\mathbb{A}^1_{\bar{K}} \setminus \{0\})^2$ with equation

$$x + x^{-1} + y + y^{-1} + 1 = 0\,. \tag{3}$$

Thus to determine all toric grids $\mathbb{T}_{m,n}$ admitting a nonzero binary harmonic function is the same as to determine all bi-torsion orders of points on the cubic (3), see [Za$_1$].

4. *Linear cellular automata on abelian groups.* In the present paper we consider similar problems for general linear cellular automata on abelian groups. Recall that the theory of cellular automata aims, in particular, to provide a model for inter-cellular or inter-molecular interactions. One can regard a linear cellular automaton on a group, or rather on the Caley graph of a group, as a discrete dynamical system generated by a convolution operator with kernel concentrated in a nearest neighborhood of the neutral element [MOW].

In more detail, suppose we are given a collection (a colony) of 'cells' placed at the vertices of a locally finite graph Γ. This determines the relation 'neighbors' for cells; neighbors can interact. Each cell can be in one of n cyclically ordered states; thus the state of the whole collection at a moment t is codified by a function $f_t : \Gamma \to \mathbb{Z}/n\mathbb{Z}$. In the subsequent portions of time, the cells simultaneously change their states. According to a certain local rule, the new state of a cell depends on the previous states of the given cell and of all its neighbors.

To define a cellular automaton, say, σ on Γ means to fix at each vertex v of Γ such a local rule, which does not depend on t. Such a collection of local rules determines a discrete dynamical system $\sigma : f_t \longmapsto f_{t+1}$. Usually the edges $[v_0, v_1], \ldots, [v_0, v_s]$ at v_0 are ordered. So the local rule at v_0 is a function $\phi_{v_0} : (\mathbb{Z}/n\mathbb{Z})^{s+1} \to \mathbb{Z}/n\mathbb{Z}$, and

$$f_{t+1}(v_0) = \phi_{v_0}\left(f_t(v_0), f_t(v_1), \ldots, f_t(v_s)\right)\,. \tag{4}$$

In the case where the graph Γ is homogeneous under a group action on Γ, it is natural to assume that the family of local rules is as well homogeneous. Consider for instance the Caley graph Γ of a finitely generated group G with a generating set $\{g_1, \ldots, g_s\}$. Given a local rule ϕ_e for the neutral element $e \in G$, we can define ϕ_g for any vertex $g \in G$ as the shift of ϕ_e by g.

For an additive cellular automaton the local rule ϕ_e is a linear function. Consequently such an automaton is generated by a convolution operator $\Delta_a : f \longmapsto f * a$ on G with kernel a supported in the nearest neighborhood of e. This kernel is just the coefficient function of ϕ_e. The evolution equation (4) can be written in this case as a heat equation

$$\partial_t(f_t) := f_{t+1} - f_t = \Delta_{a'}(f_t), \qquad \text{where} \qquad a' = a - \delta_e\,.$$

Here we restrict to additive cellular automata on lattices or toric grids, viewed as the Caley graphs of finitely generated free abelian groups and finite abelian groups, respectively. In contrast with the classical setting, we allow distant interactions. So we deal with general convolution operators.

5. *Convolution operators on lattices and Chebyshev-Dickson systems.* For a field K of characteristic $p > 0$ and for a group G we let $\mathcal{F}(G, K)$ and $\mathcal{F}^0(G, K)$ denote the vector space of all functions $f : G \to K$, of all those with finite support, respectively. We consider the convolution

$$* : \mathcal{F}(G, K) \times \mathcal{F}^0(G, K) \ni (f, a) \longmapsto f * a \in \mathcal{F}(G, K)\,,$$

where

$$f * a(g) = \sum_{h \in G} f(h)a(h^{-1}g) \qquad \forall g \in G\,.$$

Fixing a we get the convolution operator $\Delta_a : f \longmapsto f * a$ acting on the space $\mathcal{F}(G, K)$. All such operators form a K-algebra $\mathrm{Conv}_K(G)$ with $\Delta_{a_1} \circ \Delta_{a_2} = \Delta_{a_2 * a_1}$.

6. For a subgroup $H \subseteq G$ we let $\Delta_a|H$ denote the restriction of Δ_a to the subspace $\mathcal{F}_H(G, K) \subseteq \mathcal{F}(G, K)$ of all H-periodic functions. Clearly $\mathcal{F}_H(G, K)$ is of finite dimension whenever H is of finite index in G, and $\dim \mathcal{F}_H(G, K) = [G : H]$.

In the sequel $G = \Lambda$ will be a lattice i.e., a free abelian group of finite rank. We let $\mathcal{L}$ denote the set of all finite index sublattices in Λ. Ordered by inclusion, $\mathcal{L}$ can be regarded as an ordered graph. Given a function $a \in \mathcal{F}^0(\Lambda, K)$ with finite support, we consider the spectra and the spectral multiplicities of $\Delta_a|\Lambda'$, $\Lambda' \in \mathcal{L}$, in the algebraic closure $\bar{K}$ of K. In particular we consider the function

$$d(a, \Lambda') = \dim \ker (\Delta_a|\Lambda'), \qquad \Lambda' \in \mathcal{L}\,. \tag{5}$$

The family of characteristic polynomials

$$\mathrm{CharPoly}_{a,\Lambda'} = \mathrm{CharPoly}(\Delta_a|\Lambda'), \qquad \Lambda' \in \mathcal{L},$$

will be called a Chebyshev-Dickson system. Recall that the nth Dickson polynomial $D_n(x, \alpha)$ over a finite field F is the unique polynomial verifying the identity $D_n(x + \alpha/x, \alpha) = x^n + \alpha^n/x^n$, where $\alpha \in F$. Whereas for $F = \mathrm{GF}(2)$, the nth Chebyshev-Dickson polynomial of the first kind is $T_n(x) = D_n(x, 1)$. We recover the latter one as $T_n = \mathrm{CharPoly}_{a,\Lambda'}$ when taking $K = \mathrm{GF}(2)$, $a = a^+ - \delta_e$, $\Lambda = \mathbb{Z}$ and $\Lambda' = n\mathbb{Z}$.

The classical Chebyshev-Dickson system (T_n) possesses a number of interesting properties [LMT]. It forms a commutative semigroup under composition that is, $T_n \circ T_m = T_{mn}$. Furthermore T_m divides T_n if $m \mid n$, moreover, $\gcd(T_m, T_n) = T_{\gcd(m,n)}$, etc. The composition property is not stable under shifts in the argument, and so does not hold in our more general setting. However, the divisibility property does hold. Namely we have the following

Theorem 0.1. $\mathrm{CharPoly}_{a,\Lambda''}$ *divides* $\mathrm{CharPoly}_{a,\Lambda'}$ *whenever* $\Lambda' \subset \Lambda''$. *Moreover, if the index of* Λ' *in* Λ'' *equals* p^α *then* $\mathrm{CharPoly}_{a,\Lambda'} = (\mathrm{CharPoly}_{a,\Lambda''})^{p^\alpha}$.

The second assertion allows to restrict to the subgraph $\mathcal{L}^0 \subseteq \mathcal{L}$ of all sublattices $\Lambda' \subseteq \Lambda$ with indices coprime with p. For a sublattice $\Lambda' \in \mathcal{L}^0$, the dimension $d(a, \Lambda')$ of the kernel of Δ_a equals the multiplicity of the zero root of the polynomial $\mathrm{CharPoly}_{a,\Lambda'}$.

7. *Systems of convolution equations.* We let $K = \mathrm{GF}(p)$ be the Galois field of characteristic $p > 0$, $\bar{K}$ be the algebraic closure of K and $\bar{K}^\times$ be the multiplicative group of $\bar{K}$. We fix a cortege $\bar{a} = (a_1, \ldots, a_t)$ consisting of functions $a_j : \Lambda \to \bar{K}$ with bounded supports.

Given a lattice Λ we consider the system of convolution equations

$$\Delta_{a_j}(f) := f * a_j = 0, \ \ j = 1, \ldots, t, \qquad \text{where} \quad f : \Lambda \to \bar{K}\,. \tag{6}$$

We let $d(\bar{a}, \bar{n})$ denote the number of linearly independent $\bar{n}$-periodic solutions of (6). We call these solutions $\bar{a}$-harmonic.

We let

$$\mathrm{CharPoly}_{\bar{a},\Lambda'} = \gcd\left(\mathrm{CharPoly}_{a_j,\Lambda'} \ : j = 1, \ldots, t\right),$$

$$\ker(\Delta_{\bar{a}}|\Lambda') = \bigcap_{j=1}^{t} \ker\left(\Delta_{a_j}|\Lambda'\right) \qquad \text{and} \qquad d(\bar{a}, \Lambda') = \dim \ker(\Delta_{\bar{a}}|\Lambda')\,.$$

Thus $\mathrm{CharPoly}_{\bar{a},\Lambda'}(\lambda) = 0$ if and only if there exists a nonzero Λ'-periodic eigenfunction $f \in \mathcal{F}_{\Lambda'}(\Lambda, \bar{K})$ of $\Delta_{\bar{a}}$ with

$$\Delta_{a_j}(f) = \lambda \cdot f \quad \forall j = 1, \ldots, t\,.$$

The set of zeros of $\mathrm{CharPoly}_{\bar{a},\Lambda'}$ counted with multiplicities is called the spectrum of $\Delta_{\bar{a}}|\Lambda'$, and is denoted by $\mathrm{spec}\,(\Delta_{\bar{a}}|\Lambda')$. The set of spectra forms a graph Ξ ordered by inclusion. Due to the divisibility property in Chebyshev-Dickson systems, the map $\mathrm{spec} : \mathcal{L} \to \Xi$ of ordered graphs is monotonous.

8. *Symbolic variety.* Given a basis $\mathcal{V} = (v_1, \ldots, v_s)$ of Λ we can identify Λ with $\mathbb{Z}^s$. To a function $a : \Lambda = \mathbb{Z}^s \to \bar{K}$ with bounded support we associate the Laurent polynomial

$$\sigma_a = \sum_{u=(u_1,\ldots,u_s)\in\mathbb{Z}^s} a(u)x^{-u} \in \bar{K}[x_1, x_1^{-1}, \ldots, x_s, x_s^{-1}], \ \ j = 1, \ldots, t.$$

A cortege $\bar{a} = (a_1, \ldots, a_n)$ determines an affine algebraic variety

$$\Sigma_{\bar{a}} = \{\sigma_{a_j} = 0 \,:\, j = 1, \ldots, t\}$$

in the torus $(\bar{K}^\times)^s$. We call $\Sigma_{\bar{a}}$ the symbolic variety associated with the system (6).

The logarithm of the Weil zeta function counts the points on $\Sigma_{\bar{a}}$ over the Galois fields $\mathrm{GF}(q)$, where $q = p^r$, $r \geq 0$. Whereas our purpose is to count, for every multi-index $\bar{n} = (n_1, \ldots, n_s) \in \mathbb{N}^s$ with all n_i coprime to p, the number

$$d_{\bar{a}}(\bar{n}) = \mathrm{card}\,(\Sigma_{\bar{a},\bar{n}})$$

of $\bar{n}$-torsion points on the symbolic variety $\Sigma_{\bar{a}}$, where

$$\Sigma_{\bar{a},\bar{n}} = \{\xi = (\xi_1, \ldots, \xi_s) \in \Sigma_{\bar{a}} \,:\, \xi_i^{n_i} = 1,\ i = 1, \ldots, s\}\,.$$

Due to Theorem 0.2(a) below this same quantity arises in the spectral problem:

$$d_{\bar{a}}(\bar{n}) = d(\bar{a}, \bar{n})\,.$$

9. *Harmonic characters.* We let $\mathrm{Char}(\Lambda, \bar{K}^\times)$ denote the set of all $\bar{K}^\times$-valued characters of Λ, that is of all homomorphisms $\chi : \Lambda \to \bar{K}^\times$. A character χ is called $\bar{a}$-harmonic if the function $\chi : \Lambda \to \bar{K}$ is; $\mathrm{Char}_{\bar{a}-\mathrm{harm}}(\Lambda, \bar{K}^\times)$ stands for the set of all $\bar{a}$-harmonic characters of Λ.

We denote by $\mathbb{N}_{\mathrm{co}(p)}$ the set of all naturals coprime with p. Given a multi-index $\bar{n} \in \mathbb{N}^s_{\mathrm{co}(p)}$ we consider the finite subgroup of the torus $(\bar{K}^\times)^s$

$$\mu_{\bar{n}} := \{\xi = (\xi_1, \ldots, \xi_s) \in (\bar{K}^\times)^s \,:\, \xi_i^{n_i} = 1,\ \ i = 1, \ldots, s\}\,.$$

Fixing a basis $\mathcal{V} = (v_1, \ldots, v_s)$ of Λ we consider the product sublattices

$$\Lambda' = \Lambda_{\bar{n},\mathcal{V}} = \sum_{i=1}^{s} n_i \mathbb{Z} v_i \subseteq \Lambda\,.$$

For such a sublattice $\Lambda' \subseteq \Lambda$, the Fourier transform provides a natural one to one correspondence between the set of all $\bar{a}$-harmonic characters of the quotient group $G = \Lambda/\Lambda'$ and the set of points on the symbolic variety with multi-torsion order dividing $\bar{n}$. Namely the following hold.

Theorem 0.2.

(a) *For any sublattice $L' \in \mathcal{L}^0$, the subspace* $\ker(\Delta_{\bar{a}}|\Lambda')$ *possesses an orthonormal basis of $\bar{a}$-harmonic characters. In particular there are*

$$d(\bar{a}, \Lambda') = \mathrm{mult}_{\lambda=0}\,(\mathrm{CharPoly}_{\bar{a},\Lambda'})$$

such characters.

(b) *Fixing a basis $\mathcal{V}$ of Λ provides a natural bijection* $\mathrm{Char}(\Lambda, \bar{K}^\times) \xrightarrow{\cong} (\bar{K}^\times)^s$. *This bijection restricts to*

$$\mathrm{Char}_{\bar{a}-\mathrm{harm}}(\Lambda, \bar{K}^\times) \xrightarrow{\cong} \Sigma_{\bar{a}} \subseteq (\bar{K}^\times)^s\,.$$

Moreover $\forall \bar{n} \in \mathbb{N}^s_{\mathrm{co}(p)}$ it further restricts to

$$\mathrm{Char}_{\bar{a}-\mathrm{harm}}(\Lambda/\Lambda_{\bar{n},\mathcal{V}}, \bar{K}^\times) \xrightarrow{\cong} \Sigma_{\bar{a},\bar{n}} := \Sigma_{\bar{a}} \cap \mu_{\bar{n}}\,.$$

In Section 1 we prove the second part of Theorem 0.1, see Corollary 1.5. The first parts of Theorems 0.1 and 0.2 are proven in Section 2, see Corollary 2.2 and Theorem 2.4, respectively. We also deduce an expression of polynomials in a Chebyshev-Dickson system via iterated resultants, see Proposition 2.11.

In Section 3 we prove Theorem 0.2(b) (see Proposition 3.1). Besides, we provide in Theorem 3.2 a dynamical criterion for existence of a nonzero periodic $\bar{a}$-harmonic function with a given pluri-period. Or what is the same, for existence of a point on the symbolic variety[2] with a given multi-torsion order. In addition we provide a formula for the orthogonal projection onto the space of all $\bar{a}$-harmonic functions.

In the final Section 4 we study similar problems over finite fields. Assuming p-freeness, in Theorem 4.4 we show that any $\bar{a}$-harmonic function with values in the original Galois field GF(p) is a linear combination of traces of harmonic characters.

The interested reader will find some additional information and many concrete examples in [Za$_1$] and in the preliminary version [Za$_2$] of this paper. In particular, he will find there a discussion concerning the partnership graph; especially Zagier's observation on finiteness of every connected component of this graph in the case of the σ^+-automaton over the Galois field GF(2) on a plane lattice.

The author is grateful to Don Zagier for clarifying discussions, in particular for the idea of processing in the present generality. Our thanks also to Vladimir Berkovich for a kind assistance.

1. Sylow p-subgroups and Chebyshev-Dickson systems

The Dickson polynomials $D_n(x,a) \in \mathbb{Z}[x,a]$ ($E_n(x,a) \in \mathbb{Z}[x,a]$, respectively) of degree n of the first (second) kind can be characterized by the identities [LMT, ACZ]:

$$D_n(\mu_1+\mu_2,\mu_1\mu_2) = \mu_1^n+\mu_2^n \quad \text{resp.} \quad E_n(\mu_1+\mu_2,\mu_1\mu_2) = \mu_1^{n+1}-\mu_2^{n+1}/(\mu_1-\mu_2)\,.$$

Being reduced modulo a prime p, they also satisfy the relations [LMT, BZ]:

$$D_{p^\alpha m} = (D_m)^{p^\alpha} \qquad \text{resp.} \qquad E_{p^\alpha m} = (E_m)^{p^\alpha}\,.$$

In this section we show that analogous relations hold for any Chebyshev-Dickson polynomial system $\mathrm{CharPoly}_a : \mathcal{L} \to K[x]$ (see Corollary 1.5). Here K denotes a field of characteristic $p > 0$, Λ a lattice, $a \in \mathcal{F}^0(\Lambda, K)$ a K-valued function on Λ with bounded support and $\mathcal{L}$ the ordered graph of all finite index sublattices $\Lambda' \subseteq \Lambda$, see §6 in the Introduction. This allows to recover $\mathrm{CharPoly}_a$ by its restriction to the subset $\mathcal{L}^0 \subseteq \mathcal{L}$ of all sublattices with indices coprime to p.

---

[2]Which can be an arbitrary affine algebraic variety in a torus.

For a finite group G and a subset $A \subseteq G$ we let $\delta_A = \sum_{u\in A} \delta_u$ denote the characteristic function of A, where δ_u stands for the delta-function on G concentrated on $u \in G$. For a function $a = \sum_{u\in G} a(u)\delta_u$ on G we let

$$|a|A| = \sum_{u\in A} a(u), \quad \mathrm{CharPoly}_{a,G} = \det(\Delta_a - \lambda \cdot 1) \quad \text{and} \quad d(a,G) = \dim\ker(\Delta_a)\,.$$

The following Pushforward Lemma is simple, so we leave the proof to the reader.

Lemma 1.1. *For a normal subgroup $H \subseteq G$ and for any $a \in \mathcal{F}^0(G,K)$, the function*

$$a_*(v+H) = |a|(v+H)| = \sum_{v'\in H} a(v+v') \tag{7}$$

is a unique function in $\mathcal{F}^0(G/H,K)$ satisfying $a_* \circ \pi = a * \delta_H$, where $\pi : G \twoheadrightarrow G/H$ is the canonical surjection. Moreover, there is a commutative diagram

$$\begin{array}{ccc} \mathcal{F}_H(G,K) & \xrightarrow{\Delta_a} & \mathcal{F}_H(G,K) \\ \downarrow\cong & & \downarrow\cong \\ \mathcal{F}(G/H,K) & \xrightarrow{\Delta_{a_*}} & \mathcal{F}(G/H,K)\,, \end{array}$$

where

$$\mathcal{F}_H(G,K) = \{f \in \mathcal{F}(G,K) \,:\, \tau_h(f) = f \quad \forall h \in H\} \quad \textit{and} \quad \tau_h(f)(g) = f(gh)\,.$$

The convolution algebra $\mathrm{Conv}_K(G)$ consists of all operators on the space $\mathcal{F}^0(L,K)$ commuting with shifts. Moreover the shifts $(\tau_u \;:\; u \in G)$ generate $\mathrm{Conv}_K(G)$ as a K-vector space. Indeed $\forall a \in \mathcal{F}^0(G,K)$ one has

$$\Delta_a = \sum_{g\in G} a(g)\tau_{g^{-1}}\,. \tag{8}$$

Notice that $\tau_{g^{-1}} = \Delta_{\delta_g}$ and $a = \Delta_a(\delta_e)$, where $e \in G$ denotes the neutral element.

For an endomorphism $A \in \mathrm{End}(\mathbb{A}^n_K)$ we let $A = S_A + N_A$ be the Jordan decomposition, where S_A is semisimple, N_A nilpotent and S_A, N_A commute. It is defined over the algebraic closure $\bar{K}$ of K.

Proposition 1.2. *Let $G = F \times H$ be a direct product of two abelian groups, where $H = \bigoplus_{i=1}^n \mathbb{Z}/p^{r_i}\mathbb{Z}$. Then for any $a \in \mathcal{F}^0(G,K)$ the following hold.*

(a) *$S_{\Delta_a} = S_{\Delta_{a_*}} \otimes 1_H$, where $a_* \in \mathcal{F}(F,K)$ is as in (7) above.*
(b) *$\mathrm{CharPoly}_{a,G} = (\mathrm{CharPoly}_{a_*,F})^{\mathrm{ord}\,H}$.*

Consequently there exists a nonzero a-harmonic function on G if and only if there is a nonzero a_*-harmonic function on F.

Proof. (a) follows by induction on n, once it is established for $n = 1$. Letting $H = \mathbb{Z}/p^r\mathbb{Z}$ we will show that

$$\Delta_a^{p^r} = \Delta_{a_*}^{p^r} \otimes 1_H \,. \tag{9}$$

To this end, decomposing $u \in G$ as $u = u' + u''$ with $u' \in F$ and $u'' \in H$, we obtain

$$\Delta_a^{p^r} = \sum_{u \in G} [a(u)]^{p^r} (\tau_{-u})^{p^r} = \sum_{u' \in F} \left(\sum_{u'' \in H} [a(u' + u'')]^{p^r} \right) (\tau_{-u'})^{p^r}$$

$$= \left(\sum_{u' \in F} a_*(u')\tau_{-u'} \right)^{p^r} \otimes 1_H = \Delta_{a_*}^{p^r} \otimes 1_H \,.$$

This proves (9). By virtue of (9) we get

$$S_{\Delta_a}^{p^r} + N_{\Delta_a}^{p^r} = S_{\Delta_{a*}}^{p^r} \otimes 1_H + N_{\Delta_{a*}}^{p^r} \otimes 1_H \,.$$

By the uniqueness of the Jordan decomposition we have $S_{\Delta_a}^{p^r} = S_{\Delta_{a_*}}^{p^r} \otimes 1_H$ and so (a) follows. Now (b) is immediate from (a). □

From Proposition 1.2 we deduce the following corollary.

Corollary 1.3. *If* $G = \bigoplus_{i=1}^k \mathbb{Z}/p^{r_i}\mathbb{Z}$, *where* $p = \operatorname{char} K$, *then*

$$\operatorname{CharPoly}_{a,G} = (x - |a|)^{\operatorname{ord} G} \,.$$

In particular there exists an a-harmonic function on G if and only if the constant function 1 on G is a-harmonic, if and only if $|a| = 0$.

Remark 1.4. Letting $K = \mathrm{GF}(2)$, $G = \mathbb{Z}/4\mathbb{Z}$ and $a = a^+ = \delta_{\bar{0}} + \delta_{\bar{1}} + \delta_{-\bar{1}} \in \mathcal{F}(G, K)$ we obtain $\operatorname{CharPoly}_{a,G} = (x+1)^4$. Hence the algebraic multiplicity of the spectral value $\lambda = 1$ of Δ_a equals 4. Although $S_{\Delta_a} = 1_G$, the nilpotent part N_{Δ_a} is present and $\dim \ker(\Delta_a + 1_G) = 2$.

Given $\bar{a} = (a_1, \ldots, a_t) \in (\mathcal{F}^0(G, K))^t$ we let as before

$$\ker(\Delta_{\bar{a}}) = \bigcap_{j=1}^t \ker(\Delta_{a_j}), \qquad d(\bar{a}, G) = \dim \ker(\Delta_{\bar{a}})$$

and

$$\operatorname{CharPoly}_{\bar{a},G} = \operatorname{CharPoly}(\Delta_{\bar{a}}) = \gcd\left(\operatorname{CharPoly}(\Delta_{a_j}) \,:\, j = 1, \ldots, t\right) \,.$$

For a subgroup $H \subseteq G$ we let

$$\operatorname{CharPoly}_{\bar{a}_*, G/H} = \operatorname{CharPoly}(\Delta_{\bar{a}} | \mathcal{F}_H(G, K)) \,,$$

where $\bar{a}_* = ((a_1)_*, \ldots, (a_n)_*) \in \mathcal{F}(G/H, K)$.

Corollary 1.5. *Let Λ be a lattice and $\Lambda' \subseteq \Lambda$ a sublattice of index $p^\alpha q$, where $q \not\equiv 0 \mod p$. Then there exists a unique intermediate sublattice Λ'' of index q in Λ, where $\Lambda' \subseteq \Lambda'' \subseteq \Lambda$. Moreover $\forall \bar{a} \in (\mathcal{F}^0(\Lambda, K))^t$ one has*

$$\mathrm{CharPoly}_{\bar{a},\Lambda'} = (\mathrm{CharPoly}_{\bar{a},\Lambda''})^{p^\alpha} . \tag{10}$$

Proof. It is enough to show (10) for $t = 1$; then it follows easily for any $t \geq 1$. So letting $t = 1$ and $a_1 = a$ we decompose

$$G = \Lambda/\Lambda' = F \oplus G(p) ,$$

where $G(p)$ is the Sylow p-subgroup of G and $\operatorname{ord} F = q$. We let $\Lambda'' = \pi'^{-1}(G(p))$, where $\pi' : \Lambda \twoheadrightarrow G$, so that $\Lambda/\Lambda'' \cong F$. Due to the Pushforward Lemma 1.1,

$$\mathrm{CharPoly}_{a,\Lambda'} = \mathrm{CharPoly}_{\pi'_* a,G} \quad \text{and} \quad \mathrm{CharPoly}_{a,\Lambda''} = \mathrm{CharPoly}_{\pi''_* a,F} , \tag{11}$$

where $\pi'' : \Lambda \twoheadrightarrow F$. Now (10) follows from (11) in view of Proposition 1.2(b). $\square$

2. Chebyshev-Dickson systems in the p-free case

2.1. Naive Fourier transform on convolution algebras

For a finite group G there are natural isomorphisms

$$(\mathcal{F}(G,K), *) \xrightarrow{\varphi} \mathrm{Conv}_K (G) \xrightarrow{\psi} K[G] ,$$

where $\varphi : a \longmapsto \Delta_a$, $\mathrm{Conv}_K (G)$ is the convolution algebra and $K[G]$ is the group algebra of G over K. For instance [MOW], the group ring of a finite abelian group

$$G = \bigoplus_{i=1}^{s} \mathbb{Z}/n_i\mathbb{Z}$$

is the truncated polynomial ring

$$K[G] = \bigotimes_{i=1}^{s} K[x_i]/(x_i^{n_i} - 1) .$$

The ideals of $\mathcal{F}(G,K)$ are called convolution ideals. In particular, for any subgroup $H \subseteq G$, the translation invariant subspace

$$\mathcal{F}_H(G,K) = \{f \in \mathcal{F}(G,K) \,:\, \tau_h(f) = f \quad \forall h \in H\}$$

is a convolution ideal.

The composition $\psi \circ \varphi : \mathcal{F}(G,K) \to K[G]$, $a \longmapsto \tilde{a}$, provides a naive Fourier transform, which sends Δ_a to the operator of multiplication by $\tilde{a}$ in $K[G]$ and $\ker(\Delta_a)$ to the annihilator ideal $\mathrm{Ann}(\tilde{a})$, where $(\tilde{a}) \subseteq K[G]$ is the principal ideal generated by $\tilde{a}$. Thus G possesses a nonzero a-harmonic function if and only if $\tilde{a} \in K[G]$ is a zero divisor.

The next result follows immediately from the Burnside Theorem. Alternatively it can be deduced using the Fourier transform, see below.

Lemma 2.1. *For any finite abelian group G of order coprime to p the following hold.*

(a) *$\mathcal{F}(G,\bar{K})$ admits a decomposition into a direct sum of one-dimensional $\mathrm{Conv}_{\bar{K}}(G)$-submodules generated by characters:*

$$\mathcal{F}(G,\bar{K}) = \bigoplus_{g^\vee \in G^\vee} (g^\vee)\,.$$

(b) *Any convolution ideal $I \subseteq \mathcal{F}(G,\bar{K})$ is principal, generated by the sum of characters contained in I. Furthermore there is a decomposition*

$$\mathcal{F}(G,\bar{K}) = I \oplus \mathrm{Ann}(I)\,.$$

In particular, for any subgroup $H \subseteq G$ one has

$$\mathcal{F}(G,\bar{K}) = \mathcal{F}_H(G,\bar{K}) \oplus \mathrm{Ann}(\mathcal{F}_H(G,\bar{K}))\,,$$

where

$$\mathcal{F}_H(G,\bar{K}) = \sum_{H \subseteq \ker(g^\vee)} (g^\vee)\,.$$

2.2. Dual group and Fourier transform

In the sequel we let $K = \mathrm{GF}(p^r)$. Thus the multiplicative group $\bar{K}^\times$ is a torsion group, with torsion orders coprime with p. We let G be a finite abelian group of order coprime with p, and $\mathbb{N}_{\mathrm{co}(p)}$ be the set of all positive integers coprime to p.

The field $\bar{K}$ contains all roots of unity with orders dividing $\mathrm{ord}\, G$. Hence the characters of G can be realized as homomorphisms $G \to \bar{K}^\times$. This defines a natural embedding $G^\vee \hookrightarrow \mathcal{F}(G,\bar{K})$.

The Fourier transform $F : \mathcal{F}(G,\bar{K}) \to \mathcal{F}(G^\vee,\bar{K})$ is defined as usual [Nic] via

$$F : f \longmapsto \widehat{f}, \quad \text{where} \quad \widehat{f}(g^\vee) = \sum_{g\in G} f(g) g^\vee(g), \quad g^\vee \in G^\vee\,. \tag{12}$$

Its inverse $F^{-1} : \mathcal{F}(G^\vee,\bar{K}) \to \mathcal{F}(G,\bar{K})$ can be defined via

$$F^{-1} : \varphi \longmapsto \widehat{\varphi}, \quad \text{where} \quad \widehat{\varphi}(g) = \frac{1}{\mathrm{ord}\,(G)} \sum_{g^\vee \in G^\vee} \varphi(g^\vee) g^\vee(g^{-1}), \quad g \in G\,.$$

With this notation $\hat{\hat{f}} = f$ and $\hat{\hat{\varphi}} = \varphi$, which does not lead to a confusion as we never exploit the Fourier transform on the dual group $G^\vee$.

Furthermore F sends the convolution in the ring $\mathcal{F}(G,\bar{K})$ into the pointwise multiplication on $\mathcal{F}(G^\vee,\bar{K})$ giving an isomorphism of $\bar{K}$-algebras. The convolution operator Δ_a is sent to the operator of multiplication by $\hat{a}$. The Fourier transform of a character is proportional to a delta-function. So up to a scalar factor, any character $g^\vee \in G^\vee$ is a convolution idempotent, and any convolution idempotent of $(\mathcal{F}(G,\bar{K}),*)$ is proportional to a sum of characters.

In the $\bar{K}$-vector space $\mathcal{F}(G,\bar{K})$ we consider the non-degenerate symmetric bilinear form

$$\langle f_1, f_2\rangle_1 = \frac{1}{\mathrm{ord}\,(G)} \sum_{g\in G} f_1(g) f_2(g^{-1}) = \frac{1}{\mathrm{ord}\,(G)} f_1 * f_2(e)\,.$$

Its Fourier dual is the following bilinear form in $\mathcal{F}(G^\vee, \bar{K})$:

$$\langle \varphi_1, \varphi_2 \rangle_2 = \frac{1}{(\mathrm{ord}\,(G))^2} \sum_{g^\vee \in G^\vee} \varphi_1(g^\vee)\varphi_2(g^\vee)\,.$$

Indeed we have $\langle \widehat{f_1}, \widehat{f_2} \rangle_2 = \langle f_1, f_2 \rangle_1$. The characters $(g^\vee : g^\vee \in G^\vee)$ form an orthonormal basis in $\mathcal{F}(G, \bar{K})$ with respect to the form $\langle \cdot, \cdot \rangle_1$.

2.3. Divisibility in Chebyshev-Dickson systems

In this section G denotes a finite abelian group of order coprime to p. For a t-tuple $\bar{a} = (a_1, \ldots a_t) \in (\mathcal{F}^0(G, \bar{K}))^t$ we let as before $\ker(\Delta_{\bar{a}}) = \bigcap_{j=1}^t \ker(\Delta_{a_j})$ and

$$\mathrm{CharPoly}_{\bar{a},G} = \mathrm{CharPoly}(\Delta_{\bar{a}}) = \gcd\left(\mathrm{CharPoly}(\Delta_{a_j}) \,:\, j = 1, \ldots, t\right)\,.$$

By the Pushforward Lemma 1.1, for a subgroup $H \subseteq G$ we have

$$\mathrm{CharPoly}_{\bar{a}_*, G/H} = \mathrm{CharPoly}(\Delta_{\bar{a}} | \mathcal{F}_H(G, \bar{K}))\,.$$

We let

$$V(a_j) = \hat{a}_j^{-1}(0) \subseteq G^\vee \quad \text{and} \quad V(\bar{a}) = \bigcap_{j=1}^t V(a_j) \subseteq G^\vee\,.$$

We also let $\mathrm{Char}_{\bar{a}-\mathrm{harm}}(G, \bar{K}^\times)$ denote the set of all $\bar{a}$-harmonic characters of G. The following corollary is straightforward from Lemma 2.1.

Corollary 2.2. (a) *For any $a \in \mathcal{F}^0(G, \bar{K})$, the characters form an orthonormal basis $\mathcal{F}(G, \bar{K})$ of eigenfunctions of Δ_a, the matrix of Δ_a in this basis is diagonal and so*

$$\mathrm{CharPoly}(\Delta_a) = \prod_{g^\vee \in G^\vee} (x - \hat{a}(g^\vee))\,.$$

Consequently $\mathrm{spec}(\Delta_a) = \hat{a}(G^\vee) \subseteq \bar{K}$, $\quad \ker(\Delta_a) = \left(\widehat{\delta_{V(a)}}\right)$ *and*

$$d(a, G) = \mathrm{card}\, V(a) = \mathrm{mult}_0 \mathrm{CharPoly}(\Delta_a)\,.$$

(b) *Similarly for any $\bar{a} = (a_1, \ldots, a_t) \in (\mathcal{F}^0(G, \bar{K}))^t$, the $\bar{a}$-harmonic characters form an orthonormal basis in* $\ker(\Delta_{\bar{a}})$,

$$\ker(\Delta_{\bar{a}}) = \left(\widehat{\delta_{V(\bar{a})}}\right) \qquad \text{and} \quad d(\bar{a}, G) = \mathrm{card}\, V(\bar{a})\,.$$

Moreover there is a bijection $V(\bar{a}) \cong \mathrm{Char}_{\bar{a}-\mathrm{harm}}(G, \bar{K}^\times)$. Consequently the group G admits a nonzero $\bar{a}$-harmonic function if and only if it admits an $\bar{a}$-harmonic character.

A convolution ideal $I \subseteq \mathcal{F}(G, \bar{K})$ and the annihilator ideal $\mathrm{Ann}(I)$ being $\Delta_{\bar{a}}$-stable (see Lemma 2.1), the Pushforward Lemma 1.1 yields the following result.

Proposition 2.3. *For any convolution ideal $I \subseteq \mathcal{F}(G, \bar{K})$, any subgroup $H \subseteq G$ and any $\bar{a} \in (\mathcal{F}^0(G, \bar{K}))^t$ one has*

$$\mathrm{CharPoly}(\Delta_{\bar{a}}|I) \mid \mathrm{CharPoly}(\Delta_{\bar{a}}) \quad \textit{and} \quad \mathrm{CharPoly}_{\bar{a}_*, G/H} \mid \mathrm{CharPoly}_{\bar{a},G}\,. \qquad (13)$$

Now we can readily deduce the divisibility property in Chebyshev-Dickson systems disregarding the assumption of p-freeness.

Theorem 2.4. *For any* $\Lambda_1, \Lambda_2 \in \mathcal{L}$ *we have:*

$$\mathrm{CharPoly}_{\bar{a},\Lambda_1+\Lambda_2} \mid \gcd(\mathrm{CharPoly}_{\bar{a},\Lambda_1}, \mathrm{CharPoly}_{\bar{a},\Lambda_2})$$

and

$$\mathrm{lcm}(\mathrm{CharPoly}_{\bar{a},\Lambda_1}, \mathrm{CharPoly}_{\bar{a},\Lambda_2}) \mid \mathrm{CharPoly}_{\bar{a},\Lambda_1\cap\Lambda_2}\,.$$

Consequently $\mathrm{CharPoly}_{\bar{a},\Lambda_2} \mid \mathrm{CharPoly}_{\bar{a},\Lambda_1}$ *whenever* $\Lambda_1 \subseteq \Lambda_2$.

Proof. It suffices to show the latter assertion, then the former ones follow immediately. For a chain of finite index sublattices $\Lambda' \subseteq \Lambda'' \subseteq \Lambda$ we let $G = \Lambda/\Lambda'$, $H = \Lambda''/\Lambda'$ so that $G/H = \Lambda/\Lambda''$, and

$$\bar{a}' = \pi'_*\bar{a} \in \mathcal{F}(G, \bar{K}), \qquad \bar{a}'' = \pi''_*\bar{a} \in \mathcal{F}(G/H, \bar{K})\,,$$

where $\pi : G \twoheadrightarrow G/H$, $\pi' : \Lambda \twoheadrightarrow G$ and $\pi'' = \pi \circ \pi' : \Lambda \twoheadrightarrow G/H$.

We assume first that the index of Λ' in Λ is coprime with p and so $\Lambda', \Lambda'' \in \mathcal{L}^0$. By the Pushforward Lemma 1.1 we obtain

$$\mathrm{CharPoly}_{\bar{a},\Lambda'} = \mathrm{CharPoly}_{\bar{a}',G} \quad \text{and} \quad \mathrm{CharPoly}_{\bar{a},\Lambda''} = \mathrm{CharPoly}_{\bar{a}'',G/H}\,.$$

Hence by (13)

$$\mathrm{CharPoly}_{\bar{a},\Lambda''} \mid \mathrm{CharPoly}_{\bar{a},\Lambda'}\,. \tag{14}$$

In the general case, assuming that $\Lambda_1 \subseteq \Lambda_2$ we consider the decomposition $G_1 = \Lambda/\Lambda_1 = F \oplus G_1(p)$, where $G_1(p) \subseteq G_1$ is the Sylow p-subgroup. Letting $G_2 = \Lambda_2/\Lambda_1 \subseteq G_1$ we obtain $G_2(p) = G_1(p) \cap G_2$. For the sublattices $\Lambda''_i = \pi^{-1}(G_i(p)) \subseteq \Lambda$, $i = 1, 2$, where $\pi : \Lambda \twoheadrightarrow G_1$, we have $\Lambda''_i \supseteq \Lambda_i$ and $\Lambda''_1, \Lambda''_2 \in \mathcal{L}^0$. Since also $\Lambda''_1 \subseteq \Lambda''_2$, by virtue of (14) we obtain $\mathrm{CharPoly}_{a,\Lambda'_2} \mid \mathrm{CharPoly}_{a,\Lambda'_1}$. By (10) $\mathrm{CharPoly}_{a,\Lambda_i} = (\mathrm{CharPoly}_{a,\Lambda''_i})^{p^{\alpha_i}}$, $i = 1, 2$, where $\alpha_2 \le \alpha_1$. Now the result follows. □

Remark 2.5. Letting $\Lambda'' = \Lambda$ we deduce that $(x - |a|) \mid \mathrm{CharPoly}_{a,\Lambda'}$ $\forall \Lambda' \in \mathcal{L}$, $\forall a \in \mathcal{F}^0(\Lambda, \bar{K})$. We note that the eigenspace in $\mathcal{F}(\Lambda, \bar{K})$, which corresponds to the eigenvalue $|a|$ of Δ_a, contains the constant function 1; cf. Corollary 1.3.

Examples 2.6. 1. Letting $G = G_1 \times G_2$ and $a = a_1 \otimes a_2 \in \mathcal{F}(G, \bar{K})$, where $a_i \in \mathcal{F}(G_i, \bar{K})$, $i = 1, 2$, we obtain

$$\mathrm{CharPoly}(\Delta_a) = \prod_{i,j} (x - \lambda_i \mu_j)\,,$$

where $\lambda_1, \ldots, \lambda_{\mathrm{ord}\,(G_1)}$ and $\mu_1, \ldots, \mu_{\mathrm{ord}\,(G_2)}$ denote the eigenvalues of Δ_{a_1} and Δ_{a_2}, respectively.

2. Similarly, letting $a = a_1 \otimes 1 \oplus 1 \otimes a_2 \in \mathcal{F}(G, \bar{K})$, where again $G = G_1 \times G_2$, we obtain

$$\mathrm{CharPoly}(\Delta_a) = \prod_{i,j} (x - (\lambda_i + \mu_j))\,.$$

The latter formula applies in particular to the graph Laplacians with kernels $a_i^- = a_{G_i}^+ - \delta_0$, $i = 1, 2$ and $a^- = a_G^+ - \delta_0$, respectively, where a^+ is the star-function as in (2). In the characteristic 2 case cf. Bacher's Lemma 2.10(a) in [Za$_1$].

2.4. Symbol of a convolution operator

Definition 2.7. Fixing a lattice Λ of rank s, a basis $\mathcal{V} = (v_1, \ldots, v_s)$ of Λ and an n-tuple $\bar{n} = (n_1, \ldots, n_s)$, where $n_i \in \mathbb{N}$, we let

$$\Lambda_{\bar{n},\mathcal{V}} = \sum_{i=1}^{s} n_i \mathbb{Z} v_i \cong \bigoplus_{i=1}^{s} n_i \mathbb{Z}$$

be the product sublattice of Λ generated by $n_1 v_1, \ldots, n_s v_s$. There is an isomorphism of K-algebras

$$\sigma_{\mathcal{V}} : \mathrm{Conv}_K(\Lambda) \xrightarrow{\cong} K[x_1, x_1^{-1}, \ldots, x_s, x_s^{-1}], \qquad \Delta_a \longmapsto \sigma_{a,\mathcal{V}},$$

which associates to a convolution operator Δ_a on Λ its $\mathcal{V}$-symbol, that is the Laurent polynomial in s variables

$$\sigma_{a,\mathcal{V}} = \sum_{v=\sum_{i=1}^{s} \alpha_i e_i \in \Lambda} a(v) x^{-\alpha(v)} = \sum_{v=\sum_{i=1}^{s} \alpha_i e_i \in \Lambda} a(v) x_1^{-\alpha_1} \cdot \ldots \cdot x_s^{-\alpha_s}. \tag{15}$$

The inverse $\sigma_{\mathcal{V}}^{-1}$ is given by

$$x_i^{-1} \longmapsto \tau_{v_i} \qquad \text{and} \qquad x_i \longmapsto \tau_{-v_i}, \qquad i = 1, \ldots, s.$$

The algebraic hypersurface in the s-torus

$$\Sigma_{a,\mathcal{V}} = \sigma_{a,\mathcal{V}}^{-1}(0) \subseteq (K^\times)^s \tag{16}$$

associate with Δ_a will be called the symbolic hypersurface. Similarly, for a sequence of convolution operators $\Delta_{\bar{a}} = (\Delta_{a_j} : j = 1, \ldots, t)$ its symbolic variety is the affine subvariety in the torus

$$\Sigma_{\bar{a},\mathcal{V}} = \bigcap_{j=1}^{t} \sigma_{a_j,\mathcal{V}}^{-1}(0) \subseteq (K^\times)^s. \tag{17}$$

Example 2.8. (see [Za$_1$]) For $K = \mathrm{GF}(2)$, $\Lambda = \mathbb{Z}^2$, $\mathcal{V} = (e_1, e_2)$ and $a = a^+$ the symbolic hypersurface $\Sigma_{a^+,\mathcal{V}}$ is the elliptic cubic (3) in $(\bar{K}^\times)^2$. Alternatively, this cubic can be given by the equation

$$x^2 y + x y^2 + xy + x + y = 0.$$

For a finite abelian group $G = \mathbb{Z}_{\bar{n}} = \bigoplus \mathbb{Z}/n_i\mathbb{Z}$, where $\bar{n} = (n_1, \ldots, n_s) \in \mathbb{N}^s_{\mathrm{co}(p)}$, we let $\mathcal{U} = (e_1, \ldots, e_s)$ denote the standard base of G. We let also

$$\mu_{\bar{n}} = \bigoplus_{i=1}^{s} \mu_{n_i} \subseteq (\bar{K}^\times)^s,$$

where $\mu_n \subseteq \bar{K}^\times$ stands for the cyclic group of nth roots of unity. The correspondence

$$g^\vee \longmapsto (g^\vee(e_1), \ldots, g^\vee(e_1))$$

yields an isomorphism

$$\varphi : G^\vee \xrightarrow{\cong} \mu_{\bar{n}} .$$

This isomorphism relates the symbol of a convolution operator Δ_a and the Fourier transform $\hat{a}$ of its kernel.

Lemma 2.9. *For any* $a \in \mathcal{F}(G, \bar{K})$ *we have*

$$\hat{a} = (\sigma_{a,\mathcal{V}} |\mu_{\bar{n}}) \circ \varphi .$$

Consequently

$$\mathrm{CharPoly}_{a,\bar{n},\mathcal{V}} := \mathrm{CharPoly}_{a,\Lambda_{\bar{n}},\mathcal{V}} = \prod_{\xi \in \mu_{\bar{n}}} (x - \sigma_{a,\mathcal{V}}(\xi)) . \tag{18}$$

Proof. For any $g \in G, g^\vee \in G^\vee$, letting $\xi_i = g^\vee(e_i)$, $i = 1, \ldots, s$, by (8) and (15) we obtain:

$$\hat{a}(g) \cdot g^\vee(g) = \Delta_a(g^\vee)(g) = \left(\sum_{v \in G} a(v)\tau_{-v} \right) (g^\vee)(g) = \sum_{v \in G} a(v)\tau_{-v}(g^\vee)(g)$$

$$= \sum_{v=\sum_{j=1}^{s} \alpha_j e_j \in G} a(v) g^\vee(g - v) = \sigma_{a,\mathcal{V}}(\xi_1, \ldots, \xi_s) \cdot g^\vee(g) = \sigma_{a,\mathcal{V}}(\xi) ,$$

where $\xi = (\xi_1, \ldots, \xi_s) \in \mu_{\bar{n}}$. Indeed

$$g^\vee(g - v) = g^\vee(g)g^\vee(-v) = g^\vee(g)g^\vee\left(-\sum_{j=1}^{s} \alpha_j e_j \right) = g^\vee(g) \prod_{j=1}^{s} \xi_i^{-\alpha_i} .$$

The equality (18) follows now from Proposition 2.2. □

2.5. Chebyshev-Dickson systems and iterated resultants

Definition 2.10. We consider a Laurent polynomial $\Omega = P/y^\alpha$, where $P \in \bar{K}[y_1, \ldots, y_s]$ is a polynomial coprime with $y^\alpha = y_1^{\alpha_1} \cdot \ldots \cdot y_s^{\alpha_s}$. Given a multi-index $\bar{n} = (n_1, \ldots, n_s) \in \mathbb{N}^s$, we define recursively the iterated resultant $\mathrm{res}_{\bar{n}}(\Omega) := Q_s \in \bar{K}[x]$ via

$$Q_0(x, y_1, \ldots, y_s) = y^\alpha x - P(y_1, \ldots, y_s)$$

and

$$Q_i(x, y_{i+1}, \ldots, y_s) = \mathrm{res}_{y_i} (Q_{i-1}(x, y_i, \ldots, y_s), y_i^{n_i} - 1) \in \bar{K}[x, y_{i+1}, \ldots, y_s],$$

$i = 1, \ldots, s$. In detail

$$\mathrm{res}_{\bar{n}}(\Omega) = \mathrm{res}_{y_s} (\ldots \mathrm{res}_{y_1} (y_1^{\alpha_1} \ldots y_s^{\alpha_s} x - P(y_1, \ldots, y_s), y_1^{n_1} - 1) , \ldots, y_s^{n_s} - 1) .$$

Clearly $\lambda = \Omega(\xi) = P(\xi)/\xi^\alpha$ for some $\xi \in \mu_{\bar{n}}$ if and only if $\mathrm{res}_{\bar{n}}(\Omega)(\lambda) = 0$.

Given a lattice Λ of rank s, a basis $\mathcal{V}$ of Λ and a multi-index $\bar{n} \in \mathbb{N}^s_{\mathrm{co}(p)}$, we let as before $\Lambda_{\bar{n},\mathcal{V}}$ denote the product sublattice $\bigoplus n_i \mathbb{Z} v_i$ of Λ. In the p-free case we deduce from Lemma 2.9 the following expression for the Chebyshev-Dickson polynomial $\mathrm{CharPoly}_{a,\bar{n},\mathcal{V}}$ in (18) as the multivariate iterated resultant of the symbol $\sigma_{a,\mathcal{V}}$ in (15). We leave the details to the reader.

Proposition 2.11. *In the notation as above, the characteristic polynomial of the restriction $\Delta_a|\Lambda_{\bar{n},\mathcal{V}}$, where $a \in \mathcal{F}(\Lambda, \bar{K})$ and $\bar{n} \in \mathbb{N}^s_{\mathrm{co}(p)}$, can be expressed as follows:*

$$\mathrm{CharPoly}_{a,\bar{n},\mathcal{V}} = \mathrm{res}_{\bar{n}}(\sigma_{a,\mathcal{V}}) .$$

Cf. [HMP, 3.3] for an alternative expression of the characteristic polynomials of σ^+-automata on multi-dimensional grids in terms of iterated resultants.

Example 2.12. Using the above proposition we derive the following expression for the classical Chebyshev-Dickson polynomials T_n of the first kind over the Galois field GF(2):

$$T_n(x) = \mathrm{res}_y(y^2 + xy + 1, y^n + 1) .$$

3. Counting points on symbolic variety

3.1. Harmonic characters as points on symbolic variety

We let as before $K = \mathrm{GF}(p^r)$. Given $\bar{a} = (a_1, \ldots, a_t) \in (\mathcal{F}^0(\Lambda, \bar{K}))^t$, we establish in Proposition 3.1 below a natural bijection between the set of points on the symbolic variety $\Sigma_{\bar{a},\mathcal{V}}$ in (17) and the set $\mathrm{Char}_{\bar{a}-\mathrm{harm}}(\Lambda, \bar{K}^\times)$ of all $\bar{a}$-harmonic characters of Λ.

Since $\bar{K}^\times$ is a torsion group, given a basis $\mathcal{V} = (v_1, \ldots, v_s)$ of Λ, any character $g^\vee \in \mathrm{Char}(\Lambda, \bar{K}^\times)$ is $(\bar{n}, \mathcal{V})$-periodic for $\bar{n} = (n_1, \ldots, n_s)$, where $n_i = \mathrm{ord}\,(g^\vee(v_i)) \in \mathbb{N}_{\mathrm{co}(p)}$, $i = 1, \ldots, s$. Letting $G = G_{\bar{n},\mathcal{V}} = \Lambda/\Lambda_{\bar{n},\mathcal{V}}$ we have $g^\vee = h^\vee \circ \pi$ for a character $h^\vee \in G^\vee$, where $\pi : \Lambda \twoheadrightarrow G$. By virtue of the Pushforward Lemma 1.1, $g^\vee$ is $\bar{a}$-harmonic if and only if $h^\vee$ is $\bar{a}_*$-harmonic. Consequently

$$\mathrm{Char}_{\bar{a}-\mathrm{harm}}(\Lambda, \bar{K}^\times) = \bigcup_{\bar{n} \in \mathbb{N}^s_{\mathrm{co}(p)}} (G_{\bar{n},\mathcal{V}})^\vee_{\bar{a}_*-\mathrm{harm}} .$$

For any $\bar{a} = (a_1, \ldots, a_t) \in (\mathcal{F}^0(\Lambda, \bar{K}))^t$ and $\bar{n} = (n_1, \ldots, n_s) \in \mathbb{N}^s_{\mathrm{co}(p)}$, we consider the following over-determined system of algebraic equations, cf. (15):

$$\sigma_{a_j,\mathcal{V}}(x_1, \ldots, x_s) = 0, \quad x_i^{n_i} = 1, \quad i = 1, \ldots, s, \;\; j = 1, \ldots, t . \tag{19}$$

We let $\Sigma_{\bar{a},\bar{n},\mathcal{V}} = \Sigma_{\bar{a},\mathcal{V}} \cap \mu_{\bar{n}}$ denote the set of all solutions of (19), or in other words the set of all points on the symbolic variety $\Sigma_{\bar{a},\mathcal{V}}$ in (15) whose multi-torsion orders divide $\bar{n} = (n_1, \ldots, n_s)$. The following result yields (b) of Theorem 0.2 in the Introduction.

Proposition 3.1. *Given a basis $\mathcal{V}$ of Λ, the natural bijection*

$$\mathrm{Char}(\Lambda, \bar{K}^\times) \xrightarrow{\cong} (\bar{K}^\times)^s$$

restricts to

$$\mathrm{Char}_{\bar{a}-\mathrm{harm}}(\Lambda, \bar{K}^\times) \xrightarrow{\cong} \Sigma_{\bar{a},\mathcal{V}}$$

and further yields the bijections

$$\Sigma_{\bar{a},\bar{n},\mathcal{V}} = \Sigma_{\bar{a},\mathcal{V}} \cap \mu_{\bar{n}} \xrightarrow{\cong} (G_{\bar{n},\mathcal{V}})^\vee_{\bar{a}_*-\mathrm{harm}} \xrightarrow{\cong} V(\bar{a}_*) , \tag{20}$$

where $\bar a_* = \pi_*\bar a$ *is the pushforward of* $\bar a$ *under the canonical surjection* $\pi : \Lambda \twoheadrightarrow G_{\bar n,\mathcal V} = \Lambda/\Lambda_{\bar n,\mathcal V}$. *Consequently*

$$d(\bar a, \Lambda_{\bar n,\mathcal V}) = \operatorname{card} V(\bar a_*) = \operatorname{card} \Sigma_{\bar a,\bar n,\mathcal V}\,.$$

Proof. For a character $h^\vee \in G^\vee_{\bar n,\mathcal V}$, letting $g^\vee = h^\vee \circ \pi \in \operatorname{Char}(\Lambda, \bar K^\times)$ and $\xi_i = g^\vee(v_i) \in \bar K^\times$ we obtain $\xi_i^{n_i} = 1\ \forall i = 1,\dots,s$ (indeed $n_i v_i \in \Lambda_{\bar n,\mathcal V}\ \forall i$). Moreover $h^\vee \in (G_{\bar n,\mathcal V})^\vee_{\bar a_*-\mathrm{harm}}$ if and only if $\forall j = 1,\dots,t$,

$$h^\vee * (a_j)_* = 0 \iff g^\vee * a_j = 0 \iff g^\vee * \left(\sum_{v\in L} a_j(v)\delta_v\right) = 0$$

$$\iff \sum_{v=\sum_{i=1}^s \alpha_i v_i \in L} a_j(v)(g^\vee)^{-1}(v) = 0 \iff \sigma_{a_j,\mathcal V}(\xi_1,\dots,\xi_s) = 0\,,$$

and so $\xi = (\xi_1,\dots,\xi_s) \in \bigcap_{j=1}^t \Sigma_{a_j,\mathcal V} = \Sigma_{\bar a,\mathcal V}$. Vice versa, given a solution $\xi = (\xi_1,\dots,\xi_s) \in (\bar K^\times)^s$ of (19), letting $g^\vee(v_i) = \xi_i$ defines an $(\bar n,\mathcal V)$-periodic character $g^\vee \in \operatorname{Char}(\Lambda, \bar K^\times)$. By the same argument as above, $g^\vee$ and the pushforward character $h^\vee = \pi_*(g^\vee) \in (G_{\bar n,\mathcal V})^\vee$ are $\bar a$- and $\bar a_*$-harmonic, respectively. The correspondence $\xi = (\xi_1,\dots,\xi_s) \longleftrightarrow h^\vee$ yields the first bijection in (20). As for the second one, see 2.2(b). □

3.2. Criteria of harmonicity

We let as before $K = \mathrm{GF}(p^r)$. Given a basis $\mathcal V$ of Λ, a sequence $\bar a \in (\mathcal F^0(\Lambda,\bar K))^t$ and a multi-index $\bar n \in \mathbb N^s_{\mathrm{co}(p)}$, we let $G = \Lambda/\Lambda_{\bar n,\mathcal V}$ and $\bar a_* = \pi_*\bar a$, where $\pi : \Lambda \twoheadrightarrow G$. We fix a minimal $q_0 = q(\bar a_*) = p^{r_0}$ $(r_0 > 0)$ such that $\widehat a_j(G^\vee) \subseteq \mathrm{GF}(q_0)\ \forall j = 1,\dots,t$. The preceding results lead to the following criteria.

Theorem 3.2.

(a) *With the notation as above, the following conditions are equivalent.*
   (i) *There exists a nonzero* $(\bar n,\mathcal V)$*-periodic*[3] $\bar a$*-harmonic function on* Λ.
   (ii) $V(\bar a_*) := \bigcap_{j=1}^t V(a_{j*}) \neq \emptyset$.
   (iii) *The system (19) has a solution* $\xi = (\xi_1,\dots,\xi_s) \in \Sigma_{\bar a,\bar n,\mathcal V} = \Sigma_{\bar a,\mathcal V} \cap \mu_{\bar n}$.

(b) *Furthermore* $\ker(\Delta_{\bar a_*}) \subseteq (\mathcal F(G,\bar K), *)$ *coincides with the principal convolution ideal generated by the function* $\prod_{j=1}^t \left(\delta_e - \Delta^{q_0-1}_{(a_j)_*}(\delta_e)\right)$, *and*

$$\prod_{j=1}^t \left(1_G - \Delta^{q_0-1}_{(a_j)_*}\right) : \mathcal F(G,\bar K) \twoheadrightarrow \ker(\Delta_{\bar a_*})$$

is an orthogonal projection.

(c) *For* $t = 1$ *and* $a_1 = a$, (i)–(iii) *are equivalent to every one of the following conditions:*
   (iv) $(\Delta_{a_*})^{q_0-1}(\delta_e) \neq \delta_e$ *or, equivalently,* $(\Delta_{a_*})^{q_0-1} \neq 1_G$.
   (v) *The sequence* $\left(\Delta^k_{a_*}(\delta_e)\right)_{k\geq 0} \subseteq \mathcal F(G,\bar K)$ *is not periodic.*

---

[3] I.e., stable under the shifts by elements of $\Lambda_{\bar n,\mathcal V}$.

Proof. The equivalences (i)$\Longleftrightarrow$(ii)$\Longleftrightarrow$(iii) follow immediately from 2.2 and 3.1. Since the function $\widehat{(a_j)_*} \in \mathcal{F}(G^\vee, \bar{K})$ takes values in the field $\mathrm{GF}(q_0)$ we have $\delta_{V(\widehat{(a_j)_*})} = 1 - \widehat{(a_j)_*}^{\,q_0-1}$. For every $j = 1, \ldots, t$ the Fourier transform sends $1_G - \Delta_{(a_j)_*}^{q_0-1}$ to the operator of multiplication by $\delta_{V((a_j)_*)}$ in $\mathcal{F}(G^\vee, \bar{K})$, which coincides with the orthogonal projection onto the corresponding principal ideal. This yields (b) and (c). □

Remarks 3.3. 1. For $t = 1$ and $a = a_1$ we have $\widehat{a_*}^{\,q_0} = \widehat{a_*}$ and so $\Delta_{a_*}^{q_0+1}(\delta_e) = \Delta_{a_*}(\delta_e) = a_*$. Consequently the truncated sequence $\left(\Delta_{a_*}^k(\delta_e)\right)_{k\geq 1}$ (which starts with a_*) is periodic with period l dividing $q_0 - 1$. Whereas the sequence in (iv) (which starts with δ_e) is periodic if and only if Λ does not admit a nonzero a-harmonic $(\bar{n}, \mathcal{V})$-periodic function. In the latter case Δ_{a_*} is invertible of finite order in the group $\mathrm{Aut}\mathcal{F}(G, \bar{K})$.

2. For $K = \mathrm{GF}(2)$ and $G = \mathbb{Z}/n\mathbb{Z}$, according to [J, 1.1.7] or [MOW], we have

$$K[G]^\times = (K[x]/(x^n - 1))^\times \cong \mathbb{Z}/\nu\mathbb{Z}\,,$$

where

$$\nu = \nu(n) = 2^n \prod_{d|n} \left(1 - \frac{1}{2^{f(d)}}\right)^{g(d)}$$

with $f(n) = \mathrm{ord}_n(2) = \min\{j : 2^j \equiv 1 \mod n\}$ and $g(n) = \varphi(n)/f(n)$. Here φ stands for the Euler totient function. We recall that G admits a nonzero a^+-harmonic function if and only if $n \equiv 0 \mod 3$. Otherwise the minimal period l as in (1) above coincides with the order of $\tilde{a}^+$ in the cyclic group $\mathbb{Z}/\nu\mathbb{Z}$, so $l \mid \nu$.

3. For $K = \mathrm{GF}(2)$, $t = 1$ and $a = a^+$, $\Delta_{a_*}^{q_0-1} : \mathcal{F}(G, \bar{K}) \twoheadrightarrow (\ker(\Delta_{a_*}))^\perp$ is the orthogonal projection onto the space $(\ker(\Delta_{a_*}))^\perp$ of all winning patterns of the 'Lights Out' game on the toric grid $G = \mathbb{Z}_{\bar{n}}$; see [Za$_1$, §2.8] or §2 in the Introduction.

4. Convolution equations over finite fields

We fix a Galois field $K = \mathrm{GF}(q)$ with $q = p^r$ and a finite abelian group G of order coprime with p. Let $\Delta_{\bar{a}} = (\Delta_{a_1}, \ldots, \Delta_{a_t})$ be a system of convolution operators with kernel $\bar{a} \in (\mathcal{F}^0(G, K))^t$. Clearly the dimension of the space of solutions $\ker(\Delta_{\bar{a}})$ is the same in $\mathcal{F}(G, K)$ and in $\mathcal{F}(G, \bar{K})$. We show in Theorem 4.4 below that, moreover, the former subspace can be recovered by taking traces of $\bar{a}$-harmonic $\bar{K}$-valued characters.

We let $\phi_q : \xi \longmapsto \xi^q$ denote the Frobenius automorphism of $\bar{K} = \overline{\mathrm{GF}(q)}$. By the same letter we denote the induced action $\phi_q : f \longmapsto f^{\phi_q} = f^q$ on the function spaces $\mathcal{F}(G, \bar{K})$ and $\mathcal{F}(G^\vee, \bar{K})$, respectively.

The rings of invariants

$$[\mathcal{F}(G, \bar{K})]^{\phi_q} = \mathcal{F}(G, K) \quad \text{and} \quad [\mathcal{F}(G^\vee, \bar{K})]^{\phi_q} = \mathcal{F}(G^\vee, K)$$

do not correspond to each other under the Fourier transform. The restriction $D_q = \phi_q|G^\vee$ to the image of $G^\vee \hookrightarrow \mathcal{F}(G,\bar{K})$ is just the multiplication by q in the abelian group $G^\vee$. We keep again the same symbol D_q for the induced automorphism of the function space $\mathcal{F}(G^\vee,\bar{K})$. The latter one being different from ϕ_q, we let α_q denote the automorphism $\phi_q \circ (D_q)^{-1}$ of $\mathcal{F}(G^\vee,\bar{K})$ which measures this difference. In the next simple lemma (cf. [J]) we show that α_q is the Fourier dual of the Frobenius automorphism acting on $\mathcal{F}(G,\bar{K})$.

Lemma 4.1. *The automorphism $\alpha_q \in \operatorname{Aut}(\mathcal{F}(G^\vee,\bar{K}))$ is the Fourier dual of $\phi_q \in \operatorname{Aut}(\mathcal{F}(G,\bar{K}))$. Hence the Fourier image $F(\mathcal{F}(G,K))$ coincides with the subalgebra $(\mathcal{F}(G^\vee,\bar{K}))^{\alpha_q} \subseteq \mathcal{F}(G^\vee,\bar{K})$ of α_q-invariants.*

Proof. For any $f \in \mathcal{F}(G,\bar{K})$ we have

$$\left(\hat{f}\right)^{\phi_q} = \widehat{f^{\phi_q}} \circ D_q\,.$$

Indeed, for any $g^\vee \in G^\vee$,

$$\left(\hat{f}(g^\vee)\right)^q = \left(\sum_{v\in G} f(v)g^\vee(v)\right)^q = \sum_{v\in G} f^{\phi_q}(v)(g^\vee)^{\phi_q}(v) = \widehat{f^{\phi_q}}((g^\vee)^{\phi_q})\,.$$

Therefore

$$f \in \mathcal{F}(G,K) \iff f = f^{\phi_q} \iff \hat{f} = \widehat{f^{\phi_q}} \iff \hat{f}\circ D_q = (\hat{f})^{\phi_q} \iff \hat{f} = \alpha(\hat{f})\,,$$

as stated. □

Now one can easily deduce the following fact.

Corollary 4.2. *For any $\bar{a} \in (\mathcal{F}(G,K))^t$, the locus $V(\bar{a}) \subset G^\vee$ of $\bar{a}$-harmonic characters is D_q-stable.*

This leads to a direct sum decomposition of the space of solutions, see 4.3(b) below. For a function $f \in \mathcal{F}(G,\bar{K})$ we let $GF(q(f))$, where $q(f) = q^{r(f)}$, denote the minimal subfield of $\bar{K}$ generated by K and by the image $f(G)$. The trace of f is

$$\operatorname{Tr}(f) = \operatorname{Tr}_{GF(q(f)):GF(q)}(f) = f + f^q + \ldots + f^{q^{r(f)-1}} \in \mathcal{F}(G,K)\,.$$

Proposition 4.3. *For any $\bar{a} \in (\mathcal{F}(G,K))^t$ the following hold.*

(a) *There is a bijection between the set of traces $\operatorname{Tr}(g^\vee) \in \mathcal{F}(G,K)$ of all $\bar{a}$-harmonic characters $g^\vee \in V(\bar{a})$ and the orbit space $V(\bar{a})/\langle D_q\rangle$ of the cyclic group $\langle D_q\rangle$ acting on $V(\bar{a})$.*

(b) *Given a set of representatives $g_1^\vee,\ldots,g_m^\vee$ of the $\langle D_q\rangle$-orbits on $V(\bar{a})$, there is a decomposition into orthogonal direct sum of convolution ideals*

$$\ker(\Delta_{\bar{a}}) = \bigoplus_{i=1}^{m} (\operatorname{Tr}(g_i^\vee)) \subseteq \mathcal{F}(G,K)\,.$$

(c) *For any* $g^\vee \in G^\vee$ *one has*

$$g^\vee = \frac{1}{\mathrm{ord}(G)} \sum_{g\in G} g^\vee(g^{-1}) h_g, \qquad \textit{where} \quad h_g(x) = \mathrm{Tr}(g^\vee(gx)) = \tau_g\left(\mathrm{Tr}(g^\vee(x))\right).$$

Proof. By virtue of 4.2 for any character $g^\vee \in V(\bar{a})$ we have

$$h = \mathrm{Tr}(g^\vee) = g^\vee + (g^\vee)^q + \ldots + (g^\vee)^{q^{r(g^\vee)-1}} \in \ker(\Delta_{\bar{a}}) \cap \mathcal{F}(G,K).$$

Letting

$$O(g^\vee) = \{g^\vee, (g^\vee)^q, \ldots, (g^\vee)^{q^{r(g^\vee)-1}}\}$$

be the orbit of $g^\vee$ under the action of the cyclic group $\langle D_q \rangle$ on $V(\bar{a})$, one can easily deduce that $\mathrm{card}\,(O(g^\vee)) = r(g^\vee)$ and, by (12),

$$\widehat{h} = \mathrm{ord}(G) \sum_{i=0}^{r(g^\vee)-1} \delta_{(g^\vee)^{-q^i}} = \mathrm{ord}(G)\delta_{O((g^\vee)^{-1})}.$$

Now $h \longleftrightarrow O((g^\vee)^{-1})$ is the correspondence required in (a). In turn (a) implies (b). Whereas (c) follows by using the orthogonality relations for characters. Indeed by virtue of (12), for any $x \in G$ one has

$$\sum_{g\in G} g^\vee(g^{-1}) h_g(x) = \sum_{g\in G} g^\vee(g^{-1}) \sum_{i=0}^{r(g^\vee)-1} (g^\vee)^{q^i}(gx)$$

$$= \sum_{i=0}^{r(g^\vee)-1} \left(\sum_{g\in G} (g^\vee)^{-q^i}(g^{-1}) g^\vee(g^{-1}) \right) (g^\vee)^{q^i}(x) = \sum_{i=0}^{r(g^\vee)-1} \widehat{(g^\vee)^{-q^i}}(g^\vee)(g^\vee)^{q^i}(x)$$

$$= \mathrm{ord}(G) \sum_{i=0}^{r(g^\vee)-1} \delta_{(g^\vee)^{q^i}}(g^\vee)(g^\vee)^{q^i}(x) = \mathrm{ord}(G) g^\vee(x).$$

□

The following result is straightforward from 3.1 and 4.3(c).

Theorem 4.4. (a) *For a finite abelian group G of order coprime to p and for any $\bar{a} \in (\mathcal{F}(G,K))^t$, where $K = \mathrm{GF}(p^r)$, the kernel* $\ker(\Delta_{\bar{a}})$ *is spanned over K by the shifts of traces of $\bar{a}$-harmonic characters $g^\vee \in G^\vee_{\bar{a}-\mathrm{harm}}$.*

(b) *Similarly, for any sublattice $\Lambda' \subseteq \Lambda$ of finite index coprime to p and for any $\bar{a} \in (\mathcal{F}(\Lambda,K))^t$, the kernel* $\ker(\Delta_{\bar{a}}|\mathcal{F}_{\Lambda'}(\Lambda,K))$ *is spanned over K by the shifts of traces of $\bar{a}$-harmonic characters $g^\vee \in$* $\mathrm{Char}_{\bar{a}-\mathrm{harm}}(\Lambda, \bar{K}^\times)$ *with* $\Lambda' \subseteq \ker(g^\vee)$.

References

[ACZ] S.S. Abhyankar, S.D. Cohen, M.E. Zieve, *Bivariate factorizations connecting Dickson polynomials and Galois theory.* Trans. Amer. Math. Soc. 352 (2000), 2871–2887.

[BR] R. Barua and S. Ramakrishnan, *σ-game, σ^+-game and two-dimensional additive cellular automata*, Theoret. Comput. Sci. **154** (1996), 349–366.

[BZ] M. Bhargava and M. E. Zieve, *Factoring Dickson polynomials over finite fields*, Finite Fields Appl. **5** (1999), 103–111.

[GKW] J. Goldwasser, W. Klostermeyer and H. Ware, *Fibonacci Polynomials and Parity Domination in Grid Graphs*, Graphs and Combinatorics **18** (2002), 271–283.

[HMP] M. Hunziker, A. Machiavelo and J. Park, *Chebyshev polynomials over finite fields and reversibility of σ-automata on square grids*, Theoret. Comput. Sci. **320** (2004), 465–483.

[J] D. Jungnickel, *Finite fields. Structure and arithmetics*, Bibliographisches Institut, Mannheim, 1993.

[LMT] R. Lidl, G.L. Mullen, G. Turnwald, *Dickson polynomials.* John Wiley & Sons, Inc., New York, 1993.

[MOW] O. Martin, A.M. Odlyzko and S. Wolfram, *Algebraic properties of cellular automata*, Comm. Math. Phys. **93** (1984), 219–258.

[Nic] P.J. Nicholson, *Algebraic theory of finite Fourier transforms*, J. Comput. System Sci. **5** (1971), 524–547.

[SB] P. Sarkar and R. Barua, *Multidimensional σ-automata, π-polynomials and generalised S-matrices*, Theoret. Comput. Sci. **197** (1998), 111–138.

[Su] K. Sutner, *σ-automata and Chebyshev-polynomials*, Theoret. Comput. Sci. **230** (2000), 49–73.

[Za_1] M. Zaidenberg, *Periodic binary harmonic functions on lattices*, preprint MPIM-2006-14, 33p.; ArXiv math-ph/0608027. To appear in: Adv. Appl. Math.

[Za_2] M. Zaidenberg, *Convolution equations on lattices: periodic solutions with values in a prime characteristic field*, preprint Max-Planck-Institute für Mathematik, Bonn, MPIM2006-67, 30p.; ArXiv math-ph/0606070.

Mikhail Zaidenberg
Université Grenoble I
Institut Fourier
UMR 5582 CNRS-UJF
BP 74
F-38402 St. Martin d'Hères cédex
France
e-mail: `zaidenbe@ujf-grenoble.fr`